30 *Years*

Previous Solved Papers

GATE 2025

Instrumentation Engineering

- Fully solved with explanations
- Analysis of previous papers
- Topicwise presentation
- Thoroughly revised & updated

MADE EASY PUBLICATIONS

MADE EASY Publications Pvt. Ltd.

Corporate Office: 44-A/4, Kalu Sarai (Near Hauz Khas Metro Station), New Delhi-110016 | **Ph. :** 9021300500

Email : infomep@madeeasy.in | **Web :** www.madeeasypublications.org

GATE - 2025

Instrumentation Engineering

Topicwise Previous GATE Solved Papers (1995-2024)

Editions

1st Edition	:	2011
2nd Edition	:	2012
3rd Edition	:	2013
4th Edition	:	2014
5th Edition	:	2015
6th Edition	:	2016
7th Edition	:	2017
8th Edition	:	2018
9th Edition	:	2019
10th Edition	:	2020
11th Edition	:	2021
12th Edition	:	2022
13th Edition	:	2023
14th Edition	**:**	**2024**

Preface

B. Singh (Ex. IES)

Over the period of time the GATE examination has become more challenging due to increasing number of candidates. Though every candidate has ability to succeed but competitive environment, in-depth knowledge, quality guidance and good source of study is required to achieve high level goals.

The new edition of **GATE 2025 Solved Papers : Instrumentation Engineering** has been fully revised, updated and edited. The whole book has been divided into topicwise sections.

At the beginning of each subject, analysis of previous papers are given to improve the understanding of subject.

I have true desire to serve student community by way of providing good source of study and quality guidance. I hope this book will be proved an important tool to succeed in GATE examination. Any suggestions from the readers for the improvement of this book are most welcome.

B. Singh (Ex. IES)
Chairman and Managing Director
MADE EASY Group

GATE-2025

Instrumentation Engineering

CONTENTS

1. Engineering Mathematics .... 1 - 50
2. Electrical Circuits & Electrical Machines .... 51 - 99
3. Sensors and Industrial Instrumentation .... 100 - 164
4. Analog Electronics .... 165 - 239
5. Signal & Systems and Communication Systems ... 240 - 293
6. Control Systems and Process Control .... 294 - 358
7. Digital Electronics and Microprocessors .... 359 - 415
8. Measurements .... 416 - 476
9. Optical Instrumentation .... 477 - 513
10. Electricity & Magnetism .... 514 - 518
11. General Aptitude .... 519 - 555

Engineering Mathematics

UNIT

I

CONTENTS

1. Linear Algebra 3
2. Differential Equations 14
3. Probability & Statistics 19
4. Numerical Methods 27
5. Calculus 31
6. Vector Calculus 39
7. Complex Numbers 43

Engineering Mathematics

Syllabus

Linear Algebra : Matrix algebra, systems of linear equations, consistency and rank, Eigen value and Eigen vectors.

Calculus : Mean value theorems, theorems of integral calculus, partial derivatives, maxima and minima, multiple integrals, Fourier series, vector identities, line, surface and volume integrals, Stokes, Gauss and Green's theorems.

Differential equations : First order equation (linear and nonlinear), second order linear differential equations with constant coefficients, method of variation of parameters, Cauchy's and Euler's equations, initial and boundary value problems, solution of partial differential equations: variable separable method.

Analysis of complex variables : Analytic functions, Cauchy's integral theorem and integral formula, Taylor's and Laurent's series, residue theorem, solution of integrals.

Probability and Statistics : Sampling theorems, conditional probability, mean, median, mode, standard deviation and variance; random variables: discrete and continuous distributions: normal, Poisson and binomial distributions.

Numerical Methods : Matrix inversion, solutions of non-linear algebraic equations, iterative methods for solving differential equations, numerical integration, regression and correlation analysis.

Analysis of Previous GATE Papers

Exam Year	1 Mark Ques.	2 Marks Ques.	Total Marks
1995	–	–	–
1996	–	–	–
1997	–	–	–
1998	1	–	**1**
1999	–	–	–
2000	–	–	–
2001	–	–	–
2002	–	1	**2**
2003	–	1	**2**
2004	–	–	–
2005	–	2	**4**
2006	1	4	**9**
2007	2	4	**10**
2008	2	7	**16**
2009	–	2	**4**

Exam Year	1 Mark Ques.	2 Marks Ques.	Total Marks
2010	2	2	**6**
2011	2	3	**8**
2012	4	6	**16**
2013	3	5	**13**
2014	3	4	**11**
2015	3	5	**13**
2016	3	3	**9**
2017	4	3	**10**
2018	4	4	**12**
2019	4	3	**10**
2020	5	5	**15**
2021	5	3	**11**
2022	4	4	**12**
2023	4	4	**12**
2024	5	4	**13**

1 Linear Algebra

1.1 The rank of matrix $A = \begin{bmatrix} 1 & 2 & 3 \\ 3 & 4 & 5 \\ 4 & 6 & 8 \end{bmatrix}$ is

(a) 0 (b) 1
(c) 2 (d) 3

[2000 : 1 M]

1.2 The necessary condition to diagonalize a matrix is that
(a) its all eigen values should be distinct
(b) its eigen vectors should be independent
(c) its eigen value should be real
(d) the matrix is non-singular

[2000 : 1 M]

1.3 A system of equations represented by $AX = 0$, where X is a column vector of unknowns and A is matrix containing coefficients, has a nontrivial solution when A is
(a) nonsingular
(b) singular
(c) symmetric
(d) Hermetian

[2003 : 1 M]

1.4 Let A be a 3 × 3 matrix with rank 2. Then $AX = 0$ has
(a) only the trival solution $X = 0$
(b) one independent solution
(c) two independent solutions
(d) three independent solutions

[2005 : 1 M]

1.5 Identify which one of the following is an eigenvector of the matrix $A = \begin{bmatrix} 1 & 0 \\ -1 & -2 \end{bmatrix}$
(a) $[-1 \quad 1]^T$ (b) $[3 \quad -1]^T$
(c) $[1 \quad -1]^T$ (d) $[-2 \quad 1]^T$

[2005 : 1 M]

1.6 For a given 2 × 2 matrix A, it is observed that,

$$A\begin{bmatrix} 1 \\ -1 \end{bmatrix} = -\begin{bmatrix} 1 \\ -1 \end{bmatrix} \text{ and } A\begin{bmatrix} 1 \\ -2 \end{bmatrix} = -2\begin{bmatrix} 1 \\ -2 \end{bmatrix}$$

Then matrix A is

(a) $A = \begin{bmatrix} 2 & 1 \\ -1 & -1 \end{bmatrix}\begin{bmatrix} -1 & 0 \\ 0 & -2 \end{bmatrix}\begin{bmatrix} 1 & 1 \\ -1 & -2 \end{bmatrix}$

(b) $A = \begin{bmatrix} 1 & 1 \\ -1 & -2 \end{bmatrix}\begin{bmatrix} 1 & 0 \\ 0 & 2 \end{bmatrix}\begin{bmatrix} 2 & 1 \\ -1 & -1 \end{bmatrix}$

(c) $A = \begin{bmatrix} 1 & 1 \\ -1 & -2 \end{bmatrix}\begin{bmatrix} -1 & 0 \\ 0 & -2 \end{bmatrix}\begin{bmatrix} 2 & 1 \\ -1 & -1 \end{bmatrix}$

(d) $A = \begin{bmatrix} 0 & -2 \\ 1 & -3 \end{bmatrix}$

[2006 : 2 M]

1.7 Let A be an $n \times n$ real matrix such that $A^2 = I$ and y be an n-dimensional vector.
Then the linear system of equations $Ax = y$ has
(a) no solution
(b) a unique solution
(c) more than one but finitely many independent solutions
(d) infinitely many independent solutions

[2007 : 1 M]

1.8 Let $A = [a_{ij}]$, $1 \le i, j \le n$, with $n \ge 3$ and $a_{ij} = i.j$.
Then the rank of A is
(a) 0 (b) 1
(c) $n - 1$ (d) n

[2007 : 2 M]

1.9 Let $P \neq 0$ be a 3 × 3 real matrix. There exist linearly independent vectors x and y such that $PX = 0$ and $PY = 0$. The dimension to the range space of P is
(a) 0 (b) 1
(c) 2 (d) 3

[2009 : 1 M]

1.10 The eigen values of a (2 × 2) matrix X are –2 and –3. The eigen values of the matrix $(X + I)(X + 5I)$ are
(a) –3, –4 (b) –1, –2
(c) –1, –3 (d) –2, –4

[2009 : 2 M]

1.11 The matrix $P=\begin{bmatrix}0&0&1\\1&0&0\\0&1&0\end{bmatrix}$ rotates a vector about the axis $\begin{bmatrix}1\\1\\1\end{bmatrix}$ by angle of

(a) 30° (b) 60°
(c) 90° (d) 120°

[2009 : 2 M]

1.12 A real $n \times n$ matrix $A = \{a_{ij}\}$ is defined as follows: $a_{ij} = i$, if $i = j$, otherwise 0

The summation of all n eigen values of A is

(a) $n(n+1)/2$ (b) $n(n-1)/2$
(c) $\frac{n(n+1)(2n+1)}{6}$ (d) n^2

[2010 : 1 M]

1.13 X and Y are non-zero square matrices of size $n \times n$. If $XY = 0_{n\times n}$ then

(a) $|X| = 0$ and $|Y| \neq 0$
(b) $|X| \neq 0$ and $|Y| = 0$
(c) $|X| = 0$ and $|Y| = 0$
(d) $|X| \neq 0$ and $|Y| \neq 0$

[2010 : 2 M]

1.14 The matrix $M=\begin{bmatrix}-2&2&-3\\2&1&-6\\-1&-2&0\end{bmatrix}$ has eigen values −3, −3, 5. An eigen vector corresponding to the eigen value 5 is $[1\ \ 2\ \ -1]^T$. One of the eigen vectors of the matrix M^3 is

(a) $[1\ \ 8\ \ -1]^T$ (b) $[1\ \ 2\ \ -1]^T$
(c) $[1\ \ \sqrt[3]{2}\ \ -1]^T$ (d) $[1\ \ 1\ \ -1]^T$

[2011 : 1 M]

1.15 The series $\sum_{m=0}^{\infty} \frac{1}{4^m}(x-1)^{2m}$ converges for

(a) $-2 < X < 2$ (b) $-1 < X < 3$
(c) $-3 < X < 1$ (d) $X < 3$

[2011 : 2 M]

1.16 Given that: $A=\begin{bmatrix}-5&-3\\2&0\end{bmatrix}$ and $I=\begin{bmatrix}1&0\\0&1\end{bmatrix}$, the value A^3 is

(a) $15A + 12I$ (b) $19A + 30I$
(c) $17A + 15I$ (d) $17A + 21I$

[2012 : 2 M]

1.17 The dimension of the null space of the matrix $\begin{bmatrix}0&1&1\\1&-1&0\\-1&0&-1\end{bmatrix}$ is

(a) 0 (b) 1
(c) 2 (d) 3 [2013 : 1 M]

1.18 One pair of eigen vectors corresponding to the two eigen values of the matrix $\begin{bmatrix}0&-1\\1&0\end{bmatrix}$ is

(a) $\begin{bmatrix}1\\-j\end{bmatrix},\begin{bmatrix}j\\-1\end{bmatrix}$ (b) $\begin{bmatrix}0\\1\end{bmatrix},\begin{bmatrix}-1\\0\end{bmatrix}$
(c) $\begin{bmatrix}1\\j\end{bmatrix},\begin{bmatrix}0\\1\end{bmatrix}$ (d) $\begin{bmatrix}1\\j\end{bmatrix},\begin{bmatrix}j\\1\end{bmatrix}$

[2013 : 2 M]

1.19 Given:

$x(t) = 3\sin(1000\pi t)$ and $y(t) = 5\cos\left(1000\pi t + \frac{\pi}{4}\right)$

The X-Y plot will be

(a) a circle
(b) a multi-loop closed curve
(c) a hyperbola
(d) an ellipse [2014 : 1 M]

1.20 A scalar valued function is defined as $f(\boldsymbol{X}) = \boldsymbol{X}^T\boldsymbol{A}\boldsymbol{X} + \boldsymbol{b}^T\boldsymbol{X} + c$, where $\boldsymbol{A}$ is a symmetric positive definite matrix with dimension $n \times n$; $\boldsymbol{b}$ and $\boldsymbol{x}$ are vectors of dimension $n \times 1$. The minimum value of $f(\boldsymbol{X})$ will occur when $\boldsymbol{X}$ equals

(a) $(\boldsymbol{A}^T\boldsymbol{A})^{-1}\boldsymbol{b}$ (b) $-(\boldsymbol{A}^T\boldsymbol{A})^{-1}\boldsymbol{b}$
(c) $-\left(\frac{\boldsymbol{A}^{-1}\boldsymbol{b}}{2}\right)$ (d) $\frac{\boldsymbol{A}^{-1}\boldsymbol{b}}{2}$

[2014 : 2 M]

1.21 For the matrix $\boldsymbol{A}$ satisfying the equation given below, the eigen values are

$$[A]\begin{bmatrix}1&2&3\\7&8&9\\4&5&6\end{bmatrix}=\begin{bmatrix}1&2&3\\4&5&6\\7&8&9\end{bmatrix}$$

(a) $(1, -j, j)$ (b) $(1, 1, 0)$
(c) $(1, 1, -1)$ (d) $(1, 0, 0)$

[2014 : 2 M]

1.22 Let A be an $n \times n$ matrix with rank $r(0 < r < n)$. Then $AX = 0$ has p independent solutions, where p is
(a) r (b) n
(c) $n - r$ (d) $n + r$

[2015 : 1 M]

1.23 A straight line of the form $y = mx + c$ passes through the origin and the point $(x, y) = (2, 6)$. The value of m is________.

[2016 : 1 M]

1.24 Consider the matrix $A = \begin{pmatrix} 2 & 1 & 1 \\ 2 & 3 & 4 \\ -1 & -1 & -2 \end{pmatrix}$ whose eigen values are 1, –1 and 3. Then Trace of $(A^3 - 3A^2)$ is_____.

[2016 : 2 M]

1.25 The eigen values of the matrix $A = \begin{bmatrix} 1 & -1 & 5 \\ 0 & 5 & 6 \\ 0 & -6 & 5 \end{bmatrix}$ are
(a) $-1, 5, 6$ (b) $1, -5 \pm j6$
(c) $1, 5 \pm j6$ (d) $1, 5, 5$

[2017 : 1 M]

1.26 The figure shows a shape ABC and its mirror image $A_1 B_1 C_1$ across the horizontal axis (X-axis). The coordinate transformation matrix that maps ABC to $A_1B_1C_1$ is

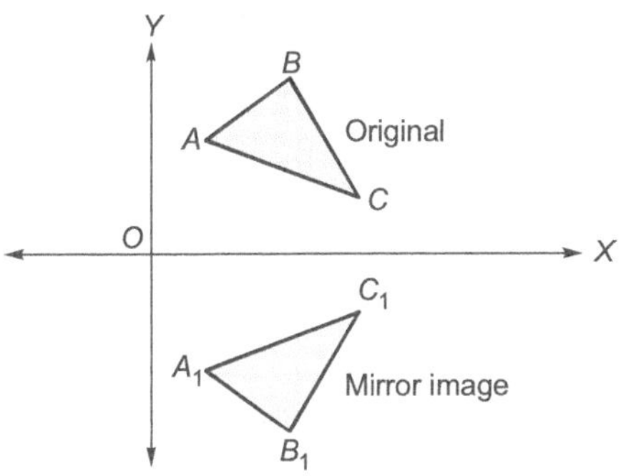

(a) $\begin{bmatrix} 0 & 1 \\ 1 & 0 \end{bmatrix}$ (b) $\begin{bmatrix} 0 & 1 \\ -1 & 0 \end{bmatrix}$

(c) $\begin{bmatrix} -1 & 0 \\ 0 & 1 \end{bmatrix}$ (d) $\begin{bmatrix} 1 & 0 \\ 0 & -1 \end{bmatrix}$

[2017 : 1 M]

1.27 If $\mathbf{v}$ is a non-zero vector of dimension 3×1, then the matrix $A = \mathbf{v}\mathbf{v}^T$ has rank = ________

[2017 : 1 M]

1.28 Let N be a 3 by 3 matrix with real number entries. The matrix N is such that $N^2 = 0$. The eigen values of N are
(a) 0, 0, 0 (b) 0, 0, 1
(c) 0, 1, 1 (d) 1, 1, 1

[2018 : 1 M]

1.29 Consider two functions $f(x) = (x - 2)^2$ and $g(x) = 2x - 1$, where x is real. The smallest value of x for which $f(x) = g(x)$ is ________.

[2018 : 1 M]

1.30 Consider the following system of linear equations:

$$3x + 2ky = -2$$
$$kx + 6y = 2$$

Here, x and y are the unknown and k is a real constant. The value of k for which there are infinite number of solutions is
(a) 3 (b) 1
(c) –3 (d) –6

[2018 : 2 M]

1.31 A 3×3 matrix has eigen values 1, 2 and 5. The determinant of the matrix is________.

[2019 : 1 M]

1.32 The curve $y = f(x)$ is such that the tangent to the curve at every point (x, y) has a Y-axis intercept c, given by $c = -y$. Then $f(x)$ is proportional to
(a) x^{-1} (b) x^2
(c) x^3 (d) x^4 [2019 : 2 M]

1.33 A set of linear equations is given in the form $Ax = b$, where A is a 2×4 matrix with real number entries and $b \neq 0$. Will it be possible to solve for x and obtain a unique solution by multiplying both left and right sides of the equation by A^T (the super script T denotes the transpose and inverting the matrix A^TA? Answer is _____.
(a) Yes, can obtain a unique solution provided the matrix A is will conditioned.
(b) Yes, it is always possible to get a unique solution for any 2×4 matrix A.
(c) Yes, can obtain a unique solution provided the matrix A^TA is well conditional.
(d) No, it is not possible to get a unique solution for any 2×4 matrix A. [2020 : 1 M]

1.34 Consider the matrix $M=\begin{bmatrix}1 & -1 & 0\\ 1 & -2 & 1\\ 0 & -1 & 1\end{bmatrix}$. One of the eigen vectors of M is

(a) $\begin{bmatrix}-1\\ 1\\ -1\end{bmatrix}$ (b) $\begin{bmatrix}1\\ 1\\ 1\end{bmatrix}$

(c) $\begin{bmatrix}1\\ -1\\ 1\end{bmatrix}$ (d) $\begin{bmatrix}1\\ 1\\ -1\end{bmatrix}$

[2020 : 2 M]

1.35 Given $A=\begin{bmatrix}2 & 5\\ 0 & 3\end{bmatrix}$, the value of the determinant

$|A^4-5A^3+6A^2+2I|=$ ______.

[2021 : 2 M]

1.36 Consider the rows vectors $v=[1, 0]$ and $w=[2, 0]$. The rank of the matrix $M=2v^Tv+3w^Tw$, where the superscript T denotes the transpose, is

(a) 3 (b) 2

(c) 4 (d) 1

[2021 : 1 M]

1.37 The determinant of the matrix M shown below is ______.

$$M=\begin{bmatrix}1 & 2 & 0 & 0\\ 3 & 4 & 0 & 0\\ 0 & 0 & 4 & 3\\ 0 & 0 & 2 & 1\end{bmatrix}$$

[2021 : 1 M]

1.38 Given $M=\begin{bmatrix}2 & 3 & 7\\ 6 & 4 & 7\\ 4 & 6 & 14\end{bmatrix}$, which of the following statement(s) is/are correct?

(a) The rank of M is 2

(b) The rank of M is 3

(c) The rows of M are linearly independent

(d) The determinant of M is 0

[2022 : 1 M]

1.39 The matrix $A=\begin{bmatrix}4 & 3\\ 9 & -2\end{bmatrix}$ has eigenvalues – 5 and 7. The eigenvector(s) is/are ______.

(a) $\begin{bmatrix}1\\ 1\end{bmatrix}$ (b) $\begin{bmatrix}3\\ 4\end{bmatrix}$

(c) $\begin{bmatrix}2\\ -6\end{bmatrix}$ (d) $\begin{bmatrix}2\\ 8\end{bmatrix}$

[2022 : 2 M]

1.40 Choose solution set S corresponding to the systems of two equations

$$x-2y+z=0$$
$$x-z=0$$

Note: . R denotes the set of real numbers

(a) $S=\left\{\alpha\begin{pmatrix}1\\ 1\\ 1\end{pmatrix}\middle|\alpha\in R\right\}$

(b) $S=\left\{\alpha\begin{pmatrix}1\\ 1\\ 1\end{pmatrix}+\beta\begin{pmatrix}1\\ 0\\ 1\end{pmatrix}\middle|\alpha,\beta\in R\right\}$

(c) $S=\left\{\alpha\begin{pmatrix}1\\ 1\\ 1\end{pmatrix}+\beta\begin{pmatrix}2\\ 1\\ 2\end{pmatrix}\middle|\alpha,\beta\in R\right\}$

(d) $S=\left\{\alpha\begin{pmatrix}1\\ 0\\ 1\end{pmatrix}\middle|\alpha\in R\right\}$

[2023 : 1 M]

1.41 The rank of the matrix A given below is one. The ratio $\frac{\alpha}{\beta}$ is ______ (rounded off to the nearest integer).

$$A=\begin{bmatrix}1 & 4\\ -3 & \alpha\\ \beta & 6\end{bmatrix}$$

[2023 : 2 M]

1.42 A matrix M is constructed by stacking three column vectors v_1, v_2, v_3 as $M=[v_1\ \ v_2\ \ v_3]$. Choose the set of vectors from the following options such that rank $(M)=3$.

(a) $v_1=\begin{bmatrix}1\\ 0\\ 1\end{bmatrix}, v_2=\begin{bmatrix}0\\ -1\\ 0\end{bmatrix}, v_3=\begin{bmatrix}1\\ -1\\ 1\end{bmatrix}$

(b) $v_1=\begin{bmatrix}1\\ 1\\ 1\end{bmatrix}, v_2=\begin{bmatrix}-1\\ 0\\ 1\end{bmatrix}, v_3=\begin{bmatrix}0\\ 0\\ 0\end{bmatrix}$

(c) $v_1 = \begin{bmatrix}1\\0\\1\end{bmatrix}, v_2 = \begin{bmatrix}-1\\0\\1\end{bmatrix}, v_3 = \begin{bmatrix}1\\-1\\1\end{bmatrix}$

(d) $v_1 = \begin{bmatrix}1\\1\\1\end{bmatrix}, v_2 = \begin{bmatrix}-1\\1\\-1\end{bmatrix}, v_3 = \begin{bmatrix}0\\-1\\0\end{bmatrix}$

[2024 : 1 M]

1.43 A 3 × 3 matrix P with all real elements has eigenvalues $\frac{1}{4}$, 1, and –2. The value of $[P^{-1}]$ is _______ (rounded off to nearest integer)

[2024 : 1 M]

Answers Linear Algebra

1.1 (c)	1.2 (a)	1.3 (b)	1.4 (b)	1.5 (b)	1.6 (c)	1.7 (b)
1.8 (b)	1.9 (*)	1.10 (a)	1.11 (*)	1.12 (a)	1.13 (c)	1.14 (b)
1.15 (b)	1.16 (b)	1.17 (b)	1.18 (a, d)	1.19 (d)	1.20 (c)	1.21 (c)
1.22 (c)	1.23 (3)	1.24 (–6)	1.25 (c)	1.26 (d)	1.27 (1)	1.28 (a)
1.29 (1)	1.30 (c)	1.31 (10)	1.32 (b)	1.33 (d)	1.34 (b)	1.35 (4)
1.36 (d)	1.37 (4)	1.38 (a, d)	1.39 (a, c)	1.40 (a)	1.41 (–8)	1.42 (c)
1.43 (–2)						

Explanations Linear Algebra

1.1 (c)

$$\begin{bmatrix}1&2&3\\3&4&5\\4&6&8\end{bmatrix} R_2-3R_1, R_3-4R_1$$

$$\begin{bmatrix}1&2&3\\0&-2&-4\\0&-2&-4\end{bmatrix} R_3-R_2$$

$$\begin{bmatrix}1&2&3\\0&-2&-4\\0&0&0\end{bmatrix}$$

Number of non-zero rows = 2

∴ Rank of A = 2

1.2 (a)

By known theorem.

A square matrix is diagonalizable if it has distinct eigen values.

1.3 (b)

$AX = 0$ means system of homogenous equations. Which has only trivial solutions if $|A| \neq 0$ i.e. A is non singular. For non trivial solutions $|A| = 0$ i.e. A must be singular.

1.4 (b)

If r is the rank of matrix A and $n \times n$ is the order of matrix then we shall have $(n - r)$ linearly independent non-trival infinite. Any linear combination of these $(n - r)$ solutions will also be a solution of $AX = 0$.

1.5 (b)

$$A = \begin{bmatrix}1&0\\-1&-2\end{bmatrix}$$

Characteristic equation of A

$$|A-\lambda I| = 0$$

$$\begin{vmatrix}1-\lambda&0\\-1&-2-\lambda\end{vmatrix} = 0$$

$$-(1-\lambda)(2+\lambda) = 0$$

$$\Rightarrow \lambda = 1, -2$$

Put $\lambda = 1$ in $[A-\lambda I]\bar{X} = 0$

$$\begin{bmatrix}0&0\\-1&-3\end{bmatrix}\begin{bmatrix}X_1\\X_2\end{bmatrix} = 0$$

$$\Rightarrow X_1 + 3X_2 = 0$$

Solution is $X_2 = -k$

and $X_1 = +3k$

$$\bar{X}_1 = [3 \quad -1]^T$$

Since option (b) in is same ratio of X_1 to X_2

∴ $[3 \quad -1]^T$ is an eigen vector.

1.6 (c)

Let, 2 × 2 matrix, $A = \begin{bmatrix} a & c \\ b & d \end{bmatrix}$

$$A\begin{bmatrix} 1 \\ -1 \end{bmatrix} = -\begin{bmatrix} 1 \\ -1 \end{bmatrix}$$

$$\Rightarrow \quad \begin{bmatrix} a & c \\ b & d \end{bmatrix}\begin{bmatrix} 1 \\ -1 \end{bmatrix} = \begin{bmatrix} -1 \\ 1 \end{bmatrix}$$

$$\Rightarrow \quad a - c = -1 \quad \text{...(i)}$$

$$b - d = 1 \quad \text{...(ii)}$$

and $$A\begin{bmatrix} 1 \\ -2 \end{bmatrix} = -2\begin{bmatrix} 1 \\ -2 \end{bmatrix}$$

$$\Rightarrow \quad \begin{bmatrix} a & c \\ b & d \end{bmatrix}\begin{bmatrix} 1 \\ -2 \end{bmatrix} = \begin{bmatrix} -2 \\ 4 \end{bmatrix}$$

$$a - 2c = -2 \quad \text{...(iii)}$$

$$b - 2d = 4 \quad \text{...(iv)}$$

From equation (i) and (iii)

$$c = 1 \text{ and } a = 0$$

From equation (ii) and (iv)

$$d = -3 \text{ and } b = -2$$

$$\therefore \quad [A] = \begin{bmatrix} 0 & 1 \\ -2 & -3 \end{bmatrix}$$

On simplification of option (c)

$$A = \begin{bmatrix} 1 & 1 \\ -1 & -2 \end{bmatrix}\begin{bmatrix} -1 & 0 \\ 0 & -2 \end{bmatrix}\begin{bmatrix} 2 & 1 \\ -1 & -1 \end{bmatrix}$$

$$= \begin{bmatrix} -1 & -2 \\ 1 & 4 \end{bmatrix}\begin{bmatrix} 2 & 1 \\ -1 & -1 \end{bmatrix} = \begin{bmatrix} 0 & 1 \\ -2 & -3 \end{bmatrix}$$

1.7 (b)

Given, $A^2 = I$

$$|A^2| = |I|$$

$$|A| \cdot |A| = 1$$

$$|A| = \pm 1$$

So, $|A| \neq 0$, so system of equations $AX = Y$ is consistent, and has unique solution given by $X = A^{-1}Y$.

1.8 (b)

$$A = \begin{bmatrix} 1\cdot1 & 1\cdot2 & 1\cdot3 & \dots & 1\cdot n \\ 2\cdot1 & 2\cdot2 & 2\cdot3 & \dots & 2\cdot n \\ 3\cdot1 & 3\cdot2 & 3\cdot3 & \dots & 3\cdot n \\ \vdots & \vdots & \vdots & \vdots & \vdots \\ n\cdot1 & n\cdot2 & n\cdot3 & \dots & n\cdot n \end{bmatrix}$$

All row are the multiple of first row

$$\therefore \quad A = \begin{bmatrix} 1\cdot1 & 1\cdot2 & 1\cdot3 & \dots & 1\cdot n \\ 0 & 0 & 0 & \dots & 0 \\ 0 & 0 & 0 & \dots & 0 \\ \vdots & \vdots & \vdots & \vdots & \vdots \\ 0 & 0 & 0 & \dots & 0 \end{bmatrix}$$

Rank of A = 1.

1.10 (a)

X has eigen values –2, and –3.

So, taking $X = \begin{bmatrix} -2 & 0 \\ 0 & -3 \end{bmatrix}$, since additional information is not given, so we can take this value.

$$X + I = \begin{bmatrix} -1 & 0 \\ 0 & -2 \end{bmatrix}$$

$$(X + 5I) = \begin{bmatrix} 3 & 0 \\ 0 & 2 \end{bmatrix}$$

$$(X + I)(X + 5I) = \begin{bmatrix} -3 & 0 \\ 0 & -4 \end{bmatrix}$$

So, eigen values are –3 and –4.

Alternate method:

Let, $A = (X + I)(X + 5I) = X^2 + 6X + 5I$ ...(i)

Since, eigen value of $X^2 + 6X + 5I$ is $\lambda^2 + 6\lambda + 5$

where, λ is eigen value of X.

So, substituting values,

$\lambda = -2, \quad \lambda_1 = (-2)^2 - 6 \times 2 + 5 = -3$

$\lambda = -3, \quad \lambda_2 = (-3)^2 - 6 \times 3 + 5 = -4$

So eigen values of

$A = (X + I)(X + 5I)$ are –3 and –4

1.12 (a)

$$a_{ij} = \begin{cases} i; & i = j \\ 0; & i \neq j \end{cases}$$

$$\Rightarrow \quad [A] = \begin{bmatrix} 1 & 0 & 0 & \dots & 0 \\ 0 & 2 & 0 & \dots & 0 \\ 0 & 0 & 3 & \dots & 0 \\ \vdots & \vdots & \vdots & \vdots & \vdots \\ 0 & 0 & 0 & 0 & n \end{bmatrix}$$

Summation of all n eigen value of A

= Trace of A

= sum of diagonal elements

= 1 + 2 + ... + n

$$= \frac{n}{2}(n+1)$$

1.13 (c)

$$[X]_{n\times n}[Y]_{n\times n} = [0]_{n\times n}$$

$$\det([X][Y]) = \det([Y][X]) = \det[X]\det[Y]$$

$\therefore$ $\det[X]\det[Y] = 0$

$\Rightarrow$ $\det[X] = 0$ or $\det[Y] = 0$

or both are zero but it is not necessary that one of them is non-zero.

1.14 (b)

If $\lambda_1, \lambda_2, \lambda_3......\lambda_n$ are the eigen values of matrix A, then $\lambda_1^k, \lambda_2^k, ...\lambda_n^k$ will be the eigen values of A^k.

But the A and A^k will have same eigen vector.

Example:

Let a matrix

$$A = \begin{bmatrix} 4 & 3 \\ 2 & 5 \end{bmatrix}$$

$$A^2 = \begin{bmatrix} 4 & 3 \\ 2 & 5 \end{bmatrix}\begin{bmatrix} 4 & 3 \\ 2 & 5 \end{bmatrix} = \begin{bmatrix} 22 & 27 \\ 18 & 31 \end{bmatrix}$$

eigen values of $A = 2, 7$

eigen values of $A^2 = 4, 49$

eigen vector of matrix A corresponding to eigen value 2 is $[3, \ -2]^T$

eigen vector of matrix A^2 corresponding to eigen value 4 is $[3, \ -2]^T$.

1.15 (b)

$$\sum_{m=0}^{\infty} \frac{1}{4^m}(X-1)^{2m} = \sum_{m=0}^{\infty} \frac{1}{2^{2m}}(x-1)^{2m}$$

$$= \sum_{m=0}^{\infty}\left(\frac{X-1}{2}\right)^{2m}$$

$$= 1 + \left(\frac{X-1}{2}\right)^2 + \left(\frac{X-1}{2}\right)^4 + ...\infty$$

This is a geometric progression it will converges

if $r = \left(\frac{X-1}{2}\right)^2 < 1$

$\Rightarrow$ $\frac{(X-1)^2}{4} < 1$

$\Rightarrow$ $(X-1)^2 < 4$

$\Rightarrow$ $-2 < (X-1) < 2$

$\Rightarrow$ $-1 < X < 3$

1.16 (b)

Given, $A = \begin{bmatrix} -5 & -3 \\ 2 & 0 \end{bmatrix}$

The characteristic equation is

$$|A - \lambda I| = 0$$

$\Rightarrow$ $\begin{vmatrix} -5-\lambda & -3 \\ 2 & -\lambda \end{vmatrix} = 0$

$\Rightarrow$ $\lambda(\lambda + 5) + 6 = 0$

$\Rightarrow$ $\lambda^2 + 5\lambda + 6 = 0$

$\Rightarrow$ $\lambda^3 + 5\lambda^2 + 6\lambda = 0$

$\Rightarrow$ $\lambda^3 + 5(-5\lambda - 6) + 6\lambda = 0$

$\Rightarrow$ $\lambda^3 = 25\lambda + 30 - 6\lambda = 19\lambda + 30$

$\therefore$ $A^3 = 19A + 30I$

1.17 (b)

$$A = \begin{bmatrix} 0 & 1 & 1 \\ 1 & -1 & 0 \\ -1 & 0 & -1 \end{bmatrix}_{3\times 3}$$

Order of matrix = 3

Rank = 2

$\therefore$ dimension of null space of $A = 3 - 2 = 1$.

1.18 (a, d)

Eigen values are

$$|A - \lambda I| = 0$$

$$\left|\begin{bmatrix} 0 & -1 \\ 1 & 0 \end{bmatrix} - \begin{bmatrix} \lambda & 0 \\ 0 & \lambda \end{bmatrix}\right| = 0$$

$$\begin{vmatrix} -\lambda & -1 \\ 1 & -\lambda \end{vmatrix} = 0$$

$$\lambda^2 + 1 = 0$$

$$\lambda^2 = -1$$

$\therefore$ $\lambda = \pm i$

to find eigen vector,

$$\lambda = +i$$

$$\begin{bmatrix} -i & -1 \\ 1 & -i \end{bmatrix}\begin{bmatrix} X_1 \\ X_2 \end{bmatrix} = \begin{bmatrix} 0 \\ 0 \end{bmatrix}$$

$\therefore$ $-iX_1 - X_2 = 0$

and $X_1 - iX_2 = 0$

clearly,

$\begin{bmatrix} X_1 \\ X_2 \end{bmatrix} = \begin{bmatrix} 1 \\ -j \end{bmatrix}$ and $\begin{bmatrix} j \\ 1 \end{bmatrix}$, satisfy

$$\lambda = -i$$

$$\begin{bmatrix} i & -1 \\ 1 & i \end{bmatrix}\begin{bmatrix} X_1 \\ X_2 \end{bmatrix} = \begin{bmatrix} 0 \\ 0 \end{bmatrix}$$

$iX_1 - X_2 = 0$ and $X_1 + iX_2 = 0$

Clearly, $\begin{bmatrix} X_1 \\ X_2 \end{bmatrix} = \begin{bmatrix} j \\ -1 \end{bmatrix}$ and $\begin{bmatrix} 1 \\ j \end{bmatrix}$, satisfy

Thus, the two eigen value of the given matrix are

$\begin{bmatrix} 1 \\ -j \end{bmatrix}, \begin{bmatrix} j \\ -1 \end{bmatrix}$ or $\begin{bmatrix} 1 \\ j \end{bmatrix}, \begin{bmatrix} j \\ 1 \end{bmatrix}$.

1.19 (d)

$$x(t) = 3\sin(1000\pi t)$$

$$y(t) = 5\cos\left(1000\pi t + \frac{\pi}{4}\right)$$

at $t = 0$,

$$x(0) = 0$$

$$y(0) = 5\cos\frac{\pi}{4} = \frac{5}{\sqrt{2}}$$

at $t = \frac{1}{2}$,

$$x\left(\frac{1}{2}\right) = \sin\left(1000\frac{\pi}{2}\right) = 0$$

$$y\left(\frac{1}{2}\right) = 5\cos\left(500\pi + \frac{\pi}{4}\right) = \frac{5}{\sqrt{2}}$$

at $t = \frac{1}{4000}$,

$$x\left(\frac{1}{4000}\right) = 3\sin\frac{\pi}{4} = \frac{3}{\sqrt{2}}$$

$$y\left(\frac{1}{4000}\right) = 5\cos\left(\frac{\pi}{4} + \frac{\pi}{4}\right) = 0$$

Thus, this represents the equation of ellipse.

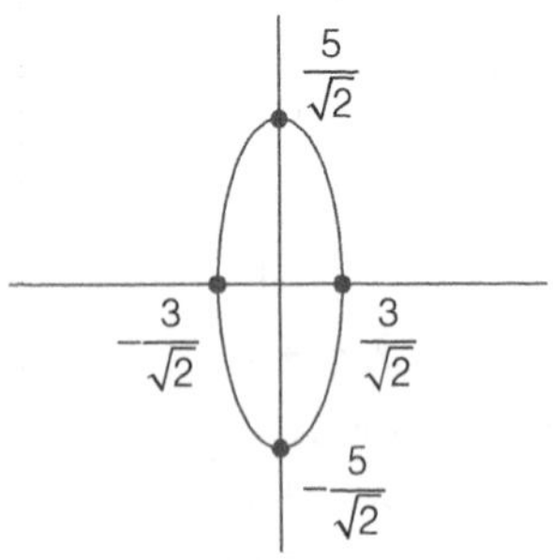

1.20 (c)

Let $n = 2$ and $A = \begin{bmatrix} p & q \\ q & r \end{bmatrix}$

$$X = \begin{bmatrix} x \\ y \end{bmatrix} \quad B = \begin{bmatrix} b_1 \\ b_2 \end{bmatrix}$$

then, $f(x) = X^T AX + B^T X + C$

$$= (px^2 + 2qxy + ry^2) + (b_1 x + b_2 y) + C$$

$$\frac{\partial f}{\partial x} = 2px + 2qy + b_1 = 0 \quad \text{..(i)}$$

$$\frac{\partial f}{\partial y} = 2qx + 2ry + b_2 = 0 \quad \text{...(ii)}$$

From equations (i) and (ii)

$$2\begin{bmatrix} p & q \\ q & r \end{bmatrix}\begin{bmatrix} x \\ y \end{bmatrix} = -\begin{bmatrix} b_1 \\ b_2 \end{bmatrix}$$

i.e. $\quad 2AX = -B$

$$\Rightarrow \quad AX = -\frac{B}{2}$$

$$\therefore X = -\frac{A^{-1}B}{2} \quad [\because A \text{ is +ve definitely} \Rightarrow |A| \geq 0]$$

$$r = \frac{\partial^2 f}{\partial x^2} = 2p \; ; \; s = \frac{\partial^2 f}{\partial x \partial y} = 2q$$

$$t = \frac{\partial^2 f}{\partial y^2} = 2r$$

$$rt - s^2 = 4(pr - q^2) > 0$$

$$r = 2p > 0 \quad [\because A \text{ is +ve definitely}]$$

$\therefore$ At $X = -\frac{A^{-1}B}{2}$

We get minimum of $f(x)$.

Similarly we can verify for $n > 2$ also.

1.21 (c)

$$[A] = \begin{bmatrix} 1 & 2 & 3 \\ 4 & 5 & 6 \\ 7 & 8 & 9 \end{bmatrix}\begin{bmatrix} 1 & 2 & 3 \\ 7 & 8 & 9 \\ 4 & 5 & 6 \end{bmatrix}^{-1} = \begin{bmatrix} -1 & 0 & 0 \\ 0 & 1 & 0 \\ 0 & 0 & 1 \end{bmatrix}$$

Eigen values

$$[A - \lambda I] = \begin{bmatrix} (-1-\lambda) & 0 & 0 \\ 0 & (1-\lambda) & 0 \\ 0 & 0 & (\lambda) \end{bmatrix} = 0$$

$$(-1-\lambda)\left[(1-\lambda)^2\right] = 0$$

$$-1 - \lambda = 0 \qquad (1-\lambda)^2 = 0$$

$$\boxed{\lambda = -1} \qquad \boxed{\lambda = 1, 1}$$

$$\therefore \quad (\lambda = -1, 1, 1)$$

1.22 (c)

Given, $\quad AX = 0$

$$\rho(A_{n \times n}) = r\,(0 < r < n)$$

$p =$ Number of independent solutions = nullity

We know that

$$\text{rank} + \text{nullity} = n$$

$$r + p = n$$

$$p = n - r$$

1.23 (3)

Equation of straight line,

$$y = mx + c$$

passing through (0, 0)

$$0 = 0 + c \Rightarrow c = 0$$

$$y = mx$$

Passing through (2, 6)

$$\therefore \quad 6 = 2m$$

$$\therefore \quad m = 3$$

1.24 (–6)

Eigen values of given matrix A are 1, –1, 3

Eigen values of A^3 are	1,	–1,	27
Eigen values of $3A^2$ are	3,	3,	27
Eigen values of $A^3 - 3A^2$ are	–2,	–4,	0

trace of $A^3 - 3A^2 = -2 - 4 + 0 = -6$

1.25 (c)

Characteristics equation is $|A - \lambda I| = 0$

$$\begin{vmatrix} 1-\lambda & -1 & 5 \\ 0 & 5-\lambda & 6 \\ 0 & -6 & 5-\lambda \end{vmatrix} = 0$$

$$(1 - \lambda)((5 - \lambda)^2 + 36) + 1(0 - 0) + 5(0 - 0) = 0$$

$$(1 - \lambda)(\lambda^2 - 10\lambda + 61) = 0$$

$$\lambda = 1,$$

$$\lambda = \frac{10 \pm \sqrt{100 - 244}}{2} = \frac{10 \pm 12i}{2} = 5 \pm 6i$$

$$\lambda = 1,\ 5 \pm 6i$$

1.26 (d)

Coordinate transformation matrix

$$= \begin{bmatrix} \cos\theta & \sin\theta \\ -\sin\theta & \cos\theta \end{bmatrix}$$

Orthogonal $\quad \therefore \ \theta = 90°$

Coordinate transformation matrix of mirror image

$$= \begin{bmatrix} \sin\theta & \cos\theta \\ \cos\theta & -\sin\theta \end{bmatrix}$$

$$= \begin{bmatrix} \sin 90° & \cos 90° \\ \cos 90° & -\sin 90° \end{bmatrix} = \begin{bmatrix} 1 & 0 \\ 0 & -1 \end{bmatrix}$$

1.27 (1)

Since V is non-zero vector of dimension 3×1

Therefore, $\quad \rho(A) \le \min\{\rho(V), \rho(V^T)\}$

$$\le \min\{1, 1\}$$

$$\le 1$$

Since V is non-zero. Hence $\rho(A) = 1$

1.28 (a)

$$N^2 = 0$$

Let Eigen values of N are $\lambda_1, \lambda_2, \lambda_3$;

Eigen values of N^2 are $\lambda^2_1, \lambda^2_2, \lambda^2_3$

But $\quad N^2 = 0$

$$\Rightarrow \quad \lambda_1^2 = 0,\ \lambda_2^2 = 0,\ \lambda_3^2 = 0$$

$\therefore$ Eigen values of N are 0, 0, 0.

1.29 (1)

$$(x - 2)^2 = 2x - 1$$

$$x^2 - 6x + 5 = 0$$

$$x = 1, 5$$

Smallest value of x is 1.

1.30 (c)

Put $k = -3$ in options,

$$3x + 2ky = -2 \Rightarrow 3x - 6y = -2$$

$$kx + 6y = 2 \Rightarrow 3x - 6y = -2$$

Both the equations are reducing in the same form, thus the system will have infinite number of solutions.

1.31 (10)

Product of eigen values $= |A|$

$$= (1)(2)(5) = 10$$

1.32 (b)

$$y = mx + c = \frac{dy}{dx}x + c$$

(given that Y intercept is $-y$)

$$y = \frac{dy}{dx}x - y$$

$$\partial y = \frac{dy}{dx}x$$

$$\frac{dy}{\partial y} = \frac{dx}{x}$$

$$\Rightarrow \quad \frac{1}{2}\ln y = \ln x + \ln C$$

$$\ln y^{1/2} = \ln(xC)$$

$$y = x^2C^2$$

$$\therefore \quad y \propto x^2$$

1.33 (d)

Matrix A has rank 2

Matrix A^T has rank 2

$\Rightarrow AA^T$ has rank 2

$\therefore \quad \left|AA^T{}_{4\times4}\right| = 0$

Hence system cannot have unique solution.

1.34 (b)

$\because \; C_1 + C_2 + C_3 = 0 \Rightarrow \lambda = 0$

We know that, $MX = \lambda X \quad \Rightarrow MX = 0$

$$\begin{bmatrix} 1 & -1 & 0 \\ 1 & -2 & 1 \\ 0 & -1 & 1 \end{bmatrix}\begin{bmatrix} x \\ y \\ z \end{bmatrix} = \begin{bmatrix} 0 \\ 0 \\ 0 \end{bmatrix}$$

Consider two equations

$x - y - 0\cdot z = 0$

$x - 2y + z = 0$

Solving $\quad \dfrac{x}{-1} = \dfrac{y}{-1} = \dfrac{z}{-1} = K$

$$\Rightarrow \quad \begin{bmatrix} x \\ y \\ z \end{bmatrix} = K_1\begin{bmatrix} 1 \\ 1 \\ 1 \end{bmatrix}$$

Hence one of the eigen vector is $\begin{bmatrix} 1 \\ 1 \\ 1 \end{bmatrix}$

1.35 (4)

Characteristic Eq. of A is $|A - \lambda I| = 0$

$\Rightarrow \quad (\lambda - 2)(\lambda - 3) = 0 \quad \Rightarrow \quad \lambda^2 - 5\lambda + 6 = 0$

By C.H theorem replace $\lambda \to A$ in characteristic equation

$A^2 - 5A + 6I = 0 \qquad \ldots(1)$

Now,

$|A^4 - 5A^3 + 6A^2 + 2I| = |A^2(A^2 - 5A + 6I) + 2I|$

$= |0 + 2I| = |2I| = 2^2|I| = 4|I| = 4 \times 1 = 4$

1.36 (d)

$$A = 2\begin{bmatrix} 1 \\ 0 \end{bmatrix}[1 \;\; 0] + 3\begin{bmatrix} 2 \\ 0 \end{bmatrix}[2 \;\; 0]$$

$$= 2\begin{bmatrix} 1 & 0 \\ 0 & 0 \end{bmatrix} + 3\begin{bmatrix} 4 & 0 \\ 0 & 0 \end{bmatrix}$$

$$= \begin{bmatrix} 2 & 0 \\ 0 & 0 \end{bmatrix} + \begin{bmatrix} 12 & 0 \\ 0 & 0 \end{bmatrix} = \begin{bmatrix} 14 & 0 \\ 0 & 0 \end{bmatrix}$$

$\rho(A) = 1$

1.37 (4)

Expanding along R_1 (Row 1)

$$|A| = +(1)\begin{vmatrix} 4 & 0 & 0 \\ 0 & 4 & 3 \\ 0 & 2 & 1 \end{vmatrix} - (2)\begin{vmatrix} 3 & 0 & 0 \\ 0 & 4 & 3 \\ 0 & 2 & 1 \end{vmatrix}$$

$= (1)[4(4 - 6)] - 2[3(4 - 6)] = 4$

1.38 (a, d)

Given matrix, $\quad M = \begin{bmatrix} 2 & 3 & 7 \\ 6 & 4 & 7 \\ 4 & 6 & 14 \end{bmatrix}$

Applying $R_3 \to R_3 - 2R_1$

$$M = \begin{bmatrix} 2 & 3 & 7 \\ 6 & 4 & 7 \\ 0 & 0 & 0 \end{bmatrix}$$

Rank of matrix is 2.

Rank of matrix is less than 3.

Hence, determinant is zero.

1.39 (a, c)

Eigen values given,

$\lambda = -5, 7$

Now for eigen values,

$\lambda_1 = -5$

$[A - \lambda_1 I][x] = 0$

$$\begin{bmatrix} 4-(-5) & 3 \\ 9 & -2-(-5) \end{bmatrix}\begin{bmatrix} x_1 \\ x_2 \end{bmatrix} = \begin{bmatrix} 0 \\ 0 \end{bmatrix}$$

$$\begin{bmatrix} 9 & 3 \\ 9 & 3 \end{bmatrix}\begin{bmatrix} x_1 \\ x_2 \end{bmatrix} = \begin{bmatrix} 0 \\ 0 \end{bmatrix}$$

$9x_1 + 3x_2 = 0$

$x_2 = -3x_1$

Let $x_1 = K$, then$x_2 = -3K$

So, eigen vector for $\lambda_1 = 5$,

$$= K\begin{bmatrix} 1 \\ -3 \end{bmatrix}$$

For eigen value $\lambda_2 = 7$

$$\begin{bmatrix} 4-7 & 3 \\ 9 & -2-7 \end{bmatrix}\begin{bmatrix} x_1 \\ x_2 \end{bmatrix} = \begin{bmatrix} 0 \\ 0 \end{bmatrix}$$

$$\begin{bmatrix} -3 & 3 \\ 9 & -9 \end{bmatrix}\begin{bmatrix} x_1 \\ x_2 \end{bmatrix} = \begin{bmatrix} 0 \\ 0 \end{bmatrix}$$

$-3x_1 + 3x_2 = 0$

$x_1 = x_2$

So, eigen vector is $K\begin{bmatrix} 1 \\ 1 \end{bmatrix}$.

1.40 (a)

Given :$x - 2y + z = 0$

$x - z = 0$

Solution is given by solving also equations.

	x	y	z	
−2	1	1	−2	
0	−1	1	0	

$$\frac{x}{2} = \frac{y}{2} = \frac{z}{2} = K$$

$\Rightarrow$ $x = K,\ y = K,\ z = K$

$$\therefore \quad \begin{bmatrix} x \\ y \\ z \end{bmatrix} = K\begin{bmatrix} 1 \\ 1 \\ 1 \end{bmatrix},\ K \in R$$

1.41 (−8)

Given : $\rho(A_{3\times 2}) = 1$

So, $C_2 = KC_1$

$\Rightarrow$ $4 = K(1)$

$\Rightarrow$ $K = 4$

Also, $\alpha = -3K$ and $6 = K\beta$

$$\therefore \quad \frac{\alpha}{\beta} = -\frac{1}{2}K^2 = -\frac{1}{2}\times 4^2 = -8$$

1.42 (c)

We have,

$$M = [v_1 \ \ v_2 \ \ v_3]$$

By options, when we put,

$$v_1 = \begin{bmatrix} 1 \\ 0 \\ 1 \end{bmatrix} \quad v_2 = \begin{bmatrix} -1 \\ 0 \\ 1 \end{bmatrix} \quad v_3 = \begin{bmatrix} 1 \\ -1 \\ 1 \end{bmatrix}$$

$$|M| = \begin{vmatrix} 1 & -1 & 1 \\ 0 & 0 & -1 \\ 1 & 1 & 1 \end{vmatrix}$$

$$|M| = 1(0 + 1) + 1(1) + 1(0)$$

$$|M| = 2$$

Though $|M| \neq 2$. Hence, Rank of $M = 3$.

Hence, option (c) is correct.

1.43 (−2)

Matrix P has eigen value

$$\lambda = \frac{1}{4}, 1, -2$$

$|P^{-1}|$

Since $= |P| =$ Product of eigen values

$$= \frac{1}{4}\times 1\times(-2) = \frac{-1}{2}$$

$$|P^{-1}) = \frac{1}{|P|} = -2$$

■■■

2 Differential Equations

2.1 The general solution of the differential equation $(D^2 - 4D + 4)y = 0$, is of the form (given $D = d/dx$), and C_1 and C_2 are constants

(a) C_1e^{2x} (b) $C_1e^{2x} + C_2e^{-2x}$

(c) $C_1e^{2x} + C_2xe^{2x}$ (d) $C_1e^{2x} + C_2xe^{-2x}$

[2005 : 2 M]

2.2 The curves, for which the curvature ρ at any point is equal to $\cos^3\theta$, where θ is the angle made by the tangent at that point with the positive direction of the x-axis, are (given $\rho = y''/[1 + (y')^2]^{3/2}$, where y' and y'' are the first and second derivatives of y with respect to x)

(a) circles (b) parabolas

(c) ellipses (d) hyperbolas

[2005 : 2 M]

2.3 The solution of the integral equation

$$y(t) = t\exp(t) - 2\exp(t)\int_0^t \exp(-\tau)\ y(\tau)\,d\tau \text{ is}$$

(a) $\frac{1}{2}(\exp(t) - \exp(-t))$

(b) $\frac{1}{2}(\exp(t) + \exp(-t))$

(c) $\frac{(\exp(t) + \exp(-t))}{(\exp(t) + \exp(-t))}$

(d) $\frac{(\exp(t) + \exp(t))}{(\exp(-t) - \exp(t))}$

[2006 : 2 M]

2.4 For a initial value problem

$$\ddot{y} + 2\dot{y} + 101y = 10.4e^x,\ y(0) = 1.1$$

and $\dot{y}(0) = -0.9$.

Various solutions are written in the following groups. Match the type of solution with the correct expression.

Group-1

P. General solution of homogeneous equations

Q. Particular integral

R. Total solution satisfying boundary conditions

Group-2

1. $0.1\,e^x$
2. $e^{-x}(A\cos 10x + B\sin 10x)$
3. $e^{-x}\cos 10x + 0.1e^x$

(a) P-2, Q-1, R-3 (b) P-1, Q-3, R-2

(c) P-1, Q-2, R-3 (d) P-3, Q-2, R-1

[2006 : 2 M]

2.5 The boundary-value problem $y'' + \lambda y = 0$, $y(0) = y(\pi) = 0$ will have non-zero solutions if and only if the values of λ are

(a) 0, ± 1, ± 2,........ (b) 1, 2, 3,........

(c) 1, 4, 9,........ (d) 1, 9, 25,........

[2007 : 2 M]

2.6 Consider the differential equation $\frac{dy}{dx} = 1 + y^2$.

Which one of the following can be a particular solution of this differential equation?

(a) $y = \tan(x + 3)$ (b) $y = \tan x + 3$

(c) $x = \tan(y + 3)$ (d) $x = \tan y + 3$

[2008 : 2 M]

2.7 Consider the differential equation $\ddot{y} + 2\dot{y} + y = 0$ with boundary conditions $y(0) = 1$, $y(1) = 0$. The value of $y(2)$ is

(a) –1 (b) $-e^{-1}$

(c) $-e^{-3}$ (d) $-e^{-2}$

[2011 : 2 M]

2.8 With initial condition $x(1) = 0.5$, the solution of the differential equation $t\frac{dx}{dt} + x = t$ is

(a) $x = t - \frac{1}{2}$ (b) $x = t^2 - \frac{1}{2}$

(c) $x = \frac{t^2}{2}$ (d) $x = \frac{t}{2}$

[2012 : 1 M]

2.9 Consider the differential equation,

$$\frac{d^2 y(t)}{dt^2} + 2\frac{dy(t)}{dt} + y(t) = \delta(t) \text{ with}$$

$$y(t)|_{t=0^-} = -2 \text{ and } \left.\frac{dy}{dt}\right|_{t=0^-} = 0.$$

The numerical value of $\left.\frac{dy}{dt}\right|_{t=0^+}$ is

(a) –2 (b) –1
(c) 0 (d) 1

[2012 : 2 M]

2.10 The type of the partial differential equation

$\frac{\partial f}{\partial t} = \frac{\partial^2 f}{\partial x^2}$ is

(a) Parabolic (b) Elliptic
(c) Hyperbolic (d) Non-linear

[2013 : 1 M]

2.11 The maximum value of the solution $y(t)$ of the differential equation $y(t) + \ddot{y}(t) = 0$ with initial conditions $\dot{y}(0) = 1$ and $y(0) = 1$, for $t \geq 0$ is

(a) 1 (b) 2
(c) π (d) $\sqrt{2}$

[2013 : 2 M]

2.12 The figure shows the plot of y as a function of x

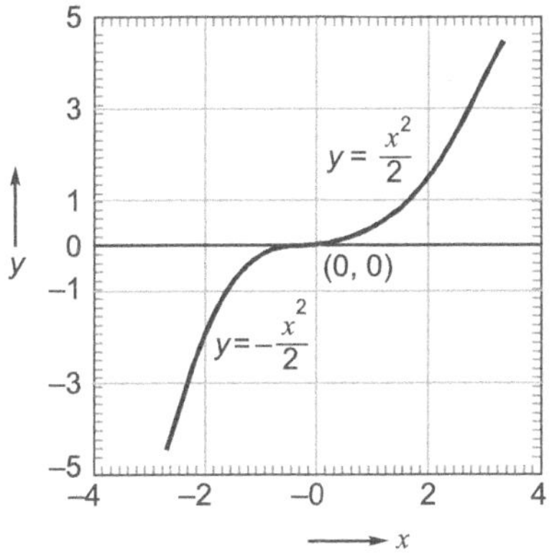

The function shown is the solution of the differential equation (assuming all initial conditions to be zero) is

(a) $\frac{d^2y}{dx^2} = 1$ (b) $\frac{dy}{dx} = x$
(c) $\frac{dy}{dx} = -x$ (d) $\frac{dy}{dx} = |x|$

[2014 : 1 M]

2.13 Consider the differential equation $\frac{dx}{dt} = \sin(x)$, with the initial condition $x(0) = 0$. The solution to this ordinary differential equation is _____.

(a) $x(t) = 0$
(b) $x(t) = \sin(t)$
(c) $x(t) = \sin(t) - \cos(t)$
(d) $x(t) = \cos(t)$ [2020 : 2 M]

2.14 Consider the differential equation

$$\frac{dy}{dx} + y\ln(y) = 0$$

If $y(0) = e$, then $y(1)$ is _______.

(a) e^e (b) e^{-e}
(c) $e^{(1/e)}$ (d) $e^{(-1/e)}$ [2022 : 2 M]

2.15 The solution $x(t)$, $t \geq 0$, to the differential equation $\ddot{x} = -k\dot{x}, k > 0$ with initial conditions $x(0) = 1$ and $\dot{x}(0) = 0$ is

(a) $x(t) = 2e^{-kt} + 2kt - 1$
(b) $x(t) = 2e^{-kt} + 1$
(c) $x(t) = 1$
(d) $x(t) = 2e^{-kt} - kt - 1$ [2023 : 1 M]

2.16 The solution of an ordinary differential equation $y''' + 3y'' + 3y' + y = 30e^{-t}$ is
$y(t) = (c_0 + c_1 t - c_2 t^2 + c_3 t^3)e^{-t}$
Given $y(0) = 3$, $y'(0) = -3$ and $y''(0) = -47$, the value of $(c_0 + c_1 + c_2 + c_3)$ is _______ (rounded off to nearest integer)
Note: $y''' = d^3y/dt^3$, $y'' = d^2y/dt^2$, $y' = dy/dt$ and c_0, c_1, c_2, c_3 are constants.

[2024 : 2 M]

Answers Differential Equations

2.1 (c)	2.2 (b)	2.3 (b)	2.4 (a)	2.5 (c)	2.6 (a)	2.7 (d)
2.8 (d)	2.9 (d)	2.10 (a)	2.11 (d)	2.12 (d)	2.13 (a)	2.14 (c)
2.15 (c)	2.16 (30)					

Explanations Differential Equations

2.1 (c)

$$D^2 - 4D + 4 = 0$$
$$(D-2)^2 = 0$$
$$D = 2, 2$$
$$\therefore \quad y = (C_1 + C_2 x)e^{2x} = C_1 e^{2x} + C_2 x e^{2x}$$

2.2 (b)

$$\rho = \frac{y''}{\left[1+(y')^2\right]^{3/2}}$$

We know, $\frac{dy}{dx} = y' = \tan\theta$

$$\rho = \frac{y''}{(1+\tan^2\theta)^{3/2}} = \frac{y''}{\sec^3\theta}$$
$$\cos^3\theta = \cos^3\theta\, y''$$
$$\Rightarrow \quad \frac{d^2y}{dx^2} = 1 \Rightarrow \frac{dy}{dx} = x + c$$
$$\Rightarrow \quad y = Ax^2 + cx + B$$

Hence the curve is parabola.

2.3 (b)

$$y(t) = t\cdot e^t - 2e^t\int_0^t e^{-\tau}y(\tau)d\tau \quad ...(i)$$

Differentiating the above equation

$$\frac{dy(t)}{dt} = t\cdot e^t + e^t - [2e^t\cdot e^{-t}\cdot y(t) + 2e^t\int_0^t e^{-\tau}y(\tau)dt]$$

From equation (i)

$$\frac{dy(t)}{dt} = te^t + e^t - 2y(t) + y(t) - te^t$$
$$\Rightarrow \quad \frac{dy(t)}{dt} + y(t) = e^t$$

This is Leibnitz linear equation

$\therefore$ Integrating Factor I.F. $= e^{\int dt} = e^t$

and the solution is

$$y(t)\cdot e^t = \int e^t\cdot e^t dt + c$$
$$y(e^t) = \frac{e^{2t}}{2} + c$$
$$y = \frac{e^t}{2} + ce^{-t} = \frac{1}{2}\left[e^t + e^{-t}\right] \text{ for } c = \frac{1}{2}$$

2.4 (a)

Auxiliary equation is

$$D^2 + 2D + 101 = 0$$
$$\Rightarrow \quad D = \frac{-2\pm\sqrt{4-404}}{2} = -1\pm j10$$
$$\therefore \quad \text{C.F.} = e^{-x}(C_1\cos 10x + C_2\sin 10x)$$
$$\text{P.I.} = \frac{1}{D^2+2D+101}10.4e^x$$
$$= \frac{1}{1+2+101}\times 10.4e^x = 0.1\, e^x$$

Complete solution is

$$y = e^{-x}(C_1\cos 10x + C_2\sin 10x) + 0.1e^x \quad ...(i)$$

Given, $y(0) = 1.1$

$$1.1 = e^0(C_1\cos 0^\circ + C_2\sin 0^\circ) + 0.1e$$
$$1.1 = C_1 + 0.1$$
$$\Rightarrow \quad C_1 = 1$$

From equation (i),

$$\dot{y} = e^{-x}(-10C_1\sin 10x + 10C_2\cos 10x) - e^{-x}(C_1\cos 10x + C_2\sin 10x) + 0.1e^x$$
$$\dot{y}(0) = -0.9 = 10C_2 - C_1 + 0.1$$
$$\Rightarrow \quad C_2 = 0$$

2.5 (c)

$$y'' + \lambda y = 0$$

Auxiliary equation, $(D^2 + \lambda) = 0$

$$D = \pm j\sqrt{\lambda}$$

solution is $y = A\cos\sqrt{\lambda}\cdot x + B\sin\sqrt{\lambda}\cdot x$

given, $y(0) = 0 \quad \Rightarrow A = 0$

and $y(\pi) = 0$

$$0 = B\sin\sqrt{\lambda}\cdot\pi$$

$\because B \neq 0$, therefore $\sqrt{\lambda}\cdot\pi = \pm n\pi$, where n is integer

$$\sqrt{\lambda} = \pm n$$
$$\lambda = n^2 = 1, 4, 9.......$$

2.6 (a)

$$\frac{dy}{dx} = (1+y)^2 \Rightarrow \frac{dy}{1+y^2} = dx$$

On integrating both sides

$$\int\frac{dy}{1+y^2} = \int dx$$
$$\tan^{-1}y = x + c$$
$$y = \tan(x + c)$$

2.7 (d)

$$\ddot{y} + 2\dot{y} + y = 0$$

Auxiliary equation is

$$D^2 + 2D + 1 = 0$$

$\therefore \quad D = -1, -1$

Here both roots are common

$\therefore$ complete solution is

$$y = (c_1 x + c_2)e^{-x}$$

given, $y(0) = 1$

$\therefore \quad 1 = c_2 \cdot e^0$

$\Rightarrow \quad c_2 = 1$

and $y(1) = 0$

$\therefore \quad 0 = (c_1 + 1)e^{-1}$

$\Rightarrow \quad c_1 = -1$

$\therefore$ Complete solution is

$$y = (1 - x)e^{-x}$$

$\therefore \quad y(2) = (1 - 2)e^{-2} = -e^{-2}$

2.8 (d)

Given, differential equation is

$$t\frac{dx}{dt} + x = t \Rightarrow \frac{dx}{dt} + \left(\frac{1}{t}\right)x = 1$$

The above equation is a linear differential equation

$$\therefore \quad \text{I.F} = e^{\int \frac{1}{t}dt = e^{\ln t}} = t$$

Solution is

$$x(\text{I.F}) = \int 1 \cdot (I.F) \cdot dt$$

$$xt = \int t\,dt \quad ; \quad xt = \frac{t^2}{2} + c$$

At $t = 1, x = 0.5$

$$\frac{1}{2} = \frac{1}{2} + c \Rightarrow c = 0$$

$$\therefore \quad x = \frac{t}{2}$$

2.9 (d)

$$\frac{d^2y(t)}{dt^2} + 2\frac{dy(t)}{dt} + y(t) = \delta(t)$$

$\Rightarrow s^2 Y(s) - sy(0) - y'(0) + 2[sY(s) - y(0)] + Y(s) = 1$

$\Rightarrow Y(s)[s^2 + 2s + 1] - (s \times -2) - y'(0) - (2 \times -2) = 1$

$\Rightarrow Y(s)[s^2 + 2s + 1] = -2s - 3$

$$\Rightarrow \quad Y(s) = \frac{-2s - 3}{s^2 + 2s + 1} = \frac{-2}{s+1} - \frac{1}{(s+1)^2}$$

$$\Rightarrow \quad y(t) = -2e^{-t} - te^{-t}$$

$$\frac{dy(t)}{dt} = 2e^{-t} - (-t\,e^{-t} + e^{-t}) = e^{-t} + te^{-t}$$

$$\left.\frac{dy(t)}{dt}\right|_{t=0^+} = 1 + 0 = 1$$

2.11 (d)

$$y(t) + \ddot{y}(t) = 0$$

$1 + D^2 = 0$

$\therefore \quad D = \pm i$

$\therefore \quad y = C_1 e^{ix} + C_2 e^{-ix} = A\cos x + B\sin x$

$y(0) = 1$

$\therefore \quad 1 = A \times 1 + B \times 0$

$A = 1$

$\dot{y} = -A\sin x + B\cos x$

$\dot{y}(0) = 1$

$\therefore \quad 1 = -A \times 0 + B \times 1$

$\therefore \quad B = 1$

So, $y = \cos x + \sin x$

for maxima,

$y' = -\sin x + \cos x = 0$

$\therefore \sin x = \cos x$

$\therefore \quad x = 45°$

$y'' = -\cos x - \sin x$

$\therefore \quad y'' < 0$ for $x = 45°$ $\therefore$ maxima

$y(\text{max}) = \cos 45° + \sin 45°$

$$= \frac{1}{\sqrt{2}} + \frac{1}{\sqrt{2}} = \frac{2}{\sqrt{2}} = \sqrt{2}$$

2.12 (d)

$$\frac{dy}{dx} = x \text{ for } x \geq 0 = -x \text{ for } x < 0$$

Integrating both sides, we get

$$\int \frac{dy}{dx} = \int x\,dx \text{ for } x \geq 0 = \int -x\,dx \text{ for } x < 0$$

$$\Rightarrow \quad y = \frac{x^2}{2} \text{ for } x \geq 0 = -\frac{x^2}{2} \text{ for } x < 0$$

Alternate solution:

$$\text{From plot,} \quad y = \begin{cases} -\dfrac{x^2}{2} & x < 0 \\ \dfrac{x^2}{2} & x \geq 0 \end{cases}$$

So, for $x < 0$,

$$y = -\frac{x^2}{2}$$

$$\Rightarrow \quad \frac{dy}{dx} = -\frac{2x}{2} = -x \quad \text{...(i)}$$

For $x \geq 0$,

$$y = \frac{x^2}{2}$$

$$\Rightarrow \quad \frac{dy}{dx} = \frac{2x}{2} = x \quad \text{...(ii)}$$

From equations (i) and (ii)

$$\frac{dy}{dx} = \begin{cases} -x & x < 0 \\ x & x \geq 0 \end{cases}$$

$$\Rightarrow \quad \frac{dy}{dx} = |x|$$

2.13 (a)

$$\frac{dx}{dt} = \sin(x)$$

Using Variable Separable Method:

$$\int \frac{dx}{\sin x} = \int dt$$

$$\log[\text{cosec } x - \cot x] = t + c$$

$$\text{cosec } x - \cot x = e^{t+c}$$

$$\frac{1-\cos x}{\sin x} = e^{t+c}$$

$$\tan\frac{x}{2} = e^{t+c}$$

$$\frac{x}{2} = \tan^{-1}[e^{t+c}]$$

Put $t = 0$, $x = 0$

$$0 = \tan^{-1}[e^{0+c}]$$

$$\Rightarrow \quad c = -\infty$$

$\therefore$ solution is $\frac{x}{2} = \tan^{-1}[e^{t-\infty}]$

but as $t \geq 0$ and finite

We get, $\frac{x}{2} = \tan^{-1}[e^{-\infty}]$

$$\frac{x}{2} = 0 \Rightarrow x(t) = 0$$

2.14 (c)

Given differential equation :

$$\frac{dy}{dx} + y\ln y = 0$$

$$\frac{dy}{dx} = -y \ln y$$

$$\frac{dy}{y\ln y} = -dx$$

Integrating both sides

$$\int \frac{dy}{y\ln y} = -\int dx$$

$$\ln|\ln y| = -x + C$$

$$\ln y = Ke^{-x} \qquad [\text{Let } K = e^C]$$

Given : $y(0) = e^1$

So, $\ln(e^1) = Ke^{-0}$

$$K = 1$$

So, $\ln y = e^{-x}$

For $x = 1$,

$$\ln y(1) = e^{-1} = \frac{1}{e}$$

$$y(1) = e^{1/e}$$

2.15 (c)

Differential equation : $\frac{d^2x}{dt^2} + k\frac{dx}{dt} = 0,\ k > 0$

$(D_2 + KD)\,x = 0$ is a homogeneous linear differential equation

Solution is $x = x_{CF}$

given by $f(m) = m^2 + km = 0$

$$\Rightarrow \quad m = 0, -k$$

$$\therefore \quad x = x_{CF} = C_1e^{0t} + C_2e^{-kt}$$

Put $t = 0$, $x = 1$

We get $C_1 + C_2 = 1$

$$\frac{dx}{dt} = -kC_2e^{-kt}$$

Put $t = 0$, $\frac{dx}{dt} = 0$

We get $C_2 = 0 \Rightarrow C_1 = 1$

2.16 (30)

$$y''' + 3y'' + 3y' + y = 30e^{-t}$$

Solution is given

$$y(t) = (c_0 - c_1t - c_2t^2 + c_3t^3)e^{-t}$$

Put $t = 0$, $y = 3 \Rightarrow 3 = c_0$

$$y'(t) = (-c_1, -2c_2t + 3c_3t^2)e^{-t} - y(t)$$

Put $t = 0$, $y' = -3$

$$3 = -c, -3 \Rightarrow c_1 = 0$$

$$y'' = (-2c_2 + 6c_3t)e^{-t} - y'(t)$$
$$- y'(t) + y(t)$$

Put $t = 0$, $y' = -3$, $y = 3$, $y'' = -47$

$$-47 = -2c_2 - 6 + 3$$

$$c_2 = \frac{44}{2} = 22$$

$$y_{PI} = \frac{30e^{-t}}{(D+1)^3} = 3$$

$$= 30t^3 \frac{e^{-t}}{6} = 5t^3e^{-t}$$

$$c_3 = 5$$

$$\therefore c_0 + c_1 + c_2 + c_3 = 0 + 3 + 22 + 5 = 30$$

■■■

3 Probability & Statistics

3.1 The probability that there are 53 Sundays in a randomly chosen leap year is

(a) $\frac{1}{7}$ (b) $\frac{1}{14}$

(c) $\frac{1}{28}$ (d) $\frac{2}{7}$

[2005 : 1 M]

3.2 Probability density function $p(x)$ of a random variable x is as shown below. The value of α is

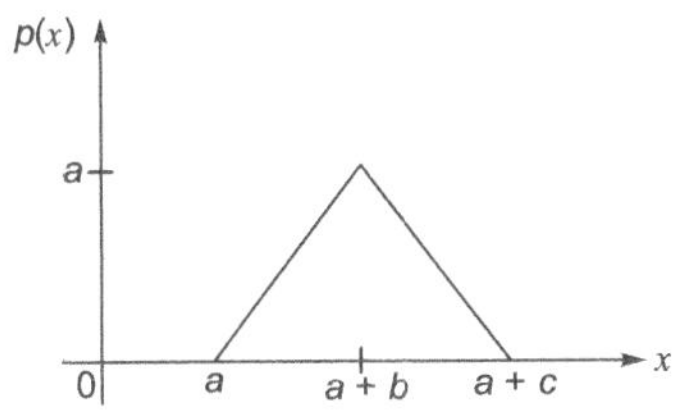

(a) $\frac{2}{c}$ (b) $\frac{1}{c}$

(c) $\frac{2}{(b+c)}$ (d) $\frac{1}{(b+c)}$

[2005 : 1 M]

3.3 Which of the following integrals provides a measure of the rapidity of change in a random variable $f(t)$?

(a) $\lim_{T\to\infty} \frac{1}{2T} \int_{-T}^{+T} f(t)\, dt$

(b) $\lim_{T\to\infty} \frac{1}{2T} \int_{-T}^{+T} f^2(t)\, dt$

(c) $\left[\lim_{T\to\infty} \frac{1}{2T} \int_{-T}^{+T} f^2(t)\, dt\right]^{1/2}$

(d) $\lim_{T\to\infty} \frac{1}{2T} \int_{-T}^{+T} f(t)\, f(t+\tau)\, dt$

[2006 : 2 M]

3.4 You have gone to a cyber-cafe with a friend. You found that the cyber-cafe has only three terminals. All terminals are unoccupied. You and your friend have to make a random choice of selecting a terminal. What is the probability that both of you will NOT select the same terminal?

(a) 1/9 (b) 1/3

(c) 2/3 (d) 1

[2006 : 1 M]

3.5 Two dices are rolled simultaneously. The probability that the sum of digits on the top surface of the two dices is even, is

(a) 0.5 (b) 0.25

(c) 0.167 (d) 0.125

[2006 : 2 M]

3.6 Assume that the duration in minutes of a telephone conversion follows the expo-nential distribution $f(x) = \frac{1}{5} e^{-x/5}, x \geq 0$. The probability that the conversion will exceed five minutes is

(a) $\frac{1}{e}$ (b) $1 - \frac{1}{e}$

(c) $\frac{1}{e^2}$ (d) $1 - \frac{1}{e^2}$

[2007 : 2 M]

3.7 Consider a Gaussian distributed random variable with zero mean and standard deviation σ. The value of its cumulative distribution function at the origin will be

(a) 0 (b) 0.5

(c) 1 (d) 10σ

[2008 : 2 M]

3.8 A random variable is uniformly distributed over the interval 2 to 10. Its variance will be

(a) $\frac{16}{3}$ (b) 6

(c) $\frac{256}{9}$ (d) 36

[2008 : 2 M]

3.9 A screening test is carried out to detect a certain disease. It is found that 12% of the positive reports and 15% of the negative reports are incorrect. Assuming that the probability of a person getting a positive report is 0.01, the probability that a person tested gets in incorrect report is

(a) 0.0027 (b) 0.0173

(c) 0.1497 (d) 0.2100

[2009 : 2 M]

3.10 The diameters of 10000 ball bearings were measured. The mean diameter and standard deviation were found to be 10 mm and 0.05 mm respectively. Assuming Gaussian distribution of measurements, it can be expected that the number of measurement more than 10.15 mm will be

(a) 230 (b) 115
(c) 15 (d) 2 [2010 : 1 M]

3.11 The box 1 contains chips numbered 3, 6, 9, 12 and 15. The box 2 contains chips numbered 6, 11, 16, 21 and 26. Two chips, one from each box, are drawn at random. The numbers written on these chips are multiplied. The probability for the product to be an even number is

(a) $\frac{6}{25}$ (b) $\frac{2}{5}$
(c) $\frac{3}{5}$ (d) $\frac{19}{25}$

[2011 : 2 M]

3.12 A fair coin is tossed till a head appears for the first time. The probability that the number of required tosses are odd, are

(a) 1/3 (b) 1/2
(c) 2/3 (d) 3/4

[2012 : 2 M]

3.13 A continuous random variable X has a probability density $f(x) = e^{-x}$, $0 < x < \infty$. Then $P\{X > 1\}$ is

(a) 0.368 (b) 0.5
(c) 0.632 (d) 1.0

[2013 : 1 M]

3.14 Given that x is a random variable in the range $[0, \infty]$ with a probability density function $\frac{e^{-\frac{x}{2}}}{K}$, the value of the constant K is _____.

[2014 : 1 M]

3.15 The figure shows the schematic of a production process with machines A, B and C. An input job needs to be pre-processed either by A or by B before it is fed to C, from which the final finished product comes out. The probabilities of failure of the machines are given as

$$P_A = 0.15, \quad P_B = 0.05, \quad P_C = 0.1.$$

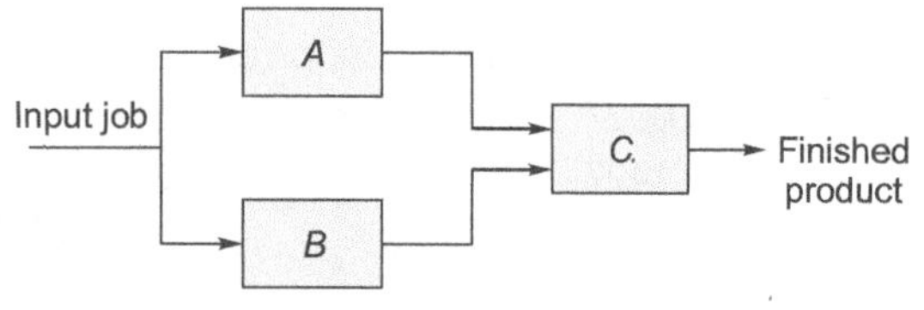

Assuming independence of failures of the machines, the probability that a given job is successfully processed (up to the third decimal place) is _____.

[2014 : 1 M]

3.16 The probability that a thermistor randomly picked up from a production unit is defective is 0.1. The probability that out of 10 thermistors randomly picked up, 3 are defective is

(a) 0.001 (b) 0.057
(c) 0.107 (d) 0.3

[2015 : 2 M]

3.17 An urn contains 5 red and 7 green balls. A ball is drawn at random and its colour is noted. The ball is placed back into the urn along with another ball of the same colour. The probability of getting a red ball in the next draw is

(a) $\frac{65}{156}$ (b) $\frac{67}{156}$
(c) $\frac{79}{156}$ (d) $\frac{89}{156}$

[2016 : 2 M]

3.18 The probability that a communication system will have high fidelity is 0.81. The probability that the system will have both high fidelity and high selectivity is 0.18. The probability that a given system with high fidelity will have high selectivity is

(a) 0.181 (b) 0.191
(c) 0.222 (d) 0.826

[2017 : 2 M]

3.19 Considers a sequence of tossing of a fair coin where the outcomes of tosses are independent. The probability of getting the head for the third time in the fifth toss is

(a) $\frac{5}{16}$ (b) $\frac{3}{16}$
(c) $\frac{3}{5}$ (d) $\frac{9}{16}$

[2018 : 1 M]

3.20 Two bags A and B have equal number of balls. Bag A has 20% red balls and 80% green balls. Bag B has 30% red balls. 60% green balls and 10% yellow balls. Contents of Bags A and B are mixed thoroughly and a ball is randomly picked from the mixture. What is the chance that the ball picked is red?

(a) 20% (b) 25%
(c) 30% (d) 40%

[2018 : 2 M]

3.21 A box has 8 red balls and 8 green balls. Two balls are drawn randomly in succession from the box *without replacement*. The probability that the first ball drawn is red *and* the second ball drawn is green is

(a) $\frac{4}{15}$ (b) $\frac{7}{16}$

(c) $\frac{1}{2}$ (d) $\frac{8}{15}$ [2019 : 1 M]

3.22 The function $p(x)$ is given by $p(x) = \frac{A}{x^{\mu}}$, where A and μ are constants with $\mu > 1$ and $1 \le x < \infty$ and $p(x) = 0$ for $-\infty < x < 1$. For $p(x)$ to be a probability density function, the value of A should be equal to

(a) $\mu - 1$ (b) $\mu + 1$

(c) $\frac{1}{(\mu - 1)}$ (d) $\frac{1}{(\mu + 1)}$

[2019 : 2 M]

3.23 A player throws a ball at a basket kept at distance. The probability that the ball falls into the basket in a single attempt is 0.1. The player attempts to throw the ball twice. Considering each attempt to be independent, the probability that this player puts the ball into the basket only in the second attempt (rounded of to two decimal places) is ____.

[2020 : 1 M]

3.24 Consider two identical bags B_1 and B_2 each containing 10 balls of identical shapes and sizes. Bag B_1 contains 7 Red and 3 Green balls, while bag B_2 contains 3 Red and 7 Green balls. A bag is picked at random and a ball is drawn from it, which was found to be Red. The probability that the Red ball came from bag B_1 (rounded off to one decimal place) is ____.

[2020 : 2 M]

3.25 Consider that X and Y are independent continuous valued random variables with uniform PDF given by $X \sim U(2, 3)$ and $Y \sim U(1, 4)$. Then $P(Y \le X)$ is equal to ____ (rounded off to two decimal places).

[2021 : 2 M]

3.26 A bag contains six red balls and four blue balls. If three balls are drawn in succession without replacement, the probability that the second and third balls drawn are red is ________ (round off to two decimal places).

[2022 : 1 M]

3.27 X is a discrete random variable which takes values 0, 1 and 2. The probabilities are $P(X = 0) = 0.25$ and $P(X = 1) = 0.5$. With $E[\ .\ ]$ denoting the expectation operator, the value of $E[X] - [X^2]$ is _______ (rounded off to one decimal place).

[2023 : 1 M]

3.28 How many five-digit numbers can be formed using the integers 3, 4, 5 and 6 with exactly one digit appearing twice?

[2023 : 2 M]

3.29 Indian Premier League has divided the sixteen cricket teams into two equal pools: Pool-A and Pool-B. Four teams of Pool-A have blue logo jerseys while the rest four have red logo jerseys. Five teams of Pool-B have blue logo jerseys while the rest three have red logo jerseys.
If one team from each pool reaches the final, the probability that one team has a blue logo jersey and another has a red logo jersey is ________ (rounded off to one decimal place).

[2024 : 1 M]

3.30 A random variable X has a probability density function

$$f_X(x) = \begin{cases} e^{-x}, & x \ge 0 \\ 0, & \text{otherwise} \end{cases}$$

The probability of $X > 2$ is __________ (rounded off to three decimal places).

[2024 : 2 M]

■■■

Answers Probability & Statistics

3.1 (d)	3.2 (a)	3.3 (c)	3.4 (c)	3.5 (a)	3.6 (a)	3.7 (b)
3.8 (a)	3.9 (c)	3.10 (c)	3.11 (d)	3.12 (c)	3.13 (a)	3.14 (2)
3.15 (0.893)	3.16 (b)	3.17 (a)	3.18 (c)	3.19 (b)	3.20 (b)	3.21 (a)
3.22 (a)	3.23 (0.09)	3.24 (0.7)	3.25 (0.5)	3.26 (0.33)	3.27 (–0.5)	3.28 (240)
3.29 (0.5)	3.30 (0.135)					

Explanations Probability & Statistics

3.1 (d)

Number of days in a leap year

= 366 days

Number of weeks in a leap year

$$= \frac{366}{7} = 52\,\text{weeks} + 2\text{ days}$$

i.e. in any leap year there are minimum 52 Sundays probability that out of two days there is one Sunday $= \frac{2}{7}$.

3.2 (a)

We know area under the probability density curve = 1.

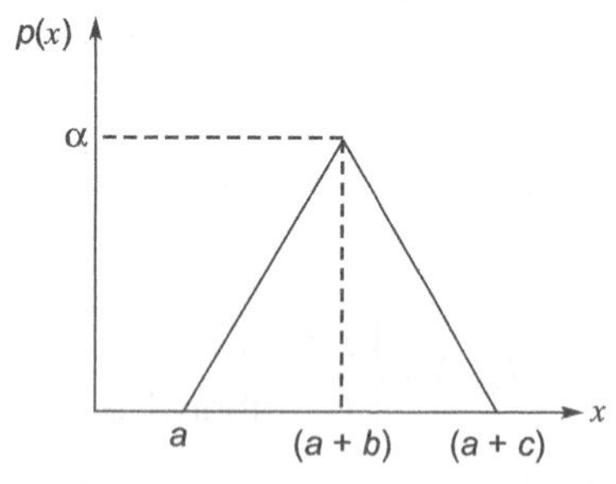

$$\therefore \quad \text{Area} = \frac{1}{2}\alpha[a+c-a] = \frac{1}{2}\alpha c$$

$$\because \quad \frac{1}{2}\alpha c = 1$$

$$\therefore \quad \alpha = \frac{2}{c}$$

3.3 (c)

Standard deviation is a measure of rapidity of change in a random variable.

It can be express as

$$\left[\lim_{T\to\infty} \frac{1}{2T}\int_{-T}^{T} f^2(t)\,dt\right]^{1/2}$$

3.4 (c)

Probability that both friend select same terminal

$$P(E) = \frac{1}{3}$$

Probability of not occurring the event

$$= 1 - P(E) = 1 - \frac{1}{3} = \frac{2}{3}$$

3.5 (a)

Probability that the sum of digits of two dices is even is same either both dices shows even numbers or odd numbers on the top of the surface.

$$\therefore \quad P(E) = P(E_1) + P(E_2)$$

where,

$P(E_1)$ = probability of occurring even number on both the dices

$P(E_2)$ = probability of occurring odd number on both the dices

$$P(E_1) = \frac{3}{6}\times\frac{3}{6} = \frac{1}{4} \quad \text{and} \quad P(E_2) = \frac{3}{6}\times\frac{3}{6} = \frac{1}{4}$$

$$\therefore \quad P(E) = \frac{1}{4}+\frac{1}{4} = \frac{1}{2}$$

3.6 (a)

Given: $f(x) = \frac{1}{5}e^{-x/5}$, $x \ge 0$

Comparing the general function

$$f(x) = \begin{cases} \lambda e^{-\lambda x}; & x \ge 0 \\ 0; & x < 0 \end{cases} \Rightarrow \lambda = \frac{1}{5}$$

The cumulative distributive function

$$f(a) = 1 - e^{-\lambda a}$$

$$\Rightarrow \quad P(E > 5) = 1 - f(5)$$

$$= 1 - (1 - e^{-\lambda.5})$$

$$= e^{-\frac{1}{5}\times 5} = e^{-1} = 1/e$$

Alternate solution:

$$P(x>5) = \int_5^\infty f(x)\,dx = \int_5^\infty \frac{1}{5}e^{-x/5}dx$$

$$= \frac{1}{5}\int_5^\infty e^{-x/5}dx$$

$$= \frac{1}{5}\left[-5e^{-x/5}\Big|_5^\infty\right] = -1\left[e^{-\infty} - e^{-5/5}\right]$$

$$= e^{-1} = \frac{1}{e}$$

3.7 (b)

Cumulative distribution function

$$= \int_{-\infty}^{x} (\text{Probability density function})\,dx$$

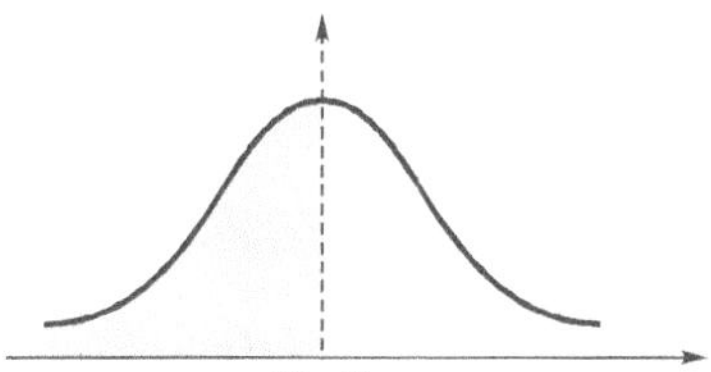

Since,

$$\int_{-\infty}^{\infty} f(x)\,dx = 1 \text{ [complete area of normal curve] ...(i)}$$

Also, cumulative distribution function at origin, is (shaded area),

$$\int_{-\infty}^{0} f(x)\,dx = \frac{1}{2} \text{ [using (i)]}$$

Alternate solution:

$$f_x(x) = \frac{1}{\sqrt{2\pi\sigma_x^2}}e^{-\frac{(x-\mu_x)^2}{2\sigma_x^2}}$$

If mean $\mu_x = 0$

$$f_x(x) = \frac{1}{\sqrt{2\pi\sigma_x^2}}e^{-\frac{x^2}{2\sigma_x^2}}$$

$$\int_{-\infty}^{\infty} f_x(x)\,dx = 1$$

Since mean is zero therefore,

$$\int_{-\infty}^{0} f_x(x)\,dx = \int_0^\infty f_x(x) = \frac{1}{2}$$

and $$F_x(x) = \int_{-\infty}^{x^+} f_x(x)\,dx$$

$$F_X(0) = \int_{-\infty}^{0^+} f_x(x)\,dx = \frac{1}{2} = 0.5$$

3.8 (a)

For uniform distribution

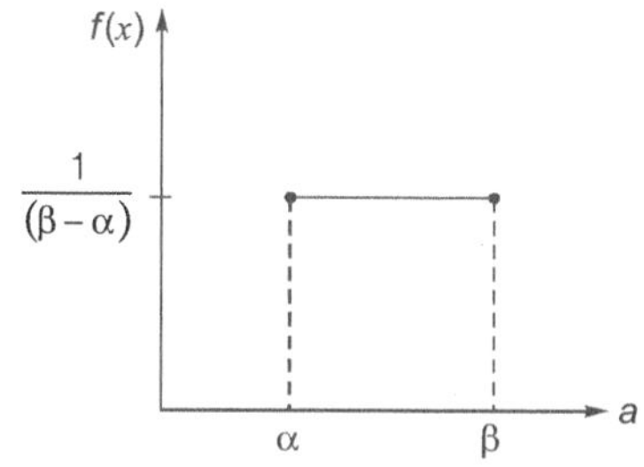

Here α = 2 and β = 10

variance;

$$V(x) = \left(\frac{\beta-\alpha}{12}\right)^2 = \frac{(10-2)^2}{12} = \frac{64}{12} = \frac{16}{3}$$

3.9 (c)

Let total sample = 100

out of which $x \rightarrow$ positive reports

then $(100 - x) \rightarrow$ –ve reports

Probability of +ve reports

$$= 0.01 = \frac{x}{100} \Rightarrow x = 1$$

so probability of getting incorrect report is

$$P(A) = \frac{0.12x + 0.15(100-x)}{100}$$

$$= \frac{15 - 0.03x}{100}$$

$$= \frac{14.97}{100} = 0.1497 \qquad (\because x = 1)$$

Alternate solution:

Tree diagram:

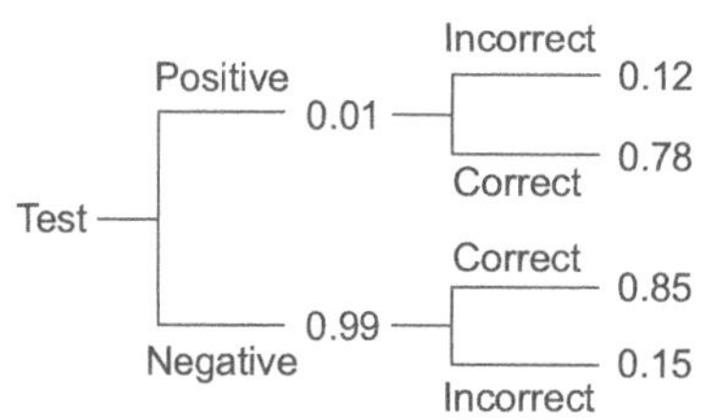

So, *P*(incorrect report)

= *P*(positive incorrect) + *P*(negative incorrect)

= 0.01 × 0.12 + 0.99 × 0.15 = 0.1497

3.10 (c)

$$\sigma = 0.05 \text{ mm}$$
$$\mu_x = 10 \text{ mm}$$

Then probability,

$$P(X > 10.15) = Q\left(\frac{x-\mu_x}{\sigma_x}\right)$$

where, $x = 10.15$ mm

$$\Rightarrow \quad P = Q\left(\frac{10.15-10}{0.05}\right) = Q(3)$$

$$= \frac{1}{\sqrt{2\pi\times 3^2}} e^{-x^2/2} = \frac{1}{\sqrt{18\pi}} e^{-4.5}$$

$$= 1.477 \times 10^{-3}$$

So, number of measurement more than 10.15 mm

$$N = P \times \text{Total number of measurement}$$

$$= 1.477 \times 10^{-3} \times 10^4 = 14.77$$

$$N \simeq 15$$

Note: $Q(x) \simeq \frac{1}{\sqrt{2\pi x^2}} \times e^{-x^2/2}$

Alternate solution:

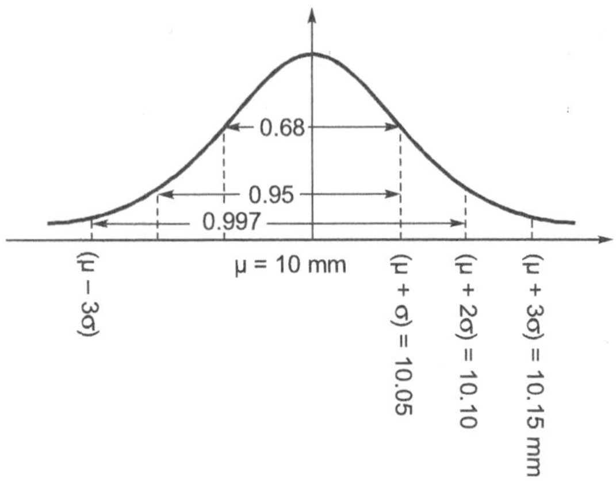

For measurement more than 10.15 mm the deviation is 0.15 mm i.e. 3σ.

For deviation beyond ±3σ limits expectation is,

$$E_{\pm 3\sigma} = 10000\,[1 - 0.997] = 30$$

So, for measurement 10.15 mm i.e. beyond 3σ limits,

$$E_{+3\sigma} = \frac{E \pm 3\sigma}{2} = 15$$

3.11 (d)

Total number of ways of selection of two chips (one from each box)

$$= {}^5C_1 \times {}^5C_1$$

The number of ways such that the product of these two chip numbers to be odd

$$= {}^3C_1 \times {}^2C_1 = 3 \times 2 = 6$$

The probability for the product to be an even number = 1 – probability for the product to be an odd number

$$= 1 - \frac{6}{25} = \frac{19}{25}$$

Alternate solution:

$B_1 \rightarrow 3, 6, 9, 12, 15$; $B_2 \rightarrow 6, 11, 16, 21, 26$

So, probability of getting even product is probability of even from B_1 × probability of odd from B_2 + probability of odd from B_1 × probability of even from B_2 + probability of even from both B_1 and B_2.

$$\Rightarrow \quad P(E) = \frac{2}{5}\times\frac{2}{5} + \frac{3}{5}\times\frac{3}{5} + \frac{2}{5}\times\frac{3}{5} = \frac{19}{25}$$

3.12 (c)

$$P(E) = \text{Probability of occurring head in odd number of tosses}$$

$$= P(H) + P(TTH) + \ldots$$

$$= \frac{1}{2} + \frac{1}{2}\times\frac{1}{2}\times\frac{1}{2} + \ldots + \infty$$

$$= \frac{\frac{1}{2}}{1-\frac{1}{4}} = \frac{1}{2}\times\frac{4}{3} = \frac{2}{3}$$

3.13 (a)

$$P = \int_1^\infty f(x)\,dx = \int_1^\infty e^{-x}dx$$

$$= e^{-x}\Big|_\infty^1 = e^{-1} = 0.368$$

3.14 (2)

$$\int_{-\infty}^{+\infty} \frac{e^{-x/2}}{K} = 1$$

$$\frac{1}{K}\int_0^{+\infty} e^{-x/2} = 1 \qquad \text{[in range } (0, \infty)]$$

$$\therefore \quad \frac{-2}{K} e^{-x/2}\Big|_0^\infty = 1$$

$$\frac{-2}{K}(e^{-\infty} - e^0) = 1$$

$$\therefore \quad K = -2(0-1) = 2$$

3.15 (0.893)

$$P(A) = 0.85,\ P(B) = 0.95,\ P(C) = 0.9$$

P[Job to be completed successfully]

= P[(Machine *A* or Machine *B* success) and (machine *C* success)]

$= P((A \cup B) \cap C)$

$= P(A \cap C) + P(B \cap C) - P(A \cap B \cap C)$

$= (0.85)(0.9) + (0.95)(0.9) - (0.85)(0.95)(0.9)$

$= 0.89325 \simeq 0.893$

3.16 (b)

$$\text{Probability} = {}^{10}C_3\,(0.1)^3\,(0.9)^7 = 0.057$$

3.17 (a)

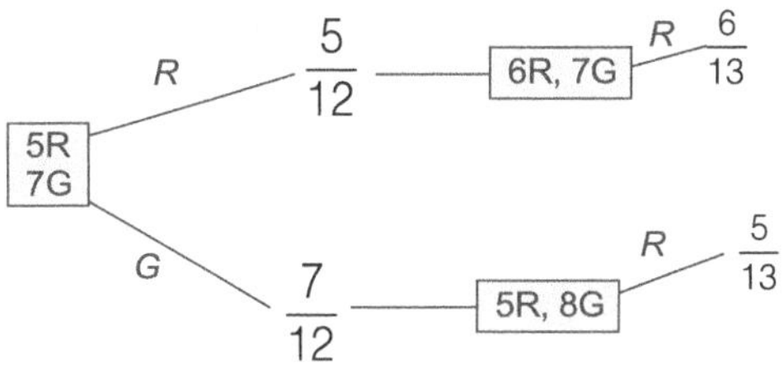

$$P(\text{Red}) = \frac{5}{12} \times \frac{6}{13} + \frac{7}{12} \times \frac{5}{13} = \frac{65}{156}$$

3.18 (c)

$$P(F) = 0.81$$
$$P(F \cap S) = 0.18$$

By conditional probability

$$P(S/F) = \frac{P(S \cap F)}{P(F)} = \frac{0.18}{0.81} = 0.222$$

3.19 (b)

Probability of getting head in a toss = $p = \frac{1}{2}$

Probability of getting tail in a toss = $q = \frac{1}{2}$

Required probability

$$= \begin{pmatrix} \text{Probability of getting} \\ \text{2 heads in first 4 tosses} \end{pmatrix} \times \begin{pmatrix} \text{Probability of getting} \\ \text{head in } 5^{th} \text{ toss} \end{pmatrix}$$

$$= ({}^4C_2\, p^2 q^2) \times \left(\frac{1}{2}\right)$$

$$= 6 \times \frac{1}{4} \times \frac{1}{4} \times \frac{1}{2} = \frac{6}{32} = \frac{3}{16}$$

3.20 (b)

Required probability = 5/20 = 0.25 or 25%

3.21 (a)

First is red and second is green $= \frac{8}{16} \times \frac{8}{15} = \frac{4}{15}$

3.22 (a)

$$\int_{-\infty}^{\infty} f(x)\,dx = 1$$

$$\int_{-\infty}^{1} f(x)\,dx + \int_{1}^{\infty} f(x)\,dx = 1$$

$$0 + \int_{1}^{\infty} \frac{A}{x^{\mu}}\,dx = 1$$

$$A\left(\frac{x^{-\mu+1}}{-\mu+1}\right)\Bigg|_1^{\infty} = 1 \quad ; \quad \frac{A(0-1)}{1-\mu} = 1$$

$$A = \mu - 1$$

3.23 (0.09)

Given, the probability that the ball faces into the basket in a single attempt = 0.1

i.e., probability that the base does not fall into basket = 1 − 0.1 = 0.9

Given each attempt is independent.

∴ Probability that the ball falls into basket only in second attempt

= P(does not fall) × P(falls) = 0.9 × 0.1 = 0.09

3.24 (0.7)

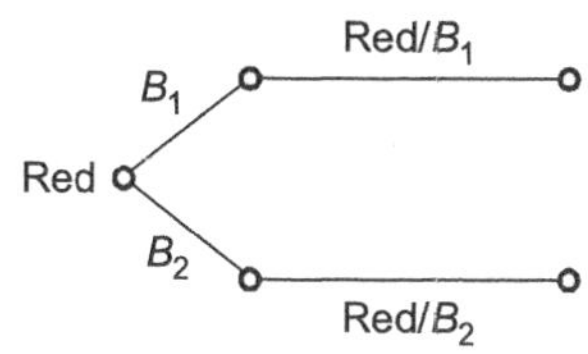

$$P(B_1) = \frac{1}{2}; \quad P(B_2) = \frac{1}{2}$$

$$P\left(\frac{\text{Red}}{B_1}\right) = \frac{7}{10}; \quad P\left(\frac{\text{Red}}{B_2}\right) = \frac{3}{10}$$

By Baye's Theorem

$$P\left(\frac{B_1}{\text{Red}}\right) = \frac{P(B_1 \cap \text{Red})}{P(\text{Red})}$$

$$= \frac{P(B_1).P\left(\frac{\text{Red}}{B_1}\right)}{P(B_1).P\left(\frac{\text{Red}}{B_1}\right) + P(B_2)P\left(\frac{\text{Red}}{B_1}\right)}$$

$$= \frac{\frac{1}{2} \times \frac{7}{10}}{\left(\frac{1}{2} \times \frac{7}{10}\right) + \left(\frac{1}{2} \times \frac{3}{10}\right)} = \frac{\frac{7}{20}}{\frac{1}{2}} = \frac{7}{10} = 0.7$$

3.25 (0.5)

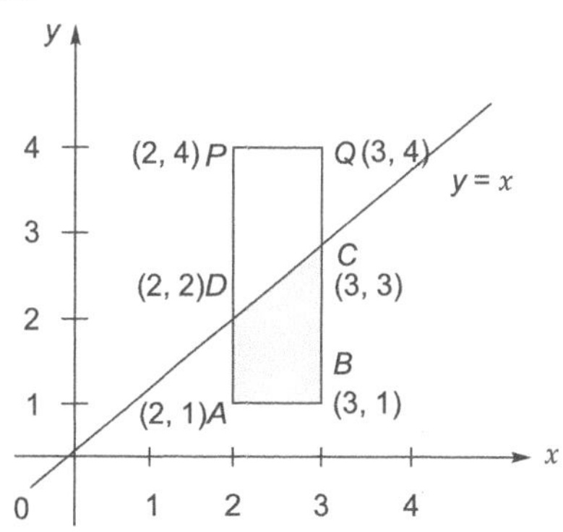

Total area $= AB \times AP$

$= 1 \times 3 = 3$

Favourable Area (F_{av})

$=$ Area of $ABCD$ (i.e., Trapezium)

$$= \frac{1}{2}(\text{sum of parallel sides}) \times (\text{Distance between them})$$

$$= \frac{1}{2}(AD + BC) \times (AB) = \frac{1}{2}(1+2) \times 1 = \frac{3}{2}$$

$$P(y \leq n) = \frac{F_{av.}\text{ area}}{\text{Total area}} = \frac{3/2}{3} = \frac{3}{6} = 0.5$$

3.26 (0.33)

Number of red balls = 6

Number of blue balls = 4

Balls are drawn three times. So, probability of getting red color balls during 2nd and 3rd time

$$P(R) = \frac{6 \times 5 \times 4 + 4 \times 6 \times 5}{10 \times 9 \times 8}$$

$$P(R) = \frac{240}{720} = \frac{1}{3} = 0.333$$

3.27 (–0.5)

x is discrete Random Variable

As we know,

$$\sum_{i=1}^{n} P(x = x_i) = 1$$

$\Rightarrow 0.25 + 0.5 + k = 1$

$\Rightarrow k = 1 - 0.75$

$k = 0.25$

$$\therefore \quad E(x) = \sum_i x_i P_i$$

$= 0 \times 0.25 + 1 \times 0.5 + 2 \times 0.25 = 0.5 + 0.5$

$E(x) = 1$

$$E(x^2) = \sum_i x_i^2 P_i$$

$= 0^2 \times 0.25 + 1^2 \times 0.5 + 2^2 \times 0.25$

$= 0 + 0.5 + 1$

$E(x^2) = 1.5$

$\therefore \quad E(x) - E(x^2) = 1 - 1.5 = -0.5$

3.28 (240)

We have 5 digits out of which 2 are identical.

$$\text{Number of ways} = \frac{5!}{2!} = \frac{120}{2} = 60$$

Out of four digit exactly one digit appearing twice.

Number of ways $= {}^4C_1 = 4$

Total 5-digit number $= 4 \times 60 = 240$

3.29 (0.5)

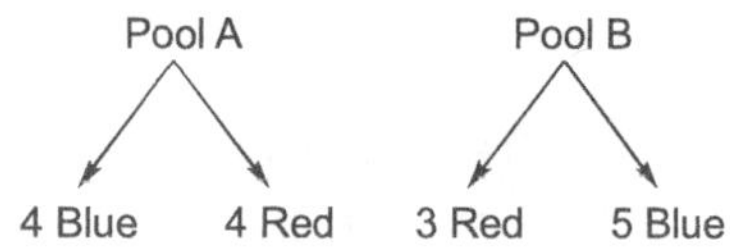

Probability $\rightarrow \frac{4}{8} \times \frac{3}{8} \times \frac{4}{8} \times \frac{5}{8}$

Pool-A Pool-B Pool-A Pool-B

Blue Red Red Blue

$$= \frac{12}{64} + \frac{20}{64} = \frac{32}{64} = \frac{1}{2} = 0.5$$

3.30 (0.135)

Probability of $X > 2$ is given as $\int_2^{\infty} e^{-x} dx$

$$P(X > 2) = \int_2^{\infty} e^{-x} dx$$

$$P(X > 2) = -e^{-x}\Big|_2^{\infty}$$

$= -e^{-\infty} + e^{-2}$

$= 0.135$

■■■

4 Numerical Methods

4.1 A linear ordinary differential equation is given as

$$\frac{d^2y}{dt^2} + 3\frac{dy}{dt} + 2y = \delta(t)$$

where $\delta(t)$ is an impulse input. The solution is found by Euler's forward-difference method that uses an integration step h. What is suitable value of h?

(a) 2.0 (b) 1.5
(c) 1.0 (d) 0.2

[2006 : 1 M]

4.2 For k = 0, 1, 2,............., the steps of Newton-Raphson method for solving a non-linear equation is given as

$$x_{k+1} = \frac{2}{3}x_k + \frac{5}{3}x_k^{-2}$$

Starting from a suitable initial choice as k tends to ∞, the iterate x_k tends to

(a) 1.7099 (b) 2.2361
(c) 3.1251 (d) 5.0000

[2006 : 2 M]

4.3 The polynomial $p(x) = x^5 + x + 2$ has

(a) all real roots
(b) 3 real and 2 complex roots
(c) 1 real and 4 complex roots
(d) all complex roots [2007 : 2 M]

4.4 Identify the Newton-Raphson iteration scheme for finding the square root of 2.

(a) $x_{n+1} = \frac{1}{2}\left(x_n + \frac{2}{x_n}\right)$

(b) $x_{n+1} = \frac{1}{2}\left(x_n - \frac{2}{x_n}\right)$

(c) $x_{n+1} = \frac{2}{x_n}$

(d) $x_{n+1} = \sqrt{2 + x_n}$ [2007 : 2 M]

4.5 The differential equation $\frac{dx}{dt} = \frac{4-x}{\tau}$, with $x(0) = 0$ and the constant $\tau > 0$, is to be numerically integrated using the forward Euler method with a constant integration time step T. The maximum value of T such that the numerical solution of x converges is

(a) $\tau/4$ (b) $\tau/2$
(c) τ (d) 2τ

[2009 : 2 M]

4.6 The velocity v(in m/s) of a moving mass, starting from rest, given as $dv/dt = v + t$. Using Euler forward difference method (also known as Cauchy-Euler method) with a step size of 0.1 s, the velocity at 0.2 s evaluate to

(a) 0.01 m/s (b) 0.1 m/s
(c) 0.2 m/s (d) 1 m/s

[2010 : 2 M]

4.7 The extremum (minimum or maximum) point of a function $f(x)$ is to be determined by solving $\frac{df(x)}{dx} = 0$ using the Newton-Raphson method. Let $f(x) = x^3 - 6x$ and $x_0 = 1$ be the initial guess of x. The value of x after two iterations (x_2) is

(a) 0.0141 (b) 1.4142
(c) 1.4167 (d) 1.5000

[2011 : 2 M]

4.8 While numerically solving the differential equation $\frac{dy}{dx} + 2xy^2 = 0$, y(0) = 1 using Euler's predictor-corrector (improved Euler-Cauchy) with a step size of 0.2, the value of y after the first step is

(a) 1.00 (b) 1.03
(c) 0.97 (d) 0.96

[2013 : 2 M]

4.9 The iteration step in order to solve for the cube roots of a given number N using the Newton-Raphson's method is

(a) $x_{k+1} = x_k + \frac{1}{3}(N - x_k^3)$

(b) $x_{k+1} = \frac{1}{3}\left(2x_k + \frac{N}{x_k^2}\right)$

(c) $x_{k+1} = x_k - \frac{1}{3}(N - x_k^3)$

(d) $x_{k+1} = \frac{1}{3}\left(2x_k - \frac{N}{x_k^2}\right)$ [2014 : 2 M]

4.10 The following table lists an nth order polynomial $f(x) = a_n x^n + a_{n-1}x^{n-1} + ... a_1 x + a_0$ and the forward difference evaluated at equally spaced values of x. The order of the polynomial is

x	$f(x)$	Δf	$\Delta^2 f$	$\Delta^3 f$
–0.4	1.7648	–0.2965	0.089	–0.03
–0.3	1.4683	–0.2075	0.059	–0.0228
–0.2	1.2608	–0.1485	0.0362	–0.0156
–0.1	1.1123	–0.1123	0.0206	–0.0084
0	1	–0.0917	0.0122	–0.0012
0.1	0.9083	–0.0795	0.011	0.006
0.2	0.8288	–0.0685	0.017	0.0132

(a) 1 (b) 2
(c) 3 (d) 4 [2017 : 2 M]

4.11 Consider the recursive equation

$$X_{n+1} = X_n - h(F(X_n) - X_n),$$

with initial condition $X_0 = 1$ and h > 0 being a very small valued scalar. This recursion numerically solves the ordinary differential equation _____.

(a) $\dot{X} = -F(X),\ X(0) = 1$
(b) $\dot{X} = F(X),\ X(0) = 1$
(c) $\dot{X} = F(X) + X,\ X(0) = 1$
(d) $\dot{X} = -F(X) + X,\ X(0) = 1$

[2020 : 1 M]

4.12 The Newton-Raphson method is applied to determine the solution of $f(x) = 0$ where $f(x) = x - \cos(x)$. If the initial guess of the solution is $x_0 = 0$, the value of the next approximation x_1 is ________

(round off to two decimal places)

[2022 : 1 M]

4.13 Consider a system given by the following first order differential equation:

$$\frac{dy}{dt} = y + 2t - t^2$$

where, $y(0) = 1$ and $0 < t < \infty$. Using a step size $h = 0.1$ for the improved Euler method, the value of $y(t)$ at $t = 0.1$ is ________ (rounded off to two decimal places).

[2024 : 1 M]

■■■

Answers Numerical Methods

4.1 (d)	4.2 (a)	4.3 (c)	4.4 (a)	4.5 (d)	4.6 (a)	4.7 (c)
4.8 (d)	4.9 (b)	4.10 (d)	4.11 (d)	4.12 (1.00)	4.13 (1.1145)	

Explanations Numerical Methods

4.1 (d)

For better approximation step size should be as small as possible.

So, $h = 0.2$

4.2 (a)

Given, $x_{k+1} = \frac{2}{3}x_k + \frac{5}{3}x_k^{-2}$

as $k \to \infty$ series converges.

$\therefore \quad x_{k+1} = x_k = \alpha$ = root of equation

$\Rightarrow \quad \alpha = \frac{2}{3}\alpha + \frac{5}{3}\alpha^{-2}$

$\Rightarrow \quad \frac{\alpha}{3} = \frac{5}{3}\frac{1}{\alpha^2}$

$\Rightarrow \quad \alpha^3 = 5$

$\alpha = 5^{1/3} = 1.7099$

$x_k = 1.7099$

4.3 (c)

$P(x) = x^5 + x + 2$

It has a real root at $x = -1$

$\Rightarrow \quad (x^4 - x^3 + x^2 - x + 2)(x + 1) = 0$

$\Rightarrow \quad (x^4 - x^3 + x^2 - x + 2)$ will give other 4 roots

$x^3(x - 1) + x(x - 1) = -2$

$(x^3 + x)(x - 1) = -2$

$x(x^2 + 1)(x - 1) = -2$

Since $(x^2 + 1)$ will be always positive for any real value of x. So for x to be real and satisfying above equation one of the expression x or $(x - 1)$ has to negative.

For $x > 1$, both will be +ve and for $x < 0$ both will be negative so not possible.

Only possible value of x is $0 < x < 1$.

Since $|x| < 1$, so $1 + x^2 < 2$ and both x and $(x - 1)$ are less than 1 so this product of less than 2 and less than 1 can never have magnitude equal to 2 so real roots are not possible.

So, for above fourth order equation 4 complex roots will be there.

Alternate solution:

Descartes rule:

Number of real positive roots in $f(x) \le$ Number of sign change in $f(x)$

Number of real negative roots in $f(x) \le$ Number of sign change in $f(-x)$

$\Rightarrow$ Number of real roots in $f(x) \le$ Number of sign changes in $f(x)$ + Number of sign changes in $f(-x)$ ...(i)

Now, $f(x) = x^5 + x + 2$

$\Rightarrow$ Number of real roots in $f(x) \le 1$

By observation, $x = -1$ is one real root.

So, real root = 1 and 4 imaginary roots.

4.4 (a)

To find square root of 2

Let the equation $f(x) = x^2 - 2 = 0$

Newton-Raphson iteration equation is

$$x_{n+1} = x_n - \frac{f(x_n)}{f'(x_n)}$$

$$f(x_n) = x_n^2 - 2$$

$$f'(x_n) = 2x_n$$

$$\therefore \quad x_{(n+1)} = x_n - \left(\frac{x_n^2 - 2}{2x_n}\right) = \frac{x_n^2 + 2}{2x_n}$$

$$x_{(n+1)} = \frac{1}{2}\left[x_n + \frac{2}{x_n}\right]$$

4.5 (d)

$$\frac{dx}{dt} = \frac{4-x}{\tau}$$

$$f(x, y) = \frac{4-x}{\tau}$$

$$x_{j+1} = x_j + h_f(x_j, y_j) = x_j + h\left(\frac{4-x_i}{\tau}\right)$$

$$= \left(1 - \frac{h}{\tau}\right)x_j + \frac{4h}{\tau}$$

for stability, $\left|1 - \frac{h}{\tau}\right| < 1$

$$\therefore \quad -1 \le 1 - \frac{h}{\tau} \le 1$$

$$-1 \le 1 - \frac{\Delta T}{\tau} \le 1$$

$$\therefore \quad \Delta T < 2\tau$$

4.6 (a)

$$v_{i+1} = v_i + h[v_i + t_i]$$

$$v_0 = 0 \text{ (at } t_0 = 0)$$

$$\therefore \quad v_1 = v_0 + h[v_0 + t_0] = 0$$

$$v_2 = v_1 + h[v_1 + t_1]$$

$$= 0 + 0.1[0 + 0.1] = 0.01 \text{ m/sec}$$

4.7 (c)

$$g(x) = \frac{df(x)}{dx} = 3x^2 - 6$$

$$g'(x) = 6x$$

Newton Raphson iteration equation is

$$x_1 = x_0 - \left[\frac{g(x_0)}{g'(x_0)}\right]$$

$$= 1 - \left[\frac{3 \times (1)^2 - 6}{6 \times 1}\right] = 1 + \frac{3}{6} = 1.5$$

$$x_2 = x_1 - \left[\frac{g(x_1)}{g'(x_1)}\right]$$

$$= 1.5 - \left[\frac{3 \times (1.5)^2 - 6}{6 \times 1.5}\right] = 1.5 - \frac{0.75}{9} = 1.4167$$

4.8 (d)

$$\frac{dy}{dx} + 2xy^2 = 0$$

$$\therefore \quad \frac{dy}{dx} = -2xy^2$$

after one iteration

$$y_1^* = y_0 + h[-2x_0 y_0^2]$$

$$= 1 + 0.2[-2 \times 0 \times 1^2] = 1 + 0 = 1$$

$$y_1 = y_0 + \frac{1}{2} \times (0.2)[-2x_0 y_0^2 - 2x_1 y_1^*]$$

$$= 1 + 0.1[-(2 \times 0 \times 1^2) - (2 \times 0.2 \times 1)]$$

$$= 1 + 0.1[-0 - 0.4] = 1 - 0.04 = 0.96$$

4.9 (b)

$$x = N^{1/3}, \quad \therefore x^3 = N$$

$$f(x) = x^3 - N$$

$$f'(x) = 3x^2 - 0 = 3x^2$$

$$x_{k+1} = x_k - \frac{f(x_k)}{f'(x_k)}$$

(By Newton Raphson method)

$$= x_k - \frac{(x_k^3 - N)}{3x_k^2} = \frac{3x_k^3 - x_k^3 + N}{3x_k^2}$$

$$= \frac{2x_k^3 + N}{3x_k^2} = \frac{1}{3}\left[2x_k + \frac{N}{x_k^2}\right]$$

4.10 (d)

In the following table by calculating $\Delta^4 f$ we get 7.2×10^{-3} for all the differences. Which is constant for all values. Therefore the order of the polynomial is 4.

4.11 (d)

Given recursive equation $X_{n+1} = X_n - h[F(X_n) - X_n]$ can be written at

$$X_{n+1} = X_n + h[X_n - F(X_n)]$$

Clearly it is Euler forward numerical solution for the differential equation.

$$\frac{dX}{dt} = X - F(X)$$

4.12 (1.00)

According to Newton-Raphson method,

$$x_{n+1} = x_n - \frac{f(x_n)}{f'(x_n)}$$

$$= 0 - \left[\frac{0-1}{1}\right] = 1$$

4.13 (1.1145)

By using improved Euler's method (2nd order Runge-Kutta method)

$$\frac{dy}{dt} = y + 2t - t^2$$

$$f(t, y) = y + 2t - t^2$$

Given $y(0) = 1 \Rightarrow y_0 = 1 \quad t_0 = 1$

Given, $h = 0.1$

then $t_i = t_0 + h = 0 + 0.1 = 0.1$

and $y_1 = y_0 + k$ ...(1)

where, $k = \dfrac{k_1 + k_2}{2}$ ...(2)

Now, $k_1 = hf(t_0, y_0) = 0.1[1 + 2(0) - 0^2]$

$= 0.1$

$k_2 = hf(t_0 + h, y_0 + k_1)$

$= 0.1 f(0.1, 1.1)$

$= 0.1[1.1 + 2(0.1) - 0.1^2]$

$k_2 = 0.129$

from equation (2)

$$k = \frac{k_1 + k_2}{2} = 0.1145$$

Now-from equation (1)

$y_1 = y_0 + k = 1 + 0.1145$

$= 1.1145$

$y(0.1) = 1.11$

■■■

5 Calculus

5.1 $\lim_{x \to 0} \frac{1}{10} \frac{1-e^{-j5x}}{1-e^{-jx}} =$ ____.

(a) 0 (b) 1.1
(c) 0.5 (d) 1

[1999 : 1 M]

5.2 $\lim_{x \to \pi/4} \frac{\sin 2\left(x - \frac{\pi}{4}\right)}{x - \frac{\pi}{4}} =$ ____.

(a) 0 (b) 1/2
(c) 1 (d) 2

[2001 : 1 M]

5.3 If $f = a_0 x^n + a_1 x^{n-1} y + ... + a_{n-1} x y^{n-1} + a_n y^n$, where, a_i (i = 0 to n) are constants, then

$x\frac{\partial f}{\partial x} + y\frac{\partial f}{\partial y}$ is

(a) $\frac{f}{n}$ (b) $\frac{n}{f}$

(c) nf (d) $n\sqrt{f}$

[2005 : 1 M]

5.4 The value of the integral $\int_{-1}^{1} \frac{1}{x^2} dx$ is

(a) 2 (b) does not exist
(c) –2 (d) ∞

[2005 : 1 M]

5.5 Given a real-valued continuous function $f(t)$ defined over [0, 1], $\lim_{t \to 0} \frac{1}{t} \int_0^t f(x) dx$ is

(a) ∞ (b) 0
(c) $f(1)$ (d) $f(0)$

[2005 : 2 M]

5.6 The plot of a function $f(x)$ is shown in the following figure. A possible expression for the function $f(x)$ is

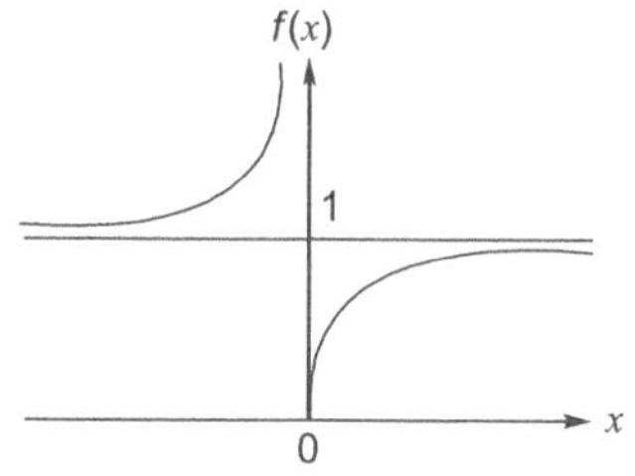

(a) $\exp(|x|)$ (b) $\exp\left(\frac{-1}{x}\right)$

(c) $\exp(-1)$ (d) $\exp\left(\frac{1}{x}\right)$

[2005 : 2 M]

5.7 The function $y = \sin\phi$, ($\phi > 0$) is approximated as $y = \phi$, where ϕ is in radian. The maximum value of ϕ for which the error due to the approximation is within ± 2% is

(a) 0.1 rad (b) 0.2 rad
(c) 0.3 rad (d) 0.4 rad

[2006 : 2 M]

5.8 For real x the maximum value of $\frac{e^{\sin x}}{e^{\cos x}}$ is

(a) 1 (b) e

(c) $e^{\sqrt{2}}$ (d) ∞

[2007 : 2 M]

5.9 Consider the function $f(x) = |x|^3$, where x is real. Then the function $f(x)$ at $x = 0$ is

(a) continuous but not differentiable
(b) once differentiable but not twice
(c) twice differentiable but not thrice
(d) thrice differentiable

[2007 : 2 M]

5.10 The value of the integral $\int_0^\infty \int_0^\infty e^{-x^2} e^{-y^2} dx\, dy$ is

(a) $\sqrt{\frac{\pi}{2}}$ (b) $\sqrt{\pi}$

(c) π (d) $\pi/4$

[2007 : 2 M]

5.11 Given $y = x^2 + 2x + 10$, the value of $\left.\frac{dy}{dx}\right|_{x=1}$ is equal to
(a) 0 (b) 4
(c) 12 (d) 13
[2008 : 1 M]

5.12 $\lim_{x \to 0} \frac{\sin x}{x}$ is
(a) indeterminate (b) 0
(c) 1 (d) ∞
[2008 : 1 M]

5.13 The expression $e^{-\ln x}$ for $x > 0$ is equal to
(a) $-x$ (b) x
(c) x^{-1} (d) $-x^{-1}$
[2008 : 2 M]

5.14 Consider the function $y = x^2 - 6x + 9$. The maximum value of y obtained when x varies over the interval 2 to 5 is
(a) 1 (b) 3
(c) 4 (d) 9
[2008 : 2 M]

5.15 It is known that two roots of the nonlinear equation $x^3 - 6x^2 + 11x - 6 =$ are 1 and 3. The third root will be
(a) j (b) $-j$
(c) 2 (d) 4
[2008 : 2 M]

5.16 One of the roots of the equation $x^3 = j$, where j is the positive square root of –1, is
(a) j (b) $\frac{\sqrt{3}}{2} + j\frac{1}{2}$
(c) $\frac{\sqrt{3}}{2} - j\frac{1}{2}$ (d) $-\frac{\sqrt{3}}{2} - j\frac{1}{2}$
[2009 : 2 M]

5.17 The infinite series
$f(x) = x - \frac{x^3}{3!} + \frac{x^5}{5!} - \frac{x^7}{7!} \ldots \infty$ converges to
(a) $\cos(x)$ (b) $\sin(x)$
(c) $\sinh(x)$ (d) e^x [2010 : 1 M]

5.18 The maximum value of
$f(x) = x^3 - 9x^2 + 24x + 5$ in the interval [1, 6] is
(a) 21 (b) 25
(c) 41 (d) 46
[2012 : 2 M]

5.19 The double integral $\int_0^a \int_0^y f(x, y)\, dx\, dy$ is equivalent to
(a) $\int_0^x \int_0^y f(x, y)\, dx\, dy$ (b) $\int_0^a \int_x^y f(x, y)\, dx\, dy$
(c) $\int_0^a \int_x^a f(x, y)\, dy\, dx$ (d) $\int_0^a \int_0^a f(x, y)\, dx\, dy$
[2015 : 1 M]

5.20 $\lim_{n \to \infty} \left(\sqrt{n^2 + n} - \sqrt{n^2 + 1}\right)$ is ______.
[2016 : 1 M]

5.21 Let $f: [-1, 1] \to R$, where $f(x) = 2x^3 - x^4 - 10$. The minimum value of $f(x)$ is______.
[2016 : 2 M]

5.22 Consider the following equations
$$\frac{\partial V(x,y)}{\partial x} = px^2 + y^2 + 2xy$$
$$\frac{\partial V(x,y)}{\partial y} = x^2 + qy^2 + 2xy$$
where p and q are constants. $V(x, y)$ that satisfies the above equations is
(a) $p\frac{x^3}{3} + q\frac{y^3}{3} + 2xy + 6$
(b) $p\frac{x^3}{3} + q\frac{y^3}{3} + 5$
(c) $p\frac{x^3}{3} + q\frac{y^3}{3} + x^2y + xy^2 + xy$
(d) $p\frac{x^3}{3} + q\frac{y^3}{3} + x^2y + xy^2$
[2018 : 2 M]

5.23 Consider the function $f(x, y) = x^2 + y^2$. The minimum value of the function attains on the line $x + y = 1$ (rounded off to two decimal places) is _____.
[2020 : 2 M]

5.24 A straight line drawn on an x-y plane intercepts the x-axis at –0.5 and the y-axis at 1. The equation that describes this line is _____.
(a) $y = x - 0.5$ (b) $y = 2x + 1$
(c) $y = 0.5x - 1$ (d) $y = -0.5x + 1$
[2020 : 2 M]

5.25 Consider the sequence, $x_n = 0.5x_{n-1} + 1$, $n = 1, 2, ...$ with $x_0 = 0$. Then $\lim_{n\to\infty} x_n$ is

(a) 2 (b) 1

(c) 0 (d) ∞

[2021 : 1 M]

5.26 Consider the function $f(x) = -x^2 + 10x + 100$. The minimum value of the function in the interval [5, 10] is ______.

[2021 : 1 M]

5.27 The global minimum of $x3e^{-|x|}$ for $x \in (-\infty, \infty)$ occurs at x = ________

(round off to one decimal place) [2022 : 1 M]

5.28 What is $\lim_{x\to\infty} f(x)$, where $f(x) = x\sin\frac{1}{x}$?

(a) 0 (b) 1

(c) ∞ (d) Limit does not exist

[2023 : 1 M]

5.29 Consider the real-valued function $g(x) = \max\{(x-2)^2, -2x+7\}$, where $x \in (-\infty, \infty)$. The minimum value attained by $g(x)$ is ______. (rounded off to one decimal place).

[2023 : 2 M]

5.30 Let C be the closed curve in the xy-plane, traversed in the counterclockwise direction along the boundary of the rectangle with vertices at (0, 0), (2, 0), (2, 1), (0, 1). The value of the line integral $\oint_C (-e^y dx + e^x dy)$ is

(a) $e^2 + 2e - 3$ (b) $e^2 - 2e - 3$

(c) $e^2 + e - 1$ (d) $e^2 + e + 1$

[2024 : 2 M]

■■■

Answers Calculus

5.1	(c)	5.2	(d)	5.3	(c)	5.4	(d)	5.5	(d)	5.6	(b)	5.7	(b)
5.8	(c)	5.9	(c)	5.10	(d)	5.11	(b)	5.12	(c)	5.13	(c)	5.14	(c)
5.15	(c)	5.16	(b)	5.17	(b)	5.18	(c)	5.19	(c)	5.20	(0.5)	5.21	(–13)
5.22	(d)	5.23	(0.5)	5.24	(b)	5.25	(a)	5.26	(100)	5.27	(–3.0)	5.28	(a)
5.29	(1)	5.30	(a)										

Explanations Calculus

5.1 (c)

The limit is in the form of $\left(\frac{0}{0}\right)$

By L-Hospital's rule,

$$\lim_{x\to 0} \frac{1}{10}\cdot\frac{5j}{j}\frac{e^{-j5x}}{e^{-jx}} = \frac{5}{10} = 0.5$$

5.2 (d)

$$\lim_{x-\frac{\pi}{4}\to 0} \frac{\sin 2\left(x-\frac{\pi}{4}\right)}{\left(x-\frac{\pi}{4}\right)} = \lim_{x-\frac{\pi}{4}\to 0} \frac{\sin 2\left(x-\frac{\pi}{4}\right)}{2\left(x-\frac{\pi}{4}\right)}\cdot 2$$

$$= 1 \times 2 = 2$$

5.3 (c)

$f = a_0x^n + a_1x^{n-1}y + ... + a_{n-1}{}^{xy}y - 1^{+} a_ny^n$

f is a homogenous function of variable x and y.

$\therefore$ by Euler's theorem on homogenous function

$$x\frac{\partial f}{\partial x} + y\frac{\partial f}{\partial y} = nf$$

5.4 (d)

$\frac{1}{x^2}$ is discontinuous at $x = 0$, but limit exist at $x = 0$

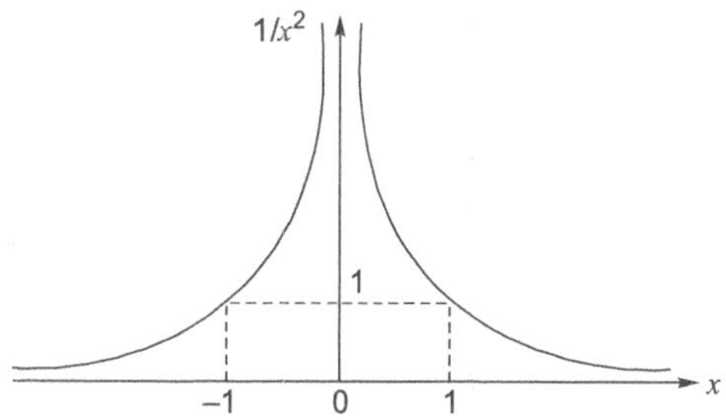

so integral $\int_{-1}^{1} \frac{1}{x^2} dx = \infty$

as clear from the curve area under curve is ∞.

5.5 (d)

Let, $$I = \lim_{t \to 0} \frac{\int_0^t f(x)dx}{t}$$

By applying *L*-hospital's rule

$$I = \lim_{t \to 0} \frac{f(t)1}{1} = f(0)$$

5.6 (b)

$$f(x) = \exp\left(-\frac{1}{x}\right)$$

when $x \to -\infty$

$$f(x) = \exp\left(\frac{-1}{-\infty}\right) \to 1$$

when $x \to 0^-$

$$f(x) = \exp\left(\frac{-1}{0^-}\right) \to \infty$$

when $x \to 0^+$

$$f(x) = \exp\left(\frac{-1}{0^+}\right) \to 0$$

and when $x \to +\infty$

$$f(x) = \exp\left(\frac{-1}{+\infty}\right) \to 1$$

∴ The possible expression for $f(x)$ is $\exp\left(-\frac{1}{x}\right)$.

5.7 (b)

Error, $$e = \frac{\phi - \sin\phi}{\phi}$$

for error to be minimum $\frac{de}{d\phi} = 0$

$$\therefore \quad \frac{de}{d\phi} = \frac{\phi(1-\cos\phi) - (\phi - \sin\phi)}{\phi^2} = 0$$

$$\Rightarrow \quad -\phi\cos\phi + \sin\phi = 0$$

$$\Rightarrow \quad \tan\phi = \phi$$

$$\Rightarrow \quad \sin\phi = \frac{\phi}{\sqrt{1+\phi^2}}$$

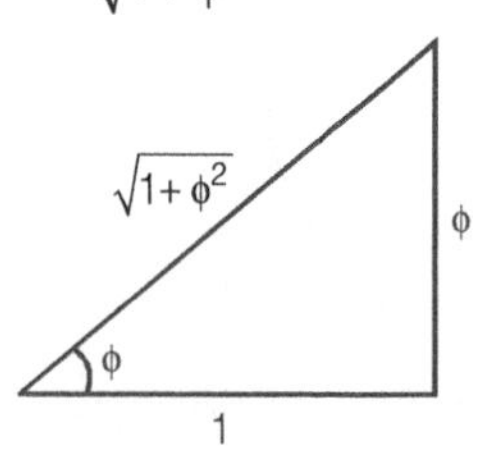

Given,

$$\therefore \quad \text{error} = 2\%$$

$$\frac{\phi - \sin\phi}{\phi} = \frac{2}{100}$$

$$\Rightarrow \quad \phi - \frac{\phi}{\sqrt{1+\phi^2}} = \frac{\phi}{50}$$

$$\Rightarrow \quad \frac{\phi}{\sqrt{1+\phi^2}} = \frac{49}{50}\phi$$

$$\Rightarrow \quad 1 + \phi^2 = \left(\frac{50}{49}\right)^2$$

$$\Rightarrow \quad \phi^2 = 0.0412 \quad \Rightarrow \phi = 0.203 \text{ rad}$$

5.8 (c)

Let $$y = \frac{e^{\sin x}}{e^{\cos x}}$$

$$\Rightarrow \quad y = e^{(\sin x - \cos x)}$$

$$\because \quad (\sin x - \cos x) = \sqrt{2}\sin\left(x - \frac{\pi}{4}\right)$$

$$\Rightarrow \quad y = e^{\sqrt{2}\sin\left(x - \frac{\pi}{4}\right)}$$

∴ Maximum value of $y = e^{\sqrt{2}}$.

5.9 (c)

It is differentiable once.

For checking 2nd order differentiation

$$\text{R.H.D.} = \lim_{h \to 0} \frac{f'(0+h) - f'(0)}{h}$$

$$= \lim_{h \to 0} \frac{3h^2}{h} = 0$$

$$\text{L.H.D.} = \lim_{h \to 0} \frac{f'(0-h) - f'(0)}{-h}$$

$$= \lim_{h \to 0} \frac{3h^2}{-h} = 0$$

So, second order differentiation exist and

$$\frac{d^2f}{dx^2} = 6|x| \text{ at } x = 0$$

for checking 3rd order differentiation

$$\text{R.H.D.} = \lim_{h \to 0} \frac{f''(0+h) - f''(0)}{h}$$

$$= \lim_{h \to 0} \frac{6h}{h} = 6$$

$$\text{L.H.D.} = \lim_{h \to 0} \frac{f''(0-h) - f''(0)}{-h}$$

$$= \lim_{h\to 0} \frac{6h}{-h} = -6$$

$\because$ R.H.D. $\neq$ L.H.D., so not differentiable thrice.

5.10 (d)

$$\int_0^\infty \int_0^\infty e^{-x^2} \cdot e^{-y^2} \cdot dx\,dy$$

$$= \int_0^\infty e^{-x^2} dx \cdot \int_0^\infty e^{-y^2} dy = I \quad \text{...(i)}$$

$$\frac{1}{\sqrt{2\pi\sigma_x^2}} \times \left[e^{-\left[\frac{(x-\mu_x)^2}{2\sigma_x^2}\right]} \right]$$

is a P.D.F. of Gaussian random variable.

Here mean $\mu_x = 0$, $\sigma_x^2 = \frac{1}{2}$

and we know that,

$$\int_{-\infty}^{\infty} f_x(x)\,dx = 1$$

$$\int_{-\infty}^{\infty} f_x(x)\,dx = \frac{1}{2} \text{ (if } \mu_x = 0)$$

so, $$\int_0^\infty e^{-x^2} dx = \frac{1}{\sqrt{2\pi\sigma_x^2}} \times \sqrt{2\pi\sigma_x^2} \int_0^\infty e^{-x^2} dx$$

$$= \sqrt{2\pi \times \frac{1}{2}} \times \frac{1}{\sqrt{\pi}} \int_0^\infty e^{-x^2} dx = \sqrt{\pi} \times \frac{1}{2} = \frac{\sqrt{\pi}}{2}$$

so, $$I = \frac{\sqrt{\pi}}{2} \times \frac{\sqrt{\pi}}{2} = \frac{\pi}{4}$$

Alternate solution:

Let, $$I = \int_0^\infty \int_0^\infty e^{-x^2} \cdot e^{-y^2} dx\,dy$$

[Here e^{-x^2} is given in integral so gamma function can be used to solve this]

Let, $x^2 \to t$

$\Rightarrow$ $2x\,dx \to dt$

$\Rightarrow$ $$dx = \frac{dt}{2\sqrt{5}}$$

$\Rightarrow$ $$I = \frac{1}{2} \int_0^\infty \int_0^\infty e^{-t}\, t^{-1/2}\, dt\,dy$$

$\Rightarrow$ $$I = \frac{1}{2} \int_0^\infty \int_0^\infty e^{-t}\, e^{-1+1/2}\, dt\,dy$$

Also, $$\sqrt{n} = \int_0^\infty e^{-t}\, t^{n-1}\, dt$$

Comparing, $n = \frac{1}{2}$

$\Rightarrow$ $$I = \frac{1}{2} \int_0^\infty e^{-y^2} \sqrt{\frac{1}{2}}\, dy = \frac{\sqrt{\pi}}{2} \int_0^\infty e^{-y^2}\, dy$$

Similarly, $$\int_0^\infty e^{-y^2} dy = \frac{\sqrt{\pi}}{2}$$

$\Rightarrow$ $$I = \frac{\sqrt{\pi}}{2} \cdot \frac{\sqrt{\pi}}{2} = \frac{\pi}{4}$$

5.11 (b)

$$y = x^2 + 2x + 10$$

$$\left.\frac{dy}{dx}\right|_{x=1} = (2x+2)\big|_{x=1} = 2 + 2 = 4$$

5.12 (c)

We know,

$$\sin x = x - \frac{x^3}{3!} + \frac{x^5}{5!} - \frac{x^7}{7!} + \ldots\ldots$$

$$\therefore \quad \frac{\sin x}{x} = 1 - \frac{x^2}{3!} + \frac{x^4}{5!} - \frac{x^6}{7!} + \ldots\ldots$$

$$\Rightarrow \lim_{x\to 0} \frac{\sin x}{x} = 1 - 0 + 0 - 0 = 1$$

Alternate solution:

$$\lim_{x\to\infty} \frac{\sin x}{x} \left[\frac{0}{0} \text{ form}\right]$$

Applying L'hospital rule,

$$= \lim_{x\to 0} \frac{\cos x}{1} = 1$$

5.13 (c)

$$y = e^{-\ln x} = e^{\ln(x^{-1})} = e^{\ln\left(\frac{1}{x}\right)} = \frac{1}{x}$$

5.14 (c)

$$y = x^2 - 6x + 9 = (x-3)^2$$

$$y(2) = 1$$

$$y(5) = 4$$

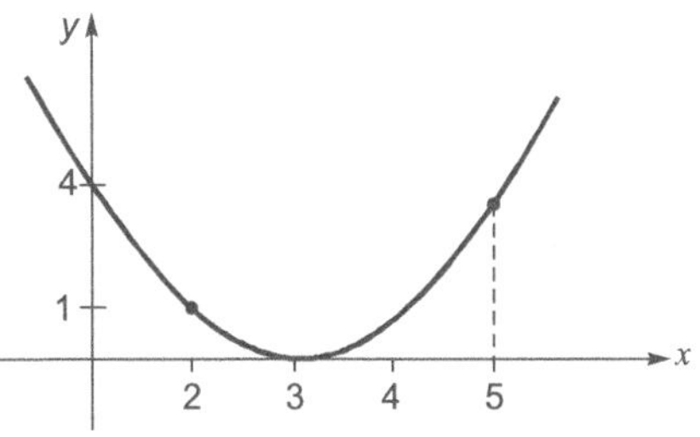

$\therefore$ Maximum value of y over the interval 2 to 5 will be at $x = 4$.

5.15 (c)

Two roots are 1 and 3

i.e. $(x^3 - 6x^2 + 11x - 6) = f(x)(x-1)(x-3)$

$$f(x) = \frac{x^3 - 6x^2 + 11x - 6}{x^2 - 4x + 3} = (x-2)$$

$\therefore$ The third root will be $x = 2$.

Alternate solution:

$f(x) = x^3 - 6x^2 + 11x - 6 = (x-1)(x-3)(x-\alpha)$

where, α is third root,

$$-6 = -1 \times -3 \times -\alpha$$

$$\Rightarrow \quad -6 = -3\alpha$$

$$\Rightarrow \quad \alpha = 2$$

Note: One can directly check option (c) and (d) options as (a) and (b) cannot be correct roots as imaginary roots occur in pair for real equations.

5.16 (b)

$$x^3 = j = e^{j\frac{\pi}{2}}$$

$$x = e^{\frac{j\pi}{6}}$$

$$x = \cos\left(2n\pi + \frac{\pi}{6}\right) + j\sin\left(2n\pi + \frac{\pi}{6}\right)$$

$(n = 0, 1, 2)$

For $n = 0$

$$x_1 = \cos\frac{\pi}{6} + j\sin\frac{\pi}{6} = \frac{\sqrt{3}}{2} + j\frac{1}{2}$$

5.17 (b)

Taylor's series

$$f(x) = f(0) + xf'(0) + \frac{x^2}{2!}f''(0) + ...\infty$$

if $\quad f(x) = \sin x$

then,

$$f(x) = x - \frac{x^3}{3!} + \frac{x^5}{5!} - \frac{x^7}{7!} + ...\infty$$

5.18 (c)

$$f(x) = x^3 - 9x^2 + 24x + 5$$

For maximum or minimum value of $f(x)$

$$f'(x) = 0$$

$$3x^2 - 18x + 24 = 0$$

$$x^3 - 6x + 8 = 0$$

$$x = 2, 4$$

$$f''(x) = 6x - 18$$

$$f''(2) < 0 \text{ and } f''(4) > 0$$

$\therefore$ at $\quad x = 2$

There will be a local maxima,

$$f(2) = 2^3 - 9 \times 2^2 + 24 \times 2 + 5 = 25$$

at end points,

$$f(1) = 1^3 - 9 \times 1^2 + 2^4 \times 1 + 5 = 21$$

$$f(6) = 6^3 - 9 \times 6^2 + 24 \times 6 + 5$$

$$= 216 - 324 + 144 + 5 = 41$$

$\therefore$ The maximum value of $f(x)$ in interval [1, 6] is 41.

5.19 (c)

$$I = \int_0^a \int_0^y f(x, y)\, dx\, dy$$

Limit of x:

Lower limit $x = 0$

Upper limit $x = y$

Limit of y:

Lower limit $y = 0$

Upper limit $y = a$

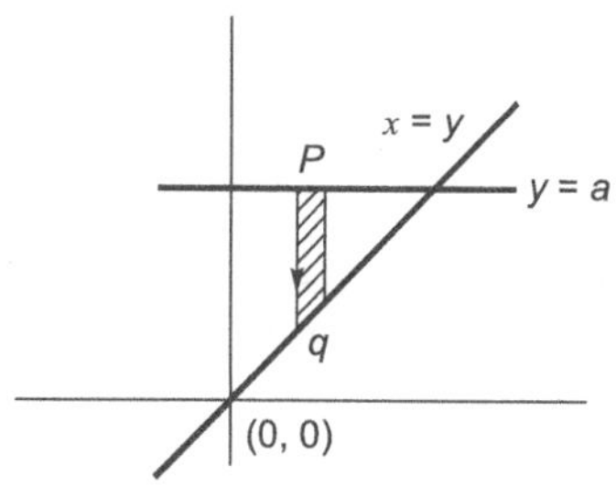

By change of order of integration limit of y

Limit of y:

Lower limit $y = x$

Upper limit $y = a$

Limit of x:

Lower limit $x = 0$

Upper limit $x = a$

So, $I = \int_0^a \int_x^a f(x, y)\, dy\, dx$

5.20 (0.5)

$$\underset{n \to \infty}{Lt} \sqrt{n^2 + n} - \sqrt{n^2 + 1}$$

$$= \underset{n \to \infty}{Lt} \frac{\sqrt{n^2 + n} - \sqrt{n^2 + 1}}{\left(\sqrt{n^2 + n} + \sqrt{n^2 + 1}\right)}\left(\sqrt{n^2 + n} + \sqrt{n^2 + 1}\right)$$

$$= \underset{n \to \infty}{Lt} \frac{n^2 + n - n^2 - 1}{\sqrt{n^2 + n} + \sqrt{n^2 + 1}}$$

$$= \underset{n \to \infty}{Lt} \frac{n - 1}{n\sqrt{1 + \frac{1}{n}} + n\sqrt{1 + \frac{1}{n^2}}}$$

$$= \underset{n\to\infty}{Lt} \frac{n\left(1-\frac{1}{n}\right)}{n\left(\sqrt{1+\frac{1}{n}}+\sqrt{1+\frac{1}{n^2}}\right)} = \frac{1}{1+1} = \frac{1}{2}$$

5.21 (–13)

$f(x) = 2x^3 - x^4 - 10$ in $[-1, 1]$

$f'(x) = 6x^2 - 4x^3$

for minima and maxima

$f'(x) = 0$

$6x^2 - 4x^3 = 0$

$2x^2(3-2x) = 0$

$x = 0, 0, \frac{3}{2}$

$f''(x) = 12x - 12x^2$

for $x = 0$, $f''(0) = 0$

for $x = \frac{3}{2}$ $f''\left(\frac{3}{2}\right) = 18 - 27 = -9 < 0$ maxima

at $x = -1$ $f(-1) = -2 - 1 - 10 = -13$

at $x = 1$ $f(1) = 2 - 1 - 10 = -9$

At $x = -1$, function attains global minimum value with $f(x)_{min} = -13$.

5.22 (d)

From trial and error method, we can found that, option (d) is correct.

5.23 (0.5)

$f(x, y) = x^2 + y^2$ with respect to line $x + y = 1$

from constraint $x + y = 1 \Rightarrow y = 1 - x$

Then $f(x, y) = x^2 + (1 - x)^2$

$= x^2 + 1 + x^2 - 2x$

$f(x, y) = \phi(x) = 2x^2 - 2x + 1$ to be minimum

find stationary point

$\phi'(x) = 4x - 2 = 0$

$\Rightarrow x = 1/2$

Then, $y = 1 - x = 1 - \frac{1}{2} = \frac{1}{2}$

$\therefore$ stationary point is $\left(\frac{1}{2}, \frac{1}{2}\right)$

and $\phi''(x) = 4 > 0$

$\Rightarrow f(x, y)$ is minimum at point $\left(\frac{1}{2}, \frac{1}{2}\right)$

Hence, minimum value of $f(x, y) = x^2 + y^2$

$$= \left(\frac{1}{2}\right)^2 + \left(\frac{1}{2}\right)^2 = \frac{1}{4} + \frac{1}{4} = \frac{1}{2}$$

5.24 (b)

Given line x-intercept is $-\frac{1}{2}$ and y-intercept is 1

Then the line equation is $\frac{x}{-\frac{1}{2}} + \frac{y}{1} = 1$

$\Rightarrow -2x + y = 1$

$\Rightarrow y = 2x + 1$

5.25 (a)

$$x_n = \frac{x_{n-1}}{2} + 1$$

$$\Rightarrow x_1 = 1, x_2 = \frac{3}{2}, x_3 = \frac{7}{4}, x_4 = \frac{15}{8}, x_5 = \frac{31}{16}$$

Sequence is $1, \frac{3}{2}, \frac{7}{2^2}, \frac{15}{2^3}, \frac{31}{2^4}, \ldots, \frac{2^n - 1}{2^{n-1}}, \ldots$

i.e., $x_n = \frac{2^n - 1}{2^{n-1}} = \frac{2^n - 1}{2^n/2} = 2\left[1 - \frac{1}{2^n}\right]$

$$\lim_{n\to\infty} x_n = \lim_{n\to\infty} 2\left(1 - \frac{1}{2^n}\right) = 2\left(1 - \frac{1}{\infty}\right) = 2$$

5.26 (100)

$f(x) = -x^2 + 10x + 100$

$f'(x) = -2x + 10$

Stationary point $f'(x) = 0 \Rightarrow x = 5$

$f''(x) = -2 < 0$

So, at $x = 5$ function has local maxima.

Minimum value of function = min[$f(5)$, $f(10)$]

$f(5) = -25 + 50 + 100 = 125$

$f(10) = -100 + 100 + 100 = 100$

Hence, minimum value = 100.

5.27 (–3.0)

Given function, $y = x^3 e^{-|x|}$

For $x < 0$, $y = x^3 e^x$

$\frac{dy}{dx} = 3x^2e^x + x^3e^x = 0$

$3x^2e^x = -x^3e^x$

$x = -3$

$\frac{d^2y}{dx^2} = 3x^2e^x + x^3e^x + 6xe^x + 3x^2e^x$

$= e^x(x^3 + 6x^2 + 6x)$

$= e^{-3}(-27 + 54 - 18)$

$= 9e^{-3} > 0$

Hence, at $x = -3$ global minimum occurs.

5.28 (a)

$$\lim_{x\to 0} x\sin\frac{1}{x} = \lim_{x\to 0}\frac{\sin\frac{1}{x}}{\frac{1}{x}}$$

$$\lim_{y\to\infty}\frac{\sin y}{y} = \frac{\text{finite}}{\infty} = 0$$

5.29 (1)

$$g(x) = \max\{(x-2)^2, -2x+7\}$$

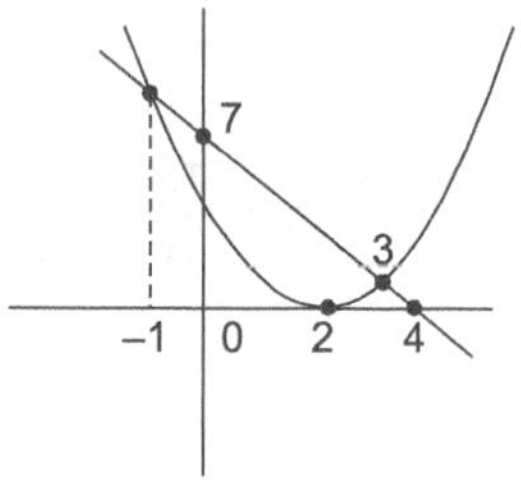

Then $g(x)$

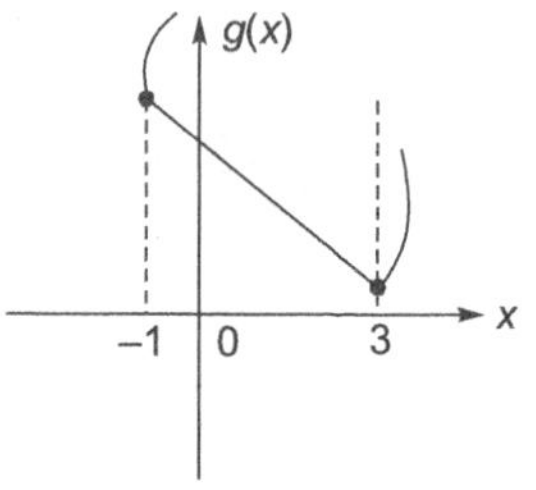

From the figure:

Minimum value of $g(x)$ is at

$$x = 3$$

$$\therefore \quad g_{min}(x) = g(3)$$
$$= (3-2)^2 \text{ or } (-2x+7)_{x=3}$$
$$= 1$$

5.30 (a)

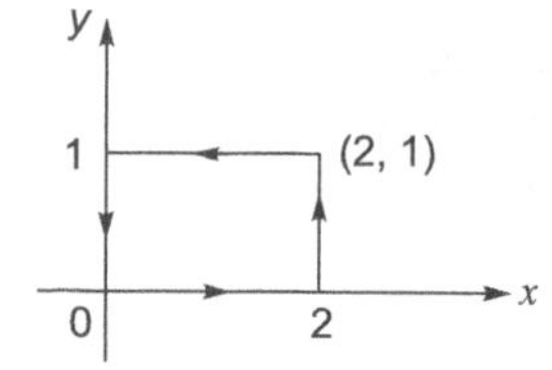

$$\oint_c (-e^y dx + e^x dy)$$

$$= \int_{x=0}^{2}\int_{y=0}^{1} (e^x y + e^y)_0^1 \, dy dx$$

$$= \int_0^2 (e^x y + e^y)_0^1 dx$$

$$= \int_0^2 (e^x + e - 1) dx$$

$$= \left(e^x + ex - x\right)_0^2 = (e^2 + 2e - 2) - (1 + 0 - 0)$$
$$= e^2 + 2e - 3$$

■■■

6 Vector Calculus

6.1 Vectors $\vec{x}_1 = 5i - 2j + k$ and $\vec{x}_2 = 4i - j + 2k$ are transformed into $\vec{y}_1 = A\vec{x}_1$ and $\vec{y}_2 = A\vec{x}_2$ using an orthogonal matrix $A = \frac{1}{3}\begin{bmatrix} 2 & 1 & 2 \\ -2 & 2 & 1 \\ 1 & 2 & -2 \end{bmatrix}$.

The cosine of the angle between $\vec{y}_1$ and $\vec{y}_2$ is

(a) $\frac{8}{\sqrt{70}}$ (b) $\frac{\sqrt{3}}{2}$

(c) $\frac{1}{\sqrt{70}}$ (d) $\frac{1}{2}$

[2004 : 2 M]

6.2 If a vector $\vec{R}(t)$ has a constant magnitude, then

(a) $\vec{R} \cdot \frac{d\vec{R}}{dt} = 0$ (b) $\vec{R} \times \frac{d\vec{R}}{dt} = 0$

(c) $\vec{R} \cdot \vec{R} = \frac{d\vec{R}}{dt}$ (d) $\vec{R} \times \vec{R} = \frac{d\vec{R}}{dt}$

[2005 : 1 M]

6.3 A scalar field is given by $f = x^{2/3} + y^{2/3}$, where x and y are the Cartesian coordinates. The derivative of f along the line $y = x$ directed away from the origin, at the point (8, 8) is

(a) $\frac{\sqrt{2}}{3}$ (b) $\frac{\sqrt{3}}{2}$

(c) $\frac{2}{\sqrt{3}}$ (d) $\frac{3}{\sqrt{2}}$

[2005 : 2 M]

6.4 A sphere of unit radius is centered at the origin. The unit normal at a point (x, y, z) on the surface of the sphere is the vector

(a) (x, y, z) (b) $\left(\frac{1}{\sqrt{3}}, \frac{1}{\sqrt{3}}, \frac{1}{\sqrt{3}}\right)$

(c) $\left(\frac{x}{\sqrt{3}}, \frac{y}{\sqrt{3}}, \frac{z}{\sqrt{3}}\right)$ (d) $\left(\frac{x}{\sqrt{2}}, \frac{y}{\sqrt{2}}, \frac{z}{\sqrt{2}}\right)$

[2009 : 1 M]

6.5 The direction of vector A is radially outward from the origin, with $|A| = kr^n$ where $r^2 = x^2 + y^2 + z^2$ and k is a constant. The value of n for which $\nabla \cdot A = 0$ is

(a) -2 (b) 2

(c) 1 (d) 0 [2012 : 2 M]

6.6 For a vector E, which one of the following statements is **NOT TRUE**?

(a) If $\nabla \cdot E = 0$, E is called solenoidal.

(b) If $\nabla \times E = 0$, E is called conservative.

(c) If $\nabla \times E = 0$, E is called irrotational.

(d) If $\nabla \cdot E = 0$, E is called irrotational.

[2013 : 1 M]

6.7 A vector is defined as

$$f = y\hat{i} + x\hat{j} + z\hat{k}$$

where $\hat{i}, \hat{j}$ and $\hat{k}$ are unit vectors in Cartesian (x, y, z) coordinate system.

The surface integral $\oiint f \cdot ds$ over the closed surface S of a cube with vertices having the following coordinates: (0, 0, 0), (1, 0, 0), (1, 1, 0), (0, 1, 0), (0, 0, 1), (1, 0, 1), (1, 1, 1), (0, 1, 1) is __.

[2014 : 1 M]

6.8 The magnitude of the directional derivative of the function $f(x, y) = x^2 + 3y^2$ in a direction normal to the circle $x^2 + y^2 = 2$, at the point (1, 1), is ____.

(a) $4\sqrt{2}$ (b) $5\sqrt{2}$

(c) $7\sqrt{2}$ (d) $9\sqrt{2}$

[2015 : 1 M]

6.9 The vector that is NOT perpendicular to the vectors $(i + j + k)$ and $(i + 2j + 3k)$ is _____.

(a) $(i - 2j + k)$ (b) $(-i + 2j - k)$

(c) $(0i + 0j + 0k)$ (d) $(4i + 3j + 5k)$

[2016 : 1 M]

6.10 The angle between two vectors $X_1 = [2\ 6\ 14]^T$ and $X_2 = [-12\ 8\ 16]^T$ in radian is ________.

[2017 : 2 M]

6.11 Given $\vec{F} = (x^2 - 2y)\vec{i} - 4yz\vec{j} + 4xz^2\vec{k}$, the value of the line integral $\int_c \vec{F} \cdot d\vec{l}$ along the straight line c from (0, 0, 0) to (1, 1, 1) is

(a) $\frac{3}{16}$ (b) 0

(c) $\frac{-5}{12}$ (d) –1

[2018 : 2 M]

6.12 The vector function $\vec{A}$ is given by $\vec{A} = \vec{\nabla}u$, where $u(x, y)$ is a scalar function, Then $|\vec{\nabla} \times \vec{A}|$

(a) –1 (b) 0

(c) 1 (d) ∞

[2019 : 1 M]

6.13 $\vec{a}, \vec{b}, \vec{c}$ are three orthogonal vectors, Given that $\vec{a} = \hat{i} + 2\hat{j} + 5\hat{k}$ and $\vec{b} = \hat{i} + 2\hat{j} - \hat{k}$, the vector $\vec{c}$ is parallel to

(a) $\hat{i} + 2\hat{j} + 3\hat{k}$ (b) $2\hat{i} + \hat{j}$

(c) $2\hat{i} - \hat{j}$ (d) $4\hat{k}$ [2019 : 1 M]

6.14 The unit vectors along the mutually perpendicular x, y and z axes are $\hat{i}, \hat{j}$ and $\hat{k}$ respectively. Consider the plane $z = 0$ and two vector $\vec{a}$ and $\vec{b}$ on that plane such that $\vec{a} \neq \alpha\vec{b}$ for any scalar α. A vector perpendicular to both $\vec{a}$ and $\vec{b}$ is____.

(a) $\hat{i} - \hat{j}$ (b) $-\hat{j}$

(c) $\hat{k}$ (d) $\hat{i}$

[2020 : 1 M]

Answers Vector Calculus

6.1 (a)	6.2 (a)	6.3 (a)	6.4 (a)	6.5 (a)	6.6 (d)	6.7 (1)
6.8 (a)	6.9 (d)	6.10 (0.723)	6.11 (d)	6.12 (b)	6.13 (c)	6.14 (c)

Explanations Vector Calculus

6.1 (a)

Given:

$$\vec{x}_1 = [5 \ -2 \ 1]^T$$

$$\vec{x}_2 = [4 \ -1 \ 2]^T$$

$$A = \frac{1}{3}\begin{bmatrix} 2 & 1 & 2 \\ -2 & 2 & 1 \\ 1 & 2 & -2 \end{bmatrix}$$

$$\vec{y}_1 = A\vec{x}_1$$

$$\therefore \quad \vec{y}_1 = \frac{1}{3}\begin{bmatrix} 2 & 1 & 2 \\ -2 & 2 & 1 \\ 1 & 2 & -2 \end{bmatrix}\begin{bmatrix} 5 \\ -2 \\ 1 \end{bmatrix} = \frac{1}{3}\begin{bmatrix} 10 \\ -13 \\ -1 \end{bmatrix}$$

$$\text{and } \vec{y}_2 = \frac{1}{3}\begin{bmatrix} 2 & 1 & 2 \\ -2 & 2 & 1 \\ 1 & 2 & -2 \end{bmatrix}\begin{bmatrix} 4 \\ -1 \\ 2 \end{bmatrix} = \frac{1}{3}\begin{bmatrix} 11 \\ -8 \\ -2 \end{bmatrix}$$

Cosine of angle between

$$\vec{y}_1 \text{ and } \vec{y}_2 = \frac{\vec{y}_1 \cdot \vec{y}_2}{|\vec{y}_1||\vec{y}_2|}$$

$$= \frac{\frac{1}{9}(110 + 104 + 2)}{\left(\frac{1}{3}\sqrt{100 + 169 + 1}\right)\left(\frac{1}{3}\sqrt{121 + 64 + 4}\right)}$$

$$= \frac{216}{27\sqrt{70}} = \frac{8}{\sqrt{70}}$$

6.2 (a)

Let $\vec{R}(t) = x(t)\hat{i} + y(t)\hat{j} + z(t)\hat{k}$

$$\frac{d\vec{R}(t)}{dt} = \frac{dx(t)}{dt}\hat{i} + \frac{dy(t)}{dt}\hat{j} + \frac{dz(t)}{dt}\hat{k}$$

$$\vec{R}(t) \cdot \frac{d\vec{R}(t)}{dt} = x(t)\frac{dx(t)}{dt} + y(t)\frac{dy(t)}{dt} + z(t)\frac{dz(t)}{dt} \quad \text{...(i)}$$

as $\left|\vec{R}(t)\right| = \text{Constant} = R$

$$\therefore \quad \{x(t)\}^2 + \{y(t)\}^2 + \{z(t)\}^2 = R^2 \quad \text{...(ii)}$$

By differentiating equation (ii)

$$2x(t)\frac{dx(t)}{dt} + 2y(t)\frac{dy(t)}{dt} + 2z(t)\frac{dz(t)}{dt} = 0$$

$$\therefore \quad x(t)\frac{dx(t)}{dt} + y(t)\frac{dy(t)}{dt} + z(t)\frac{dz(t)}{dt} = 0$$

$$\therefore \quad \vec{R}(t)\cdot\frac{d\vec{R}(t)}{dt} = 0$$

6.3 (a)

$$f = x^{2/3} + y^{2/3}$$

Gradient of $f(x, y) = \frac{\partial f}{\partial x}\hat{i} + \frac{\partial f}{\partial y}\hat{j}$

At point (8, 8),

$$\text{grad} f(x,y)\Big|_{(8,8)} = \left(\frac{2}{3}\cdot x^{-1/3}\,\hat{i} + \frac{2}{3}\cdot y^{-1/3}\hat{j}\right)\Bigg|_{(8,8)}$$

$$= \frac{2}{3}(8)^{-1/3}\hat{i} + \frac{2}{3}(8)^{-1/3}\hat{j}$$

$$= \frac{1}{3}\hat{i} + \frac{1}{3}\hat{j}$$

And the magnitude

$$= \sqrt{\left(\frac{1}{3}\right)^2 + \left(\frac{1}{3}\right)^2} = \frac{1}{3}\cdot\sqrt{2}$$

6.4 (a)

Unit normal at point $P(x, y\ z)$ on the surface of sphere

$$= \frac{x\hat{i} + y\hat{j} + z\hat{k}}{\sqrt{x^2 + y^2 + z^2}}$$

Radius of sphere $= \sqrt{x^2 + y^2 + z^2} = 1$

$\therefore$ unit normal $= (x\hat{i} + y\hat{j} + z\hat{k}) = (x, y, z)$

6.5 (a)

$$\vec{A} = kr^n\,\hat{i}_r$$

$$\nabla\cdot\vec{A} = \frac{1}{r^2}\frac{\partial}{\partial r}(r^2\cdot kr^n) = \frac{1}{r^2}\cdot\frac{\partial}{\partial r}(kr^{n+2})$$

$$= (n+2)\,k\cdot\frac{r^{n+1}}{r^2} = (n+2)k\,r^{n-1}$$

For $\nabla\cdot\vec{A} = 0$

$n + 2 = 0$

$\Rightarrow \quad n = -2$

6.6 (d)

Option (d) is not true as irrotational vector has cross product as zero. Thus for vector to be irrotational $\nabla\times E = 0$.

6.7 (1)

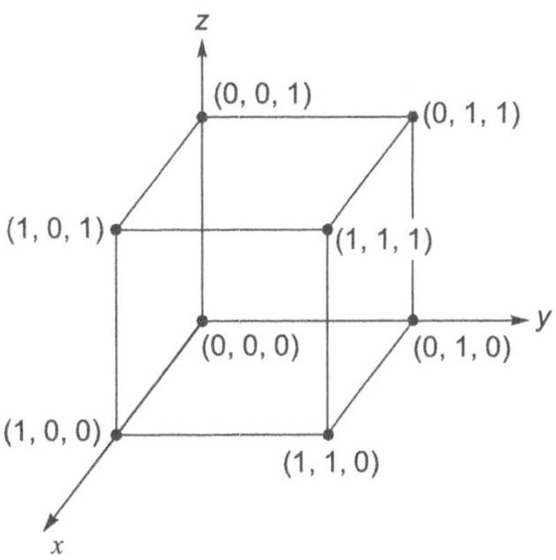

$$\vec{F} = y\hat{i} + x\hat{j} + z\hat{k}$$

$$\nabla\cdot\vec{F} = \frac{\partial}{\partial x}(y) + \frac{\partial}{\partial y}(x) + \frac{\partial}{\partial z}(z)$$

$$= 0 + 0 + 1 = 1$$

By Gauss divergence theorem,

$$\int_S \vec{F}\cdot\hat{n}ds = \int_V \nabla\cdot\vec{F}dv = \int_V 1\,dV = V = 1$$

Where V is volume of the cube (since given cube is unit cube)

$$V = l^3 = (1)^3 = 1$$

6.8 (a)

$f(x, y) = x^2 + 3y^2$

$\phi = x^2 + y^2 - 2$

and point $P \Rightarrow (1, 1)$

Normal to the surface,

$$\nabla\phi = \hat{i}\frac{\partial\phi}{\partial x} + \hat{j}\frac{\partial\phi}{\partial y} = 2x\hat{i} + 2y\hat{j}$$

$$\nabla\phi\big|_{\text{at }P(1,1)} = 2\hat{i} + 2\hat{j}$$

the normal vector is $\vec{a} = 2\hat{i} + 2\hat{j}$

Magnitude of directional derivative of f along $\vec{a}$ at (1, 1) is $\Rightarrow \nabla\cdot f\cdot\hat{a}$

$$\nabla f = \hat{i}\frac{\partial f}{\partial x} + \hat{j}\frac{\partial f}{\partial y} = 2x\hat{i} + 6y\hat{j}$$

$$\nabla f\big|_{(1,1)} = 2\hat{i} + 6\hat{j}$$

$$|\vec{a}| = \sqrt{4+4} = 2\sqrt{2}$$

$$\hat{a} = \frac{2\hat{i} + 2\hat{j}}{2\sqrt{2}} = \frac{\hat{i} + \hat{j}}{\sqrt{2}}$$

$\therefore$ Magnitude of directional derivative

$$= (2\hat{i} + 6\hat{j})\left(\frac{\hat{i}+\hat{j}}{\sqrt{2}}\right)$$

$$= \frac{2+6}{\sqrt{2}} = \frac{8}{\sqrt{2}} = 4\sqrt{2}$$

6.9 (d)

We know that if $\vec{a}$ and $\vec{b}$ are perpendicular

then $\qquad \vec{a}\cdot\vec{b} = 0$

options (a), (b), (c) are perpendicular.

options (d) is not perpendicular.

6.10 (0.723)

$$\bar{x}_1 = 2\bar{i} + 6\bar{j} + 14\bar{k}$$

$$\bar{x}_2 = -12\bar{i} + 8\bar{j} + 16\bar{k}$$

$$\cos\theta = \frac{\bar{a}\cdot\bar{b}}{|\bar{a}||\bar{b}|}$$

$$= \frac{(2\bar{i}+6\bar{j}+14\bar{k})(-12\bar{i}+8\bar{j}+16\bar{k})}{\sqrt{4+36+196}\sqrt{144+64+256}}$$

$$= \frac{-24+48+224}{(15.36)(21.54)} = \frac{248}{330.8544} = 0.7495$$

$$\theta = \cos^{-1}(0.7495) = 41.45°$$

$$= 0.723 \text{ radians}$$

6.11 (d)

$$x = y = z$$

$$\Rightarrow \qquad dx = dy = dz$$

$$\int_C \vec{F}\cdot d\vec{l} = \int_C \left(x^2 - 2y\right)dx - 4yzdy + 4xz^2dz$$

$$= \int_0^1 \left(x^2 - 2x - 4x^2 + 4x^3\right)dx$$

$$= -3\left[\frac{x^3}{3}\right]_0^1 - 2\left[\frac{x^2}{2}\right]_0^1 + 4\left[\frac{x^4}{4}\right]_0^1$$

$$= -1 - 1 + 1 = -1$$

6.12 (b)

Given that $u(x, y)$ is a scaler point function,

and $\qquad \bar{A} = \nabla u$

$$\nabla\times\bar{A} = \nabla\times(\nabla u) = 0$$

Vector identity Curl (gradϕ) = 0

6.13 (c)

$\bar{a}, \bar{b}, \bar{c}$ are mutually perpendicular

$\therefore\quad \bar{c}$ is parallel is $\bar{a}\times\bar{b}$

$$\bar{a}\times\bar{b} = \begin{vmatrix} \bar{i} & \bar{j} & \bar{k} \\ 1 & 2 & 5 \\ 1 & 2 & -1 \end{vmatrix}$$

$$= \hat{i}(-12) - \hat{j}(-6) + 0\hat{k}$$

$$= -6\left(2\hat{i} - \hat{j}\right)$$

6.14 (c)

Given z = 0 and $\bar{a} \neq \alpha\bar{b}$

So, it is clear that both $\bar{a}, \bar{b}$ are ine xy-plane.

$\therefore\quad$ The vector perpendicular to both $\bar{a}, \bar{b}$ will be in the direction of z-axis.

■■■

7 Complex Numbers

7.1 The real part of the complex number $z = x + iy$ is given by

(a) $\text{Re}(z) = z - z^*$ (b) $\text{Re}(z) = \dfrac{z - z^*}{2}$

(c) $\text{Re}(z) = \dfrac{z + z^*}{2}$ (d) $\text{Re}(z) = z + z^*$

[1994 : 1 M]

7.2 $\cos\phi$ can be presented as

(a) $\dfrac{e^{i\phi} - e^{-i\phi}}{2}$ (b) $\dfrac{e^{i\phi} - e^{-i\phi}}{2i}$

(c) $\dfrac{e^{i\phi} + e^{-i\phi}}{i}$ (d) $\dfrac{e^{i\phi} + e^{-i\phi}}{2}$

[1994 : 1 M]

7.3 The complex number $z = x + iy$ which satisfy the equation $|z + 1| = 1$ lie on

(a) a circle with (1, 0) as the centre and radius 1
(b) a circle with (–1, 0) as the centre and radius 1
(c) y – axis
(d) x – axis

[1997 : 2 M]

7.4 Consider the circle $|z - 5 - 5i| = 2$ in the complex plane (x, y) with $z = x + iy$. The minimum distance from the origin to the circle is

(a) $5\sqrt{2} - 2$ (b) $\sqrt{54}$

(c) $\sqrt{34}$ (d) $5\sqrt{2}$

[2005 : 2 M]

7.5 Let $z^3 = \bar{z}$, where z is a complex number not equal to zero. Then z is a solution of

(a) $z^2 = 1$ (b) $z^3 = 1$
(c) $z^4 = 1$ (d) $z^9 = 1$

[2005 : 2 M]

7.6 The value of the integral of the complex function

$$f(s) = \frac{3s + 4}{(s + 1)(s + 2)}$$

along the path $|s| = 3$ is

(a) $2\pi j$ (b) $4\pi j$
(c) $6\pi j$ (d) $8\pi j$

[2006 : 2 M]

7.7 Let $j = \sqrt{-1}$. Then one value of j^j is

(a) $\sqrt{j}$ (b) -1
(c) $\pi/2$ (d) $e^{-\pi/2}$

[2007 : 1 M]

7.8 For the function $\dfrac{\sin z}{z^3}$ of a complex variable z, the point $z = 0$ is

(a) a pole of order 3 (b) a pole of order 2
(c) a pole of order 1 (d) not a singularity

[2007 : 2 M]

7.9 A complex variable $Z = x + j0.1$ has its real part x varying in the range $-\infty$ to $+\infty$. Which one of the following is the locus (shown in thick lines) of $1/Z$ in the complex plane?

(a)

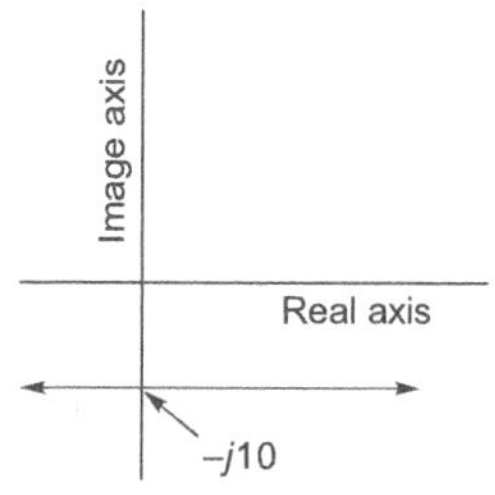

(b)

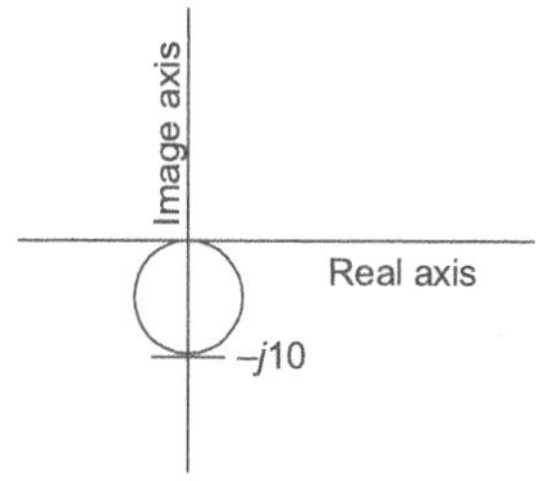

(c)

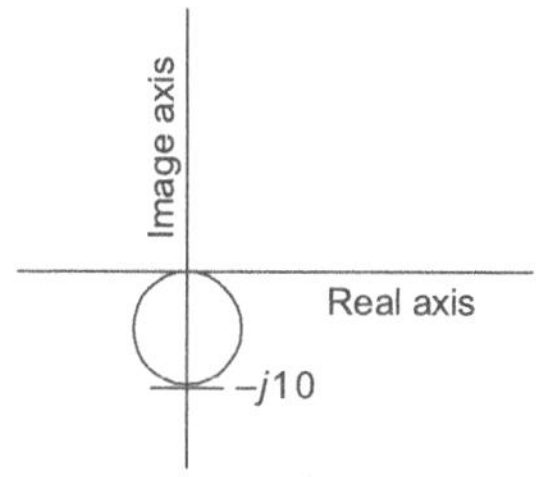

(d)

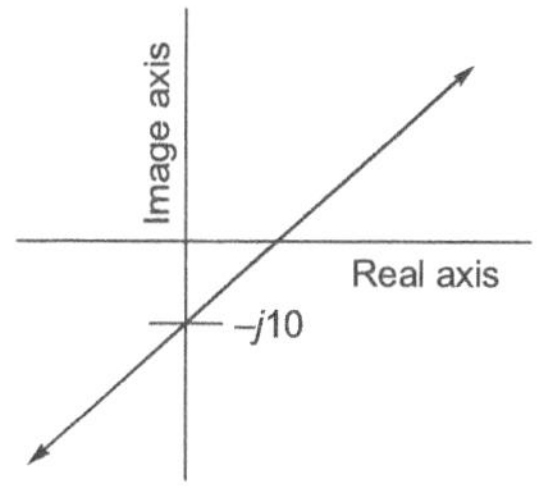

[2008 : 2 M]

7.10 If $z = x + jy$, where x and y are real, the value of $\left|e^{jz}\right|$ is

(a) 1 (b) $e\sqrt{x^2 + y^2}$
(c) e^y (d) e^{-y}

[2009 : 1 M]

7.11 The value of $\oint \frac{\sin z}{z} dz$, where the counter of integration is a simple closed curve around the origin, is

(a) 0 (b) $2\pi j$
(c) ∞ (d) $1/(2\pi j)$

[2009 : 1 M]

7.12 Contour C in the adjoining figure is described by $x^2 + y^2 = 16$. The value $\oint_C \frac{z^2+8}{0.5z - 1.5j} dz$ is

(Note: $j = \sqrt{-1}$)

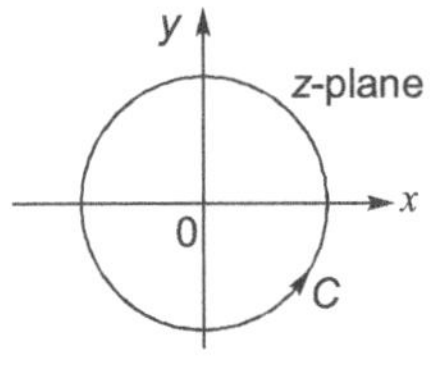

(a) $-2\pi j$ (b) $2\pi j$
(c) $4\pi j$ (d) $-4\pi j$

[2010 : 1 M]

7.13 The contour integral $\oint_C e^{1/z} dz$ with C as the counter-clockwise unit circle in the z-plane is equal to

(a) 0 (b) 2π
(c) $2\pi\sqrt{-1}$ (d) ∞

[2011 : 1 M]

7.14 If $x = \sqrt{-1}$, then the value of x^x is

(a) $e^{-\pi/2}$ (b) $e^{\pi/2}$
(c) x (d) 1

[2012 : 1 M]

7.15 Given: $f(z) = \frac{1}{z+1} - \frac{2}{z+3}$. If C is a counter clockwise path in the z-plane such that $|z+1| = 1$, the value of $\frac{1}{2\pi j}\oint_C f(z)\, dz$ is

(a) –2 (b) –1
(c) 1 (d) 2

[2012 : 1 M]

7.16 The complex function tanh(s) is analytic over a region of the imaginary axis of the complex s-plane if the following is TRUE everywhere in the region for all integers n

(a) $\text{Re}(s) = 0$ (b) $Im(s) \neq n\pi$
(c) $Im(s) \neq \frac{n\pi}{3}$ (d) $Im(s) \neq \frac{(2n+1)\pi}{2}$

[2013 : 1 M]

7.17 The value of $\oint \frac{1}{z^2} dz$, where the contour is the unit circle traversed clockwise, is

(a) $-2\pi i$ (b) 0
(c) $2\pi i$ (d) $4\pi i$

[2015 : 1 M]

7.18 In the neighborhood of $z = 1$, the function $f(z)$ has a power series expansion of the form
$f(z) = 1 + (1 - z) + (1 - z)^2 + \ldots\ldots\ldots$
Then $f(z)$ is

(a) $\frac{1}{z}$ (b) $\frac{-1}{z-2}$
(c) $\frac{z-1}{z+2}$ (d) $\frac{1}{2z-1}$

[2016 : 1 M]

7.19 The value of the integral $\frac{1}{2\pi j}\int_C \frac{z^2+1}{z^2-1} dz$ where z is a complex number and C is a unit circle with center at $1 + 0j$ in the complex plane is_______.

[2016 : 2 M]

7.20 Let $z = x + jy$ where $j = \sqrt{-1}$. Then $\overline{\cos z} =$

(a) $\cos z$ (b) $\cos \bar{z}$
(c) $\sin z$ (d) $\sin \bar{z}$

[2017 : 1 M]

7.21 Let $f_1(z) = z^2$ and $f_2(z) = \bar{z}$ be two complex variable function. Here $\bar{z}$ is the complex conjugate of z. Choose the correct answer.

(a) Both $f_1(z)$ and $f_2(z)$ are analytic
(b) Only $f_1(z)$ is analytic
(c) Only $f_2(z)$ is analytic
(d) Both $f_1(z)$ and $f_2(z)$ are not analytic

[2018 : 1 M]

7.22 A complex function $f(z) = u(x, y) + iv(x, y)$ and its complex conjugate, $f^*(z) = u(x, y) - iv(x, y)$ are both analytic in the entire complex plane, where $z = x + iy$ and $i = \sqrt{-1}$. The function f is then given by

(a) $f(z) = x + iy$ (b) $f(z) = x^2 - y^2 + i2xy$
(c) $f(z) = \text{constant}$ (d) $f(z) = x^2 + y^2$

[2019 : 2 M]

7.23 Let $f(x) = \frac{1}{z+a}, a > 0$. The value of the integral $\oint f(x)dx$ over a circle C with centre $(-a, 0)$ and radius $R > 0$ evaluated in the anti-clockwise direction is _____.

(a) 0 (b) $2\pi i$
(c) $-2\pi i$ (d) $4\pi i$

[2020 : 1 M]

7.24 Let $f(z) = \frac{1}{z^2 + 6z + 9}$ defined in the complex plane. The integral $\oint_c f(z)dz$ over the contour of a circle c with center at the origin and unit radius is ______.

[2021 : 1 M]

7.25 $f(z) = (z-1)^{-1} - 1 + (z-1) - (z-1)^2 + \ldots$ is the series expansion of

(a) $\frac{1}{(z-1)^2}$ for $|z-1| < 1$

(b) $\frac{1}{z(z-1)}$ for $|z-1| < 1$

(c) $\frac{-1}{z(z-1)}$ for $|z-1| < 1$

(d) $\frac{-1}{(z-1)}$ for $|z-1| < 1$

[2021 : 2 M]

7.26 For the complex number $Z = \frac{a+jb}{a-jb}$, where $a > 0$ and $b > 0$. Which of the following statement(s) is/are true?

(a) The phase is $2\tan^{-1}\frac{b}{a}$

(b) The phase is $\tan^{-1}\frac{2b}{a}$

(c) The magnitude is 1

(d) The magnitude is $\sqrt{\frac{a^2+b^2}{a^2-b^2}}$

[2022 : 2 M]

7.27 Consider thefunction $f(z) = \frac{1}{(z+1)(z+2)(z+3)}$.

The residue of $f(z)$ at $z = -1$, is ______

[2022 : 2 M]

7.28 Let $f(z) = j\frac{1-z}{1+z}$, where z denotes a complex number and j denotes $\sqrt{-1}$. The inverse function $f^{-1}(z)$ maps the real axis to the________.

(a) unit circle with centre at the origin
(b) unit circle with centre not at the origin
(c) imaginary axis
(d) real axis

[2023 : 2 M]

7.29 Let $z = x + iy$ be a complex variable and $\bar{z}$ be its complex conjugate. The equation $\bar{z}^2 + z^2 = 2$ represents a

(a) parabola (b) hyperbola
(c) ellipse (d) circle

[2024 : 1 M]

7.30 The complex functions $f(z) = u(x, y) + iv(x, y)$ and $\overline{f(z)} = u(x,y) - iv(x,y)$ are both analytic in a given domain. Choose the correct option(s) from the following.

(a) $\frac{\partial u}{\partial x} = \frac{\partial v}{\partial y} = 0$ (b) $\frac{\partial u}{\partial y} = -\frac{\partial v}{\partial x} \neq 0$

(c) $\frac{df(z)}{dz} = 0$ (d) $\frac{df(z)}{dz} \neq 0$

[2024 : 2 M]

■■■

Answers Complex Numbers

7.1	(c)	7.2	(d)	7.3	(b)	7.4	(a)	7.5	(c)	7.6	(c)	7.7	(d)
7.8	(b)	7.9	(b)	7.10	(d)	7.11	(a)	7.12	(d)	7.13	(c)	7.14	(a)
7.15	(c)	7.16	(d)	7.17	(b)	7.18	(a)	7.19	(1)	7.20	(b)	7.21	(b)
7.22	(c)	7.23	(b)	7.24	(0)	7.25	(b)	7.26	(a, c)	7.27	(0.50)	7.28	(a)
7.29	(b)	7.30	(a, c)										

Explanations Complex Numbers

7.1 (c)

$$z = x + iy$$
$$z^* = x - iy$$
$$z + z^* = 2x$$
$$\therefore \quad x = \frac{z+z^*}{2}$$
$$\therefore \quad x = \text{Re}\, z = \frac{z+z^*}{2}$$

7.2 (d)

$$e^{i\phi} = \cos\phi + i\sin\phi$$
$$e^{-i\phi} = \cos\phi - i\sin\phi$$
$$e^{i\phi} + e^{-i\phi} = 2\cos\phi$$
$$\cos\phi = \frac{e^{i\phi}+e^{-i\phi}}{2}$$

7.3 (b)

$$|z+1| = 1$$
$$|x+iy+1| = 1$$
$$|(x+1+iy)| = 1$$
$$(x+1)^2 + y^2 = 1$$

Centre (–1, 0) radius 1.

7.4 (a)

$$|z-(5+5i)| = 2$$

Radius of the circle is 2 and centre is at $5 + 5i$

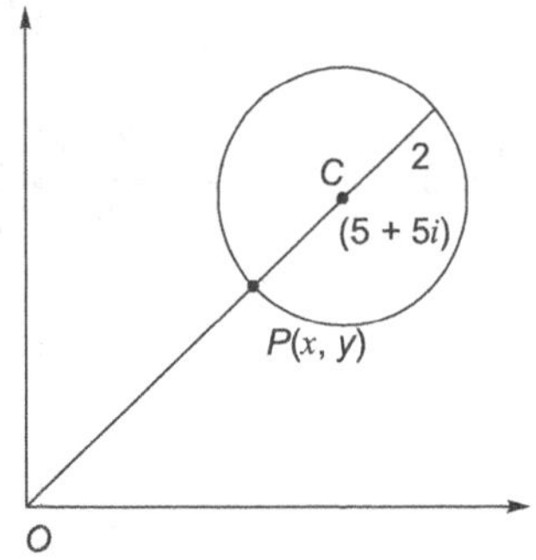

For distance to be min. The point P will be on the line passes though origin and centre of the circle.

$\therefore$ Slope of line OP = Slope of line OC

$$\frac{y}{x} = \frac{5}{5}$$
$$\Rightarrow \quad x = y$$
$$(PC)^2 = (\text{radius of the circle})^2$$
$$(5-x)^2 + (5-y)^2 = 4$$
$$(5-x)^2 = 2$$
$$\therefore \quad x = 5-\sqrt{2}$$
$$OP = \sqrt{2}\cdot x = 5\sqrt{2}-2$$

7.5 (c)

$$z^3 = \bar{z}$$

multiply both the sides by z, we get

$$\Rightarrow \quad z^4 = \bar{z}\cdot z$$
$$z^4 = |z|^2$$
$$|z|^4 e^{j4\theta} = |z|^2 \quad (\text{where } \theta \text{ is angle of z})$$

since $|z|^2$ is a real quantity so in order to satisfy above equation $e^{j4\theta}$ has to be real quantity = 1

and $\theta = \frac{n\pi}{2}$,

(where $n = \pm 1 + 2.....$)

so, $|z|^2 = 1$

taking $\theta = \pi/2$

$$z = |z|e^{j\pi/2}$$
$$z^4 = |z|^4 e^{j2\pi}$$
$$\boxed{z^4 = 1}$$

7.6 (c)

$$\text{Given: } f(s) = \frac{3s+4}{(s+1)(s+2)}$$
$$= \frac{1}{s+1} + \frac{2}{s+2} = f_1(s) + f_2(s)$$

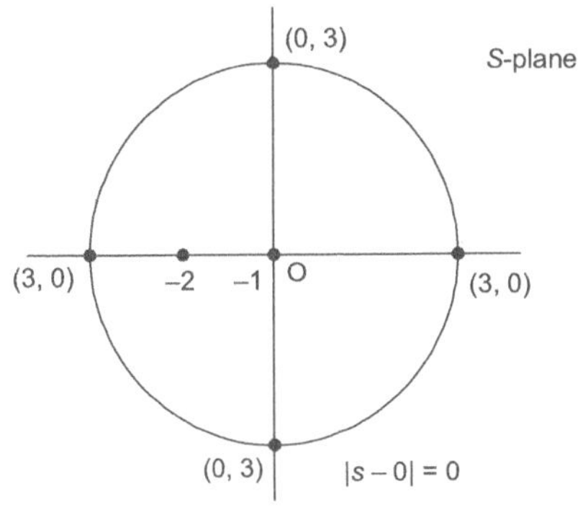

$$\oint_c \frac{3s+4}{(s+1)(s+2)}ds = \oint_c \frac{1}{(s+1)}ds + \oint_c \frac{2}{(s+2)}ds$$

$$= 2\pi j + 2\pi j(2) = 6\pi j$$

7.7 (d)

Let, $y = j^j$

$y = (e^{j\pi/2})^j$

$\Rightarrow \quad ln\ y = jln(e^{j\pi/2}) = j \cdot j\frac{\pi}{2}$

$ln\ y = -\frac{\pi}{2}$

$y = e^{-\pi/2}$

7.8 (b)

$$f(x) = \frac{\sin z}{z^3} = \frac{1}{z^3}\left[z - \frac{z^3}{3!} + \frac{z^5}{5!} - \frac{z^7}{7!} + ...\right]$$

$$= \frac{1}{z^2} - \frac{1}{3!} + \frac{z^2}{5!} - \frac{z^4}{7!} + ...$$

$\therefore$ The function $f(x)$ has the pole of order 2.

7.9 (b)

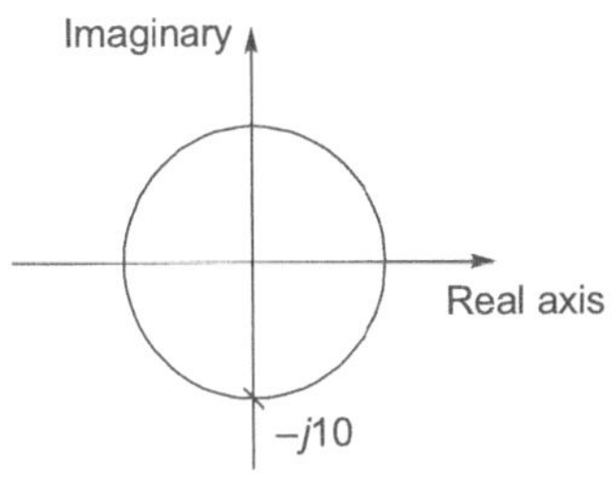

$Z = x + j0.1$

$$\Rightarrow \quad z^{-1} = \frac{1}{x + j0.1} = \frac{x - j0.1}{x^2 + (0.1)^2}$$

$$= \frac{x}{x^2 + (0.1)^2} - \frac{j0.1}{x^2 + (0.1)^2} = u + iv$$

$$\Rightarrow \quad u = \frac{x}{x^2 + (0.1)^2} \quad ...(i)$$

$$v = \frac{-0.1}{x^2 + (0.1)^2}$$

$(i)^2 + (ii)^2$,

$$u^2 + v^2 = \frac{x^2 + (0.1)^2}{\left[x^2 + (0.1)^2\right]^2} = \frac{1}{x^2 + (0.1)^2}$$

$$\Rightarrow u^2 + v^2 = \left[\frac{0.1}{x^2 + (0.1)^2}\right]10$$

Using equation (ii),

$u^2 + v^2 = -10\ V$

$\Rightarrow\ u^2 + v^2 + 10\ V = 0$

$\Rightarrow\ u^2 + (v + 5)^2 = 25$

$\Rightarrow$ Required locus

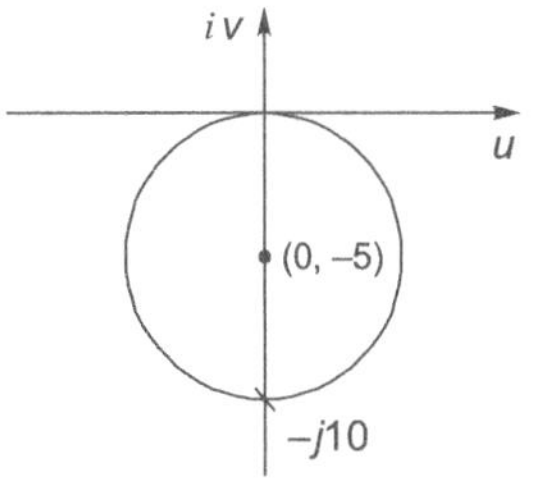

7.10 (d)

$f = e^{jz} = \cos z + j \sin z$

$= \cos (x + jy) + j \sin (x + iy)$

$= \cos x \cdot \cos (jy) - \sin i \cdot \sin(iy)$

$+ j\{\sin \cdot \cos (iy) + \cos \cdot \sin(iy)\}$

We know,

$\sin i\theta = i \sin h\theta$

$\cos i\theta = \cos h\theta$

$\therefore \quad f = \cos x \cos hy - j \sin x \sin hy$

$+ j \sin x \cos hy - \cos x \sin hy$

$= \cos x (\cos hy - \sin hy)$

$+ j \sin x (\cos hy - \sin hy)$

$\cos hy - \sin hy = e^{-y}$

$f = e^{-y} (\cos x + j \sin x)$

$|f| = e^{-y} \times 1 = e^{-y}$

Alternate solution:

$$|e^{+jz}| = |e^{j(x+iy)}| = |e^{(jx-y)}|$$

$$= |e^{jx} \cdot e^{-y}| = |e^{jx}| \cdot |e^{-y}|$$

$\because \quad |e^{jx}| = 1 \qquad$ [where x is real]

$\Rightarrow \quad |e^{jz}| = e^{-y}$

7.11 (a)

$$f(z) = \frac{\sin z}{z} \text{ has no poles as}$$

$$= \frac{1}{z}\left[z - \frac{z^3}{\underline{|3}} + \frac{z^5}{\underline{|5}} - \ldots\ldots\right]$$

$$= 1 - \frac{z^2}{\underline{|3}} + \frac{z^4}{\underline{|5}} - \ldots\ldots$$

So, no poles lies inside or on the given contour.
So, according to Cauchy integral theorem

$$\oint \frac{\sin z}{z} dz = 0$$

7.12 (d)

$$I = \oint_c \frac{z^2+8}{0.5z-1.5j} dz = \oint_c \frac{2z^2+16}{z-3j} dz$$

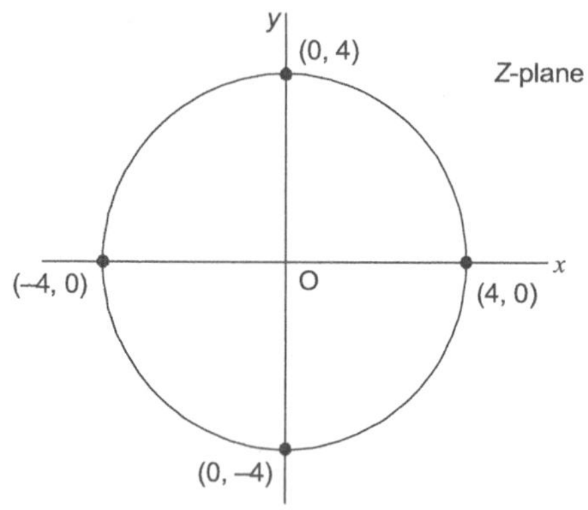

From Cauchy's integral formula

$$\oint_c \frac{f(\tau)dz}{z-a} = 2\pi j f(a) = 2\pi j[2(3j)^2 + 16]$$

$$= 2\pi j[-18 + 16] = -4\pi j$$

Alternate solution:

Using Cauchy residue theorem,

$$\oint_c \frac{z^2+8}{0.5z-1.5j} dz$$

$$= 2\pi i\left[\text{Sum of residue of } \frac{z^2+8}{0.5z-1.5j} \text{ at poles inside } c\right]$$

$$\underset{z\to 3j}{\text{Res}} \frac{z^2+8}{0.5z-1.5j} = \underset{z\to 3j}{\text{Res}} \frac{2z^2+16}{z-3j}$$

$$= \lim_{z\to 3j} \frac{(z-3j)(2z^2+16)}{(z-3j)} = 2(3j)^2 + 16 = -2$$

which is inside $c : x^2 + y^2 = 16$

$$\Rightarrow \oint_c \frac{z^2+8}{0.5z-1.5j} dz = 2\pi i[-2] = -4\pi i$$

7.13 (c)

$$f(z) = e^{1/z} = 1 + \frac{1}{z} + \frac{1}{2!\,z^2} + \frac{1}{3!\,z^3} + \ldots\infty$$

$$= 1 + z^{-1} + \frac{z^{-2}}{2!} + \frac{z^{-3}}{3!} + \ldots\infty$$

$f(z)$ has an isolated essential singularity at $z = 0$ and $z = 0$ is inside unit circle on complex plane

∴ Residue of $f(z)$ at $z = 1$ (coefficient of $1/z$)

$$\oint_C e^{1/z} dz = 2\pi i \text{ (Residue at } z = 0)$$

$$= 2\pi i \cdot 1 = 2\pi i$$

7.14 (a)

Given: $x = \sqrt{-1} = j = \left(\cos\frac{\pi}{2} + j\sin\frac{\pi}{2}\right) = e^{j\frac{\pi}{2}}$

Let, $x^x = y$

$$y = (j)^j = \left(e^{j\frac{\pi}{2}}\right)^j = e^{j^2\frac{\pi}{2}} = e^{-\frac{\pi}{2}}$$

7.15 (c)

$$f(z) = \left(\frac{1}{z+1} - \frac{2}{z+3}\right)$$

$|z + 1| = 1$ is a circle having centre at $z = -1$ and radius 1.

This will include the pole at $z = -1$ only.

Residue of function at $z = -1$ is

$$R = \lim_{z\to -1}(z+1)f(z)$$

$$R = \lim_{z\to -1}\left[1 - \frac{2(z+1)}{z+3}\right] = 1$$

So, $\oint_c f(z)dz = 2\pi j$[sum of residues at poles inside and at curve]

$$= 2\pi j[1]$$

$$\frac{1}{2\pi j}\oint_c f(z)dz = 1$$

7.16 (d)

$$\tanh s = \frac{e^s - e^{-s}}{e^s + e^{-s}}$$

it is analytic if $e^s + e^{-s} \neq 0$

∴ $e^s \neq e^{-s}$

$e^{2s} \neq -1$

$$s \neq \frac{i(2n+1)}{2}\pi$$

$$\therefore \quad Im(s) \neq \frac{(2n+1)}{2}\pi$$

7.17 (b)

Given, $\oint \frac{1}{z^2} dz$ where C is the unit circle. By Cauchy's residue theorem

$$\oint \frac{1}{z^2} dz = 2\pi i \text{ [sum of residues]}$$

$\because \frac{1}{z^2}$ is NOT analytical at $z = 0$

So, $z = 0$ is the pole of order 2.

$$\text{So, residue at } z = 0 = \frac{1}{\underline{2-1|}}\left[\frac{d}{dz} z^2 . \frac{1}{z^2}\right]_{z=0}$$

$$= 0$$

$$\text{So, } \oint \frac{1}{z^2} dz = 2\pi i[0] = 0$$

7.18 (a)

$$f(z) = 1 + (1-z) + (1-z)^2 + \ldots$$

$$= \frac{1}{1-(1-z)} = \frac{1}{1-1+z} = \frac{1}{z}$$

7.19 (1)

$$\frac{1}{2\pi j}\int_C \frac{z^2+1}{(z^2-1)} dz = \frac{1}{2\pi j}\int_C \frac{z^2+1}{(z-1)(z+1)} dz$$

Poles are at $z = 1, -1$

Given, circle is $|z-1| = 1$

pole $z = 1$ lies inside C

pole $z = -1$ lies outside C

Res $f(z)$ at $z = 1$ is

$$= \underset{z\to 1}{Lt}(z-1)\frac{z^2+1}{(z-1)(z+1)} = \frac{2}{2} = 1$$

Res $f(z)$ at $z = -1$ is $= 0$

By Cauchy's residue theorem

$$\frac{1}{2\pi j}\int_C \frac{z^2+1}{z^2-1} dz = \frac{1}{2\pi j} \times 2\pi j(1+0) = 1$$

7.20 (b)

$$\overline{\cos z}$$

$$= \overline{\cos(x+iy)} = \overline{\cos x \cos iy - \sin x \sin iy}$$

$$= \overline{\cos x \cosh y - i \sin x \sinh y}$$

$$= \cos x \cosh y + i \sin x \sinh y$$

$$= \cos x \cos iy + \sin x \sin iy = \cos(x - iy) = \cos \bar{z}$$

7.21 (b)

Only $f_1(z) = z^2$ is analytic

$f_2(z) = \bar{z}$ is not analytic

7.22 (c)

$$f(z) = u + iv$$

$$f^*(z) = u - iv$$

Both are analytic when $f(z)$ is constant.

7.23 (b)

Given $f(z) = \frac{1}{z+a}, a > 0$ and singularity of $f(z)$ is $z = -a$

$$\oint_c f(z)dz = \oint_c \frac{1}{z+a} dz, \text{ where 'c' : } |z + a| = R$$

singularity $(z = -a)$ lies inside 'c'

$\therefore$ By Cauchy integral formula

$$\oint_c \frac{1}{z-(-a)} dz = 2\pi i(1) = 2\pi i$$

7.24 (0)

$$f(z) = \frac{1}{z^2+6z+9} = \frac{1}{(z+3)^2}$$

Poles of $f(z)$, $(z + 3)^2 = 0$

$\Rightarrow$ $z = -3$ (Double pole)

$C : |z| = 1$

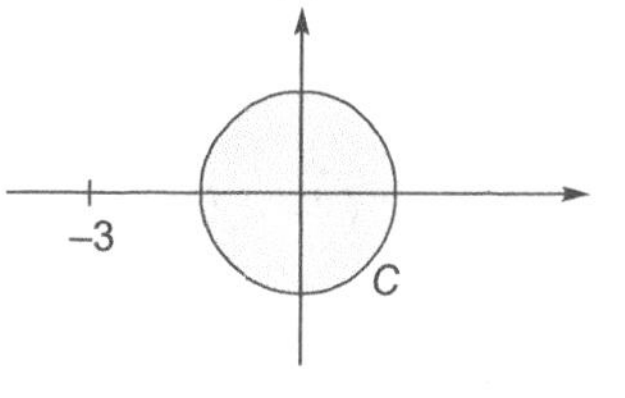

$\because$ Pole lies outside 'C' so by CIT, $\oint_C f(z)dz = 0$

7.25 (b)

$$f(z) = (z-1)^{-1}\left[1 - \frac{1}{(z-1)^{-1}} + \frac{(z-1)}{(z-1)^{-1}} - \frac{(z-1)^2}{(z-1)^{-1}} + \ldots\right]$$

$$= (z-1)^{-1}\left[1-(z-1)+(z-1)^2-(z-1)^3+\ldots\right]$$

We know that,

$(1 + x)^{-1} = 1 - x + x^2 - x^3 + \ldots$; for $|x| < 1$

$$= \frac{1}{(z-1)}[1+(z-1)]^{-1}; \text{ for } |z - 1| < 1$$

$$= \frac{1}{(z-1)}\left[z^{-1}\right]; \text{ for } |z - 1| < 1$$

$$= \frac{1}{z(z-1)}; \text{ for } |z - 1| < 1$$

7.26 (a, c)

For given complex number

$$Z = \frac{a+jb}{a-jb},\ a > 0,\ b > 0$$

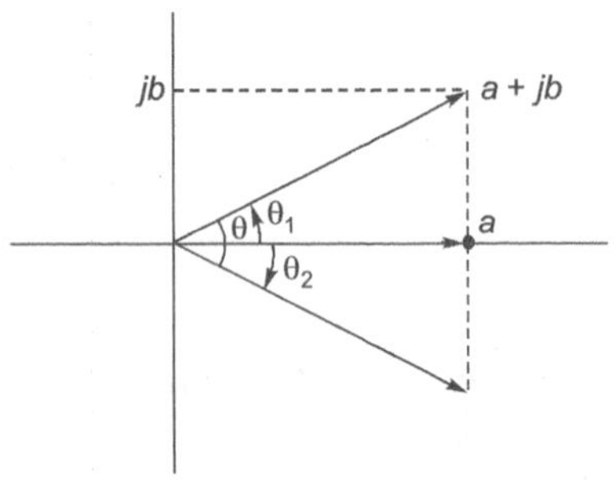

$$\text{Magnitude}, |Z| = \frac{\sqrt{a^2+b^2}}{\sqrt{a^2+b^2}} = 1$$

$$\text{Phase angle}, \theta = \theta_1 - (\theta_2)$$

$$\theta = \tan^{-1}\left(\frac{b}{a}\right) - \left(-\tan^{-1}\left(\frac{b}{a}\right)\right)$$

$$\theta = 2\tan^{-1}\left(\frac{b}{a}\right)$$

7.27 (0.50)

Given function,

$$f(z) = \frac{1}{(z+1)(z+2)(z+3)}$$

Residue at $z = -1$

$$\text{Res}\ (z = -1) = \lim_{z\to -1}(z+1)f(z)$$

$$= \lim_{z\to -1}\frac{1}{(z+2)(z+3)} = \frac{1}{(-1+2)(-1+3)}$$

$$\text{Res}\ (z = -1) = 0.5$$

7.28 (a)

$$\text{Given : } w = u + iv = \frac{i(1-z)}{(1+z)}$$

$$= \frac{i - ix - i^2y}{1+x+iy} = \frac{i(1-x)+y}{(1+x)+iy}\times\frac{(1+x)-iy}{(1+x)-iy}$$

$$u + iv = \frac{y(1-x)+y(1+x)}{(1+x)^2+y^2} + i\frac{(1-x^2)-y^2}{(1+x)^2+y^2}$$

To find image of real axis is w-plane.

We take $V = 0$

$$\Rightarrow \frac{(1-x^2)-y^2}{(1+x)^2+y^2} = 0$$

$$\Rightarrow 1 - x^2 - y^2 = 0$$

$$\Rightarrow x^2 + y^2 = 1$$

$\therefore$ Under the mapping of $f^{-1}(z)$ real axis in w-plane maps to unit circle in z-plane.

7.29 (b)

$$z = x + iy;\quad \bar{z} = x - iy$$

$$\bar{z}^2 + z^2 = 2$$

$$(x - iy)^2 + (x + iy)^2 = 2$$

$$x^2 + (iy)^2 - 2xyi + x^2 + (iy)^2 + 2xyi = 2$$

$$x^2 - y^2 + x^2 - y^2 = 2 \Rightarrow x^2 - y^2 = 1 \quad \text{....Hyperbola}$$

7.30 (a, c)

$\because f(z)$ is analytic,

$\Rightarrow$ C-R equation holds

$$\Rightarrow \quad \frac{\partial u}{\partial y} = \frac{-\partial v}{\partial x} \text{ and } \frac{\partial u}{\partial x} = \frac{\partial v}{\partial y}$$

$$\because \quad \overline{f(z)} = u - iv \text{ also analytic}$$

$\Rightarrow$ C-R equation holds

$$\text{Then} \quad \frac{\partial u}{\partial x} = \frac{-\partial v}{\partial y},\ \frac{\partial u}{\partial y} = \frac{\partial v}{\partial x}$$

$$\frac{\partial u}{\partial x} = \frac{\partial v}{\partial y} = \frac{-\partial v}{\partial y}$$

$$\Rightarrow \quad 2\frac{\partial v'}{\partial y} = 0 \Rightarrow \frac{\partial v}{\partial y} = 0$$

$$\therefore \quad \frac{\partial u}{\partial x} = \frac{\partial u}{\partial y} = 0 \Rightarrow \text{(a) is true}$$

$$\text{Similarly} \quad \frac{\partial u}{\partial y} = \frac{+\partial v}{\partial x} = 0$$

$$\text{WKT} \quad f'(z) = \frac{\partial u}{\partial x} + i\frac{\partial v}{\partial x}$$

$$= 0 + i0$$

$$= 0$$

$\therefore$ (c) is true

■■■

Electrical Circuits & Electrical Machines

UNIT

II

CONTENTS

1. Network Laws and Network Theorems 53
2. RL, RC, RLC Circuit and Transients 69
3. Two-Port Network 92
4. Electrical Machines 96

Electrical Circuits & Electrical Machines

Syllabus

Voltage and current sources : independent, dependent, ideal and practical; v-i relationships of resistor, inductor, mutual inductance and capacitor; transient analysis of RLC circuits with dc excitation.

Kirchhoff's laws, mesh and nodal analysis, superposition, Thevenin, Norton, maximum power transfer and reciprocity theorems.

Peak-, average- and rms values of ac quantities; apparent-, active- and reactive powers; phasor analysis, impedance and admittance; series and parallel resonance, locus diagrams, realization of basic filters with R, L and C elements. transient analysis of RLC circuits with ac excitation.

One-port and two-port networks, driving point impedance and admittance, open-, and short circuit parameters.

Single phase transformer : equivalent circuit, phasor diagram, open circuit and short circuit tests, regulation and efficiency; Three phase induction motors: principle of operation, types, performance, torque-speed characteristics, no-load and blocked rotor tests, equivalent circuit, starting and speed control; Types of losses and efficiency calculations of electric machines.

Analysis of Previous GATE Papers

Exam Year	1 Mark Ques.	2 Marks Ques.	Total Marks
1995	–	–	–
1996	–	–	–
1997	–	–	–
1998	1	–	**1**
1999	–	–	–
2000	–	–	–
2001	–	–	–
2002	–	1	**2**
2003	–	1	**2**
2004	–	–	–
2005	–	2	**4**
2006	1	4	**9**
2007	2	4	**10**
2008	2	7	**16**
2009	–	2	**4**

Exam Year	1 Mark Ques.	2 Marks Ques.	Total Marks
2010	2	2	**6**
2011	2	2	**6**
2012	4	6	**16**
2013	3	5	**13**
2014	3	4	**11**
2015	3	4	**11**
2016	3	3	**9**
2017	3	4	**11**
2018	3	4	**11**
2019	5	2	**9**
2020	2	3	**8**
2021	1	5	**11**
2022	2	4	**10**
2023	2	5	**12**
2024	3	4	**11**

1 Network Laws and Network Theorems

1.1 The output resistance across the terminals 1 and 2 of the DC bridge in figure is

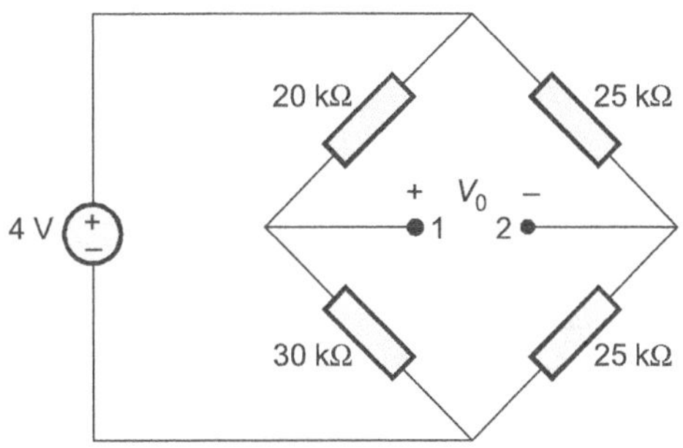

(a) 12.5 kΩ (b) 24.5 kΩ
(c) 25.0 kΩ (d) 100 kΩ

[2003 : 2 M]

1.2 The root-mean-square value of a voltage waveform consisting of a superimposition of 2 V dc and a 4 V peak-to-peak square wave is

(a) 2 V (b) $\sqrt{6}$ V

(c) $\sqrt{8}$ V (d) $\sqrt{12}$ V

[2006 : 1 M]

1.3 A metal wire has a uniform cross-section A, length l, and resistance R between its two end points. It is uniformly stretched so that its length becomes αl. The new resistance is

(a) αR (b) $\alpha^2 R$

(c) $\sqrt{\alpha} R$ (d) $e^{\alpha} R$

[2006 : 2 M]

1.4 In full sunlight, a solar cell has a short circuit current of 75 mA and a current of 70 mA for a terminal voltage of 0.6 V with a given load. The Thevenin resistance of the solar cell is

(a) 8 Ω (b) 8.6 Ω
(c) 120 Ω (d) 240 Ω

[2007 : 1 M]

1.5 The power supplied by the dc voltage source in the circuit shown below is

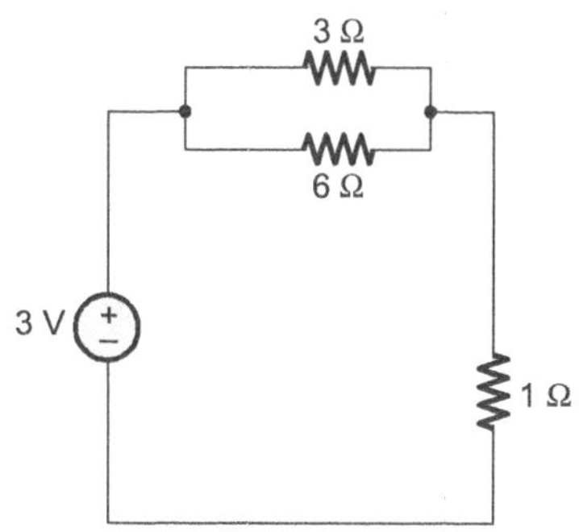

(a) 0 W (b) 1.0 W
(c) 2.5 W (d) 3.0 W

[2008 : 1 M]

1.6 The current I supplied by the dc voltage source in the circuit shown below is

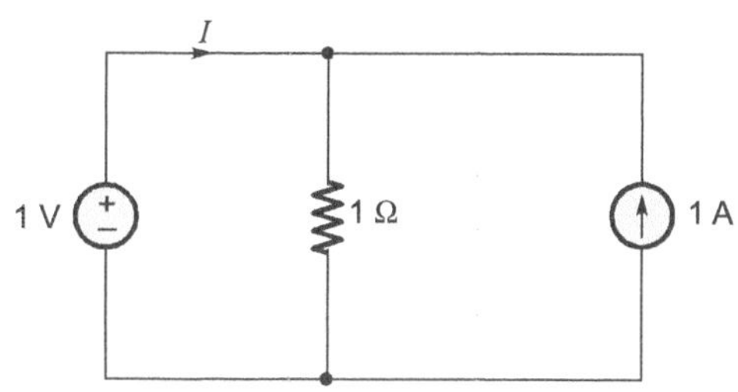

(a) 0 A (b) 0.5 A
(c) 1 A (d) 2 A

[2008 : 1 M]

1.7 In the circuit shown below the maximum power that can be transferred to the load Z_L is

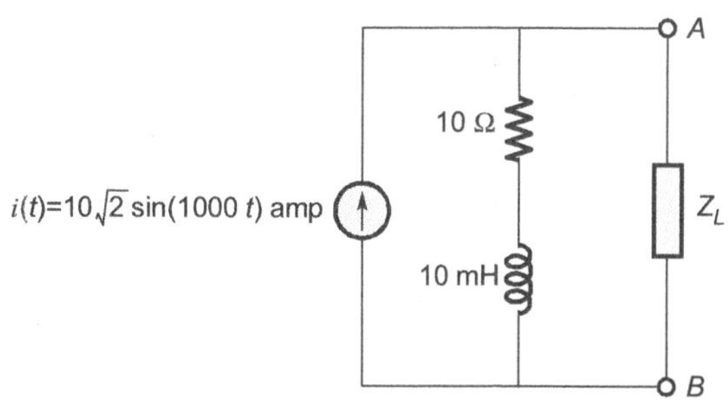

(a) 250 W (b) 500 W
(c) 1000 W (d) 2000 W

[2008 : 2 M]

1.8 In the circuit shown below the average power consumed by the 1 Ω resistor is

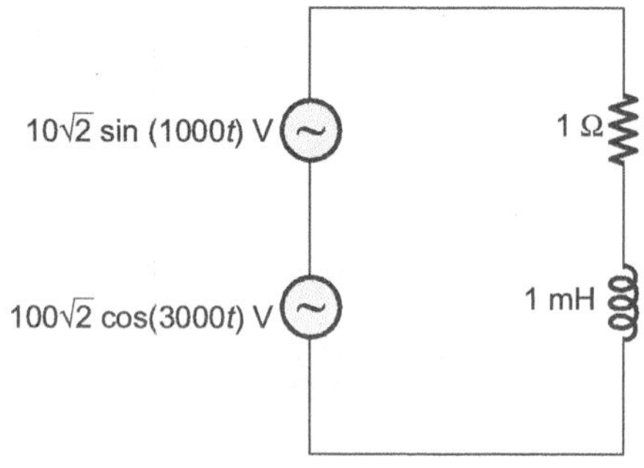

(a) 50 W (b) 1050 W
(c) 5000 W (d) 10100 W

[2008 : 2 M]

1.9 Which one of the following equations is valid for the circuit shown below?

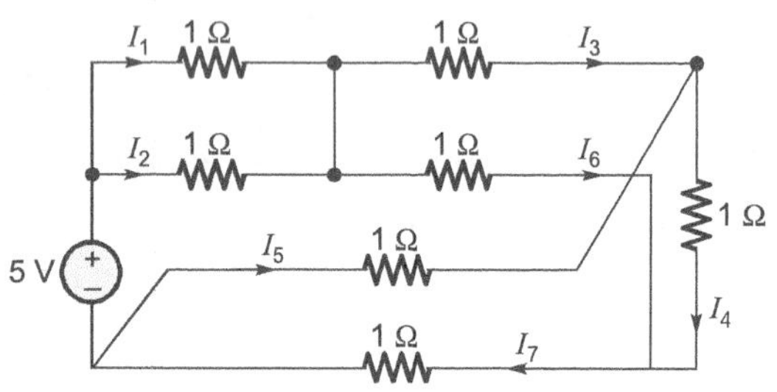

(a) $I_3 + I_5 - I_6 + I_7 = 0$
(b) $I_3 - I_5 + I_6 + I_7 = 0$
(c) $I_3 + I_5 + I_6 + I_7 = 0$
(d) $I_3 + I_5 + I_6 - I_7 = 0$

[2008 : 2 M]

1.10 The source network S is connected to the load network L as shown by dashed lines. The power transferred from S to L would be maximum when R_L is

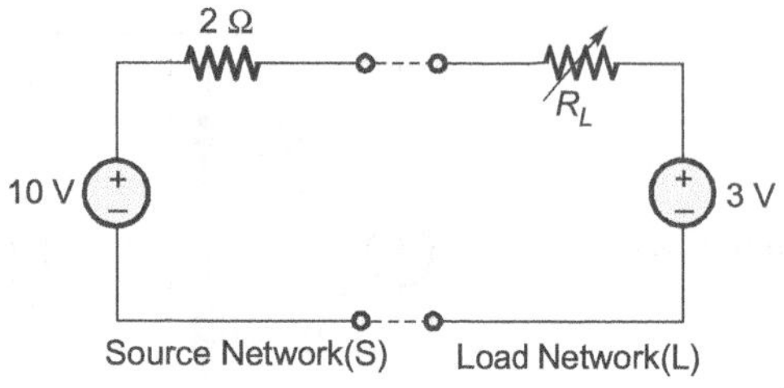

(a) 0 Ω (b) 0.6 Ω
(c) 0.8 Ω (d) 2 Ω

[2009 : 2 M]

1.11 The root mean squared value of $x(t) = 3 + 2\sin(t)\cos(2t)$ is

(a) $\sqrt{3}$ (b) $\sqrt{8}$
(c) $\sqrt{10}$ (d) $\sqrt{11}$

[2009 : 2 M]

1.12 In the dc circuit shown in the adjoining figure, the node voltage V_2 at steady state is

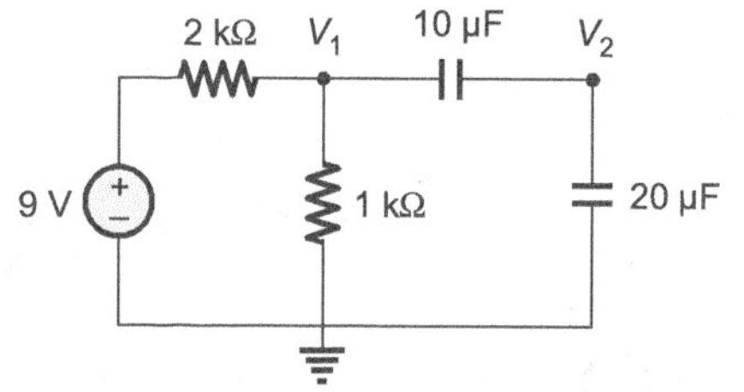

(a) 0 V (b) 1 V
(c) 2 V (d) 3 V

[2010 : 1 M]

1.13 A 100 W, 1 Ω resistor and a 800 W, 2 Ω resistor are connected in series. The maximum dc voltage that can be applied continuously to the series circuit without exceeding the power limit of any of the resistor is

(a) 90 V (b) 0 V
(c) 45 V (d) 40 V

[2010 : 1 M]

1.14 The current I shown in the circuit given below is equal to

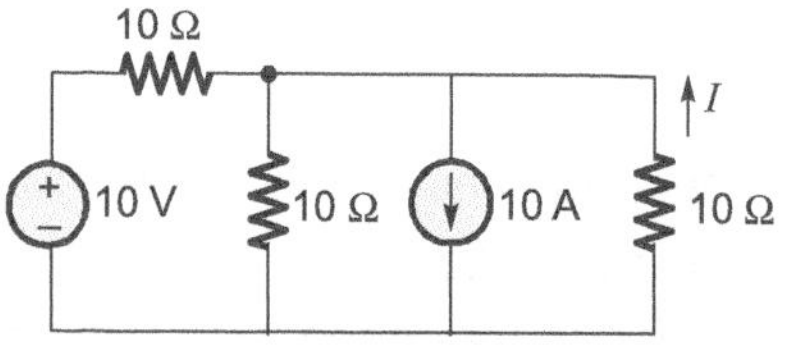

(a) 3A (b) 3.67 A
(c) 6 A (d) 9 A

[2011 : 1 M]

1.15 In the circuit shown below, the current through the inductor is

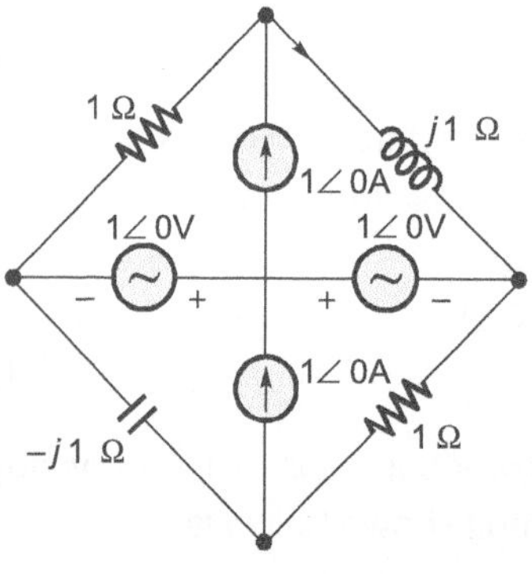

(a) $\frac{2}{1+j}A$ (b) $\frac{-1}{1+j}A$

(c) $\frac{1}{1+j}A$ (d) 0 A

[2012 : 1 M]

1.16 The impedance looking into nodes 1 and 2 in the given circuit is

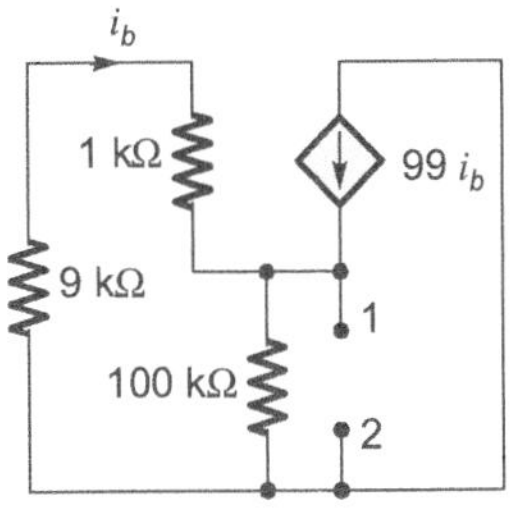

(a) 50 Ω (b) 100 Ω
(c) 5 kΩ (d) 10.1 kΩ

[2012 : 1 M]

1.17 If $V_A - V_B = 6$ V, then $V_C - V_D$ is

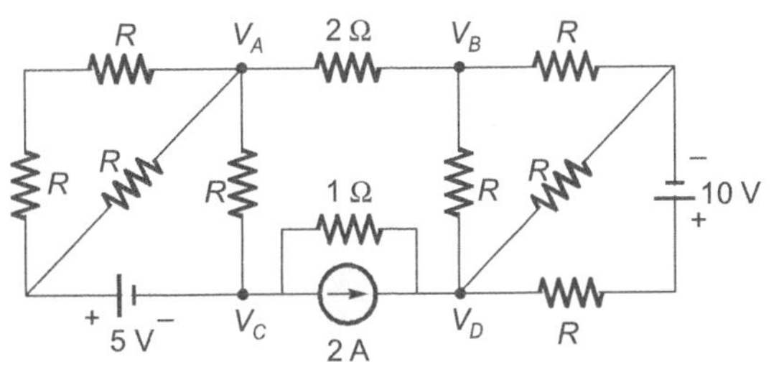

(a) –5 V (b) 2 V
(c) 3 V (d) 6 V

[2012 : 2 M]

1.18 Assuming both the voltage sources are in phase, the value of R for which maximum power is transferred from circuit A to circuit B is

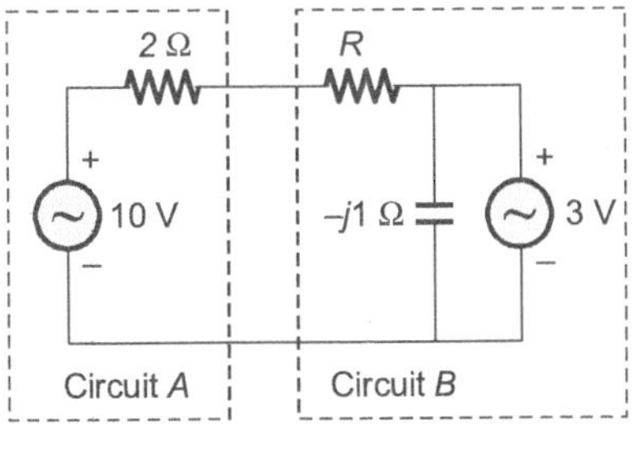

(a) 0.8 Ω (b) 1.4 Ω
(c) 2 Ω (d) 2.8 Ω

[2012 : 2 M]

1.19 A source $v_s(t) = V \cos 100\,\pi t$ has an internal impedance of $4 + j3$ Ω. If a purely resistive load connected to this source has to extract the maximum power out of the source, its value in Ω should be

(a) 3 (b) 4
(c) 5 (d) 7

[2013 : 1 M]

1.20 The transfer function $V_2(s)/V_1(s)$ of the circuit shown below is

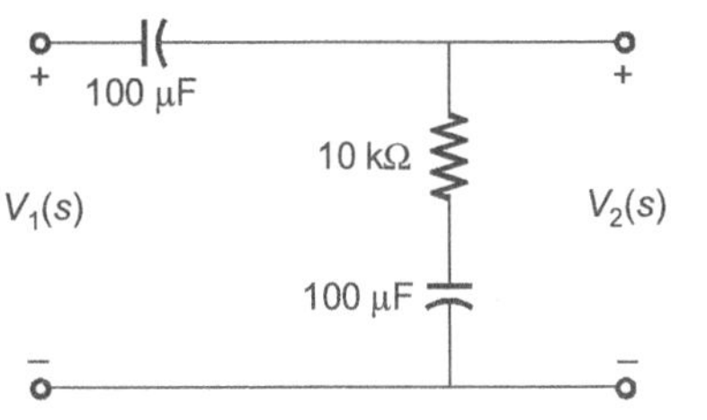

(a) $\frac{0.5s+1}{s+1}$ (b) $\frac{3s+6}{s+2}$

(c) $\frac{s+2}{s+1}$ (d) $\frac{s+1}{s+2}$

[2013 : 1 M]

1.21 Consider a delta connection of resistors and its equivalent star connection as shown. If all elements of the delta connection are scaled by a factor k, $k > 0$, the elements of the corresponding star equivalent will be scaled by a factor of

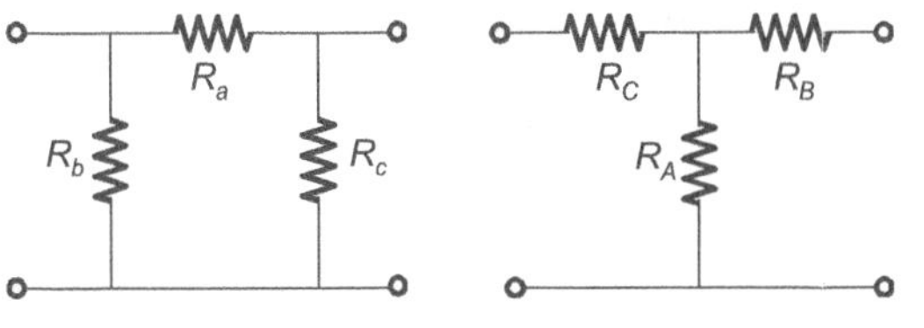

(a) k^2 (b) k
(c) $1/k$ (d) $\sqrt{k}$

[2013 : 1 M]

1.22 The following arrangement consists of an ideal transformer and an attenuator, which attenuates by a factor of 0.8. An ac voltage $V_{WX1} = 100$ V is applied across WX to get an open circuit voltage V_{YZ1} across YZ Next, an ac voltage $Y_{YZ2} = 100$ V is applied across YZ to get an open circuit voltage V_{WX2} across WX. Then V_{YZ1}/V_{WX1}, V_{WX2}/V_{YZ2} are respectively.

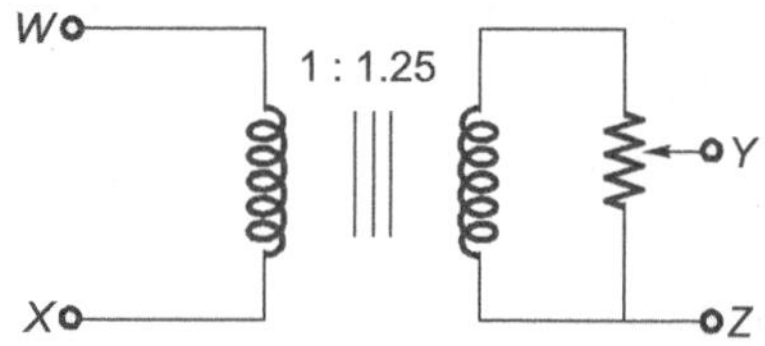

(a) $\frac{125}{100}$ and $\frac{80}{100}$ (b) $\frac{100}{100}$ and $\frac{80}{100}$

(c) $\frac{100}{100}$ and $\frac{100}{100}$ (d) $\frac{80}{100}$ and $\frac{80}{100}$

[2013 : 2 M]

1.23 In the circuit shown below, if the source voltage $V_S = 100\angle 53.13°$ Volts, then the Thevenin's equivalent voltage in Volts as seen by the load resistance R_L is

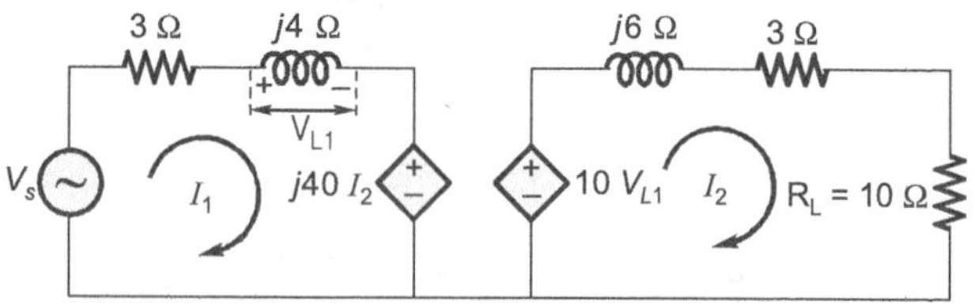

(a) $100\angle 90°$ (b) $800\angle 0°$

(c) $800\angle 90°$ (d) $100\angle 60°$

[2013 : 2 M]

1.24 Time domain expressions for the voltage $v_1(t)$ and $v_2(t)$ are given as

$v_1(t) = V_m \sin(10t - 130°)$

and $v_2(t) = V_m \cos(10t + 10°)$.

Which one of the following statements is TRUE?

(a) $v_1(t)$ leads $v_2(t)$ by 130°

(b) $v_1(t)$ lags $v_2(t)$ by 130°

(c) $v_1(t)$ lags $v_2(t)$ by −130°

(d) $v_1(t)$ leads $v_2(t)$ by −130°

[2014 : 1 M]

1.25 The circuit shown in the figure contains a dependent current source between A and B terminals. The Thevenin's equivalent resistance in kΩ between the terminals C and D is ______.

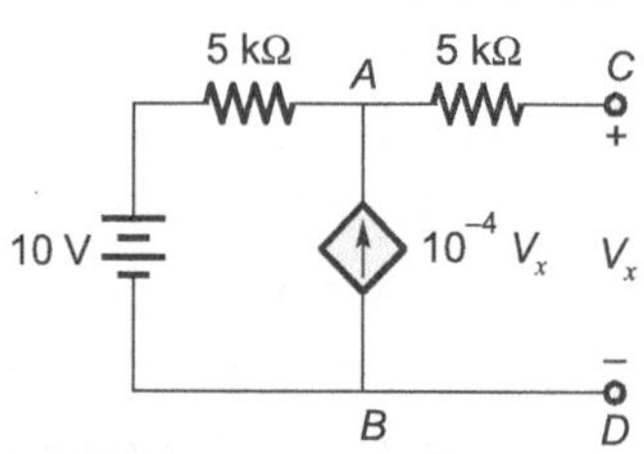

[2014 : 2 M]

1.26 A load resistor R_L is connected to a battery of voltage E with internal resistance R_i through a resistance R_s as shown in the figure. For fixed values of R_L and R_i, the value of $R_s(\geq 0)$ for maximum power transfer to R_L is

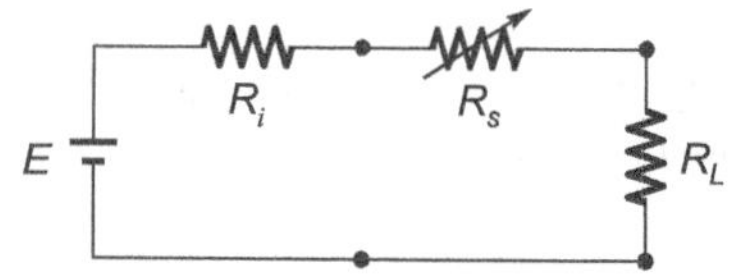

(a) 0 (b) $R_L - R_i$

(c) R_L (d) $R_L + R_i$

[2015 : 1 M]

1.27 The linear I-V characteristic of 2-terminal non-ideal DC sources X and Y are shown in the figure. If the sources are connected to a 1 Ω resistor as shown, the current through the resistor in amperes is ______ A.

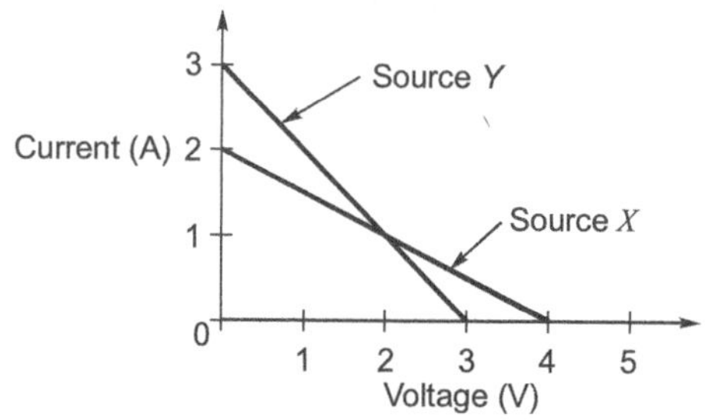

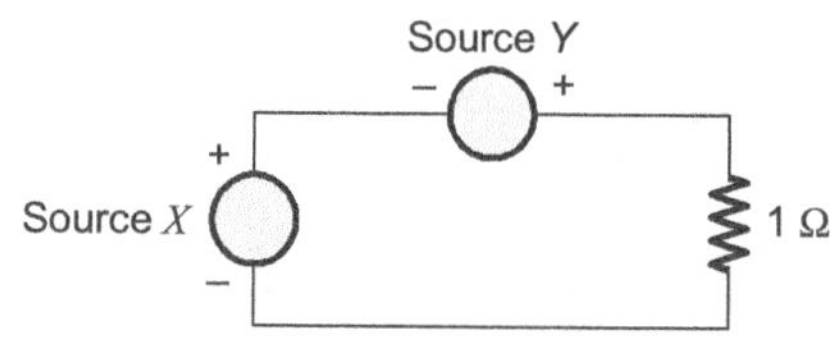

[2015 : 2 M]

1.28 Consider the circuits shown in the figure. The magnitude of the ratio of the currents, i.e. $|I_1/I_2|$, is ______.

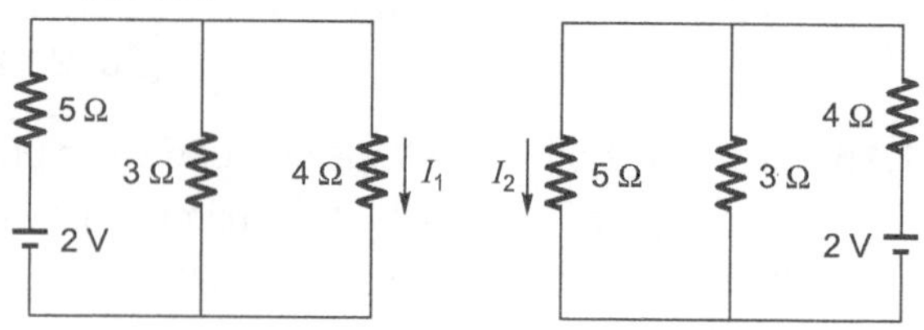

[2015 : 2 M]

1.29 The current in amperes through the resistor R in the circuit shown in the figure is ____ A.

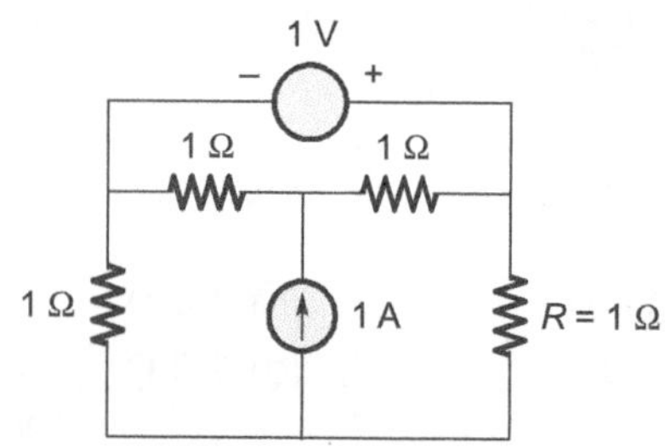

[2015 : 2 M]

1.30 Three currents i_1, i_2 and i_3 meet at a node as shown in the figure below. If $i_1 = 3\cos(\omega t)$ ampere, $i_2 = 4\sin(\omega t)$ ampere and $i_3 = I_3\cos(\omega t + \theta)$ ampere, the value of I_3 in ampere is______.

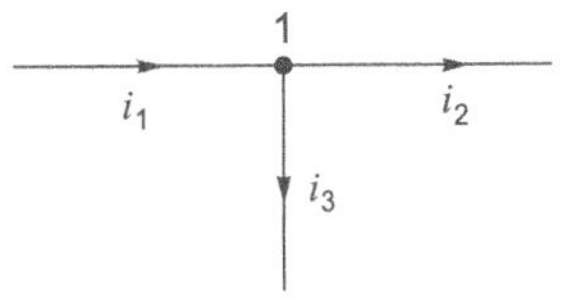

[2016 : 1 M]

1.31 The current I_X in the circuit given below in milliampere is______.

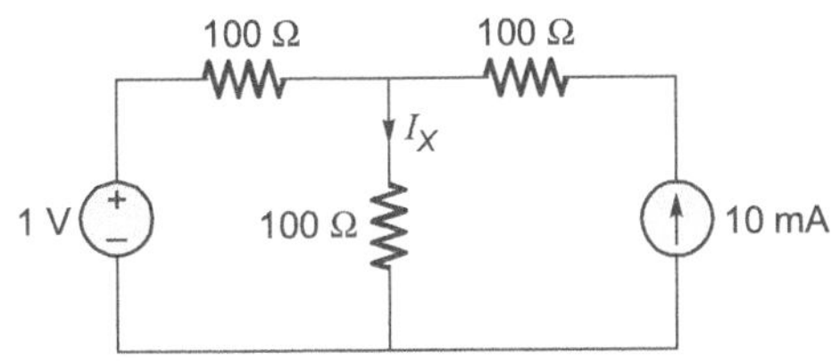

[2016 : 2 M]

1.32 A circuit consisting of dependent and independent source is shown in the figure. If the voltage at Node-1 is –1 V, then the voltage at Node-2 is ______ V.

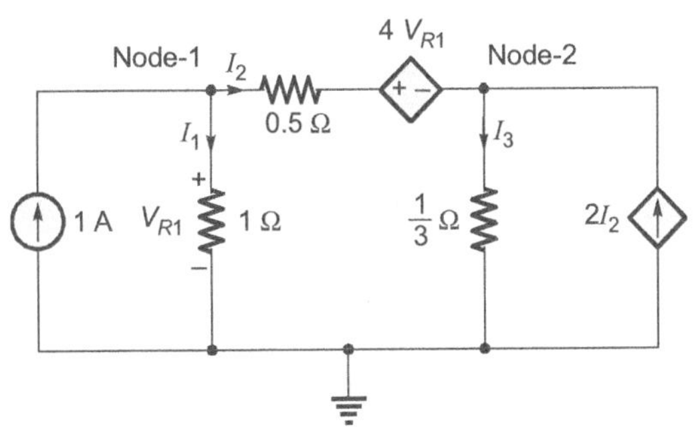

[2017 : 1 M]

1.33 The Thevenin equivalent circuit representation across terminals p-q of the circuit shown in the figure is a

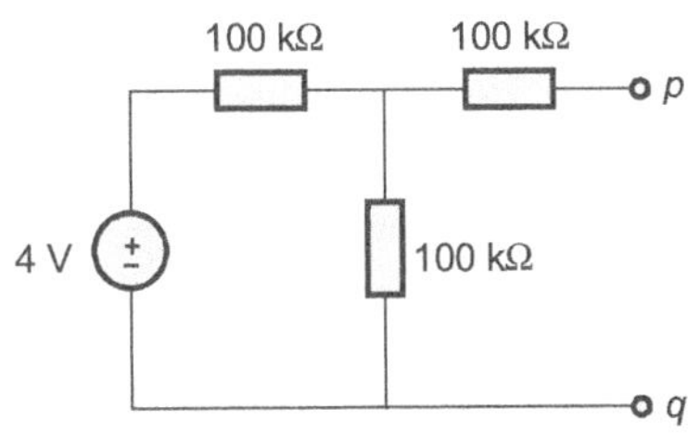

(a) 1 V source in series with 150 kΩ
(b) 1 V source in parallel with 100 kΩ
(c) 2 V source in series with 150 kΩ
(d) 2 V source in parallel with 200 kΩ

[2018 : 1 M]

1.34 In the given circuit the mesh current I_1, I_2 and I_3 are

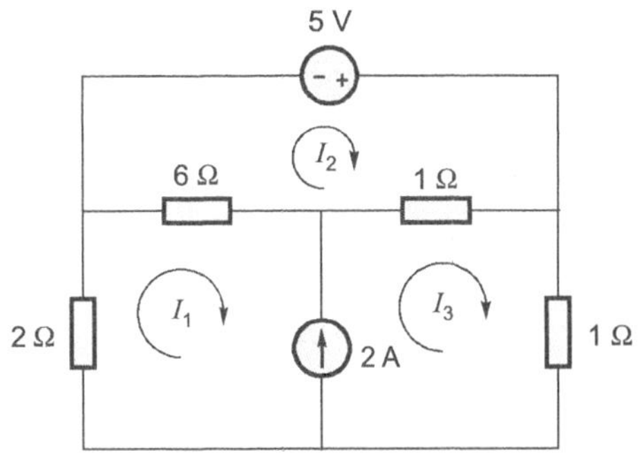

(a) $I_1 = 1$ A, $I_2 = 2$ A and $I_3 = 3$ A
(b) $I_1 = 2$ A, $I_2 = 3$ A and $I_3 = 4$ A
(c) $I_1 = 3$ A, $I_2 = 4$ A and $I_3 = 5$ A
(d) $I_1 = 4$ A, $I_2 = 5$ A and $I_3 = 6$ A

[2018 : 2 M]

1.35 In the given circuit, superposition is applied. When V_2 is set to 0 V, the current I_2 is –6 A. When V_1 is set to 0 V, the current I_1 is +6 A. Current I_3 (in A) when both sources are applied will be (up to two decimal places) ______.

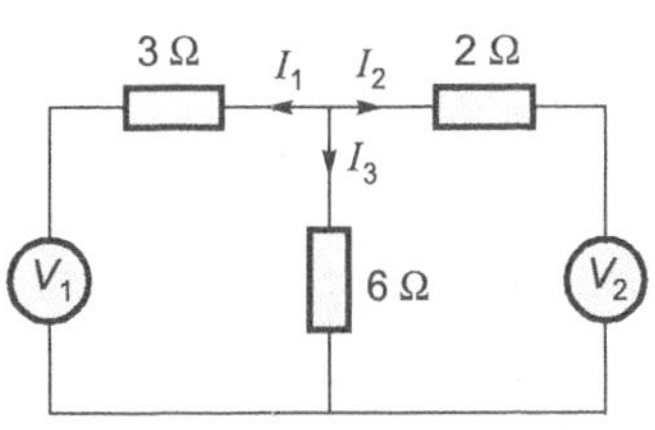

[2018 : 2 M]

1.36 In the circuit shown below, maximum power is transferred to the load resistance R_L, when R_L = ________Ω.

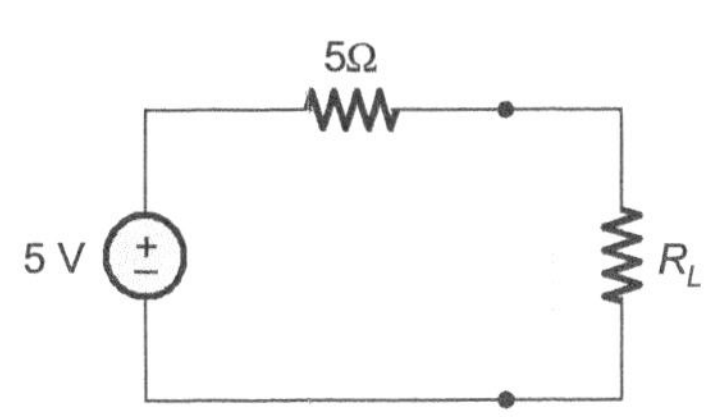

[2019 : 1 M]

1.37 Consider a circuit comprising only resistors with constant resistance and ideal independent DC voltage sources. If all the resistances are scaled down by a factor 10, and all source voltages are scaled up by a factor 10, the power dissipated in the circuit scales up by a factor of______.

[2019 : 1 M]

1.38 Three 400 Ω resistors are connected in delta and powered by a 400 V (rms), 50 Hz, balanced, symmetrical *R-Y-B* sequence, three-phase three-wire mains. The rms value of the line current (in amperes, rounded off to one decimal place) is _____.

[2020 : 1 M]

1.39 In the circuit shown below, the safe maximum value for the current I is _____.

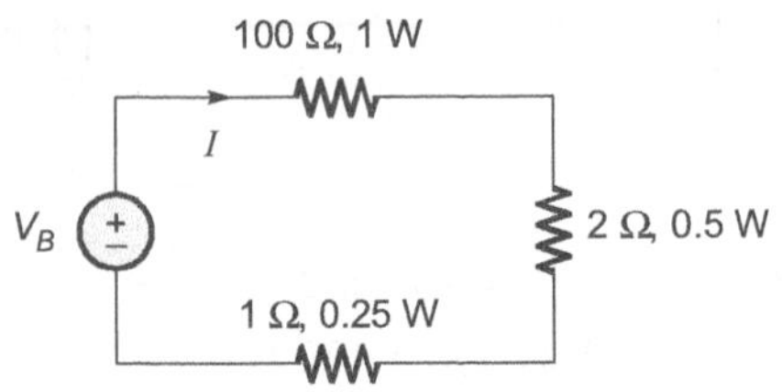

(a) 0.5 A (b) 0.1 A
(c) 1.0 A (d) 0.05 A

[2020 : 1 M]

1.40 I_1, I_2 and I_3 in the figure below are mesh currents. The correct set of mesh equations for these currents, in matrix form, is _____.

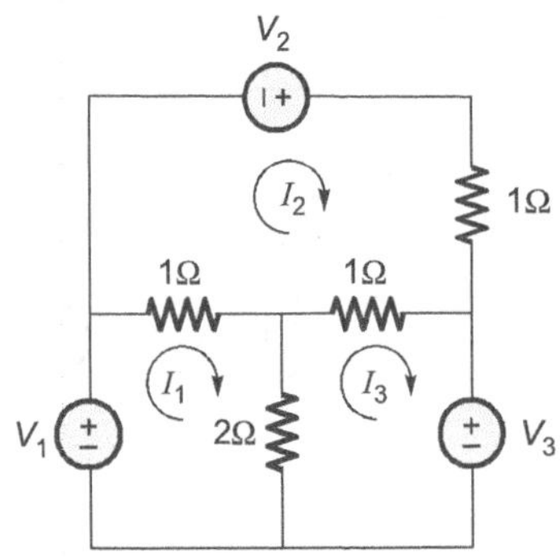

(a) $\begin{bmatrix} 1 & -1 & -2 \\ -1 & 2 & -1 \\ -2 & -1 & 3 \end{bmatrix}\begin{bmatrix} I_1 \\ I_2 \\ I_3 \end{bmatrix} = \begin{bmatrix} V_1 \\ V_2 \\ V_3 \end{bmatrix}$

(b) $\begin{bmatrix} -3 & -1 & -2 \\ -1 & 3 & -1 \\ -2 & -1 & 3 \end{bmatrix}\begin{bmatrix} I_1 \\ I_2 \\ I_3 \end{bmatrix} = \begin{bmatrix} V_1 \\ V_2 \\ -V_3 \end{bmatrix}$

(c) $\begin{bmatrix} 3 & -1 & -2 \\ -1 & 3 & -1 \\ -2 & -1 & 3 \end{bmatrix}\begin{bmatrix} I_1 \\ I_2 \\ I_3 \end{bmatrix} = \begin{bmatrix} V_1 \\ V_2 \\ -V_3 \end{bmatrix}$

(d) $\begin{bmatrix} 3 & -1 & -2 \\ -1 & 3 & -1 \\ -2 & -1 & -3 \end{bmatrix}\begin{bmatrix} I_1 \\ I_2 \\ I_3 \end{bmatrix} = \begin{bmatrix} V_1 \\ V_2 \\ V_3 \end{bmatrix}$

[2020 : 2 M]

1.41 In the circuit shown, the rms value of the voltage across the 100 Ω, resistor (in volts) _____.

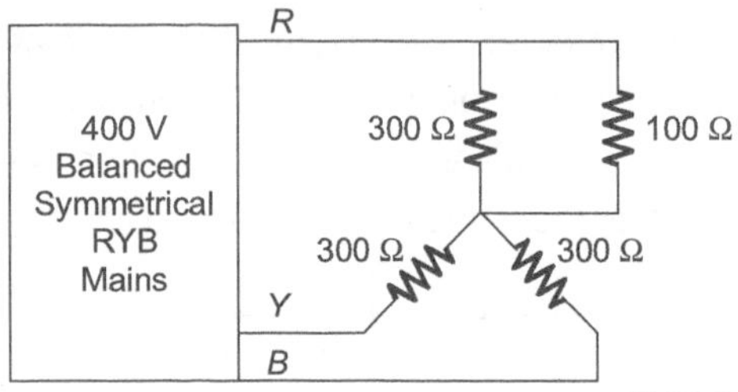

[2020 : 2 M]

1.42 Given Circuit *A* with currents I_1 and I_2 as shown, the current I_3 in Circuit B (in amperes), is _______(round off to one decimal place)

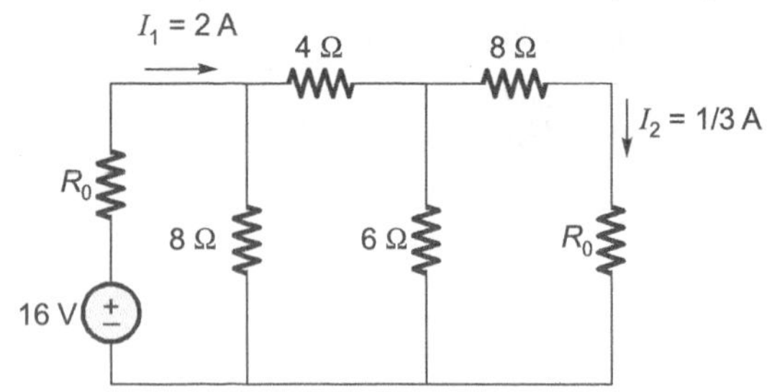

Circuit A

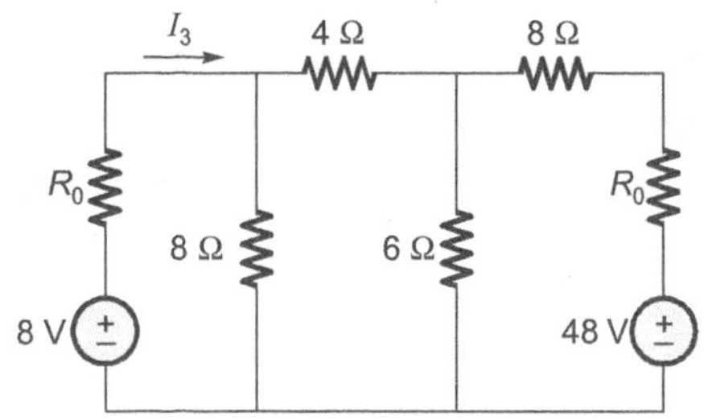

Circuit B

[2022 : 2 M]

1.43 A 1 kHz sine-wave generator having an internal resistance of 50 Ω generates an open-circuit voltage of 10 $V_{p\text{-}p}$. When a capacitor is connected across the output terminals, the voltage drops to 8 $V_{p\text{-}p}$. The capacitance of the capacitor (in microfarads) is ______ (round off to two decimal places)

[2022 : 2 M]

1.44 In the circuit given, the Thevenin equivalent resistance R_{th} across the terminals '*a*' and '*b*' is _______ Ω (rounded off to one decimal place).

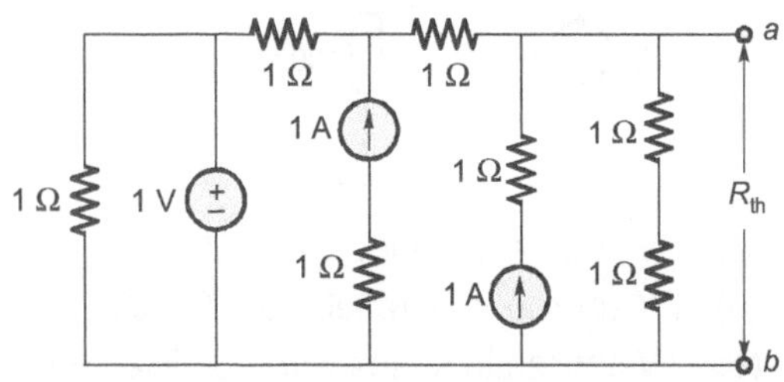

[2023 : 1 M]

1.45 The voltage applied and the current drawn by a circuit are

$v(t) = 95 + 200\cos(120\pi t) + 90\cos(360\pi t - 60°)$V

$i(t) = 4\cos(120\pi t - 60°) + 1.5\cos(240\pi t - 75°)$A

The average power absorbed by the circuit is ________W (rounded off to nearest integer).

[2024 : 1 M]

1.46 The current $i(t)$ drawn by a circuit is given as

$i(t) = 4 + 30\cos(t) - 20\sin(t) + 15\cos(3t) - 10\sin(3t)$ A

The root-mean-square value of $i(t)$ is ________ A (rounded off to one decimal place).

[2024 : 1 M]

Answers Network Laws & Network Theorems

1.1	(b)	1.2	(c)	1.3	(b)	1.4	(c)	1.5	(d)	1.6	(a)	1.7	(b)
1.8	(b)	1.9	(d)	1.10	(c)	1.11	(c)	1.12	(b)	1.13	(c)	1.14	(a)
1.15	(c)	1.16	(a)	1.17	(a)	1.18	(a)	1.19	(c)	1.20	(d)	1.21	(b)
1.22	(b)	1.23	(c)	1.24	(a)	1.25	(20)	1.26	(a)	1.27	(1.75)	1.28	(1)
1.29	(1)	1.30	(5)	1.31	(10)	1.32	(2)	1.33	(c)	1.34	(a)	1.35	(1)
1.36	(5)	1.37	(1000)	1.38	(1.7)	1.39	(b)	1.40	(c)	1.41	(115.47)	1.42	(0.0)
1.43	(2.39)	1.44	(1)	1.45	(200)	1.46	(28.78)						

Explanations Network Laws & Network Theorems

1.1 (b)

We can find the Thevenin resistance between terminals 1 and 2, by short circuiting the battery of 4 V.

$$R_{12} = 30 || 20 + 25 || 25$$

$$= \frac{30 \times 20}{30+20} + \frac{25 \times 25}{25+25} = \frac{600}{50} + \frac{625}{50}$$

$$= 12 + 12.5 = 24.5 \text{ k}\Omega$$

1.2 (c)

Given function,

$$y(t) = \begin{cases} 2+a & ; \ 0 < t < \dfrac{T}{2} \\ 2-a & ; \ \dfrac{T}{2} < t < T \end{cases}$$

here T is the period of square wave and a is peak value of square wave.

Given, $a = 2$

$$\therefore \quad y(t) = \begin{cases} 4 & ; \ 0 < t < \dfrac{T}{2} \\ 0 & ; \ \dfrac{T}{2} < t < T \end{cases}$$

RMS value of any periodic function $y(t)$

$$= \sqrt{\frac{1}{T}\int_0^T y^2(t)\,dt}$$

$$\therefore \quad y_{rms} = \sqrt{\frac{1}{T}\left[\int_0^{T/2}(4)^2 dt + \int_{T/2}^{T}(0)^2 dt\right]}$$

$$= \sqrt{\frac{1}{T}\cdot 16 \cdot \frac{T}{2}} = \sqrt{8} \text{ V}$$

1.3 (b)

We know, $R = \rho\dfrac{l}{A}$

as ρ is the property of the material, so it will be constant for same material.

For constant volume

$$R = kl^2$$

$$\therefore \quad R' = k(\alpha l)^2 = \alpha^2 \cdot (kl)^2 = \alpha^2 R$$

1.4 (c)

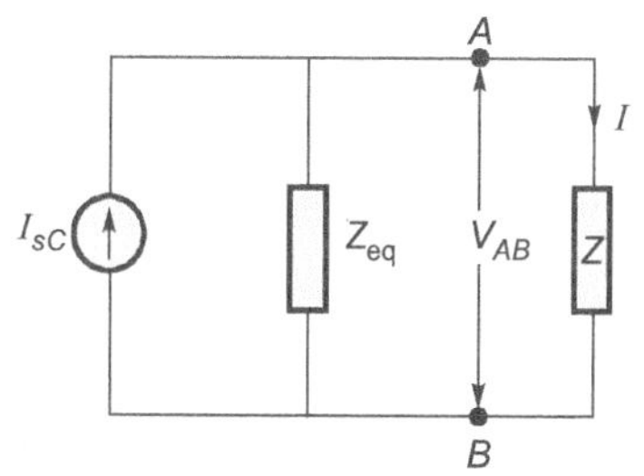

Let the given load is Z.

Then the current,

$$I = \frac{Z_{eq}}{Z + Z_{eq}} \times I_{SC} \qquad \text{...(i)}$$

Given, $I = 70$ mA, $I_{SC} = 75$ mA, $V_{AB} = 0.6$ V

From equation (i),

$$70 = \frac{Z_{eq.}}{Z + Z_{eq}} \times 75$$

$$\Rightarrow \quad \frac{75}{70} = 1 + \frac{Z}{Z_{eq.}}$$

$$\Rightarrow \quad \boxed{Z_{eq} = 14 \times Z}$$

and $\quad V_{AB} = Z \times I$

$$0.6 = Z \times 70 \times 10^{-3}$$

$$\Rightarrow \quad Z = \frac{600}{70}$$

$$\therefore \quad Z_{eq} = 14 \times \frac{600}{70} = 120\ \Omega$$

1.5 (d)

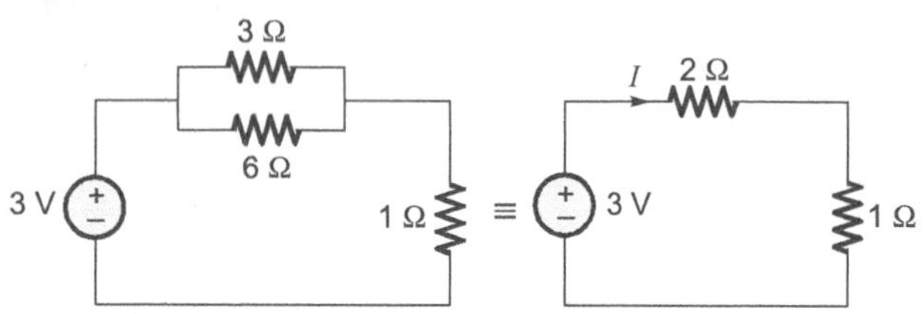

Current I in the circuit

$$I = \frac{3}{2+1} = 1\text{ A}$$

∵ Power supplied by the d.c. voltage source

= Power absorbed by the load.

∴ Power supplied = $I^2 \cdot R_{eq}$

$= I^2 \times 3 = 3.0$ W

1.6 (a)

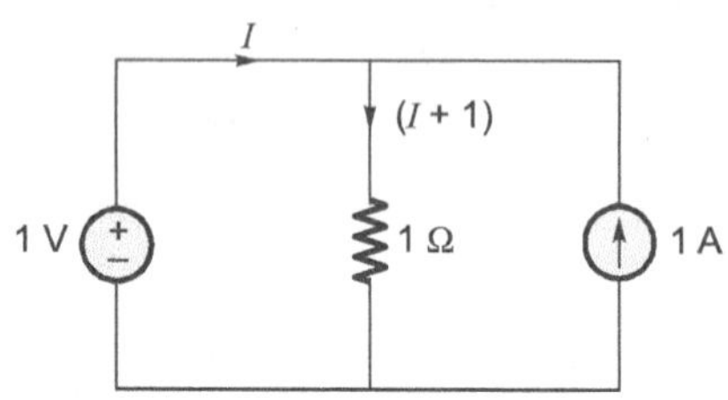

By applying KVL in 1st loop

$$1 = I + 1$$

$$I = 0\text{ A}$$

1.7 (b)

$$i(t) = 10\sqrt{2}\sin(1000t)\,A$$

$$\omega = 1000$$

$$\therefore \quad j\omega L = j10^3 \times 10 \times 10^{-3} = j10\ \Omega$$

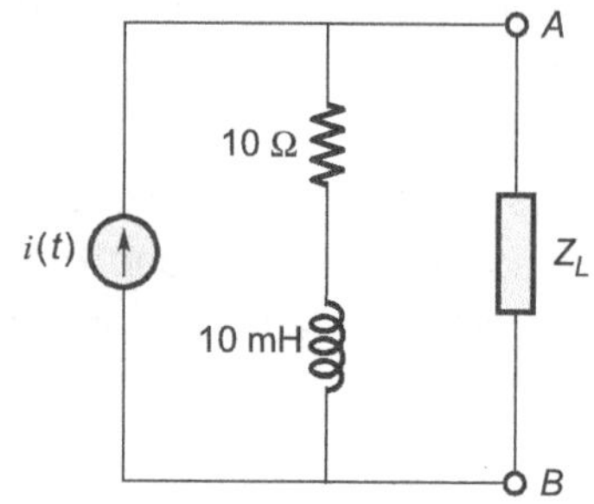

Thevenin resistance across AB

$$Z_{Th} = (10 + j10)$$

For maximum power transfer

Load impedance = Z_{Th}^*

$\Rightarrow Z_L = (10 - j10)$

r.m.s value of source current

$$I_{rms} = \frac{10\sqrt{2}}{\sqrt{2}} = 10\text{ A}$$

∴ Current in branch AB (or load current)

$$I_L = \frac{Z_{Th}}{Z_{Th} + Z_L} \times I_{rms}$$

$$= \frac{(10 + j10)}{(10 + j10 + 10 - j10)} \times 10$$

$$I_L = \frac{10 + j10}{20} \times 10 = (5 + j5)$$

$$I_L = 5\sqrt{2}\angle 45°\text{A}$$

Only resistive part of load impedance is responsible for power loss.

∴ Power transferred to load

$= |I_L|^2 \cdot$ Real part of Z_L

$= (5\sqrt{2})^2 \times 10 = 50 \times 10$

$= 500$ W

1.8 (b)

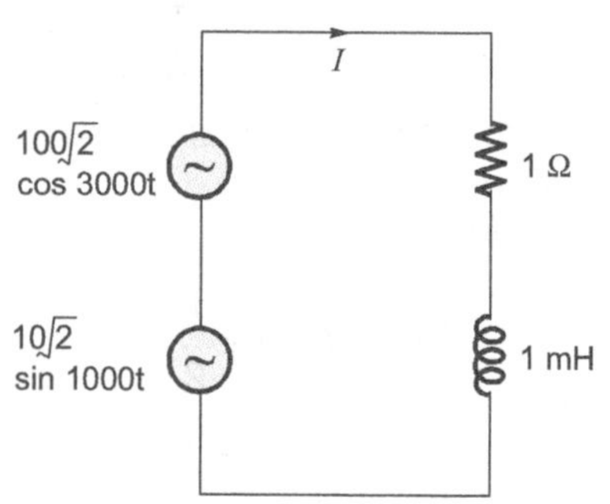

Current,

$$I = \frac{100\sqrt{2}\cos 3000t}{1 + j\omega_1 L} + \frac{10\sqrt{2}\sin 1000t}{1 + j\omega_2 L}$$

$$I = \frac{100\sqrt{2}\cos 3000t}{1 + j\times 3000\times 10^{-3}} + \frac{10\sqrt{2}\sin 1000t}{1+j\times 1000\times 10^{-3}}$$

$$= \frac{100\sqrt{2}\cos 3000t}{1+j3} + \frac{10\sqrt{2}\sin 1000t}{1+j1}$$

$$= \frac{100\sqrt{2}\cos 3000t}{\sqrt{10}\angle\phi_1} + \frac{10\sin(1000t)\times\sqrt{2}}{\sqrt{2}\angle\phi_1}$$

where,

$\phi_1 = \tan^{-1}(3)$ and $\phi_2 = \tan^{-1}(1)$

So, $I = \frac{100\sqrt{2}}{\sqrt{10}}\cos(3000t - \phi_1) + \frac{10\sin(1000t)\times\sqrt{2}}{\sqrt{2}\angle\phi_2}$

$\Rightarrow$ So, RMS of I is

$$I_{rms} = \sqrt{\frac{1}{2}\left[\frac{10000\times 2}{10} + 100\right]} = \sqrt{1050}$$

So, power dissipated is

$$P = I_{rms}^2 \times R = 1050\times 1 = 1050 \text{ Watt}$$

1.9 (d)

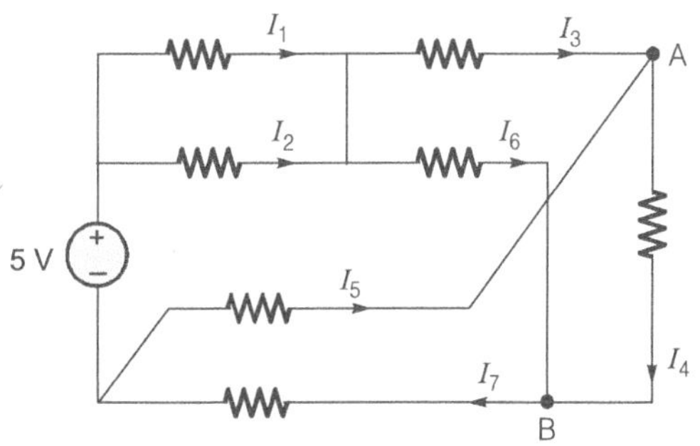

By applying KCL at node A,

$$I_3 + I_5 = I_4 \qquad \text{...(i)}$$

and at node B,

$$I_4 + I_6 = I_7$$

$$\Rightarrow \qquad I_4 = I_7 - I_6 \qquad \text{...(ii)}$$

From equation (i) and (ii)

$$I_3 + I_5 = I_7 - I_6$$

$$I_3 + I_5 + I_6 - I_7 = 0$$

1.10 (c)

Current, $I = \frac{7}{2+R_L}$

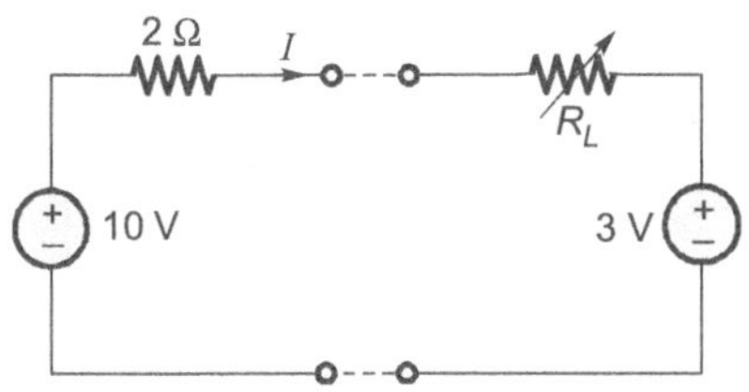

Power transferred to load,

$$= \left(\frac{7}{2+R_L}\right)^2 R_L + 3\times\frac{7}{2+R_L}$$

$$= \frac{49R_L}{(2+R_L)^2} + \frac{21}{(2+R_L)}$$

$$= \frac{49R_L + 21R_L + 42}{(2+R_L)^2}$$

$$= \frac{70R_L + 42}{(2+R_L)^2} = \frac{14(5R_L+3)}{(2+R_L)^2}$$

For maximum power transfer

$$\frac{dP}{dR_L} = 0$$

$$\Rightarrow (R_L + 2)^2 \times 5 - (5R_L + 3)\times 2(R_L + 2) = 0$$

$$\Rightarrow 5R^2_L + 20R_L + 20 - 2(5R^2_L + 13R_L + 6) = 0$$

$$\Rightarrow \quad -5R^2_L - 6R_L + 8 = 0$$

$$\Rightarrow \quad 5R^2_L + 6R_L - 8 = 0$$

$$\Rightarrow \quad R_L = 0.8 \text{ and } -2$$

Hence, $R_L = 0.8\ \Omega$

1.11 (c)

$$x(t) = 3 + 2\sin t\cdot\cos 2t$$

$$= 3 + \sin 3t - \sin t$$

If $y(t) = a_0 + (a_1\cos\omega t + a_2\cos 2\omega t + ...) + (b_1\sin\omega t + b_2\sin 2\omega t + ...)$

then,

$$y_{rms} = \sqrt{a_0^2 + \frac{1}{2}(a_1^2 + a_2^2 + ...) + \frac{1}{2}(b_1^2 + b_2^2 + ...)}$$

$\therefore$ Rms value of $x(t) = \sqrt{(3)^2 + \frac{1}{2}(1 + (-1)^2)}$

$$= \sqrt{9 + \frac{1}{2}(2)} = \sqrt{10}$$

1.12 (b)

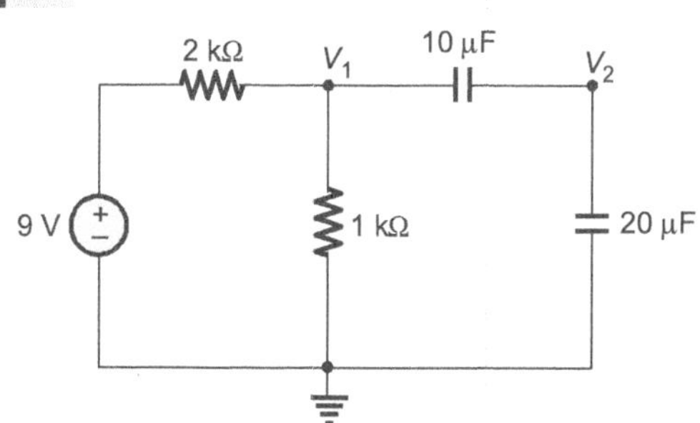

By applying voltage divider rule.

The voltage,

$$V_1 = \frac{1}{1+2}\times 9\text{ V} = 3\text{ V}$$

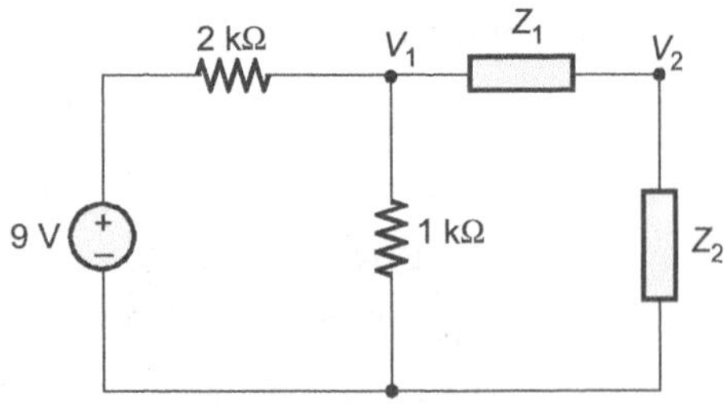

$$Z_1 = \frac{1}{j\omega \times 10 \times 10^{-6}}$$

and $$Z_2 = \frac{1}{j\omega \times 20 \times 10^{-6}}$$

$\Rightarrow$ $$Z_1 = 2Z_2$$

Again by applying voltage divider rule,

$$V_2 = \frac{Z_2}{Z_1 + Z_2} \times 3\text{V}$$

$$= \frac{Z_2}{2Z_2 + Z_2} \times 3\text{V} = 1\text{ V}$$

1.13 (c)

Resistor 1 : 100 Ω, 1 W

Resistor 2 : 800 Ω, 2 W

maximum current that resistor can withstand

$$I_1 = \sqrt{\frac{1}{100}} = \frac{1}{10}\text{A}$$

Similarly,

$$I_2 = \sqrt{\frac{2}{800}} = \frac{1}{20}\text{A}$$

if these two resistors are connected in series.

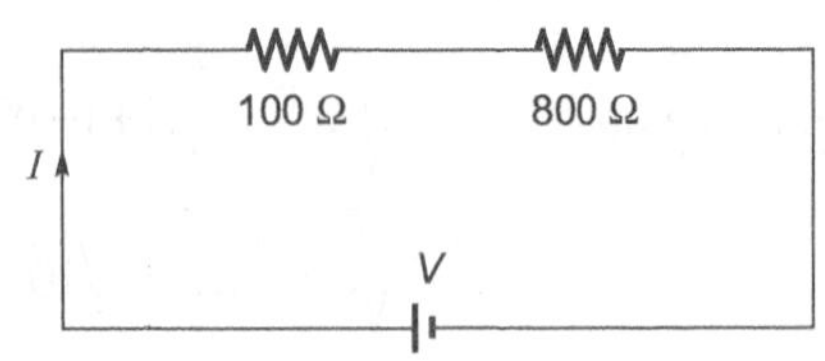

Then maximum value of $I = \dfrac{1}{20}$

$\therefore$ $$V = I(100 + 800)$$

$$= \frac{1}{20}(900) = 45\text{ volt}$$

1.14 (a)

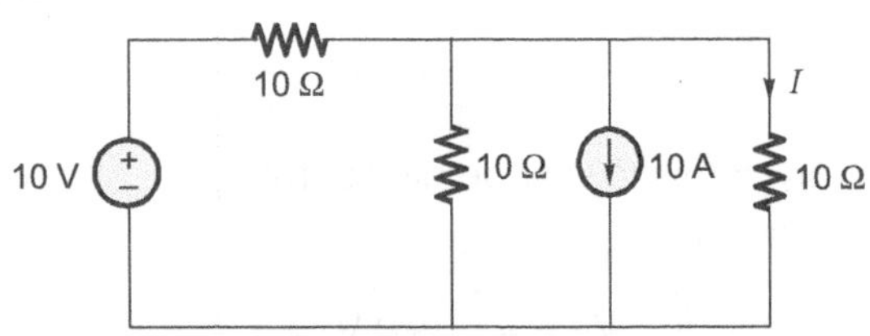

By source transformation,

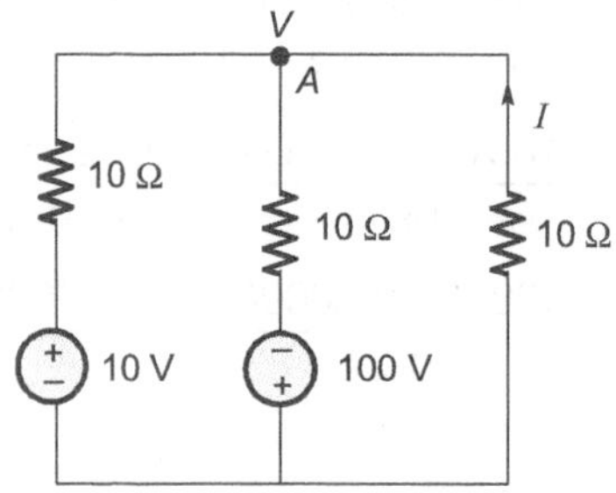

Let V is the voltage at node A.

By applying KCL at node A

$$\frac{V-10}{10} + \frac{V+100}{10} + \frac{V}{10} = 0$$

$$3\text{ V} + 90 = 0$$

$$V = -30\text{ V} \quad \text{and} \quad V = -I \times 10$$

$\therefore$ $$-30 = -I \times 10 \quad \Rightarrow I = 3\text{ A}$$

1.15 (c)

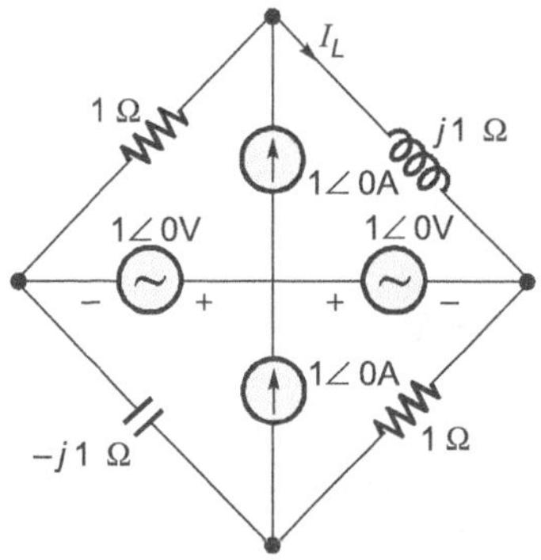

By applying current division rule in upper part of the circuit.

$$I_L = \frac{1}{1+j} \times 1\angle 0 = \frac{1}{1+j}$$

1.16 (a)

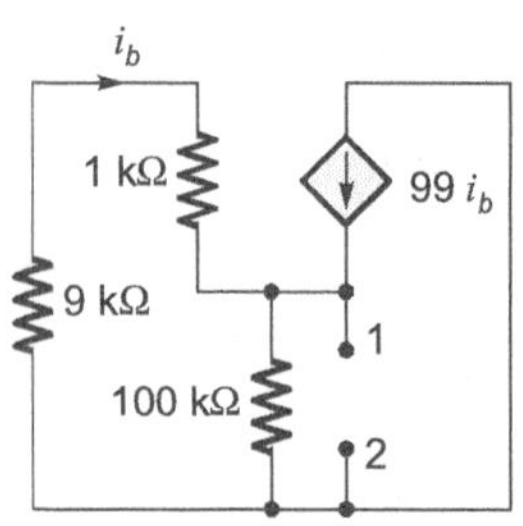

$\Downarrow$

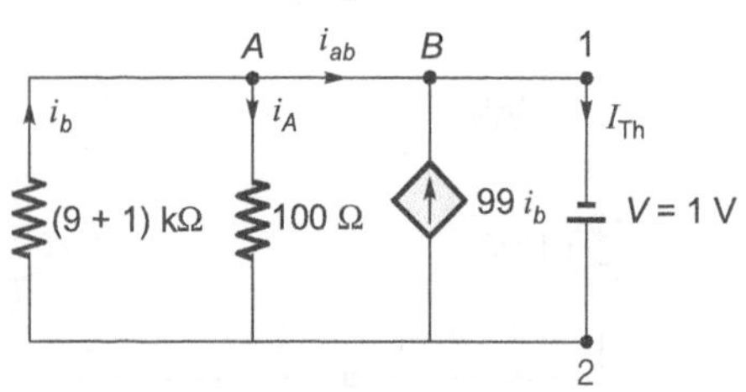

To find Thevenin impedance across node 1 and 2. Connect a 1 V source and find the current through voltage source.

Then, $Z_{Th} = \frac{1}{I_{Th}}$

By applying KCL at node B and A

$$i_{AB} + 99i_b = I_{Th}$$
$$i_b = i_A + i_{AB}$$
$$\Rightarrow \quad i_b - i_A + 99i_b = I_{Th}$$
$$\Rightarrow \quad 100\, i_b - i_A = I_{Th} \quad \text{...(i)}$$

By applying KVL in outer loop

$$10 \times 10^3\, i_b = 1$$
$$i_b = 10^{-4}\text{ A}$$

and $10 \times 10^3\, i_b = -100\, i_A$

$$\Rightarrow \quad i_A = -100\, i_b$$

∴ From equation (i)

$$100\, i_b + 100\, i_b = I_{Th}$$
$$\Rightarrow \quad i_{Th} = 200\, i_b$$
$$= 200 \times 10^{-4} = 0.02$$

$$\therefore \quad Z_{Th} = \frac{1}{I_{Th}} = \frac{1}{0.02} = 50\ \Omega$$

1.17 (a)

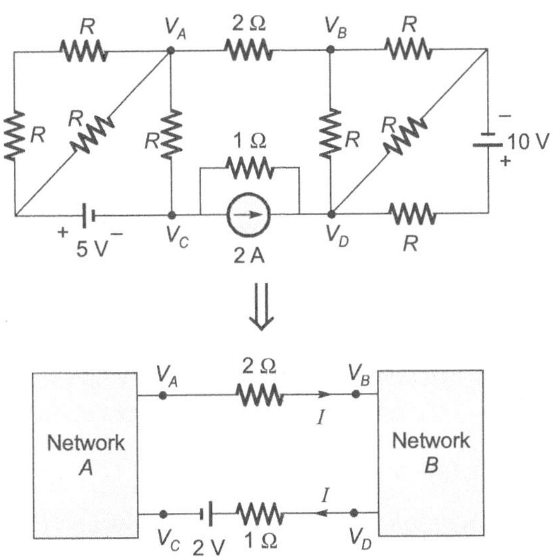

$$V_A - V_B = 2I$$
$$\Rightarrow \quad 2I = 6$$
$$\Rightarrow \quad I = 3\text{A}$$
$$V_c + 2 + 1 \times I = V_D$$
$$V_C - V_D = -2 - 3$$
$$= -5\text{ V}$$

1.18 (a)

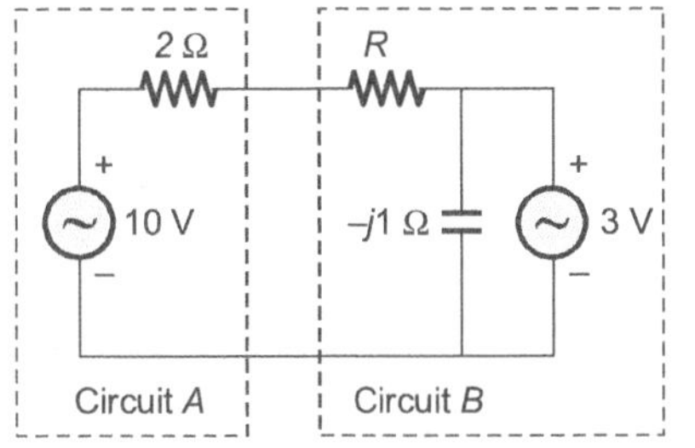

Thevenin equivalent circuit

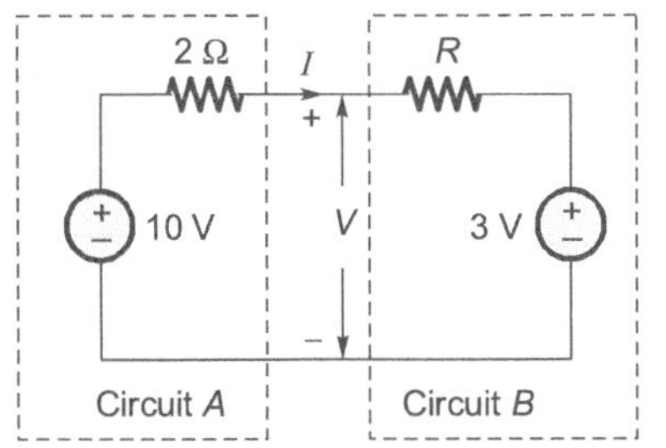

$$I = \frac{7}{R+2}$$

and $V = 10 - 2I$

$$= 10 - \frac{14}{R+2} = \frac{10R+6}{R+2}$$

Power transferred from circuit A to circuit B

$$P = VI = \frac{10R+6}{R+2} \times \frac{7}{R+2}$$

For P to be maximum $\frac{dP}{dR} = 0$

$$(R+2)^2(10) - (10R+6) \times 2(R+2) = 0$$
$$5R^2 + 20R + 20 = 10R^2 + 26R + 12$$
$$5R^2 + 6R = 8$$
$$\Rightarrow \quad R = 0.8\ \Omega$$

1.19 (c)

For pure resistive load to extract the maximum power

$$R_L = \sqrt{R_s^2 + X_s^2} = \sqrt{4^2 + 3^2}$$
$$R_L = 5\ \Omega$$

1.20 (d)

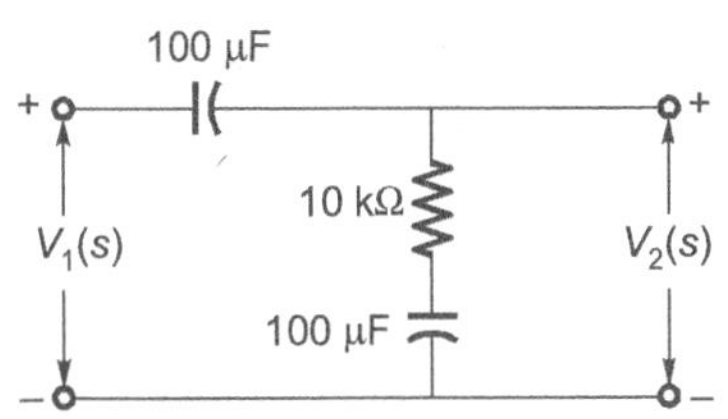

$$\frac{V_2(s)}{V_1(s)} = \frac{R+\frac{1}{Cs}}{\frac{1}{Cs}+R+\frac{1}{Cs}} = \frac{1+RCs}{2+RCs}$$

$$= \frac{1+10\times10^3\times100\times10^{-6}s}{2+10\times10^3\times100\times10^{-6}s} = \frac{1+s}{2+s}$$

1.21 (b)

$$R_A = \frac{R_b R_c}{R_a+R_b+R_c}$$

$$R'_a = kR_a$$

$$R'_b = kR_b$$

$$R'_c = kR_c$$

$$R'_A = \frac{kR_b\cdot kR_c}{kR_a+kR_b+kR_c} = \frac{k^2R_bR_c}{k(R_A+R_b+R_c)}$$

$$= k\times\frac{R_bR_c}{R_a+R_b+R_c}$$

$$R'_A = kR_A$$

1.22 (b)

$$V_{YZ_1} = 100\times1.25\times0.8 = 100\text{ V}$$

In second case when 100 V is applied at YZ terminals, this whole 100 V will appear across the secondary winding.

Hence,

$$V_{Wx_2} = \frac{100}{1.25} = 80\text{ V}$$

$$\Rightarrow \quad \frac{Y_{YZ_1}}{Y_{wx_1}} = \frac{100}{100}, \frac{V_{wx_2}}{V_{YZ_2}} = \frac{80}{100}$$

1.23 (c)

To find V_{Th}, open circuit the load voltage R_L then

$I_2 = 0$

$j40\,I_2 = 0$

$$V_{L_1} = \frac{V_S\cdot(j4)}{3+j4} = \frac{100\angle53.13^\circ}{5\angle53.13^\circ}\times4\angle90^\circ$$

$$V_{L_1} = 80\angle90^\circ$$

$$V_{Th} = 10V_{L_1}+I_2j6+I_25$$

$$V_{Th} = 10\times80\angle90^\circ+0\times j6+0\times5$$

$$V_{Th} = 800\angle90^\circ+0+0$$

$$V_{Th} = 800\angle90^\circ$$

1.24 (a)

$$v_1(t) = V_m\sin(10t-130^\circ)$$

$$= V_m\cos(10t-130^\circ-90^\circ)$$

$$= V_m\cos(10t-220^\circ)$$

$$= V_m\cos(10t-220^\circ+360^\circ)$$

$$= V_m\cos(10t+140^\circ)$$

$$v_2(t) = V_m\cos(10t+10^\circ)$$

$\therefore$ Clearly $v_1(t)$ is leading $v_2(t)$ by $140^\circ - 10^\circ$

$= 130^\circ$

1.25 (20)

Thevenin equivalent circuit is made by connecting 1 A current source at the output and short circuiting the battery.

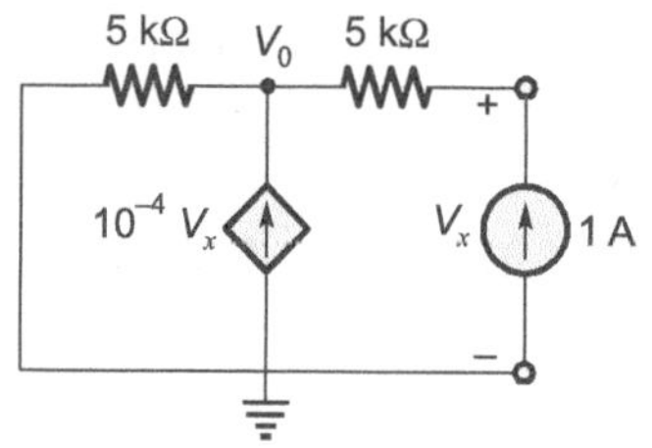

At node, V_0

$$\frac{V_o}{5000} = 10^{-4}V_x+1 \quad \text{...(i)}$$

Also, $\quad V_x = 5000+V_0 \quad$ ...(ii)

Putting equation (ii) in (i), we get

$$\frac{V_x-5000}{5000} = 10^{-4}V_x+1$$

$$\therefore \quad \frac{2V_x}{10000} = \frac{V_x}{10000}+2$$

$$\therefore \quad V_x = 20000$$

$$R_{Th} = \frac{V_x}{1A} = \frac{20000}{1} = 20\text{ k}\Omega$$

1.26 (a)

- For the above network maximum power transfer theorem can not be applied since load resistor is constant and other resistance is variable.
- In the above network power dissipation in the R_L is maximum for minimum value of R_s so that current in the R_L will be maximum hence $R_s = 0$.

1.27 (1.75)

The transfer characteristics curve and circuit of a non ideal source is given as

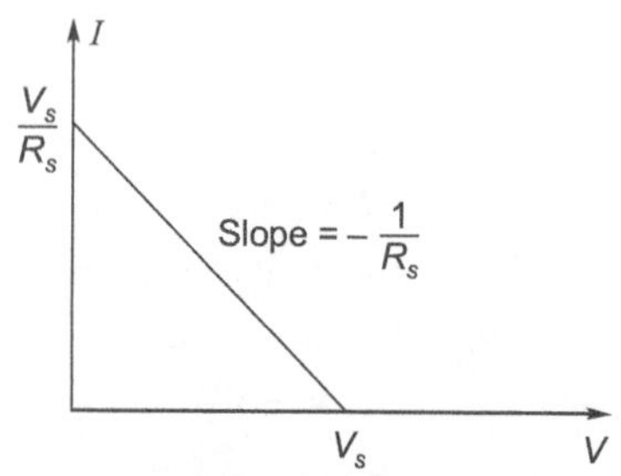

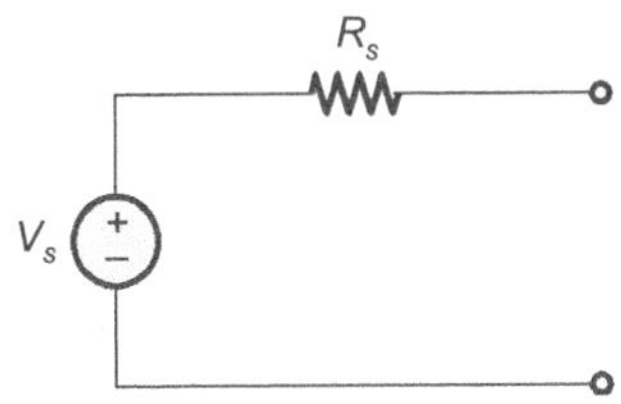

Comparing is with the given graph

X: $V_x = 4$ V, $R_x = 2\ \Omega$

Y: $V_Y = 3$ V, $R_x = 1\ \Omega$

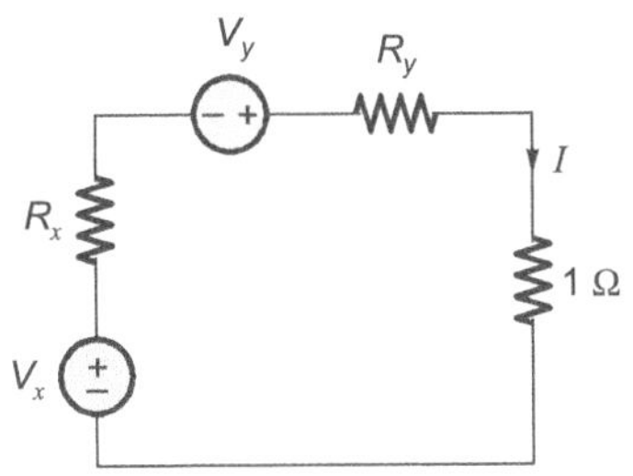

$$\therefore \quad I = \frac{4+3}{2+1+1} = 1.75 \text{ A}$$

1.28 (1)

By Reciprocity theorem, $I_1 = I_2$

So, $\left|\frac{I_1}{I_2}\right| = 1$

1.29 (1)

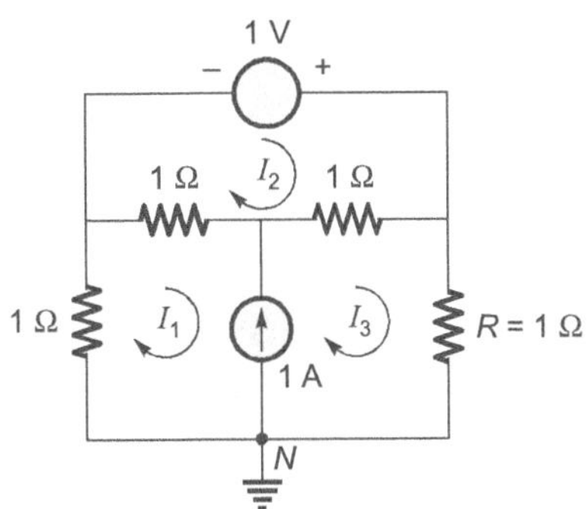

Mesh, 1 and 3 form a supermesh

$$2I_1 - I_2 + 2I_3 - I_2 = 0$$

$$2I_1 - 2I_2 + 2I_3 = 0 \quad \ldots\text{(i)}$$

at point N applying KCL, we get

$$I_1 - I_3 = -1 \quad \ldots\text{(ii)}$$

In mesh 2

$$-1 + 2I_2 - I_3 - I_1 = 0$$

$$-I_1 + 2I_2 - I_3 = 1 \quad \ldots\text{(iii)}$$

Solving equations (i), (ii) and (iii)

Current through R, i.e. $I_3 = 1$ A

Also, since the direction of current is not specified in the question so, it could also be −1 A.

1.30 (5)

Apply KCL at NODE - 1

$$i_1 = i_2 + i_3$$

$$i_3 = i_1 - i_2$$

$$= 3\cos\omega t - 4\sin\omega t$$

$$= \sqrt{3^2+4^2}\cos(\omega t + \phi)$$

$$= 5\cos(\omega t + \phi)$$

$$\therefore \quad I_3 = 5 \text{ Amp.}$$

1.31 (10)

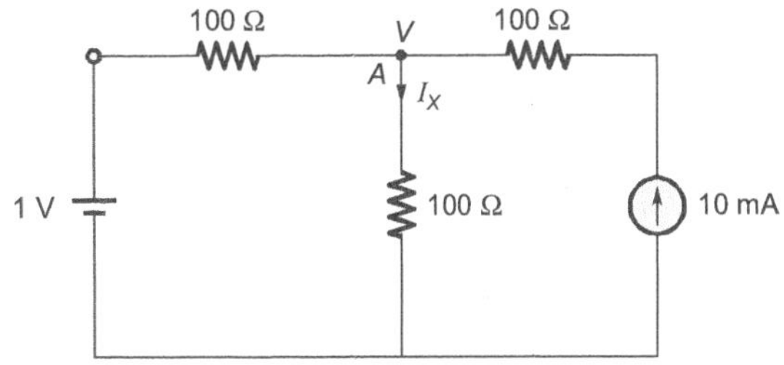

Applying Node equation at Node A

$$\frac{V-1}{100} + \frac{V}{100} = 10 \times 10^{-3}$$

$$2V - 1 = 1$$

$$V = 1$$

$$I_x = \frac{V}{100} = \frac{1}{100} = 10 \text{ mA}$$

1.32 (2)

By KCL at node 2

$$2I_2 = \frac{V_2}{1/3} + \frac{V_2 + 4V_{R_1} - V_1}{0.5} \quad \ldots\text{(i)}$$

$$V_{R1} = -1 \quad \ldots\text{(ii)}$$

and $$I_2 = \frac{V_1 - 4V_{R_1} - V_2}{0.5} \quad \ldots\text{(iii)}$$

Sub equation (ii) and (iii) in equation (i),

$$V_2 = 2 \text{ V}$$

1.33 (c)

Case I: (V_{OC})

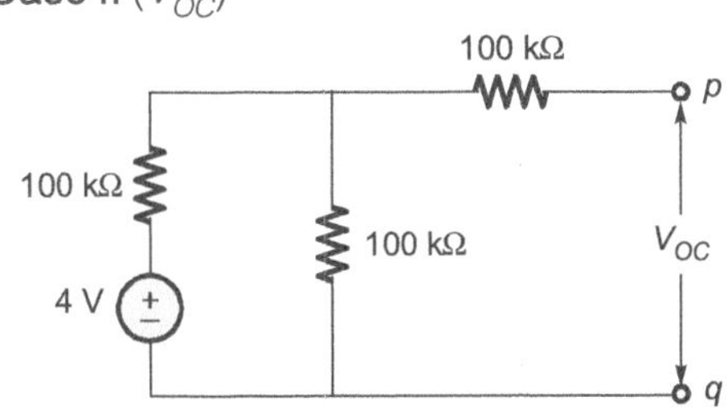

$$V_{OC} = 4 \times \frac{100}{100+100} = 2 \text{ V}$$

Case II: (R_{Th})

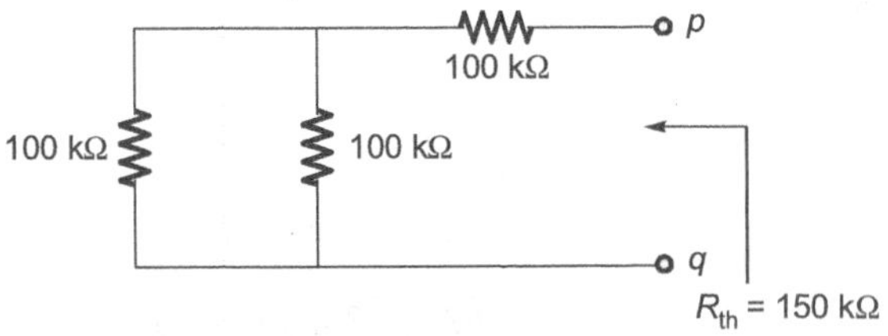

$$R_{Th} = 150\ \text{k}\Omega$$

1.34 (a)

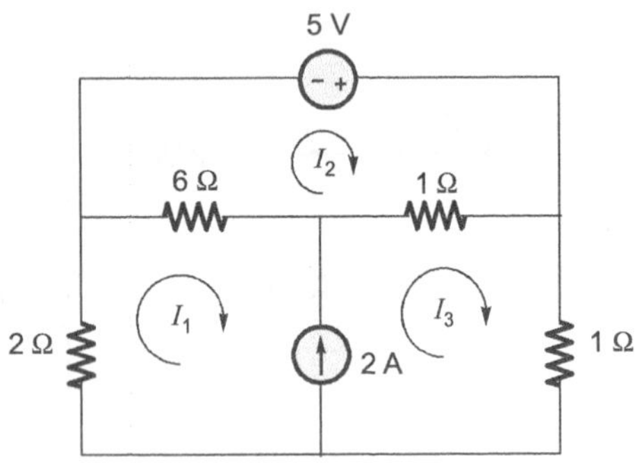

$$8I_1 - 6I_2 + 2I_3 - I_2 = 0 \quad \text{... (i)}$$

$$I_3 - I_1 = 2\ \text{A} \quad \text{... (ii)}$$

$$-6I_1 + 7I_2 - I_3 = 5 \quad \text{... (iii)}$$

By solving above equations,

$I_1 = 1$ A, $I_2 = 2$ A and $I_3 = 3$ A

1.35 1 (0.95 to 1.05)

Case I:

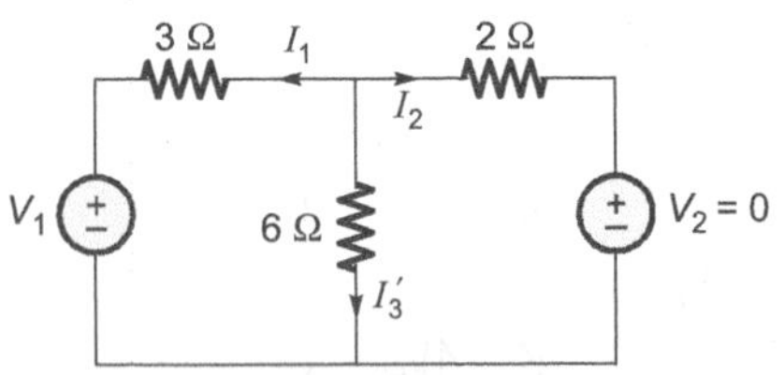

$$I_2 = -6\ \text{A}$$

Apply current division and KCL we get

$$I_3' = -2\ \text{A}$$

Case II:

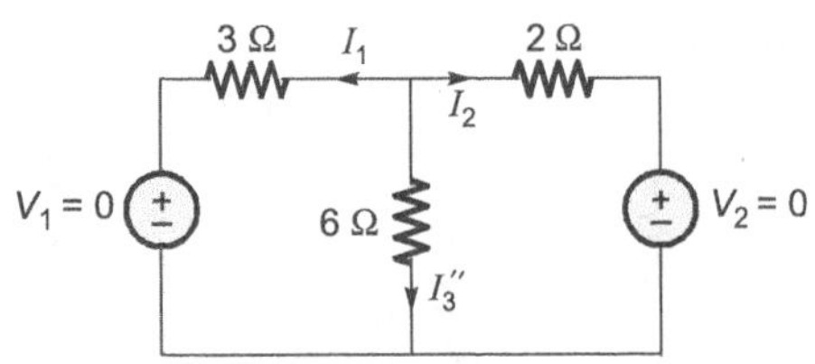

$$I_1 = 6\ \text{A}$$

Apply current division we get

$$I_3'' = 3\ \text{A}$$

$$I_3 = I_3' + I_3'' = -2\ \text{A} + 3\ \text{A} = 1\ \text{A}$$

1.36 (5)

According to maximum power transfer theorem, the maximum power will be transferred to the load when,

$$R_L = R_S$$

$$\therefore \quad R_L = 5\ \Omega$$

1.37 (1000)

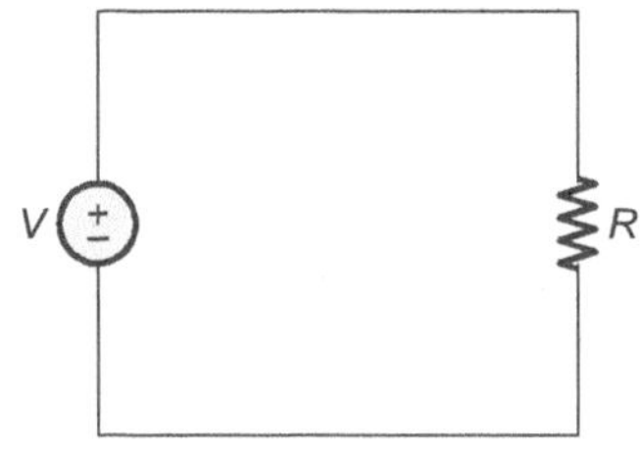

$$P = \frac{V^2}{R}$$

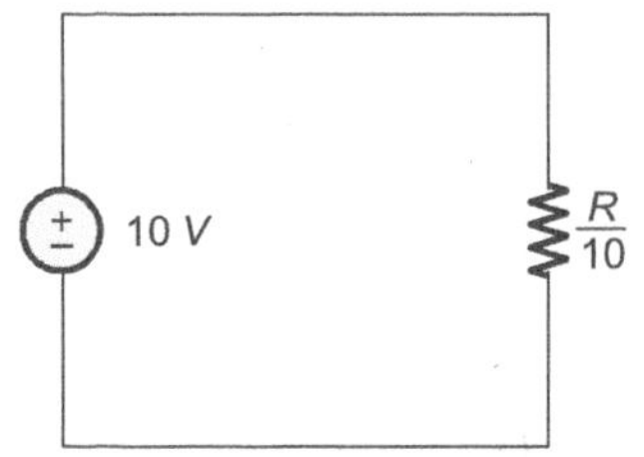

$$P' = \frac{(10V)^2}{R/10} = 1000\frac{V^2}{R}$$

$$\therefore \quad P' = 1000P$$

$\therefore$ Power dissipated scales up by a factor of 1000.

1.38 (1.7)

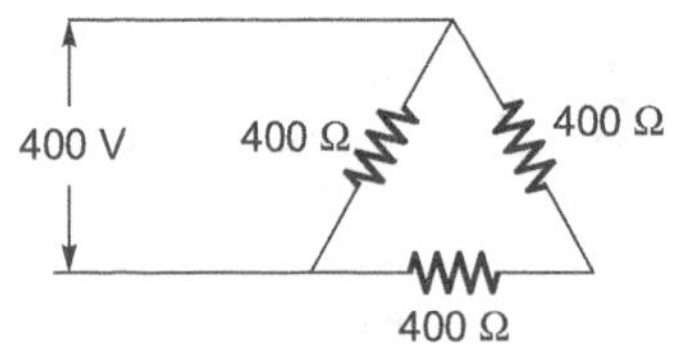

$$I_{Ph} = \frac{V_{Ph}}{R_{Ph}} = \frac{400}{400} = 1\text{A}$$

$$I_{Line} = \sqrt{3}\, I_{Ph} = \sqrt{3} \times 1 = 1.732\ \text{A}$$

1.39 (b)

Maximum current for 100 Ω, 1 W;

$$I = \sqrt{\frac{P}{R}} = 0.1\ \text{A}$$

Maximum current for 2 Ω, 0.5 W;

$$I = \sqrt{\frac{0.5}{2}} = 0.5\ \text{A}$$

Maximum current for 1 Ω, 0.25 W;

$$I = \sqrt{0.25} = 0.5 \text{ A}$$

The safe maximum value of current I which can flow through series resistances is 0.1 A.

1.40 (c)

From inspection

$$3I_1 - I_2 - 2I_3 = V_1 \quad ...(1)$$
$$-I_1 + 3I_2 - I_3 = V_2 \quad ...(2)$$
$$-2I_1 - I_2 + 3I_3 = -V_3 \quad ...(3)$$

$$\begin{bmatrix} 3 & -1 & -2 \\ -1 & 3 & -1 \\ -2 & -1 & 3 \end{bmatrix}\begin{bmatrix} I_1 \\ I_2 \\ I_3 \end{bmatrix} = \begin{bmatrix} V_1 \\ V_2 \\ -V_3 \end{bmatrix}$$

1.41 (115.47)

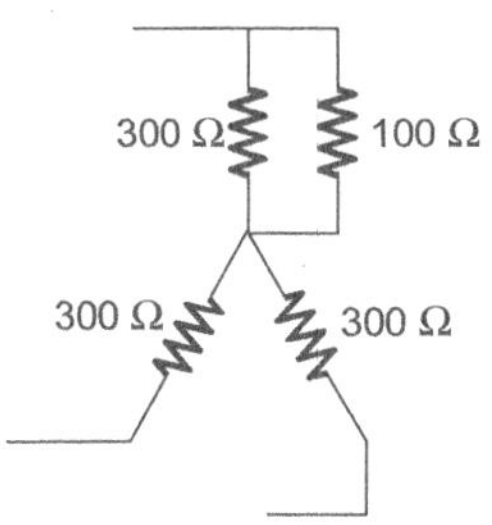

$R = 300 \,||\, 100$
$R = 75\,\Omega$

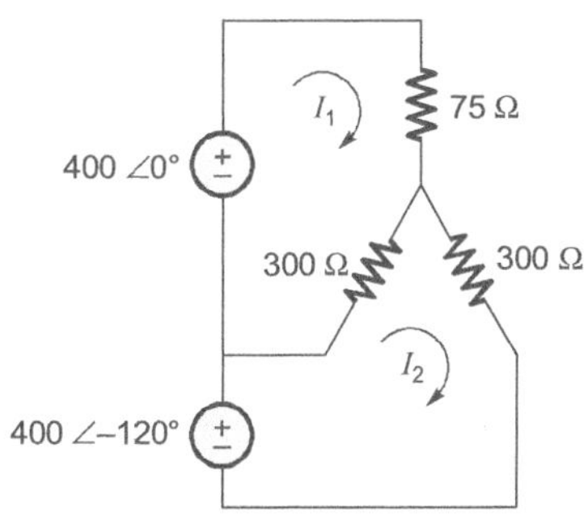

Apply KVL in each loop

$$375\, I_1 - 300\, I_2 = 400\angle 0° \quad ...(1)$$
$$-300\, I_1 + 600\, I_2 = 400\angle{-120°} \quad ...(2)$$

Solving (1) and (2),

$$I_1 = 1.5396\angle{-30°}$$
$$V = 75 \times I_1 \Rightarrow 115.47\angle{-30°}$$

The voltage across 100 Ω resistor is 115.47 V.

1.42 (0.0)

Case-I: Taking 8 V only

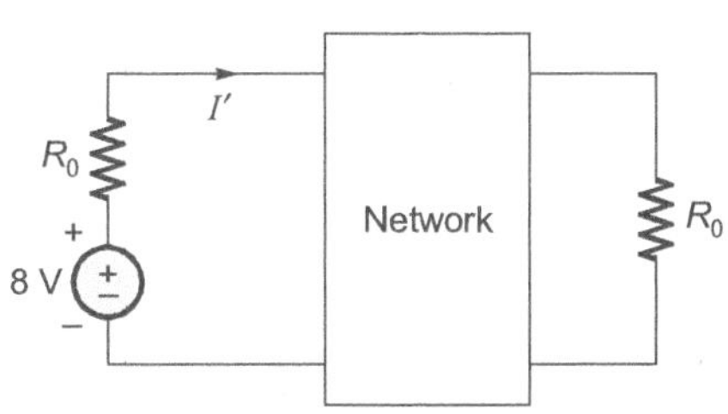

$$I' = 2\left(\frac{8}{16}\right) = 1 \text{ Amp}$$

Case-II: Taking 48 V only

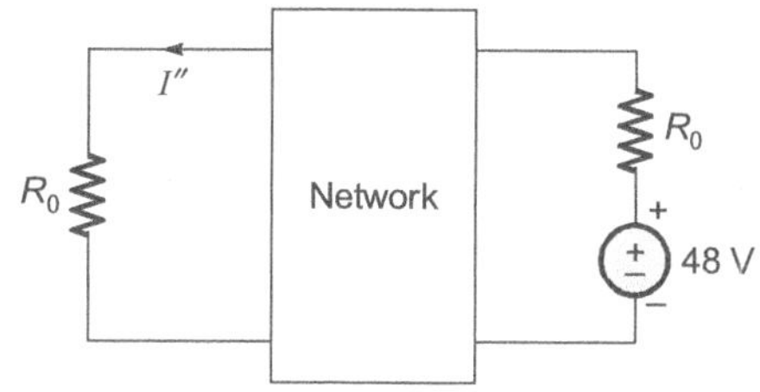

$$I'' = \left(\frac{1}{3}\right)\left(\frac{48}{16}\right) = 1 \text{ Amp}$$

So, $\quad I = I' - I'' = 1 - 1 = 0$ Amp

So current-I is zero (0).

1.43 (2.39)

Given data: $V_{in} = 10V_{P-P}$; $V_o = 8V_{P-P}$

The frequency,

$$f = 1 \text{ kHz}$$

We know, $\quad V_o = \dfrac{X_C}{\sqrt{X_C^2 + R^2}} V_{in}$

$$8 = \frac{X_C}{\sqrt{X_C^2 + 50^2}} \times 10$$

$$0.8 = \frac{X_C}{\sqrt{X_C^2 + 50^2}}$$

$$X_C^2 + 50^2 - 1.5625X_C^2 = 0$$
$$X_C^2 - 1.5625X_C^2 + 2500 = 0$$

$$X_C = \frac{200}{3}\Omega$$

$$C = \frac{1}{2\pi f X_C} = \frac{1}{2\pi \times 10^3 \times \frac{200}{3}}$$

$$= 2.387\ \mu\text{F}$$

So, value of capacitance,

$$C = 2.387\ \mu\text{F}.$$

1.44 (1)

For Thevenin resistance, we need to deactivate all the independent source.

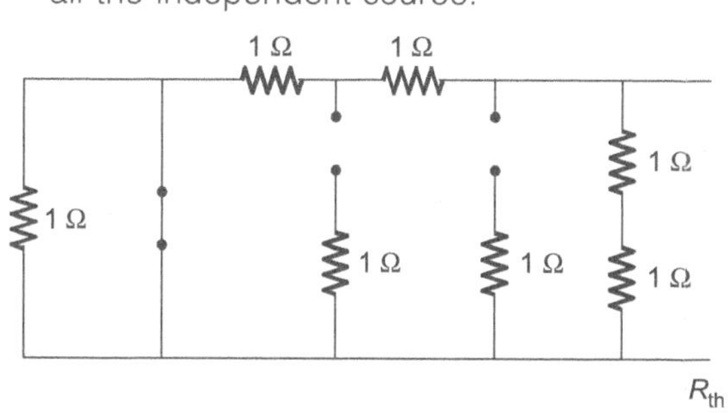

$$R_{th} = (1+1)\,||\,(1+1) = 2\,||\,2$$

$$= \frac{2\times 2}{2+2} = \frac{4}{4} = 1\,\Omega$$

$$R_{th} = 1\ \Omega$$

1.45 (200)

Given:

$v(t) = 95 + 200\cos(120\pi t) + 90\cos(360\pi t - 60°)$ V

$i(t) = 4\cos(120\pi t - 60°) + 1.5\cos(240\pi t - 75°)$A

$$P_{avg} = \frac{200\times 4}{2}\cos 60° + 0°$$

$$P_{avg} = 200\text{ W}$$

1.46 (28.78)

Given $\quad i(t) = 4 + 30\cos t - 20\sin t + 15\cos 3t - 10\sin 3t$

$$I_{rms} = \sqrt{4^2 + \frac{1}{2}(900+400+225+100)}$$

$$= 28.78\text{ A}$$

■■■

2 RL, RC, RLC Circuit and Transients

2.1 In the given below figure the stored energy of the capacitor is

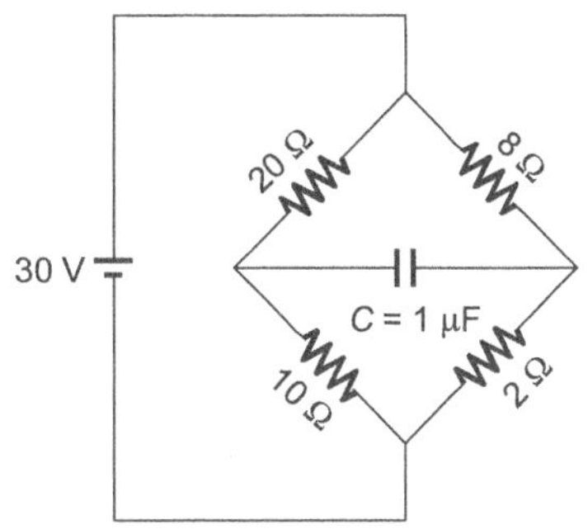

(a) 8 μJ (b) 30 μJ
(c) 450 μJ (d) 900 μJ

[1998 : 1 M]

2.2 In the circuit shown in the figure, the switch S is closed at time $t = 0$. The supply current at $t = 0^+$ and the capacitor voltage at $t \to \infty$ are, respectively

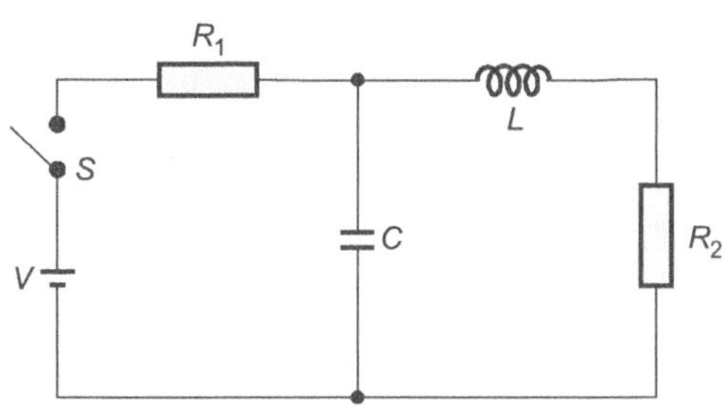

(a) 0 and V

(b) $\dfrac{V}{R_1}$ and $\dfrac{VR_2}{R_1 + R_2}$

(c) $\dfrac{V}{R_1 + R_2}$ and $\dfrac{VR_2}{R_1 + R_2}$

(d) $\dfrac{V}{R_1}$ and V

[2005 : 2 M]

2.3 Figure shows a circuit which has a coil of resistance R and inductance L. At resonance, the Q-factor of the coil is given by

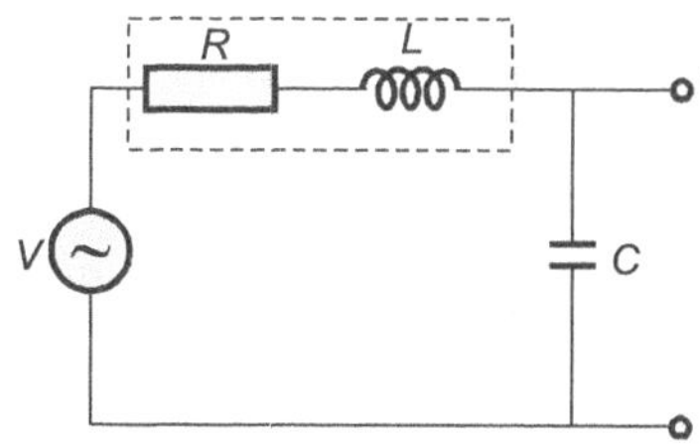

(a) $\left(\dfrac{V - V_0}{V}\right)$ (b) $\dfrac{V_0}{V}$

(c) $\left(\dfrac{V - V_0}{V_0}\right)$ (d) $\dfrac{V}{V_0}$

[2005 : 2 M]

2.4 In the circuit shown in the following figure, the input voltage $v_i(t)$ is constant at 2 V for time >> 1 s and then it changes to 1 V. The output voltage, $v_0(t)$, 2 s after the change will be

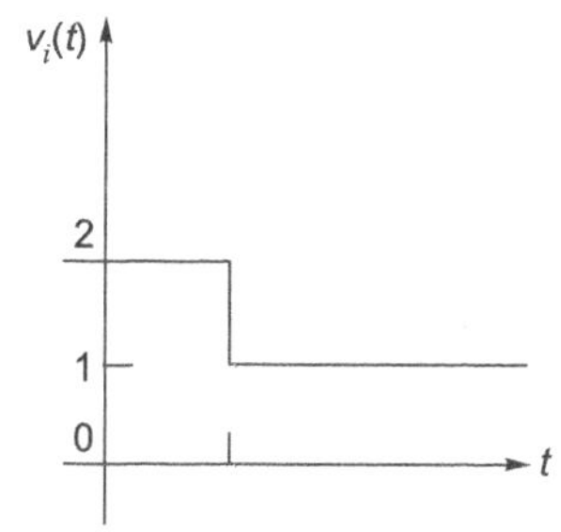

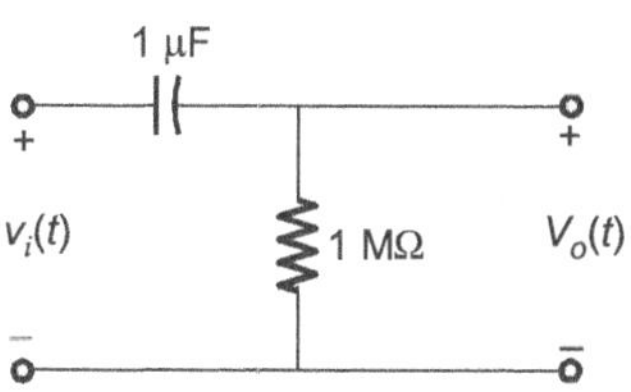

(a) –exp (–2) V (b) –1 + exp (–2) V
(c) exp (–2) V (d) 1 – exp (–2) V

[2006 : 2 M]

Statement for Linked Answer Questions (2.5 to 2.6):

In the circuit shown below, the current through the PMMC meter is assumed to be zero. The ideal switch toggles between position 1 and position 2. For each position, it is connected for time $T/2$. Assume $R_4C_x << T/2$ and $R_2C_x >> T/2$.

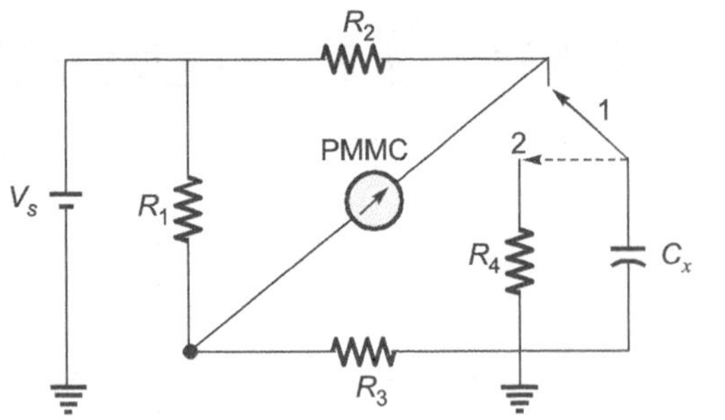

2.5 The capacitor C_x can be charged up to the maximum voltage

(a) $\frac{V_s T}{2R_2 C_x}$ (b) $\frac{V_s T}{2R_4 C_x}$

(c) $\frac{2V_s R_2 C_x}{T}$ (d) $\frac{2V_s R_4 C_x}{T}$

[2006 : 2 M]

2.6 In the bridge circuit shown in the figure above, if the PMMC meter shows null deflection, then the value C_x is

(a) $\frac{T}{R_4}$ (b) $\frac{TR_3}{(R_1+R_2)R_2}$

(c) $\frac{T(R_3+R_1)}{4R_2(R_3-R_1)}$ (d) $\frac{TR_3}{2R_2(R_1-R_3)}$

[2006 : 2 M]

2.7 In the circuit shown in the following figure, the current through the 1 Ω resistor is

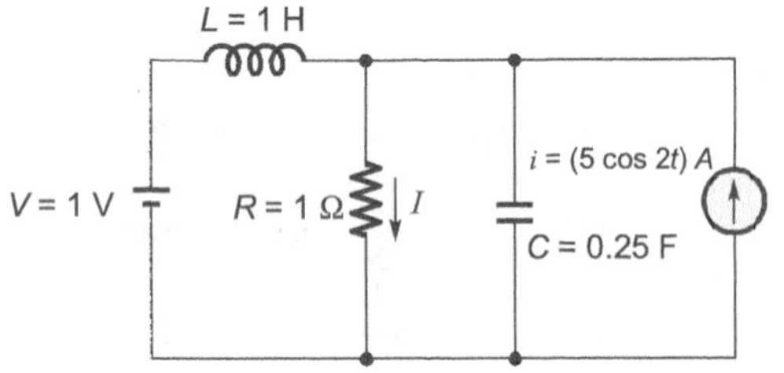

(a) $(1 + 5\cos 2t)$ A (b) $(5 + \cos 2t)$ A
(c) $(1 - 5\cos 2t)$ A (d) 6 A

[2007 : 2 M]

2.8 In the circuit shown in the following figure, the switch is kept closed for a long time and then opened at $t = 0$.
The values of the current just before opening the switch ($t = 0^-$) and after closing the switch ($t = 0^+$) are, respectively.

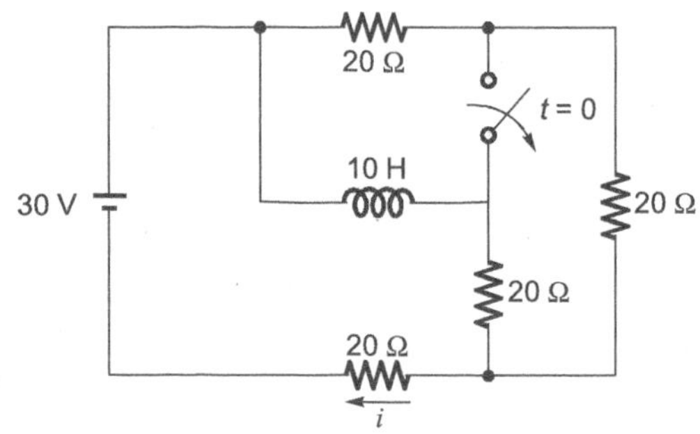

(a) $\frac{3}{4}$ A and 1 A (b) $\frac{7}{6}$ A and $\frac{5}{2}$ A

(c) 1 A and $\frac{7}{6}$ A (d) 1 A and 1 A

[2007 : 2 M]

2.9 In the circuit shown in the figure, the input signal is $v_i(t) = 5 + 3\cos\omega t$.
The steady state output is expressed as $v_o(t) = P + Q\cos(\omega t - \phi)$. If $\omega CR = 2$, the values of P and Q are

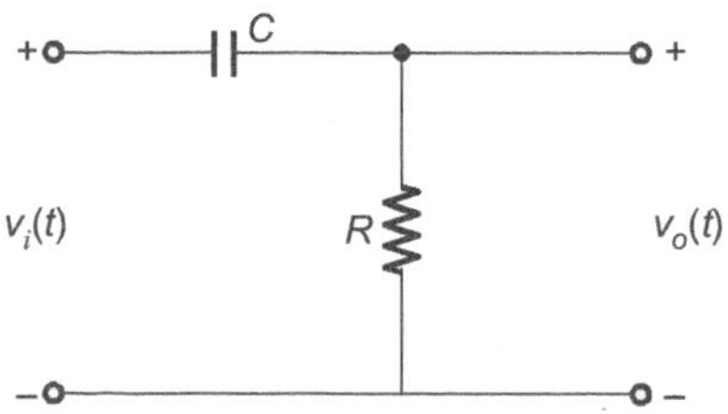

(a) $P = 0$ and $Q = \frac{6}{\sqrt{5}}$

(b) $P = 0$ and $Q = \frac{3}{\sqrt{5}}$

(c) $P = 5$ and $Q = \frac{6}{\sqrt{5}}$

(d) $P = 5$ and $Q = 3$

[2007 : 2 M]

2.10 For the circuit shown below the steady-state current is

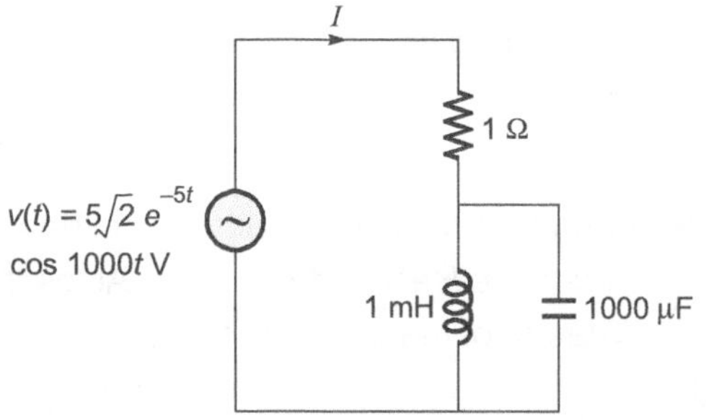

(a) 0 A

(b) $5\sqrt{2}\cos(1000t)$ A

(c) $5\sqrt{2}\cos\left(1000t - \frac{\pi}{4}\right)$ A

(d) $5\sqrt{2}$ A

[2008 : 2 M]

2.11 For the circuit shown below the voltage across the capacitor is

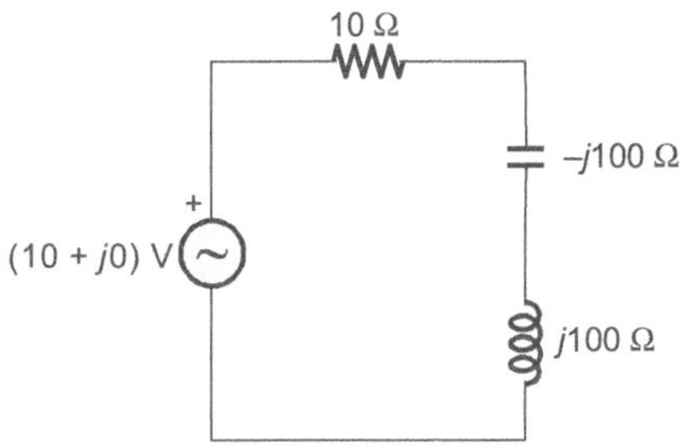

(a) (10 + j0) V (b) (100 + j0) V

(c) (0 + j100) V (d) (0 – j100) V

[2008 : 2 M]

Statement for Linked Answer Question (2.12 and 2.13):

In the circuit shown below the steady-state is reached with the switch *K* open. Subsequently the switch is closed at time $t = 0$.

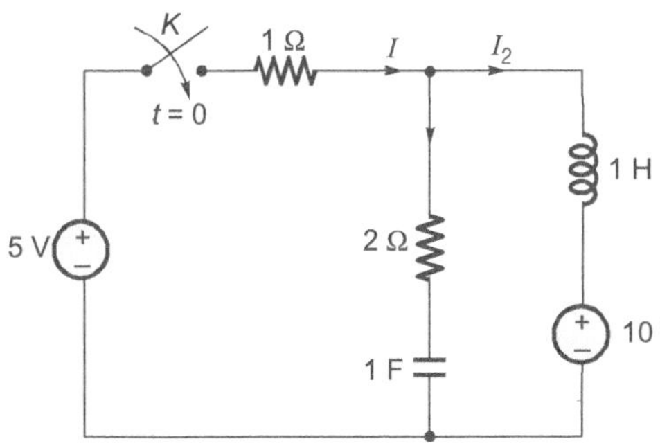

2.12 At time $t = 0^+$, current I_1 is

(a) $-\frac{5}{3}$ A (b) 0 A

(c) $\frac{5}{3}$ A (d) ∞ A

[2008 : 2 M]

2.13 At time $t = 0^+$, $\frac{dI_2}{dt}$ is

(a) –5 A/s (b) $-\frac{10}{3}$ A/s

(c) 0 A/s (d) 5 A/s

[2008 : 2 M]

Statement for Linked Answer Questions (2.14 & 2.15):

A coil having an inductance (L) of 10 mH and resistance R is connected in series with an ideal 100 μF capacitor (C). When excited by a voltage source of value of $10\sqrt{2}\cos(1000t)$ V, the series RLC circuit draws 20 W of power.

2.14 The value of the coil resistance R is

(a) 1 Ω (b) 2 Ω

(c) 4 Ω (d) 5 Ω

[2010 : 2 M]

2.15 The Q-factor of the coil at an angular frequency of 1000 rad/s is

(a) 1 (b) 2

(c) 4 (d) 5

[2010 : 2 M]

2.16 In the circuit shown below, the switch, initially at position 1 for a long time, is changed to position 2 at $t = 0$.

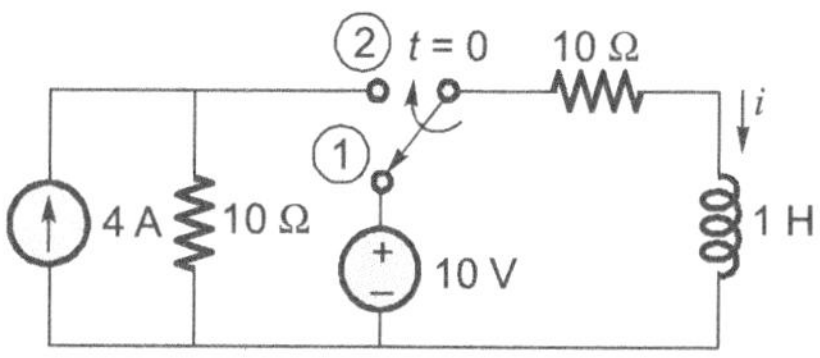

The current i through the inductor for $t \geq 0$ is

(a) $I - e^{-20t}$A (b) $I + e^{-20t}$A

(c) $I + 2e^{-20t}$A (d) $2 - e^{-20t}$A

[2011 : 1 M]

Common data Question for (2.17 and 2.18):

Consider the circuit shown below:

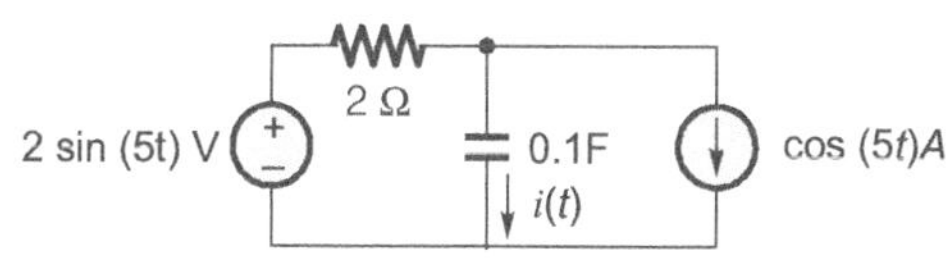

2.17 The current $i(t)$ through the capacitor is

(a) sin (5t) A (b) cos (5t) A

(c) sin (5t – 45°) A (d) 1 A

[2011 : 2 M]

2.18 The average total power delivered by the two sources is

(a) 1 W (b) 0.5 W

(c) 2 W (d) 4 W

[2011 : 2 M]

2.19 The average power delivered to an impedance $(4 - j3)\,\Omega$ by a current $5\cos(100\pi t + 100)$ A is

(a) 44.2 W (b) 50 W
(c) 62.5 W (d) 125 W

[2012 : 1 M]

2.20 In the following figure, C_1 and C_2 are ideal capacitors. C_1 has been charged to 12 V before the ideal switch S is closed at $t = 0$. The current $i(t)$ for all t is

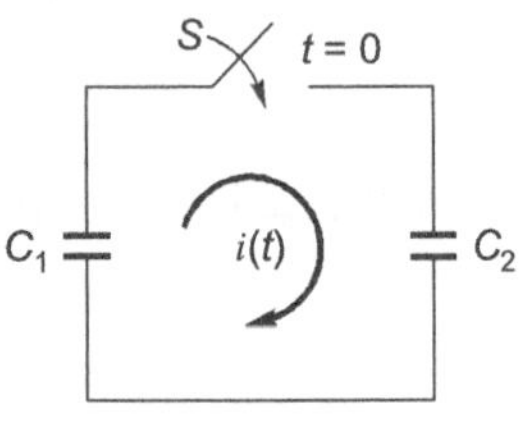

(a) zero
(b) a step function
(c) an exponentially decaying function
(d) an impulse function

[2012 : 1 M]

Statement for Linked Answer (2.21 and 2.22):

In the circuit shown, the three voltmeter readings are $V_1 = 220$ V, $V_2 = 122$ V, $V_3 = 136$ V.

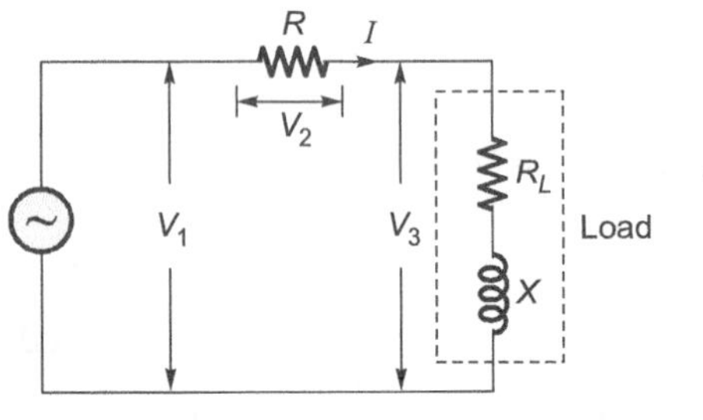

2.21 The power factor of the load is

(a) 0.45 (b) 0.50
(c) 0.55 (d) 0.60

[2012 : 2 M]

2.22 If $R_L = 5\,\Omega$, the approximate power consumption in the load is

(a) 700 W (b) 750 W
(c) 800 W (d) 850 W

[2012 : 2 M]

2.23 Two magnetically uncoupled inductive coils have Q factors q_1 and q_2 at the chosen operating frequency. Their respective resistances are R_1 and R_2. When connected in series, the effective Q factor of the series combination at the same operating frequency is

(a) $q_1 + q_2$
(b) $(1/q_1) + (1/q_2)$
(c) $(q_1R_1 + q_2R_2)/(R_1 + R_2)$
(d) $(q_1R_2 + q_2R_1)/(R_1 + R_2)$

[2013 : 2 M]

2.24 Three capacitor C_1, C_2 and C_3 whose values are 10 μF, 5 μF, and 2 μF respectively, have breakdown voltages of 10 V, 5 V and 2 V respectively. For the interconnection shown below, the maximum safe voltage in volts that can be applied across the combination, and the corresponding total charge in μC stored in the effective capacitance across the terminals are, respectively

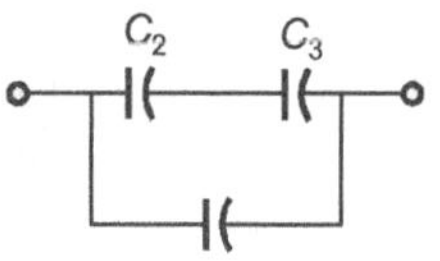

(a) 2.8 and 36 (b) 7 and 119
(c) 2.8 and 32 (d) 7 and 80

[2013 : 2 M]

2.25 The circuit shown in figure was at steady state for $t < 0$ with the switch at position 'A'. The switch is thrown to position 'B' at time $t = 0$. The voltage V(volts) across the 10 Ω resistor at time $t = 0^+$ is _______ V.

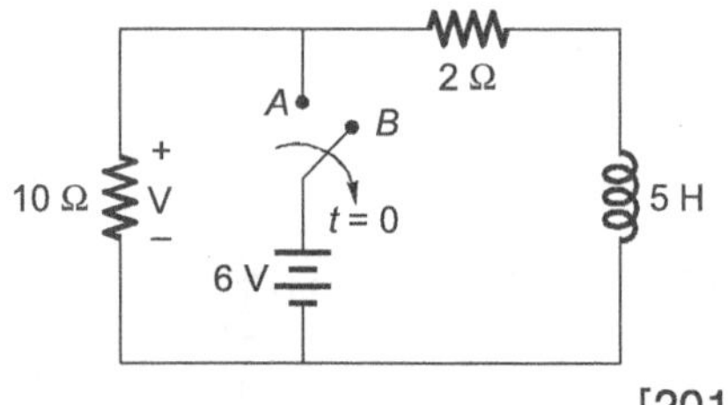

[2014 : 1 M]

2.26 The average real power in watts delivered to a load impedance $Z_L = (4 - j2)$ by an ideal current source $i(t) = 4\sin(\omega t + 20°)A$ is ____.

[2014 : 1 M]

2.27 In the circuit shown in the figure, initially the capacitor is uncharged. The switch 'S' is closed at $t = 0$. Two milliseconds after the switch is closed, the current through the capacitor (in mA) is ____.

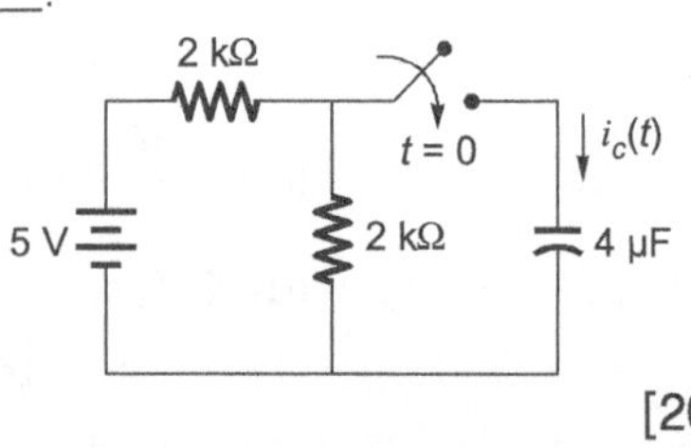

[2014 : 2 M]

2.28 A capacitor 'C' is to be connected across the terminals 'A' and 'B' as shown in the figure so that the power factor of the parallel combination becomes unity. The value of the capacitance required in μF is______.

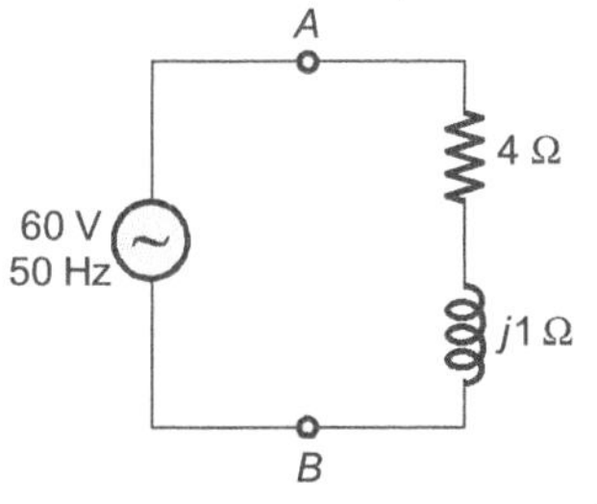

[2014 : 2 M]

2.29 In the microprocessor controlled measurement scheme shown in the figure, R_x is the unknown resistance to be measured, while R_{ref} and C_{ref} are known. C_{ref} is charged from voltage V_L to V_H (by a constant DC voltage source V_S), once through R_{ref} in T_{ref} seconds and then discharged to V_L. It is again charged from voltage V_L to V_H through R_x in T_x seconds.

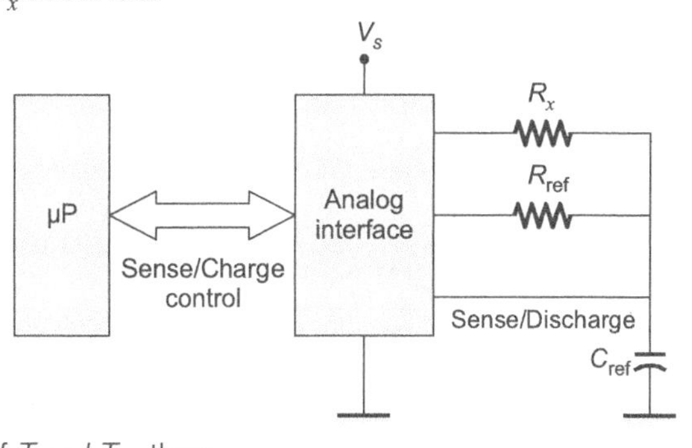

If $T_x = kT_{ref}$ then

(a) $R_x = kR_{ref}\left(1-\frac{V_L}{V_H}\right)$

(b) $R_x = kR_{ref}\ln\left(\frac{V_H}{V_L}\right)$

(c) $R_x = kR_{ref}$

(d) $R_x = R_{ref}\ln k$

[2014 : 2 M]

2.30 The capacitor shown in the figure is initially charged to +10 V. The switch closes at time $t = 0$. Then the value of $v_c(t)$ in volts at time $t = 10$ ms is ______ V.

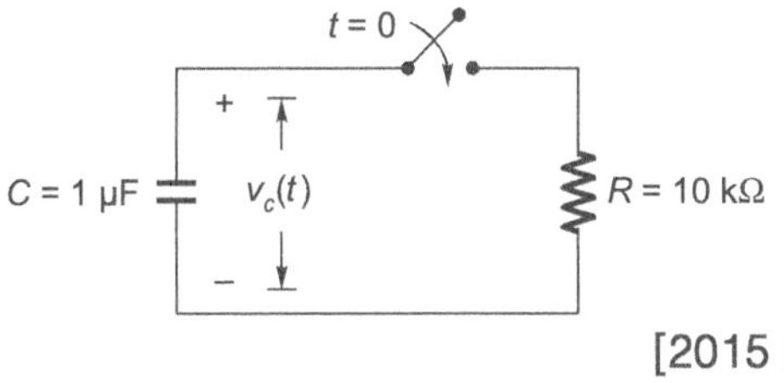

[2015 : 1 M]

2.31 The circuit shown in the figure is in series resonance at frequency f_c Hz. The value of V_c in volts is ______ V.

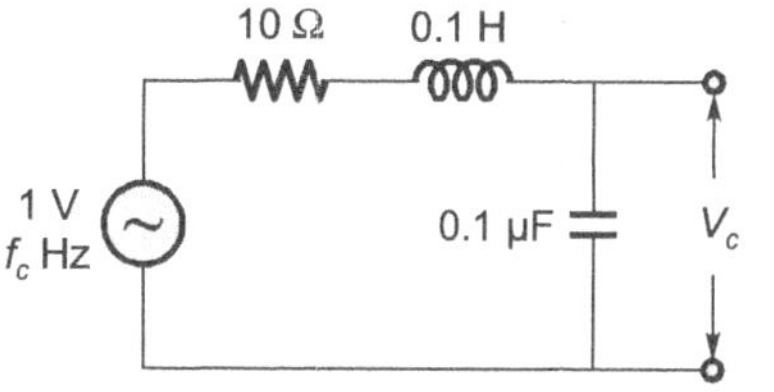

[2015 : 2 M]

2.32 A current $i(t)$ shown in the figure below is passed through a 1 F capacitor that had zero initial charge. The voltage across the capacitor for $t > 2$s in volt is______.

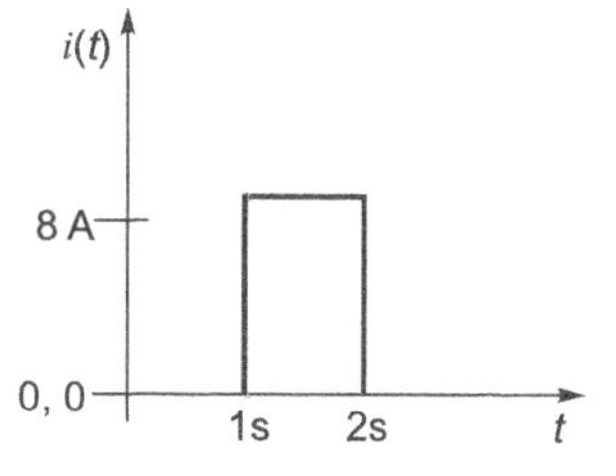

[2016 : 1 M]

2.33 In the circuit shown below:

$(v_1 + v_2) = [1 \sin(2\pi\ 10000t) + 1 \sin(2\pi\ 30000t)]$ V

The RMS value of the current through the resistor R will be minimum if the value of the capacitor C in microfarad is______.

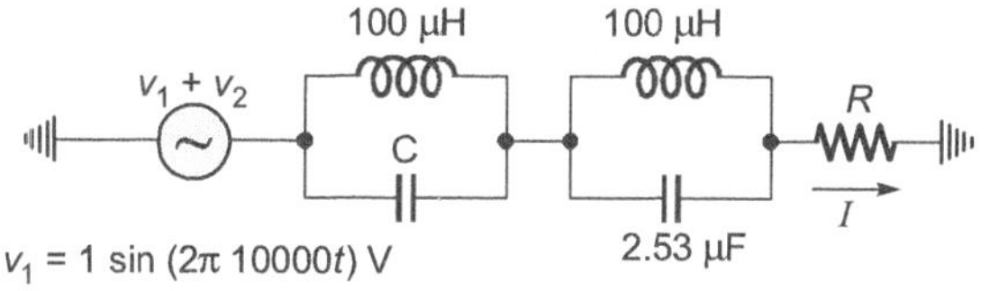

[2016 : 1 M]

2.34 In the circuit shown below, $V_S = 101\angle 0$ V, $R = 10\ \Omega$ and $\omega L = 100\ \Omega$. The current I_S is in phase with V_S. The magnitude of I_S in milliampere is_______.

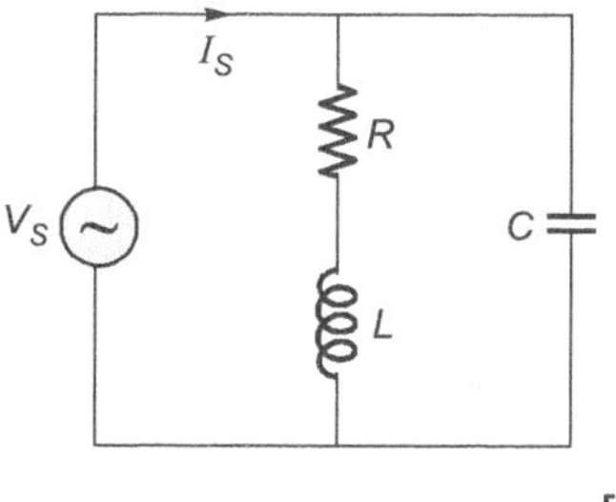

[2016 : 2 M]

2.35 The voltage $v(t)$ shown below is applied to the given circuit. $v(t) = 3$ V for $t < 0$ and $v(t) = 6$ V for $t > 0$. The value of current $i(t)$ at $t = 1$ s, in ampere is________.

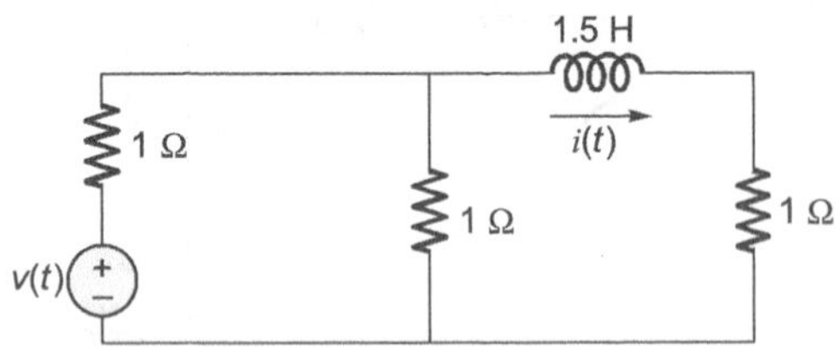

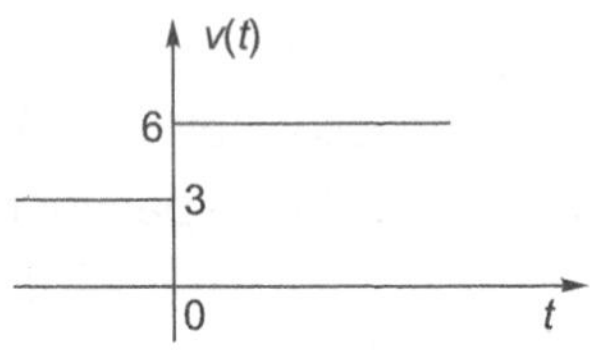

[2016 : 2 M]

2.36 A series R-L-C circuit is excited with a 50 V, 50 Hz sinusoidal source. The voltages across the resistance and the capacitance are shown in the figure. The voltage across the inductor (V_L) is _______ V.

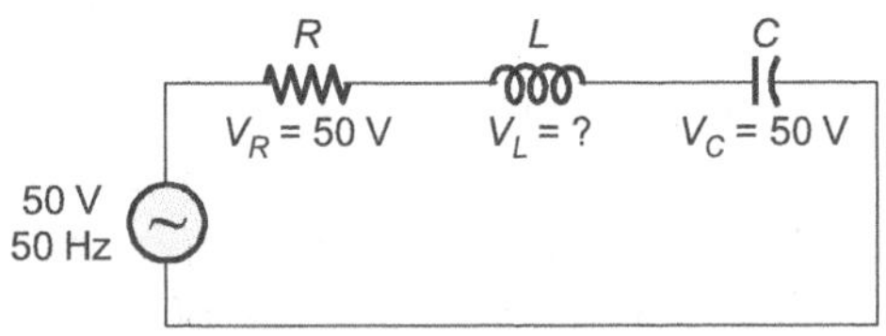

[2017 : 1 M]

2.37 The current response of a series R-L circuit to a unit step voltage is given in the table. The value of L is _______ H.

t in s	0	0.25	0.5	0.75	1.0	...	∞
$i(t)$ in A	0	0.197	0.316	0.388	0.432	...	0.5

[2017 : 2 M]

2.38 In the circuit diagram, shown in the figure, S_1 was closed and S_2 was open for a very long time. At $t = 0$, S_1 is opened and S_2 is closed. The voltage across the capacitor, in volts, at $t = 5$ μs is _____.

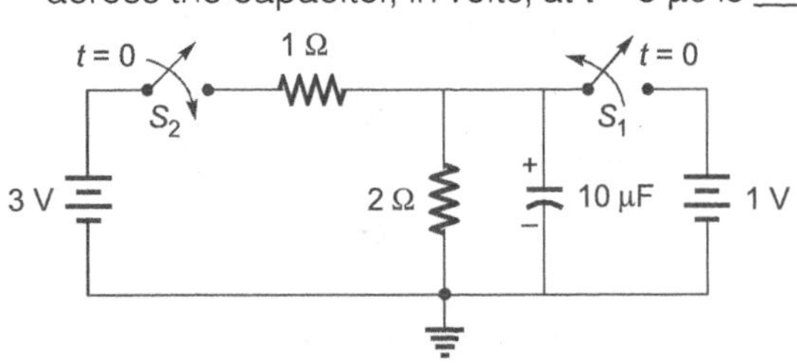

[2017 : 2 M]

2.39 A series R-L-C circuit is excited with an A.C voltage source. The quality factor (Q) of the circuit is given as $Q = 30$. The amplitude of current in ampere at upper half-power frequency will be _______.

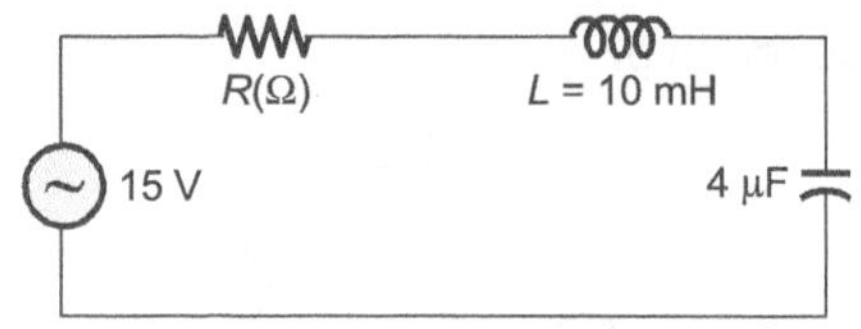

[2017 : 2 M]

2.40 For the circuit, shown in the figure, the total real power delivered by the source to the loads is _____ kW.

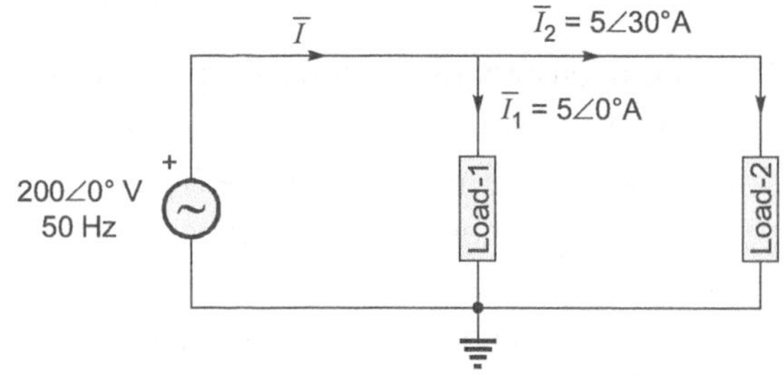

[2017 : 2 M]

2.41 A series R-C circuit is excited by a 1∠0 V sinusoidal ac voltage source. The locus diagram of the phasor current $I = (x + jy)$ A. When C is varied, while keeping R fixed, is

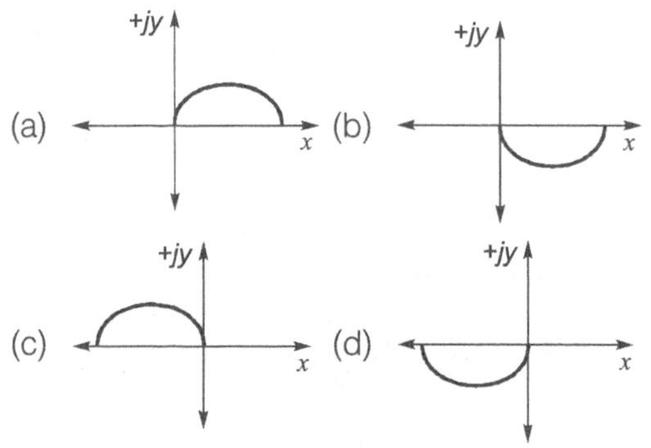

[2018 : 1 M]

2.42 A coil having an impedance of (10 + j100) Ω is connected in parallel to a variable capacitor as shown in figure. Keeping the excitation frequency unchanged, the value of the capacitor is changed so that parallel resonance occurs.. The impedance across terminals p-q at resonance (in Ω) is

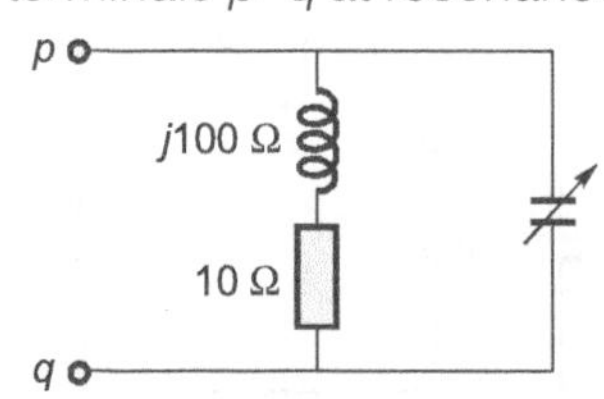

[2018 : 1 M]

2.43 In the figure, an RLC load is supplied by a 230 V, 50 Hz single phase source. The magnitude of the reactive power (in VAr) supplied by the source is ________.

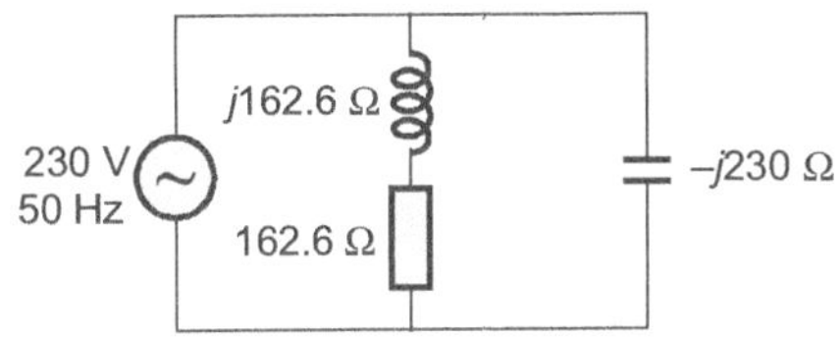

[2018 : 2 M]

2.44 Assuming ideal opamp, the RMS voltage (in mV) in the output V_0 only due to the 230 V, 50 Hz interference is (up to one decimal place) ________.

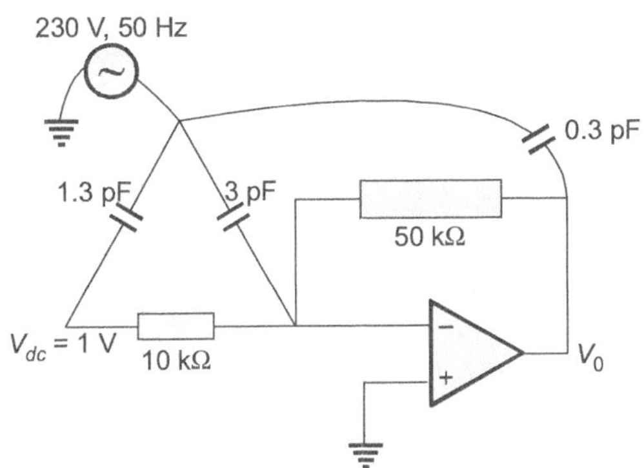

[2018 : 2 M]

2.45 If each of the values of inductance, capacitance and resistance of a series LCR circuit are doubled, the Q-factor of the circuit would

(a) reduce by a factor $\sqrt{2}$

(b) reduce by a factor 2

(c) increase by a factor $\sqrt{2}$

(d) increase by a factor 2

[2019 : 1 M]

2.46 In the circuit shown below, initially the switch S_1 is open, the capacitor $C1$ has a charge of 6 coulomb, and the capacitor $C2$ has 0 coulomb. After S_1 is closed, the charge on $C2$ in steady state is________Coulomb.

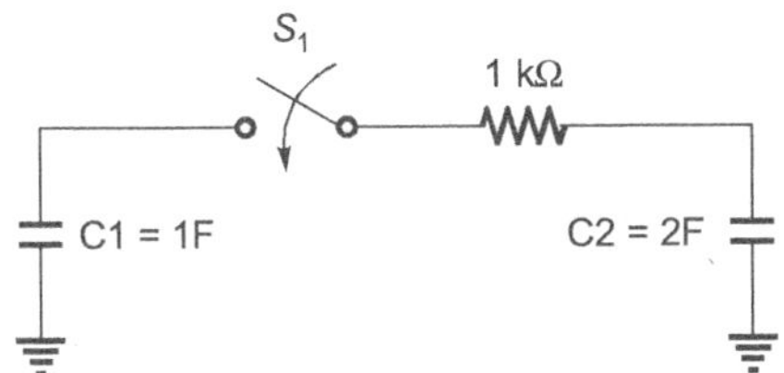

[2019 : 1 M]

2.47 In the circuit shown below, the angular frequency ω at which the current is in phase with the voltage is ________ rad/s.

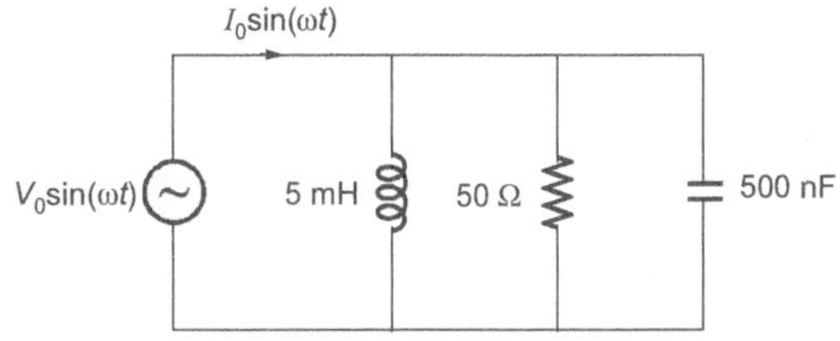

[2019 : 2 M]

2.48 In the circuit shown below, a step input voltage of magnitude 5 V is applied at node A at time $t = 0$. If the capacitor has no charge for $t \leq 0$, the voltage at node P at $t = 6$ μs is ________ V.

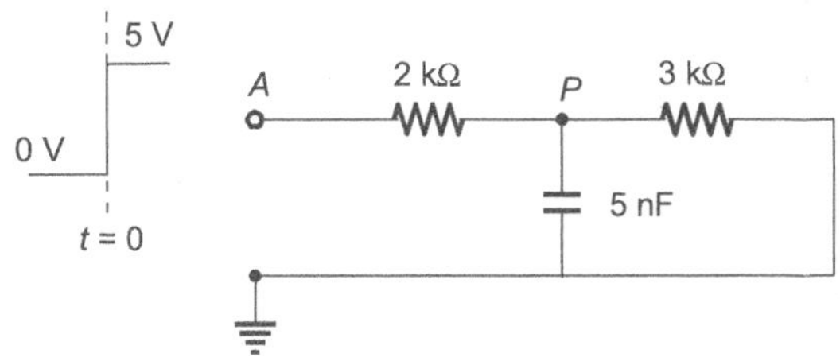

[2019 : 2 M]

2.49 The rms value of the phasor current I in the circuit shown (in amperes) is _____.

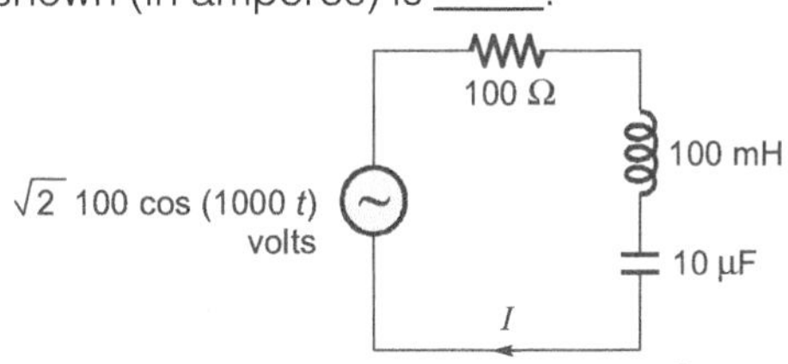

[2020 : 2 M]

2.50 A single-phase transformer has a magnetizing inductance of 250 mH and a core loss resistance of 300 Ω, referred to primary side, When excited with a 230 V, 50 Hz sinusoidal supply at the primary, the power factor of the input current drawn, with secondary on open circuit, is _____. (rounded off to two decimal places).

[2021 : 1 M]

2.51 An air cored coil having a winding resistance of 10 Ω. is connected in series with a variable capacitor C_x. The series circuit is excited by a 10 V sinusoidal voltage source of angular frequency 1000 rad/s. As the value of the capacitor is varied, a maximum voltage of 30 V was observed across it. Neglecting skin-effect, the value of the inductance of the coil in millihenry is _____.

[2021 : 2 M]

2.52 In the bandpass filter circuit shown, $R_0 = 50\ \Omega$, $L_0 = 1$ mH, $C_0 = 10$ nF. The Q factor of the filter is ______ (round off to two decimal places)

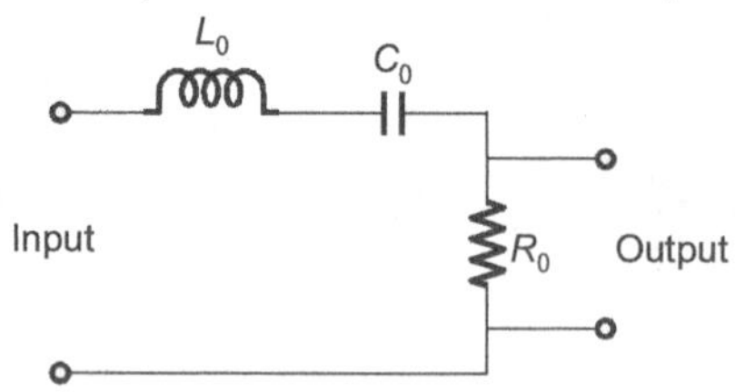

[2022 : 1 M]

2.53 In the circuit shown, the load is driven by a sinusoidal ac voltage source $V_1 = 100\angle 0^0$ V at 50 Hz. Given $R_1 = 20\ \Omega$, $C_1 = \left(\frac{1000}{\pi}\right)\mu F$, $L_1 = \left(\frac{20}{\pi}\right)$mH and $R_2 = 4\ \Omega$, the power factor is ______ (round off to one decimal place)

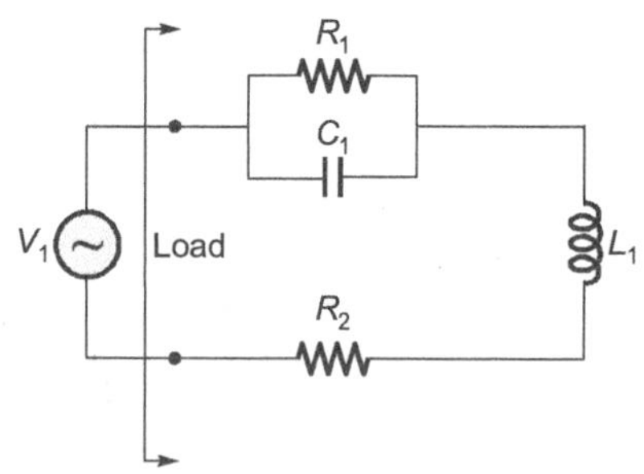

[2022 : 2 M]

2.54 In the circuit shown, the capacitance $C_0 = 10\ \mu F$ and inductance $L_0 = 1$ mH and the diode is ideal. The capacitor is initially charged to 10 V and the current in the inductor is initially zero. If the switch is closed at $t = 0$ s, the voltage $V_c(t)$ (in volts) across the capacitor at $t = 0.5$s is ______ (round off to one decimal place)

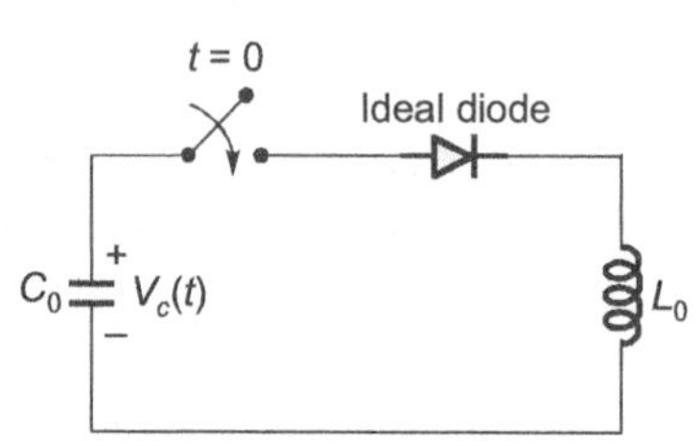

[2022 : 2 M]

2.55 In the circuit shown, $\omega = 100\ \pi$ rad/s, $R_1 = R_2 = 2.2\ \Omega$ and $L = 7$ mH. The capacitance C for which Y_{in} is purely real is ______ mF. (rounded off to two decimal places)

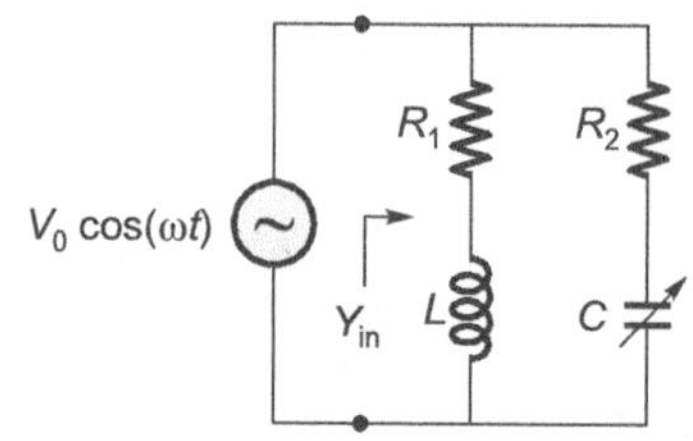

[2023 : 2 M]

2.56 The R-L circuit with $R = 10$ kΩ and $L = 1$ mH is excited by a step current $I_0u(t)$. At $t = 0^-$, there is a current $I_L = I_0/5$ flowing through the inductor. The minimum time taken for the current through the inductor to reach 99% of its final value is ______ μs (rounded off to two decimal places).

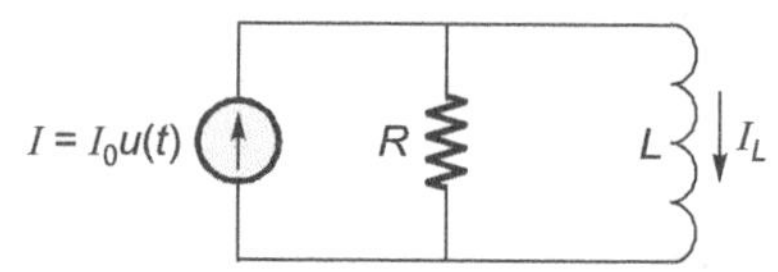

[2023 : 2 M]

2.57 A sinusoidal current of $i_1(t) = 1\sin(200\pi t)$ mA is flowing through a 4 H inductor which is mutually coupled to another 5 H inductor carrying

$$i_2(t) = 2\sin(200\pi t)\text{ mA}$$

as shown in the figure. The coupling coefficient between the inductors is 0.6. The peak energy stored in the circuit is ______ μJ (rounded off to two decimal places).

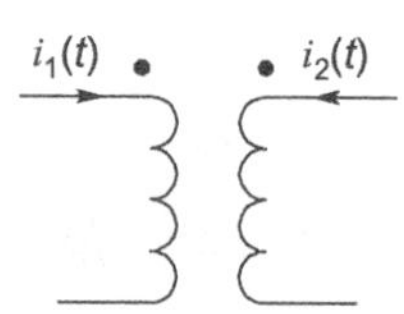

[2023 : 2 M]

2.58 The voltage source $V_s = 10\sqrt{2}\sin(20000\pi t)$ V has an internal resistance of 50 Ω. The RMS value of the current through R is ______ mA (rounded off to one decimal place).

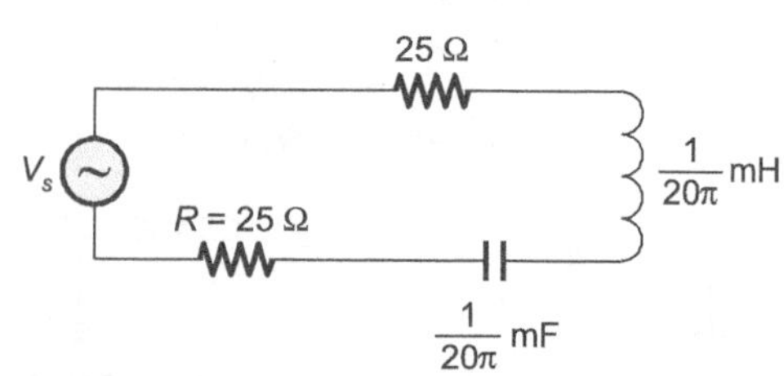

[2023 : 2 M]

2.59 The switch in the following figure has been closed for a long time ($t < 0$). It is opened at $t = 0$ seconds. The value of d_{vc}/dt at $t = 0^+$ is ________ V/s (rounded off to nearest integer).

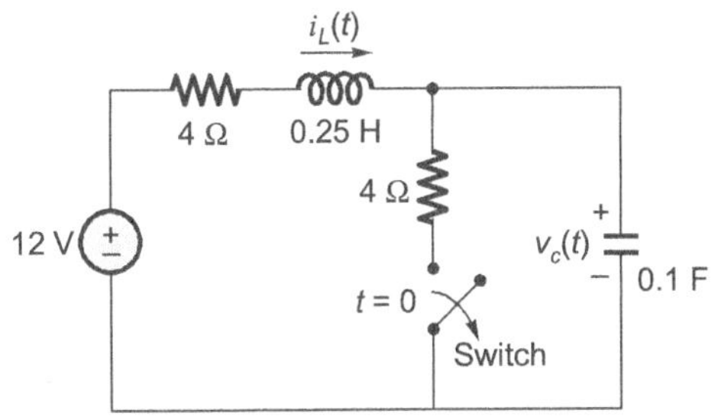

[2024 : 1 M]

2.60 Two magnetically coupled coils, when connected in series-aiding configuration, have a total inductance of 500 mH. When connected in series-opposing configuration, the coils have a total inductance of 300 mH. If the self-inductance of both the coils are equal, then the coupling coefficient is ________ (rounded off to two decimal places).

[2024 : 2 M]

■■■

Answers RL, RC, RLC Circuit & Transients

2.1	(a)	2.2	(b)	2.3	(b)	2.4	(a)	2.5	(a)	2.6	(*)	2.7	(a)
2.8	(c)	2.9	(a)	2.10	(a)	2.11	(d)	2.12	(a)	2.13	(b)	2.14	(d)
2.15	(b)	2.16	(d)	2.17	(a)	2.18	(a)	2.19	(b)	2.20	(d)	2.21	(a)
2.22	(b)	2.23	(c)	2.24	(c)	2.25	(30)	2.26	(32)	2.27	(1.51)	2.28	(187.33)
2.29	(c)	2.30	(3.678)	2.31	(100)	2.32	(8)	2.33	(0.28)	2.34	(100)	2.35	(1.632)
2.36	(50)	2.37	(1)	2.38	(1.52)	2.39	(6.36)	2.40	(1.866)	2.41	(a)	2.42	(1010)
2.43	(67.36)	2.44	(10.84)	2.45	(b)	2.46	(4)	2.47	(20000)	2.48	(1.896)	2.49	(1)
2.50	(0.25)	2.51	(30)	2.52	(6.32)	2.53	(0.8)	2.54	(–10.0)	2.55	(1.45)	2.56	(0.44)
2.57	(17.37)	2.58	(100)	2.59	(15)	2.60	(0.25)						

Explanations RL, RC, RLC Circuit & Transients

2.1 (a)

Voltage across capacitor, V

$$= \left[\frac{10}{10+20} - \frac{2}{2+8}\right] \times 30$$

$$= 10 - 6 = 4 \text{ V}$$

Energy stored in the capacitor

$$= \frac{1}{2}CV^2$$

$$= \frac{1}{2} \times 1 \times 10^{-6} \times 4^2$$

$$= 8 \, \mu\text{J}$$

2.2 (b)

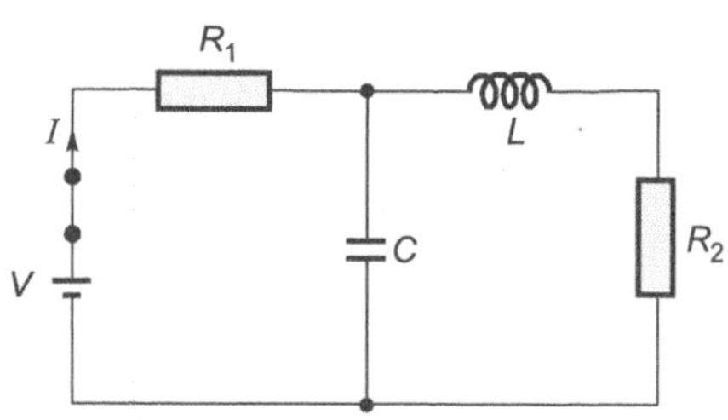

At $t = 0^+$ capacitor will short hence the supply current, $I = \dfrac{V}{R_1}$

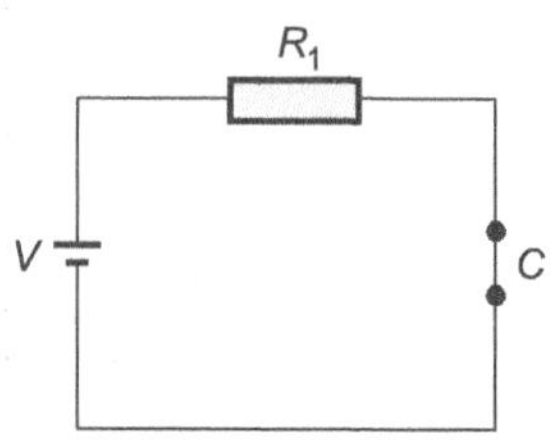

as $t \to \infty$ capacitor will open and inductor will short

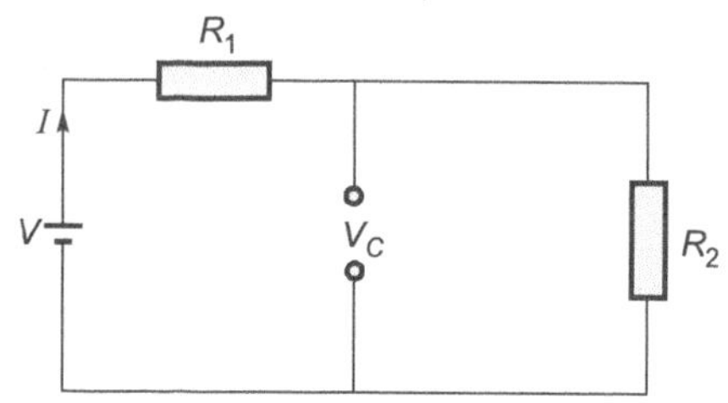

$$\therefore \quad I = \frac{V}{R_1 + R_2}$$

and $\quad V_C = I R_2$

$$\Rightarrow \quad V_C = \frac{R_2}{R_1 + R_2} \cdot V$$

2.3 (b)

$$\text{Quality factor } (Q) = \frac{X_C}{R} = \frac{I X_C}{I R} = \frac{V_0}{V}$$

2.4 (a)

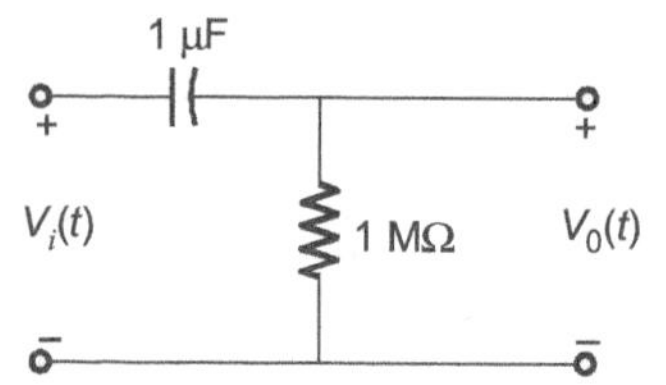

Time constant = RC = 1 sec.

The input voltage is applied for long time (>> time constant τ) .

$\therefore$ The capacitor will be fully charged.

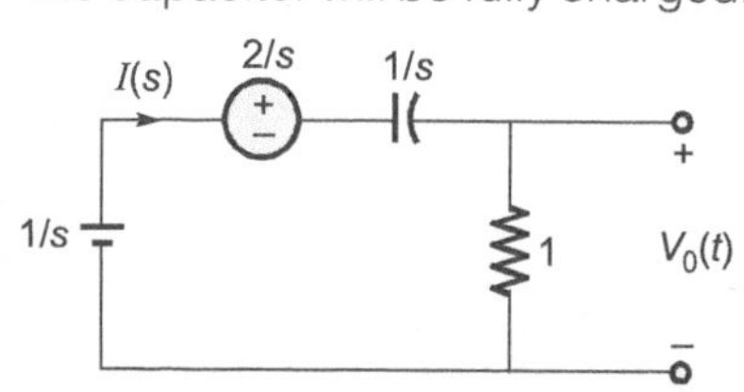

Applying KVL,

$$\frac{1}{s}-\frac{2}{s} = \left(1+\frac{1}{s}\right)I(s)$$

$$\Rightarrow \quad -\frac{1}{s} = \frac{s+1}{s}\cdot I(s)$$

$$\Rightarrow \quad I(s) = -\frac{1}{s+1}$$

taking Laplace inverse

$$\Rightarrow \quad I(t) = -e^{-t} \Rightarrow V_0(t) = -1\cdot e^{-t}$$

$$\therefore \quad V_0(2) = -e^{-2}\text{ V}$$

2.5 (a)

When switch is at position 1 and current through PMMC is zero then modified circuit will be as

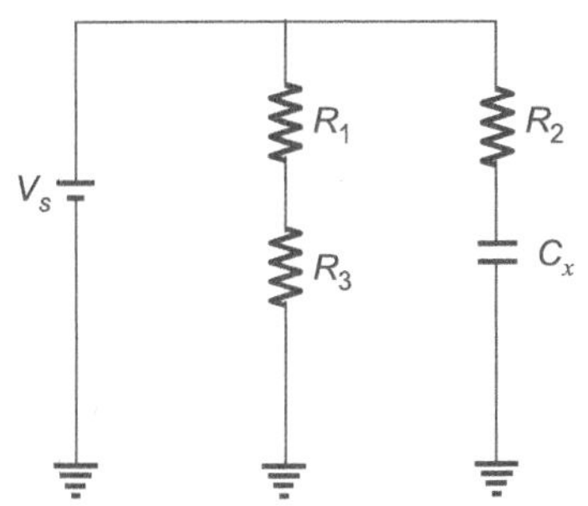

time constant $\tau = R_{eq}\,C$

net resistance across C will be find as

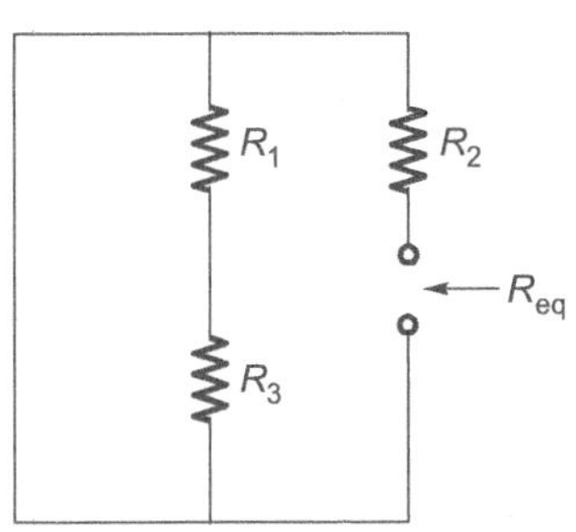

$$R_{eq} = R_2$$

therefore, $\tau = R_2\,C_x$

Now, $V_\infty = V_s$

$$V(0) = 0$$

So, charging equation for capacitor will be

$$V_c(t) = V_\infty - [V_\infty - V(0)]\,e^{-t/\tau}$$

$$= V_s - [V_s - 0]\,e^{-t/R_2C_x}$$

$$= V_s[1-e^{-t/R_2C_x}]$$

as time increases, voltage across capacitor increases, so at $t = T/2$, we get maximum voltage during charging

so, $V_c = V_s[1-e^{-T/2\,R_2C_x}]$

$\because \quad T/2 << R_2C_x$

so, $\dfrac{T}{2R_2C_x} << 1$

so for e^{-x}, we can take linear approximation as $= 1-x$

where, $x = \dfrac{T}{2R_2C_x}$

Now, $V_c = V_s\left[1-\left(1-\dfrac{T}{2R_2C_x}\right)\right]$

$$= \frac{V_sT}{2R_2C_x}$$

during discharging, $R_4C_x << T/2$, so it will discharge rapidly at position 2.

2.6 (*)

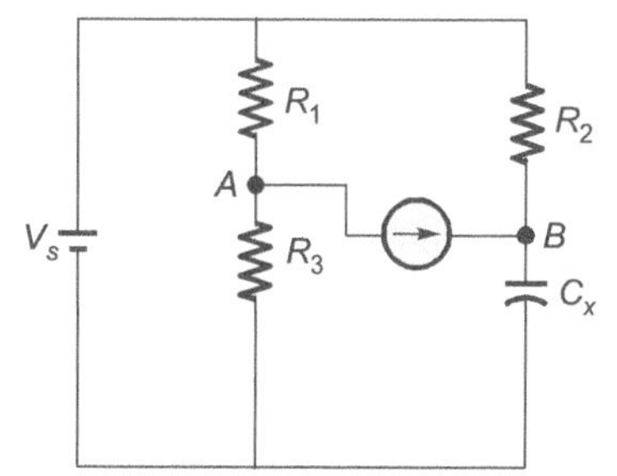

Bridge circuit will be as (at position 1)

Voltage at B at $t = T/2$ is

$$V_B = V_C = \frac{V_sT}{2R_2C_x} \quad \text{...(i)}$$

and voltage at A is

$$V_A = \frac{R_3}{R_1+R_3}\times V_s \quad \text{...(ii)}$$

for null deflection both V_A and V_B must be equal

so, $\dfrac{R_3}{R_1+R_3}V_s = \dfrac{V_sT}{2R_2C_x}$

$$\boxed{C_x = \frac{T(R_1+R_3)}{2R_2R_3}}$$

* = No option is matching.

2.7 (a)

Case-I:

When ac current source is open circuited.

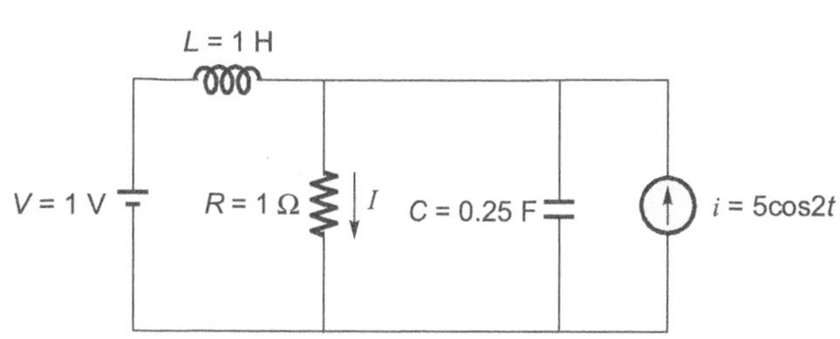

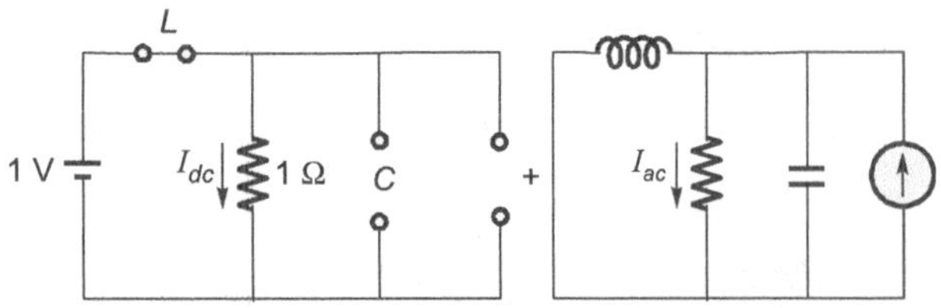

The inductor will short-circuit and capacitor will open-circuit for dc voltage source

$$I_{dc} = \frac{1}{1} = 1\text{A}$$

Case-II:

When dc voltage source is short-circuited.

$$X_L = j\omega L = j \times 2 \times 1 = j2$$

$$X_C = \frac{1}{j\omega C} = \frac{1}{j\times 2\times 0.25} = -j2$$

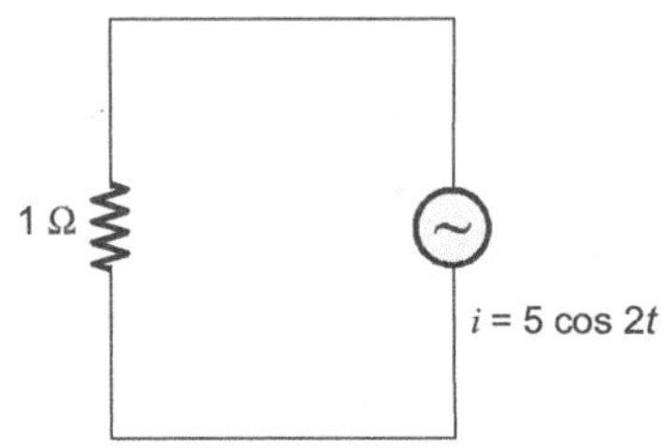

equivalent impedance of inductor and capacitor

$$Z_{eq} = \frac{j2\times(-1)}{j2\times(-j2)} \to \infty$$

$$\therefore \quad I_{ac} = 5\cos 2t$$

Total current through 1 Ω resistor

$$I = I_{dc} + I_{ac}$$
$$= (1 + 5\cos 2t)\text{ A}$$

2.8 (c)

at $t = 0^-$

The inductor will be short,

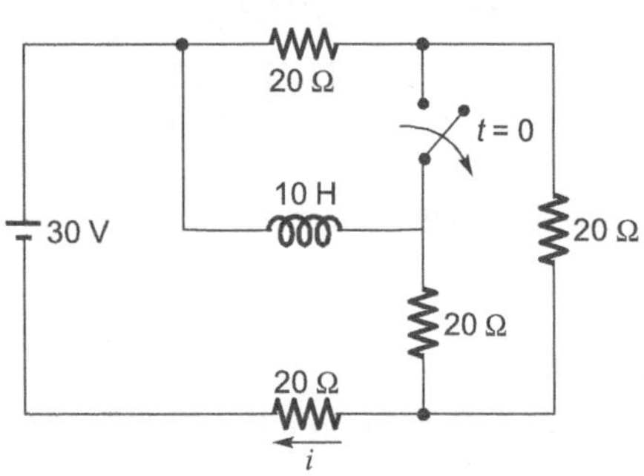

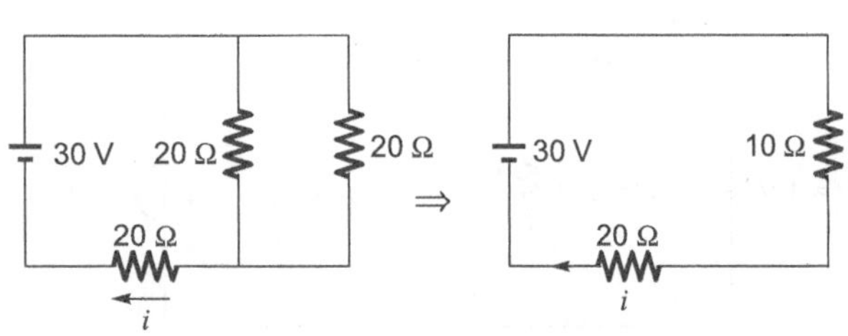

$$\therefore \quad i = \frac{30}{20+10}\text{ A} = 1\text{ A}$$

and current through inductor at $t = 0^+$

$$I_L = 1\text{ A}$$

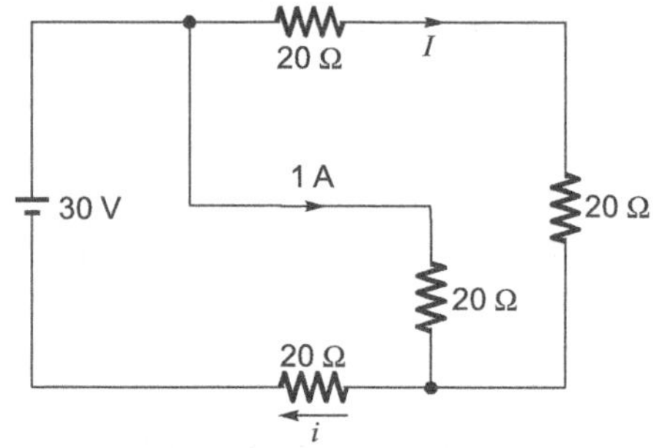

From the figure,

Applying KVL in outer loop.

$$30 = (20 + 20)I + 20i \qquad \text{...(i)}$$

As the current through the inductor does not change instantaneously. The current the inductor will be 1 A at $t = 0^+$

$$i = (I + 1)$$
$$30 = 40I + 20I + 20$$
$$I = \frac{10}{60} = \frac{1}{6}$$

Hence current,

$$i = I + 1 = \frac{1}{6} + 1 = \frac{7}{6}\text{ A}$$

2.9 (a)

Input signal,

$$v_i(t) = 5 + 3\cos\omega t$$
$$v_i(t) = V_1 + V_2(t)$$

By applying superposition theorem

Case-I

When V_1 is applied

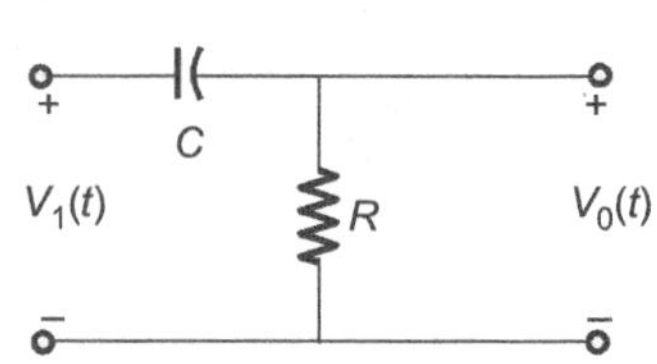

$$V_{01}(t) = V_1\, e^{-t/RC}$$

$\therefore$ Steady state output $V_{01}(t) = 0$

Case-II

When $V_2(t)$ is applied

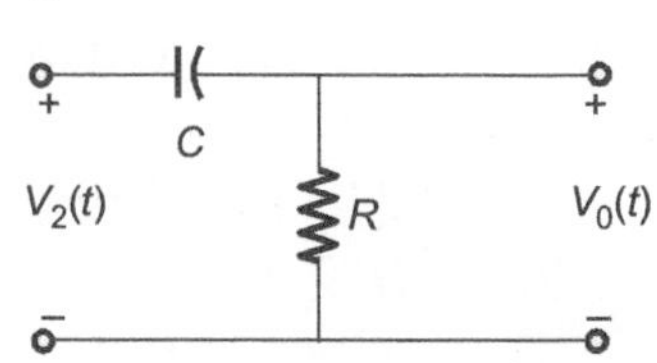

$$\because \quad Z = \sqrt{R^2 + \frac{1}{\omega^2 C^2}}$$

$$\therefore V_{02}(t) = \frac{V_2(t)}{Z\angle\phi} \cdot R = \frac{3\cos(\omega t - \phi)}{\sqrt{1+R^2\omega^2C^2}} \cdot R\omega C$$

as $\omega RC = 2$

$$\therefore V_{02}(t) = \frac{6\cos(\omega t-\phi)}{\sqrt{1+4}} = \frac{6}{\sqrt{5}}\cos(\omega t - \phi)$$

$\therefore$ Steady state output $V_0(t)$

$$= V_{01}(t) + V_{02}(t)$$

$$= 0 + \frac{6}{\sqrt{5}}\cos(\omega t - \phi)$$

$$\therefore \quad P = 0 \text{ and } Q = \frac{6}{\sqrt{5}}$$

2.10 (a)

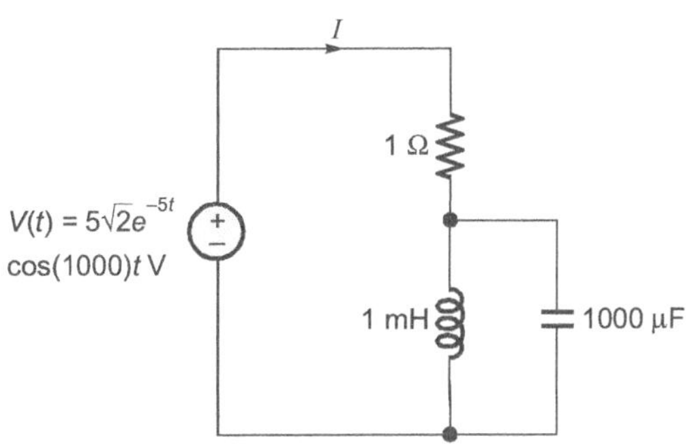

$$Z_c = \frac{1}{j\omega C} = \frac{1}{j \times 1000 \times 10^{-3}} = -j$$

$$Z_L = j\omega L = j \times 1000 \times 10^{-3}$$

$$= j$$

$$\therefore \quad Z_{eq} = 1 + \frac{-j \times j}{-j + j}$$

$$Z_{eq} \to \infty$$

The current in the given circuit will be zero.

2.11 (d)

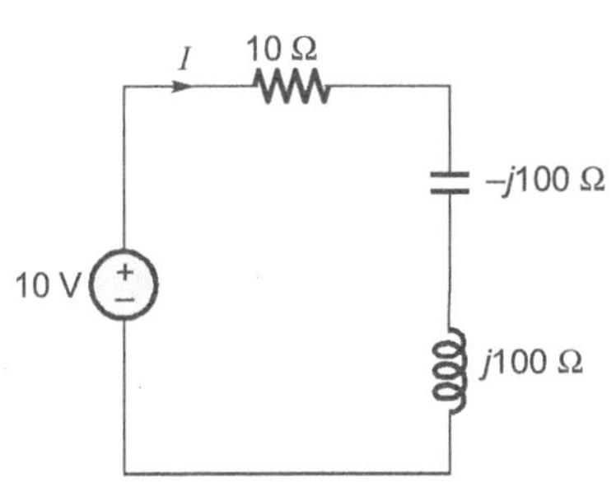

Total impedance of circuit

$$Z_T = 10 + (-j100) + j100 = 10\ \Omega$$

$$\text{Current, } I = \frac{10}{10} = 1\text{ A}$$

Hence, voltage across capacitor

$$V_C = 1 \times -j100 = -j100\text{ V}$$

2.12 (a)

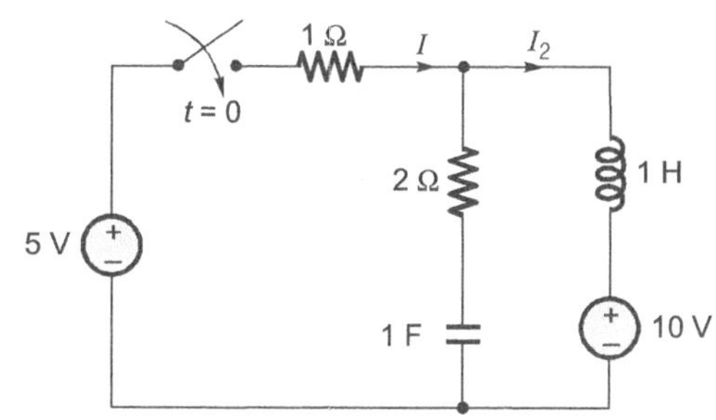

As the steady-state is reached at $t = 0^-$

The capacitor will be open and inductor will be short and current through inductor and voltage across capacitor will not change after switch is closed.

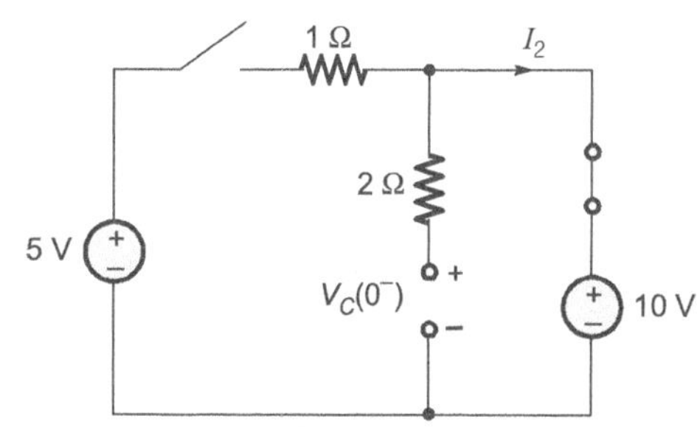

(at $t = 0^-$)

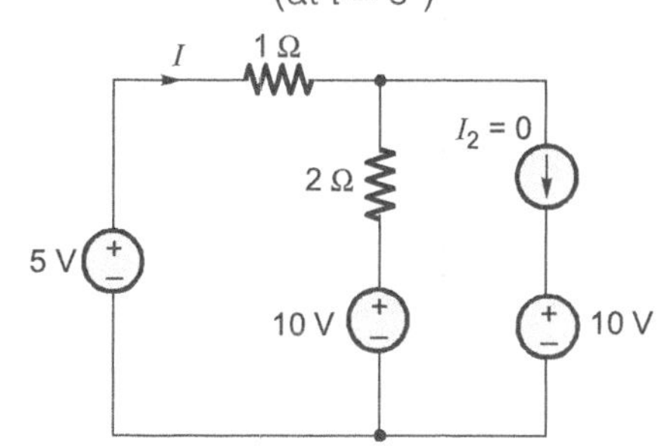

(at $t = 0^+$)

at $t = 0^-$

$$I_2(0^-) = 0\text{ A}$$

and $\quad V_C(0^-) = 10\text{ V}$

at $t = 0^+$

apply KVL in Ist loop

$$5 - I - 2I - 10 = 0$$

$$\Rightarrow \quad 3I = -5$$

$$I = -\frac{5}{3}\text{A}$$

2.13 (b)

Applying KVL loop II

$$V_L + 2\times\frac{5}{3} - 10 + 10 = 0$$

$$V_L = -\frac{10}{3} = L\frac{dI_2(0^+)}{dt}$$

$$\Rightarrow \quad \boxed{\frac{dI_2(0^+)}{dt} = -\frac{10}{3}\text{ A/s}}$$

2.14 (d)

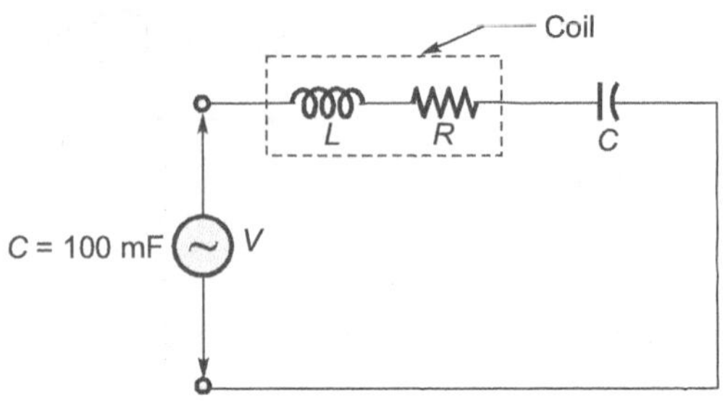

Given:

$L = 10$ mH, $C = 100$ μF

$$V = 10\sqrt{2}\cos(1000t)$$

$$V_{rms} = \frac{10\sqrt{2}}{\sqrt{2}} = 10\text{ V}$$

$$X_L = j\omega L = j \times 1000 \times 10^{-2} = j10\ \Omega$$

$$X_C = \frac{1}{j\omega C} = \frac{1}{j\times1000\times10^{-4}} = -j10\ \Omega$$

Total impedance of the circuit

$$Z_T = R + j10 - j10 = R$$

Current, $I = \dfrac{V}{R}$

Power drawn $= I^2R$

$$\Rightarrow \quad I^2R = 20$$

$$\Rightarrow \quad \frac{V^2}{R} = 20$$

$$\Rightarrow \quad \frac{10^2}{R} = 20 \Rightarrow R = 5\ \Omega$$

2.15 (b)

$$Q\text{-factor} = \frac{\omega_0 L}{R} = \frac{1000\times10\times10^{-3}}{5} = 2$$

2.16 (d)

When switch is at position 1,

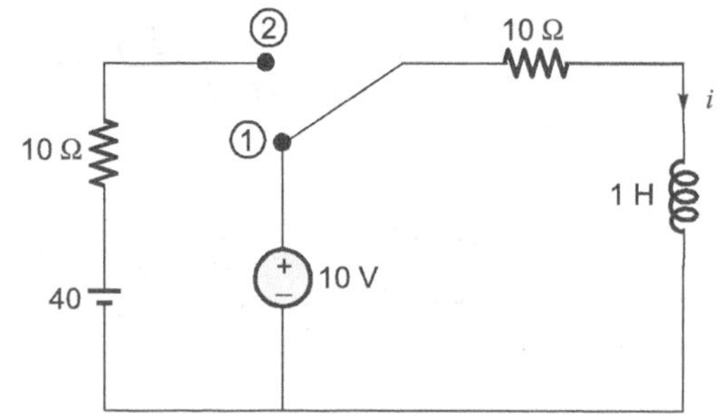

for long time, inductor behave as short circuit

$$\therefore \quad i = \frac{10}{10} = 1\text{ A}$$

when switch at position 2,

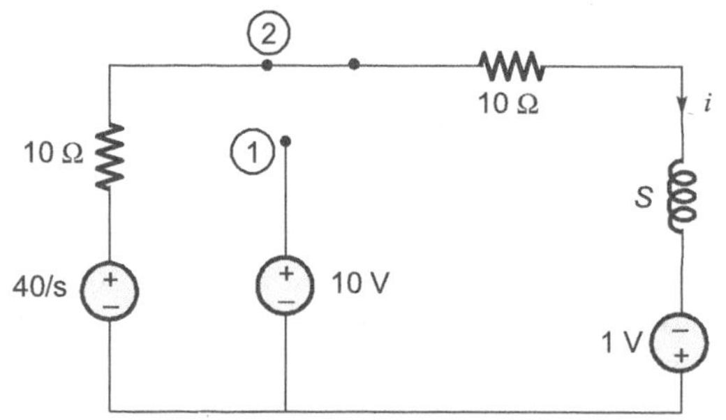

By applying KVL,

$$\frac{40}{s} - 10I(s) - 10I(s) - sI(s) + 1 = 0$$

$$\Rightarrow \quad (20 + s)\,I(s) = \frac{40}{s} + 1 = \frac{40+s}{s}$$

$$I(s) = \frac{s+40}{s(s+20)} = \frac{2}{s} - \frac{1}{s+20}$$

$$i(t) = (2 - e^{-20t})\text{ A}$$

2.17 (a)

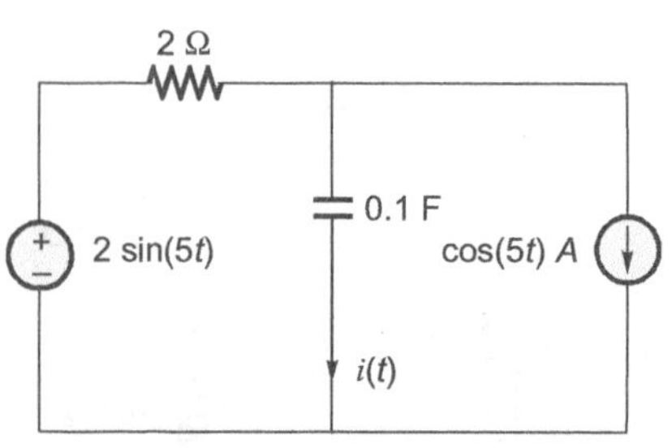

By source transformation,

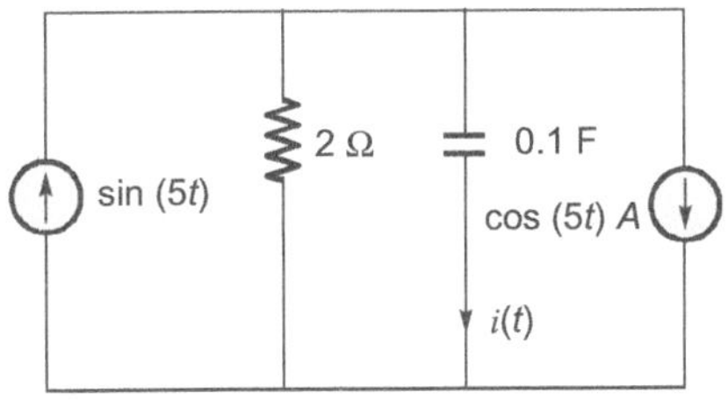

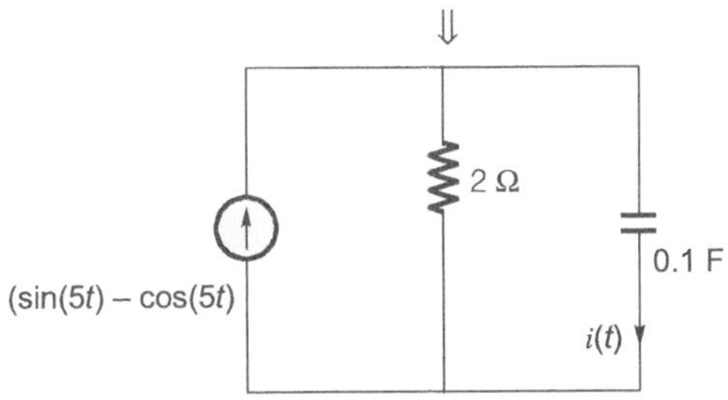

By source transformation,

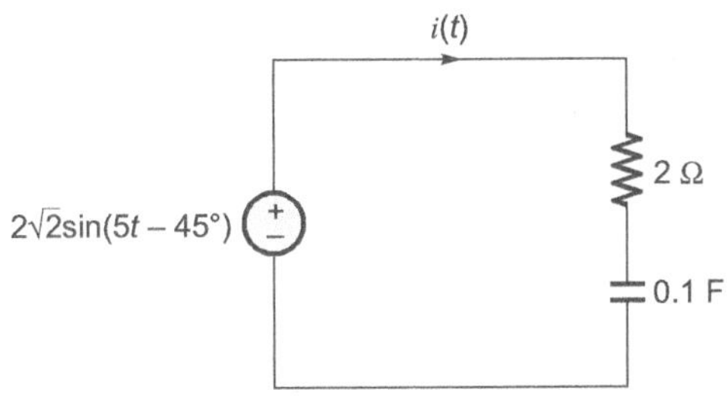

$$Z = \sqrt{R^2 + \frac{1}{(C\omega)^2}} = \sqrt{4 + \frac{1}{(0.1\times 5)^2}}$$

$$= 2\sqrt{2}$$

$$i = \frac{2\sqrt{2}\sin(5t - 45°)}{2\sqrt{2}\angle(-45)} = \sin 5t \text{ A}$$

2.18 (a)

$$i(t) = \sin 5t$$

$$i_{rms} = \frac{1}{\sqrt{2}} \text{ A}$$

Power delivered = Power dissipated

$$= i_{rms}^2 R = \frac{1}{2}\times 2 = 1 \text{ W}$$

2.19 (b)

$$i = 5\cos(100\pi t + 100°)\text{A}$$
$$= 5\angle 100°\text{A}$$
$$Z = (4 - j3)\Omega$$
$$= 5\angle{-36.87°}\Omega$$
$$v = iZ = 25\angle 63.13° \text{ V}$$

The average power is

$$P = \frac{1}{2}V_m I_m \cos\phi$$

$$= \frac{1}{2}\times 25\times 5\times\cos(36.87°)$$

$$= \frac{1}{2}\times 25\times 5\times\frac{4}{5} = 50 \text{ W}$$

2.20 (d)

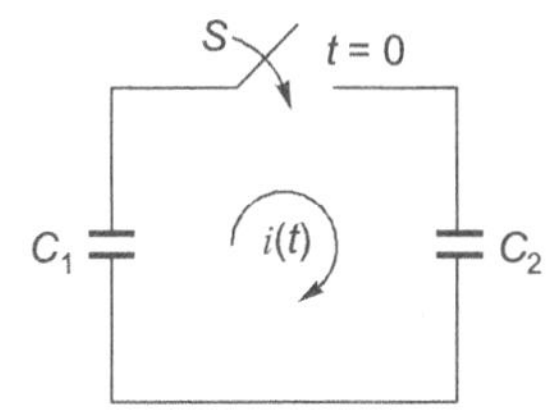

Circuit is s-domain

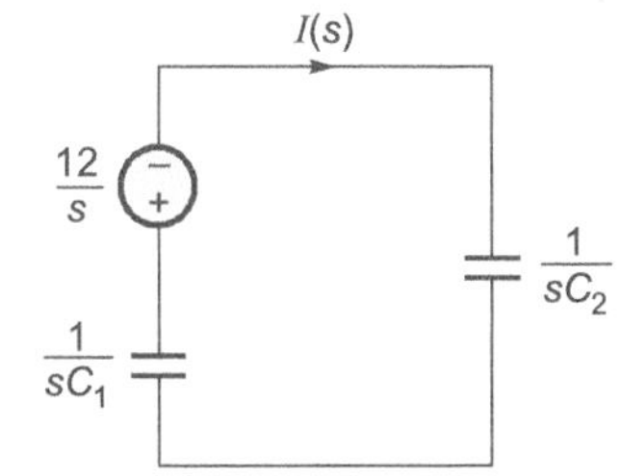

By applying kVL,

$$\frac{12}{s} + \frac{I(s)}{s}\left(\frac{1}{C_1} + \frac{1}{C_2}\right) = 0$$

$$I(s) = -\frac{12C_1C_2}{C_1 + C_2} = k \text{ (constant)}$$

$\Rightarrow i(t) = k\delta(t)$

$\therefore$ Current $i(t)$ is an impulse function.

2.21 (a)

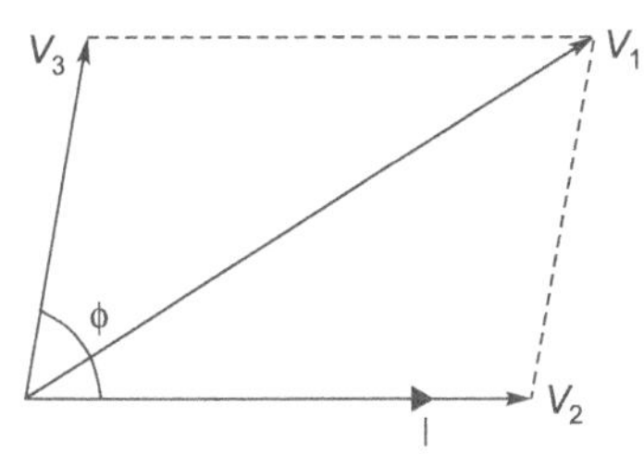

$$V_1^2 = V_2^2 + V_3^2 + 2V_2V_3\cos\phi$$

$$(220)^2 = (122)^2 + (136)^2 + 2\times 122 \times 136\times\cos\phi$$

$\Rightarrow \cos\phi = 0.45$

2.22 (b)

Given, $R_L = 5\ \Omega$

$$\therefore \cos\phi = \frac{R_L}{|Z|}$$

$$\Rightarrow |Z| = \frac{5}{0.45} = 11.11$$

Power consumed by load

$$P_L = \left(\frac{V_3}{|Z|}\right)^2 R_L$$

$$= \left(\frac{136}{11.11}\right)^2 \times 5 = 749.1$$

$$\simeq 750 \text{ W}$$

2.23 (c)

$$q_1 = \frac{\omega L_1}{R_1}$$

$$q_2 = \frac{\omega L_2}{R_2}$$

$$\omega L_1 = q_1 R_1$$

$$\omega L_2 = q_2 R_2$$

Coils are connected in series.

So, $q.R = \omega L_1 + \omega L_2 = q_1 R_1 + q_2 R_2$

$$q = \frac{q_1 R_1 + q_2 R_2}{R}$$

$$R = R_1 + R_2$$

$$q = \frac{q_1 R_1 + q_2 R_2}{R_1 + R_2}$$

2.24 (c)

$$Q = CV$$

$$Q_1 = C_1 V_1 = 10 \times 10^{-6} \times 10$$

$$= 100 \mu C$$

$$Q_2 = C_2 V_2 = 5 \times 10^{-6} \times 5$$

$$= 25 \mu C$$

$$Q_3 = C_3 V_3 = 2 \times 10^{-6} \times 2$$

$$= 4 \mu C$$

Capacitors C_2 and C_3 are in series.

In series charge is same.

So, the maximum charge on C_2 and C_3 will be minimum of (Q_2, Q_3) = min(25 μC, 4 μC) = 4μC = Q_{23}.

In series the equivalent capacitance of C_2 and C_3 is

$$C_{23} = \frac{C_2 C_3}{C_2 + C_3} = \frac{5 \times 2}{5+2} = \frac{10}{7} \mu F$$

So, the equivalent voltage

$$V_{23} = \frac{Q_{23}}{C_{23}} = \frac{4 \times 10^{-6}}{\frac{10}{7} \times 10^{-6}} = \frac{28}{10} = 2.8 \, V$$

In parallel, the voltage is same.

$$V_1 = V_{23} = 2.8 \text{ V}$$

Charge in capacitor C_1

$$Q_1 = C_1 V_1$$

$$= 10 \times 10^{-6} \times 2.8 = 28 \mu C$$

In parallel, the total charge

$$Q = Q_1 + Q_{23}$$

$$Q = 4 + 28$$

$$Q = 32 \mu C$$

2.25 (30)

at, $t = 0^-$

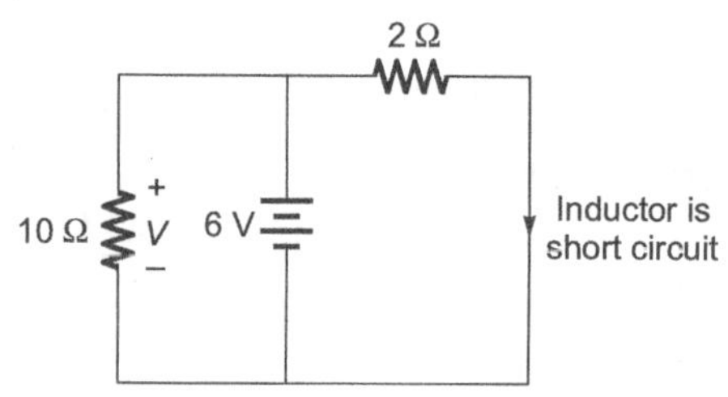

$$i = \frac{6}{2} = 3 \text{ A}$$

at $t = 0^+$, current across the 10 Ω resistor will be,

∴ $V = -10 \times 3 = -30$ V

2.26 (32)

Average real power delivered to inductor or to capacitor is zero.

∴ Average real power delivered to load Z_L, is

$$P_{over} = (I_{rms})^2 R = \left(\frac{4}{\sqrt{2}}\right)^2 \times 4 = 32 \text{ Watt}$$

2.27 1.51 (1.48 to 1.53)

At $t = 0^+$, $V_c = 0$ V

At $t = \infty$, $V_c = \frac{2}{2+2} \times 5 = 2.5$ V

$$V_c = V_c(\infty) + [V_c(0^+) - V_c(\infty)] \exp(-t/\tau)$$

($\because$ RC time constant $(\tau) = (2k \parallel 2k) \times 4 \times 10^{-6}$

$$= 4 \times 10^{-3}$$

$$\therefore \quad v_c(t) = 2.5 - 2.5 e^{\frac{-t}{4 \times 10^3}}$$

$$i_c = C \frac{dV_c}{dt}$$

$= 4 \times 10^{-6} \times (-2.5) \times (-250) \, e^{-250 \, t}$

$= 2.5 \times 10^{-3} \, e^{-250 \, t}$

at $t = 2$ msec.

$$i_c = 2.5 \times 10^{-3} \times e^{-250 \times 2/1000}$$

$$i_c = 1.51 \text{ mA}$$

2.28 **187.33 (187.0 to 187.50)**

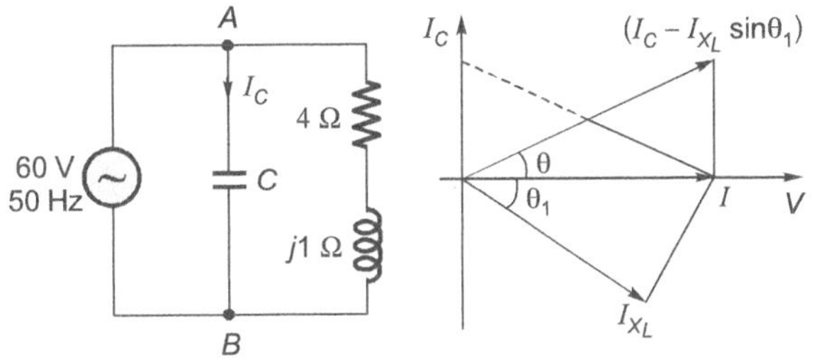

For unity power factor $\theta = 0°$

$\therefore \quad \cos\theta = 1$

$$\tan\theta = \frac{I_C - I_{X_L}\sin\theta_1}{I} = 0$$

$$I_C = I_{X_L}\sin\theta_1$$

$$\frac{V_c}{X_C} = \frac{V_c}{X_L}\sin\theta_1$$

$$\omega C = \frac{1}{\sqrt{(\omega L)^2 + R^2}} \times \frac{\omega L}{\sqrt{(\omega L)^2 + R^2}}$$

$$C = \frac{1}{\sqrt{1^2 + 4^2} \times \sqrt{1^2 + 4^2} \times 2\pi \times 50}$$

$$= 187.33\ \mu\text{F}$$

2.29 **(c)**

$$v_c(t) = v_c(\infty) + \left[v_c(0^+) - v_c(\infty)\right]e^{-t/RC}$$

$$V_H = V_S + (V_L - V_S)\, e^{-T_{\text{ref}}/R_{\text{ref}}C_{\text{ref}}} \quad \text{...(i)}$$

$$V_H = V_S + (V_L - V_S)\, e^{-T_x/R_x C_{\text{ref}}} \quad \text{...(ii)}$$

Comparing equations (i) and (ii),

$$-\frac{T_{\text{ref}}}{R_{\text{ref}}C_{\text{ref}}} = -\frac{T_x}{R_x C_{\text{ref}}}$$

for $(T_x = kT_{\text{ref}})$

$$R_x = k\,R_{\text{ref}}$$

2.30 **3.678 (3.6 to 3.7)**

$$v_c(t) = V_o e^{-t/\tau}$$

$$= 10e^{-100t} \quad \left[\because \tau = RC = \frac{1}{100}\right]$$

$\therefore\ v_c(\text{at 10 msec}) = 10e^{(-100 \times 10 \times 10^{-3})}$

$$= 3.678\ \text{V}$$

2.31 **100 (98 to 102)**

Under series resonance condition the voltage across L, C is Q times of the supply voltage.

$$V_c = QV$$

$$Q = \frac{1}{R}\sqrt{\frac{L}{C}}$$

$$= \frac{1}{10}\sqrt{\frac{0.1}{0.1 \times 10^{-6}}} = 100$$

$\therefore \quad V_C = QV$

$$= 100 \times 1 = 100\ \text{V}$$

2.32 **(8)**

$$\because \quad i_c = C\frac{dV_c}{dt}$$

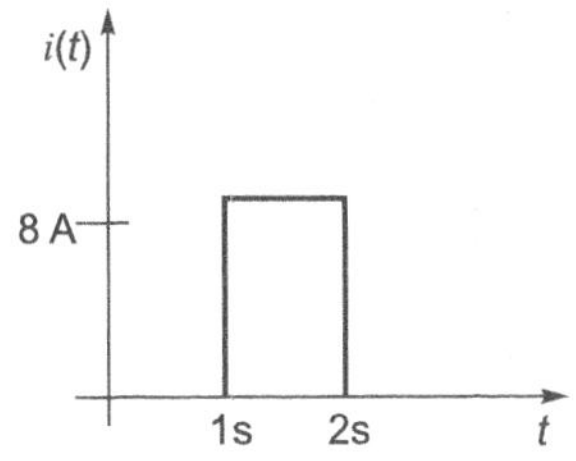

$$V = \frac{1}{C}\int_{-\infty}^{t} i_c(t)dt$$

$$C = 1\ \text{F}$$

$$V_C = \int_{-\infty}^{t} i_c(t)dt$$

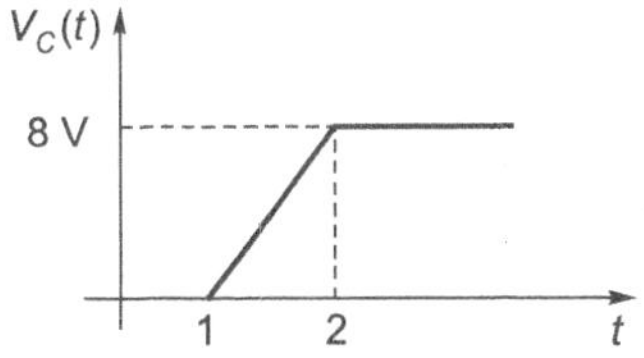

for $t > 0$; $\quad V_C = 8$ V

2.33 **0.28 (0.280 to 0.283)**

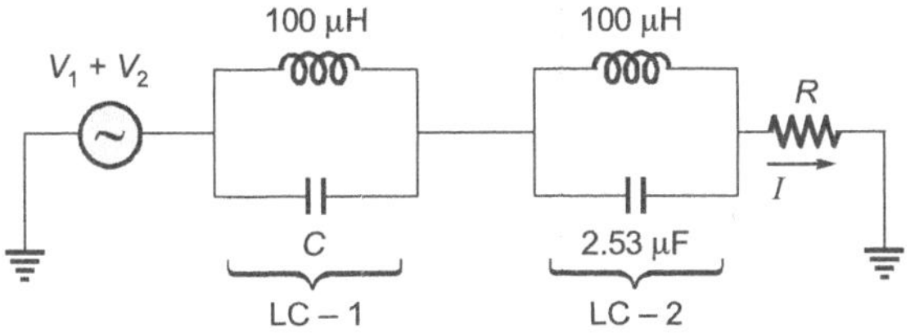

$$V_1 = 1\sin(2\pi\ 10000\ t)\ \text{V}$$

$$V_2 = 1\sin(2\pi\ 30000\ t)\ \text{V}$$

$$\therefore \quad f_1 = 10000\ \text{Hz}$$

$$f_2 = 30000\ \text{Hz} = 3\ f_1$$

For LC – 2

$$\omega_0 = \frac{1}{\sqrt{LC}} \Rightarrow f_0 = \frac{1}{2\pi\sqrt{LC}}$$

$$f_0 = \frac{1}{2\pi\sqrt{100 \times 10^{-6} \times 2.53 \times 10^{-6}}} \simeq 10000\ \text{Hz}$$

So, for $\quad f_1 = f_0\big|_{LC-2} \Rightarrow$ Resonance

It act as open so $I = 0$

for $LC-1$ to be open circuit i.e., for resonance

$$(3f_1) = \frac{1}{2\pi\sqrt{LC}}$$

$$C = \frac{1}{(2\pi)^2 \times (3f_1)^2 \times L}$$

$$= \frac{1}{(2\pi)^2 \times (30{,}000)^2 \times 100 \times 10^{-6}}$$

$$= 0.28\ \mu F$$

2.34 100 (99.5 to 100.5)

According to statement given circuit is at resonance

$$Y = \frac{1}{R + jX_L} + \frac{1}{-jX_C}$$

$$Y = \frac{R - jX_L}{R^2 + X_L^2} + \frac{j}{X_C}$$

$\because$ At resonance imaginary part is zero

$$I_s = V_s \cdot \text{Real}\{Y\}$$

$$= 101 \cdot \left[\frac{10}{10^2 + 100^2}\right]$$

$$= \frac{101 \times 10}{10100} = 0.1\ A = 100\ mA$$

2.35 1.632 (1.55 to 1.65)

At $t = 0^-$ (steady state)

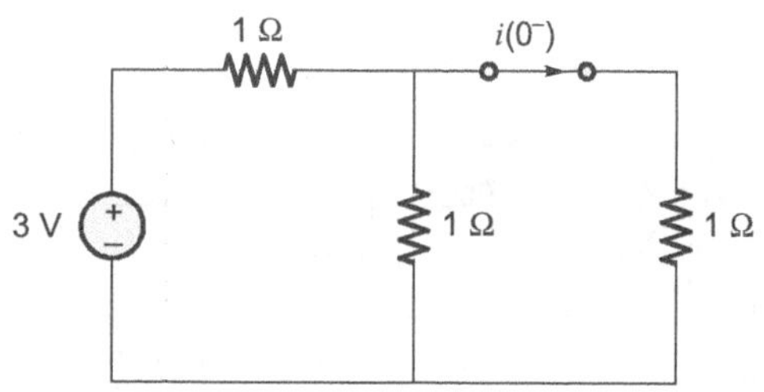

$$i(0)^- = \frac{3}{1.5} \times \frac{1}{2} = 1\ A$$

$$\therefore \quad i(0^+) = 1\ A$$

at $t = \infty$

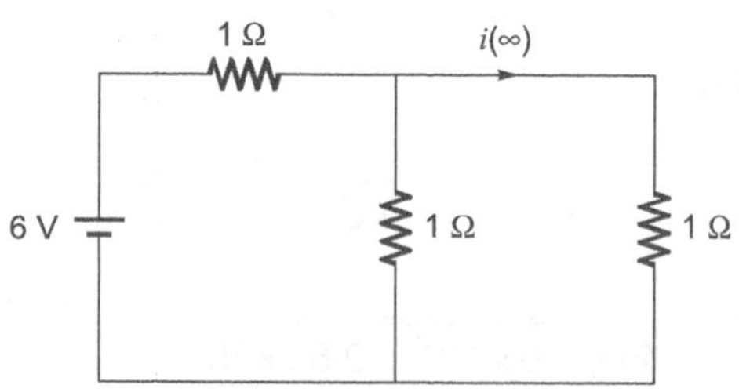

$$i(\infty) = \frac{6}{1.5} \times \frac{1}{2} = 2\ A$$

Time constant

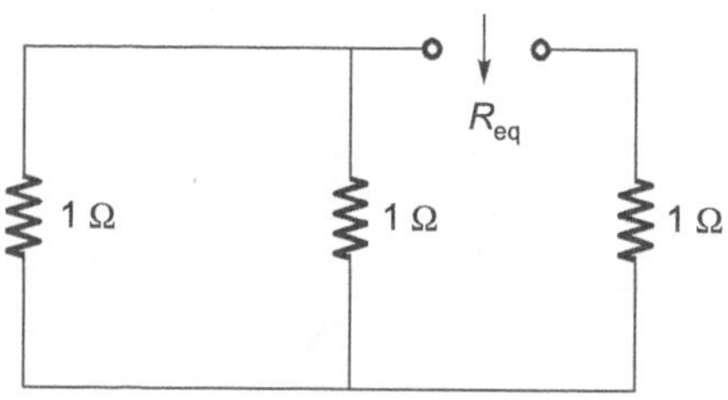

$$R_{eq} = \frac{1 \times 1}{1 + 1} + 1 = 1.5\ \Omega$$

$$L = 1.5\ H$$

$$\tau = \frac{L}{R_{eq}} = 1\ \text{sec}$$

$$i(t) = \left[i(0^+) - i(\infty)\right]e^{-t/\tau} + i(\infty)$$

$$= [1 - 2]e^{-t} + 2$$

$$i(1) = 2 - 1e^{-1/1}$$

$$i(1) = 1.632\ A$$

2.36 (50)

$$V = \sqrt{V_R^2 + (V_L - V_C)^2}$$

$$50 = \sqrt{50^2 + (V_L - 50)^2}$$

$$V_L = 50$$

2.37 1 (0.99 to 1)

$$i(t) = \frac{V}{R}\left[1 - e^{-\frac{Rt}{L}}\right]$$

From given data at $t = \infty$

$$\frac{V}{R} = 0.5$$

$$\because \quad V = 1$$

$$R = 2\ \Omega$$

$$\text{at} \quad t = 0.25$$

$$0.197 = 0.5\left[1 - e^{-\frac{2(0.25)}{L}}\right]$$

$$L = 1\ H$$

2.38 1.52 (1.47 to 1.55)

At $t = 0^-$

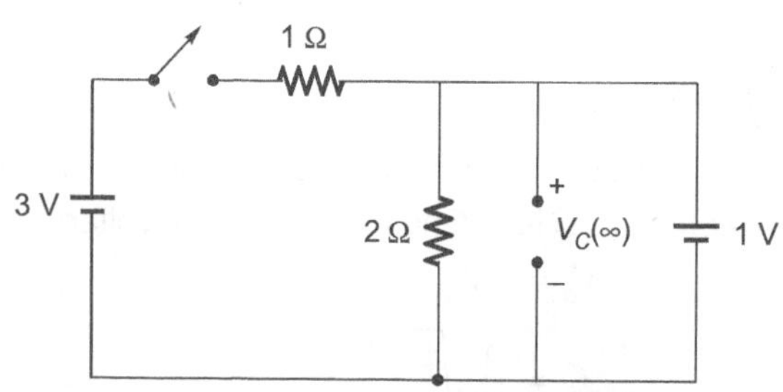

$$V_C(0^-) = 0$$

At $t = \infty$

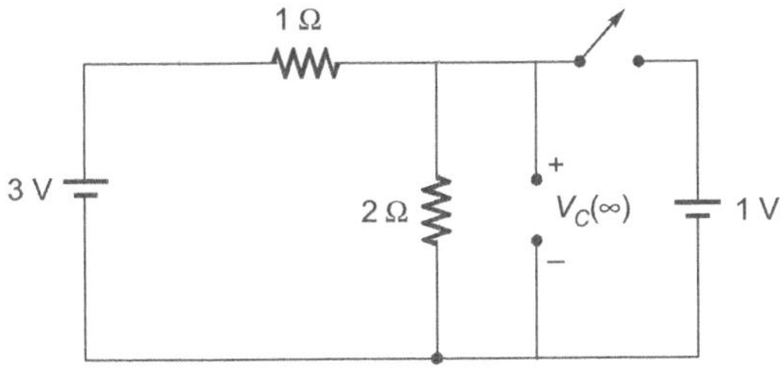

$$V_C(\infty) = 3\frac{2}{2+1} = 2\text{ V}$$

$$V_C(t) = [V_C(0^+) - V_C(\infty)]e^{-\frac{t}{RC}} + V_C(\infty)$$

$$= [1-2]e^{-\frac{t}{\frac{2}{3}\times10\times10^{-6}}} + 2$$

at $t = 5\ \mu\text{sec}$

$$V_C = 1.52\text{ Volts}$$

$C = \infty$	$C \downarrow$	
$X_C = 0$	$X_C \uparrow$	$C = 0$
$Z = R$	$Z \uparrow$	$X_C = \infty$
$I = V/R$	$I \downarrow$	$Z = \infty$
$\theta = 0$	$\theta = \tan^{-1}\left(\frac{-X_C}{R}\right)\uparrow$	$I = 0$

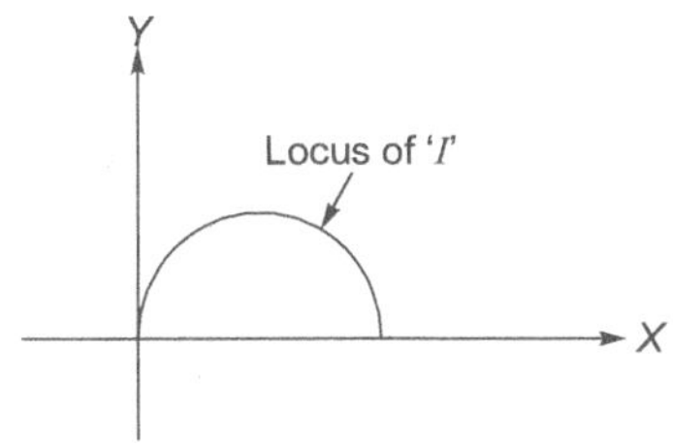

2.39 6.36 (6 to 7)

$$\omega_0 = \frac{1}{\sqrt{LC}} = \frac{1}{\sqrt{10\times10^{-3}\times4\times10^{-6}}}$$

$$= 5000\text{ rad/sec}$$

$$Q = \frac{\omega_0 L}{R}$$

$$30 = \frac{5000(10\times10^{-3})}{R}$$

$$R = \frac{5}{3}\ \Omega$$

at $\omega_0 \rightarrow \quad I = \frac{V}{R} = \frac{15}{5/3} = 9\text{ V}$

at $\omega_2 \rightarrow \quad \frac{I}{\sqrt{2}} = \frac{9}{\sqrt{2}} = 6.36\text{ A}$

2.40 1.866 (1.75 to 1.96)

$$\bar{I} = \bar{I}_1 + \bar{I}_2 = 5\angle0° + 5\angle30°$$

$$I = 9.659\angle14.94$$

$$P = VI\cos\theta$$

$$= 200 \times 9.659 \times \cos(14.94)$$

$$= 1.866\text{ kW}$$

2.41 (a)

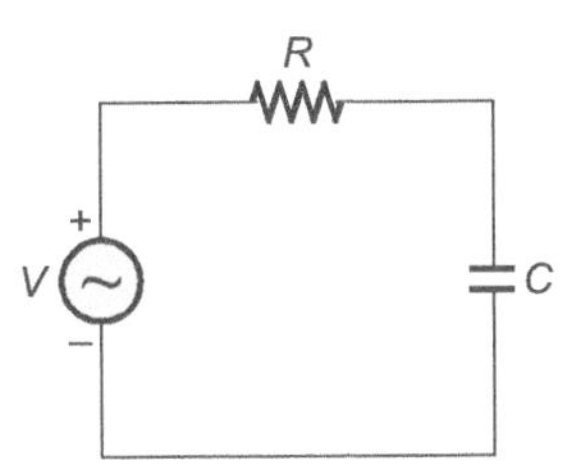

2.42 1010 (1009 to 1011)

At resonance,

$$I_m(Y_{eq}) = 0$$

At resonance,

$$\text{Re}(Y_{eq}) = G = \frac{R}{R^2 + X_L^2} = \frac{10}{10^2 + 100^2}$$

$$Z = \frac{1}{G} = \frac{10100}{10} = 1010\ \Omega$$

2.43 67.36 (66 to 68)

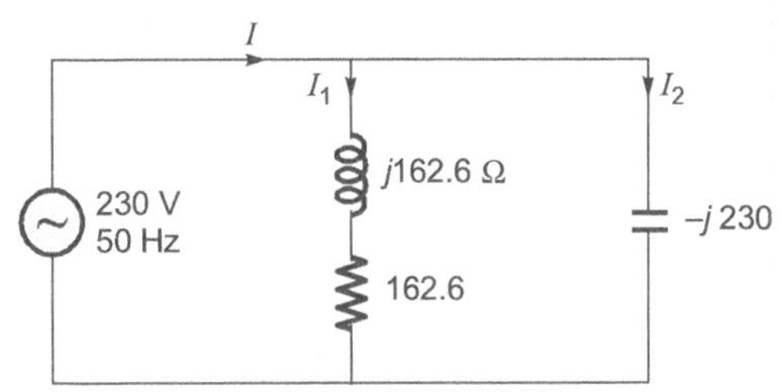

$$I_1 = \frac{230}{162.6 + j162.6} = 1\angle-45°$$

$$I_2 = \frac{230}{230\angle-90°} = 1\angle90°$$

$$I = I_1 + I_2 = \frac{1}{\sqrt{2}} - j\frac{1}{\sqrt{2}} + j1$$

$$I = 0.707 + j0.293$$

$$= 0.765\angle22.51°$$

$$Q = VI\sin\theta$$

$$= 230(0.765)\sin(22.51)$$

$$= 67.36\text{ VAR}$$

2.44 10.84 (10.6 to 11.0)

When voltage at the output due to only 230 V, 50 Hz supply is to be calculated, the DC supply should be short circuited as shown below.

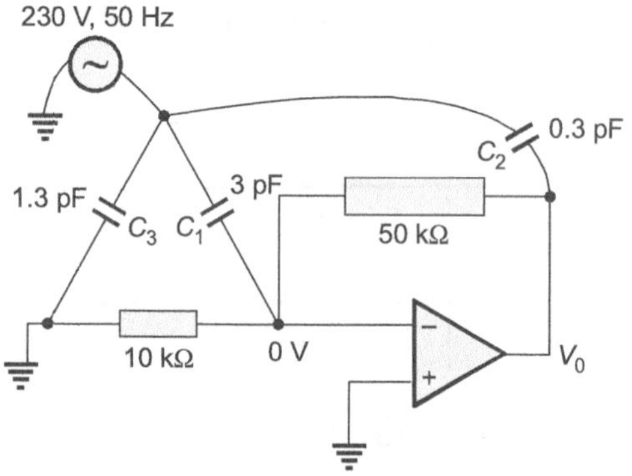

C_3 and 10 kΩ does not play any role in the value of V_0. So, the reduced equivalent circuit can be given as shown below.

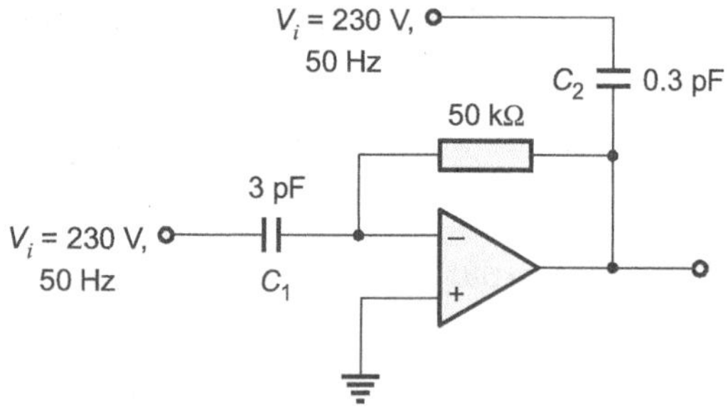

$$V_0 = -\frac{50\text{ k}\Omega}{1/j\omega C_1} V_i = -j\omega C_1 (50\text{ k}) V_i$$

$$V_{0(\text{rms})} = (2\pi \times 50) \times 3 \times 10^{-12} \times 50 \times 10^3 \times V_{i(\text{rms})}$$

$$= 2\pi \times 75 \times 10^{-7} \times 230\text{ V}$$

$$= \frac{345\pi}{100}\text{ mV} = 10.84\text{ mV}$$

2.45 (b)

$$Q = \frac{1}{R}\sqrt{\frac{L}{C}}$$

and as R, L and C are doubled.

$$\therefore \quad Q' = \frac{1}{2R}\sqrt{\frac{2L}{2C}} = \frac{1}{2}\left[\frac{1}{R}\sqrt{\frac{L}{C}}\right]$$

$$Q' = \frac{1}{2}Q$$

2.46 (4)

$$V_{C1}(0^-) = \frac{6}{1} = 6\text{ V and } V_{C2}(0^-) = 0\text{ V}$$

For $t > 0$

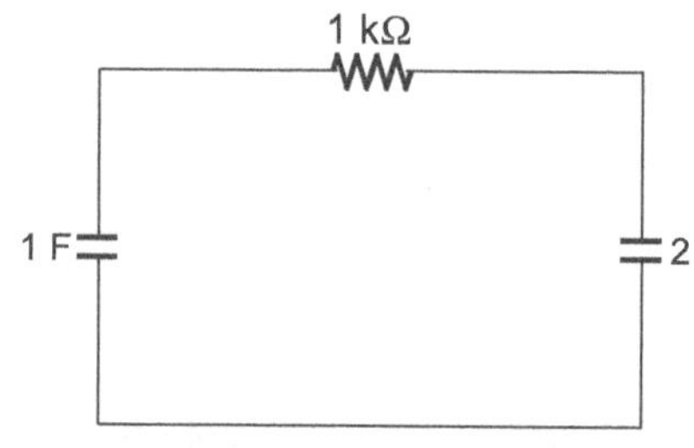

Transforming the circuit to the Laplace domain,

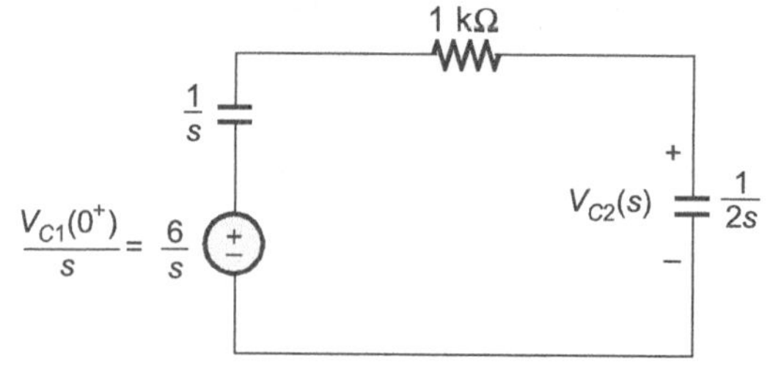

$$V_{C2}(s) = \frac{\frac{6}{s}}{\frac{1}{s} + 1k + \frac{1}{2s}} \times \frac{1}{2s} = \frac{6}{s[2 + 2sk + 1]}$$

Voltage across the capacitor C2 in steady state,

$$V_{C2}(\infty) = \lim_{s\to 0} sV_{C2}(s) = \lim_{s\to 0}\frac{6}{[2 + 2sk + 1]} = \frac{6}{3}$$

$$V_{C2}(\infty) = 2\text{ V}$$

$\therefore$ The charge on C_2 in steady state, $Q = C_2 V_{C2}(\infty)$

$$Q = (2)(2) = 4C$$

Alternate method :

$$Q_1(0^-) = 6C$$

$$V_{C_1}(0^-) = \frac{Q_1(0^-)}{C} = \frac{6}{1} = 6V$$

The initial voltage across the capacitor C_1 is 6V and capacitor C_2 is 0V i.e. $V_{C_2}(0^-) = 0V$

In steady state the voltage across two capacitors are equal i.e.

$$V_{C_1}(\infty) = V_{C_2}(\infty) = \frac{V_{C_1}(0^-)C_1 + V_{C_2}(0^-)C_2}{C_1 + C_2}$$

$$V_{C_1}(\infty) = V_{C_2}(\infty)$$

$$\Rightarrow \quad \frac{(6\times 1) + 0}{1 + 2} = 2V$$

The steady state voltage across capacitor C_2 is 2 V

and change $Q_2 = C_2V_2 = 2 \times 2 = 4C$

2.47 (20000)

At resonance the source voltage and source current will be in phase.

$$\therefore\ \omega = \frac{1}{\sqrt{LC}} = \frac{1}{\sqrt{5\times10^{-3}\times500\times10^{-9}}} \text{ rad/sec}$$

$$= \frac{1}{\sqrt{25\times10^{-10}}} = \frac{100\times10^{3}}{5} = 20\text{ k}$$

$$\omega = 20000 \text{ rad/sec}$$

2.48 (1.896) 1.87 to 1.91

$$V_C(0^-) = 0 \text{ V}$$

As $V_C(0^+) = V_C(0^-)$

$\therefore$ $V_C(0^+) = 0$ V

For $t > 0$

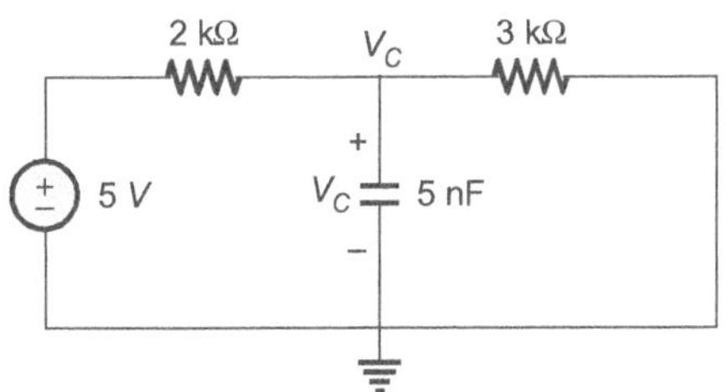

$$V_C(t) = V_C(\infty) + [V_C(0^+) - V_C(\infty)]\, e^{-t/\tau} \quad \text{... (i)}$$

$$\tau = R_{eq}.C$$

$$R_{eq} = 2k \parallel 3k = \frac{(2k)(3k)}{5k} = \frac{6}{5}\text{k}\Omega$$

$$\therefore\ \tau = \left(\frac{6}{5}\text{k}\Omega\right)(5\text{ nf}) = 6\times10^{-6}\text{sec}$$

At $t = \infty$,

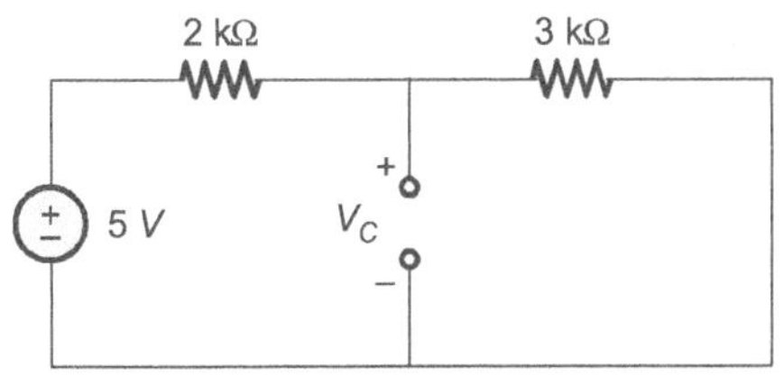

$$V_C(\infty) = \frac{5}{2k+3k}\times 3k = 3\text{ V}$$

Using equation (i)

$$V_C(t) = 3 + [0-3]e^{-t/6\times10^{-6}}$$

$\therefore$ Voltage at node P at $t = 6$ μs,

$$V_C(6\ \mu\text{s}) = 3\left[1-e^{-6\times10^{-6}/6\times10^{-6}}\right] = 3\left[1-e^{-1}\right]$$

$$V_C(6\ \mu\text{s}) = 1.896 \text{ V}$$

2.49 (1)

$$R = 100\ \Omega$$

$$X_L = \omega L = 100\ \Omega$$

$$X_C = \frac{1}{\omega C} = 100\ \Omega$$

$$Z = 100 + j(100 - 100)$$

$$|Z| = 100\ \Omega$$

$$I = \frac{V_{rms}}{Z} = \frac{(\sqrt{2}\times100)/\sqrt{2}}{100} \Rightarrow 1\text{ A}$$

2.50 (0.25)

$$L = 250 \times 10^{-3}\text{ H};\quad R_0 = 300\ \Omega$$

$$B_L = \frac{1}{2\pi fL} = \frac{1}{2\pi(50)250\times10^{-3}}$$

$$= 0.012732$$

$$G_0 = \frac{1}{300} = 3.33\times10^{-3}$$

$$\cos\theta = \frac{G_0}{\sqrt{G_0^2 + B_L^2}} = 0.25 \text{ lag}$$

2.51 (30)

As capacitor's varied maximum value of voltage occurs when it is in resonance condition.

At resonance condition,

$$I = \frac{V}{R} = \frac{10}{10} = 1\text{ A}$$

$$V_L = V_C = 30\text{ V}$$

$$I\omega L = 30\text{ V}$$

$$\omega L = 30$$

$$L = \frac{30}{1000} = 30\text{ mH}$$

2.52 (6.32)

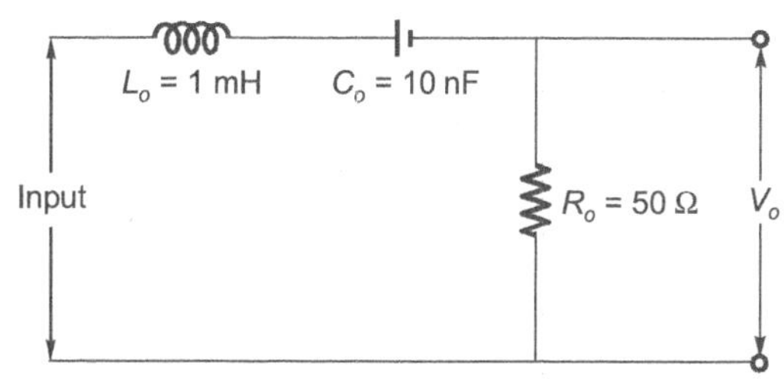

The quality factor,

$$Q = \frac{1}{R}\sqrt{\frac{L}{C}} = \frac{1}{50}\sqrt{\frac{1}{10\times10^{-6}}} = 6.32$$

2.53 (0.8)

$$\vec{Z}_{in} = R_2 + j\omega L_1 + \frac{R_1}{1+j\omega R_1 C_1}$$

$$\vec{Z}_{in} = 4 + j100\pi \times \frac{20}{\pi} \times 10^{-3} + \frac{20}{1+j100\pi \times 20 \times \frac{10^{-3}}{\pi}}$$

$$= 4 + j2 + \frac{20}{1+j2} = 4 + j2 + \frac{20(1-j2)}{5} = 4 + j2 + 4 - j8$$

$$\vec{Z}_{in} = (8 - j6)\ \Omega$$

$$\cos\phi = \frac{8}{\sqrt{8^2+6^2}} = 0.80 \text{ leading}$$

2.54 (–10.0)

$$V_c(0) = V_o = 10 \text{ V}$$

Capacitor voltage,

$$V_c(t) = V_o \cos \omega_o t$$

$$i_c(t) = i_L(t) = -c\frac{dV_c}{dt}$$

$$= -cV_o(-\sin \omega_{ot}) \times \omega_o$$

$$i_L(t) = V_o\sqrt{\frac{C}{L}}\sin\omega_o t$$

at $\omega_o t = \pi, i_L(\pi) = 0$

$$V_c(\pi) = V_o \cos(\pi) = -V_o = -10 \text{ V}$$

So, this time capacitor plates are charged opposite to its initial.

So, after $\omega_o t = \pi$,

$$t = \frac{\pi}{\omega_o} = \pi\sqrt{L_C}$$

$t = 10^{-4}\pi$ sec capacitor

Voltage is always –10 V.

2.55 (1.45)

Given : $R_1 = R_2 = 2.2\ \Omega$

$L = 7$ mH, $\omega = 100\pi$ rad/sec

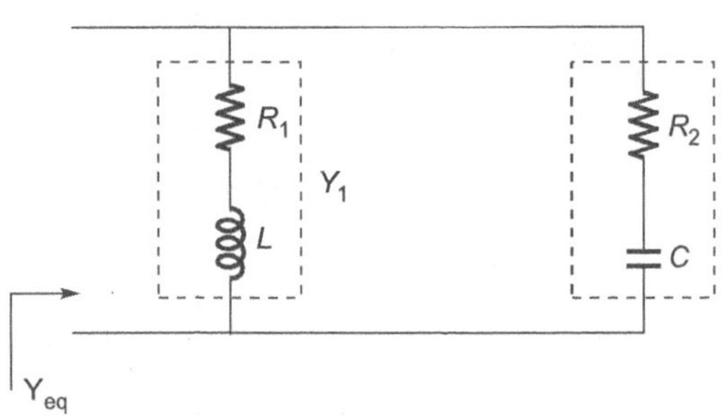

$$Y_{eq} = Y_1 + Y_2$$

$$= \frac{1}{R_1 + j\omega L} + \frac{1}{R_2 - \frac{j}{\omega C}}$$

$$= \frac{(R_1 - j\omega L)}{\sqrt{R_1^2 + (\omega L)^2}} + \frac{R_2 + \frac{j}{\omega C}}{\sqrt{R_2^2 + \left(\frac{1}{\omega C}\right)^2}}$$

For resonant frequency,

$$I_m\{Y_{eq}\} = 0$$

$$\frac{-j\omega L}{\sqrt{R_1^2+(\omega L)^2}} + \frac{\frac{j}{\omega C}}{\sqrt{R_2^2+\left(\frac{1}{\omega C}\right)^2}} = 0$$

$$\omega L\sqrt{R_2^2+\left(\frac{1}{\omega C}\right)^2} = \frac{1}{\omega C}\sqrt{R_1^2+(\omega L)^2}$$

$$\omega^2 L^2\left(R_2^2 + \frac{1}{\omega^2 C^2}\right) = \frac{1}{\omega^2 C^2}(R_1^2 + \omega^2 L^2)$$

$$\omega^2 L^2 R_2^2 + \frac{L^2}{C^2} = \frac{R_1^2}{\omega^2 C^2} + \frac{L^2}{C^2}$$

$$\omega^2 L^2 R_2^2 = \frac{R_1^2}{\omega^2 C^2}$$

$$C^2 = \frac{R_1^2}{R_2^2}\cdot\frac{1}{\omega^4 L^2}$$

$$C = \frac{R_1}{R_2}\cdot\frac{1}{\omega^2 L}$$

$$= \frac{2.2}{2.2} \times \frac{1}{(100\pi)^2 \times 7 \times 10^{-3}}$$

$$C = 1.45 \times 10^{-3} \text{ F}$$

2.56 (0.44)

The current equation for inductor

$$i_L(t) = i_L(\infty) + [i_L(0^+) - i_L(\infty)]e^{-t/\tau}$$

$$i_L(0^+) = \frac{I_o}{5}$$

$$i_L(\infty) = I_o$$

$$i_L(t) = I_o + \left[\frac{I_o}{5} - I_o\right]e^{-t/\tau} = I_o - \frac{4I_o}{5}e^{-t/\tau}$$

The inductor current reaches to 99% of its final value.

i.e. $i_L(t) = 0.99\, I_o$

$$0.99\, I_o = I_o\left[1 - \frac{4}{5}e^{-t/\tau}\right]$$

$$\frac{4}{5}e^{-t/\tau} = 0.01$$

$$t = -\tau \ln\left(\frac{0.05}{4}\right) = 4.382\tau$$

$$\tau = \frac{L}{R} = \frac{1\times10^{-3}}{10\times10^{3}} = 10^{-7}\text{ sec}$$

$$t = 4.382 \times 10^{-7}\text{ sec} = 0.4382\ \mu\text{s}$$

2.57 (17.37)

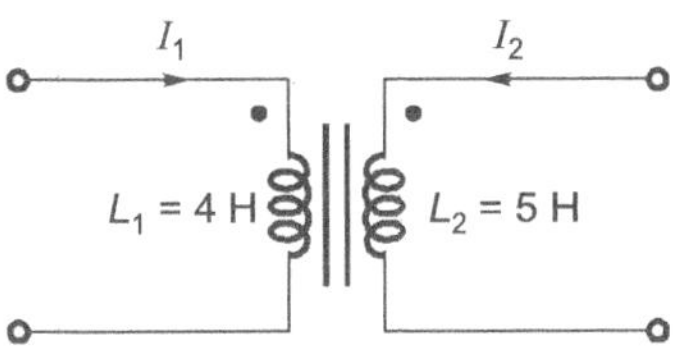

$$i_1(t) = 1\sin(200\ \pi t)\text{ mA}$$
$$i_2(t) = 2\sin(200\ \pi t)\text{ mA}$$
$$M = K_1\sqrt{L_1L_2}$$
$$M = 0.6\sqrt{4\times5} = 2.683\text{ H}$$

$$W_{max} = \frac{1}{2}L_1 I_{1max}^2 + \frac{1}{2}L_2 I_{2max}^2 + MI_{1max}I_{2max}$$

$$= \frac{1}{4}(4)1^2\times10^{-6} + \frac{1}{2}(5)(2)^2\times10^{-6} + 2.683(1)(2)10^{-6}$$

$$W_{max} = 17.366\ \mu\text{J}$$

2.58 (100)

Given, $V_s = 10\sqrt{2}\sin(20000\pi t)$ V

$$\text{Current, } I = \frac{V_s}{50+25+25+j\omega L+\frac{1}{j\omega C}}$$

$$= \frac{V_s}{100 + j20000\pi\times\frac{10^{-3}}{20\pi} + \frac{1}{j20000\pi\times\frac{10^{-3}}{20\pi}}}$$

$$= \frac{V_s}{100+j1-j1} = \frac{V_s}{100}$$

$$I = \frac{10\sqrt{2}}{100}\sin(20000\pi t)$$

$$\therefore \quad I_{RMS} = \frac{\frac{10\sqrt{2}}{100}}{\sqrt{2}} = 0.1 = 100\text{ mA}$$

2.59 (15)

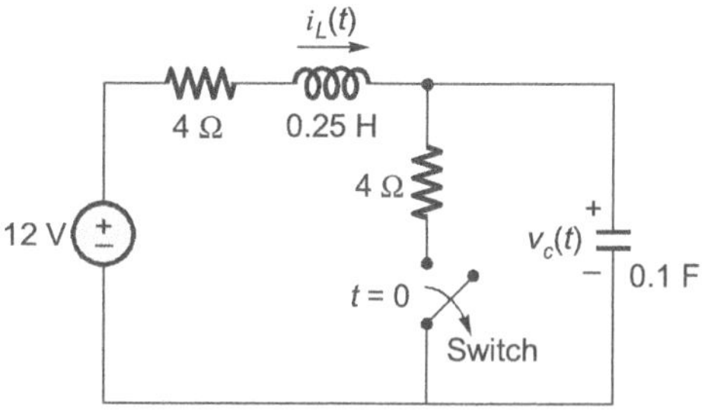

$t = 0^-$ s : closed steady state

Capacitor : OC

Inductor : SC

$$V_C = 12\text{ V}\times\frac{4}{8} = 6\text{ V and } I_L = 1.5\text{ A}$$

$t = 0^+ \rightarrow$ circuit

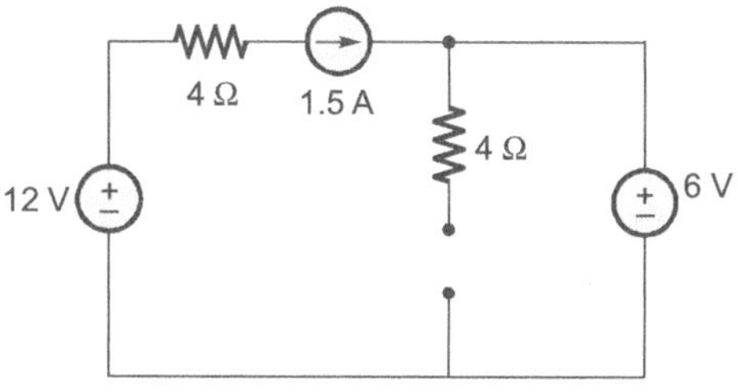

$$I_C = 1.5 = C\cdot\left.\frac{d_{VC}}{dt}\right|_{t=0^+}$$

$$\left.\frac{d_{VC}}{dt}\right|_{t=0^+} = \frac{1.5}{0.1} = 15\text{ V/sec}$$

2.60 (0.25)

Series aiding

$$L_{eq} = L_1 + L_2 + 2M$$
$$L_1 = L_2 = L$$
$$L_{eq} = 2L + 2M$$
$$500\text{ m} = 2L + 2M \quad ...(1)$$

Series opposing

$$L_{eq} = L_1 + L_2 - 2M$$
$$L_1 = L_2 = L$$
$$300\text{ m} = 2L - 2M \quad ...(2)$$

Solve eq. (1) and eq. (2)

$$L = 200\text{ mH};\ M = 50\text{ mH}$$

$$K = \frac{M}{\sqrt{L_1L_2}} = 0.25$$

3 Two-Port Network

3.1 The DC voltage gain $\frac{V_o}{V_i}$ in the following circuit is given by

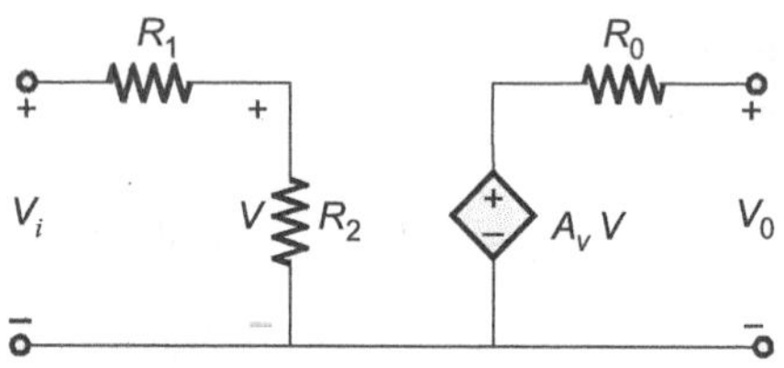

(a) $A_v \frac{R_2}{R_1 + R_2}$
(b) $A_v \frac{R_1}{R_1 + R_2}$
(c) $A_v \frac{R_2}{R_1 + R_2} + R_0$
(d) A_v

[2007 : 1 M]

3.2 Consider the coupled circuit shown below:

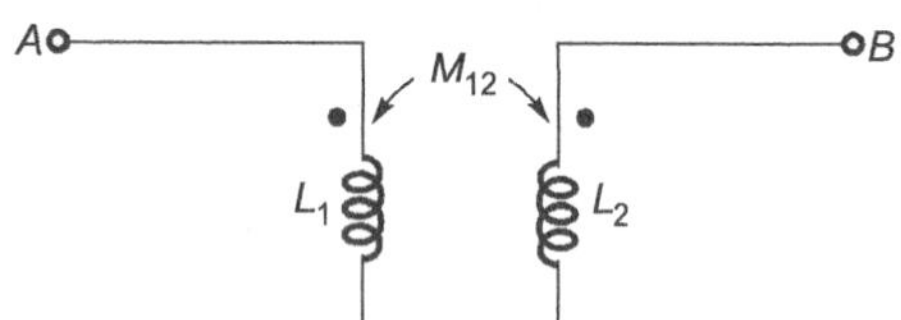

At angular frequency ω, this circuit can be represented by the equivalent *T*-network, shown below:

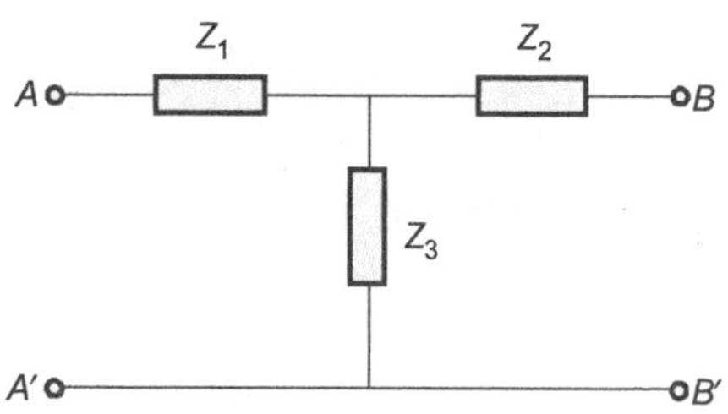

Indicate the correct set of expressions for the impedance of the *T*-network.

(a) $Z_1 = j\omega(L_1 - M_{12})$
$Z_2 = j\omega(L_2 - M_{12})$
$Z_3 = j\omega M_{12}$

(b) $Z_1 = j\omega(L_1 + M_{12})$
$Z_2 = j\omega(L_2 + M_{12})$
$Z_3 = j\omega M_{12}$

(c) $Z_1 = j\omega L_1$
$Z_2 = j\omega L_2$
$Z_3 = -j\omega M_{12}$

(d) $Z_1 = j\omega(L_1 - M_{12})$
$Z_2 = j\omega(L_2 - M_{12})$
$Z_3 = j\omega(L_1 + L_2 + M_{12})$

[2007 : 2 M]

3.3 For the circuit shown below the input resistance $R_{11} = \left.\frac{V_1}{I_1}\right|_{I_2=0}$ is

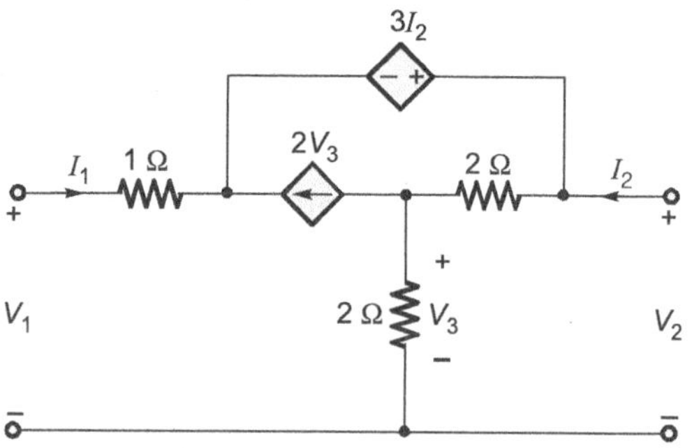

(a) –3 Ω
(b) 2 Ω
(c) 3 Ω
(d) 13 Ω

[2008 : 2 M]

Common Data for (3.4 and 3.5):

With 10 V dc connected at port *A* in the linear nonreciprocal two-port network shown below, the following were observed:

(i) 1 Ω connected at port *B* draws a current of 3 A
(ii) 2.5 Ω connected at port *B* draws a current of 2 A

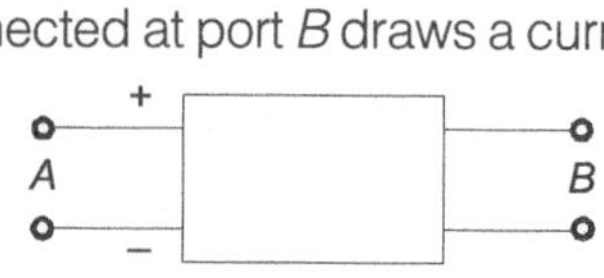

3.4 For the same network with 6 V dc connected at port *A*, 1 Ω connected port *B* draws 7/3 A. If 8 V dc is connected to port *A*, the open circuit voltage at port *B* is

(a) 6 V
(b) 7 V
(c) 8 V
(d) 9 V

[2012 : 2 M]

3.5 With 10 V dc connected at port A, the current drawn by 7 Ω connected at port B is

(a) 3/7 A (b) 5/7 A
(c) 1 A (d) 9/7 A

[2012 : 2 M]

3.6 Considering the transformer to be ideal, the transmission parameter 'A' of the 2-port network shown in the figure below is

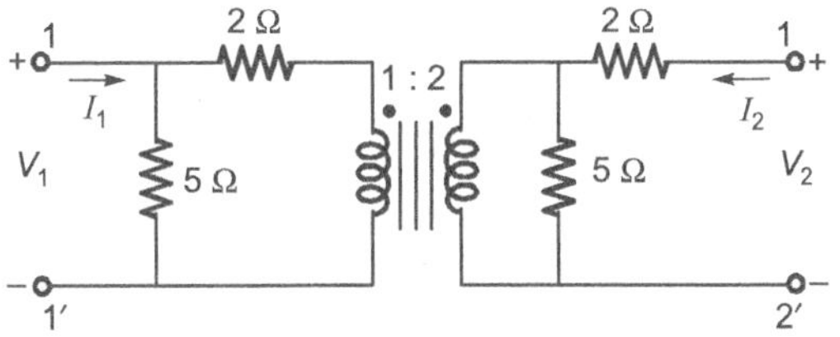

(a) 1.3 (b) 1.4
(c) 0.5 (d) 2.0

[2013 : 2 M]

3.7 The output of the ideal transformer with the polarities and dots shown in the figure is given by

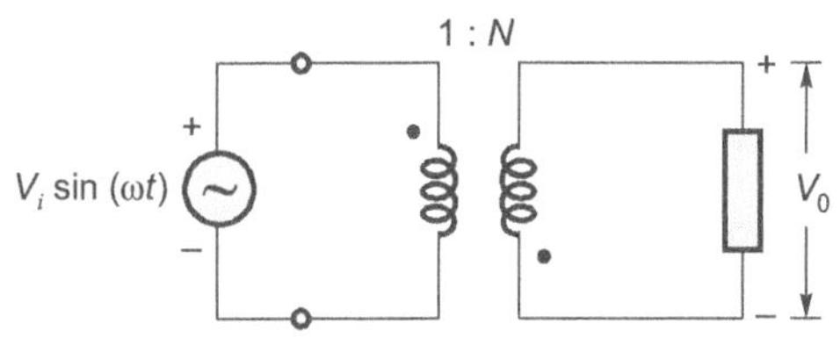

(a) $NV_i \sin\omega t$ (b) $-NV_i \sin\omega t$

(c) $\frac{1}{N}V_i \sin\omega t$ (d) $-\frac{1}{N}V_i \sin\omega t$

[2015 : 1 M]

3.8 The connection of two 2-port networks is shown in the figure. The ABCD parameters of N_1 and N_2 networks are given as

$$\begin{bmatrix} A & B \\ C & D \end{bmatrix}_{N_1} = \begin{bmatrix} 1 & 5 \\ 0 & 1 \end{bmatrix} \text{and} \begin{bmatrix} A & B \\ C & D \end{bmatrix}_{N_2} = \begin{bmatrix} 1 & 0 \\ 0.2 & 1 \end{bmatrix}$$

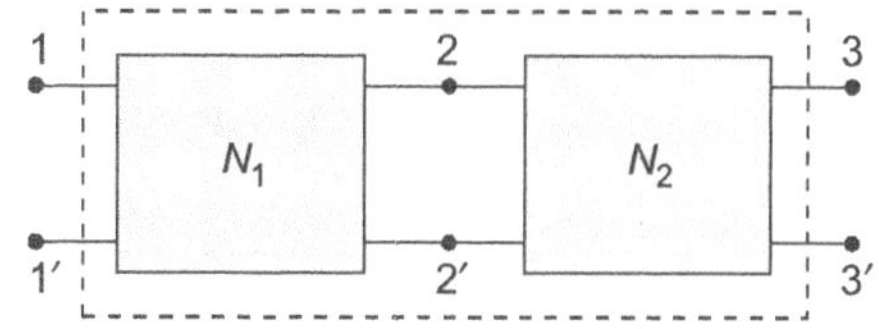

The $ABCD$ parameters of the combined 2-port network are

(a) $\begin{bmatrix} 2 & 5 \\ 0.2 & 1 \end{bmatrix}$ (b) $\begin{bmatrix} 1 & 2 \\ 0.5 & 1 \end{bmatrix}$

(c) $\begin{bmatrix} 5 & 2 \\ 0.5 & 1 \end{bmatrix}$ (d) $\begin{bmatrix} 1 & 2 \\ 0.5 & 5 \end{bmatrix}$

[2017 : 1 M]

3.9 In the Figures (a) and (b) shown below, the transformers are identical and ideal, except that the transformer in Figure (b) is centre-tapped. Assuming ideal diodes, the ratio of the root-mean-square (RMS) voltage across the resistor R in Figure (a) to that in Figure (b) is

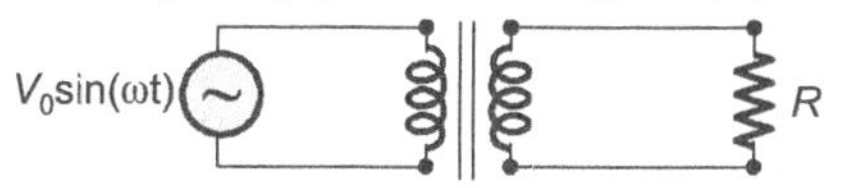

Fig (a)

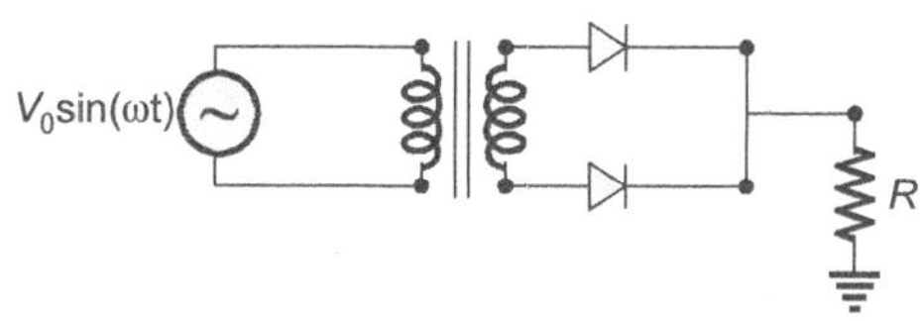

Fig (b)

(a) $\sqrt{2}:1$ (b) 2 : 1

(c) $2\sqrt{2}:1$ (d) 4 : 1

[2019 : 1 M]

■■■

Answers Two-Port Network

3.1 (a) 3.2 (a) 3.3 (d) 3.4 (b) 3.5 (c) 3.6 (a) 3.7 (b)

3.8 (a) 3.9. (b)

Explanations Two-Port Network

3.1 (a)

By voltage division rule voltage V across R_2

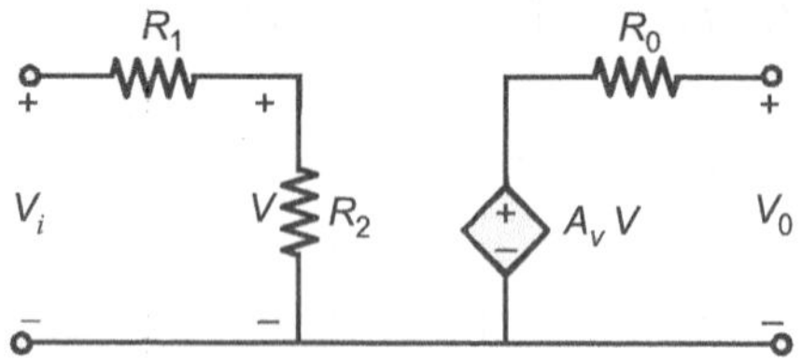

$$V = \frac{R_2}{R_1 + R_2} \cdot V_i$$

and output voltage $V_0 = A_v V$

$$\therefore \quad V_0 = A_v \cdot \frac{R_2}{R_1 + R_2} \cdot V_i$$

$\therefore$ DC voltage gain,

$$\frac{V_0}{V_i} = \frac{R_2}{R_1 + R_2} A_v$$

3.2 (a)

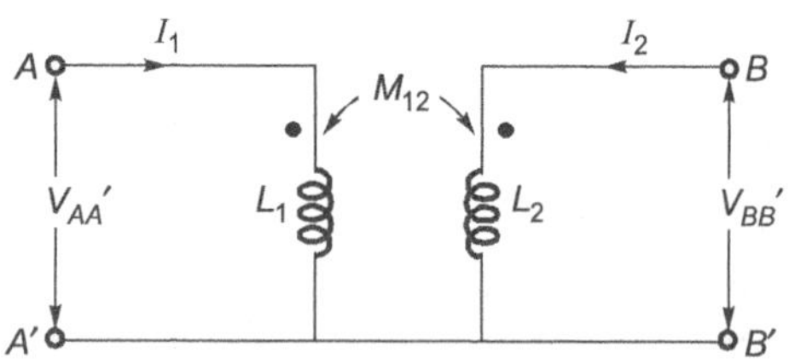

Applying KVL,

$$V_{AA}' = j\omega L_1 I_1 + j\omega M_{12} I_2 \quad \text{...(i)}$$

$$V_{BB}' = j\omega M_{12} I_1 + j\omega L_2 I_2 \quad \text{...(ii)}$$

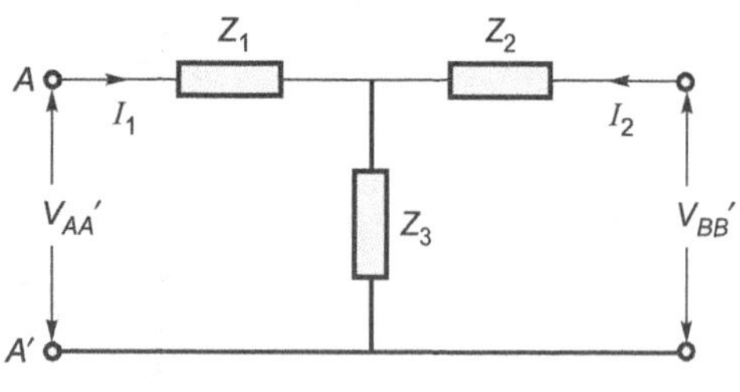

Applying KVL,

$$V_{AA}' = (Z_1 + Z_3)I_1 + Z_3 I_2 \quad \text{...(iii)}$$

$$V_{BB}' = Z_3 I_1 + (Z_2 + Z_3)I_2 \quad \text{...(iv)}$$

By comparing equation (i) with equation (iii) and (ii) with equation (iv) we get,

$$Z_1 + Z_3 = j\omega L_1$$

$$Z_2 + Z_3 = j\omega L_2$$

$$Z_3 = j\omega M_{12}$$

$$\Rightarrow \quad Z_1 = j\omega(L_1 - M_{12}),$$

$$\Rightarrow \quad Z_2 = j\omega(L_2 - M_{12})$$

$$\Rightarrow \quad Z_3 = j\omega M_{12}$$

3.3 (d)

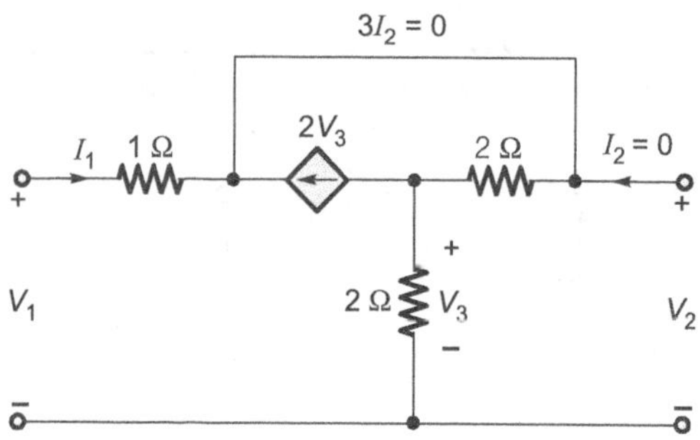

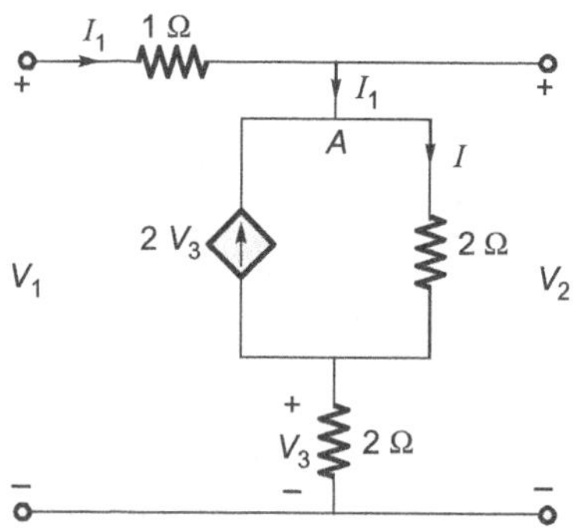

By applying KCL at point A,

$$I_1 + 2V_3 = I$$

and by applying KVL in outer loop

$$V_1 = I_1 + 2(I_1 + 2V_3) + 2I_1$$

$$V_1 = 5I_1 + 4V_3 \quad \text{...(i)}$$

$$\because \quad V_3 = 2I_1$$

$$\therefore \quad V_1 = 5I_1 + 8I_1$$

$$\frac{V_1}{I_1} = 13\ \Omega$$

3.4 (b)

(i) $V_1 = 10$ V, $V_2 = 3$ V, $I_2 = -3$A

$$V_1 = AV_2 - BI_2$$

$$10 = 3A + 3B \quad \text{...(1)}$$

(ii) $V_2 = 5$ V, $I_2 = -2$A

$$10 = 5A + 2B \quad \text{...(2)}$$

$$A = \frac{10}{9} \quad \text{...(3)}$$

$$B = \frac{20}{9} \quad ...(4)$$

Given,
$$V_1 = 8\text{ V}, (V_2)_{OC} = ?, I_2 = 0$$
$$V_1 = AV_2 - BI_2$$
$$8 = A(V_2)_{OC} - 0$$
$$(V_2)_{OC} = \frac{8}{A} = \frac{8}{10/9} = 7.2\text{ V}$$

3.5 (c)

$$V_1 = 10\text{ V}, V_2 = (-7I_2)$$
$$V_1 = AV_2 - BI_2$$
$$10 = -7I_2A - BI_2$$
$$10 = -\frac{70}{9}I_2 - \frac{20}{9}I_2$$
$$I_2 = -1\text{A}$$

–ve sign signifies that current is drawn.

Alternate Solution:

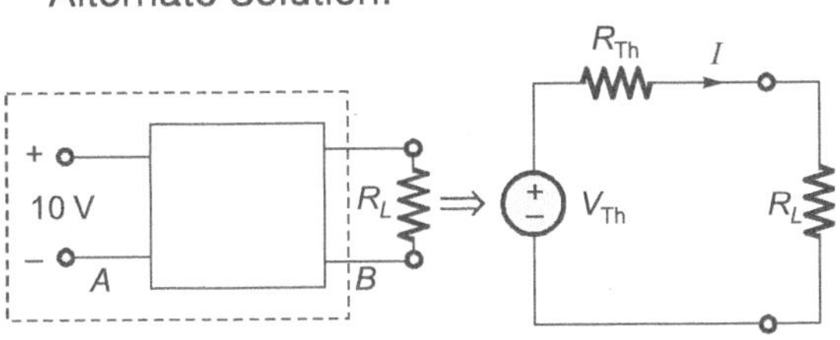

(i) $$3 = \frac{V_{Th}}{1 + R_{Th}}$$

(ii) $$2 = \frac{V_{Th}}{2.5 + R_{Th}}$$

$$\Rightarrow \frac{3}{2} = \frac{R_{Th} + 2.5}{R_{Th} + 1}$$
$$\Rightarrow 2R_{Th} + 5 = 3R_{Th} + 3$$
$$\Rightarrow R_{Th} = 2\ \Omega \text{ and } V_{Th} = 9\text{ V}$$
$$I = \frac{9}{7+2} = 1\text{ Amp.}$$

3.6 (a)

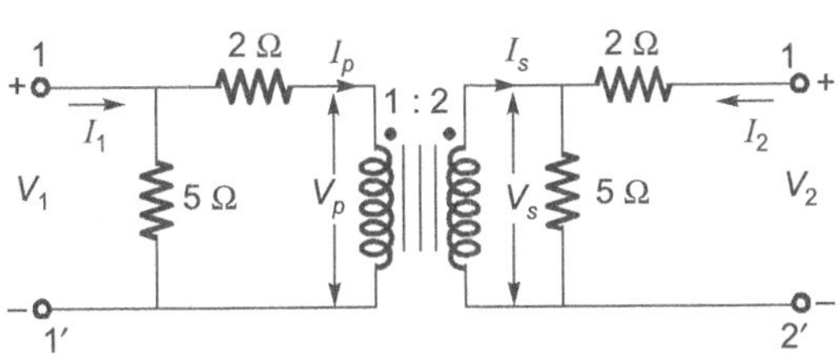

$$I_p = 2I_s \text{ and } V_s = 2V_p \quad ...(i)$$
$$\because \quad A = \left.\frac{V_1}{V_2}\right|_{I_2=0}$$
$\therefore$ At $I_2 = 0$
$$V_2 = V_s = 5I_s = 2.5I_p \quad ...(ii)$$
$$V_1 = V_p + 2I_p$$

from equation (i),
$$V_1 = \frac{V_s}{2} + 2I_p$$
from equation (ii),
$$V_1 = \frac{5}{4}I_p + 2I_p$$
$$V_1 = 3.25\, I_p \quad ...(iii)$$
from equation (ii) and (iiii)
$$A = \frac{V_1}{V_2} = \frac{3.25\, I_p}{2.5\, I_p} = 1.3$$

3.7 (b)

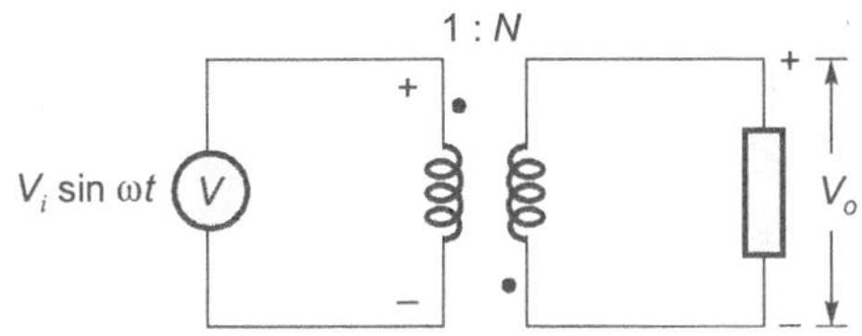

$$\frac{V_2}{V_1} = \frac{N_2}{N_1}$$
$$V_2 = V_i \sin\omega t\left(\frac{N}{1}\right) = N\, V_i \sin\omega t$$
$$V_- = -V_2 = -NV_i \sin\omega t$$

3.8 (a)

$$\begin{bmatrix} A & B \\ C & D \end{bmatrix} = \begin{bmatrix} 1 & 5 \\ 0 & 1 \end{bmatrix}\begin{bmatrix} 1 & 0 \\ 0.2 & 1 \end{bmatrix}$$
$$\begin{bmatrix} A & B \\ C & D \end{bmatrix} = \begin{bmatrix} 2 & 5 \\ 0.2 & 1 \end{bmatrix}$$

3.9 (b)

Fig (a)

The resultant output V_{01} is

 $\longrightarrow$ $V_{01\,rms}$ across $R = \dfrac{V_m}{\sqrt{2}}$

fig (b)

The resultant output V_{02} is

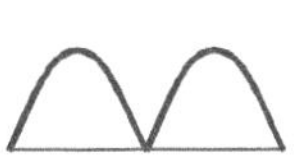 $\longrightarrow$ $V_{02\,rms}$ across $R = \dfrac{V_m/2}{\sqrt{2}}$

$\dfrac{V_m}{2}$ due to centertap

The ratio $\dfrac{V_{01}}{V_{02}}$ is

$$\frac{V_m/\sqrt{2}}{V_m/2\sqrt{2}} = 2 : 1$$

■■■

4 Electrical Machines

4.1 A toroid made of CRGO has an inner diameter of 10 cm and an outer diameter of 14 cm. The thickness of the toroid is 2 cm, 200 turns of copper wire is wound on the core. $\mu_0 = 4\pi \times 10^{-7}$ H/m and μ_R of CRGO is 3000. When a current of 5 mA flows through the winding, the flux density in the core in millitesla is _____.

[2021 : 2 M]

4.2 A single-phase transformer has maximum efficiency of 98 %. The core losses are 80 W and the equivalent winding resistance as seen from the primary side is 0.5 Ω. The rated current on the primary side is 25 A. The percentage of the rated input current at which the maximum efficiency occurs is

(a) 80.5%
(b) 50.6%
(c) 35.7%
(d) 100%

[2021 : 2 M]

4.3 A slip-ring induction motor is expected to be started by adding extra resistance in the rotor circuit. The benefit that is derived by adding extra resistance in the rotor circuit in comparison to the rotor being shorted is

(a) The power factor at start will be lower.
(b) The starting current is higher.
(c) The starting torque would be higher.
(d) The losses at starting would be lower.

[2021 : 2 M]

4.4 A household fan consumes 60 W and draws a current of 0.3125 A (rms) when connected to a 230 V (rms) AC, 50 Hz single phase mains. The reactive power drawn by the fan in VAr is (rounded off to the nearest integer).

[2021 : 2 M]

4.5 A 440 V, 8 kW, 4-pole, 50 Hz, star-connected induction motor has a full load slip of 0.04. The rotor speed (in rpm) at full load is ________ (round off to one decimal place)

[2022 : 1 M]

4.6 The no-load steady-state output voltage of a DC shunt generator is 200 V when it is driven in the clockwise direction at its rated speed. If the same machine is driven at the rated speed but in the opposite direction, the steady-state output voltage will be ________V (rounded off to the nearest integer).

[2023 : 1 M]

4.7 A short-circuit test is conducted on a single-phase transformer by shorting its secondary. The frequency of input voltage is 1 kHz. The corresponding wattmeter reading, primary current and primary voltage are 8 W, 2 A and 6 V respectively. Assume that the no-load losses and the no-load currents are negligible, and the core has linear magnetic characteristics. Keeping the secondary shorted, the primary is connected to a 2 V (RMS), 1 kHz sinusoidal source in series with a $\frac{1}{2\pi\sqrt{5}}$ mF capacitor. The primary current (RMS) will be _____ A (rounded off to two decimal places).

[2023 : 2 M]

4.8 An ADC has a full scale voltage of 1.4 V, resolution of 200 mV, and produces binary output data. The input signal of the ADC has a bandwidth of 500 MHz, and it samples the data at the Nyquist rate. The parallel data output is converted to a serial bit stream using a parallel-to-serial converter. The data rate at the output of the parallel-to-serial converter is _____ Gbps (rounded off to nearest integer).

[2024 : 2 M]

4.9 A 50 kVA transformer has an efficiency of 95% at full load and unity power factor. Assume the core losses are negligible. The efficiency of the transformer at 75% of the full load and 0.8 power factor is ________% (rounded off to one decimal place).

[2024 : 2 M]

4.10 A three-phase squirrel-cage induction motor has a starting torque of 100% of the full load torque and a maximum torque of 300% of the full load torque. Neglecting the stator impedance, the slip at the maximum torque is ________% (rounded off to two decimal places).

[2024 : 2 M]

■■■

Answers Electric Machines

4.1 (10)	4.2 (b)	4.3 (c)	4.4 (39.56)	4.5 (1440.0)	4.6 (0)	4.7 (1)
4.8 (3)	4.9 (95.30)	4.10 (17.15)				

Explanations Electric Machines

4.1 (10)

$$\phi = \frac{MMF}{S} = \frac{NI}{\frac{l}{a\mu_0\mu_r}} = \frac{NI}{l}(a\mu_0\mu_r)$$

$$B = \frac{\phi}{a} = \frac{\frac{NI}{l}(a\mu_0\mu_r)}{a}$$

$$B = \frac{NI\mu_0\mu_r}{l}$$

$$B = \frac{200(5\times10^{-3})(4\pi\times10^{-7})(3000)}{2\pi\left(\frac{12\times10^{-2}}{2}\right)}$$

B = 10 millitesla

4.2 (b)

At maximum efficiency,

Core loss = Copper losses

Core loss = 80 W

Copper loss at rated current

$= (25)^2 \times 0.5 = 312.5$ W

When maximum efficiency occurred,

$$I^2_m R = 80$$

$$I^2_m = \frac{80}{0.5} = 160$$

$$I_m = 12.649 \text{ A}$$

$$\% \text{ of rate of current } = \frac{12.649}{25}\times100 = 50.60\%$$

4.3 (c)

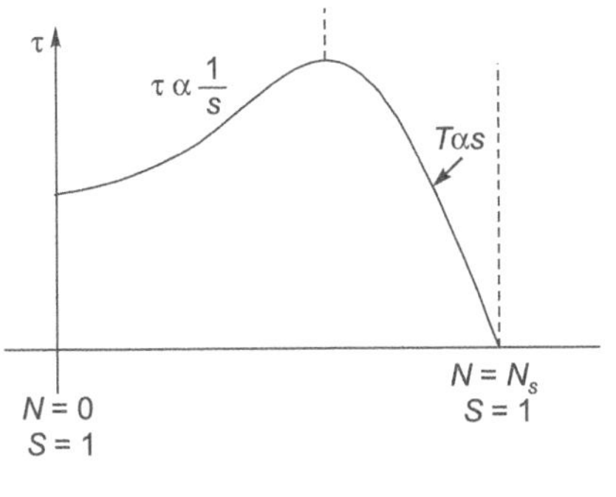

For power factor $\cos\theta_2 = \frac{R_2}{Z_2} = \frac{R_2}{R_2 + jsX_2}$

At starting $(\cos\theta_2)_{\text{starting}} = \frac{R_2}{R_2 + X_2}$

Order of R_2/phase = 0.1 to 0.2Ω

Order of X_2/phase = 1.5Ω to 2Ω

$$R_2^2 <<< X_2^2$$

So $R_2^2 \rightarrow$ neglected

$$\cos\theta_2 = \frac{R_2}{X_2}$$

As $R_2\uparrow$ so power factor increases.

Hence option (a) is wrong.

$$T_s = \frac{180}{2\pi N_s}\times\frac{E_2^2 R_2}{R_2^2 + X_2^2}$$

So, R_2^2 term is neglected

$$T_s = \frac{180}{2\pi N_s}\frac{E_2^2 R_2}{X_2^2},$$

So, starting torque increases as it is directly proportional to rotor resistance.

4.4 (39.56)

$$P_{out} = 60 \text{ W}$$
$$I = 0.3125 \text{ A}$$
$$V = 230 \text{ V}$$
$$VI\cos\phi = P_{out}$$
$$\cos\phi = 0.8347 \qquad \left[\because \cos\phi = \frac{P_{out}}{VI}\right]$$
$$\phi = 33.406$$
$$Q = P\tan\phi$$
$$Q = 39.56 \text{ VAr}$$

4.5 (1440.0)

The slip of motor,

$$s = 0.04$$

The synchronous speed,

$$N_s = 120 \times \frac{f}{P}$$

$$N_s = 120 \times \frac{50}{4} = 1500 \text{ rpm}$$

So,
$$N_r = N_s(1 - s)$$
$$= 1500(1 - 0.04)$$
$$N_r = 1440 \text{ rpm}$$

4.6 (0)

As due to change in direction polarity change, which cancel out residual field.

Alternate:

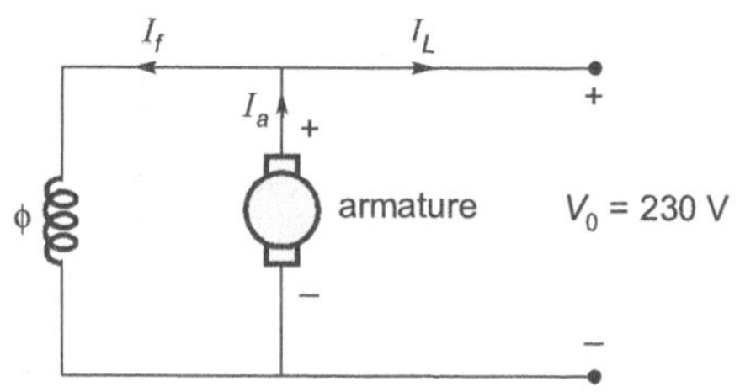

Using Fleming right hand rule.

(a) Initially in clockwise direction

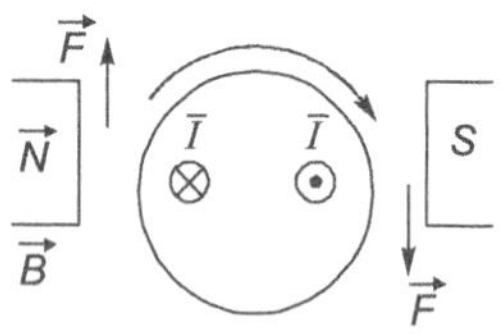

Now its characteristic curve B and H

(b) Now its direction is anticlockwise achieve this firstly its speed must be reduced to 0 but at 0 speed, the ϕ_f reaches residual flux ϕ_{res}.

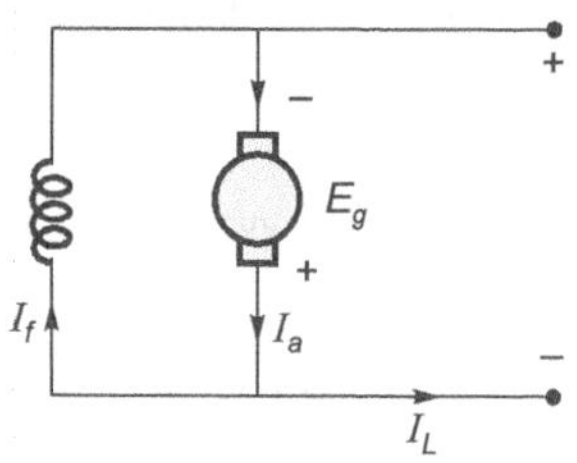

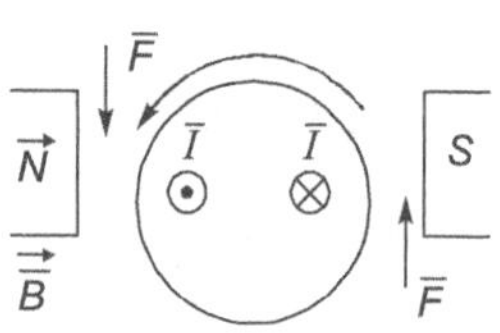

Now I_f direction reversed so slowly the residual flux of ϕ_f becomes '0'.

$$E_g \propto \phi \cdot N$$
$$E_g = 0 \text{ (where } \phi = 0)$$

Steady-state output voltage will be 0 V.

4.7 (1)

$$P_m = I^2 R_{eq}$$
$$8 = 2^2 \times R_{eq}$$
$$R_{eq} = 2\ \Omega$$
$$Z_{eq} = \frac{6}{2} = 3\ \Omega$$
$$X_{eq} = \sqrt{3^2 - 2^2} = \sqrt{5}\ \Omega$$

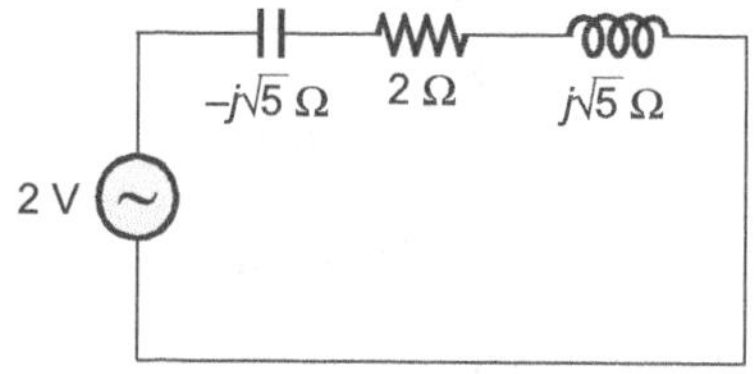

$$C = \frac{1}{2\pi\sqrt{5}} \text{mF},$$

$$X_C = -j\frac{1}{2\pi \times 1 \times 10^3 \times \frac{1}{2\pi\sqrt{5}} \times 10^{-3}} = -j(\sqrt{5})$$

$$I_{primary} = \frac{2}{2} = 1 \text{ A}$$

4.8 (3)

Full scale voltage → V_{FS} = 1.4 V

Resolution = 200 mV

Bandwidth → f_m = 500 MHz

$$\text{Resolution} = \frac{V_{FS}}{2^n - 1}$$

$$200 \text{ m} = \frac{1.4}{2^n - 1}$$

$$2^n - 1 = \frac{1.4}{200 \times 10^{-3}}$$

$$2^n - 1 = 7 \rightarrow 2^n = 8;\ n = 3$$

Sampling rate = N.R

$= 2f_m = 1000$ MHz

Data Rate → $R_b = nf_s$

$= 3 \times 1000$ m

= 3000 Mbps

= 3 Gbps

4.9 (95.30)

$$\text{Efficiency} = \frac{\text{Output}}{\text{Input}} = \frac{\text{Output}}{\text{Output} + \text{Losses}}$$

$$0.95 = \frac{\text{output}}{\text{output + losses}}$$

$$\text{losses} = \frac{0.05 \times \text{output}}{0.95}$$

$$= \frac{0.05 \times 50 \times 10^3}{0.95} = 2631.58 \text{ W}$$

It is given that assume core losses are negligible so

Losses = copper loss

% Efficiency new

$$= \frac{50 \times 0.75 \times 10^3 \times 0.8}{50 \times 0.75 \times 10^3 \times 0.8 + \frac{9}{16} \times 2631.58} \times 100$$

$$= \frac{30000}{30000 + 1480.26} \times 100$$

$= 95.30\%$

4.10 (17.15)

Starting torque = Full load torque

Maximum torque = 3 × Full load torque

$$\frac{T_{starting}}{T_{max}} = \frac{2}{\frac{S_{T_{max}}}{S_{starting}} + \frac{S_{starting}}{S_{T_{max}}}}$$

$$\frac{1}{3} = \frac{2}{\frac{S_{T_{max}}}{1} + \frac{1}{S_{T_{max}}}} = \frac{2S_{T_{max}}}{S_{T_{max}}^2 + 1}$$

$$S_{T_{max}}^2 + 1 = 6S_{T_{max}}$$

$$S_{T_{max}}^2 - 6S_{T_{max}} + 1 = 0$$

$$S_{T_{max}} = 0.1715$$

$$\%S_{T_{max}} = 17.15$$

■■■

Sensors and Industrial Instrumentation

UNIT III

CONTENTS

1. Basics of Transducers 102
2. Resistive and Inductive Transducers 104
3. Capacitive and Piezoelectric Transducers 118
4. Pressure, Flow and Temperature Measurement 131
5. Velocity, Acceleration, Force, Torque & Seismic Measurement 154
6. Miscellaneous 160

Sensors and Industrial Instrumentation

Syllabus

Resistive-, capacitive-, inductive-, piezoelectric-, Hall effect sensors and associated signal conditioning circuits; transducers for industrial instrumentation: displacement (linear and angular), velocity, acceleration, force, torque, vibration, shock, pressure (including low pressure), flow (variable head, variable area, electromagnetic, ultrasonic, turbine and open channel flow meters) temperature (thermocouple, bolometer, RTD (3/4 wire), thermistor, pyrometer and semiconductor); liquid level, pH, conductivity and viscosity measurement. 4-20 mA two-wire transmitter.

Analysis of Previous GATE Papers

Exam Year	1 Mark Ques.	2 Marks Ques.	Total Marks
1995	–	2	**4**
1996	6	2	**10**
1997	4	2	**8**
1998	3	1	**5**
1999	2	4	**10**
2000	1	3	**7**
2001	4	3	**10**
2002	5	3	**11**
2003	6	11	**28**
2004	5	5	**15**
2005	3	7	**17**
2006	3	5	**13**
2007	1	6	**13**
2008	3	6	**15**
2009	7	–	**7**

Exam Year	1 Mark Ques.	2 Marks Ques.	Total Marks
2010	5	–	**5**
2011	4	3	**10**
2012	2	5	**12**
2013	2	6	**14**
2014	4	4	**12**
2015	5	2	**9**
2016	1	4	**9**
2017	2	5	**12**
2018	2	2	**6**
2019	3	3	**9**
2020	2	3	**8**
2021	4	3	**10**
2022	3	–	**3**
2023	3	2	**7**
2024	4	4	**12**

1 Basics of Transducers

1.1 Which one of the following is a derived unit (not a fundamental unit) in SI system?
(a) Candela (b) Coulomb
(c) Kelvin (d) mol.
[2003 : 1 M]

1.2 Different types of transducers and signal conditioning circuits are given below. Match the transducer with the appropriate signal conditioning circuit commonly used with it.
P. Thermocouple
Q. Strain gage
R. Piezoelectric sensor
S. L.V.D.T.
1. DC-Bridge
2. Phase sensitive detector
3. Charge amplifier
4. Cold junction compensation
(a) P-2, Q-3, R-4, S-1 (b) P-1, Q-4, R-2, S-3
(c) P-4, Q-1, R-2, S-3 (d) P-4, Q-1, R-3, S-2
[2003 : 1 M]

1.3 A zero error in a Vernier caliper is termed as
(a) accidental error (b) systematic error
(c) interference error (d) random error
[2004 : 1 M]

1.4 Match the essential amplifier characteristics with the sensing applications given below:
Amplifier Characteristics
P. Charge amplifier with very low bias current and high input impedance.
Q. Voltage amplifier with low bias current and very high input impedance.
R. Voltage amplifier with very high CMRR.
Sensing Application
L. Strain gauge in unipolar DC Wheatstone bridge.
M. Glass electrode pH sensor.
N. Piezoelectric sensor for measurement of static force.
(a) P-L, Q-M, R-N (b) P-M, Q-N, R-L
(c) P-N, Q-L, R-M (d) P-N, Q-M, R-L
[2007 : 1 M]

1.5 A pneumatic nozzle-flapper system is conventionally used to convert ________
(a) Small changes in flapper's velocity to large changes in output temperature.
(b) Small changes in flapper's displacement to large changes in output temperature.
(c) Small changes in flapper's velocity to large changes in output pressure.
(d) Small changes in flapper's displacement to large changes in output pressure.
[2022 : 1 M]

1.6 A Hall sensor is based on the principle of _____.
(a) Photoelectric effect (b) Seebeck effect
(c) Piezoelectric effect (d) Lorentz force
[2022 : 1 M]

1.7 The two secondaries of a linear variable differential transformer (LVDT) showed a magnitude of 2 V (RMS) for zero displacement position of the core. It is noted that the phase of one of the secondaries has a deviation of one degree from the expected phase. Other than this deviation, the LVDT is ideal. If the differential output sensitivity of the LVDT is 1 mV (RMS)/1 μm, the output for zero displacement is ________ μm (rounded off to one decimal place). [2023 : 2 M]

1.8 A linear transducer is calibrated for the ranges shown in the figure. The gain of the transducer is ______ mA/°C (rounded off to two decimal places).

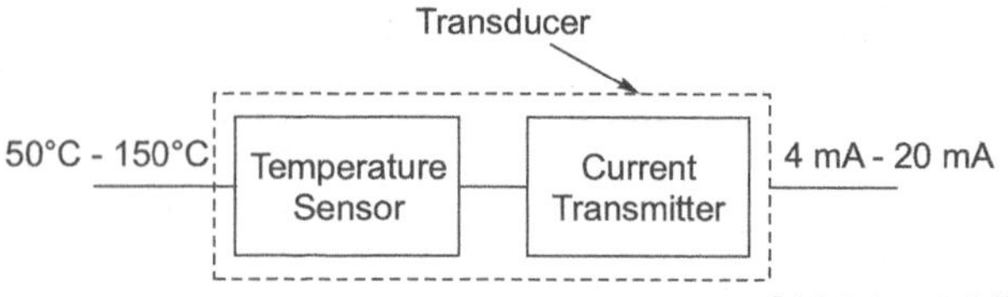

[2024 : 1 M]

■■■■

Answers Basics of Transducers

1.1 (b) 1.2 (d) 1.3 (b) 1.4 (c) 1.5 (d) 1.6 (d) 1.7 (34.9)

1.8 (0.16)

Explanations Basics of Transducers

1.1 (b)

Fundamental units is a set of those units of physical quantities from which every other unit can be derived.

eg. Candela – Fundamental unit of light
Kelvin – Fundamental unit of temp.
Mol. – Fundamental unit of amount of substance

Derived units are those which are derived from the fundamental units.

eg. Coulomb = 1 Ampere × 1 Second
= 1 Farad × 1 Volt

1.2 (d)

Thermocouple: When reference junction is not at 0°C then cold junction compensation is done in thermocouples.

Strain Gauge: Change in resistance due to applied strain on strain gauge is measured using DC bridge.

Piezoelectric sensor: A charge amplifier is used signal conditioning circuit of piezoelectric sensor.

L.V.D.T: To identify positive and negative displacements phase sensitive detectors are used.

1.5 (d)

The nozzle and flapper mechanism is a displacement type detector which converts mechanical movement into pressure signal.

1.6 (d)

Hall effect works on the principle of Lorentz force.

1.7 (34.9)

For an LVDT,

Output voltage = $\overline{E_{S1}} - \overline{E_{s2}}$

Given there is deviation of 1° and rms value is 2 V.

Voltage at zero displacement

$$V_0 = 2\angle 0° - 2\angle 1°$$
$$V_0 = 2 - 2\cos 1 - j2\sin 1$$
$$V_0 = 3.04609 \times 10^{-4} - j[349.04812 \times 10^{-4}]$$
$$V_0 = 3.04609 \times 10^{-4} - j[349.04812 \times 10^{-4}]$$
$$V_0 = 349.061411 \times 10^{-4} \angle -89.5$$
$$|V_0| = 3.49061411 \times 10^{-2}$$

So given sensitivity is 1 mV/1 μm

Output voltage = $|V_0| \times S$

$$\text{displacement} = \frac{3.49061411 \times 10^{-2}}{1 \times 10^{-3} / 10^{-6}\text{m}}$$

displacement (μm) = 3.49061411×10^{1} μm
= 34.906 μm

1.8 (0.16)

Linear transducer output,

$$V_0 = KV_{in} + C$$

where $K \rightarrow$ gain
$C \rightarrow$ constant

$$V_0 = K\,V_{in} + C$$

$\Rightarrow$ 4 mA = K(50°C) + C ...(1)

$\Rightarrow$ 20 mA = K(150°C) + C ...(2)

(2) – (1) $\Rightarrow$ 16 mA = K(150 – 50)

$$K = \frac{16\text{ mA}}{100°C} = 0.16\text{ mA/°C}$$
$$= 0.16$$

■■■

2 Resistive and Inductive Transducers

2.1 The two secondary coils of an LVDT have wrongly been connected as shown in figure. Then the input-output relationship would be

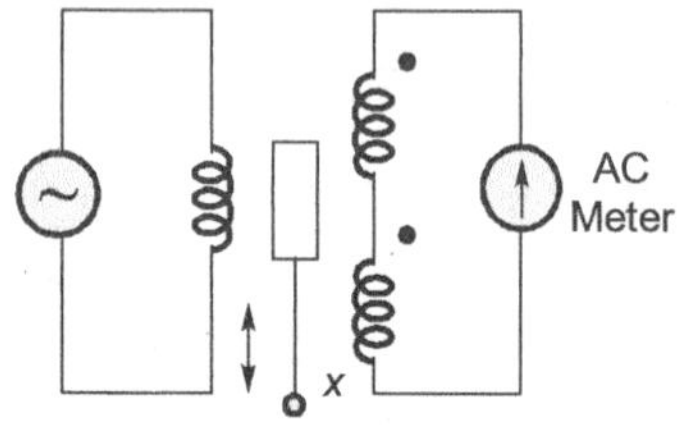

(a) 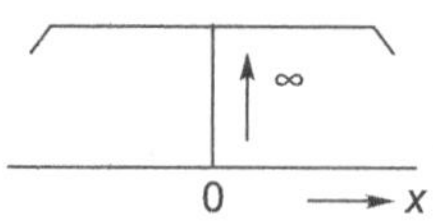

(b) 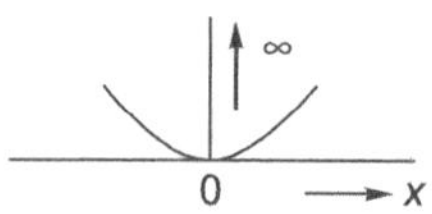

(c) 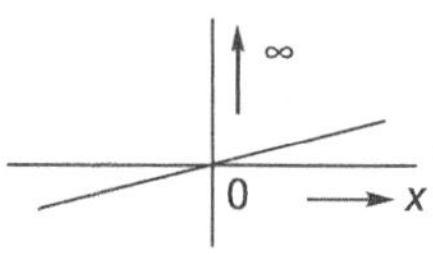

(d) 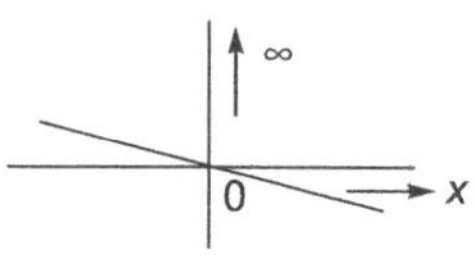

[1998 : 1 M]

2.2 A strain gauge has gauge factor $G = -100$. The type of the strain gauge is
(a) Unbounded metal type
(b) Bonded metal foil type
(c) *p*-type semiconductor
(d) *n*-type semiconductor [1998 : 1 M]

2.3 The resistance temperature characteristics of a temperature transducer is shown in fig. The transducer is,

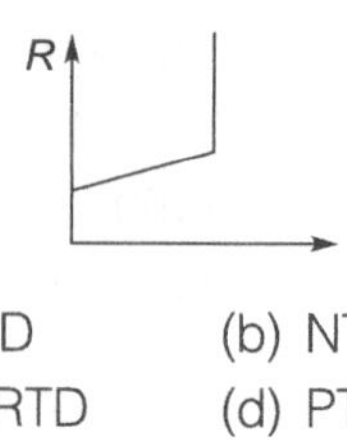

(a) Nickel RTD (b) NTC thermistor
(c) Platinum RTD (d) PTC thermistor
[2000 : 1 M]

2.4 Among the bonded metal strain gauges, the foil type is more popular than the wire type because,
(a) Error due to transverse strain is much less in foil type
(b) Gauge factor is much higher in foil type
(c) The foil type is much more insensitive to temperature variation.
(d) None of these [2000 : 2 M]

2.5 Linear variable differential transformer has
(a) two primary coils connected in phase and a secondary coil
(b) two primary coils connected in opposition and a secondary coil
(c) one primary coil and two secondary coils connected in phase
(d) one primary coil and two secondary coils connected in opposition
[2001 : 1 M]

2.6 A strain gauge is attached to a bar of length 20 cm which is subjected to a tensile force. The nominal resistance of strain gauge is 100 Ω. The changes in resistance and elongation in the bar measured are 0.35 Ω and 0.2 mm, respectively. The gauge factor of the strain gauge is
(a) 2 (b) 3.5
(c) 10 (d) 100 [2001 : 2 M]

2.7 All metal resistive strain gages have a gauge factor (GF) nearly 2.5 because
(a) Young's modulus is the same for all metals and alloys
(b) Poisson's ratio is the same for all metals and alloys
(c) The conductivity of the material changes with applied strain in the elastic region
(d) The conductivity of the material is independent of the applied strain.
[2002 : 1 M]

2.8 An RTD is assumed to be represented in the range 0°C to 100°C by a linear model as
$$R_T = R_0(1 + 0.004\,T)$$
where T is the temperature in °C. $R_0 = 100\ \Omega$ with a variation of ±2 Ω. The true model of the RTD is
$$R_T = R_0(1 + 0.004\,T + 6 \times 10^{-7}\,T^2)$$
The worst-case error magnitude in the resistance value that will be introduced, if the linear model is used over a range of 0 to 100°C is
(a) 2 Ω (b) 2.8 Ω
(c) 3.0 Ω (d) 3.4 Ω
[2003 : 2 M]

2.9 A resistance R_T of a resistance transducer as $R_T = R(1 - Kx)$, where K is the constant of transformation and x the input quantity being sensed. The resistor is connected to an ideal, opamp signal conditioning circuit of figure. The value of R is 100 Ω, K is 0.004 and x is 75. The output V_0 is

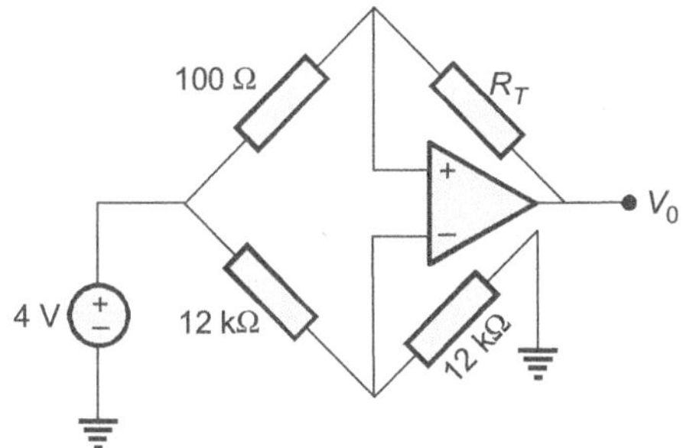

(a) –400 mV (b) –4 mV
(c) –300 mV (d) +600 mV

[2003: 2 M]

2.10 The excitation frequency of an LVDT is 2 kHz. The maximum frequency of displacement should be limited to

(a) 200 Hz (b) 1.5 kHz
(c) 2 kHz (d) 2.5 kHz

[2004 : 1 M]

2.11 A strain gauge of resistance 120 Ω and gauge factor 2.0 is at zero strain condition. A 200 kΩ fixed resistance is connected in parallel with it. Then the combination will represent an equivalent strain of

(a) +290 μm/m (b) zero
(c) –123.8 μm/m (d) –300 μm/m

[2004: 2 M]

Common Data for Questions (2.12 and 2.13):

A bonded metal strain gauge has the following specifications:

Gauge factor = 2.0
Unstrained resistance = 120 Ω at 0°C
Maximum operating temperature = 150°C
Temperature coefficient of resistance = $(2 \times 10^{-5})°C^{-1}$
Maximum tensile and compressive strain = 2×10^{-2}

2.12 The temperature-induced resistance error at the maximum operating temperature expressed as the percentage of resistance change due to maximum tensile strain only, assuming self-heating effects to be negligible, is

(a) 0.0625 (b) 0.625
(c) 6.26 (d) 7.5

[2005 : 2 M]

2.13 Two strain gauges, with the specifications given above are mounted on a structural member in a push-pull configuration, connected in a bridges along with two standard resistances, whose output is fed to an amplifier as shown in the figure. The output voltage V_0 per unit battery voltage, under maximum loading, is

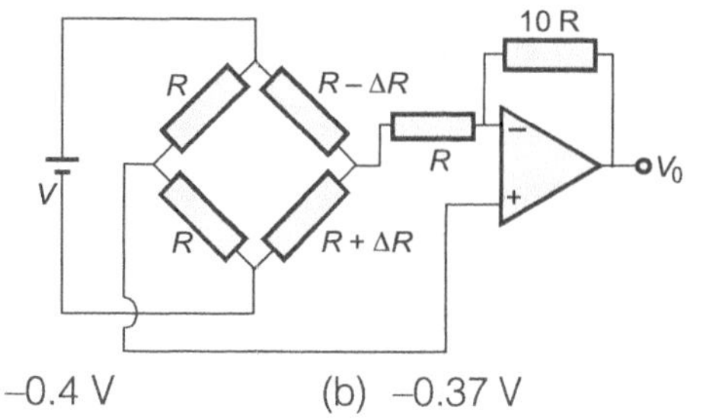

(a) –0.4 V (b) –0.37 V
(c) 0.37 V (d) 0.4 V

[2005 : 2 M]

2.14 The secondary induced voltages of a LVDT, shown in the following figure, at null position are $\bar{V}_1 = 1.0\ V \angle 0°, \bar{V}_2 = 1.0\ V \angle 10°$ respectively. Then the null voltage of the LVDT is

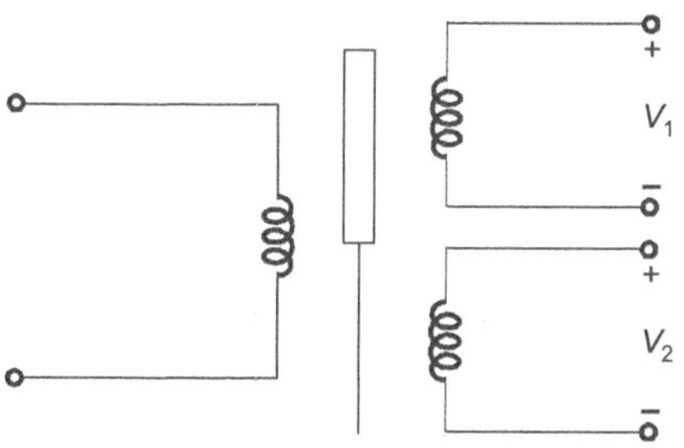

(a) 0 V (b) 0.014 V
(c) 0.174 V (d) 2 V

[2006 : 2 M]

2.15 A linear variable differential transformer (LVDT) is

(a) a displacement transducer
(b) an impedance matching transformer
(c) a differential temperature sensor
(d) an auto transformer

[2008 : 1 M]

2.16 A strain gauge has a nominal resistance of 600 Ω and a gauge factor of 2.5. The strain gauge is connected in a dc bridge with three other resistance of 600 Ω each. The bridge is excited by a 4 V battery. If the strain gauge is subjected to a strain of 100 μm/m, the magnitude of the bridge output will be

(a) 0 V (b) 250 μV
(c) 500 μV (d) 750 μV

[2008: 2 M]

2.17 Four strain gauges are fixed on a cylindrical shaft to measure torque, as shown in the figure.
A correct way to place these gauges in the bridge is

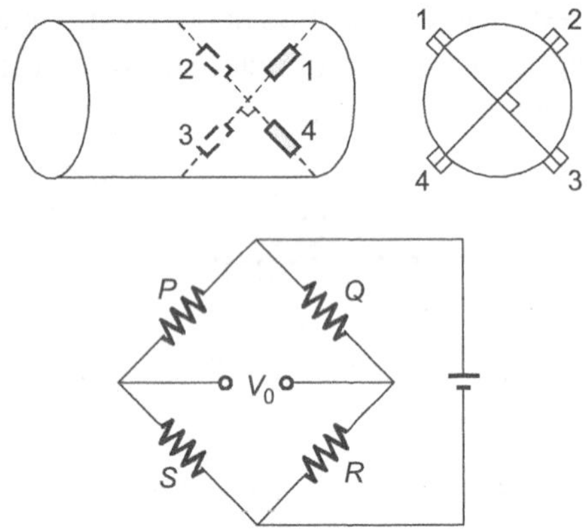

(a) P-1, Q-2, R-3, S-4
(b) P-1, Q-3, R-2, S-4
(c) P-3, Q-1, R-2, S-4
(d) P-2, Q-1, R-3, S-4

[2009 : 1 M]

2.18 An LVDT is supplied with a sinusoidal voltage of amplitude 5 V and frequency 1 kHz. The output is connected to an ac voltmeter. The reading of the voltmeter is 1 V for a displacement of 1 mm from the null position. When the displacement is 1 mm in the opposite direction from the null position, the reading of the voltmeter is

(a) –1 V (b) –0.2 V
(c) 1 V (d) 5 V

[2009 : 1 M]

2.19 Poisson's ratio for a metal is 0.35. Neglecting piezo-resistance effect, the gage factor of a strain gage made of this metal is

(a) 0.65 (b) 1
(c) 1.35 (d) 1.70

[2010 : 1 M]

2.20 The primary and secondary of an LVDT (stroke length ±50 mm) are connected to 3 kHz sinusoidal source and ideal semiconductor diode bridge-based phase sensitive demodulator circuit. The core of the LVDT remains static at 15 mm above the ideal null position. The frequency of the voltage observed at the input of the low-pass filter is

(a) 1 kHz (b) 1.5 kHz
(c) 3 kHz (d) 6 kHz

[2011 : 1 M]

2.21 A strain gauge is attached on a cantilever beam as shown. If the base of the cantilever vibrates according to the equation $x(t) = \sin \omega_1 t + \sin \omega_2 t$, where 2 rad/s < ω_1, ω_2 < 3 rad/s, then the output of the strain gauge is proportional to

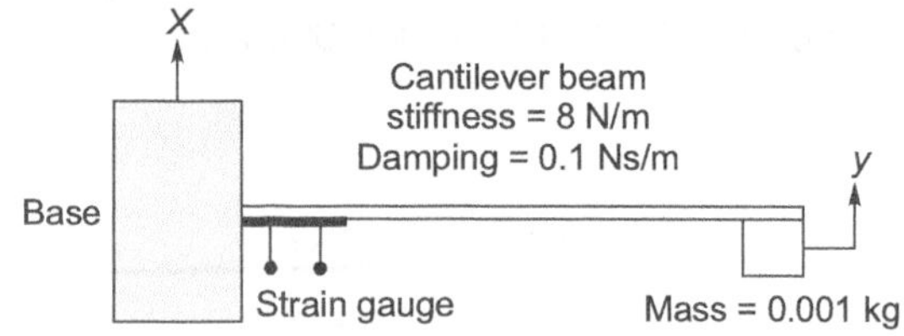

(a) x (b) $\frac{dx}{dt}$
(c) $\frac{d^2x}{dt^2}$ (d) $\frac{d(x-y)}{dt}$

[2012 : 1 M]

Linked Answer Questions (2.22 and 2.23):

A differential amplifier with signal terminals *X*, *Y*, *Z* is connected as shown in figure (a) below for CMRR measurement where the differential amplifier has an additional constant offset voltage in the output. The observations obtained are: when $V_i = 2$ V, $V_0 = 3$ mV, and when $V_i = 3$ V, $V_0 = 4$ mV.

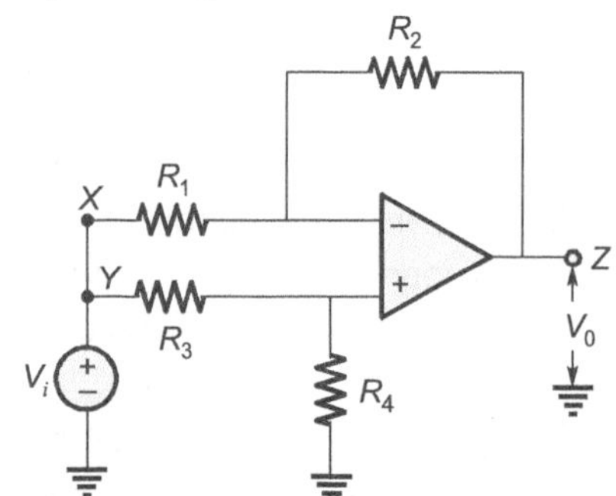

Fig. (a)

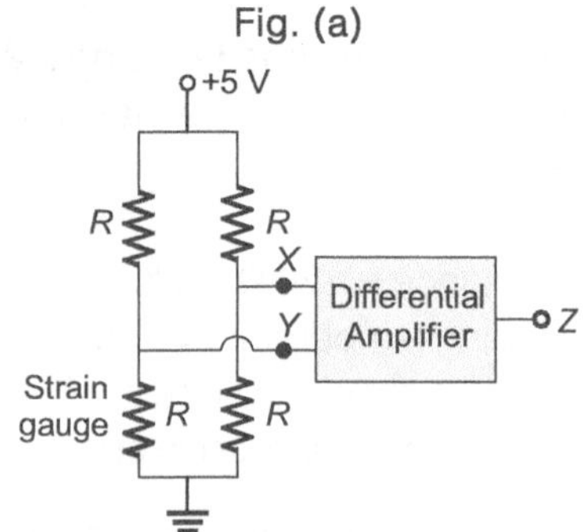

Fig. (b)

2.22 Assuming its differential gain to be 10 and the op-amp to be otherwise ideal, the CMRR is

(a) 10^2 (b) 10^3
(c) 10^4 (d) 10^5

[2013: 2 M]

2.23 The differential amplifier is connected as shown in figure (b) above to a single strain gage bridge. Let the strain gage resistance vary around its no-load resistance R by $\pm 1\%$. Assume the input impedance of the amplifier to be high compared to the equivalent source resistance of the bridge, and the common mode characteristic to be as obtained above. The output voltage in mV varies approximately from

(a) +128 to −128 (b) +128 to −122
(c) +122 to −122 (d) +99 to −101

[2013: 2 M]

2.24 Semiconductor strain gauges typically have much higher gauge factors than those of metallic strain gauges, primarily due to

(a) higher temperature sensitivity
(b) higher Poisson's ratio
(c) higher piezoresistive coefficient
(d) higher magnetostrictive coefficient

[2014 : 1 M]

2.25 A thermistor has a resistance of 1 kΩ at temperature 298 K and 465 Ω at temperature 316 K. The temperature sensitivity in K^{-1} [i.e. $(1/R)(dR/dT)$, where R is the resistance at the temperature T(in K)], of this thermistor at 316 K is ______.

[2014 : 2 M]

2.26 A p-type semiconductor strain gauge has a nominal resistance of 1000 Ω and a gauge factor of +200 at 25°C. The resistance of the strain gauge in ohms when subjected to a strain of $+10^{-4}$ m/m at the same temperature is ____ Ω. [2015 : 1 M]

2.27 In the strain gauge bridge circuit given below, $R_1 = R_3 = R(1-x)$ and $R_2 = R_4 = R(1+x)$, where R is 350 Ω. The voltage sources v_s and v_n represent the dc excitation and the undesired noise/interference, respectively. The value of capacitor C in microfarad that is required to ensure that the output across a and b is low-pass filtered with a cutoff frequency of 150 Hz is______.

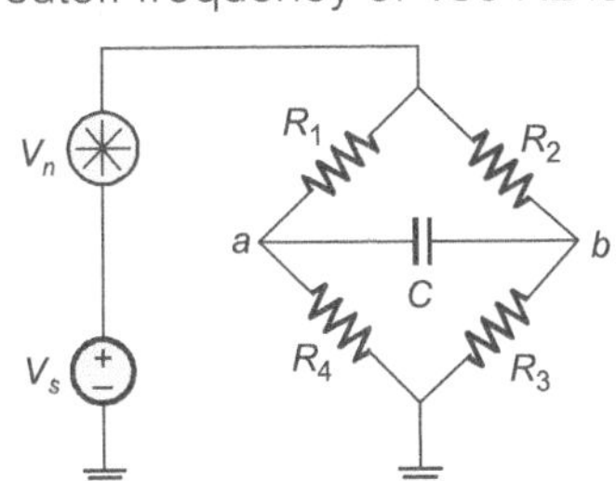

[2016 : 2 M]

2.28 In the circuit below, the opamp is ideal and the sensor is an RTD whose resistance $R_\theta = 1000(1 + 0.004\theta)$ Ω, where θ is temperature in °C. The output sensitivity in mV/°C is______.

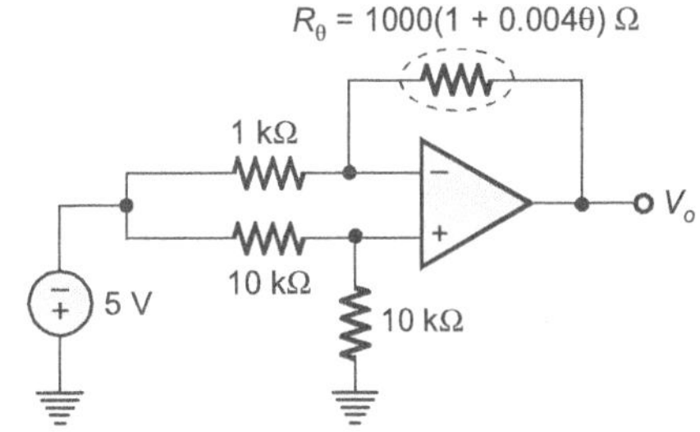

[2016 : 2 M]

2.29 A resistance temperature detector (RTD) is connected to a circuit, as shown in the figure. Assume the opamp to be ideal. If V_0 = +2.0 V, then the value of x is ________.

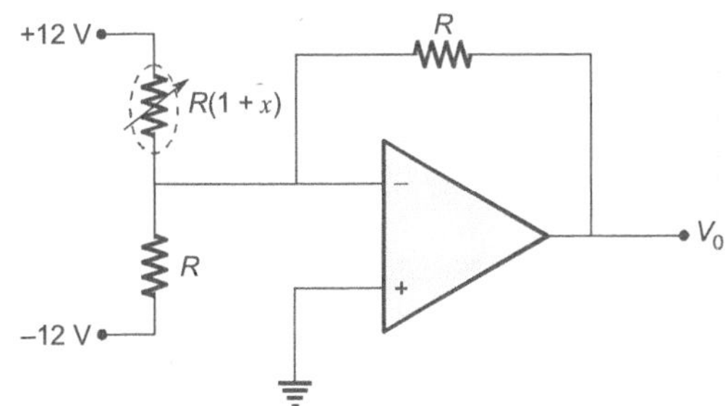

[2017 : 2 M]

2.30 The unbalanced voltage of the Wheatstone bridge, shown in the figure is measured using a digital voltmeter having infinite input impedance and a resolution of 0.1 mV. If R = 1000 Ω, then the minimum value of ΔR in Ω to create a detectable unbalanced voltage is _______.

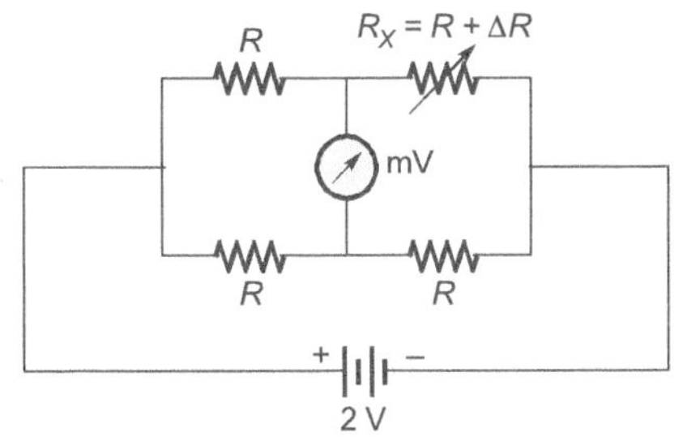

[2017 : 2 M]

2.31 The junction semiconductor temperature sensor shown in the figure is used to measure the temperature of hot air. The output voltage V_0 is 2.1 V. The current output of the sensor is given by $I = T$ μA where T is the temperature in K. Assuming the opamp to be ideal, the temperature of the hot air in °C is approximately _______.

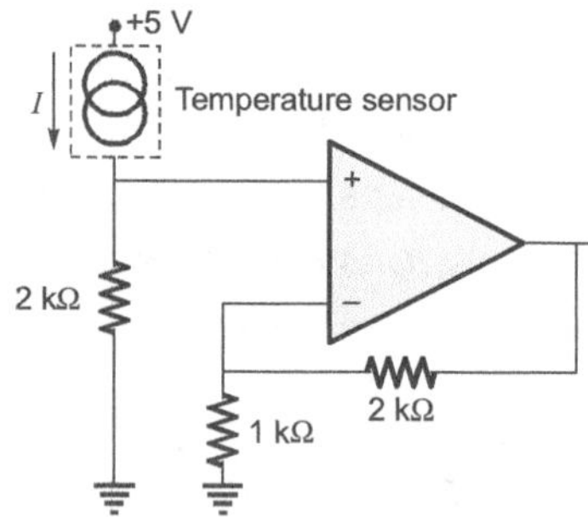

[2017 : 2 M]

2.32 A 1000 Ω strain gauge (R_g) has a gauge factor of 2.5. It is connected in the bridge as shown in figure. The strain gauge is subjected with a positive strain of 400 μm/m. The output V_0 (in mV) of the bridge is (up to two decimal places) _____.

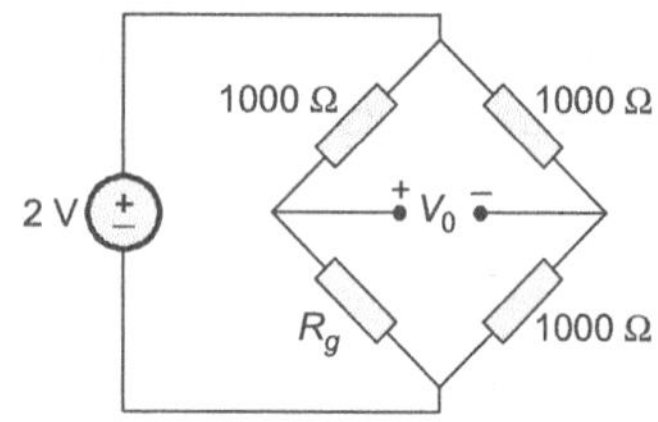

[2018 : 2 M]

2.33 Four identical resistive strain gauges with gauge factor of 2.0 are used in a Wheatstone bridge as shown in the figure below. Only one of the strain gauges R_{SENSE} changes its resistance due to strain. If the output voltage V_{out} is measured to be 1 mV, the magnitude of strain, in units of microstrain is

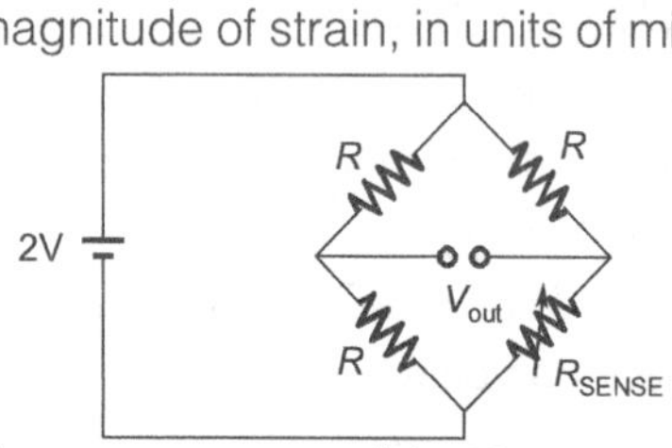

(a) 1
(b) 10
(c) 100
(d) 1000

[2019 : 2 M]

2.34 Four strain gauges in a Wheatstone bridge configuration are connected to an instrumentation amplifier as shown in the figure. From the choices given below, the preferred value for the common mode rejection ratio (CMRR) of the amplifier, in dB, would be

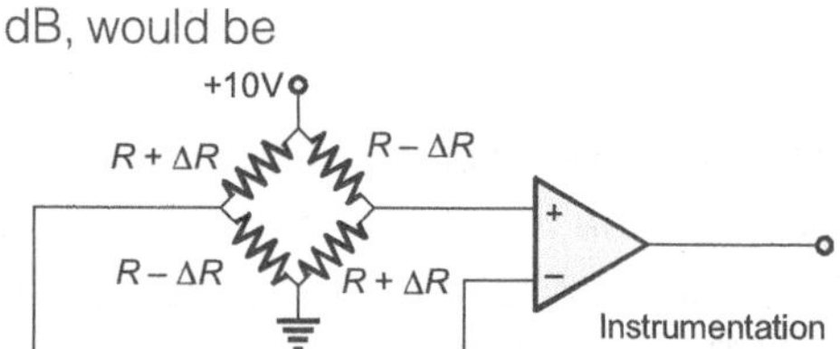

(a) −20
(b) 0
(c) 3
(d) 100

[2019 : 1 M]

2.35 A metallic strain gauge of resistance R_x with a gauge factor of 2 is bonded to a structure made of a metal with modulus of elasticity of 200 GN/m². The value of R_x is 1 kΩ when no stress is applied. R_x is a part of a quarter bridge with three identical fixed resistors of 1 kΩ each. The bridge is excited from a DC voltage of 4 V. The structure is subjected to a stress of 100 MN/m². Magnitude of the output of the bridge (in mV, rounded off to two decimal places) is _____.

[2020 : 2 M]

2.36 The mutual inductances between the primary coil and the secondary coils of a linear variable differential transformer (LVDT) shown in the figure are M_1 and M_2. Assume that the self-inductances L_{S1} and L_{S2} remain constant and are independent of x. When $x = 0$, $M_1 = M_2 = M_0$. When x is in the range ± 10 mm, M_1 and M_2 change linearly with x. At x = +10 mm or −10 mm, the change in the magnitude of M_1 and M_2 is 0.25 M_0. For a particular displacement $x = D$, the voltage across the detector becomes zero when $|V_2| = 1.25\,|V_1|$. The value of D (in mm, rounded off to one decimal place) is _____.

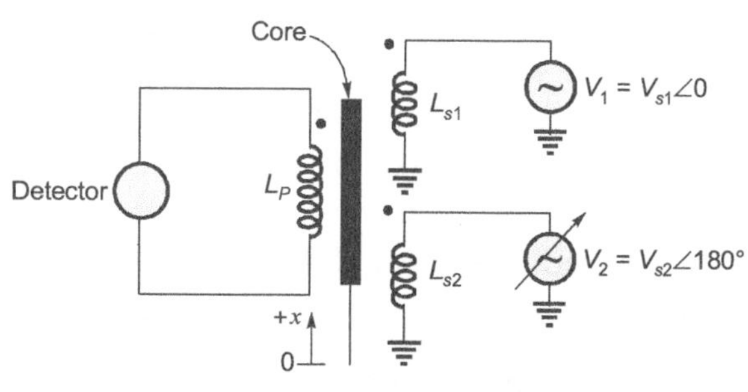

[2020 : 2 M]

2.37 Pair the quantities (p) to (s) with the measuring devices (i) to (iv).

(i) Linear Variable Differential Transformer (LVDT)	(p) Torque
(ii) Thermistor	(q) Pressure
(iii) Strain gauge	(r) Linear position
(iv) Diaphragm	(s) Temperature

(a) (i) - (r), (ii) - (s), (iii) - (q), (iv) - (p)
(b) (i) - (p), (ii) - (s), (iii) - (r), (iv) - (q)
(c) (i) - (r), (ii) - (s), (iii) - (p), (iv) - (q)
(d) (i) - (q), (ii) - (s), (iii) - (p), (iv) - (r)

[2023 : 1 M]

2.38 Four strain gauges R_A, R_B, R_C and R_D, each with nominal resistance R, are connected in a bridge configuration. When a force is applied, R_A and R_D increase by ΔR and R_B and R_C decrease by ΔR as shown. A potentiometer with total resistance R_V is connected as shown. If R = 100 Ω, and ΔR = 1 Ω, the minimum value of resistance R_V required to balance the bridge is ______Ω (rounded off to two decimal places).

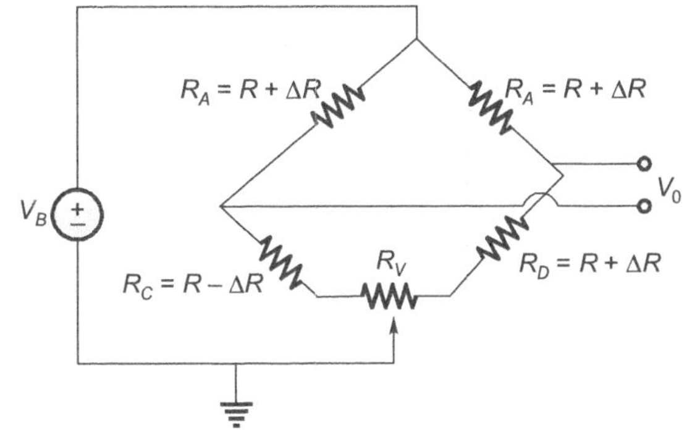

[2023 : 2 M]

Answers Resistive & Inductive Transducers

2.1 (a)	2.2 (d)	2.3 (d)	2.4 (a)	2.5 (d)	2.6 (b)	2.7 (d)
2.8 (d)	2.9 (d)	2.10 (a)	2.11 (d)	2.12 (d)	2.13 (c)	2.14 (c)
2.15 (a)	2.16 (b)	2.17 (a)	2.18 (c)	2.19 (d)	2.20 (c)	2.21 (c)
2.22 (c)	2.23 (b)	2.24 (c)	2.25 (–0.040)	2.26 (1020)	2.27 (3.03)	2.28 (10)
2.29 (0.2)	2.30 (0.2)	2.31 (77)	2.32 (0.5)	2.33 (d)	2.34 (d)	2.35 (1)
2.36 (10)	2.37 (c)	2.38 (–3.96)				

Explanations Resistive & Inductive Transducers

2.1 (a)

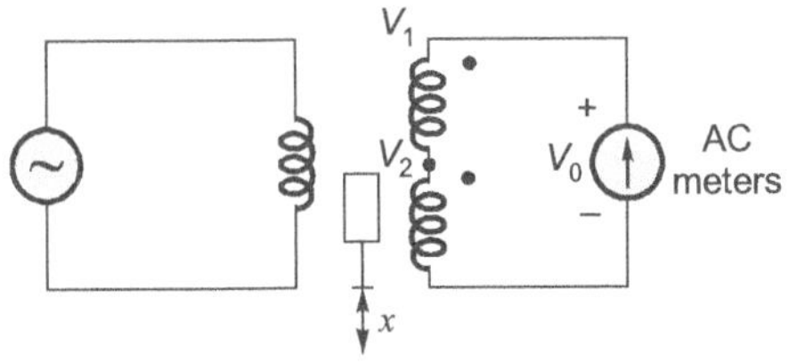

Since, secondary coils are arranged is series addition. So, V_0 = constant.

2.2 (d)

Semiconductor strain gauge have very high gauge factor and for

p-type ⇒ Gauge factor is positive.

n-type ⇒ Gauge factor is negative.

2.3 (d)

There is abrupt increase in resistance of transducer after a certain temperature T. This is an attribute of self heating effect which is seen in thermistors. Also resistance of transducer increases with temperature so it has positive temperature coefficient.

Note: Self heating effect is not present in RTD.

2.5 (d)

LVDT has one primary coils connected in phase opposition with two secondary coil.

2.6 (b)

$$R = 100\ \Omega$$
$$\Delta R = 0.35\ \Omega$$
$$L = 20\text{ cm}$$
$$\Delta L = 0.2\text{ mm} = 0.02\text{ cm}$$

$$\text{G.F.} = \frac{\Delta R/R}{\Delta L/L} = \frac{\frac{0.35}{100}}{\frac{0.02}{20}} = 3.5$$

2.7 (d)

$$\text{Gauge factor} = 1 + 2r + \frac{\Delta\rho/\rho}{\varepsilon}$$

Since conductivity (resistivity) of metal is independent of applied strain. So change in conductivity (change in resistivity) is zero.

So, for metals,

Gauge factor = $1 + 2r \approx 2.5$

2.8 (d)

True value at 100°C,

$$R_{100°C} = R_0(1 + 0.004\,T + 6 \times 10^{-7}\,T^2)$$
$$= 100(1+0.004\times100+6\times10^{-7}\times100^2)$$
$$= 140.6\ \Omega$$

Value of $R_0 = 100 \pm 2\Omega = 98, 102\ \Omega$

$$R_{Ia} = R_0(1 + 0.004\text{ T})$$
$$= 98\,(1 + 0.004 \times 100) = 137.2\ \Omega$$
$$R_{Ib} = R_0(1 + 0.004\text{ T})$$
$$= 102(1 + 0.004 \times 100) = 142.8$$

$\therefore$ Worst error = (140.6 – 137.2) Ω = 3.4 Ω

2.9 (d)

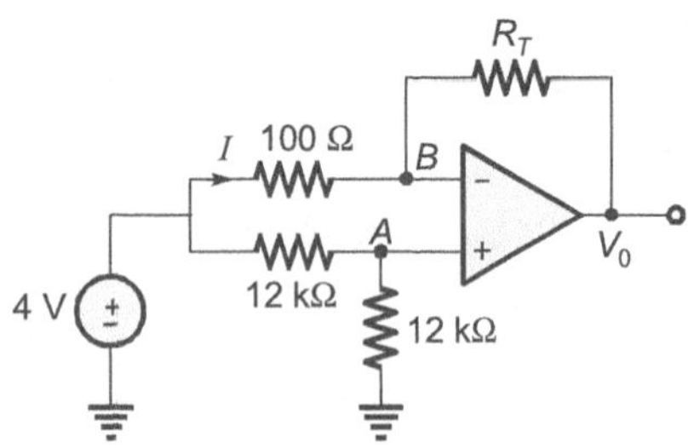

From figure,

$$V_A = \frac{4\times12}{12+12} = 2\text{ V} \quad \text{...(i)}$$

Also using virtual ground,

$$V_A = V_B \quad \text{...(ii)}$$
$$\Rightarrow \quad V_B = 2\text{ V}$$

Given,

$$R_T = R(1 - kx)$$

for, $k = 0.004$, $x = 75$

$$R_T = 100\,(1 - 0.004 \times 75) = 70\ \Omega$$

From network,

$$I = \frac{4-V_B}{100} = \frac{4-2}{100} = 20\text{ mV}$$

Also,

$$\frac{V_B - V_0}{R_T} = I$$
$$\Rightarrow \quad \frac{2-V_0}{70} = 20 \times 10^{-3}$$
$$\Rightarrow \quad V_0 = 2 - 1.4 = 0.6\text{ V} = 600\text{ mV}$$

2.10 (a)

For accurate filtering and demodulation of the signal a frequency ratio of 10 : 1 is followed:

$\therefore$ Maximum frequency of displacement should be limited to

$$= \frac{2\text{kHz}}{10} = \frac{2000}{10} = 200\text{ Hz}$$

2.11 (d)

Net equivalent resistance

$$= 120\ \Omega \parallel 200\text{ k}\Omega$$
$$= 119.92\ \Omega$$

Change in resistance,

$$\Delta R = -0.08$$
$$\because \quad \frac{\Delta R}{R} = G_F \in$$

Change in per unit resistance for strain gauge

$$\in = \frac{\Delta R}{R\,G_F} = -\frac{0.08}{120\times2} = -300\ \mu\text{m/m}$$

2.12 (d)

Due to temperature, the resistance of the gauge changes to

$$R_T = R_0\,(1 + \alpha\,\Delta t)$$
$$= 120\,(1 + 2 \times 10^{-5} \times 150)$$
$$= 120\,(1 + 0.003) = 120.36$$

Change in resistance due to temperature variation is

$$(\Delta R)_T = R_T - R_0$$
$$= 120.36 - 120 = 0.36$$
$$\therefore \quad \frac{(\Delta R)_T}{R} = \frac{0.36}{120} \quad \text{...(i)}$$

Resistance error due to the tensile strain only is

$$\frac{\Delta R}{R} = G\in \quad \text{...(ii)}$$

$$\text{Percentage resistance change} = \frac{0.36/120}{G\times\in}\times100$$
$$= \frac{0.36}{2\times2\times10^{-2}\times120}\times100$$
$$= 0.075 \times 100 = 7.5$$

2.13 (c)

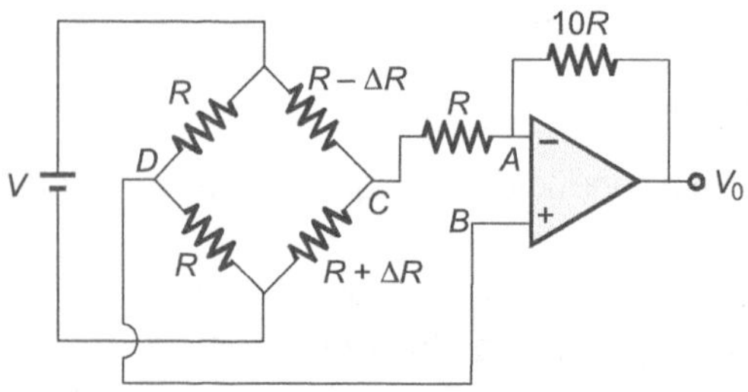

For maximum loading, $\varepsilon = 2 \times 10^{-2}$

$\Rightarrow \Delta R = GF\cdot\varepsilon\cdot R = 2 \times (2 \times 10^{-2}) \times 120 = 4.8\ \Omega$

For push-pull configuration are strain gauge resistance decreases by ΔR and other resistance increases by ΔR.

So, $\quad R + \Delta R = 120 + 4.8$

$$= 124.8\ \Omega$$

and $R - \Delta R = 120 - 4.8 = 115.2\ \Omega$

From network,

$$V_D = \frac{V \cdot R}{R + R} = \frac{V}{2}$$

Also, $V_D = V_B$

and $V_B = V_A$ (using virtual ground)

$$\Rightarrow \quad V_A = \frac{V}{2} \qquad \text{...(i)}$$

From network,

$$V_C = \frac{V(R + \Delta R)}{(R + \Delta R) + (R - \Delta R)}$$

$$= \frac{V(R + \Delta R)}{2R} = 0.52 \qquad \text{...(ii)}$$

Also, $$I = \frac{V_C - V_A}{R} = \frac{0.52\text{ V} - 0.5\text{ V}}{120}$$ (using (i) and (ii))

$$\Rightarrow \quad I = \frac{V}{6000}\text{ A}$$

From network,

$$V_0 = V_A - (10R)I$$

$$\Rightarrow \quad V_0 = \frac{V}{2} - (10 \times 120)\frac{V}{6000}$$

$$= \frac{V}{2} - \frac{V}{5} = 0.3\text{ V}$$

2.14 (c)

The null voltage of the above shown LVDT is given by

$$\text{Null voltage} = \bar{V}_1 - \bar{V}_2$$

$$= 1.0\angle 0° - 1.0\angle 10°$$

$$= 1(\cos 0° + i\sin 0°) - 10(\cos 10° + i\sin 10°)$$

$$= 1(1 + i0) - 1(0.9848 + i\,0.1736)$$

$$= 0.0152 - i\,0.1736$$

Magnitude of null voltage

$$= \sqrt{(0.0152)^2 + (0.1736)^2}$$

$$= \sqrt{0.03036} = 0.1742$$

2.15 (a)

LVDT is a displacement transducer.

2.16 (b)

Clearly it is the case of a quarter bridge,

G.F = 2.5

$\varepsilon = 1\ \mu\text{m/m} = 10^{-6}$ m/m

$E = 4$ V

$$V_0 = \frac{E}{4} G_f \cdot \varepsilon \quad \text{(for quarter bridge)}$$

$$= \frac{4}{4} \times 2.5 \times 10^{-6} = 250\ \mu\text{V}$$

2.17 (a)

The gauge 1 and 3 must be placed in the opposite arms and so is the case of gauges 2 and 4. So, the option (a) is correct.

2.18 (c)

From the output characteristics of LVDT as shown below, the output voltage is 1 V.

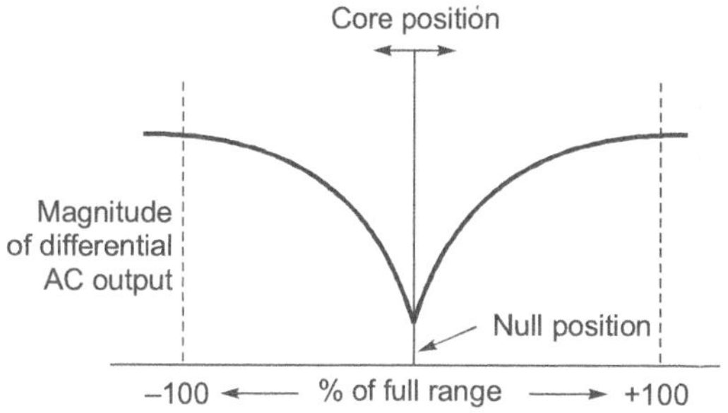

Since output of LVDT is sinusoidal, so meter used to measure output voltage is RMS measuring meter and RMS value is always greater than zero.

$$E_0 = |E_1 - E_2| = 1\text{ V}$$

2.19 (d)

G.F = 1 + 2ν, neglecting the piezo-resistive effect

= 1 + 2 × 0.35

= 1 + 0.70 = 1.70

2.20 (c)

Phase sensitive demodulator circuit do not change signal frequency.

So frequency of output voltage = frequency of input voltage = 3 kHz.

2.21 (c)

Spring stiffness (k) = 8 N/m.

Mass (m) = 0.001 kg

So, the Natural frequency $(\omega_n) = \sqrt{\frac{k}{m}}$

$$= \sqrt{\frac{8}{0.001}} = \sqrt{8000}$$

= 89.442 rad/sec

and given that

$$x(t) = \sin\omega_1 t + \sin\omega_2 t$$

where $\omega_1 > 2$ rad/sec and $\omega_2 < 3$ rad/sec

For every $\omega_1 > 2$ rad/sec and $\omega_2 < 3$ rad/sec; then frequency of $x(t)$ will be less than natural frequency.

So, the pickup is used to Measure "Acceleration".

output of strain gauge is directly proportional to $\frac{d^2x}{dt^2}$.

2.22 (c)

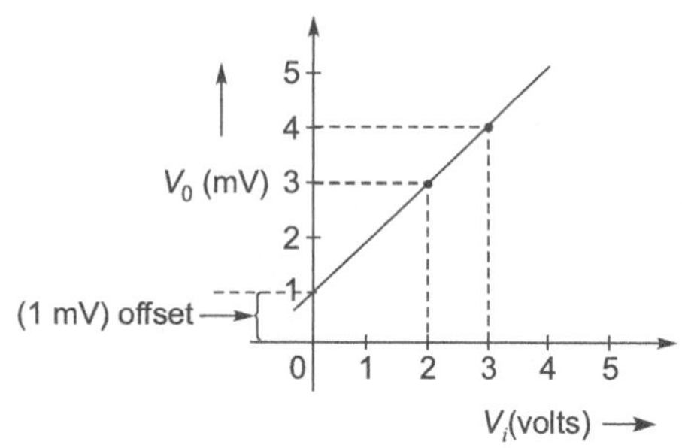

$\therefore$ $A_c = 10^{-3}$, $A_d = 10$ (given),

Offset voltage = 1 mV

$$\text{CMRR} = \frac{A_d}{A_c} = \frac{10}{10^{-3}} = 10^4$$

2.23 (b)

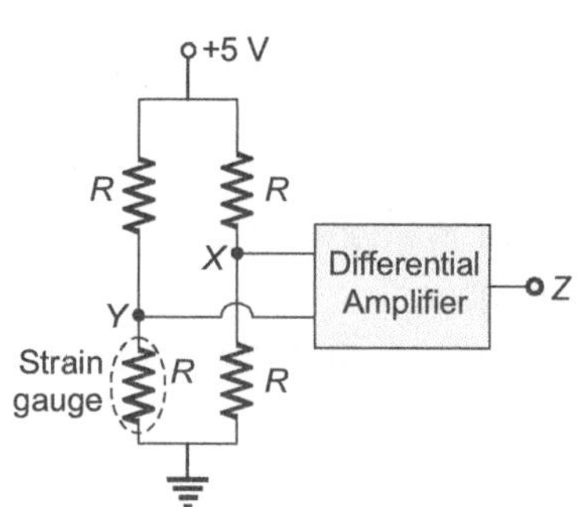

For variation in R as +1%

$$V_x = \frac{R}{R+R}\times 5 = \frac{5}{2}$$

$$V_y = \frac{1.01R}{R+1.01R}\times 5 = \left(\frac{1.01}{2.01}\right)\times 5$$

$$V_z = A_dV_d + A_cV_c + \text{offset voltage}$$

$$= 10(V_y - V_x) + 10^{-3}\frac{(V_y+V_x)}{2} + 1 \text{ mV}$$

$$= +128 \text{ mV}$$

For variation in R as –1%

$$V_x = \frac{R}{R+R}\times 5 = \left(\frac{1}{2}\right)5$$

$$V_y = \frac{0.99R}{R+0.99R} = \left(\frac{0.99}{1.99}\right)5$$

$$V_z = A_dV_d + A_cV_c + \text{offset voltage}$$

$$= -122 \text{ mV}$$

2.24 (c)

S.C. strain gauges are used when a very high gauge factor is required. They have a gauge factor 50 times as wire strain gauge.

S.C. strain gauges depend for their action upon the piezoresistive effect; change in value of resistance due to change in resistivity.

2.25 –0.040 (–0.042 to –0.038)

For thermistor resistance R at temperature T is,

$$R = R_0 \exp\left[\beta\left(\frac{1}{T} - \frac{1}{T_0}\right)\right] \quad \text{...(i)}$$

we need to find $\frac{1}{R}\frac{dR}{dT}$,

Differentiating (i) w.r.t. T,

$$\frac{dR}{dT} = R_0 \exp\left[\frac{1}{T} - \frac{1}{T_0}\right] \cdot \frac{\beta}{-T^2}$$

$$\Rightarrow \quad \frac{dR}{dT} = R \cdot \frac{\beta}{-T^2}$$

$$\Rightarrow \quad \frac{1}{R}\frac{dR}{dt} = \frac{-\beta}{T^2} \quad \text{...(ii)}$$

Given that, $R_{298} = 1000\ \Omega$ and $R_{316} = 465\ \Omega$

$$\Rightarrow \quad 465 = 1000\exp\left[\beta\left[\frac{1}{316} - \frac{1}{298}\right]\right]$$

$$\Rightarrow \quad \beta = \frac{\ln(465/1000)}{\left[\frac{1}{316} - \frac{1}{298}\right]} = 4005.89 \text{ K} \quad \text{...(iii)}$$

Using (ii) and (iii), temperature sensitivity at 316 K is,

$$\left(\frac{1}{R}\frac{dR}{dT}\right)_{T=316\text{k}} = \frac{-4005.89}{(316)^2} = -0.04$$

Note: You can observe huge change in resistance of thermistor when temperature changes. Also $R_{298\text{ K}} = 1000\ \Omega$ and $R_{316\text{ K}} = 465\ \Omega$ i.e., as temperature increases, resistance decreases i.e., thermister has NTC negative sign in answer (–0.04) shows NTC.

2.26 (1020)

$$\text{G.F.} = \frac{\Delta R/R}{\Delta L/L}$$

$$\therefore \quad +200 = \frac{\Delta R}{1000 \times 10^{-4}}$$

$$\therefore \quad \Delta R = 20$$

$\therefore$ Resistance of gauge = 1000 + 20

= 1020 Ω

2.27 3.03 (2.9 to 3.1)

Let Thevenin equivalent voltage across terminal $ab = V_{Th}$

and Thevenin equivalent resistance across terminal $ab = R_{Th}$

$$\therefore \quad R_{Th} = \frac{R_1R_4}{R_1+R_4} + \frac{R_2R_3}{R_2+R_3}$$

$$= \frac{2R^2(1-x^2)}{[R(1-x)+R(1+x)]}$$

$$= R(1-x^2)$$

as $x << 1 \Rightarrow x^2 << 1$

$$\therefore \quad R_{Th} \approx R$$

Draw equivalent circuit

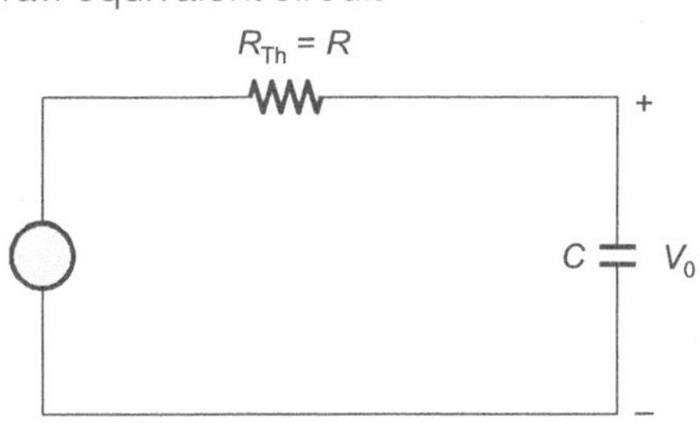

$$\therefore \quad \frac{V_0}{V_{Th}} = \frac{1/CS}{R+1/CS} = \frac{1}{RCS+1} = \frac{1}{RCj\omega+1}$$

According to given condition,

at f = 150 Hz (3-dB cutoff frequency)

$$\left|\frac{V_0}{V_{Th}}\right| = \frac{1}{\sqrt{2}}$$

$$\frac{1}{(RC\omega)^2+1} = \frac{1}{2}$$

$$\therefore \quad f = \frac{1}{2\pi RC}$$

$$C = \frac{1}{2\times\pi\times350\times150} = 3.03\,\mu\text{F}$$

Method 2:

$$f_{cutoff} = \frac{1}{2\pi R_{eq}C} \quad \text{...(i)}$$

where R_{eq} is the Thevenin's resistance of network seen by capacitor.

≡

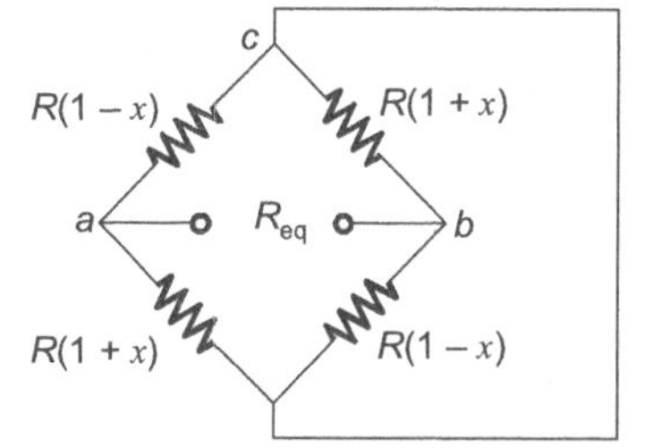

≡

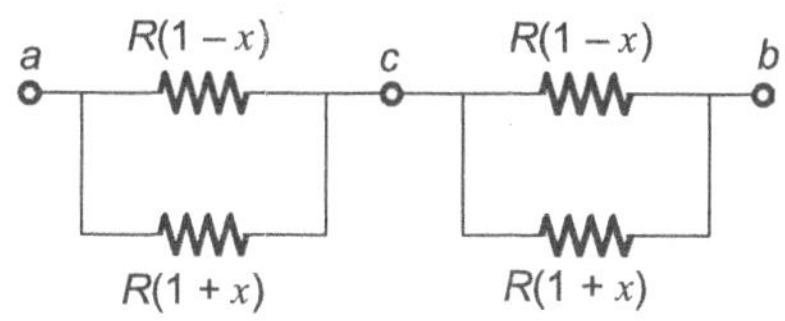

≡

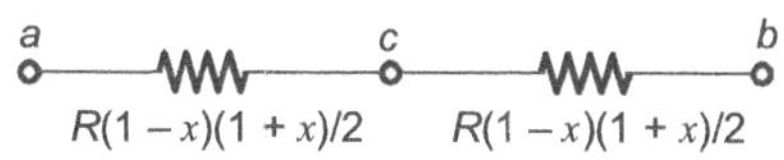

$$\Rightarrow \quad R_{eq} = \frac{R(1-x)(1+x)}{2} + \frac{R(1-x)(1+x)}{2}$$

$$= R(1-x^2)$$

For small variations,

$$x << 1$$

$$\Rightarrow \quad R_{eq} \approx R = 350\ \Omega \quad \text{...(ii)}$$

using (i) and (ii),

$$150 = \frac{1}{2\pi\cdot 350\cdot C}$$

$$\Rightarrow \quad C = 3.03\,\mu\text{F}$$

2.28 (10)

Redraw the circuit,

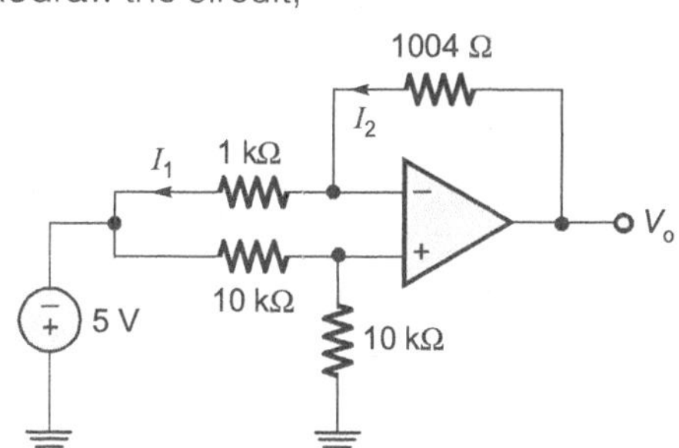

$$R_\theta = 1000\,[1 + 0.004\,\theta]$$

for 1° C rise in temperature

$$R_\theta = 1000\,[1 + 0.004]$$

$$= 1004\ \Omega$$

Voltage at Node A by potential divider

$$V_A = \frac{(-5)10}{10+10} = -2.5\ \text{V}$$

Using virtual ground

$$V_A = V_B = -2.5\text{ V}$$

Apply KCL at Node B

$$I_1 = I_2$$

$$\frac{V_B-(-5)}{1000} = \frac{V_0 - V_B}{1004}$$

$$V_0 = 10\text{ mV}$$

$$\text{Sensitivity} = 10\text{ mV/°C}$$

2.29 (0.2)

Given that:

V_0 =+2 V and opamp is ideal.

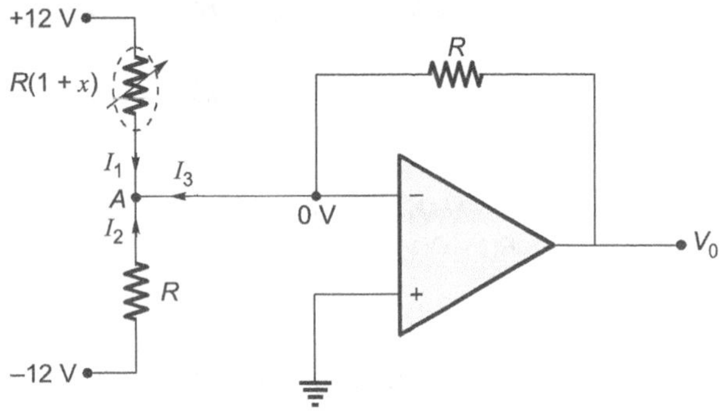

KCL at node A

$$I_1 + I_2 + I_3 = 0$$

$$\frac{12-0}{R(1+x)} + \frac{-12-0}{R} + \frac{V_0-0}{R} = 0$$

$$\frac{12}{1+x} - 12 + 2 = 0$$

$$\frac{12}{1+x} = 10$$

$$\Rightarrow \quad 1 + x = 12$$

$$x = 0.2$$

2.30 (0.2)

Method I

Given bridge is the quarter bridge arrangement.

For quarter bridge arrangement

$$V_0 = \frac{V_S}{4}\frac{\Delta R}{R}$$

$$\Delta R = \frac{V_0 \times 4 \times R}{V_S} = 0.2\ \Omega$$

$$= \frac{0.1\times 10^{-3}\times 4\times 1000}{2}$$

$$= 0.2\ \Omega$$

Method II

Redraw the bridge we have

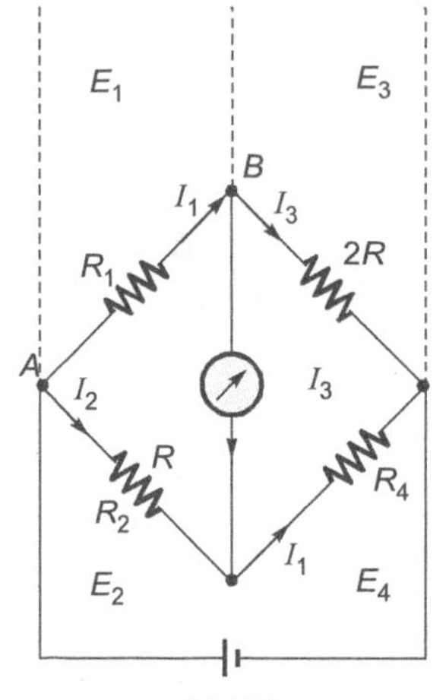

Since the voltmeter is heaving an infinite input impedance $I_D = 0$, and $I_1 = I_3$ and $I_2 = I_4$ the voltmeter ready E_0 is given by

$$E_0 = E_3 - E_4 = I_3R_3 - I_4R_4$$

But,

$$I_1 = I_3 = \frac{E}{R_1 + R_3}$$

$$I_2 = I_4 = \frac{E}{R_2+R_4}$$

i.e.

$$E_0 = \frac{ER_3}{R_1+R_3} - \frac{ER_4}{R_2+R_4}$$

$$= E\left\{\frac{R_3(R_2+R_4) - R_4(R_1+R_3)}{(R_1+R_3)(R_2+R_4)}\right\}$$

$$= E\left\{\frac{R_2R_3 + R_3R_4 - R_1R_4 - R_4R_3}{(R_1+R_3)(R_2+R_4)}\right\}$$

$$= E\left\{\frac{R_2R_3 - R_1R_4}{(R_1+R_3)(R_2+R_4)}\right\}$$

But given, $R_1 = R_2 = R_4 = R$ and

$$R_3 = (R + \Delta R)$$

$\therefore$Voltmeter ray $= E_0 + \Delta E_0$

and as $E_0 = 0$

$$= E\left\{\frac{R(R+\Delta R) - R^2}{(R_1 + R + \Delta R)2R}\right\}$$

$$E_0 + \Delta E_0 = E\left\{\frac{R^2 + R\Delta R - R^2}{(2R+\Delta R)2R}\right\}$$

$$\Delta E_0 = E\left\{\frac{\Delta R}{4R + 2\Delta R}\right\} = E\left\{\frac{\Delta R/R}{4 + 2\frac{\Delta R}{R}}\right\}$$

$$\because \quad 4 >>> 2\frac{\Delta R}{R}$$

$$\therefore \quad \Delta E_0 = \frac{E}{4}\cdot\frac{\Delta R}{R}$$

Here, $\Delta E_0 = 0.1 \times 10^{-3}$ V

$E = 2$ V

and $R = 1000\ \Omega$

$$\therefore \quad 0.1 \times 10^{-3} = \frac{2}{4}\times\frac{\Delta R}{1000}$$

$$\Delta R = \frac{4\times 0.1\times 10^{-3}\times 1000}{2} = 0.2\ \Omega$$

2.31 (77)

Given that the constant output of sensor $I = T\,\mu$A (T is kelvin).

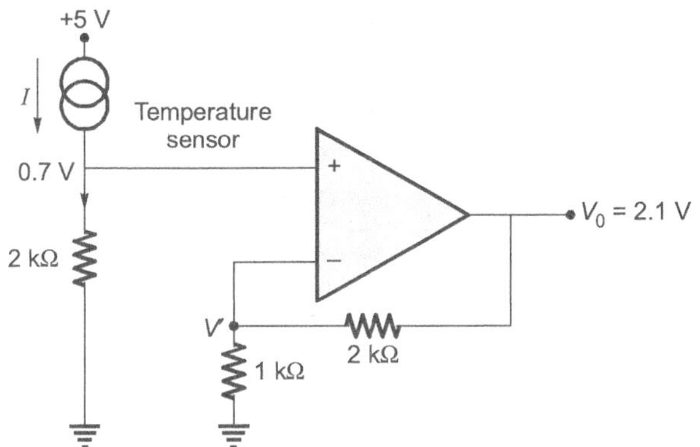

$$V' = 2.1\times\frac{1\text{k}\Omega}{1\text{k}\Omega + 2\text{k}\Omega} = 2.1\times\frac{1}{3} = 0.7 \text{ V}$$

And from ideal "Op-amp" Configuration

$$V' = V_- = V_+ = 0.7 \text{ V}$$

And the voltage across the 2 kΩ resistor (follow the sensor) is

$$0.7 \text{ V} = I \times 2 \times 10^3$$

$$\frac{0.7\,V}{2\times 10^{-3}} = I$$

$$I = 0.35 \text{ mA}$$

$$\Rightarrow \quad T \times 10^{-6} = 0.35 \times 10^{-3}$$

$$T = 350 \text{ kelvin}$$

$$T = 350 - 273$$

$$= 77°\text{C}$$

$I = \frac{5-0}{2\text{ k}\Omega} = 2.5$ mA is wrong as voltage drop across the sensor is not zero.

2.32 0.5 (0.48 to 0.51)

From the given "Quarter Bridge"

Output voltage of the bridge is

$$V_0 = \frac{V_s}{4R_b}\cdot\Delta R \qquad [\text{where } V_s = 2 \text{ V}]$$

$$V_o = \frac{2}{4R_b}\cdot R_o G_f \cdot \epsilon \quad [\Delta R = R\cdot G_f \epsilon]$$

$$V_o = \frac{2}{4}\times\frac{1000}{1000}\times 2.5\times 400\times 10^{-6}$$

$$V_o = 0.5 \text{ mV}$$

2.33 (d)

The given resistance bridge is of Quarter bridge. So the output voltage of the bridge is

$$V_{OB} = \frac{V_s}{4}\times\frac{\Delta R}{R_0} \qquad \text{...(1)}$$

$$\text{In strain gauge, } G_f = \frac{\left(\frac{\Delta R}{R_0}\right)}{\epsilon} \qquad \text{...(2)}$$

where, V_s = Supply voltage ΔR = change in resistance, R_o = Initial resistance of the gauge and ϵ = strain,

From eq. (1) and (2),

$$V_{OB} = \frac{V_s}{4}\times G_f\,\epsilon$$

$$1 \times 10^{-3} = \frac{2}{4}\times 2\times\epsilon$$

$$\epsilon = 1 \times 10^{-3} \text{ units}$$

$$= 1000 \text{ micro strain}$$

So, option (d) is correct answer.

2.35 (1)

Given, The initial resistance of strain gauge (R_x) is

$$R_x = 1 \text{ k}\Omega$$

Young's Modulus of Elasticity (E) = 200×10^9 N/m²

Applied stress = 100×10^6 N/m²

Quarter bridge configuration

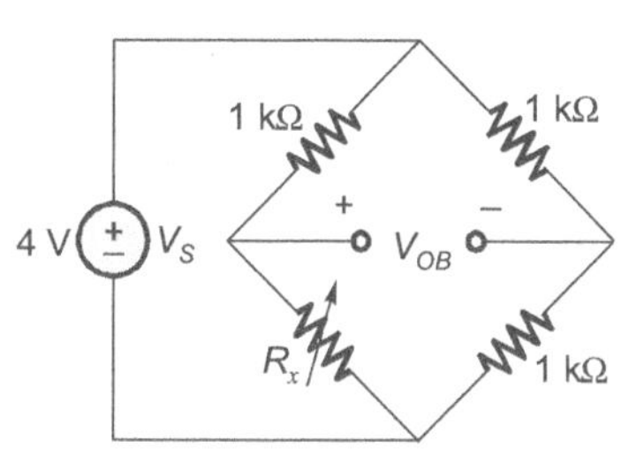

Output voltage of quarter bridge for small strain

$$V_{OB} = \frac{V_S}{4}\times\frac{\Delta R_x}{R_x}$$

$$\Delta R_x = R_x G_f \epsilon$$

(R_x = No load resistance; G_f = Gauge factor; $\in$ = Applied strain)

$$V_{OB} = \frac{V_S}{4} \times G_f \in$$
$$= \frac{4}{4} \times 2 \times \left[\frac{\text{Stress}}{\text{Youngs Modulus}}\right]$$
$$= 2 \times \frac{100 \times 10^6}{200 \times 10^9} = 1 \times 10^{-3} \text{ V}$$
$$= 1 \text{ mV}$$

2.36 (10)

Given that,

Self-inductances L_{s1} and L_{s2} remain constant and are independent of x.

At $x = 0$, $M_1 = M_2 = M_0$

M_1 and M_2 change linearly with x. This means

$$M_1 = M_0 + k_1 x$$
$$M_2 = M_0 + k_2 x$$

at $x = +10$ mm

$$M_1 = M_0 + k_1 \times 10 \times 10^{-3}$$
$$M_2 = M_0 + k_2 \times 10 \times 10^{-3}$$
$$M_1 - M_2 = (k_1 - k_2) \times 10 \times 10^{-3} = 0.25\ M_0$$
$$(k_1 - k_2) = 25\ M_0 \quad \text{...(i)}$$

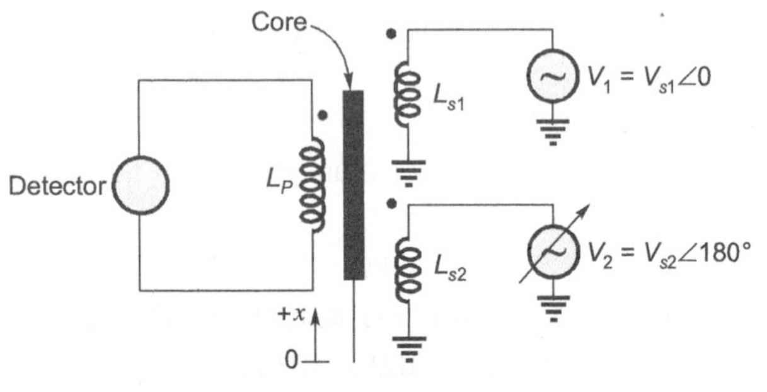

Voltage across the detector (V_D) = Voltage developed across the primary coil (I_P)

$$V_D = V_{LP}$$

Voltage developed across the primary coil depends on the primary current and current in the secondary coils.

So, $V_D = V_{LP}$

$$= L_P \frac{di_P}{dt} + M_1 \frac{di_{s_1}}{dt} - M_2 \frac{di_{s_2}}{dt} \quad \text{...(ii)}$$

($\because$ i_{s1} and i_{s2} are out of phase)

According to the given information detector output voltage (V_D) = 0 at $x = D$ and there can't be any primary current (i_p)

So, equation (ii) can be written as

$$M_1 \frac{di_{s_1}}{dt} - M_2 \frac{di_{s_2}}{dt} = 0 \quad \text{...(iii)}$$

From the diagram, voltages across secondary coils can be written as

$$V_1 = L_{s_1} \frac{di_{s_1}}{dt}$$
$$V_2 = L_{s_2} \frac{di_{s_2}}{dt}$$

as L_{s_1}, L_{s_2} are not changing and initial voltages across both should be zero.

So, $L_{s_1} = L_{s_2} = L_s$

$$V_1 = L_s \frac{di_{s_1}}{dt}$$

and $$V_2 = L_s \frac{di_{s_2}}{dt} \quad \text{... (iv)}$$

Consider equation (iii) xL_s

Then, $$M_1 L_s \frac{di_{s_1}}{dt} - M_2 L_s \frac{di_{s_2}}{dt} = 0$$

$$M_1 V_1 - M_2 V_2 = 0 \quad \text{...(v)}$$

at $x = D$

$$M_1 = M_0 + k_1 D$$
$$M_2 = M_0 + k_2 D$$
$$V_2 = 1.25\ V_1$$

From equation (v)

$$(M_0 + k_1 D)V_1 - (M_0 + k_2 D) \times 1.25 V_1 = 0$$
$$(k_1 - k_2)D = 0.25\ M_0$$
$$D = \frac{0.25 M_0}{k_1 - k_2} = \frac{0.25 M_0}{25 M_0}$$
$$= 10 \times 10^{-3} \text{ (From eq.(i))}$$
$$D = 10 \text{ mm}$$

2.38 (-3.96)

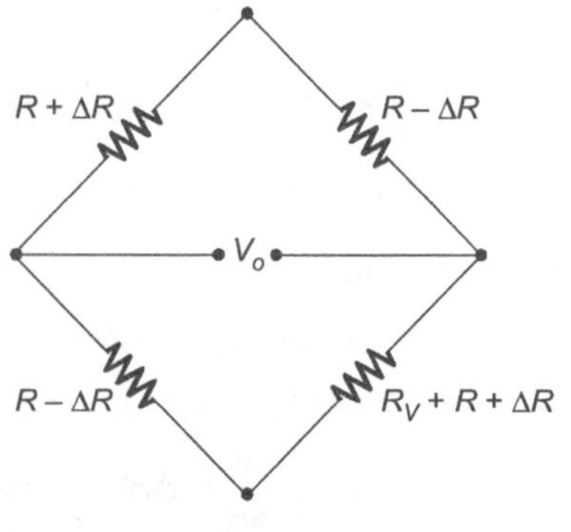

Given that R = 100 Ω, ΔR = 1 Ω

$$V_o = E\left[\frac{R+\Delta R}{R+\Delta R+R-\Delta R} - \frac{R-\Delta R}{R-\Delta R+(R_V+R+\Delta R)}\right]$$

$$V_o = E\left[\frac{R+\Delta R}{2R} - \frac{R-\Delta R}{2R+R_V}\right]$$

For a balanced bridge ⇒ $V_b = 0$

$$0 = E\left[\frac{R+\Delta R}{2R} - \frac{R-\Delta R}{2R+R_V}\right]$$

$$\frac{R+\Delta R}{2R} = \frac{R-\Delta R}{2R+R_V} \Rightarrow 2R + R_V$$

$$= 2R\left[\frac{R-\Delta R}{R+\Delta R}\right]$$

$$R_V = 2R\left[\frac{R-\Delta R}{R+\Delta R} - 1\right] = 2R\left[\frac{R-\Delta R-R-\Delta R}{R+\Delta R}\right]$$

$$= 2R\left[-\frac{2\Delta R}{R+\Delta R}\right] = \frac{-4R\Delta R}{R\left[1+\frac{\Delta R}{R}\right]} = -\frac{4\Delta R}{1+\frac{\Delta R}{R}}$$

$$\left(\frac{\Delta R}{R} \text{ neglected}\right)$$

$= -4\Delta R \Rightarrow$ minimum value

$$R_V = \frac{-4\times 1}{1+\frac{1}{100}} = \frac{-4}{1.01} = -3.96\ \Omega$$

■■■

3 Capacitive and Piezoelectric Transducers

3.1 A piezoelectric transducer is directly connected through a cable to an electronic voltmeter. The minimum operating frequency of measurement is 1000 Hz. If the connecting cable length is doubled, the new minimum operating frequency is
(a) 500 Hz
(b) 1000 Hz
(c) 2000 Hz
(d) 4000 Hz

[1996 : 2 M]

3.2 A piezoelectric type 50 MPa full-scale pressure sensor with built-in electronics has a sensitivity of 100 mV per MPa. If this sensor is subjected to a static pressure of 10 MPa then its output will be
(a) 0 mV (b) 100 mV
(c) $1000/\sqrt{2}$ mV (d) 1000 mV

[2002 : 1 M]

3.3 A quartz crystal (Young's modulus, $E = 9 \times 10^{10}$ N/m²) with piezoelectric properties has a diameter of 10 mm and thickness of 2 mm. Its voltage sensitivity constant is 4500 V/μm. If the voltage output is 127.3 V, the applied load is approximately.
(a) 100 N (b) 200 N
(c) 127.3 N (d) 6.4 N

[2002 : 2 M]

3.4 The expression for the capacitance (*C* in pF) of a parallel plate capacitor is given by : $C = 6.94 \times 10^{-3} (d^2/S)$. The diameter (*d*) of each plate is 20 mm and the spacing between the plates (S) is 0.25 mm. The displacement sensitivity of the capacitor is approximately.
(a) 44.4 pF/mm
(b) –44.4 pF/mm
(c) 11.1 pF/mm
(d) –11.1 pF/mm

[2002 : 2 M]

3.5 A signal conditioning circuit suitable for a push-pull type capacitive transducer is given in figure. The output V_0 is

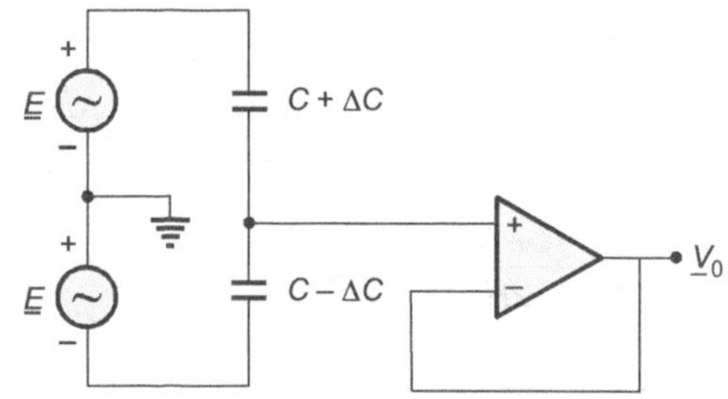

(a) $\underline{V}_0 = \underline{E}\dfrac{C}{\Delta C}$

(b) $\underline{V}_0 = \underline{E}\dfrac{\Delta C}{C}$

(c) $\underline{V}_0 = \underline{E}\dfrac{C - \Delta C}{C + \Delta C}$

(d) $\underline{V}_0 = \underline{E}\dfrac{C + \Delta C}{C - \Delta C}$

[2003 : 2 M]

3.6 A Piezoelectric transducer with a sensitivity of 2.0 pC/N having a capacitance of 1600 pF and a leakage resistance of 10^{12} Ω is connected to a charge amplifier as shown in figure. If a force of 0.1 sin10*t* N is applied to the transducer, the output-amplitude of the charge amplifier is

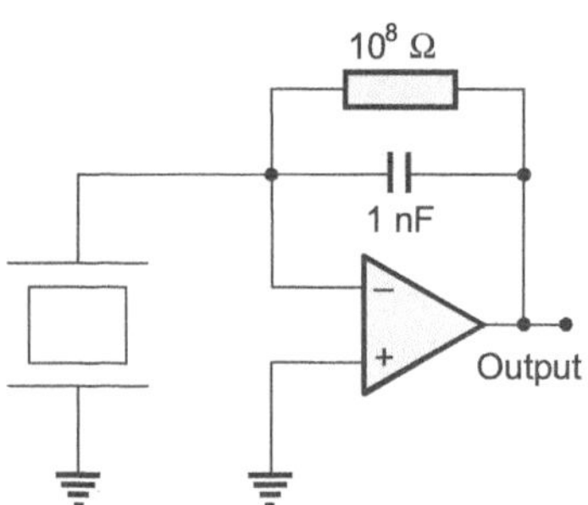

(a) 0.141 V (b) 0.232 mV
(c) 1.414 mV (d) 1.732 mV

[2003 : 2 M]

3.7 To avoid the effect of fringing, the potential of the guard ring of a capacitance transducer, shown in figure is to be made equal to

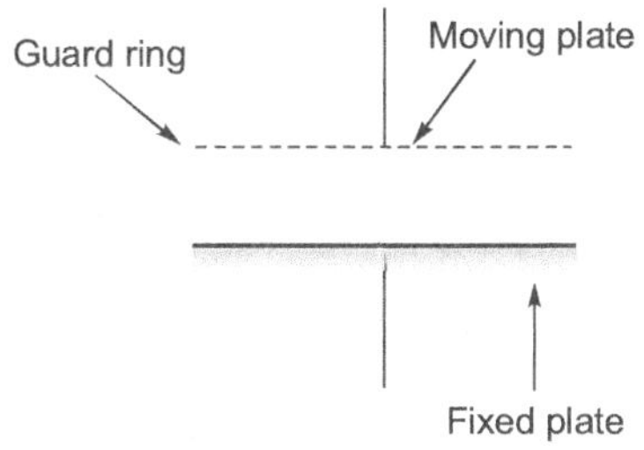

(a) the ground potential
(b) half of the excitation voltage
(c) the voltage of the fixed plate
(d) the voltage of the moving plate

[2004 : 1 M]

3.8 A quartz peizo-electric type pressure sensor has a built-in charge amplifier. The sensor has a sensitivity of 1 μV/Pa. It is subjected to a constant pressure of 120 kPa. The output of the transducer at steady state is

(a) 0 mV (b) 100 μV
(c) 120 μV (d) 120 mV

[2004 : 1 M]

3.9 A variable capacitance angular velocity pickup is made with two concentrically mounted parallel semicircular plates with a small separation between them as shown in figure. A dc voltage is connected across the terminals of the capacitance sensor as shown. For a constant angular velocity ω, the waveform of the current $i(t)$ will be

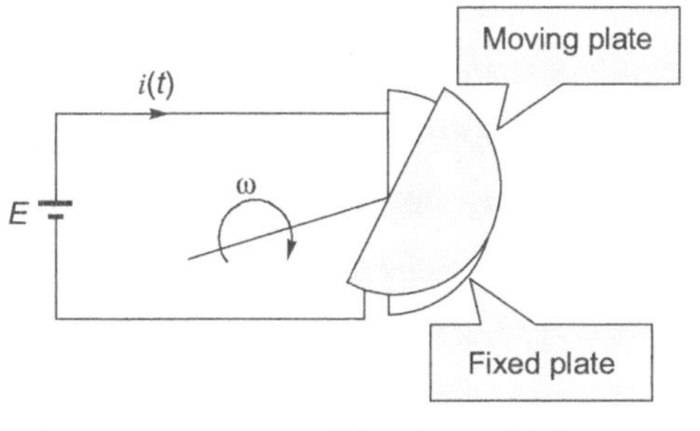

(a) dc (b) sinusoidal
(c) triangular (d) square

[2004 : 2 M]

Statement for Linked Answer Questions (3.10 and 3.11):

A capacitive type displacement transducer consists of two triangular plates, placed side by side, with negligible gap in between them and a rectangular plate moving laterally with an uniform air gap of 1 mm between the fixed plates and the moving plate. The schematic diagram is shown below with appropriate dimension.

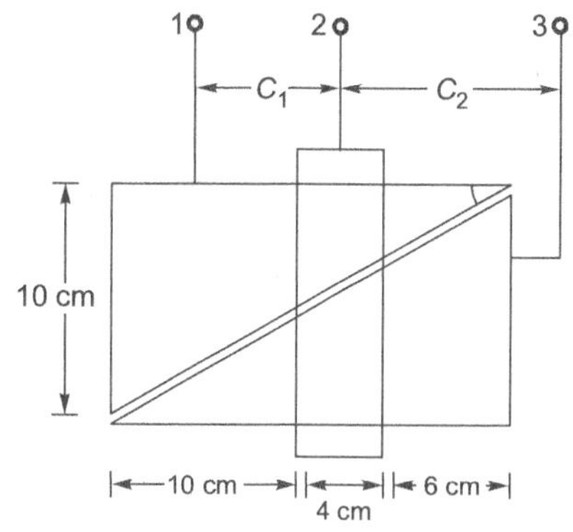

3.10 With the position of the moving plate shown in the figure above, the values of the capacitances C_1 and C_2 thus formed are

(a) $C_1 = 2.14$ pF, $C_2 = 1.29$ pF
(b) $C_1 = 14.17$ pF, $C_2 = 21.25$ pF
(c) $C_1 = 339$ pF, $C_2 = 212$ pF
(d) $C_1 = 18.16$ nF, $C_2 = 23.2$ nF

[2006 : 2 M]

3.11 The above sensor is incorporated in a capacitance measuring circuit as shown in the following figure. Assuming ideal op amp, the output voltage under the conditions mentioned above is

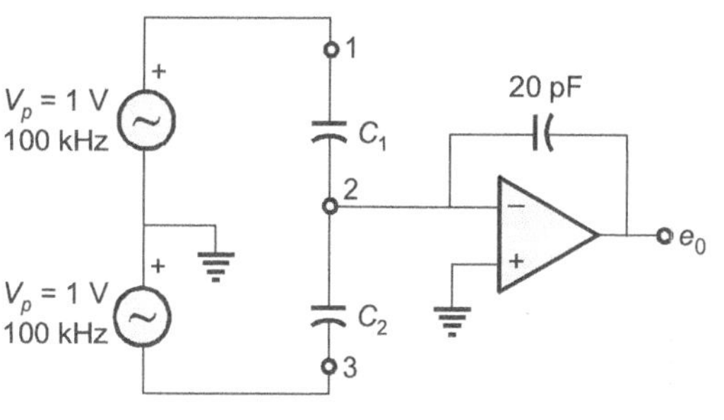

(a) 0.012 V (b) 0.354 V
(c) 1.23 V (d) 2.541 V

[2006 : 2 M]

3.12 A variable air gap type capacitor consists of two parallel plates : a fixed plate and a moving plate at a distance x. If a potential V is applied across the two plates, then force of attraction between in the plates is related to x as

(a) $F \propto x^2$ (b) $F \propto \frac{1}{x^2}$
(c) $F \propto \frac{1}{x}$ (d) $F \propto x$

[2006 : 2 M]

Statement for Linked Answer Questions (3.13 to 3.14):

A push-pull capacitive displacement transducer is interfaced to a differential amplifier and an ADC as shown in the figure below.

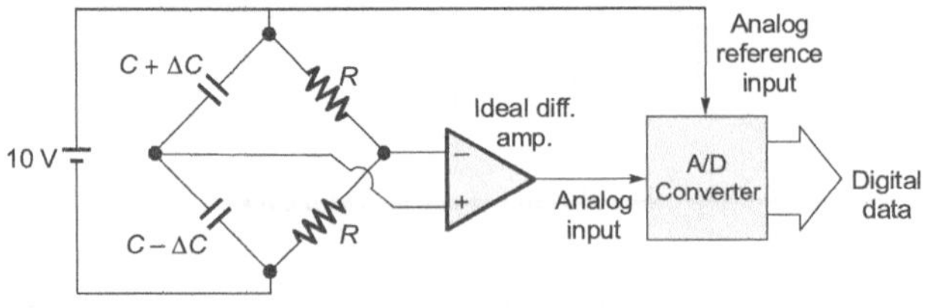

Note that the bridge supply and the analog reference input for the ADC are derived from the same 10 V DC source.

3.13 The change in capacitance for full scale displacement is ±5% for the capacitors. The gain of the differential amplifier for utilization of the full range of the ADC (which is ±10 V) is

(a) 10 (b) 20

(c) 30 (d) 40

[2007 : 2 M]

3.14 If the supply voltage to the bridge decreases by 5%, the sensitivity of the measurement system

(a) decreases by 5% (b) does not change

(c) increases by 5% (d) increases by 200%

[2007 : 2 M]

3.15 For signal conditioning of a piezoelectric type transducer we require

(a) a charge amplifier

(b) a differential amplifier

(c) an instrumentation amplifier

(d) a transconductance amplifier

[2008 : 1 M]

3.16 A piezoelectric type accelerometer has a sensitivity of 100 mV/g. The transducer is subjected to a constant acceleration of 5 g. The steady state output of the transducer will be

(a) 0 V (b) 100 mV

(c) 0.5 V (d) 5 V

[2008 : 2 M]

3.17 To reduce the effect of fringing in a capacitive type transducer,

(a) the transducer is shielded and the shield is kept at ground potential

(b) a guard ring is provided and it is kept at ground potential

(c) the transducer is shielded and the shield is kept at the same potential as the moving plate

(d) a guard ring is provided and it is kept at the same potential as the moving plate

[2008 : 2 M]

Common Data for Questions (3.18 and 3.19):

The figure shows an arrangement for measuring small angular displacements in a vertical plane. A non-conducting tube of length $2l$ and rectangular cross-section (width w, height d) is bent along an arc of a circle with radius $R >> d$, centered at P. Four electrode plates of length l and width w are plated to form two curved parallel plate capacitors C_1 and C_2 with a negligible gap between them. The tube contains water with an air bubble of rectangular cross-section (with w, height d) and length $l/4$. The capacitors are connected in a bridge circuit as shown in the figure, where the bridge has ac excitation v_i. Angular displacement $\Delta\theta$ occurs about the point P.

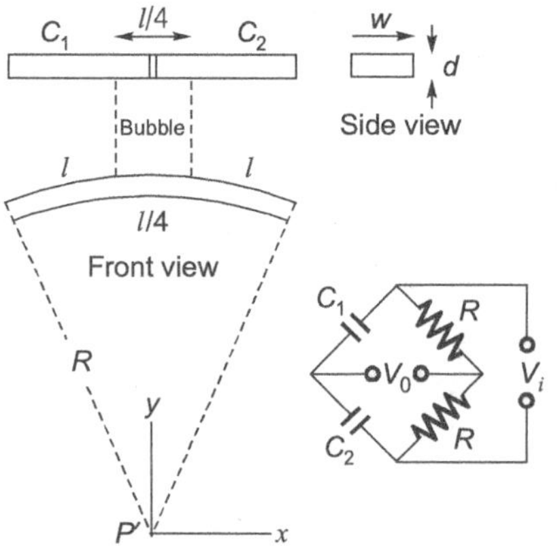

3.18 The range of angular displacement (in radians) this system can measure is

(a) $-\frac{l}{8R}$ to $+\frac{l}{8R}$ (b) $-\frac{l}{4R}$ to $+\frac{l}{4R}$

(c) $-\frac{3l}{4R}$ to $+\frac{l}{4R}$ (d) $-\frac{l}{R}$ to $+\frac{l}{R}$

[2009 : 1 M]

3.19 The sensitivity $\frac{v_0/v_1}{\Delta\theta}$ is

(a) inversely proportional to R and l

(b) inversely proportional to R and directly proportional to l

(c) directly proportional to R and l

(d) directly proportional to R and inversely proportional to l

[2009 : 1 M]

3.20 A differential push-pull type capacitive displacement sensor (nominal capacitance $C_0 = 0.01\ \mu F$) is connected in two adjacent arms of an ac bridge in such a way that the output voltage of the bridge is independent of the frequency of the supply voltage. Supply to the bridge is 1 V at 1 kHz. and two equal resistances ($R = 3.9\ k\Omega$) are placed in the other two arms of the bridge. The bridge sensitivity is

(a) 0.001 mV/pF (b) 0.05 mV/pF
(c) 0.1 mV/pF (d) 0.5 mV/pF

[2011 : 2 M]

3.21 A capacitive motion transducer circuit is shown. The gap d between the parallel plates of the capacitor is varied as

$$d(t) = 10^{-3}[1 + 0.1 \sin(1000\pi t)]\text{m}.$$

If the value of the capacitance is 2 pF at $t = 0$ ms, the output voltage V_o at $t = 2$ ms is

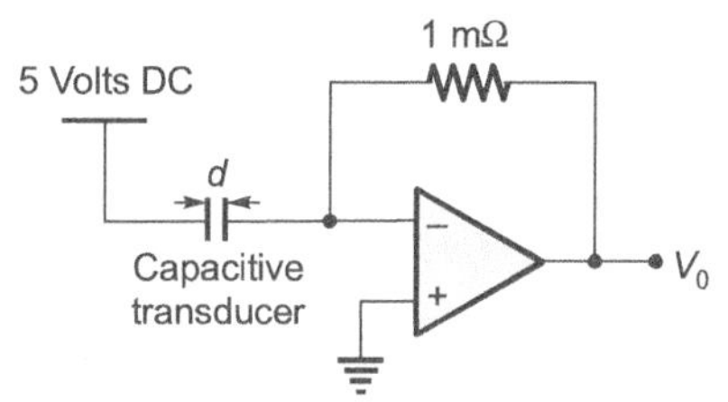

(a) $\frac{\pi}{2}$ mV (b) π mV
(c) 2π mV (d) 4π mV

[2012 : 1 M]

3.22 A dynamometer arm makes contact with the piezoelectric load cell as shown. The g-constant of the piezoelectric material is 50×10^{-3} Vm/N and the surface area of the load cell is 4 cm^2. If a torque $\tau = 20$ Nm is applied to the dynamometer, the output voltage V_o of the load cell is

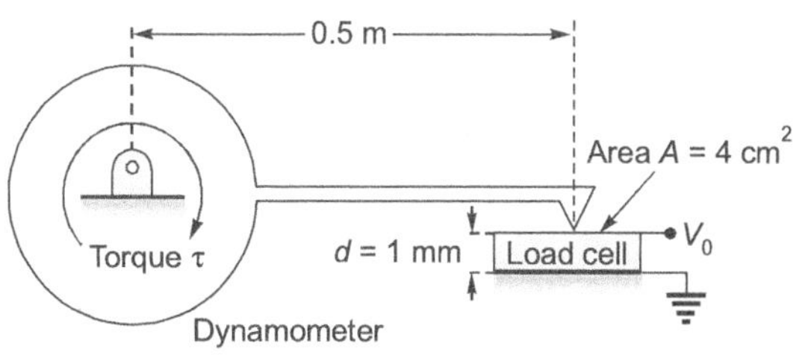

(a) 4 V (b) 5 V
(c) 10 V (d) 16 V

[2012 : 2 M]

Common Data Questions (3.23 and 3.24):

A piezoelectric force sensor, connected by a cable to a voltage amplifier, has the following parameters: Crystal properties: Stiffness 10^9 N/m, Damping ratio 0.01, Natural frequency 10^5 rad/s, Force-to-charge sensitivity 10^{-9} C/N, Capacitance 10^{-9} F with its loss angle assumed negligible.

Cable properties: Capacitance 2×10^{-9} F with its resistance assumed negligible.

Amplifier properties: Input impedance 1 MΩ, Bandwidth 1 MHz, Gain 3.

3.23 The maximum frequency of a force signal in Hz below the natural frequency within in useful midband range of measurement, for which the gain amplitude is less than 1.05, approximately is

(a) 35 (b) 350
(c) 3500 (d) 16×10^3

[2013 : 2 M]

3.24 The minimum frequency of a force signal in Hz within its useful mid-band range of measurement, for which the gain amplitude is more than 0.95, approximately is

(a) 16 (b) 160
(c) 1600 (d) 16×10^3

[2013 : 2 M]

3.25 A barium titanate piezoelectric crystal with $d_{33} = 150$ pC/N, $C_{crystal} = 25$ pF and $R_{crystal} = 10^{10}\ \Omega$ is used to measure the amplitude of a step force. The voltage output is measured using a digital voltmeter with input impedance $10^{13}\ \Omega$ connected across the crystal. All other capacitances and resistances may be neglected. A step force of 2 N is applied from direction "3" on the crystal. The time in milliseconds within which the voltmeter should sample the crystal output voltage so that the drop from the peak value is no more than 0.12 V is _____.

[2014 : 2 M]

3.26 A piezo-electric type pressure sensor has a sensitivity of 1 mV/kPa and a bandwidth of 300 Hz to 300 kHz. For a constant (dc) pressure of 100 kPa, the steady state output of the sensor in millivolt is _____.

[2016 : 1 M]

3.27 A piezoelectric pressure sensor has a bandpass characteristic with cut-off frequencies of 0.1 Hz and 1 MHz and a sensitivity of 100 mV/kPa. The sensor is subjected to a static constant pressure of 100 kPa. Its steady-state output will be

(a) 0 V (b) 0.1 V
(c) 1 V (d) 10 V

[2018 : 1 M]

3.28 A differential capacitive sensor with a distance between the extreme plates 100 mm is shown in figure below. The difference voltage $\Delta V = V_1 - V_2$, where V_1 and V_2 are the rms values, for a downward displacement of 10 mm of the intermediate plate from the central position, in volts, is

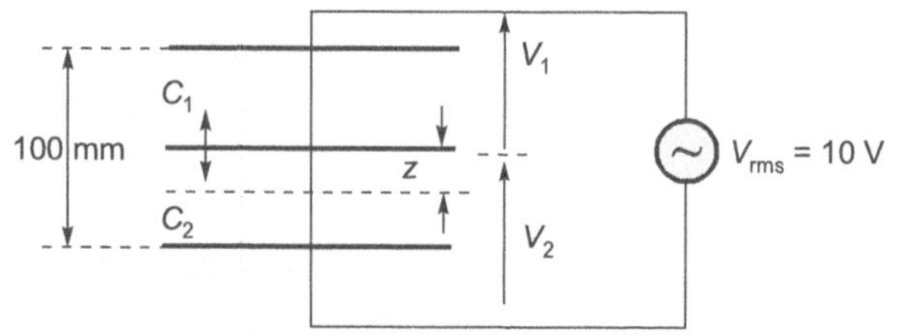

(a) 0.9 (b) 1.0
(c) 1.1 (d) 2

[2019 : 2 M]

3.29 A Piezoelectric transducer with sensitivity of 30 mV/kPa is intended to be used in the range of 0 kPa to 100 kPa. The readout circuit has peak noise amplitude of 0.3 mV arid measured signals over the full pressure range are encoded with 10 bits. The smallest pressure that produces a non-zero output, in units of Pa, is approximately

(a) 10 (b) 100
(c) 240 (d) 300

[2019 : 2 M]

3.30 The capacitance C_x of a capacitive type sensor is $(1000\,x)$ pF, where x is the input to the sensor. As shown in the figure, the sensor is excited by a violate $10 \sin(100\,\pi t)$ V. The other terminal of the sensor is tied to the input of a high input impedance amplifier through a shielded cable, with shield connected to ground. The cable capacitance is 100 pF. The peak of the voltage V_A at the input of the amplifier when $x = 0.1$ (in volts) is _____.

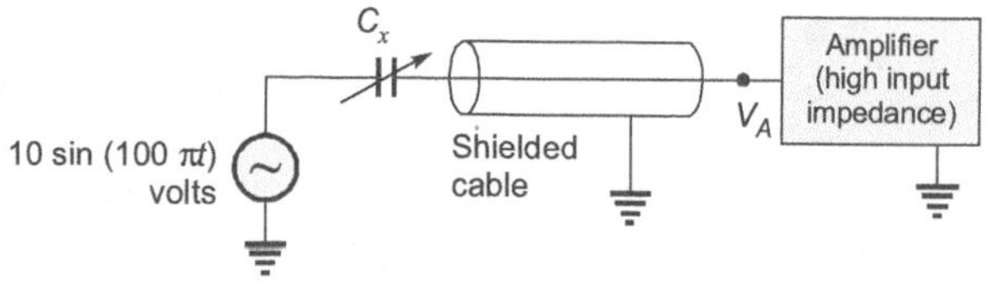

[2020 : 1 M]

3.31 The circuit shown uses ideal opamp powered from a supply $V_{CC} = 5$ V. If the charge q_p generated by the piezoelectric sensor is of the form $q_p = 0.1 \sin(10000\,pt)$ mC, the peak detector output after 10 cycles of q_p (in volts, rounded off to one decimal place) is _____.

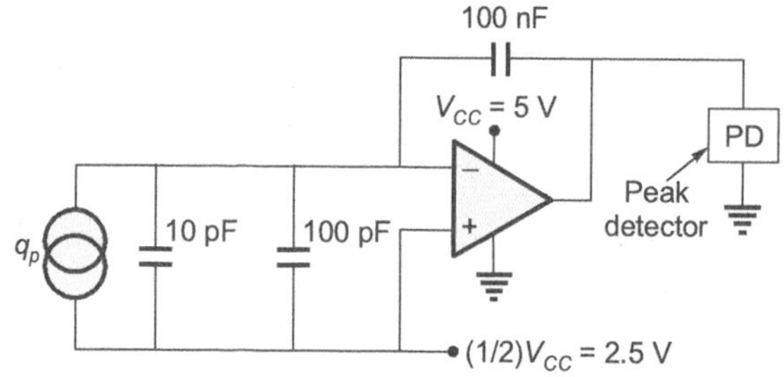

[2020 : 2 M]

3.32 A piezoresistive pressure sensor has a sensitivity of 1 (mV/V) kPa. The sensor is excited with a DC supply of 10 V and the output is read using a $3\frac{1}{2}$ digit 200 mV full-scale digital multimeter. The resolution of the measurement set-up, in pascal is _____.

[2021 : 1 M]

Answers Capacitive & Piezoelectric Transducers

3.1 (a)	3.2 (a)	3.3 (a)	3.4 (b)	3.5 (b)	3.6 (a)	3.7 (a)
3.8 (a)	3.9 (a)	3.10 (b)	3.11 (b)	3.12 (b)	3.13 (d)	3.14 (b)
3.15 (a)	3.16 (a)	3.17 (b)	3.18 (a)	3.19 (d)	3.20 (b)	3.21 (b)
3.22 (b)	3.23 (c)	3.24 (b)	3.25 (2.49)	3.26 (0)	3.27 (a)	3.28 (d)
3.29 (b)	3.30 (5)	3.31 (3.5)	3.32 (10)			

Explanations Capacitive & Piezoelectric Transducers

3.1 (a)

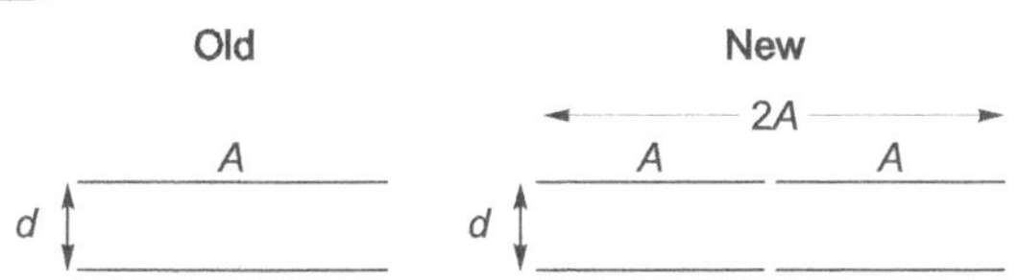

$$C = \frac{A\varepsilon_0}{d}$$

$$\therefore \quad C \propto A$$

$$\therefore \quad C_{new} = 2\,C_{old}$$

$$X_c = \frac{1}{\omega C}$$

$$\therefore \quad \omega = \frac{1}{X_c C}$$

$$\therefore \quad f \propto \frac{1}{C}$$

$$\frac{1}{f_{new}} = 2\frac{1}{f_{old}}$$

$$f_{new} = \frac{f_{old}}{2} = \frac{1000}{2} = 500 \text{ Hz}$$

3.2 (a)

A piezoelectric type transducer only responds to the dynamic inputs. For static inputs, the output of the piezoelectric sensor is zero.

3.3 (a)

$$E = 9 \times 10^{10} \text{ N/m}^2$$

$$V = gtp = gE(\Delta t)$$

$$\Delta t = \frac{V}{gE} = \frac{127.3 \text{ V}}{4500 \text{ V/}\mu\text{m}}$$

$$= 0.02828\ \mu\text{m}$$

$$F = EA\left(\frac{\Delta t}{t}\right)$$

$$= 9\times10^{10} \times \frac{0.02828\times10^{-6}}{2\times10^{-3}} \times \frac{\pi(10\times10^{-3})^2}{4}$$

$$F = 99.49 \text{ N} \approx 100 \text{ N}$$

3.4 (b)

$$C = 6.94\times10^{-3}\left(\frac{d^2}{S}\right)$$

$$\text{Sensitivity} = \frac{\partial C}{\partial S} = -6.94\times10^{-3}\frac{d^2}{S^2}$$

$$= -6.94\times10^{-3}\frac{(20\times10^{-3})^2}{(0.25\times10^{-3})^2}$$

$$= -44.4 \text{ pF/mm}$$

Note: Negative sign indicates that as displacement increases then capacitance decreases.

3.5 (b)

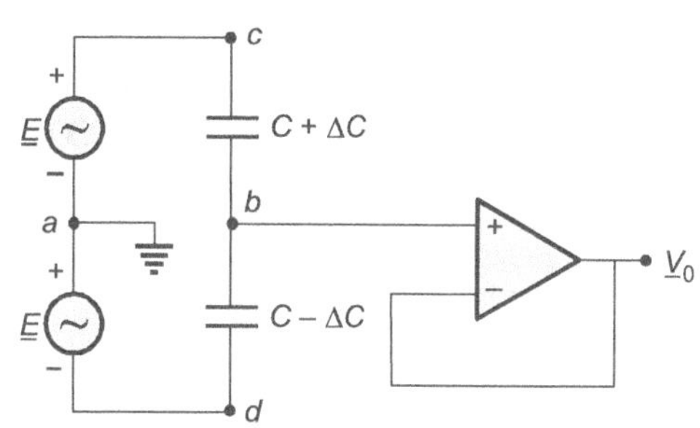

$$V_{cd} = 2E$$

$$V_0 = V_{ba}$$

$$V_{bd} = \frac{C+\Delta C}{C+\Delta C+C-\Delta C}\times 2E$$

$$V_{ba} = V_{bd} - E$$

$$= \frac{C+\Delta C}{2C}\times 2E - E$$

$$= \left(\frac{C+\Delta C}{C} - 1\right)E = \frac{\Delta C}{C}E$$

3.6 (a)

Charge, $Q = dF = 2 \times 10^{-12} \times 0.1 \sin 10t$
$= 0.2 \times 10^{-12} \sin 10t$
$= 0.2 \sin 10t (pC)$

Voltage developed (V) =

$$\frac{Q}{C_p} = \frac{0.2}{1600}\sin(10t)V$$
$$= 0.125 \sin 10t (\text{mV})$$

$R_F \Rightarrow$ Resistance of feedback circuit

$$R_F = 10^8 || X_C$$
$$X_C = \frac{1}{\omega C} = \frac{1}{10 \times 10^{-9}} = 10^8\ \Omega$$
$$\therefore \quad R_F = 10^8 || 10^8$$
$$= 0.5 \times 10^8\ \Omega$$

$R_1 \Rightarrow$ Resistance of the inverting terminal of the op-amp

$$C_p = 1600\text{ pF} = 1600 \times 10^{-12}\text{ F}$$
$$X_{C_p} = \frac{1}{\omega C_p} = \frac{1}{10 \times 1600 \times 10^{-12}}$$
$$= \frac{10^{12}}{16000}$$
$$R_L = 10^{12}$$
$$R_1 = R_L || X_{C_p} = 10^{12} || \frac{10^{12}}{16000}$$
$$= \frac{10^{12} \times \frac{10^{12}}{16000}}{10^{12} + \frac{10^{12}}{16000}} \simeq \frac{10^{12}}{16000}$$
$$V_{out} = -\frac{R_F}{R_1}(V)$$
$$|V_{out}| = \frac{R_F}{R_1}(V)$$
$$V_{out} = \frac{0.5 \times 10^8}{10^{12}} \times 16000 \times 0.125 \times 10^{-3}$$
$$= 0.1 \times 10 - 3 = 0.1\text{ mV}$$

$\therefore$ V_{out} is nearest to 0.141 mV.

3.7 (a)

To avoid the effect of fringing, the potential is to be made equal to the ground potential.

3.8 (a)

For constant pressure applied the output of a piezoelectric sensor at steady state is zero volts.

3.9 (a)

$$\because \quad i(t) = \frac{dQ}{dt}$$
$$Q = CE$$
$$i(t) = \frac{d}{dt}(C \times E)$$
$$= E\frac{d}{dt}\left(\frac{\in A}{d}\right) = \frac{E\in}{d}\frac{d}{dt}\left(\frac{\theta r^2}{2}\right) = \frac{E \in r^2}{d\ 2}\frac{d\theta}{dt}$$
$$\because \quad \theta = \omega t$$
$$\frac{d\theta}{dt} = \omega$$
$$i(t) = \frac{E \in r^2}{d \times 2}\omega$$
$$\therefore \quad i(t) \propto \omega$$
$$\because \quad \omega \rightarrow \text{constant}$$
$$\therefore \quad i(t) \rightarrow \text{constant}$$

3.10 (b)

$$C_1 = \frac{A\varepsilon}{d}, C_2 = \frac{A_2\,\varepsilon}{d}$$

we have, $d = 1$ mm

Calculating A_1:

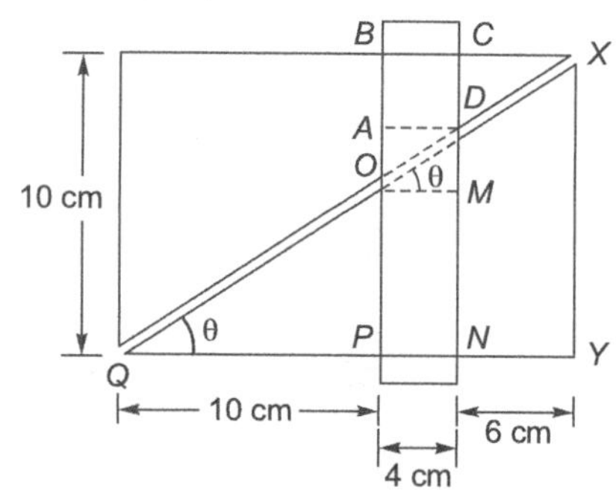

A_1 = Area of $ABCD$ + Area of OAD

In $\Delta\, QXY$,

$$\tan\theta = \frac{XY}{QY} = \frac{10}{20}$$
$$\therefore \quad \tan\theta = \frac{1}{2}$$

In ΔOQP,

$$\tan\theta = \frac{OP}{QP}$$
$$\frac{1}{2} = \frac{OP}{10}$$
$$OP = 5\text{ cm}$$

In ΔOMD,

$$\tan\theta = \frac{DM}{OM}$$

$$\frac{1}{2} = \frac{DM}{4}$$
$$DM = 2 \text{ cm}$$
$$DM = AO = 2 \text{ cm}$$
$$BP = AB + OA + PO$$
$$10 = AB + 2 + 5$$
$$\therefore \quad AB = 3 \text{ cm}$$
$$A_1 = \text{Area of } ABCD + \text{Area of } \Delta OAD$$
$$= 3 \times 4 + \frac{1}{2} \times 2 \times 4$$
$$= 12 + 4 = 16 \text{ cm}^2$$
$$A_2 = \text{Area of strip BCNP} - A_1$$
$$= 10 \times 4 - 16 = 40 - 16 = 24 \text{ cm}^2$$
$$C_1 = \frac{A_1 \varepsilon}{d} = \frac{16 \times 8.84 \times 10^{-12}}{10^{-3}} = 14.17 \text{ pF}$$
$$C_2 = \frac{A_2 \varepsilon}{d} = \frac{24 \times 8.84 \times 10^{-12}}{10^{-3}} = 21.25 \text{ pF}$$

3.11 (b)

In the given circuit,

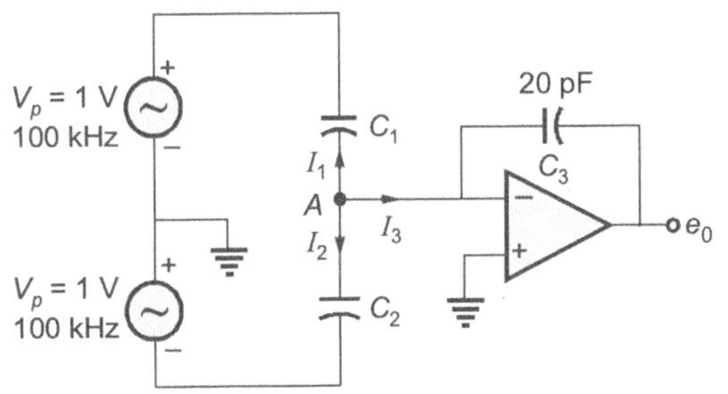

$$V_+ = 0 \text{ V}$$

Using virtual ground,

$$V_+ = V_- = V_A = 0 \text{ V}$$

Apply KCL at node A,

$$I_1 + I_2 + I_3 = 0$$
$$\frac{V_A - V_P}{1/j\omega C_1} + \frac{V_A + V_P}{1/j\omega C_2} + \frac{V_A - e_0}{1/j\omega C_3} = 0$$

Put $V_A = 0$ V,

$$V_P = 1 \text{ V}$$
$$e_0 = \left[\frac{C_2 - C_1}{C_3}\right] 1 \text{ V}$$
$$= \left[\frac{21.25 \times 10^{-12} - 14.17 \times 10^{-12}}{20 \times 10^{-12}}\right] \times 1$$
$$= 0.354 \text{ V}$$

3.12 (b)

If voltage = V then electric field E between the plates,

$$E = \frac{V}{x}$$

Therefore force of attraction

$$F = qE = CVE = \frac{\in AV}{x} \times \frac{V}{x}$$
$$F = \frac{\in AV^2}{x^2}$$

here $\in$, A and V are constant,

so, $$F \propto \frac{1}{x^2}$$

3.13 (d)

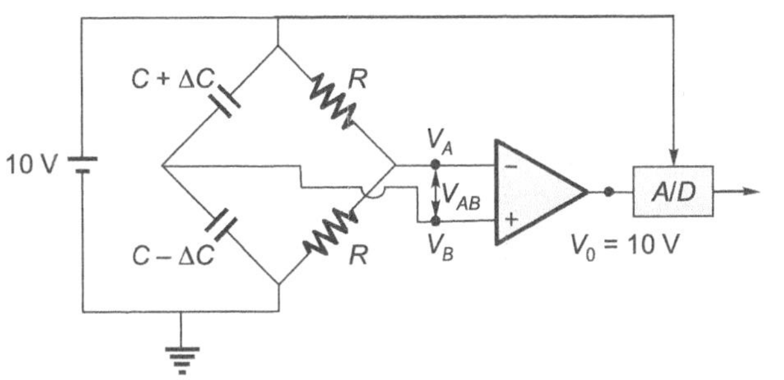

Let gain of amplifier is A_D then,

$$A_D = \frac{V_0}{V_{BA}} \quad \text{...(i)}$$

Given, $V_0 = 10$ V ...(ii)

Also, $V_{BA} = V_B - V_A$ ...(iii)

From network,

$$V_A = \frac{10R}{R + R} = 5 \text{ V} \quad \text{...(iv)}$$
$$V_B = \frac{10 \cdot \frac{1}{C - \Delta C}}{\frac{1}{C - \Delta C} + \frac{1}{C + \Delta C}}$$
$$= \frac{10 \cdot \frac{1}{0.95}}{\frac{1}{0.95} + \frac{1}{1.05}} = 5.25 \text{ V} \quad \text{...(v)}$$

Using (iii), (iv) and (v),

$$V_{BA} = 5.25 - 5 = 0.25 \text{ V} \quad \text{...(vi)}$$

Using (i), (ii) and (vi)

$$A_D = \frac{10}{0.25} = 40$$

3.14 (b)

If the supply voltage to the bridge is reduced by 5% then the reference voltage to A/D converter also reduces by 5%

So;

$$V_{in} = 10\text{ V} - 0.05 \times 10 = 9.5\text{ V}$$

V_{ref} to A/D converter = 9.5 V

Then at full scale displacement

$$\frac{\Delta C}{C} = 0.005$$

Bridge output voltage,

$$V_{OB} = \frac{V_{in}}{2} \cdot \frac{\Delta C}{C} \quad \text{[Push - Pull arrangement]}$$

$$V_{OB} = \frac{9.5}{2} \times 0.05 = 0.2375$$

Analog input to A/D converter

$$(V_{in})_{ADC} = \text{Gain of amplifier} \times V_{OB}$$
$$= 40 \times 0.2375 = 9.5\text{ V}$$

So, at full scale input displacement analog input to A/D converter is equals to analog reference input, then digital display will be full scale.

This means there is no change in sensitivity.

3.16 (a)

At steady state, for constant pressure applied the output voltage generated by a piezoelectric crystal is zero volts.

3.18 (a)

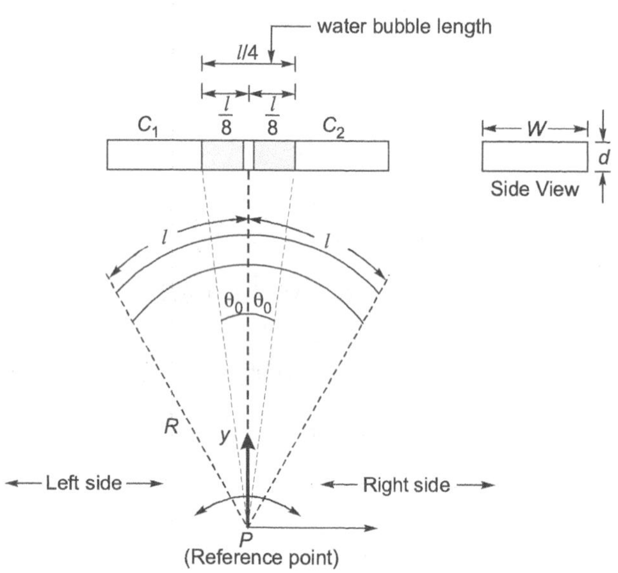

Initially (at reference point P)

	C_1	C_2
Air occupied length	$\frac{7l}{8}$	$\frac{7l}{8}$
Water occupied length	$\frac{l}{8}$	$\frac{l}{8}$

So, $C_1 = C_2 = C_0$

where,

$$C_0 = \frac{\epsilon_0 \epsilon_{air}\left(\frac{7l}{8}\right)W + \epsilon_0 \epsilon_{Water}\left(\frac{l}{8}\right)W}{d}$$

$$C_0 = \frac{\epsilon_0 W}{d}\left[\epsilon_{air}\left(\frac{7l}{8}\right) + \epsilon_{Water}\left(\frac{l}{8}\right)\right] \quad \text{... (i)}$$

Initially, Angular displacement, $\theta = \theta_0 = \frac{l}{8R}$

Suppose,

If we angularly displace arrangement right side then slowly bubble moves (Displaces) left side as shown below.

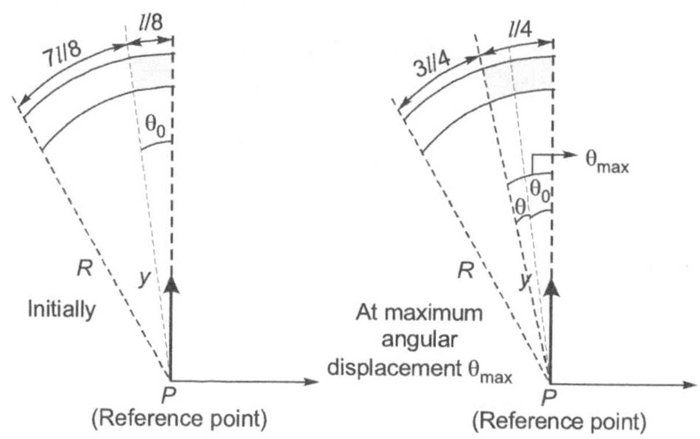

For C_1.

Initially, $\theta_i = \theta_0 = \frac{l}{8R}$

at maximum $\theta_{final} = \theta_{max} = \frac{l}{4R}$

So, detectable, $\theta = \theta_{max} - \theta_i = \frac{l}{4R} - \frac{l}{8R} = \frac{l}{8R}$

This will be same even if we displace the arrangement vertically left side also. So the correct answer is $\frac{l}{8R}$ left side to $\frac{l}{8R}$ right side.

with respect to position '*P*' $-\frac{l}{8R}$ to $+\frac{l}{8R}$

Note: Once the full bubble enters anyone side, there after capacitance won't change.

3.19 (d)

From the given bridge configuration; We can understand that it is push - pull arrangement

$$V_0 = V_i \times \frac{\Delta C}{2C_0}$$

$$\frac{V_0}{V_i} = \frac{\Delta C}{2C_0} \quad \text{... (ii)}$$

In general,

$$C = \frac{\in A}{d} = \frac{\in}{d}(l \times W)$$

$$C' = \frac{\in}{d}(l \pm \Delta l)W$$

$$\Delta C = C' - C = \pm\frac{\in}{d}Wl\left(\frac{\Delta l}{l}\right)$$

$$= \pm C_0 \cdot \frac{\Delta l}{l} \quad \text{... (iii)}$$

From (ii) and (iii)

$$\frac{V_0}{V_i} = \frac{1}{2}\cdot\frac{\Delta l}{l}R\cdot\Delta\theta = \frac{1}{2}\cdot\frac{R\cdot\Delta\theta}{l}$$

$$\Rightarrow \left(\frac{V_0/V_i}{\Delta\theta}\right) = \frac{1}{2}\times\frac{R}{l}$$

3.20 (b)

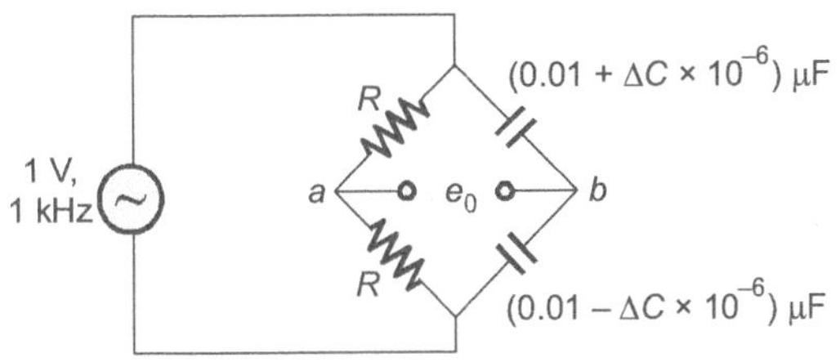

Taking ΔC of the order of pF (10^{-12} F)

$$V_a = \frac{R}{R+R}\times 1 = \frac{1}{2} = 0.5 \text{ V}$$

$$V_b = \frac{0.01+\Delta C\times 10^{-6}}{0.01+\Delta C\times 10^{-6}+0.01-\Delta C\times 10^{-6}}\times 1$$

$$= \frac{0.01+\Delta C\times 10^{-6}}{0.02}$$

$$V_0 = V_b - V_a$$

$$= \frac{0.01+\Delta C\times 10^{-6}-0.5\times 0.02}{0.02}$$

$$= \frac{\Delta C\times 10^{-6}}{0.02} = 50\times 10^{-6}\,\Delta C$$

$$\therefore \frac{V_0}{\Delta C} = 50\times 10^{-6}\text{ V/pF} = 0.05\text{ mV/pF}$$

3.21 (b)

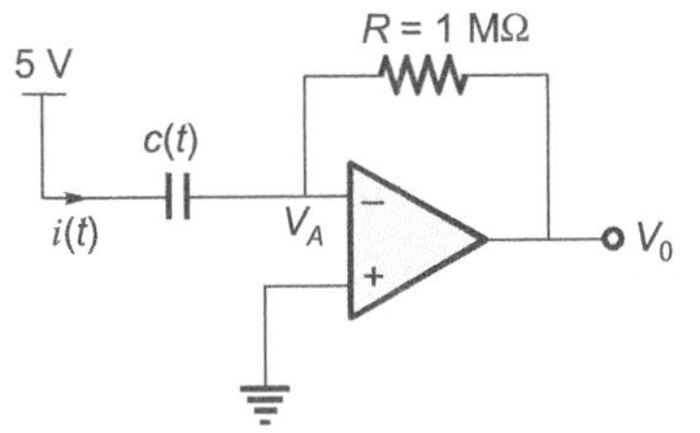

We need to find V_0 at t = 2 ms, so we will try to find $i(t)$ first followed by V_0.

We know,

$$i(t) = \frac{dQ}{dt} = \frac{d(CV_c)}{dt} \quad \text{...(i)}$$

Since $V_A = 0$

$$\Rightarrow \quad V_C = 5\text{ V} \quad \text{...(ii)}$$

Also, $$c(t) = \frac{\varepsilon_0 A}{d(t)} \quad \text{...(iii)}$$

Given, $c(0) = 2$ pF

$$\Rightarrow \quad 2\text{ pF} = \frac{\varepsilon_0 A}{10^{-3}[1+0.1\sin(1000\pi\cdot 0)]\text{ m}}$$

$$\Rightarrow \quad \varepsilon_0 A = 2\times 10^{-3}\text{ pF-m}$$

using equation (iii),

$$c(t) = \frac{2\times 10^{-3}}{10^{-3}[1+0.1\sin(1000\pi t)]}\text{pF}$$

$$= \frac{2}{1+0.1\sin(1000\pi t)}\text{pF} \quad \text{...(iv)}$$

Using (i), (ii) and (iv),

$$i(t) = \frac{10d\left[\dfrac{1}{1+0.1\sin 1000\pi t}\right]}{dt}\times 10^{-12}\text{ A}$$

$$= \frac{-1\times 100\pi\cos 1000\pi t}{(1+0.1\sin 1000\pi t)^2}\times 10^{-12}\text{ A}$$

for t = 2 m sec

$$i(2) = \frac{-\pi\cdot 1}{[1+0.1\sin 2\pi]^2}\times 10^{-9}\text{ A}$$

$$\Rightarrow \quad i(2) = -\pi\times 10^{-9}\text{ A}$$

Now, $$i(2) = \frac{V_A - V_0(2)}{1\times 10^6}$$

$$\Rightarrow -\pi\times 10^{-9} = (0 - V_0(2))\times 10^{-6}$$

$$\Rightarrow \quad V_0(2) = \pi\text{ mV}$$

Note: Even when DC is applied to capacitor here, we cannot replace it by an open circuit because capacitor has time varying capacitance.

i.e., $$i = \frac{d(cV_c)}{dt} = 0 \quad \{\text{if } C, V_c \to \text{constant}\}$$

$$\neq 0 \quad \{\text{if either of } C \text{ or } V_c \text{ is time varying}\}$$

3.22 (b)

$$T = Fr$$

$$F = \frac{T}{r} = \frac{20}{0.5} = 40\text{ N}$$

$$t = 1\text{ mm} \quad \text{(given)}$$

Voltage output of the piezoelectric

$$= g\,t\,P$$

$$= gt\left(\frac{F}{A}\right)$$

$$= 50\times10^{-3}\times1\times10^{-3}\times\frac{40}{4\times10^{-4}}$$

$$= 5 \text{ volts}$$

3.23 (c)

Electrical equivalent circuit for given data is shown below.

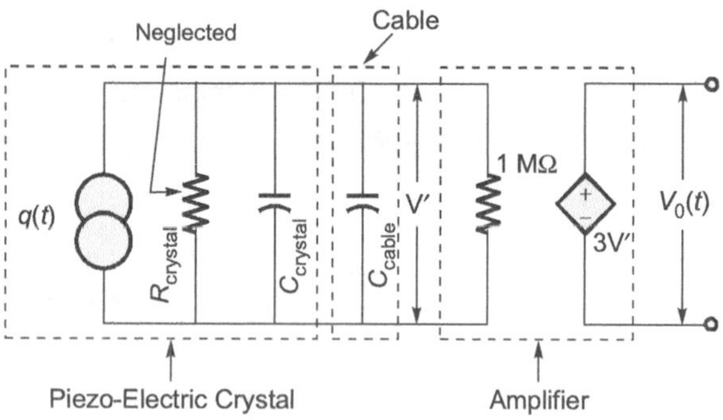

From the equivalent diagram

$$V_0(s) = 3V'(s) \quad \text{... (i)}$$

From Piezo-electric Crystal Theory

Keep, $R = R_{\text{amplifier}} = 1\times10^6\ \Omega$

$$C = C_{\text{crystal}} \| C_{\text{cable}} = 3\times10^{-9}\text{F}$$

Then,

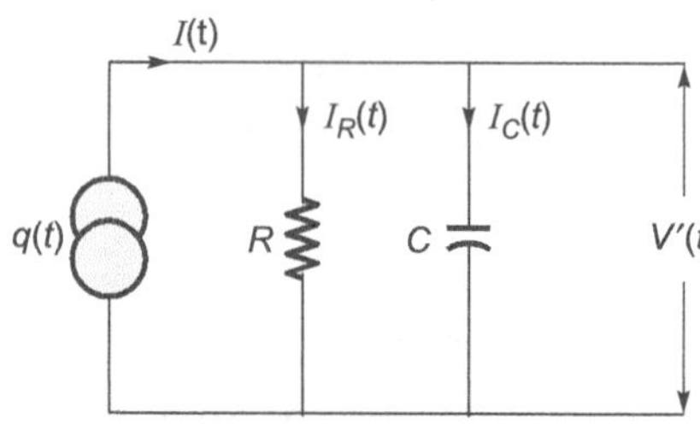

Then, from circuit theory concepts

$$I(t) = I_R(t) + I_C(t)$$

$$V'(t) = R(I_R(t))$$

$$\Rightarrow \quad = R\times(I(t) - I_C(t))$$

$$\Rightarrow \quad = R\left(\frac{dq(t)}{dt} - C\cdot\frac{dV'(t)}{dt}\right)$$

Apply Laplace transform on both side

$$\Rightarrow \quad V'(s) = R(SQ(s) - SCV'(s))$$

$$\Rightarrow \quad \frac{V'(s)}{Q(s)} = \frac{1}{C}\frac{\cdot\tau s}{1+\tau s} \quad \text{...(ii)}$$

(where, $\tau = RC$)

But dynamic model of Piezo-electric crystal is

$$\frac{Q(s)}{F(s)} = \frac{d.\omega_n^2}{S^2 + 2\xi\omega_n S + \omega_n^2} \quad \text{...(iii)}$$

From (ii) and (iii),

$$\frac{V'(s)}{F(s)} = \frac{V'(s)}{Q(s)}\times\frac{Q(s)}{F(s)}$$

$$= \frac{d}{c}\frac{\tau s}{(1+\tau S)}\times\frac{\omega_n^2}{S^2 + 2\xi\omega_n^2 S + \omega_n^2} \quad \text{...(iv)}$$

From (i) and (iv)

$$\frac{V'(s)}{F(s)} = \frac{V_0(s)}{V'(s)}\times\frac{V'(s)}{F(s)}$$

$$\Rightarrow \frac{V_0(s)}{F(s)} = \frac{3\times\frac{d}{c}\times\tau\times\omega_n^2\times S}{(1+\tau S)(S^2 + 2\xi\omega_n^2 + \omega_n^2)}$$

From the given data

$$d = 10^{-9}\ \text{C/N}$$

$$\tau = RC = 3\times10^{-3} = 0.003 \text{ sec.}$$

$$\omega_n = 10^5 \text{ rad/sec}$$

$$\xi = 0.01$$

$$|G(s)| = \left|\frac{V_0(s)}{F(s)}\right| < 1.05$$

Replace $s = j\omega$

$$\Rightarrow \quad |G(j\omega)| = \left|\frac{V_0(j\omega)}{F(j\omega)}\right| < 1.05$$

$$\Rightarrow \quad \frac{3\times\frac{d}{c}\times\tau\times\omega_n^2\times\omega}{\sqrt{1+(\tau\omega)^2}\sqrt{(\omega_n^2-\omega^2)^2 + 4\xi\omega_n^2\omega^2}} < 1.05$$

$$\Rightarrow \quad \frac{3\times10^{-3}\times10^{10}\times\omega}{\sqrt{1+9\times10^{-6}\omega^2}\sqrt{(10^{10}-\omega^2)^2 + 4\times0.01\times10^{10}\omega^2}} < 1.05 \quad \text{...(v)}$$

⇒ If we solve for 'ω' from equation (5); then we will get $\omega \cong 3500$ rad/sec.

3.24 (b)

$$|G(s)| = \left|\frac{V_0(s)}{F(s)}\right| > 0.95$$

$$\Rightarrow \quad \frac{3\times\frac{d}{c}\times\tau\times\omega_n^2\times\omega}{\sqrt{1+(\tau\omega)^2}\sqrt{(\omega_n^2-\omega^2)^2 + 4\xi\omega_n^2\omega^2}} > 0.95$$

$$\Rightarrow \quad \frac{3\times10^{-3}\times10^{10}\times\omega}{\sqrt{1+9\times10^{-6}\omega^2}\sqrt{(10^{10}-\omega^2)^2 + 4\times0.01\times10^{10}\omega^2}} > 0.95$$

⇒ If we solve for 'ω' then we will get $\omega \cong 160$ rad/s.

3.25 2.49 (2.48 to 2.52)

Equivalent circuit of piezoelectric crystal.

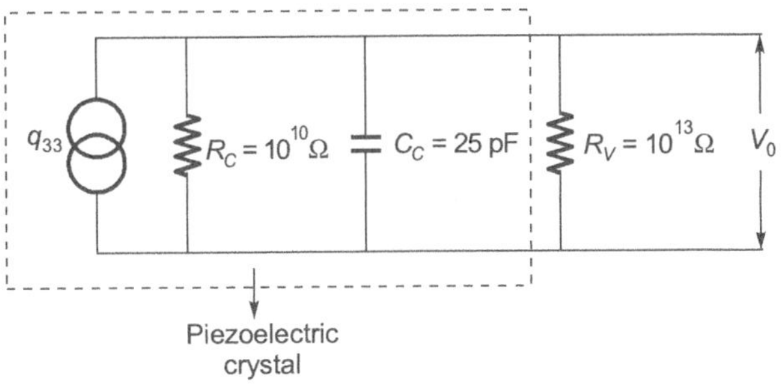

Given, circuit is an RC circuit whose output voltage V_0 is,

$$V_0(t) = V_{in}\left[e^{-t/\tau} - 1\right] \quad \text{...(i)}$$

Now, time constant, $\tau = R_{eq}C$

$$R_{eq} = 10^{10} \,||\, 10^{13} \approx 10^{10}\ \Omega$$

$$\Rightarrow \quad \tau = 10^{10} \times 25 \times 10^{-12} \text{ sec} = 250 \text{ msec} \quad \text{...(ii)}$$

Also, $$V_{in} = \frac{dF}{C} = \frac{150 \times 10^{-12} \times 2}{25 \times 10^{-12}} = 12 \quad \text{...(iii)}$$

Using (i), (ii) and (iii),

$$V_0(t) = 12\left[e^{\frac{-t}{250 \times 10^{-3}}} - 1\right]$$

For $V_0(t) = 0.12$ V

$$\Rightarrow \quad 0.12 = 12\left[e^{\frac{-t}{250 \times 10^{-3}}} - 1\right]$$

$$\Rightarrow \quad 1.01 = e^{\frac{-t}{250 \times 10^{-3}}}$$

$$\Rightarrow \quad t = 2.4876 \text{ msec} \approx 2.49 \text{ msec}$$

3.26 (0)

For constant (DC) pressure steady state output of piezoelectric crystal = 0 mV

3.27 (a)

The given piezo-electric pressure sensor has bandwidth "0.1 Hz to 1 MHz".

As the given input pressure is "static-input" means "0 Hz" signal. The output voltage at steady state = 0 V

3.28 (d)

The given figure in the question can be redrawn as

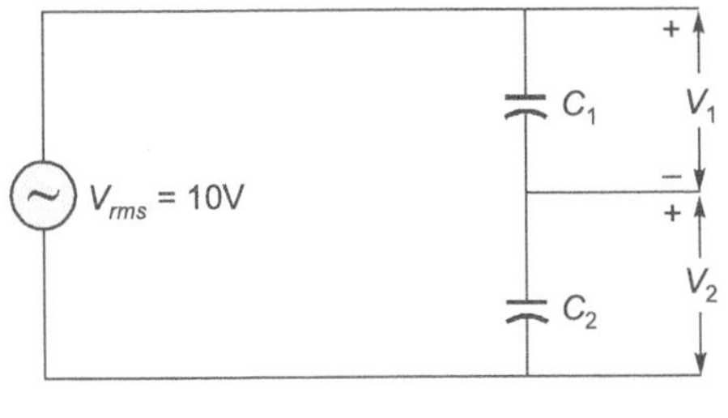

$$V_0 = 10\left[\frac{\frac{\epsilon A}{d-z} - \frac{\epsilon A}{d+z}}{\frac{\epsilon A}{d+z} + \frac{\epsilon A}{d-z}}\right]$$

$$= 10\left[\frac{d+z-d+z}{d+z+d-z}\right] = 10 \times \frac{z}{d}$$

$$= 10 \times \frac{10 \text{ mm}}{50 \text{ mm}} = 2\text{V}$$

So, option (d) is correct answer.

3.29 (b)

The overall block diagram of the above measurement system is given by

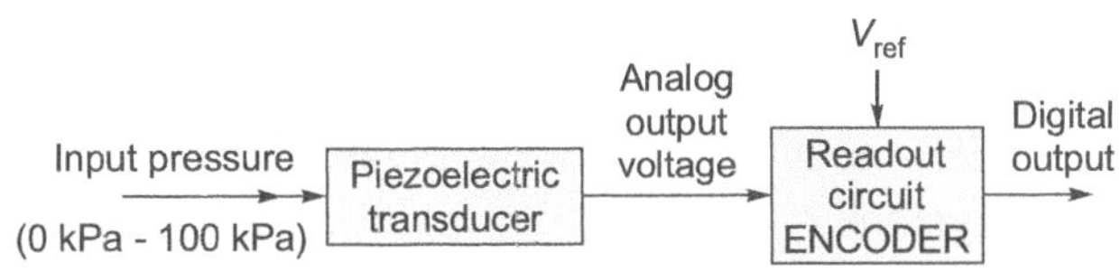

- Generally, encoder is a setup/device, which converts Analog voltage Signal into Digital code.
- According to the question, readout circuit has encoder of 10 bits. This means the resolution of encoder is given by

$$\frac{V_{ref}}{2^n - 1} \quad \text{...(i)}$$

[Where 'n' represent number of bits]

- Piezoelectric transducer converts the input pressure to voltage by follows.

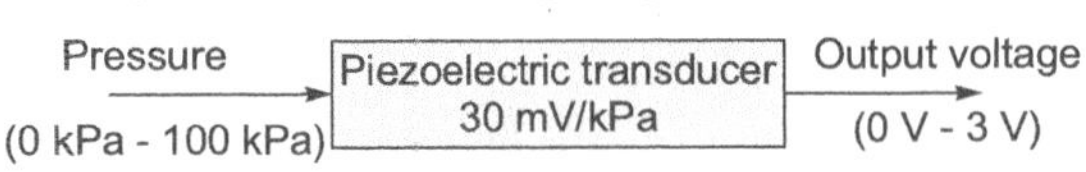

- The output voltage of Piezoelectric transducer is the input Analog voltage to Readout/Encoder circuit.
- To cover full pressure range,

  V_{ref} = Maximum output voltage of Piezoelectric transducer + Peak noise amplitude

  $$V_{ref} = 3 \text{ V} + 0.3 \text{ mV} = 3.0003 \text{ V} \quad \text{...(ii)}$$

  So, from equation (i) and (ii)

  Resolution of encoder

  $$= \frac{3.0003}{2^{10} - 1} = 0.00293 \text{ V}$$

- The minimum input voltage, the readout requires is 0.00293 V. To get non zero output voltage, Piezoelectric transducer should provide minimum output voltage of 0.00293 V.

- As we already knows that Piezoelectric transducer converts
  1 kPa → 30 mV.
  ? ← 0.00293 V
  The minimum pressure that can produce non zero output voltage

$$= \frac{0.00293 \times 1 \times 10^3}{30 \times 10^{-3}} = 97.66 \simeq 100 \text{ Pa}$$

3.30 (5)

The capacitance of the capacitive sensor

$C_x = (1000x)$ Pf

x value is = 0.1

$C_x = (1000 \times 0.1)$ Pf = 100 Pf

Cable capacitance = 100 Pf

Electrical equivalent of given measurement system.

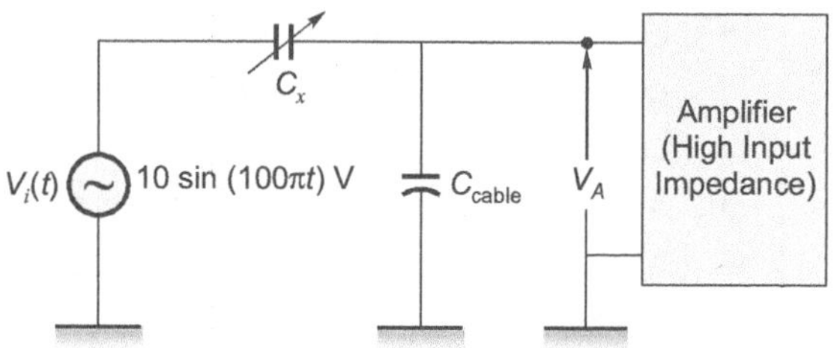

As the amplifier has high input impedance there is no loading effect.
So,Peak value of

$$V_A = V_i \times \frac{\frac{1}{j\omega C_{cable}}}{\frac{1}{j\omega C_{cable}} + \frac{1}{j\omega C_x}}$$

$$= 10 \times \frac{C_x}{C_x + C_{cable}}$$

$$= 10 \times \frac{100 \times 10^{-12}}{(100+100) \times 10^{-12}} = 5 \text{ V}$$

3.31 (3.5)

The potential at $V_+ = V_- = 2.5$ V

because of virtual ground current drawn by opamp is 0 from q_p

So, peak detector output $V_p(t) = 2.5 - V_c(t)$

$$2.5 - V_p(t) = \frac{1}{C} q_p(t)$$

$$2.5 - V_p(t) = \frac{1}{100 \times 10^{-9}} \times 0.1 \sin(10^4 \pi t) \times 10^{-6}$$

$$2.5 - V_p(t) = \sin(10^4 \pi t)$$

$$\therefore \quad V_p(t) = 2.5 - \sin(10^4 \pi t)$$

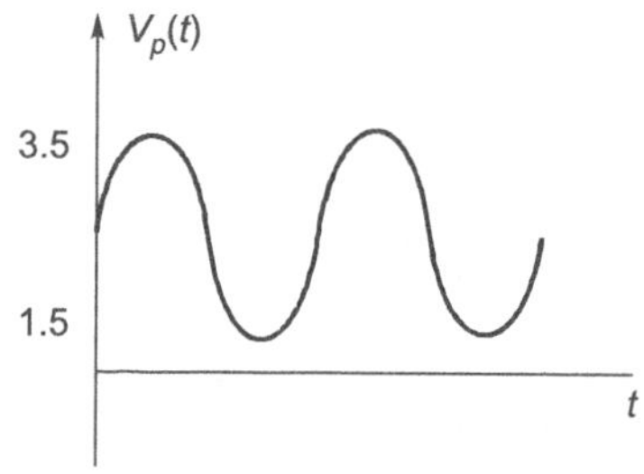

$V_p(t) = 3.5$ V

After 10 cycles, peak value detected by peak detector should be 3.5 Volts.

3.32 (10)

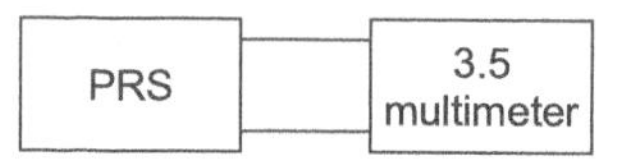

Total counts = 2000 counts

Maximum voltage = 200 mV

$$\text{Sensitivity} = \frac{\frac{1 \text{mV}}{\text{V}}}{1 \text{kPascal}} \quad \text{... (i)}$$

Input voltage = 10 V

Full scale output,

$V_0 = 200$ mV

$$\text{Also, Sensitivity} = \frac{\frac{200 \text{mV}}{10 \text{ V}}}{P_{in}} \quad \text{... (ii)}$$

From (i) and (ii),

$$\frac{\frac{1 \text{mV}}{\text{V}}}{1 \text{kPascal}} = \frac{\frac{200 \text{ mV}}{10 \text{ V}}}{P_{in}}$$

$P_{in} = 20$ k Pascal

$$\text{Resolution} = \frac{20 \text{ k pascal}}{2000 \text{ counts}} = 10 \text{ pascal}$$

■■■

4 Pressure, Flow and Temperature Measurement

4.1 An example of a variable area device for measuring flow is

(a) Flow nozzle (b) Orifice meter
(c) Venturimeter (d) Rotameter

[1996 : 1 M]

4.2 The primary standard for calibrating vacuum is

(a) McLeod gauge
(b) Dead weight tester
(c) Thermocouple gauge
(d) Knudsen gauge

[1996 : 1 M]

4.3 A pressure gauge used to measure vacuum indicates a gauge pressure of 5 kPa. If the atmospheric pressure is 100 kPa, the absolute pressure is

(a) 105 kPa (b) 0.05 kPa
(c) 95 kPa (d) 20 kPa

[1996 : 1 M]

4.4 Match the device with its working principle

List-I
(a) Bourdon gauge
(b) Pirani gauge
(c) Semiconductor strain gauge
(d) Bimetallic strip

List-II
(i) Thermal conductivity
(ii) Thermal coefficient of expansion
(iii) Piezo-resistive effect
(iv) Photo-electric effect
(v) Deformation of element

[1996 : 2 M]

4.5 A Pitot static tube is used to measure the airflow rate in a square tube. If P_d is the differential pressure, the flow rate is proportional to

(a) $\sqrt{P_d}$ (b) P_d
(c) $\dfrac{1}{\sqrt{P_d}}$ (d) P_d^2

[1997 : 1 M]

4.6 A McLeod gauge has a volume of the bulb equal to 100 cm^3 and a capillary of diameter 1 mm. Calculate the pressure indicated by a reading of 3 cms. What error would result if the capillary volume is assumed to be negligible compared to the volume of the bulb.

[1997 : 1 M]

4.7 Match the following instruments with the measurements:

List-I	List-II
(a) Rotameter	(i) Rotational speed
(b) Tachometer	(ii) Roughness
(c) Tracer	(iii) Temperature
(d) Pyrometer	(iv) Flow rate

[1997 : 2 M]

4.8 Match the type of the thermocouple with its characteristics.

List-I
(a) Type E Chromel-Constantan
(b) Type J Fe-Alumel
(c) Type T Cu-Constantan
(d) Type S Pt-PtRh

List-II
(i) Low cost
(ii) Good stability
(iii) High precision
(iv) High output

[1997 : 2 M]

4.9 Choose the thermocouple that can measure a temperature in the range of 1300°C to 1400°C.

(a) Iron-constantan
(b) Copper-constantan
(c) Platinum-platinum rhodium
(d) Chromel-alumel

[1998 : 1 M]

4.10 Match the pressure measuring devices with the pressure levels in torr.

List-I	List-II
(a) Bourdon tube	(i) 10^3
(b) Ionization gage	(ii) 10^{-10}
(c) Water manometer	(iii) 1.0
(d) Pirani	(iv) 10^{-5}

[1998 : 2 M]

4.11 The emf-temperature data for a thermocouple with reference junction at 0°C is a follows:

Temperature (°C)	20	180	200
emf (mV)	1.2	11.8	13.5

The emf developed in mV with junctions at 200°C is

(a) 11.8 (b) 12.3
(c) 13.0 (d) 13.5

[1999 : 1 M]

4.12 An elastic transducer is used to measure pressure in a vessel and it indicates a pressure of 3.2 bar. Atmospheric pressure is 1.01 bar. The absolute pressure in the vessel in bar is

(a) 1.01 (b) 2.19
(c) 3.20 (d) 4.21

[1999 : 2 M]

4.13 Which of the following gauges can measure the lowest vacuum pressure?

(a) McLeod gauge
(b) Pirani gauge
(c) Ionisation gauge

[2000 : 2 M]

4.14 Which of the following restrictors has the highest discharge coefficient?

(a) Orifice plate
(b) Flow nozzle
(c) Venturi tube

[2000 : 2 M]

4.15 For flow measurement, a rotameter can be installed in a pipeline

(a) horizontally with flow inlet in a specific direction
(b) horizontally with flow inlet in any direction
(c) vertically with flow inlet at the bottom and the outlet at the top
(d) vertically with flow inlet at the top and outlet at the bottom

[2001 : 1 M]

4.16 An example of a positive displacement flow meter is

(a) orifice meter
(b) rotary vane type meter
(c) turbine type meter
(d) ultrasonic flow meters

[2001 : 1 M]

4.17 A thermocouple is suddenly immersed in a medium of high temperature. The approximate time taken by the thermocouple to reach 98% of the steady-state value is

(a) equal to the time constant of the thermocouple
(b) equal to twice the value of the time constant of the thermocouple
(c) equal to four times the value of the time constant of the thermocouple
(d) independent of the time constant

[2002 : 1 M]

4.18 A Pirani gage sensor is used to measure pressure of the order of

(a) 10 MPa (b) 1 MPa
(c) 100 Pa (d) 1 Pa

[2002 : 1 M]

4.19 In a rotameter, used for measuring flow rate of a fluid, P_a : pressure above the float, P_b : pressure below the float, A : area of float, V: volume of the float, d_1 : density of float material, d_2 : density of the fluid, g : acceleration due to gravity. The following equation describes the equilibrium of the float.

(a) $(P_b - P_a)A = Vg(d_2 - d_1)$
(b) $(P_a + P_b)A = Vg(d_2 + d_1)$
(c) $(P_a - P_b)A = Vg(d_2 - d_1)$
(d) $(P_b - P_a)A = Vg(d_2 + d_1)$

[2002 : 2 M]

4.20 A Pirani gage sensor is used to measure pressure of the order of

(a) 10 MPa (b) 1 MPa
(c) 100 Pa (d) 1 Pa

[2003 : 1 M]

4.21 Which of the following meters has the lowest pressure drop for a given range of flow?

(a) Orifice meter (b) Venturi meter
(c) Flow Nozzle (d) Rotameter

[2003 : 1 M]

4.22 A temperature-measuring instrument is modeled as a first order system with a time constant of 5s. The sensor of the instrument is placed inside an oil bath whose temperature has a sinusoidal variation with amplitude of 10°C and a period of 20s around an average temperature of 200°C. The sinusoidal component at the output of the instrument will have amplitude.

(a) 0°C (b) 5.37°C
(c) 8.57°C (d) 10°C

[2003 : 2 M]

Data given below should be used for answering Questions (4.23 and 4.24).

The table provides the thermo-emf sensitivity of five materials with reference to platinum around 273 K.

Material	Constantan	Nickel	Copper	Iron	Nichrome
Sensitivity (μVK^{-1})	–35	–25	+6	+18.5	+25

4.23 The thermocouple pair that gives the maximum sensitivity around 273 K is
(a) Platinum-Constantan
(b) Nichrome-Constantan
(c) Nickel-Constantan
(d) Copper-Nickel

[2003 : 2 M]

4.24 Two copper-constantan thermocouples are connected such that the two constantan wires are joined together. The two copper wires are connected to the input of a low noise chopper stabilized differential amplifier having a gain of 1000. One of the thermocouple junctions is immersed in a flask containing ice and water in equal proportion. The other thermocouple is at a temperature T. If the output of the amplifier is 2.050 V, the temperature T is
(a) 250°C (b) 102.5°C
(c) 51.25°C (d) 50°C

[2003 : 2 M]

4.25 An N-type semiconductor strain gage has a nominal resistance of 1000 Ω and gage factor of –100. The resistance of the gage, when a compressive strain of 100 μm/m is applied, is
(a) 900 Ω (b) 990 Ω
(c) 1010 Ω (d) 1100 Ω

[2003 : 2 M]

4.26 For a steady flow of liquid, the float of the rotameter will remain in a particular position when
(a) drag force is balanced by weight of the float and buoyancy force
(b) weight of the float is balanced by the drag force and buoyancy force
(c) buoyancy force is balanced by the drag force and weight of the float
(d) drag force and buoyancy together is slightly greater than the weight of the float

[2003 : 2 M]

4.27 A thermocouple can be considered as a single capacity system with a time constant of 100 ms and unity gain. It is immersed in a heat bath whose temperature is varying at 1°C/s from 0°C. When the actual temperature of the heat bath is 70°C, the reading of the thermocouple will be
(a) 70.1°C (b) 71°C
(c) 69.9°C (d) 69°C

[2004 : 1 M]

4.28 An iron-constantan thermocouple is to be used to measure temperature between 0 and 300°C. With reference junction at 0°C, the emf induced at 100°C and 300°C are

$$E_{100} = 5268\ \mu V,\ E_{300} = 16325\ \mu V$$

Then the non-linearity at 100°C as a percentage of full scale is
(a) –3.23% (b) –1.06%
(c) –0.96% (d) +3.23%

[2004 : 2 M]

4.29 An electromagnetic flowmeter is used to measure the average flow rate of a liquid in a pipe of diameter 50 mm. The flux density in the liquid has a peak value of 0.1 T. The output from the flowmeter is taken to an amplifier of gain 1000 and input impedance of 1 MΩ. If the impedance of the liquid between the electrodes is 200 kΩ, then the peak value of the amplifier output voltage for an average flow velocity of 20 mm/s is
(a) 1.365 V (b) 1 V
(c) 0.1 V (d) 0.083 V

[2004 : 2 M]

4.30 In a variable-area flowmeter (rotameter), the inlet tube diameter is the same as flow rate of a fluid is Q_1, the float stands at a height of 70 mm from the inlet and when the volume flow rate of the fluid is Q_2, the float moves to a height of 20 mm from the inlet. The ratio of the flow rates Q_1/Q_2 is
(a) 3.5 (b) 0.29
(c) 0.082 (d) 12.3

[2005 : 1 M]

4.31 A mercury barometer reads h mm Hg with the temperature of the mercury at T°C. The barometer reading corrected for the standard temperature 0°C with β denoting the volumetric expansion coefficient of mercury in 0°C, is
(a) $\dfrac{h}{1+\beta T}$ (b) $h(\beta + T)$
(c) $h(1 + \beta T)$ (d) $h(\beta - T)$

[2005 : 2 M]

4.32 Figure given below shows an ultrasound flowmeter with the transducers separated by a distance L = 25 mm with the line joining the transducers making an angle 60° with the direction of the flow. The transit time difference between upstream and downstream measurements is 10 ns with the sound velocity in the medium being 1000 m/s. Assuming that the size of the transducers is very small as compared to the diameter of the pipe, the flow velocity is

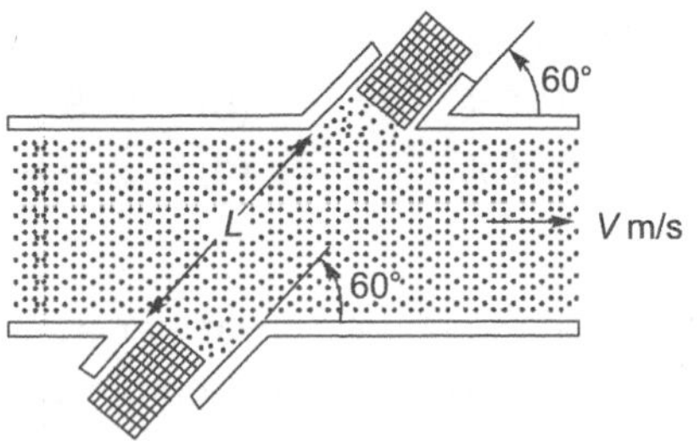

(a) 0.2 m/s (b) 2 m/s
(c) 0.4 m/s (d) 40 m/s

[2005 : 2 M]

Common Data for Questions (4.33 to 4.34):

The temperature e.m.f. characteristics of the iron constantan and copper-constantan junctions at 0°C are given by E_{IC} = 53 T μV and E_{CC} = 43T μV, over an appropriate temperature range, where T is in 0°C. It is known that AD 590 is an IC temperature sensor which can drive a current of (273 + T) mA at T°C into an appropriate load resistance.

4.33 An iron-constantan thermocouple has its measuring junction in a bath whose temperature can vary between 100°C-200°C. The reference junction is maintained at ambient temperature. If the emf developed is to correspond to that with the reference junction at 0°C, the maximum error, expressed as a percentage of the emf developed, due to ambient temperature variation in the range 0°C-45°C, is

(a) 8.182 (b) 22.5
(c) 45 (d) 81.82

[2005 : 2 M]

4.34 A thermocouple is made using a copper and an iron wire. Its measuring junction is kept at a temperature of 100°C and the reference junction is kept at the ambient temperature of 30°C. The emf developed would be

(a) 700 μV (b) 1000 μV
(c) 3010 μV (d) 5300 μV

[2005 : 2 M]

4.35 A voltage generated using a AD590 sensor is used to compensate for ambient temperature error for the iron-constantan thermocouple in the circuit shown in the figure such that the output voltage V is proportional to the measuring junction temperature T_m. The values of R_L and E for exact compensation, assuming ideal components, are

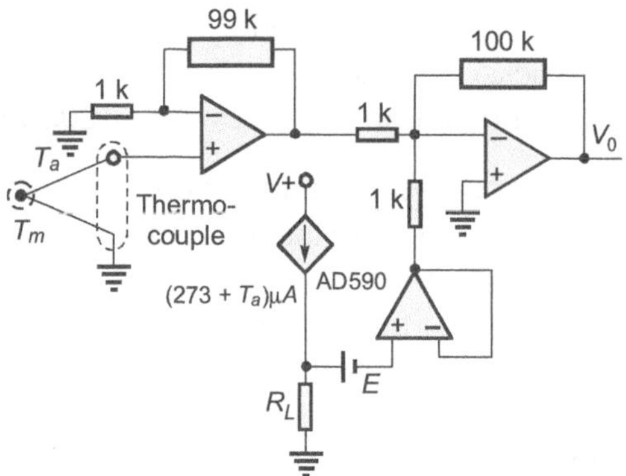

(a) 1 kΩ, 273 mV (b) 4.3 kΩ, 1.174 V
(c) 5.3 kΩ, 1.447 V (d) 1 kΩ, 303 mV

[2005 : 2 M]

4.36 "Vena Contracta" is the cross-section where the flow area is a minimum for a restriction type flowmeter. For an orifice meter, if d is the diameter of the orifice opening, then the area of the vena contracta is approximately

(a) $\frac{\pi d^2}{4}$ (b) $\frac{0.99\pi d^2}{4}$

(c) $\frac{0.8\pi d^2}{4}$ (d) $\frac{0.6\pi d^2}{4}$

[2006 : 1 M]

4.37 A type J (iron-constantan) thermocouple has a voltage sensitivity of 55 μV/°C. A digital voltmeter (DVM) is used to measure the voltage under the condition shown in the following figure.
Given that: T_1 = 300°C, T_2 = 100°C and $T_3 = T_4$ = 20°C, the meter will indicate a voltage of

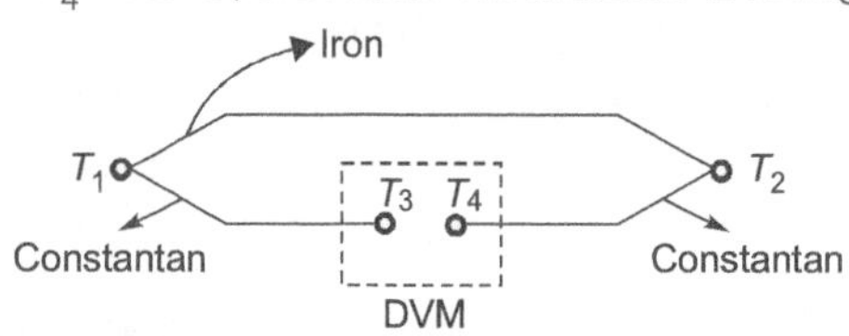

(a) 11.0 mV (b) 15.4 mV
(c) 16.5 mV (d) 17.6 mV

[2006 : 1 M]

4.38 A Pirani gage measures vacuum pressure and works on the principle of
(a) change in ionizing potential
(b) change in thermal conductivity
(c) deformation of elastic body
(d) change in self-inductance

[2006 : 1 M]

4.39 A well of cross-sectional area a_w is connected to an inclined tube of cross-sectional area a_t to form a differential pressure gauge as shown in the figure below. When $p_1 = p_2$ the common liquid level is denoted by A. When $p_1 > p_2$, the liquid level in the well is depressed to B, and the level in the tube rises by l along its length such that difference in the tube and well levels is h_d.
The angle of inclination θ of the tube with the horizontal is

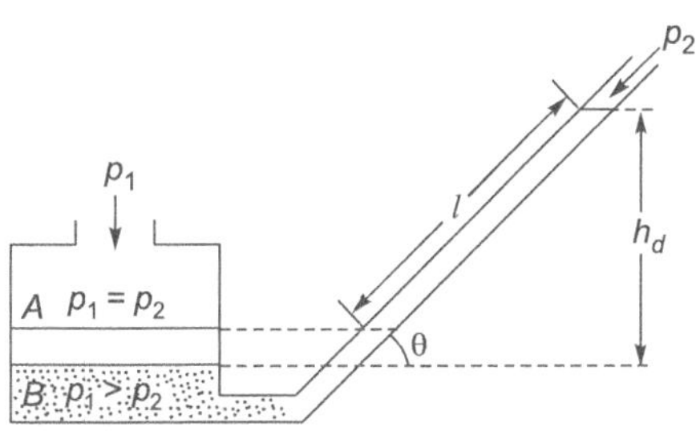

(a) $\sin^{-1}\left[\frac{l}{h_d} - \frac{a_w}{a_t}\right]$ (b) $\sin^{-1}\left[\frac{h_d}{l} + \frac{a_t}{a_w}\right]$

(c) $\sin^{-1}\left[\frac{h_d}{l} - \frac{a_t}{a_w}\right]$ (d) $\sin^{-1}\left[\frac{h_d}{l} + \frac{a_w}{a_t}\right]$

[2007 : 2 M]

4.40 The temperature being sensed by a negative temperature coefficient (NTC) type thermistor is linearly increasing. Its resistance will
(a) linearly increase with temperature
(b) exponentially increase with temperature
(c) linearly decrease with temperature
(d) exponentially decrease with temperature

[2008 : 1 M]

4.41 A pair of identical thermocouples is employed for measuring the temperature of a specimen as shown below. The characteristic of the thermocouples is tabulated below. The reference junction is at 2°C. The meter reads 48 μV. The correct temperature of the specimen is

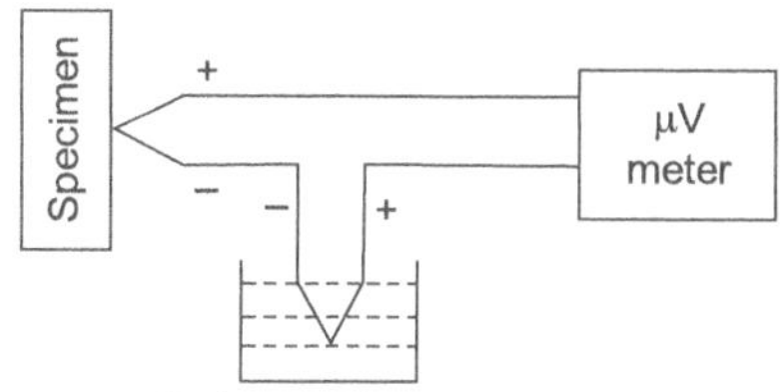

Temp.(°C)	0	10	20	30	40	50	60	70	80	90
Output(μV)	35	45	55	65	75	85	95	105	115	125

(a) 13°C (b) 46°C
(c) 48°C (d) 50°C

[2008 : 2 M]

4.42 The figure shows the cross-sectional diagram of an orifice flow meter with an orifice radius R. Point 'a' is 30 R upstream while points 'b' and 'c' are 0.8 R and 30 R downstream from the orifice respectively. The pressure at points a, b and c are $P_a P_b$, and P_c respectively. Then

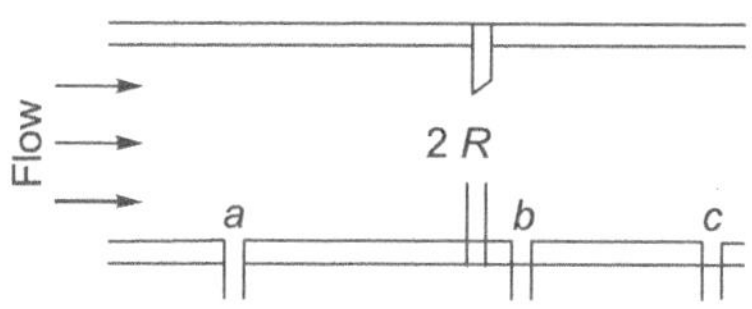

(a) $P_c > P_b > P_a$ (b) $P_b > P_c > P_a$
(c) $P_a > P_b > P_c$ (d) $P_a > P_c > P_b$

[2009 : 1 M]

4.43 Match the following:

P. Radiation Pyrometer — W. Angular velocity measurement
Q. Dall tube — X. Vacuum pressure measurement
R. Pirani gauge — Y. Flow measurement
S. Gyroscope — Z. Temperature measurement

(a) P-Z, Q-W, R-X, S-Y
(b) P-Z, Q-Y, R-X, S-W
(c) P-W, Q-X, R-Y, S-Z
(d) P-Z, Q-X, R-W, S-Y

[2010 : 1 M]

4.44 A thermocouple based temperature measurement system is shown in the adjoining figure. Relevant thermocouple emf data (in mV) is given below. The cold junction is kept at 0°C. The temperature is 30°C in the other parts of the system. The emf V is measured to be 26.74 mV. The temperature of the hot liquid is

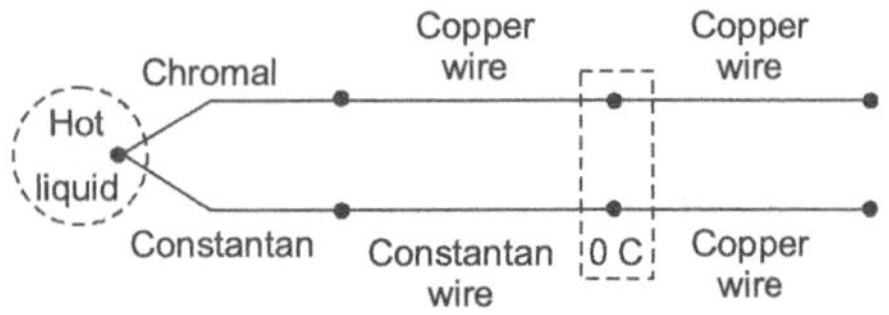

Temp.	emf of Chromel-Constantan	emf of Copper-Constantan
10°C	0.591	0.391
20°C	1.192	0.789
30°C	1.801	1.196
370°C	26.549	19.027
380°C	27.345	19.638

(a) 370.0°C (b) 372.4°C
(c) 376.6°C (d) 380.0°C

[2010 : 1 M]

4.45 A differential pressure transmitter is used to measure the flow rate in a pipe. Due to ageing, the sensitivity of the pressure transmitter is reduced by 5%. All other aspects of the flow meter remaining constant, change in the sensitivity of the flow measurement is

(a) 10.0% (b) 5.0%
(c) 2.5% (d) 2.2%

[2010 : 1 M]

4.46 For a copper-constantan (Type T) thermocouple, the junction potential E(in μV) at θ°C is given by
$E = 38.74\theta + 3.3 \times 10^{-2}\theta^2 + 2.07 \times 10^{-4}\theta^3 - 2.2 \times 10^{-6}\theta^4$ + higher order terms, assuming the cold junction compensation. The sensitivity of thermocouple at 100°C is approximately

(a) 45.35 μV/°C (b) 42.75 μV/°C
(c) 38.74 μV/°C (d) 0.06 μV/°C

[2011 : 1 M]

4.47 The temperature of a furnace is monitored at a distance of 50 m away. The temperature transmitter has a range of 0-500°C and provides 4–20 mA current output. The measured temperature and the output current have a straight line relationship with positive slope. The temperature is determined from the voltage measured across a resistance of 500 Ω in the current loop. If the voltage measured across the resistance is 4 V. The temperature of the furnace is

(a) 100 °C (b) 125 °C
(c) 150 °C (d) 200 °C

[2011 : 1 M]

4.48 In a flapper-nozzle displacement transducer, the values of the following parameters are given: Diameter of the orifice = 0.2 mm, Diameter of the nozzle = 0.8 mm, Supply pressure = 1.4×10^2 kPa (gauge), Ambient pressure = 0 (gauge). The maximum value of the sensitivity is

(a) 4.0 MPa/mm (b) 5.6 MPa/mm
(c) 6.4 MPa/mm (d) 7.3 MPa/mm

[2011 : 2 M]

4.49 A turbine flowmeter is rotating at 72 rpm. The flux ψ linked to the nearby magnet and coil assembly is given by $\psi(\theta) = 3 + \cos(4\theta)$ mWb, where θ is the angular position (in radian). The amplitude and frequency of the output voltage signal, respectively, are

(a) 4 mV and 45.8 Hz
(b) 30.2 mV and 4.8 Hz
(c) 30.2 mV and 30.2 Hz
(d) 288 mV and 45.8 Hz

[2011 : 2 M]

4.50 Water (density: 1000 kgm^{-3}) stored in a cylindrical drum of diameter 1 m is emptied through a horizontal pipe of diameter 0.05 m. A pitot-static tube is placed inside the pipe facing the flow. At the time when the difference between the stagnation and static pressures measured by the pitot-static tubes is 10 kPa, the rate of reduction in water level in the drum is,

(a) $\frac{1}{200\sqrt{5}}$ ms^{-1} (b) $\frac{1}{75\sqrt{10}}$ ms^{-1}

(c) $\frac{1}{50\sqrt{10}}$ ms^{-1} (d) $\frac{1}{40\sqrt{5}}$ ms^{-1}

[2012 : 2 M]

4.51 A U-tube manometer of tube diameter D is filled with a liquid of zero viscosity. If the volume of the liquid filled is V, the natural frequency of oscillations in the liquid level about its mean position, due to small perturbations, is

(a) $\frac{D}{2\sqrt{2\pi}}\sqrt{\frac{g}{V}}$ (b) $\frac{2\sqrt{2}}{\sqrt{\pi}}\frac{\sqrt{gV}}{D^2}$

(c) $\frac{1}{2\sqrt{\pi}}\frac{\sqrt{gD}}{V^{1/3}}$ (d) $\frac{1}{\sqrt{\pi}}\sqrt{\frac{g}{D}}$

[2012 : 2 M]

4.52 The differential pressure transmitter of a flow meter using a venturi tube reads 2.5×10^5 Pa for a flow rate of 0.5 m^3/s. The approximate flow rate in m^3/s for a differential pressure 0.9×10^5 Pa is

(a) 0.30 (b) 0.18
(c) 0.83 (d) 0.60

[2013 : 1 M]

Common Data Questions (4.53 and 4.54):

A tungsten wire used in a constant current hot wire anemometer has the following parameters:
Resistance at 0°C is 10 Ω, Surface area is 10^{-4} m^2, Linear temperature coefficient of resistance of the tungsten wire is 4.8×10^{-3}/°C, Convective heat transfer coefficient is 25.2 W/m^2/°C, flowing air temperature is 30°C, wire current is 100 mA, mass-specific heat products is 2.5×10^{-5} J/°C.

4.53 The thermal time constant of the hot wire under flowing air condition in ms is

(a) 24.5 (b) 12.25
(c) 6.125 (d) 3.0625

[2013 : 2 M]

4.54 At steady state, the resistance of the wire in Ω is

(a) 10.000 (b) 10.144
(c) 12.152 (d) 14.128

[2013 : 2 M]

4.55 The pressure and velocity at the throat of a venture tube, measuring the flow of a liquid, are related to the upstream pressure and velocity, respectively, as follows:

(a) pressure is lower but velocity is higher
(b) pressure is higher but velocity is lower
(c) both pressure and velocity are lower
(d) pressure and velocity are identical

[2014 : 1 M]

4.56 For a rotameter, which one of the following statements is TRUE?

(a) The weight of the float is balanced by the buoyancy and the drag force acting on the float.
(b) The velocity of the fluid remains constant for all positions of the float.
(c) The measurement of volume flow rate of gas is not possible.
(d) The volume flow rate is insensitive to changes in density of the fluid.

[2014 : 1 M]

4.57 A thermopile is constructed using 10 junctions of Chromel-Constantan (sensitivity 60 μV/°C for each junction) connected in series. The output is fed to an amplifier having an infinite input impedance and a gain of 10. The output from the amplifier is acquired using a 10-bit *ADC*, with reference voltage of 5 V. The resolution of this system in units of °C is ______.

[2014 : 2 M]

4.58 A transit ultrasonic flowmeter uses a pair of ultrasonic transducers placed at 45° angle, as shown in the figure.

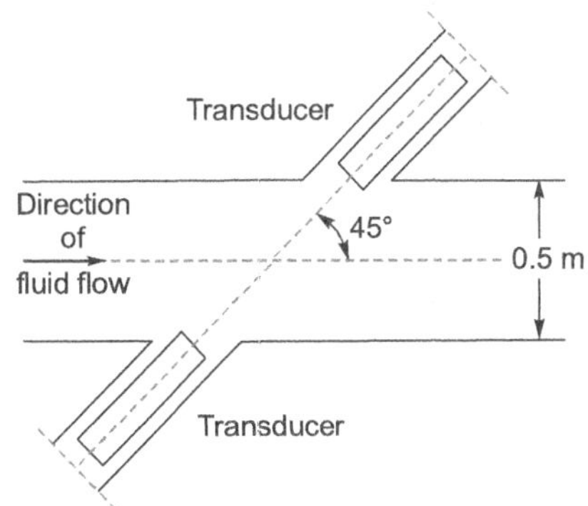

The inner diameter of the pipe is 0.5 m. The differential transit time is directly measured using a clock of frequency 5 MHz. The velocity of the fluid is small compared to the velocity of sound in the static fluid, which is 1500 m/s and the size of the crystals is negligible compared to the diameter of the pipe. The minimum change in fluid velocity (m/s) that can be measured using this system is _____.

[2014 : 2 M]

4.59 Liquid flow rate is measured using

(a) a Pirani gauge (b) a pyrometer
(c) an orifice plate (d) a Bourdon tube

[2015 : 1 M]

4.60 The Seebeck coefficients, in μV/°C, for copper, constantan and iron, with respect to platinum, are 1.9, –38.3 and 13.3, respectively. The magnitude of the thermo emf *E* developed in the circuit shown in the figure, in millivolts is _____ mV.

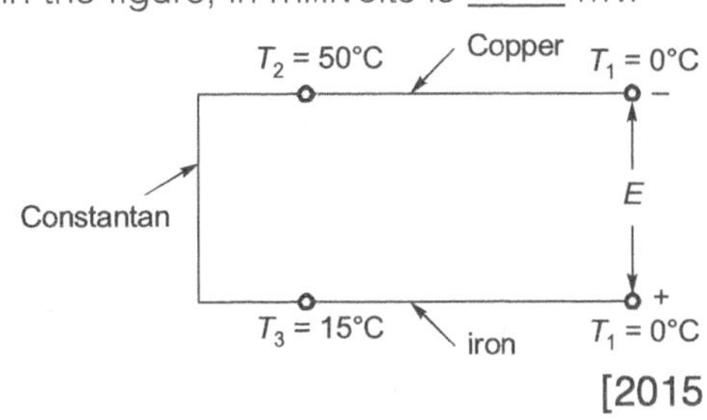

[2015 : 2 M]

4.61 In the figure shown, R_T represents a resistance temperature device (*RTD*), whose characteristic is given by $R_T = R_0(1 + \alpha T)$, where $R_0 = 100\ \Omega$, $\alpha = 0.0039°C^{-1}$ and T denotes the temperature in °C. Assuming the op-amp to be ideal, the value of V_o in volts when $T = 100°C$, is _____ V.

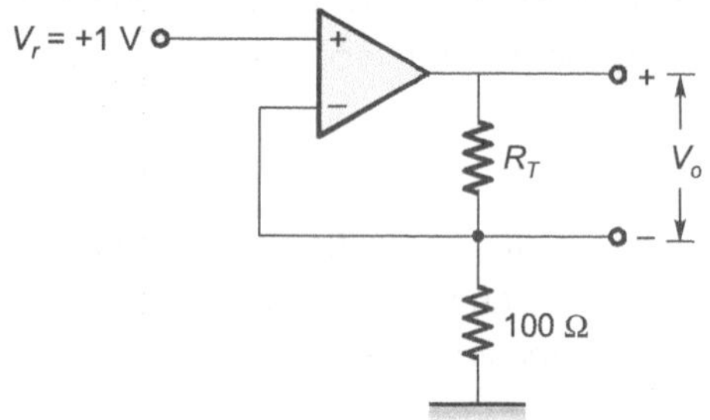

[2015 : 2 M]

4.62 The velocity of flow of water (density 1000 kg/m^3) in a horizontal pipe is measured using the PITOT tube shown below. The fluid in the U-tube manometer is mercury with a density of 13534 kg/m^3. Assume $g = 9.81$ m/s^2. If the height difference (h) is measured as 94.1 mm, the velocity of flow of water in m/s is_________.

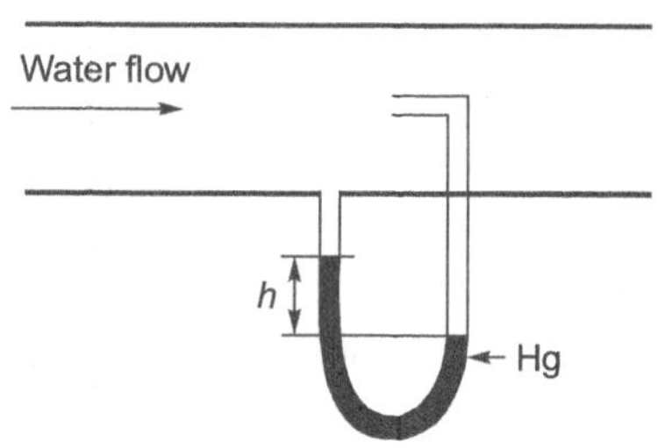

[2016 : 2 M]

4.63 The pressure drop across an orifice plate for a particular flow rate is 5 kg/m^2. If the flow rate is doubled (within the operating range of the orifice), the corresponding pressure drop in kg/m^2 is

(a) 2.5 (b) 5.0
(c) 20.0 (d) 25.0

[2017 : 1 M]

4.64 The most suitable pressure gauge to measure pressure in the range of 10^{-4} to 10^{-3} torr is

(a) Bellows (b) Barometer
(c) Strain gauge (d) Pirani gauge

[2017 : 1 M]

4.65 The hot junction of a bare thermocouple, initially at room temperature (30°C), is suddenly dipped in molten metal at t = 0s. The cold junction is kept at room temperature. The thermocouple can be modeled as a first-order instrument with time constant of 1.0s and a static sensitivity of 10 μV/°C. If the voltage, measured across the thermocouple indicates 10.0 mV at t = 1.0 s, then the temperature of the molten metal in °C is _____.

[2017 : 2 M]

4.66 The magnetic flux density of an electromagnetic flowmeter is 100 mWb/m^2. The electrodes are wall-mounted inside the pipe having a diameter of 0.25 m. A voltage of 1 V is generated when a conducting fluid is passed through the flowmeter. The volumetric flow rate of the fluid in m^3/s is _____.

[2017 : 2 M]

4.67 As shown in the figure, temperature θ is measured using a *K* type thermocouple. It has a sensitivity of 40 μV/°C. The gain (G) of the ideal instrumentation amplifier is 25. If the output V_0 is 96 mV, then the value of θ (in °C) is _______.

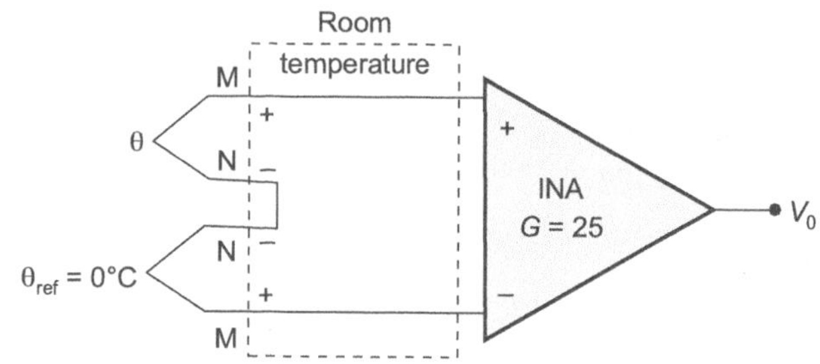

[2018 : 1 M]

4.68 The average velocity v of flow of clear water in a 100 cm (inner) diameter tube is measured using the ultrasonic flow meter as shown in the figure. The angle θ is 45°. The measured transit times are t_1 = 0.9950 ms and t_2 = 1.0000 ms. The velocity v (in m/s) in the pipe is (up to one decimal place) _______.

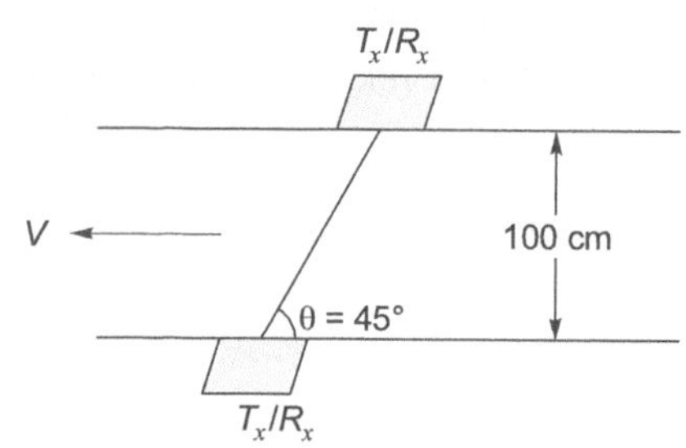

[2018 : 2 M]

4.69 A pitot-static tube is used to estimate the velocity of an incompressible fluid of density 1 kg/m^3. If the pressure difference measured by the tube is 200 N/m^2, the velocity of the fluid, assuming the pitot-tube coefficient to be 1.0, is_______m/s.

[2019 : 1 M]

4.70 Thermocouples measure temperature based on
(a) Photoelectric effect
(b) Seebeck effect
(c) Hall effect
(d) Thermal expansion [2019 : 1 M]

4.71 For the full bridge made of linear strain gauges with gauge factor 2 as shown in the diagram. $R_1 = R_2 = R_3 = R_4 = 100\ \Omega$ at 0°C and strain is 0. The temperature coefficient of resistance of the strain gauges used is 0.005 per °C. All strain gages are made of same material and exposed to same temperature. While measuring a strain of 0.01 at a temperature of 50°C, the output V_o in millivolt is _____ (rounded off to two decimal places).

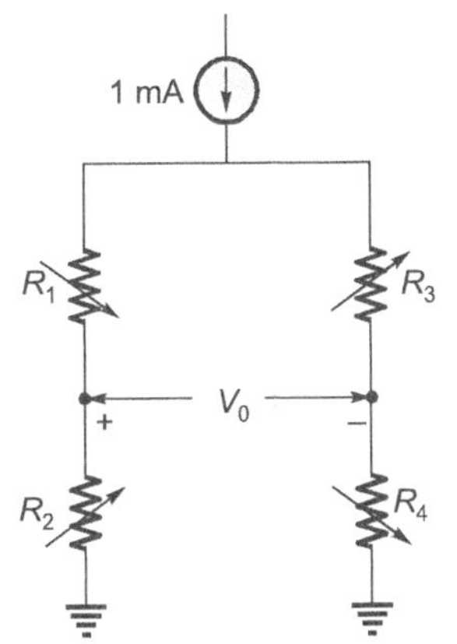

[2021 : 2 M]

4.72 Input-Output characteristic of a temperature sensor is exponential for a
(a) Mercury thermometer
(b) Thermocouple
(c) Resistive Temperature Device (RTD)
(d) Thermistor

[2021 : 1 M]

4.73 A strain gage having nominal resistance of 1000 Ω has a gage factor of 2.5. If The strain applied to the gage is 100 μm/m, its resistance in ohm will change to _____ (rounded off to two decimal places).

[2021 : 1 M]

4.74 Given: Density of mercury is 13600 kg/m^3 and acceleration due to gravity is 9.81 m/s^2. Atmospheric pressure is 101 kPa. In a mercury U-tube manometer, the difference between the heights of the liquid in the U-tube is 1 cm. The differential pressure being measured in pascal is (rounded off to the nearest integer).

[2021 : 1 M]

4.75 A J-type thermocouple has an output voltage $V_\theta = (13650 + 50\,\theta_x)\ \mu V$. where θ_x is the junction temperature in Celsius (°C). The thermocouple is used with reference junction compensation, as shown in the figure. The Instrumentation amplifier used has a gain $G = 20$. If θ_{Ref} is 1°C. for an input θ_x of 100 °C. the output V_o of the instrumentation amplifier in millivolt is:

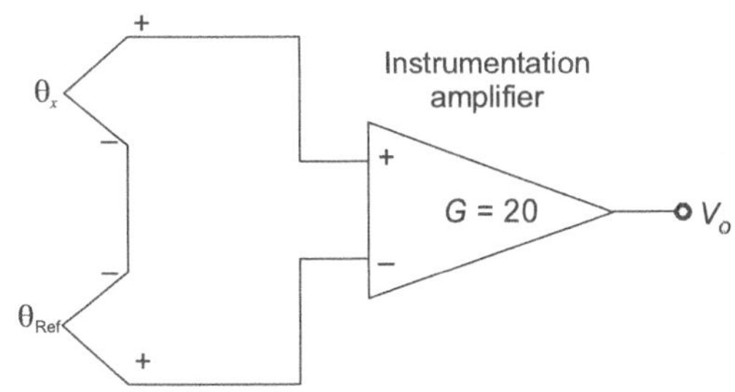

(a) 100 mV (b) 98 mV
(c) 99 mV (d) 101 mV

[2021 : 2 M]

4.76 The figure shows a Chromel-Alumel thermocouple, where the junction A is held at temperature T_A, and a thermal emf E_1 is measured using an ideal voltmeter between the open ends B1 and B2, both held at temperature T_B. Two identical copper wires are introduced between B1-C1 and B2-C2 as shown in the figure. When C1 and C2 are held at temperature T_C, the voltmeter reads a thermal emf E_2. Then, _______

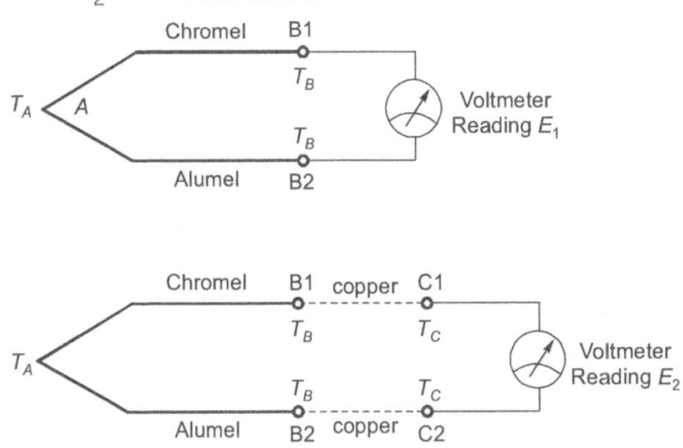

(a) $E_1 < E_2$ (b) $E_1 > E_2$
(c) $E_1 = 2E_2$ (d) $E_1 = E_2$

[2022 : 1 M]

4.77 Which of the following flow meters offers the lowest resistance to the flow?
(a) Turbine flow meter
(b) Orifice flow meter*
(c) Venturi meter
(d) Electromagnetic flow meter

[2023 : 1 M]

4.78 The pressure drop across a control valve is constant. The control valve with inherent characteristic has decreasing sensitivity. If x represents the fraction of maximum stem position of the control valve, then the function $f(x)$ representing the fraction of maximum flow is

(a) a^{x-1}, where a is constant
(b) $\sqrt{x}$
(c) x
(d) x^2

[2024 : 1 M]

4.79 A linear potentiometer (0 – 10 kΩ) is used to measure the water level as shown in the figure. The resistance between A and C varies linearly from 0 to 10 kΩ for a change in water level from 0 to 20 cm. The sensor is excited using a DC voltage source, V_S = 10 V with an internal resistance, R_S = 200 Ω. If V_{out} = 5 V, the water level is ________ cm (rounded off to one decimal place).

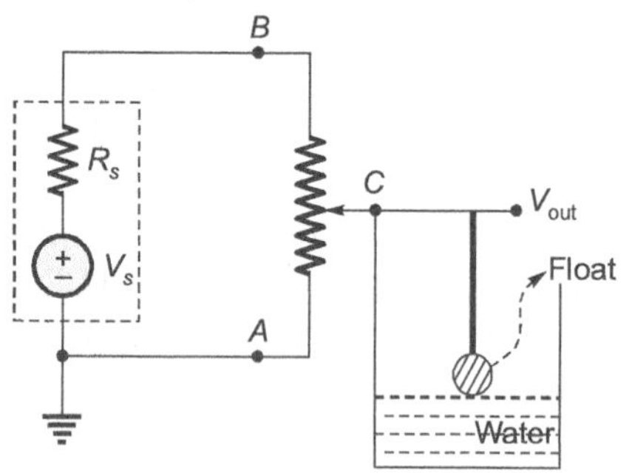

[2024 : 1 M]

4.80 The measured values from a flow instrument, whose range is between 0 and 2 flow units, are shown in the histogram. The systematic error (bias) and the maximum error (in flow units), respectively are

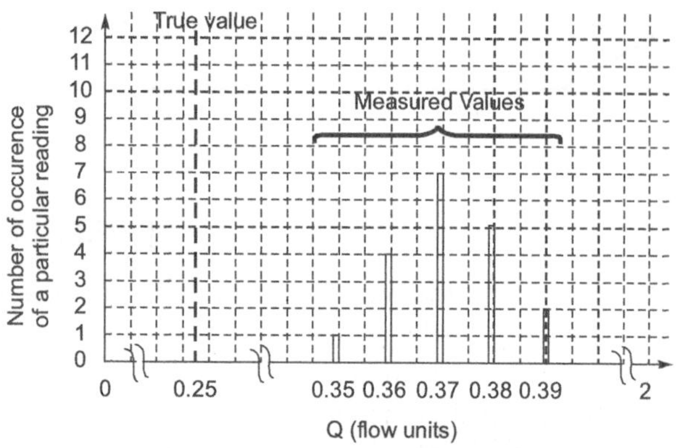

(a) 0.12 and 0.14 (b) 0.01 and 0.10
(c) 0.10 and 0.14 (d) 0.04 and 0.12

[2024 : 2 M]

4.81 Consider an ultrasonic measurement system shown in the figure. The ultrasonic transmitter (T) sends a continuous wave signal $x(t) = \cos(2\pi f_1 t)$ volts towards an object whose vibration is modeled as $m(t) = 0.5\sin(2\pi f_2 t)$ volts. Neglecting the phase shift due to any other effect, the received signal at the receiver (R) is $y(t) = \cos(2\pi f_1 t + \beta\cos(2\pi f_2 t))$ volts.

Assuming the frequency sensitivity factor as 500 Hz/volt, f_1 = 40 kHz, f_2 = 1 kHz, the modulation index (β) and the frequency deviation in $y(t)$, respectively, are

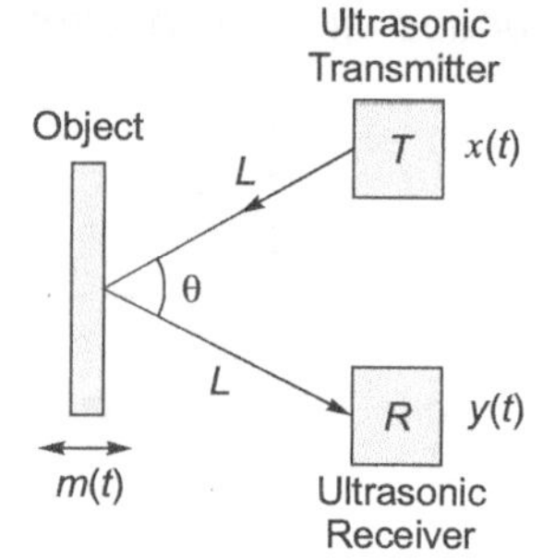

(a) 0.25 and ±250 Hz (b) 0.5 and ±500 Hz
(c) 1 and ±1000 Hz (d) 0.75 and ±1000 Hz

[2024 : 2 M]

4.82 The readings recorded from a 20-psig pressure gauge are given in the Table. The regression line obtained for the data is $y = 0.04x + 10.32$. The regression coefficient of determination, R^2 = ________ (rounded off to three decimal places).

x	1	2	3	4	5
y(psig)	10.3	10.5	10.4	10.5	10.5

[2024 : 2 M]

■■■

Answers Pressure, Flow and Temperature Measurement

4.1	(d)	4.2	(a)	4.3	(a)	4.4	(Sol.)	4.5	(a)	4.6	(2)	4.7	(Sol.)
4.8	Sol.	4.9	(c)	4.10	Sol.	4.11	(d)	4.12	(d)	4.13	(c)	4.14	(c)
4.15	(c)	4.16	(b)	4.17	(c)	4.18	(d)	4.19	(a)	4.20	(d)	4.21	(d)
4.22	(b)	4.23	(b)	4.24	(d)	4.25	(c)	4.26	(b)	4.27	(c)	4.28	(b)
4.29	(d)	4.30	(a)	4.31	(c)	4.32	(c)	4.33	(b)	4.34	(a)	4.35	(c)
4.36	(c)	4.37	(a)	4.38	(b)	4.39	(c)	4.40	(d)	4.41	(d)	4.42	(d)
4.43	(b)	4.44	(d)	4.45	(b)	4.46	(b)	4.47	(b)	4.48	(b)	4.49	(b)
4.50	(d)	4.51	(a)	4.52	(a)	4.53	(b)	4.54	(d)	4.55	(a)	4.56	(a)
4.57	(0.814)	4.58	(0.45)	4.59	(c)	4.60	(1.236)	4.61	(1.39)	4.62	(4.81)	4.63	(c)
4.64	(d)	4.65	(1612.28)	4.66	(1.96)	4.67	(96)	4.68	(5.02)	4.69	(20)	4.70	(b)
4.71	(2.5)	4.72	(d)	4.73	(1000.25)	4.74	(1334.16)	4.75	(c)	4.76	(d)	4.77	(d)
4.78	(b)	4.79	(10.2)	4.80	(a)	4.81	(a)	4.82	(0.5)				

Explanations Pressure, Flow and Temperature Measurement

4.1 (d)

Rotameter is a constant pressure drop variable area flowmeter.

4.2 (a)

The pressure measured by the McLeod gauge is

$$P = \frac{h^2 A_t}{V}$$

$A_t \Rightarrow$ tube area

$V \Rightarrow$ volume of bulb of the tube

Hence, it is evident that the unknown pressure is calculated in term of the physical dimensions of the gauge and hence this gauge can be used as standard for vacuum measurement.

4.3 (a)

Gauge pressure = absolute pressure- atmosphere pressure

$\therefore$ Absolute pressure = 5 + 100 = 105 kPa

4.4 Sol.

(a)-(v), (b)-(i), (c)-(iii), (d)-(ii)

4.5 (a)

Using Bernoulli's equation.

Volumetric flow rate, $Q \propto \sqrt{P_d}$.

This equation holds true for every shape of tube.

4.6 (2)

$$\text{Pressure indicated} = \frac{h^2 A_t}{V - A_t h}$$

$$= \frac{(0.03)^2\left(\frac{\pi}{4}(10^{-3})^2\right)}{100\times10^{-6} - \frac{\pi}{4}\times10^{-6}\times0.03}$$

$$= \frac{7.065\times10^{-4}}{99.97} = 7.067\ \mu\text{m}$$

if capillary value is assumed to be negligible

$$P = \frac{A_t h^2}{V} = \frac{\frac{\pi}{4}\times10^{-6}\times(0.03)^2}{100\times10^{-6}}$$

$$= 7.065\times10^{-6}$$

$$\text{error} = (7.067 - 7.065)\ \mu\text{m}$$

$$= 2\times10^{-3}\ \mu\text{m} = 2\ \text{nm}$$

4.7 Sol.

(a)-(iv), (b)-(i), (c)-(ii), (d)-(iii)

4.8 Sol.

(a)-(iv), (b)-(ii), (c)-(i), (d)-(iii)

4.10 Sol.

(a)-(i), (b)-(ii), (c)-(iii), (d)-(iv)

4.11 (d)

The measuring junction is at 200°C

∴ output = 13.5 mV

the reference junction in at 0°C

∴ output voltage = (13.5 – 0) = 13.5 mV

4.12 (d)

Absolute pressure = 3.2 + 1.01 = 4.21 bar.

4.13 (c)

Pirani gauge → upto 10^{-4} Torr

Mcleod gauge → upto 10^{-6} Torr

Ionisation vacuum gauge → upto 10^{-9} Torr

So, Ionisation vacuum gauge can measure lowest vacuum pressure.

Note: Vacuum pressure is pressure having a value below atmospheric pressure.

4.14 (c)

	Flow meter type	Discharge coefficient, C_d
1.	Orifice	0.56 to 0.68
2.	Flow nozzle	0.92 to 0.98
3.	Venturi tube	0.95 to 0.995

4.15 (c)

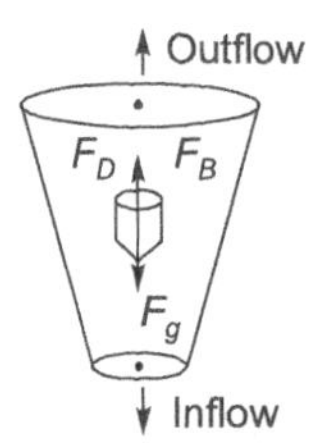

Force balance equation of flow is

$$F_g = F_B + F_D$$

is valid only vertical pipeline with flow inlet at bottom.

In any other configuration, above equation do not held.

In horizontal pipe, float will flow in direction of flow.

In vertical pipe with in flow at top, float will not be static and will move downwards as,

$$(F_D + F_g) > F_B$$

4.16 (b)

Positive displacement meter measures fixed volume of fluid at every cycle and thus measures volumetric flow rate.

4.17 (c)

By thermometer formulae,

$$\theta(t) = \theta_f(1 - e^{-t/\tau})$$

$$0.98\,\theta_f = \theta_f(1 - e^{-t/\tau})$$

$$0.98 = 1 - e^{-t/\tau}$$

$$e^{-t/\tau} = 0.02$$

$$-\frac{t}{\tau} = -4$$

$$\therefore \quad \boxed{t = 4\tau}$$

4.18 (d)

Pirani gauge range is 10^{-2} to 10^{-4} torr ≈ 1 Pa to 0.01 Pa.

4.19 (a)

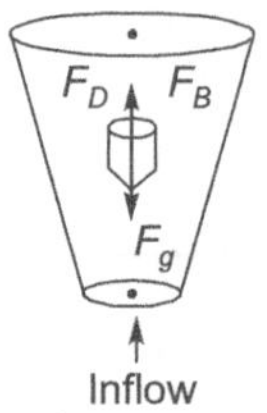

Force balance equation for float is,

$$F_g = F_B + F_D$$

where, $F_g \rightarrow$ gravitational force

$F_B \rightarrow$ Buoyant force

$F_D \rightarrow$ Drag force

$$F_g = Vgd_1 \quad \text{...(i)}$$

$$F_B = Vgd_2 \quad \text{...(ii)}$$

$$F_D = (P_b - P_a)A \quad \text{...(iii)}$$

So, using force balancing equation and (i), (ii) and (iii).

$$(P_b - P_a)A = Vg(d_1 - d_2)$$

$$\Rightarrow (P_a - P_b)A = Vg(d_2 - d_1)$$

4.21 (d)

Rotameters are the constant pressure drop, and variable area type flow meters, all other meters like orifice meter, venturimeter and flow nozzle are differential pressure meters.

4.22 (b)

The frequently response of a first order system when it is subjected to the sinusoidal input is given as:

$$\frac{A_0}{A_i} = \frac{1}{\sqrt{(1+\omega\tau)^2}}$$

Given, $\tau = 5$ sec ; $\omega = \frac{2\pi}{T} = \frac{2\pi}{20} = \frac{\pi}{10}$; $A_i = 10°C$

$$\therefore \quad A_0 = \frac{10}{\sqrt{1+\left(\frac{\pi}{10}\times 5\right)^2}} = \frac{10}{\sqrt{1+\left(\frac{3.14}{2}\right)^2}} = \frac{10}{\sqrt{1+2.46}} = \frac{10}{1.86}$$

$$= 5.37°$$

4.23 (b)

	Thermocouple pair	Sensitivity
A	Platinum - Constantan	0 – (–35) = 35 μV/K
B	Nichrome - Constantan	25 – (–35) = 60 μV/K
C	Nickle - Constantan	–25 – (–35) = 10 μV/K
D	Copper - Nickle	6 – (–25) = 31 μV/K

So, maximum sensitivity is for Nichrome - Constantan thermocouple.

4.24 (d)

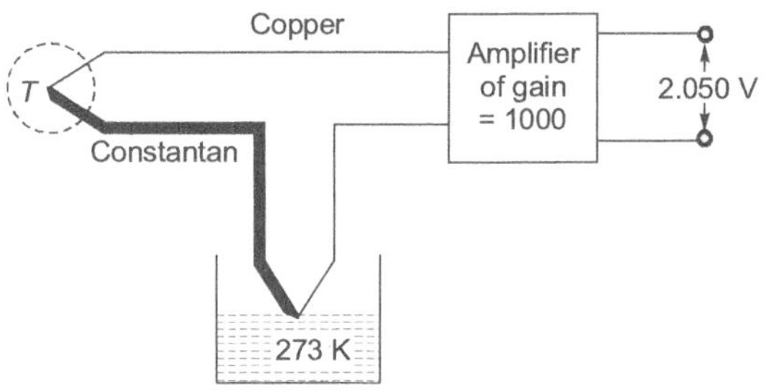

Sensitivity of copper

$$= 6 - (-35) = 41\ \mu V/K$$

$$41 \times 10^{-6} \times 1000 \times (T - 273) = 2.050$$

$$41 \times 10^{-3}(T - 273) = 2.050$$

$$(T - 273) = \frac{2.050}{41 \times 10^{-3}} = 0.05 \times 10^3$$

$$(T - 273) = 50$$

$$\therefore \quad T = 273 + 50\text{ K} = 50°\text{C}$$

4.25 (c)

$$G_f = \frac{\Delta R/R}{\epsilon}$$

Given that,

$$G_f = -100$$

$$R = 1000\ \Omega$$

$$\epsilon = -100\ \mu m/m = -100 \times 10^{-6}\ m/m$$

$$-100 = \frac{\Delta R/1000}{-100 \times 10^{-6}}$$

$$\Delta R = 10$$

∴ Resistance of gauge

$$= R + \Delta R = 1000 + 10$$

$$= 1010\ \Omega$$

Note: It can be observed that for semiconductor strain gauges, sensitivity is very high i.e., for 100 μm/m strain, change in resistance is 10 Ω.

4.26 (b)

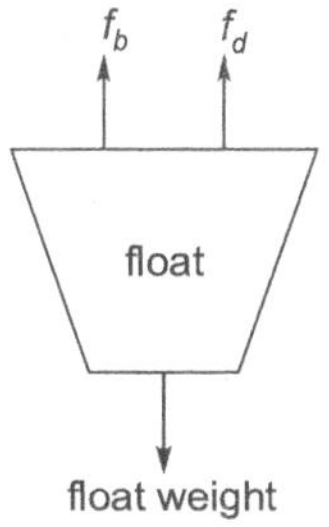

Float weight = drag force + buoyancy force

4.27 (c)

Error is measurement when a ramp of magnitude 'k' is applied is

$$e_{ss} = k\tau$$

where, τ = time constant

$$e_{ss} = 1\frac{°\text{C}}{\text{sec}} \times 100 \times 10^{-3}\text{ sec} = 0.1\ °\text{C}$$

∴ the reading of the thermocouple will be

$$= 70° - 0.1°\text{C} = 69.9°\text{C}$$

4.28 (b)

$$E_{100} = 5268\ \mu V$$

$$E_{300} = 16325\ \mu V$$

By the linear relationship

$$E_{100}\text{ must be} = \frac{E_{300}}{3}$$

$$\therefore \quad E_{100} = \frac{16325}{3} = 5441.66\ \mu V$$

$$\%\text{ age error} = \frac{5268 - 5441.66}{16325} \times 100 = -1.06\%$$

4.29 (d)

$$v = 60\text{ mm/s}$$

For $e = Bdv$

Here, $B = 0.1$ T

$$d = 50 \times 10^{-3}\text{ m}$$

$$v = 20\text{ mm/sec} = 20 \times 10^{-3}\text{ m/sec.}$$

$$e = 0.1 \times 50 \times 10^{-3} \times 20 \times 10^{-3}$$

$$= 100 \times 10^{-6} = 100\ \mu V$$

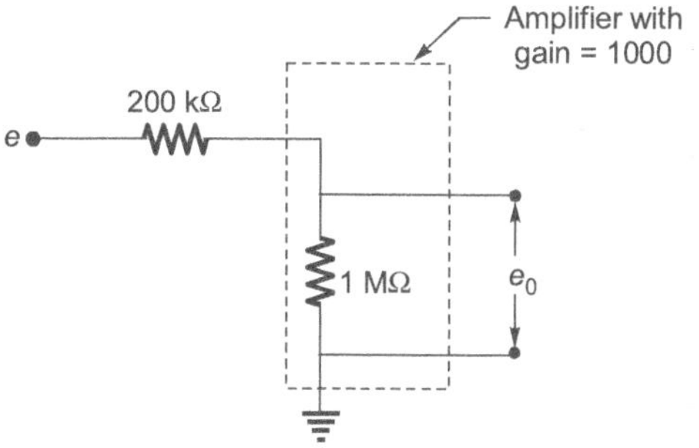

∴ Clearly by voltage divider rule we have

$$e_0 = \frac{1\text{M}\Omega}{1\text{M}\Omega + 200\text{ k}\Omega} e \times \text{gain}$$

$$= \frac{1\text{M}\Omega}{1.2\text{M}\Omega} \times e \times \text{gain}$$

$$= \frac{1}{1.2} \times (100 \times 10^{-6}) \times 1000$$

$$= 0.083\text{ volts}$$

4.30 (a)

Since the scale of a rotameter is linear, so the rising of the float in the tube depends directly on the applied flow

$\therefore \quad Q \propto x$

$\therefore \quad \frac{Q_1}{Q_2} = \frac{70}{20} = 3.5$

4.31 (c)

The correction applied = $h\beta T$

$\therefore$ The reading of barometer = $h + h\beta T$

$= h(1 + \beta T)$

4.32 (c)

As se can see the above situation it is similar to the Case-II as we discussed in the text

$$\Delta t = \frac{2L\cos\theta}{C^2}v$$

$$\therefore \quad v = \frac{(\Delta t)C^2}{2L\cos\theta}$$

$$= \frac{10\times10^{-9}\times(1000)^2}{2\times25\times10^{-3}\times\cos60°} = 0.4 \text{ m/sec}$$

Method-2:

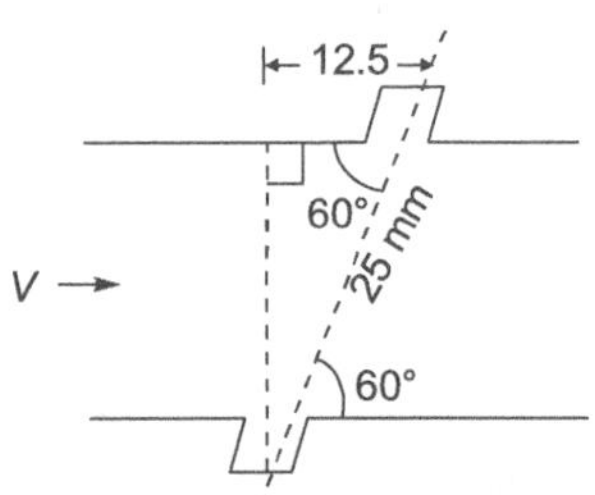

Let, t_1 be upstream signal time and t_2 be downstream signal time.

For upstream transmission,

12.5 mm = $(V + 1000)t_1$

$$\Rightarrow t_1 = \frac{12.5}{V+1000} \quad ...(i)$$

For downstream transmission,

12.5 mm = $(1000 - V)t_2$

$$\Rightarrow t_2 = \frac{12.5}{1000-V} \quad ...(ii)$$

$$\Rightarrow \Delta t = \frac{12.5(2\text{ V})}{(1000)^2 - V^2}$$

$(1000)^2 >> V^2$

$$\Rightarrow \Delta t = \frac{2\times12.5\times V}{(1000)^2}$$

$$\Rightarrow V = \frac{10\times10^{-9}\times(1000)^2}{2\times12.5\times10^{-3}} = 0.4 \text{ m/sec}$$

4.33 (b)

e.m.f. developed when reference junction in at 0°C $[E_{R0}]$ for junction at 200°C

$= 200 \times 53\ \mu V = 10600\ \mu V$

e.m.f. developed when reference is at 45°C $[E_{R45}]$ for junction at 200°C

$= 200 \times 53 - 53 \times 45$

$= 10600 - 2385$

$= 8215\ \mu V$

$$\%\text{age error} = \frac{E_{R0} - E_{R45}}{E_{R0}} \times 100$$

$$= \frac{10600 - 8215}{10600} \times 100$$

$$= \frac{2385}{106} = 22.5$$

4.34 (a)

$E_{IC\,(\text{iron-constantan})} = 53 \times T\ \mu V$

$E_{CC(\text{Copper-constantan})} = 43 \times T\ \mu V$

$\therefore \quad E_{(ICo)\,(\text{iron-copper})} = 53T - 43T = 10\ T\ \mu V$

Output when the measuring junction is kept at 100°C and the reference junction is at 30°C is

e.m.f. developed = $10 \times 100 - 10 \times 30$

$= 1000 - 300$

$= 700\ \mu V$

4.35 (c)

The given circuit can be redrawn as:

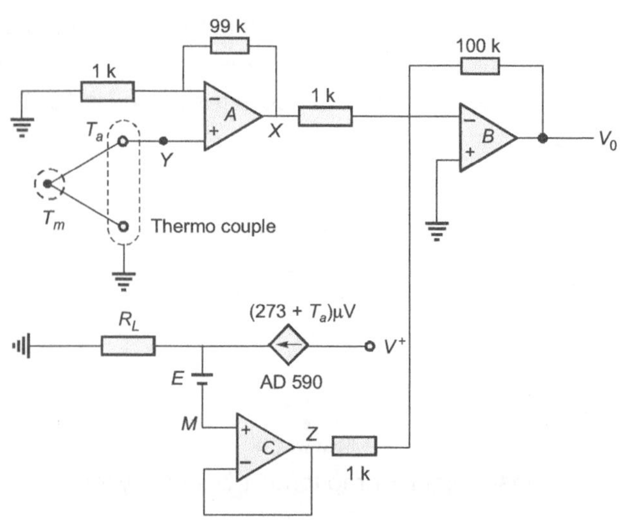

- Clearly it can be observed from the above circuit diagram that the op-amp B is behaving as a summer circuit, with two arms as the input, (i.e. arm A and arm C).
- The thermocouple is placed in the circuit A, measuring temperature (T_m) and the reference junction at the ambient temperature T_a, for output to be error free, T_a must be kept 0°C but here it has some finite temperature. So we have to compensate for this error by using a compensation circuit, i.e. (Circuit C) here.

Let the T_m be at 100°C, and T_a be at 30°C

∴ the emf produced by the thermocouple at point Y

$$= (100 - 30)\, 53\ \mu V$$
$$= 3710\ \mu V \quad (\text{for } T_a = 30°C)$$

The op-amp A is configured in the non-inverting configuration.

$$\therefore \quad E_x = \left(1 + \frac{99}{1}\right) E_Y = 100 \times 3710\ \mu V$$
$$= 100 \times 3710\ \mu V = 371\ mV$$

But to have the zero error. T_a must be at 0°C.

So, E_Y for $T_a = 0°C$ is,

$$E_Y = (100 - 0)\, 53 \times 10^{-3}\ \mu V$$
$$= 5300\ \mu V$$

and $$E_x = \left(1 + \frac{99}{1}\right) 5300\ \mu V = 530\ mV$$

$$\therefore \quad \text{error} = (E_x)_{T_a = 0°} - (E_x)_{T_a = 30°}$$
$$= 530 - 371\ mV = 159\ mV$$

This error must be compensated by the output of the op-amp C.

For op-amp C, all the current coming from AD 590 sensor will pass through the resistor R_L, (Since no current) goes inside the op-amp).

$$\therefore \quad E_M = \text{Voltage across } R_L - E$$
$$= (273 + 30) \times 10^{-6} \times R_L - E$$
$$= 303 \times 10^{-6} \times R_L - E \quad \text{...(i)}$$

op-amp 'C' is used in the voltage follower configuration.

$$\therefore \quad E_Z = E_M$$

and in order to provide temperature compensation.

(E_Z = error voltage i.e. = 159 mV) ...(ii)

Putting equation (ii) in equation (i) we get,

$$303 \times 10^{-6} \times R_L - E = 159 \times 10^{-3}$$

By hit and trial clearly option (C) satisfies the above equation.

Hence, $R_L = 5.3\ k\Omega$,

and $E = 1.447$ V

4.37 (a)

Difference in the temperature between the hot and the cold junction is

$$\Delta T = 300°C - 100°C$$
$$= 200°C$$

The temperatures T_3 and T_4 do not have any effects since the metal is same (i.e. constantan). [According to first law, the application of heat on a single homogenous metal do not produce any electric current].

Sensitivity = 55 μV/°C

∴ for 1°C output = 55 μV

for 200°C output = 55 × 200 μV

= 110 × 100 μV = 11.0 mV

4.39 (c)

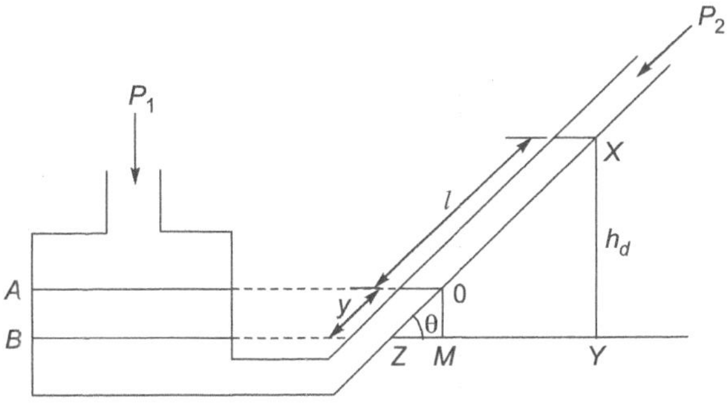

By the application of pressure P_1 on the well, the depression is observed from point A to B and the liquid rises in the tube.

Volume of liquid depressed in the well = Volume of liquid raised in the tube

⇓ ⇓

$$(AB)\, a_w = l \times at$$
$$AB = \frac{a_t}{a_w} l$$

In the triangle OMZ.

$OM = AB$ (as can be observed from the geometry clearly)

$$\therefore \quad OM = \frac{a_t}{a_w} l$$

$$\sin\theta = \frac{\text{Perpendicular}}{\text{Hypotenuse}} = \frac{OM}{y}$$

$$\therefore \quad y = \frac{OM}{\sin\theta} = \frac{a_t}{a_w} \frac{l}{(\sin\theta)}$$

In triangle ZYX,

$$ZX = y + l = \frac{a_t l}{a_w \sin\theta} + l$$

$$\sin\theta = \frac{XY}{ZX} = \frac{h_d}{ZX}$$

$$\therefore \quad h_d = \sin\theta(ZX)$$

$$= \sin\theta\left[\frac{a_t l}{a_w \sin\theta} + l\right] = \frac{a_t l}{a_w} + l\sin\theta$$

$$l\sin\theta = \frac{h_d}{l} - \frac{a_t}{a_w}$$

$$\therefore \quad \theta = \sin^{-1}\left[\frac{h_d}{l} - \frac{a_t}{a_w}\right]$$

4.40 (d)

For thermister,

$$R_1 = R_0 e^{\beta\left[\frac{1}{T_1} - \frac{1}{T_0}\right]}$$

So, as temperature varies linearly, resistance decreases exponentially.

4.41 (d)

From table,
Sensitivity of thermocouple is

$$= \frac{125 - 35}{90 - 0}\mu V/°C = 1\mu V/°C$$

Also for thermocouple,

$$V = a(\Delta T)$$

$$\Rightarrow \quad 48 = 1[T - 2]$$

$$\Rightarrow \quad T = 50°C$$

4.42 (d)

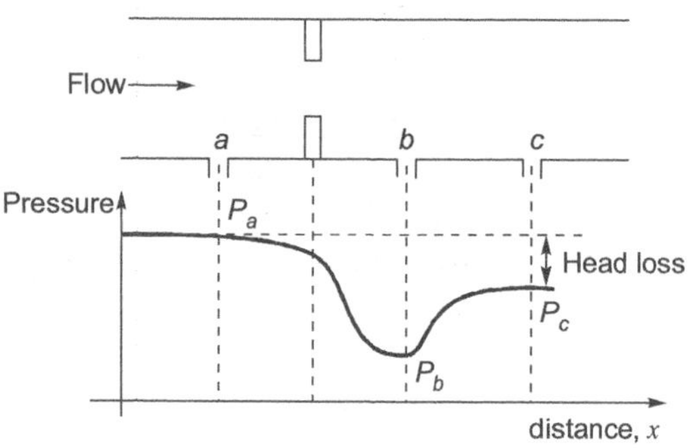

$$P_a > P_c > P_b$$

4.44 (d)

The e.m.f. produced at the point (*B*) and (*C*) will be zero. Since it is a combination of the similar metals [in order to generate e.m.f. the junction must be made up of dissimilar metals]. Point (*A*) is at 30°C, from the given data and is a combination of chromel and copper.

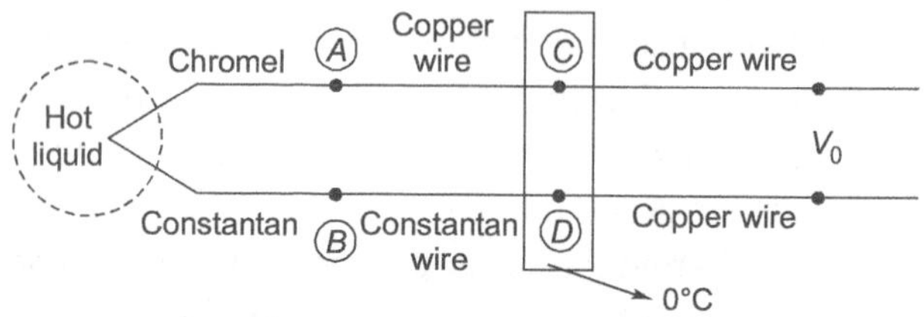

At 30°C

$$V_{(\text{Chromel-copper})} = V_{(\text{Chromel-constantan})} - V_{(\text{Copper-constantan})}$$

$$= 1.801 - 1.196 = 0.605 \text{ mV}$$

Now, $\quad V_0 = V_{\text{Hot junction}} - 0.605$

$$V_{\text{Hot junction}} = 26.74 + 0.605 = 27.345 \text{ mV}$$

It can be clearly seen from the figure that the Hot junction is made up of Chromel-constantan. Looking in the table for Chromel-constantan junction the voltage output of 27.345 μV corresponds to 380°C.

Note: The effect of the junction (*D*) is not taken or is taken as 0 μV as its value is not given, hence it is neglected in the above calculations.

4.45 (b)

Correct explanation:

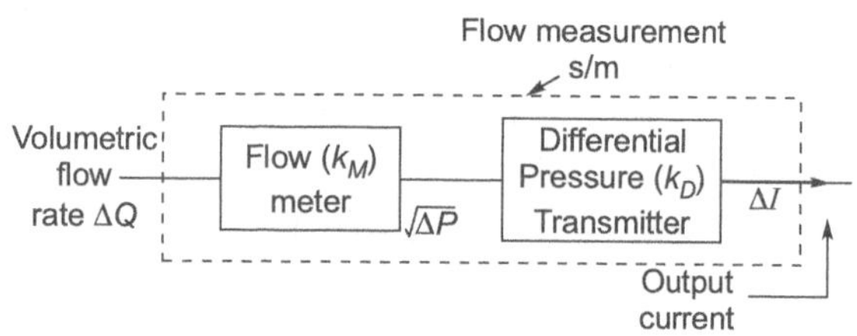

Sensitivity of overall measurement system (K_{MS})
= Sensitivity of flow meter (k_{FM}) × sensitivity of differential pressure transmitter $(k_D)_P$

$$\Rightarrow \quad k_{MS} = k_{FM} \times k_{DP}$$

Case-1

$$(k_{MS})_1 = (k_{FM})_1 \times (k_{DP})_1$$

Case-2

If sensitivity of differential transmitter is reduced by 5% then.

$$(k_{DP})_2 = 0.95(k_{DP})_1;\ (k_{FM})_1 = (k_{FM})_2$$

$$(k_{MS})_2 = (k_{FM})_2 \times (k_{DP})_2$$

$$(k_{MS})_2 = (k_{FM1}) \times 0.95(k_{DP})_1$$

$$(k_{MS})_2 = 0.95(k_{FM})_1 (k_{DP})_1$$

$$(k_{MS})_2 = 0.95(k_{MS})_1$$

This means sensitivity of measurement system reduces by 5%.

4.46 (b)

Sensitivity here$(s) = \dfrac{dE}{d\theta}$

$= \dfrac{d}{d\theta}(38.74\theta + 3.3\times10^{-2}\theta + 2.07\times 10^{-4}\theta^3 - 2.2\times10^{-6}\theta^4)$

$= (38.74 + 6.6 \times 10^{-2} + 6.21 \times 10^{-4}\theta^2 - 8.8 \times 10^{-6}\theta^3)\ \mu V$

$= (38.74 + 6.6 \times 10^{-2} + 6.21 \times 10^{-4}\theta^2 - 8.8 \times 10^{-6}\theta^3)\ \mu V$

$S|_{\theta=100c} = 37.74 + 6.6 + 6.21 - 8.8$

$= 42.75\ \mu V / °C$

4.47 (b)

$V = 4\text{ V} \Rightarrow I = \dfrac{V}{R} = \dfrac{4}{500} = 8\text{mA}$

4 – 20 mA correspond to 0.500°C.

4 mA ⇒ 0°C

20 mA ⇒ 500°C

$I = Mt + C$

$4\text{ mA} = M(0) + C$ ...(i)

$20\text{ mA} = 500\,M + C$ ...(ii)

$\boxed{C = 4\text{ mA}}$

$500\text{ M} = 20\text{ mA} - 4\text{ mA} = 16\text{ mA}$

$M = \dfrac{16}{500} = 0.032$

$\therefore\ I = 0.032 \times T + 4\text{ mA}$

$8\text{ mA} = 0.032 \times T + 4\text{ mA}$

$\therefore\ 0.032\,T = 4\text{ mA}$

$\therefore\ \boxed{T = 125°C}$

4.48 (b)

Given,

Diameter of orifice = 0.2 mm

⇒ Radius of orifice, $r_0 = \dfrac{0.2}{2} = 0.1\text{mm}$

and diameter of nozzle = 0.8 mm

⇒ radius of nozzle, $r_n = \dfrac{0.8}{2} = 0.4\text{ mm}$

Now,

Maximum sensitivity $= \dfrac{P_{rn}}{r_0^2}$

$= \dfrac{1.4\times10^2 \times 0.4\times10^{-3}}{(0.1\times10^{-3})^2}\text{ kPa/m} = 5.6\text{ MPa/mm}$

4.49 (b)

Given, flux linkage, $\Psi(\theta) = 3 + \cos(4\theta)$ mWb

Also, according to Faraday's law of electromagnetic induction.

$|E(\theta)| = \left|\dfrac{d\Psi(\theta)}{dt}\right| = 4(\sin 4\theta)\dfrac{d\theta}{dt}$

$= 4\sin(4\theta)$ WmV

Now, frequency of output voltage

$= 4 \times 72$ rpm

$= \dfrac{288}{60}$ rpm or 4.8 Hz

Magnitude of output voltage

$= \dfrac{4\times 2\pi\times 72}{60} = 30.2\text{ mV}$

4.50 (d)

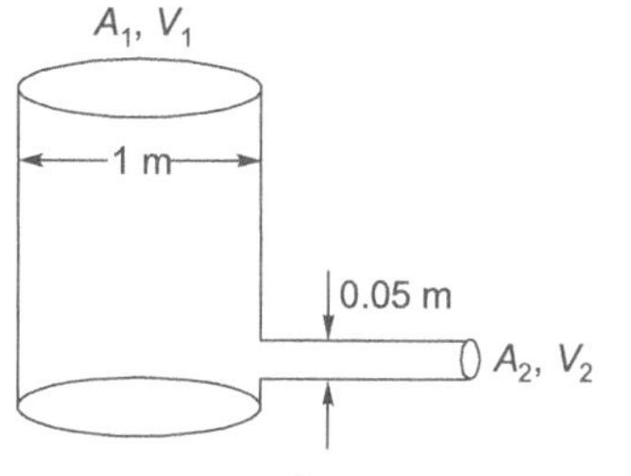

$A_1 = \pi r_1^2 = \pi\times\left(\dfrac{1}{2}\right)^2 m^2 = \dfrac{\pi}{4} m^2$

$A_2 = \pi r_2^2 = \pi\times\left(\dfrac{0.05}{2}\right)^2$

$= \pi\times\left(\dfrac{5}{200}\right)^2 = \dfrac{\pi\times 25}{4\times10^4} m^2$

$V_2 = \sqrt{\dfrac{2g}{w}(\Delta P)}$

where, w = Specific weight of water = ρg

ΔP = Pressure drop cross the Pitot tube

by, equation of continuity of flow

$A_1V_1 = A_2V_2$

$\therefore\ \dfrac{\pi}{4}\times V_1 = \dfrac{25\pi}{4\times10^4}\times\sqrt{\dfrac{2\times g}{1000\times g}\times 10\times 10^3}$

$V_1 = \dfrac{25}{10^4}\times\sqrt{20}$

$= \dfrac{1}{400}\times\sqrt{20}\times\dfrac{\sqrt{20}}{\sqrt{20}} = \dfrac{1}{400}\times\dfrac{20}{\sqrt{20}}$

$= \dfrac{1}{20\sqrt{20}} = \dfrac{1}{20\times 2\sqrt{5}} = \dfrac{1}{40\sqrt{5}}\text{ msec}^{-1}$

Method 2:

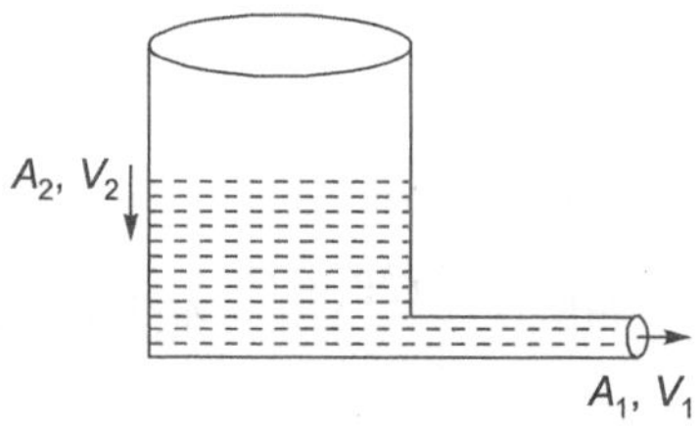

Difference between stagnation pressure and static pressure is due to kinetic energy so,

$$p_{stagnation} - p_{static} = 10 \text{ kPa} = \frac{1}{2}\rho V_1^2$$

$$\Rightarrow 10 \times 10^3 = \frac{1}{2} \times 1000 \times V_1^2$$

$$\Rightarrow \quad V_1 = 2\sqrt{5} \text{ m/sec}$$

We need to find rate of reduction in water level in the drum i.e., V_2. So using equation of continuity,

$$A_1 V_1 = A_2 V_2$$

$$\Rightarrow 2\sqrt{5} \times \pi\left(\frac{0.05}{2}\right)^2 = \pi \times \left(\frac{1}{2}\right)^2 \times V_2$$

$$\Rightarrow \quad V_2 = \frac{1}{40\sqrt{5}} \text{ m/sec}$$

4.51 (a)

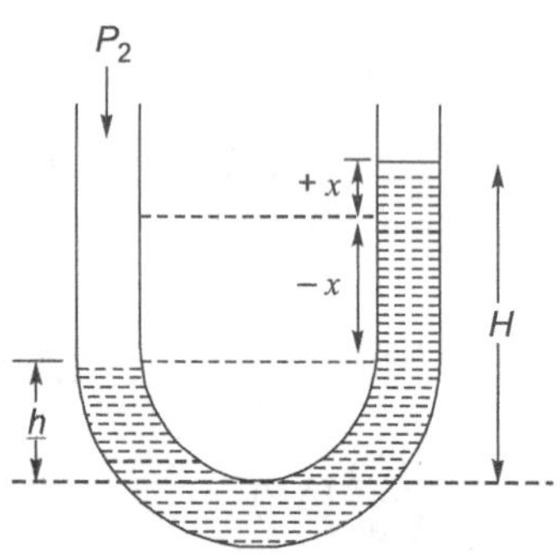

The case given is as shown in the figure.

Total force exerted on the liquid

$$= \rho\, a\, g\, (H\text{-}h)$$

$$= 2\, \rho\, a\, g\, x$$

where a = area of the tube (m^2)

x = displacement of the liquid (m)

p = applied pressure

v = total volume of manometer fluid (m^3)

Force due to friction $= B\dfrac{dx}{dt} = 0$ (given)

as here, friction (viscosity) are zero.

Force acting due to inertia $= \rho v \dfrac{d^2x}{dt^2}$.

Total force due to pressure P

= total force on the liquid + force due to inertia + force due to viscous friction (= 0)

$$Pa = \rho v \frac{d^2x}{dt^2} + 2\pi a g x$$

taking Laplace transform

$$P(s) = \frac{\rho v}{a} s^2 X(s) + 2\rho g X(s)$$

$$G(s) = \frac{X(s)}{P(s)}$$

normalized transfer function is

$$G(s) = 2\rho a g \frac{X(s)}{P(s)} = \frac{2ag/V}{s^2 + \dfrac{2ag}{V}}$$

$$s^2 + \frac{2ag}{V} = 0$$

Replacing s by $(j\omega)$ we get

$$(j\omega)^2 + \frac{2ag}{V} = 0$$

$$-\omega^2 = -\frac{2ag}{V}$$

$$\omega = \sqrt{\frac{2ag}{V}}$$

$$a = \pi\left(\frac{D}{2}\right)^2 = \frac{\pi D^2}{4}$$

$$\therefore \quad \omega = \sqrt{2\frac{\pi D^2}{4}\frac{g}{V}} = \sqrt{\frac{\pi D^2 g}{2V}} = D\sqrt{\frac{\pi g}{2V}}$$

$$f = \frac{1}{2\pi} D\sqrt{\frac{\pi g}{2V}}$$

$$\boxed{f = \frac{D}{2\sqrt{2\pi}}\sqrt{\frac{g}{V}}}$$

Method-2 (Dimensional analysis)

It can be observed from options that dimensions of all options are differents so we can go for dimensional analysis.

Diameter, $D \rightarrow [L]$

Volume, $V \rightarrow [L^3]$

Acceleration due to gravity, $g \rightarrow$ m/sec$^2 \rightarrow [LT^{-2}]$

Frequency $\rightarrow [T^{-1}]$

$$\frac{D}{2\sqrt{2\pi}}\sqrt{\frac{g}{V}} = L\sqrt{\frac{LT^{-2}}{L^3}} = [T^{-1}]$$

Hence option (a) is correct answer.

Note: You should go for dimensional analysis only after checking whether dimensions in options are different or not.

4.52 (a)

In case of differential pressure measurement

$Q(\text{flow rate}) \propto \sqrt{(\Delta P)}$

$$\frac{Q_1}{Q_2} = \frac{\sqrt{\Delta P_1}}{\sqrt{\Delta P_2}}$$

$$\therefore \quad \frac{0.5}{Q_2} = \frac{\sqrt{2.5\times 10^5}}{\sqrt{0.9\times 10^5}}$$

$$Q_2 = \frac{\sqrt{0.3}}{\sqrt{2.5}}\times 0.5 = 0.3 \text{ m}^3/\text{sec}$$

4.53 (b)

- Resistance of tungsten wire at 0°C = 10 Ω

  $R_0 = 10\ \Omega$
- Surface Area, $A = 10^{-4}$ m².
- Coefficient of resistance of the wire

  $\alpha = 4.8 \times 10^{-3}$ /°C
- Convective heat transfer coefficient

  $U = 25.2$ W/m²/°C
- Temperature of air (T) = 30°C
- Current passing through wire (I_0) = 100 mA.

- Mass-specific heat product (mC)

  $= 2.5 \times 10^{-5}$ J/°C

  To get the time constant, first we should model the hot wire.
- By applying conservation of heat energy to hot wire.

  (Input electrical energy) – (Output heat energy loss) = Accumulation of energy.

$$i^2R_T - U(V)\,(T - T_F) = MC\cdot\frac{dT}{dt} \qquad \dots (1)$$

Where $U(v)$ is the convective heat transfer coefficient between sensor and fluid. If I_0 and R_{T0}, T_0 and V_0 represent steady equilibrium conditions, then

$$i_0^2 R_0 - U(V_0)\,A(T_0 - T_F) = 0 \qquad \dots (2)$$

Defining Δi, ΔR_T, ΔV and ΔT to be small deviations from the above equilibrium values, we have

$$i = i_0 + \Delta i$$
$$R_T = R_{T0} + \Delta R_T$$
$$T = T_0 + \Delta T$$
$$U(v) = U(v_0) + \sigma\Delta v \qquad \dots (3)$$

in equation (3) $\sigma = \left(\dfrac{\partial U}{\partial v}\right)_{v_0}$. that means the rate of change of U with respect to v, evaluated at equilibrium velocity v_0.

From (1) and (3) we can write as

$$(i_0 + \Delta i)^2(R_{T0} + \Delta R_T) - [U(v_0) + \sigma\Delta v]A(T_0 + \Delta T - T_F)$$
$$= MC\frac{d}{dt}(T_0 + \Delta T) \qquad \dots (4)$$

neglecting all terms involving the product of two small quantities gives

$$(i_0^2 + 2i_0\Delta i)\,R_{T0} + i_0^2\Delta R_T - U(v_0)A\,(T_0 - T_F) - U(v_o)$$
$$A\Delta T - \sigma A(T_0 - T_F)\,\Delta v = MC\frac{d\Delta T}{dt} \qquad \dots (5)$$

Subtracting (2) from (5) gives,

$$2i_0\,R_{T0}\,\Delta i + i_0^2R_T - U(v_0)A\,\Delta T - \sigma A(T_0 - T_F) - \Delta v$$
$$= MC\frac{d\Delta T}{dt} \qquad \dots (6)$$

ΔT can be eliminated by setting

$$K_T = \frac{\Delta R_T}{\Delta T};\ \text{i.e. } \Delta T = \left(\frac{1}{K_T}\right)\Delta R_T \text{ where '}K_T\text{' is}$$

the slope of the element resistance - temperature characteristics, thus

$$\left(\frac{U(v_0)A}{K_T} - i_0^2\right)\Delta R_T + \frac{MC}{K_T}\frac{d\Delta R_T}{dt}$$
$$= 2i_0R_{T0}\Delta i - \sigma A(T_0 - T_F)\Delta U$$

which reduces to

$$\Delta R_T + \tau_v\frac{d\Delta R_T}{dt} = K_v\Delta i - K_v\,\Delta v \qquad \dots (7)$$

where
$$\tau_v = \frac{MC}{U(v_0)A - i_0^2K_T} \qquad \dots (8)$$

$$K_I = \frac{2K_Ti_0R_{T_0}}{U(v_0)A_T - i_0^2R_{T_0}}$$

$$K_v = \frac{K_I\sigma A(T_0 - T_F)}{U(v_0)A - i_0^2K_T}$$

Apply Laplace transform to equation (7),

$$\Delta R_T(s) = \left(\frac{K_I}{1+\tau_v s}\right)\Delta I(s) - \left(\frac{K_v}{1+\tau_v s}\right)\Delta v(s)$$

and we know that for metallic wire

$$R_T = R_0\,(1 + \alpha\Delta T)$$
$$\Rightarrow \quad R_T - R_0 = \alpha R_0\Delta T$$
$$\Delta R_T = \alpha R_0\Delta T$$
$$\frac{\Delta R_T}{\Delta T} = \frac{dR_T}{dT} = \alpha R_0 = K_T$$

From given data and (8)

Time constant,

$$\tau_v = \frac{2.5\times 10^{-5}}{25.2\times 10^{-4} - 10^{-2}\times 4.8\times 10^{-3}\times 10} = 12.254 \text{ s}$$

4.54 (d)

Method-1:

At steady state from equation (7)

$$\Delta R_T = k_i \Delta i - k_v \Delta v$$

then final steady state resistance

$$R_T = R_{T0} + \Delta R_T$$

Method-2:

Heat generated in the wire = $I^2 R_w$

I = Current through the wire (A)

R_w = Resistance of wire (Ω)

$$\text{Heat dissipated due to convection} = hA(\theta_W - \theta_f)$$

h = coefficient of heat transfer (W/m^2 °C)

A = heat transfer area (m^2)

θ_w = temperature of wire (°C)

θ_f = temperature of flowing fluid (°C)

For equilibrium condition, we can write the energy balance for the hot wire as,

$$I^2 R_w = hA(\theta_w - \theta_f) \quad \text{...(i)}$$

$$R_w = R_0[1 + \alpha(\theta_w - \theta_0)] \quad \text{...(ii)}$$

Putting equation (ii) in (i) we get,

$$I^2 R_0[1 + \alpha(\theta_w - \theta_0)] = hA(\theta_w - \theta_f)$$

$$(100 \times 10^{-3})^2 \times 10\,[1 + 4.8 \times 10^{-3}\,(\theta_w - 0)] = 25.2 \times 10^{-4}\,(\theta_w - 30°)$$

$$1 + 4.8 \times 10^{-3}\,\theta_w = 25.2 \times 10^{-3}\,(\theta_w - 30°)$$

$$20.4 \times 10^{-3}\,\theta_w = 1.756$$

$$\theta_w = \frac{1.756 \times 10^3}{20.4} = 86.07\ °\text{C}$$

R_w at 86.07 °C is,

$$R_w = R_0[1 + \alpha(\theta_w - 0)]$$

$$R_w = 10[1 + 4.8 \times 10^{-3}\,(86.07 - 0)]$$

$$= 10[1 + 0.413136]$$

$$= 10[1.413136] = 14.131\ \Omega$$

So, the nearest answer is option (d).

4.55 (a)

Pressure and velocity profile of a venturi tube are:

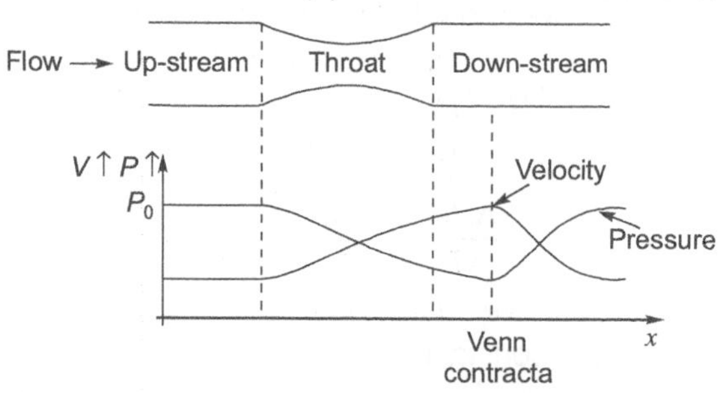

So,

$$P_{\text{throat}} < P_{\text{upstream}}$$

$$V_{\text{throat}} > V_{\text{upstream}}$$

4.56 (a)

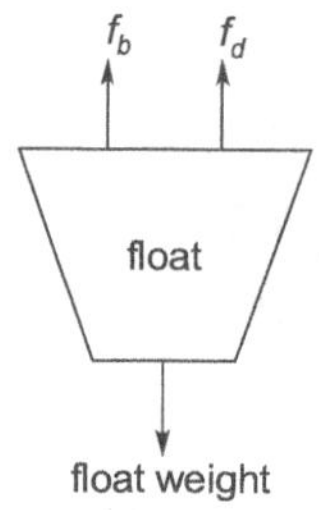

Float weight = drag force + buoyancy force

4.57 0.814 (0.800 to 0.833)

Output of thermopile

$$= 60 \times 10 \times 10\ \mu\text{V/°C}$$

$$= 6\ \text{mV/°C}$$

Resolution of ADC

$$= \frac{5}{2^{10} - 1} = \frac{5}{1023} = 4.88\ \text{mV}$$

Resolution of circuit in °C, is

$$6\ \text{mV} \longrightarrow 1°\text{C}$$

$$4.88\ \text{mV} \longrightarrow \frac{1}{6} \times 4.88$$

$$= 0.814°\text{C}$$

4.58 (0.45)

$$L = \frac{0.5}{\sin 45°} = 0.707$$

Distance between transmitter and receiver is

$(2 \times 0.354) = 0.707$ m

Change fluid velocity Δv

$$\Delta t = \frac{2d\,(\Delta v)\cos\theta}{C^2}, \quad \text{as } C >> 0$$

$$\frac{1}{5 \times 10^6} = \frac{2 \times 707 \times (\Delta v) \cos 45°}{(1500)^2}$$

$$\therefore \quad \Delta v = 0.45\ \text{m/sec}$$

4.59 (c)

- Pirani gauge is used to measure vacuum pressure.
- Pyrometer is used to measure high temperature.
- Orifice plate is used to measure liquid flow rate.
- Bourdon tube is used to measure pressure/force.

4.60 1.236 (1.2 to 1.3)

$$S_{\text{Cu-constantan}} = 1.9 - (-38.3) = 40.2\ \mu\text{V/°C}$$

$$V_{\text{Cu-constantan}} = 40.2 \times 50 = 2010\ \mu\text{V} = 2.01\ \text{mV}$$

$$S_{\text{Fe-constantan}} = 13.3 - (-38.3) = 51.6\ \mu\text{V/°C}$$

$$V_{\text{Fe-constantan}} = 51.6 \times 15 = 0.774\ \text{mV}$$

$$V_{\text{out}} = 2.01 - 0.774 = 1.236\ \text{mV}$$

4.61 1.39 (1.3 to 1.4)

$R_T = R_o[1 + \alpha T]$

$= 100[1 + 0.0039 \times 100] = 139\ \Omega$

It is a non-inverting amplifier

$\therefore$ Its output $= \left(1+\frac{R_F}{R_1}\right)V_i = \left(1+\frac{139}{100}\right)\times 1$

$= 2.39$ V

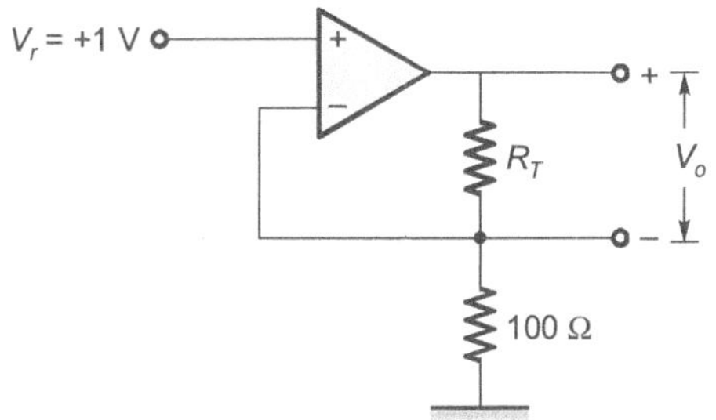

$V_C = V_B = V_A = 1$ V

V_o = Output of op-amp $- V_C$

$= 2.39 - 1 = 1.39$ V

Method 2:

$R_{100} = 100(1 + 0.0039 \times 100) = 139\ \Omega$

Using virtual ground concept,

$V_a = 1$ V

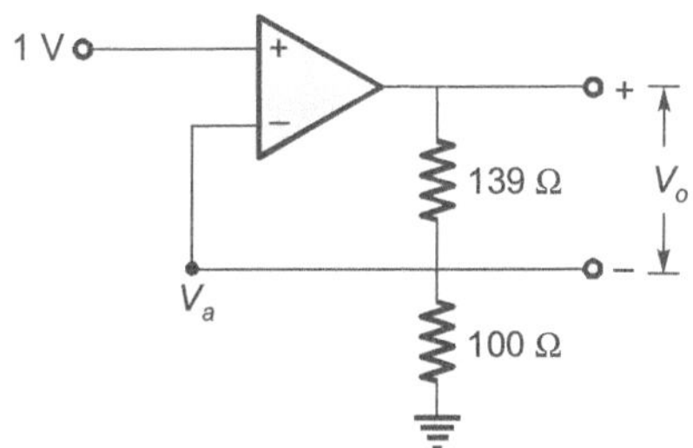

$\Rightarrow \quad I = \frac{1}{100} = 10$ mA

Now, $V_0 = 139 \times 10$ mA $= 1.39$ V

4.62 4.81 (4.8 to 5.2)

Differential pressure, $\Delta P = (\rho_m - \rho_W)gh$

ρ_m = Density of mercury = 13534 kg/m^3

ρ_W = Density of water = 1000 kg/m^3

For Pitot - Static tube

Velocity of flow of water (v) $= \sqrt{\frac{2\Delta P}{\rho_W}}$ m/sec.

$$v = \sqrt{\frac{2(\rho_m - \rho_W)gh}{\rho_W}}$$

$$= \sqrt{\frac{2(13534-1000)}{1000}\times 9.81\times 94.1\times 10^{-3}}$$

$$= \sqrt{23.140} = 4.81 \text{ m/sec}$$

4.63 (c)

In orifice meter the relation between flowrate and differential pressure is, $Q \,\alpha\, \sqrt{\Delta P}$

['Q' is volumetric flow rate, 'Δp' Differential pressure]

So, $\frac{Q_2}{Q_1} = \sqrt{\frac{(\Delta P)_2}{(\Delta P)_1}}$

$\Rightarrow \left(\frac{Q_2}{Q_1}\right)^2 = \frac{\Delta P_2}{\Delta P_1}$

$\Rightarrow \Delta P_2 = \Delta P_1\left[\frac{Q_2}{Q_1}\right]^2$

Given, $Q_2 = 2Q_1$

$\Rightarrow \Delta P_2 = \Delta P_1 \times 4$

$= 5$ kg/m$^2 \times 4 = 20$ kg/m^2

4.64 (d)

- Given pressure range is 10^{-4} to 10^{-3} torr this means it is low pressure range.
- Out of given choices "Pirani gauge" is the suitable pressure gauge which can measure the given range of pressure.

4.65 1612.28 (1605 to 1618)

At $t = 1$ sec, thermocouple voltage equation is,

$V = \alpha\Delta T$

$10 \times 10^{-3} = 10 \times 10^{-6}[T_1 - 30°]$

$\Rightarrow T_1 = 1030°$C

Now we have calculated temperature of thermocouple's hot junction at $t = 1$ sec.

We need to find temperature of molten metal which is steady state temperature of hot junction.

Since thermocouple is modelled as first order instrument so,

$T(t) = T_\infty + [T_D - T_\infty]\, e^{-t/\tau}$

at $t = 1$ sec

$1030 = T_\infty + [30 - T_\infty]e^{-1/1}$

$\Rightarrow T_\infty = 1612°$C

4.66 1.96 (1.9 to 2.0)

Given that:

Magnetic flux density (B)

$= 100 \times 10^{-3}$ Wb/m^2.

Diameter (D) $= 0.25$ m

Generated voltage (e_0) $= 1$ V

From the concept of "Electro magnetic flowmeter".

We can write that, $e_0 = B \times D \times v$

Where, v = Mean velocity of conductive fluid

$$v = \frac{e_0}{BD} \quad \text{... (i)}$$

But we know that

$$Q = Av \quad \text{... (ii)}$$

From equation (i) and (ii),

$$Q = A \times \frac{e_0}{BD} = \pi \frac{D^2}{4} \times \frac{e_0}{BD} = \frac{\pi \cdot D \cdot e_0}{4B} = \frac{3.14 \times 0.23 \times 1}{4 \times 100 \times 10^{-3}}$$

$$= 1.96 \text{ m}^3/\text{sec}$$

4.67 96 (95 to 97)

From the given figure,

Output of thermocouple, $V_{OT} = e_{M/N}(\theta - \theta_{Ref})$

where, $e_{M/N}$ is sensitivity of thermocouple.

Total output voltage after instrumentation amplifier is

$$V_o = A_I \times V_{OT}$$

where, A_I is gain of instrumentation amplifier

$$96 \times 10^{-3} = 25 \times e_{M/N}(\theta - 0)$$

$$\theta = \frac{96 \times 10^{-3}}{25 \times 40 \times 10^{-6}} = 96°\text{C}$$

4.68 5.02 (3.5 to 5.7)

From the given figure and Information

$$\text{Transit time, } t_1 = 0.995 \times 10^{-3} = \frac{L}{C + v\cos\theta}$$

$$\text{Transit time, } t_2 = 1 \times 10^{-3} = \frac{L}{C - v\cos\theta}$$

Where 'L' is the length of the line joining both the transducers.

'C' is velocity of ultra sound in water.

'v' is velocity of fluid in pipe.

'θ' is the given angle 45°.

From the given figure,

$$\sin\theta = \frac{D}{L} \; ; \; L = \frac{D}{\sin\theta}$$

$$\text{So, } C + v\cos\theta = \frac{L}{0.995 \times 10^{-3}} \quad \text{... (i)}$$

$$C - v\cos\theta = \frac{L}{1 \times 10^{-3}} \quad \text{... (ii)}$$

(i) – (ii)

$$2v\cos\theta = 1000 \cdot L\left[\frac{1}{0.995} - 1\right]$$

$$2 \times v \times \cos 45 = 1000 \times \frac{100 \times 10^{-2}}{\sin 45}\left[\frac{1}{0.995} - 1\right]$$

$$\left(\because L = \frac{D}{\sin\theta}\right)$$

$$2 \times v \times \frac{1}{\sqrt{2}} = 10^3 \times \sqrt{2} \times (0.00502)$$

$$v = 5.02 \text{ m/sec}$$

4.69 (20)

Given that, Density of the fluid, $\rho = 1$ kg/m^3

If the pressure difference measured,

$$\Delta p = 200 \text{ N/m}^2$$

The velocity of the fluid measured by pitot static tube is

$$v = \sqrt{\frac{2\Delta p}{\rho}} = \sqrt{\frac{2 \times 200}{1}} = 20 \text{ m/s}$$

4.71 (2.5)

Given data : $G_F = 2$ At $T = 0°$C, $\epsilon = 0$

$$R_1 = R_2 = R_3 = R_4 = 100\ \Omega$$

$$\alpha = 0.005/°\text{C}$$

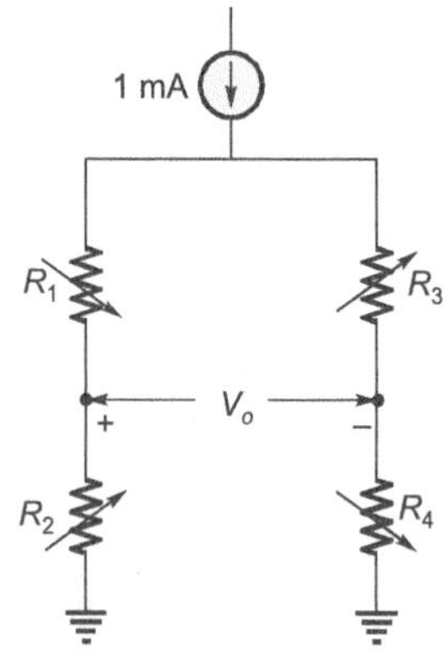

At $T = 50°$C, $\epsilon = 0.01$ and $V_o = 1$

$$R = R_o(1 + \alpha\Delta T)$$

$$= 100(1 + 0.005 \times 50)$$

$$R = 125\ \Omega$$

Resistance change due to ϵ,

$$R = 125 \pm \Delta R_\epsilon$$

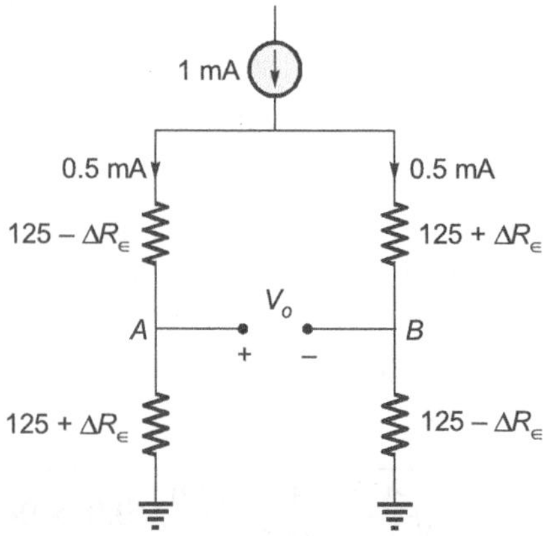

$V_o = 0.5 \text{ mA}(125 + \Delta R_\epsilon - 125 + \Delta R_\epsilon)$

$V_o = \Delta R_\epsilon$ mV

$V_o = R\epsilon G_F \text{ mV} = 125 \times 0.01 \times 2 \text{ mV} = 2.5 \text{ mV}$

4.73 (1000.25)

For strain gauge $= \frac{\delta R}{R} = G_f \times \in$ $(\in = \text{strain})$

$$\delta R = 1000 \times 2.5 \times 100 \times 10^{-6} = 0.25\Omega$$

Resistance of the strain gauge changes to $1000 + 0.25 = 1000.25\Omega$

4.74 (1334.16)

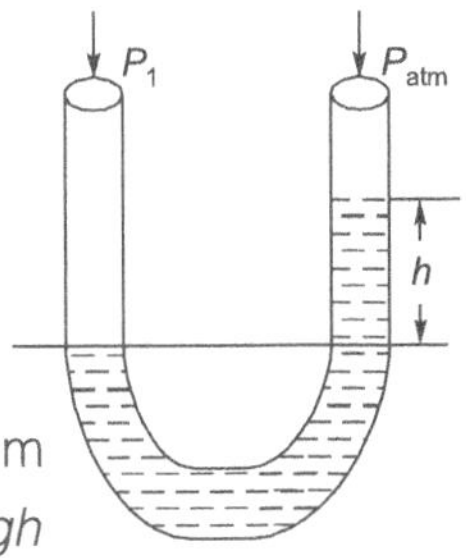

Density of monometric fluid (ρ) = 13,600 kg/m^3

Acceleration due to gravity is (g) = 9.81 m/s^2

Height indicated (h) = 1 cm

$$P_1 = P_{atm} + \rho g h$$

$$P_1 - P_{atm} = \rho g h$$

Differential pressure

$$\Delta P = P_1 - P_{atm} = \rho g h$$

$$\Delta P = 13600 \times 9.81 \times 1 \times 10^{-2} = 1334.16 \text{ Pa}$$

4.75 (c)

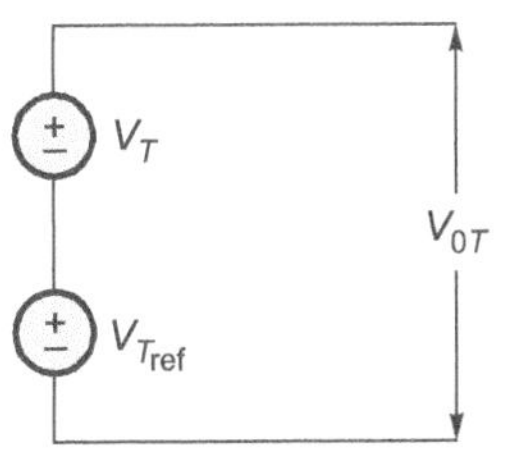

$$V_{0T} = V_T - V_{Tref} = [13650 + 50\theta_x] - [13650 + 50\theta_{ref}]$$

$$= 50\,[\theta_x - \theta_{ref}] = 50\,[100 - 1] = 4950$$

$$V_0 = 20 \times V_{0T} = 4950 \times 20\ \mu\text{V} = 99 \text{ mV}$$

4.76 (d)

In the given problem, since copper wires are identical, therefore, the reading of voltmeter will not change at all.

4.77 (d)

Electromagnetic flow meter offers lowest resistance to the flow.

4.79 (10.2)

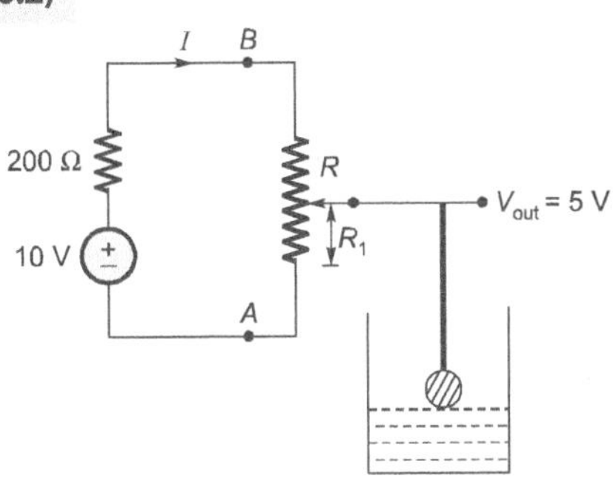

$R \rightarrow (0 - 10\ \text{k}\Omega)$

level $\rightarrow$ (0 to 20 cm)

$$I = \frac{10}{10.2\ \text{k}\Omega}$$

$$R_1 = \frac{5\ V_{out}}{I} = \frac{5}{10/10.2} = 5.1\ \text{k}\Omega$$

Now, $10\ \text{k}\Omega \xrightarrow{\text{length}} 20 \text{ cm}$

$$5.1\ \text{k}\Omega \longrightarrow \frac{20}{10} \times 5.1 = 10.2 \text{ cm}$$

4.81 (a)

$$x(t) = \cos 2\pi f_1 t \rightarrow \text{carrier signal}$$

$$m(t) = 0.5 \sin 2\pi f_2 t$$

$$y(t) = \cos[2\pi f_1 t + \beta \cos 2\pi f_2 t]$$

$\rightarrow$ FM signal;

Given, $K_f = 500 \frac{\text{Hz}}{\text{volt}}$

$$f_1 = f_c = 40 \text{ kHz}$$

$$f_2 = f_m = 1 \text{ kHz}$$

$$\beta = \frac{K_f A_m}{f_m} = \frac{500 \times 0.5}{1000} = 0.25$$

Freq. deviation $\rightarrow \Delta f = K_f A_m = 250$ Hz

4.82 (0.5)

$$R^2 = 1 - \frac{\Sigma[y - (ax + b)]^2}{\Sigma[y - \bar{y}]^2}$$

$$\bar{y} = \frac{\Sigma y}{n} = 10.44$$

$$\Sigma(y - \bar{y})^2 = (10.3 - 10.44)^2 + 3(10.5 - 10.44)^2 + (10.4 - 10.44)^2 = 0.032$$

$$y_i = ax_i + b$$

$X =$	1	2	3	4	5
$aX + b = y_i =$	10.36	10.4	10.44	10.48	10.52

$$\Sigma[y - (ax_i + b)]^2 = [10.3 - 10.36]^2 + [10.5 - 10.4]^2 + [10.4 - 10.44]^2 + [10.5 - 10.48]^2 + [10.5 - 10.52]^2 = 0.016$$

$$R^2 = 1 - \frac{0.016}{0.032} = 0.5$$

■■■

5 Velocity, Acceleration, Force, Torque & Seismic Measurement

5.1 A weight of 4.9 N rests on a rough horizontal floor that has a coefficient of friction of 0.5. If $g = 9.8$ m/s^2, the smallest horizontal force that gives the weight an acceleration of 4 m/s^2 is approximately

(a) 2.0 N (b) 2.5 N
(c) 4.5 N (d) 19.6 N

[1995 : 2 M]

5.2 With the stroboscope, synchronism has been obtained at the flashing rates of 25 and 20 pulses per second. The speed of rotation is

(a) 0.6 rev/min. (b) 50 rev/min.
(c) 100 rev/min. (d) 6000 rev/min.

[1995 : 2 M]

5.3 A cantilever is used as a primary sensing element for the measurement of force. The measured deflection is 1 mm. Suppose each of the dimensions, length, width and thickness of the cantilever are doubled for the same material and force, then the deflection will be

(a) 0.125 mm (b) 0.25 mm
(c) 0.5 mm (d) 2 mm

[1996 : 1 M]

5.4 The natural frequency of a seismic mass type pickup is f_n. The pick-up will give the correct motion information, if the exciting frequency f_e is

(a) f_n (b) $\frac{1}{\sqrt{2}} f_n$
(c) $>> f_n$ (d) $<< f_n$

[1997 : 1 M]

5.5 Light from a mains powered fluorescent lamp is falling on a gear rotating clockwise. The gear wheel has one tooth painted white. The white tooth appears to be rotating in the anti-clockwise direction with a speed of 60 rpm. If the mains frequency is 50 Hz, the true speed of the gear wheel is

(a) 2940 r.p.m (b) 3060 r.p.m
(c) 5940 r.p.m (d) 5960 r.p.m

[1997 : 1 M]

5.6 A seismic vibration sensor, having natural frequency ω_n and damping ratio ζ, used for measuring amplitude of vibration with frequency ω, should have

(a) $\omega_n << \omega$ and ξ much greater than 1
(b) $\omega_n = \omega$ and ξ slightly less than 1
(c) $\omega_n >> \omega$ and ξ slightly less than 1
(d) $\omega_n <<, \omega$ and ξ slightly less than 1

[2001 : 2 M]

5.7 A seismic type of transducer has a damping constant of 10. The transducer is designed to measure

(a) Acceleration (b) Velocity
(c) Displacement (d) Force

[2002 : 1 M]

5.8 A shaft encoder attached to a dc motor has a sensitivity of 500 pulses per revolution. A frequency meter connected to the output of the encoder indicates the frequency to be 5500 Hz. The speed of the motor in RPM is

(a) 110 (b) 220
(c) 550 (d) 660

[2003 : 1 M]

5.9 In a seismic acceleration sensor, with a displacement transducer as the secondary element, decreasing the mass while maintaining all other parameters of the sensor unchanged will

(a) reduce both natural frequency and steady-state sensitivity.
(b) increase both natural frequency and steady-state sensitivity.
(c) increase natural frequency but reduce steady-state sensitivity.
(d) increase natural frequency without affecting steady-state sensitivity.

[2003 : 2 M]

5.10 A cantilever-type micro-accelerometer is designed by scaling down each dimension of a macro-accelerometer by a factor of 100. If the natural frequency of the macro-accelerometer is ω, then the natural frequency of the micro-accelerometer will be

(a) $100\,\omega$ (b) $10\,\omega$
(c) $0.1\,\omega$ (d) $0.01\,\omega$

[2004 : 2 M]

5.11 In a flyball-type angular speed magnitude measuring device, the centrifugal force, given $10^{-6}\omega^2$, where ω is the angular speed in rad/s, is transmitted to a spring with the non-linear characteristic $F = 16\pi^2 x^2$, where F is the applied force and x is the spring compression in meters. The sensitivity of the device in mm/s is

(a) $\frac{1}{\pi}$ (b) $\frac{1}{2\pi}$

(c) $\frac{1}{3\pi}$ (d) $\frac{1}{4\pi}$

[2005 : 1 M]

5.12 An accelerometer has a seismic mass of 1000 gram and natural frequency of 1 kHz. With g the acceleration due to gravity, the static sensitivity of the accelerometer in (nm/g) is

(a) 2.58 nm/g (b) 9.81 nm/g

(c) 7.28×10^2 nm/g (d) 6.28 × nm/g

[2006 : 2 M]

5.13 The accompanying figure shows a VERTICAL venturimeter with upward water flow. When the measured static pressure difference, $p_1 - p_2$, between the inlet and the throat is 30 kPa, the flow rate is found to be 50 litres per second. Assume that the coefficient of discharge remains the same. When $p_1 - p_2 = 20$ kPa, the flow rate, in litres per second, is

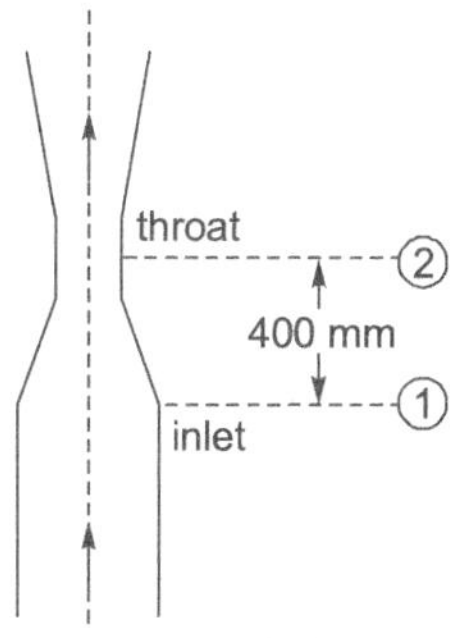

(a) 33.3 (b) 39.3

(c) 40.8 (d) 54.2

[2007 : 2 M]

5.14 The speed of a gear having 60 teeth is measured using a proximity sensor. The output of the proximity sensor is fed to a counter with a gating time of 1s. The counter indicates a value of 120. The speed at which the gear is rotating is

(a) 60 rpm (b) 120 rpm

(c) 600 rpm (d) 1200 rpm

[2008 : 2 M]

5.15 The torque in a rotating shaft is measured using strain gauges. The strain gauges, must be positioned on the shaft such that axes of the strain gauges are at

(a) 0° with respect to the axis of the shaft

(b) 30° with respect to the axis of the shaft

(c) 45° with respect to the axis of the shaft

(d) 90° with respect to the axis of the shaft

[2008 : 2 M]

5.16 A stroboscopic system is used for measuring the speed of a rotating shaft. The shaft has one target mark on it. The maximum strobe rate at which synchronism is achieved is r_1 flashes per minute. The next lower flash rate at which synchronism is achieved is r_2 flashes per minute. The speed of the shaft in rpm is

(a) $\frac{r_1 r_2}{r_1 - r_2}$ (b) $\frac{r_1 r_2}{r_1 + r_2}$

(c) $\frac{r_2^2}{r_1 + r_2}$ (d) $\frac{r_1^2}{r_1 + r_2}$

[2009 : 1 M]

5.17 The seismic mass of an accelerometer oscillates sinusoidally at 100 Hz with a maximum displacement of 10 mm from its mean position. The peak acceleration of the seismic mass is

(a) 3947.84 m/s^2 (b) 3141.50 m/s^2

(c) 314.15 m/s^2 (d) 100.00 m/s^2

[2010 : 1 M]

Common Data for (5.18 and 5.19):

The deflection profile $y(x)$ of a cantilever beam due to application of a point force F (in Newton), as a function of distance x from its base, is given by

$$y(x) = 0.001F\,x^2\left(1 - \frac{x}{3}\right)\text{ m.}$$

The angular deformation θ at the end of the cantilever is measured by reflecting a laser beam off a mirror M as shown in the figure.

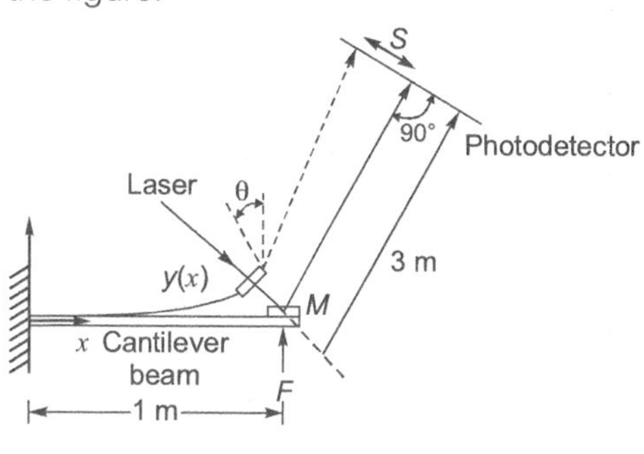

5.18 The translation S of the spot of laser light on the photodetector when a force of $F = 1$ N is applied to the cantilever is

(a) 1 mm (b) 3 mm
(c) 6 mm (d) 12 mm

[2012 : 2 M]

5.19 If linear variable differential transformers (LVDTs) are mounted at $x = \frac{1}{2}$m and $x = \frac{1}{4}$m on the cantilever to measure the effect of time varying forces, the ratio of their output is

(a) 12/7 (b) 40/11
(c) 176/23 (d) 112/15

[2012 : 2 M]

5.20 An accelerometer has input range of 0 to 10g, natural frequency 30 Hz and mass 0.001 kg. The range of the secondary displacement transducer in mm required to cover the input range is

(a) 0 to 2.76 (b) 0 to 9.81
(c) 0 to 11.20 (d) 0 to 52.10

[2013 : 1 M]

5.21 The torque transmitted by a cylindrical shaft is to be measured by using two strain gauges. The angles for mounting the strain gauges relative to the axis of the shaft for maximum sensitivity are

(a) ±45° (b) ±60°
(c) ±90° (d) ±180°

[2015 : 1 M]

5.22 A mass-spring damper system with force as input and displacement of the mass as output has a transfer function

$$G(s) = \frac{1}{(s^2 + 24s + 900)}$$

A force input $F(t) = 10\sin(70t)$ Newtons is applied at time $t = 0$ s. A beam from an optical stroboscope is focused on the mass. In steady state, the strobe frequency in hertz at which the mass appears to be stationary is

(a) $\frac{5}{\pi}$ (b) $\frac{15}{\pi}$
(c) $\frac{35}{\pi}$ (d) $\frac{50}{\pi}$

[2015 : 1 M]

5.23 The relationship between the force $f(t)$ and the displacement $x(t)$ of a spring-mass system (with a mass M, viscous damping D and spring constant K) is

$$M\frac{d^2x(t)}{dt^2} + D\frac{dx(t)}{dt} + Kx(t) = f(t).$$

$X(s)$ and $F(s)$ are the Laplace transforms of $x(t)$ and $f(t)$ respectively. With $M = 0.1$, $D = 2$, $K = 10$ in appropriate units, the transfer function $G(s) = \frac{X(s)}{F(s)}$ is

(a) $\frac{10}{s^2 + 20s + 100}$ (b) $s^2 + 20s + 100$
(c) $\frac{10s^2}{s^2 + 20s + 100}$ (d) $\frac{s}{s^2 + 20s + 100}$

[2016 : 2 M]

5.24 A laser pulse is sent from ground level to the bottom of a concrete water tank at normal incidence. The tank is filled with water up to 2 m below the ground level. The reflected pulse from the bottom of the tank travels back and hits the detector. The round-trip time elapsed between sending the laser pulse, the pulse hitting the bottom of the tank reflecting back and sensed by the detector is 100 ns. The depth of the tank from ground level marked as x in metre is _____.
[Refractive index of water $n_{water} = 1.3$ and velocity of light in air $c_{air} = 3 \times 10^8$ m/s]

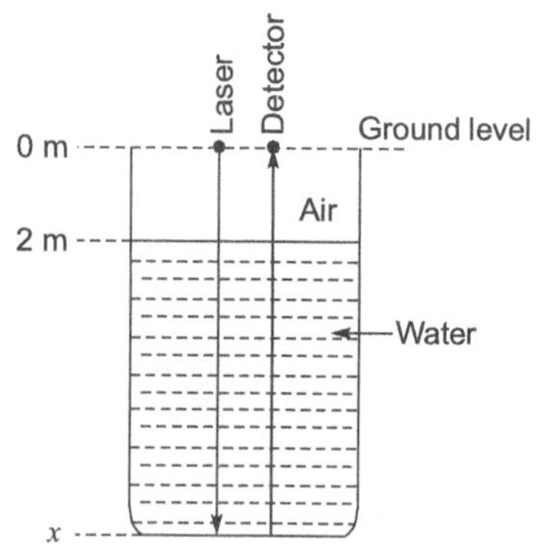

(a) 11 (b) 10
(c) 9 (d) 12

[2021 : 2 M]

5.25 The force per unit length between two infinitely long parallel conductors, with a gap of 2 cm between them is 10 μN/m. When the gap is doubled, the force per unit length will be _________ μN/m (rounded off to one decimal place).

[2023 : 1 M]

■■■

Answers Velocity, Acceleration, Force, Torque & Seismic Measurement

5.1 (c)	5.2 (d)	5.3 (c)	5.4 (c)	5.5 (a)	5.6 (d)	5.7 (a)
5.8 (d)	5.9 (c)	5.10 (b)	5.11 (d)	5.12 (b)	5.13 (c)	5.14 (b)
5.15 (c)	5.16 (a)	5.17 (a)	5.18 (b)	5.19 (b)	5.20 (a)	5.21 (a)
5.22 (c)	5.23 (a)	5.24 (d)	5.25 (5.0)			

Explanations Velocity, Acceleration, Force, Torque & Seismic Measurement

5.1 (c)

$$F = \mu mg + ma$$

$$4.9 = mg$$

$$m = \frac{4.9}{g} = \frac{4.9}{9.8} = 0.5\,\text{kg}$$

$$\therefore \quad F = 0.5 \times 0.5 \times 9.8 + 0.5 \times 4$$

$$= 2.45 + 2 = 4.45 \approx 4.5\text{ N}$$

5.2 (d)

$$\text{Speed of rotation} = \frac{f_1 f_2}{(f_1 - f_2)}$$

$$= \frac{25 \times 20}{25 - 20} = 100\text{ rev/sec}$$

$$= 6000\text{ rev/min}$$

5.3 (c)

$$x = 1\text{ mm}$$

$$x = \frac{Fl^3}{3EI} = \frac{Fl^3}{3E\left(\frac{1}{12}bt^3\right)}$$

$$x' = \frac{F(2l)^3}{3E\frac{1}{12}[2b)(2t)^3]}$$

$$= \frac{1}{2}x = \frac{1}{2} \times 1\text{mm} = 0.5\text{ mm}$$

5.4 (c)

For the seismic transducer to work in the displacement mode, the frequency ratio $\frac{\omega_e}{\omega_n} >> 1$

$$\therefore \quad \frac{2\pi f_e}{2\pi f_n} >> 1$$

$$\therefore \quad f_e >> f_n$$

5.5 (a)

True speed = (60 × 50) – 60 = 2940 rpm

5.6 (d)

For the seismic transducer to work in the displacement mode (i.e. for measuring amplitude) high frequency ratio is required.

i.e. $\frac{\omega_1}{\omega_n} >> 1$,

i.e. $\omega_n << \omega$

and $\xi \approx 1$

5.7 (a)

A seismic transducer works in the acceleration mode for high value of damping constants.

For seismic transducer to work in displacement mode, the damping constant in nearly equal to unity.

5.8 (d)

Sensitivity = 500 pulses/revolution frequency observed by frequency meter

= 5500 Hz

= 5500 pulses/sec.

∴ Number of revolution/sec.

$$= \frac{5500}{500} = 11\text{ rev/sec.}$$

$$\therefore \text{Speed motor} = 11\frac{\text{rev.}}{\text{sec.}}$$

$$= 11 \times 60\frac{\text{rev.}}{\text{min.}} = 660\text{ rev/min.}$$

5.9 (c)

For Seismic accelerometer,

$$\text{Sensitivity} = \frac{x}{a} = \frac{M}{k}$$

$\Rightarrow$ Sensitivity $\propto$ Mass

Also,

$$\text{Natural frequency} = \sqrt{\frac{k}{M}}$$

⇒ As mass decreases
⇒ Sensitivity decreases
⇒ Natural frequency increases

5.10 (b)

$$x = \frac{Fl^3}{3EI}$$

$$\therefore \quad F = \frac{x(3EI)}{l^3}$$

but, $F = kx$

$$\therefore \quad k = \frac{3EI}{l^3} \qquad (k = \text{stiffness})$$

$$\omega = \sqrt{\frac{k}{J}} = \sqrt{\frac{3EI}{l^3}} = \sqrt{\frac{3E\times\frac{1}{12}bt^3}{l^3}}$$

Now,

$b \to 100\,b$
$t \to 100\,t$
$l \to 100\,l$

$$\therefore \quad \omega' = \sqrt{\frac{3E\frac{1}{12}\times(100b)\,(100)^3t^3}{(100)^3l^3}} = 10\,\omega$$

5.11 (d)

Balancing forces we get,

$$10^{-6}\,\omega^2 = 16\pi^2 x^2$$

$$\therefore \quad \frac{x^2}{\omega^2} = \frac{10^{-6}}{16\pi^2}$$

$$\therefore \quad \text{Sensitivity}\left(\frac{x}{\omega}\right) = \frac{10^{-3}}{4\pi}\,\text{m/sec.}$$

$$= \frac{1}{4\pi}\,\text{mm/sec.}$$

5.12 (b)

$$g = 9.81\text{ m/sec}^2$$

$$f = 1\text{ kHz} = 10^3\text{/sec}$$

$$\frac{g}{f^2} = \frac{9.81}{10^6} = 9.81\times10^{-6}\text{ m}$$

$$\text{Now resistivity in } (nm/g) = \frac{g/f^2}{W}$$

W = weight = 1000 gm

$$\therefore \quad S = \frac{9.81\times10^{-6}}{1000\text{gm}}$$

$$= 9.81\times10^{-9}\text{ m/g} = 9.81\text{ nm/g}$$

5.13 (c)

Since, $Q \propto \sqrt{\Delta P}$ (in venturimeter)

$$\text{Hence,} \quad \frac{50}{Q} = \sqrt{\frac{30}{20}}$$

$$Q = 40.8\text{ litres/sec}$$

5.14 (b)

$$\text{Speed} = \frac{\text{Puses recorded per second}}{\text{Number of teeths on the rotor}}\text{rps}$$

$$= \frac{120}{60}\text{rps} = 2\text{ rps}$$

$$= 2\times60\text{ rpm} = 120\text{ rpm}$$

Hence, (b) option is correct.

5.15 (c)

Strain gauges are positioned as 45° w.r.t. axis of shaft so as to strain it in direction of major axis and hence ensure maximum sensitivity.

5.17 (a)

Peak acceleration $= \omega^2 x_m$

$= (2\pi\times100)^2\times10\times10^{-3} = 3947.84\text{ m/sec}^2$

5.18 (b)

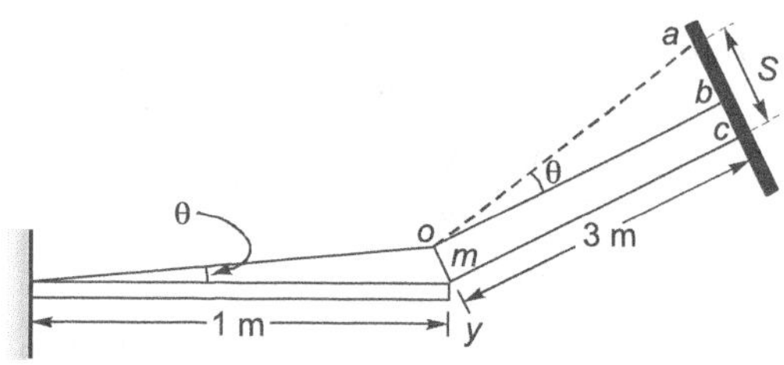

$$F = 1\text{ N}$$

$$x = 1\text{ m}$$

$$\therefore \quad y = 0.001\,Fx^2\left(1-\frac{x}{3}\right)m$$

$$= 0.001\times1\times1\left(1-\frac{1}{3}\right)$$

$$= 0.001\times\frac{2}{3} = \frac{2}{3}\text{mm}$$

$$om = y = bc$$

$$bc = \frac{2}{3}\text{mm}$$

in triangle oab

$$\theta = \frac{ab}{ob}$$

$$\therefore \quad ab = \theta(ob)$$

$$\text{Also,} \quad \theta = \frac{y}{x}$$

$\therefore \qquad \theta = y \qquad [\text{for } x = 1 \text{ m}]$

$$ab = \frac{2}{3} \times 10^{-3} \times 3 = 2 \times 10^{-3} \text{ m}$$

$$s = bc + ab$$

$$= \left(\frac{2}{3} + 2\right) \times 10^{-3} \text{ m} = 2.7 \text{ mm}$$

$$\therefore \quad s \approx 3 \text{ mm}$$

5.19 (b)

Output of *LVDT* ∝ deflection,

$$\therefore \frac{\text{o/p of } LVDT \text{ at}\left(x = \frac{1}{2}\text{m}\right)}{\text{o/p of } LVDT \text{ at}\left(x = \frac{1}{4}\text{m}\right)} = \frac{0.001\, F\left(\frac{1}{2}\right)^2\left[1 - \frac{1}{2 \times 3}\right]}{0.001\, F\left(\frac{1}{4}\right)^2\left[1 - \frac{1}{4 \times 3}\right]}$$

$$= \frac{\frac{1}{4} \times \frac{5}{6}}{\frac{1}{16} \times \frac{11}{12}} = \frac{40}{11}$$

5.20 (a)

$$\text{Acceleration} = \omega^2 x$$

Case-I:

At acceleration = 0

$$0 = \omega^2 x$$

$$\therefore \quad x = 0 \text{ mm}$$

Case-II:

At acceleration = 10g

$$10g = \omega^2 x = (2\pi f)^2 x$$

$$= (2\pi \times 30)^2 x$$

$$\therefore \quad x = \frac{10 \times 9.8}{(2\pi \times 30)^2} = 2.76 \text{ mm}$$

5.21 (a)

For maximum sensitivity the angles for measuring the strain gauges is $\pm\pi/4$ or $\pm 45°$.

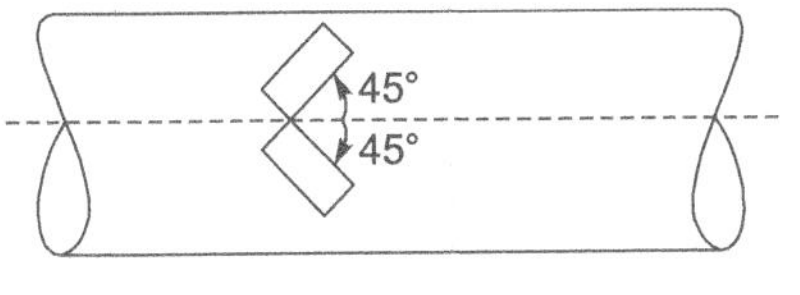

5.22 (c)

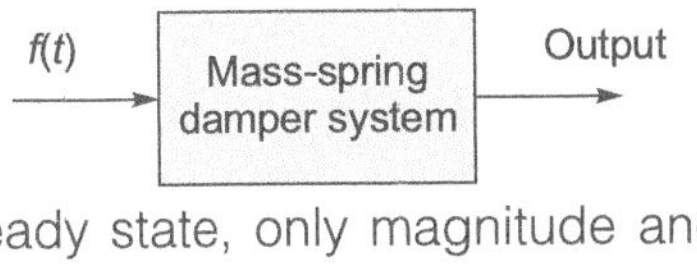

At steady state, only magnitude and phase of input change frequency does not change.

$$\omega = 70$$

$$f = \frac{70}{2\pi} = \frac{35}{\pi}$$

5.23 (a)

$$M\frac{d^2x(t)}{dt^2} + D\frac{dx(t)}{dt} + Kx(t) = f(t)$$

Taking Laplace transform on both sides

$$Ms^2 X(s) + Ds\, X(s) + K\, X(s) = F(s)$$

$$\frac{X(s)}{F(s)} = \frac{1}{Ms^2 + Ds + K}$$

By substituting, *M*, *D* and *K* values

$$\frac{X(s)}{F(s)} = \frac{1}{0.1s^2 + 2s + 10} = \frac{10}{s^2 + 20s + 100}$$

5.24 (d)

$$t_{\text{journey}} = t_1 + t_2 + t_3 + t_4$$

$$= \frac{2}{c} + 2t_2 + \frac{2}{c}$$

$$\left[t_2 = t_3 = \frac{h}{v} = \frac{1.3Ch}{c}\right]$$

$$= \frac{4}{c} + \frac{2 \times 1.3h}{c}$$

$$100\text{ns} = \frac{4 + 2.6h}{c}$$

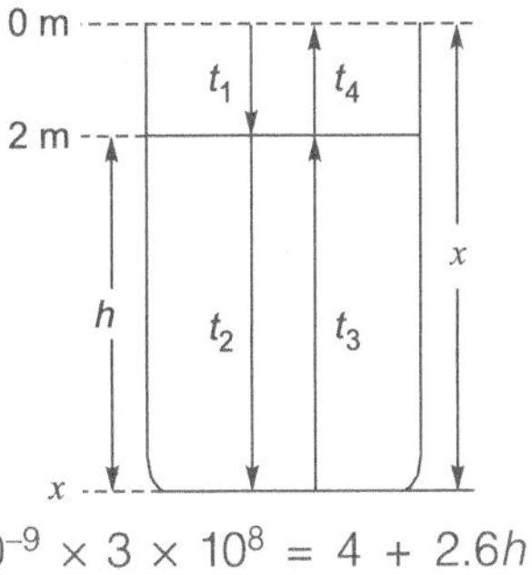

$$100 \times 10^{-9} \times 3 \times 10^8 = 4 + 2.6h$$

$$30 - 4 = 2.6h$$

$$h = \frac{26}{2.6} = 10$$

$$\therefore \quad x = (10 + 2)m = 12 \text{ m}$$

5.25 (5.0)

We know that,

Force is inversely proportional to '*d*'

$$F \propto \frac{1}{d}$$

$$\frac{F_1}{F_2} = \frac{d_2}{d_1}$$

$$\frac{10}{F_2} = \frac{4 \text{ cm}}{2 \text{ cm}}$$

$$F_2 = \frac{10}{2} = 5\ \mu\text{N/m}$$

■■■

6 Miscellaneous

6.1 Assuming uniform pressure over the supported area, the friction moment in an axial bearing of a shaft of diameter d for an axial load of F (with coefficient of friction μ) is given by

(a) $\frac{4}{3}F\mu d$ (b) $\frac{1}{3}F\mu d$

(c) $\frac{2}{3}F\mu d$ (d) $F\mu d$

[1996 : 1 M]

6.2 Measurement of viscosity involves measuring

(a) Frictional force (b) Coriolis's force
(c) Centrifugal force (d) Buoyant force

[1996 : 1 M]

6.3 The Sound Pressure Level (SPL) measured in open space (free field), at a distance of 6 m from a noise source is 80 dB. At a distance of 60 m, the SPL is

(a) 80 dB (b) 60 dB
(c) 8 dB (d) 1.34 dB

[1999 : 1 M]

6.4 Identify the correct set of matches from the following

List-I
(a) LVDT
(b) Piezoelectric crystal
(c) pH probe
(d) Thermocouple

List-II
(i) Reference junction compensation
(ii) Charge amplifier
(iii) Phase sensitive detector
(iv) High input impedance amplifier

Codes:
(a) (a) – (i), (b) – (iii), (c) – (iv), (d) – (ii)
(b) (a) – (iii), (b) – (ii), (c) – (iv), (d) – (i)
(c) (a) – (iv), (b) – (i), (c) – (ii), (d) – (i)
(d) None of the above

[1999 : 2 M]

6.5 Identify the correct set of matches from the following:

List-I
(a) Pirani gauge
(b) Venturimeter
(c) Radiation pyrometer
(d) Load cell

List-II
(i) Temperature measurement
(ii) Vacuum pressure measurement
(iii) Force measurement
(iv) Flow measurement

Codes:
(a) (a) – (ii), (b) – (i), (c) – (iv), (d) – (i)
(b) (a) – (iv), (b) – (iii), (c) – (ii), (d) – (i)
(c) (a) – (ii), (b) – (iv), (c) – (i), (d) – (iii)
(d) None of the above

[1999 : 2 M]

6.6 A photo conductive transducer works on the principle that when a light beam strikes the material

(a) its resistance decreases, which is sensed by an external circuit.
(b) the barrier between transparent metal layer and a semiconductor material, a voltage is generated.
(c) the barrier between transparent metal layer and a semiconductor material, a current is generated in the external circuit.
(d) the cathode, it releases electrons, which are attracted towards anode, thereby producing electric current in the external circuit.

[1999 : 2 M]

6.7 Value of pH of a solution is 4. If indicates that concentration of hydrogen ions is

(a) 10^{-4} g/litre and the solution is acidic
(b) 10^{-4} g/litre and the solution is alkalite
(c) 10^{-4} mg/litre and the solution is acidic
(d) 10^{-4} mg/litre and the solution is alkalite

[2001 : 1 M]

6.8 In the context of transducers, identify the correct matches:

List-I
(a) mean free path
(b) humidity
(c) heat transfer coefficient
(d) intensity of radiation

List-II
(p) optical pyrometer
(q) Knudsen gauge
(r) sling psychrometer
(s) hot wire anemometer

Codes:
(a) (a) – (p), (b) – (q), (c) – (r), (d) – (s)
(b) (a) – (q), (b) – (p), (c) – (s), (d) – (r)
(c) (a) – (r), (b) – (s), (c) – (p), (d) – (q)
(d) (a) – (q), (b) – (r), (c) – (s), (d) – (p)

[2001 : 2 M]

6.9 A viscous damper consists of a sliding piston and a cylinder filled with oil of kinematic viscosity 5×10^{-5} m^2/s. A damping force of 20 N is applied on the piston and the steady state velocity reached is 10 mm/s. The damping coefficient of the damper is
(a) 2 Ns/mm (b) 4 Ns/mm
(c) 10 Ns/mm (d) 20 Ns/mm

[2003 : 1 M]

6.10 Measurement of large external radius is possible with the help of an angle piece and a screw gauge as shown in figure. The radius R can be expressed from the measurement of H and θ as

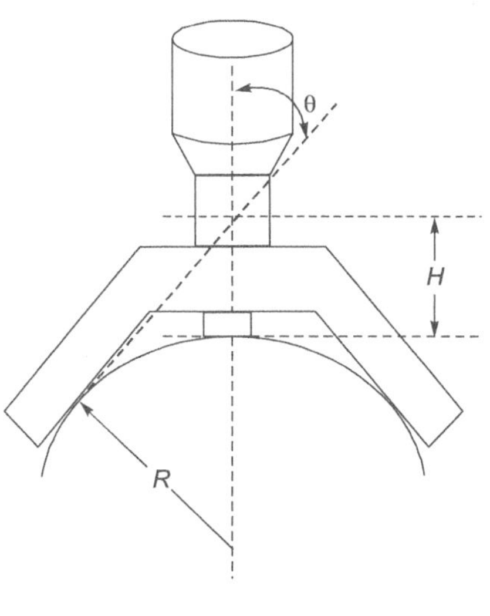

(a) $\frac{H\sin\theta}{1-\sin\theta}$ (b) $\frac{H\cos\theta}{1-\cos\theta}$

(c) $\frac{H\tan\theta}{1+\sin\theta}$ (d) $\frac{H\tan\theta}{1-\cos\theta}$

[2003 : 2 M]

6.11 In a falling-ball viscometer, the ball attains terminal velocities of 0.01 m/s for oil A and 0.002 m/s for oil B. Assuming the oils have the same density and oil A has kinematic viscosity of 5×10^{-3} m^2/s, the kinematic viscosity of oil B in m^2/s is
(a) 15×10^{-3} (b) 20×10^{-3}
(c) 25×10^{-3} (d) 30×10^{-3}

[2005 : 1 M]

6.12 In a laminar flow experiment, Fluid A is pumped through a straight tube and the volumetric flow rate and pressure drop per unit length are recorded. In a second straight tube having twice the internal diameter of the first one, Fluid B records the same pressure drop per unit length at the same volumetric flow rate. Assuming fully developed flow conditions in the tubes, the ratio of the dynamic viscosity of Fluid B to that of Fluid A is
(a) 16 (b) 32
(c) 64 (d) 128

[2007 : 2 M]

6.13 A pH electrode, being used at 25°C, has a source resistance of 10^{10} Ω. The electrode obeys the Nernst equation perfectly. The electronic voltmeter, with which the potential is being measured, has an input impedance of 10^{11} Ω and a gain of 100. If the pH of the analyte changes from 6.5 to 7.8, the change in voltage observed in the voltmeter is
(a) less than 6.8 V
(b) between 6.8 V and 7.19 V
(c) between 7.2 V and 7.49 V
(d) greater than 7.49 V [2007 : 2 M]

6.14 Assuming complete disassociation, the pH of a 2 mH solution of H_2SO_4 is closest to
(a) 0 (b) 3.7
(c) 2.4 (d) 2.7 [2009 : 1 M]

6.15 The conventional way of expressing vibration is in terms of
(a) Richter scale
(b) acceleration due to gravity
(c) speed of sound
(d) atmospheric pressure [2011 : 1 M]

6.16 A pH electrode obeys Nernst equation and is being operated at 25°C. The change in the open circuit voltage in millivolts across the electrode for a pH change from 6 to 8 is ___. [2014 : 1 M]

6.17 The voltage (E_0) developed across a glass electrode for pH measurement is related to the temperature (T) by the relation

(a) $E_0 \propto \frac{1}{T^2}$ (b) $E_0 \propto \frac{1}{T}$

(c) $E_0 \propto T$ (d) $E_0 \propto T^2$

[2015 : 1 M]

6.18 If I is the current flowing through a Hall effect sensor and B is the magnetic flux density perpendicular to the direction of the current (in the plane of the Hall effect sensor), the Hall voltage generated is _____.

(a) Directly proportional to both I and B

(b) Inversely proportional to I and directly proportional to B

(c) Directly proportional to I and inversely proportional to B

(d) Inversely proportional to both I and B

[2020 : 1 M]

6.19 Match the following measuring devices with their principle of measurement.

Measurement Device

(P) Optical pyrometer

(Q) Thermocouple

(R) Strain gauge

(S) Linear variable differential transformer

Principle of Measurement

(I) Variation in mutual inductance

(II) Change in resistance

(III) Wavelength of radiated energy

(IV) Electromotive force generated by two dissimilar metals

(a) (P) – (III); (Q) – (IV); (R) – (II); (S) – (I)

(b) (P) – (IV); (Q) – (III); (R) – (II); (S) – (I)

(c) (P) – (III); (Q) – (I); (R) – (IV); (S) – (II)

(d) (P) – (II); (Q) – (IV); (R) – (I); (S) – (III)

[2024 : 1 M]

6.20 A shielded cable with C_{stray} = 20 pF and R_{wire} = 10 Ω is used to connect the inductive sensors as shown in the figure. The RMS value of V_{out} is ____V (rounded off to two decimal places).

Note: Assume all components are ideal, and sensors are not magnetically coupled.

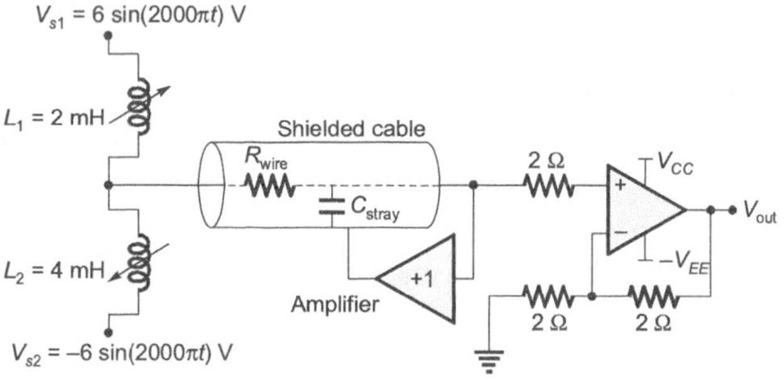

[2024 : 2 M]

■■■

Answers Miscellaneous

6.1 (d)	6.2 (a)	6.3 (b)	6.4 (b)	6.5 (c)	6.6 (d)	6.7 (a)
6.8 (d)	6.9 (a)	6.10 (a)	6.11 (c)	6.12 (a)	6.13 (b)	6.14 (c)
6.15 (b)	6.16 Sol.	6.17 (c)	6.18 (a)	6.19 (a)	6.20 (2.83)	

Explanations Miscellaneous

6.1 (d)

The pressure is uniform so, frictional moment $= F \mu d$.

6.3 (b)

$$\text{Sound level } (dB) \propto \frac{1}{\text{distance}}$$

$2 \log(A_v) = 80$

$\therefore \quad A_v = 10^4$

for 6 m distance sound level $= 10^4$

$$\therefore \quad \text{for 60 m sound level} = \frac{10^4}{10} = 10^3$$

Sound level in $(dB) = 20 \log 10^3$

for 60 m = 60 dB

6.7 (a)

$$\text{pH} = -\log [H^+]$$
$$4 = -\log [H^+]$$
$$\therefore \quad [H^+] = 10^{-4} \text{ gm/litre}$$

Since, pH < 7 therefore ACIDIC

$$\therefore \quad \text{pH} = -\log_{10} 10^{-4} = 4$$

6.9 (a)

The damping coefficient of the damper (D) is given as:

$$D = \frac{\text{Force}}{\text{Velocity}}$$

$$\frac{20\text{N}}{10 \text{ mm/sec.}} = 2 \text{ Ns/mm}$$

6.10 (a)

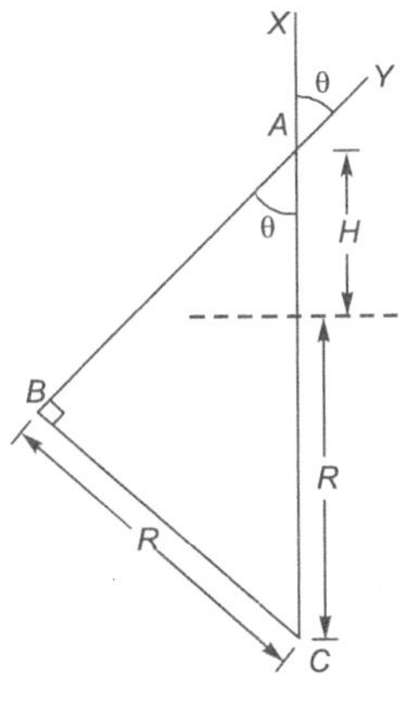

R is the Radius of the circle.

In $\Delta\, ABC$, $\angle B = 90°$

Since, $\angle XAY = \theta$, $\therefore$ $\angle BAC$ is also θ

i.e., ($\angle XAY = \angle BAC$) (Vertically opposite angles).

In ΔABC,

$$\sin\theta = \frac{\text{perpendicular}}{\text{hypotinuse}} = \frac{R}{R+H}$$

$$(R + H) \sin\theta = R$$
$$R \sin\theta + H \sin\theta = R$$
$$R - R \sin\theta = H \sin\theta$$
$$R(1 - \sin\theta) = H \sin\theta$$

$$\therefore \quad \boxed{R = \frac{H \sin\theta}{(1 - \sin\theta)}}$$

6.11 (c)

In case of falling sphere viscometer,

$$\text{kinematic viscosity } (\nu) \propto \frac{1}{\text{terminal velocity } (v_T)}$$

$$\therefore \quad \frac{\nu_A}{\nu_B} = \frac{(v_T)_B}{(v_T)_A}$$

$$\nu_B = \frac{\nu_A (V_T)_A}{(V_T)_B} = \frac{5 \times 10^{-3} \times 0.01}{0.002}$$

$$= 25 \times 10^{-3} \text{ m}^2\text{/sec}$$

6.12 (a)

$$Q_A = \frac{\pi r_A^4 (P_1 - P_2)}{8 \mu L}$$

Since, $(P_1 - P_2)$ is constant, and all other variables are also made constant in the second experiment. So,

$$\frac{\pi (P_1 - P_2)}{8 \mu L} = k \text{ (say)}$$

$$\therefore \quad Q_A = k r_A^4 \quad \text{...(i)}$$

and also, $\quad Q_B = k r_B^4$

Put, $\quad r_B = 2 r_A$

$$\therefore \quad Q_B = k(2\, r_A)^4 = 16 k r_A^4 \quad \text{...(ii)}$$

Dividing (ii) by (i) we get,

$$\frac{Q_B}{Q_A} = \frac{16 k r_A^4}{k r_A^4} = 16$$

6.13 (b)

$$\text{pH change} = 7.8 - 6.5 = 1.3$$
$$E_o = 59.1 \text{ mV/pH} \times 1.3 \text{ ph} \times 100$$
$$= 7.683 \text{ V}$$

Voltmeter reading,

$$V_L = \frac{V_o}{R_o + R_L} \times R_L = 6.98 \text{ V}$$

6.14 (c)

$$H_2SO_4 \rightarrow 2H^+ + SO_4^{-2}$$

$$pH = -\log[2 \times H^+]$$

$$pH = \log_{10}[2 \times 2 \times 10^{-3}]$$

$$= 2.4$$

6.16 Sol.

Sensitivity of an ideal pH probe is, 59.2 mV/pH and pH is changing from 6 to 8

$$= 8 - 6 = 2$$

∴ Output change of pH probe

$$= 59.2 \times 2 = 118.4 \text{ mV}$$

6.17 (c)

Voltage E_o developed across the glass electrode for pH measurement is directly proportional to the temperature (T).

$$E_o \propto T$$

For pH measurement, nearest equation is given by

$$E_o = -\frac{RT}{nf} \ln \frac{C_o}{C_i}$$

6.18 (a)

The Hall voltage developed across the Hall sensor is given by

$$V_H = K_H \frac{IB}{t}$$

So, $V_H \propto I;\quad V_H \propto B;\quad V_H \propto \frac{1}{t}$

■■■

Analog Electronics

UNIT IV

CONTENTS

1. Diode Circuits, Zener Diode as a Circuit Element 167
2. BJT, JFET, MOSFET, their Biasing & Frequency Response 176
3. Op-Amp and Instrumentation Amplifier 189
4. Wave Shaping Circuits and Filters 222
5. Miscellaneous 234

Analog Electronics

Syllabus

Characteristics and applications of diode, Zener diode, BJT and MOSFET; small signal analysis of transistor circuits, feedback amplifiers. Characteristics of ideal and practical operational amplifiers; applications of opamps: adder, subtractor, integrator, differentiator, difference amplifier, instrumentation amplifier, precision rectifier, active filters, oscillators, signal generators, voltage controlled oscillators and phase locked loop, sources and effects of noise and interference in electronic circuits.

Analysis of Previous GATE Papers

Exam Year	1 Mark Ques.	2 Marks Ques.	Total Marks
1995	5	2	**9**
1996	2	–	**2**
1997	3	–	**3**
1998	4	–	**4**
1999	5	1	**7**
2000	3	–	**3**
2001	3	–	**3**
2002	1	–	**1**
2003	4	7	**18**
2004	3	8	**19**
2005	2	8	**18**
2006	2	8	**18**
2007	2	13	**28**
2008	3	4	**11**
2009	1	5	**11**

Exam Year	1 Mark Ques.	2 Marks Ques.	Total Marks
2010	1	4	**9**
2011	1	6	**13**
2012	1	2	**5**
2013	4	4	**12**
2014	3	5	**13**
2015	3	5	**13**
2016	4	3	**10**
2017	5	4	**13**
2018	2	3	**8**
2019	1	3	**7**
2020	3	3	**9**
2021	2	3	**8**
2022	2	6	**14**
2023	3	5	**13**
2024	3	7	**17**

1 Diode Circuits, Zener Diode as a Circuit Element

1.1 In the circuit shown in figure, V_i is 4 V. Assuming the diodes to be ideal, V_0 is

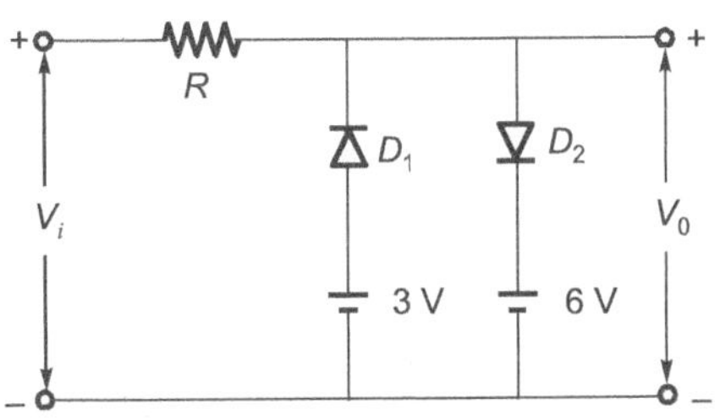

(a) 3 V (b) 4 V
(c) 4.5 V (d) 6 V

[1999 : 1 M]

1.2 The input voltage, V_i is $4 + 3 \sin \omega t$. Assuming all elements to be ideal. The average of the output voltage V_0 in figure is

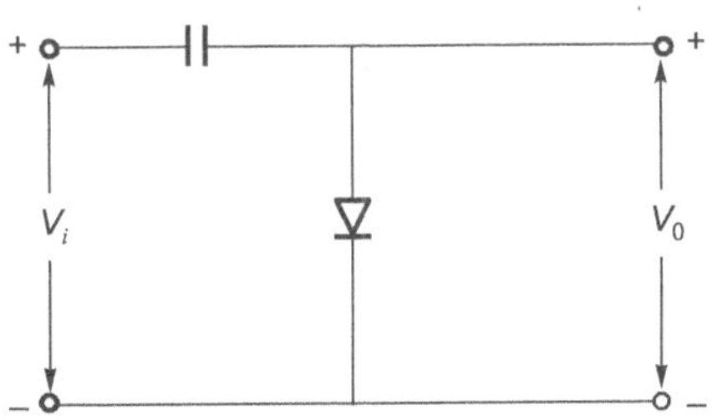

(a) –3 V (b) +3 V
(c) –7 V (d) +7 V

[1999 : 1 M]

1.3 For a sinusoidal input of 50 V amplitude, the circuit shown in figure can be used as

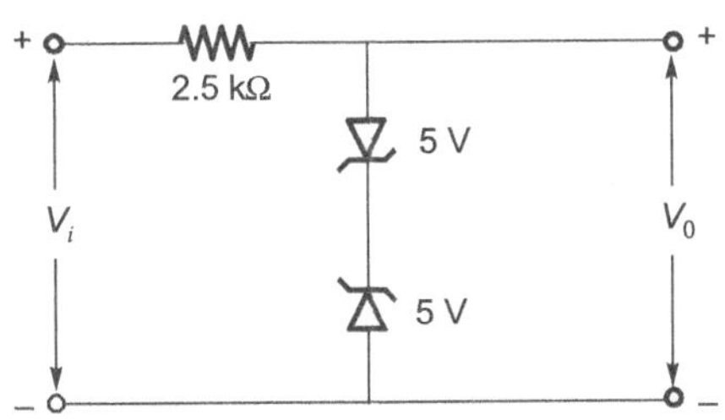

(a) Regulated DC power supply
(b) Square Wave generator
(c) Half wave rectifier
(d) Full wave rectifier

[2000 : 1 M]

1.4 A forward-biased silicon diode when carrying negligible current, has a voltage drop of 0.64 V. When the current is 1 A it dissipates 1 W. The ON-resistance of the diode is

(a) 0.36 Ω (b) 0.64 Ω
(c) 0.74 Ω (d) 1.0 Ω

[2003 : 2 M]

1.5 The 5 V Zener diode in figure is ideal and the ammeter (A), of full-scale 1 mA, has an internal resistance of 100 Ω. The circuit shown, with terminal 1 positive, functions as a

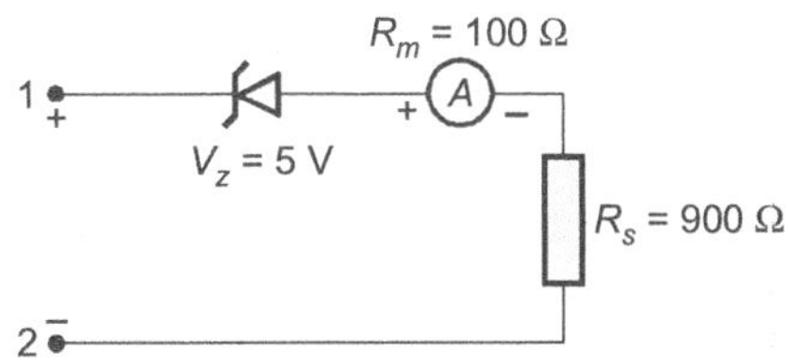

(a) 0 – 1 V DC voltmeter
(b) 0 – 1 mA DC ammeter
(c) 0 – 6 V DC voltmeter
(d) 0 – 5 V AC voltmeter

[2003 : 2 M]

1.6 Assuming ideal diode characteristics, the input/output voltage relationship for the circuit shown in figure is

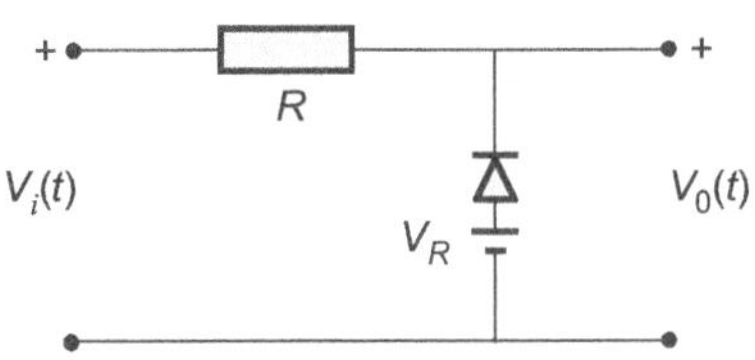

(a) $v_o(t) = v_i(t)$, for all $v_i(t)$
(b) $v_o(t) = v_i(t)$, for $v_i(t) > v_R$
$= 0$, otherwise
(c) $v_o(t) = v_i(t)$, for $v_i(t) < v_R$
$= v_R$, otherwise
(d) $v_o(t) = v_i(t)$, for $v_i(t) > v_R$
$= v_R$, otherwise

[2004 : 1 M]

1.7 The device shown in figure is a

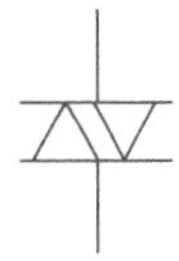

(a) Variac (b) Triac
(c) Quadrac (d) Diac

[2004 : 1 M]

1.8 For the circuit shown in the figure, the diode D is ideal. The power dissipated by the 300 Ω resistor is

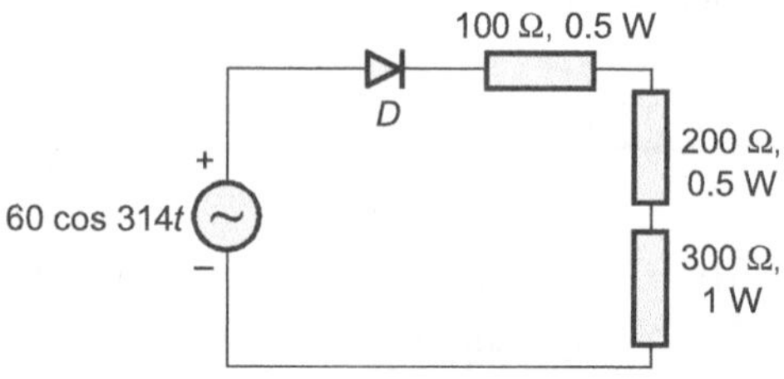

(a) 0.25 W (b) 0.50 W
(c) 0.75 W (d) 1.00 W

[2004 : 2 M]

1.9 The peak value of the output voltage V_0 across the capacitor shown in the figure for a 230 : 9 transformer and a 230 V, 50 Hz, input, assuming 0.7 V diode drop and an ideal transformer, is

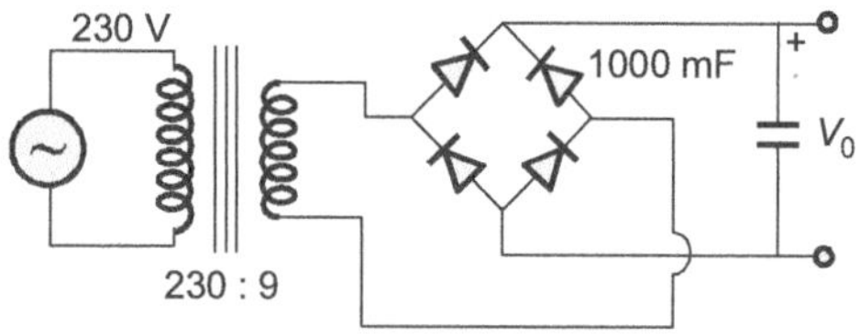

(a) 12.73 (b) 11.33
(c) 7.6 (d) 9.0

[2005 : 1 M]

1.10 When light falls on the photodiode shown in the following circuit, the reverse saturation current of the photodiode changes from 100 μA to 200 μA.

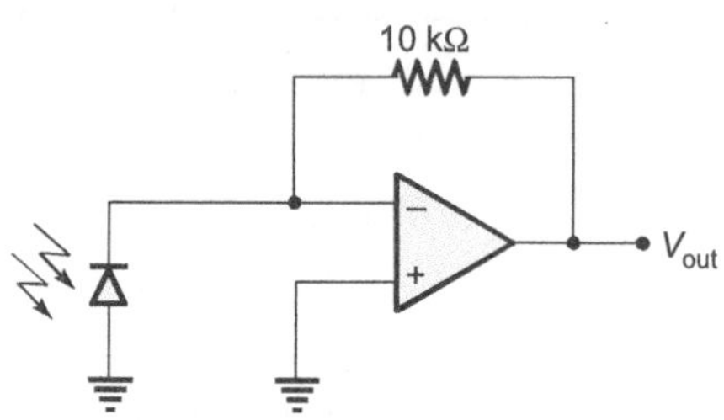

Assuming the op-amp to be ideal, the output voltage, V_{out} of the circuit

(a) does not change
(b) changes from 1 V to 2 V
(c) changes from 2 V to 1 V
(d) changes from –1 V to –2 V

[2007 : 1 M]

1.11 In the circuit shown below, the ideality factor η of the diode is unity and the voltage drop across it is 0.7 V. The dynamic resistance of the diode at room temperature is approximately

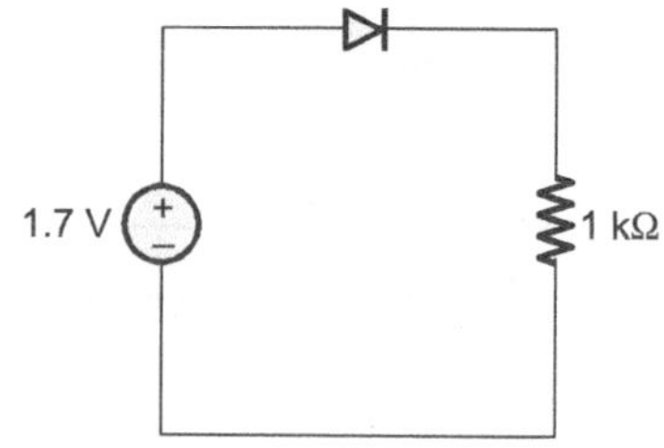

(a) 15 Ω (b) 25 Ω
(c) 50 Ω (d) 700 Ω

[2008 : 1 M]

1.12 Assuming Zener diode D_1 has current-voltage characteristics as shown below on the right and forward voltage drop of diode D_2 is 0.7 V, the voltage V_0 in the circuit shown below is

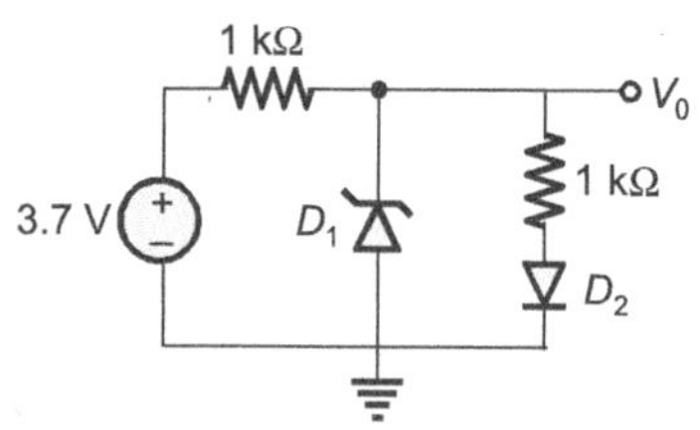

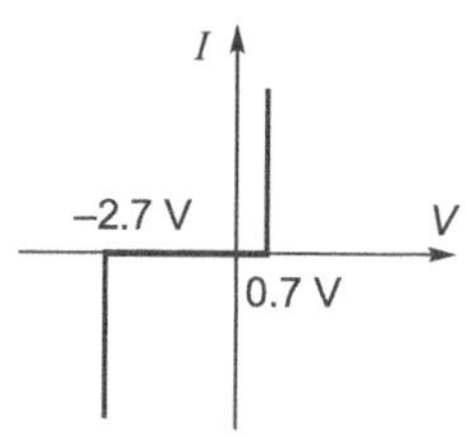

(a) 3.7 V (b) 2.7 V
(c) 2.2 V (d) 0 V

[2011 : 2 M]

1.13 The i-v characteristics of the diode in the circuit given below are:

$$i = \begin{cases} \dfrac{v-0.7}{500} A, & v \geq 0.7\text{ V} \\ 0\ A, & v < 0.7\text{ V} \end{cases}$$

The current in the circuit is

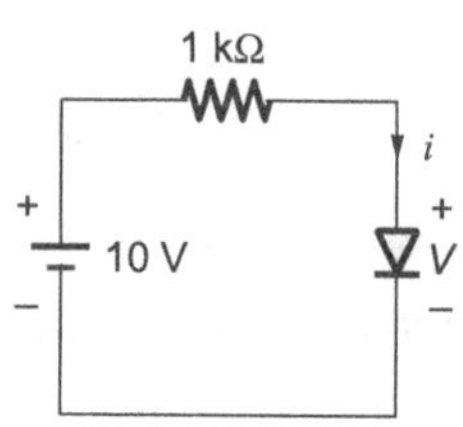

(a) 10 mA (b) 9.3 mA
(c) 6.67 mA (d) 6.2 mA

[2012 : 1 M]

1.14 For the circuit shown below, the knee current of the ideal Zener diode is 10 mA. To maintain 5 V across R_L, the minimum value of the load resistor R_L in Ω and the minimum power rating of the Zener diode in mW, respectively, are

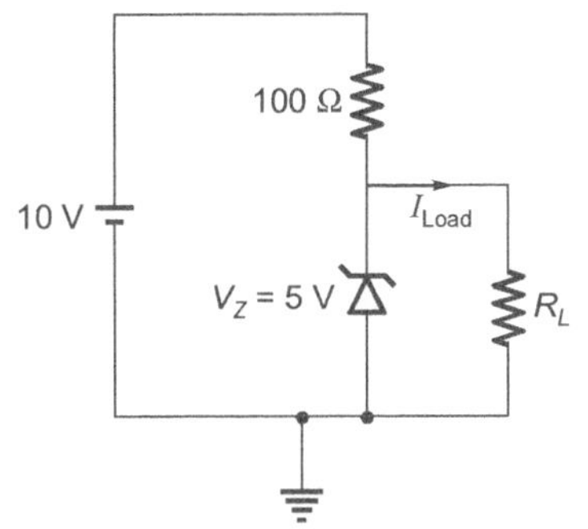

(a) 125 and 125 (b) 125 and 250
(c) 250 and 125 (d) 250 and 250

[2013 : 2 M]

1.15 A voltage 1000 sin ωt volts is applied across YZ. Assuming ideal diodes, the voltage measured across WX in volts, is

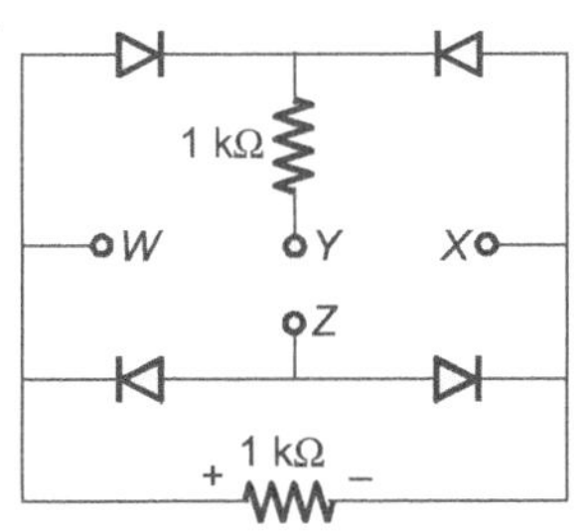

(a) sin ωt
(b) (sin ωt + | sin ωt|)/2
(c) (sin ωt – | sin ωt|)/2
(d) 0 for all t

[2013 : 2 M]

1.16 For the circuit shown in the figure assume ideal diodes with zero forward resistance and zero forward voltage drop. The current through the diode D_2 in mA is ____.

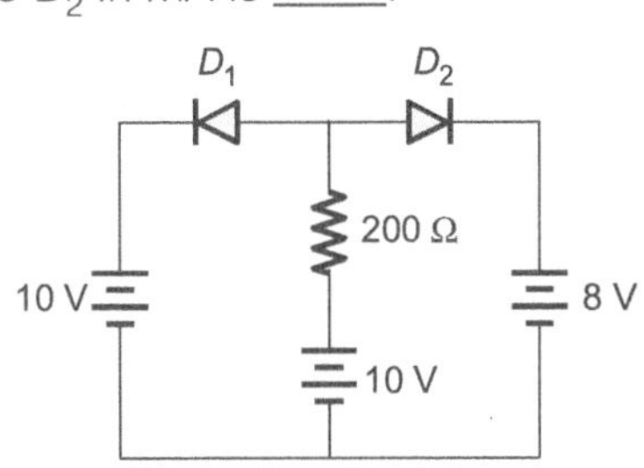

[2014 : 1 M]

1.17 The figure shows a half-wave rectifier circuit with input voltage $v(t)$ = 10sin(100 πt) volts. Assuming ideal diode characteristic with zero forward voltage drop and zero reverse current, the average power consumed in watts by the load resistance R_L is ______ W.

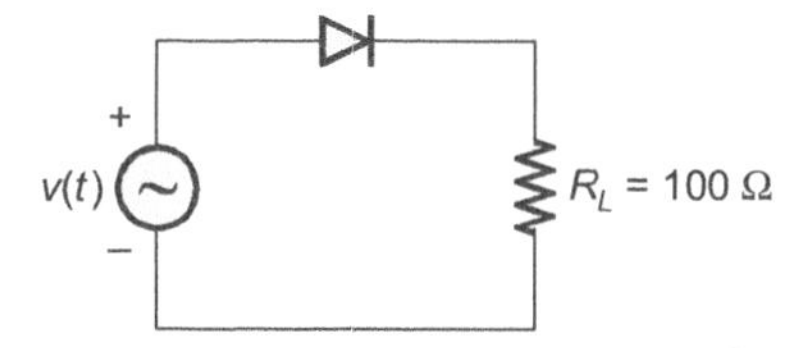

[2015 : 1 M]

1.18 The diode D used in the circuit below is ideal. The voltage drop V_{ab} across the 1 kΩ resistor in volt is______.

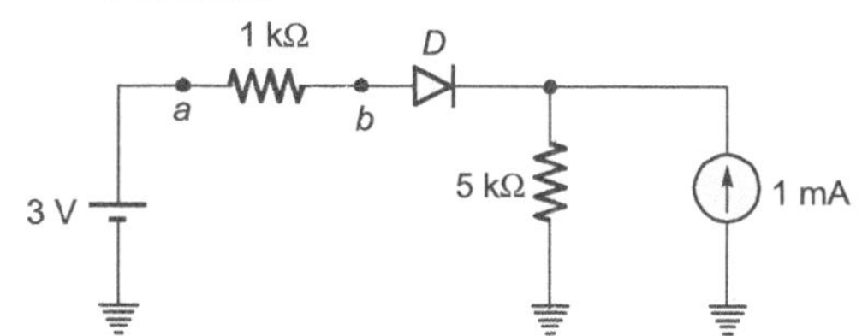

[2016 : 1 M]

1.19 The silicon diode, shown in the figure, has a barrier potential of 0.7 V. There will be no forward current flow through the diode. If V_{dc}, in volts, is greater than

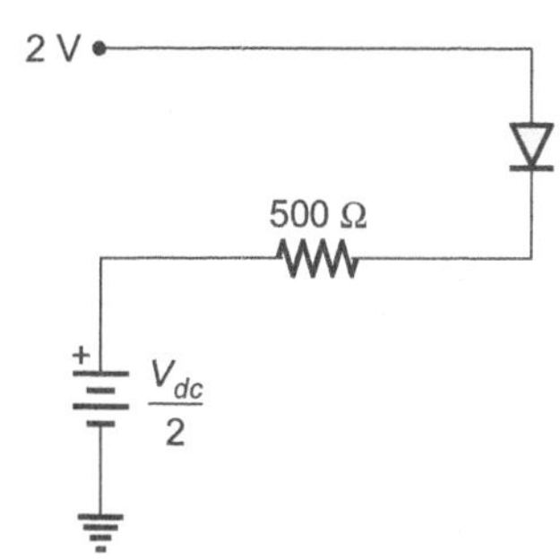

(a) 0.7 (b) 1.3
(c) 1.8 (d) 2.6

[2017 : 1 M]

1.20 The diodes given in the circuit are ideal. At t = 60 ms, V_{pq} (in Volts) is ________.

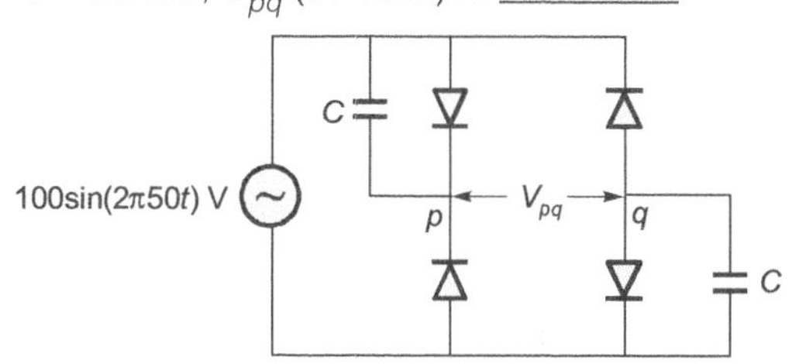

[2018 : 1 M]

1.21 If the diodes in the circuit shown are ideal and the breakdown voltage V_z of the Zener diode is 5 V, the power dissipated in the 100 Ω resistor (in watts) is _____.

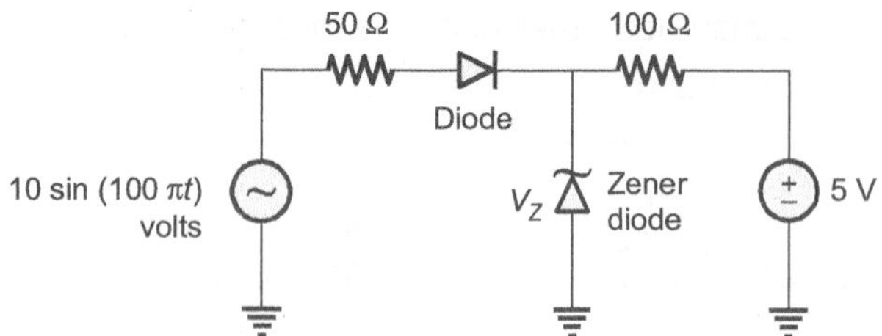

(a) 225/100 (b) 1
(c) 0 (d) 25/100

[2020 : 1 M]

1.22 Assume the diodes in the circuit shown are ideal. The current I_x flowing through the 3 kΩ resistor (in mA, rounded off to one decimal place) is _____.

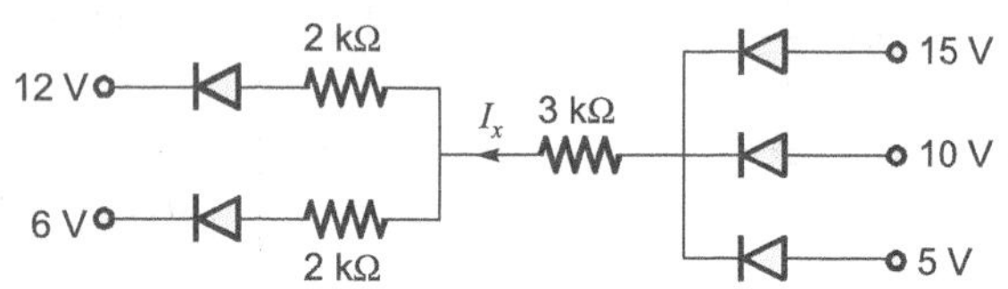

[2020 : 2 M]

1.23 A Zener diode is used as a 4 V voltage regulator in the circuit shown. Given that the diode requires a minimum current of 4 mA for voltage regulation, the maximum current (in milliamperes) permitted to flow through the load R_L is _______ (round off to one decimal place)

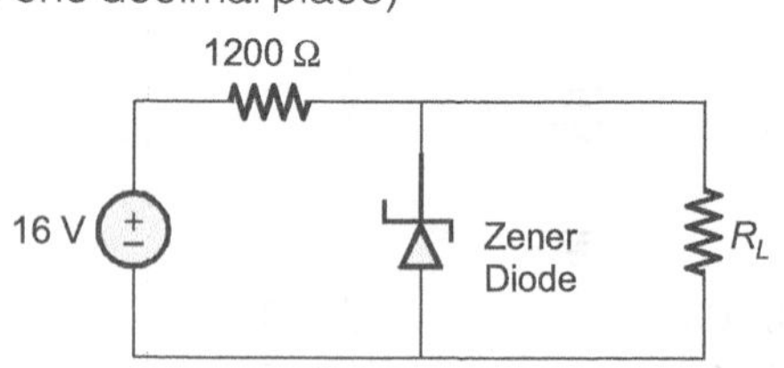

[2022 : 1 M]

1.24 A signal $V_{in}(t)$ shown is applied from t = 0 ms to t = 6 ms to the circuit shown. Given the initial voltage across the capacitor is 0.3 V, and that the diode is ideal, the open circuit voltage $V_{out}(t)$ at t = 5 ms is _______.

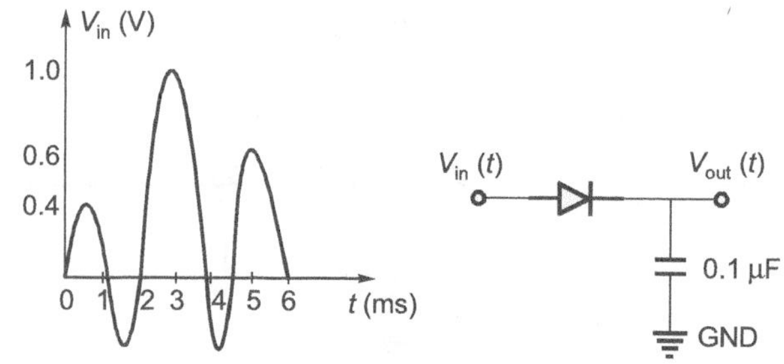

(a) 0.3 V (b) 0.6 V
(c) 0.7 V (d) 1.0 V

[2022 : 2 M]

1.25 A light emitting diode (LED) emits light when it is _____ biased. A photodiode provides maximum sensitivity to light when it is _____ biased.

(a) forward, forward (b) forward, reverse
(c) reverse, reverse (d) reverse, forward

[2023 : 1 M]

1.26 The diode in the circuit is ideal. The current source $i_s(t) = \pi \sin(3000\pi\, t)$ mA. The magnitude of the average current flowing through the resistor R is _____ mA (rounded off to two decimal places).

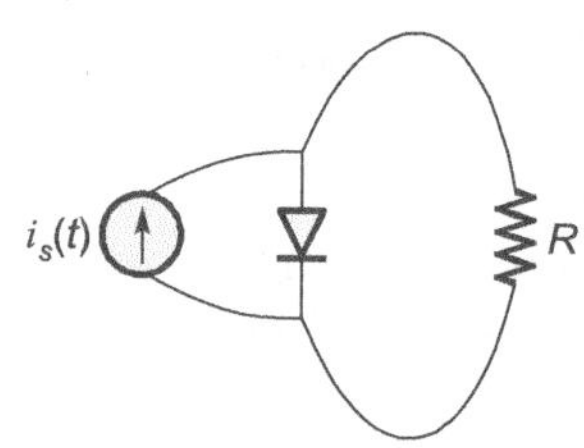

[2023 : 1 M]

1.27 In a p-i-n photodiode, a pulse of light containing 8×10^{12} incident photons at wavelength λ_0 = 1.55 μm gives rise to an average 4×10^{12} electrons collected at the terminals of the device. The quantum efficiency of the photodiode at this wavelength is _______%.

(a) 50 (b) 54.2
(c) 62.5 (d) 80

[2023 : 2 M]

■■■

Answers Diode Circuits, Zener Diode as a Circuit Element

1.1 (b)	1.2 (a)	1.3 (b)	1.4 (d)	1.5 (c)	1.6 (d)	1.7 (d)
1.8 (c)	1.9 (b)	1.10 (b)	1.11 (b)	1.12 (c)	1.13 (d)	1.14 (b)
1.15 (d)	1.16 (10)	1.17 (0.25)	1.18 (0)	1.19 (d)	1.20 (200)	1.21 (c)
1.22 (1.8)	1.23 (6.0)	1.24 (d)	1.25 (b)	1.26 (1)	1.27 (a)	

Explanations Diode Circuits, Zener Diode as a Circuit Element

1.1 (b)

for $V_i = 4$ V
Diode D_1 and D_2 are off
so, $V_0 = V_i = 4$ V

1.2 (a)

Once capacitor is fully charged to +7 V, it remains charged as diode is reverse biased, hence the circuit reduces to

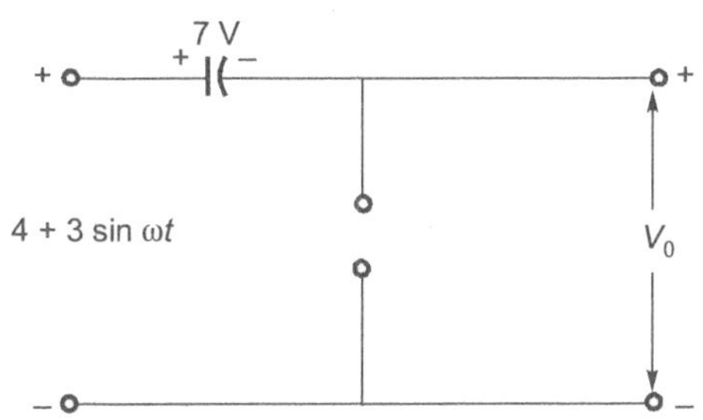

since average value of sinωt is zero
so average value of the output voltage V_0 is 4 − 7 = $V_0 \Rightarrow V_0 = -3$ V

1.3 (b)

When $V_i > 5$ V, $V_0 = 5$ V
and $V_i < -5$ V, $V_0 = -5$ V
output waveform,

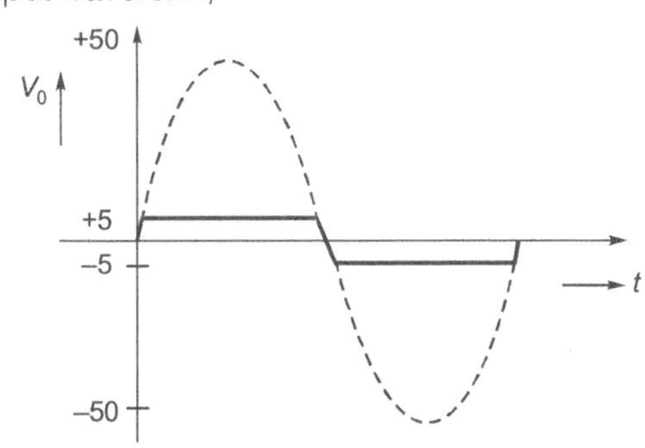

Hence the most appropriate answer is (b).

1.4 (d)

$$R = \frac{dV}{dI}$$

$$P = (dI)^2 \times R_{ON}$$

$$\therefore \quad R_{ON} = \frac{P}{(dI)^2} = \frac{1}{1} = 1.0\ \Omega$$

1.5 (c)

Zener diode gets on when input will be greater than 5 V and full scale current of ammeter is 1 mA. So, maximum input that it can measure

$$= 5 + 1 \times 10^{-3}[100 + 900]$$
$$= 6\text{ V}$$

So, this circuit can measure voltages ranging from 5 V → 6 V from the options, none of them is correct. But we can take option (c).

1.6 (d)

When $V_i(t) \le V_R$, diode D gets on and output $V_0 = V_R$.
When $V_i(t) > V_R$, diode D gets off and output $V_0 = V_i(t)$.

1.7 (d)

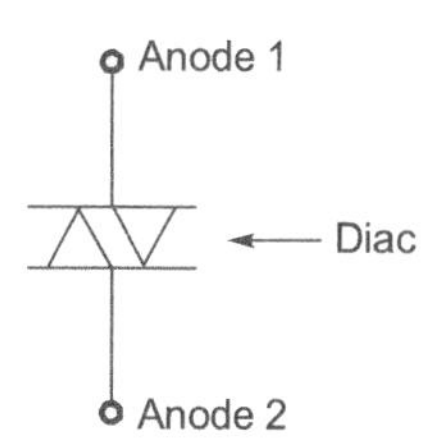

1.8 (c)

Diode conduct for positive half cycle
so, $P = I^2 R$
Power dissipated in 300 Ω resister

$$P = \frac{1}{T}\int_0^{T/2}\left(\frac{V_m \cos\omega t}{600}\right)^2 \cdot 300\, d\omega t$$

$$= 0.75 \text{ watts}$$

1.9 (b)

We know that, for a transformer

$$\frac{V_1}{V_2} = \frac{n_1}{n_2}$$

$$\frac{230\,V}{V_2} = \frac{230}{9}$$

$$\Rightarrow \quad V_2 = 9\text{ V}$$

So, in the secondary coil induced rms voltage $V_{rms} = 9$ V.

$$V_m(V_p) = \sqrt{2}\,V_{rms}$$

$$= \sqrt{2}\times 9 = 12.73$$

At a time two diode is forward biased in one half cycle, so the peak voltage across capacitor

$$= 12.73 - 1.4\text{ V}$$

$$= 11.33\text{ V}$$

1.10 (b)

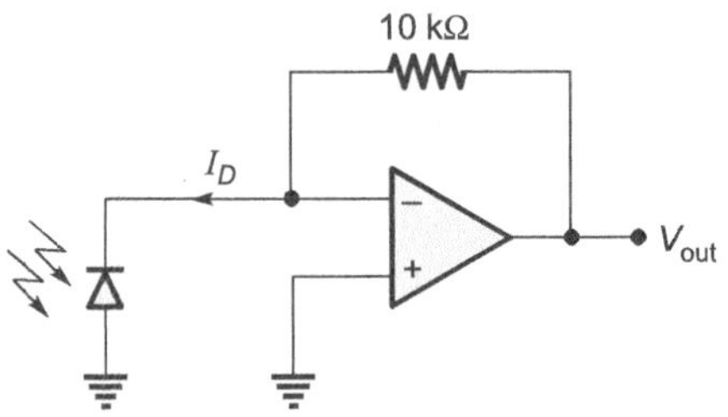

When, $I_D = 100\ \mu$A

$$V_0 = I_D R_f$$

$$= +100\times 10^{-6}\times 10\times 10^3 = +1\text{ V}$$

and when $I_D = 200\ \mu$A

$$V_0 = I_D R_f = +200\times 10^{-6}\times 10\times 10^3 = +2\text{ V}$$

1.11 (b)

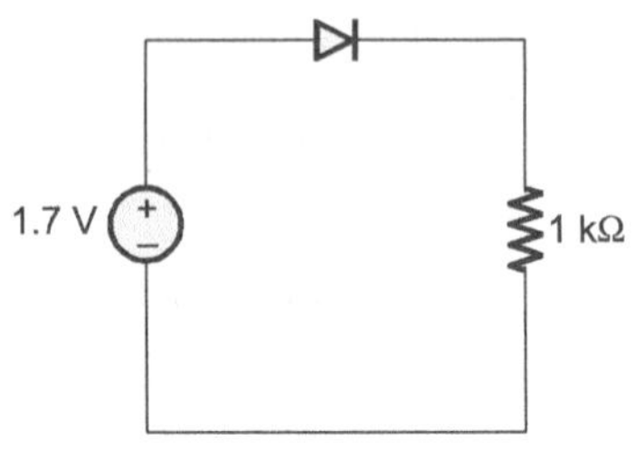

Now the equivalent circuit is

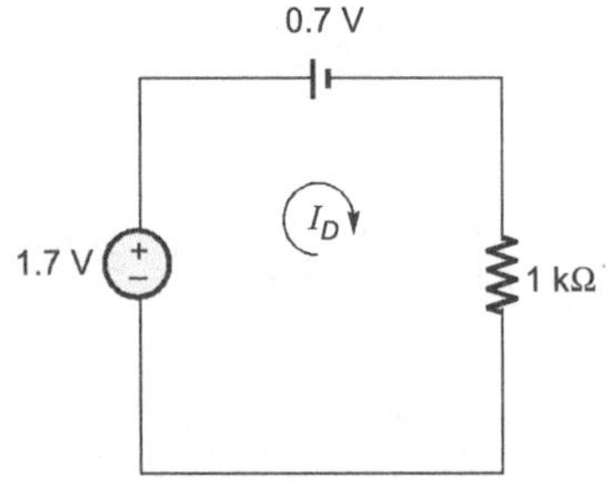

Apply KVL in the above circuit

$$1.7 - 0.7 - 1\,k\cdot I_D = 0$$

$$I_D = 1\text{ mA}$$

And we know that the dynamic resistance is

$$r_d = \frac{\eta V_T}{I_D} = \frac{1\times 26\text{ mV}}{1\text{mA}} = 26\ \Omega \cong 25\ \Omega$$

1.12 (c)

Zener will be O.C.

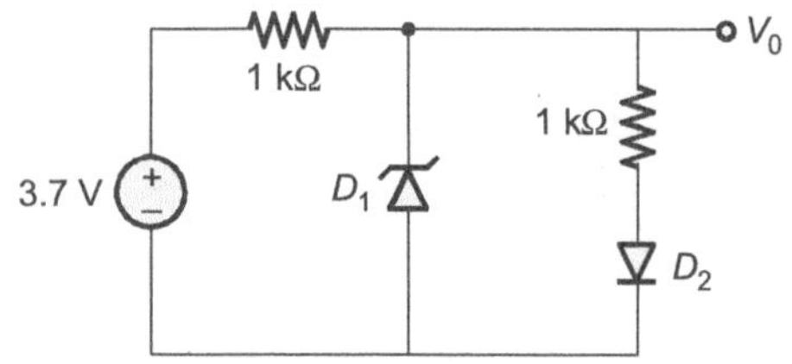

$$\frac{V_0 - 3.7}{1\text{k}\Omega} + \frac{V_0 - 0.7}{1\text{k}\Omega} = 0 \qquad (\because V = V_0)$$

$$2V_0 = 4.4$$

$$\therefore \quad V_0 = 2.2\text{ V}$$

1.13 (d)

By applying kVL in loop

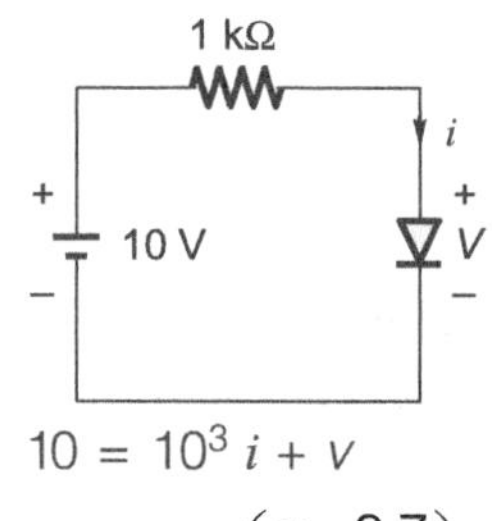

$$10 = 10^3\,i + v$$

$$10 = 10^3\left(\frac{v - 0.7}{500}\right) + v$$

$$10 = 2v - 1.4 + v$$

$$3v = 11.4$$

$$\Rightarrow \quad v = 3.8$$

$$\Rightarrow \quad i = (10 - 3.8)\times 10^{-3} = 6.2\text{ mA}$$

1.14 (b)

Voltage across 100 Ω resistance

$$V_{100} = 10 - 5 = 5\text{ V}$$

Current through 100 Ω resistance

$$I_{100} = \frac{V_{100}}{100} = \frac{5}{100} = 50\text{ mA}$$

Voltage across load R_L

$$V_{RL} = 5\text{ V} = I_{Lmax}\cdot R_{min}$$

$$R_{min} = \frac{5}{I_{Lmax}}$$

$$I = I_{zk} + I_{Lmax} = 50\text{ mA}$$

$$I_{Lmax} = 50 - I_{Zk} = 50 - 10 = 40\text{ mA}$$

$$R_{min} = \frac{5}{40 \times 10^{-3}} = \frac{5000}{40} = \frac{500}{4}$$

$$R_{min} = 125\ \Omega$$

Minimum power rating of Zener diode

$$P_z = V_z I_{zmax}$$

$$I_{zmax} = 50\text{ mA}$$

$$P_z = 5 \times 50 = 250\text{ mW}$$

1.15 (d)

Case 1:

When V_{YZ} is positive.

Then all the four diodes are reverse bias

So, $V_w = V_x = 0$

$V_{wx} = 0\text{ V}$

Case 2:

When V_{YZ} is negative

then all the four diodes are forward bias

So, all the diodes will be short circuit

$$V_w = V_x$$

$$V_{wx} = V_w - V_x = V_w - V_w = 0$$

$$V_{wx} = 0$$

So, $V_{wx} = 0$ for all t

1.16 (10)

Closely, diode D_1 is reversed biased and diode D_2 is forward biased,

(as voltage at point between D_1 and D_2 is $V < 10$ V)

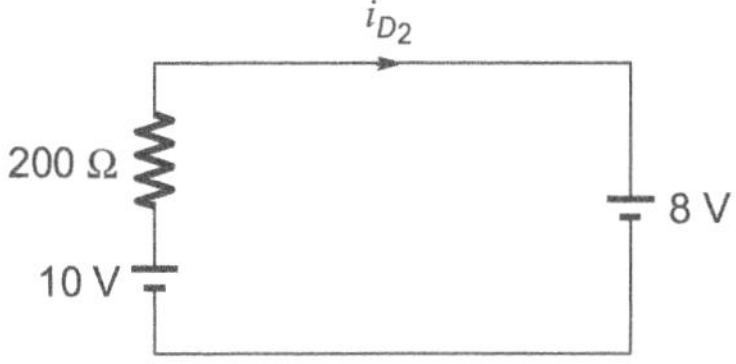

$$i_{D_2} = \frac{10-8}{200} = 10\text{ mA}$$

1.17 0.25 (0.24 to 0.26)

It is a case of half wave rectifier.

∴ Output voltage is

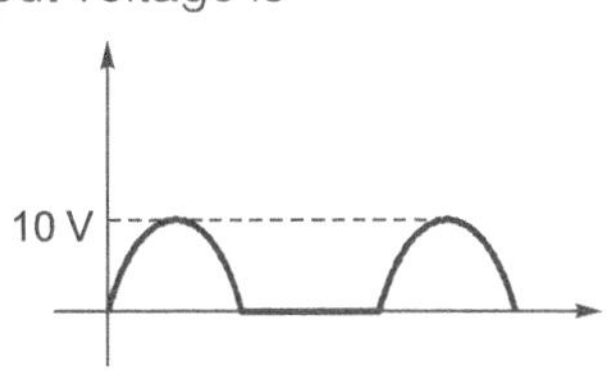

$$P_{avg} = \frac{V_{rms}^2}{R} = \frac{(10/2)^2}{100}$$

$= 0.25$ Watt

1.18 (0)

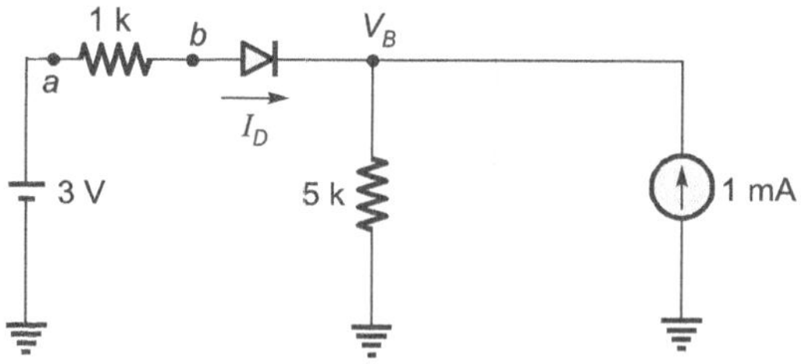

Let the diode is ON.

The diode is replaced by short circuit

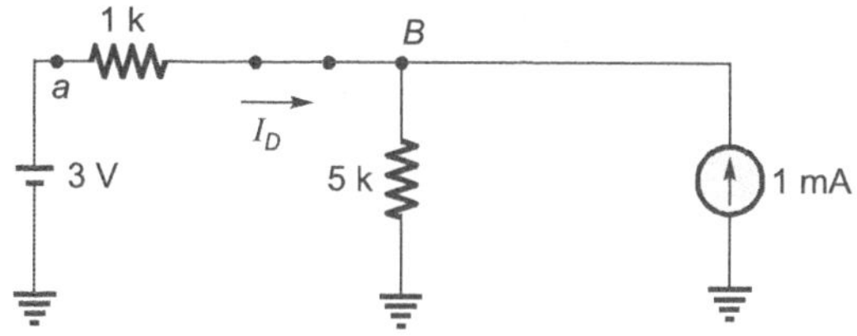

KCL at node B:

$$\frac{V_B - 3}{1\text{k}\Omega} + \frac{V_B}{5\text{ k}\Omega} - 1 \times 10^{-3} = 0$$

$$\Rightarrow \quad V_B = \frac{10}{3} = 3.33\text{ V}$$

$$\Rightarrow \quad I_D = \frac{3 - 3.33}{1 \times 10^3} = -\text{ve not possible}$$

⇒ Diode is off.

So the current passing through diode = 0

$$\Rightarrow \quad V_{ab} = V_{1\text{ K}} = 0\text{ Volts}$$

1.19 (d)

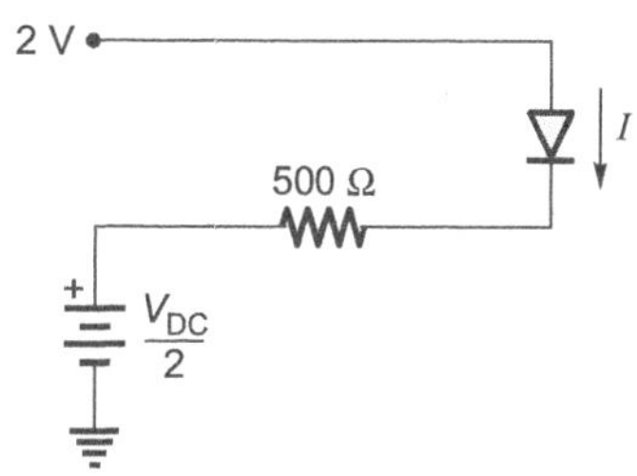

$$I = \frac{2 - 0.7 - \frac{V_{DC}}{2}}{500} = 0$$

$$V_{DC} = 2.6\text{ V}$$

∴ If $V_{DC} > 2.6$ V then diode will become off.

1.20 (200)

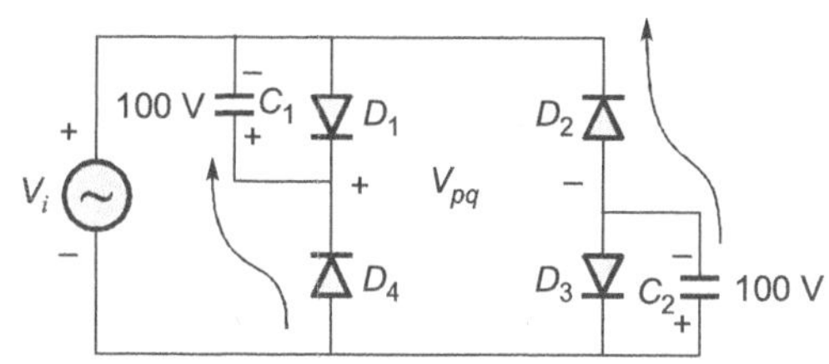

During +ve cycle: D_2 and D_4 are OFF and no current flows.

During -ve cycle:

C_1 charges through D_4 upto 100 V

C_2 charges through D_2 upto 100 V

KVL:

$$-V_i - 100 + V_{pq} - 100 = 0$$

$$V_{pq} = 200 + V_i$$

At $t = 60$ msec, $V_i = 0$

$$V_{pq} = 200 \text{ V}$$

1.21 (c)

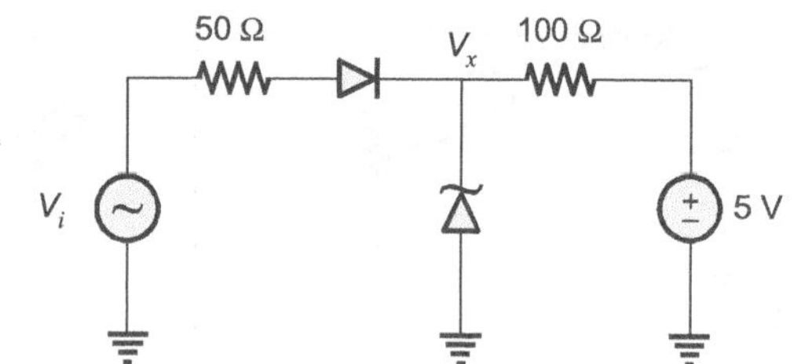

Let both diodes be OFF $\Rightarrow V_x = 5$ V

If $V_i < 5$ V then both diodes are OFF

$\therefore$ current through 100 Ω remains zero.

If $V_i > 5$ V then normal diode becomes ON and zener diode undergoes breakdown.

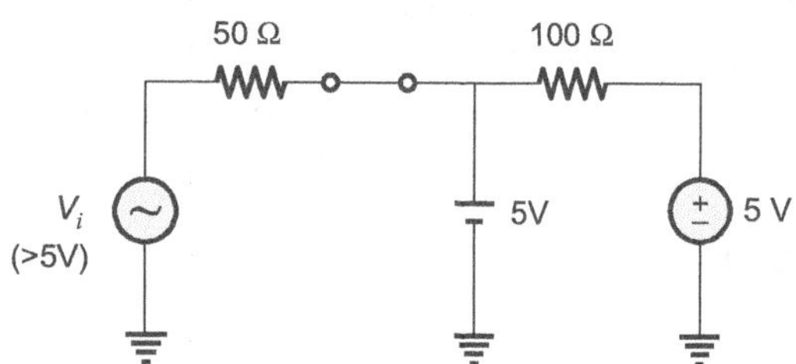

Now also current through 100 Ω is zero. Hence power dissipation in 100 Ω resistor is zero.

1.22 (1.8)

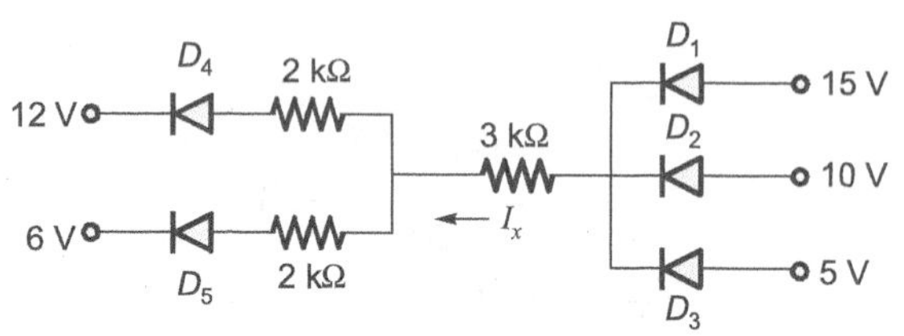

D_1, D_2, D_3 are in common cathode connection. Only D_1 can conduct D_2 and D_3 remain OFF. Let D_4, D_5 be ON.

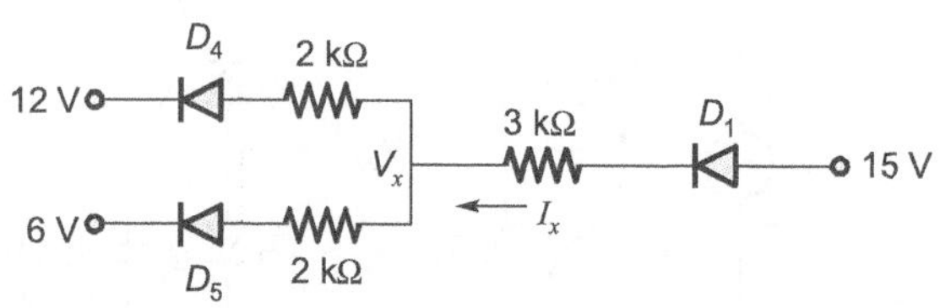

KCL at V_x Node:

$$I_x = I_4 + I_5$$

$$\frac{15 - V_x}{3} = \frac{V_x - 12}{2} + \frac{V_x - 6}{2} = \frac{2V_x - 18}{2}$$

$$\frac{15 - V_x}{3} = V_x - 9 \Rightarrow V_x = 10.5 \text{ V}$$

$\because V_x < 12$ V $\Rightarrow D_4$ must be OFF

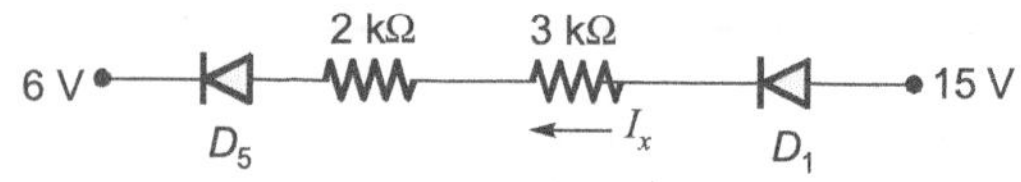

$$I_x = \frac{15 - 6}{2 + 3} = 1.8 \text{ mA}$$

1.23 (6.0)

$$I_Z = I - I_L$$

$$I_{Z,min} = I - I_{L,max}$$

$$I_{L,max} = \frac{16 - 4}{1.2} - 4 = 6 \text{ mA}$$

1.24 (d)

Given circuit is positive peak detector, so capacitor will detect the peak of input and capacitor voltage will remains at peak value of input voltage.

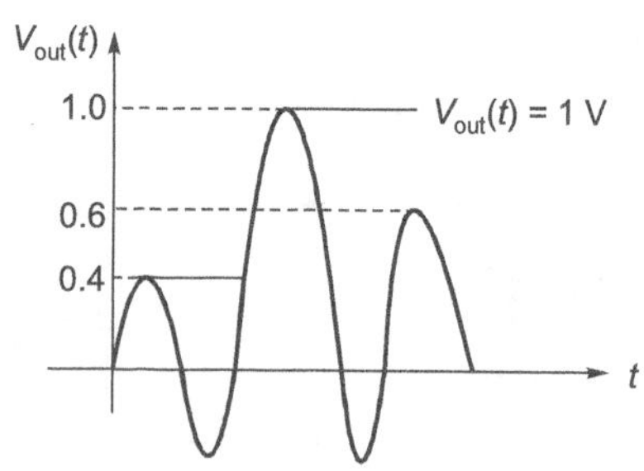

1.25 (b)

LED emits light when it is **forward** biased. Photodiode provides maximum sensitivity to light when it is operated in **reverse** bias.

1.26 (1)

If i_s is positive : Diode is ON.

$$i_R = 0$$

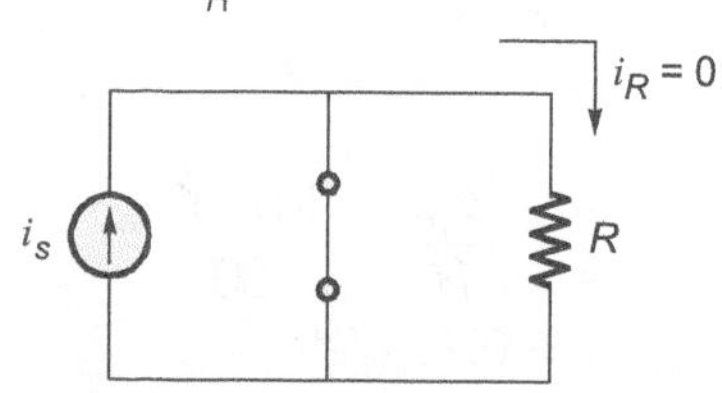

If i_s is negative : Diode is OFF.

$$i_R = i_s = \pi \sin 300\pi t \text{ mA}$$

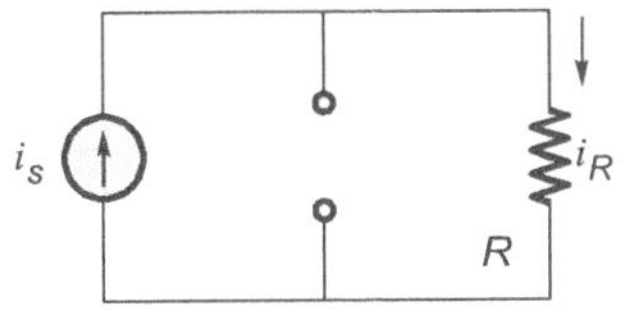

Average current through the resistor R,

$$i_R = \frac{\text{Peak value}}{\pi} = -\frac{\pi}{\pi} = -1\text{mA}$$

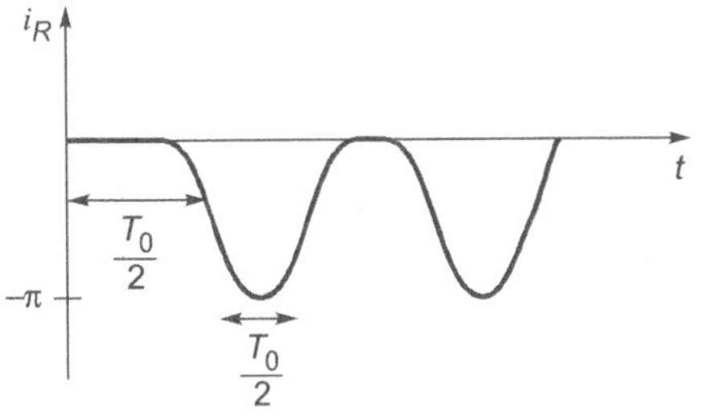

Magnitude of average current through the resistor R,

$$i_R = 1 \text{ mA}$$

1.27 (a)

Incident photons = 8×10^{12}

Electron collected/generated = 4×10^{12}

Quantum efficiency of photodiode

$$= \frac{\text{Number of electron generated}}{\text{Number of photon incident}}$$

$$= \frac{4 \times 10^{12}}{8 \times 10^{12}} = \frac{4}{8} = \frac{1}{2} = 0.5 = 50\%$$

■■■

2 BJT, JFET, MOSFET, their Biasing & Frequency Response

2.1 The circuit shown in figure is said to be in common _______ configuration.

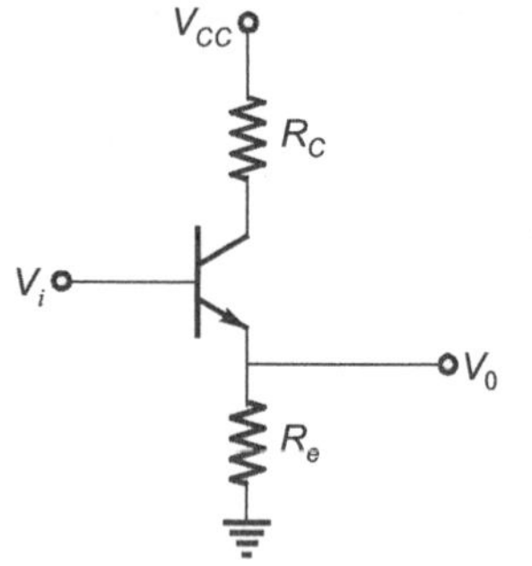

[1995 : 1 M]

2.2 As a switching device, JFET has a very low offset Voltage in the ON state as compared to BJT. (TRUE / FALSE).

[1995 : 1 M]

2.3 A common emitter transistor amplifier using a collector load of 1 kΩ and operating at room temperature with I_C = 1 mA, gives a voltage gain of nearly equal to

(a) 25 (b) 40
(c) 250 (d) 1000 [1997 : 1 M]

2.4 For a silicon BJT shown in figure find R_B to establish V_{CE} = 2 V, Assume V_{BE} = 0.7 V

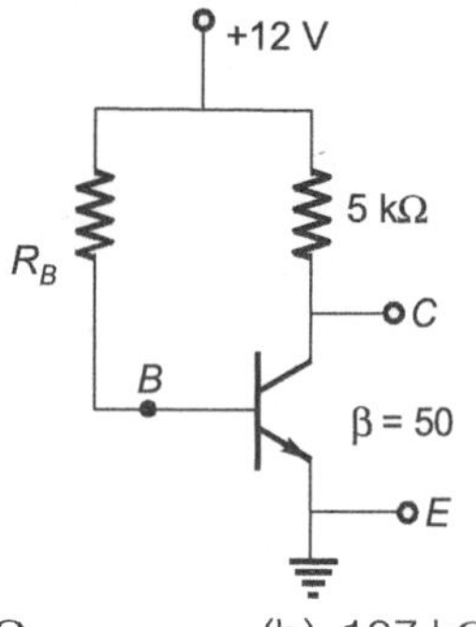

(a) 283 kΩ (b) 107 kΩ
(c) 200 kΩ (d) 242 kΩ

[1998 : 1 M]

2.5 Match the figures with the devices names

List-I

A. 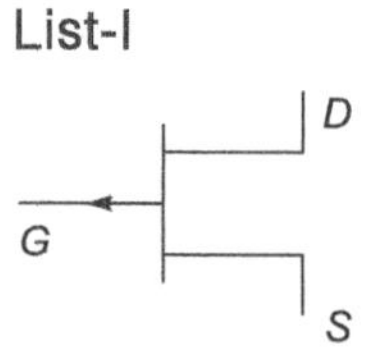

B. 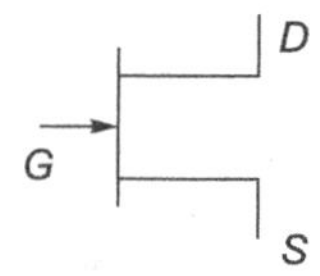

C. 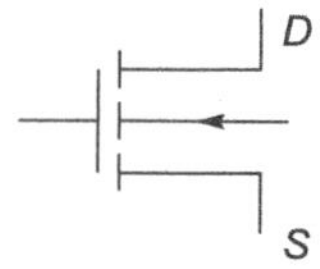

D. 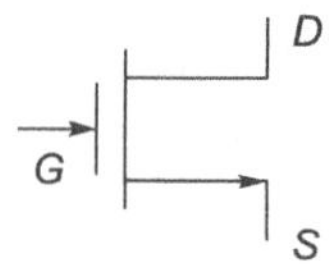

List-II

1. Depletion mode MOSFET
2. P-channel JFET
3. N-channel JFET
4. Enhancement Mode MOSFET

Codes:

	A	B	C	D
(a)	2	3	4	1
(b)	1	2	3	4
(c)	1	3	4	2
(d)	3	1	2	4

[1998 : 1 M]

2.6 In the JFET amplifier circuit shown in figure, the signal outputs V_1 and V_2 are related as

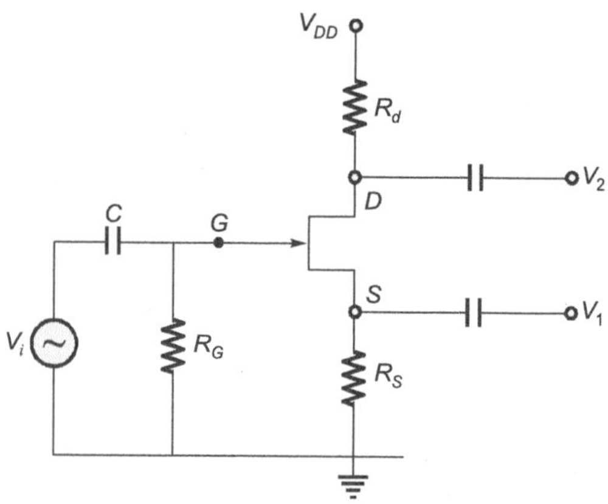

(a) $V_2 = \frac{R_d}{R_s} V_1$ (b) $V_2 = \frac{-R_d}{R_s} V_1$

(c) $V_2 = \frac{R_s}{R_d} V_1$ (d) $V_2 = -\frac{R_s}{R_d} V_1$

[1999 : 1 M]

2.7 A unity gain buffer amplifier has a bandwidth of 1 MHz. The output voltage of the amplifier for an input of 2 V sinusoid of frequency 1 MHz will be

(a) 2 V (b) $2\sqrt{2}$ V

(c) $\frac{2}{\sqrt{2}}$ V (d) $\frac{1}{\sqrt{2}}$ V [1999 : 2 M]

2.8 If both the junctions of a transistor are forward biased it will be in

(a) saturation mode (b) active mode

(c) cut-off mode (d) inverse active mode

[2000 : 1 M]

2.9 A transistor amplifier circuit is shown in figure. The quiescent collector current, rounded off to first decimal, is

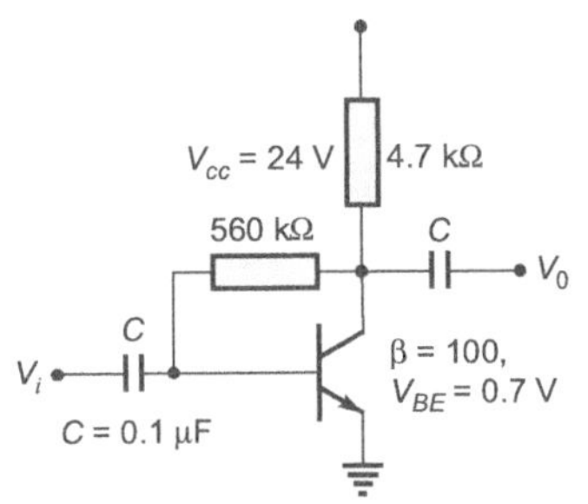

(a) 2.6 mA (b) 2.3 mA

(c) 2.1 mA (d) 2.0 mA [2003 : 2 M]

2.10 The parameters of the JEET in figure are: g_m = 1 mA/V, r_d = 15 kΩ. Neglecting the effect of the capacitor for A.C. analysis, the small signal A.C. voltage gain for the circuit is

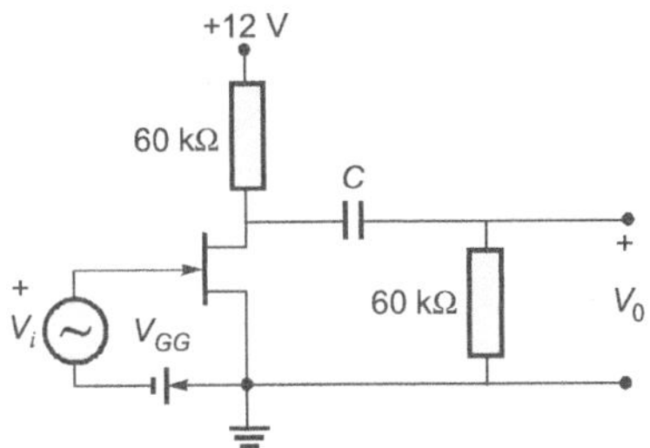

(a) –30 (b) –10

(c) +4 (d) +60 [2004 : 2 M]

2.11 For the circuit shown in the figure, I_{DSQ} (in mA) and V_{GSQ} (in V) are related through $2I_{DSQ} = (4 + V_{GSQ})^2$.

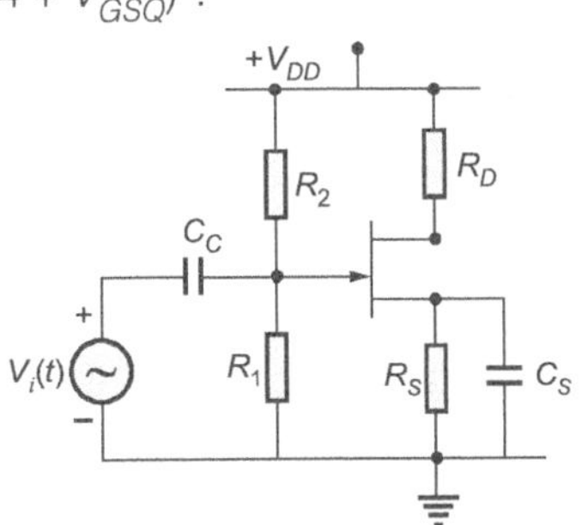

The following data is given:
V_{DD} = 15 V, R_1 = 1.0 MΩ, R_2 = 6.5 MΩ, R_D = 2.0 kΩ, R_S = 1.0 kΩ, I_{DSS} = 8 mA.
The value of I_{DSQ}, assuming the gate current is negligible, is approximately equal to

(a) 5 mA (b) 2.0 mA

(c) 2.3 μA (d) 3.4 mA [2005 : 2 M]

2.12 An amplifier circuit is shown below. Assume that the transistor works in active region. The low frequency small-signal parameters for the transistor are g_m = 20 mS, β_0 = 50, $r_0 = \infty$, r_b = 0. What is the voltage gain, $A_V = \left(\frac{V_0}{V_i}\right)$, of the amplifier?

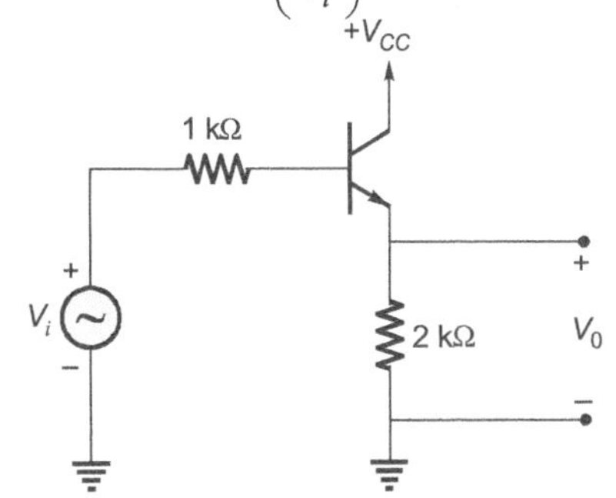

(a) 0.967 (b) 0.976

(c) 0.983 (d) 0.998 [2006 : 2 M]

2.13 The biasing circuit of a silicon transistor is shown below. If β = 80, then what is V_{CE} for the transistor?

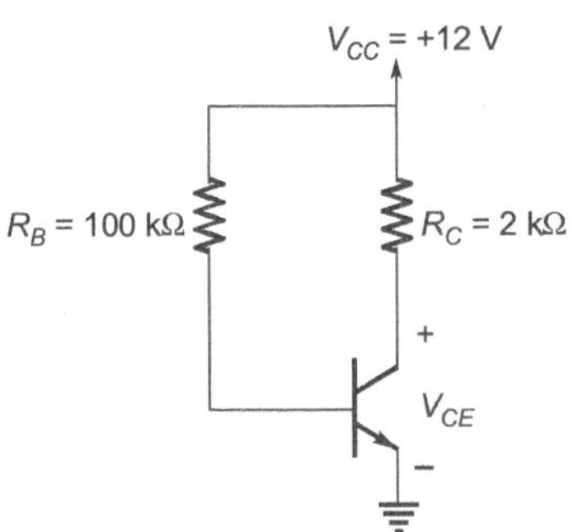

(a) –6.08 V (b) 0.2 V

(c) 1.2 V (d) 6.08 V [2006 : 2 M]

2.14 A FET source follower is shown in the figure below:

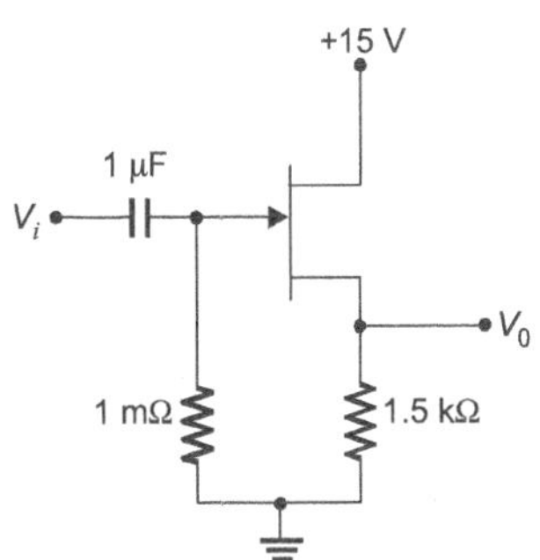

The nature of feedback in this circuit is
(a) positive current
(b) negative current
(c) positive voltage
(d) negative voltage [2007 : 2 M]

2.15 In the circuit shown below, $V_{BE} = 0.7$ V.
The β of the transistor and V_{CE} are, respectively

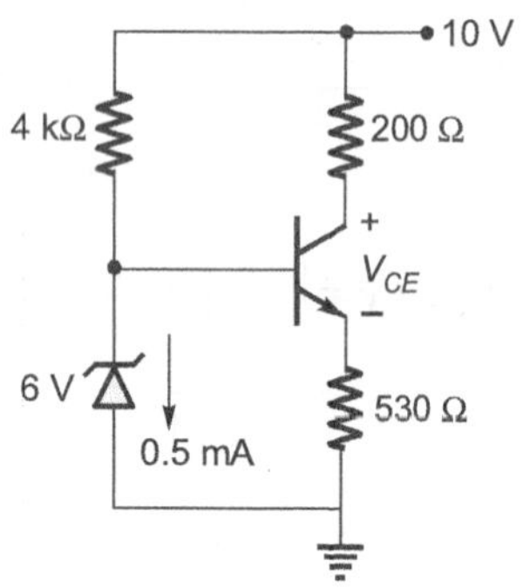

(a) 19 and 2.8 V (b) 19 and 4.7 V
(c) 38 and 2.8 V (d) 38 and 4.7 V
[2007 : 2 M]

2.16 The three transistors in the circuit shown below are identical, with $V_{BE} = 0.7$ V and β = 100.

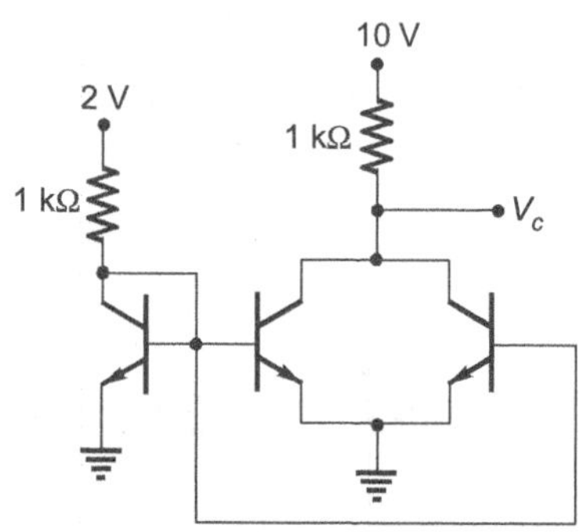

The voltage V_c is
(a) 0.2 V (b) 2 V
(c) 7.4 V (d) 10 V [2007 : 2 M]

2.17 For a single stage BJT common base amplifier,
(a) current gain as well as voltage gain can be greater than unity
(b) current gain can be greater than unity but voltage gain is always less than unity
(c) voltage gain can be greater than unity but current gain is always less than unity
(d) current gain as well as voltage gain is always less than unity [2008 : 1 M]

2.18 In the amplifier circuit shown below, assume $V_{BE} = 0.7$ V and the β of the transistor and the values of C_1 and C_2 are extremely high. If the amplifier is designed such that at the quiescent point its $V_{CE} = V_{CC}/2$, where V_{CC} is the power supply voltage, its small signal voltage gain $\left|\frac{V_{out}}{V_{in}}\right|$ will be

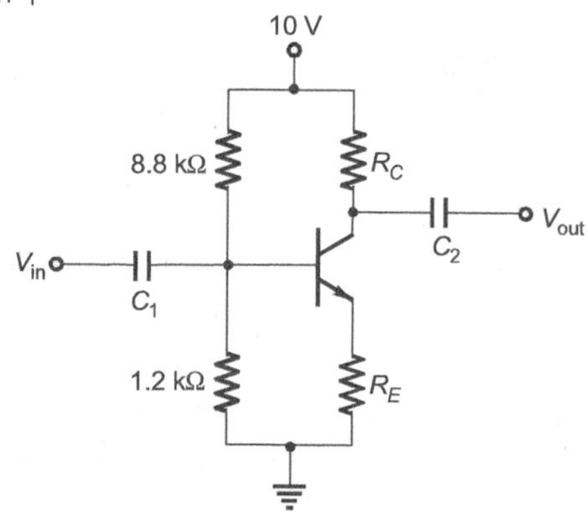

(a) 3.75 (b) 4.5
(c) 9 (d) 19 [2008 : 2 M]

Common Data Questions (2.19 and 2.20):
The circuit shown in the figure uses three identical transistors with $V_{BE} = 0.7$ and β = 100.
Given: $R_1 = R_2 = R_3 = 1$ kW

$$\frac{kT}{q_e} = 25 \text{ mV}$$

The collector current of transistor Q_3 is 2 mA.

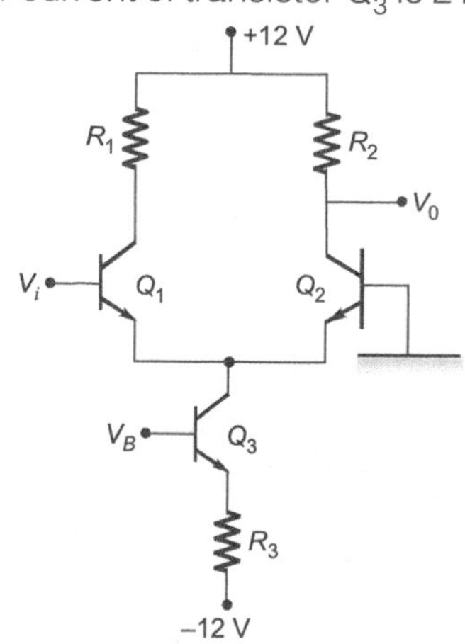

2.19 The bias voltage V_B at the base of the transistor Q_3 is approximately
(a) −9.3 V (b) −10.0 V
(c) −10.3 V (d) −11.0 V [2009 : 2 M]

2.20 The small-signal voltage gain of the circuit is
(a) −20 (b) −40
(c) 20 (d) 40 [2009 : 2 M]

2.21 The matched transistors Q_1 and Q_2 shown in the adjoining figure have β = 100. Assuming the base-emitter voltage to be 0.7 V, the collector-emitter voltage V_2 of the transistor Q_2 is

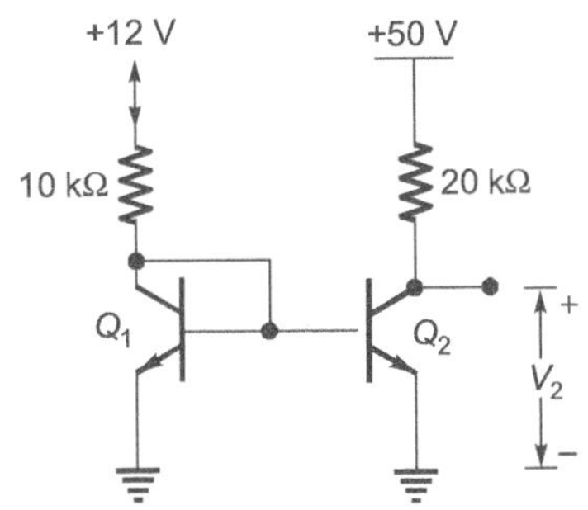

(a) 33.9 V (b) 27.8 V
(c) 16.2 V (d) 0.7 V

[2010 : 2 M]

2.22 The amplifier shown below as a voltage gain of –2.5, an input resistance of 10 kΩ and a lower 3 dB cut-off frequency of 20 Hz. Which on the following statements is **TRUE** when the emitter resistance R_E is doubled?

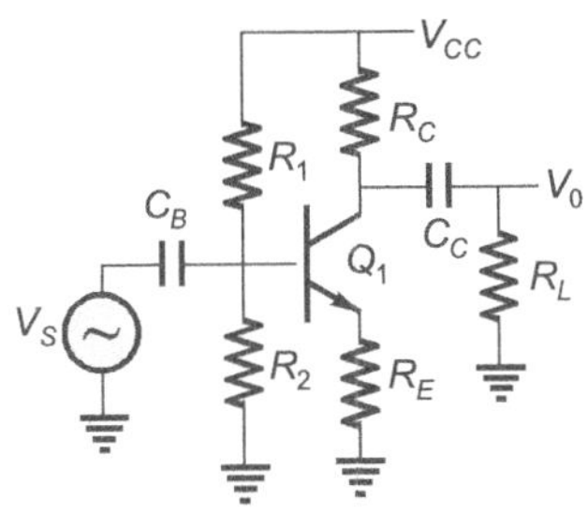

(a) Magnitude of voltage gain will decrease
(b) Input resistance will decrease
(c) Collector bias current will increase
(d) Lower 3-*dB* cut-off frequency will increase

[2011 : 1 M]

2.23 The voltage gain A_v of the circuit shown below is

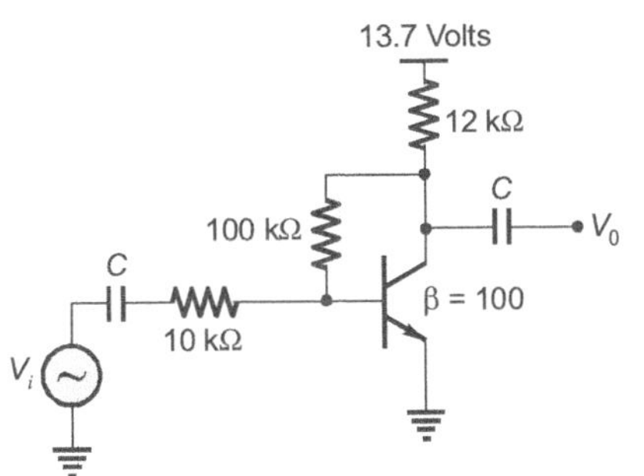

(a) $|A_v| \approx 200$ (b) $|A_v| \approx 100$
(c) $|A_v| \approx 20$ (d) $|A_v| \approx 10$

[2012 : 2 M]

2.24 In the transistor circuit as shown below, the value of resistance R_E in kΩ is approximately.

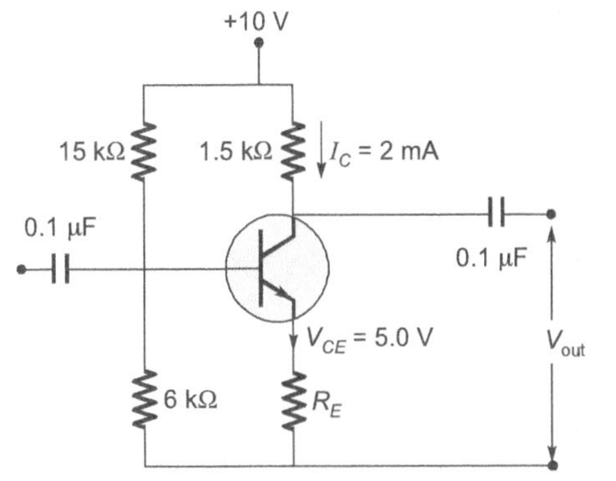

(a) 1.0 (b) 1.5
(c) 2.0 (d) 2.5

[2013 : 1 M]

2.25 For the circuit shown in the figure, the transistor has β = 40, V_{BE} = 0.7 V, and the voltage across the Zener diode is 15 V. The current (in mA) through the Zener diode is ____.

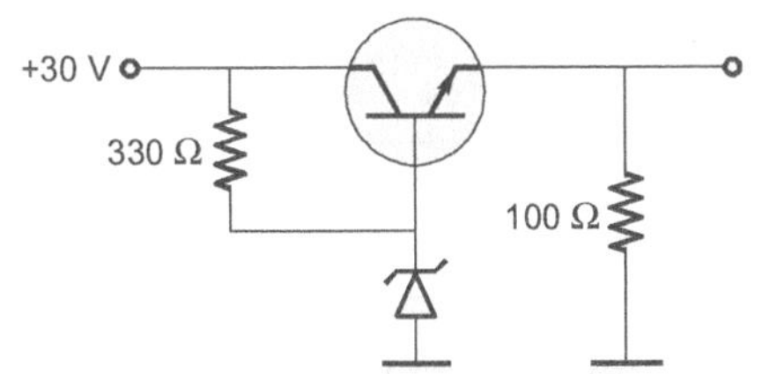

[2014 : 2 M]

2.26 In the figure, transistors T_1 and T_2 have identical characteristics. $V_{CE(sat)}$ of transistor T_3 is 0.1 V. The voltage V_1 is high enough to put T_3 in saturation. Voltage V_{BE} of transistors T_1, T_2 and T_3 is 0.7 V. The value of $(V_1 - V_2)$ in V is ____.

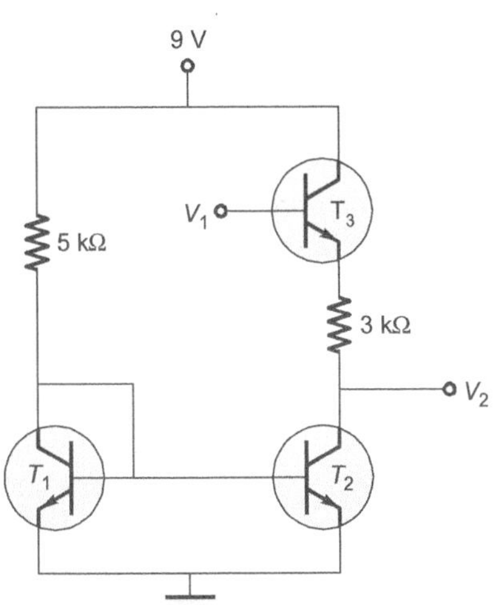

[2014 : 2 M]

2.27 In the circuit shown in the figure, it is found that V_{BE} = 0.7 V and V_E = 0 V. If β_{DC} = 99 for the transistor, then the value of R_B in kilo ohms is ___ kΩ.

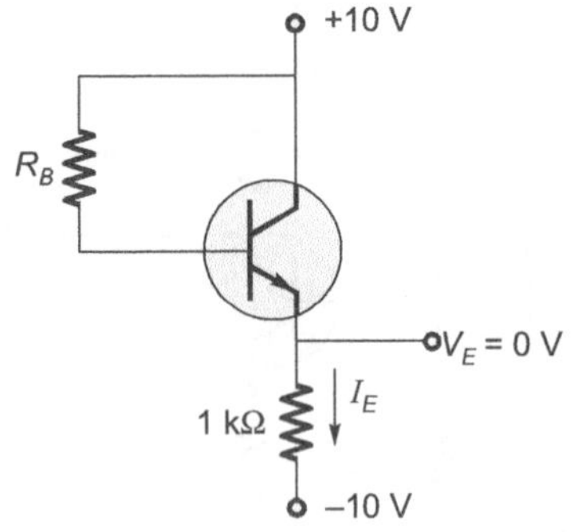

[2015 : 2 M]

2.28 In the circuit, shown in the figure, the MOSFET is operating in the saturation zone. The characteristics of the MOSFET is given by $I_D = \frac{1}{2}(V_{GS} - 1)^2$ mA, where V_{GS} is in V. If V_s = +5 V, then the value of R_s in kΩ is _____.

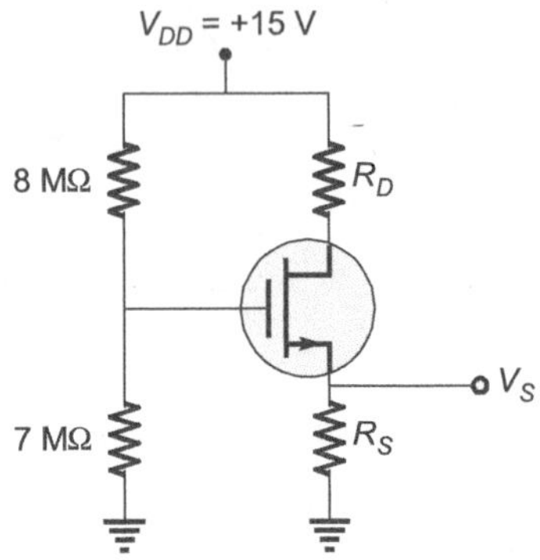

[2017 : 2 M]

2.29 A voltage amplifier is constructed using enhancement mode MOSFETs labeled M_1, M_2, M_3 and M_4 in the figure below. M_1, M_2 and M_4 are *n*-channel MOSFETs and M_3 is a *p*-channel MOSFET. All MOSFETs operate in saturation mode and channel length modulation can be ignored. The low frequency, small signal input and output voltages are v_{in} and v_{out} respectively and the dc power supply voltage is V_{DD}. All n-channel MOSFETs have identical transconductance g_{mn} while the p-channel MOSFET has transconductance g_{mp}. The expressions for the low frequency small signal voltage gain v_{out}/v_{in} is

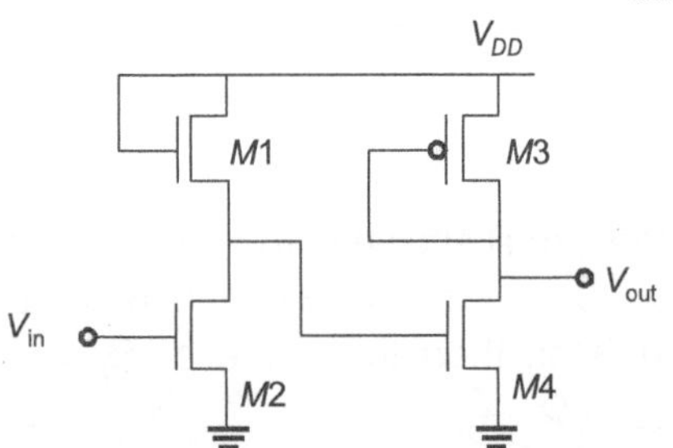

(a) $\frac{-g_{mn}}{g_{mp}}$ (b) $-g_{mn}(g_{mn} + g_{mp})^{-1}$

(c) $\frac{+g_{mn}}{g_{mp}}$ (d) $g_{mn}(g_{mn} + g_{mp})^{-1}$

[2019 : 2 M]

2.30 In the circuit shown below, all transistors are n-channel enhancement mode MOSFETs. They are identical and are biased to operate in saturation mode. Ignoring channel length modulation, the output voltage V_{out} is__________V.

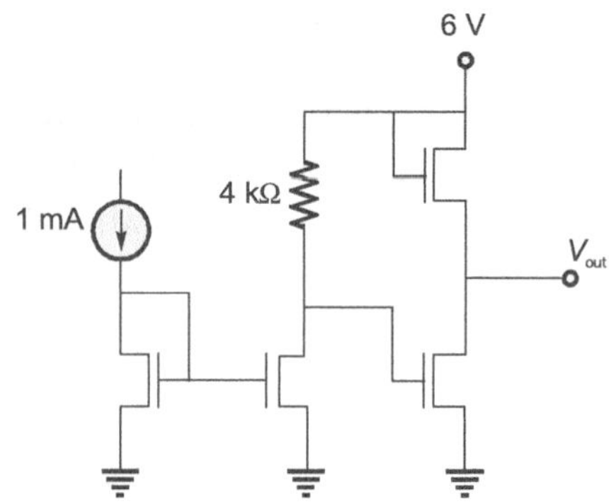

[2019 : 2 M]

2.31 All the transistors used in the circuit are matched and have a current gain β of 20. Neglecting the Early effect, the current I_{04} in milliampere is _____.

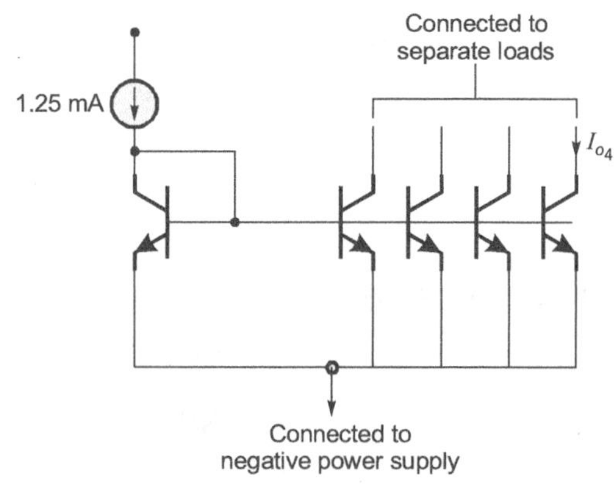

[2021 : 2 M]

2.32 The transistor Q_1 has a current gain β_1 = 99 and the transistor Q_2 has a current gain β_2 = 49. The current I_{B2} in microampere is _____.

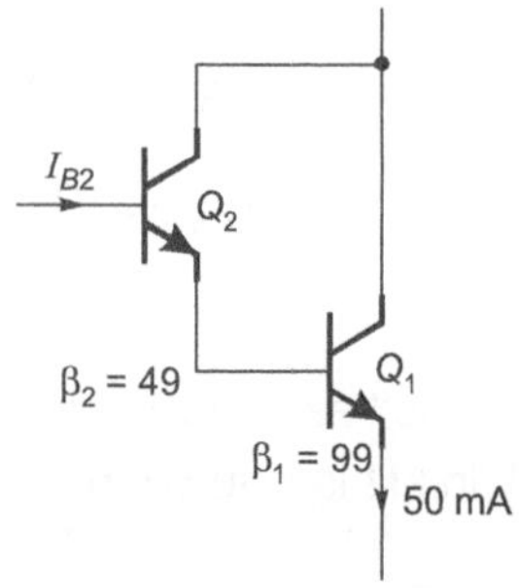

[2021 : 1 M]

2.33 In the small signal circuit shown, the enhancement mode n-channel MOSFET is biased in saturation with transconductance g_m. If channel length modulation is ignored, the small signal impedance looking into the node P is given by ________.

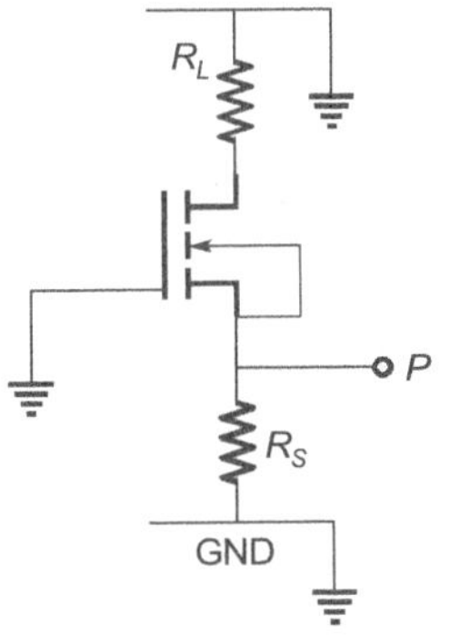

(a) $R_S \parallel R_L \parallel g_m^{-1}$ (b) $R_s \parallel g_m^{-1}$

(c) $(R_S + R_L) \parallel g_m^{-1}$ (d) $\dfrac{R_L g_m}{1+R_S g_m}(R_L \parallel g_m^{-1})$

[2022 : 2 M]

2.34 In the small signal circuit shown, the enhancement mode n-channel MOSFET is biased in saturation with a transconductance g_m. A small signal low-frequency voltage v_d injected at the supply terminal results in a small signal voltage fluctuation v_0 at the output. If the channel length modulation of the MOSFET is ignored, the small signal gain v_0/v_d is given by ________

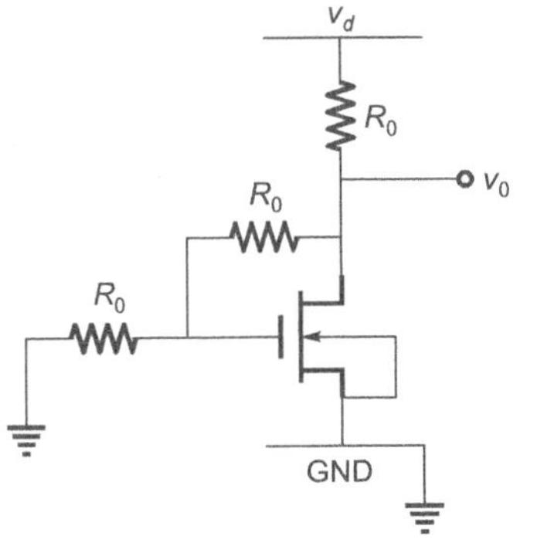

(a) $\dfrac{-g_m R_0}{1+g_m R_0}$ (b) $(g_m R_0 + 1)^{-1}$

(c) $\dfrac{-g_m R_0}{1+2g_m R_0}$ (d) $\left(\dfrac{g_m R_0}{2}+\dfrac{3}{2}\right)^{-1}$

[2022 : 2 M]

2.35 In the circuit shown, $R_1 = 100\ \text{k}\Omega$ and $R_2 = 1\ \text{k}\Omega$. If the base-to-emitter voltage of the npn BJT is 0.7 V and the collector-to-emitter voltage is 5.2 V, the β (current gain) of the BJT is ________ (round off to two decimal places)

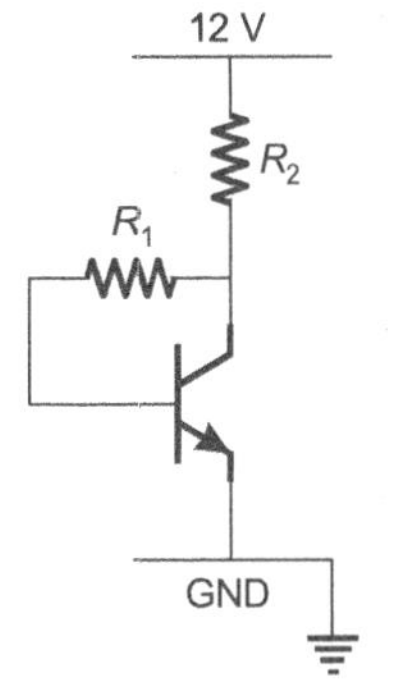

[2022 : 2 M]

2.36 The figure below shows a feedback amplifier constructed using an nMOS transistor. Assume that $\mu_n C_{ox} = 1\ \text{mA/V}^2$, threshold voltage $V_T = 1$ V and $W/L = 2$. The bias voltage at the drain terminal is 4 V. The capacitors C_∞ offer zero impedance at the signal frequency. The ratio V_{out}/V_{in} is ____ (rounded off to two decimal places).

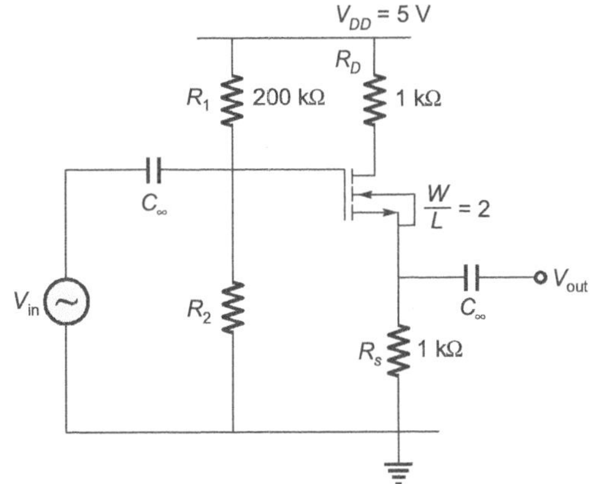

[2023 : 2 M]

2.37 Figure shows an amplifier using an NMOS transistor. Assume that the transistor is in saturation with device parameters, $\mu_n C_{ox} = 250\ \mu\text{A/V}^2$, threshold voltage $V_T = 0.65$ V and W/L = 4. Ignore the channel length modulation effect. The drain current of the transistor at the operating point is ________ μA (rounded off to nearest integer).

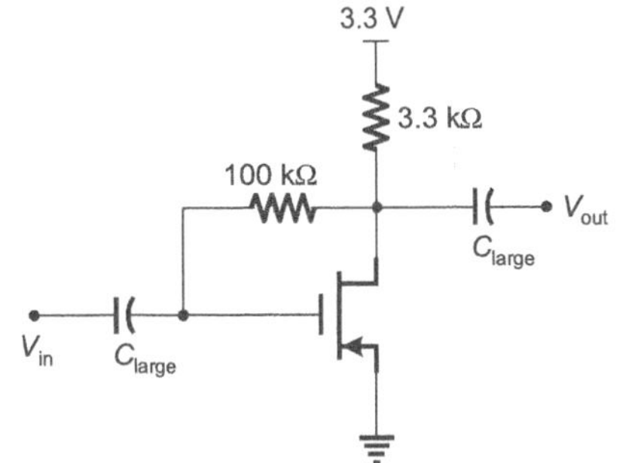

[2024 : 1 M]

■■■

Answers BJT, JFET, MOSFET, their Biasing & Frequency Response

2.1 (Sol.)	2.2 (True)	2.3 (b)	2.4 (a)	2.5 (a)	2.6 (b)	2.7 (a)
2.8 (a)	2.9 (b)	2.10 (b)	2.11 (d)	2.12 (d)	2.13 (b)	2.14 (d)
2.15 (a)	2.16 (c)	2.17 (c)	2.18 (c)	2.19 (a)	2.20 (b)	2.21 (b)
2.22 (a)	2.23 (d)	2.24 (a)	2.25 (41.96)	2.26 (5.58)	2.27 (93)	2.28 (10)
2.29 (c)	2.30 (4)	2.31 (1)	2.32 (10)	2.33 (b)	2.34 (d)	2.35 (150.11)
3.36 (0.67)	2.37 (500)					

Explanations BJT, JFET, MOSFET, their Biasing & Frequency Response

2.1 Sol.

Circuit shown below is a common collector configuration.

2.3 (b)

For common emitter transistor amplifier,

$$\text{Gain } A_V = \frac{V_{out}}{V_{in}} = -\frac{R_C}{r_e}$$

where, $r_e = \frac{V_T}{I_C}$

V_T = Voltage at room temperature = 25 mV

$$r_e = \frac{25\text{ mV}}{1\text{mA}} = 25\ \Omega$$

$$A_V = -\frac{R_C}{r_e} = -\frac{1000}{25} = -40$$

2.4 (a)

By applying KVL to the output loop,

$12 - I_C \times 5 \times 10^3 - V_{CE} = 0$

$$\Rightarrow \quad I_C = \frac{12-2}{5\times10^3} = 2\text{ mA}$$

assuming transistor is in active region

$\therefore \quad I_C = \beta I_B$

applying KVL to input loop

$12 - I_B \times R_B - V_{BE} = 0$

$$\Rightarrow \quad R_B = \frac{(12-0.7)\beta}{I_C} = 282.5\text{ k}\Omega$$

Since, $R_B > R_C$

$\therefore \quad V_B < V_C \Rightarrow V_{CB} > 0$

hence transistors is in active region, so our assumption is right, and the required value of R_B is 282.5 kΩ.

2.6 (b)

From the given amplifier circuit

$$V_2 = -I_D R_d \qquad \text{...(i)}$$

and $$V_1 = I_D R_s \qquad \text{...(ii)}$$

from equation (i) and (ii),

$$\therefore \quad \frac{V_2}{V_1} = -\frac{R_d}{R_s} \Rightarrow V_2 = -\frac{R_d}{R_s}V_1$$

2.7 (a)

Since Gain × Bandwidth = constant for a amplifier so for unity gain buffer

Output = Input × Gain = 2 × 1 = 2 V

$\Rightarrow \quad V_0 = 2$ V

2.8 (a)

Condition for saturation mode is the forward biased of both the junction.

2.9 (b)

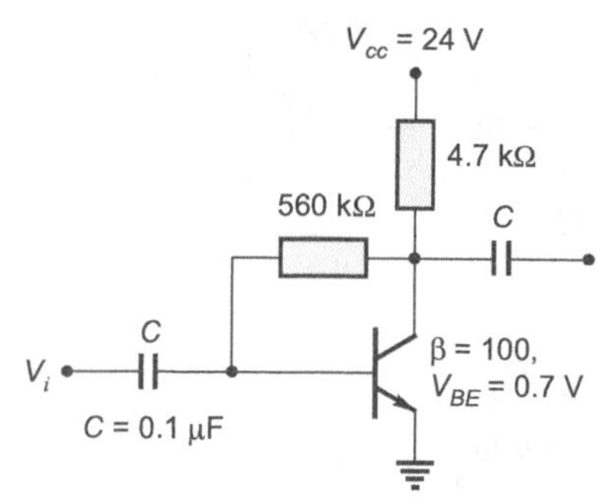

Apply KVL to the collector-base-emitter circuit we get

$V_{CC} - (I_C + I_B)R_C - I_B R_B - V_{BE} = 0$

$24 - (I_C + I_B)\,4.7 - I_B \times 560 - 0.7 = 0$

$24 - (100 + 1)\,I_B 4.7 - I_B 560 - 0.7 = 0$

$(\because I_C = \beta I_B)$

$$I_B = \frac{23.3}{560 + 101\times4.7} = 0.0225\text{ mA}$$

So, $I_{CQ} = \beta I_B$

$= 100 \times 0.0225\text{ mA} = 2.25\text{ mA} \simeq 2.3\text{ mA}$

2.10 (b)

Small AC voltage gain:

$$A_V = -g_m R_D$$

$$R_D = (60\ k\Omega)||(60\ k\Omega)||r_d$$

$$= (30\ k\Omega)||(15\ k\Omega) \quad (\because r_d = 15\ k\Omega)$$

$$= \frac{30 \times 15}{45} = 10\ k\Omega$$

So, $A_V = 1\ mA/V \times 10\ k\Omega = -10$

2.11 (d)

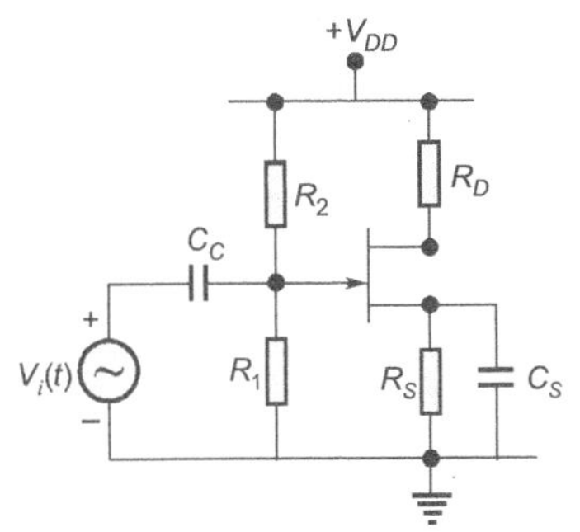

$$\text{Voltage across } (R_1) = \frac{R_1}{R_1 + R_2} \cdot V_{DD}$$

$$= \frac{1}{1+6.5} \times 15 = \frac{15}{7.5} = 2\ V$$

$\therefore$ $$2 = V_{GSQ} + R_S \times I_{DSQ}$$

$$= V_{GSQ} + I_{DSQ}$$

$$(\because R_s = 1\ k\Omega)$$

$\therefore$ $$I_{DSQ} = 2 - V_{GSQ}$$

$$2I_{DSQ} = 4 - 2\ V_{GSQ} \quad \text{...(i)}$$

Given, $$2I_{DSQ} = (4 + V_{GSQ})^2 \quad \text{...(ii)}$$

Solving equations (i) and (ii), we get

$$4 - 2\ V_{GSQ} = (4 + V_{GSQ})^2$$

$$V_{GSQ}^2 + 10\ V_{GSQ} + 12 = 0$$

$\therefore$ $$V_{GSQ} = -1.39\ V$$

$$\boxed{|V_{GSQ}| = 1.39\ V}$$

Put the value of V_{GSQ} in equation (ii), we get

$$2I_{DSQ} = (4 - 1.39)^2 = 6.8121\ mA$$

$\therefore$ $$I_{DSQ} = 3.4\ mA$$

2.12 (d)

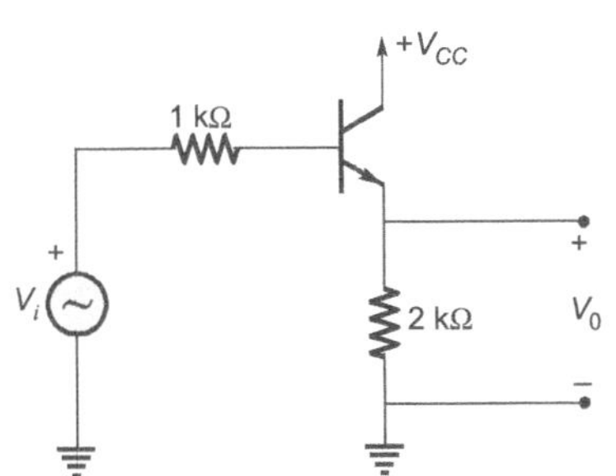

The small signal equivalent circuit of the above circuit diagram is

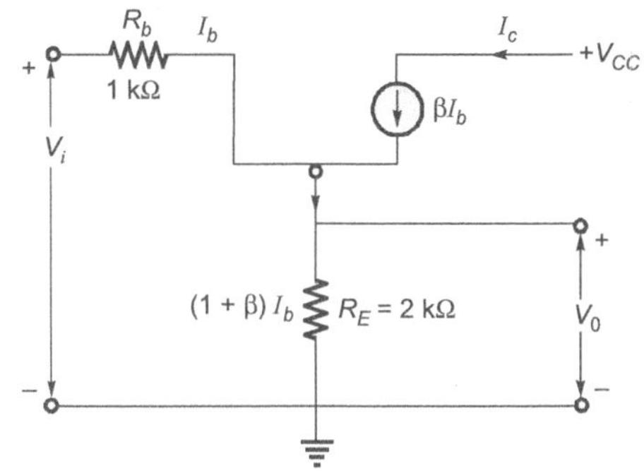

$$V_0 = R_E(1+\beta)I_b$$

$$V_i = R_b I_b + R_E(1+\beta)I_b$$

Voltage gain (A_V) =

$$\frac{V_0}{V_i} = \frac{R_E(1+\beta)I_b}{R_b I_b + R_E(1+\beta)I_b} = \frac{R_E(1+\beta)}{R_b + R_E(1+\beta)}$$

$$= \frac{2 \times 51}{1 + 2 \times 51} = \frac{102}{103} = 0.990$$

2.13 (b)

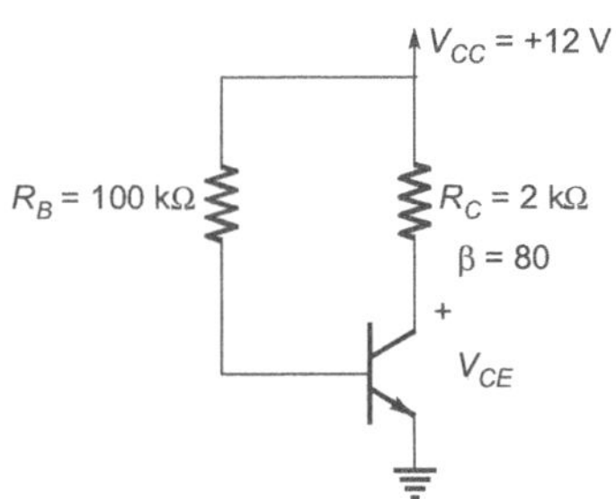

V_{OC} across base-emitter junction = 12 V

Since, $V_{OC} > V_{BEactive}$.

Therefore transistor will be on, it may be in active region or saturation region.

Let us assume that this is in active region then applying KVL in base-emitter loop.

We have

$$12 - 100 \times 10^3 I_B - 0.7 = 0$$

$$I_B = 0.113\ mA$$

$\therefore$ $$I_C = \beta I_B = 80 \times 0.113\ mA = 9.04\ mA$$

$\therefore$ $$V_{CE} = V_{CC} - I_C R_C$$

$$= 12 - 2 \times 9.04 = -6.08\ V$$

Since V_{CE} must be greater than 0.2 V for active region, so our assumption is wrong and transistor is in saturation region.

$\therefore$ $$V_{CE} = 0.2\ V$$

2.14 (d)

It is –ve voltage feedback with $\beta = 1$.

2.15 (a)

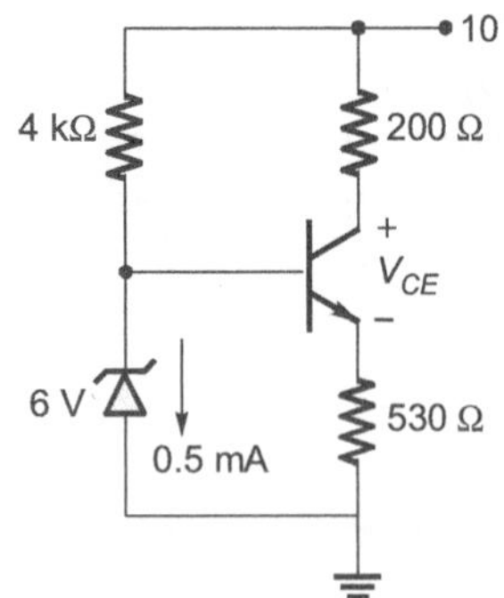

The current through base resistor is

$$I_b' = \frac{(10-6)V}{4\,\text{k}\Omega} = 1\text{mA}$$

Now, 0.5 mA of current passes through zener diode hence effective base current is

$$I_b = 0.5 \text{ mA}$$

Apply KVL to the base circuit,

$V_Z - V_{BE} - I_E R_E = 0$

$6 - 0.7 - (\beta + 1)I_b \times 530 = 0$

$5.3 - (\beta + 1)\, 0.5 \times 10^{-3} \times 530 = 0$

$5.3 - 0.265\,(\beta + 1)$

$$\therefore \quad \beta + 1 = \frac{5.3}{0.265} = 20\,;\ \therefore \beta = 19$$

Now, apply KVL to the collector to emitter circuit

$V_{CC} - I_C R_C - V_{CE} - I_E R_E = 0$

$$\begin{aligned} V_{CE} &= V_{CC} - I_C R_C - I_E R_E \\ V_{CE} &= 10 - I_C 200 - 530\, I_E \\ &= 10 - \beta I_b 200 - 530 \times (\beta + 1)\, I_b \\ &= 10 - 19 \times 0.5 \times 10^{-3} \times 200 \\ &\quad -530 \times 20 \times 0.5 \times 10^{-3} \\ &= 10 - 1.9 - 5.3 = 2.8 \text{ volts} \end{aligned}$$

2.16 (c)

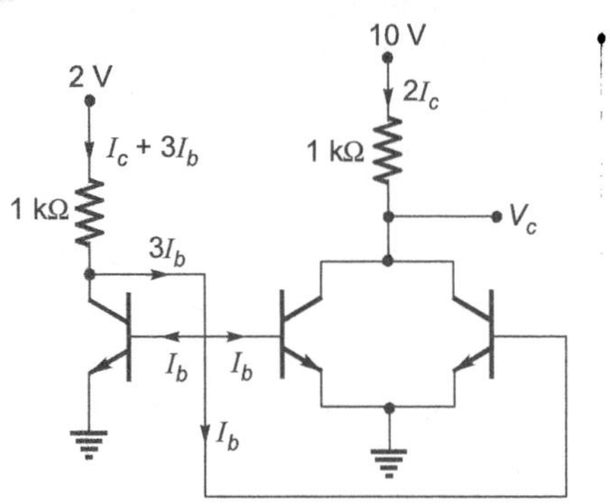

Using KVL in the left hand transistor,

$2 - (I_C + 3\,I_b) \times 1 - V_{BE} = 0$

$$2 - 0.7 = (\beta + 3)I_b \qquad (\because I_c = \beta I_b)$$

$$\therefore \quad I_b = \frac{1.3}{103} \qquad (\because \beta = 100 \text{ given})$$

Now again apply KVL to the collector resistor of the right and transistor

$V_{CC} - 2I_C \times 1 - V_C = 0$

$$\therefore \; V_C = 10 - 2 \times \beta I_b = 10 - 2 \times 100 \times \frac{1.3}{103}$$

$$\therefore \; V_C = (10 - 2.52)\, V = 7.4 \text{ Volts} = 7.4 \text{ Volts}$$

2.17 (c)

$$A_V > 1 \text{ and } A_I = \frac{\beta}{\beta + 1} < 1$$

2.18 (c)

D.C. analysis,

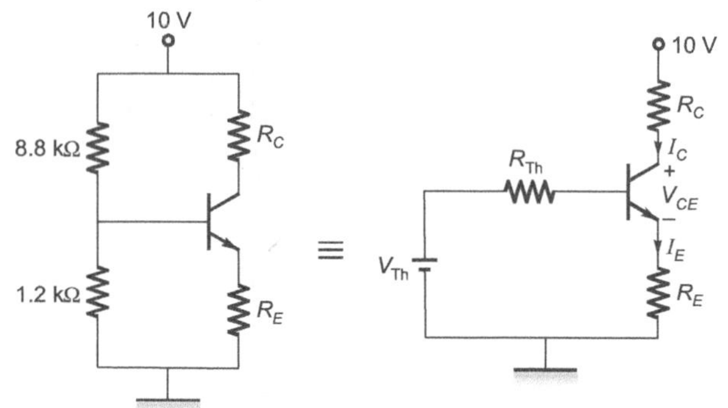

$$\text{where,} \quad V_{Th} = \frac{10 \times 1.2}{1.2 + 8.8} = 1.2 \text{ V}$$

$$R_{Th} = \frac{1.2 \times 8.8}{1.2 + 8.8} = 1.056 \text{ V}$$

$\because$ β is very large,

$I_B \approx 0$ [No voltage drop across R_{Th}]

$I_C = I_E$

Apply KVL in input loop,

$$V_{Th} = I_B R_{Th} + V_{BE} + I_E R_E$$

$$1.2 = 0 + 0.7 + I_C R_E$$

$$I_C = \frac{0.5}{R_E} \qquad \text{...(i)}$$

Apply KVL in output loop,

$$-10 + I_C R_C + \frac{10}{2} + I_E R_E = 0$$

$$5 = I_C(R_C + R_E)$$

From equation (i),

$$5 = \frac{0.5}{R_E}[R_C + R_E]$$

$$\Rightarrow \quad \frac{R_C}{R_E} = 9 \qquad \text{...(ii)}$$

AC equivalent small is signal re model,

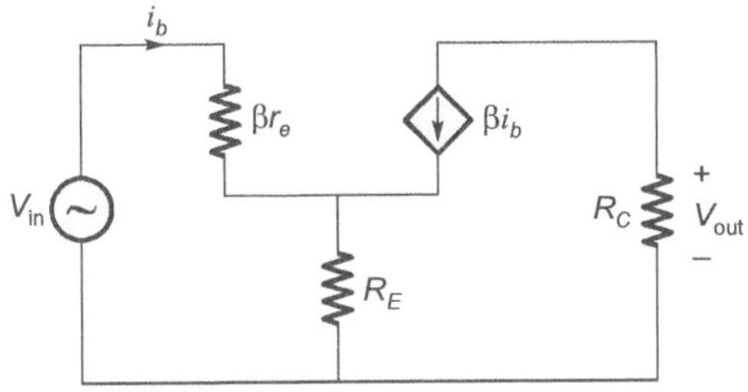

$$V_{out} = -\beta\, i_b R_C$$

$$V_{in} = i_b\beta r_e + (i_b + \beta i_b)\, R_E$$

$$= i_b[\beta r_e + (1+\beta)\, R_E]$$

$\because \quad 1 + \beta \approx \beta$

$$\cong \beta i_b[r_e + R_E]$$

$$\left|\frac{V_{out}}{V_{in}}\right| = \frac{\beta i_b R_C}{\beta i_b[r_e + R_E]} = \frac{R_C}{r_e + R_E}$$

$\because \quad r_e << R_E$

$$\left|\frac{V_{out}}{V_{in}}\right| = \frac{R_C}{R_E}$$

From equation (ii), $\left|\frac{V_{out}}{V_{in}}\right| = \frac{R_C}{R_E} = 9$

2.19 (a)

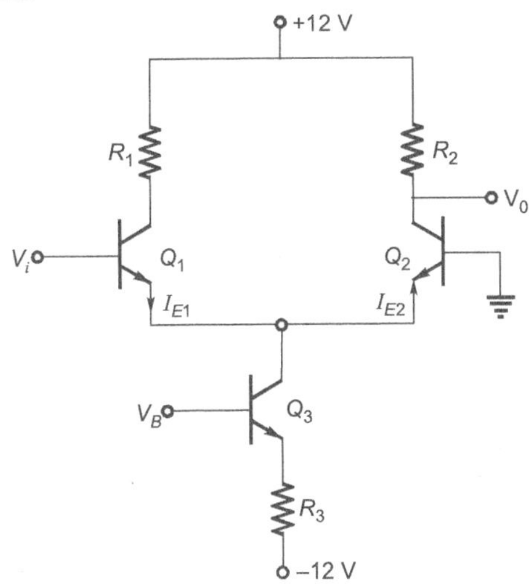

$$I_{E1} = I_{E2}$$

($\because$ β is same and $R_1 = R_2 = 1\ k\Omega$)

$$I_{C3} = I_{E1} + I_{E2} = 2\, I_{E1}$$

$$I_{E3} = I_{C3} + I_{B3} = \left(1 + \frac{1}{\beta}\right) I_{C3}$$

Given, $I_{C3} = 2$ mA

$$I_{E3} = \left(1 + \frac{1}{100}\right) \times 2\text{ mA}$$

$$= \frac{101}{100} \times 2\text{ mA} = \frac{202}{100} \times 1\text{mA}$$

$$V_B - V_{BE} - I_{E3} R_{E3} = -12$$

$\therefore \quad V_B = -12 + V_{BE} + I_{E3} R_{E3}$

$$= -12 + \frac{202}{100} \times 1 + 0.7 = -9.28\text{ V} \simeq -9.3\text{ V}$$

2.20 (b)

When output is taken between one collector and ground then $A_d = -\frac{1}{2} g_m R_C$ (i.e. singly ended gain).

$\Rightarrow$ and if take the output between two collectors, then differential gain,

$$A_d = -g_m R_C$$

$\Rightarrow$ $R_C \rightarrow$ collector resistance = 1 kΩ = 1000 Ω

$$g_m = \frac{I_C}{V_T} = \frac{2\text{ mA}}{25\text{ mV}} = \frac{2}{25}\text{ A/V}$$

$\Rightarrow$ And here output is taken between one collector and ground, therefore gain,

$$A_d = -\frac{1}{2} g_m R_C = -\frac{1}{2} \times \frac{2}{25} \times 1000 = -40$$

2.21 (b)

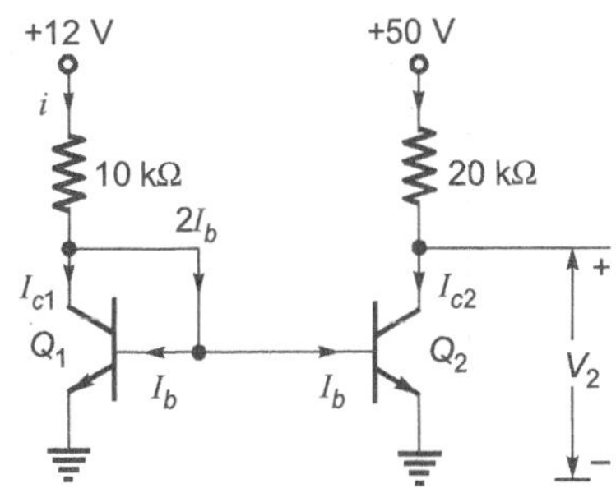

$$12 - i \times 10 \times 10^3 - 0.7 = 0$$

Now, $\quad i = \frac{12 - 0.7}{10 \times 10^3} = 1.13$ mA

$$i = I_{c1} + 2\,I_b$$

$$i = \beta\, I_b + 2\,I_b = (\beta + 2) I_b$$

$$1.13 = 102\, I_b$$

$\therefore \quad I_b = \frac{1.13}{102}\text{ mA} = 0.01107\text{ mA}$

$\therefore \quad I_c = 100\, I_b = 1.1078\text{ mA}$

For matched transistor

$$I_{c1} = I_{c2}$$

($\because$ β and I_b is same for both the transistors)

Now apply KVL in the outer port of Q_2 transistor

$$50 - I_{c2} \times 20 \times 10^3 - V_2 = 0$$

$\therefore \quad V_2 = 50 - 1.1078 \times 10^{-3} \times 20 \times 10^3 = 27.84$ V

2.22 (a)

R_E doubled, then –ve feedback increases so voltage gain decreases.

2.23 (d)

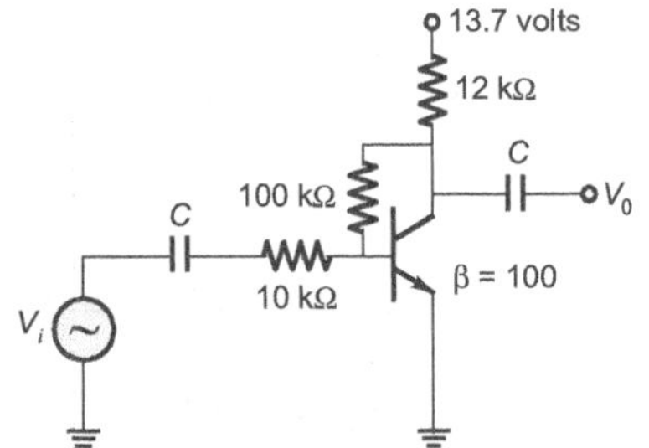

Equivalent A.C. model will be

taking, $h_{ie} = 1\text{ k}\Omega$; $h_{fe} = \beta = 100$

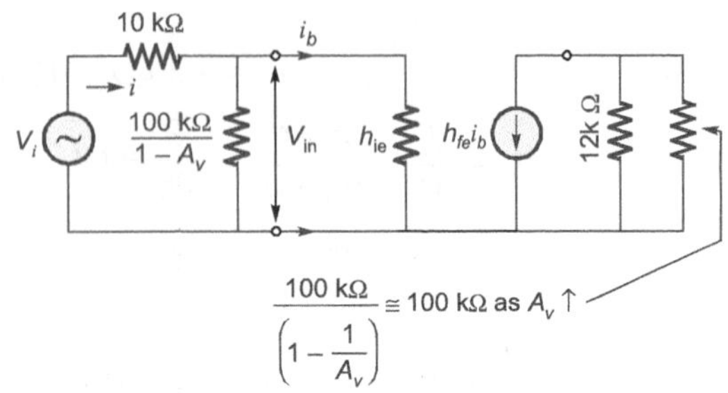

$$A_V = \frac{V_o}{V_{in}} = \frac{-h_{fe}i_b \times (100 \parallel 12)\text{k}\Omega}{h_{ie} \times i_b}$$

$$A_V = \frac{-100 \times 10.71\text{k}\Omega}{1\text{k}\Omega} = -1071.42$$

$$V_i = h_{ie}\, i_b + 10^4\, i$$

$$= 10^3\, i_b + 10^4 \left[i_b + \frac{h_{ie} \times i_b}{\left(\frac{100\text{ k}\Omega}{1-A_v}\right)} \right]$$

$$V_i = 10^3 i_b + i_b[10^4 + (1072.42) \times 100]$$

$$A_{VS} = \frac{V_0}{V_i} = \frac{-h_{fe}\, i_b \times 10.7142 \times 10^3}{i_b[10^3 + 10^4 + 10^2 \times 1072.42]}$$

$$A_{VS} = \frac{-100 \times 10.7142 \times 10^3}{(10^3 + 10^4 + 10^2 \times 1072.42)}$$

$$A_{VS} = -9.06$$

$$|A_{VS}| = 9.06 \simeq 10$$

2.24 (a)

The transistor is operating in the saturation mode here.

Also, $I_C \approx I_E$, as β not given so β assumed to be very high

$$\frac{10-5}{1.5+R_E} = 2$$

$$\therefore \quad \frac{5}{2} = 1.5 + R_E$$

$$\therefore \quad R_E = 2.5 - 1.5 = 1\text{ k}\Omega$$

2.25 41.96 (40 to 43)

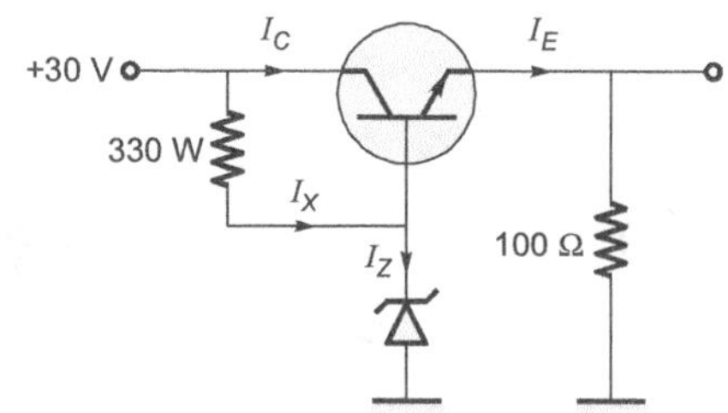

$$I_X = \frac{30-15}{330} = 45.45\text{ mA}$$

$$I_E = \frac{15-0.7}{100} = 143\text{ mA} = I_C + I_B = \beta I_B + I_B = (\beta + 1)I_B$$

$$\therefore \quad I_C = \frac{143}{40+1} = 3.48\text{ mA}$$

$$I_Z = I_X - I_B = 45.45 - 3.48 = 41.96\text{ mA}$$

2.26 5.58 (5.5 to 5.8)

For transistor T, $I_C + 2\,I_B = \frac{9 - V_{BE}}{5000} = 1.66\text{ mA}$

$$I_C = \beta I_B \qquad \therefore \quad I_c + 2\frac{I_c}{\beta} = 1.66$$

$$I_C = \frac{1.66}{1+\frac{2}{100}} \qquad (\text{for } \beta = 100)$$

$$= 1.62\text{ mA}$$

I_c in same for T_1 and T_2, since they are identical.

$$\therefore \quad V_1 - V_2 = V_{BE} + (3 \times 1.62)$$

$$= 0.7 + (3 \times 1.62) = 5.58\text{ V}$$

2.27 93 (92 to 94)

$$\because \quad V_E = 0\text{ V}$$

$$\therefore \quad I_E(1\text{ k}\Omega) - 10 = 0$$

$$I_E = \frac{10}{1000} = 10\text{ mA}$$

$$I_B = \frac{I_E}{\beta+1} = \frac{10}{1+99} = 0.1\text{mA}$$

KVL in Input Loop, $-10 + I_B R_B + 0.7 = 0$

$$\therefore \quad R_B = \frac{10-0.7}{0.1 \times 10^{-3}} = 93\text{ k}\Omega$$

2.28 (10)

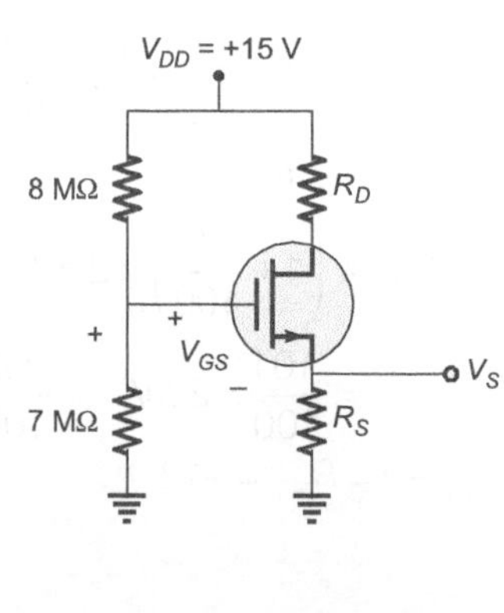

$$V_G = \frac{15 \times 7M}{8M + 7M} = 7\text{ V}$$

$$V_{GS} = V_G - V_S = 7 - 5 = 2\text{ V}$$

$$I_D = \frac{1}{2}(V_{GS} - 1)^2 \text{ mA} = 0.5 \text{ mA}$$

$$V_S = I_D R_S$$

$$\Rightarrow \quad R_S = \frac{5}{0.5 \times 10^{-3}} = 10\text{ k}\Omega$$

2.29 (c)

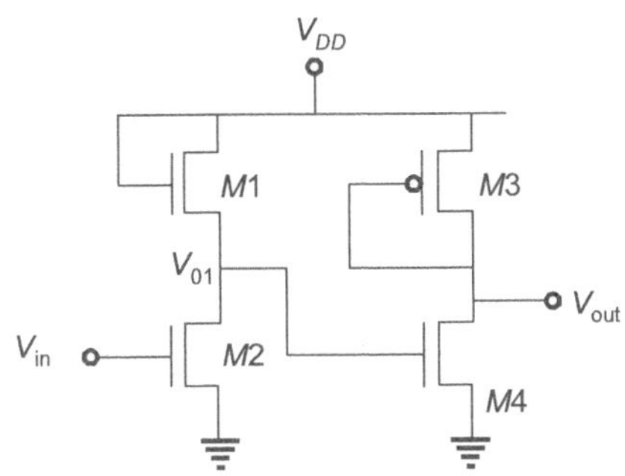

$$V_{01} = A_x V_{in} = -g_{mn} \times \frac{1}{g_{mn}} \times V_{in} = -V_{in}$$

$$V_0 = A_y V_{01} = -g_{mn} \times \frac{1}{g_{mp}} \times V_{01}$$

$$= -g_{mn} \times \frac{1}{g_{mp}} \times (-V_{in}) = +\frac{g_{mn}}{g_{mp}} \times V_{in}$$

$$\frac{V_0}{V_{in}} = \frac{g_{mn}}{g_{mp}}$$

2.30 (4)

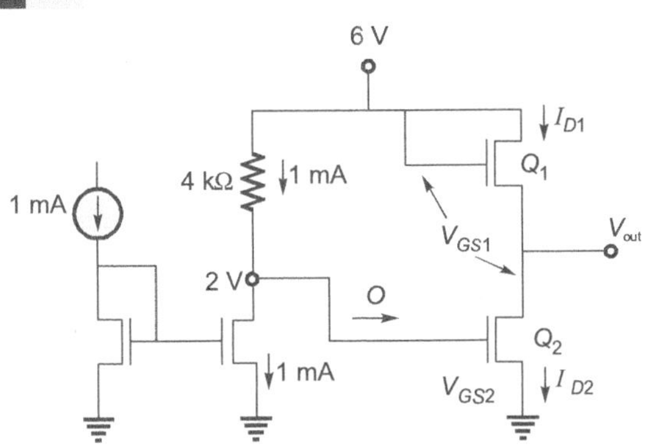

Current mirror

Q_1 and Q_2 are in current saturation mode

$$-6 + \left(1 \times 10^{-3} \times 4 \times 10^3\right) + V_{GS_2} = 0$$

$$V_{GS_2} = 2\text{V}$$

$$I_{D_1} = I_{D_2}$$

$$\frac{1}{2}k_n\left(V_{GS_1} - V_T\right)^2 = \frac{1}{2}k_n\left(V_{GS_2} - V_T\right)^2$$

$$V_{GS_1} = V_{GS_2} = 2V$$

$$-6 + V_{GS_1} + V_0 = 0$$

$$V_0 = 6 - V_{GS_1}$$

$$V_0 = 6 - 2 = 4\text{V}$$

2.31 (1)

$$\beta = 20$$

All transistors are identical and their V_{BE} values are perfectly equal. Hence, their currents will be equal.

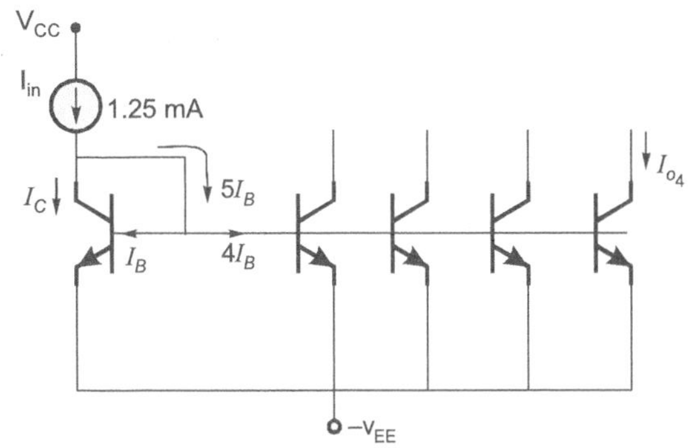

By KCL, $\quad I_{\text{In}} = I_C + 5I_B = I_C + \dfrac{5I_C}{\beta}$

$$I_C = \frac{I_{\text{in}}}{1 + \frac{5}{\beta}}$$

$$I_{o4} = \frac{I_{\text{in}}}{1 + \frac{5}{\beta}} = \frac{1.25}{1 + \frac{5}{20}} = 1\text{mA}$$

2.32 (10)

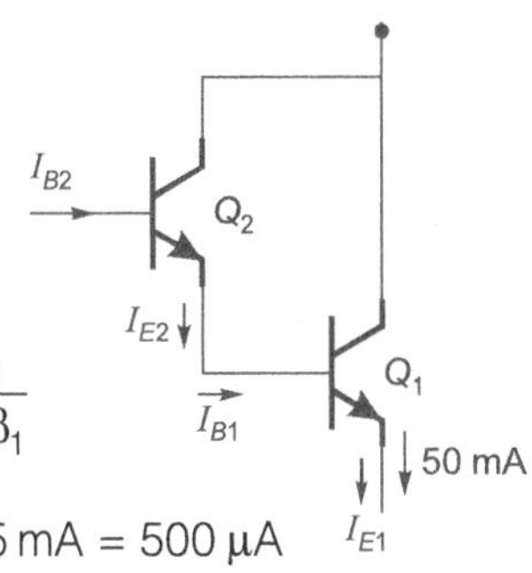

$$\beta_1 = 99$$

$$\beta_2 = 49$$

$$I_{E1} = 50 \text{ mA}$$

$$I_{E2} = I_{B1} = \frac{I_{E1}}{1 + \beta_1}$$

$$\Rightarrow \quad I_{E2} = \frac{50}{100} = 0.5\text{ mA} = 500\ \mu\text{A}$$

$$I_{B2} = \frac{I_{E2}}{1 + \beta_2} = \frac{500}{50}$$

$$\Rightarrow \quad I_{B2} = 10\ \mu\text{A}$$

2.33 (b)

Resistance seen at point P in small signal analysis is

$$R_P = (g_m^{-1} \parallel R_s)$$

2.34 (d)

For given circuit AC model is

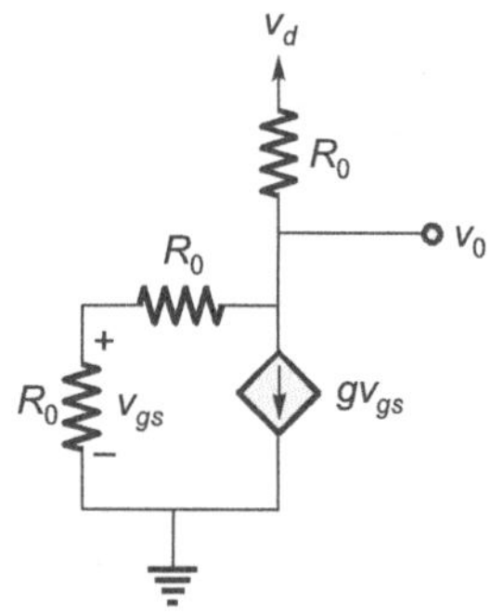

$$V_{gs} = \frac{R_0}{R_0 + R_0} \times V_0 = \frac{V_0}{2}$$

Applying KCL at node V_o

$$\frac{V_d - V_0}{R_0} = \frac{V_0}{2R_0} + gV_{gs} = \frac{V_0}{2R_0} + g\frac{V_0}{2}$$

$$\frac{V_d}{R_0} = V_0\left(\frac{3}{2R_0} + \frac{g}{2}\right)$$

$$\Rightarrow \quad V_d = V_0\left(\frac{3}{2} + \frac{1}{2}g_m R_0\right)$$

$$\Rightarrow \quad \frac{V_0}{V_d} = \left(\frac{3}{2} + \frac{1}{2}g_m R_0\right)^{-1}$$

2.35 (150.11)

Applying KVL in base emitter loop

$-5.2 + I_B R_1 + 0.7 = 0$

$$I_B = \frac{4.5}{100} = 0.045 \text{ mA}$$

$$I_C + I_B = (1 + \beta)I_B = \frac{12 - 5.2}{1}$$

$$(1 + \beta) = 151.11$$

Current gain, $\beta = 150.11$

2.36 (0.67)

Given : $V_D = 4$ V, $\frac{W}{L} = 2$, $V_T = 1$ V

$\mu_n C_{ox} = 1$ mA/V^2

Given amplifier is common drain amplifier.

Now, $$g_m = \mu_n C_{ox}\frac{W}{L}(V_{GS} - V_T)$$

$$V_{GS} = V_G - V_S$$

$$V_{GS} = \frac{5 \times 300}{500} - I_D R_S$$

$$V_{GS} = 3 - I_D R_S$$

Also, $$I_D = \frac{5 - V_D}{R_D} = \frac{5 - 4}{1} = 1 \text{ mA}$$

$\therefore$ $$V_{GS} = 3 - 1$$

$$V_{GS} = 2 \text{ Volt}$$

Hence, $$g_m = 1 \times 10^{-3} \times 2[2 - 1]$$

$$g_m = 2 \text{ mA/V}$$

$\therefore$ For common drain amplifier,

$$\frac{V_{out}}{V_{in}} = \frac{g_m R_s}{1 + g_m R_s} \quad (\text{Taking } r_{ds} = \infty)$$

$$= \frac{2 \times 1}{1 + 2 \times 1}$$

$$\frac{V_{out}}{V_{in}} = \frac{2}{3} \text{ V/V}$$

2.37 (500)

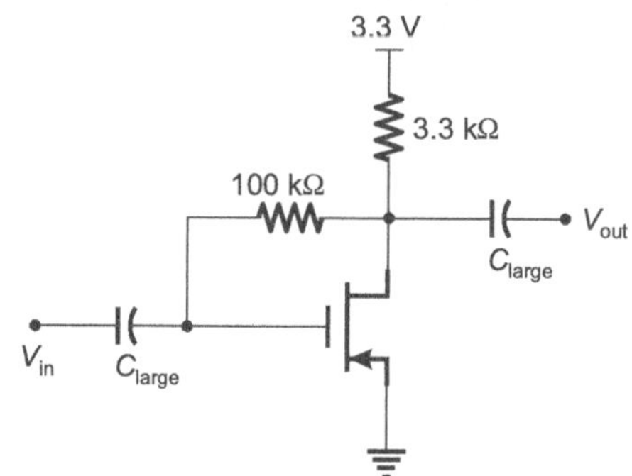

DC analysis

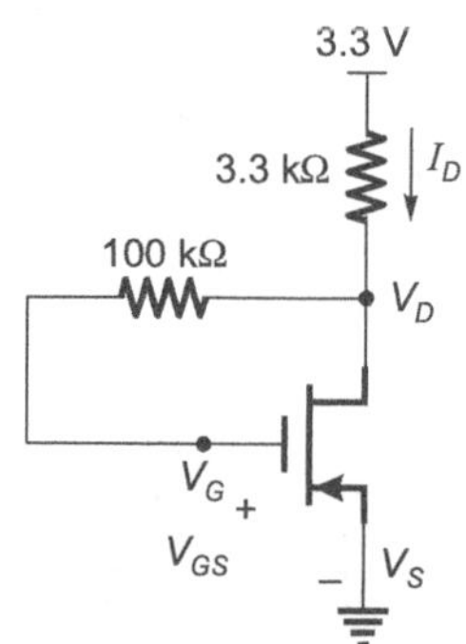

from the circuit:

$$V_D = V_G \text{ (saturation region)}$$

$$I_D = \frac{1}{2}\mu_n C_{ox}\frac{W}{L}(V_{GS} - V_T)^2$$

$$= \frac{1}{2} \times 250 \times 4(3.3 - 3.3I_d - 0.64)^2$$

$$2I_D = (2.65 - 3.3I_D)^2$$

$$10.89I_D^2 - 19.49I_D + 7.0225 = 0$$

$$I_D = 0.5 \text{ mA}$$

$$I_D = 500\ \mu\text{A}$$

■■■

3 Op-Amp and Instrumentation Amplifier

3.1 The output voltage V_0 of the circuit shown in the figure is

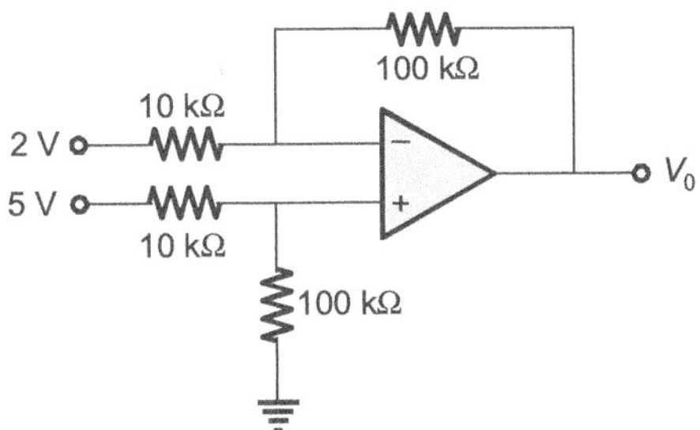

(a) – 20 V (b) 20 V
(c) – 30 V (d) 30 V [1995 : 1 M]

3.2 For the sample and hold circuit shown in figure both the op-amps have input bias current of 1 nA. The maximum leakage current through the switch is 0.5 nA. The worst case drop rate during the hold operation is ____ mV/m sec.

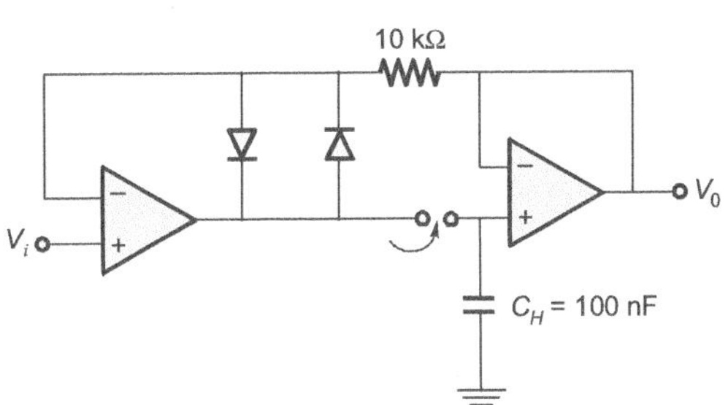

[1995 : 1 M]

3.3 The Common Mode Rejection Ratio (CMRR) of an op-amp with a common mode gain of 0.01 and a differential mode gain of 10^5 is _____ dB

[1995 : 1 M]

3.4 In the circuit shown in figure, if e_i = sin t, the voltage V_0 is

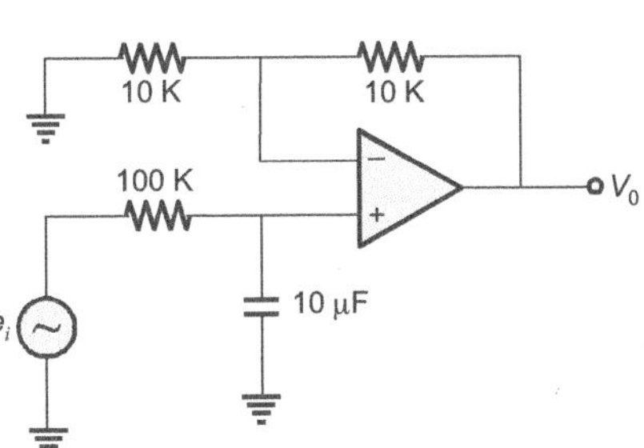

(a) $\sqrt{2}\sin\left(t-\frac{\pi}{4}\right)$ (b) $\sqrt{2}\sin\left(t+\frac{\pi}{4}\right)$
(c) $\frac{1}{\sqrt{2}}\sin\left(t-\frac{\pi}{4}\right)$ (d) $\frac{1}{\sqrt{2}}\sin\left(t+\frac{\pi}{4}\right)$

[1996 : 1 M]

3.5 The differential input resistance of the circuit shown in figure is

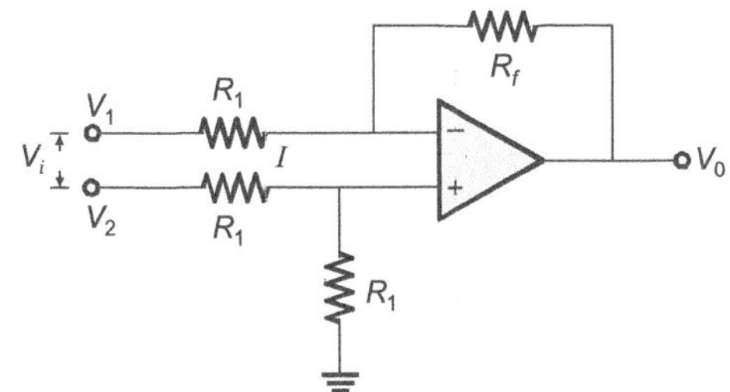

(a) R_1 (b) $R_1/2$
(c) $2R_1$ (d) R_f [1996 : 1 M]

3.6 The operational amplifier use a differential input stage with a constant current source mainly to obtain
(a) Very low common mode gain
(b) Very high differential gain
(c) Very low input noise
(d) Very high input resistance [1997 : 1 M]

3.7 In the op-amp circuit shown in figure the output voltage V_0 is equal to

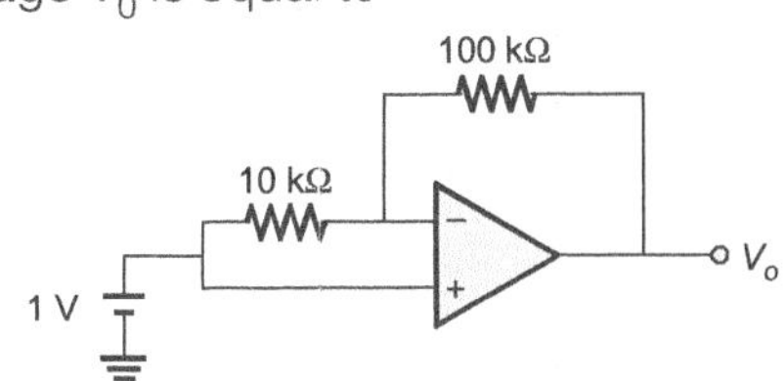

(a) 11 V (b) –10 V
(c) 9 V (d) 1 V [1997 : 1 M]

3.8 Pick up the correct relationship for the circuit shown in figure

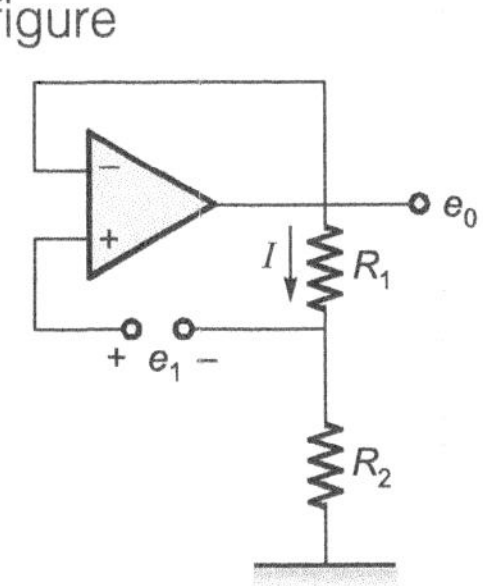

(a) $e_0 = e_1$ (b) $e_0 = \frac{R_2}{R_1} e_1$

(c) $I = \frac{e_1}{R_2}$ (d) $I = \frac{e_1}{R_1}$ [1999 : 1 M]

3.9 The op-amp in the amplifier circuit shown in figure has an offset voltage of 10 mV and it is ideal otherwise. If V_i is zero, the output voltage V_0 is

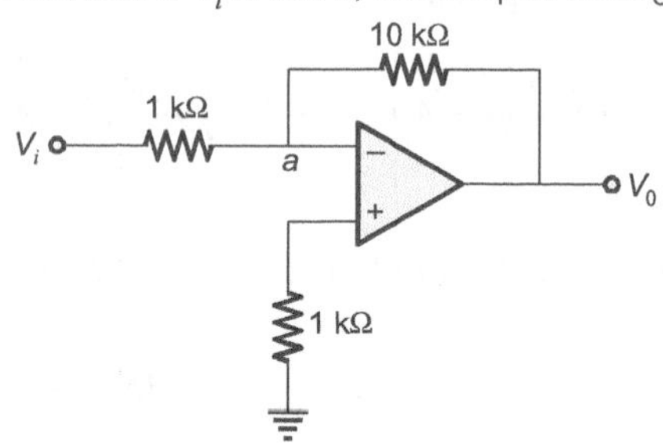

(a) 0 (b) 10 mV
(c) 100 mV (d) 110 mV [1999 : 1 M]

3.10 An op-amp with a slew rate of 1 V/μsec has been used to build an amplifier of gain +10. If the input to the amplifier is a sinusoidal voltage having peak amplitude of 1 V, the maximum allowable frequency of the input signal for undistorted output is

(a) 830 Hz (b) 15.92 kHz
(c) 31.84 kHz (d) 1 MHz

[2000 : 1 M]

3.11 A sample and hold circuit have two buffers one at input and other at the output. The primary requirements for the buffers are

(a) The input buffer should have high slew rate and output buffer should have low bias current.
(b) The input buffer should have low bias current and output buffer should have high slew rate.
(c) both the buffers should have low bias currents.
(d) both the buffers should have high slew rate.

[2001 : 1 M]

3.12 In the figure shown input offset voltage of the operational amplifier is 2 mV. The output DC-error voltage is

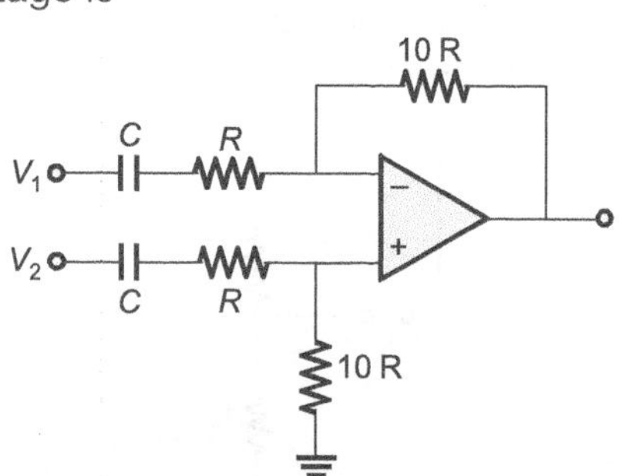

(a) 0 (b) 2 mV
(c) 11 mV (d) 22 mV

[2001 : 1 M]

3.13 An amplifier of gain 10 with a gain BW product of 1 MHz and slew rate of 0.1 V/μsec is fed with a 10 kHz symmetrical square wave of ±1 V amplitude. Its output will be

(a) ± 10 V amplitude square wave
(b) ± 2.5 V amplitude square wave
(c) ± 10 V amplitude triangular wave
(d) ± 2.5 V amplitude triangular wave

[2002 : 1 M]

3.14 The output of an opamp whose input is a 2.5 MHz square wave is shown in figure. The slew rate of the opamp is

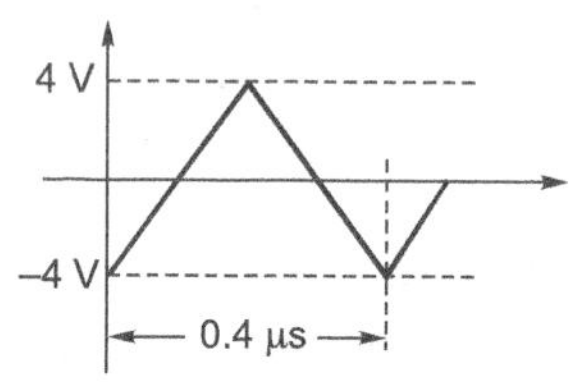

(a) 0.8 V/μs (b) 8.0 V/μs
(c) 20.0 V/μs (d) 40.0 V/μs

[2003 : 1 M]

3.15 The opamp and the 1 mA current source in the circuit of figure are ideal. The output of the op-amp is

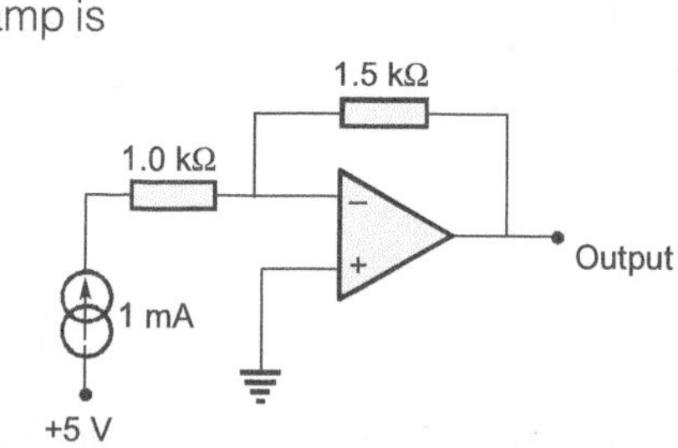

(a) –1.5 mA (b) –1.5 V
(c) –7.5 V (d) +1.5 V [2003 : 1 M]

3.16 The opamp used in the inverting amplifier shown in figure has an equivalent input offset voltage V_{ios} of 5 mV. The output offset voltage is

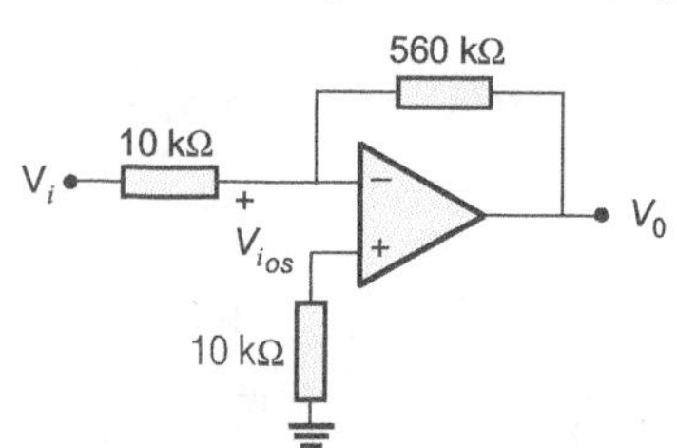

(a) 5 mV (b) 280 mV
(c) 285 mV (d) 560 mV [2003 : 2 M]

3.17 In the circuit shown in figure, the opamps used are ideal. The output V_0 is

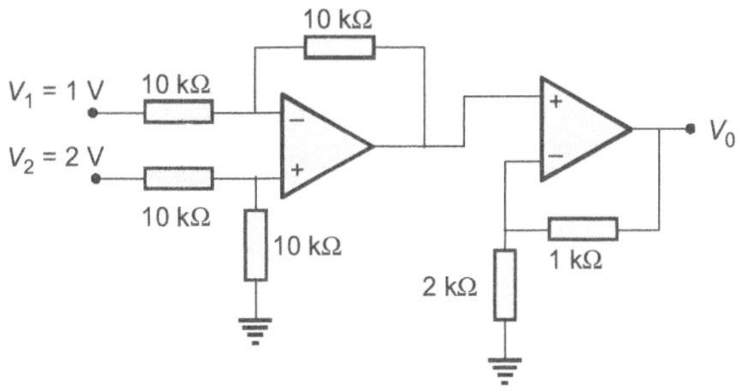

(a) 3.0 V (b) 1.5 V
(c) 1.0 V (d) 0.5 V [2003 : 2 M]

3.18 The output of the opamp in the circuit of figure is

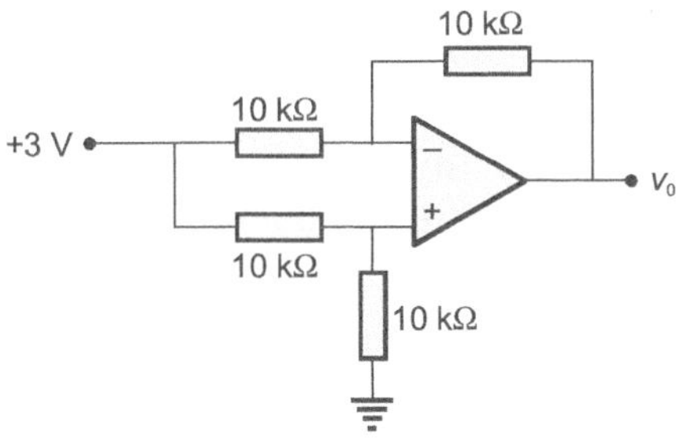

(a) 0 V (b) –3 V
(c) +1.5 V (d) +3 V [2004 : 1 M]

3.19 The value of V_0 in the circuit, shown in figure is

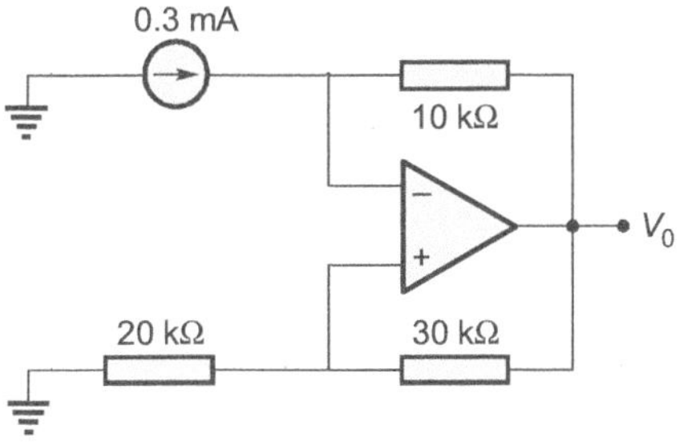

(a) –5 V (b) –3 V
(c) +3 V (d) +5 V [2004 : 2 M]

3.20 The gain $\left(\frac{V_O}{V_i}\right)$ of the amplifier circuit shown in figure is

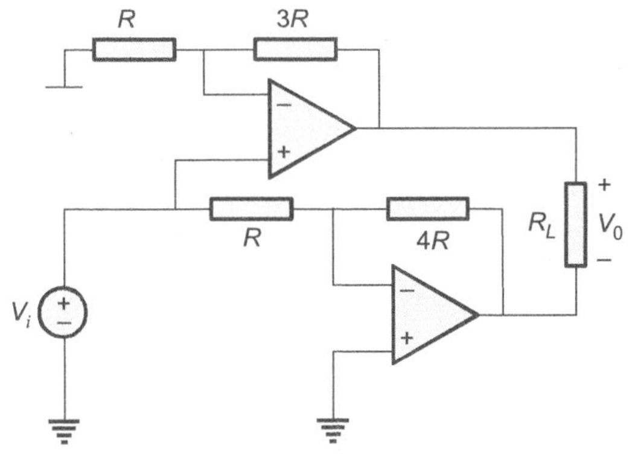

(a) 8 (b) 4
(c) –4 (d) $\frac{3R_L}{R}$

[2004 : 2 M]

3.21 The output voltage V_o in the circuit in figure is

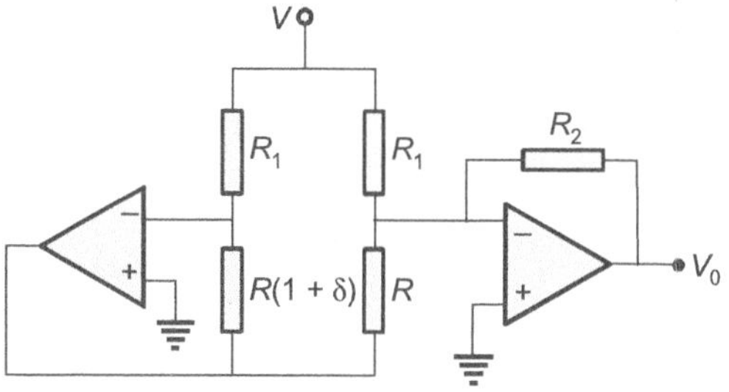

(a) $\frac{R_2}{R}V\delta$ (b) $\frac{R_2}{R_1}V\delta$
(c) $\frac{R_2}{R_1\delta}V$ (d) $\frac{R_2}{R(1+\delta)}V$

[2004 : 2 M]

3.22 V_1 and V_2 are the input voltages of an instrumentation amplifier. The output of the instrumentation amplifier is found to be 100 $(V_1 - V_2)$ + 10^{-4} $(V_1 + V_2)$. The gain and the common mode rejection ratio (CMRR) of the instrumentation amplifier respectively are

(a) (50, 60 dB) (b) (50, 120 dB)
(c) (100, 60 dB) (d) (100, 120 dB)

[2004 : 2 M]

3.23 In the circuit shown in the given figure the input voltage $V_{in}(t)$ is given by 2 sin(100 πt). For R_L in the range 0.5 kΩ to 1.5 kΩ, the current through R_L is

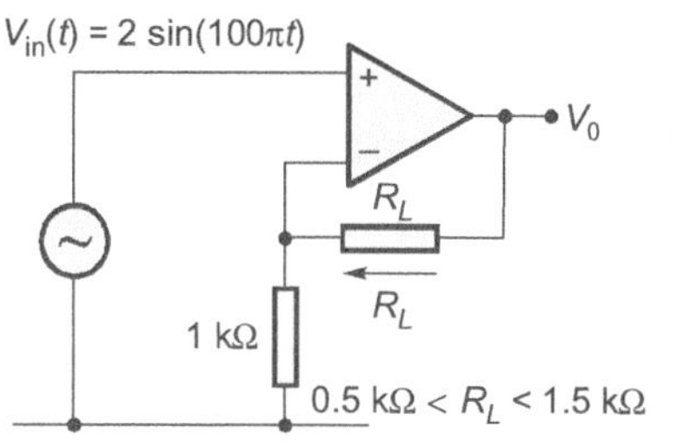

(a) +2 sin (100 πt) mA
(b) –2 sin (100 πt) mA
(c) +0.5 sin (100 πt) mA
(d) 1.5 sin (100 πt) mA

[2005 : 1 M]

3.24 In the circuit shown in the figure the input voltage V_i is a symmetrical saw-tooth wave of average value zero, positive slope and peak-to-peak value

20 V. The average value of the output, assuming an ideal operational amplifier with peak-to-peak symmetrical swing of 30 V, is

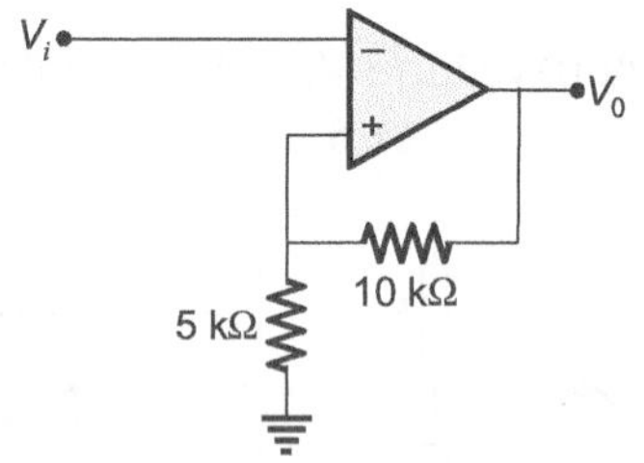

(a) 5 V (b) 10 V
(c) –5 V (d) 7.5 V

[2005 : 2 M]

3.25 In the instrumentation amplifier shown in the figure if the switch SW is changed from position A to B, the values of the amplifier gain G before and after changing the switch respectively are

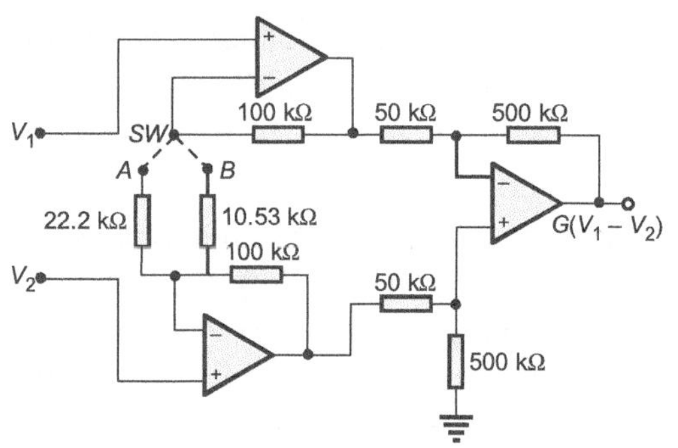

(a) 45, 95 (b) 50, 100
(c) 100, 200 (d) 90, 180

[2005 : 2 M]

3.26 In the circuit shown in the figure, assuming ideal diode characteristics with zero forward resistance and 0.7 V forward drop, the average value of V_0 when the input waveform is as shown, is

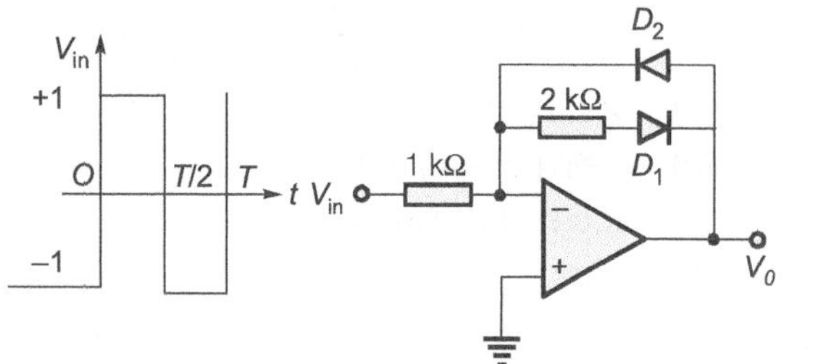

(a) –0.7 V (b) –1.0 V
(c) –2.0 V (d) –2.7 V

[2005 : 2 M]

3.27 If the value of the resistance R in the following figure is increased by 50%, then voltage gain of the amplifier shown in the figure will change by

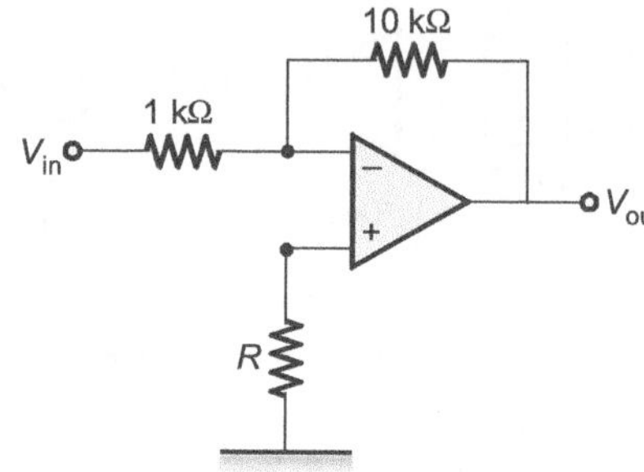

(a) 50% (b) 5%
(c) –50% (d) negligible amount

[2006 : 1 M]

3.28 When the switch S_2 is closed the gain of the programmable gain amplifier shown in the following figure is

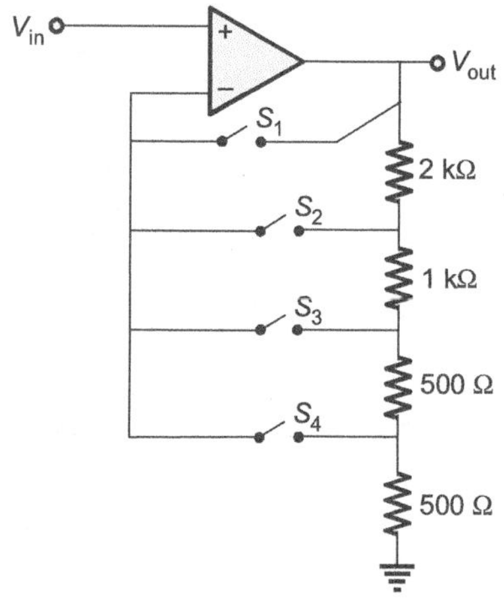

(a) 0.5 (b) 2
(c) 4 (d) 8 [2006 : 1 M]

3.29 In the circuit shown in the following figure, the opamp has input bias current I_b < 10 nA, and input offset voltage V_{i0} < 1 mV. The maximum dc error in the output voltage is

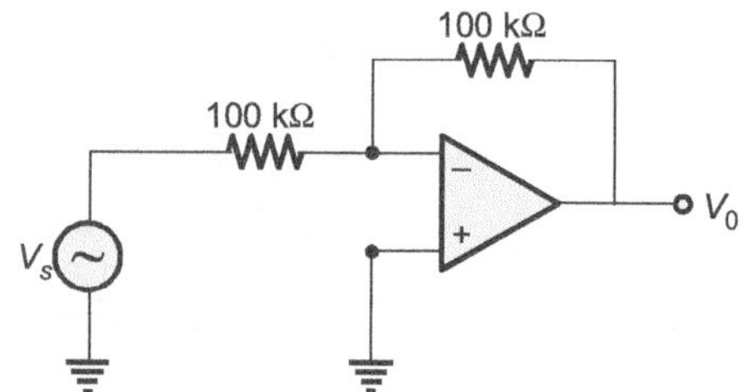

(a) 1.0 mV (b) 2.0 mV
(c) 2.5 mV (d) 3.0 mV

[2006 : 2 M]

3.30 The potential difference between the input terminals of an opamp may be treated to be nearly zero, if
(a) the two supply voltages are balanced
(b) the output voltage is not saturated
(c) the opamp is used in a circuit having negative feedback
(d) there is a dc bias path between each of the terminals and the circuit ground
[2006 : 2 M]

3.31 A dual opamp instrumentation amplifier is shown below. The expression for the output of the amplifier is given by

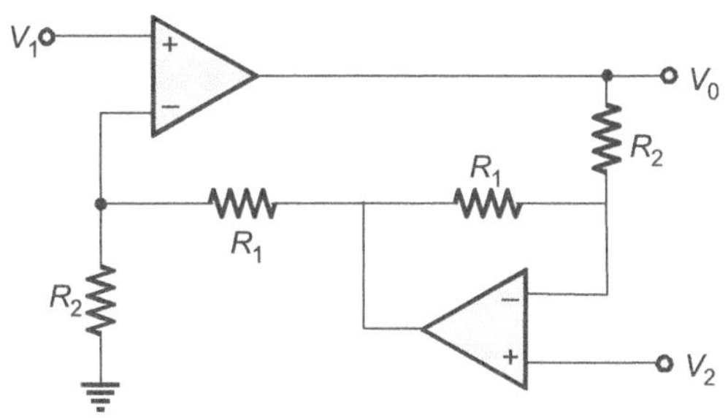

(a) $v_0 = \left(1 + \frac{R_2}{R_1}\right)(v_2 - v_1)$

(b) $v_0 = \left(1 + \frac{2R_2}{R_1}\right)(v_2 - v_1)$

(c) $v_0 = \frac{2R_2}{R_1}(v_2 - v_1)$

(d) $v_0 = \left(1 + \frac{2R_1}{R_2}\right)(v_2 - v_1)$ [2006 : 2 M]

3.32 Consider the linear circuit with and ideal op-amp shown in the figure shown below:

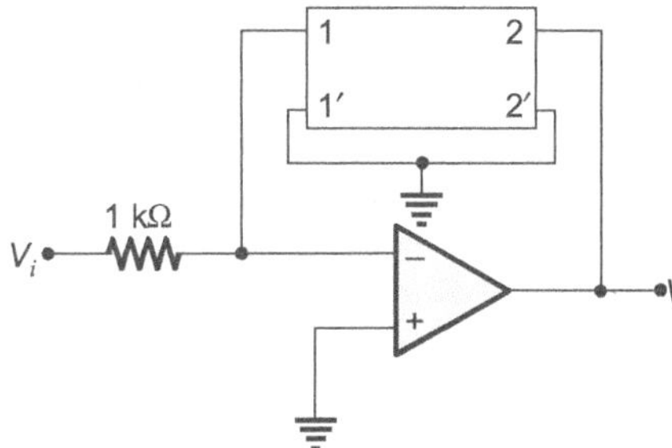

The Z-parameters of the two port feedback network are:
$Z_{11} = Z_{22} = 11$ kΩ, and
$Z_{12} = Z_{21} = 1$ kΩ.
The gain of the amplifier is
(a) +110 (b) +11
(c) –1 (d) –120
[2007 : 2 M]

3.33 The figure shows a single op-amp differential amplifier circuit.

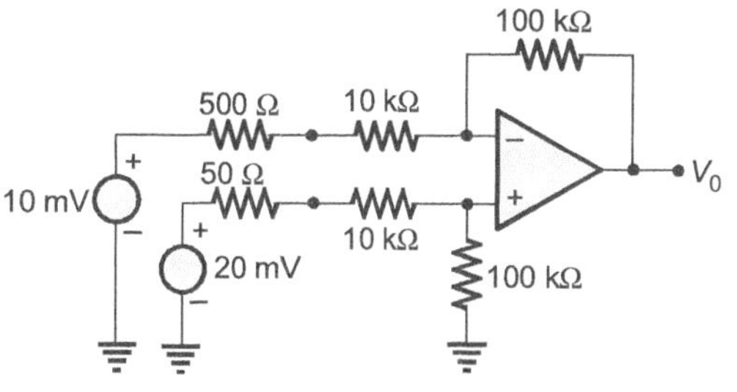

Which one of the following statements about the output is correct?
(a) $V_0 \leq 95$ mV
(b) 95 mV < $V_0 \leq 98$ mV
(c) 98 mV < $V_0 \leq 101$ mV
(d) $V_0 > 101$ mV
[2007 : 2 M]

Common Data Questions (3.34 to 3.35):
Consider the op-amp circuit shown in the figure below:

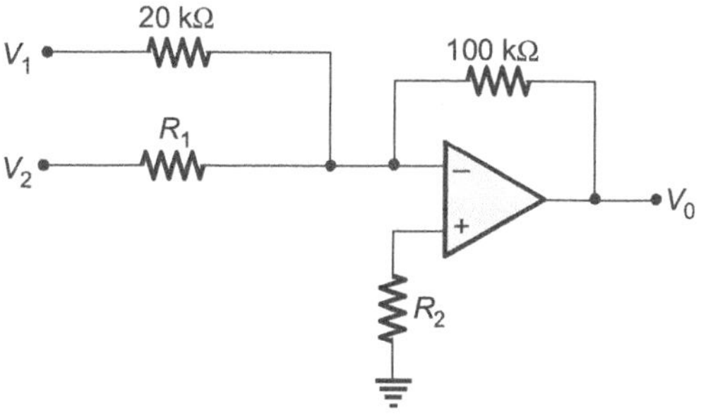

3.34 If $V_1 = 0.2$ V, $V_2 = 0.6$ V and $V_0 = -7$ V and the op-amp is ideal, the value of R_1 is
(a) 5 kΩ (b) 10 kΩ
(c) 15 kΩ (d) 20 kΩ [2007 : 2 M]

3.35 Let $V_1 = V_2 = V_c \sin 2\pi f t$ and $R_1 = 20$ kΩ. The op-amp has a slew rate of 0.5 V/μs with its other parameters being ideal. The values of V_C and f for which the amplifier output will have no distortion are, respectively
(a) 0.1 V and 300 kHz
(b) 0.5 V and 300 kHz
(c) 0.1 V and 30 kHz
(d) 0.5 V and 30 kHz [2007 : 2 M]

3.36 Let $V_1 = V_2 = 0$ and $R_1 = 20$ kΩ. Assume that the op-amp is ideal except for a non-zero input bias current. What is the value of R_2 for the output voltage of the op-amp to be zero?

(a) 2.2 kΩ (b) 9.1 kΩ
(c) 20 kΩ (d) 100 kΩ

[2007 : 2 M]

3.37 A differential amplifier shown below has a differential mode gain of 100 and a CMRR of 40 dB. If V_1 = 0.55 V and V_2 = 0.45 V, output is

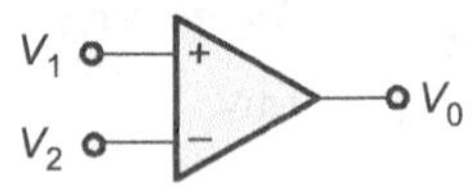

(a) 10 V (b) 10.5 V
(c) 11 V (d) 15 V [2007 : 2 M]

3.38 An ideal op-amp has the characteristics of an ideal
(a) voltage controlled voltage source
(b) voltage controlled current source
(c) current controlled voltage source
(d) current controlled current source

[2008 : 1 M]

3.39 For the op-amp circuit shown below, v_0 is approximately equal to

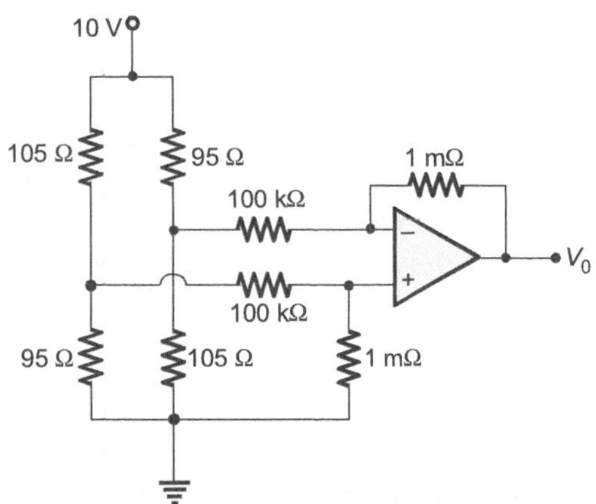

(a) –10 V (b) –5 V
(c) +5 V (d) +10 V [2008 : 2 M]

3.40 In the circuit shown, the Zener diode has ideal characteristics and a breakdown voltage of 3.2 V. The output voltage V_0 for an input voltage V_0 = +1 V is closest to

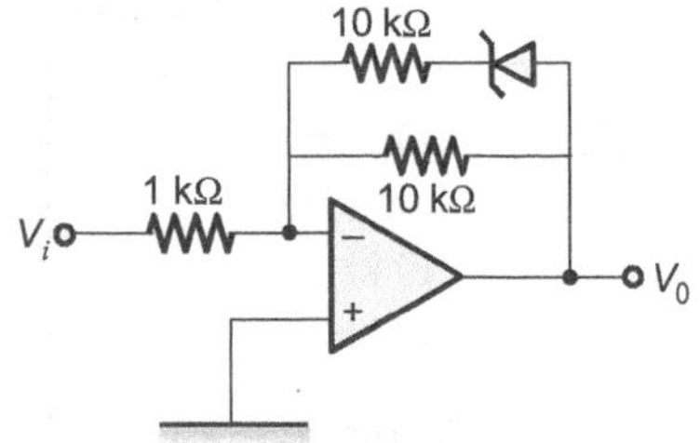

(a) –10 V (b) –6.6 V
(c) –5 V (d) –3.2 V

[2009 : 2 M]

3.41 The input resistance of the circuit shown in the figure, assuming an ideal op-amp is

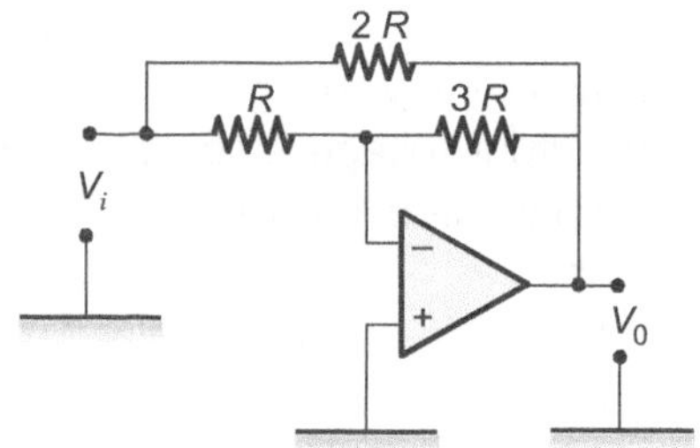

(a) $\frac{R}{3}$ (b) $\frac{2R}{3}$
(c) R (d) $\frac{4R}{3}$ [2009 : 2 M]

3.42 In the ideal op-amp circuit given in the adjoining figure, the value of R_f is varied from 1 kΩ to 100 kΩ. The gain $G = (v_0/v_i)$

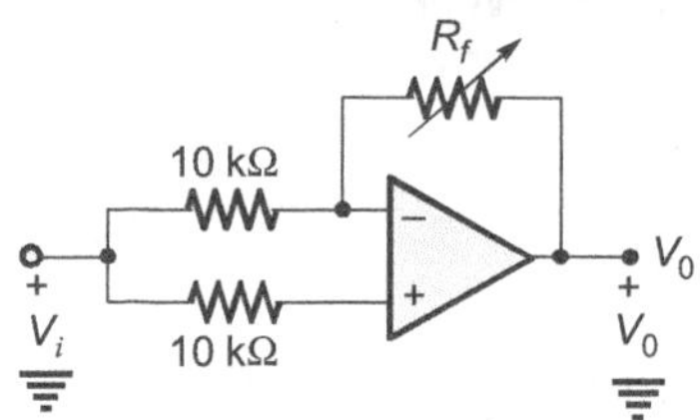

(a) remain constant at +1
(b) remain constant at –1
(c) vary as $-R_f/$ 10,000)
(d) vary as (1 + R_f/10,000) [2010 : 1 M]

Common Data Questions (3.43 and 3.44):

A differential amplifier is constructed using an ideal op-amp as shown in the adjoining figure. The values of R_1 and R_2 are 47 kΩ and 470 kΩ respectively.

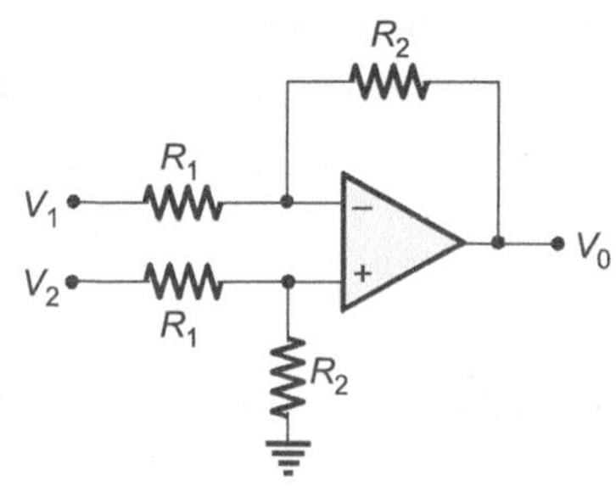

3.43 The input impedances seen looking into the terminals V_1 and V_2, with respect to ground, respectively are
(a) 47 kΩ and 43 kΩ
(b) 47 kΩ and 47 kΩ
(c) 47 kΩ and 517 kΩ
(d) 517 kΩ and 517 kΩ

[2010 : 2 M]

3.44 V_1 and V_2 are connected to voltage sources having an open circuit output of +1 V each and internal resistances of 13 kΩ and 3 kΩ respectively. The output voltage V_0 is

(a) 0 V (b) 0.15 V
(c) 1.5 V (d) 10 V

[2010 : 2 M]

3.45 Assuming base-emitter voltage of 0.7 V and β = 99 of transistor Q_1, the output voltage V_0 in the ideal opamp circuit shown below is

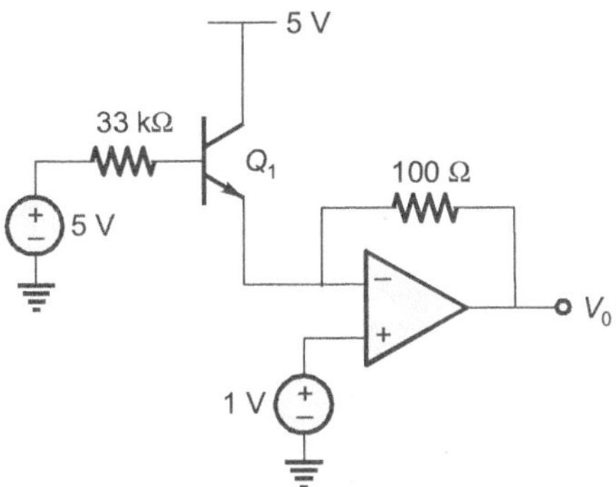

(a) –1 V (b) –1/3.3 V
(c) 0 V (d) 2 V [2011 : 2 M]

3.46 The value of V_0 of the series regulator shown below is

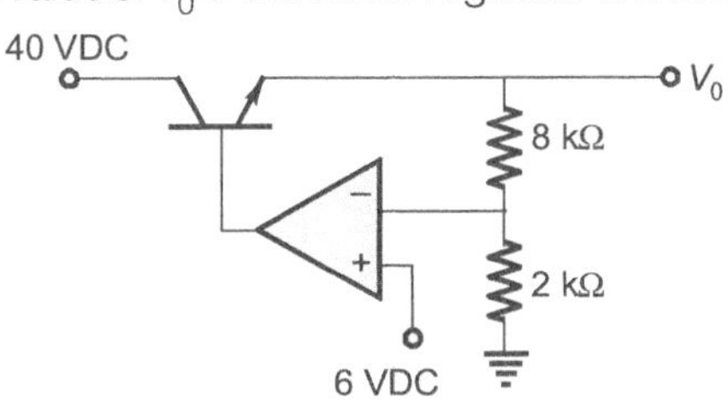

(a) 24 V (b) 28 V
(c) 30 V (d) 32 V [2011 : 2 M]

3.47 The operational amplifier shown in the circuit below has a slew rate of 0.8 Volts/μs. The input signal is 0.25 sin(ω*t*). The maximum frequency of input in kHz for which there is no distortion in the output is

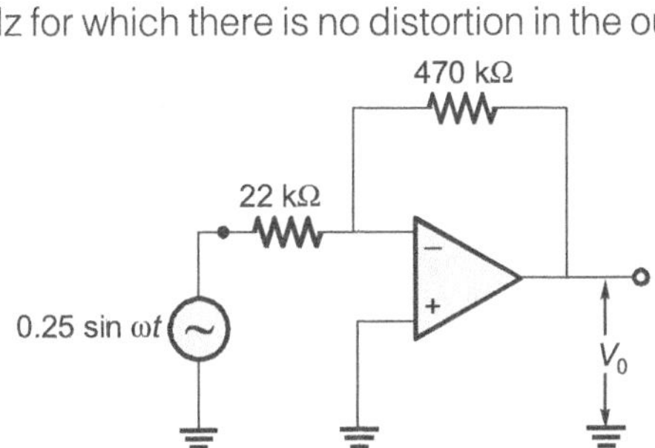

(a) 23.84 (b) 25.0
(c) 50.0 (d) 46.60 [2013 : 1 M]

3.48 In the circuit shown below what is the output voltage (V_{out}) in volts if a silicon transistor *Q* and an ideal op-amp are used?

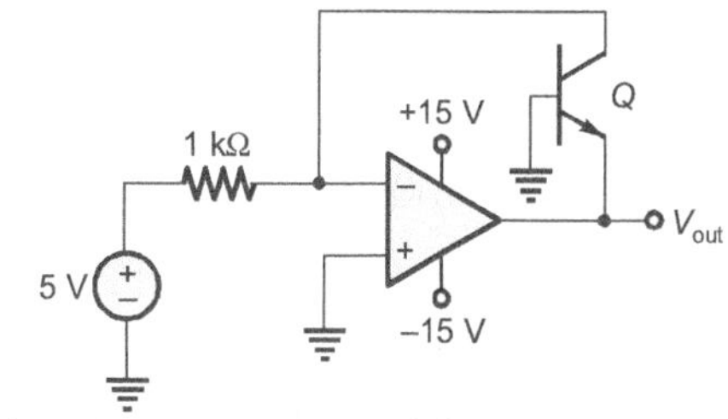

(a) –15 (b) –0.7
(c) +0.7 (d) +15 [2013 : 1 M]

3.49 In the circuit shown below the op-amps are ideal. The V_{out} in volts is

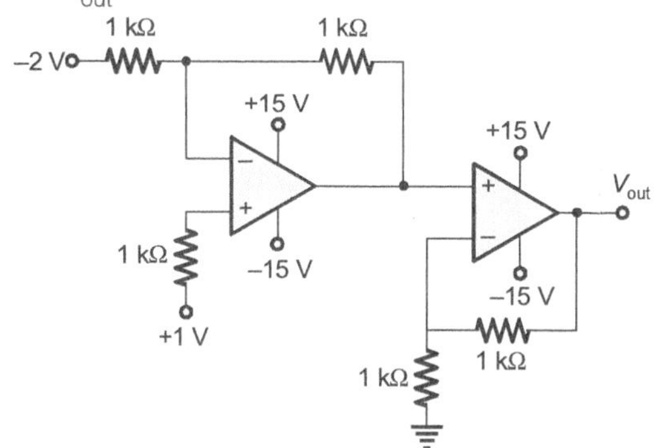

(a) 4 (b) 6
(c) 8 (d) 10 [2013 : 2 M]

3.50 For the op-amp shown in the figure, the bias currents are I_{b1} = 450 nA and I_{b2} = 350 nA. The values of the input bias current (I_B) and the input offset current (I_f) are

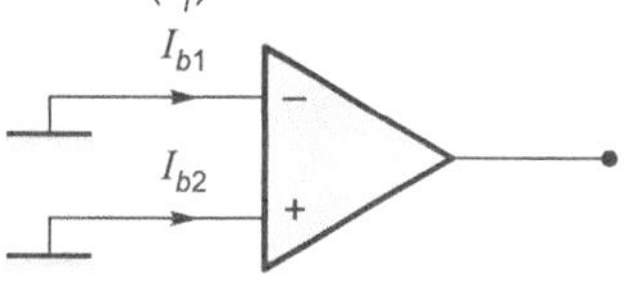

(a) I_B = 800 nA, I_f = 50 nA
(b) I_B = 800 nA, I_f = 100 nA
(c) I_B = 400 nA, I_f = 50 nA
(d) I_B = 400 nA, I_f = 100 nA

[2014 : 1 M]

3.51 The amplifier in the figure has gain of –10 and input resistance of 50 kΩ. The values of R_i and R_f are

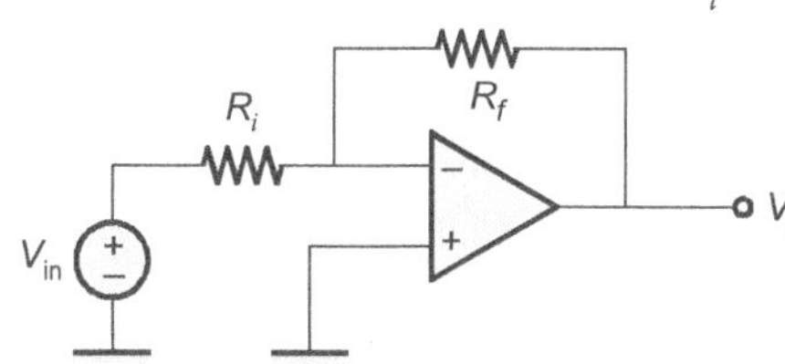

(a) R_i = 500 kΩ, R_f = 50 kΩ
(b) R_i = 50 kΩ, R_f = 500 kΩ
(c) R_i = 5 kΩ, R_f = 10 kΩ
(d) R_i = 50 kΩ, R_f = 200 kΩ [2014 : 1 M]

3.52 Assuming an ideal op-amp in linear range of operation, the magnitude of the transfer impedance $\frac{V_o}{i}$ in MΩ of the current to voltage converter shown in the figure is ____.

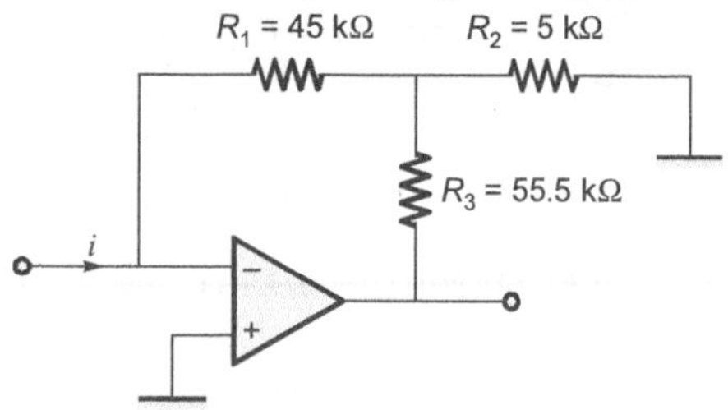

[2014 : 2 M]

3.53 In the circuit given below, the opamp is ideal. The output voltage V_0 in volt is______.

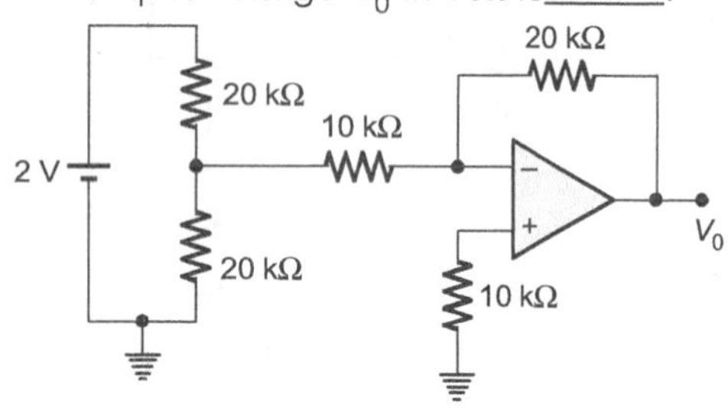

[2016 : 1 M]

3.54 In the circuit given below, the diodes D_1 and D_2 have a forward voltage drop of 0.6 V. The opamp used is ideal. The magnitude of the negative peak value of the output V_0 in volt is______.

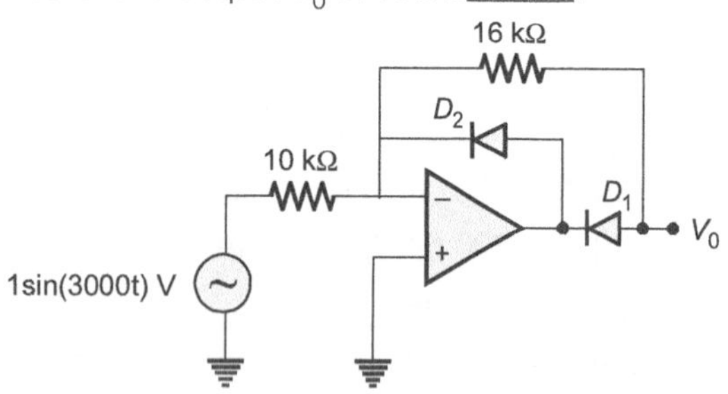

[2016 : 1 M]

3.55 In the circuit given below, each input terminal of the opamp draws a bias current of 10 nA. The effect due to these input bias currents on the output voltage V_0 will be zero, if the value of R chosen in kilo-ohm is_______.

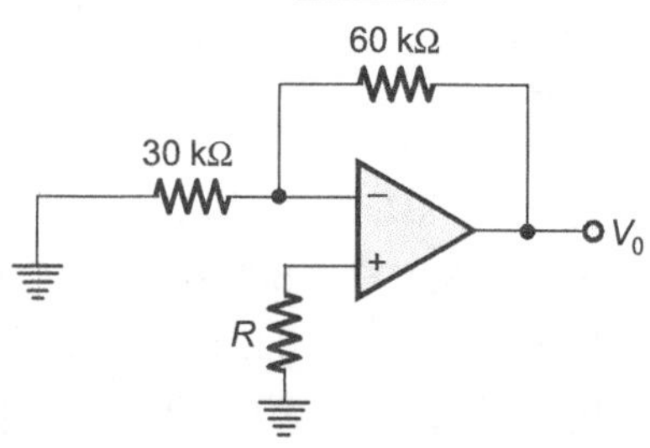

[2016 : 1 M]

3.56 An ideal opamp is used to realize a difference amplifier circuit given below having a gain of 10. If x = 0.025, the CMRR of the circuit in dB is______.

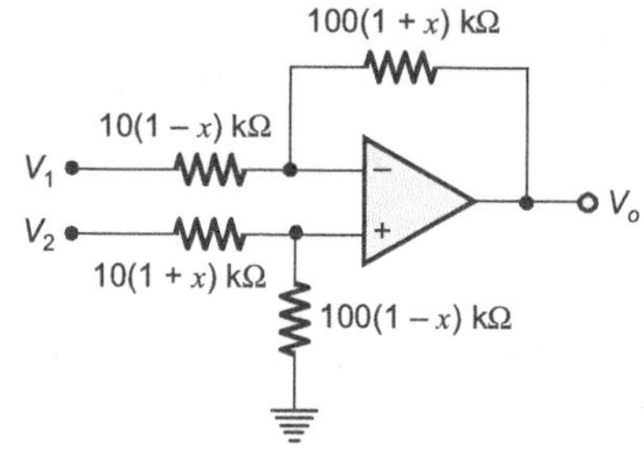

[2016 : 2 M]

3.57 In the circuit given below, the opamp is ideal. The input v_x is a sinusoid. To ensure $v_y = v_x$, the value of C_N in Pico farad is_____.

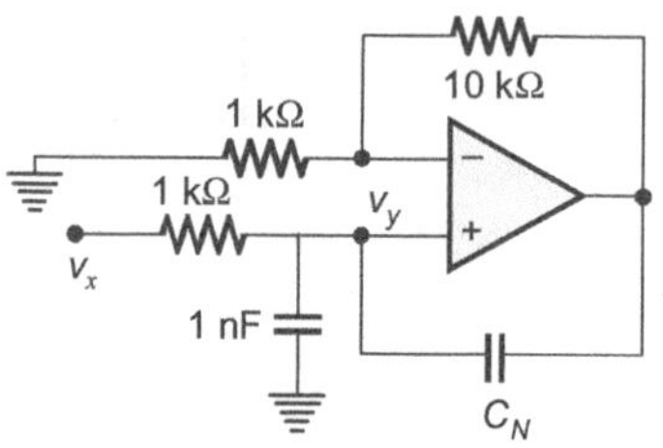

[2016 : 2 M]

3.58 In the circuit given below, the opamp is ideal. The value of current I_L in microampere is____.

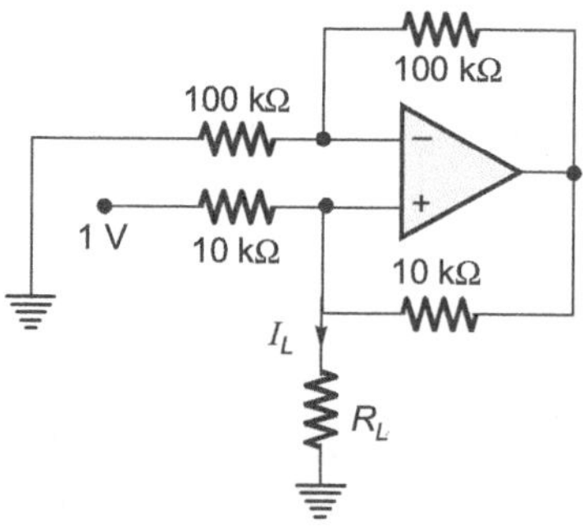

[2016 : 2 M]

3.59 The output V_0 shown in the figure, in volt, is close to

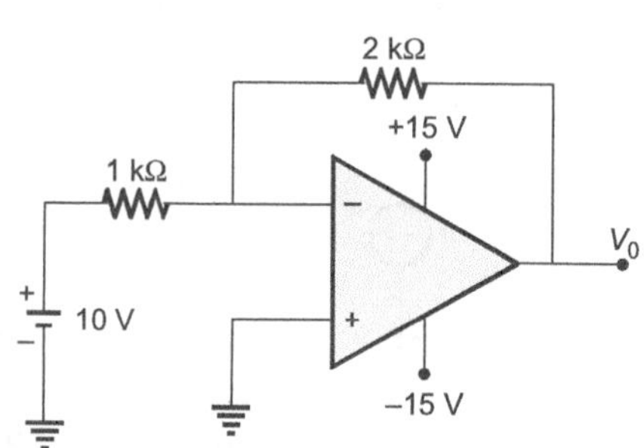

(a) –20 (b) –15
(c) –5 (d) 0

[2017 : 1 M]

3.60 The differential amplifier, shown in the figure, has a differential gain of $A_d = 100$ and common mode gain of $A_c = 0.1$. If $V_1 = 5.01$ V and $V_2 = 5.00$ V, then V_0 in volt (up to one decimal place) is _____.

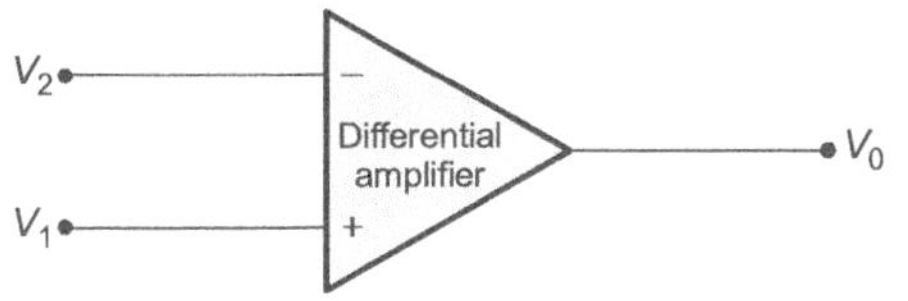

[2017 : 1 M]

3.61 The two-input voltage multiplier, shown in the figure has a scaling factor of 1 and produces voltage output. If $V_1 = +15$ V and $V_2 = +3$ V, the value of the V_0 in volt is _______.

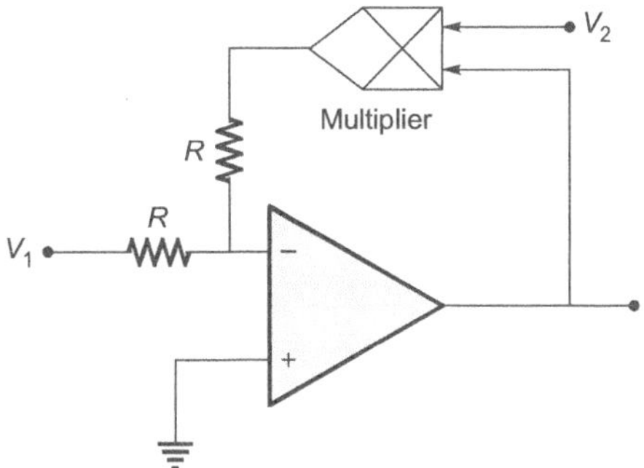

[2017 : 2 M]

3.62 In the given circuit, assume that the opamp is ideal and the transistor has a β of 20. The current I_0 (in μA) flowing through, the load Z_L is .

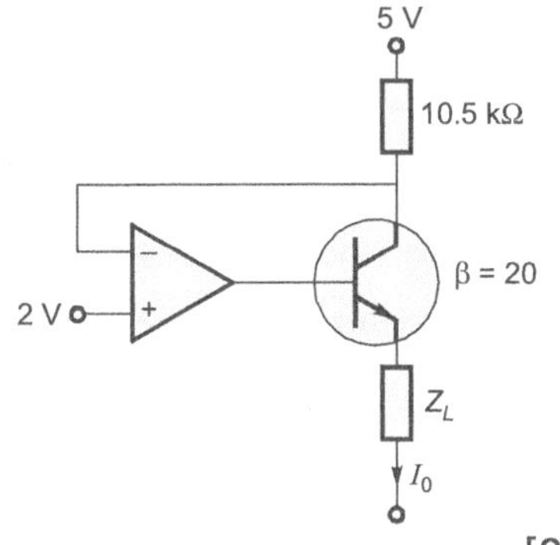

[2018 : 1 M]

3.63 The circuit given use ideal opamps. The current I (in μA) drawn from the source v_s is (up to two decimal places) _________.

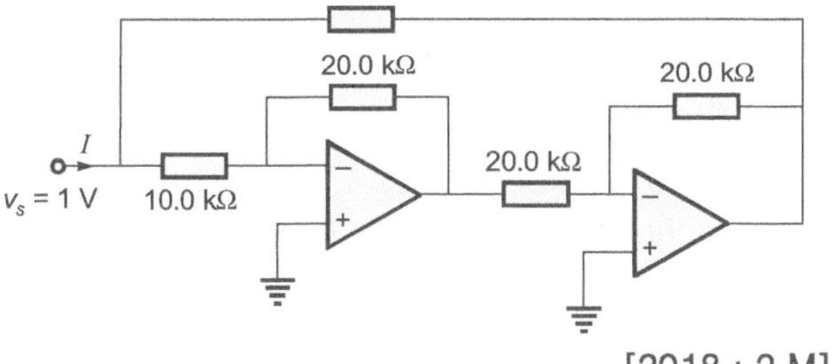

[2018 : 2 M]

3.64 An opamp that is powered from a ± 5 V supply is used to build a non-inverting amplifier having a gain of 15. The slew rate of the opamp is 0.5×10^6 V/s. For a sinusoidal input with amplitude of 0.3 V, the maximum frequency (in kHz) up to which it can be operated without any distortion is (up to one decimal place) _______.

[2018 : 2 M]

3.65 In the circuit shown below, the input voltage V_{in} is positive. The current (I)-voltage (V) characteristics of the diode can be assumed to be $I = I_0 e^{V/V_T}$ under the forward bias condition, where V_T is the thermal voltage and I_0 is the reverse saturation current. Assuming an ideal op-amp, the output voltage V_{out} of the circuit is proportional to

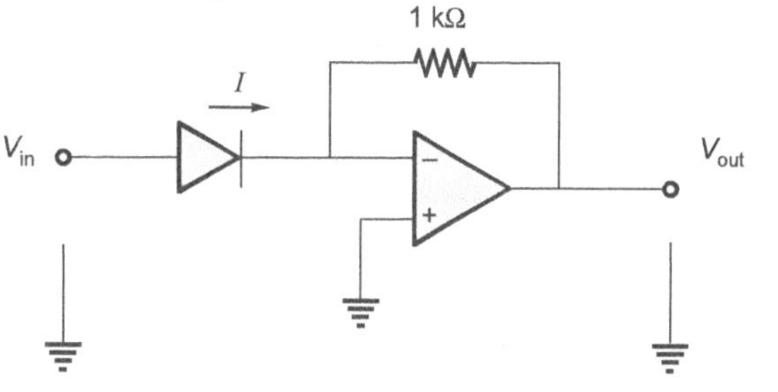

(a) $\log_e\left(\dfrac{V_{in}}{V_T}\right)$ (b) $2\ V_{in}$

(c) e^{V_{in}/V_T} (d) V_{in}^2

[2019 : 1 M]

3.66 In the circuit shown below all OPAMPS are ideal. The current $I = 0$ A when the resistance R = ______kΩ.

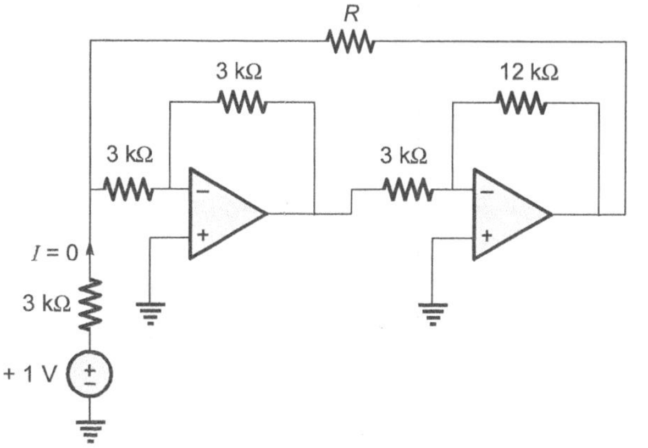

[2019 : 2 M]

3.67 Assume that the opamp in the circuit shown is ideal. The value of $\dfrac{V_x}{I_x}$ (in kΩ) is _____.

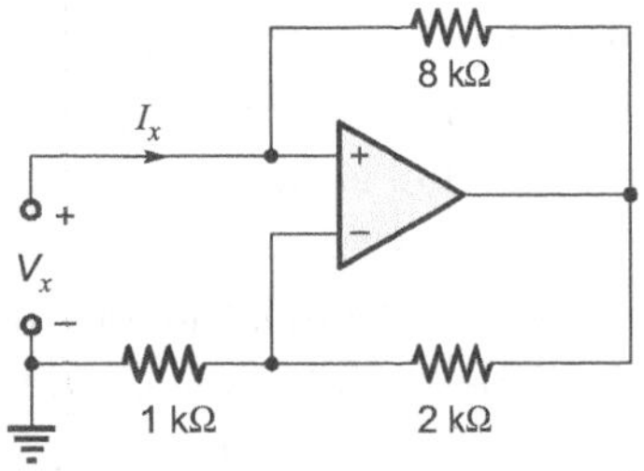

[2020 : 1 M]

3.68 Assuming ideal opamps, the output voltage at V_1 in the figure shown (in volts) is _____.

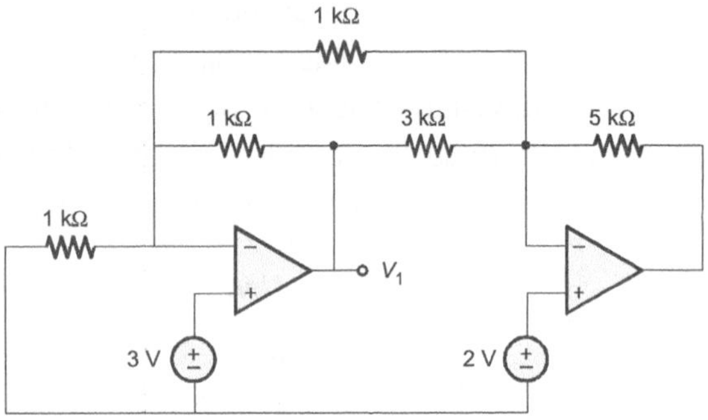

[2020 : 1 M]

3.69 If the opamps in the circuit shown are ideal and V_x = 0.5 mV, the steady state value of V_0 (in volts, rounded off to two decimal places) is _____.

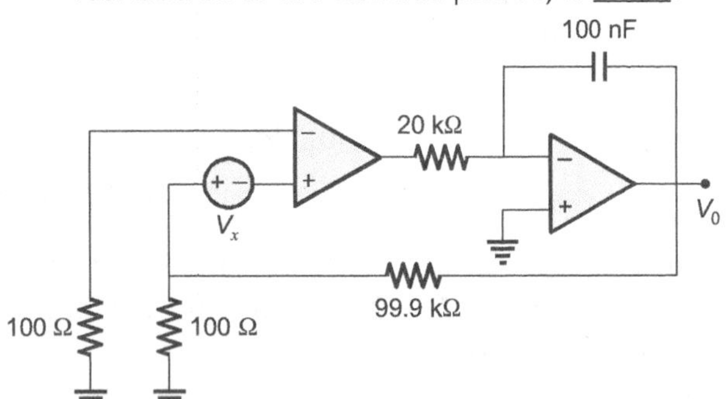

[2020 : 2 M]

3.70 A circuit consisting of capacitors, DC voltage source and an amplifier having a voltage gain $G = -5$ is shown in the figure. The effective capacitance across the nodes A and B (in μF, rounded off to one decimal place) is _____.

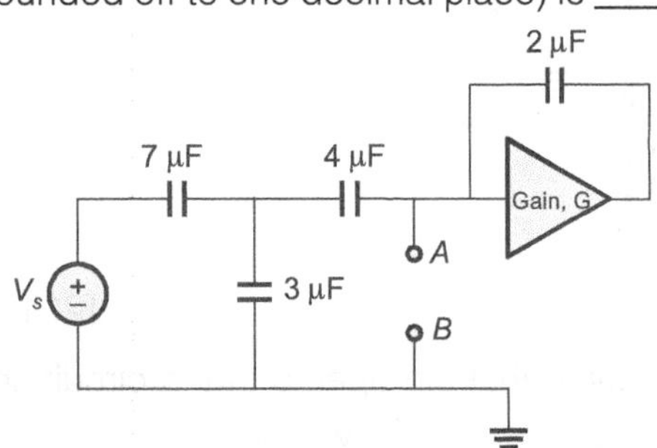

[2020 : 2 M]

3.71 The output V_o of the ideal OpAmp used in the circuit shown below is 5 V. Then the value of resistor R_L in kilo ohm (kΩ) is

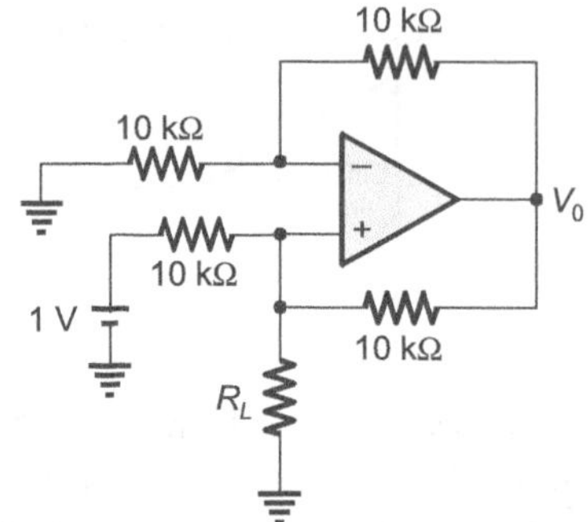

(a) 2.5
(b) 25
(c) 5
(d) 50

[2021 : 2 M]

3.72 The diode used in the circuit has a fixed voltage drop of 0.6 V when forward biased, the signal v_S is given to the ideal OpAmp as shown. When v_S is at its positive peak, the output (v_{OA}) of the OpAmp in volts is _____.

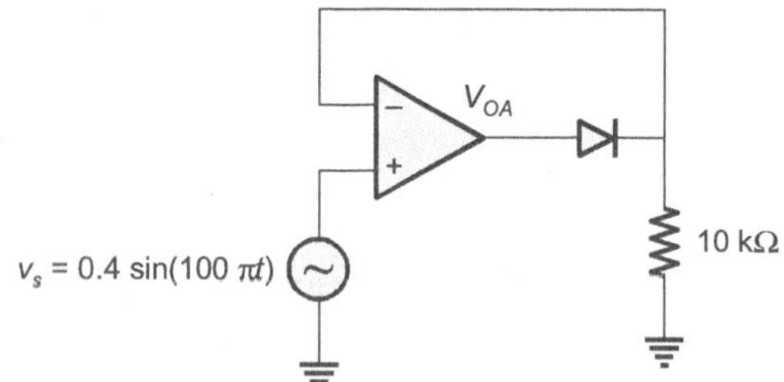

[2021 : 1 M]

3.73 The circuit shown below uses an ideal OpAmp. Output V_o in volt is _____ (rounded off to one decimal place).

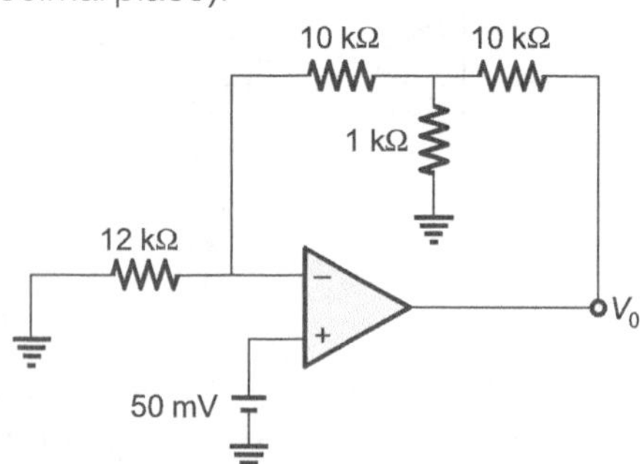

[2021 : 2 M]

3.74 An OPAMP has a gain of 10^4, an input impedance of 10 MΩ and an output impedance of 100 Ω. The OPAMP is used in unity-gain feedback configuration in a voltage buffer circuit. The closed-loop output impedance of the OPAMP (in milliohms) in the circuit is _______ (round off to one decimal place)

[2022 : 1 M]

3.75 The circuit shown is driven by a sinusoidal input voltage, V_{in}, resulting in the output voltage, V_{out}. The frequency (in kilohertz) at which the voltage gain is 0 dB is ________ (round off to two decimal places)

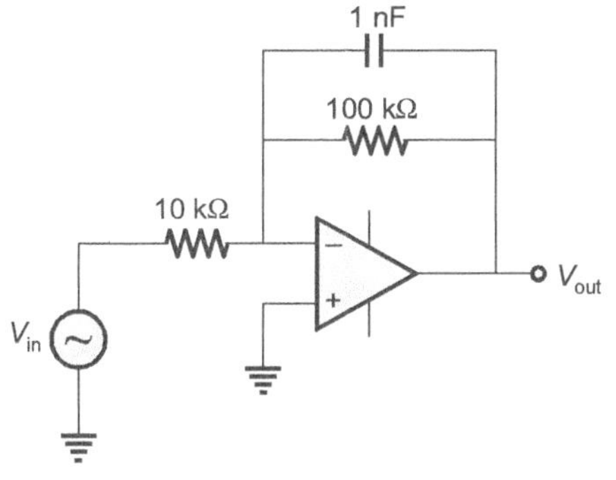

[2022 : 2 M]

3.76 In the circuit shown, the switch is initially closed. It is opened at $t = 0$s and remains open thereafter. The time (in milliseconds) at which the output voltage V_{out} becomes LOW is ________ (round off to three decimal places)

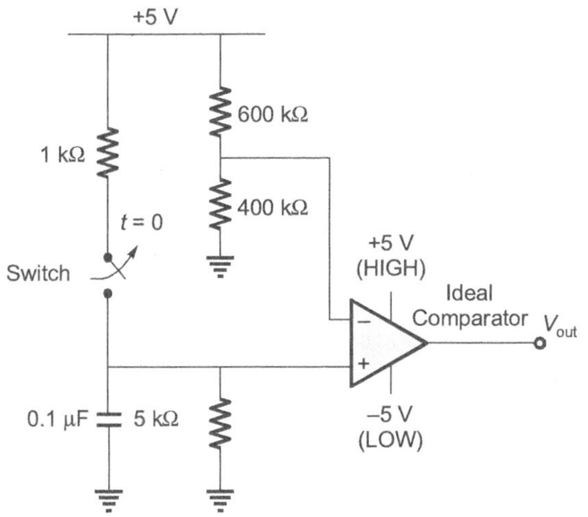

[2022 : 2 M]

3.77 The opamp in the circuit shown is ideal, except that it has an input bias current of 1 nA and an input offset voltage of 10 μV. The resulting worst-case output voltage will be ± ________ μV (rounded off to the nearest integer).

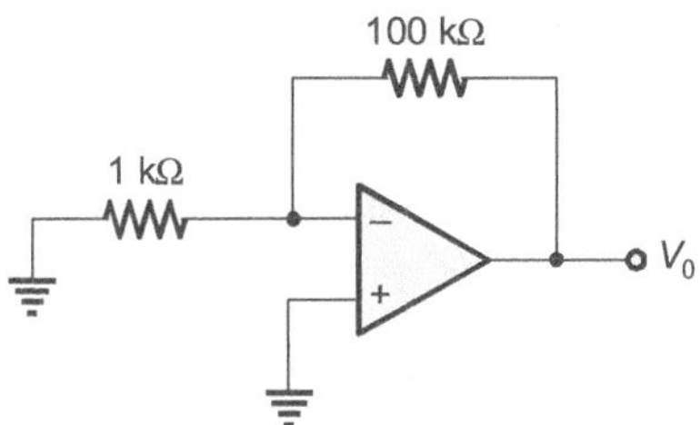

[2023 : 1 M]

3.78 The opamps in the circuit are ideal. The input signals are $V_{S1} = 3 + 0.10\sin(300t)$ V and $V_{S2} = -2 + 0.11\sin(300t)$ V.

The average value of the voltage V_0 is ______V (rounded off to two decimal places).

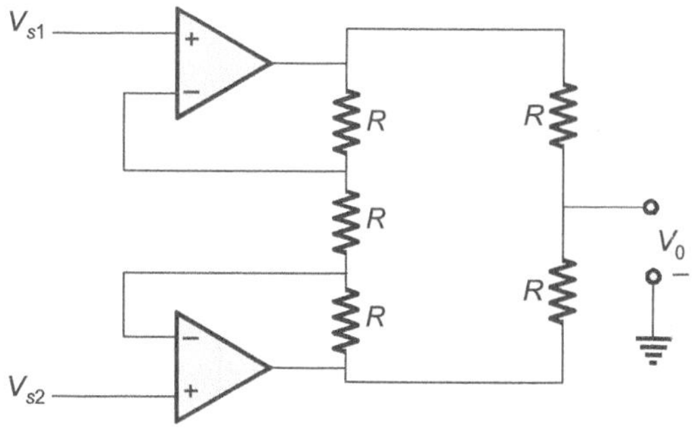

[2023 : 2 M]

3.79 In the circuit shown, the input voltage V_{in} = 100 mV. The switch and the opamp are ideal. At time $t = 0$, the initial charge stored in the 10 nF capacitor is 1 nC, with the polarity as indicated in the figure. The switch S is controlled using a 1 kHz square-wave voltage signal V_s as shown. Whenever V_s is 'High', S is in position '1' and when V_s is 'Low', S is in position '2'.

At t = 20 ms, the magnitude of the voltage V_0 will be ________ mV (rounded off to the nearest integer).

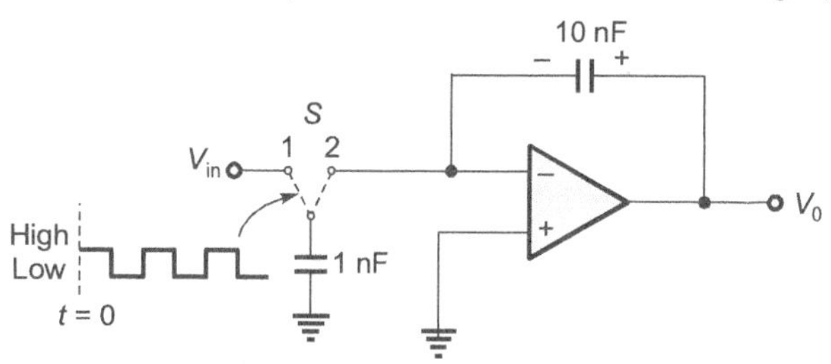

[2023 : 2 M]

3.80 In the circuit shown, assuming an ideal opamp, the value of the output voltage V_0 = ____ V (rounded off to one decimal place).

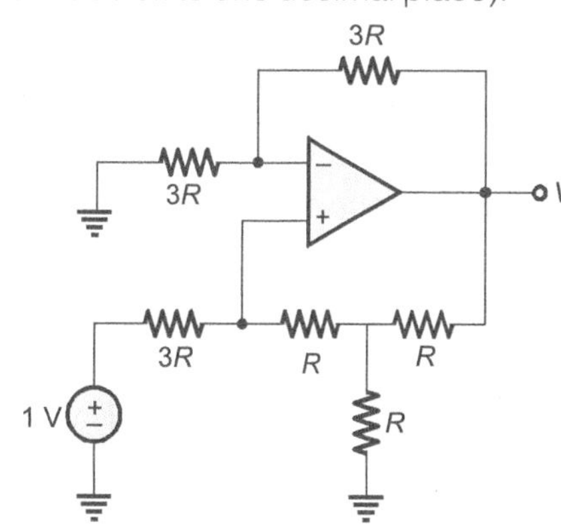

[2023 : 2 M]

3.81 Consider the capacitive sensor circuit and its output voltage shown in the figure. The circuit is switched ON at $t = 0$. Assuming the opamp to be ideal, the frequency of the output voltage V_0 is ________ kHz (rounded off to two decimal places).

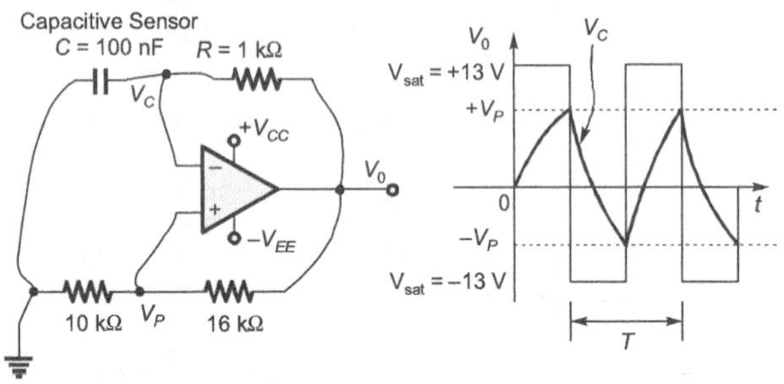

[2024 : 2 M]

3.82 In the figure shown, the diode current is given by $I_D = I_S e^{\frac{\alpha V_D}{T}}$. V_D is the diode voltage in volts, T is the absolute temperature in Kelvin, $\alpha = 1.16 \times 10^4$ K/V, and $I_S = 10^{-15}$ A is the saturation current. The dc current source, opamp and the resistors are ideal, and are assumed to be temperature independent. The change in the output voltage (V_{out}) per Kelvin change in temperature is ________ mV (rounded off to one decimal place).

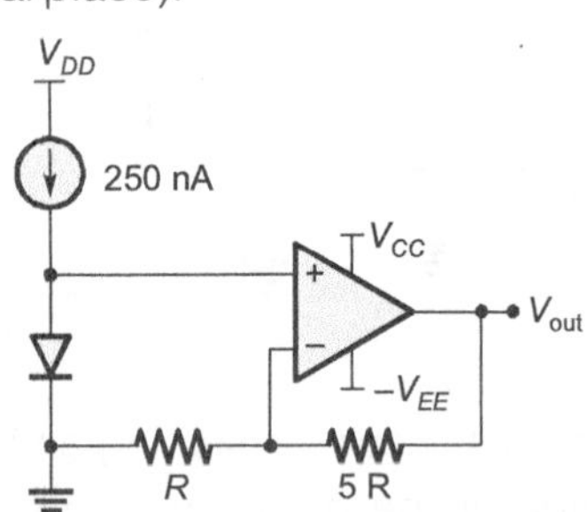

[2024 : 2 M]

3.83 In the circuit shown, assume the opamp is ideal and the initial charge on the capacitor is zero. The output voltage at time t = 2 ms is _____ V (rounded off to one decimal place).

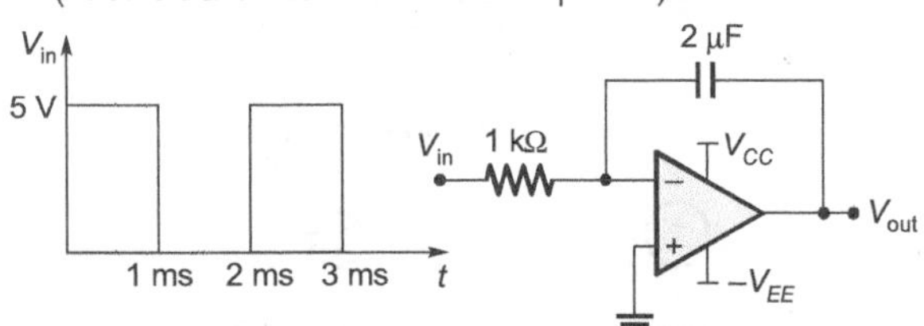

[2024 : 2 M]

3.84 In the figure shown, S_W is a switch whose position changes from 1 to 0 when V_C changes from logic HIGH to LOW and vice versa. The bandwidth of the permanent magnet moving coil (PMMC) type voltmeter is 1 Hz. If $V_{sense} = 2 \sin(4000\pi t)$ V and $V_{ref} = 4 \sin(2000\pi t)$ V, the voltmeter reading is _______V (rounded off to nearest integer).

Note: Assume all components are ideal.

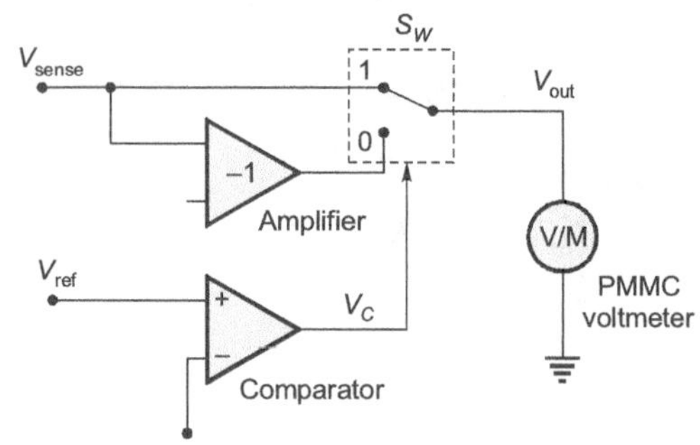

[2024 : 2 M]

3.85 Consider the circuit shown in the figure.

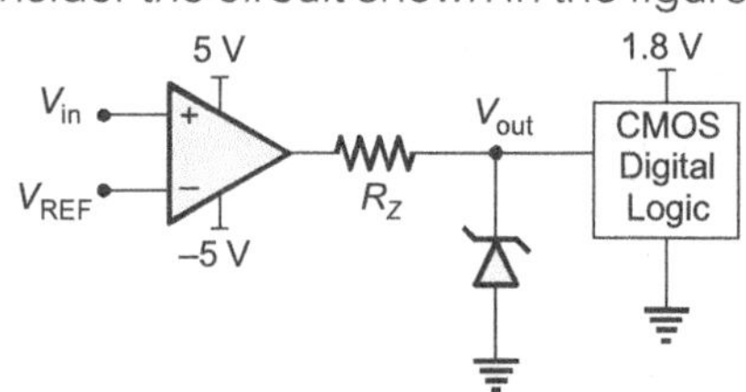

The CMOS digital logic circuit has infinite input impedance. Assume the opamp is ideal. A 1.8 V Zener diode with a minimum Zener current of 2 mA is used. The corresponding maximum value of resistance R_Z is ________ kΩ. (rounded off to one decimal place).

[2024 : 1 M]

3.86 Consider the readout circuit of a piezoelectric sensor shown in the figure.

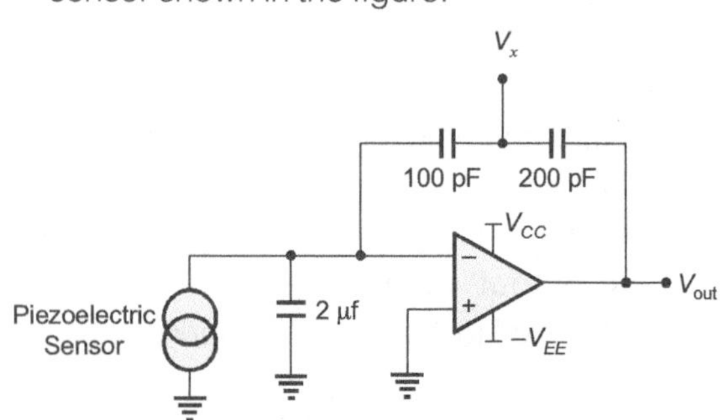

When the piezoelectric sensor generates a charge q_p, the resulting change in voltage V_x is −2 V. Then the corresponding change in the voltage V_{out} is _______ V (rounded off to nearest integer).

Note: Assume all components are ideal.

[2024 : 1 M]

3.87 In the figure shown, both the opamps A_1 and A_2 are ideal, except that the opamp A_1 has an offset voltage (V_{os}) of 1 mV. For V_{in} = 0 V, the values of the output voltages V_{out1} and V_{out2}, respectively, are

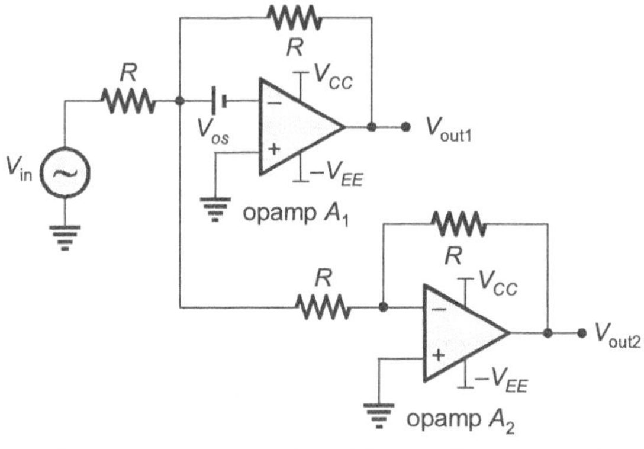

(a) 3 mV and –1 mV (b) 1 mV and 0 mV
(c) 1 mV and –1 mV (d) 2 mV and 0 mV

[2024 : 2 M]

3.88 In the figure shown, $R = 1\ k\Omega$ and $C = 0.1\ \mu F$. For a dc gain of –10, the 3 dB cut-off frequency (rounded off to one decimal place) is
Assume the opamp is ideal.

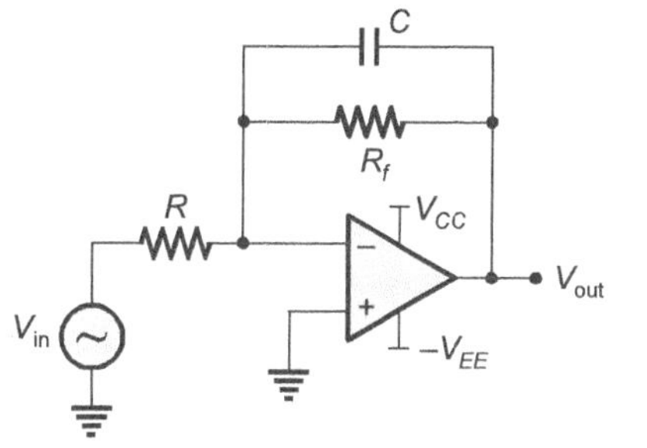

(a) 159.1 Hz (b) 1591.5 Hz
(c) 1750.7 Hz (d) 175.0 Hz

[2024 : 2 M]

3.89 In the figure shown, $R = 4.5\ k\Omega$, $\Delta R = 1.5\ k\Omega$, and INA is assumed to be ideal. The equivalent resistance between A and B is ____ kΩ (rounded off to nearest integer).

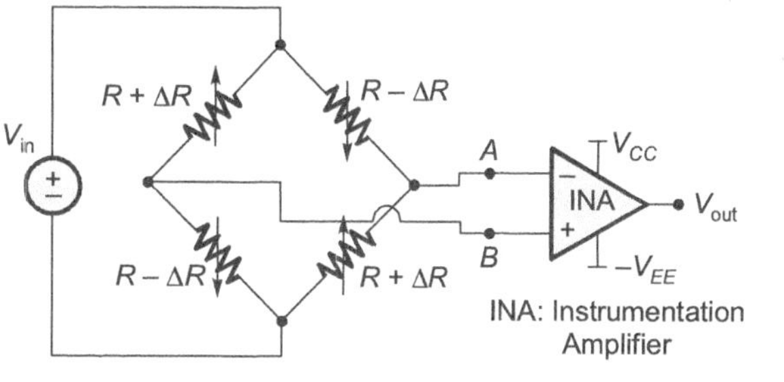

[2024 : 2 M]

Answers Op-Amp and Instrumentation Amplifier

3.1 (d)	3.2 (0.15)	3.3 (140)	3.4 (a)	3.5 (c)	3.6 (b)	3.7 (d)
3.8 (d)	3.9 (d)	3.10 (b)	3.11 (c)	3.12 (b)	3.13 (c)	3.14 (d)
3.15 (b)	3.16 (c)	3.17 (b)	3.18 (a)	3.19 (a)	3.20 (a)	3.21 (b)
3.22 (d)	3.23 (a)	3.24 (d)	3.25 (c)	3.26 (b)	3.27 (d)	3.28 (b)
3.29 (d)	3.30 (c)	3.31 (a)	3.32 (d)	3.33 (b)	3.34 (b)	3.35 (c)
3.36 (b)	3.37 (b)	3.38 (a)	3.39 (b)	3.40 (b)	3.41 (a)	3.42 (a)
3.43 (c)	3.44 (b)	3.45 (c)	3.46 (c)	3.47 (a)	3.48 (b)	3.49 (c)
3.50 (d)	3.51 (b)	3.52 (0.6)	3.53 (–1)	3.54 (1.6)	3.55 (20)	3.56 (40.83)
3.57 (100)	3.58 (100)	3.59 (b)	3.60 (1.5)	3.61 (–5)	3.62 (0)	3.63 (0.99)
3.64 (17.69)	3.65 (c)	3.66 (9)	3.67 (–4)	3.68 (7)	3.69 (0.50)	3.70 (14.86)
3.71 (b)	3.72 (1)	3.73 (1.1)	3.74 (10.0)	3.75 (15.84)	3.76 (0.367)	3.77 (1110)
3.78 (0.5)	3.79 (100)	3.80 (2)	3.81 (6.17)	3.82 (10)	3.83 (–2.5)	3.84 (0)
3.85 (1.6)	3.86 (–3)	3.87 (a)	3.88 (a)	3.89 (4)		

Explanations Op-Amp and Instrumentation Amplifier

3.1 (d)

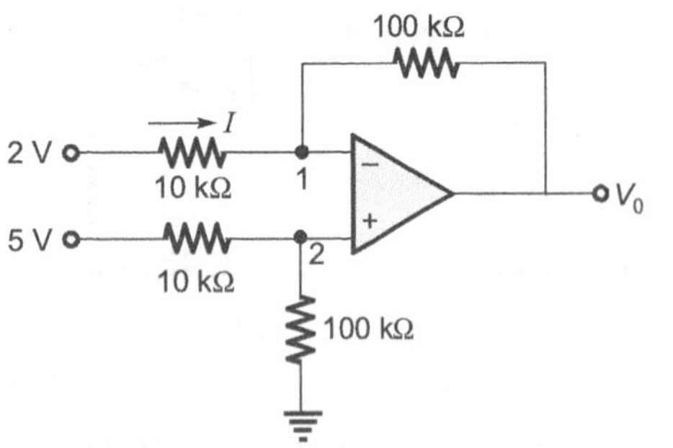

Given circuit is

$$V_1 = V_2$$

(Because of very high input impedance of the op-amp)

also, $V_2 = \frac{100}{110} \times 5 = \frac{500}{110} = \frac{50}{11}$ V

$$V_0 = -\frac{100}{10} \times 2V + \left(1 + \frac{100}{10}\right) V_2$$

$$= -20 + \left(11 + \frac{50}{11}\right) = -20 + 50 = 30 \text{ V}$$

3.2 (0.15)

Redraw the circuit during hold op-amp

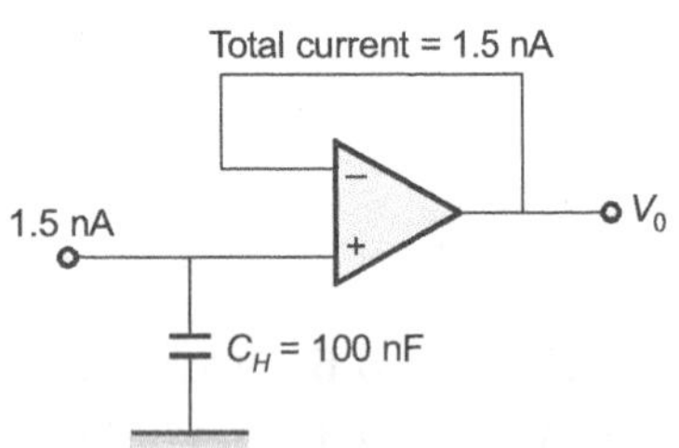

During holding,

$$C_H \frac{dV_c}{dt} = 1.5 \text{ nA}$$

$$\frac{dV_c}{dt} = \frac{1.5 \times 10^{-9}}{C} = \frac{1.5 \times 10^{-9}}{100 \times 10^{-9}}$$

$$= 0.15 \text{ V/sec} = 0.15 \text{ mV/m-sec}$$

3.3 (140)

Given, $A_{CM} = 0.01$,

$A_d = 10^6$

$$\text{CMRR} = \left|\frac{A_d}{A_{CM}}\right| = \frac{10^5}{0.01} = 10^7$$

$$(\text{CMRR})_{dB} = 20 \log_{10} 10^7 = 140 \text{ dB}$$

3.4 (a)

Given circuit is

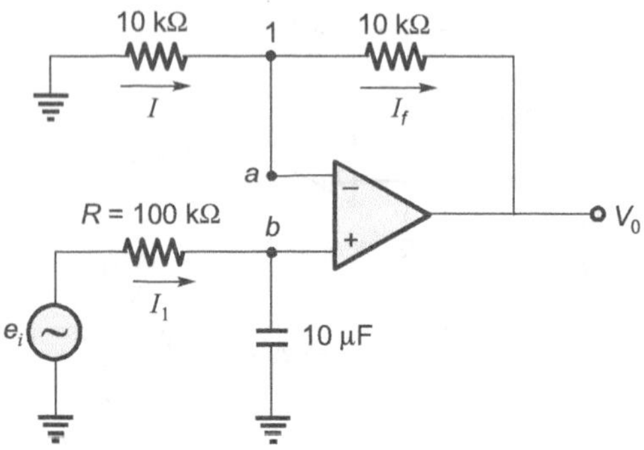

KCL at node 1

$$\frac{0 - V_a}{10} = \frac{V_a - V_0}{10} \Rightarrow \boxed{V_0 = 2V_a} \quad \text{...(i)}$$

KCL at node b,

$$\frac{e_i - V_b}{R} = \frac{V_b - 0}{1/sC}$$

$$\Rightarrow \quad V_b = \left(\frac{e_i}{1 + sCR}\right)$$

$$\Rightarrow \quad V_0 = \frac{2e_i}{1 + sCR}$$

($\because V_b = V_a$, By virtual ground)

$$V_0 = e_i\left(\frac{2}{1 + sCR}\right)$$

$\therefore$ transfer function is

$$H(s) = \frac{2}{1 + sCR}$$

$$|H(s)| = \frac{2}{\sqrt{1 + \omega^2 C^2 R^2}}$$

for $\omega = 1$, $C = 10\ \mu$F and $R = 100$ K

$$\omega CR = 1 \times 10 \times 10^{-6} \times 100 \times 10^3 = 1$$

$$= \frac{2}{\sqrt{1+1}} = \frac{2}{\sqrt{2}} = \sqrt{2}$$

$$\angle H(s) = -\tan^{-1} \omega CR = -\tan^{-1}(1) = 45°$$

$$\therefore \quad H(s) = \sqrt{2} \angle -45°$$

$$\Rightarrow \quad \boxed{V_0 = \sqrt{2} \sin\left(t - \frac{\pi}{4}\right)}$$

Hence option (a) is correct.

3.5 (c)

For the given circuit,

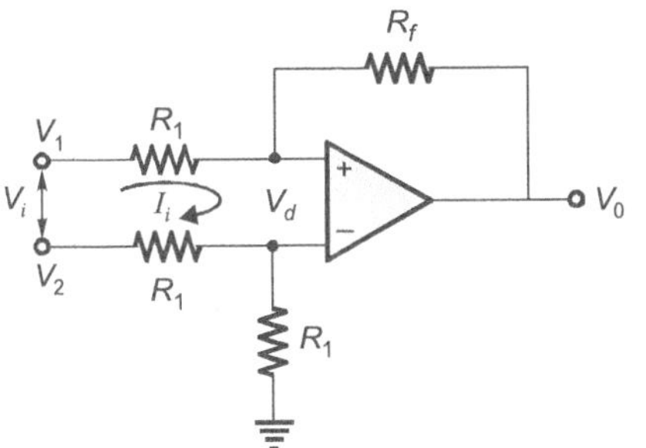

input resistance is V_i/I_i

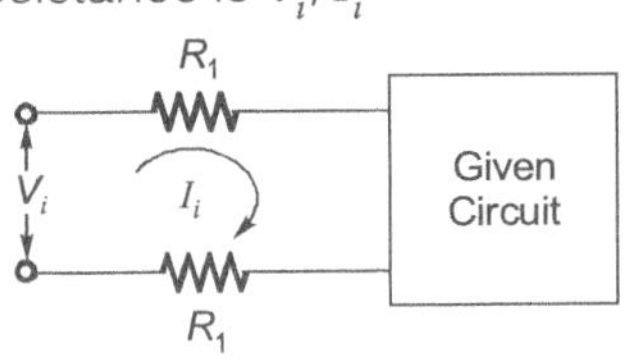

$$\Rightarrow \quad \boxed{\frac{V_i}{I_i} = 2R_1}$$

hence option (c) is correct.

3.6 (b)

The op-amp use a differential input stage with a constant current source to obtain very high differential gain.

3.7 (d)

The given op-amp circuit can be redrawn as

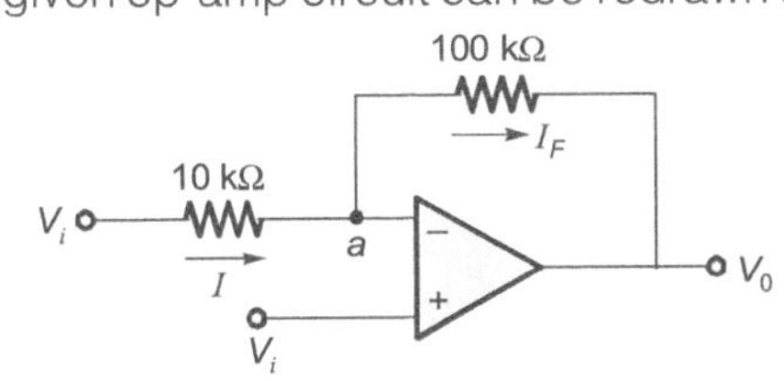

Since, $V_a = 1$ V

so current, $I = I_F = 0$,

hence there is no voltage drop across 100 kΩ resistor.

So output voltage, $V_0 = V_a = 1\,\text{V}$

3.8 (d)

Given circuit is

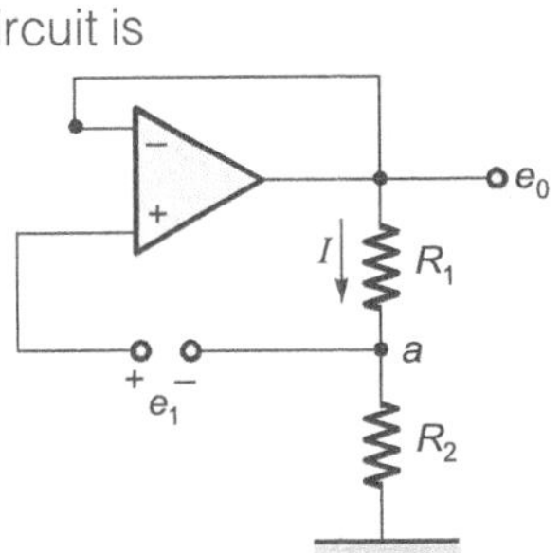

from the given circuit,

$$I = \frac{e_0}{R_1 + R_2} \qquad \text{...(i)}$$

Also, $\quad e_1 = e_0 - V_a$

$$V_a = \frac{e_0 \times R_2}{R_1 + R_2}$$

$$\Rightarrow \quad e_1 = e_0 - \frac{e_0 R_2}{R_1 + R_2} = e_0\left[\frac{R_1 + R_2 - R_2}{R_1 + R_2}\right]$$

$$= \frac{e_0 R_1}{R_1 + R_2} \qquad \text{...(ii)}$$

$$I = \frac{e_0}{R_1 + R_2} = \frac{e_1 (R_1 + R_2)}{R_1} \times \frac{1}{(R_1 + R_2)}$$

$$\Rightarrow \quad \frac{I}{e_1} = \frac{I}{R_1} \text{ or } I = \frac{e_1}{R_1}$$

Hence, option (d) is correct.

3.9 (d)

For $V_i = 0$, the output voltage of op-amp is

$$V_0 = -\left(\frac{10}{1} \times V_i\right) + \left(1 + \frac{10}{1}\right) V_{offset}$$

$$= -(10 \times 0) + (11 \times 10) = 110 \text{ mV}$$

3.10 (b)

Gain $A_V = \frac{V_0}{V_i} = 10$, slew rate $\left(\frac{dV}{dt}\right)_{max} = 10^6$

$$\Rightarrow \quad V_0 = 10\,V_i = 10 \sin \omega t$$

also, $\quad \frac{dV_0}{dt} = 10\,\omega \cos \omega t$

$$\text{slew rate} = \left(\frac{dV_0}{dt}\right)_{max} = (10\omega\cos\omega t) = 10\,\omega$$

$$= 2\pi f \times 10 = 10^6$$

$$\Rightarrow \quad f = \frac{10^6}{20\pi} = 15.92 \text{ kHz}$$

Hence option (b) is correct.

3.11 (c)

Both the buffers should have low bias current to get the DC error output voltage low.

3.12 (b)

For output DC error voltage V_1 and V_2 are connected at ground and capacitor acts as open. Redrawn the circuit,

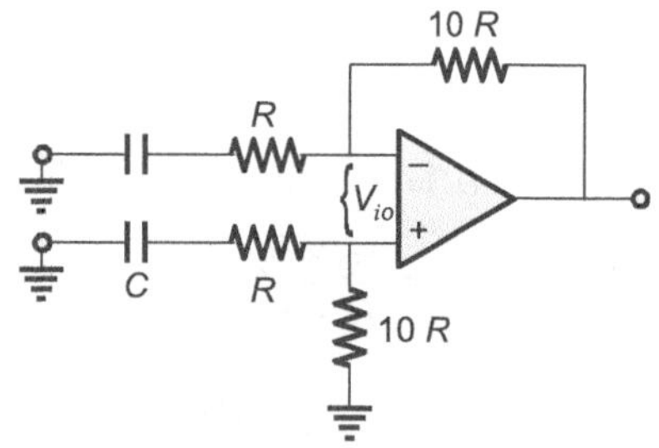

Output DC error voltage = $V_{io} = \left[1+\frac{R_F}{R_1}\right]$

$\because$ Capacitor act as open for DC voltage. Therefore, R_1 = 1 kΩ becomes ∞ so its gain = 1

$$V_{\text{off set}} = V_{io} \times 1 = 2 \text{ mV}$$

3.13 (c)

Input waveform,

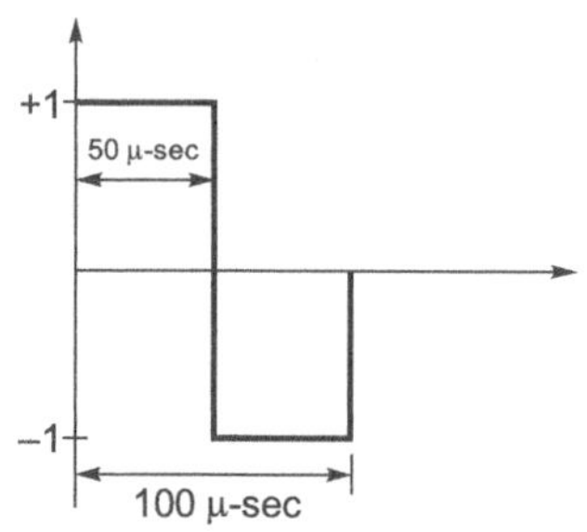

Slew rate = 0.1 V/μ-sec

so, in 50 μ-sec output can swing only 5 volt.

Output will never be able to reach amplitude of ± 10 V.

So, output will be ± 2.5 V amplitude triangular wave.

3.14 (d)

Slew rate: Slew rate is defined as maximum rate at which amplifier output can change in volts per microsecond ($V/\mu s$)

$$\text{Slew rate} = \frac{\Delta V_0}{\Delta t}\left(\frac{V}{\mu s}\right) = \frac{8}{0.2} = 40.0 \text{ V}/\mu\text{s}$$

3.15 (b)

$$V_- = V_+ = 0 \text{ volt}$$

($\because$ V is virtually grounded)

So, $V_0 = (-1.5 \times 10^3 \times 1.0 \times 10^{-3})$ V
$= -1.5$ V

3.16 (c)

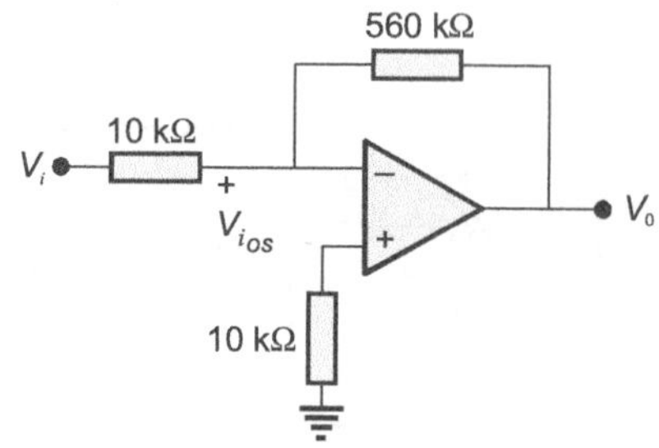

$$V_{0(\text{offset})} = V_{IO}\left(\frac{R_1+R_f}{R_1}\right) = 5 \text{ mV}\left(1+\frac{R_f}{R_1}\right)$$

$$= 5 \text{ mV}\left(1+\frac{560}{10}\right)$$

$$= 57 \times 5 \text{ mV} = 285 \text{ mV}$$

3.17 (b)

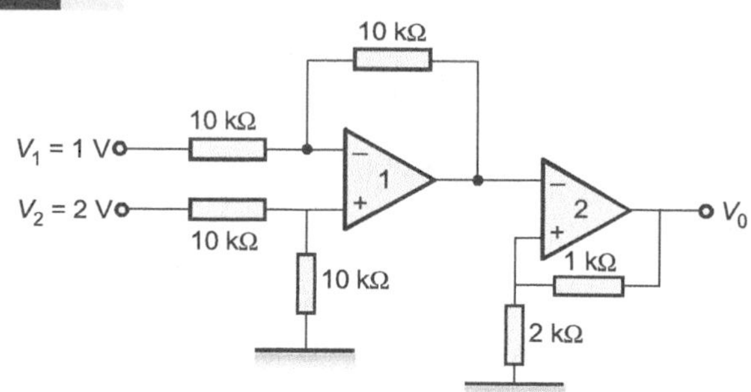

Apply superposition theorem,

V_1 = 1 V and V_2 = 0 V

then output of op-amp-1 is

$$V_{01} = -\frac{R_f}{R_1}\cdot V_{in} = -\frac{1}{1}\times 1 = -1 \text{ V}$$

The output of op-amp-2 is

$$V'_{01} = V_{01}\times\left(1+\frac{R_f}{R_1}\right) = -1\times\left(1+\frac{1}{2}\right) = -\frac{3}{2}V$$

$$= -1.5 \text{ V}$$

Now V_1 = 0 V and V_2 = 2 V

output of op-amp1 is

$$V_+ = \frac{10}{20}\times 2 = 1 \text{ V}$$

$$V_{02} = 1\times\left(1+\frac{10}{10}\right) = 2 \text{ V}$$

Output of op-amp-2 is

$$V'_{02} = 2\times\left(1+\frac{R_f}{R_1}\right) = 2\times\left(1+\frac{1}{2}\right) = 3V$$

$$= 3 \text{ V}$$

$$\therefore \quad V_0 = V'_{01} + V'_{02}$$

$$= -1.5 \text{ V} + 3 \text{ V} = 1.5 \text{ V}$$

3.18 (a)

Voltage at non-inverting terminal,

$$V_+ = \frac{10}{10+10}\times 3 = 1.5\text{ V}$$

Using KCL at inverting terminal, we have

$$\frac{3-V_-}{10K} = \frac{V_- - V_0}{10K}$$

or

$$V_0 = 2V_- - 3$$
$$= 2\times 1.5 - 3 \quad (\because V_+ = V_-)$$
$$= 0\text{ V}$$

3.19 (a)

$$V_+ = V_- = V_0 + (10\text{ k}\Omega)\times(0.3\text{ mA}) = V_0 + 3$$

Now, using KCL at non-inverting terminal

$$\frac{V_+ - 0}{20\text{K}} + \frac{V_+ - V_0}{30\text{K}} = 0$$

$$\frac{V_0+3}{2} + \frac{V_0+3-V_0}{3} = 0$$

$$V_0 = -5\text{ V}$$

3.20 (a)

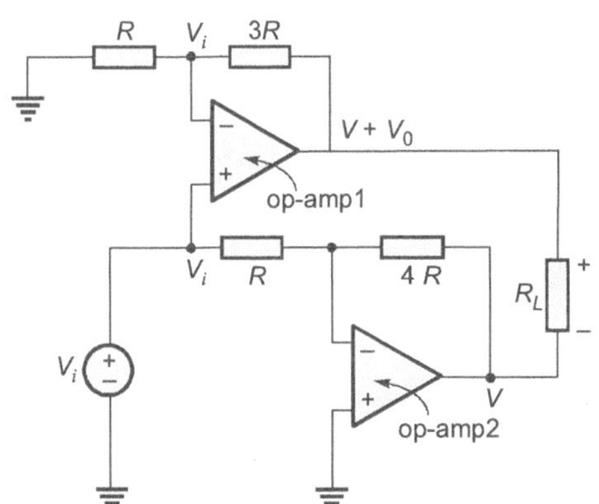

Using KCL at inverting terminal of opamp, we have

$$\frac{V_i - 0}{R} + \frac{V_i - (V+V_0)}{3R} = 0 \quad \text{...(i)}$$

Again use KCL at inverting node of opamp2, We have

$$\frac{0-V_i}{R} + \frac{0-V}{4R} = 0 \quad \text{...(ii)}$$

$$V = -4V_i$$

Put the value of V in equation (i), we get

$$\frac{V_i}{R} + \frac{V_i + 4V_i - V_0}{3R} = 0$$

$$3V_i + V_i + 4V_i - V_0 = 0$$

$$8V_i - V_0 = 0$$

$$\boxed{\frac{V_0}{V_i} = 8}$$

3.21 (b)

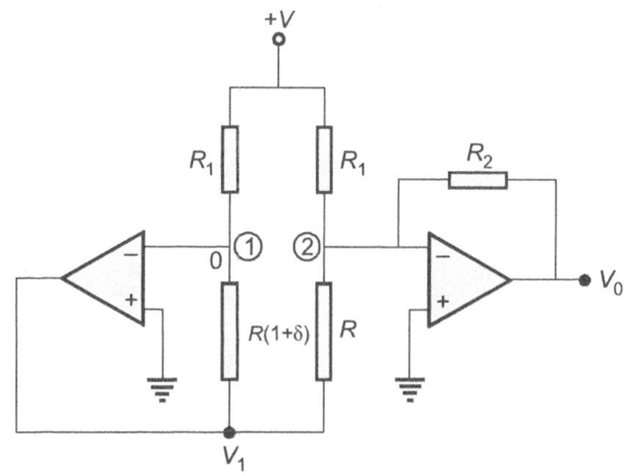

$V_+ = V_- = 0$ (due to virtual ground concept) KCL at node (1)

$$\frac{0-V}{R_1} + \frac{0-V_1}{R(1+\delta)} = 0 \quad \text{...(i)}$$

$$\frac{V_1}{R} = \frac{-V}{R_1}(1+\delta)$$

KCL at node (2)

$$\frac{0-V}{R_1} + \frac{0-V_1}{R} + \frac{0-V_0}{R_2} = 0 \quad \text{...(ii)}$$

From equation (i) and (ii) we get

$$V_0 = -R_2\left[\frac{V}{R_1} + \frac{V_1}{R}\right]$$

$$= -R_2\left[\frac{V}{R_1} - \frac{V}{R_1}(1+\delta)\right]$$

$$\boxed{V_o = \frac{R_2}{R_1}V\delta}$$

3.22 (d)

For an op-amp the output is

$$V_0 = A_dV_d + A_cV_c \quad \text{...(i)}$$

where, $V_d = V_1 - V_2$

and $V_c = \dfrac{V_1+V_2}{2}$

Given output is

$$V_0 = 100(V_1 - V_2) + 10^{-4}(V_1 + V_2) \quad \text{...(ii)}$$

Compare equation (i) and (ii) we get

$$A_d = 100$$
$$A_c = 2\times 10^{-4}$$

$$\text{CMRR} = \frac{A_d}{A_c} = \frac{100}{2\times 10^{-4}} = 0.5\times 10^6$$

$$\text{CMRR (dB)} = 20\log(0.5\times 10^6)$$
$$= 114\text{ dB}$$

The gain of op-amp is the gain in difference mode only i.e. $A_d = 100$.

So, the nearest answer is option (d).

3.23 (a)

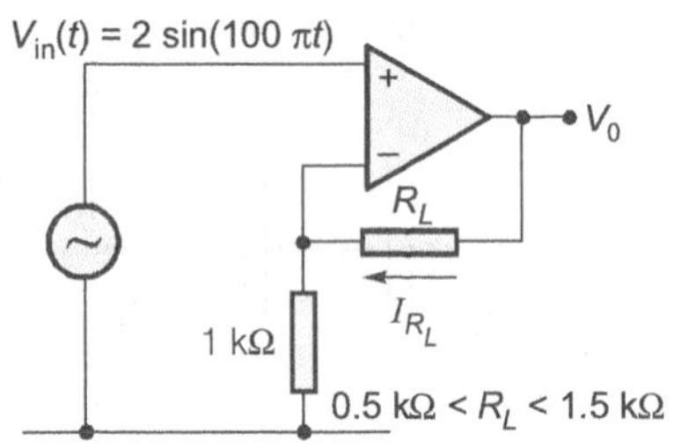

$$V_0 = \left(1+\frac{R_f}{R_1}\right)V_{in}$$

$$= \left(1+\frac{R_L}{1\text{k}\Omega}\right)\times 2\sin(10\pi t)$$

$$I_{R_L} = \frac{V_0}{R_{eq}} = \frac{\left(\frac{1\text{k}\Omega + R_L}{1\text{k}\Omega}\right)\times 2\sin(100\pi t)}{(R_L + 1\text{k}\Omega)}$$

$$\therefore \quad I_{R_L} = 2\sin(100\pi t)\text{ mA}$$

3.24 (d)

Given, saw tooth wave is as

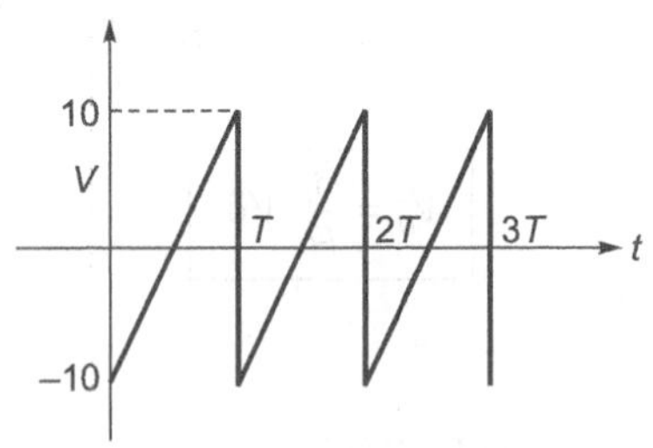

so equation for one period

$$y = \frac{20}{T}t - 10 \text{ (for first period)}$$

for the circuit given

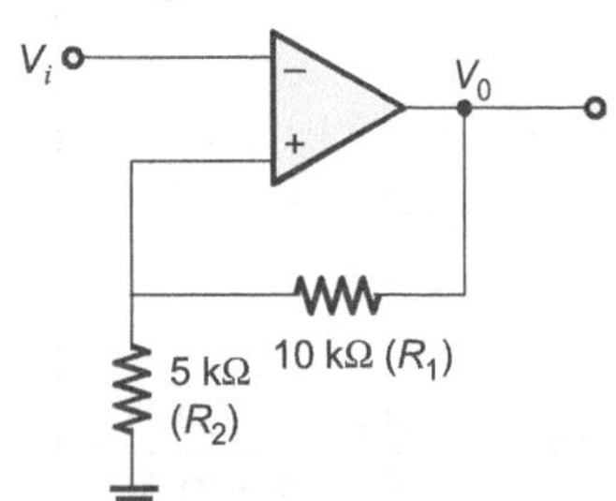

since output is symmetrical and peak to peak = 30 V

So, $+V_{sat} = 15\text{ V}$

$-V_{sat} = -15\text{ V}$

Now, $$V_{ut} = \frac{R_2}{R_1+R_2}\times +V_{sat} = \frac{5}{15}\times 15 = 5\text{ V}$$

lower threshold voltage

$$V_{lt} = \frac{R_2}{R_2+R_1}\times(-V_{sat}) = -5\text{ V}$$

so time at which input will be +5 V

$$5 = \frac{20}{T}t - 10 \rightarrow t = \frac{3}{4}T$$

Now output waveform

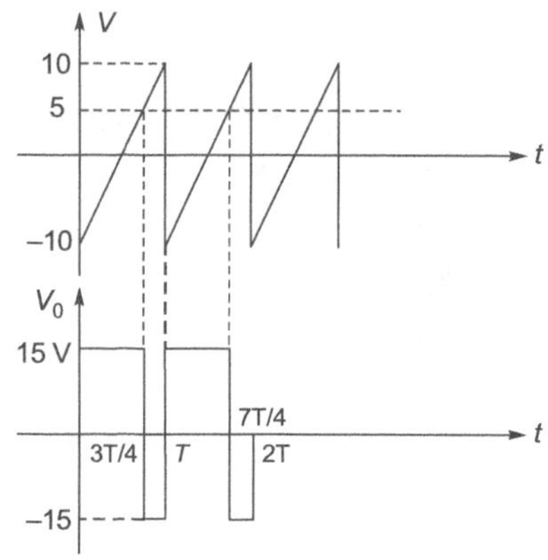

So average value of output

$$V_{avg} = \frac{15\times 3T/4 + (-15)\times T/4}{T}$$

$$= \frac{45}{4} - \frac{15}{4} = \frac{30}{4} = 7.5\text{ V}$$

3.25 (c)

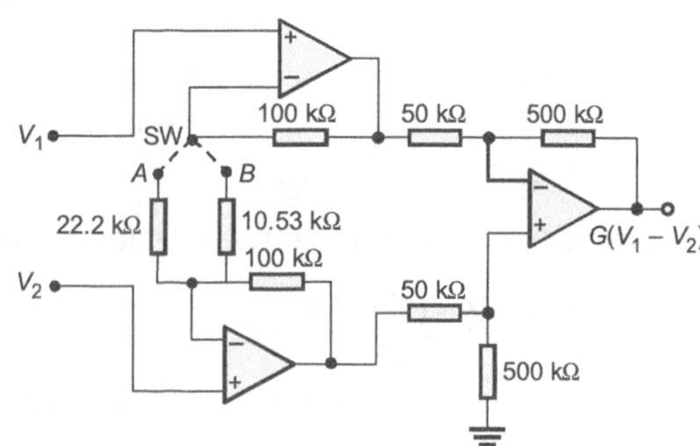

The gain of the instrumentation amplifier is given as

$$G = \frac{V_{out}}{V_2 - V_1} = \left(1+\frac{2R_1}{R_{gain}}\right)\frac{R_3}{R_2} \qquad \text{...(i)}$$

Here, $R_1 = 100\text{ k}\Omega$

$R_{gain} = 22.2\text{ k}\Omega$

and $R_2 = 50\text{ k}\Omega$

At position '*A*' the gain is

$$G = \left(1+\frac{2\times 100}{22.2}\right)\times\frac{500}{50} = 100$$

At position 'B' the R_{gain} = 10.53 kΩ

So now gain is

$$G = \left(1+\frac{2\times100}{10.53}\right)\times\frac{500}{50} = 200$$

3.26 (b)

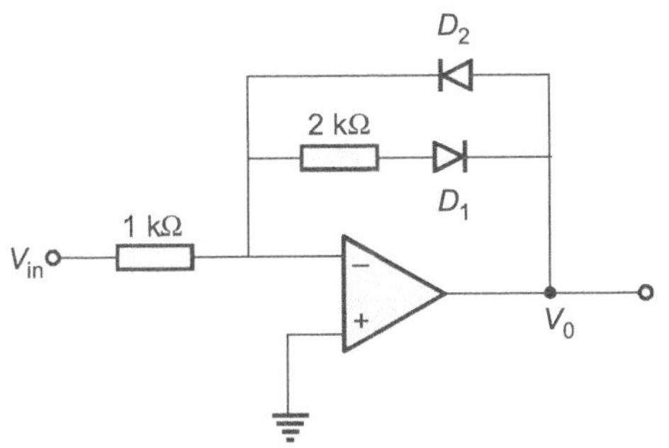

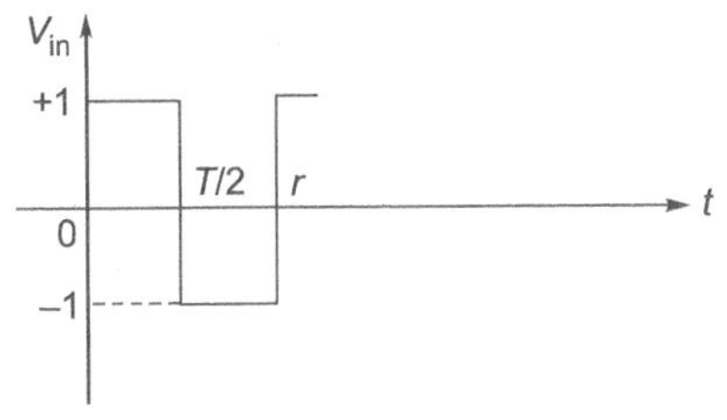

For +ve half cycle V_{in} = 1 V, then diode D_2 is reverse bias and diode D_1 is forward biased, then circuit becomes

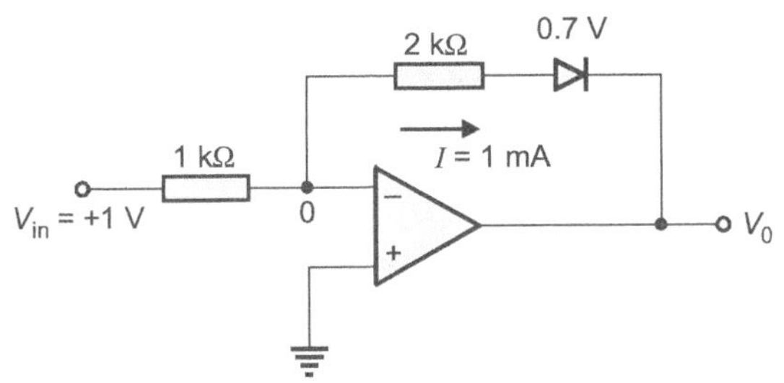

Then current,

$$1 - 1\text{ k}\Omega \times I = 0$$

$$\therefore \quad I = 1\text{ mA}$$

Then apply KVL,

$$0 - 2\times10^3\times10^{-3} - 0.7 = V_0$$

$$\therefore \quad V_0 = -2.7\text{ V}$$

For –ve half cycle V_{in} = –1 V, then diode D_2 is forward biased and diode D_1 is reverse bias and circuit becomes

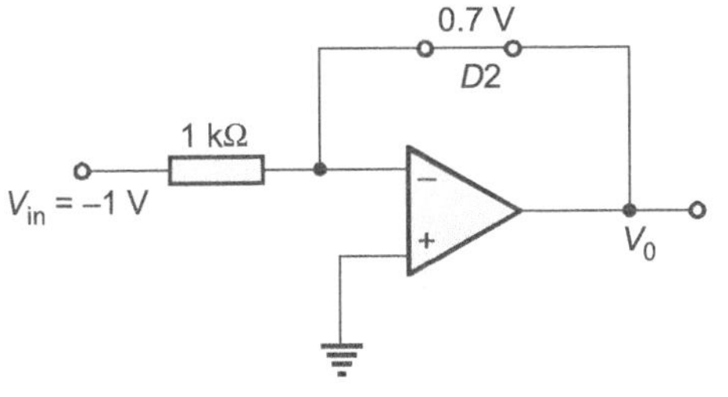

$$0 + 0.7 = V_0$$

$$\therefore \quad V_0 = 0.7\text{ V}$$

and the output waveform is like as given below:

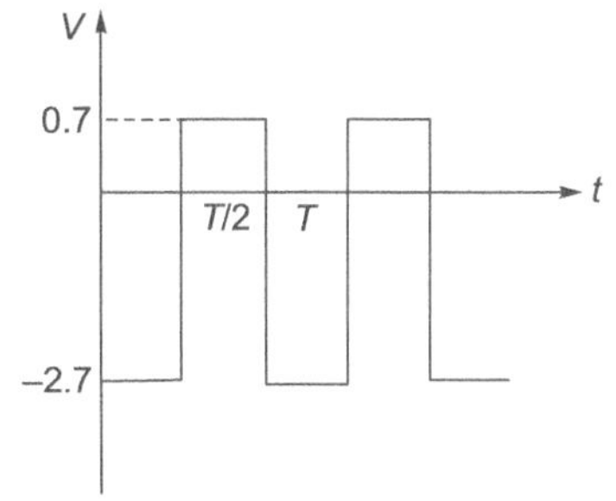

$$V_0(\text{avg}) = \frac{0.7+(-2.7)}{2} = -1\text{ V}$$

3.27 (d)

$$\frac{0-V_0}{10\text{ k}\Omega} = \frac{V_{in}-0}{1\text{k}\Omega}$$

$$\frac{V_0}{V_{in}} = -10 \text{ (independent of } R)$$

3.28 (b)

When switch S_2 is closed the circuit becomes:

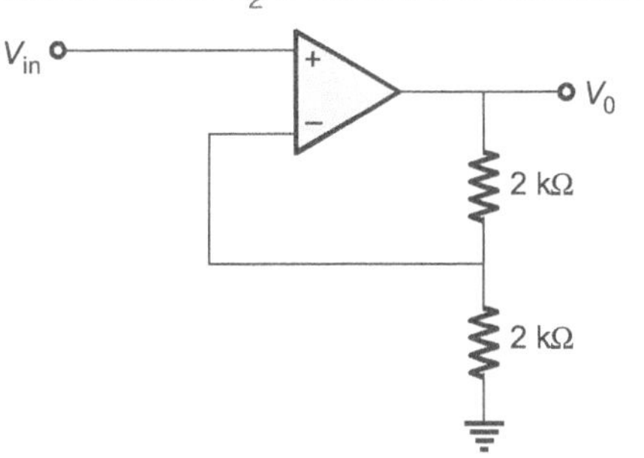

$$V_0 = \left(1+\frac{R_f}{R_1}\right)V_{in} = \left(1+\frac{2}{2}\right)V_{in}$$

$$\therefore \quad \frac{V_0}{V_{in}} = 2$$

3.29 (d)

DC output voltage that is output offset voltage

$$= \left(1+\frac{R_f}{R_1}\right)V_{i0} + R_f I_b$$

So maximum offset (dc voltage) at the output will be

$$V_{00} = \left(1+\frac{100}{100}\right)1\text{ mV} + \frac{10\times10^{-9}\times100\times10^{3}\text{mV}}{10^{-3}}$$

$$V_{00} = 2\text{ mV} + 1\text{ mV}$$

$$V_{00} = 3\text{ mV}$$

3.30 (c)

When OPAMP is used in –ve feedback then voltage difference between the terminals is treated as zero.

3.31 (a)

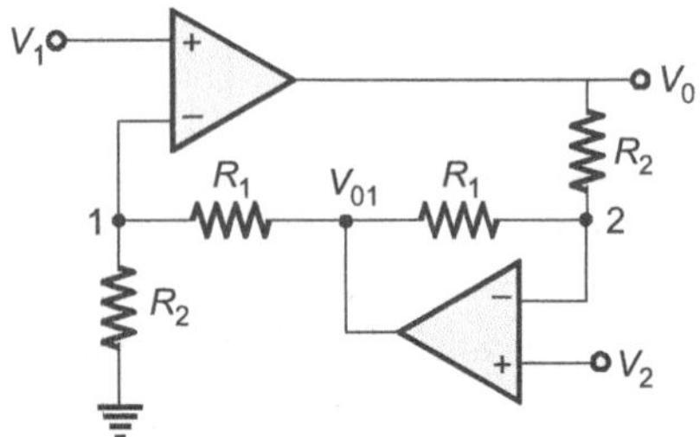

Voltage at node 1 = V_1
(Virtual ground concept)

and $$V_1 = \frac{R_2}{(R_1+R_2)}V_{01}$$

or $$V_{01} = \left(1+\frac{R_1}{R_2}\right)V_1 \quad \text{...(i)}$$

voltage at node 2 = V_2
writing KCL at node 2 we have

$$\frac{V_0-V_2}{R_2} = \frac{V_2-V_{01}}{R_1}$$

$$\frac{V_0}{R_2} = V_2\left[\frac{1}{R_2}+\frac{1}{R_1}\right]-\frac{1}{R_1}\left[1+\frac{R_1}{R_2}\right]V_1$$

$$V_0 = \frac{V_2[R_1+R_2]}{R_1}-\frac{R_2}{R_1}\left[1+\frac{R_1}{R_2}\right]V_1$$

$$= (V_2-V_1)\left(1+\frac{R_2}{R_1}\right)$$

3.32 (d)

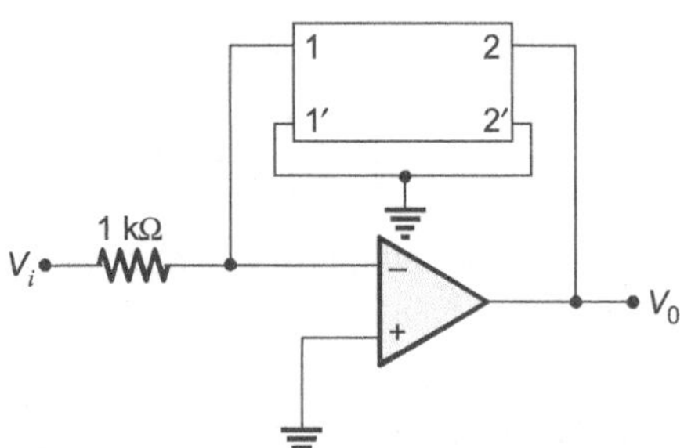

$$Z = \begin{bmatrix} 11 & 1 \\ 1 & 11 \end{bmatrix} \text{ (All parameters are in k}\Omega)$$

$$\Delta Z = 120 \text{ k}\Omega$$

Now the circuit becomes

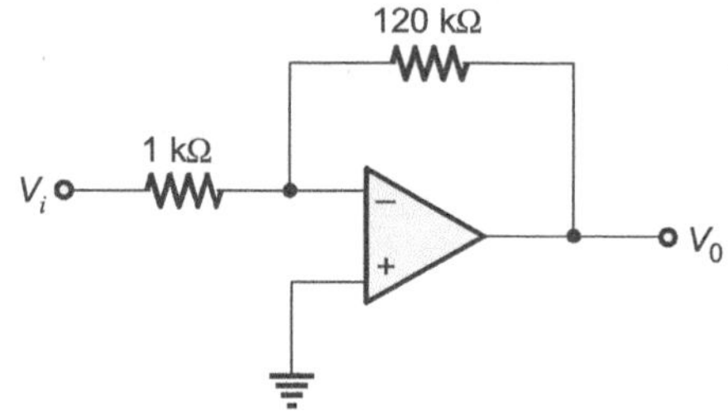

$$\text{Gain} = \frac{V_0}{V_i} = -\frac{R_f}{R_1} = -\frac{120}{1} = -120$$

3.33 (b)

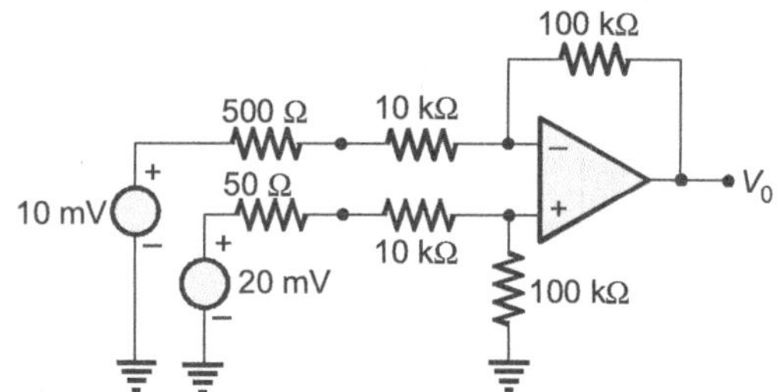

Apply KCL at V_+ node

$$\frac{V_+}{100}+\frac{V_+-20}{10.05} = 0$$

$$\frac{V_+}{100}+\frac{V_+}{10.05}-\frac{20}{10.05} = 0$$

$$\frac{V_+}{100}+\frac{V_+}{10.05} = 1.99$$

$$(0.01+0.0995)\,V_+ = 1.99$$

$$\therefore \quad V_+ = \frac{1.99}{0.109} = 18.25$$

Now, apply KCL at V_- node

$$\frac{V_- - V_0}{100}+\frac{V_- - 10}{10.5} = 0$$

For an ideal op-amp $V_+ = V_-$

$$\frac{18.25-V_0}{100}+\frac{18.25-10}{10.5} = 0$$

$$V_0 = 100(0.1825+0.7857)$$

$$= 96.86 \text{ mV}$$

3.34 (b)

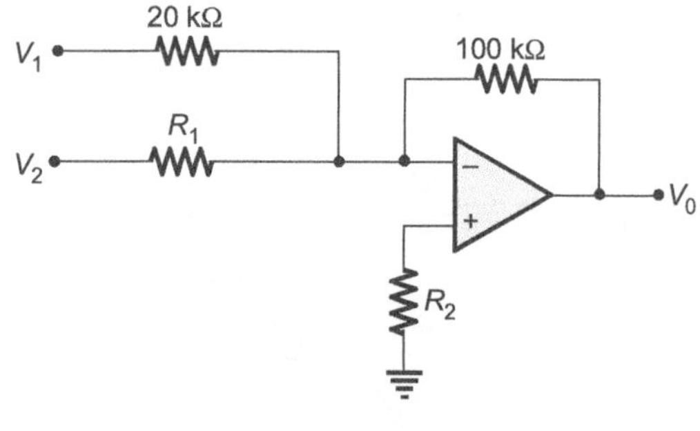

Apply KCL at the inverting node,

$$\frac{0-V_1}{20}+\frac{0-V_2}{R_1}+\frac{0-V_0}{100}=0$$

$$V_0 = -100\left[\frac{V_1}{20}+\frac{V_2}{R_1}\right]$$

Put, $V_0 = -7$ V, $V_1 = 0.2$ V and $V_2 = 0.6$ V

$$-7 = -100\left[\frac{0.2}{20}+\frac{0.6}{R_1}\right]$$

$$0.07 = 0.01+\frac{0.6}{R_1}$$

$$\therefore \quad R_1 = 10\text{ k}\Omega$$

3.35 (c)

$$V_0 = -100\, V_c \sin(2\pi f)\left(\frac{1}{20}+\frac{1}{20}\right)$$

$$= -10\, V_c \sin(2\pi ft)$$

$$\therefore \left|\frac{dV}{dt}\right|_{max} = |-10\, V_c(2\pi f)\cos(2\pi ft)|_{max}$$

$$= 20\,\pi f\, V_c$$

To avoid distortion,

$$\text{slew rate} \geq \left|\frac{dV_0}{dt}\right|_{max}$$

$$0.5\times10^6 \geq 20\pi f\, V_c$$

$$f\, V_c \leq \frac{0.5\times10^6}{20\pi}$$

$$f\, V_c \leq 7957.75$$

So when $f = 30$ kHz and $V_c = 0.1$ V

then, $f\, V_c = 3000 < 7957.74$

So option (c) is correct.

3.36 (b)

In order to avoid the effect of bias current, resistance at +ve terminal must be equal to dc resistance seen from –ve terminal by replacing all the sources by their internal resistances. Therefore,

$$R = 20\text{ k}\Omega\,||\,20\text{ k}\Omega\,||\,100\text{ k}\Omega$$

$$= 10\text{ k}\Omega\,||\,100\text{ k}\Omega$$

$$= 9.09\text{ k}\Omega = 9.1\text{ k}\Omega$$

3.37 (b)

$$\text{CMRR} = 20\log_{10}\frac{A_d}{A_{cm}}$$

$$40 = 20\log_{10}\frac{A_d}{A_{cm}}$$

$$\frac{A_d}{A_{cm}} = 100$$

$$\therefore \quad A_{cm} = \frac{A_d}{100} = \frac{100}{100} = 1$$

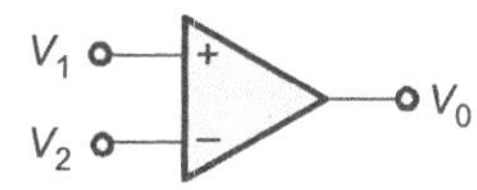

$$V_0 = A_d(V_1 - V_2) + A_{cm}\left(\frac{V_1+V_2}{2}\right)$$

$$= 100(0.55-0.45)+1\frac{(0.55+0.45)}{2}$$

$$= 10 + 0.5 = 10.5\text{ V}$$

3.38 (a)

It is VCVS.

3.39 (b)

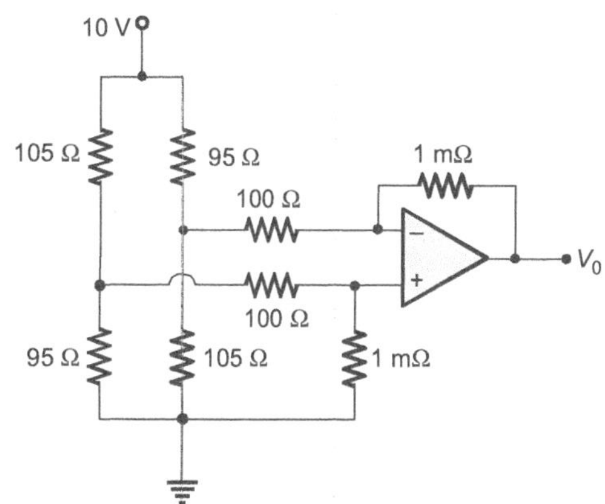

Break this figure in two parts

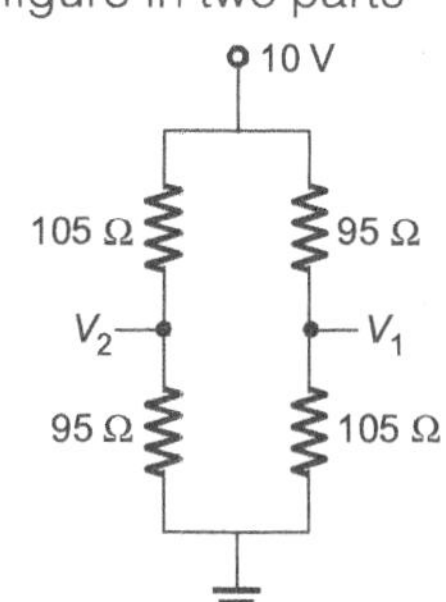

Fig. (a)

Break this figure in two parts

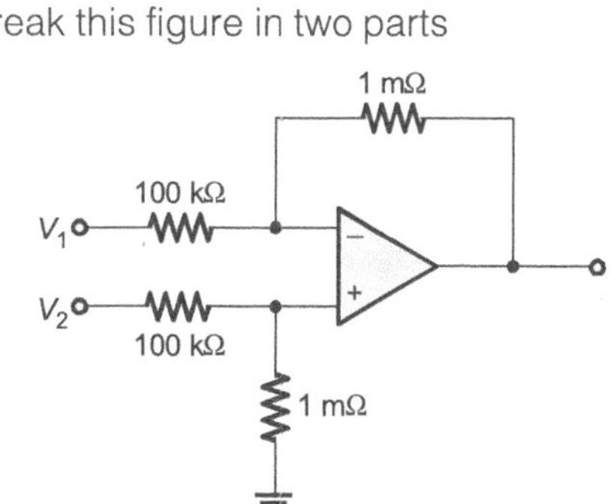

Fig. (b)

From figure (a) by voltage division rule

$$V_1 = \frac{95}{200} \times 10 = 4.75 \text{ V}$$

and $$V_2 = \frac{105}{200} \times 10 = 5.25 \text{ V}$$

Due to symmetry,

$$V_0 = A_v(V_2 - V_1)$$

$$A_v = -\frac{R_f}{R_1} = -\frac{10^6}{10^5} = -10$$

$$\therefore \quad V_0 = -10(5.25 - 4.75) = -5 \text{ V}$$

3.40 (b)

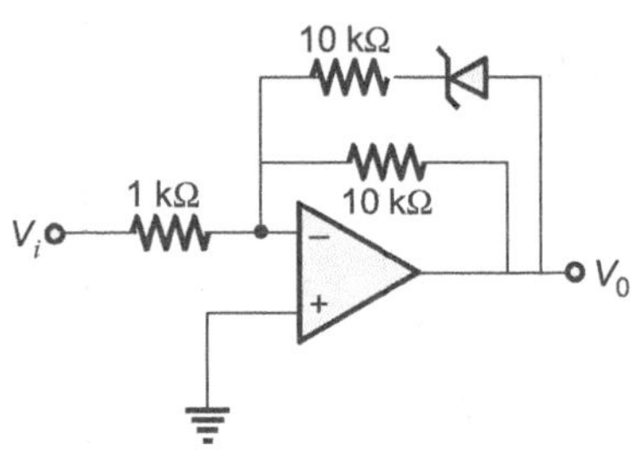

$$V_z = 3.2 \text{ V}$$

Now first open the Zener diode then circuit will be

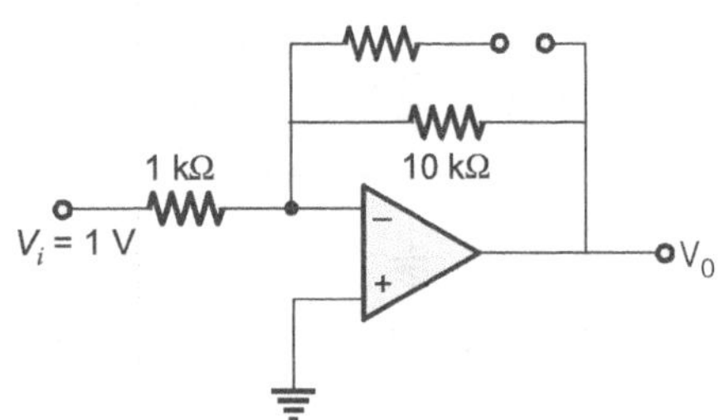

$$V_0 = -\frac{R_f}{R_1} V_i = -10 \text{ V}$$

$$V_{OC} = 10 \text{ V} > V_z$$

So Zener will be on and circuit will be as

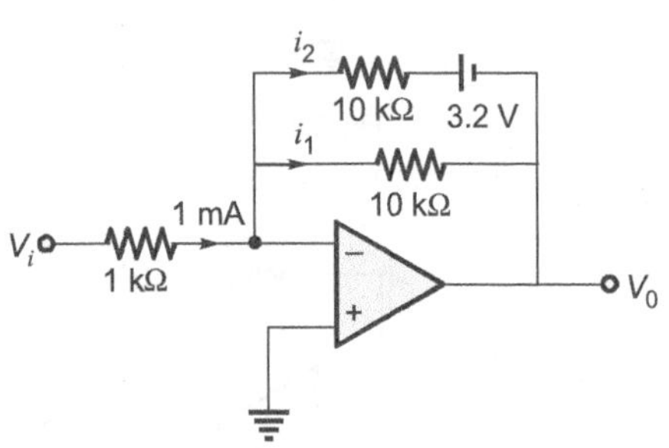

V_0 will lie between –10 and –3.2 V.

Because Zener will be on then current 1 mA will be divided in two parts and output will be

$$-10 \text{ V} < V_0 < -3.2 \text{ V}$$

If output V_0 = –5 V

then, i_1 = 0.5 mA

i_2 = 0.18 mA

so, $i_1 + i_2 \neq i$

(hence, this is not possible)

When, V_0 = –6.6 V

then, i_1 = 0.66 mA

$$i_2 = \frac{6.6 - 3.2}{10} = 0.34 \text{ mA}$$

here, $i_1 + i_2 = i$,

3.41 (a)

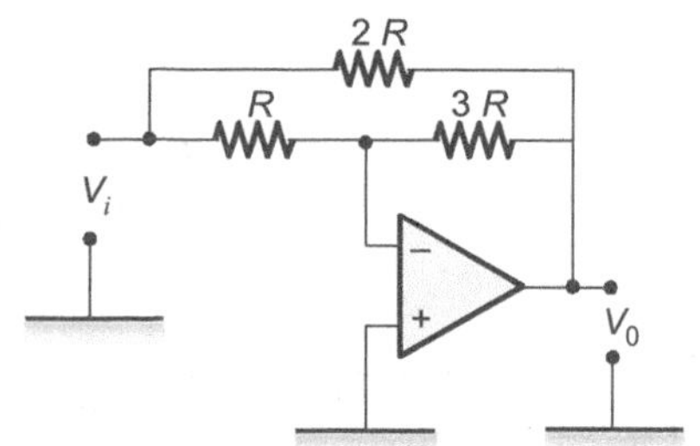

Now convert ∇ network to Y network and circuit will become as given below.

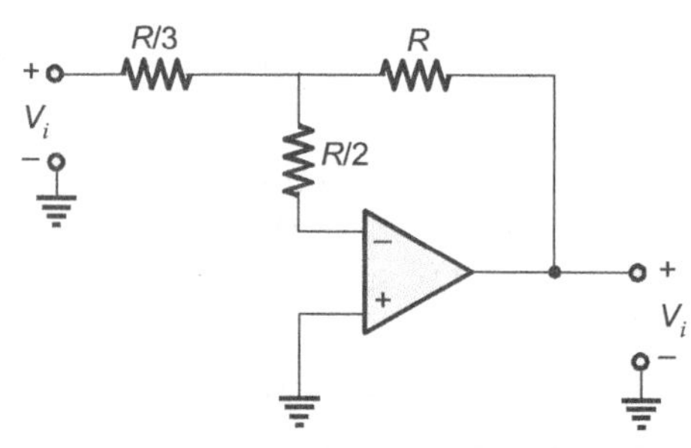

Because no current will flow through $R/2$ resister, so the input resistance will be $\boxed{R_{in} = \frac{R}{3}\,\Omega}$.

3.42 (a)

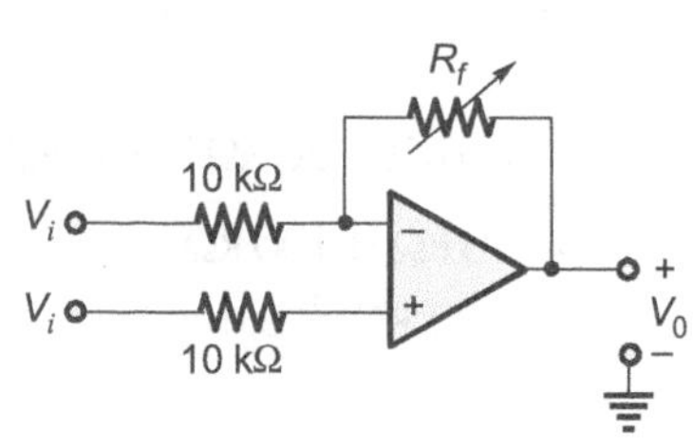

$$\text{Gain} = \frac{V_0}{V_i} = \left(1 + \frac{R_f}{R_1}\right) + \left(-\frac{R_f}{R_1}\right)$$

$$= 1 + \frac{R_f}{R_1} - \frac{R_f}{R_1} = +1$$

3.43 (c)

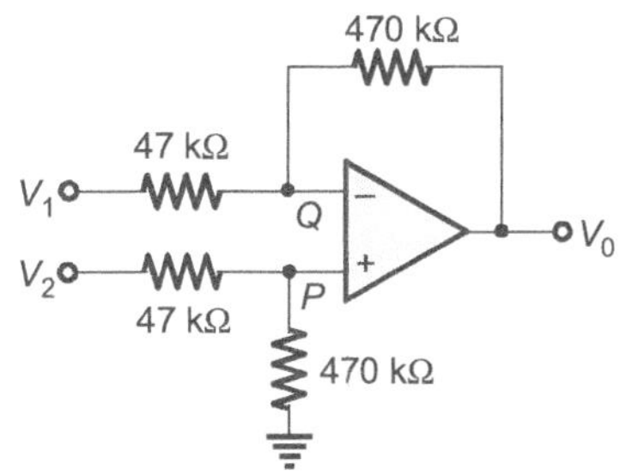

Input impedance seen looking into the terminals V_1 and V_2 with respect to ground.

Apply superposition theorem put $V_2 = 0$, So voltage at $p = 0$.

By virtual ground concept

voltage at $Q = 0$

So, input impedance seen from V_1 with respect to ground is $R_1 = 47\ k\Omega$.

Now, put $V_1 = 0$

The input impedance from V_2 with respect to ground $= R_1 + R_2$

$= (47 + 470)\ k\Omega = 517\ k\Omega$

3.44 (b)

Now the effective circuit is

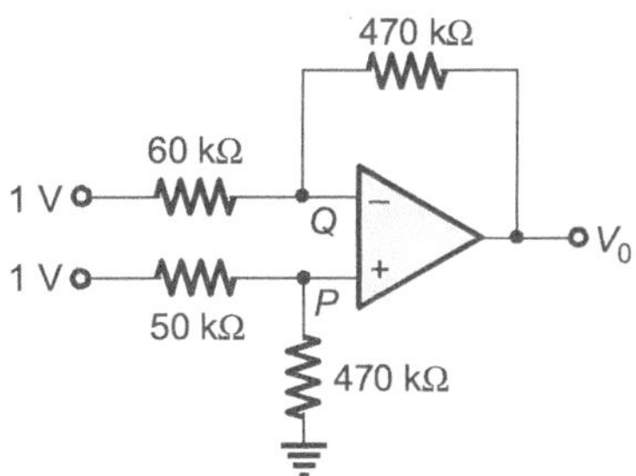

$$V_0 = 1\left[\frac{470}{470+50}\right]\left[1+\frac{470}{60}\right] - 1\left[\frac{470}{60}\right] = 0.15\ V$$

3.45 (c)

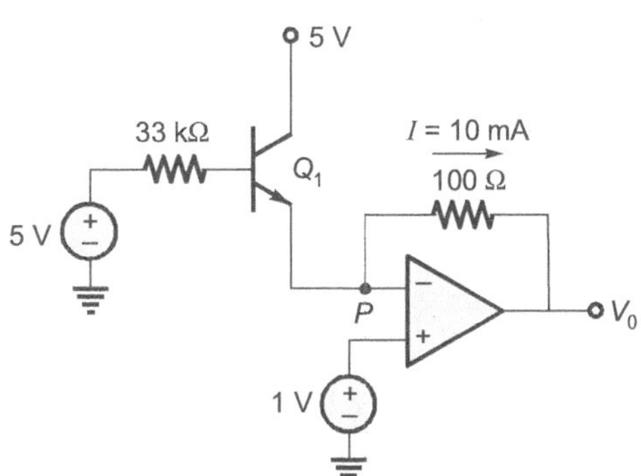

Due to virtual concept the voltage at point $P = V_P = 1\ V$.

Now, apply KVL to the base-emitter circuit of transistor Q_1

$5 - 33 \times 10^3 \times I_b - 0.7\ V - 1 = 0$

$$\therefore \quad I_b = \frac{3.3}{33}\ mA = 0.1\ mA$$

$$I_E = (1 + \beta)\, I_0$$

$$= (1 + 99) \times 0.1\ mA = 10\ mA$$

$$\therefore \quad V_0 = V_P - 10^{-2} \times 100$$

$$= 1 - 1 = 0\ V$$

3.46 (c)

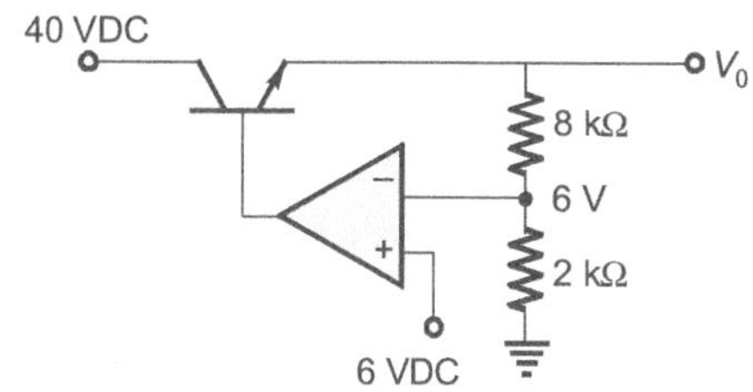

Due to virtual ground concept

$$V_+ = V_- = 6\ V$$

$$\frac{2}{10} \times V_0 = 6$$

$$\therefore \quad V_0 = 30\ V$$

3.47 (a)

$$V_0 = -\frac{470}{22} \times (0.25 \sin\omega t)$$

for no distortion in the output

$$\frac{dV_0}{dt} \le \text{slew rate}$$

$$-\frac{470}{22} \times 0.25 \frac{d}{dt}(\sin\omega t) \le \frac{0.8}{10^{-6}}\ V/\text{sec}$$

$$\therefore \quad -\frac{470}{22} \times 0.25\ \omega \cos\omega t \le \frac{0.8}{10^{-6}}$$

$$\omega \le \frac{0.8}{10^{-6}} \times \frac{22}{470} \times \frac{1}{0.25} \qquad [\because |\cos\omega t| = 1]$$

$$2\pi f \le 0.1497 \times 10^6$$

$$f \le \frac{0.1497}{2\pi} \times 10^6\ Hz$$

$$f \le 0.02385 \times 10^6\ Hz$$

$$\therefore \quad f \le 23.85\ kHz$$

Thus, the maximum frequency of input for distortionless output is 23.85 kHz.

3.48 (b)

Due to virtual short,

$$V_c = V_B = 0\ V$$

(collector voltage of transistor Q)

and given that,

Base voltage of transistor Q

$$V_B = 0\text{ V}$$

So, $V_C = 0\text{ V}$

It mean collector to base of transistor Q is short circuited.

If any junction of transistor is short circuited then the junction acts as reverse bias. So, the C-B junction is reverse bias.

The given op-amp is an inverting configuration which have positive voltage as input. So the output voltage of op-amp will be negative voltage.

$$V_{out} = \text{Negative voltage}$$

Emitter voltage of transistor

$$V_E = \text{Negative voltage}$$

given transistor is n-p-n transistor and

$$V_B = 0\text{ V}$$

$$V_E = \text{Negative voltage}$$

So, the E-B junction will be forward bias. Thus, the transistor is in active region and will behave as closed switch.

So, $V_{BE} = 0.7\text{ V}$ (for silicon transistor)

$$V_B - V_E = 0.7\text{ V}$$

$$V_E = V_B - 0.7$$

$$V_E = 0 - 0.7$$

$$V_E = -0.7\text{ V}$$

3.49 (c)

Output of first op-amp

$$V_{out1} = -2\left[-\frac{1}{1}\right] + \left[1+\frac{1}{1}\right]$$

$$= 2 + 1(2) = 4\text{ V}$$

$$V_{out} = 4\left(1+\frac{1}{1}\right) = 4(2) = 8\text{ V}$$

3.50 (d)

$$I_B = \frac{1}{2}(I_{b_1} + I_{b_2})$$

$$= \frac{1}{2}(450 + 350) = 400\text{ nA}$$

$$I_f = I_{b_1} - I_{b_2} = (450 - 350)\text{ nA} = 100\text{ nA}$$

3.51 (b)

Gain of inverting amplifier,

$$A_v = -\frac{R_f}{R_i} = -10$$

$\therefore$ $R_f = 10\,R_i$

$\therefore$ $R_f = 500\text{ k}\Omega,$

$R_i = 50\text{ k}\Omega$

3.52 (0.6)

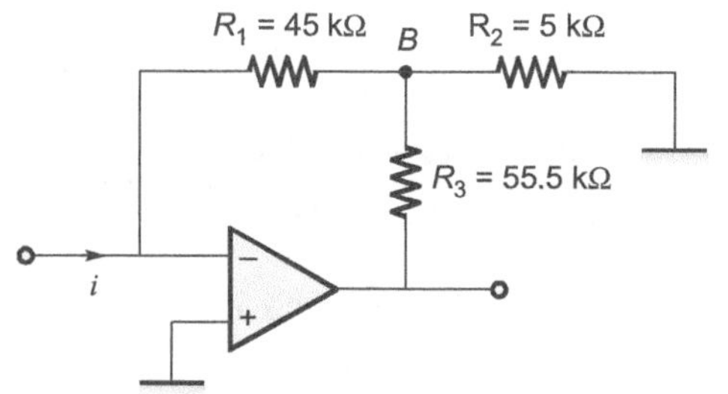

$$V_A = 0 \text{ (By virtual ground)}$$

$$\frac{V_B - 0}{45} + \frac{V_B - V_o}{55.5} + \frac{V_B - 0}{5} = 0$$

$$i = \frac{-(V_B - 0)}{45}$$

$\therefore$ $V_B = -45\,i$

$$\frac{-45i}{45} + \frac{(-45i) - V_o}{55.5} + \frac{(-45i)}{5} = 0$$

$$-i - \frac{45i + V_o}{55.5} - 9i = 0$$

$$10i = -\frac{45i + V_o}{55.5}$$

$$600i = V_o$$

$$\frac{V_o}{i} = 600\text{ k}\Omega = 0.6\text{ M}\Omega$$

3.53 (-1)

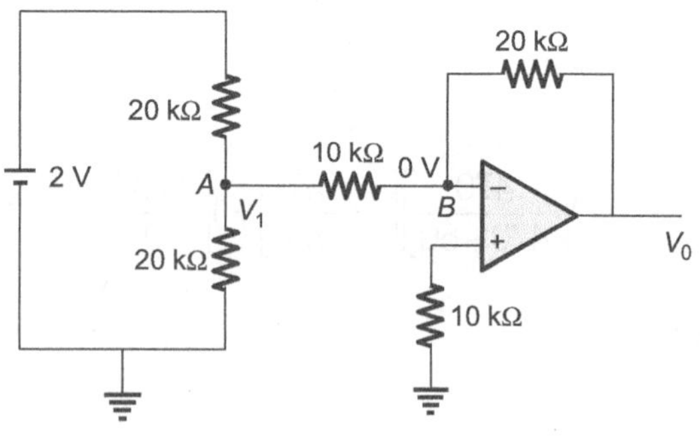

KCL at node A:

$$\frac{V_1 - 2}{20\text{ k}\Omega} + \frac{V_1}{20\text{ k}\Omega} + \frac{V_1 - 0}{10} = 0$$

$\Rightarrow$ $V_1 = 0.5$

Apply KCL at node B

$$\frac{V_1 - 0}{10\text{ k}\Omega} = \frac{0 - V_0}{20\text{ k}\Omega}$$

$$V_0 = -\frac{20\text{k}\Omega}{10\text{k}\Omega}V_1 = -1\text{ Volt}$$

3.54 1.6 (1.55 to 1.65)

During positive half cycle

when $v_i > 0.6$ V

$\Rightarrow$ D_1-ON and D_2-OFF

Then equivalent circuit

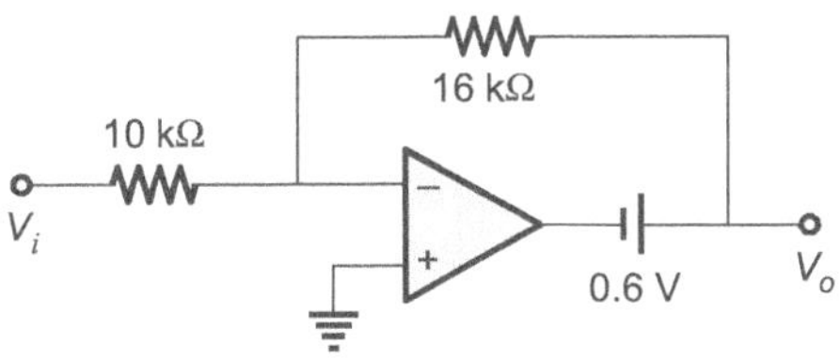

$$V_0 = -v_i\left[\frac{16}{10}\right] = -1.6\,v_i$$

$$= -1.6 \sin(3000\,t)$$

For negative half cycle when $v_i < -0.6$ V

$\Rightarrow$ D_1-OFF & D_2-ON

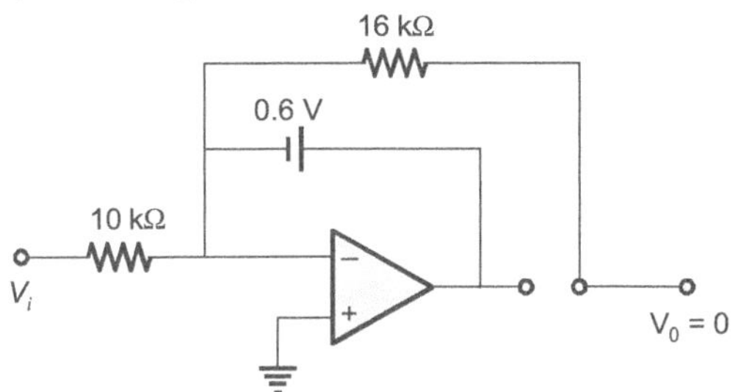

Then magnitude of the negative peak value of the output V_0 is 1.6 V.

3.55 (20)

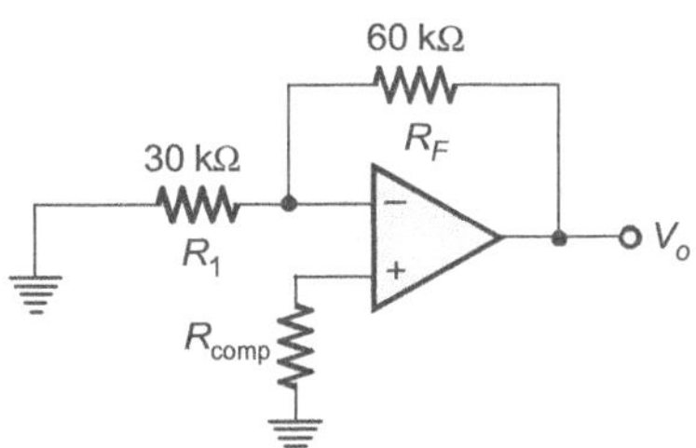

$$R_{comp} = \frac{R_F \times R_1}{R_F + R_1} = \frac{30 \times 60}{30 + 60} = 20\text{ k}\Omega$$

3.56 40.83 (40 to 41)

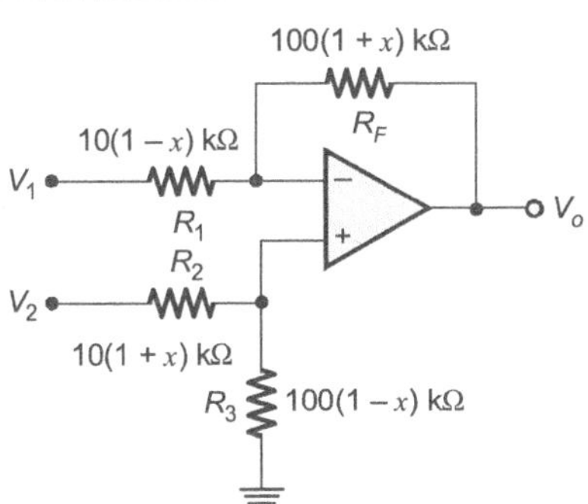

Here, $R_F = 10\,R_2$

$R_3 = 10\,R_1$

$$\text{CMRR} = -\frac{1}{2}\left[\frac{(R_1+R_F)R_3 + R_F(R_2+R_3)}{(R_1+R_F)R_3 - R_F(R_2+R_3)}\right]$$

$$= -\frac{1}{2}\left[\frac{(R_1+10R_2)10R_1 + 10R_2(R_2+10R_1)}{(R_1+10R_2)10R_1 - 10R_2(R_2+10R_1)}\right]$$

$$= -\frac{1}{2}\left[\frac{10R_1^2 + 100R_1R_2 + 10R_2^2 + 100R_1R_2}{10R_1^2 + 100R_1R_2 - 10R_2^2 - 100R_1R_2}\right]$$

$$= -\frac{1}{2}\left[\frac{R_1^2 + R_2^2 + 20R_1R_2}{R_1^2 - R_2^2}\right]$$

$$= -\frac{1}{2}\left[\frac{1(1+x^2) + 20(1-x^2)}{-4x}\right] = 110$$

CMRR (DB) = 20 $\log_{10}$ 110 = 40.83 dB

3.57 (100)

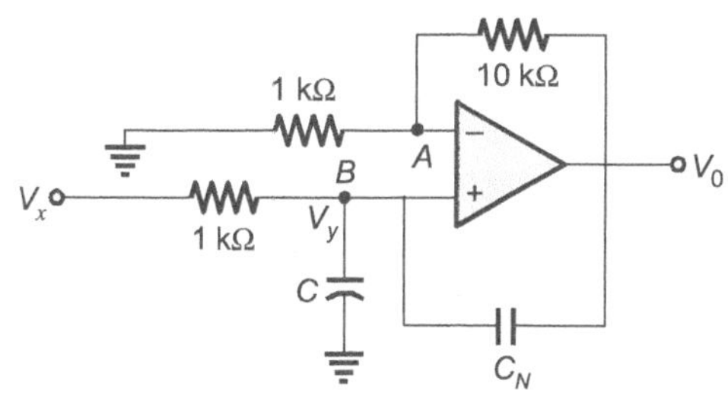

Apply KCL at node A

$$\frac{0 - V_a}{1\text{k}\Omega} = \frac{V_a - V_0}{10\text{ k}\Omega}$$

$$\Rightarrow \quad V_a = \frac{V_0}{11}$$

Using virtual ground concept

$$V_A = V_B = V_y = \frac{V_0}{11}$$

Apply KCL at Node B

$$\frac{V_y - V_x}{1\text{k}\Omega} + \frac{V_y}{1/CS} + \frac{V_y - V_0}{1/C_NS} = 0$$

$\because$ $V_x = V_y$ [Given]

$$V_y(CS) + (V_y - V_0)(C_NS) = 0$$

$$\frac{V_0}{11}(CS) + \left(\frac{-10}{11}\right)V_0C_NS = 0$$

$$C_N = 0.1\,\text{C}$$

$$= 0.1\text{ nF} = 100\text{ pF}$$

3.58 (100)

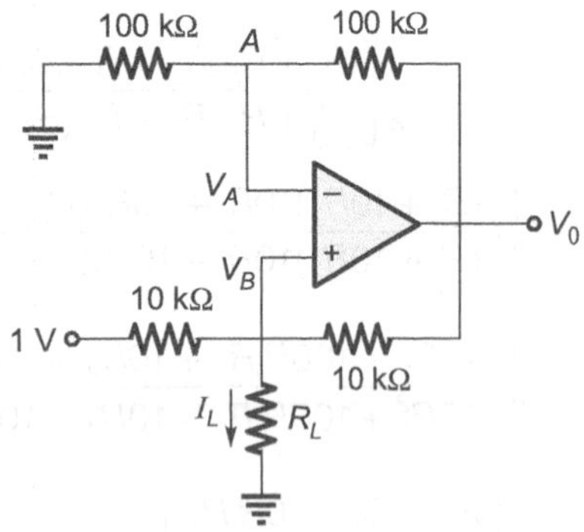

Apply KCL at Node A

$$\frac{0 - V_a}{100\text{ k}\Omega} = \frac{V_a - V_0}{100\text{ k}\Omega}$$

$$\Rightarrow \quad V_a = \frac{V_0}{2}$$

Using virtual ground concept

$$V_a = V_B = \frac{V_0}{2}$$

Apply KCL at Node B

$$\frac{V_B - 1}{10\text{ k}\Omega} + \frac{V_B - V_0}{10\text{ k}\Omega} + I_L = 0$$

$$\Rightarrow \quad I_L = 0.1\text{ mA} = 100\,\mu\text{A}$$

3.59 (b)

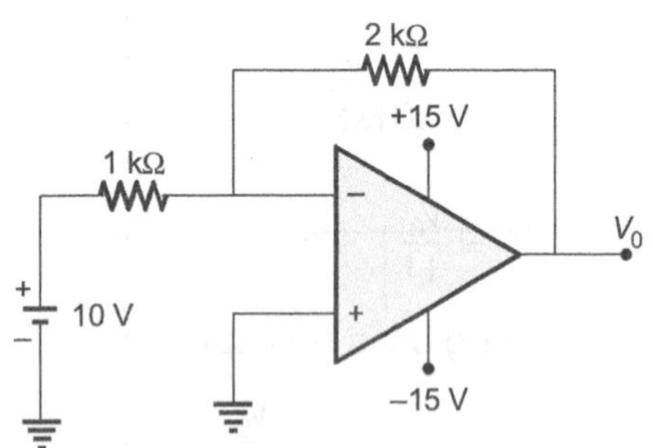

$$V_0 = -\frac{R_F}{R_1}V_i = -\frac{2\text{ k}\Omega}{1\text{k}\Omega}\times 10 = -20\text{ V}$$

Not possible because V_0 is saturated,

$$\therefore \quad V_0 = -V_{sat} = -15\text{ V}$$

3.60 1.5 (1.4 to 1.6)

$$V_0 = A_d(V_1 - V_2) + A_C\left(\frac{V_1 + V_2}{2}\right)$$

$$= 100(5.01 - 5) + 0.1\left(\frac{5.01 + 5}{2}\right) = 1.5005\text{ V}$$

3.61 (−5)

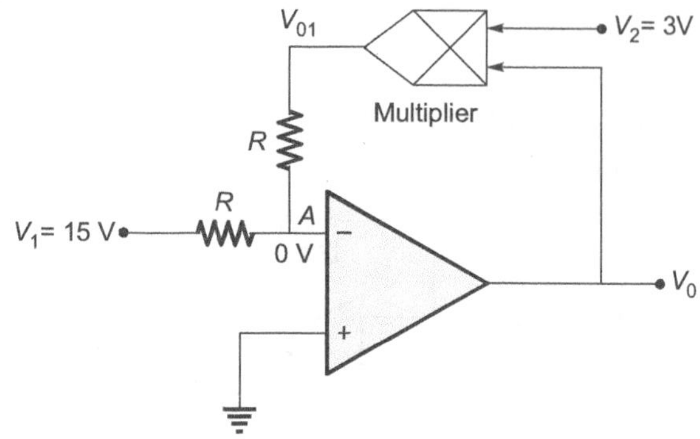

Scaling factor $\Rightarrow V_R = 1$

KCL at Node A:

$$\frac{0 - V}{R} + \frac{0 - V_{01}}{R} = 0$$

$$\Rightarrow \quad V_{01} = -V_1 = -15\text{ V}$$

Also, $$V_{01} = \frac{V_0 \times V_1}{V_R} = \frac{V_0 \times 3}{1}$$

$$\Rightarrow \quad V_0 = -\frac{V_{01}}{3} = -\frac{15}{3} = -5\text{ V}$$

3.62 (0)

The overall feedback to the op-amp is positive and op-amp output goes to $-V_{sat}$.

Hence, BJT does not conduct.

So, $I_B = I_C = I_E = 0$

$I_o = 0$

3.63 0.99 (0.98 to 1.00)

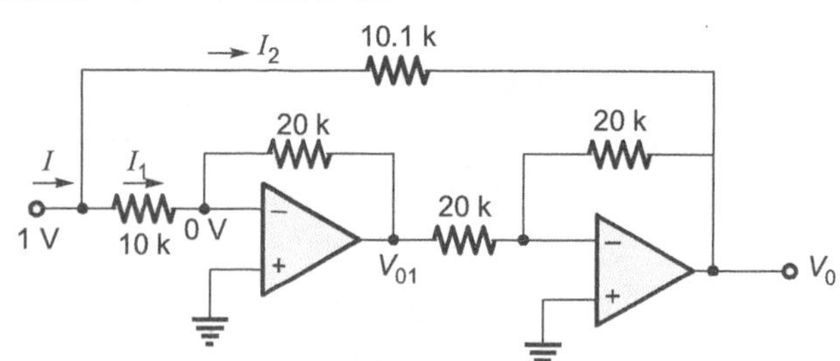

$$V_{01} = \frac{-20\text{k}}{10\text{k}}\times 1\text{V} = -2\text{V}$$

$$V_0 = \frac{-20\text{k}}{20\text{k}}\times V_{01} = 2\text{V}$$

$$I_1 = \frac{1-0}{10\text{k}} = 0.1\text{mA}$$

$$I_2 = \frac{1-2}{10.1\text{k}} = -0.099\text{ mA}$$

$$I = I_1 + I_2 = 0.1\text{ mA} - 0.099\text{ mA} = 0.99\,\mu\text{A}$$

3.64 **17.69 (17.5 to 17.9)**

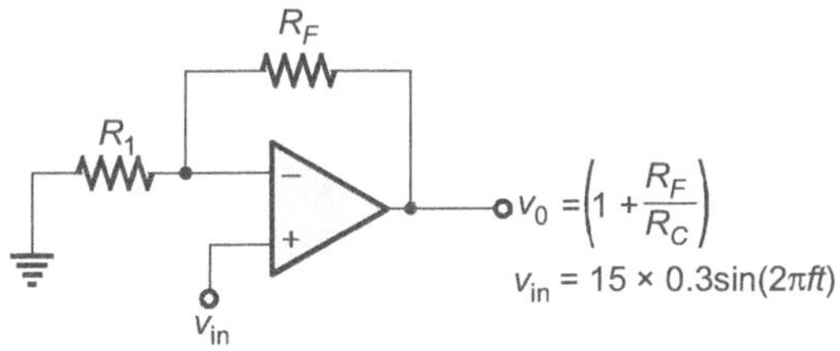

$$\text{Slew rate,}\quad SR = \left|\frac{\partial V_0}{\partial t}\right|_{max} = A \cdot V_m \cdot 2\pi f$$

$$0.5 \times 10^6 = 15 \times 0.3 \times 2\pi f$$

$$\Rightarrow \quad f_{max} = \frac{0.5 \times 10^6}{9\pi} = 17.69 \text{ kHz}$$

3.65 **(c)**

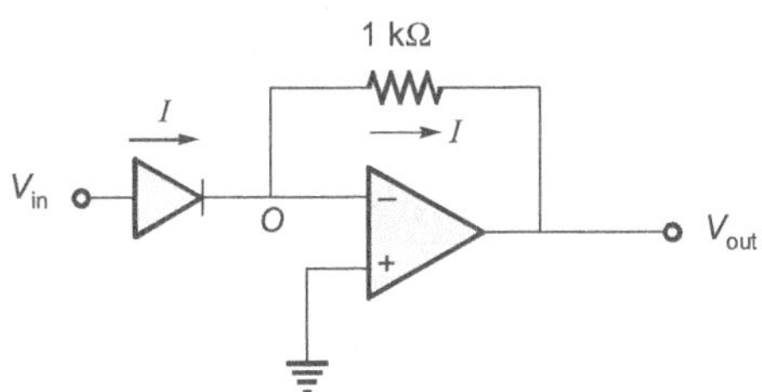

$$V^+ = V^- = 0 \text{ (Virtual ground)}$$

$$I = I_0 e^{V_D/V_T} = I_0 e^{V_{in}/V_T}$$

$$0 - V_{out} = I \times 1 \text{ k}\Omega$$

$$V_{out} = -I \times 1 \text{ k}\Omega$$

$$V_{out} = -I \times 1 \text{ k}\Omega \; e^{V_{in}/V_T}$$

$$V_{out} \propto e^{V_{in}/V_T}$$

This circuit is an exponential amplifier.

3.66 **(9) 9.0 to 9.1**

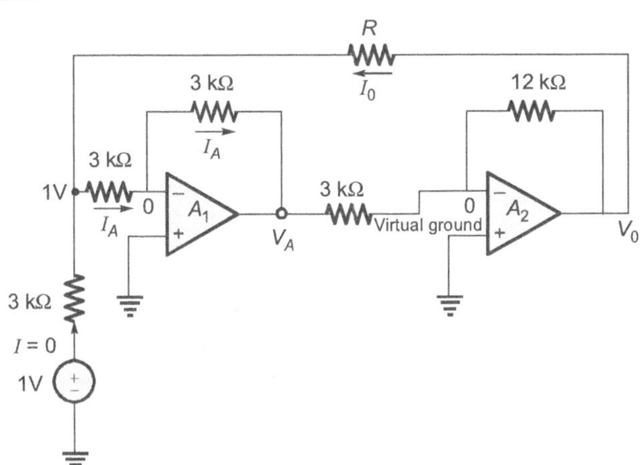

Opamp A_1

KCLat V^-

$$\frac{1V - 0}{3k} = \frac{0 - V_A}{3k}$$

$$V_A = -1 \text{ V}$$

Opamp A_2

KCL at V^-

$$\frac{V_A - 0}{3k} = \frac{0 - V_0}{12k}$$

$$-\frac{1}{3k} = -\frac{V_0}{12k}$$

$$V_0 = 4 \text{ V}$$

KCL at 1 V

$$I_0 + I = I_A$$

$$I_0 = I_A = \frac{1V - 0}{3k} = \frac{1}{3} \text{mA}$$

$$R = \frac{V_0 - 1V}{I_0} = \frac{4V - 1V}{\frac{1}{3}\text{mA}} = \frac{3}{\frac{1}{3}\text{mA}} = 9 \text{ k}\Omega$$

$$R = 9 \text{ k}\Omega$$

3.67 **(–4)**

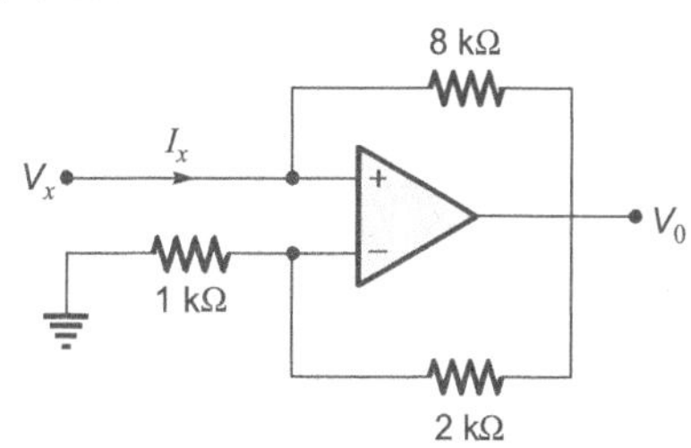

$$V^+ = V^-$$

$$V_x = \frac{V_0 \times 1}{1 + 2}$$

$$\Rightarrow \quad V_0 = 3V_x \quad \text{...(1)}$$

$$I_x = \frac{V_x - V_0}{8} = \frac{V_x - 3V_x}{8} = \frac{-V_x}{4}$$

$$\Rightarrow \quad \frac{V_x}{I_x} = -4 \text{ k}\Omega$$

3.68 **(7)**

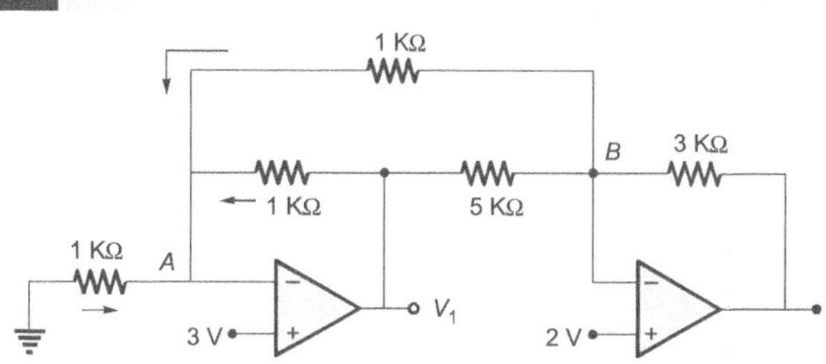

$$V_A = 3 \text{ V}; \; V_B = 2 \text{ V}$$

KCL at Node A:

$$\frac{0 - 3}{1} + \frac{V_1 - 3}{1} + \frac{2 - 3}{1} = 0$$

$$\Rightarrow \quad V_1 = 7 \text{ V}$$

3.69 (0.50)

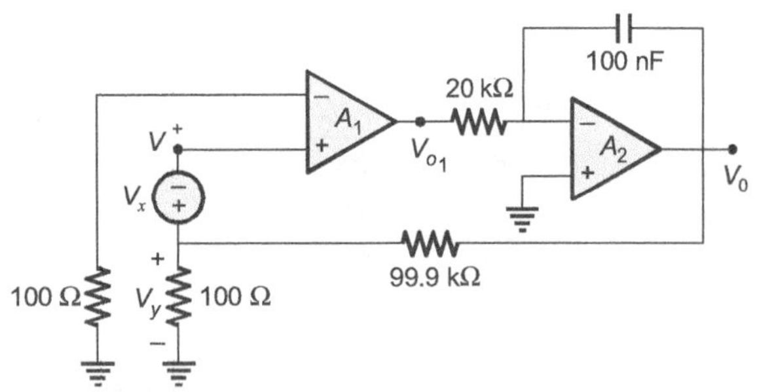

Given: $V_x = 0.5$ mV

For op-amp A_1:

$$V^- = 0$$

$$V^+ = -V_x + V_y = -0.5\times10^{-3} + \frac{V_o \times 0.1}{0.1+99.1}$$

$$= (-0.5 + V_o) \times 10^{-3} \text{ V}$$

Let $V_o = 0$ at $t = 0 \Rightarrow V^+ = -0.5$ mV

$\Rightarrow$ Output of A_1 should be $-V_{sat}$

Op-amp A_2 is integrator

$$V_o = -\frac{1}{RC}\int_t V_{o_1}\,dt = \frac{-1}{20\times10^3\times10^{-7}}\int_t -V_{sat}\,dt$$

$$V_o = 500 \times V_{sat} \times t$$

If $t \uparrow$ then $V_o \uparrow \Rightarrow V_y \uparrow \Rightarrow V^+ \downarrow$

If V_y becomes +0.5 mV, then

$$V_y = \frac{V_o}{1000}$$

$$\Rightarrow \quad V_o = 1000\ V_y = 0.5 \text{ V}$$

If V_y becomes slightly greater than 0.5 mV then V^+ becomes slightly positive.

Hence V_{o_1} becomes $+V_{sat}$.

$$\Rightarrow \quad V_o = -500 \times V_{sat} \times t$$

i.e. V_o should becomes a negative ramp.

As $V_o \downarrow \Rightarrow V_y \downarrow \Rightarrow V^+ \downarrow$

If V_y again becomes 0.5 mV or slightly less then V^+ becomes negative.

$\Rightarrow V_{o_1}$ becomes $-V_{sat}$ and V_o comes back to 0.5 V.

Thus output V_o is forced to remain constant at 0.5 V.

3.70 (14.86)

Apply Miller's theorem to $C = 2$ μF

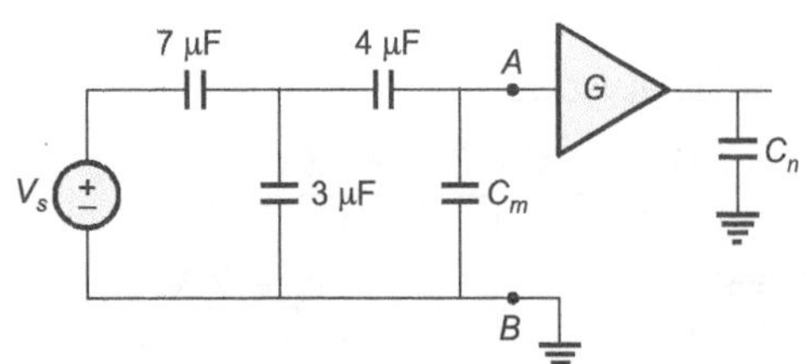

$$C_m = C(1 + A_V) = 2 \times (1 + 5) = 12 \text{ μF}$$

Disable V_S

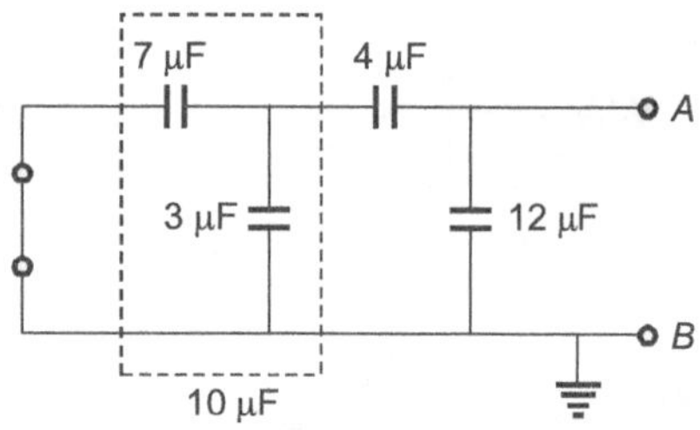

$$C_{AB} = \frac{10\times4}{10+4} + 12 = 14.86 \text{ μF}$$

3.71 (b)

Given, $V_o = 5$ V

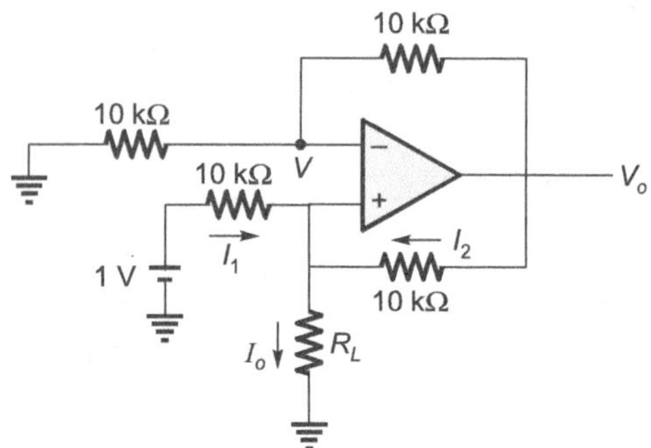

$$V = \frac{V_o\times10}{10+10} = \frac{V_o}{2} = 2.5V$$

Using virtual short property,

$$V^+ = V^- = 2.5 \text{ V}$$

by KCL, $I_1 + I_2 = I_o$

$$\frac{1-2.5}{10} + \frac{5-2.5}{10} = \frac{2.5}{R_L}$$

$$\Rightarrow \quad R_L = 25 \text{ kΩ}$$

3.72 (1)

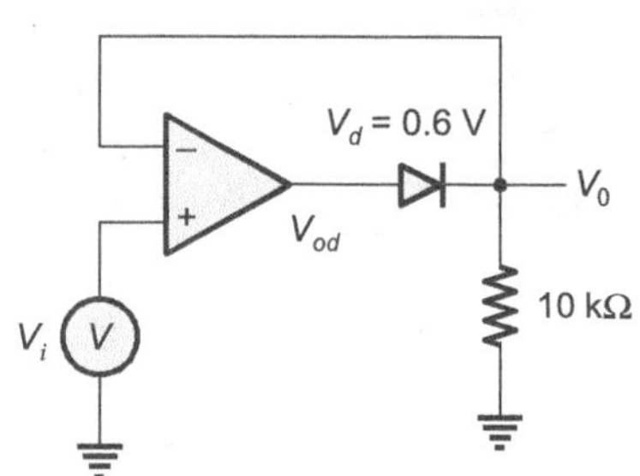

Given circuit is called precision diode.

$V_i = 0.4 \sin \omega t$

Let diode be OFF.

Then $V^- = V_0 = 0$ and $V^+ = V_i$;

If $V^+ < V^-$ i.e., $V_i < 0$ then $V_{od} = -V_{sat}$ and diode remains OFF $\Rightarrow V_0 = 0$

If $V^+ > V^-$ i.e., $V_i > 0$ then V_{od} becomes positive
Now diode becomes ON and Op-Amp will be closed-loop.

$$\Rightarrow \quad V^- = V^+ \quad \Rightarrow \quad V_0 = V_i$$

If V_i has peak value i.e. $V_i = 0.4$ V

$$\Rightarrow \quad V_0 = 0.4 \text{ V}$$

Now $V_{od} - V_0 = V_d$

$$\Rightarrow \quad V_{od} = V_d + V_0 = 0.6 + 0.4$$

$$\Rightarrow \quad V_{od} = 1 \text{ V}$$

3.73 (1.1)

Using virtual s/c property,

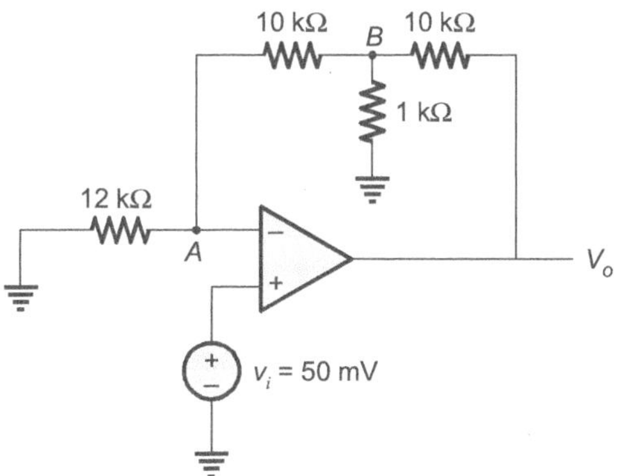

$$V^- = V^+$$

$$\Rightarrow \quad V_A = V_i = 50 \text{ mV}$$

by KCL at A: $\dfrac{0-V_A}{12} + \dfrac{V_B - V_A}{10} = 0$

$$\Rightarrow \quad V_B = \frac{11V_A}{6} = \frac{11V_i}{6} \quad \ldots \text{(i)}$$

by KCL at B:

$$\frac{V_A - V_B}{10} + \frac{0 - V_B}{1} + \frac{V_o - V_B}{10} = 0$$

$$\Rightarrow \quad V_A - V_B - 10V_B + V_0 - V_B = 0$$

$$V_0 = 12V_B - V_A = 12 \times \frac{11V_i}{6} - V_i$$

$$= 21V_i = 21 \times 50 \text{ mV}$$

$$= 1050 \text{ mV or } 1.05 \text{ V}$$

Answer upto 1 decimal place = 1.1 V

3.74 (10.0)

Output impedance of unity-gain feedback

$$R_{of} = \frac{R_0}{1 + A_{OL}\beta} = \frac{100}{1 + 10^4 \times 1} = \frac{100}{10^4} = 10 \text{ m}\Omega$$

3.75 (15.84)

As we know that,

$$\frac{V_o}{V_i} = -\frac{Z_2}{Z_1}$$

$$Z_1 = 10 \text{ k}\Omega$$

$$Z_2 = R \| \frac{1}{j\omega C} = \frac{R}{1 + j\omega CR}, \quad \text{where, } R = 100 \text{ k}\Omega, \; C = 1\text{nF}$$

$$\left|\frac{V_o}{V_i}\right| = 0 \text{ dB}$$

$$\Rightarrow \quad \left|\frac{V_o}{V_i}\right| = 1$$

$$\Rightarrow \quad \left|\frac{Z_2}{Z_1}\right| = 1$$

$$\Rightarrow \quad \left|\frac{R}{(1 + j\omega CR) \times 10}\right| = 1$$

$$\Rightarrow \quad 10 = \sqrt{1 + \omega^2 C^2 K^2}$$

$$\Rightarrow \quad \omega CR = \sqrt{99}$$

$$\Rightarrow \quad f = \frac{\sqrt{99}}{2\pi CR} = \frac{\sqrt{99}}{2\pi \times 100 \text{ k}\Omega \times 1\text{n}}$$

$$= 15.84 \text{ kHz}$$

3.76 (0.367)

For $t < 0$, capacitor open circuited :

$$V_c(0) = \frac{5}{5+1} \times 5 = \frac{25}{6} \text{V}$$

At $t = 0$; switch opened.

Capacitor starts discharging through 5 kΩ resistor.

$$V_c(t) = V_c(0)e^{-t/\tau} \text{ V} \qquad (t > 0)$$

$$V_c(t) = \frac{25}{6}e^{-t/\tau}$$

Let at $t = t_1$; comparator switches to $-V_{sat}$ (–5 V) and at that instant,

$$V_c(t) = 2 \text{ V}$$

$$2 = \frac{25}{6}e^{-t/\tau}$$

On solving,

$$t = \tau \ln\left[\frac{25}{12}\right] = 0.1 \times 5 \ln\left[\frac{25}{12}\right] \text{ msec} = 0.367 \text{ msec}$$

3.77 (1110)

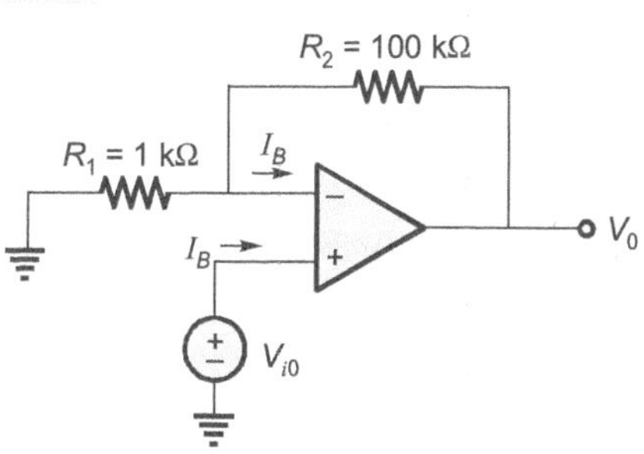

KCL at inverting terminal:

$$\frac{0 - V_{i0}}{R_1} + \frac{V_0 - V_{i0}}{R_2} = I_B$$

$$\frac{V_0}{R_2} = I_B + \left(\frac{1}{R_1} + \frac{1}{R_2}\right)V_{i0}$$

$$V_0 = I_B R_2 + \left(1 + \frac{R_2}{R_1}\right)V_{i0}$$

$$V_0 = 10^{-9} \times 100 \times 10^3 + \left(1 + \frac{100}{1}\right) \times 10 \times 10^{-6}$$

$$V_0 = 1110\ \mu\text{V}$$

3.78 (0.5)

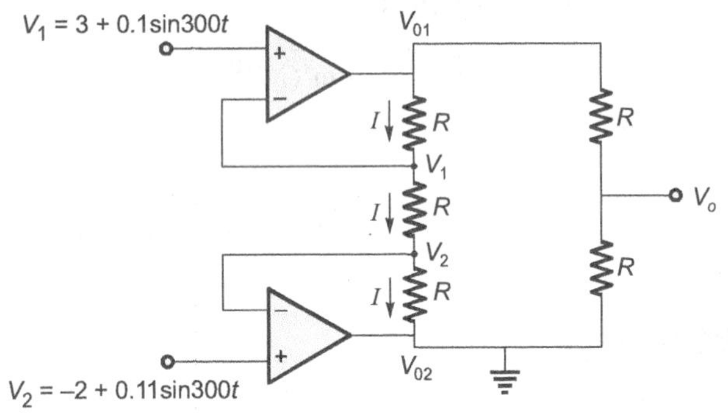

$$V_{o_1} = IR + V_1$$

$$= \frac{(V_1 - V_2)}{R}R + V_1 = 2V_1 - V_2 \quad \text{...(i)}$$

$$V_{o_2} = V_2 - IR = V_2 - \frac{(V_1 - V_2)}{R} \times R$$

$$= 2V_2 - V_1 \quad \text{...(ii)}$$

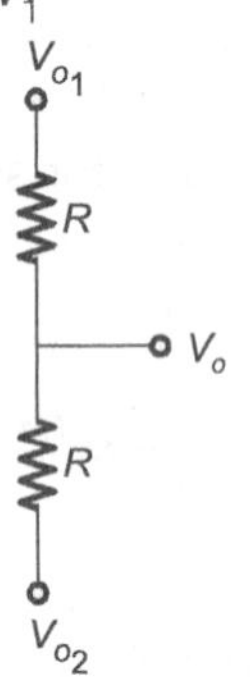

$$V_o = \frac{V_{o_1} \times R}{R + R} + \frac{V_{o_2} \times R}{R + R} = \frac{V_{o_1} + V_{o_2}}{2}$$

$$= \frac{2V_1 - V_2 + 2V_2 - V_1}{2} = \frac{V_1 + V_2}{2}$$

$$V_o = \frac{V_1 + V_2}{2}$$

$$= \frac{3 + 0.10\sin 300t - 2 + 0.11\sin 300t}{2}$$

$$= \frac{1 + 0.21\sin 300t}{2} = \frac{1}{2} + \frac{0.21}{2}\sin 300t$$

$$= 0.5 + 0.105\sin 300t$$

$$V_{o(\text{avg})} = 0.5\text{ V}$$

3.79 (100)

For $t \le 0$

Charge stored in C_2 is 1 nC

$$V_{C2} = \frac{Q}{C_2} = \frac{1\text{nC}}{10\text{ nF}} = 0.1\text{V}$$

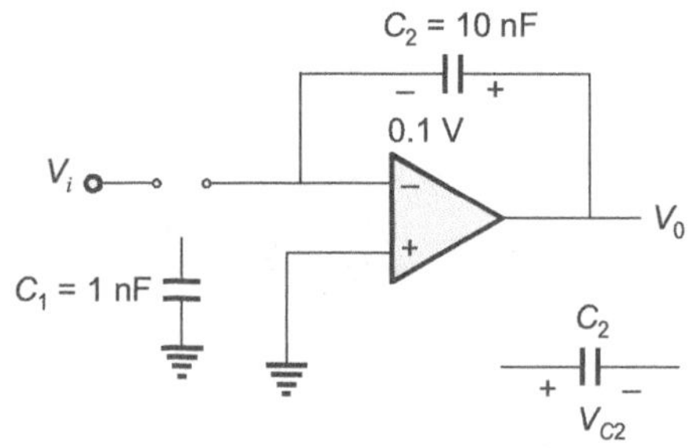

Considering given polarity,

$$V_{C2} = -0.1\text{ V}$$

$$T_0 = \frac{1}{f_0} = 1\text{msec}$$

For $0 < t < 0.5$ msec

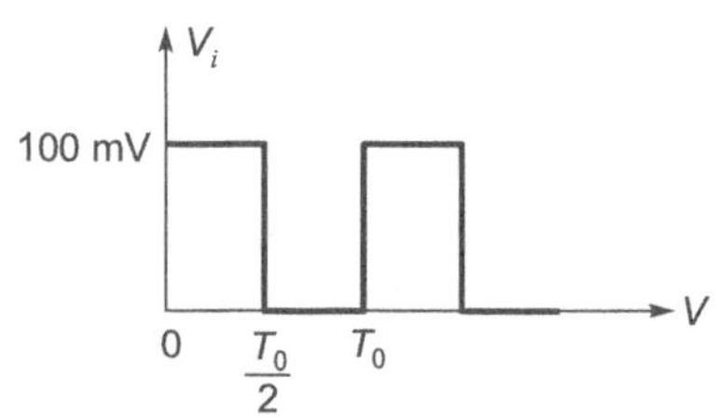

C_1 is connected to V_i

C_1 charges upto 100 mV

For 0.5 msec $< t <$ 1 msec

C_1 is connected to inverting terminal.

Consider V^- at virtual ground.

C_1 should instantaneously discharge to 0.

$$\Delta V_{C1} = 100\text{ mV} - 0 = 100\text{ mV}$$

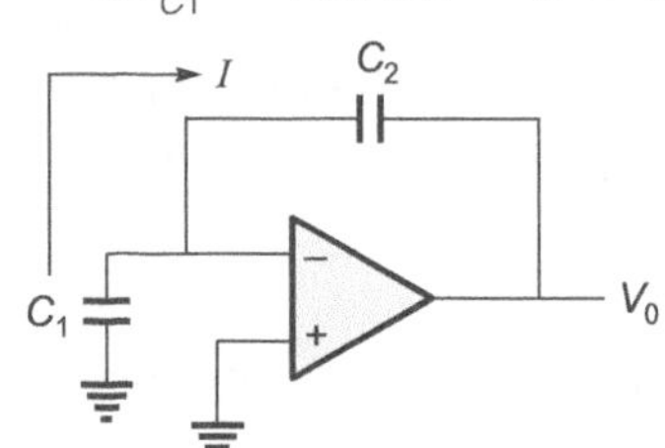

C_1 discharges through C_2

Charge lost by C_1 = Charge supplied to C_2

$$C_1 \times \Delta V_{C1} = C_2 \times \Delta V_{C2}$$

$$\Delta V_{C2} =$$

$$\frac{C_1}{C_2} \times \Delta V_{C1} = \frac{1\text{nF}}{10\text{ nF}} \times 100\text{ mV} = 0.01\text{V}$$

In one cycle of input, voltage across C_2 changes by 0.01 V.

At t = 20 msec : Input has 20 cycles

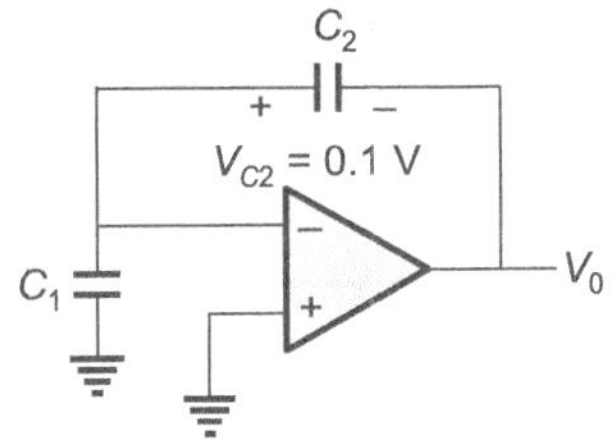

$\Rightarrow$Change in $V_{C2} = 20 \times 0.01 = 0.2$ V

$$V_{C2}|_{t=20\text{ ms}} = V_{C2}|_{t=0} + \Delta V_{C2}$$
$$= -0.1 + 0.2$$
$$= 0.1 \text{ V}$$
$$V^- - V_0 = 0.1 \text{ V}$$
$$0 - V_0 = 0.1 \text{ V}$$
$$\Rightarrow \quad V_0 = -0.1 \text{ V}$$
$$|V_0| = 0.1 \text{ V}$$

3.80 (2)

Assuming op-amp is not saturated.

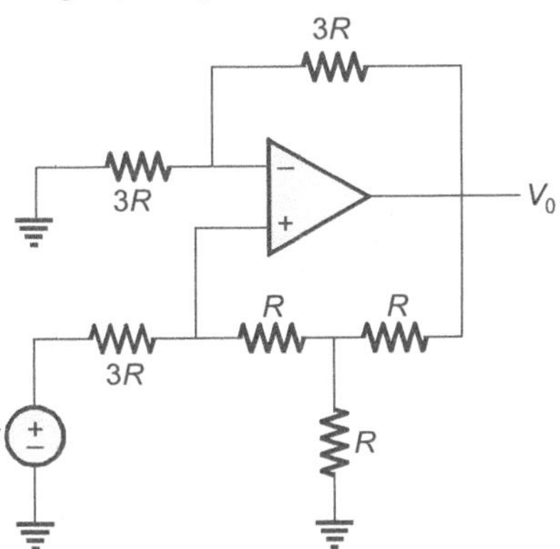

$$\frac{V_A - 1}{4R} + \frac{V_A}{R} + \frac{V_A - V_o}{R} = 0$$
$$\frac{V_A - 1}{4} + V_A + V_A - V_o = 0$$
$$9V_A - 1 - 4V_o = 0$$
$$V_A = \frac{1}{9} + \frac{4}{9}V_o$$

$\therefore$ Current flowing through $3R\ \Omega$.

$$i = \frac{1 - V_A}{4R} = \frac{1}{4R}\left[1 - \left(\frac{1}{9} + \frac{4V_o}{9}\right)\right]$$

$\therefore$ Voltage at non-inverting terminal

$$V_+ = 1 - \frac{1}{4R} \times 3R\left[1 - \left(\frac{1}{9} + \frac{4V_o}{9}\right)\right]$$
$$V_+ = 1 - \frac{3}{4}\left[\frac{8}{9} - \frac{4V_o}{9}\right]$$
$$V_+ = 1 - \frac{2}{3} + \frac{V_o}{3} = \frac{1}{3} + \frac{V_o}{3}$$

Using virtual short concept

$$V_+ = V_- = \frac{1}{3} + \frac{V_o}{3}$$

Applying KCL at inverting node

$$\frac{0 - V_-}{3R} = \frac{V_- - V_o}{3R}$$
$$-V_- = V_- - V_o$$
$$V_o = 2V_-$$
$$V_o = 2\left[\frac{1}{3} + \frac{V_o}{3}\right]$$
$$3V_o = 2 + 2V_o$$
$$V_o = 2 \text{ Volt}$$

3.81 (6.17)

Given circuit is an Astable multivibrator

$$T = 2\tau \ln\left(\frac{1+\beta}{1-\beta}\right)$$

where $\tau = RC = 10^3 \times 100 \times 10^{-9}$
$= 10^{-4}$ sec

$$\beta = \frac{10}{10+16} = \frac{10}{26} = \frac{5}{13}$$

$$T = 2 \times 10^{-4} \ln\left(\frac{1+5/13}{1-5/13}\right) = 2 \times 10^{-4} \ln\left(\frac{18}{8}\right)$$

$$\text{freq. of oscillation } f = \frac{1}{T} = \frac{10^4}{2\ln\left(\frac{18}{8}\right)} \text{Hz}$$

$$f = \frac{5}{\ln\left(\frac{18}{8}\right)} \text{kHz} = 6.1657 \text{ kHz} = 6.17 \text{ kHz}$$

3.82 (10)

$$I_D = I_s e^{\alpha V_D / T}$$

$T \to$ Kelvin; $\alpha = 1.16 \times 10^4$ K/V; $I_S = 10^{-15}$ A

if $T \to$ change

$V_{out} \to$ change = ?

$$V_{out} = \left(1 + \frac{5R}{R}\right)V_D = 6V_D \qquad \text{...(1)}$$

$$I_D = I_S e^{\alpha V_D / V_T} \Rightarrow \frac{I_D}{I_S} = e^{\alpha V_D / T}$$

$$\frac{\alpha V_D}{T} = \ln\left(\frac{I_D}{I_S}\right) = \ln\left(\frac{250 \text{ nA}}{10^{-15} \text{ A}}\right)$$

$$\frac{\alpha V_D}{T} = 19.3369$$

$$V_D = \frac{T}{\alpha} \times 19.3369$$

$$\therefore \quad V_{out} = 6V_D = 6 \times \frac{T}{\alpha} \times 19.3369$$

$$\frac{dV_{out}}{dT} = \frac{6}{\alpha} \times 19.3369 \frac{\text{Volt}}{\text{Kelvin}}$$

$$= 0.01000 \frac{\text{Volt}}{\text{Kelvin}} = 10.00 \frac{\text{mV}}{\text{Kelvin}} = 10.0$$

3.83 (–2.5)

$$\frac{V_{in} - 0}{R} = 0 + I_C$$

$$\frac{V_{in}}{R} = C\frac{dV_C}{dt} = C\frac{d}{dt}(0 - V_{out})$$

$$-\frac{V_{in}}{RC} = \frac{dV_{out}}{dt}$$

$$V_{out} = \frac{-1}{RC}\int V_{in}\, dt$$

$$V_{out} = \frac{-1}{10^3 \times 2 \times 10^{-6}}\int V_{in} dt$$

$$V_{out} = \frac{-10^3}{2}\int V_{in} dt \text{ Volt}$$

$0 < t < 1$ ms :

$$V_{in} = 5 \text{ volt}$$

$$V_{out} = -\frac{10^3}{2} \times 5(t)$$

at $t = 1$ ms

$$V_{out} = -10^3 \times 2.5 \times 1 \times 10^{-3}$$
$$= -2.5 \text{ volt}$$

1 ms $< t <$ 2 ms:

$$V_{in} = 0 \text{ volt}$$
$$V_{out} = -2.5 \text{ volt}$$

$\therefore$ at $t = 2$ ms, $V_{out} = -2.5$ volt

3.84 (0)

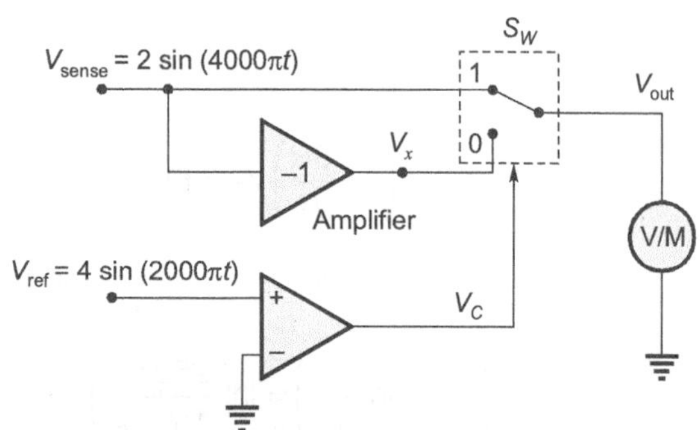

$$\omega_{sense} = 4000\,\pi \Rightarrow f_{sense} = 2000 \text{ Hz}$$

$$\omega_{ref} = 2000\,\pi \Rightarrow f_{ref} = 1000 \text{ Hz}$$

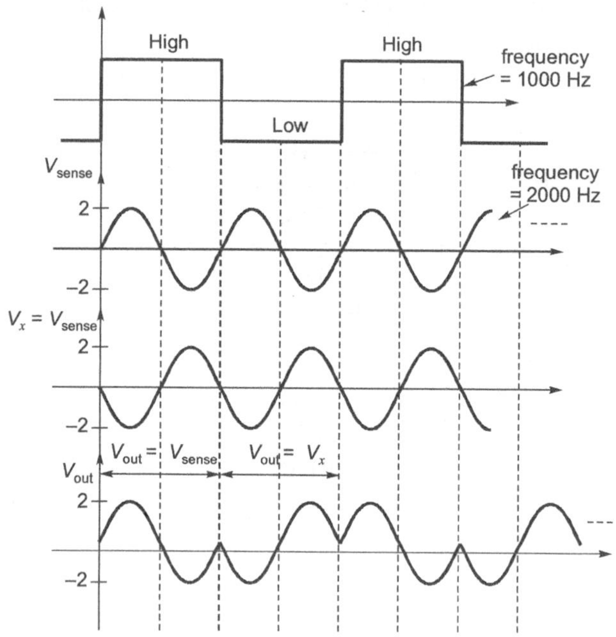

Average value of V_{out} = 0 volt

Vout has fundamental frequency

= 1000 Hz = 1 kHz

Since B.W of PMMC is 1 Hz.

Hence, $\quad V_{out} = 0$ V

3.85 (1.6)

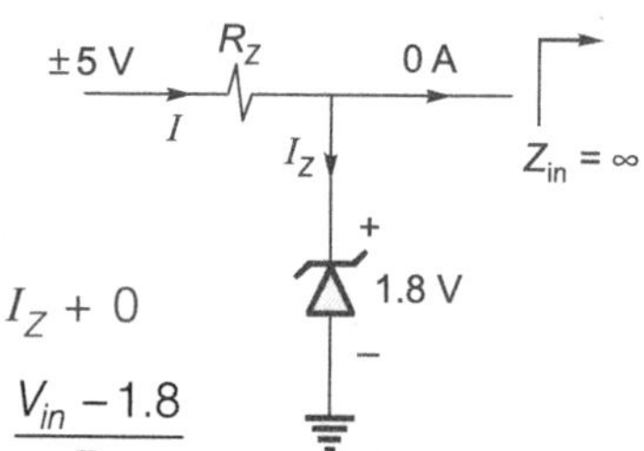

$I_{Z\,min}$ = 2 mA

V_Z = 1.8 volt

$R_{Z\,(max)}$ = ?

$$I = I_Z + 0$$

$$\Rightarrow \quad I = I_Z = \frac{V_{in} - 1.8}{R_Z}$$

$$R_Z = \frac{V_{in} - 1.8}{I_Z}$$

$$R_{Z\,(max)} = \frac{5 - 1.8}{2\,\text{mA}} = 1.6\,\text{k}\Omega$$

3.86 (–3)

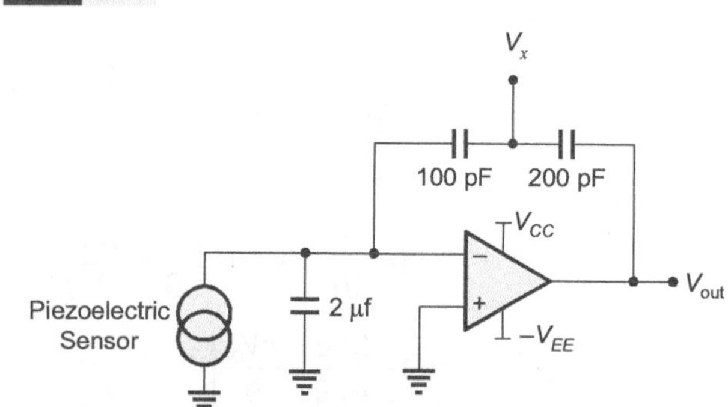

From the circuit,

$$V_x = \frac{200}{200+100} \times V_{out}$$

$$V_x = \frac{2}{3} V_{out}$$

$$\frac{dV_x}{dQ_p} = \frac{2}{3} \times \frac{dV_{out}}{dQ_p}$$

Given, $$\frac{dV_x}{dQ_p} = -2 \text{ V}$$

$$-2 = \frac{2}{3} \times \frac{dV_{out}}{dQ_p}$$

$$\frac{dV_{out}}{dQ_p} = -3 \text{ V}$$

3.87 (a)

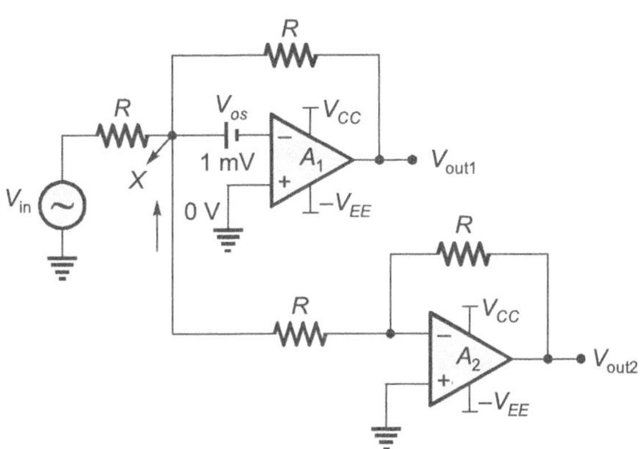

⇒ applying KCL at node X—

$$\frac{1}{2} + \frac{1 - V_{01}}{R} + \frac{1-0}{R} = 0$$

$$V_{01} = 3 \text{ mV}$$

and $$V_{out2} = 1 \times \left(\frac{-R}{R}\right) = -1 \text{mV}$$

3.88 (a)

DC Gain = –10

3 dB BW = ?

$C = 1\ \mu\text{F}$

$$\frac{V_0}{V_{in}} = \frac{-\left(R_f \parallel \frac{1}{sc}\right)}{R} = \frac{-R_f}{R(1+scR_f)}$$

$$\frac{-R_f}{R} = -10 \quad R_f = 10\ R$$

$$3 \text{ dB BW} = \frac{1}{2\pi c R_f} = \frac{1}{2\pi \times 10^{-7} \times 10^4}$$

$$= 159.15 \text{ Hz}$$

3.89 (4)

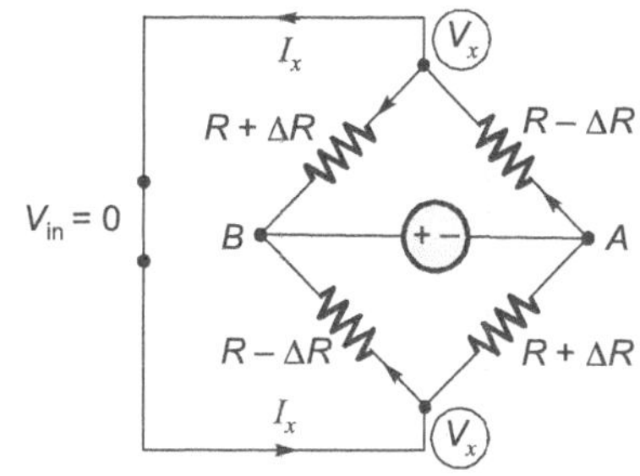

$$R_{th} = \frac{V}{I} = ?$$

$$R_{th} = \frac{V_A - V_B}{I} = \frac{V}{I} = ?$$

KCL: $$I = \frac{V_A - V_x}{(R-\Delta R)} + \frac{V_A - V_x}{(R+\Delta R)} \quad \text{...(1)}$$

KCL: $$I_x + \frac{V_A - V_x}{R+\Delta R} = \frac{V_x - V_B}{R-\Delta R} \quad \text{...(2)}$$

KCL: $$\frac{V_A - V_x}{R-\Delta R} = I_x + \frac{V_x - V_B}{R+\Delta R} \quad \text{...(3)}$$

Eq. (1) and (2)

$$I = \frac{V_A - V_x}{R-\Delta R} + \left(\frac{V_x - V_B}{R-\Delta R} - I_x\right)$$

$$I = \frac{V_A - V_x + V_x - V_B}{R-\Delta R} - I_x$$

$$\frac{V_A - V_B}{R-\Delta R} = I + I_x \quad \text{...(a)}$$

Eqn (1) and (3)

$$I - \frac{V_A - V_x}{R+\Delta R} = I_x + \frac{V_x - V_B}{R+\Delta R}$$

$$I - I_x = \frac{V_x - V_B + V_A - V_x}{R+\Delta R}$$

$$= \frac{V_A - V_B}{R+\Delta R} \quad \text{...(b)}$$

From eqn. (a) and (b) ⇒ add eqn (a) and (b)

$$2I = \frac{V_A - V_B}{R-\Delta R} + \frac{V_A - V_B}{R+\Delta R} = V\left[\frac{1}{R-\Delta R} + \frac{1}{R+\Delta R}\right]$$

$$2I = V\left[\frac{R+\Delta R + R - \Delta R}{(R-\Delta R)(R+\Delta R)}\right] = V\frac{2R}{(R-\Delta R)(R+\Delta R)}$$

$$\frac{V}{I} = \frac{(R-\Delta R)(R+\Delta R)}{R}$$

$$= \frac{(4.5-1.5)(4.5+1.5)}{4.5} = \frac{3 \times 6}{4.5} = \frac{18}{4.5} = 4$$

Req = 4

4 Wave Shaping Circuits and Filters

4.1 An integrator circuit is shown in figure. The op-amp is of type 741 and has an input offset current i_{os} of 1 μA. C is 1 μF and R is 1 MΩ. If the input V_i is a 1 kHz square wave of 1 V peak to peak, the output V_0, under steady condition, will be

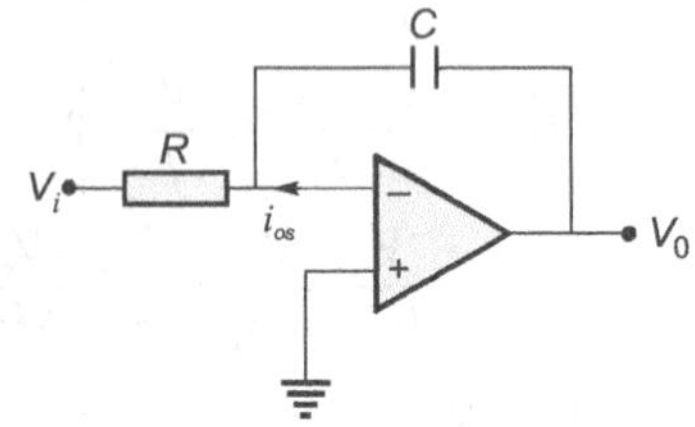

(a) a square wave of 1 V peak to peak
(b) a triangular wave of 1 V peak to peak
(c) positive supply voltage $+V_{CC}$
(d) negative supply voltage $-V_{CC}$

[2003 : 1 M]

4.2 The input-output characteristics of a Schmitt trigger is given in below figure. The noise margin of the Schmitt trigger is

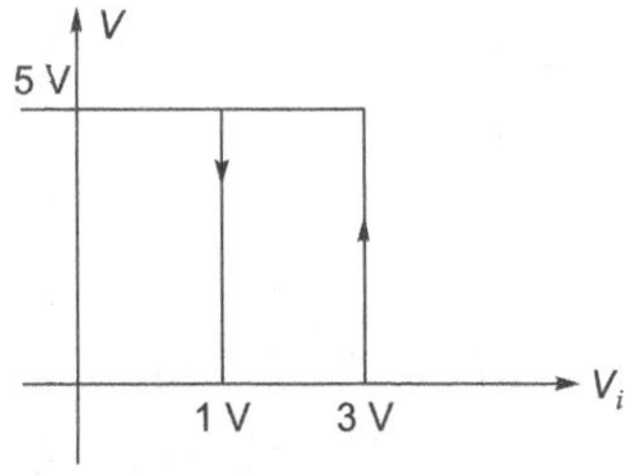

(a) 1 V (b) 2 V
(c) 3 V (d) 4 V

[2003 : 1 M]

4.3 The circuit shown in figure is that of a waveform generator. Assuming ideal devices, and ±12 V supply, the output V_o is a

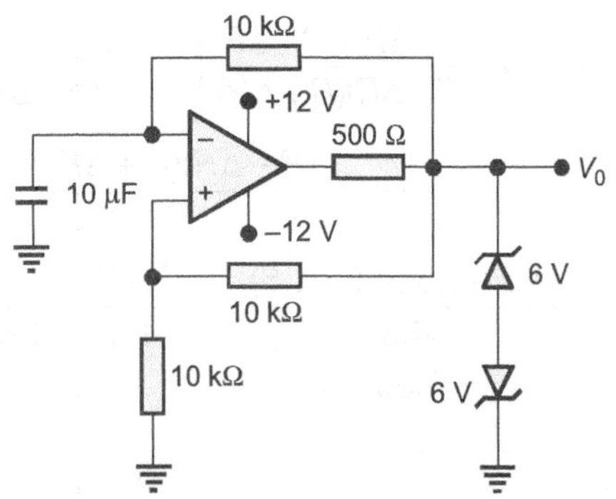

(a) triangular wave of period 120 ms and amplitude ±6 V
(b) square wave of period 60 ms and amplitude ±6 V
(c) square wave of period 120 ms and amplitude ±6 V
(d) square wave of period 60 ms and amplitude ±12 V

[2003 : 2 M]

4.4 The circuit in figure is a

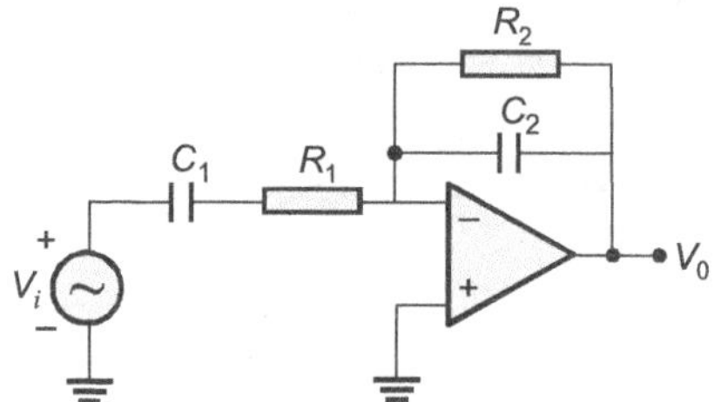

(a) band-pass filter with

lower cut-off $\omega_L = \dfrac{1}{R_1 C_1}$ and

higher cut-off $\omega_H = \dfrac{1}{R_2 C_2}$

(b) band-reject filter with

lower cut-off $\omega_L = \dfrac{1}{R_1 C_1}$ and

higher cut-off $\omega_H = \dfrac{1}{R_2 C_2}$

(c) band-pass filter with

lower cut-off $\omega_L = \dfrac{1}{R_2 C_2}$ and

higher cut-off $\omega_H = \dfrac{1}{R_1 C_1}$

(d) band-reject filter with

lower cut-off $\omega_L = \dfrac{1}{R_2 C_2}$ and

higher cut-off $\omega_H = \dfrac{1}{R_1 C_1}$

[2004 : 2 M]

4.5 Figure (a) shows a Schmitt trigger circuit and figure (b) the corresponding hysteresis characteristics. The values of V_{TL} and V_{TH} are

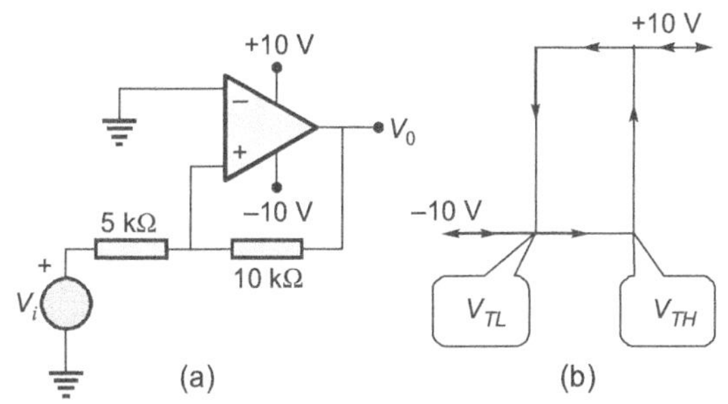

(a) $V_{TL} = -3.75$, $V_{TH} = +3.75$ V
(b) $V_{TL} = -1$ V, $V_{TH} = +5$ V
(c) $V_{TL} = -5$ V, $V_{TH} = +1$ V
(d) $V_{TL} = -5$ V, $V_{TH} = +5$ V

[2004 : 2 M]

4.6 The input-output characteristic of a Schmitt trigger has a hysteresis band of ±0.1 V. If the input voltage is 5 sin (1000 πt), the delay between the corresponding zero cross-over points of the output and input signals is

(a) 6.37 μs (b) 0.02 μs
(c) 63.7 μs (d) 2.0 μs

[2005 : 2 M]

4.7 An Astable multivibrator circuit using a 555 IC is given in the following figure. The frequency of oscillation is

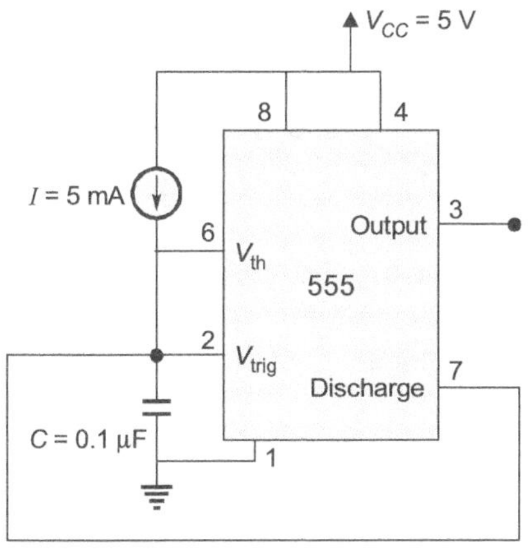

(a) 20 kHz (b) 30 kHz
(c) 40 kHz (d) 45 kHz

[2006 : 2 M]

Statement for Linked Answer Questions (4.8 and 4.9):

In the Schmitt trigger circuit shown below, the Zener diodes have V_Z (reverse saturation voltage) = 6 V and V_D (forward voltage drop) = 0.7 V.

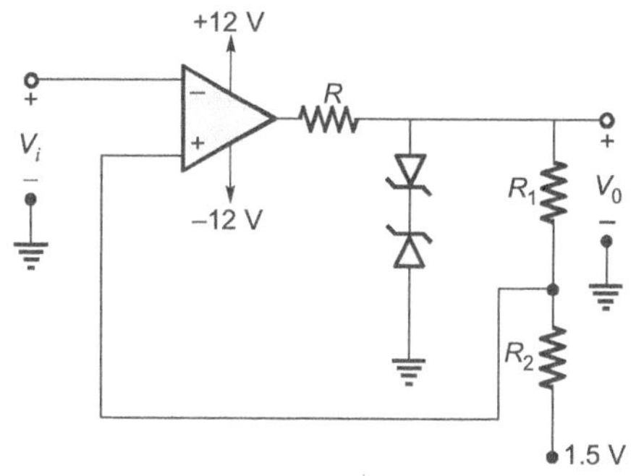

4.8 If the circuit has the input lower trip point (LTP) = 0 V, then value of R_1/R_2 is given as

(a) 0.223 (b) 2.67
(c) 4.67 (d) ∞

[2006 : 2 M]

4.9 The input upper trip point (UTP) of the Schmitt trigger is

(a) 1.5 V (b) 2.1 V
(c) 2.42 V (d) 6.7 V

[2006 : 2 M]

4.10 A 555 Astable multivibrator circuit is shown in the figure below:

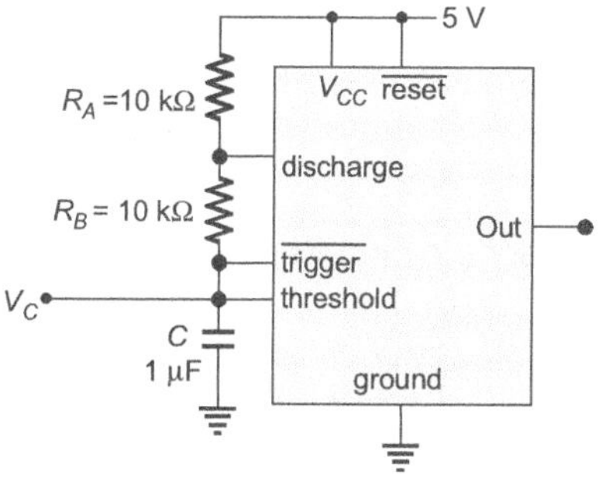

If R_B is shorted, the waveform at V_C is

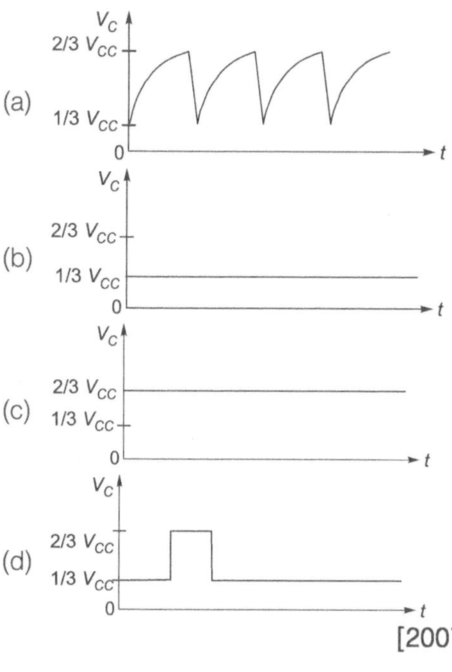

[2007 : 1 M]

4.11 The input signal shown in the figure shown above is fed to a Schmitt trigger. The signal has a square wave amplitude of 6 V *p-p*. It is corrupted by an additive high frequency noise of amplitude 8 $V_{p\text{-}p}$.

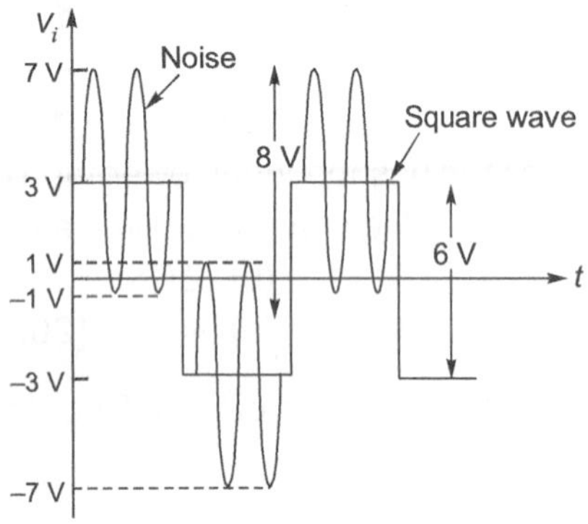

Which one of the following is an appropriate choice for the upper and lower trip points of the Schmitt trigger to recover a square wave of the same frequency from the corrupted input signal V_i ?

(a) ±8.0 V (b) ±2.0 V

(c) ± 0.5 V (d) 0 V

[2007 : 2 M]

4.12 Consider the circuit shown below:

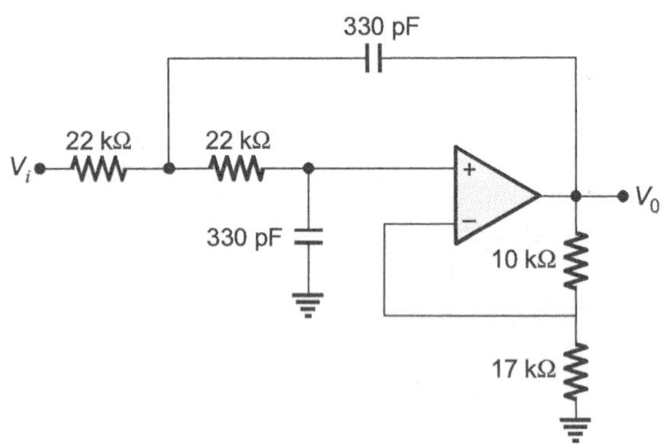

The correct frequency response of the circuit is

(a)

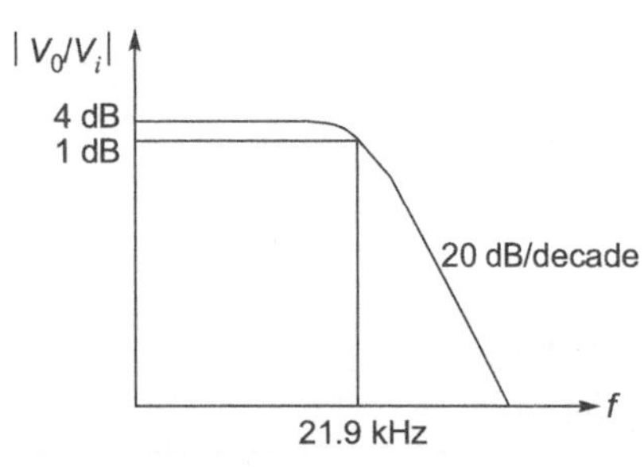

(b)

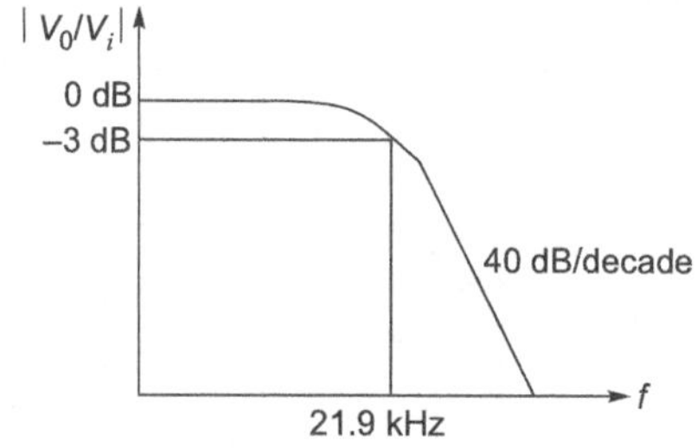

(c)

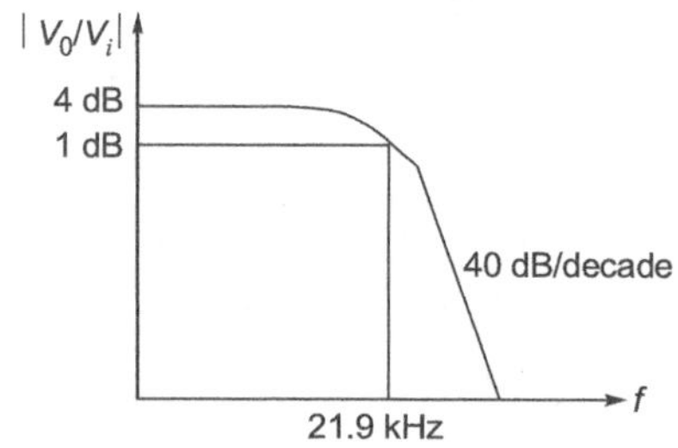

(d)

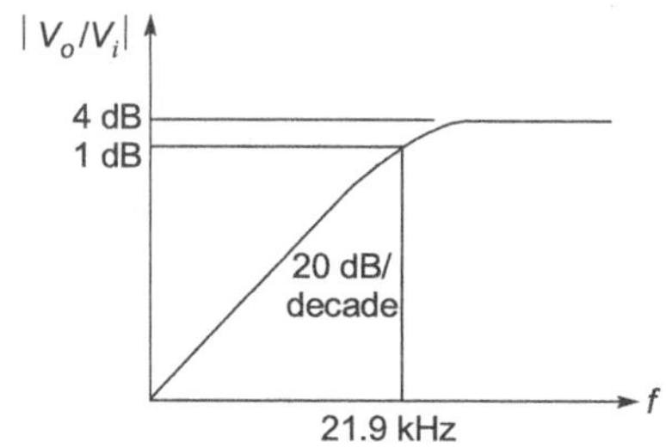

[2007 : 2 M]

4.13 In the circuit shown below the switch (S) is closed whenever the input voltage (V_{in}) is positive and open otherwise.

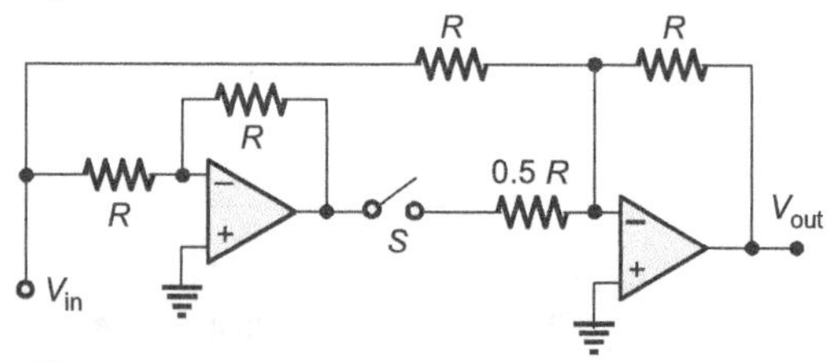

The circuit is a

(a) low-pass filter (b) level shifter

(c) modulator (d) precision rectifier

[2007 : 2 M]

4.14 Consider the triangular wave generator shown below:

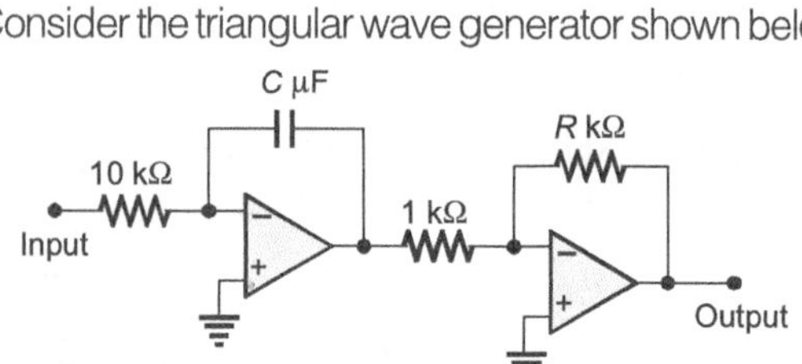

Assume that the op-amps are ideal and have ±12 V power supply. If the input is a ±5 V 50 Hz square wave of duty cycle 50%, the condition that results in a triangular wave of peak to peak amplitude 5 V and frequency 50 Hz at the output is

(a) $RC = 1$ (b) $\frac{R}{C} = 1$

(c) $\frac{R}{C} = 5$ (d) $\frac{C}{R} = 5$

[2007 : 2 M]

4.15 The op-amp circuit shown below is that of a

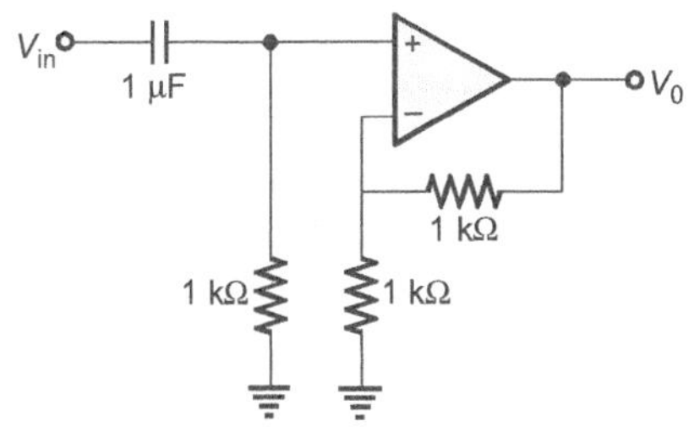

(a) low-pass filter with a maximum gain of 1
(b) low-pass filter with a maximum gain of 2
(c) high-pass filter with a maximum gain of 1
(d) high-pass filter with a maximum gain of 2

[2008 : 2 M]

4.16 In the op-amp circuit shown below the input voltage v_{in} is gradually increased from –10 V to +10 V. Assuming that the output voltage v_{out} saturates at –10 V and +10 V, v_{out} will change from

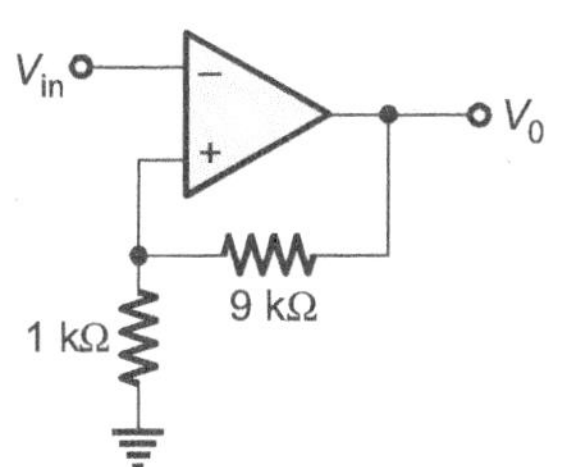

(a) –10 V to +10 V when $v_{in} = -1$ V
(b) –10 V to +10 V when $v_{in} = +1$ V
(c) +10 V to –10 V when $v_{in} = -1$ V
(d) +10 V to –10 V when $v_{in} = +1$ V

[2008 : 2 M]

4.17 The circuit shown in the figure is

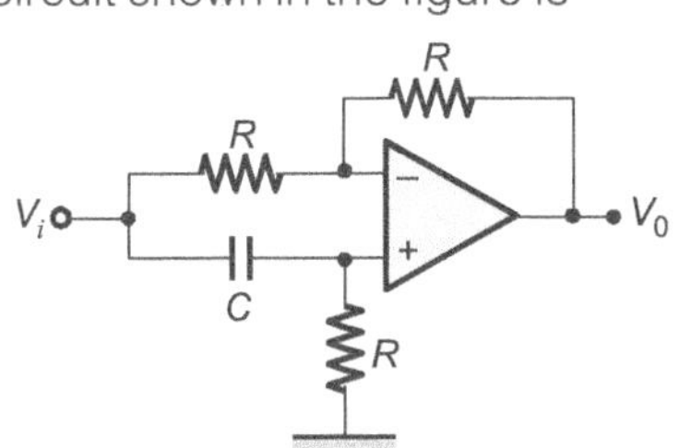

(a) an all-pass filter
(b) a band-pass filter
(c) a high-pass filter
(d) a low-pass filter

[2009 : 1 M]

4.18 In the circuit shown in the figure, the switch S has been in Position 1 for a long time. It is then moved to Position 2. Assume the Zener diodes to be ideal. The time delay between the switch moving to Position 2 and the transition in the output voltage V_0 is

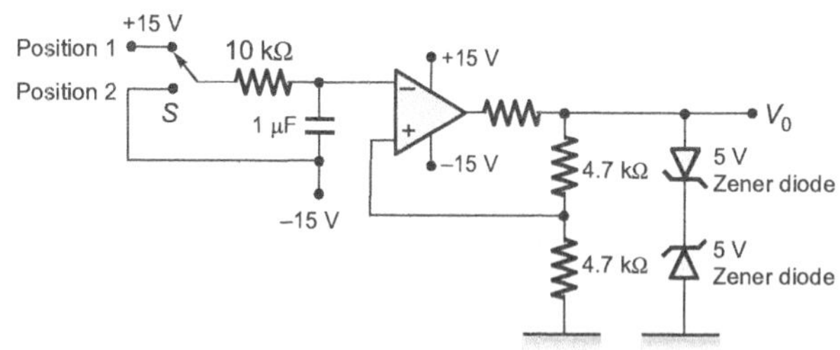

(a) 5.00 ms (b) 8.75 ms
(c) 10.00 ms (d) 13.75 ms

[2009 : 2 M]

4.19 An active filter is shown in the adjoining figure. The dc gain and the 3 dB cut-off frequency of the filter respectively, are nearly

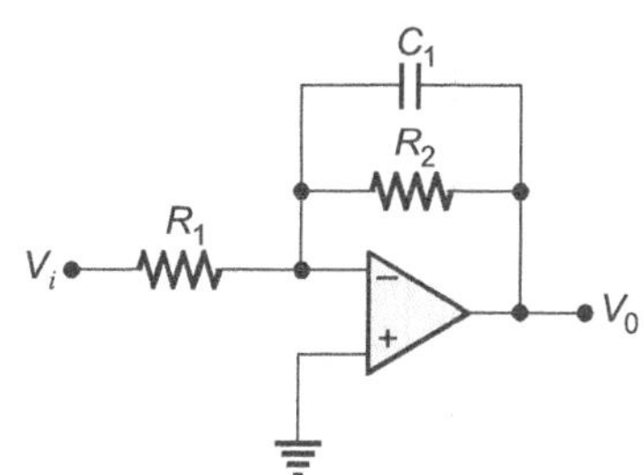

R_1 = 15.9 kΩ, R_2 = 159 kΩ, C_1 = 1.0 nF

(a) 40 dB, 3.14 kHz (b) 40 dB, 1.00 kHz
(c) 20 dB, 6.28 kHz (d) 20 dB, 1.00 kHz

[2010 : 2 M]

4.20 The ideal opamp based circuit shown below acts as a

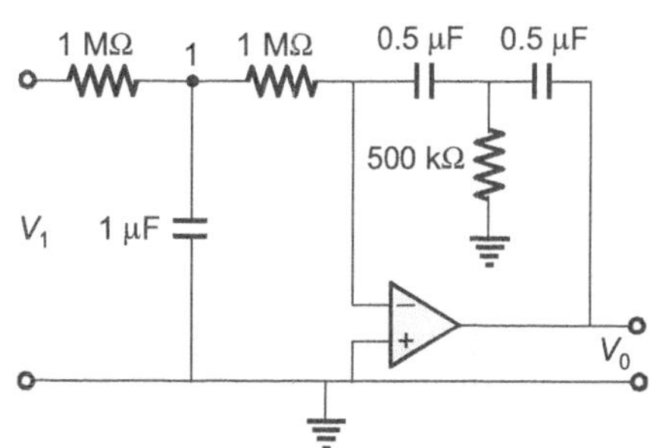

(a) low-pass filter (b) high-pass filter
(c) band-pass filter (d) band-reject filter

[2011 : 2 M]

4.21 The circuit shown is a

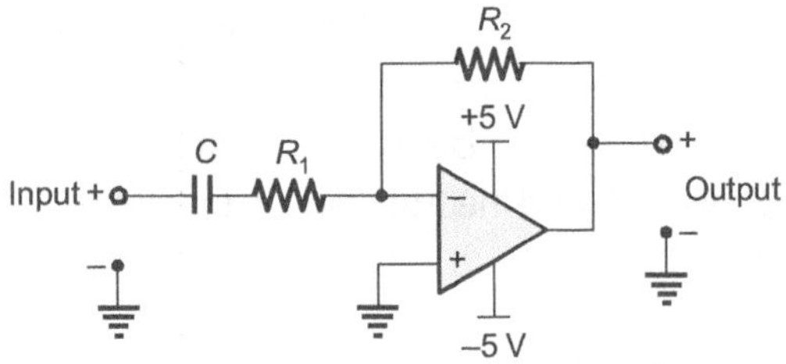

(a) low pass filter with $f_{3dB} = \frac{1}{(R_1+R_2)C}$ rad/s

(b) high pass filter with $f_{3dB} = \frac{1}{R_1 C}$ rad/s

(c) low pass filter with $f_{3dB} = \frac{1}{R_1 C}$ rad/s

(d) high pass filter with $f_{3dB} = \frac{1}{(R_1+R_2)C}$ rad/s

[2012 : 2 M]

4.22 A signal $V_i(t) = 10 + 10\sin 100\pi t + 10\sin 4000\pi t + 10\sin 100000\pi t$ is supplied to a filter circuit (shown below) made up of ideal op-amps. The least attenuated frequency component in the output will be

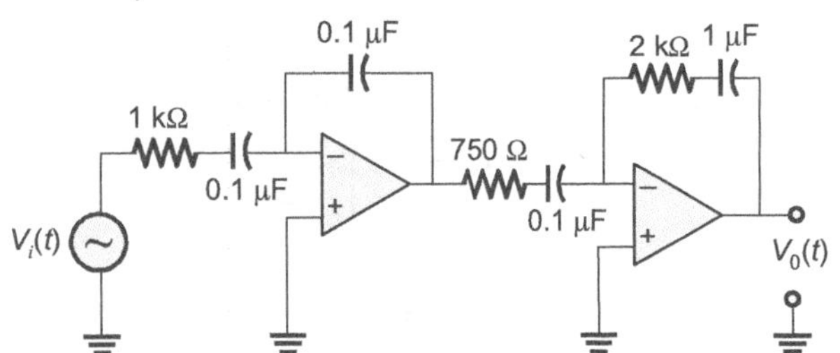

(a) 0 Hz (b) 50 Hz
(c) 2 kHz (d) 50 kHz

[2013 : 2 M]

4.23 The figures show an oscillator circuit having an ideal Schmitt trigger and its input-output characteristics. The time period (in ms) of $v_o(t)$ is ______.

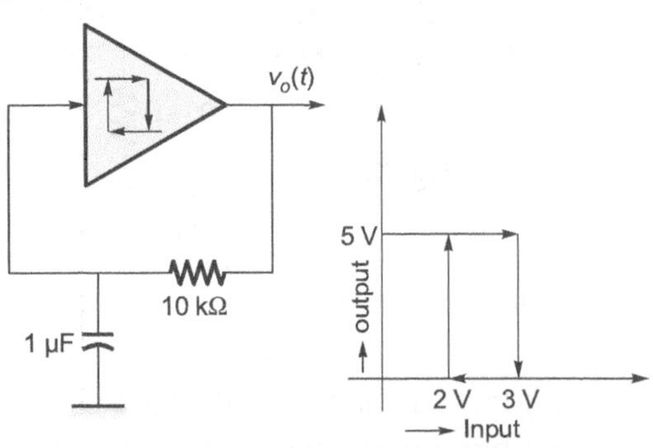

[2014 : 2 M]

4.24 The filter whose transfer function is of the form

$$G(s) = \frac{s^2 - bs + c}{s^2 + bs + c} \text{ is}$$

(a) a high-pass filter (b) a low-pass filter
(c) an all-pass filter (d) a band-reject filter

[2015 : 1 M]

4.25 The condition for oscillation in a feedback oscillator circuit is that at the frequency of oscillation, initially the loop gain is greater than unity while the total phase shift around the loop in degree is

(a) 0 (b) 90
(c) 180 (d) 270

[2017 : 1 M]

4.26 Assuming the opamp shown in the figure to be ideal, the frequency at which the magnitude of V_0 will be 95% of the magnitude of V_{in} is ______ kHz.

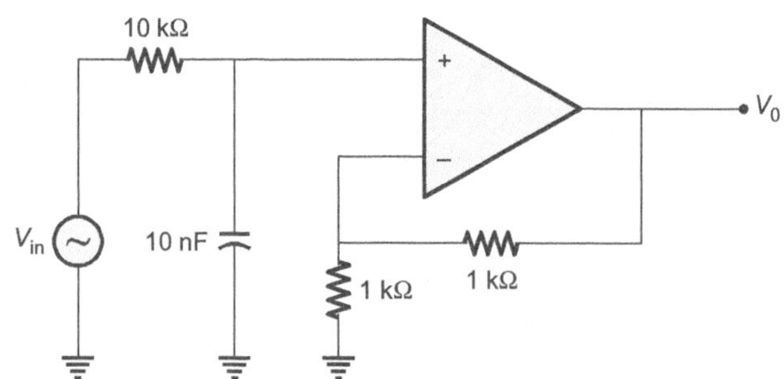

[2017 : 2 M]

4.27 The circuit of a Schmitt trigger is shown in the figure. The Zener-diode combination maintains the output between ± 7 V. The width of the hysteresis band is ______ V.

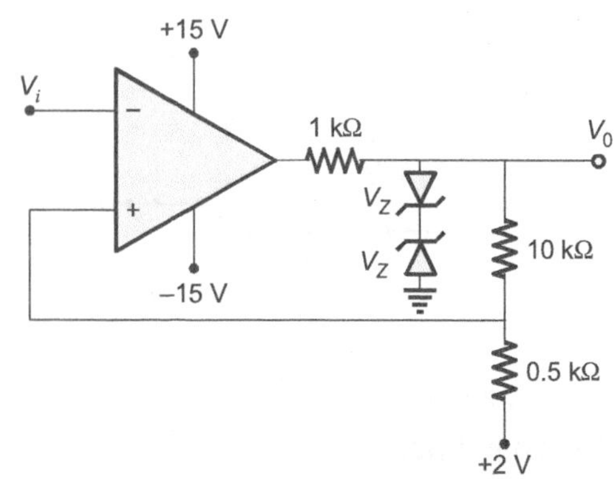

[2017 : 2 M]

4.28 For a first order lowpass filter with unity dc, gain and –3 dB corner frequency of 2000π rad/s, the transfer function $H(j\omega)$ is

(a) $\frac{1}{1000 + j\omega}$ (b) $\frac{1}{1 + j1000\omega}$

(c) $\frac{2000\pi}{2000\pi + j\omega}$ (d) $\frac{1000\omega}{1 + j1000\omega}$

[2017 : 1 M]

4.29 In the given relaxation oscillator, the opamps and the zener diodes are ideal. The frequency (in kHz) of the square wave output v_0 is ________.

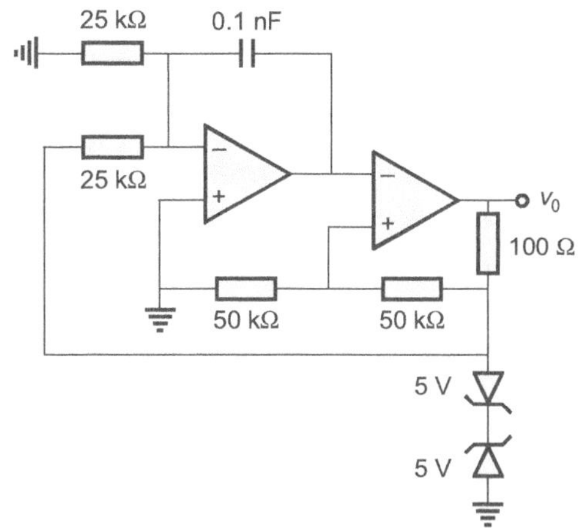

[2018 : 2 M]

Answers Wave Shaping Circuits & Filters

4.1 (b)	4.2 (b)	4.3 (c)	4.4 (a)	4.5 (d)	4.6 (a)	4.7 (b)
4.8 (c)	4.9 (c)	4.10 (a)	4.11 (b)	4.12 (c)	4.13 (d)	4.14 (b)
4.15 (d)	4.16 (d)	4.17 (a)	4.18 (a)	4.19 (d)	4.20 (c)	4.21 (b)
4.22 (c)	4.23 (8.109)	4.24 (c)	4.25 (a)	4.26 (2.95)	4.27 (0.67)	4.28 (c)
4.29 (200)						

Explanations Wave Shaping Circuits & Filters

4.1 (b)

Using KVL at the inverting terminal,

We have,

$$\frac{V_i - 0}{R} = (0 - V_0)\,Cs$$

$$V_0 = -\frac{1}{RC}\int_0^t v(t)\,dt$$

Thus, given circuit will behave as an integrator. The input of the integrator circuit is square wave, so the output will produce triangular wave with same amplitude and frequency.

4.2 (b)

Noise margin = 3 V – 1 V = 2 V

4.3 (c)

The given circuit is astable multivibrator, so output will be a square wave.

Due to two Zener diode output will not be allowed to go beyond 6 V. So output will be ±6 V.

Here $V_{ut} = V_{it} = 3$ V, so it will be a square wave of 50% duty cycle.

and $T/2 = RC\,ln\,2$

$$T = 2\,RC\,ln\,2$$
$$= 2 \times 10 \times 10^3 \times 10 \times 10^{-6}\,ln\,2$$
$$= 2 \times 10^{-1} \times 0.693 \text{ sec}$$
$$= 2 \times 100 \times 0.693 \text{ msec}$$
$$= 138.6 \text{ msec}$$

So, looking from the options (c) is correct option.

4.4 (a)

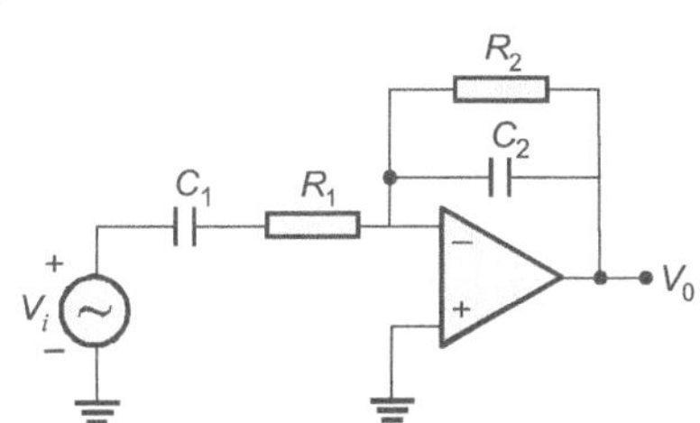

Using KCL at the inventing terminal

$$\frac{V_i - 0}{\left(R_1 + \frac{1}{C_1 S}\right)} = \frac{0 - V_0}{\frac{R_2}{(R_2 C_2 s + 1)}}$$

$$A_v = \frac{V_0}{V_i} = \frac{-R_2 C_1 S}{(R_2 C_2 S + 1)(R_1 C_1 S + 1)}$$

or $$A_v(f) = \frac{-R_2/R_1}{\left(1 - j\left(\frac{f_L}{f}\right)\right)\cdot\left(1 + j\left(\frac{f}{f_H}\right)\right)}$$

$$= [-A_L(jf) \cdot A_H(jf)]$$

where, $f_L = \frac{1}{2\pi R_1 C_1}$ and $f_H = \frac{1}{2\pi R_2 C_2}$

so, the given circuit will as a band pass filter with lower cut-off frequency $\omega_L = \frac{1}{R_1 C_1}$ and higher cut-off frequency $\omega_H = \frac{1}{R_2 C_2}$.

4.5 (d)

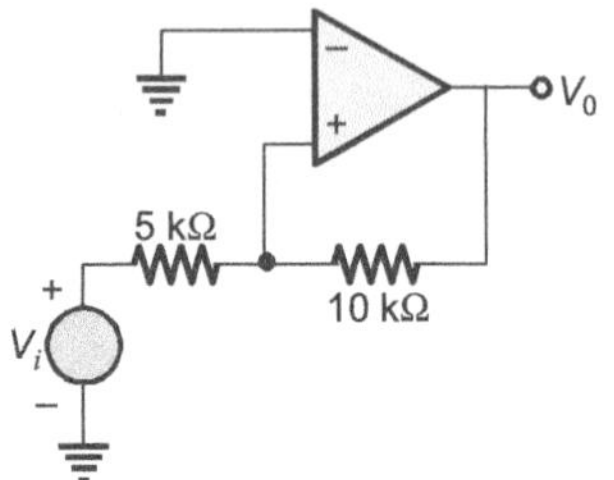

V_0 will be either at $+V_{sat}$ or $-V_{sat}$

Voltage V_1 at node 1,

$$V_1 = \frac{5}{15}V_0 + \frac{10}{15}V_i$$

when $V_1 = 0$, then $V_0 = -2V_i \Rightarrow V_i = -\frac{V_0}{2}$

so for $V_0 = +V_{sat}$

$\Rightarrow$ $V_i = V_{LT} = -5$

and for $V_0 = -V_{sat}$, $V_{UT} = 5$ V

4.6 (a)

Hysteresis loop of schmitt trigger follow the characteristic shown in below figure:

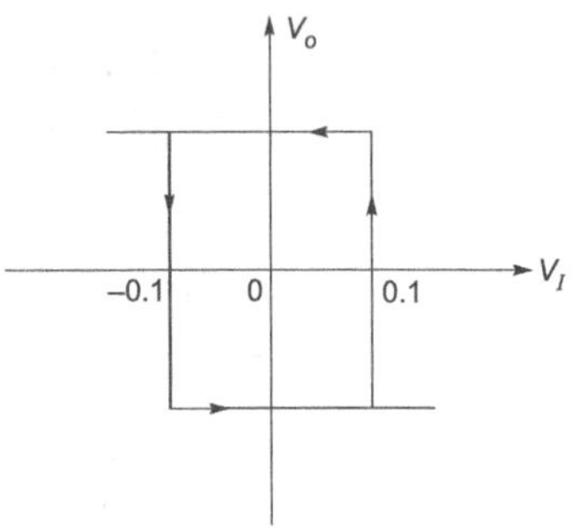

$$0.1 = 5\sin(1000\pi t)$$

$$\sin(1000\pi t) = \frac{0.1}{5}$$

$$1000\pi t = \sin^{-1}\left(\frac{0.1}{5}\right)$$

$$t = \frac{0.02 \text{ rad}}{1000\pi \text{ rad/sec}}$$

$$= 6.37 \times 10^{-6} \text{ sec} = 6.37 \ \mu\text{sec}$$

4.7 (b)

It is a sawtooth wave generator across capacitor *C*.

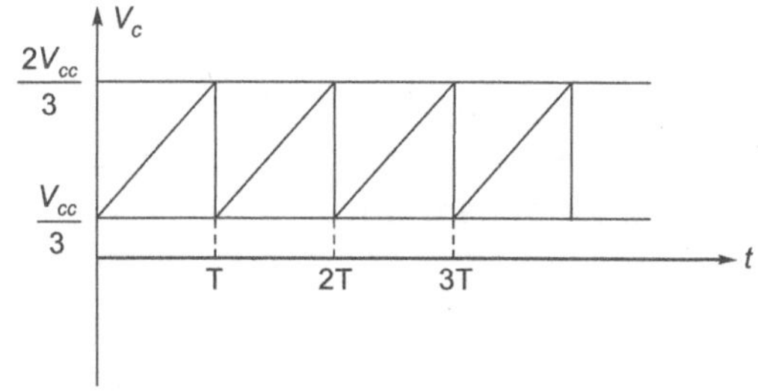

So, $$\frac{dV_c}{dt} = \frac{\frac{2V_{CC}}{3} - \frac{V_{CC}}{3}}{T} = \frac{V_{CC}}{3T}$$

$$I = C\frac{dV_c}{dt}$$

$$\frac{dV_c}{dt} = \frac{I}{C} = \frac{5\times10^{-3}}{0.1\times10^{-6}} = 5 \times 10^4$$

$$\frac{V_{CC}}{3T} = 5 \times 10^4$$

$$\Rightarrow \quad \frac{1}{T} = f = \frac{3\times5\times10^4}{5} = 30 \text{ kHz}$$

4.8 (c)

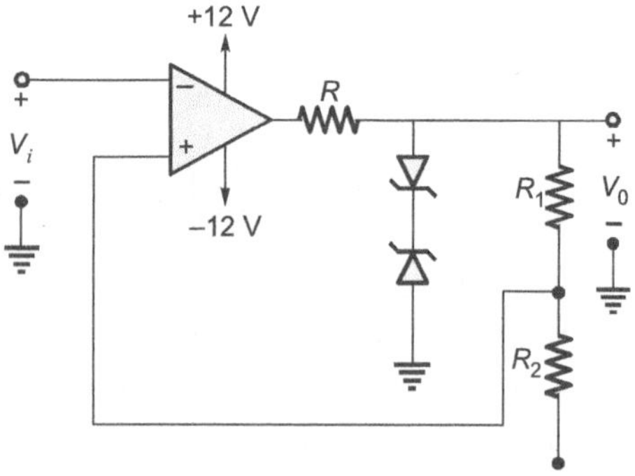

$$V_{LT} = \frac{R_2}{R_1+R_2}(V_{Z1}+V_D) + \frac{R_1}{R_1+R_2}\cdot V_R \quad \text{...(i)}$$

and

$$V_{UT} = \frac{R_2}{R_1+R_2}(V_{Z2}+V_D) + \frac{R_1}{R_1+R_2}\cdot V_R \quad \text{...(ii)}$$

So, from equation (i)

$$0 = \frac{R_2}{R_1+R_2}(6+0.7) + \frac{R_1}{R_1+R_2}\cdot(1.5)$$

$$\frac{R_1}{R_2} = \frac{6.7}{1.5} = 4.47$$

4.9 (c)

From equation (ii) we get

$$V_{UT} = \frac{R_2}{R_1+R_2}(V_{Z2}+V_D)+\frac{R_1}{R_1+R_2}V_R$$

$$= \frac{1}{1+4.47}(6+0.7)+\frac{4.47}{4.47+1}\cdot(1.5) = 2.45\text{ V}$$

4.10 (a)

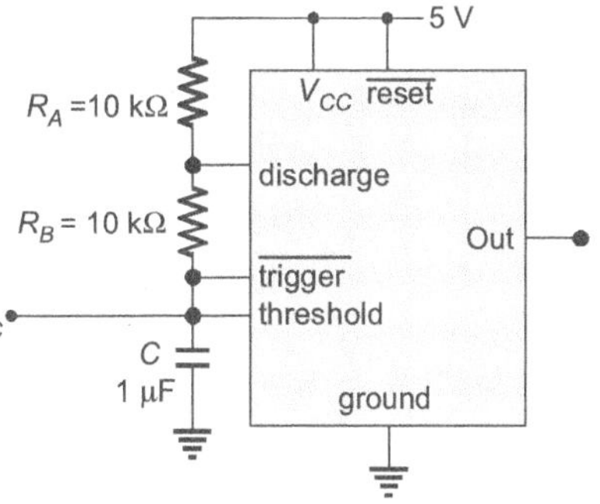

Capacitor C charges toward V_{CC} through R_A and R_B until in raises upto 2/3 V_{CC}. This voltage is the threshold voltage of pin 6, which drives comparator 1 to trigger the flip-flop so that the output at pin 3 goes low. In addition, the discharge transistor is driven on. Causing the output at pin 7 to discharge the capacitor through resistor R_B. But R_B is shorted out. So discharging time constant $R_B \cdot C$. So capacitor will instantaneously discharge.

4.11 (b)

For the positive cycle of the square wave, the maximum negative attenuation by the noise is

3 – 1 = 2 V

Thus the lower threshold for the positive cycle of the signal will be 2 V.

For the negative cycle of the square wave, the maximum positive attenuation by the noise is

–3 + 1 = –2 V

Thus the upper threshold for the negative cycle of the signal will be –2 V.

Thus, if we choose the upper and lower trip points of the Schmitt trigger as ±2 V. We will recover a square wave of the same frequency from the corrupted input signal V_i.

4.12 (c)

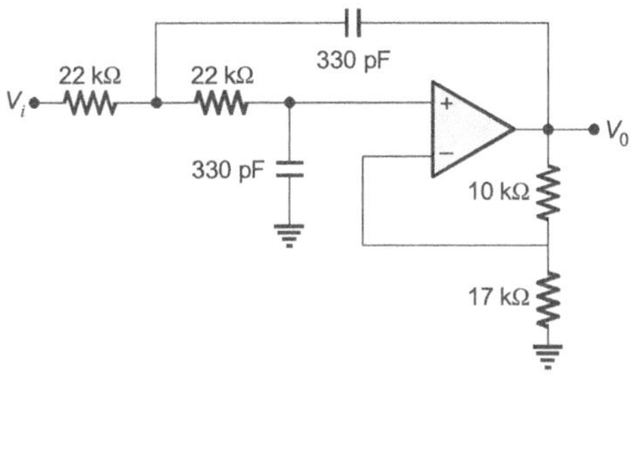

The given circuit is a second order low pass filter therefore it has a –40 dB/decade and it has 3 dB cut-off frequency as

$$f = \frac{1}{2\pi\sqrt{R_1C_1R_2C_2}}$$

$$= \frac{1}{2\pi\times 22\times 10^3\times 330\times 10^{-12}}$$

$$= 21.9\text{ kHz}$$

It has a decade from 4 dB to 1 dB in mid frequency range.

4.13 (d)

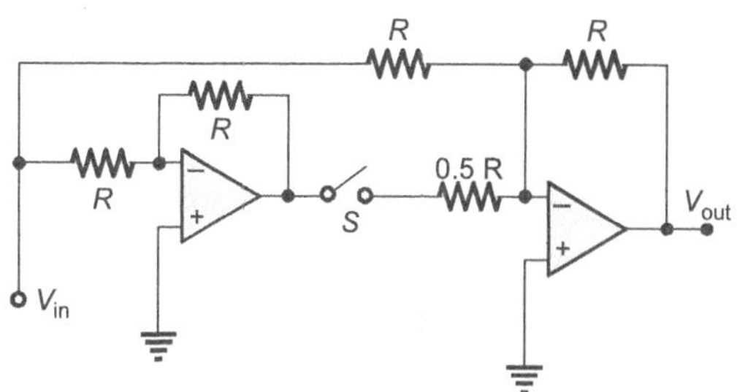

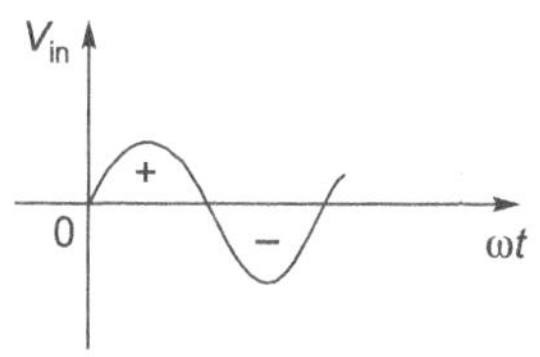

Case-I

When V_{in} = +ve half cycle then switch (*s*) is closed and circuit will become

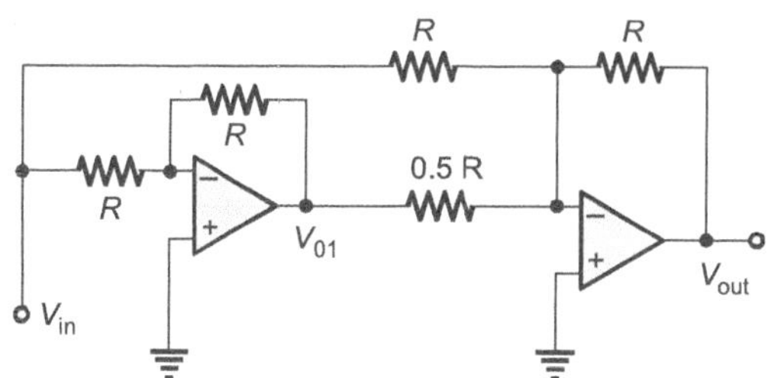

$$V_{01} = -\left(\frac{R}{R}\right)V_{in} = -V_{in}$$

$$V_{02} = -\frac{R}{R}V_{in}+\left(\frac{-R}{0.5R}\right)V_{01} = V_{in}$$

Case-II

When V_{in} = –ve half cycle then switch s is open and circuit will become.

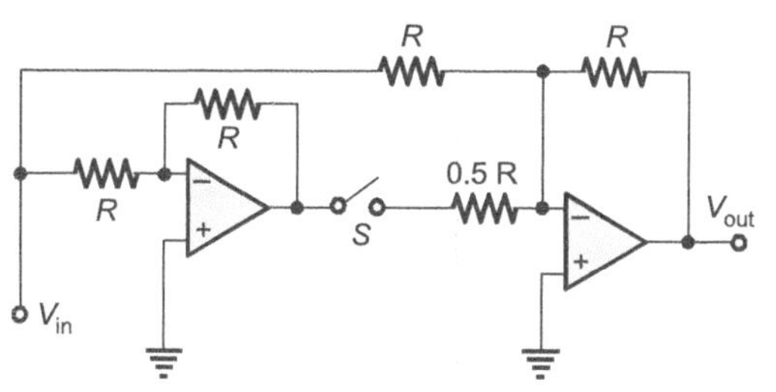

No circuit flow through left side opamp due to open switch so

$$V_{02} = -\frac{R}{R}(-V_{in}) = V_{in}$$

So, the circuit will act as a rectifier (fullwave).

4.14 (b)

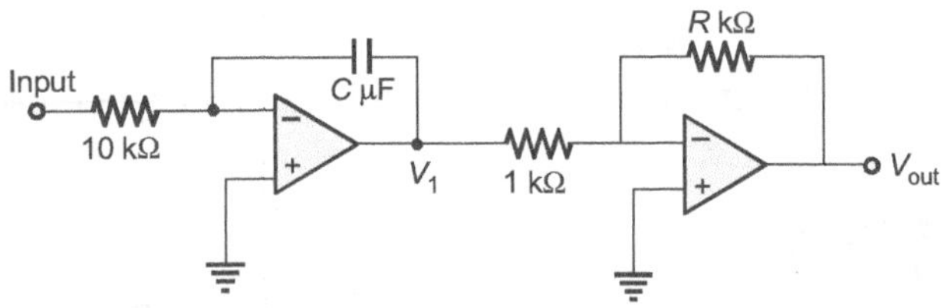

Output should be

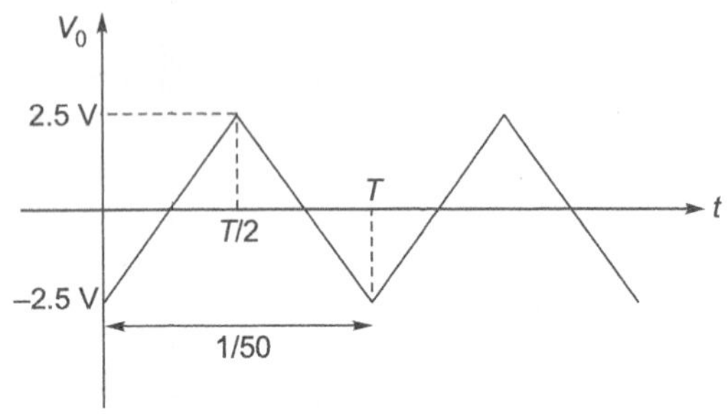

slope in +ve half is

$$\frac{dV_0}{dt} = \frac{5}{(T/2)} = \frac{10}{T}$$

$$= 500 \text{ V/sec}$$

Now, $$\frac{5}{10\times10^3} = -C\frac{dV_1}{dt}$$

$$\frac{dV_1}{dt} = \frac{-5\times10^{-4}}{C}$$

$$V_0 = -RV_1$$

$$\frac{dV_0}{dt} = -R\frac{dV_1}{dt}$$

$$= \frac{R}{C}\times5\times10^{-4}$$

$$\Rightarrow \quad \frac{R}{C}\times5\times10^{-4} = 500$$

$$\frac{R}{C} = 10^6$$

$$\frac{R}{C\times10^6} = 1$$

$$\frac{R}{C(\mu F)} = 1$$

4.15 (d)

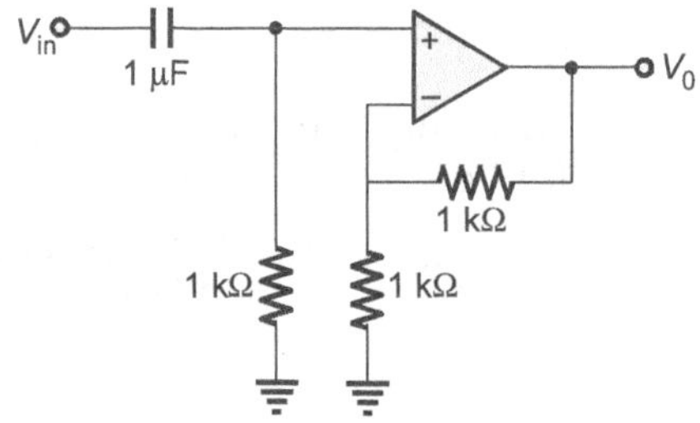

This is a non-inverting amplifier

$$\frac{V_{out}}{V_{in}} = \left(1+\frac{R_f}{R_1}\right), \text{ at high frequency}$$

$$= 1+\frac{1\text{k}\Omega}{1\text{k}\Omega}$$

$$\text{Gain} = \frac{V_{out}}{V_{in}} = 2$$

Since at the non-inverting input first element is capacitor, which will pass only high frequency, so it is a high pass filter.

4.16 (d)

It is Schmitt trigger.

Since initially $V_{in} = -10$ V, therefore output

$V_0 + V_{sat} = 10$ V

$$\therefore \quad V_+ = \frac{1}{1+9}\times10 = 1\text{ V}$$

So, when input crosses 1 V then $V_- > V_+$ and output switch to $-V_{sat}$ i.e. –10 V.

4.17 (a)

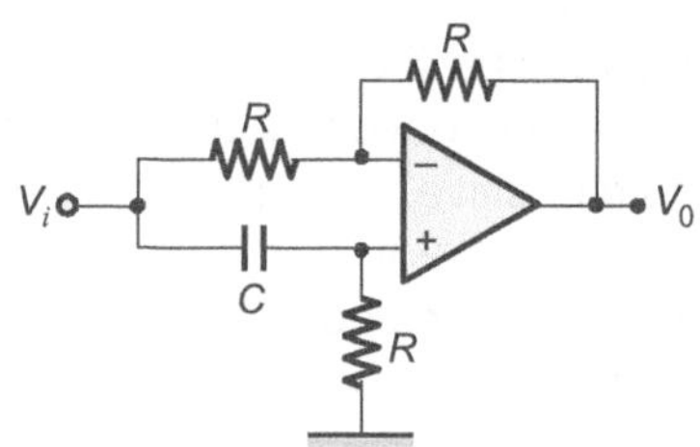

The above figure can be rearranged as below:

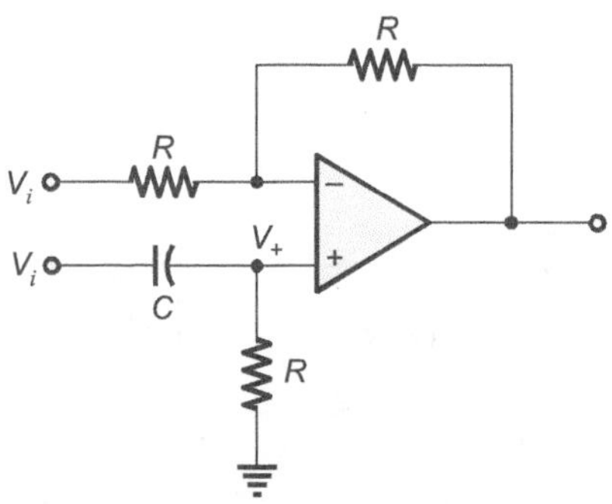

The voltage at non-inverting amplifier is

$$V_+ = \frac{R}{R+\frac{1}{Cs}} \cdot V_i$$

$$V_{01} = V_+\left[1+\frac{R_f}{R_1}\right]$$

$$= \frac{2V_iR}{R+\frac{1}{Cs}} \quad (\because R_f = R_1 = R)$$

Output voltage due to inverting input

$$V_{02} = -\frac{R_f}{R_1} \cdot V_i$$

$$= -V_i \quad (\because R_f = R_1 = R)$$

$$V_0 = V_{01} - V_{02}$$

$$= \left(\frac{2R}{R+\frac{1}{Cs}}\right)V_i - V_i = V_i\left[\frac{2R}{R+\frac{1}{Cs}} - 1\right]$$

$$= V_i\left[\frac{2R-R-\frac{1}{Cs}}{R+\frac{1}{Cs}}\right] = V_i\left[\frac{R-\frac{1}{Cs}}{R+\frac{1}{Cs}}\right]$$

$$\frac{V_0}{V_i} = \frac{s-\frac{1}{RC}}{s+\frac{1}{RC}}$$

There is one zero on right side of s-plane and one pole on left side of s-plane.
So, it is an all pass filter.

4.18 (a)

Initially, $V_c = 15$ V (switch at position 1)

and $V_0 = +5$ V

$\therefore \quad V_+ = +\frac{5}{2} = +2.5$ V

So, transition occurs when V_c reaches 2.5 V after changing the position of switch to position 2.

$$I = \frac{V_+}{1\text{k}\Omega} = V_+ \text{ mA}$$

$$= 2.5 \text{ mA}$$

$$C\frac{dV_c}{dt} = I$$

$$10^{-6}\left[\frac{15-2.5}{dt}\right] = 2.5 \times 10^{-3}$$

$$dt = 10^{-3} \times \frac{12.5}{2.5}$$

$$= 5 \times 10^{-3} = 5 \text{ msec}$$

4.19 (d)

At DC frequency, the capacitor will becomes open circuited, as now the circuit becomes

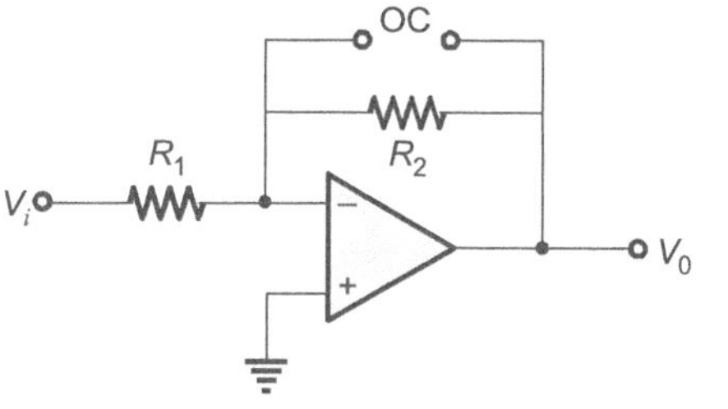

Given, $R_1 = 15.9$ kΩ

$R_2 = 159$ kΩ

and, $C_1 = 1.0$ nF

$$\text{DC gain} = -\frac{R_f}{R_1} = -\frac{159}{15.9} = -10$$

$$\text{gain (dB)} = 20 \log|\text{gain}| = 2 \log_{10} 10$$

$$\boxed{\text{DC gain(dB)} = 20 \text{ dB}}$$

3 dB cut-off frequency is

$$f_c = \frac{1}{2\pi \times R_2 \times C_1}$$

$$= \frac{1}{2\pi \times 159 \times 10^3 \times 1.0 \times 10^{-9}} \text{ Hz}$$

$$f_c = 1.0 \text{ kHz}$$

4.20 (c)

When f is very low, then all the capacitor gets open and output = 0.

When f is very high , then all the capacitors are short circuited and voltage at node 1 will be zero.

So, output V_0 also equal to zero.

This shows the characteristic of band pass filter.

4.21 (b)

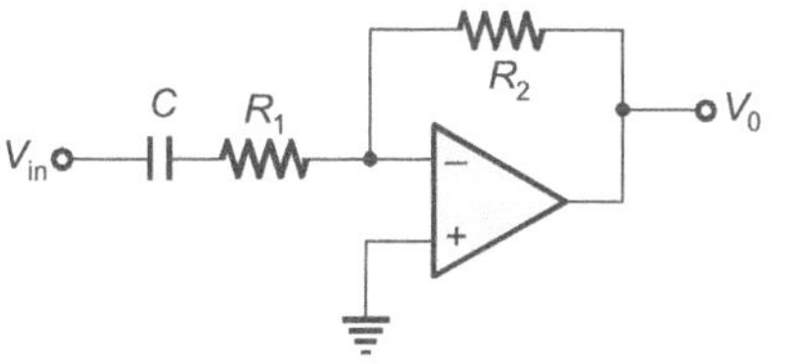

$$\frac{V_{in}-0}{R_1+\frac{1}{sC}} = \frac{0-V_o}{R_2}$$

$$\frac{V_o}{V_i} = -\left(\frac{R_2}{R_1+\frac{1}{sC}}\right)$$

It is a high pass filter with

$$f_{3dB} = \frac{1}{R_1C} \text{ rad/sec}$$

4.22 (c)

For Op-Amp A_1:

Analysis

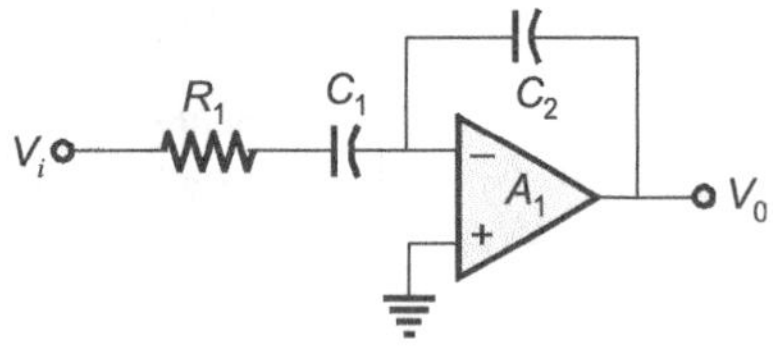

$$\text{T.F.} = \frac{V_0(s)}{V_i(s)} = \frac{-\frac{1}{sC_2}}{R_1+\frac{1}{sC_1}} = \frac{-\frac{1}{sC_2}}{\frac{1+sR_1C_1}{sC_1}}$$

Here, $C_1 = C_2$

$$= \frac{1}{1+sR_1C_1} \rightarrow \text{one pole}$$

Op-amp A_1 is lowpass response:

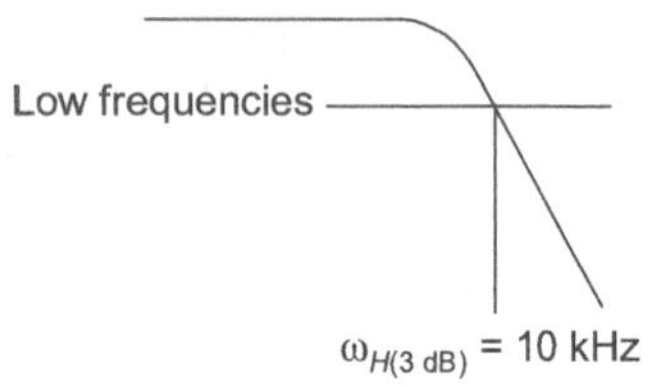

$$\omega_{H(3\ dB)} = \frac{1}{R_1C_1} = \frac{1}{1\,k\Omega\times 0.1\,\mu F} = 10\ \text{kHz}$$

Op-Amp A_2:

Analysis,

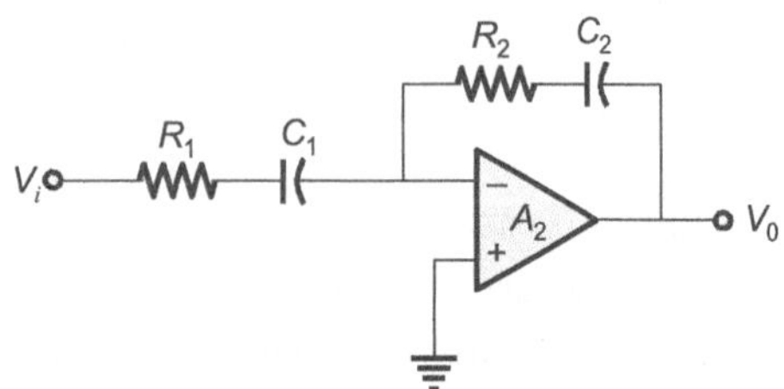

$$\text{T.F.} = \frac{V_0(s)}{V_i(s)} = -\frac{Z_2(s)}{Z_1(s)} = -\frac{R_2+\frac{1}{sC_2}}{R_1+\frac{1}{sC_1}}$$

$$= \frac{-R_2\left[1+\frac{1}{sR_2C_2}\right]}{R_1\left[1+\frac{1}{sR_1C_1}\right]} = \frac{\left[-\frac{R_2}{R_1}\right]\left[1-j\frac{\omega_{L_2 3dB}}{\omega}\right]}{\left[1-j\frac{\omega_{L_1 3dB}}{\omega}\right]}$$

$\Rightarrow$ High pass Response:

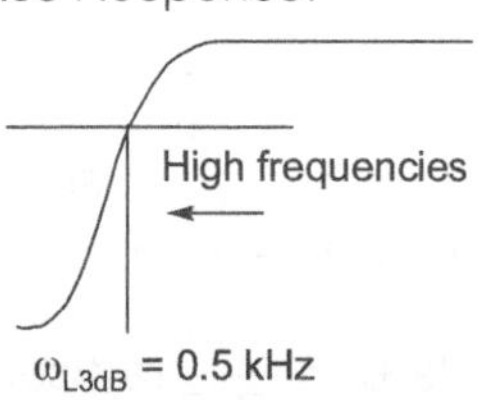

$$\omega_{L_1 3dB} = \frac{1}{R_1C_1} = \frac{1}{750\times 0.1\mu F} = \frac{1}{75\mu F} = 13.3\ \text{kHz}$$

$$\omega_{L_2 3dB} = \frac{1}{R_2C_2} = \frac{1}{2k\Omega\times 1\mu F} = 0.5\ \text{kHz}$$

$$\omega_{L_2 3dB} << \omega_{L_1 3dB}$$

Lower 3dB dominant frequency is 0.5 kHz.

Total response:

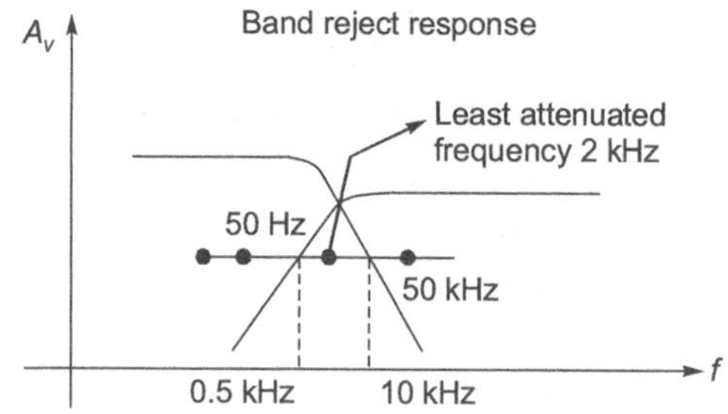

$\because$ Only option (c) is lying in this range.

4.23 8.109 (8.0 to 8.5)

$$V_{sat} = 5\ V,\quad UTP = 3\ V,\quad LTP = 2\ V$$

$$v_c(t) = V_{final} + (V_{initial} + V_{final})\,e^{-t/\tau}$$

$$\therefore\quad UTP = V_{sat} + (LTP - V_{sat})\,e^{-T_1/RC}$$

$$e^{-T_1/RC} = \left[\frac{V_{sat}-UTP}{V_{sat}-LTP}\right]$$

$$\therefore\quad \text{Total, } T = 2T_1 = 2RC\,ln\left[\frac{5-2}{5-3}\right]$$

$$\therefore\quad T = 8.109\ \text{m-sec.}$$

4.26 2.95 (2.9 to 3)

$$A_r = \left|\frac{V_0}{V_i}\right| = \frac{A}{\sqrt{1+\left(\frac{f_1}{f_c}\right)^2}}$$

Where, $A = 1+\frac{R_F}{R_1} = 2$

$$f_c = \frac{1}{2\pi RC}$$

$$= \frac{1}{2\pi\times 10\times 10^3\times 10\times 10^{-9}} = \frac{10^4}{2\pi}$$

At $f = f_1$,

$$V_0 = 0.95\ V_i \quad\Rightarrow A_V = 0.95$$

$$0.95 = \frac{2}{\sqrt{1+\left(\frac{f_1}{f_c}\right)^2}}$$

$$\Rightarrow\quad f_f = f_c\sqrt{\left(\frac{2}{0.95}\right)^2} - 1 = 2.95\ \text{kHz}$$

4.27 0.67 (0.6 to 0.7)

$$V_H = 2\ \beta V_Z = \frac{2\times 0.5k}{10k+0.5k}\times 7 = \frac{2}{3}$$

$$= 0.67\ \text{V}$$

4.28 (c)

General transfer function of 1st order LPF is

$$G(s) = \frac{1}{1+S\tau}$$

Where, $\tau = RC = \frac{1}{\omega_C} = \frac{1}{2000\pi}$

$$\therefore\quad G(j\omega) = \frac{1}{1+j\omega\left(\frac{1}{2000\pi}\right)} = \frac{2000\pi}{2000\pi + j\omega}$$

4.29 (200)

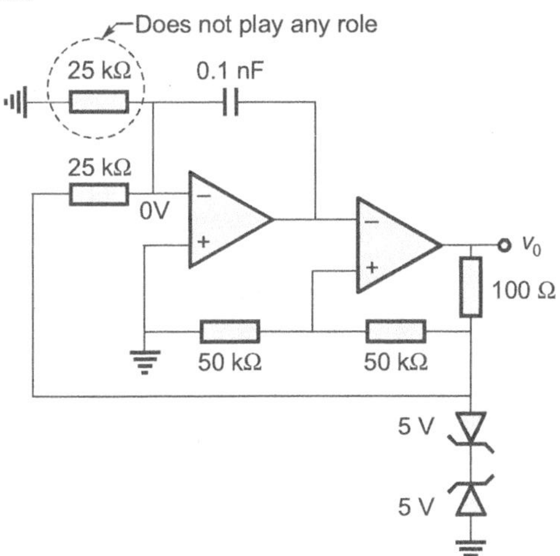

The above circuit can be redrawn as follows:

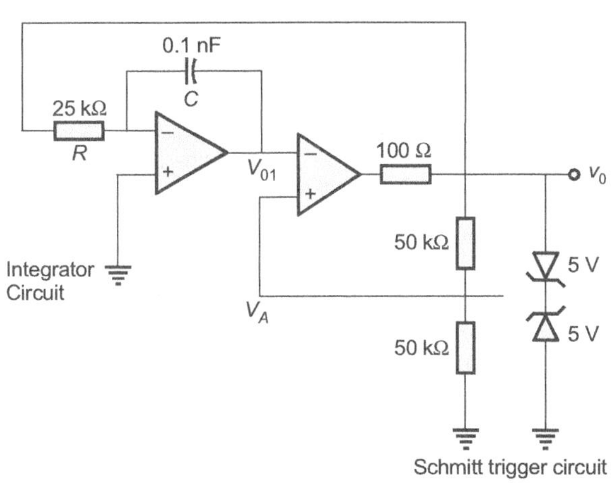

Here, $\quad V_0 = \pm 5\ \text{V}, \qquad V_A = \frac{V_0}{2} = \pm 2.5\ \text{V}$

$$V_{01} = -\frac{1}{RC}\int v_0 dt + k$$

when v_{01} exceed 2.5 V, v_0 falls from 5 V to –5 V.
when v_{01} falls below –2.5 V, v_0 falls from –5 V to 5 V.

Therefore,

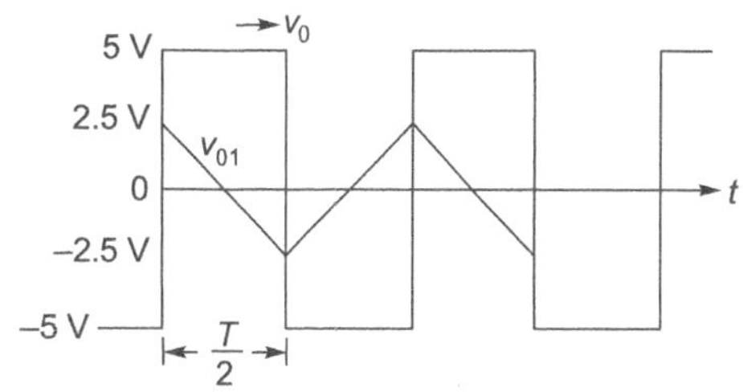

$0 < t < \frac{T}{2}$: $\quad V_0 = +5\text{V};\ V_A = 2.5\ \text{V}$

$$v_{01} = -\frac{1}{RC}\int v_0 dt + k_1 = \frac{-v_0}{RC}t + k_1$$

$$= -\frac{5}{RC}t + 2.5\ \text{V}$$

At $t = \frac{T}{2}$, $\quad v_{01} = -2.5\ \text{V}$

$$-2.5 = \frac{-5}{RC}\cdot\frac{T}{2} + 2.5$$

$$\Rightarrow \qquad T = 2RC$$

$$\Rightarrow \qquad f = \frac{1}{2RC}$$

$$= \frac{1}{2\times 25\times 10^3 \times 0.1\times 10^{-9}}$$

$$= 200\ \text{kHz}$$

Note : Actually the problem given is wrong. There should be an inventing amplifier circuit with unity gain between the two op-amp circuits. The expected circuit should be as shown below.

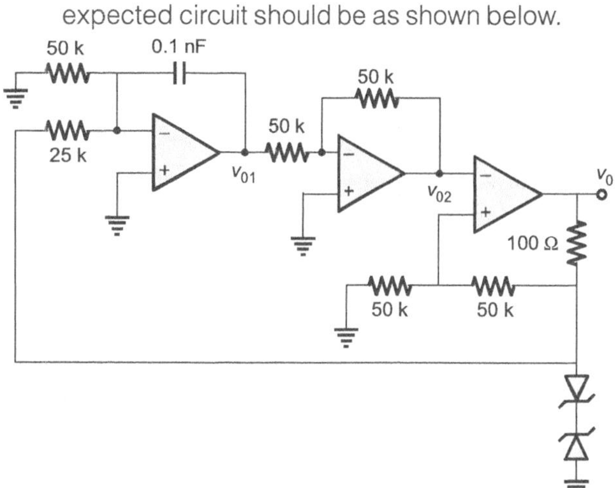

■■■

5 Miscellaneous

5.1 A class '*B*' push pull complementary symmetry amplifier uses
(a) Two npn transistors
(b) One pnp and one npn transistors
(c) Two pnp transistors
(d) One pre-amplifier of npn transistor followed by amplifier of two transistors.
[1998 : 1 M]

5.2 For full cycle of sinusoidal input match the amplifier class with its conduction cycle.
(a) class *A* (b) class *B*
(c) class *AB* (d) class *C*
(i) one half-cycle
(ii) less than half cycle
(iii) full cycle
(iv) more than half cycle but less than full cycle
[1998 : 1 M]

5.3 The 3 dB cut-off frequency of a first order analog high pass filter is ω_c. For the signal 0.5 sin $\omega_c t$, the output will have a phase shift of

(a) $-\frac{\pi}{2}$ (b) $-\frac{\pi}{4}$

(c) $\frac{\pi}{4}$ (d) $\frac{\pi}{2}$

[2001 : 1 M]

5.4 A 50 W power amplifier has a rated output of 30 V_{rms} and a voltage gain of 40 dB. It is connected to a 40 W loudspeaker having an internal resistance of 10 Ω. Determine the maximum input voltage that can be given to the power amplifier so that neither the power amplifier nor the loudspeaker is overloaded.
(a) 40 V (b) 20 V
(c) 0.4 V (d) 0.2 V
[2003 : 2 M]

5.5 For the *RC* circuit shown in the figure, the condition for obtaining an attenuation, $|V_{out}/V_{in}|$, of 1/3 at a frequency ω rad/s is

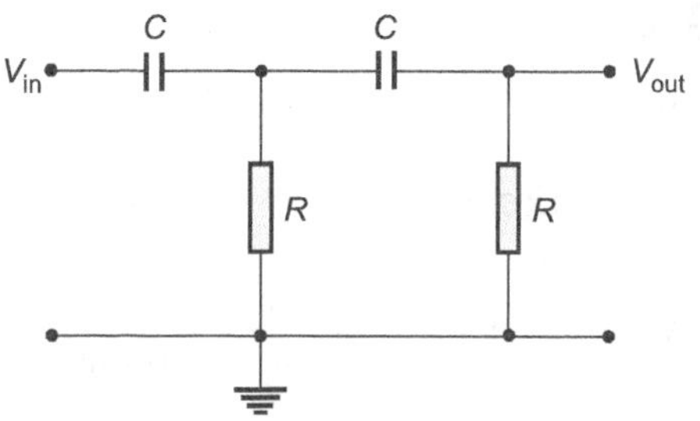

(a) $3\omega CR - 1 = 0$ (b) $2\omega CR - 1 = 0$
(c) $3\omega CR - 2 = 0$ (d) $\omega CR - 1 = 0$
[2005 : 2 M]

Statement for Linked Answer Questions (5.6 and 5.7):

Given figure shows a transistor circuit with feedback.

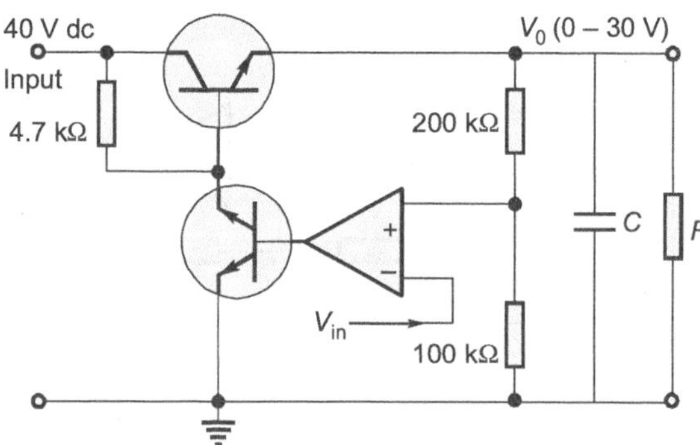

5.6 In the circuit, in order to get V_0 in the range of 0-30 V, the range of V_{in} is
(a) 0-30 V (b) 0-20 V
(c) 0-15 V (d) 0-10 V [2005 : 2 M]

5.7 If V_{in} is generated using an n-bit *DA* converter, the minimum value of *n* required, so that the value of V_0 can be set within an accuracy of less than 20 mV is,
(a) 9 (b) 10
(c) 11 (d) 12 [2005 : 2 M]

Statement for Linked Answer Questions (5.8 and 5.9):
*M*1, *M*2 and *M*3 in the circuit shown below are matched *N*-channel enhancement mode MOSFETs operating in saturation mode, forward voltage drop of each diode is 0.7 V, reverse leakage current of each diode is negligible and the opamp is ideal.

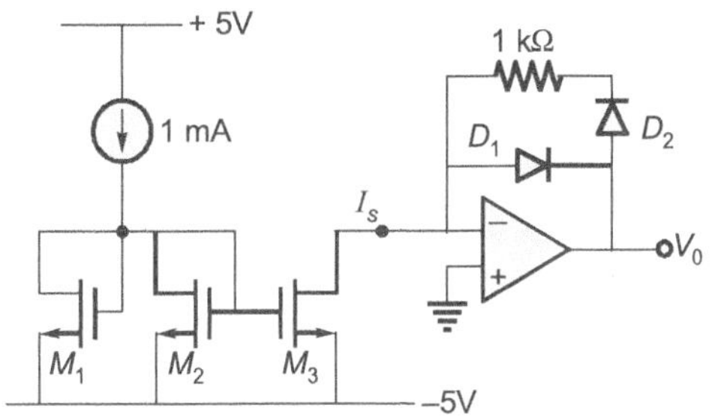

5.8 The current I_s in the circuit is

(a) –1 mA (b) 0.5 mA
(c) 1 mA (d) 2 mA [2011 : 2 M]

5.9 For the computed value of current I_s, the output voltage V_0 is

(a) 1.2 V (b) 0.7 V
(c) 0.2 V (d) –0.7 V [2011 : 2 M]

5.10 In the feedback network shown below, if the feedback factor k is increased, then the

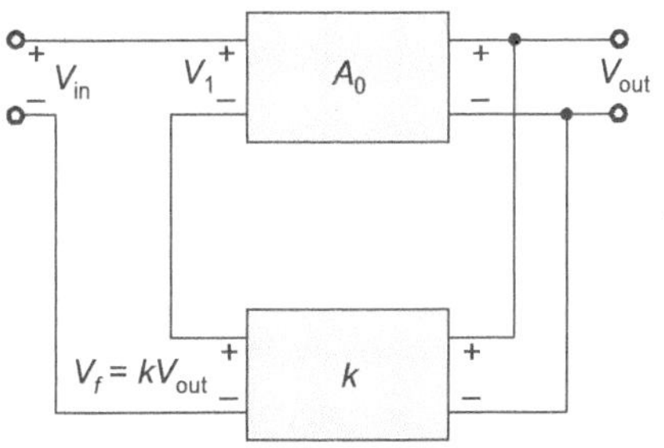

(a) input impedance increases and output impedance decreases.
(b) input impedance increases and output impedance also increases.
(c) input impedance decreases and output impedance also decreases.
(d) input impedance decreases and output impedance increases.

[2013 : 1 M]

5.11 For the given low-pass circuit shown in the figure, the cutoff frequency in Hz will be ___.

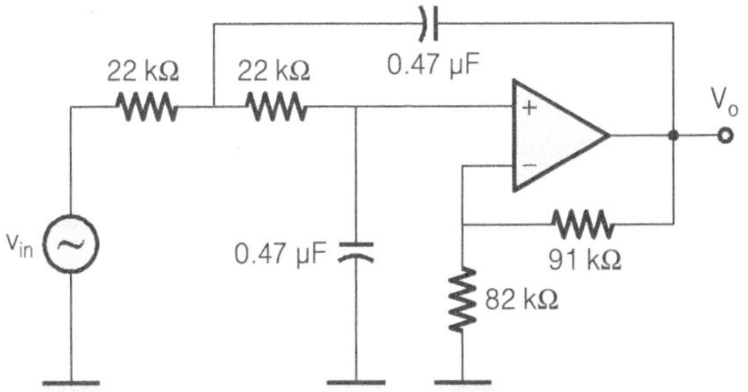

[2014 : 2 M]

5.12 A power line is coupled capacitively through various parasitic capacitances to a shielded signal line as shown in the figure. The conductive shield is grounded solidly at one end. Assume that the length of the signal wire extending beyond the shield, and the shield resistance are negligible. The magnitude of the noise voltage coupled to the signal line is

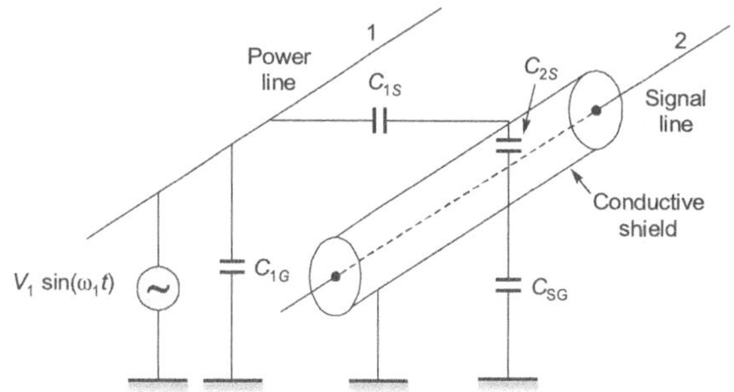

(a) directly proportional to C_{1G}
(b) inversely proportional to the power line frequency
(c) inversely proportional to C_{1S}
(d) zero

[2015 : 1 M]

5.13 For the circuit shown in the figure, the rising edge triggered D-flip flop with asynchronous reset has a clock frequency of 1 Hz. The NMOS transistor has an ON resistance of 1000 Ω and an OFF resistance of infinity. The nature of the output waveform is

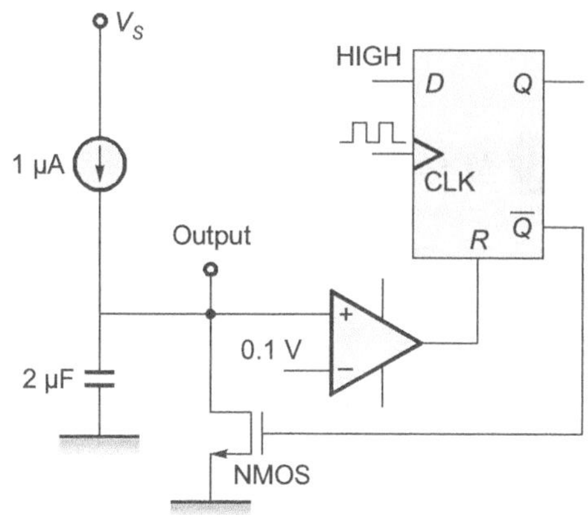

(a) 

(b) 

(c) 

(d) 

[2015 : 2 M]

5.14 In the circuit shown, the voltage source $v(t) = 15 + 0.1\sin(100t)$. The PMOS transistor is biased such that it is in saturation with its gate-source capacitance being 4 nF and its trans-conductance at the operating point being 1 mA/V. Other parasitic impedances of the MOSFET may be ignored. An external capacitor of capacitance 2 nF is connected across the PMOS transistor as shown. The input impedance in mega ohm as seen by the voltage source is _____ MΩ.

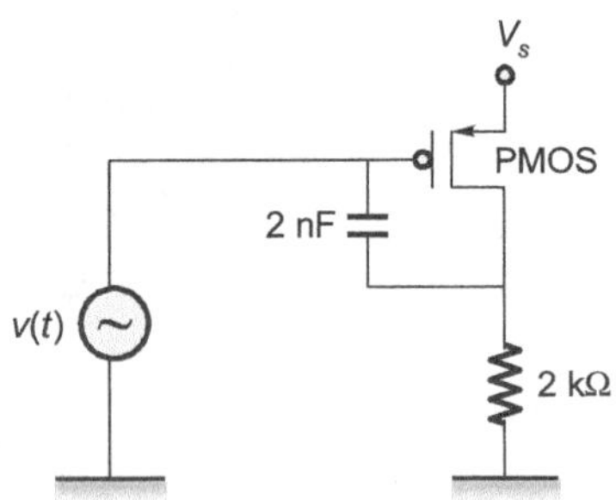

[2015 : 2 M]

5.15 In the circuit shown in the figure, both the NMOS transistors are identical with their threshold voltages being 5 V. Ignoring channel length modulation, the output voltage V_{out} in volt is _______ V.

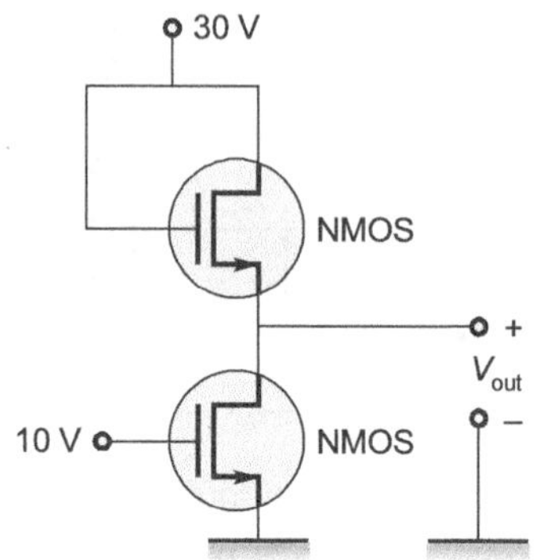

[2015 : 2 M]

5.16 An op-amp has ideal characteristics except that its open loop gain is given by the expression $A_V(s) = 10^4/(1 + 10^{-3}\,s)$. This op-amp is used in the circuit shown in the figure. The 3-dB bandwidth of the circuit, in rad/s, is

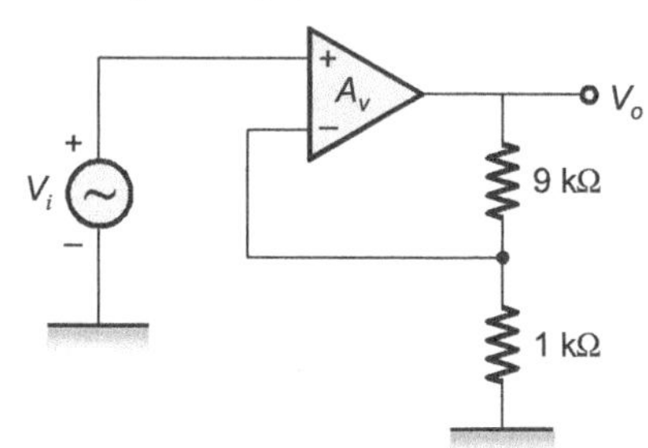

(a) 10^2
(b) 10^3
(c) 10^4
(d) 10^6

[2015 : 2 M]

Answers Miscellaneous

5.1 (b)	5.2 (Sol.)	5.3 (c)	5.4 (d)	5.5 (d)	5.6 (d)	5.7 (a)
5.8 (b)	5.9 (a)	5.10 (a)	5.11 (15.39)	5.12 (d)	5.13 (a)	5.14 (1)
5.15 (20)	5.16 (d)					

Explanations Miscellaneous

5.2 Sol.

(a)-(iii), (b)-(i), (c)-(ii), (d)-(ii)

5.3 (c)

∵ For HPF

$$\phi(\omega) = \tan^{-1}\left[\frac{\omega_H}{\omega}\right]$$

where, ω_H = 3 dB cut-off frequency
ω = Input frequency

here, $\omega_h = \omega_c$
$\omega = \omega_c$

$$\therefore \quad \phi(\omega) = \tan^{-1}\left[\frac{\omega_c}{\omega_c}\right] = \frac{\pi}{4}$$

5.4 (d)

Rated output of power amplifier is 30 V_{rms}
gain (voltage gain) in dB = 40 dB = 100
so maximum input voltage so that output can not exceed 30 V_{rms} is

$$\frac{V_{in}}{V_{out}} = \frac{1}{100} \qquad (\because \text{gain} = 100)$$

Now it is connected to loudspeaker of rated power = 40W and internal resistance $R_{in} = 10\ \Omega$.
So, maximum voltage across it can be

$$\frac{V_{rms}^2}{R_{in}} \le 40\ \text{W}$$

$$V^2_{rms} \le 40 \times R_{in}$$

$$V^2_{rms} \le 400$$

$$V_{rms} \le 20\ \text{V}$$

So, voltage across loudspeaker can not exceed 20 V.
So, maximum voltage at the output of amplifier can be less than or equal to 20 V.
So, input voltage at amplifier will be

$$V_{in} \le \frac{V_0}{100}$$

$$V_{in} \le \frac{20}{100}$$

Since 0.2 is smaller than 0.3 V, so if we keep input less than or equal to 0.2 V then both of them will not be overloaded.
So, $(V_{in})_{max} = 0.2$ V

5.5 (d)

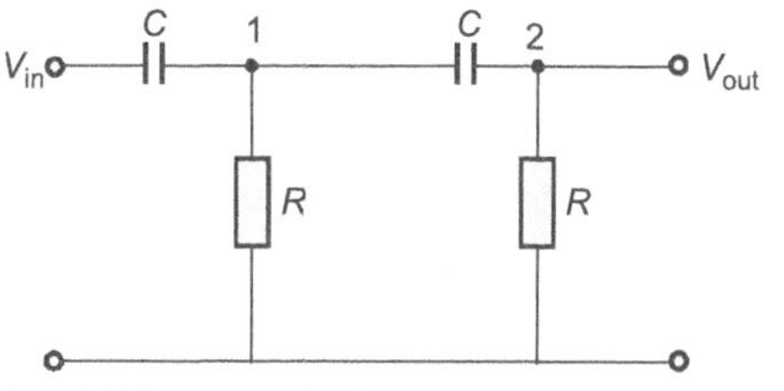

Writing KCL at node 1, we get

$$(V_1 - V_{in})sC + \frac{V_1}{R} + (V_1 - V_{out})sC = 0$$

$$\frac{V_1}{R}[2\,sCR + 1] - V_{in}\cdot sC - V_{out}\cdot sC = 0 \qquad \ldots(i)$$

Writing KCL at node 2, we get

$$(V_{out} - V_1)sC + \frac{V_{out}}{R} = 0 \qquad \ldots(ii)$$

From equation (1) and (2), we have

$$\frac{(2\,sCR+1)(sCR+1)}{sCR^2}V_{out} - V_{in}\,sC - V_{out}\,sC = 0$$

$$\frac{[2(sCR)^2 + 3sCR + 1 - (sCR)^2]}{sCR^2}V_{out} = V_{in}\,sC$$

$$\frac{V_{out}}{V_{in}} = \frac{(sCR)^2}{(sCR)^2 + 3sCR + 1}$$

when, $\omega = \dfrac{1}{RC}$

$$\omega CR - 1 = 0$$

then, $$\frac{V_{out}}{V_{in}} = \frac{-1}{3j} = \frac{j}{3}$$

$$\left|\frac{V_{out}}{V_{in}}\right| = \frac{1}{3}$$

5.6 (d)

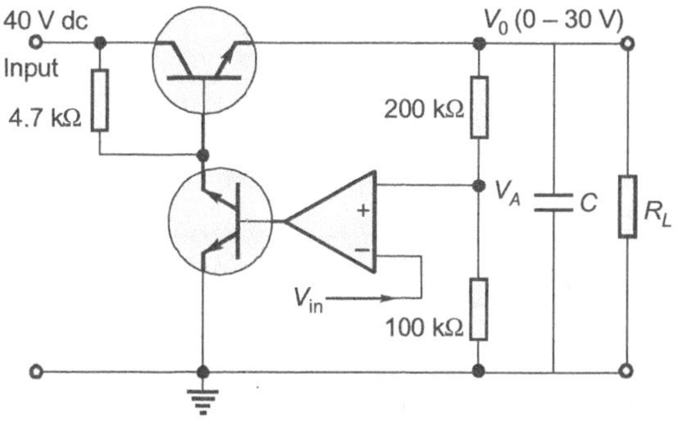

By virtual ground concept by op-amp,

$$V_{in} = V_A$$

$$V_A = \frac{100}{100+200} \times 30$$

$$= \frac{100}{300} \times 30 = 10 \text{ V}$$

$$\therefore \quad V_{in} = 0 \text{ to } 10 \text{ V}$$

5.7 (a)

Maximum error occur in a *DAC* is $\pm\Delta$.

Where, Δ = Step size

$$\Delta = \frac{V_{FS}}{2^n - 1}$$

where, V_{FS} = Full scale voltage

n = Number of bits

$$\therefore \quad 20 \times 10^{-3} = \frac{10}{2^n - 1}$$

$$2^n - 1 = 500$$

$$\Rightarrow \quad 2^n = 500 + 1 = 501$$

so, the minimum value of $n = 9$

5.8 (b)

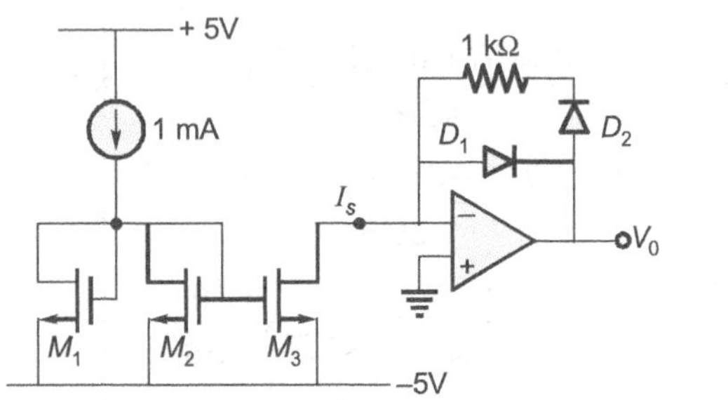

In the given circuit diagram, the MOSFETs M_1 and M_2 are matched.

Due to symmetry,

$$I_{D1} = I_{D2} = 0.5 \text{ mA}$$

and MOSFETs M_2 and M_3 are the current mirror to each other

$$\therefore \quad I_{D2} = I_{D3} = 0.5 \text{ mA}$$

ultimately $\boxed{I_{D3} = I_s = 0.5 \text{ mA}}$

5.9 (a)

Now as per the direction of I_s circuit will be

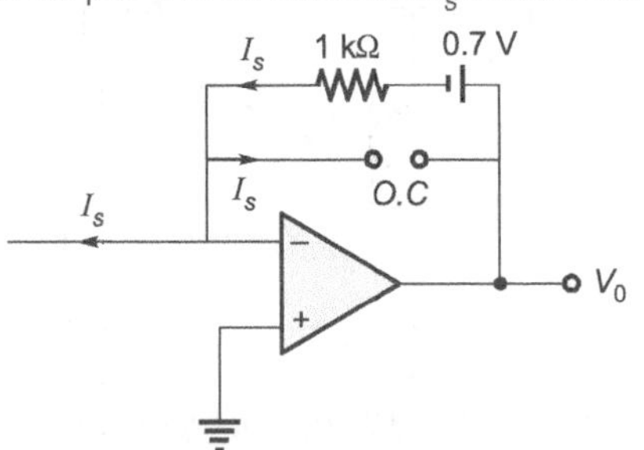

$$V_0 = 0.7 \text{ V} + I_s \times 10^3$$
$$= 0.7 + 0.5 \times 10^{-3} \times 10^3$$
$$= 1.2 \text{ V}$$

5.10 (a)

The given configuration is a voltage-series feedback configuration.

So, the input impedance increases

$$R_{if} = R_i(1 + A_o k)$$

and, the output impedance decreases

$$R_{of} = \frac{R_o}{1 + A_o k}$$

5.11 15.39 (15 to 16)

$$f = \frac{1}{2\pi RC} = \frac{1}{2 \times 22 \times 10^3 \times 0.47 \times 10^6}$$
$$= 15.39 \text{ Hz}$$

5.13 (a)

- Clock is positive edge trigger and input to the D-FF is constant at logic-1.
- Reset pin of FF is active high and it can function independent to the clock.
- Reset will be triggered when V^+ terminal voltage just crosses 0.1 V.
- N-MOS behave as open switch (off state, $R = \infty$) when $\bar{Q} = 0$ and it will behave as a resistance of 1000 Ω when $\bar{Q} = 1$ (as ON resistance in specified).
- Now for analysis, assume capacitor is initially uncharged and *a* positive edge is triggered to D-FF, as its input in high. Then ($Q = 1$ and $\bar{Q} = 0$).
- When $\bar{Q} = 0$, N-MOS is off, assuming the input resistance of comparator infinite, then the source current can flow only through the capacitor. Now we have to see how much time the capacitor will take to reach 0.1 V. So that $V^+ > V^-$ and R will change from 0 to 1 and as the reset terminal be triggered the state of FF changes i.e. ($Q = 0, \bar{Q} = 1$) and N-MOS gets ON. As the integration of step is ramp, so if we integrate the constant current the voltage signal will be ramp.

  Capacitor voltage equation,

$$V_C = \frac{1}{C}\int_0^t I_C dt$$

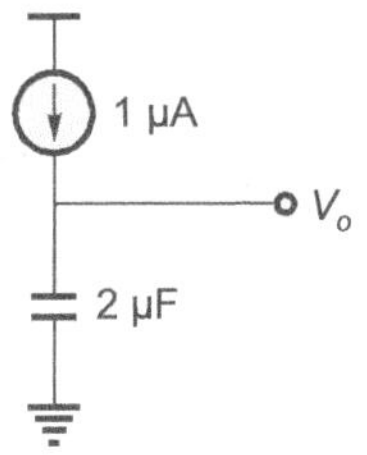

$$0.1\text{ V} = \frac{1}{2\times10^{-6}}\int_0^t 1\times10^{-6}dt$$

$$0.1\text{ V} = \frac{t}{2} \Rightarrow t = 0.2\text{ sec} = 200\text{ msec}$$

- So at $t = 0.2$ sec, N-MOS is ON. Now the circuit is equivalent to a current source connected to a parallel combination of RC with non zero initial voltage of capacitor (0.1 V).
- Now the capacitor will discharge from its initial voltage with time constant $\tau = RC = 2$ ms and at 10τ it will reach steady state i.e. at 20 msec. The nature of discharge is exponential decay.

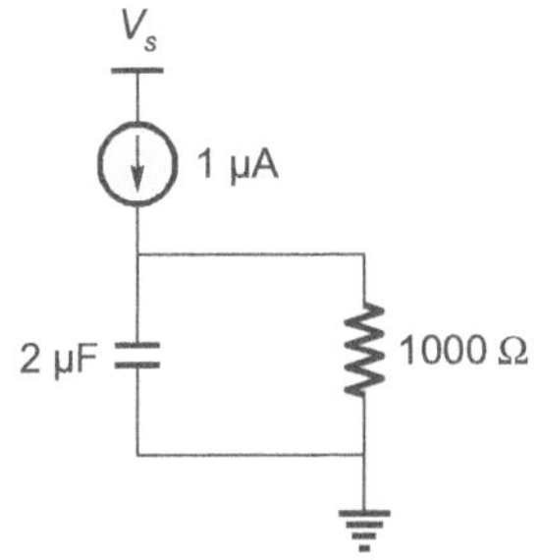

- So we can conclude from the above analysis that capacitor takes 200 msec to charge upto 0.1 V and 20 msec to discharge to the steady state value.
- So if the +ve edge of the clock appeared at $t = 0$ then at 200 msec it completed its charging and at 220 msec it completed its discharging. This continues in each clock cycle.
- The following waveform gives further clarity as $f_{clk} = 1$ Hz, $T_{clk} = 1000$ msec.

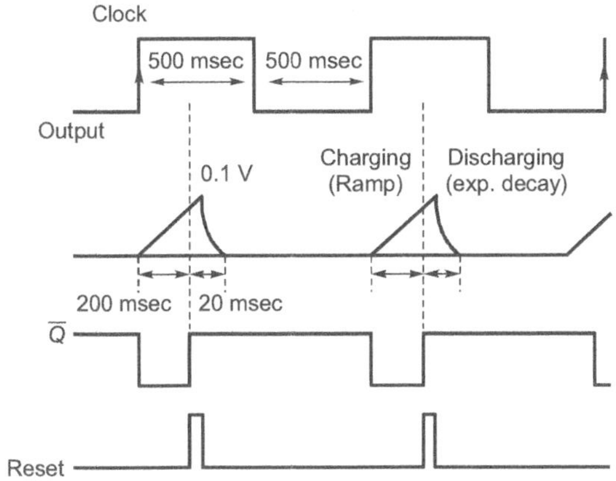

5.14 (1)

For DC component of voltage source the impedance will be infinity.

For AC component of voltage source the impedance can be calculated using the small signal model of the circuit.

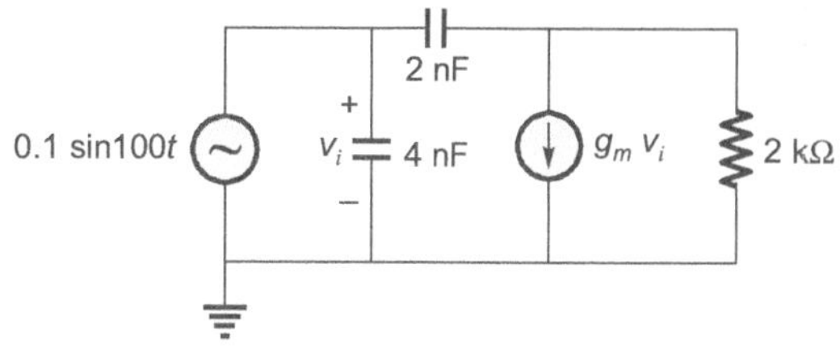

gain of the amplifier is

$$-g_m R_L = (-1\times10^{-3}\times2\times10^3) = -2$$

Applying Miller theorem, the total input capacitance will be 10 nF, thus the input impedance is

$$Z = \frac{1}{\omega C} = \frac{1}{100\times10\times10^{-9}} = 1\text{ M}\Omega$$

5.15 (20)

$$V_{GS1} = V_{GS2} \quad \text{(both are identical transistor)}$$

$$30 - V_{out} = 10 - 0$$

$$V_{out} = 30 - 10 = 20\text{ V}$$

5.16 (d)

Closed loop transfer

$$A_{CL} = \frac{A_{OL}}{1 + A_{OL}\beta}$$

Here, $$A_{CL} = \frac{10^4}{1+10^{-3}s}$$

$$\beta = \frac{R_i}{R_i + R_E} = \frac{1}{9+1} = 0.1$$

$$\therefore\ A_{CL} = \frac{\dfrac{10^4}{1+10^{-3}s}}{1+\dfrac{10^4\times0.1}{1+10^{-3}s}} = \frac{10^4}{1+10^{-3}s+10^3}$$

$$= \frac{10}{10^{-3}+10^{-6}s+1} \approx \frac{10}{1+\dfrac{s}{10^6}}$$

Comparing it with the standard first order transfer

i.e. $\dfrac{K1}{1+s\tau}$.

$$\tau = \frac{1}{10^6}$$

$$\text{B.W} = \frac{1}{\tau} = \frac{1}{1/10^6} = 10^6\text{ rad/sec.}$$

■■■

Signal & Systems and Communication Systems

UNIT V

CONTENTS

1. Continuous Time LTI System 242
2. Discrete Time LTI Systems 250
3. Fourier Series 255
4. Fourier Transform 257
5. Laplace Transform 261
6. Z-Transform 265
7. Discrete Time Fourier Transform (D.T.F.T.) 270
8. Discrete Fourier Transform (D.F.T.) and Frequency Response 271
9. Sampling 274
10. Digital Filters 277
11. Communication Systems 285

Signal & Systems and Communication Systems

Syllabus

Periodic, aperiodic and impulse signals; Laplace, Fourier and z-transforms; transfer function, frequency response of first and second order linear time invariant systems, impulse response of systems; convolution, correlation. Discrete time system: impulse response, frequency response, pulse transfer function; DFT and FFT; basics of IIR and FIR filters.

Amplitude- and frequency modulation and demodulation; Shannon's sampling theorem, pulse code modulation; frequency and time division multiplexing, amplitude-, phase-, frequency-, quadrature amplitude, pulse shift keying for digital modulation.

Analysis of Previous GATE Papers

Exam Year	1 Mark Ques.	2 Marks Ques.	Total Marks
1995	–	–	–
1996	–	2	**4**
1997	–	1	**2**
1998	1	–	**1**
1999	1	1	**3**
2000	1	–	**1**
2001	1	1	**3**
2002	1	–	**1**
2003	3	9	**21**
2004	3	8	**19**
2005	3	3	**9**
2006	–	8	**16**
2007	2	3	**8**
2008	3	7	**17**
2009	2	4	**10**

Exam Year	1 Mark Ques.	2 Marks Ques.	Total Marks
2010	6	3	**12**
2011	7	3	**13**
2012	4	3	**10**
2013	7	3	**13**
2014	4	2	**8**
2015	2	6	**14**
2016	4	4	**12**
2017	5	5	**15**
2018	5	4	**13**
2019	2	5	**12**
2020	3	2	**7**
2021	4	3	**10**
2022	6	3	**12**
2023	3	3	**9**
2024	4	3	**10**

1 Continuous Time LTI System

1.1 Given,

$x(t) * x(t) = t \exp(-2t)\, u(t)$

the function $x(t)$ is

(a) $\exp(-2t)\, u(t)$ (b) $\exp(-t)\, u(t)$
(c) $t\exp(-t)\, u(t)$ (d) $0.5 \exp(-t)\, u(t)$

[2006 : 2 M]

1.2 The Fourier series for a periodic signal is given as

$x(t) = \cos(1.2\pi t) + \cos(2\pi t) + \cos(2.8\pi t)$.

The fundamental frequency of the signal is

(a) 0.2 Hz (b) 0.6 Hz
(c) 1.0 Hz (d) 1.4 Hz

[2006 : 2 M]

1.3 Consider the periodic signal

$x(t) = (1 + 0.5 \cos 40\pi t) \cos 200\pi t$

where t is in seconds. Its fundamental frequency in Hz is

(a) 20 (b) 40
(c) 100 (d) 200

[2007 : 1 M]

1.4 The signal $x(t)$ and $h(t)$ shown in the figures are convolved to yield $y(t)$.

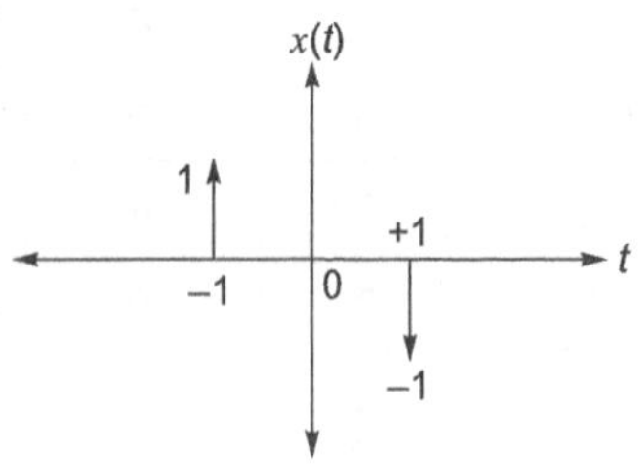

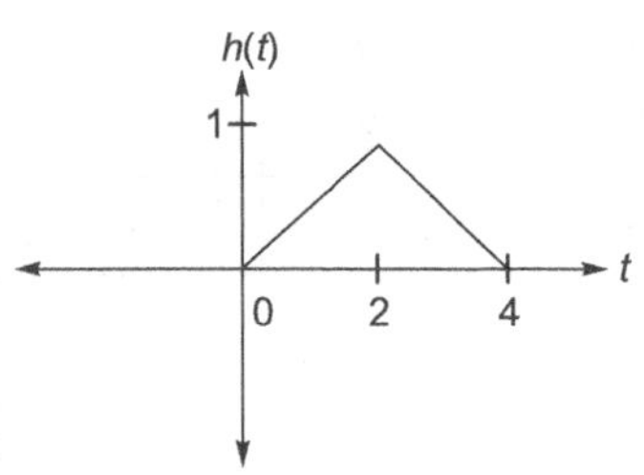

Which one of the following figures represent the output $y(t)$?

(a)

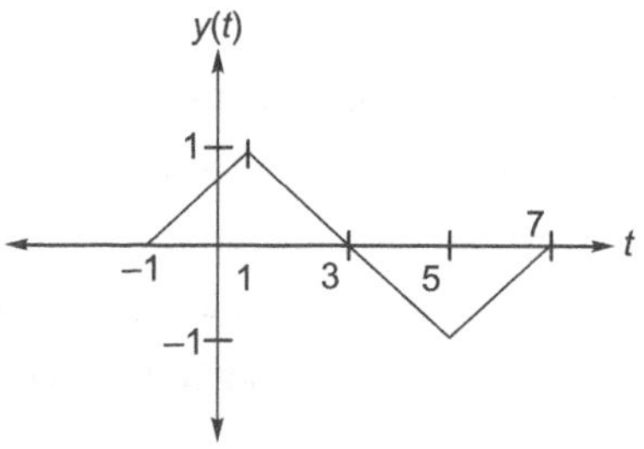

(b)

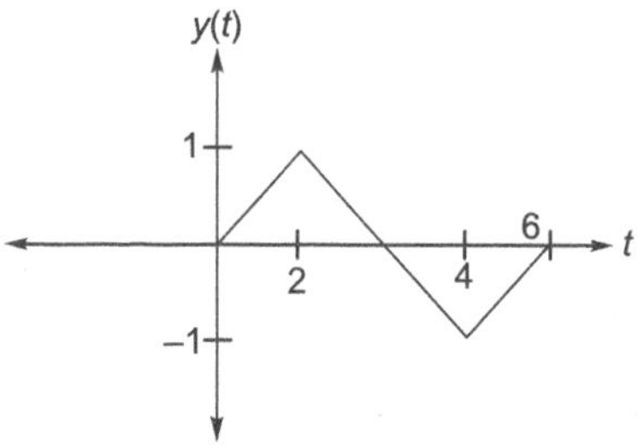

(c)

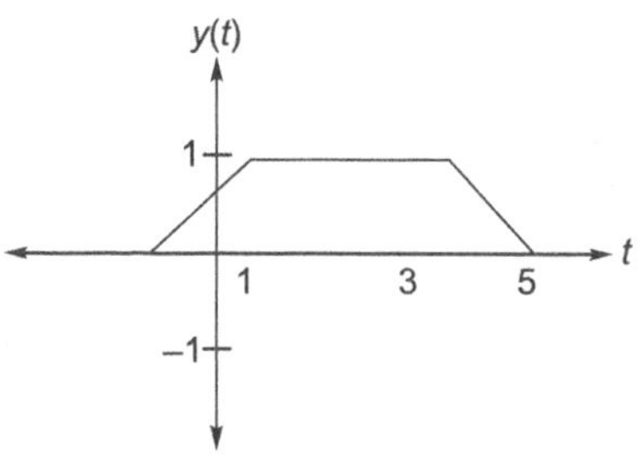

(d)

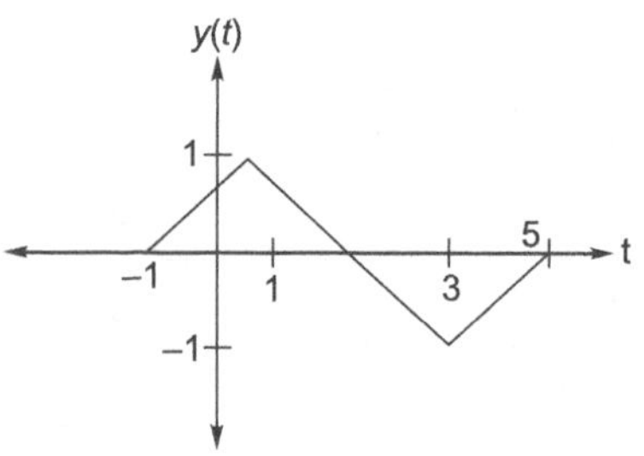

[2007 : 2 M]

1.5 The step response of a linear time invariant system is $y(t) = 5\, e^{-10t}\, u(t)$, where $u(t)$ is the unit step function. If the output of the system corresponding to an impulse input $\delta(t)$ is $h(t)$, then $h(t)$ is

(a) $-50\, e^{-10t}\, u(t)$
(b) $5\, e^{-10t}\, \delta(t)$
(c) $5\, u(t) - 50\, e^{-10t}\, u(t)$
(d) $5\, \delta(t) - 50\, e^{-10t}\, u(t)$

[2008 : 2 M]

1.6 The fundamental period of $x(t) = 2\sin\pi t + 3\sin 3\pi t$, with t expressed in seconds, is

(a) 1 (b) 0.67

(c) 2 (d) 3

[2009 : 1 M]

1.7 For input $x(t)$, an ideal impulse sampling system produces the output

$$y(t) = \sum_{k=-\infty}^{0} x(kT)\,\delta(t-kT)$$

where $\delta(t)$ is the Dirac delta function.
The system is

(a) non-linear and time invariant
(b) non-linear and time varying
(c) linear and time invariant
(d) linear and time varying

[2009 : 2 M]

1.8 The integral $\int_{-\infty}^{\infty} \delta\left(t-\frac{\pi}{6}\right) 6\sin(t)\,dt$ evaluate to

(a) 6 (b) 3

(c) 1.5 (d) 0

[2010 : 1 M]

1.9 The input $x(t)$ and the corresponding output $y(t)$ of a system are related by

$$y(t) = \int_{-\infty}^{5t} x(\tau)\,d\tau$$

The system is

(a) time invariant and causal
(b) time invariant and non causal
(c) time variant and non causal
(d) time variant and causal

[2010 : 1 M]

1.10 The continuous-time signal $x(t) = \sin\omega_0 t$ is a periodic signal. However, for its discrete-time counterpart $x[n] = \sin\omega_0 n$ to be periodic, the necessary condition is

(a) $0 \le \omega_0 < 2\pi$

(b) $\frac{2\pi}{\omega_0}$ to be an integer

(c) $\frac{2\pi}{\omega_0}$ to be a ratio of integers

(d) none

[2011 : 1 M]

1.11 The integral

$$\frac{1}{\sqrt{2\pi}}\int_{-\infty}^{\infty} t^2 e^{-t^2/2}\delta(1-2t)\,dt \text{ is equal to}$$

(a) $\frac{1}{8\sqrt{2\pi}}e^{-1/8}$ (b) $\frac{1}{4\sqrt{2\pi}}e^{-1/8}$

(c) $\frac{1}{\sqrt{2\pi}}e^{-1/2}$ (d) 1

[2011 : 1 M]

1.12 Consider a system with input $x(t)$ and output $y(t)$ related as follows

$$y(t) = \frac{d}{dt}\left\{e^{-t}x(t)\right\}$$

Which one of the following statements is **TRUE?**

(a) The system is nonlinear
(b) The system is time-invariant
(c) The system is stable
(d) The system has memory

[2011 : 1 M]

1.13 The input $x(t)$ and output $y(t)$ of a system are related as $y(t) = \int_{-\infty}^{t} x(\tau)\cos(3\tau)\,d\tau$. The system is

(a) time-invariant and stable
(b) stable and not time-invariant
(c) time-invariant and not stable
(d) not time-invariant and not stable

[2012 : 2 M]

1.14 Two system with impulse responses $h_1(t)$ and $h_2(t)$ are connected in cascade. Then the overall impulse response of the cascaded system is given by

(a) product of $h_1(t)$ and $h_2(t)$
(b) sum of $h_1(t)$ and $h_2(t)$
(c) convolution of $h_1(t)$ and $h_2(t)$
(d) subtraction of $h_2(t)$ from $h_1(t)$

[2013 : 1 M]

1.15 For a periodic signal
$v(t) = 30\sin 100t + 10\cos 300t + 6\sin(500t + \pi/4)$
the fundamental frequency in rad/s is

(a) 100 (b) 300

(c) 500 (d) 1500

[2013 : 1 M]

1.16 Which one of the following statements is not true for a continuous time causal and stable LTI system?

(a) All the poles of the system must lie on the left side of the $j\omega$-axis.
(b) Zeros of the system can lie anywhere in the s-plane.
(c) All the poles must lie within $|s| = 1$.
(d) All the roots of the characteristic equation must be located on the left side of the $j\omega$-axis.

[2013 : 1 M]

1.17 Assuming zero initial condition, the response $y(t)$ of the system given below to a unit step input $u(t)$ is

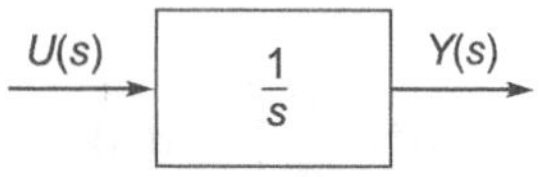

(a) $u(t)$ (b) $t\,u(t)$
(c) $\frac{t^2}{2}u(t)$ (d) $e^{-t}u(t)$

[2013 : 1 M]

1.18 The impulse response of a system is $h(t) = t\,u(t)$. For an input $u(t-1)$, the output is

(a) $\frac{t^2}{2}u(t)$ (b) $\frac{t(t-1)}{2}u(t-1)$
(c) $\frac{(t-1)^2}{2}u(t-1)$ (d) $\frac{t^2-1}{2}u(t-1)$

[2013 : 1 M]

1.19 The impulse response of a continuous time system is given by $h(t) = \delta(t-1) + \delta(t-3)$. The value of the step response at $t = 2$ is

(a) 0 (b) 1
(c) 2 (d) 3

[2013 : 2 M]

1.20 The fundamental period of the signal $x(t) = 2\cos\left(\frac{2\pi t}{3}\right) + \cos(\pi t)$, in seconds, is ___ s.

[2015 : 2 M]

1.21 A periodic signal $x(t)$ shown in the figure. The fundamental frequency of the signal $x(t)$ in Hz is _____.

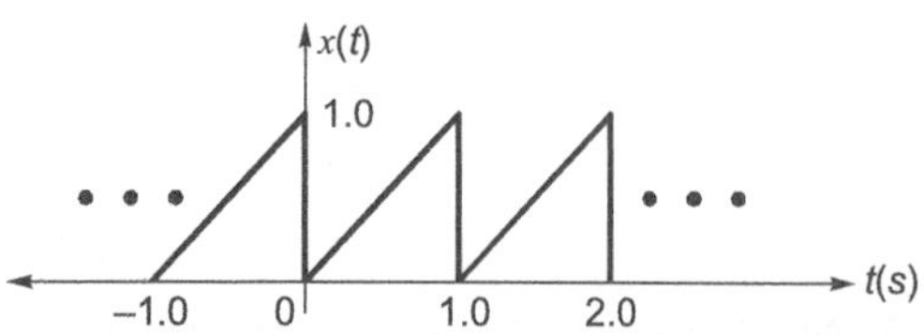

[2017 : 1 M]

1.22 Two periodic signals $x(t)$ and $y(t)$ have the same fundamental period of 3 seconds. Consider the signal $z(t) = x(-t) + y(2t+1)$. The fundamental period of $z(t)$ in seconds is

(a) 1 (b) 1.5
(c) 2 (d) 3

[2018 : 1 M]

1.23 Consider signal $x(t) = \begin{cases} 1, & |t| \le 2 \\ 0, & |t| > 2 \end{cases}$. Let $\delta(t)$ denote the unit impulse (Dirac-delta) function. The value of the integral $\int_0^5 2x(t-3)\delta(t-4)\,dt$ is

(a) 2 (b) 1
(c) 0 (d) 3

[2018 : 1 M]

1.24 The transfer function relating the input $x(t)$ to the output $y(t)$ of a system is given by $G(s) = \frac{1}{(s+3)}$.

A unit-step input is applied to the system at time $t = 0$. Assuming that $y(0) = 3$, the value of $y(t)$ at time $t = 1$ is________(Answer should be rounded off to two decimal places).

[2019 : 2 M]

1.25 The signal $\sin\left(\sqrt{2\pi t}\right)$ is

(a) periodic with period $T = \sqrt{2\pi}$
(b) periodic with period $T = 2\pi$
(c) periodic with period $T = 4\pi^2$
(d) not periodic

[2021 : 1 M]

1.26 The output of a system $y(t)$ is related to its input $x(t)$ according to the relation $y(t) = x(t)\sin(2\pi t)$. This system is _______

(a) Linear and time-variant
(b) Non-linear and time-invariant
(c) Linear and time-invariant
(d) Non-linear and time-variant

[2022 : 1 M]

1.27 A unity-gain negative-feedback control system has a loop-gain $L(s)$ given by

$$L(s) = \frac{6}{s(s-5)}$$

The closed-loop system is _______.

(a) Causal and stable
(b) Causal and unstable
(c) Non-causal and stable
(d) Non-causal and unstable

[2022 : 1 M]

1.28 Let $y(t) = x(4t)$, where $x(t)$ is a continuous-time periodic signal with fundamental period of 100 s. The fundamental period of $y(t)$ is _______ s (rounded off to the nearest integer).

[2023 : 1 M]

1.29 The impulse response of an LTI system is $h(t) = \delta(t) + 0.5\,\delta(t-4)$, where $\delta(t)$ is the continuous-time unit impulse signal. If the input signal $x(t) = \cos\left(\frac{7\pi}{4}t\right)$, the output is _______.

(a) $0.5\cos\left(\frac{7\pi}{4}t\right)$ (b) $1.5\cos\left(\frac{7\pi}{4}t\right)$

(c) $0.5\sin\left(\frac{7\pi}{4}t\right)$ (d) $1.5\sin\left(\frac{7\pi}{4}t\right)$

[2023 : 2 M]

■■■

Answers Continuous Time LTI System

1.1 (a)	1.2 (a)	1.3 (a)	1.4 (d)	1.5 (d)	1.6 (c)	1.7 (d)
1.8 (b)	1.9 (c)	1.10 (c)	1.11 (a)	1.12 (c)	1.13 (d)	1.14 (c)
1.15 (a)	1.16 (c)	1.17 (b)	1.18 (c)	1.19 (b)	1.20 (6)	1.21 (1)
1.22 (d)	1.23 (a)	1.24 (0.47)	1.25 (d)	1.26 (a)	1.27 (b)	1.28 (25)
1.29 (a)						

Explanations Continuous Time LTI System

1.1 (a)

Given,

$$x(t) * x(t) = t\exp(-2t)\,u(t)$$

taking Laplace transform on both the sides,

$$X(s)\cdot X(s) = \frac{1}{(s+2)^2}$$

$$\therefore \quad X(s) = \frac{1}{s+2}$$

taking inverse Laplace transform of $X(s)$ is

$$\therefore \quad x(t) = \exp(-2t)\,u(t)$$

1.2 (a)

Given signal is

$x(t) = \cos(1.2\pi t) + \cos(2\pi t) + \cos(2.8\pi t)$

Step-1:

$$\omega_1 = 1.2\pi$$

$$\Rightarrow \quad \frac{2\pi}{T_1} = 1.2\pi$$

$$\Rightarrow \quad T_1 = \frac{2\pi}{1.2\pi} = \frac{20}{12} = \frac{5}{3};\ \omega_2 = 2\pi$$

$$\Rightarrow \quad \frac{2\pi}{T_2} = 2\pi$$

$$\Rightarrow \quad T_2 = 1$$

$$\omega_3 = 2.8\pi$$

$$\Rightarrow \quad \frac{2\pi}{T_3} = 2.8\pi$$

$$\Rightarrow \quad T_3 = \frac{2}{2.8} = \frac{5}{7}$$

Step-2:

$$\frac{T_1}{T_2} = \frac{5}{3},\quad \frac{T_1}{T_3} = \frac{7}{3}$$

Step-3:

L.C.M. of denominators of step 2 is

L.C.M. = 3.

Step-4:

$$T = \text{L.C.M.}\{T_1, T_2, T_3\}$$
$$= 5 \text{ sec.}$$

Fundamental period = 5 sec.

Fundamental frequency of the signal

$$= \frac{1}{T} = \frac{1}{5} = 0.2 \text{ Hz.}$$

1.3 (a)

Given signal is

$$x(t) = \cos 200\pi t + \frac{1}{2}(\cos 40\pi t \times \cos 200\pi t)$$

$$= \cos 200\pi t + \frac{1}{4}\cos 240\pi t + \frac{1}{4}\cos 160\pi t$$

G.C.D. of frequency components $(\omega) = 40\pi$

$$2\pi f = 40\pi$$

$$\therefore \quad f = 20 \text{ Hz}$$

1.4 (d)

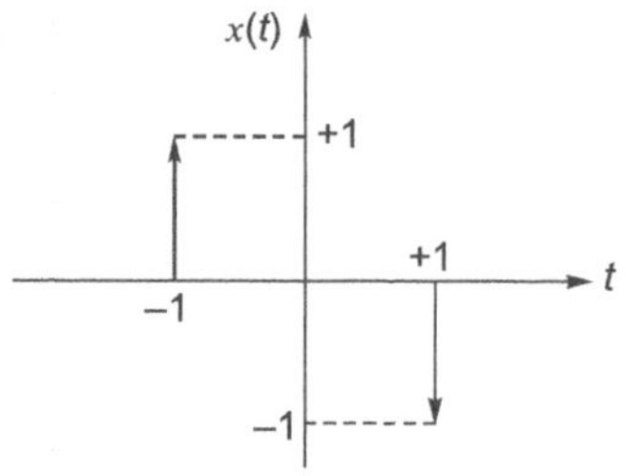

$$\Rightarrow \quad x(t) = \delta(t+1) - \delta(t-1)$$

so, $x(t) * h(t) = y(t)$

$$= h(t+1) - h(t-1)$$

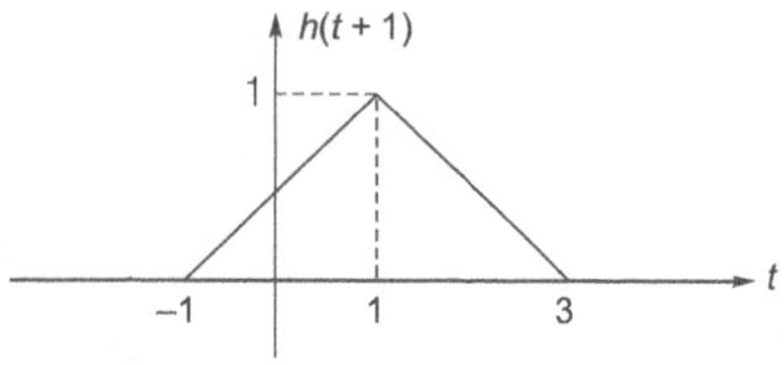

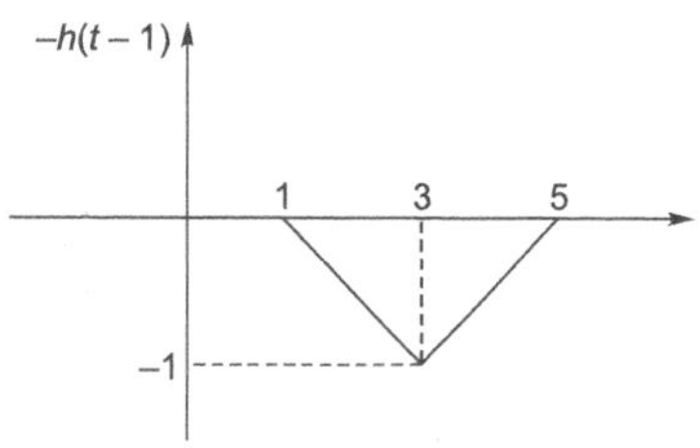

so, $\quad y(t) = h(t+1) - h(t-1)$

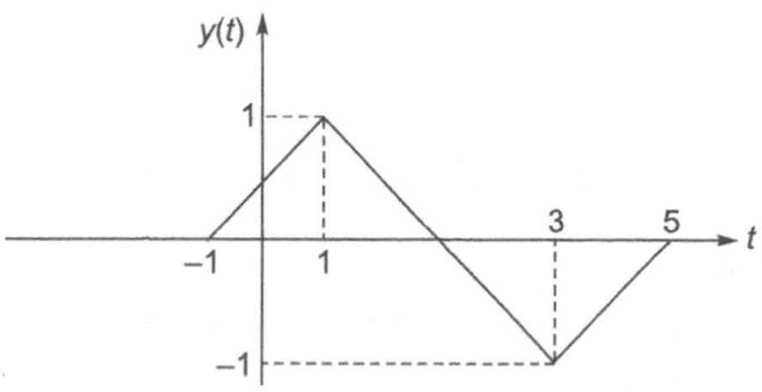

1.5 (d)

Step response $s(t) = 5\,u(t)\,e^{-10t}$

Impulse response $h(t) = \dfrac{ds(t)}{dt}$

$$\therefore \quad h(t) = \frac{ds(t)}{dt} = \frac{d\{5e^{-10t}u(t)\}}{dt}$$

$$= -50\,e^{-10t}\,u(t) + 5\,e^{-10t}\,du(t)/dt$$

$$= -50\,e^{-10t}\,u(t) + 5\,\delta(t)\,dt$$

1.6 (c)

Given signal is,

$$x(t) = 2\sin \pi t + 3\sin 3\pi t$$

$$\omega_1 = \pi$$

$$\Rightarrow \quad \frac{2\pi}{T_1} = \pi \Rightarrow T_1 = 2$$

$$\omega_2 = 3\pi$$

$$\Rightarrow \quad \frac{2\pi}{T_2} = 3\pi$$

$$\Rightarrow \quad T_2 = \frac{2}{3}$$

$$\frac{T_1}{T_2} = \frac{2}{2/3} = \frac{6}{2} = \frac{3}{1}$$

LCM of denominator of $\dfrac{T_1}{T_2} = 1$

Fundamental period

$$= (\text{LCM}) \times T_1 = 1 \times 2 = 2 \text{ sec.}$$

1.7 (d)

Given signal is

$$y(t) = \sum_{k=-\infty}^{0} x(kT)\,\delta(t-kT)$$

This is the representation of a signal $x(t)$ in weighted and sum form and it obeys the principle of superposition and homogeneity. So, this is linear but it is time varying as for $x(t-t_0)$,

$$y'(t) = \Sigma x(kT - t_0)\,\delta(t-kT)$$

and $\quad y(t-t_0) = \Sigma x(kT)\,\delta(t-t_0-kT)$,

Therefore $y(t-t_0) \neq y'(t)$, so time varying.

1.8 (b)

Given signal is

$$x(t) = \int_{-\infty}^{\infty} \delta\left(t - \frac{\pi}{6}\right) 6\sin(t)\,dt$$

By shifting property of unit impulse function

$$\int_{t_1}^{t_2} x(t)\delta(t-t_0)dt = \begin{cases} x(t_0); & t_1 < t_0 < t_2 \\ 0; & \text{elsewhere} \end{cases}$$

$$\int_{-\infty}^{\infty} \delta\left(t - \frac{\pi}{6}\right) 6\sin(t)\,dt = 6\cdot\sin\frac{\pi}{6} = 6 \times \frac{1}{2} = 3$$

Hence, correct option is (b).

1.9 (c)

$$y(t) = \int_{-\infty}^{5t} x(\tau)\,d\tau$$

For $t = 1$,

$$y(1) = \int_{-\infty}^{5} x(\tau)\,d\tau$$

here output depends on future values of $x(t)$, so system is non-causal.

Now for shifted input $x(t-t_0)$, output is

$$y'(t) = \int_{-\infty}^{5t} x(\tau - t_0)\, d\tau = \int_{-\infty}^{5t} x(\tau - t_0)\, d\tau$$

$$\tau - t_0 = \tau'$$

$$\Rightarrow \quad y'(t) = \int_{-\infty}^{(5t-t_0)} x(\tau')\, d\tau'$$

and $$y(t-t_0) = \int_{-\infty}^{5(t-t_0)} x(\tau)\, d\tau \neq y'(t)$$

So system is time variant.

1.10 (c)

$\sin \omega_0 n$ to be periodic

$$\frac{\omega_0}{2\pi} = \frac{m}{N}$$

where m and N both are integers.

$\therefore \frac{2\pi}{\omega_0}$ will be ratio of integers.

1.11 (a)

$$I = \frac{1}{\sqrt{2\pi}}\int_{-\infty}^{\infty} t^2 e^{-t^2/2} \cdot \delta(1-2t)\, dt$$

By using property

$$\int_{-\infty}^{\infty} f(t)\delta(t-t_0)\, dt = f(t_0)$$

here, $t_0 = \frac{1}{2}$

$$I = \frac{1}{\sqrt{2\pi}} \times \frac{1}{2} \times \left(\frac{1}{2}\right)^2 \times e^{-(1/2)^2/2}$$

$$= \frac{1}{\sqrt{2\pi}} \cdot \frac{1}{8} \cdot e^{-1/8}$$

1.12 (c)

$$y(t) = \frac{d}{dt}\left\{e^{-t}\, x(t)\right\} = e^{-t}\frac{dx(t)}{dt} - e^{-t}\, x(t)$$

this is a linear differential equation with variable coefficient, so system is linear but time variant.

It depends on present value of $x(t)$ so has no memory.

But for $0 \le t < \infty$ system is stable.

1.13 (d)

$$y = \int_{-\infty}^{t} x(\tau)\cos(3\tau)\, d\tau$$

$$y(t-t_0) = \int_{-\infty}^{t-t_0} x(\tau)\cos(3\tau)\, d\tau$$

$y'(t)$ for input $x(t-t_0)$ is

$$y'(t) = \int_{-\infty}^{t} x(\tau - t_0)\cos 3\tau\, d\tau$$

$$y'(t) = \int_{-\infty}^{(t-t_0)} x(\tau)\cos 3(\tau + t_0)\, d\tau$$

$y'(t) \neq y(t-t_0)$ so system is not time invariant

for input $x(\tau) = \cos(3\tau)$ bounded input

$$y(t) = \int_{-\infty}^{t} \cos^2(3\tau)\, d\tau \to \infty \text{ as } t \to \infty$$

So for bounded input, output is not bounded therefore system is not stable.

1.14 (c)

The overall impulse response $h(t)$ of the cascaded system is given by:

$$h(t) = h_1(t) * h_2(t)$$

1.15 (a)

$\omega_1 = 100$; $\omega_2 = 300$; $\omega_3 = 500$

H.C.F. of (ω_1, ω_2 and ω_3) = H.C.F. (100, 300, 500)

$\omega = 100$ rad/sec.

1.16 (c)

All poles must lie within $|z| = 1$

1.17 (b)

$$Y(s) = \frac{1}{s}U(s) = \frac{1}{s^2}$$

$$y(t) = tu(t)$$

1.18 (c)

$$h(t) = tu(t)$$

$$H(s) = \frac{1}{s^2}$$

$$\Rightarrow \quad \frac{Y(s)}{U(s)} = \frac{1}{s^2}$$

$$Y(s) = \frac{1}{s^2}\frac{e^{-s}}{s}$$

$$\Rightarrow \quad y(t) = \frac{(t-1)^2}{2}u(t-1)$$

1.19 (b)

Step response = Integration of impulse response.

$$\int \delta(t-1) = u(t-1)$$

$$\int \delta(t-3) = u(t-3)$$

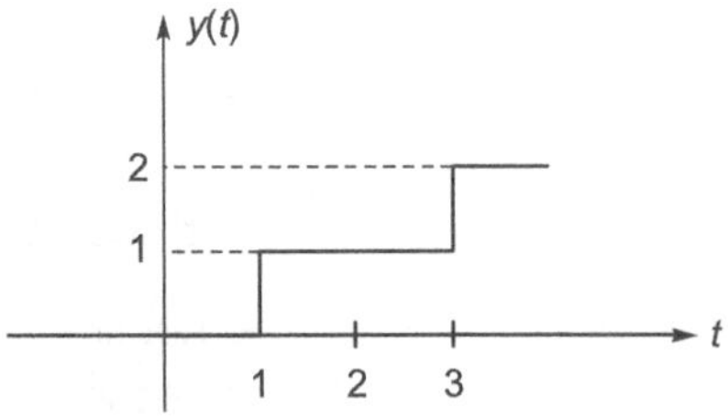

at $t = 2$

$y(t) = 1$

1.20 (6)

$$\omega_1 = \frac{2\pi}{3}, \omega_2 = \pi$$

$$\omega = \frac{\text{HCF}(2\pi, \pi)}{\text{LCM}(3, 1)} = \frac{\pi}{3}$$

$\therefore$ $T = 6$ sec.

1.21 (1)

$\because$ $T = 1$ sec.

So, $f_0 = \frac{1}{1} = 1$ Hz

1.22 (d)

$$x(-t) \rightarrow 3 \rightarrow \omega_{01} = \frac{2\pi}{3}$$

$$y(2t+1) \rightarrow \frac{3}{2} \rightarrow \omega_{02} = \frac{2\pi}{3/2} = \frac{4\pi}{3}$$

$$\omega_0 = GCD\left\{\frac{2\pi}{3}, \frac{4\pi}{3}\right\} = \frac{2\pi}{3}$$

$$T = \frac{2\pi}{2\pi/3} = 3 \text{ sec}$$

1.23 (a)

$$2\int_0^5 x(t-3)\cdot\delta(t-4)dt = 2x(1) = 2$$

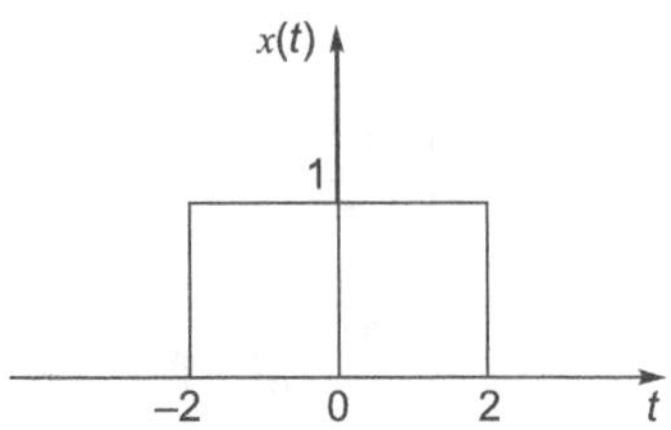

1.24 (0.47) 0.46 to 0.48

The solution can be split into two portions

$$y(t) = y_s(t) + y_n(t)$$

where, $y_s(t)$ = steady state response,

$y_n(t)$ = normal response

$\therefore$ To calculate the steady-state response, we will assume initial input to be zero.

$$\therefore \quad G(s) = \frac{1}{s+3}$$

$$\therefore \quad \frac{Y(s)}{X(s)} = \frac{1}{s+3}$$

$$(s+3)Y(s) = X(s)$$

$$sY(s) + 3Y(s) = X(s)$$

Now, given input is a step response

$$\therefore sY(s) + 3Y(s) = \frac{1}{s}$$

$$\therefore \quad Y(s) = \frac{1}{s(s+3)}$$

$$Y(s) = \frac{A}{s} + \frac{B}{s+3} = \frac{1/3}{s} + \frac{-1/3}{s+3}$$

Applying, inverse Laplace transform, we get

$$y_s(t) = \frac{1}{3}u(t) - \frac{1}{3}e^{-3t}u(t)$$

Now, to calculate the natural response, we assume that the input $x(t) = 0$

$$sy(s) - y(0) + 3y(s) = 0$$

$$(s+3)Y(s) = y(0)$$

$$y(s) = \frac{y(0)}{s+3}$$

Given $y(0) = 3$

$$\therefore \quad Y(s) = \frac{3}{s+3}$$

Taking the inverse Laplace transform, we get

$$y_n(t) = 3e^{-3t}u(t)$$

$\therefore$ Total response,

$$y(t) = y_s(t) + y_n(t)$$

$$\therefore \quad y(t) = 3e^{-3t}u(t) + \frac{1}{3}u(t) - \frac{1}{3}e^{-3t}u(t)$$

$$\therefore \quad y(t) = 0.4659 \simeq 0.47$$

1.25 (d)

$$x(t) = \sin\left(\sqrt{2\pi t}\right)$$

- For negative values of 't', sine-function will apply on imaginary values.
- For positive values of 't', sine-function will apply on real-values.
- So, there will not be repetition in the waveform. $\therefore$ We can draw waveform only in the RHS. But we cannot draw waveform in the LHS.
- Therefore, $\sin\left(\sqrt{2\pi t}\right)$ is non-periodic.

1.26 (a)

Given input-output relation as,

$$y(t) = x(t)\sin(2\pi t)$$

for $x(t) = x_1(t)$,

$$y_1(t) = x_1(t)\sin(2\pi t)$$

for $x(t) = x_2(t)$,

$$y_2(t) = x_2(t)\sin(2\pi t)$$

for input $x(t) = x_1(t) + x_2(t)$

$$y_3(t) = [x_1(t) + x_2(t)]\sin(2xt)$$

$$y_3(t) = y_1(t) + y_2(t)$$

So, it is a linear system.

Since, $\sin(2\pi t)$ has time varying. So, $y(t)$ is time variant. So, system is linear and time variant.

1.27 (b)

Loop gain of a unity feedback control system is given as

$$L(s) = \frac{6}{s(s-5)}$$

The characteristic equation

$$s^2 - 5s + 6 = 0$$

as poles lies in right side of s-plane. So, it is unstable system and system is causal. So, causal and unstable.

1.28 (25)

Let $x(t) = \sin(\omega_o t)$

We have, $T = 100$ sec

$$\omega_o = \frac{2\pi}{T} = \frac{2\pi}{100} \text{ rad/sec}$$

Now, $y(t) = x(4t) = \sin(4\omega_o t)$

$$\therefore \quad \omega' = 4\omega_o = 4 \times \frac{2\pi}{100}$$

$$\frac{2\pi}{T'} = 4 \times \frac{2\pi}{100}$$

$$T' = 25 \text{ sec}$$

1.29 (a)

We have; $h(t) = \delta(t) + 0.5\delta(t-4)$

Taking FT;

$$H(\omega) = 1 + 0.5e^{-j4\omega}$$

at $\omega = \frac{7\pi}{4}$

$$H(\omega) = 1 + 0.5e^{-j4\left(\frac{7\pi}{4}\right)}$$

$$= 1 + 0.5e^{-j7\pi}$$

$$= 0.5$$

$$\therefore \quad |H(\omega)|_{\omega = \frac{7\pi}{4}} = 0.5$$

$$\angle H(\omega)|_{\omega = \frac{7\pi}{4}} = 0$$

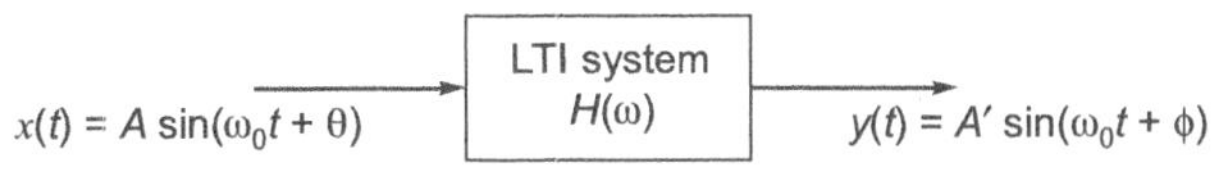

$$A' = A_0 |H(\omega)|_{\omega = \omega_0}$$

$$\phi = \theta + \angle H(\omega)|_{\omega = \omega_0}$$

$$\therefore \quad y(t) = 0.5\cos\left(\frac{7\pi}{4}t\right)$$

■■■

2 Discrete Time LTI Systems

2.1 Given $x[n] = \dfrac{\sin\omega_c n}{\pi n}$, the energy of the signal given by $\sum_{n=-\infty}^{\infty} |x[n]|^2$ is

(a) $\dfrac{\omega_c}{\pi}$ (b) $\pi\omega_c$

(c) Infinite (d) $2\pi\omega_c$

[2003 : 1 M]

2.2 Given $h[n] = [\underset{\uparrow}{1}, 2, 2]$, $f(n)$ is obtained by convolving $h[n]$ with itself and $g[n]$ by correlating $h[n]$ with itself. Which one of the following statement is true?

(a) $f[n]$ is causal and its maximum value is 9.

(b) $f[n]$ is non-causal and its maximum value is 8.

(c) $g[n]$ is causal and its maximum value is 9.

(d) $g[n]$ is non-causal and its maximum value is 9.

[2003 : 2 M]

2.3 Given $X = (a, b, c, d)$ as the input, a linear time invariant system produces an output $y = [x, x, x, x, \ldots\ldots,$ < repeated N-times >]. The impulse response of the system is

(a) $\sum_{i=0}^{N-1} \delta(n-4i)]$ (b) $u[n] - u[n-N]$

(c) $u[n] - u[n-N-1]$ (d) $\sum_{i=0}^{N-1} \delta(n-i)$

[2003 : 2 M]

2.4 The sequence $x[n]$ whose z-transform $X(z) = e^{1/z}$ is

(a) $\dfrac{1}{n!}u[n]$ (b) $\dfrac{1}{-n!}u[-n]$

(c) $(-1)^n \dfrac{1}{n!}u[n]$ (d) $\dfrac{1}{-(n+1)!}u[-n-1]$

[2003 : 2 M]

2.5 The fundamental period of the sequence $x[n] = 3\sin(1.3\pi n + 0.5\pi) + 5\sin(1.2\pi n)$ is

(a) 20 (b) $\dfrac{2\pi}{1.3\pi}$

(c) $\dfrac{2\pi}{1.2\pi}$ (d) 10

[2005 : 2 M]

2.6 Consider the discrete-time signal

$$x(n) = \left(\frac{1}{3}\right)^n u(n), \text{ where } u(n) = \begin{cases} 1, & n \geq 0 \\ 0, & n < 0 \end{cases}.$$

Define the signal

$y(n) \Rightarrow$ as $y(n) = x(-n), -\infty < n < \infty$.

Then $\sum_{n=-\infty}^{\infty} y(n)$ equals

(a) $-\dfrac{2}{3}$ (b) $\dfrac{2}{3}$

(c) $\dfrac{3}{2}$ (d) 3

[2007 : 2 M]

2.7 The fundamental period of the discrete-time signal $x[n] = e^{j\left(\frac{5\pi}{6}\right)n}$ is

(a) $\dfrac{6}{5\pi}$ (b) $\dfrac{12}{5}$

(c) 6 (d) 12

[2008 : 1 M]

2.8 Which one of the following discrete-time systems is time invariant?

(a) $y[n] = n\,x[n]$ (b) $y[n] = x[3n]$

(c) $y[n] = x[-n]$ (d) $y[n] = x[n-3]$

[2008 : 1 M]

2.9 Consider a discrete-time LTI system with input $x[n] = \delta[n] + \delta[n-1]$ and impulse response $h[n] = \delta[n] - \delta(n-1)$. The output of the system will be given by

(a) $\delta(n) - \delta[n-2]$ (b) $\delta[n] - \delta[n-1]$

(c) $\delta[n-1] + \delta[n-2]$ (d) $\delta[n] + \delta[n-1] + \delta[n-2]$

[2008 : 2 M]

2.10 Consider a discrete-time system for which the input $x[n]$ and the output $y[n]$ are related as $y[n] = x[n] - \frac{1}{3}y[n-1]$. If $y[n] = 0$ for $n < 0$ and $x[n] = \delta[n]$, then $y[n]$ can be expressed in terms of the unit step $u[n]$ as

(a) $\left(-\frac{1}{3}\right)^n u[n]$ (b) $\left(\frac{1}{3}\right)^n u[n]$

(c) $(3)^n u[n]$ (d) $(-3)^n u[n]$

[2008 : 2 M]

2.11 Let $y[n]$ denote the convolution of $h[n]$ and $g[n]$, where $h[n] = (1/2)^n u[n]$ and $g[n]$ is a causal sequence. If $y[0] = 1$ and $y[1] = 1/2$, then $g[1]$ equals

(a) 0 (b) 1/2

(c) 1 (d) 3/2

[2012 : 2 M]

2.12 The discrete-time transfer function $\frac{1-2z^{-1}}{1-0.5z^{-1}}$ is

(a) non-minimum phase and unstable.

(b) minimum phase and unstable.

(c) minimum phase and stable.

(d) non-minimum phase and stable.

[2013 : 1 M]

2.13 The impulse response of an LTI system is given as

$$h[n] = \begin{cases} \frac{\omega_c}{\pi} & n = 0 \\ \frac{\sin \omega_c n}{\pi n} & n \neq 0 \end{cases}$$

It represents an ideal

(a) non-causal, low-pass filter

(b) causal, low-pass filter

(c) non-causal, high-pass filter

(d) causal, high-pass filter [2014 : 1 M]

2.14 The signal $x[n]$ shown in the figure below is convolved with itself to get $y[n]$. The value of $y[-1]$ is____.

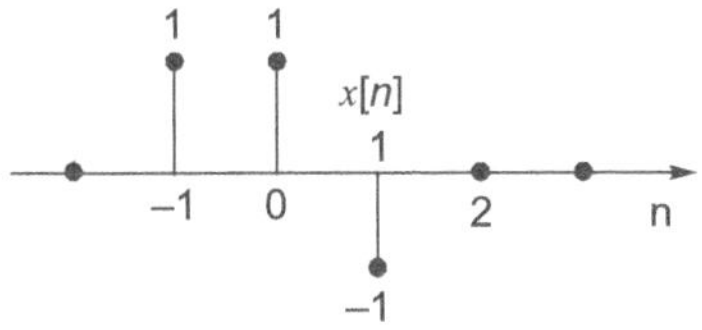

[2016 : 1 M]

2.15 The fundamental period N_0 of the discrete-time sinusoid $x[n] = \sin\left(\frac{301}{4}\pi n\right)$ is ____.

[2016 : 2 M]

2.16 A discrete-time signal $x[n] = e^{j\left(\frac{5\pi}{3}\right)n} + e^{j\left(\frac{\pi}{4}\right)n}$ is down-sampled to the signal $x_d[n]$ such that $x_d[n] = x[4n]$. The fundamental period of the down-sampled signal $x_d[n]$ is____________.

[2019 : 2 M]

2.17 Consider the signal $x[n] = \sin(2\pi n)\, u[n]$, where

$$u[n] = \begin{cases} 1 & n = 0,1,2,3,..... \\ 0 & \text{otherwise} \end{cases}$$

The period of this signal $x[n]$ is

(a) 3 (b) 4

(c) 2 (d) 1 [2020 : 1 M]

2.18 Let $g[n] = \begin{cases} 1 & n = 0 \\ 0 & n = \pm 1, \pm 2, \pm 3,..... \end{cases}$

and $h[n] = \begin{cases} 1 & n = 0,3,6,9,..... \\ 0 & \text{otherwise} \end{cases}$. Consider

$y[n] = h[n] \otimes g[n]$, where $\otimes$ denotes the convolution operator. The value of $y[2]$ is _____.

[2020 : 2 M]

2.19 The input-output relationship of an LTI system is given below:

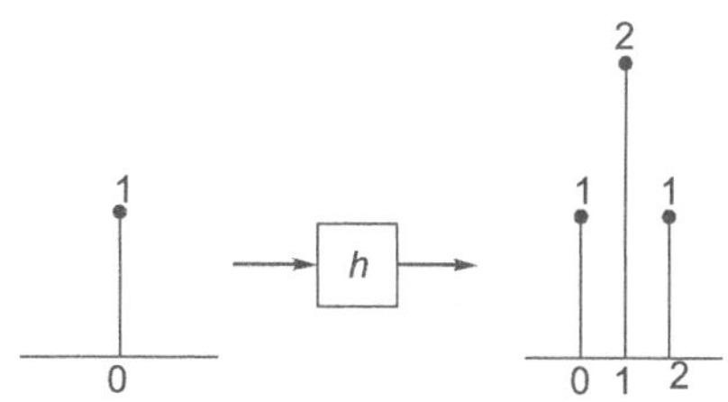

For an input $x[n]$ shown below:

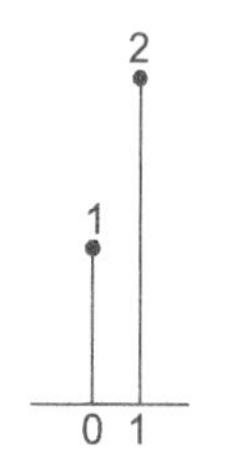

The peak value of the output when $x[n]$ passes through h is _____.

(a) 5 (b) 2

(c) 4 (d) 6 [2021 : 1 M]

2.20 Consider the discrete-time signal
$x[n] = u[-n+5] - u[n+3]$,

where $u[n] = \begin{cases} 1; & n \geq 0 \\ 0; & n < 0 \end{cases}$. The smallest n for which

$x[n] = 0$ is _______ . [2023 : 1 M]

2.21 A discrete-time sequence is given by $x[n]=[1, 2, 3, 4]$ for $0 \le n \le 3$. The zero lag auto-correlation value of $x[n]$ is

(a) 1 (b) 10
(c) 20 (d) 30 [2024 : 1 M]

Answers Discrete Time LTI Systems

2.1 (a)	2.2 (d)	2.3 (a)	2.4 (a)	2.5 (a)	2.6 (c)	2.7 (d)
2.8 (d)	2.9 (a)	2.10 (a)	2.11 (a)	2.12 (d)	2.13 (a)	2.14 (2)
2.15 (8)	2.16 (6)	2.17 (d)	2.18 (0)	2.19 (a)	2.20 (–3)	2.21 (d)

Explanations Discrete Time LTI Systems

2.1 (a)

The Fourier transform of sinc function is a gate signal shown below:

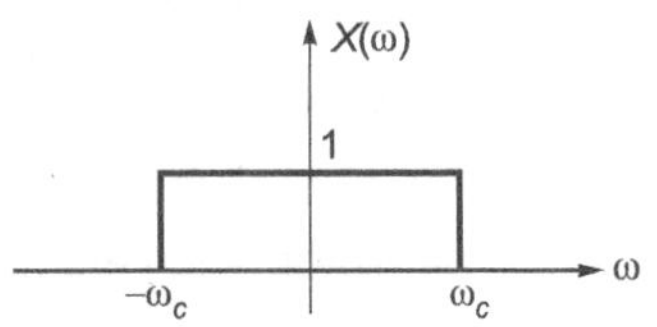

$$\text{Energy} = \frac{1}{2\pi}\int_{-\infty}^{\infty}|X(\omega)|^2\, d\omega$$

$$= \frac{1}{2\pi}\int_{-\omega_c}^{\omega_c} 1\cdot d\omega = \frac{1}{2\pi}\cdot[\omega]_{-\omega_c}^{\omega_c}$$

$$= \frac{1}{2\pi}\cdot[\omega_c - (-\omega_c)] = \frac{2\omega_c}{2\pi} = \frac{\omega_c}{\pi}$$

$$\boxed{\text{Energy} = \frac{\omega_c}{\pi}}$$

2.2 (d)

$$h[n] * h[n] = f[n] = \{\underset{\uparrow}{1}, 4, 8, 8, 4\}$$

$$g[n] = h[n] * h[-n] = \{2, 6, \underset{\uparrow}{9}, 6, 2\}$$

So $g[n]$ is non causal having maximum value 9.0

2.3 (a)

Given, $x = [a, b, c, d]$ input signal

$y = [x, \ x, \ x, \ x, \ldots\ldots N$ times] output signal

That is clearly observed for different inputs, outputs are same and periodic with fundamental period $N = 4$.

$\underbrace{a, b, c, d}_{\delta(n)} + \underbrace{a, b, c, d}_{\delta(n-4)} + \underbrace{a, b, c, d}_{\delta(n-8)} + \ldots\ldots$

In general, this can be written as

$$\sum_{i=0}^{N-1}\delta(n-4i)$$

2.4 (a)

$$e^{1/z} = \sum_{k=0}^{\infty}\frac{(1/z)^k}{k!} = \sum_{k=0}^{\infty}\frac{z^{-k}}{k!} = X(z)$$

the Z-transform of an arbitrary signal $x[n]$ is

$$X(z) = \sum_{n=-\infty}^{\infty} x[n]z^{-n}$$

by comparing

$$x[n] = \frac{1}{n!}u[n]$$

2.5 (a)

$$x[n] = 3\sin[1.3\pi n + 0.5\pi] + 5\sin[1.2\pi n]$$

using formula, $\boxed{\dfrac{\omega_0}{2\pi} = \dfrac{m}{N}}$

$$\Rightarrow \quad \frac{m_1}{N_1} = \frac{1.3\pi}{2\pi} = \frac{13}{20} \Rightarrow N_1 = 20$$

and $$\frac{m_2}{N_2} = \frac{1.2\pi}{2\pi} = \frac{12}{20} = \frac{3}{5} \Rightarrow N_2 = 5$$

$$N = \text{L.C.M}(N_1, N_2) = 20$$

2.6 (c)

Given, $$x(n) = \left(\frac{1}{3}\right)^n u(n);\ n \ge 0$$

$$y(n) = x(-n) = \left(\frac{1}{3}\right)^{-n} u(-n);\ n \le 0$$

$$\sum_{n=-\infty}^{\infty} y(n) = \sum_{n=-\infty}^{0}\left(\frac{1}{3}\right)^{-n} u(-n) = \sum_{n=-\infty}^{0}\left(\frac{1}{3}\right)^{-n}$$

$$= \frac{1}{1-\left(\frac{1}{3}\right)};\ n \le 0 = \frac{3}{2}$$

2.7 (d)

Given signal is $x[n] = e^{j\left(\frac{5\pi}{6}\right)n}$

We can write, $e^{j\left(\frac{5\pi}{6}\right)n} = \cos\left(\frac{5\pi}{6}n\right) + j\sin\left(\frac{5\pi}{6}n\right)$

$$\frac{\omega_0}{2\pi} = \frac{m}{N} \Rightarrow \frac{m}{N} = \frac{5\pi/6}{2\pi} = \frac{5}{12}$$

$\therefore$ Fundamental period $N = 12$.

2.8 (d)

By observing the options only (d) option is obeying time-shift property in input as well as output.

2.9 (a)

Given, $x[n] = \delta[n] + \delta[n-1]$

$x[n] = \{1, 1\}$

$h[n] = \delta[n] - \delta[n-1] = \{1, -1\}$

By using add and shift columnwise

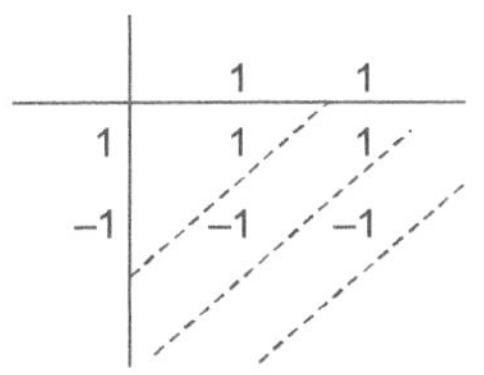

$y[n] = \{1, 0, -1\}$

$y[n] = 1 \cdot \delta[n] + 0 \cdot \delta[n-1] - 1 \cdot \delta[n-2]$

$= \delta[n] - \delta[n-2]$

2.10 (a)

Given, $y[n] = x[n] - \frac{1}{3}y[n-1]$

and $y[n] = 0;\ n < 0$

and $y[n] = \delta[n] - \frac{1}{3}y[n-1]$

Put, $n = 0$

$$y[0] = \delta[0] - \frac{1}{3}y[n-1]$$

$(\because y[n] = 0;\ n < 0$

$\therefore y[-1] = 0)$

$y[0] = 1 \quad (\because \delta[0] = 1)$

Put, $n = 1$

$$y[1] = \delta[1] - \frac{1}{3}y[0] = 0 - \frac{1}{3} \times 1 = -\frac{1}{3}$$

Put, $n = 2$

$$y[2] = \delta[2] - \frac{1}{3}y[1] = 0 - \frac{1}{3}\cdot\left(-\frac{1}{3}\right) = \left(-\frac{1}{3}\right)^2$$

In general, $y[n] = \left(-\frac{1}{3}\right)^n u(n)$

2.11 (a)

given $g[n]$ is a causal sequence

$\therefore$ $g[n]$ will be zero for $n < 0$

$$y[n] = \sum_{k=-\infty}^{\infty} g[k]\, h[n-k]$$

and $h[n] = \left(\frac{1}{2}\right)^n u[n]$

$$h[n] = \begin{cases} \left(\frac{1}{2}\right)^n & ;\ n \geq 0 \\ 0 & ;\ n < 0 \end{cases}$$

$h[0] = 1,\quad h[1] = \frac{1}{2},\quad h[2] = \frac{1}{4}$

$y[0] = g[0]\, h[0] + g[1]\, h[-1] + \cdots$

$1 = g[0] \times 1 + 0 + 0 + \cdots$

$g[0] = 1$

$y[1] = g[0]\, h[1] + g[1]\, h[0] + g[2]\, h[-1] + \cdots$

$\frac{1}{2} = 1 \cdot \frac{1}{2} + g[1] \cdot 1 + 0$

$\Rightarrow \quad g[1] = 0$

2.12 (d)

Transfer function is $\Rightarrow \dfrac{z-2}{z-0.5}$

Clearly all the zeros are on the RHS of imaginary axis so the system is non-minimum phase.

Also, the poles $z = 0.5$ is inside the unity circle thus the system is stable.

2.13 (a)

$$h(n) = \begin{cases} \dfrac{\omega_c}{\pi}; & n = 0 \\ \dfrac{\sin \omega_c n}{\pi n} & n \neq 0 \end{cases}$$

Since $h(n) \neq 0$ for $n < 0$ or negative values of n. So, the system is non-causal.

Also $\dfrac{\sin \omega_c n}{\pi n}$ represents a rectangular function with the representation as

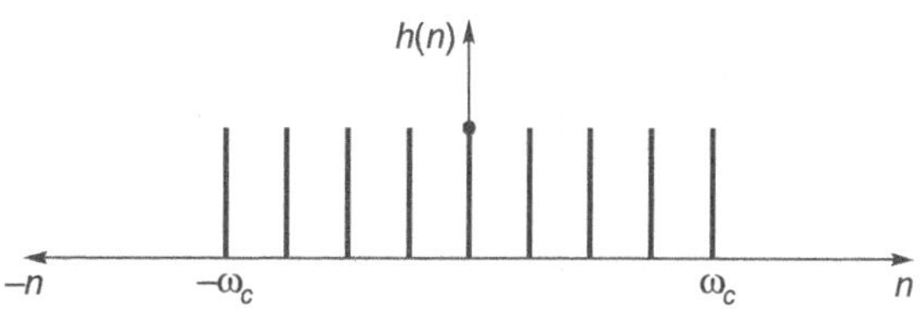

This represents the characteristics of ideal low-pass filter.

2.14 (2)

Given: $x[n] = \{1, 1, -1\}$
$\uparrow$

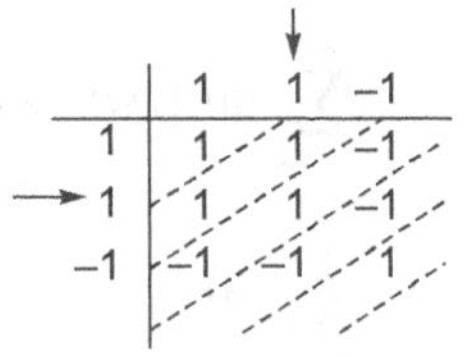

$$y[n] = x[n] \otimes x[n]$$

$$y[n] = \{1, \ 2, \ -1, \ -2, \ 1\}$$

$y[-1] = 2$ $\uparrow$

2.15 (8)

$$x[n] = \sin\left[\frac{301}{4}\pi n\right]$$

$$N_0 = \frac{2\pi}{\omega_0} = \frac{2\pi}{301\pi/4} \times m = \frac{8}{301} \times 301 = 8$$

2.16 (6)

$$x(n) = e^{j(5\pi/3)n} + e^{j(\pi/4)n}$$

$$\omega_1 = \frac{5\pi}{3} \text{ and } \omega_2 = \frac{\pi}{4}$$

GCD (ω_1, ω_2), $\omega_0 = \frac{\pi}{12}$

$$x_d(n) = x(4n)$$

Apply time scaling so,

$$\omega_d = 4\omega_0 = \frac{4\pi}{12} = \frac{\pi}{3}$$

$$\frac{\omega d}{2\pi} = \frac{m}{N} \Rightarrow \frac{\pi/3}{2\pi} = \frac{m}{N}$$

$$N = 6$$

2.17 (d)

$$x(n) = \sin 2\pi n \cdot u(n)$$

For $n = 0, x(0) = 0$

$n = 1, x(1) = \sin 2\pi = 0$

$n = 2, x(2) = \sin 4\pi = 0$

Thus, $x(n) = \{.... \ 0, 0, 0, 0, 0, 0, ...\}$
$\uparrow$

Hence, Time-period is 1.

2.18 (0)

$$g(n) = \{...., 0, 0, 1, 0, 0, ...\} = \delta(n)$$

$$h(n) = \{1, 0, 0, 1, 0, 0, 1, 0, 0, ...\}$$

Now, $y(n) = g(n) * h(n) = \delta(n) * h(n) = h(n)$

Therefore, $y(2) = h(2) = 0$

2.19 (a)

When $x[n] = \delta[n]$, then $y[n] = h[n]$

$$h[n] = \delta[n] + 2\delta[n-1] + \delta[n-2]$$

$$= \{1, 2, 1\}$$

Given, $x'[n] = \delta[n] + 2\delta[n-1]$

$$x'[n] = \{1, 2\}$$

$$y[n] = x'[n] * h[n]$$

Sum by column method:

1	2		
1	2	1	
1	2		
	2	4	
		1	2

$y[n] = \{1 \quad 4 \quad 5 \quad 2\}$
$\uparrow$

Peak value at the output is 5.

2.20 (−3)

We know that

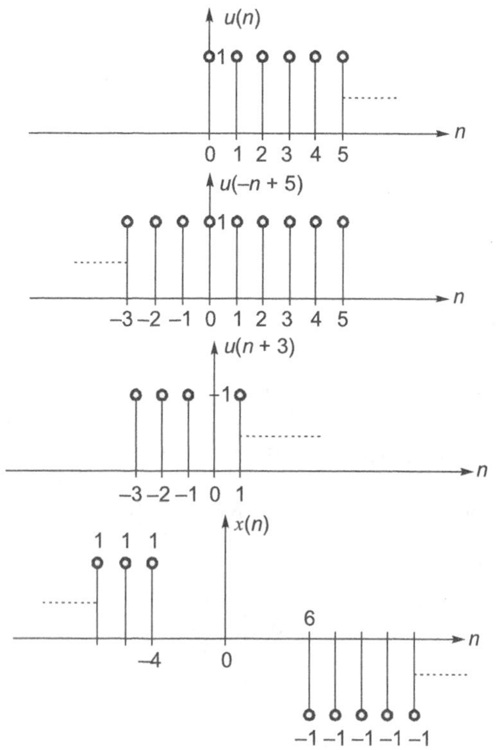

Smallest 'n' for which $x[n] = 0$

$\therefore \quad n = -3$

2.21 (d)

Zero log autocorrelation of $x(n)$

$$\text{Energy of } x(n) = \sum_{n=-\infty}^{\infty} |x(n)|^2$$

$$= 1^2 + 2^2 + 3^2 + 2^4 = 30$$

3 Fourier Series

3.1 $f(x)$, shown in the figure is represented by

$$f(x) = a_0 + \sum_{n=1}^{\infty} (a_n \cos(nx) + b_n \sin(nx)).$$

The value of a_0 is

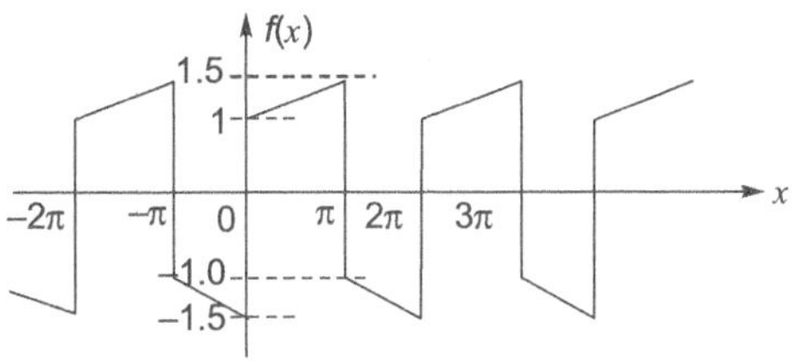

(a) 0 (b) $\pi/2$
(c) π (d) 2π [2010 : 1 M]

3.2 Consider a periodic signal $x(t)$ as shown below:

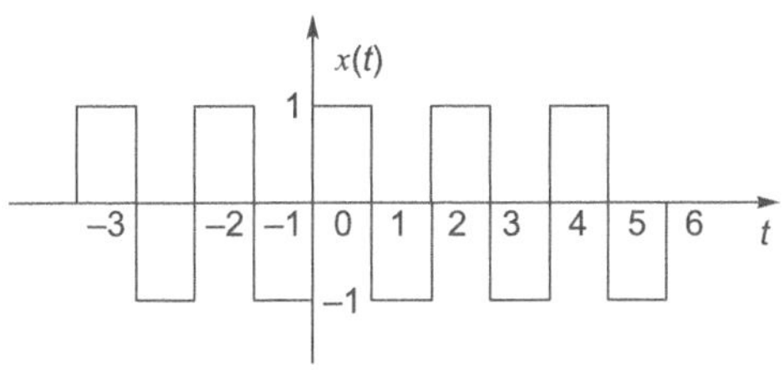

It has a Fourier series representation

$$x(t) = \sum_{k=-\infty}^{\infty} a_k e^{j(2\pi/T)kt}$$

Which one of the following statements is **TRUE?**

(a) $a_k = 0$, for k odd integer and $T = 2$
(b) $a_k = 0$, for k even integer and $T = 2$
(c) $a_k = 0$, for k even integer and $T = 4$
(d) $a_k = 0$, for k odd integer and $T = 4$

[2011 : 1 M]

3.3 For the periodic signal $x(t)$ shown below with period $T = 8s$, the power in the 10th harmonic is

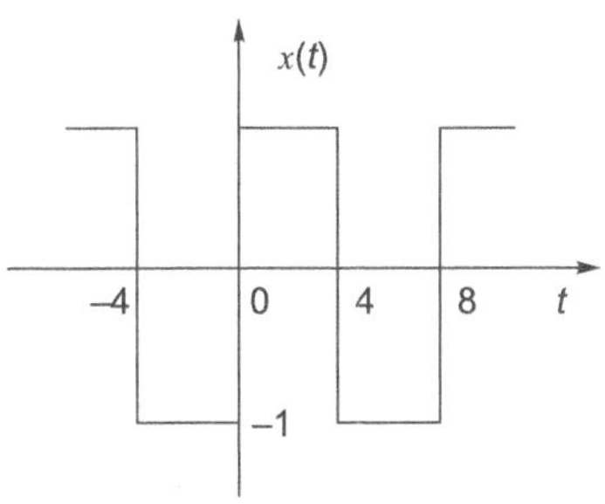

(a) 0 (b) $\frac{1}{2}\left(\frac{2}{10\pi}\right)^2$
(c) $\frac{1}{2}\left(\frac{4}{10\pi}\right)^2$ (d) $\frac{1}{2}\left(\frac{4}{5\pi}\right)^2$

[2016 : 2 M]

3.4 An ideal square wave with period of 20 ms shown in the figure, is passed through an ideal low pass filter with cut-off frequency 120 Hz. Which of the following is an accurate description of the output?

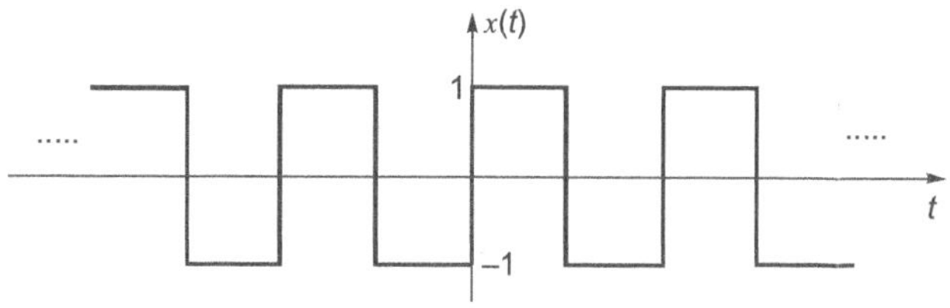

(a) Output is zero.
(b) Output consists of both 50 Hz and 100 Hz frequency components.
(c) Output is a pure sinusoid of frequency 50 Hz.
(d) Output is a square wave of fundamental frequency 50 Hz.

[2018 : 1 M]

3.5 A periodic function $f(x)$, with period 2, is defined as

$$f(x) = \begin{cases} -1-x & -1 \le x < 0 \\ 1-x & 0 < x \le 1 \end{cases}$$

The Fourier series of this function contains ________.

(a) Both cos $(n\pi x)$ and sin $(n\pi x)$ where $n = 1, 2, 3, \ldots$
(b) Only sin $(n\pi x)$ where $n = 1, 2, 3, \ldots$
(c) Only cos $(n\pi x)$ where $n = 1, 2, 3, \ldots$
(d) Only cos $(2n\pi x)$ where $n = 1, 2, 3, \ldots$

[2022 : 1 M]

■■■

Answers Fourier Series

3.1 (a) 3.2 (b) 3.3 (a) 3.4 (c) 3.5 (b)

Explanations Fourier Series

3.1 (a)

By observing the waveform, it is a half wave symmetry.

i.e. $$f(t) = -f\left(t \pm \frac{T}{2}\right)$$

So, in Fourier series a_0 is zero. So, directly we can say $a_0 = 0$. Hence (a) is correct.

3.2 (b)

The given waveform have half wave symmetry therefore a_k will be zero for even integer value of K.

Time period of given waveform is 2.

3.3 (a)

Given signal satisfies odd and half wave symmetry so it contains only odd harmonics.

$\therefore$ Power in 10th harmonic will be zero.

3.4 (c)

$$T = 20 \text{ msec}$$

Fundamental frequency,

$$f_0 = \frac{1000}{20} = 50 \text{ Hz}$$

The given signal is an odd signal and it also has half wave symmetry. So, it will have spectral components only for odd harmonics. So, the spectrum of the signal will have component of 50 Hz, 150 Hz, 250 Hz, 350 Hz, ...

When passed through an LPF of cut-off frequency 120 Hz, the output of filter has only 50 Hz component. So, option (c) is correct.

3.5 (b)

Given a periodic function with period 2 as

$$f(x) = \begin{cases} -1-x; & -1 \le x < 0 \\ +1-x; & 0 < x \le 1 \end{cases}$$

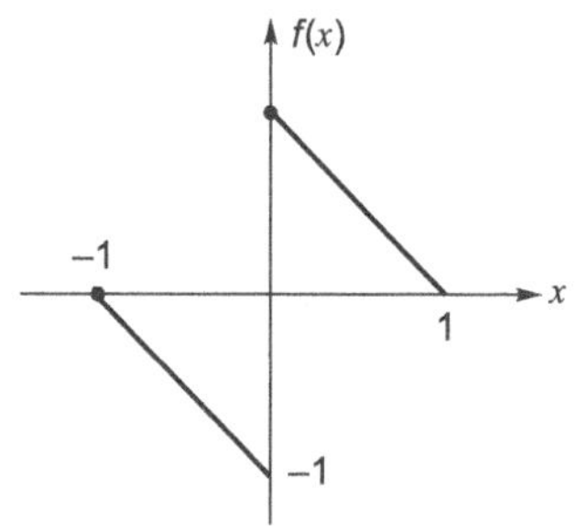

This function is odd function and $f(x)$ don't have halfwave symmetry.

So only sine terms will be present in expansion $f(x)$

$$f(x) = \sin(n\pi x)\ ; \quad n = 1, 2, 3, \ldots\infty$$

4 Fourier Transform

4.1 A real function $f(t)$ has a Fourier transform $F(\omega)$. The Fourier transform of $[f(t) - f(-t)]$ is

(a) zero (b) real

(c) real and odd (d) imaginary

[2003 : 1 M]

4.2 The continuous-time signal $x(t) = \dfrac{1}{a^2 + t^2}$ has the Fourier transform $\pi/a \exp(-a|\omega|)$. The signal $x(t)$ cosbt has Fourier transform

(a) $\dfrac{\pi}{2a}\left[\exp(-a|\omega - b|) + \exp(-a|\omega + b|)\right]$

(b) $\dfrac{\pi}{2a}\left[\exp(-a|\omega|) + \exp(-a|\omega|)\right]$

(c) $\dfrac{\pi}{a}[\exp(-a|\omega|)\cos b\omega]$

(d) $\dfrac{\pi}{2a}\left[\exp(-a|\omega - b|) - \exp(-a|\omega + b|)\right]$

[2005 : 1 M]

4.3 If the waveform, shown in the following figure, corresponds to the second derivative of a given function $f(t)$, then the Fourier transform of $f(t)$ is

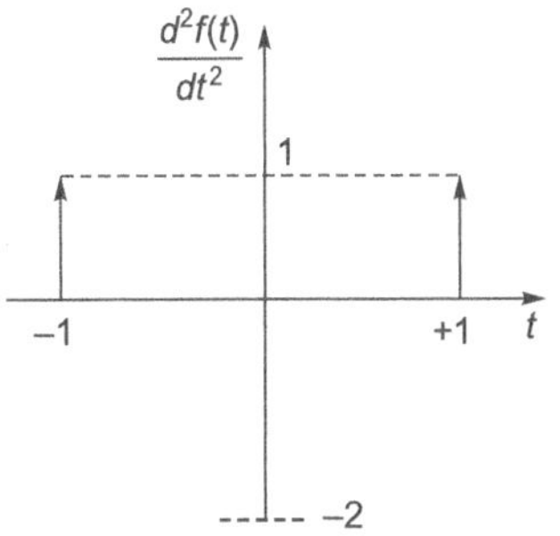

(a) $1 + \sin\omega$ (b) $1 + \cos\omega$

(c) $\dfrac{2(1-\cos\omega)}{\omega^2}$ (d) $\dfrac{2(1+\cos\omega)}{\omega^2}$

[2006 : 2 M]

4.4 The Fourier transform of a function $g(t)$ is given

$$G(\omega) = \frac{\omega^2 + 21}{\omega^2 + 9}$$

(a) $\delta(t) + 2\exp(-3|t|)$

(b) $\cos 3\omega t + 21\exp(-3t)$

(c) $\sin 3\omega t + 7\cos \omega t$

(d) $\sin 3\omega t + 21\exp(3t)$

[2006 : 2 M]

4.5 The magnitude of Fourier transform $X(\omega)$ of a function $x(t)$ is shown below in figure (a). The magnitude of Fourier transform $Y(\omega)$ of another function $y(t)$ is shown below in figure (b). The phases of $X(\omega)$ and $Y(\omega)$ are zero for all ω. The magnitude and frequency units are identical in both the figures. The function $y(t)$ can expressed in terms of $x(t)$ as

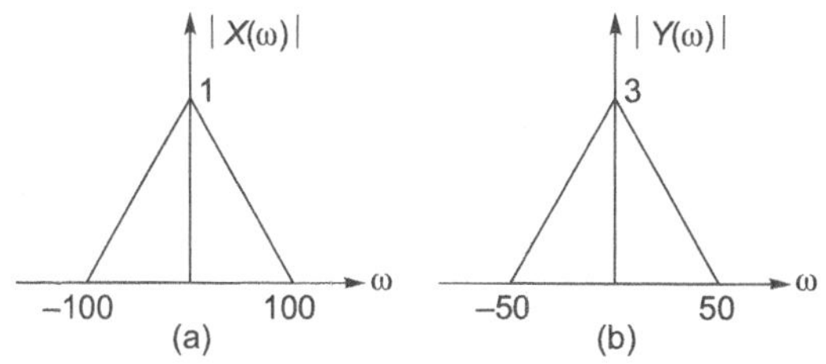

(a) $\dfrac{2}{3}x\left(\dfrac{t}{2}\right)$ (b) $\dfrac{3}{2}x(2t)$

(c) $\dfrac{2}{3}x(2t)$ (d) $\dfrac{3}{2}x\left(\dfrac{t}{2}\right)$

[2006 : 2 M]

4.6 The Fourier transform $x(t) = e^{-at}u(-t)$, when $u(t)$ is unit step function,

(a) exist for any real value of a

(b) does not exist for any real value of a

(c) exists if the real value of a is strictly negative

(d) exists if the real value of a is strictly positive

[2008 : 1 M]

4.7 Consider the signal $x(t) = \begin{cases} e^{-t}, & t \geq 0 \\ 0, & t < 0 \end{cases}$.

Let $x(\omega)$ denote the Fourier transform of this signal. The integral $\dfrac{1}{2\pi}\int_{-\infty}^{\infty} X(\omega)d\omega$ is

(a) 0 (b) 1/2

(c) 1 (d) ∞

[2011 : 1 M]

4.8 The Fourier transform of a signal $h(t)$ is $H(j\omega) = (2\cos\omega)(\sin 2\omega)/\omega$. The value of $h(0)$ is

(a) 1/4 (b) 1/2
(c) 1 (d) 2 **[2012 : 2 M]**

4.9 The Fourier Transform of a signal $x(t)$, denoted by $X(j\omega)$ is shown in the figure.

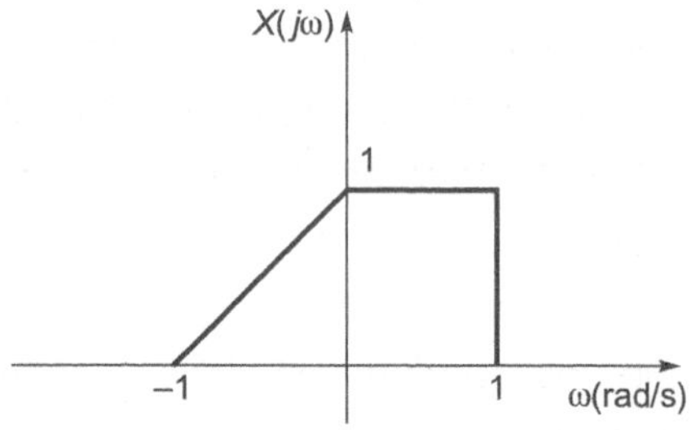

Let $y(t) = x(t) + e^{jt}x(t)$. The value of Fourier transform of $y(t)$ evaluated at the angular frequency $\omega = 0.5$ rad/s is

(a) 0.5 (b) 1
(c) 1.5 (d) 2.5 **[2018 : 2 M]**

4.10 The output of a continuous-time system $y(t)$ is related to its input $x(t)$ as $y(t) = x(t) + \frac{1}{2}x(t-1)$. If the Fourier transforms of $x(t)$ and $y(t)$ are $X(\omega)$ and $Y(\omega)$ respectively and $|X(0)|^2 = 4$. the value of $|Y(0)|^2$ is ________.

[2019 : 2 M]

4.11 Consider the signal $x(t) = e^{-|t|}$. Let $X(j\omega) = \int_{-\infty}^{\infty} x(t)e^{-j\omega t}dt$ be the Fourier transform of $x(t)$. The value of $X(j0)$ is _____.

[2020 : 1 M]

4.12 A continuous real-valued signal $x(t)$ has finite positive energy and $x(t) = 0,\ \forall\ t < 0$. From the list given below, select ALL the signals whose continuous-time Fourier transform is purely imaginary.

(a) $x(t) + x(-t)$ (b) $x(t) - x(-t)$
(c) $j(x(t) + x(-t))$ (d) $j(x(t) - x(-t))$

[2023 : 2 M]

■■■

Answers Fourier Transform

4.1	(d)	4.2	(a)	4.3	(c)	4.4	(a)	4.5	(d)	4.6	(c)	4.7	(c)
4.8	(c)	4.9	(c)	4.10	(9)	4.11	(2)	4.12	(b, c)				

Explanations Fourier Transform

4.1 (d)

The Fourier transform of

$$f(t) \xleftrightarrow{F} F(\omega) \quad \text{...(i)}$$

$$f^*(-t) \xleftrightarrow{F} F^*(\omega)$$

$$-f(-t) \xleftrightarrow{F} -F^*(\omega) \quad (\because f^*(-t) = f(-t)) \quad \text{...(ii)}$$

Add (i) and (ii) we get,

$$f(t) - f(-t) \xleftrightarrow{F} F(\omega) - F^*(\omega)$$

when $F^*(\omega)$ is conjugate of $F(\omega)$.

So, that $F(\omega) - F^*(\omega)$ is must be imaginary.

4.2 (a)

As $\cos bt = \dfrac{e^{jbt} + e^{-jbt}}{2}$

then, $x(t) \times \cos bt = \frac{1}{2}\left[x(t)\cdot e^{jbt} + x(t)e^{-jbt}\right]$

$$X(\omega) = \frac{\pi}{a}\exp(-a|\omega|)$$

By frequency shift property

using, $x(t)\cdot e^{j\omega_0 t} \leftrightarrow X(\omega - \omega_0)$

$\therefore$ $FT[x(t)\cos bt]$

$$= \frac{1}{2}\frac{\pi}{a}\left[\exp(-a|\omega - b|) + \exp(-a|\omega + b|)\right]$$

$$= \frac{\pi}{2a}\left[\exp(-a|\omega - b|) + \exp(-a|\omega + b|)\right]$$

4.3 (c)

$$\frac{d^2 f(t)}{dt^2} = \delta(t-1) - 2\delta(t) + \delta(t+1)$$

Taking Fourier transform,

$$f\left[\frac{d^2 f(t)}{dt^2}\right] = e^{-j\omega} - 2 + e^{j\omega} = 2\cos\omega - 2$$

$$F[f(t)] = \frac{2\cos\omega - 2}{(j\omega)^2} = \frac{2\cos\omega - 2}{j^2\omega^2}$$

$$\therefore \quad f(\omega) = \frac{2(1-\cos\omega)}{\omega^2} \quad (\because j^2 = -1)$$

4.4 (a)

Given, $G(\omega) = \dfrac{\omega^2 + 21}{\omega^2 + 9}$

$\Rightarrow \quad G(\omega) = 1 + 2 \cdot \dfrac{2 \times 3}{\omega^2 + (3)^2}$

$\therefore \quad g(t) = \delta(t) + 2 \cdot e^{-3|t|}$

as Fourier transform of $e^{-a|t|} = \dfrac{2a}{\omega^2 + a^2}$

4.5 (d)

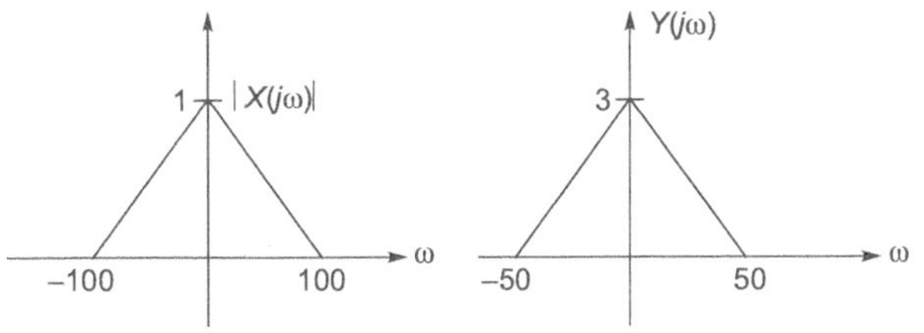

We know that expansion in frequency domain result in compression in time domain and viceversa.

Here compression is done in frequency domain, so there will be expansion in time domain by same amount.

So for $Ax(t/2) \leftrightarrow 2A\,X(2j\omega)$

$$2A = 3$$

$$A = \frac{3}{2}$$

so, $\quad y(t) = \dfrac{3}{2}x(t/2)$

4.6 (c)

The Fourier transform of $e^{bt}u(-t)$ is $\dfrac{1}{b - j\omega}$ is $b > 0$

$\therefore$ The F.T. of $e^{-at}\,u(t)$ will exist

if $\quad -a > 0$

$\Rightarrow \quad a < 0$

4.7 (c)

$$x(t) = \frac{1}{2\pi}\int_{-\infty}^{\infty} X(\omega)\, e^{-j\omega t}\, d\omega$$

putting $t = 0$ in the above expression we get,

$$x(0) = \frac{1}{2\pi}\int_{-\infty}^{\infty} X(\omega)\, e^{0}\, d\omega$$

$$\therefore \quad \frac{1}{2\pi}\int_{-\infty}^{\infty} X(\omega)\, d\omega = x(0) = e^{-0} = 1$$

4.8 (c)

$$H(j\omega) = \cos\omega \cdot \frac{2\sin 2\omega}{\omega}$$

$$= \frac{e^{j\omega} + e^{-j\omega}}{2} \cdot \frac{2\sin 2\omega}{\omega}$$

$$= \frac{1}{2}\left[e^{j\omega} \cdot \left(\frac{2\sin 2\omega}{\omega}\right) + e^{-j\omega}\left(\frac{2\sin 2\omega}{\omega}\right)\right]$$

Let, $\quad X(\omega) = \dfrac{2\sin 2\omega}{\omega}$

Then,

$$x(t) = \begin{cases} 1 & ; \quad -2 < t < 2 \\ 0 & ; \quad \text{otherwise} \end{cases}$$

By time shifting property

$$h(t) = \frac{1}{2}\begin{cases} x(t+1) & ; \quad -3 < t < -1 \\ x(t+1) + x(t-1) & ; \quad -1 < t < 1 \\ x(t-1) & ; \quad 1 < t < 3 \end{cases}$$

$$\therefore h(0) = \frac{1}{2}[x(1) + x(-1)]$$

$$= \frac{1}{2}[1 + 1] = 1$$

4.9 (c)

$$y(t) = x(t) + e^{jt}x(t)$$

$$Y(\omega) = X(\omega) + X(\omega - 1)$$

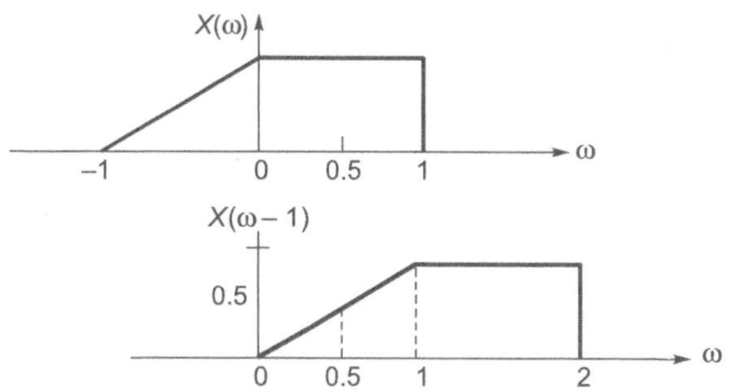

$$Y(\omega)\big|_{\omega = 0.5} = 1.5$$

4.10 (9)

$$y(t) = x(t) + \frac{1}{2}x(t-1)$$

Apply Fourier transform

$$Y(\omega) = X(\omega) + \frac{1}{2}e^{-j\omega}X(\omega)$$

$$|Y(\omega)| = |X(\omega)| + \frac{1}{2}|X(\omega)|$$

$$|Y(\omega)|^2 = |X(\omega)|^2 + \frac{1}{4}|X(\omega)|^2 + |X(\omega)|^2$$

Put $\omega = 0$

$$|Y(0)|^2 = |X(0)|^2 + \frac{1}{4}|X(0)|^2 + |X(0)|^2$$

$$|Y(0)|^2 = 9$$

4.11 (2)

$$x(t) = e^{-|t|} \rightleftharpoons X(j\omega) = \frac{2}{1+\omega^2}$$

By using area under frequency domain property,

Area of $x(t) = X(j0)$

$\Rightarrow X(j0) =$ Area of $x(t)$

$$= \int_{-\infty}^{\infty} x(t)dt$$

$$= \int_{-\infty}^{0} e^{t}dt + \int_{0}^{\infty} e^{-t}dt = 1 + 1 = 2$$

4.12 (b, c)

If $x(t)$ is real, then its Fourier transform will be complex conjugate.

i.e., $x(t) \longrightarrow X(\omega)$

$x(-t) \longrightarrow X(-\omega) = X^*(\omega)$

Option (a) $\rightarrow$ $x(t) + x(-t) \xleftrightarrow{FT} X(\omega) + X(-\omega)$

$= X(\omega) + X^*(\omega) \Rightarrow$ Real

Option (b) $\rightarrow$ $j[(x(t) + x(-t)]$

$= j[X(\omega) + X(-\omega)]$

... Taking Fourier Transform

$= j(X(\omega) + X^*(\omega))$

= Purely Imaginary

Option (c) $\rightarrow$ $x(t) - x(-t) \xleftrightarrow{F.T} X(\omega) - X(-\omega)$

$= X(\omega) - X^*(\omega) \Rightarrow$ Purely imaginary

Option (d) $\rightarrow$

$j[x(t) - x(-t)] \xleftrightarrow{F.T} j[X(\omega) - X^*(\omega)]l$

$\Rightarrow$ Purely Real

■■■

5 Laplace Transform

5.1 Identify the transfer function corresponding to an all-pass filter from the following:

(a) $\frac{1-s\tau}{1+s\tau}$ (b) $\frac{1+s\tau_1}{1+s\tau_2}$

(c) $\frac{1}{1+s\tau}$ (d) $\frac{s\tau}{1+s\tau}$

[2005 : 1 M]

5.2 Let the signal $x(t)$ have the Fourier transform $X(\omega)$.

Consider the signal $y(t) = \frac{d}{dt}[x(t-t_d)]$

(Where t_d is an arbitrary delay.)

The magnitude of the Fourier transform of $y(t)$ is given by the expression

(a) $\left|X(\omega)|\omega|\right|$ (b) $|X(\omega)|\cdot\omega$

(c) $\omega^2\cdot|X(\omega)|$ (d) $|\omega||X(\omega)|e^{-j\omega t_d}$

[2007 : 2 M]

5.3 $u(t)$ represents the unit step function. The Laplace transform of $u(t-\tau)$ is

(a) $\frac{1}{s\tau}$ (b) $\frac{1}{s-\tau}$

(c) $\frac{e^{-s\tau}}{s}$ (d) $e^{-s\tau}$

[2010 : 1 M]

5.4 The unilateral Laplace transform of $f(t)$ is $\frac{1}{s^2+s+1}$. The unilateral Laplace transform of $tf(t)$ is

(a) $-\frac{s}{(s^2+s+1)^2}$ (b) $-\frac{2s+1}{(s^2+s+1)^2}$

(c) $\frac{s}{(s^2+s+1)^2}$ (d) $\frac{2s+1}{(s^2+s+1)^2}$

[2012 : 1 M]

5.5 The Laplace Transform representation of the triangular pulse shown below is

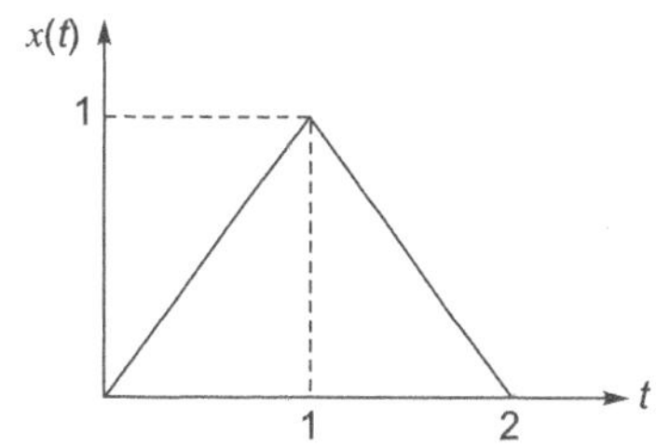

(a) $\frac{1}{s^2}[1+e^{-2s}]$ (b) $\frac{1}{s^2}[1-e^{-s}+e^{-2s}]$

(c) $\frac{1}{s^2}[1-e^{-s}+2e^{-2s}]$ (d) $\frac{1}{s^2}[1-2e^{-s}+e^{-2s}]$

[2013 : 2 M]

5.6 If $X(s)$, the Laplace transform of signal $x(t)$ is given by $X(s) = \frac{(s+2)}{(s+1)(s+3)^2}$, then the value of $x(t)$ as $t\to\infty$ is ______.

[2016 : 1 M]

5.7 A system is described by the following differential equation:

$$\frac{dy(t)}{dt} + 2y(t) = \frac{dx(t)}{dt} + x(t), x(0) = y(0) = 0$$

where $x(t)$ and $y(t)$ are the input and output variables respectively. The transfer function of the inverse system is

(a) $\frac{s+1}{s-2}$ (b) $\frac{s+2}{s+1}$

(c) $\frac{s+1}{s+2}$ (d) $\frac{s-1}{s-2}$

[2017 : 1 M]

5.8 The Laplace transform of a causal signal $y(t)$ is $Y(s) = \frac{s+2}{s+6}$. The value of the signal $y(t)$ at $t = 0.1$s is _____ units.

[2017 : 2 M]

5.9 Let u(t) denote the unit step function. The bilateral Laplace transform of the function $f(t) = e^t u(-t)$ is ______.

(a) $\frac{1}{s-1}$ with real part of $s < 1$

(b) $\frac{1}{s-1}$ with real part of $s > 1$

(c) $\frac{-1}{s-1}$ with real part of $s < 1$

(d) $\frac{-1}{s-1}$ with real part of $s > 1$

[2021 : 1 M]

5.10 Given $y(t) = e^{-3t}u(t) * u(t+3)$, where * denotes convolution operation. The value of $y(t)$ as $t \to \infty$ is ______.
(rounded off to two decimal places)

[2021 : 2 M]

5.11 The signal $x(t) = (t-1)^2 u(t-1)$, where $u(t)$ is the unit-step function, has the Laplace transform $X(s)$. The value of $X(1)$ is ______.

(a) $\frac{1}{e}$ (b) $\frac{2}{e}$

(c) $2e$ (d) e^2 [2022 : 2 M]

5.12 The Laplace transform of the continuous-time signal $x(t) = e^{-3t}u(t-5)$ is ______, where $u(t)$ denotes the continuous-time unit step signal.

(a) $\frac{e^{-5s}}{s+3}$, Real$\{s\} > -3$

(b) $\frac{e^{-5(s-3)}}{s-3}$, Real$\{s\} > -3$

(c) $\frac{e^{-5(s+3)}}{s+3}$, Real$\{s\} > -3$

(d) $\frac{e^{-5(s-3)}}{s+3}$, Real$\{s\} > -3$ [2023 : 2 M]

5.13 Laplace transform of a signal $x(t)$ is

$$X(s) = \frac{1}{s^2+13s+42}$$

Let $u(t)$ be the unit step function. Choose the signal $x(t)$ from the following options if the region of convergence is $-7 < \text{Re}\{s\} < -6$.

(a) $-e^{-6t}u(t) - e^{-7t}u(-t)$

(b) $-e^{-6t}u(-t) - e^{-7t}u(t)$

(c) $e^{-6t}u(t) - e^{-7t}u(-t)$

(d) $-e^{-6t}u(t) - e^{-7t}u(-t)$

[2024 : 2 M]

■■■

Answers Laplace Transform

5.1	(a)	5.2	(a)	5.3	(c)	5.4	(d)	5.5	(d)	5.6	(0)	5.7	(b)
5.8	(-2.195)	5.9	(c)	5.10	(0.33)	5.11	(b)	5.12	(c)	5.13	(b)		

Explanations Laplace Transform

5.1 (a)

By observing all the transfer function $\frac{1-s\tau}{1+s\tau}$ is the transfer function for all pass filter. So, the correct option is (a).

5.2 (a)

$x(t) \leftrightarrow X(\omega)$

$x(t-t_d) \leftrightarrow e^{-j\omega t_d}X(\omega)$

$$\frac{d[x(t-t_d)]}{dt} \leftrightarrow j\omega\, e^{-j\omega t_d}X(\omega)$$

$$\frac{d[x(t-t_d)]}{dt} \leftrightarrow |j\omega||X(\omega)||e^{-j\omega t_d}|$$

$$= |\omega||X(\omega)| = |X(\omega)|\omega||$$

Hence correct option is (a).

5.3 (c)

$$f(t) = u(t-\tau)$$
$$L\{f(t)\} = L\{u(t-\tau)\}$$
$$F(s) = e^{-s\tau}\cdot L\{u(t)\}$$
$$\therefore \quad F(s) = \frac{e^{-s\tau}}{s}$$

5.4 (d)

If $f(t) \xleftrightarrow{L} F(s)$

then $tf(t) \xleftrightarrow{L} -\frac{d}{ds}(F(s))$

given,

$$F(s) = \frac{1}{(s^2+s+1)}$$

$$\Rightarrow \frac{dF(s)}{ds} = \frac{(s^2+s+1)\times 0 - 1\times(2s+1)}{(s^2+s+1)^2}$$

$$= -\frac{(2s+1)}{(s^2+s+1)^2}$$

and Laplace transform of $tf(t)$ is $-\frac{d}{ds}(F(s))$

$$\therefore \mathcal{L}[tf(t)] = \frac{(2s+1)}{(s^2+s+1)^2}$$

5.5 (d)

$$x(t) = r(t) - 2r(t-1) + r(t-2)$$

$$\therefore X(s) = \frac{1}{s^2}[1-2e^{-s}+e^{-2s}]$$

5.6 (0)

$$X(s) = \frac{(s+2)}{(s+1)(s+3)^2}$$

Using final value theorem

$$\underset{t\to\infty}{\text{Lt}}\, x(t) = \underset{s\to 0}{\text{Lt}}\, sX(s)$$

$$\underset{t\to\infty}{\text{Lt}}\, x(t) = \underset{s\to 0}{\text{Lt}}\, s\frac{(s+2)}{(s+1)(s+3)^2} = 0$$

5.7 (b)

$$\frac{dy(t)}{dt} + 2y(t) = \frac{dx(t)}{dt} + x(t)$$

So, $$H(s) = \frac{Y(s)}{X(s)} = \frac{s+1}{s+2}$$

So, transfer function of inverse system.

$$H^{-1}(s) = \frac{s+2}{s+1}$$

5.8 -2.195 (-2.4 to -2.0)

$y(s) = \frac{s+2}{s+6}$ is proper converting to strictly proper

$$y(s) = 1 - \frac{4}{s+6}$$

$$y(t) = \delta(t) - 4e^{-6t}\,U(t)$$

$$y(0.1) = 0 - 4e^{-0.6}\,U(t) = -2.195$$

5.9 (c)

$$f(t) = e^{t}u(-t)$$

As we know, $e^{-t}u(t) \xrightarrow{\text{LT}} \frac{1}{s+1}, \sigma > -1$

By using time-reversal property of Laplace transform,

$$e^{t}u(-t) \xrightarrow{\text{LT}} \frac{1}{-s+1}, \quad -\sigma > -1$$

$$\therefore e^{t}u(-t) \xrightarrow{\text{LT}} \frac{-1}{s-1}, \quad \sigma < 1$$

5.10 (0.33)

$$y(t) = e^{-3t}\,u(t) * u(t+3)$$

$$Y(s) = e^{3s}\left[\frac{1}{3}\left(\frac{1}{s} - \frac{1}{s+3}\right)\right]$$

$$Y(s) = e^{3s}\,F(s) \qquad \ldots \text{(i)}$$

where, $$F(s) = \frac{1}{3}\left(\frac{1}{s} - \frac{1}{s+3}\right)$$

$$f(t) = \frac{1}{3}\left[1-e^{-3t}\right]u(t) \qquad \ldots \text{(ii)}$$

From equation (i),

$$Y(s) = e^{3s}\,F(s)$$

$$y(t) = f(t+3)$$

$$y(t) = \frac{1}{3}\left[1-e^{-3(t+3)}\right]u(t+3) \ldots.$$

From equation (ii),

Put $t = \infty, y(\infty) = \frac{1}{3}[1-0]\times 1 = \frac{1}{3}$

5.11 (b)

Given signal,

$$x(t) = (t-1)^2\,u(t-1)$$

$$x_1(t) = t^2\,u(t)$$

Taking Laplace transform,

$$X_1(s) = \frac{2}{s^2}$$

$$x_2(t) = x(t-1)$$

Taking Laplace transform,

$$X_2(s) = \frac{2e^{-s}}{s^2}$$

So, at $t = 1$,

$$X(1) = \frac{2e^{-1}}{1} = \frac{2}{e}$$

5.12 (c)

Given, $x(t) = e^{-3t}u(t-5)$

We know, $u(t) \rightarrow \frac{1}{s}$

$$u(t-5) \rightarrow \frac{1}{s}\cdot e^{-5s}$$

$$e^{-3t}u(t-5) \rightarrow \frac{1}{(s+3)} \cdot e^{-5(s+3)}$$

$$\therefore \quad X(s) = \frac{e^{-5(s+3)}}{s+3}; \qquad \text{Re}(s) > -3$$

5.13 (b)

Given $x(t) \xrightleftharpoons{L.T} X(s) = \dfrac{1}{s^2 13s + 42}$

$$\frac{1}{s^2+13s+42} = \frac{1}{(s+6)} - \frac{1}{(s+7)} \qquad -7 < 6 < -6$$

$$X(s) = \frac{1}{(s+6)} - \frac{1}{(s+7)} \qquad -7 < \text{Re}\{s\} < -6$$

Taking Inverse 1.7

$$x(t) = -e^{-6t}\, u(-t) - e^{-7t}u(t)$$

■■■

6 Z-Transform

6.1 Given $x(z) = \frac{1/2}{1-az^{-1}} + \frac{1/3}{1-bz^{-1}}$ $|a|$ and $|b| < 1$ with the ROC specified as $|a| < |z| < |b|$, x[0] of the corresponding sequence is given by

(a) 1/3 (b) 5/6
(c) 1/2 (d) 1/6

[2004 : 2 M]

6.2 A discrete-time signal, $x[n]$, suffered a distortion modeled by an LTI system with $H(z) = (1 - az^{-1})$, a is real and $|a| > 1$. The impulse response of a stable system that exactly compensates the magnitude of the distortion is

(a) $\left(\frac{1}{a}\right)^n u[n]$ (b) $-\left(\frac{1}{a}\right)^n u[-n-1]$
(c) $a^n u[n]$ (d) $a^n u[-n-1]$

[2004 : 2 M]

6.3 In the IIR filter shown below, a is a variable gain. For which of the following cases, the system will transit from stable to unstable condition

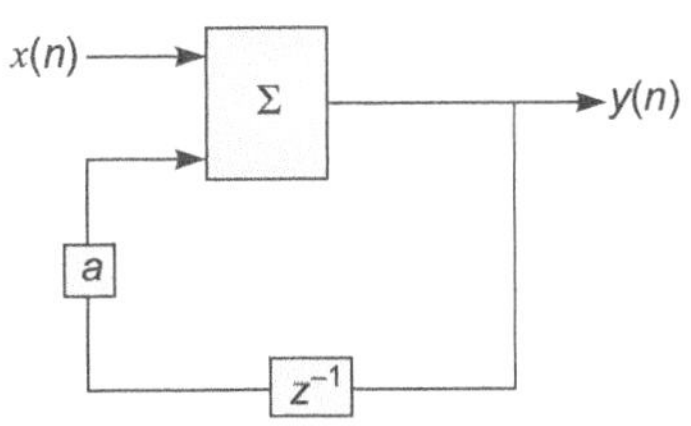

(a) 0.1 < a < 0.5 (b) 0.5 < a < 1.5
(c) 1.5 < a < 2.5 (d) 2 < a < ∞

[2004 : 2 M]

6.4 The region of convergence of the Z-transform of the discrete-time signal $x[n] = 2^n u[n]$ will be

(a) $|z| > 2$ (b) $|z| < 2$
(c) $|z| > \frac{1}{2}$ (d) $|z| < \frac{1}{2}$

[2008 : 2 M]

6.5 $H(z)$ is discrete rational transfer function. To ensure that both $H(z)$ and its inverse are stable its

(a) Poles must be inside the unit circle and zeros must be outside the unit circle.
(b) Poles and zeros must be inside the unit circle.
(c) Poles and zeros must be outside the unit circle.
(d) Poles must be outside the unit circle and the zeros should be inside the unit circle.

[2010 : 1 M]

6.6 Consider the difference equation $y[n] - \frac{1}{3}y[n-1] = x[n]$ and suppose that $x[n] = \left(\frac{1}{2}\right)^n u[n]$. Assuming the condition of initial rest, the solution for $y[n]$, $n \geq 0$ is

(a) $3\left(\frac{1}{3}\right)^n - 2\left(\frac{1}{2}\right)^n$ (b) $-2\left(\frac{1}{3}\right)^n + 3\left(\frac{1}{2}\right)^n$
(c) $\frac{2}{3}\left(\frac{1}{3}\right)^n + \frac{1}{3}\left(\frac{1}{2}\right)^n$ (d) $\frac{1}{3}\left(\frac{1}{3}\right)^n + \frac{2}{3}\left(\frac{1}{2}\right)^n$

[2011 : 2 M]

6.7 If $x[n] = (1/3)^{|n|} - (1/2)^n u[n]$, then the region of convergence (ROC) of its Z-transform in the Z-plane will be

(a) $\frac{1}{3} < |z| < 3$ (b) $\frac{1}{3} < |z| < \frac{1}{2}$
(c) $\frac{1}{2} < |z| < 3$ (d) $\frac{1}{3} < |z|$

[2012 : 1 M]

6.8 The system function of an LTI system is given by

$$H(z) = \frac{1 - \frac{1}{3}z^{-1}}{1 - \frac{1}{4}z^{-1}}$$

The above system can have stable inverse if the region of convergence of H(z) is defined as

(a) $|z| < \frac{1}{4}$ (b) $|z| < \frac{1}{12}$
(c) $|z| > \frac{1}{4}$ (d) $|z| < \frac{1}{3}$

[2014 : 1 M]

6.9 The transfer function of a digital system is given by

$$\frac{b_0}{1-z^{-1}+a_2 z^{-2}}\text{; where } a_2 \text{ is real}$$

The transfer function is BIBO stable if the value of a_2 is

(a) –1.5 (b) –0.75
(c) 0.5 (d) 1.5 [2014 : 2 M]

6.10 The z-transform of $x[n] = \alpha^{|n|}, 0 < |\alpha| < 1$, is $X(z)$. The region of convergence of $X(Z)$ is

(a) $|\alpha| < |z| < \frac{1}{|\alpha|}$ (b) $|z| > \alpha$
(c) $|z| > \frac{1}{|\alpha|}$ (d) $|z| < \min\left[|\alpha|, \frac{1}{|\alpha|}\right]$

[2015 : 2 M]

6.11 The Region of Convergence (ROC) of the Z-transform of a causal unit step discrete-time sequence is

(a) $|z| < 1$ (b) $|z| \le 1$
(c) $|z| > 1$ (d) $|z| \ge 1$ [2017 : 1 M]

6.12 Consider two discrete-time signals:
$x_1(n) = \{1, 1\}$ and $x_2(n) = \{1, 2\}$, for $n = 0, 1$.
The Z-transform of the convoluted sequence $x_1(n) * x_2(n)$ is

(a) $1 + 2z^{-1} + 3z^{-2}$ (b) $z^2 + 3z + 2$
(c) $1 + 3z^{-1} + 2z^{-2}$ (d) $z^{-2} + 3z^{-3} + 2z^{-4}$

[2017 : 2 M]

6.13 Let $y[n] = x[n] * h[n]$, where $*$ denotes convolution and $x[n]$ and $h[n]$ are two discrete time sequence. Given that the z-transform of $y[n]$ is

$$Y(z) = 2 + 3z^{-1} + z^{-2}$$

the z-transform of $p[n] = x[n] * h[n-2]$ is

(a) $2 + 3z + z^{-2}$ (b) $3z + z^{-2}$
(c) $2z^2 + 3z + 1$ (d) $2z^{-2} + 3z^{-3} + z^{-4}$

[2018 : 2 M]

6.14 The input x[n] and output $y[n]$ of a discrete-time system are related as $y[n] = \alpha y[n-1] + x[n]$. The condition on α for which the system is Bounded-Input Bounded-Output (BIBO) stable is

(a) $|\alpha| < 1$ (b) $|\alpha| = 1$
(c) $|\alpha| > 1$ (d) $|\alpha| < \frac{3}{2}$

[2019 : 1 M]

6.15 Consider the transfer function

$$H_c(s) = \frac{1}{(s+1)(s+3)}$$

Bilinear transformation with a sampling period of 0.1 s is employed to obtain the discrete-time transfer function $H_d(z)$. Then $H_d(z)$ is ________.

(a) $\frac{(1+z^{-1})^2}{(19-21z^{-1})(23-17z^{-1})}$

(b) $\frac{(1-z^{-1})^2}{(21-19z^{-1})(17-23z^{-1})}$

(c) $\frac{(1+z^{-1})^2}{(21-19z^{-1})(23-17z^{-1})}$

(d) $\frac{(1+z^{-1})^2}{(21-19z^{-1})(17-23z^{-1})}$ [2022 : 2 M]

6.16 $F(z) = \frac{1}{1-z}$, when expanded as a power series around $z = 2$, would result in $F(z) = \sum_{k=0}^{\infty} a_k(z-2)^k$ with the region of convergence (ROC) $|z-2| < 1$. The coefficients a_k, ≥ 0, are given by the expression ________ .

(a) $(-1)^k$ (b) $(-1)^{k+1}$
(c) $\left(\frac{1}{2}\right)^k$ (d) $\left(\frac{-1}{2}\right)^{k+1}$

[2023 : 1 M]

6.17 Consider a discrete-time sequence

$$x[n] = \begin{cases} (0.2)^n, & 0 \le n \le 7 \\ 0, & \text{otherwise} \end{cases}$$

The region of convergence of $X(z)$, the z-transform of $x[n]$, consists of

(a) all values of z except $z = 0.2$
(b) all values of z
(c) all values of z except $z = 0$
(d) all values of z except $z = \infty$

[2024 : 2 M]

■■■

Answers Z-Transform

6.1 (c)	6.2 (d)	6.3 (b)	6.4 (a)	6.5 (b)	6.6 (b)	6.7 (c)
6.8 (c)	6.9 (c)	6.10 (a)	6.11 (c)	6.12 (c)	6.13 (d)	6.14 (a)
6.15 (c)	6.16 (b)	6.17 (c)				

Explanations Z-Transform

6.1 (c)

$$X(z) = \frac{1/2}{(1-az^{-1})} + \frac{1/3}{1-bz^{-1}},$$

ROC $|a| < |z| < |b|$

$$\therefore \quad x[n] = \frac{1}{2}(a)^n u[n] - \frac{1}{3}(b)^n u[-n-1]$$

$$x[0] = \frac{1}{2}$$

6.2 (d)

For compensation,

$$H'(z)\,H(z) = 1$$

So, overall impulse response = $\delta(n)$
and there is no delay element (no distortion).

$$\therefore \quad H'(z) = \frac{1}{H(z)} = \frac{1}{(1-az^{-1})}$$

$$= -(a)^n u[-n-1],$$

$$(|z| < |a|)$$

(Because system $H'(z)$ has to be stable)
So unit circle must lie inside the ROC.

$$|a| > 1,$$

If ROC is $|z| < |a|$, then it will include unit circle.
Neglecting the –ve sign, option (d) is correct.

6.3 (b)

From the figure,

$$Y(z) = X(z) + az^{-1}\,Y(z)$$

$$Y[z]\,[1 - az^{-1}] = X(z)$$

Therefore system transfer function will be

$$H(z) = \frac{1}{(1-az^{-1})}$$

For system to be stable, it's impulse response's ROC must include unit circle.

$$h[n] = [a]^n u[n]$$

For ROC $|z| > |a|$
if $|a| < 1$ then system is stable
if $|a| > 1$ then system is unstable
if $0.5 < |a| < 1.5$

For $0.5 < |a| < 1 \Rightarrow$ stable
$1 < |a| < 1.5 \Rightarrow$ unstable

So, system will transit from stable to unstable state as $|a|$ become greater than 1.

6.4 (a)

Given,

$$x[n] = 2^n u(n)$$

$$x(z) = \sum_{n=-\infty}^{\infty} x[n]\,z^{-n}$$

$$= \sum_{n=-\infty}^{\infty} 2^n u[n] z^{-n} = \sum_{n=0}^{\infty} 2^n \cdot z^{-n}$$

$$= \sum_{n=0}^{\infty}\left(\frac{2}{z}\right)^n = \frac{1}{1-2z^{-1}}$$

$$ROC \Rightarrow \left|2z^{-1}\right| < 1 \Rightarrow |z| > 2$$

6.5 (b)

For $H(z)$ to be stable the poles of $H(z)$ must be inside the unit circle and for the inverse of $H(z)$ to be stable the pole of inverse of $H(z)$ also must be inside the unit circle because the pole of inverse of $H(z)$ are the zeros of $H(z)$. Hence for both $H(z)$ and its inverse to be stable poles and zeros must be inside the unit circle.

6.6 (b)

$$y[n] - \frac{1}{3}y[n-1] = x[n]$$

taking z-transform we get,

$$Y(z) - \frac{1}{3}z^{-1}Y(z) = X(z)$$

$$\frac{Y(z)}{X(z)} = \frac{1}{\left(1-\frac{1}{3}z^{-1}\right)}$$

$$Y(z) = \frac{1}{\left[1-\left(\frac{1}{3}\right)z^{-1}\right]}X(z) \quad ...(i)$$

$$x(n) = \left(\frac{1}{2}\right)^n u[n]$$

$$X(z) = \frac{1}{\left(1-\frac{1}{2}z^{-1}\right)} \quad ...(ii)$$

From (i) and (ii) we have

$$Y(z) = \frac{1}{\left[1-\left(\frac{1}{3}\right)z^{-1}\right]} \times \frac{1}{\left(1-\frac{1}{2}z^{-1}\right)}$$

$$= \frac{3}{\left(1-\frac{1}{2}z^{-1}\right)} - \frac{2}{\left(1-\frac{1}{3}z^{-1}\right)}$$

Assuming $y[n]$ to be right sided and taking inverse z-transform we get,

$$y[n] = 3\left(\frac{1}{2}\right)^n u[n] - 2\left(\frac{1}{3}\right)^n u[n]$$

So, $y[n]$ for $n \geq 0 = 3\left(\frac{1}{2}\right)^n - 2\left(\frac{1}{3}\right)^n$

6.7 (c)

Let $x_1[n] = \left(\frac{1}{3}\right)^{|n|}$

and $x_2[n] = \left(\frac{1}{2}\right)^n u[n]$

$$\Rightarrow x_1[n] = \left(\frac{1}{3}\right)^n u[n] + \left(\frac{1}{3}\right)^{-n} u[-n-1]$$

$$\left(\frac{1}{3}\right)^n u[n] \xleftrightarrow{z} \frac{1}{1-\frac{1}{3}z^{-1}}; ROC: |z| > \frac{1}{3}$$

$$\left(\frac{1}{3}\right)^{-n} u[-n-1] \xleftrightarrow{z} \frac{-1}{1-\left(\frac{1}{3}\right)^{-1} z^{-1}}; ROC: |z| < 3$$

and

$$x_2[n] = \left(\frac{1}{2}\right)^n u[n] \xleftrightarrow{z} \frac{1}{1-\frac{1}{2}z^{-1}}; ROC: |z| > \frac{1}{2}$$

$\therefore$ ROC is $\frac{1}{2} < |z| < 3$

6.8 (c)

A causal discrete time LTI system with rational transfer function is stable if and only if all the pole of the transfer function lie inside the unit circle $|z| = 1$, also ROC should include the unit are $|z| = 1$.

6.9 (c)

Transfer function of the digital system, is

$$H(z) = \frac{b_o}{1-z^{-1}+a_2 z^{-2}}, \quad a_2 \text{ is real}$$

$$= \frac{bz^2}{z^2 - z + a_2}, \quad a = -1$$

The stability of a two-pole system can be investigated by a stability criteria method called "Scho-Cohn" stability test.

In this approach there are two polynomials k_1 and k_2 which are calculated from denominator polynomial of the transfer function.

For the system to be stable,

$$|k_1| < 1, \text{ and } |k_2| < 1$$

$$k_2 = a_2 \text{ and } k_1 = \frac{a}{1+a_2}$$

thus, $\quad -1 < a_2 < 1 \quad ...(i)$

and $\quad -1 < \frac{a_1}{1+a_2} < 1 \quad ...(ii)$

these two condition (i) and (ii) are satisfied only if $a_2 = 0.5$ among the given options.

6.10 (a)

$$x[n] = \alpha^{|n|}, 0 < |\alpha| < 1$$

$$\text{So, } x[n] = \begin{cases} \alpha^n u(n) & ; n \geq 0 \\ \alpha^{-n} u(-n-1); & n < 0 \end{cases}$$

$$x[n] = \alpha^n u[n] + \alpha^{-n} u[-n-1]$$

Right sided signal	Left sided signal
ROC $\rightarrow X(z) > \alpha$	$X(z) < \frac{1}{\alpha}$
$\|z\| > \|\alpha\|$	$\|z\| < \left\|\frac{1}{\alpha}\right\|$

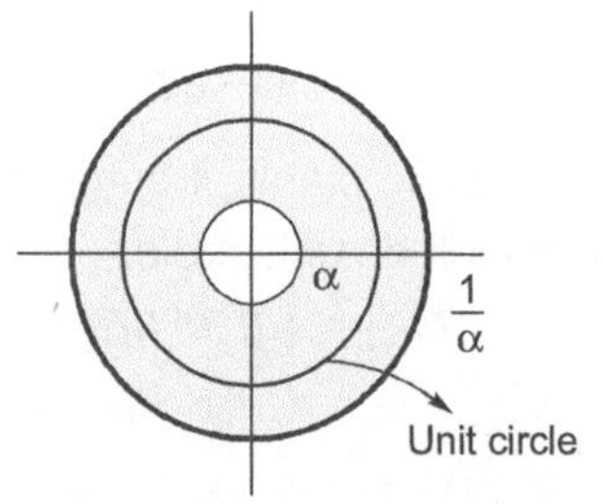

6.11 (c)

$$u[n] \xrightarrow{z} \frac{1}{1-z^{-1}} ; |z| > 1$$

6.12 (c)

$$x_1[n] = \{1, 1,\} \times x_2\{n\} = \{1, 2\}$$

So, $x[n] = x_1[n] * x_2[n] = \{1, 3, 2\}$

$$x(z) = 1 + 3z^{-1} + 2z^{-2} = -2$$

6.13 (d)

$$y[n] = x[n] * h[n]$$

$$Y(z) = 2 + 3z^{-1} + z^{-2} = X(z) \times H(z)$$

$$P(z) = z^{-2}\,[X(z) \times H(z)] = 2z^{-2} + 3z^{-3} + z^{-4}$$

6.14 (a)

Here, $H(z) = \dfrac{1}{1-\alpha z^{-1}}; |z| > |\alpha|$

For system to be stable,

$$|\alpha| < 1$$

Therefore, option (a) is correct.

6.15 (c)

$$H_c(s) = \frac{1}{(s+1)(s+3)}$$

$$T_s = 0.1 \text{ sec}$$

Bilinear transfer - transformation

$$s = \frac{2}{T}\left(\frac{1-z^{-1}}{1+z^{-1}}\right) = \frac{2}{0.1}\left(\frac{1-z^{-1}}{1+z^{-1}}\right)$$

$$s = 20\left(\frac{1-z^{-1}}{1+z^{-1}}\right)$$

So z-transform of given transfer function becomes

$$H_d(z) = \left.\frac{1}{(s+1)(s+3)}\right|_{s=20\left(\frac{1-z^{-1}}{1+z^{-1}}\right)}$$

$$H_d(z) = \frac{1}{\left[1+\dfrac{20(1-z^{-1})}{1+z^{-1}}\right]\left[3+\dfrac{20(1-z^{-1})}{1+z^{-1}}\right]}$$

$$H_d(z) = \frac{(1+z^{-1})^2}{(1+z^{-1}+20-20z^{-1})(3+3z^{-1}+20-20z^{-1})}$$

$$H_d(z) = \frac{(1+z^{-1})^2}{(21-19z^{-1})(23-17z^{-1})}$$

6.16 (b)

Power series expansion for

$$\frac{1}{1+x} = 1 - x + x^2 - x^3 + x^4 - \ldots\ldots \quad \text{when } |x| < 1$$

Now, given that,

$$F(z) = \frac{1}{1-z} = -\left[\frac{1}{1+(z-2)}\right]$$

$$= -[1 - (z-2) + (z-2)^2 - (z-2)^3 + \ldots\ldots],$$

when $|z-2| < 1$

$$= -\sum_{K=0}^{\infty} (-1)^K (z-2)^K, \quad |z-2| < 1$$

$$= \sum_{K=0}^{\infty} (-1)^{K+1} (z-2)^K, \quad |z-2| < 1$$

$$= \sum_{K=0}^{\infty} a_K (z-2)^K, \quad |z-2| < 1$$

$$\therefore \quad a_K = (-1)^{K+1}$$

6.17 (c)

Given $x[n] = \begin{cases} (0.2)^n, & 0 \le n \le 7 \\ 0, & \text{otherwise} \end{cases}$

ROC: $X(z)$

$X(z) \rightarrow$ finite duration

ROC → entire z-plane

except origin → right sided (only –ve power of z)

7 Discrete Time Fourier Transform (D.T.F.T.)

7.1 If the Fourier transform of $x[n]$ is $X(e^{j\omega})$, then the Fourier transform of $(-1)^n x[n]$ is

(a) $(-j)^{\omega} X(e^{j\omega})$ (b) $(-1)^{\omega} X(e^{j\omega})$

(c) $X\left(e^{j(\omega-\pi)}\right)$ (d) $\dfrac{d(X(e^{j\omega}))}{d\omega}$

[2004 : 1 M]

7.2 Given the discrete-time sequence

$x[n] = [2, 0, -\underset{\uparrow}{1}, -3, 4, 1, -1]$, $X(e^{j\pi})$ is

(a) 8 (b) 6π

(c) 8π (d) 6

[2005 : 1 M]

■■■

Answers Discrete Time Fourier Transform (D.T.F.T.)

7.1 (c) 7.2 (d)

Explanations Discrete Time Fourier Transform (D.T.F.T.)

7.1 (c)

$x[n] \leftrightarrow X(e^{j\omega})$

$$(-1)^n x[n] = (e^{j\pi})^n x[n] \qquad (\because e^{j\pi} = -1)$$

$$e^{j\pi n} x[n] \leftrightarrow X[e^{j(\omega-\pi)}],$$

[shifting in frequency domain]

by using the property,

$$e^{j\omega n_0} x[n] \leftrightarrow X(e^{j(\omega-\omega_0)})$$

7.2 (d)

$$x[n] = [2, 0, -\underset{\uparrow}{1}, -3, 4, 1, -1]$$

$$X(z) = 2z^{+2} + 0z^{+1} - 1z^{0} - 3z^{-1} + 4z^{-2} + 1z^{-3} - 1z^{-4}$$

$$X(e^{j\pi}) = 2e^{j2\pi} - 1 - 3e^{j(-\pi)} + 4e^{j(-2\pi)} + e^{j(-3\pi)} - e^{j(-4\pi)}$$

$$= 2 - 1 - 3 \times (-1) + 4 - 1 - 1$$

$$= 9 - 3 = 6$$

■■■

8 Discrete Fourier Transform (D.F.T.) and Frequency Response

8.1 A linear time-invariant causal system has frequency response given in polar form as:

$$\frac{1}{\sqrt{1+\omega^2}}\angle -\tan^{-1}(\omega)$$

For input $x(t) = \sin t$, the output is

(a) $\frac{1}{\sqrt{2}}\cos t$ (b) $\frac{1}{\sqrt{2}}\cos\left(t-\frac{\pi}{4}\right)$

(c) $\frac{1}{\sqrt{2}}\sin t$ (d) $\frac{1}{\sqrt{2}}\sin\left(t-\frac{\pi}{4}\right)$

[2009 : 2 M]

8.2 4-point DFT of a real discrete-time signal $x[n]$ of length 4 is given by $X[k]$, $n = 0, 1, 2, 3$ and $k = 0, 1, 2, 3$. It is given that $X[0] = 5$, $X[1] = 1 + j1$, $X[2] = 0.5$, then $X[3]$ and $x[0]$ respectively are

(a) $1 - j$, 1.875 (b) $1 - j$, 1.500

(c) $1 + j$, 1.875 (d) $0.1 - j0.1$, 1.500

[2010 : 2 M]

8.3 $X(k)$ is the Discrete Fourier Transform of a 6-point real sequence $x(n)$.
If $X(0) = 9 + j0$, $X(2) = 2 + j2$, $X(3) = 3 - j0$, $X(5) = 1 - j1$, $x(0)$ is

(a) 3 (b) 9

(c) 15 (d) 18

[2014 : 2 M]

8.4 Three DFT coefficients, out of the DFT coefficients of a five-point real sequence are given as: $X(0) = 0$, $X(1) = 1 - j1$ and $X(3) = 2 + j2$. The zeroth value of the sequence $x(n)$.

(a) 1 (b) 2

(c) 3 (d) 4

[2017 : 2 M]

8.5 For the sequence $x[n] = \{1, -1, 1, -1\}$, with $n = 0, 1, 2, 3$, the DFT is computed as

$$X(k) = \sum_{n=0}^{3} x[n]\, e^{-j\frac{2\pi}{4}nk},$$

for $k = 0, 1, 2, 3$. The value of k for which $X(k)$ is not zero is

(a) 0 (b) 1

(c) 2 (d) 3

[2018 : 2 M]

8.6 Consider the finite sequence $X = (1, 1, 1)$. The Inverse Discrete Fourier Transform (IDFT) of X is given as $(x(0), x(1), x(2))$. The value of $x(2)$ is ______.

[2020 : 2 M]

8.7 The number of complex multiplications required for computing a 16-point DFT using the decimation-in-time radix-2 FFT is ________ (in integer).

[2024 : 1 M]

8.8 The 4-point DFTs of two sequences $x[n]$ and $y[n]$ are $X[k] = [1, -j, 1, j]$ and $Y[k] = [1, 3j, 1, -3j]$, respectively. Assuming $z[n]$ represents the 4-point circular convolution of $x[n]$ and $y[n]$, the value of $z[0]$ is ________ (rounded off to nearest integer).

Note: The DFT of a N-point sequence $x[n]$ is defined as

$$X[k] = \sum_{n=0}^{N-1} x[n]e^{\frac{-j2\pi nk}{N}}$$

[2024 : 2 M]

■■■

Answers Discrete Fourier Transform (D.F.T.) & Frequency Response

8.1 (d) 8.2 (a) 8.3 (a) 8.4 (b) 8.5 (c) 8.6 (0) 8.7 (32)

8.8 (2)

Explanations Discrete Fourier Transform (D.F.T.) & Frequency Response

8.1 (d)

Given,

$$\text{Transfer function} = \frac{1}{\sqrt{1+\omega^2}}\angle -\tan^{-1}(\omega)$$

and input $x(t) = \sin t$

$\Rightarrow \quad \omega = 1$

$$\therefore \quad \text{T/F} = \frac{1}{\sqrt{1+1^2}}\angle \tan^{-1}(1) = \frac{1}{\sqrt{2}}\angle -45^\circ$$

$$\therefore \quad y(t) = \frac{1}{\sqrt{2}}\sin(t-45^\circ)$$

8.2 (a)

$$\text{IDFT } x[n] = \frac{1}{N}\sum_{k=0}^{N-1} X[k]\,e^{j2\pi nk/N}$$

$$\therefore \quad x(0) = \frac{1}{N}\sum_{k=0}^{N-1} X[k]\,e^{0} = \frac{1}{4}\sum_{k=0}^{3} X[k]$$

$$= \frac{1}{4}[X[0]+X[1]+X[2]+X[3]]$$

$$x(0) = \frac{1}{4}[5+1+j+0.5+X[3]]$$

only option (a) $X[3] = 1 - j$ and $x(0) = 1.875$ satisfies.

8.3 (a)

$$x(n) = \frac{1}{N}\sum_{k=0}^{N-1} X[k]\,e^{\left(\frac{j2\pi n}{N}\right)k}$$

Also, $X(k) = X(N-k)^*$

For $N = 6$

Given,

$X(0) = 9 + j0$, $X(1) = X(5)^* = 1 + j$

$X(2) = 2 + j2$, $X(3) = 3 - j0$,

$X(4) = X(2)^* = 2 - j2$, $X(5) = 1 - j1$

Therefore,

$$x(0) = \frac{1}{6}\sum_{K=0}^{5} X(k)$$

$$= \frac{1}{6}\begin{bmatrix} 9 + j0 + 1 + j1 + j2 + j2 \\ +3 - j0 + 2 - j2 + 1 - j1 \end{bmatrix}$$

$$= \frac{18}{6} = 3$$

8.4 (b)

Since $x[n]$ is real

So, $X[k] = X^*[N-k]$

$X[2] = X^*[3] = 2 - j2$

$X[4] = X^*[1] = 1 + j1$

$$\text{So,} \quad x[0] = \frac{1}{N}\sum_{k=0}^{4} X[k]$$

$$= \frac{4+1+2+2+1}{5} = 2 \qquad \text{...(ii)}$$

8.5 (c)

$$X[k] = \begin{bmatrix} 1 & 1 & 1 & 1 \\ 1 & -j & -1 & j \\ 1 & -1 & 1 & -1 \\ 1 & j & -1 & -j \end{bmatrix}\begin{bmatrix} 1 \\ -1 \\ 1 \\ -1 \end{bmatrix}$$

$X[k] = 0$

$X[1] = 1 + j - 1 - j = 0$

$X[2] = 1 + 1 + 1 + 1 = 4$

$X[2]$ is non-zero.

8.6 (0)

$$x(n) = \frac{1}{N}\sum_{K=0}^{N-1} X(K)e^{j\frac{2\pi}{N}Kn}$$

$$= \frac{1}{3}\sum_{K=0}^{2} X(K)e^{j\frac{2\pi}{3}Kn}$$

$[\because X(K) = \{X(0), X(1), X(2)\} = \{1,1,1\}]$

$$= \frac{1}{3}\sum_{K=0}^{2} e^{j\frac{2\pi}{3}Kn} = \frac{1}{3}\left[1+e^{j\frac{4\pi}{3}}+e^{j\frac{8\pi}{3}}\right]$$

$$= \frac{e^{j\frac{4\pi}{3}}}{3}\left[e^{j\frac{4\pi}{3}} + 1 + e^{-j\frac{4\pi}{3}}\right]$$

$$= \frac{e^{j\frac{4\pi}{3}}}{3}\left[1 + 2\cos\frac{4\pi}{3}\right]$$

$$= \frac{e^{j\frac{4\pi}{3}}}{3}\left[1 + 2\left(-\frac{1}{2}\right)\right] = 0$$

8.7 (32)

No. of complex multiplication

$$= \frac{N}{2}\log_2 N$$

$$N = 16$$

$$\text{multiplication} = \frac{16}{2}\log_2 16$$

$$= 8 \times 4 = 32$$

8.8 (2)

$$z(n) = x(n) * y(n)$$

$$Z(k) = X(k) \cdot y(k)$$

$$= \{1, -j, 1, j\} \times \{1, 3j, 1, -3j\}$$

$$= \{1, 3, 1, 3\}$$

$$z(n)\big|_{n=0} = \frac{1}{N}\sum_{K=0}^{3} Z(k) = \frac{1}{4}(1+3+1+3)$$

$$= \frac{8}{4} = 2$$

■■■

9 Sampling

9.1 A signal $x(t) = 5\cos(150\pi t - 60)$ is sampled at 200 Hz. The fundamental period of the sampled sequence $x[n]$ is

(a) $\frac{1}{200}$ (b) $\frac{2}{200}$
(c) 4 (d) 8 [2004 : 1 M]

9.2 The spectrum of a band limited signal after sampling is shown below. The value of the sampling interval is

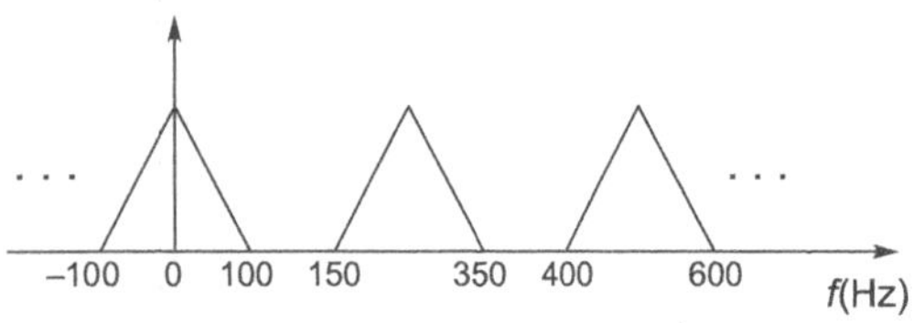

(a) 1 ms (b) 2 ms
(c) 4 ms (d) 8 ms
[2006 : 2 M]

9.3 A digital measuring instrument employs a sampling rate of 100 samples/second. The sampled input $x(n)$ is averaged using the difference equation

$$y[n] = [x(n) + x(n-1) + x(n-2) + x(n-4)/4]$$

For a step input, the maximum time taken for the output to reach the final value after the input transition is

(a) 20 ms (b) 40 ms
(c) 80 ms (d) ∞ [2006 : 2 M]

9.4 Let $x(t)$ be a continuous-time, real-valued signal band-limited to F Hz.
The Nyquist sampling rate in Hz,
For $y(t) = x(0.5\,t) + x(t) - x(2t)$ is

(a) F (b) 2F
(c) 4F (d) 8F [2007 : 1 M]

9.5 This continuous time signal $x(t) = \cos(100\pi t) + \sin(300\pi t)$ is sampled at the rate 100 Hz to get the signal

$$x_s(t) = \sum_{n=-\infty}^{\infty} x(nT_s)\delta(t - nT_s),$$

T_s = sampling period

The signal $x_s(t)$ is passed through an ideal low pass filter with cutoff frequency 100 Hz. The output of the filter is proportional to

(a) $\cos(100\pi t)$
(b) $\cos(100\pi t) + \sin(100\pi t)$
(c) $\cos(100\pi t) - \sin(100\pi t)$
(d) $\sin(100\pi t)$ [2011 : 1 M]

9.6 The highest frequency present in the signal $x(t)$ is f_{max}. The highest frequency present in the signal $y(t) = x^2(t)$ is

(a) $\frac{1}{2} f_{max}$ (b) f_{max}
(c) $2f_{max}$ (d) $4f_{max}$ [2015 : 1 M]

9.7 If a continuous time signal $x(t) = \cos(2\pi t)$ is sampled at 4 Hz, the value of the discrete time sequence $x(n) = 5$ is

(a) –0.707 (b) –1
(c) 0 (d) 1 [2017 : 1 M]

9.8 A sinusoid of 10 kHz is sampled at 15 k samples/s. The resulting signal is passed through an ideal low pass filter (LPF) with cut-off frequency of 25 kHz. The maximum frequency component at the output of the LPF (in kHz) is _____.
[2020 : 1 M]

9.9 A signal $x(t)$ is band-limited between 100 Hz and 200 Hz. A signal $y(t)$ is related to $x(t)$ as follows:

$$y(t) = x(2t - 5)$$

The statement that is always true is ________.

(a) $y(t)$ is band-limited between 50 Hz and 100 Hz
(b) $y(t)$ is band-limited between 100 Hz and 200 Hz
(c) $y(t)$ is band-limited between 200 Hz and 400 Hz
(d) $y(t)$ is not band-limited [2022 : 1 M]

9.10 The Nyquist sampling frequency for $x(t) = 10\sin^2(200\pi t)$ is _________ Hz (rounded off to nearest integer).
[2024 : 1 M]

■■■

Answers Sampling

9.1	(d)	9.2	(c)	9.3	(b)	9.4	(c)	9.5	(b)	9.6	(c)	9.7	(c)
9.8	(25)	9.9	(c)	9.10	(400)								

Explanations Sampling

9.1 (d)

The given signal is

$$x(t) = 5\cos(150\pi t - 60)$$
$$f_s = 200 \text{ Hz}$$
$$\Rightarrow \quad T_s = \frac{1}{f_s} = \frac{1}{200} \text{ sec}$$
$$x(nTs) = 5\cos(150\pi nT_s - 60)$$
$$= 5\cos\left(150\pi \times \frac{1}{200} n - 60\right)$$
$$= 5\cos\left(\frac{3\pi}{4} n - 60\right)$$

Here, $\omega_0 = \dfrac{3\pi}{4}$

$$\frac{\omega_0}{2\pi} = \frac{m}{N}$$
$$\Rightarrow \quad \frac{m}{N} = \frac{\frac{3\pi}{4}}{2\pi} = \frac{3\pi}{8\pi} = \frac{3}{8}$$

Hence, $N = 8$.

9.2 (c)

Sampling frequency $f_s = 250$ Hz,
therefore sampling period (sampling interval) =

$$1/fs = \frac{1}{250} = 4 \text{ msec}.$$

9.3 (b)

Since output y depends on input, such as no delay, delay by 1 unit, delay by 2 unit, delay by 4 unit, so it will sum all the sample after $4Ts$ (maximum delay), to get one sample of $y[n]$.

$$\therefore \quad t = 4T_s = \frac{4}{f_s} = \frac{4}{100} = 40 \text{ msec}$$

9.4 (c)

Expansion in time domain is compression in frequency domain and vice-versa. So, the maximum frequency component in given signal is 2 F Hz. And according to sampling theorem

Nyquist rate = $2f_m = 2 \times 2\text{F} = 4\text{F}$ Hz

9.5 (b)

$$x(t) = \cos(100\pi t) + \sin(300\pi t)$$

$X(f) \Rightarrow$

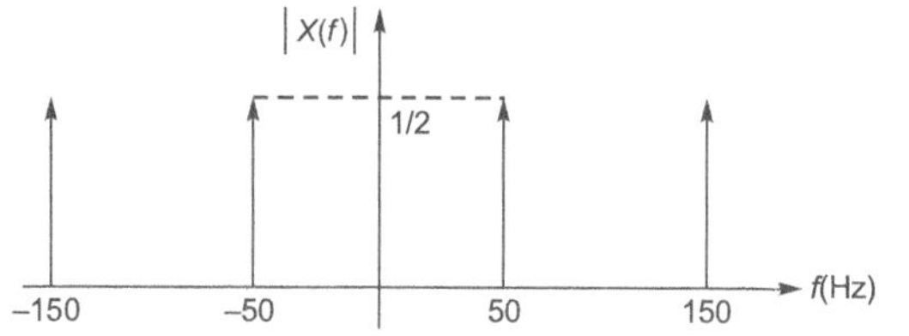

Now, $x_s(t) = \sum_{n=-\infty}^{\infty} x(nT_s)\,\delta[t - nT_s]$

$$x_s(f) = \frac{1}{T_s}\sum_{n=-\infty}^{\infty} x(f - nf_s)$$
$$f_s = 100 \text{ Hz}$$

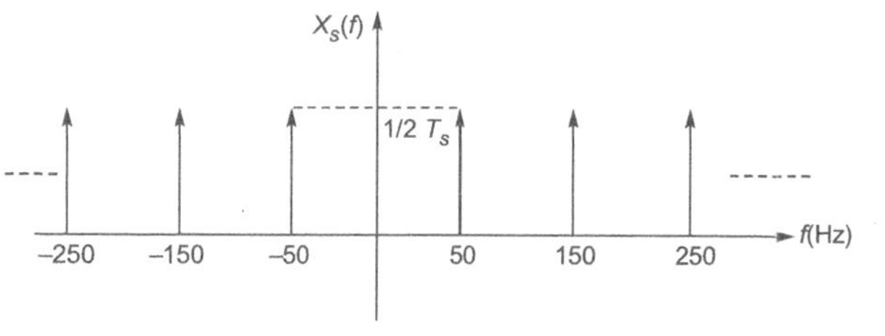

Here due to under sampling high frequency component corresponding to 150 Hz will appear as low frequency component at 50 Hz.

So when we filter it by a L.P.F. having cut-off frequency of 100 Hz, we get

$$y(t) = \cos 100\pi t + \sin 100\pi t$$

$\sin 100\pi t$ high frequency component appear as low frequency component.

9.6 (c)

By convolution property

$$x(t)\cdot x(t) = X(f) * X(f)$$

Now by limiting property of sampling frequency of signal $y(t) = 2f_{max}$

9.7 (c)

$$x(nT_s) = \cos 2\pi \cdot \frac{n}{4} = \cos\frac{\pi}{2} n = x(n)$$
$$x(n)\big|_{n=5} = \cos\frac{5\pi}{2} = 0$$

9.8 **(25)**

Given signal is sinusoidal signal having frequency
$\Rightarrow$ 10 kHz

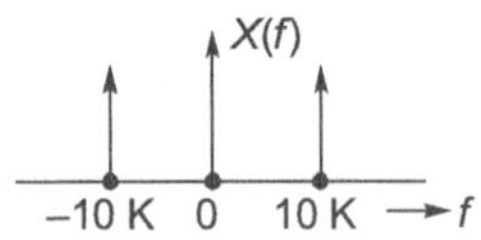

Given, $f_s = 15$ kHz

$$X_s(f) = f_s \sum_{n=-\infty}^{\infty} X(f - nf_s)$$

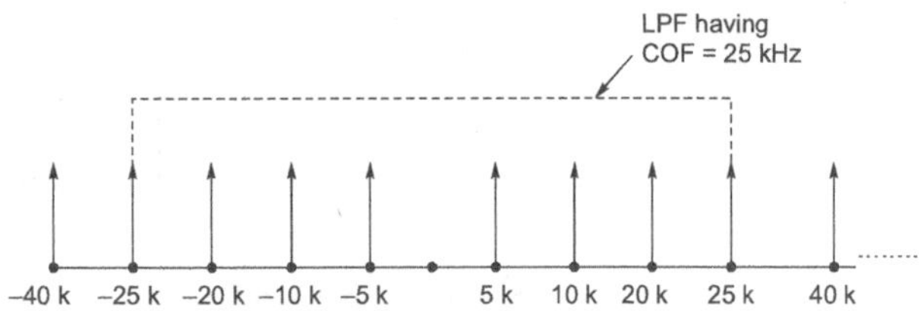

After passing through Ideal LPF having Cut-off frequency = 25 km

frequencies present at filter output are $\rightarrow$ 5 k, 10 k, 20 k, 25 k, ...

Max frequency present at filter output $\rightarrow$ 25 kHz

Method : 2

Given, $f_m = 10$ kHz

After sampling, frequencies present will be $\rightarrow$

$\pm f_m \pm nf_s$: n = 0, 1, 2, 3, 4, ...

i.e. ± 10 k, ± 10 k ± 15 k, ± 10 k ± 30 k, ...

After passing through ideal LPF having Cut-off frequency = 25 kHz, highest free component present at filter output = 25 kHz.

9.9 **(c)**

$x(t)$ is band limited to 100 Hz to 200 Hz

$$y(t) = x(2t - 5)$$

$$x(t) \rightleftharpoons X(\omega)$$

$$x(2t) \rightleftharpoons \frac{1}{2} \cdot X\left(\frac{\omega}{2}\right)$$

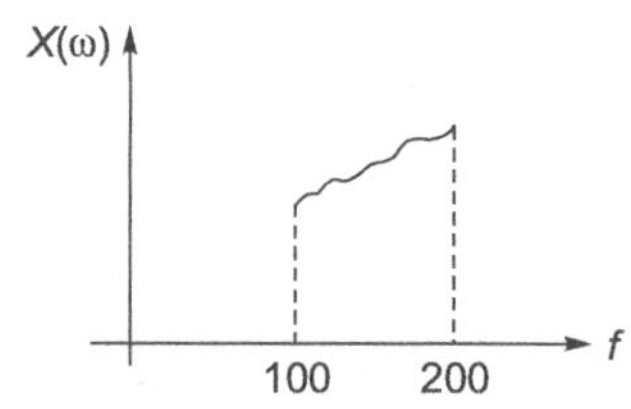

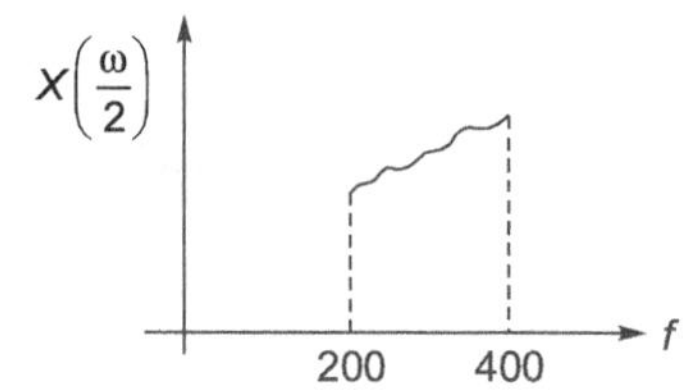

Time shifting will not change bandwidth,
So $x(2t - 5)$ will be band limited to 200 to 400.

9.10 **(400)**

Sampling freq. at $x(t) = 10 \sin^2(200\, \pi t)$

$$x(t) = \frac{1 - \cos 400\pi t}{2}$$

maximum freq. component present in $x(t)$

$$= \frac{400\pi}{2\pi} = 200 \text{ Hz}$$

$$f_{NQ} = 2f_m = 400 \text{ Hz}$$

■■■

10 Digital Filters

10.1 The bilinear transformation $\omega = \dfrac{z-1}{z+1}$

(a) maps the inside of the unit circle in the z-plane to left half of the ω-plane
(b) maps the outside of the unit circle in the z-plane to left half of the ω-plane
(c) maps the inside of the unit circle in the z-plane to right half of the ω-plane
(d) maps the outside of the unit circle in the z-plane to right half of the ω-plane

[2002 : 1 M]

10.2 A linear time invariant system with system function $H(z) = 1 + z^{-1} + z^{-2}$ is given an input signal sampled at 18 kHz. The frequency of the analog sinusoid which cannot pass through the system is

(a) $\dfrac{12}{\pi}$ kHz (b) 6 kHz
(c) 12 kHz (d) $\dfrac{6}{\pi}$ kHz

[2003 : 2 M]

10.3 A discrete-time system with input $x[n]$ output $y[n]$ relation $y[n] = x[n] - 2x[n-1] + x[n-2]$ a good approximation to a

(a) High pass filter
(b) Band-stop filter blocking $\dfrac{\pi}{8} \le |\omega| \le \dfrac{\pi}{4}$
(c) Low pass filter
(d) Band-pass filter passing $\dfrac{\pi}{8} \le |\omega| \le \dfrac{\pi}{4}$

[2003 : 2 M]

10.4 Bilinear transformation avoids the problems of aliasing connected with the use of impulse invariance through

(a) mapping the entire imaginary axis of the s-plane on the unit circle in the z-plane
(b) pre-filtering the input-signal to impose bank-limitedness
(c) mapping zeros of the left half of the s-plane inside the unit circle in the z-plane
(d) up-sampling the input signal so that the bandwidth is reduced

[2003 : 2 M]

10.5 A linear phase FIR system with impulse response real has a zero at $z = \dfrac{1}{2}e^{j\frac{\pi}{4}}$. The largest set of remaining zeros that can be obtained from the above information is

(a) $\dfrac{1}{2}e^{j\frac{\pi}{2}}, \dfrac{1}{2}e^{j\frac{3\pi}{4}}, \dfrac{1}{2}e^{-j\frac{\pi}{2}}$

(b) $2e^{-j\frac{\pi}{4}}, 2e^{-j\frac{3\pi}{4}}, \dfrac{1}{2}e^{-j\frac{\pi}{2}}$

(c) $2e^{-j\frac{\pi}{4}}, 2e^{j\frac{\pi}{4}}, \dfrac{1}{2}e^{-j\frac{\pi}{4}}$

(d) $2e^{j\frac{\pi}{4}}, \dfrac{1}{2}e^{-j\frac{\pi}{4}}$

[2003 : 2 M]

10.6 Direct form implementation of the FIR structure shown in figure (a) is also implemented in figure (b) in lattice form. The coefficient K_1 and K_2 are respectively

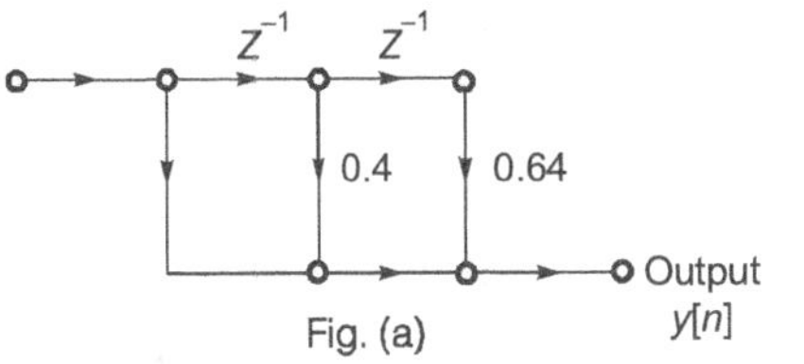

Fig. (a)

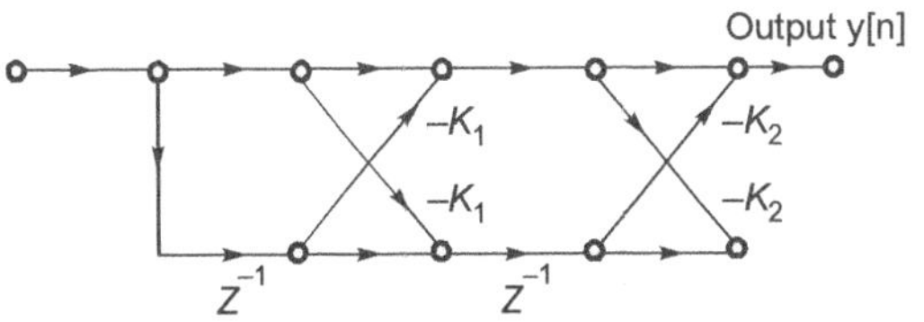

Fig. (b)

(a) 0.640 and –0.244 (b) 0.640 and 0.244
(c) 0.244 and 0.640 (d) –0.244 and –0.640

[2003 : 2 M]

10.7 A Direct Form Implementation of an LTI system with $H(z) = \frac{1}{1-0.7z^{-1}+0.13z^{-2}}$ is shown in figure

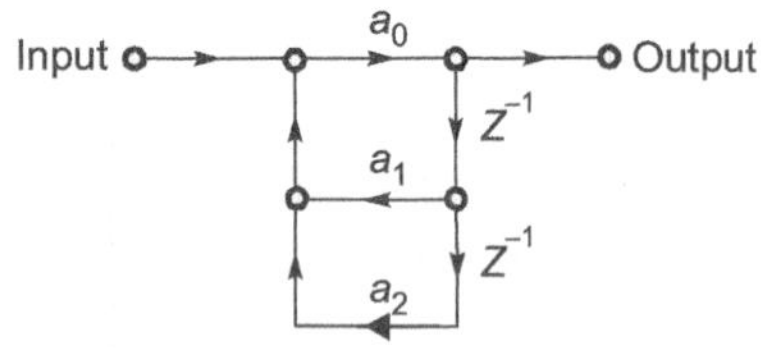

The value of a, a_1 and a_2 are respectively
(a) 1.0, 0.7 and –0.13
(b) –0.13, 0.7 and 1.0
(c) 1.0, –0.7 and 0.13
(d) 0.13, –.7 and 1.0

[2004 : 2 M]

10.8 A low-pass filter has a sampling frequency f_T = 10 kHz. The pass-band (f_p) and stop band (f_s) edge frequencies are 2 kHz and 3 kHz respectively. The frequency axis is pre-wasped before applying bilinear transformation to get new angular frequencies, $\tilde{r}_p$ and $\tilde{r}_s$. The ratio of the worked edge frequencies $\left(\frac{\tilde{r}_s}{\tilde{r}_p}\right)$ is
(a) –1.213 (b) 1.213
(c) 1.575 (d) 1.894

[2004 : 2 M]

10.9 A causal, analog system has a transfer function $H(z) = \frac{a}{s^2+a^2}$. Assuming a sampling time T seconds, the poles of the transfer function $H(z)$ for an equivalent digital system obtained using impulse invariance method are at
(a) (e^{aT}, e^{-aT}) (b) $\left(j\frac{a}{T}, -j\frac{a}{T}\right)$
(c) (e^{jaT}, e^{-jaT}) (d) $(e^{aT/2}, e^{-aT/2})$

[2004 : 2 M]

10.10 A digital notch filter with a notch frequency of 60 Hz is to be transformed into one operating at a new notch frequency of 120 Hz. The sampling frequency is 400 Hz. A low-pass to low-pass transformation $z^{-1} = \frac{1-\alpha\hat{z}}{\hat{z}-\alpha}$ is used on the first to obtain the new notch filter. Then the value of α should be
(a) –0.95 (b) 0.95
(c) –0.46 (d) 0.46

[2004 : 2 M]

10.11 An analog low-pass filter is needed with the following specifications:
Pass band: $0.9 \le |H(j\omega)| \le 1$, $0 \le \omega \le 2\pi \times 1000$
Stop band: $|H(j\omega) \le 0.2$, $|\omega| \ge 2\pi \times 1500$
The minimum order N of the Butterworth filter to approximate the above is
(a) 6 (b) 11
(c) 12 (d) 15

[2005 : 2 M]

10.12 A discrete-time system is shown in figure. The system function $H(z)$, of the network is given by

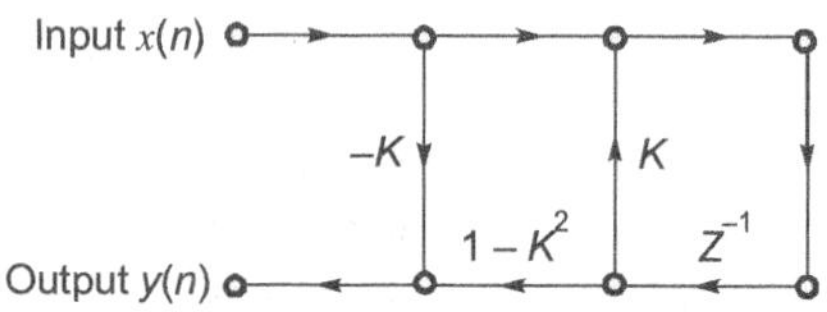

(a) $\frac{-k+z^{-1}}{1(1-k^2)z^{-1}}$ (b) $\frac{(1-k^2)+z^{-1}}{1-kz^{-1}}$
(c) $\frac{1-(1-k^2)z^{-1}}{1-kz^{-1}}$ (d) $\frac{-k+z^{-1}}{1-kz^{-1}}$

[2005 : 2 M]

10.13 A digital filter has the transfer function:

$$H(z) = \frac{z^2+1}{z^2+0.81}$$

If this filter has to reject a 50 Hz interference from the input, then the sampling frequency for the input signal should be
(a) 50 Hz (b) 100 Hz
(c) 150 Hz (d) 200 Hz

[2006 : 2 M]

10.14 An analog signal is sampled at 9 kHz. The sequence so obtained is filtered by an FIR filter with transfer function $H[z] = 1 - z^{-6}$. One of the analog frequencies for which the magnitude response of the filter is zero is
(a) 0.75 kHz (b) 1 kHz
(c) 1.5 kHz (d) 2 kHz

[2009 : 2 M]

10.15 The transfer function $H(z)$ of a fourth-order linear phase FIR system is given by

$$H(z) = (1 + 2z^{-1} + 3z^{-2})\, G(z)$$

Then $G(z)$ is

(a) $3 + 2z^{-1} + z^{-2}$ (b) $1+\frac{1}{2}z^{-1}+\frac{1}{3}z^{-2}$

(c) $\frac{1}{3}+2z^{-1}+z^{-2}$ (d) $1 + 2z + 3z^2$

[2009 : 2 M]

10.16 A digital filter having a transfer function $H(z)=\frac{P_0+P_1Z^{-1}+P_3Z^{-3}}{1+d_3Z^{-3}}$ is implemented using.

Direct form-I and direct form-II realization of IIR structure. The number of delay units required in direct form-I and direct form-II realizations are respectively,

(a) 6 and 6 (b) 6 and 3

(c) 3 and 3 (b) 3 and 2 [2010 : 1 M]

10.17 Shown below is the pole-zero plot of a digital filter

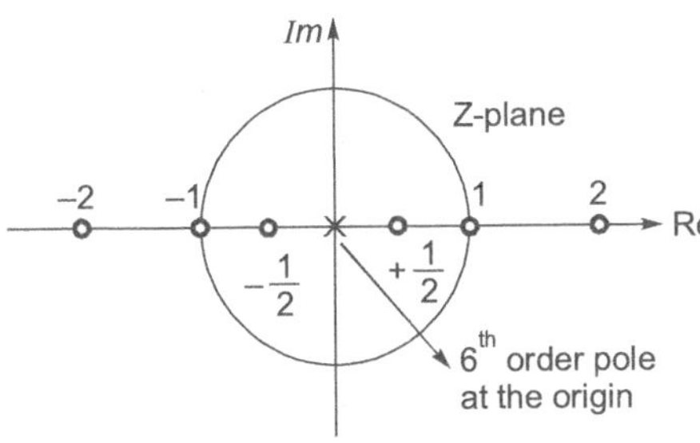

Which one of the following statements is **TRUE**?

(a) This is a low pass filter.

(b) This is a high pass filter.

(c) This is an IIR filter.

(d) This is an FIR filter. [2011 : 1 M]

10.18 A filter is represented by the signal flow graph shown in the figure. Its input is $x(t)$ and output is $y(t)$. The transfer function of the filter is

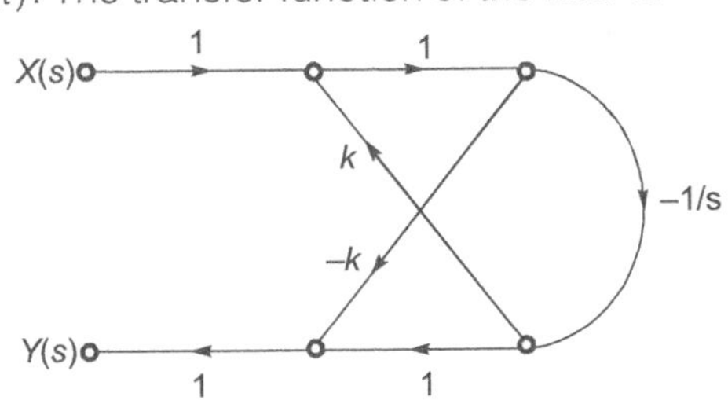

(a) $-\frac{(1+ks)}{s+k}$ (b) $\frac{1-ks}{s-k}$

(c) $\frac{-(1-ks)}{s+k}$ (d) $\frac{1-ks}{s+k}$ [2011 : 2 M]

10.19 A discrete-time $x[n]$ is obtained by sampling an analog signal at 10 kHz. The signal $x[n]$ is filtered by a system with impulse response $h[n] = 0.5\{\delta[n] + \delta[n-1]\}$. The 3 dB cut-off frequency of the filter is

(a) 1.25 kHz (b) 2.50 kHz

(c) 4.00 kHz (d) 5.00 kHz

[2014 : 1 M]

10.20 Let $3 + 4j$ be a zero of a fourth order linear-phase FIR filter. The complex number which is NOT a zero of this filter is

(a) $3-4j$ (b) $\frac{3}{25}+\frac{4}{25}j$

(c) $\frac{3}{25}-\frac{4}{25}j$ (d) $\frac{1}{3}-\frac{1}{4}j$ [2015 : 1 M]

10.21 Consider a low-pass filter module with a pass-band ripple of δ in the gain magnitude. If M such identical modules are cascaded, ignoring the loading effects, the pass-band ripple of the cascade is

(a) $1-(1-\delta)^M$ (b) δ^M

(c) $(1-\delta^M)$ (d) $(1-\delta)^M$ [2015 : 2 M]

10.22 The signal $x[n] = \sin(\pi n/6)\ (\pi n)$ is processed through a linear filter with the impulse response $h[n] = \sin(\omega_c n)/(\pi n)$ where $\omega_c > \pi/6$. The output of the filter is

(a) $\sin(2\omega_c n)/(\pi n)$ (b) $\sin(\pi n/3)/(\pi n)$

(c) $[\sin(\pi n/6)/(\pi n)]^2$ (d) $\sin(\pi n/6)/(\pi n)$

[2015 : 2 M]

10.23 The frequency response of a digital filter $H(\omega)$ has the following characteristics

Passband : $0.95 \le |H(\omega)| \le 1.05$ for $0 \le \omega \le 0.3\pi$

and Stopband : $0 \le |H(\omega)| \le 0.005$ for $0.4\pi \le \omega \le \pi$, where ω is the normalized angular frequency in rad/sample. If the analog upper cut off frequency for the passband of the above digital filter is to be 1.2 kHz, then the sampling frequency should be__________kHz.

[2019 : 2 M]

10.24 The input signal shown below

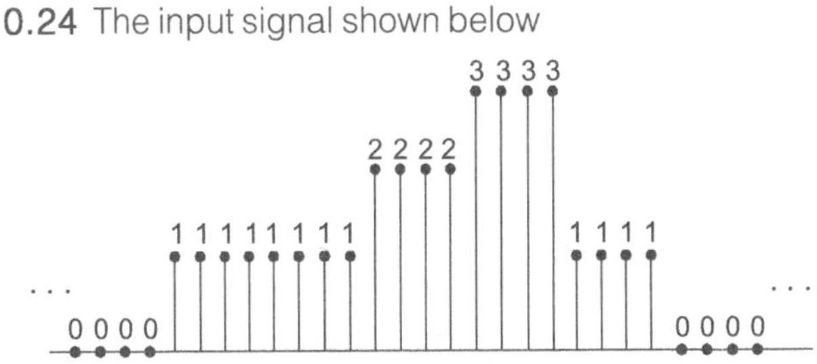

is passed through the filter with the following taps:

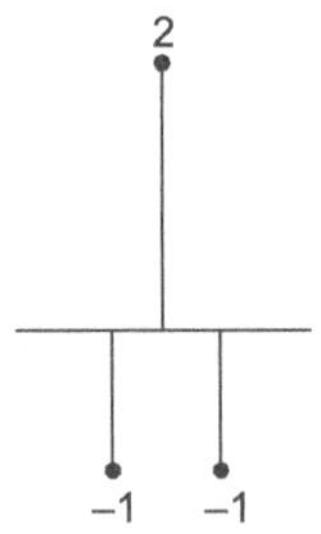

The number of non-zero output sample is ______.

[2021 : 2 M]

10.25 Consider a filter defined by the difference equation

$$y[n] - 0.5\, y[n-2] = a\, x[n-4]$$

where $x[n]$ and $y[n]$ represent the input and output, respectively. If the magnitude response of the filter at $\omega = \frac{\pi}{2}$ is $\left|H\left(\frac{\pi}{2}\right)\right| = 0.5$, the value of a is ______ (rounded off to two decimal places)

[2024 : 1 M]

■■■

Answers Digital Filters

10.1 (a)	10.2 (b)	10.3 (d)	10.4 (b)	10.5 (c)	10.6 (d)	10.7 (a)
10.8 (b)	10.9 (c)	10.10 (c)	10.11 (a)	10.12 (d)	10.13 (d)	10.14 (c)
10.15 (a)	10.16 (b)	10.17 (d)	10.18 (a)	10.19 (b)	10.20 (d)	10.21 (a)
10.22 (d)	10.23 (8)	10.24 (10)	10.25 (0.75)			

Explanations Digital Filters

10.1 (a)

The bilinear transformation $\omega = \frac{z-1}{z+1}$ maps inside the unit circle in the z-plane to left half of the ω-plane.

For z inside of the unit circle

$$0 < z < 1$$

for $0 < z < 1$

$$\omega = \frac{z-1}{z+1} = \frac{(-)\text{ve}}{(+)\text{ ve}} = (-)\text{ve}$$

$\Rightarrow$ ω-plane if lies left half of w-plane for $z < 1$.

10.3 (d)

$$y[n] = x[n] - 2x[n-1] + x[n-2]$$

taking z-transform, we get

$$Y(z) = X(z) - 2z^{-1} X[z] + z^{-2} X[z]$$

$$\frac{Y(z)}{X(z)} = H(z) = 1 - 2z^{-1} + z^{2}$$

$$\Rightarrow H(j\omega) = 1 - 2e^{-j\omega} + e^{-2j\omega}$$

$$\Rightarrow H(j\omega) = (1 - e^{-j\omega})^2$$

$$|H(j\omega)| = |(1-\cos\omega) + j\sin\omega|^2$$

$$= (\sqrt{(1-\cos\omega)^2 + \sin^2\omega})^2$$

$$= (\sqrt{1+\cos^2\omega - 2\cos\omega + \sin^2\omega})^2$$

$$= (\sqrt{2-2\cos\omega})^2$$

$$= (\sqrt{2-(1-\cos\omega)})^2$$

$$= \left(\sqrt{4\sin^2\frac{\omega}{2}}\right)^2 = (2\sin\omega/2)^2$$

So for small ω, $|H(\omega)| \to 0$

and for high $\omega(\omega \to 2\pi)$, $|H(\omega)| \to 0$

So it is characteristic of a band pass filter.

10.4 (b)

In order to avoid aliasing, we filter the high frequency (insignificant frequency component) component before sampling it.

10.5 (c)

it has a zero $z = \frac{1}{2}e^{+j\pi/4}$, since impulse response has to be real therefore corresponding to this zero there will be another zero $z = \frac{1}{2}e^{-j\pi/4}$.

So for largest set we will have zero in conjugate pairs.

10.6 (d)

From first figure (a),

$$Y[z] = X[z] + 0.4\,z^{-1}X[z] + 0.64\,z^{-2}X[z] \quad ....(i)$$

Now from figure (b),

$$Y[z] = X[z] - K_2 z^{-2}X[z] + K_1K_2 z^{-1}X[z] - K_1 z^{-1}X[z]$$

$$Y[z] = X[z] + [K_1K_2 - K_1]\,z^{-1}X[z] - K_2 z^{-2}X[z] \quad ...(ii)$$

Comparing (i) and (ii) we get

$$K_2 = -0.64$$

$$K_1K_2 - K_1 = 0.4$$

$$-0.64\,K_1 - K_1 = 0.4$$

$$K_1 = -\frac{0.4}{1.64} = -0.244$$

10.7 (a)

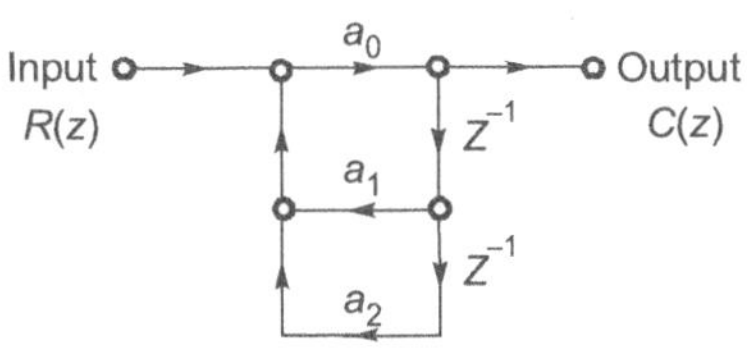

From above figure

$$C(z) = a_0[R(z) + z^{-1}\cdot a_1 C(z) + z^{-1}\cdot z^{-1}\cdot a_2\,C(z)]$$

$$\Rightarrow \quad C(z)\,[1 - a_0a_1 z^{-1} - a_0a_2 z^{-2}] = a_0R(z)$$

$$\Rightarrow \quad \frac{C(z)}{R(z)} = \frac{a_0}{1 - a_0 a_1 z^{-1} - a_0 a_2 z^{-2}}$$

as $\quad \dfrac{C(z)}{R(z)} = H(z)$

$$\Rightarrow \quad H(z) = \frac{1}{\frac{1}{a_0} - a_1 z^{-1} - a_2 z^{-2}}$$

Given, $\quad H(z) = \dfrac{1}{1 - 0.7z^{-1} + 0.13z^{-2}}$

By comparing,

$$\frac{1}{a_0} = 1 \Rightarrow a_0 = 1,\ a_1 = 0.7 \text{ and } a_2 = -0.13$$

10.8 (b)

$$f_T = 10 \text{ kHz}$$

$$T = \frac{1}{f_T} = 10^{-4} \text{ sec}$$

$$f_p = 2 \text{ kHz}$$

$$f_s = 3 \text{ kHz}$$

$$\Omega_p = 4\pi \times 10^3 \text{ rad/sec}$$

$$\Omega_s = 6\pi \times 10^3 \text{ rad/sec}$$

Now angular frequency,

$$\omega_s = 2\tan^{-1}\frac{\Omega T}{2}$$

$$= 2\tan^{-1}(3\pi \times 0.1)$$

$$\omega_s = 1.511$$

$$\omega_p = 2\tan^{-1}(2\pi \times 0.1) = 1.12$$

Now, $\quad \dfrac{\omega_s}{\omega_p} = \dfrac{1.511}{1.12}$

$$\boxed{\frac{\omega_s}{\omega_p} = 1.349}$$

10.9 (c)

$$H(s) = \frac{a}{(s^2 + a^2)}$$

$$H(t) = \sin at$$

$$H[nT] = \sin anT,$$

(where T is sampling period)

$$H[nT] = \sin aTn$$

$$H(s) = \frac{aT}{s^2 + (aT)^2}$$

for analog to digital transformation

$$z = e^s \Rightarrow \log_e z = s$$

So, $\quad H(z) = \dfrac{aT}{(lnz)^2 + (aT)^2}$

for poles $(lnz)^2 + (aT)^2 = 0$

$$(lnz)^2 = -(aT)^2$$

$$lnz = \pm jaT$$

$$z = e^{jaT}, e^{-jaT}$$

10.11 (a)

$$0.9 \le |H(j\omega)| \le 1 \qquad 0 \le \omega \le 2\pi \times 1000$$

$$|H(j\omega)| \le 0.2 \qquad |\omega| \ge 2\pi \times 1500$$

Here, $\quad \delta_1 = 0.9;\ \delta_2 = 0.2$

$$\omega_1 = 2\pi \times 1000$$

$$\omega_2 = 2\pi \times 1500$$

$$\Omega_1 = \frac{\omega_1}{T}, \text{ and } \Omega_2 = \frac{\omega_2}{T}$$

$$\therefore \quad \frac{\Omega_2}{\Omega_1} = \frac{\omega_2}{\omega_1} = \frac{1500}{1000} = 1.5$$

Now order of the filter,

$$N \geq \frac{\frac{1}{2}\log\left\{\left(\frac{1}{\delta_2^2}-1\right)\Big/\left(\frac{1}{\delta_1^2}-1\right)\right\}}{\log\left(\frac{\Omega_2}{\Omega_1}\right)}$$

$$N \geq 5.70$$

so, $N_{min} = 6$

10.12 (d)

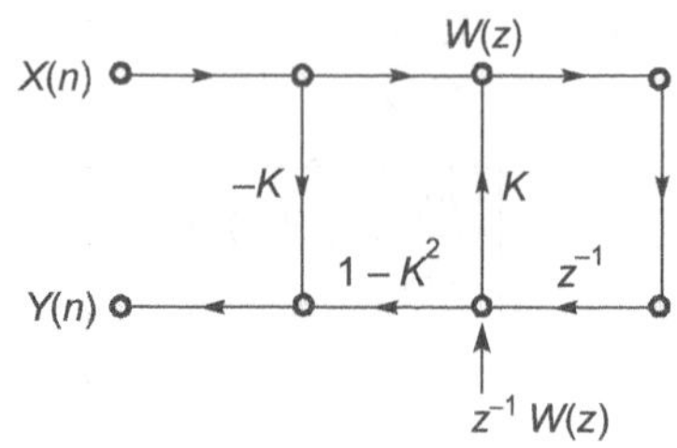

$$W(z) = X(z) + kz^{-1} W(z)$$

$$\therefore \quad W(z) = \frac{X(z)}{1-kz^{-1}}$$

$$\therefore \quad Y(z) = -k\,X(z) + (1-k^2)\,z^{-1}\,W(z)$$

$$Y(z) = -k\,X(z) + (1-k^2)z^{-1}\frac{X(z)}{(1-kz^{-1})}$$

$$\therefore \quad \frac{Y(z)}{X(z)} = -k + \frac{(1-k^2)z^{-1}}{(1-kz^{-1})}$$

$$= \frac{-k + k^2 z^{-1} + z^{-1} - k^2 z^{-1}}{1-kz^{-1}}$$

$$= \frac{(-k+z^{-1})}{(1-kz^{-1})}$$

10.14 (c)

$$H(z) = 1 - z^{-6}$$

$$H(e^{j\omega}) = 1 - e^{-6j\omega}$$

$$= (1-\cos 6\omega) + j\sin 6\omega$$

$$|H(e^{j\omega})| = \sqrt{(1-\cos 6\omega)^2 + \sin^2 6\omega}$$

$$= \sqrt{2 - 2\cos 6\omega}$$

$$= \sqrt{2\times 2\sin^2 3\omega}$$

$$|H(e^{j\omega})| = 2\sin 3\omega \quad 0 \leq \omega \leq 2\pi$$

So, $|H(e^{j\omega})| = 0$, if $3\omega = n\pi$

Now, $\omega = \frac{2\pi f}{f_s} = n\pi$

$$f = \frac{n\pi f_s}{2\pi\times 3} = \frac{n\times 9\times 10^3}{6} = 1.5\,\text{kHz} \quad (\text{for } n = 1)$$

10.15 (a)

$$H(z) = (1 + 2z^{-1} + 3z^{-2})\,G(z)$$

Let, $G(z) = (A_0 + A_1 z^{-1} + A_2 z^{-2})$

then, $H(z) = A_0 + (2A_0 + A_1)z^{-1} + (3A_0 + 2A_1 + A_2)z^{-2} + (3A_1 + 2A_2)z^{-3} + 3A_2 z^{-4}$

So, $h(0) = A_0$, $h(1) = (2A_0 + A_1)$, $h(2) = (3A_0 + 2A_1 + A_2)$

$h(3) = (3A_1 + 2A_2)$, $h(4) = 3A_2$

we know that for linear phase filter

$$h[n] = h[M-1-n]$$

for $0 \leq n \leq M-1$

where M is number of points in filter

here, $M = 5$

Now, $h(0) = h[5-1-0] = h[4]$

$\Rightarrow \quad A_0 = 3A_2$

$$h[1] = h[5-1-1] = h[3]$$

$$2A_0 + A_1 = 3A_1 + 2A_2$$

$$6A_2 + A_1 = 3A_1 + 2A_2$$

$$4A_2 = 2A_1$$

$$A_1 = 2A_2$$

So, $A_0 : A_1 : A_2 \Rightarrow 3 : 2 : 1$

10.16 (b)

$$H(z) = \frac{P_0 + P_1 z^{-1} + P_3 z^{-3}}{1 + d_3 z^{-3}}$$

$$= H_1(z)\,H_2(z)$$

where, $H_1(z) = P_0 + P_1 z^{-1} + P_3 z^{-3}$

$$H_2(z) = \frac{1}{1 + d_3 z^{-3}}$$

For direct form-I,

we realize it as

$$x(z) \rightarrow \boxed{H_1} \xrightarrow{W} \boxed{H_2} \rightarrow y$$

so it will be like this

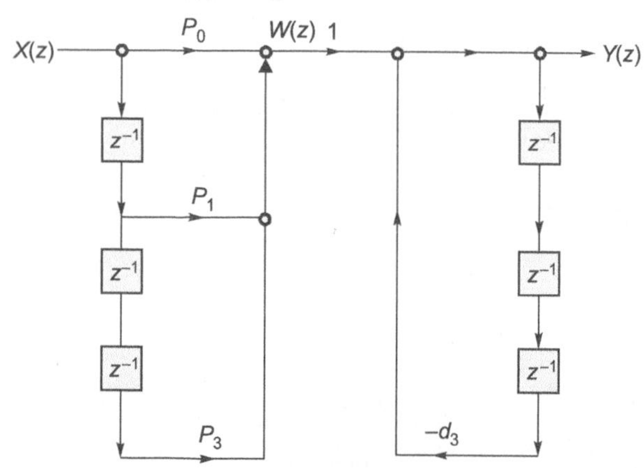

So 6 → delay elements required.

In direct form 2, $H_1(z)$ and $H_2(z)$ are interchanged.

So realization is

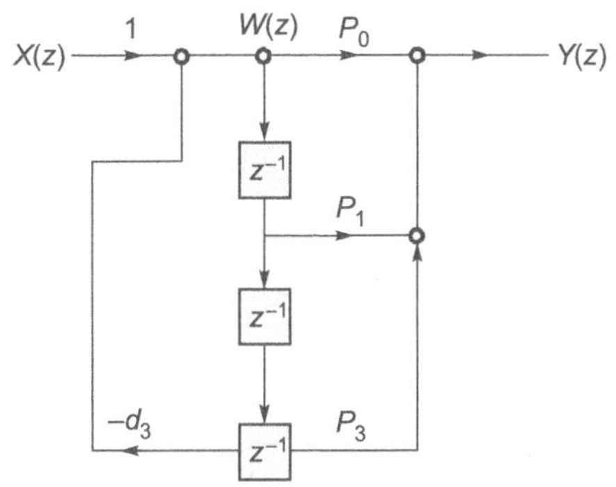

So in direct form 2, only 3 delay elements are required.

10.17 (d)

$$H(z) = \frac{(z-1)(z-2)(z+1/2)(z-1/2)(z+1)(z+2)}{z^6}$$

$$\frac{Y(z)}{X(z)} = H(z)$$

$$H(z) = \frac{(z^2-4)(z^2-1)(z^2-1/4)}{z^6}$$

$$= \frac{z^6 - \frac{21}{4}z^4 + \frac{21}{4}z^2 - 1}{z^6}$$

$$= 1 - \frac{21}{4}z^{-2} + \frac{21}{4}z^{-4} - z^{-6}$$

$$h[n] = \delta[n] - \frac{21}{4}\delta[n-2] + \frac{21}{4}\delta[n-4] - \delta[n-6]$$

So it is a FIR (finite impulse response) filter.

10.18 (a)

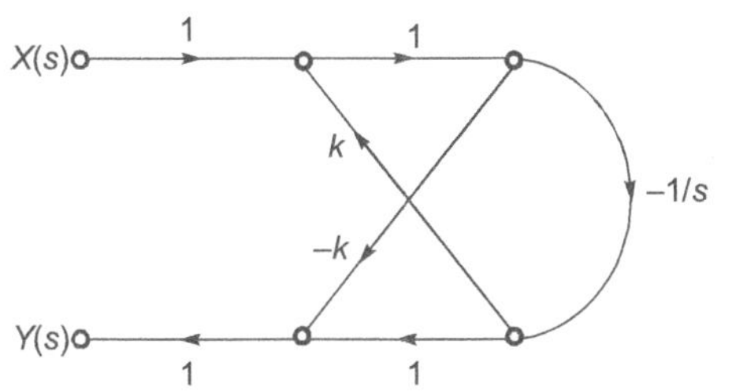

Forward paths,

$$P_1 = -\frac{1}{s}, \; P_2 = -k$$

$$\text{individual loop} = -\frac{1}{s}\cdot k$$

no two non-touching loop using mason's gain formula

$$\text{Transfer function} = \frac{\sum P_k \Delta_k}{\Delta}$$

$$= \frac{-1/s + (-k)}{1 - \left(-\frac{1}{s}\cdot k\right)} = \frac{-(1+sk)}{s\left(1+\frac{k}{s}\right)}$$

$$= \frac{-(1+sk)}{s+k}$$

10.19 (b)

$$h[n] = 0.5[\delta(n) + \delta(n-1)]$$

$$H(e^{j\Omega}) = 0.5[1 + e^{-j\Omega}]$$

$$= \frac{1}{2}[1 + \cos\Omega - j\sin\Omega]$$

$$\left|H(e^{j\omega})\right| = \frac{1}{2}\sqrt{(1+\cos\Omega)^2 + \sin^2\Omega}$$

$$= \frac{1}{2}\sqrt{1+\cos^2\Omega + 2\cos\Omega + \sin^2\Omega}$$

$$= \frac{1}{2}\sqrt{2+2\cos\Omega} = \cos\left(\frac{\Omega}{2}\right)$$

3 dB cut-off frequency,

$$= \frac{1}{\sqrt{2}}\left|H(e^{j\Omega})\right| = \frac{1}{\sqrt{2}}$$

$$\cos\frac{\Omega}{2} = \frac{1}{\sqrt{2}}$$

$$\cos\frac{\Omega}{2} = \frac{1}{\sqrt{2}}, \quad \frac{\Omega}{2} = \frac{\pi}{2}$$

$$\therefore \quad \Omega = \frac{\pi}{2}$$

2π rad/sample $\rightarrow$ 10 kHz.

So, $\frac{\pi}{2}$ rad/sample $\rightarrow \frac{10}{2\pi}\times\frac{\pi}{2} = 2.5$ kHz

10.20 (d)

In case of FIR filter, if it has one $z = 0$ as z_0.

Then the other zeros are $\frac{1}{z_o}, [z_0^*], \left[\frac{1}{z_0}\right]^*$

$$\frac{1}{z_0} \Rightarrow \left(\frac{3}{25} - j\frac{4}{25}\right)$$

$$[z_0]^* \Rightarrow (3-4j)$$

$$\left[\frac{1}{z_0}\right]^* \Rightarrow \left[\frac{3}{25} + j\frac{4}{25}\right]$$

$\therefore$ Clearly option (d) is not the zero of this FIR filter.

10.21 (a)

At ω_p (pass band cut-off frequency)

$|H(\omega)|$ is $(1-\delta)$.

If M such identical modules are cascaded then

the new $|H(\omega)|$ is $(1-\delta)^M$

$\Rightarrow$ The value of effective ripple will be

$1-(1-\delta)^M$

10.22 (d)

$$\left[\frac{\sin\alpha n}{\pi n}\right]\times\left[\frac{\sin\beta n}{\pi n}\right]=\left[\frac{\sin\gamma n}{\pi n}\right]$$

where, $\gamma = \min(\alpha, \beta)$

Output $g(n) = x(n) * h(n)$

In frequency domain they will be multiplied.

10.23 (8)

$$\omega_P = 0.3\,\pi$$
$$\omega_S = 0.4\,\pi$$
$$F_P = 1.2 \text{ kHz}$$

Since, $\omega_P = \dfrac{2\pi F_P}{F_T}$

So, sampling frequency,

$$F_T = \frac{2\pi F_P}{\omega_P}$$
$$= \frac{2\pi\times 1.2\,\text{kHz}}{0.3\pi} = 8 \text{ kHz}$$

10.24 (10)

h(n) \ x(n)	1	1	1	1	1	1	1	1	2	2	2	2	3	3	3	3	1	1	1	1
–1	–1	–1	–1	–1	–1	–1	–1	–1	–2	–2	–2	–2	–3	–3	–3	–3	–1	–1	–1	–1
2	2	2	2	2	2	2	2	2	4	4	4	4	6	6	6	6	2	2	2	2
–1	–1	–1	–1	–1	–1	–1	–1	–1	–2	–2	–2	–2	–3	–3	–3	–3	–1	–1	–1	–1

$y[n] = \{-1, 1, 0, 0, 0, 0, 0, 0, -1, 1, 0, 0, -1,$
$1, 0, 0, 2, -2, 0, 0, 1, -1\}$

Number of non-zero samples in $y[n] = 10$

10.25 (0.75)

By applying DTFT on given difference equation

$Y(e^{j\omega}) - 0.5Y(e^{j\omega})\times e^{-j2\omega} = aX(e^{j\omega})e^{-j4\omega}$

$$\Rightarrow \quad \frac{Y(e^{j\omega})}{X(e^{j\omega})} = \frac{ae^{-j4\omega}}{1-0.5e^{-j2\omega}}$$

$$\Rightarrow \quad H(e^{j\omega}) = \frac{ae^{-j4\omega}}{1-0.5e^{-j2\omega}}$$

Now, put $\omega = \pi/2$,

$$H(e^{j\pi/2}) = \frac{ae^{-j2\pi}}{1-0.5e^{-j\pi}} = \frac{a}{1+0.5}$$

Given, $|H(e^{j\pi/2})| = 0.5$

$$\Rightarrow \quad \frac{a}{1+0.5} = 0.5 \Rightarrow a = 0.5\times 1.5 = 0.75$$

■■■

11 Communication Systems

11.1 A linear phase low pass filter with a cut-off frequency 1500 Hz shifts a 50 Hz signal by 3 m-sec. It will shift a signal of 150 Hz by

(a) 3 m-sec (b) 9m-sec
(c) 1 m-sec (d) 30 m-sec

[1996 : 2 M]

11.2 The signal to noise (S/N) ratio of an amplifier developing an output of 10V and a noise voltage of 1 mV is

(a) –40 dB (b) 40 dB
(c) 80 dB (d) 100 dB

[1996 : 2 M]

11.3 The Nyquist sampling rate for the signal

$s(t) = \frac{\sin(500\pi t)}{\pi t} \times \frac{\sin(700\pi t)}{\pi t}$ is given by

(a) 400 Hz (b) 600 Hz
(c) 1200 Hz (d) 1400 Hz

[1997 : 2 M]

11.4 For an amplitude modulated signal with carrier power 500 W and modulation index 0.75, the total power is,

(a) 540.5 W (b) 640.5 W
(c) 440.5 W (d) 740.5 W

[1998 : 1 M]

11.5 A phase locked loop can be employed for demodulation of

(a) pulse amplitude modulation signal
(b) pulse code modulation signal
(c) frequency modulation signal
(d) signal side band amplitude modulation signals

[1999 : 1 M]

11.6 In frequency modulated system, the carrier and the output signals are $x_c(t) = \cos \omega_c t$ and $y(t) = \cos (\omega_c t + \sin 2t - \cos t)$ respectively. The modulating input $x_m(t)$ is proportional to

(a) $2 \cos 2t + \sin t$ (b) $\sin 2t - \cos t$
(c) $\cos 2t + \sin t$ (d) $2 \cos t\ 2t - \sin t$

[1999 : 2 M]

11.7 In an FM broadcast, the maximum frequency deviation allowed is 75 kHz, and the maximum modulation frequency is 15 kHz. The bandwidth is closest to

(a) 180 kHz (b) 60 kHz
(c) 105 kHz (d) 120 kHz

[2000 : 1 M]

11.8 A sinusoidal signal of amplitude $\sqrt{2}$ mV and frequency 5 kHz is buries in Gaussian noise with zero mean value. The noise has a uniform power spectral density of 100 pW/Hz up to a cut-off frequency of 1 MHz for this system, the signal to noise ratio in dB is

(a) 20 (b) 100
(c) – 20 (d) – 40

[2000 : 1 M]

11.9 A sample and hold circuit has two buffers, one at the input and the other at the output. The primary requirement for the buffer are

(a) the input buffer should have high slew rate and the output buffer should have low bias current.
(b) the input buffer should have low bias current and the output buffer should have high slew rate.
(c) both the buffer should have low bias currents.
(d) both the buffers should have high slew rate.

[2001 : 1 M]

11.10 For a suppressed carrier amplitude modulator (AM-SC) system, the carrier and the modulating inputs are $x_c(t) = \cos \omega_c t$ and $x_m(t) = 0.5 \sin \omega_m t$, respectively. The output of the system is proportional to

(a) $\sin (\omega_c + \omega_m)t + \sin(\omega_c - \omega_m)t$
(b) $\sin (\omega_c + \omega_m)t + \cos (\omega_c - \omega_m)t$
(c) $(1 + 0.5 \sin \omega_m t) \cos \omega_c t$
(d) $(1 - 0.5 \sin \omega_m t) \cos \omega_c t$

[2001 : 2 M]

11.11 A sinusoidal signal of frequency 1 kHz is used to produce an FM signal with a modulation index $\beta = 5$. The bandwidth (where 98% of power is contained) of the FM signal is
(a) 2 kHz (b) 3 kHz
(c) 6 kHz (d) 12 kHz
[2003 : 1 M]

11.12 The scheme shown in figure is used for the generation of wideband FM from a narrow band FM. The multiplier box multiplies the input frequency by the factor shown. The input $x(t)$ is a narrow band FM signal of carrier 100 kHz and frequency deviation of 25 Hz. The local oscillator frequency in kHz and the multiplication constant m to achieve an output $y(t)$ with a carrier of 2.0 MHz and a frequency deviation of 1.0 kHz are respectively:

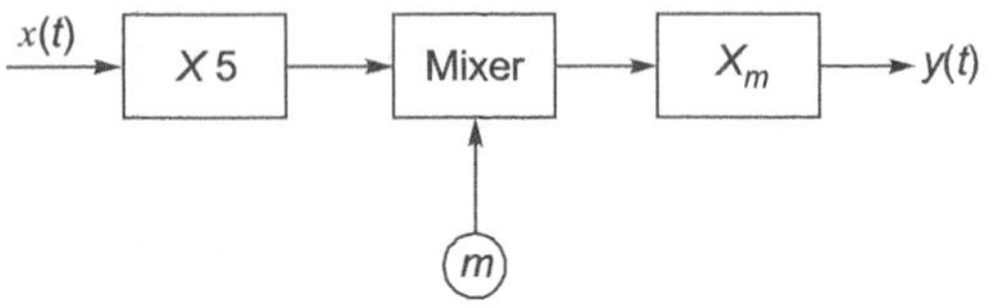

(a) 750, 4 (b) 1000, 4
(c) 750, 8 (d) 1000, 8
[2003 : 2 M]

11.13 The Nyquist rate of sampling of an analog signal $s(t)$ for alias free reconstruction is 5000 samples. For a signal $x(t) = [s(t)]^2$ the corresponding Nyquist sampling rate in sample/sec is
(a) 2500 (b) 5000
(c) 10000 (d) 25000
[2004 : 1 M]

11.14 Due to an amplitude modulation by a sine wave, if the total current in the antenna increases from 4 A to 4.8 A the depth of modulation in percentage is
(a) 93.8 (b) 80.1
(c) 40.4 (d) 20.2
[2004 : 2 M]

11.15 If the bandwidth of a low-pass signal of 3 kHz, the bandwidth of $g^2(t)$ will be
(a) $\frac{3}{2}$ kHz (b) 3 kHz
(c) 6 kHz (d) 9 kHz
[2008 : 2 M]

11.16 Consider the AM signal $s(t) = [1 + m(t)] \cos(2\pi f_c(t)$. It is given that the bandwidth of the real, low-pass message signal $m(t)$ is 2 kHz. If $f_c = 2$ MHz, the bandwidth of the band-pass signal $s(t)$ will be
(a) 2.004 MHz (b) 2 MHz
(c) 4 kHz (d) 2 kHz
[2008 : 2 M]

11.17 Ten real, band-pass message signals each of bandwidth 3 kHz, are to be frequency division multiplexed on a band-pass channel with bandwidth B kHz. If the guard band is between any two adjacent signals should be of 500 Hz width and there is no need to provide any guard band at the edge of the band-pass channel the value of B should be at least
(a) 30 (b) 34.5
(c) 35 (d) 35.5 [2008 : 2 M]

11.18 A 50% duty cycles square wave with zero mean used as a baseband signal in an ideal frequency modulator with a sinusoidal carrier of frequency ω_c. The modulated signal is given as an input to an ideal phase demodulation circuit (circuit that produces an output proportional to the difference in phase of the modulated signal from that of the carrier). The output of the circuit is
(a) a square wave
(b) a train of impulses with alternating signs
(c) a triangular wave
(d) a sinusoidal wave [2009 : 1 M]

11.19 In a pulse code modulated (PCM) signal sampled at f_s and encoded into an n-bit code, the minimum bandwidth required for faithful reconstruction is
(a) $2nf_s$ (b) nf_s
(c) $\frac{nf_s}{2}$ (d) f_s [2010 : 2 M]

11.20 A signal with frequency components 50 Hz, 100 Hz and 200 Hz only is sampled at 150 samples/sec. The ideally reconstructed signal will have frequency components of
(a) 50 Hz only
(b) 75 Hz only
(c) 50 Hz and 75 Hz
(d) 50 Hz, 70 Hz and 100 Hz
[2010 : 2 M]

11.21 A square wave (amplitude ± 10 mV, frequency 5 kHz, duty cycle 50%) is passed through an ideal low-pass filter with pass-band gain and cut-off frequency of 0 dB and 10 kHz, respectively. The filtered signal is subsequently "buried" additively into a zero-mean noise process of one-sided power-spectral density (PSD) of 25 pW Hz^{-1} up to a frequency of 2 MHz. The PSD of the noise is assumed to be zero beyond 2 MHz. The signal-to-noise ratio of the output is

(a) 0 dB (b) 0.1 dB
(c) 1.0 dB (d) 3 dB **[2011 : 2 M]**

11.22 Two independent random variables X and Y are uniformly distributed in the interval [–1, 1]. The probability that max [X, Y] is less than 1/2 is

(a) $\frac{3}{4}$ (b) $\frac{9}{16}$
(c) $\frac{1}{4}$ (d) $\frac{2}{3}$ **[2012 : 1 M]**

11.23 A band-limited signal with a maximum frequency of 5 kHz is to be sampled. According to the sampling theorem, the sampling frequency in kHz, which is not valid is

(a) 5 (b) 12
(c) 15 (d) 20 **[2013 : 1 M]**

11.24 Signals from fifteen thermocouples are multiplexed and each one is sampled once per second with a 16-bit ADC. The digital samples are converted by a parallel to serial converter to generate a serial PCM signal. This PCM signal is frequency modulated with FSK modulator with 1200 Hz as 1 and 960 Hz as 0. The minimum band allocation required for faithful reproduction of the signal by the FSK receiver without considering noise is

(a) 840 Hz to 1320 Hz
(b) 960 Hz to 1200 Hz
(c) 1080 Hz to 1320 Hz
(d) 720 Hz to 1440 Hz **[2013 : 2 M]**

11.25 A full duplex binary FSK transmission is made through a channel of bandwidth 10 kHz. In each direction of transmission the two carriers used for the two states are separated by 2 kHz. The maximum baud rate for this transmission is

(a) 2000 bps (b) 3000 bps
(c) 5000 bps (d) 10000 bps
[2014 : 1 M]

11.26 A signal is band-limited to 0 to 12 kHz. The signal spectrum is corrupted by additive noise which is band-limited to 10 to 12 kHz. Theoretically, the minimum rate in kilohertz at which the noisy signal must be sampled so that the UNCORRUPTED PART of the signal spectrum can be recovered, is ____ kHz.
[2015 : 2 M]

11.27 The probability density function of a random variable X is $p_x(x) = e^{-x}$ for $x \geq 0$ and 0 otherwise. The expected value of the function $g_x(x) = e^{3x/4}$ is _____.
[2015 : 2 M]

11.28 The signal $m(t) = \cos(\omega_m t)$ is SSB (single side-band) is modulated with a carrier $\cos(\omega_c t)$ to get $s(t)$. The signal obtained by passing $s(t)$ through an ideal envelope detector is

(a) $\cos(\omega_m t)$
(b) $\sin(\omega_m t)$
(c) $\cos(\omega_m t) + \sin(\omega_m t)$
(d) 1 **[2016 : 1 M]**

11.29 Let $s(t) = \text{rect}\left(\frac{t-3}{3}\right)$ be a signal passed through an AWGN (additive white Gaussian noise) channel with noise power spectral density (PSD) $\frac{N_0}{2}$ to get $v(t)$. If $v(t)$ is passed through a matched-filter that is matched to $s(t)$, then output signal-to noise ratio (SNR) of the matched-filter is

(a) $\frac{1}{N_0}$ (b) $\frac{2}{N_0}$
(c) $\frac{3}{N_0}$ (d) $\frac{4}{N_0}$ **[2016 : 1 M]**

11.30 The signal $m(t) = \frac{\sin(100\pi t)}{100\pi t}$ is frequency modulated (FM) with an FM modulator of frequency deviation constant of 30 kHz/V. Using Carson's rule, the approximate bandwidth of the modulated wave in kilohertz is_______.
[2016 : 2 M]

11.31 A signal $m(t)$ varies from –3.5 V to +3.5 V with an average power of 3W. The signal is quantized using a midtread type quantizer and subsequently binary encoded. With the codeword of length 3, the signal to quantization noise ratio in dB is______.
[2016 : 2 M]

11.32 An angle modulated signal with carrier frequency $\omega_c = 2\pi \times 10^6$ rad/s is given by $\phi_m(t) = \cos(\omega_c t + 5\sin(1000\pi t) + 10\sin(2000\pi t))$. The maximum deviation of the frequency in the angle modulated signal from that of the carrier is _______ kHz.

[2017 : 2 M]

11.33 In sinusoidal amplitude modulation scheme (with carrier) the modulated signal is given by

$$A_m(t) = 100\cos(\omega_c t) + 50\cos(\omega_m t)\cos(\omega_c t),$$

where ω_c is the carrier frequency and ω_m is the modulation frequency. The power carried by the sidebands in % of total power is _______%.

[2017 : 2 M]

11.34 The figure shows a phase locked loop. The output frequency is locked at f_0 = 5 kHz. The value of f_i in kHz is _______.

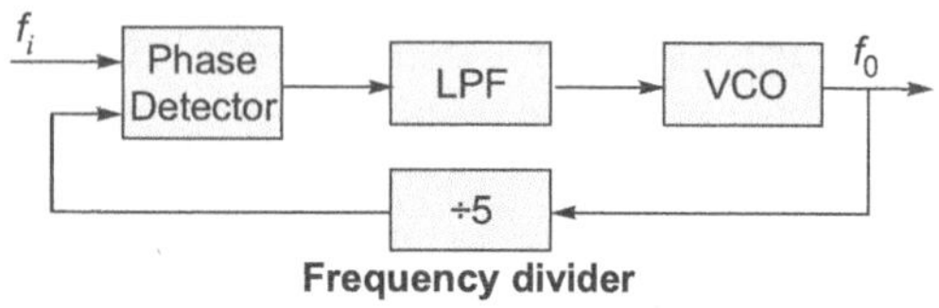

[2017 : 1 M]

11.35 X and Y are two independent random variables with variances 1 and 2, respectively. Let $Z = X - Y$. The variance of Z is

(a) 0 (b) 1

(c) 2 (d) 3

[2018 : 1 M]

11.36 The sampling rate for Compact Discs (CDs) is 44,000 samples/s. If the samples are quantized to 256 levels and binary coded, the corresponding bit rate (in bits per second) is ___.

[2018 : 2 M]

11.37 An amplitude modulated signal is shown in the figure. The modulation index is (up to one decimal place) _______.

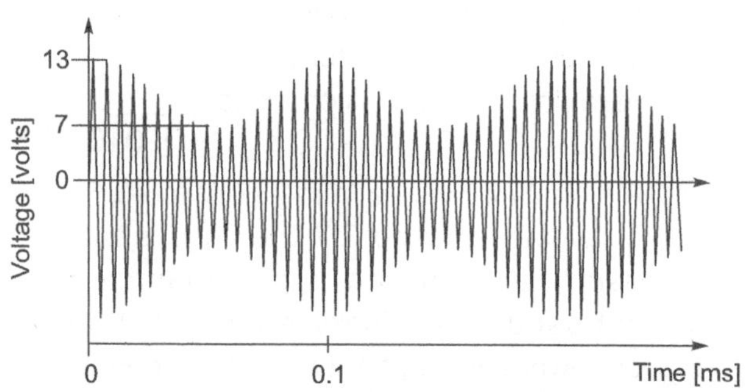

[2018 : 1 M]

11.38 A signal $x(t)$ has a bandwidth $2B$ about a carrier frequency of f_c = 2 GHz as shown in figure (a) below. In order to demodulate this signal, it is first mixed (multiplied) with a local oscillator of frequency f_{LO} = 1.5 GHz, and then passed through an ideal low-pass filter (LPF) with a cut-off frequency of 2.8 GHz. The output of the LPF is sent to a digitizer ADC with a sampling rate of 1.6 GHz as shown in figure (b) below. The maximum value of B so that the signal $x(t)$ can be reconstructed from its samples according to the Nyquist sampling theorem is_________MHz.

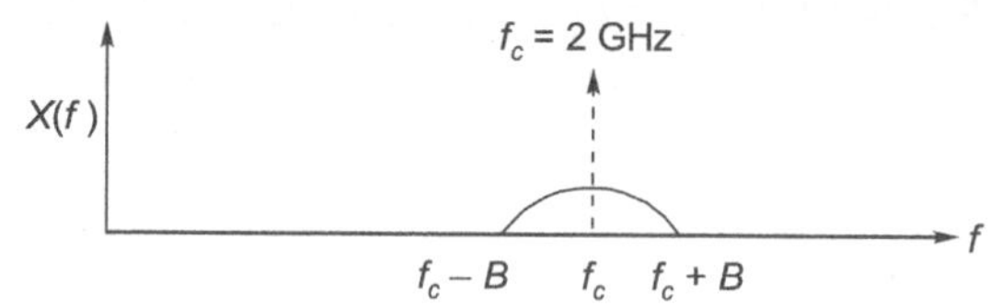

Figure (a)

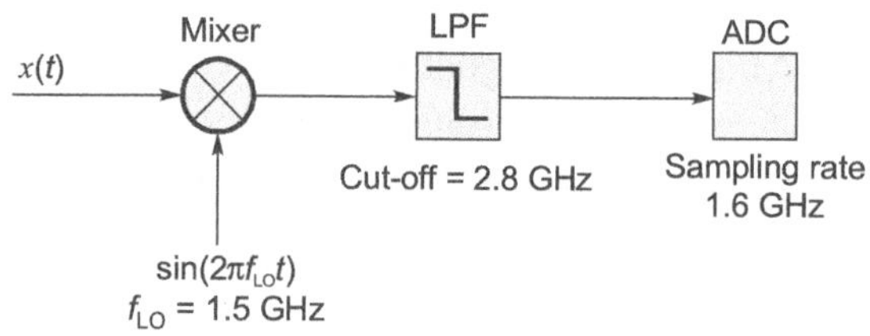

Figure (b)

[2019 : 2 M]

11.39 A signal $\cos(2\pi f_m t)$ modulates a carrier $\cos(2\pi f_c t)$ using the double-sideband-with carrier (DSBWC) scheme to yield a modulated signal $\cos(2\pi f_c t) + 0.3\cos(2\pi f_m t)\cos(2\pi f_c t)$. The modulation index is___________.

(Answer should be rounded off to one decimal place).

[2019 : 1 M]

11.40 An amplitude modulation (AM) scheme uses tone modulation, with modulation index of 0.6. The power efficiency of the AM scheme is_____% (rounded off to one decimal place).

[2021 : 1 M]

11.41 A signal having a bandwidth of 5 MHz is transmitted using the Pulse code modulation (PCM) scheme as follows. The signal is sampled at a rate of 50% above the Nyquist rate and quantized into 256 levels. The binary pulse rate of the PCM signal in Mbits per second is _______.

[2021 : 2 M]

11.42 A sinusoidal carrier wave with amplitude A_c and frequency f_c is amplitude modulated with a message signal $m(t)$ having frequency $0 < f_m << f_c$ to generate the modulated wave $s(t)$ given by

$$s(t) = A_c[1 + m(t)]\cos(2\pi f_c t)$$

The message signal that can be retrieved completely using envelope detection is ________.

(a) $m(t) = 0.5\cos(2\pi f_m t)$
(b) $m(t) = 1.5\sin(2\pi f_m t)$
(c) $m(t) = 2\sin(4\pi f_m t)$
(d) $m(t) = 2\cos(4\pi f_m t)$ [2022 : 1 M]

11.43 Consider 24 voice signals being transmitted without latency using time-division multiplexing. If each signal is sampled at 12 kHz and represented by an 8-bit word, the bit-duration (in microseconds) is ________ (round off to two decimal places).

[2022 : 1 M]

11.44 A car is moving collinearly with a laser beam emitted by a transceiver. A laser pulse emitted at $t = 0$ s is received back by the transceiver 100 ns (nanoseconds) later after reflection from the car. A second pulse emitted at $t = 0.1$ s is received back 90 ns later. Given the speed of light is 3×10^8 m/s, the average speed of the car in this interval is ________.

(a) 54 kmph, moving towards the transceiver
(b) 108 kmph, moving towards the transceiver
(c) 54 kmph, moving away from the transceiver
(d) 108 kmph, moving away from the transceiver

[2022 : 2 M]

Answers Communication Systems

11.1 (a)	11.2 (c)	11.3 (c)	11.4 (b)	11.5 (c)	11.6 (a)	11.7 (a)
11.8 (c)	11.9 (a)	11.10 (a)	11.11 (d)	11.12 (c)	11.13 (c)	11.14 (a)
11.15 (c)	11.16 (c)	11.17 (b)	11.18 (c)	11.19 (c)	11.20 (a)	11.21 (*)
11.22 (b)	11.23 (a)	11.24 (d)	11.25 (b)	11.26 (22)	11.27 (4)	11.28 (d)
11.29 (*)	11.30 (60.1)	11.31 (15.563)	11.32 (12.5)	11.33 (11.11)	11.34 (1)	11.35 (d)
11.36 (352000)	11.37 (0.30)	11.38 (300)	11.39 (0.3)	11.40 (15.25)	11.41 (120)	11.42 (a)
11.43 (0.43)	11.44 (a)					

Explanations Communication Systems

11.1 (a)

Since a linear phase low pass filter will shift a signal independent of frequency so it will shift a signal by 3 m-sec.

11.2 (c)

$$\frac{S}{N} = \frac{\text{Signal voltage}}{\text{Noise voltage}} = \frac{10\text{ V}}{1\text{ mV}}$$

$$\left(\frac{S}{N}\right)_{dB} = 20\log_{10}\frac{10}{1\times 10^{-3}}$$

$$= 20\log_{10}10^4 = 80\text{ dB}$$

11.3 (c)

$$s(t) = \frac{1}{2}\left[\frac{2\sin(500\pi t)\sin(700\pi t)}{\pi^2 t^2}\right]$$

$$s(t) = \frac{1}{2\pi^2 t^2}[\cos(700\pi t - 500\pi t) - \cos(700\pi t + 500\pi t)]$$

$$s(t) = \frac{1}{2\pi^2 t^2}[\cos(200\pi t) - \cos(1200\pi t)]$$

Maximum frequency component,

$$f_m = \frac{1200\pi}{2\pi} = 600\text{ Hz}$$

Nyquist sampling rate,

$$f_{s\,min} = 2f_m = 1200\text{ Hz}$$

11.4 (b)

Total power in A.M

Given, $P_c = 500$ W, $m = 0.75 = \dfrac{3}{4}$

$$P_t = P_c\left(1+\frac{m^2}{2}\right) = 500\left(1+\frac{1}{2}\times\frac{9}{16}\right)$$

$$= 500\left(\frac{41}{32}\right) = 640.625 \text{ W}$$

11.5 (c)

A PLL can be employed for demodulation of frequency modulated signal.

11.6 (a)

The FM signal is expressed as

$$y(t) = A_c \cos\left(\omega_c t + k_f \int_{-\infty}^{t} m(t)\,dt\right)$$

and given equation of FM is

$$y(t) = \cos(\omega_c t + \sin 2t - \cos t)$$

on comparing it with a general equation of FM

$$\Rightarrow \quad \int_{\infty}^{t} m(t)\,dt = \sin 2t - \cos t$$

differentiate above equation on both sides

$$m(t) = 2\cos 2(t) + \sin t$$

11.7 (a)

Given frequency deviation $\Delta f = 75$ kHz

Modulating frequency $f_m = 15$ kHz

$$\beta = \frac{\Delta f}{f_m} = \frac{75}{15} = 5$$

$$\text{Bandwidth} = 2(\beta+1)f_m = 2(6)\,f_m = 12\,f_m$$

$$= 180 \text{ kHz}$$

11.10 (a)

$$S_{(AMSC)}(t) = x_c(t)\cdot x_m(t)$$

$$= (\cos\omega_c t)\,0.5\,(\sin\omega_m t)$$

$$= \frac{0.5}{2}[\sin(\omega_c+\omega_m)t + \sin(\omega_m-\omega_c)t]$$

so the output is proportional to

$$\sin(\omega_c+\omega_m)t + \sin(\omega_m-\omega_c)t$$

11.11 (d)

$$f_m = 1 \text{ kHz}$$

$$\beta = 5$$

$$\text{B.W.} = 2(\beta+1)f_m$$

$$= 2(5+1)\times 1 \text{ kHz} = 12 \text{ kHz}$$

11.12 (c)

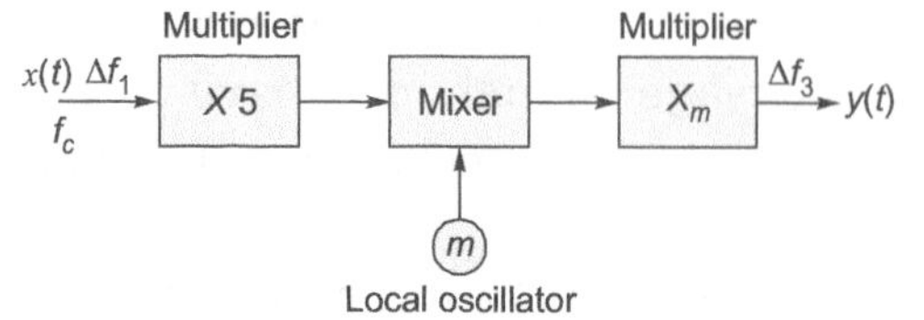

The frequency deviation of the output of first multiplier

$$\Delta f_2 = 5\,\Delta f_1 = 125 \text{ Hz}$$

and, $\quad \Delta f_3 = m\,\Delta f_2$

$$m = \frac{\Delta f_3}{\Delta f_2} = \frac{1000 \text{ Hz}}{125 \text{ Hz}} = 8$$

$$m(f_{LO} - 5\,f_c) = 2000 \text{ kHz}$$

$$\therefore \quad f_{LO} = (5f_c + 250) \text{ kHz}$$

$$= (5\times 100 + 250) \text{ kHz} = 750 \text{ kHz}$$

11.13 (c)

$s^2(t)$ has frequency component double of that of $s(t)$.

So sampling rate also become 2 times the sampling rate of $s(t)$.

$$\text{Sampling rate} = 2\times 5000$$

$$= 10000 \text{ samples/sec}$$

11.14 (a)

We know that, $\quad I_t = I_c\sqrt{1+\dfrac{m^2}{2}}$

$$4.8 = 4\sqrt{1+\frac{m^2}{2}} \Rightarrow 1.2 = \sqrt{1+\frac{m^2}{2}}$$

$$\Rightarrow \quad 1+\frac{m^2}{2} = 1.44 \quad \Rightarrow \quad \frac{m^2}{2} = 0.44$$

$$\Rightarrow \quad m = 93.8\%$$

11.15 (c)

$g^2(t)$ has frequency component double of that of $g(t)$.

So, $B = 2\times$ B.W. of $g(t) = 2\times 3$ kHz $= 6$ kHz

11.16 (c)

In case of AM modulation B.W. becomes double of that of message signal B.W.

11.17 (b)

If there are ten signals then there will be 9 guard bands.

So, the required bandwidth of the channel is

$$B = 10\times 3 \text{ kHz} + 9\times\frac{1}{2} \text{ kHz} = 34.5 \text{ kHz}$$

11.19 (c)

According to Nyquist sampling theorem for the faithful reconstruction of signal

$$nf_s = 2f_m$$

$$\therefore \quad f_m = \frac{nf_s}{2}$$

11.20 (a)

Given, $f_s = 150$ samples/sec

We know that,

$$f_s \geq 2f_m$$

$$\therefore \quad f_m < \frac{f_s}{2}$$

$$\therefore \quad f_m < \frac{150}{2} = 75 \text{ samples/sec}$$

In given signals only 50 Hz signal is satisfying Nyquist criteria, and there are no signals like 70 Hz and 75 Hz. Hence correct option is (a).

11.21 (*)

Square wave is as

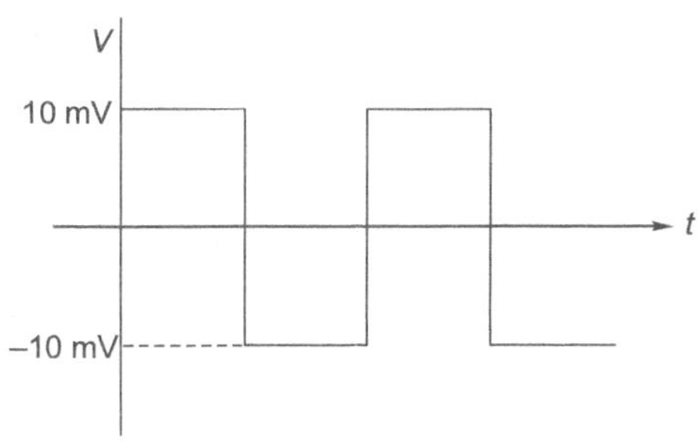

It has half wave symmetry so it will contain odd harmonics only as it is also odd signal therefore only odd harmonics of sine term will present.

Cut-off frequency of L.P.F. is 10 kHz so it will allow only to pass fundamental component at 5 kHz of amplitude $\frac{40}{\pi}$ mV (you can calculate).

So power of message

$$P_m = \frac{1}{2}\left(\frac{40 \times 10^{-3}}{\pi}\right)^2 = 81.05 \times 10^{-6} \text{ Watt}$$

Power in noise signal

$$P_N = 25 \times 10^{-12} \times 2 \times 10^6$$
$$= 50 \times 10^{-6} \text{ Watt}$$

$$(S/N) \text{ in dB} = 10\log\left(\frac{P_m}{P_N}\right) = 2.098 \text{ dB}$$

11.22 (b)

PDF of X and Y are

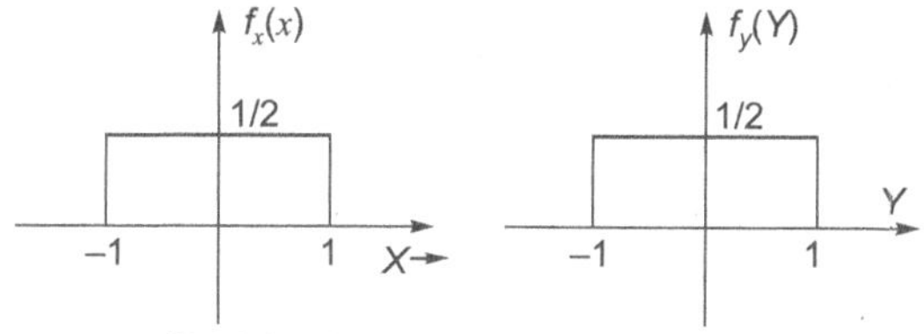

max $[X, Y]$ will be less than 1/2 if both X and Y will be less than 1/2.

So required probability is

$$p = (X < 1/2) * p(Y < 1/2)$$

$$p = \int_{-1}^{1/2} f_x(X)dX * \int_{-1}^{1/2} f_y(Y)dY = \left[\int_{-1}^{1/2} \frac{1}{2} dX\right]^2$$

$$p = \left(\frac{3}{4}\right)^2 = \frac{9}{16}$$

11.23 (a)

$$(f_s)_{min} = 2f_m$$
$$(f_s)_{min} = 2 \times 5 = 10 \text{ kHz}$$

So, $f_s \geq 10$ kHz

11.25 (b)

Since the transmission is full duplex solution 5 kHz will be allocated for each direction ($\rightleftarrows$)

$$\text{Bandwidth} = \text{Band rate} + f_2 - f_1$$

Given, $f_2 - f_1 = 2$ kHz

$$\text{Band rate} = \text{Bandwidth} - (f_2 - f_1)$$

$= 5\,k - 2\,k = 3\,k = 3000$ bps

11.27 (4)

$$p_x(x) = \begin{cases} e^{-x}; & x > 0 \\ 0; & \text{otherwise} \end{cases}$$

$$g_x(x) = e^{3x/4}$$

$$E_1(g_x(x)) = \int_0^\infty p_x(x)\, g_x(x)\, dx$$

$$= \int_0^\infty e^{-x} e^{3x/4}\, dx = \int_0^\infty e^{-x/4} dx = \left[\frac{e^{-x/4}}{-1/4}\right]_0^\infty = 4$$

11.28 (d)

SSB signal is given by

$$= A_C m(t)\cos(\omega_c t) \pm A_C \hat{m}(t)\sin(\omega_c t)$$

In phase component $= A_C m(t)$

Quadrature phase component $= A_C \hat{m}(t)$

Now, the output of envelop detector can be given by

$$|S(t)| = A_C\sqrt{m^2(t) + \hat{m}^2(t)} = \sqrt{\cos^2\theta + \sin^2\theta} = 1$$

11.29 (*)

$$\text{Output of matched filter} = \frac{E_{max}}{(N_0/2)}$$

$\because$ E_{max} of given signal = 3 Joule

$$\therefore \quad \text{Output} = \frac{3}{N_0/2} = \frac{6}{N_0}$$

11.30 60.1 (60.02 to 60.2)

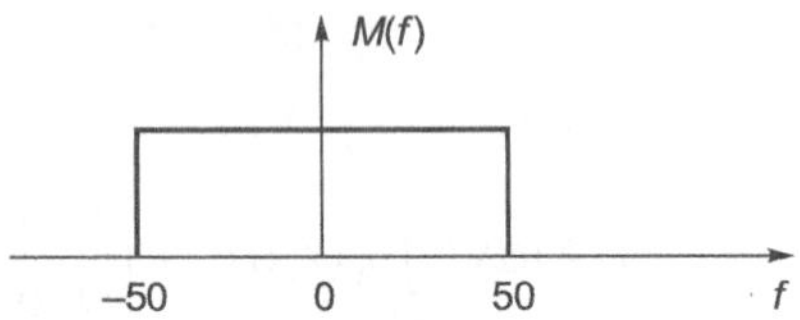

$$m(t) = \frac{\sin(100\pi t)}{100\pi t} = \text{sinc}(100t)$$

K_f = 30 kHz/Volt

$$\beta = \frac{k_f A_m}{f_m} = \frac{30 \times 10^3}{50} = \frac{3}{5}k = 600$$

$$\text{BW} = (\beta + 1)\, 2f_1 = (600 + 1) \times 100$$

BW = 60.1 kHz

11.31 (15.563)

Signal power = 3 W

$$\text{Resolution } (\Delta) = \frac{3.5 - (-3.5)}{2^3 - 1} = \frac{7}{7} = 1$$

Quantization noise power

$$N = \frac{\Delta^2}{12} = \frac{1}{12}$$

Signal to noise ratio

$$\frac{S}{N} = \frac{3}{1/12} = 36$$

$$\left(\frac{S}{N}\right)_{dB} = 10\log_{10}(36) = 15.563 \text{ dB}$$

11.32 (12.5)

Given angle modulated signal

$$\phi(t) = \cos\underbrace{(2\pi f_c t + 5\sin 1000\pi t + 10\sin 2000\pi t)}_{\phi_i(t)}$$

$$f_i = \frac{1}{2\pi}\frac{d}{dt}\phi_i(t)$$

$$= f_c + 2500 \cos 1000\pi t + 10000 \cos 2000\pi t$$

$$= f_c + K_f\, m(t)$$

Maximum frequency deviation

$$\Delta f = \max[K_f\, m(t)]$$

$$= \max[2500 \cos 1000\pi t + 10000 \cos 2000\pi t]$$

$$= 2500 + 10000$$

$$= 12.5 \text{ kHz}$$

11.33 11.11 (11 to 12.2)

Given an signal,

$$A_m(t) = 100 \cos\omega_c t + 50 \cos\omega_c t \cos\omega_m t$$

$$= 100\left[1 + \frac{1}{2}\cos\omega_c t\right]\cos\omega_c t = \frac{1}{2}$$

% of side band power in total power

$$= \frac{P_{SB}}{P_t} = \frac{\mu^2}{2 + \mu^2}$$

$$\frac{P_{SB}}{P_t} = 0.111111 = 11.11\%$$

11.34 (1)

In locked condition

$$\frac{f_0}{5} = f_i$$

$$f_i = 1 \text{ kHz}$$

11.35 (d)

$$\text{Var}(aY) = a^2\, \text{Var}(Y)$$

So, $\text{Var}(-Y) = (-1)^2\, \text{Var}(Y) = \text{var}(Y)$

Since X and Y are independent random variables,

$$\text{Var}(Z) = \text{Var}(X) + \text{Var}(-Y)$$

$$= \text{Var}(X) + \text{Var}(Y) = 1 + 2 = 3$$

11.36 (352000)

Given, f_s = 44000 samples/sec

Number of quantization levels, $L = 256$

Number of bits per sample, $n = \log_2 L = 8$

$$R_b = nf_s = 8 \times 44000$$

$$= 352000 \text{ bits/sec}$$

11.37 (0.30)

Modulation index,

$$\mu = \frac{A_{max} - A_{min}}{A_{max} + A_{min}} = \frac{13 - 7}{13 + 7} = \frac{6}{20} = 0.3$$

11.38 (300)

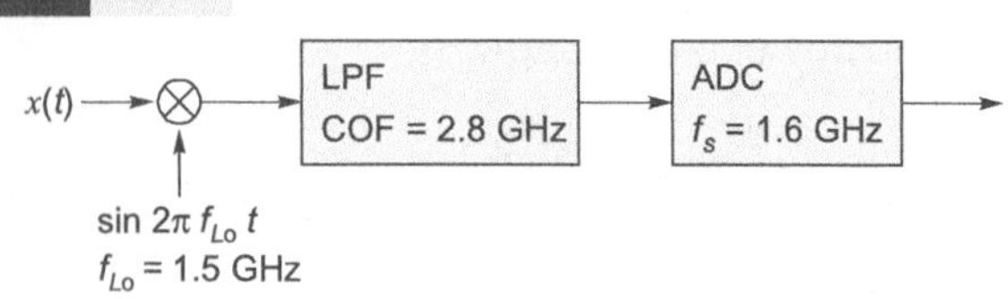

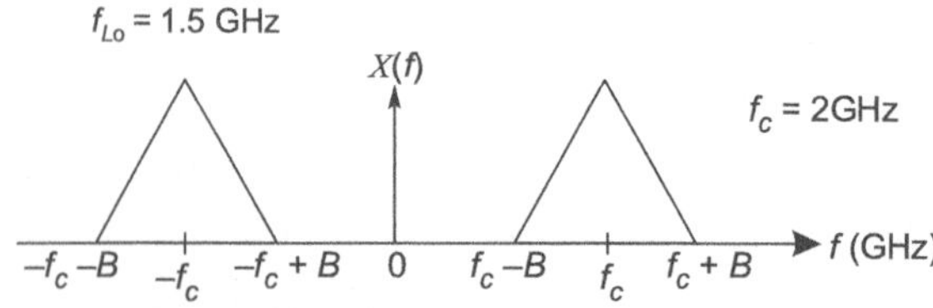

By multiplying with sinusoidal signal,

$X(f)$ shifted to right by f_{Lo} and left by f_{Lo}

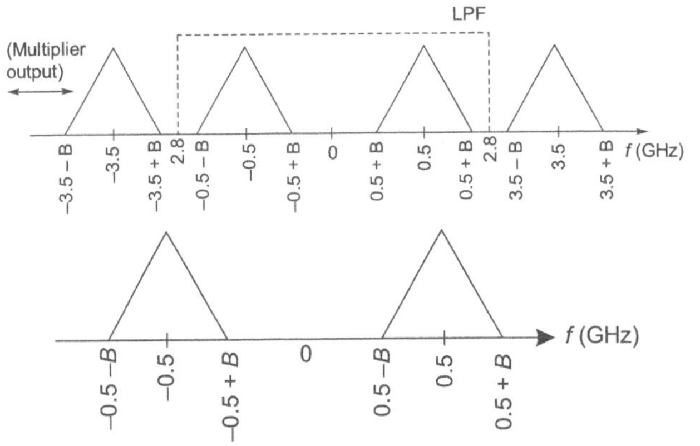

Nyquist rate = $2 \times f_{max}$

$= 2\,[0.5 + B] = 1 + 2\,B$

Given sampling rate,

$f_s = 1.6$ GHz

To recover original signal, $f_s \geq$ Nyquist rate

$1.6 \text{ GHz} \geq 1 \text{ GHz} + 2B$

$2B \leq 0.6$ GHz

$B = 0.3$ GHz

$B_{max} = 0.3 \text{ GHz} = 300$ MHz

11.39 (0.3)

Double sideband with carrier signal is given by

$$s(t) = \cos(2\pi f_c t) + 0.3\cos(2\pi f_m t)\cos(2\pi f_c t)$$

$$= \left[1 + 0.3\cos(2\pi f_m t)\right]\cos 2\pi f_c t$$

Comparing with standard amplitude modulated signal

$$s(t) = A_c[1 + \mu\cos 2\pi f_m t]\cos 2\pi f_c t$$

$$\mu = 0.3$$

11.40 (15.25)

Amplitude modulation, $\mu = 0.6$

Assuming sinusoidal message modulation efficiency,

$$\eta = \frac{\mu^2/2}{1+\frac{\mu^2}{2}} \times 100$$

$$\%\eta = \frac{(0.6)^2}{2+(0.6)^2} \times 100 = \frac{0.36}{2+0.36} \times 100$$

$= 15.25\%$

11.41 (120)

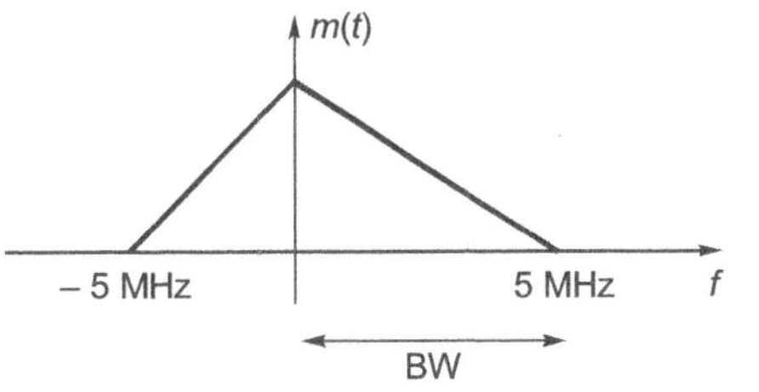

$f_m = 5$ MHz

N.R $= 2f_m = 2 \times$ BW

N.R $= 10$ MHz

$f_s = 1.5 \times \text{N.R} = 1.5 \times 10$

$$= 15 \text{ MHz} = \frac{1}{T_s}$$

$L = 256$

$L = 2^n$

So, $n = 8$ bit/sec

Bit rate, $R_b = \dfrac{n}{T_s} = 8 \times 15 = 120$ Mbit/sec

11.42 (a)

For envelope detection modulation index must be less than 1. So, option (a) is correct.

11.43 (0.43)

Bit rate $R_b = Nnf_s$

[$\because$ N = No. of signals; n = No. of bits; f_s = sampling frequency]

$$\text{Bit interval} = \frac{1}{R_b} = \frac{1}{Nnf_s}$$

$$T_b = \frac{1}{24 \times 12 \times 8 \times 10^{-3}} = 0.434\,\mu\text{sec}$$

■■■

Control Systems and Process Control

UNIT

VI

CONTENTS

1. Basics of Control System and Transfer Function 296

2. Block Diagram and Signal Flow Graph 301

3. Compensators and Controllers 305

4. Time Response Analysis 312

5. Stability Analysis 323

6. Root Locus 328

7. Frequency Response Analysis,
   Bode Plot and Nyquist Plot 334

8. State Space Analysis 348

9. Process Control 353

Control Systems and Process Control

Syllabus

Feedback principles, signal flow graphs, transient response, steady-state-errors, Bode plot, phase and gain margins, Routh and Nyquist criteria, root loci, design of lead, lag and lead-lag compensators, state-space representation of systems; time-delay systems; mechanical, hydraulic and pneumatic system components, synchro pair, servo and stepper motors, servo valves; on-off, P, PI, PID, cascade, feedforward, and ratio controllers, tuning of PID controllers and sizing of control valves.

Analysis of Previous GATE Papers

Exam Year	1 Mark Ques.	2 Marks Ques.	Total Marks
1995	4	1	**6**
1996	1	2	**5**
1997	4	–	**4**
1998	2	1	**4**
1999	5	1	**7**
2000	2	2	**6**
2001	2	1	**4**
2002	2	5	**12**
2003	4	10	**24**
2004	4	8	**20**
2005	–	10	**20**
2006	1	8	**17**
2007	2	6	**14**
2008	1	9	**19**
2009	1	6	**13**

Exam Year	1 Mark Ques.	2 Marks Ques.	Total Marks
2010	–	4	**8**
2011	1	6	**13**
2012	2	6	**14**
2013	1	4	**9**
2014	2	4	**10**
2015	1	3	**7**
2016	3	3	**9**
2017	2	4	**10**
2018	3	4	**11**
2019	3	5	**13**
2020	4	5	**14**
2021	2	3	**8**
2022	2	4	**10**
2023	3	3	**9**
2024	2	3	**8**

1 Basics of Control System and Transfer Function

1.1 $F(s) = \frac{(1-e^{sT})}{s}$ is the Laplace transform of
(a) a pulse of width T
(b) a square wave of period T
(c) a unit step delayed by T
(d) a ramp delayed by T [1997 : 1 M]

1.2 The transfer function of a system is the Laplace transform of its
(a) Square wave response
(b) Step response
(c) Ramp response
(d) Impulse response [1998 : 1 M]

1.3 Relationship between input $x(t)$ and output $y(t)$ of a system is given as

$$\frac{d^2y}{dt^2} = x(t-2) + \frac{d^2x}{dt^2}$$

The transfer function of this system is
(a) $1+\frac{e^{-2s}}{s^2}$ (b) $1+\frac{e^{2s}}{s^2}$
(c) $1 + s^2 e^{-2s}$ (d) $1 + s^2 e^{2s}$
[1999 : 1 M]

1.4 A transfer function has two zeros at infinity. Then the relation between the numerator degree (N) and the denominator degree (M) of the transfer function is
(a) $N = M + 2$ (b) $N = M - 2$
(c) $N = M + 1$ (d) $N = M - 1$
[1999 : 1 M]

1.5 The transfer function of a second order all pass filter is
(a) $\frac{s^2}{s^2+as+b}$ (b) $\frac{cs}{s^2+as+b}$
(c) $\frac{s^2+d}{s^2+as+b}$ (d) $\frac{s^2-as+b}{s^2+as+b}$
[1999 : 1 M]

1.6 The characteristic roots of the system described as $\frac{dx}{dt} = y, \frac{dy}{dt} = -x$ are at
(a) +1, +1 (b) −1, +1
(c) $-j, +j$ (d) None of these
[2000 : 2 M]

1.7 In a synchro pair, the control transmitter excites the three stator windings of the control transformer. The stator winding voltages will have
(a) equal magnitudes but different phases
(b) different magnitude and different phases
(c) equal magnitude and phases
(d) different magnitude but equal phases
[2003 : 1 M]

1.8 A three-stack variable reluctance stepper motor has 12 numbers of rotor and stator teeth on each stack. The resolution of angular rotation for the stepper motor, assuming no two stacks are excited simultaneously is
(a) 10° (b) 15°
(c) 20° (d) 30° [2003 : 1 M]

1.9 In the feedback scheme shown in figure, the time-constant of the closed loop system will be

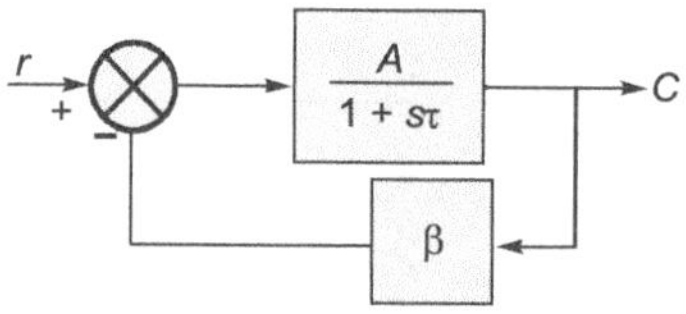

(a) $A\beta\tau$ (b) $(1 + A\beta, \tau)$
(c) τ (d) $\frac{\tau}{(1+A\beta)}$
[2003 : 1 M]

1.10 The transfer function of a second order band-pass filter, having a centre frequency of 1000 rad/s, selectivity of 100 and a gain of 0 dB at the centre frequency is
(a) $\frac{10s}{s^2+10s+10^6}$ (b) $\frac{s}{s^2+s+10^6}$
(c) $\frac{100s}{s^2+100s+10^7}$ (d) $\frac{100s}{s^2+100s+10^6}$
[2003 : 2 M]

1.11 The dominant poles of a servo system are located at $s = (-2 \pm j2)$. The damping ratio of the system is
(a) 1 (b) 0.8
(c) 0.707 (d) 0.6
[2004 : 1 M]

1.12 The torque-speed curve of a constant field armature controlled DC servomotor is shown in the figure. The armature resistance in Ω and torque constant in Nm/A of the motor respectively are

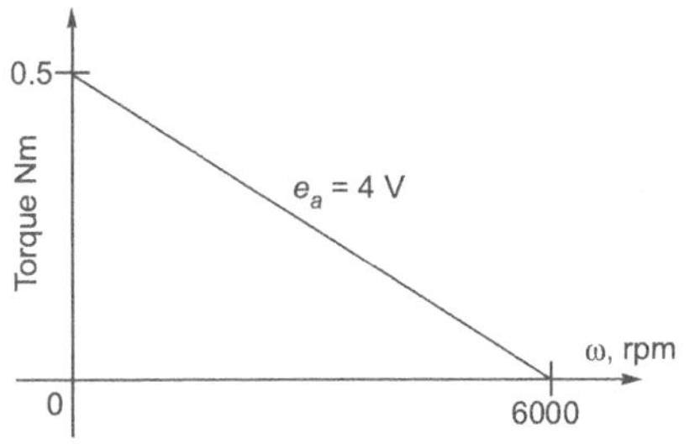

(a) (1.76, 0.68) (b) (1.76, 0.85)
(c) (2.00, 0.25) (d) (0.01, 0.81)

[2004 : 2 M]

1.13 Compared to a conventional two-phase induction motor, a two-phase servomotor has
(a) a smaller R/X ratio and a lighter rotor
(b) the same R/X ratio and a heavier rotor
(c) a larger R/X ratio and a heavier rotor
(d) a larger R/X ratio and a lighter rotor

[2006 : 1 M]

1.14 The rotor of the control transformer of a synchro pair gives a maximum voltage of 1.0 V at a particular position of the rotor of the control transmitter. The transmitter rotor is now rotated by 30° anticlockwise keeping the transformer rotor stationary. The transformer rotor voltage for this position is
(a) 1.0 V (b) 0.866 V
(c) 0.5 V (d) 0 V

[2010 : 2 M]

1.15 The transfer function of a Zero-Order-Hold system with sampling interval T is

(a) $\frac{1}{s}(1-e^{-Ts})$ (b) $\frac{1}{s}(1-e^{-Ts})^2$
(c) $\frac{1}{s}e^{-Ts}$ (d) $\frac{1}{s^2}e^{-Ts}$

[2012 : 1 M]

1.16 The overall closed loop transfer function $\frac{C(s)}{R(s)}$, represented in the figure, will be

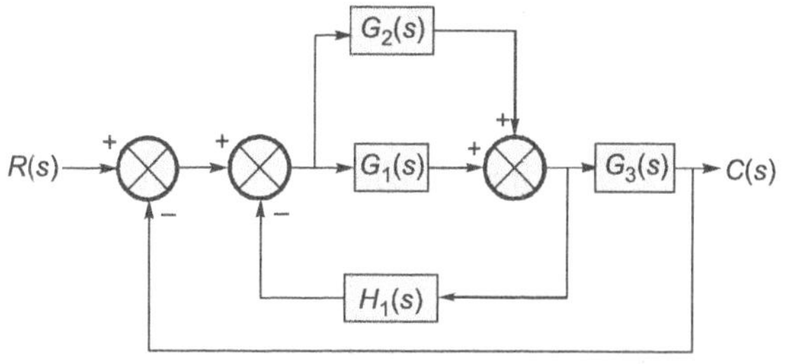

(a) $\frac{(G_1(s)+G_2(s))G_3(s)}{1+(G_1(s)+G_2(s))(H_1(s)+G_3(s))}$

(b) $\frac{(G_1(s)+G_3(s))}{1+G_1(s)H_1(s)+G_2(s)G_3(s)}$

(c) $\frac{(G_1(s)-G_2(s))H_1(s)}{1+(G_1(s)+G_3(s))(H_1(s)+G_1(s))}$

(d) $\frac{G_1(s)G_2(s)H_1(s)}{1+G_1(s)H_1(s)+G_1(s)G_3(s)}$ [2017 : 2 M]

1.17 The output $y(t)$ of a system is related to its input $x(t)$ as

$$y(t) = \int_0^t x(\tau-2)d\tau$$

where, $x(t) = 0$ and $y(t) = 0$ for $t \leq 0$. The transfer function of the system is

(a) $\frac{1}{s}$ (b) $\frac{(1-e^{-2s})}{s}$
(c) $\frac{e^{-2s}}{s}$ (d) $\frac{1}{s}-e^{-2s}$

[2019 : 1 M]

1.18 The closed loop transfer function of a control system is given by $\frac{C(s)}{R(s)} = \frac{1}{s+1}$. For the input $r(t)$ = sin t, the steady state response $c(t)$ is _____.

(a) $\frac{1}{\sqrt{2}}\cos t$ (b) $\frac{1}{\sqrt{2}}\sin\left(t+\frac{\pi}{4}\right)$
(c) 1 (d) $\frac{1}{\sqrt{2}}\sin\left(t-\frac{\pi}{4}\right)$

[2020 : 1 M]

■■■

Answers Basics of Control System and Transfer Function

1.1 (a)	1.2 (d)	1.3 (a)	1.4 (b)	1.5 (d)	1.6 (c)	1.7 (a)
1.8 (a)	1.9 (d)	1.10 (a)	1.11 (c)	1.12 (c)	1.13 (d)	1.14 (b)
1.15 (a)	1.16 (a)	1.17 (c)	1.18 (d)			

Explanations Basics of Control System and Transfer Function

1.1 (a)

$$F(s) = \frac{1-e^{-sT}}{s} = \frac{1}{s} - e^{-sT}\left(\frac{1}{s}\right)$$

$$\therefore \quad f(t) = u(t) - u(t-T)$$

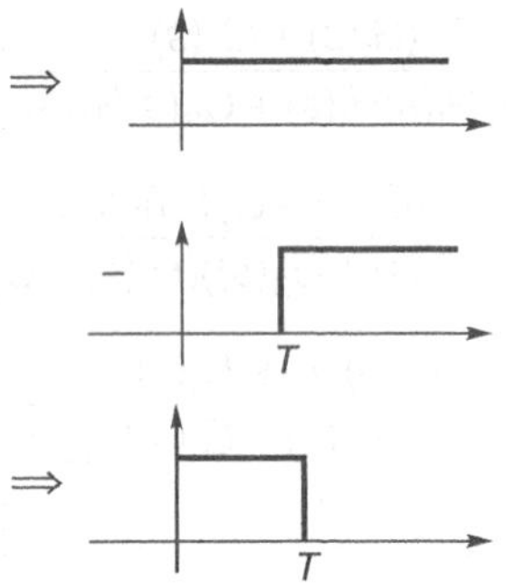

1.2 (d)

T.F. = L (Impulse response)

1.3 (a)

$$\frac{d^2y}{dt^2} = x(t-2) + \frac{d^2x}{dt^2}$$

$$s^2 Y(s) = e^{-2s} X(s) + s^2 X(s) = (e^{-2s} + s^2) X(s)$$

$$\therefore \quad \frac{Y(s)}{X(s)} = \frac{e^{-2s} + s^2}{s^2} = 1 + \frac{e^{-2s}}{s^2}$$

1.4 (b)

$\because$ Number of asymptotic root locus branch

$= M - N$

Here, $M - N = 2$

$N = M - 2$

1.5 (d)

T.F. of second order all pass filter is of the form,

$$\text{T.F.} = \frac{s^2 - as + b}{s^2 + as + b}$$

magnitude of the transfer function is unity

$$|\text{T.F.}| = 1$$

and the singularities are symmetric with respect to $(j\omega)$ axis.

1.6 (c)

$$\because \quad \frac{dx}{dt} = Y \; ; \quad \frac{dy}{dt} = -X$$

$$\frac{d^2x}{dt^2} = \frac{dy}{dt} = -X$$

Taking Laplace,

$$s^2X(s) - sx(0) = -X(s)$$

$$[s^2 + 1]\,X(s) = sx(0)$$

$$X(s) = \frac{sx(0)}{s^2+1}$$

Characteristic roots:

$$s = -j, + j$$

1.7 (a)

Equal in magnitude and have 120° of successive phase difference.

1.8 (a)

$$\text{Resolution } (\alpha) = \frac{360°}{nT}$$

Cohere, n = number of stacks

T = number of teeth

$$\therefore \quad \text{Resolution } (\alpha) = \frac{360°}{3 \times 12} = 10°$$

1.9 (d)

$$T(s) = \frac{C(s)}{R(s)} = \frac{A}{1 + s\tau + A\beta}$$

$$= \frac{\dfrac{A}{1+A\beta}}{1 + \dfrac{s\tau}{1+A\beta}} \quad \text{...(i)}$$

or $$T(s) = \frac{\alpha}{1 + \tau' s} \quad \text{...(ii)}$$

Comparing equation (i) and (ii) we get

Thus, time constant $\tau' = \dfrac{\tau}{1 + A\beta}$.

1.10 (a)

Transfer function of second order band-pass filter is

$$A_v(s) = \frac{\left(\frac{\omega_0}{Q}\right)A_0 s}{s^2 + \left(\frac{\omega_0}{Q}\right)s + \omega_0^2}$$

Given,

$$Q = 100$$
$$A_0 = 1$$
$$\omega_0 = 1000 \text{ rad/sec}$$

We get,

$$A_v = \frac{\left(\frac{1000}{100}\right)\cdot 1 \cdot s}{s^2 + \left(\frac{1000}{100}\right)s + (1000)^2} = \frac{10s}{s^2 + 10s + 10^6}$$

1.11 (c)

$$T(s) = \frac{1}{(s-(-2+j2))\,(s-(-2-j2))}$$
$$= \frac{1}{(s+2-j2)\,(s+2+j2)}$$
$$= \frac{1}{(s+2)^2+4} = \frac{1}{s^2+4s+4+4}$$
$$= \frac{1}{s^2+4s+8}$$
$$\omega_n^2 = 8$$
$$\Rightarrow \quad \omega_n = \sqrt{8}$$
$$2\xi\omega_n = 4$$
$$\therefore \quad \xi = \frac{4}{2\cdot\sqrt{8}} = \frac{2}{\sqrt{8}} = \frac{1}{\sqrt{2}} = 0.707$$

1.12 (c)

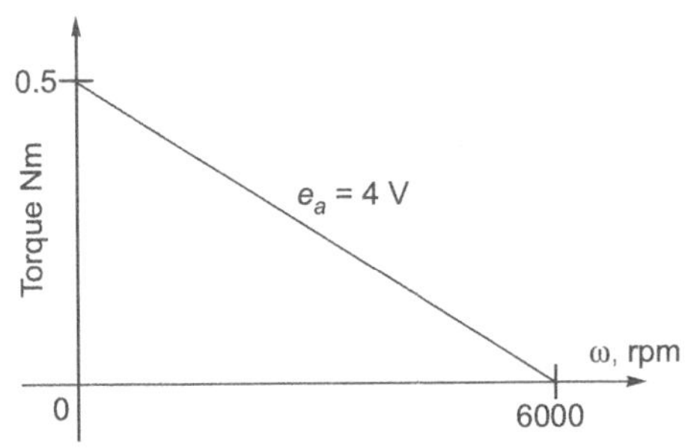

We know,

$$T_m(t) = K_m(t)\,\phi\, i_a(t)$$
$$T_m(t) = K_i\, i_a(t) \qquad (\because \phi = \text{constant})$$

Among the options only the armature resistance $R_a = 2.00\ \Omega$ will satisfy the torque equation.

$$i_a(t) = \frac{e_a}{R_a} = \frac{4}{2} = 2.0 \text{ Amp.}$$
$$K_i = \frac{T_m(t)}{i_a(t)} = \frac{0.5}{2.0} \text{ N-m/A}$$
$$= 0.25 \text{ N-m/A}$$

$$\boxed{\therefore K_i = 0.25 \text{ N-m/A}}$$

1.14 (b)

$$v_r = v_m \cos\phi$$
$$= 1.0 \cos 30$$
$$v_r = 0.866\ v$$

1.15 (a)

The transfer function of a zero-order hold system having a sampling interval T is $\frac{1}{s}(1-e^{-Ts})$.

1.16 (a)

Blocks G_1 and G_2 are in parallel

$\therefore \quad G_1 + G_2$

To the above result H_1 is in negative feedback

$$\therefore \quad \frac{(G_1+G_2)}{1+(G_1+G_2)H_1}$$

To the above result G_3 is in series

$$\therefore \quad \frac{(G_1+G_2)G_3}{1+(G_1+G_2)H_1}$$

To the above result feedback is '1'

$$\therefore \quad \frac{C(s)}{R(s)} = \frac{(G_1+G_2)\,G_3}{1+(G_1+G_2)\,[H_1+G_3]}$$
$$= \frac{(G_1(s)+G_2(s))\,G_3(s)}{1+(G_1(s)+G_2(s))\,(H_1(s)+G_3(s))}$$

1.17 (c)

Put $x(t) = \delta(t)$

$$y(t) = h(t) = \int_0^t \delta(\tau-2)dz = u(t-2)$$

So, $H(s) = \frac{e^{-2s}}{s}$

Alternatively,

$$Y(t) = \int_0^t x(\tau-2)d\tau$$

$$Y(s) = \frac{1}{s} X(s) \cdot e^{-2s}$$

$$\therefore \quad \frac{Y(s)}{X(s)} = \frac{e^{-2s}}{s}$$

1.18 (d)

$$\text{Transfer function} = \frac{1}{s+1} = G(s)$$

$$G(j\omega) = \frac{1}{j\omega+1} = \frac{1}{\sqrt{\omega^2+1}} \angle -\tan^{-1}\omega$$

$$\text{Input} = \underset{A}{1}\sin(\underset{\omega}{(1)}t + \underset{\phi_1}{0°})$$

$$\text{Output} = B\sin(\omega t + \phi_2)$$

where $B = A|G(j\omega)| = 1 \times \frac{1}{\sqrt{1+1}} = \frac{1}{\sqrt{2}}$

and $\phi_2 = \phi_1 + \angle G(j\omega) = -\tan^{-1} 1$

$= -45°$

$\therefore$ Steady state o/p

$$= \frac{1}{\sqrt{2}}\sin\left(t - \frac{\pi}{4}\right)$$

■■■

2 Block Diagram and Signal Flow Graph

2.1 For the signal flow graph show in figure, the transfer function is $C(s)/R(s)$ = ______.

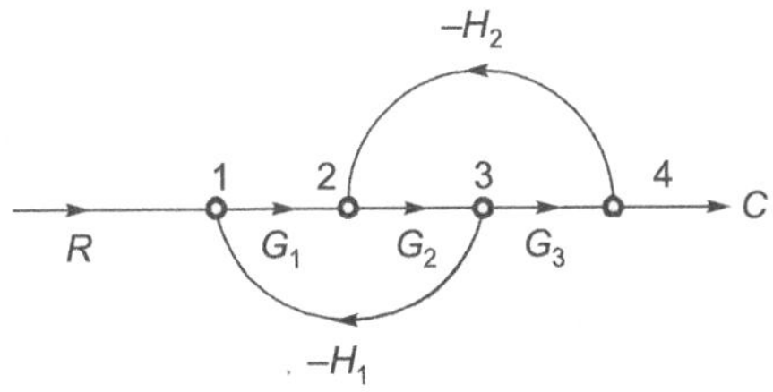

[1995 : 2 M]

2.2 Match the characteristics with the system transfer function.

(a) All pass (e) $e^{-0.5s}$

(b) Transport delay (f) $\dfrac{1-s}{1+s}$

(c) Lead compensator (g) $\dfrac{1}{s^2+2s-3}$

(d) Unstable (h) $\dfrac{s}{s^2+s+100}$

(i) $\dfrac{10(1+0.3s)}{(1+0.1s)}$

[1996 : 2 M]

2.3 The transfer function between y_2 and y_1 in figure is

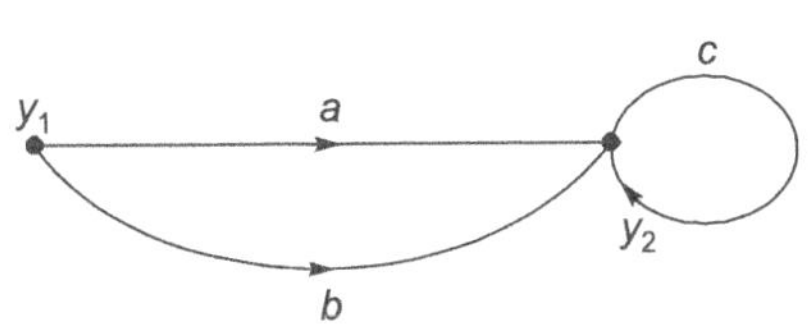

(a) $a + b$ (b) $(a + b)\,c$

(c) $\dfrac{a+b}{1-c}$ (d) $\dfrac{a+b}{1+c}$

[1999 : 1 M]

2.4 The steady state error for the feedback control system shown in figure when subjected to a unit step input is

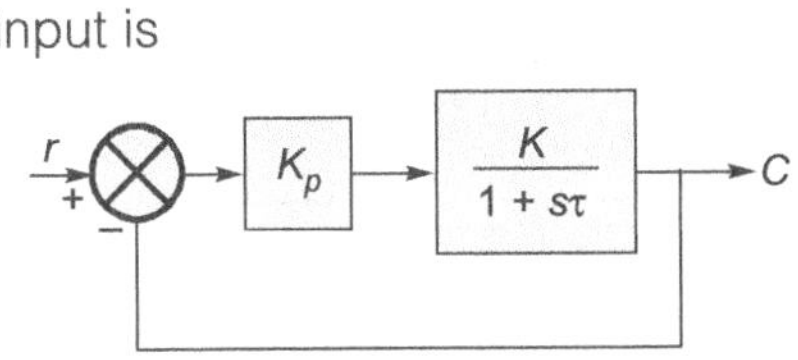

(a) zero (b) $\dfrac{1}{(1+KK_p)}$

(c) $\dfrac{KK_p}{(1+KK_p)}$ (d) Infinite

[2003 : 1 M]

2.5 The transfer function of the open loop system $G(s)$ which is represented by the signal flow graph shown in the figure below is

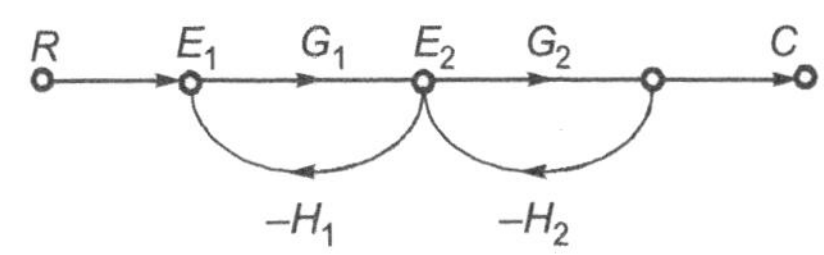

(a) $\dfrac{G_1G_2}{1+G_1H_1+G_2H_2}$

(b) $\dfrac{G_1G_2}{1-G_1H_1+G_2H_2}$

(c) $\dfrac{G_1G_2}{(1+G_1H_1)(1+G_2H_2)}$

(d) $\dfrac{G_1G_2}{1+(G_1+H_1)(G_2+H_2)}$

[2005 : 2 M]

2.6 The signal flow graph representation of a control system is shown below:

The transfer function $\dfrac{Y(s)}{R(s)}$ is computed as

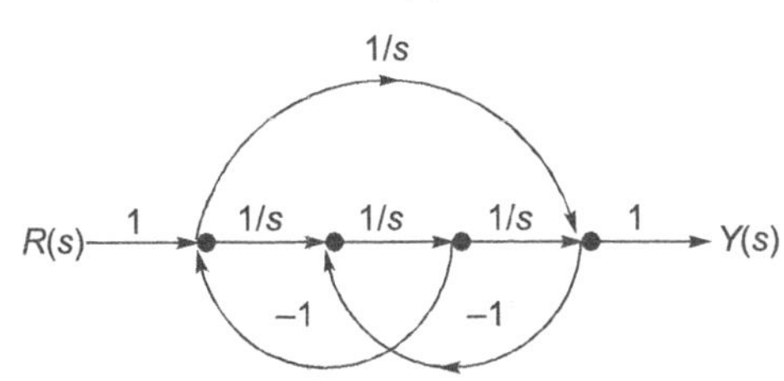

(a) $\dfrac{1}{s}$ (b) $\dfrac{s^2+1}{s(s^2+2)}$

(c) $\dfrac{s(s^2+1)}{s^2+2}$ (d) $1-\dfrac{1}{s}$

[2006 : 2 M]

2.7 A feedback control system with high gain K, is shown in the figure below:

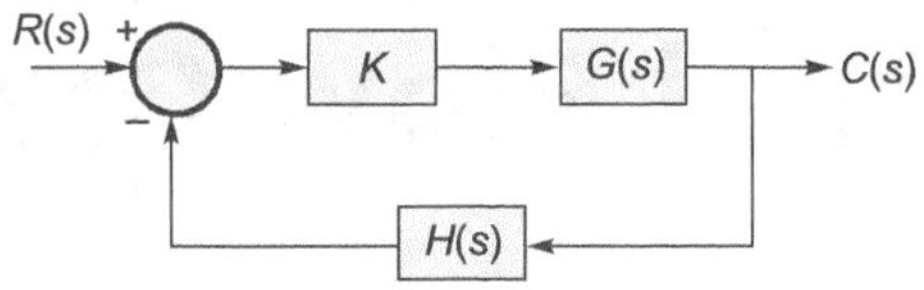

Then the closed loop transfer function is

(a) sensitive to perturbations in $G(s)$ and $H(s)$
(b) sensitive to perturbations in $G(s)$ but not to perturbations in $H(s)$
(c) sensitive to perturbations in $H(s)$ but not to perturbations in $G(s)$
(d) insensitive to perturbations in $G(s)$ and $H(s)$

[2007 : 1 M]

2.8 The signal flow graph of a system is given below:

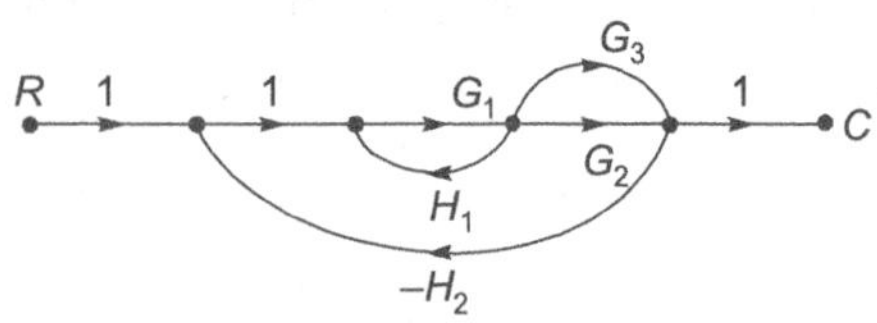

The transfer function (C/R) of the system is

(a) $(G_1G_2 + G_1G_3)/(1 + G_1G_2H_2)$
(b) $(G_1G_2 + G_1G_3)/(1 - G_1H_1 + G_1G_2H_2)$
(c) $(G_1G_2 + G_1G_3)/(1 - G_1H_1 + G_1G_2H_2 + G_1G_3H_2)$
(d) $(G_1G_2 + G_1G_3)/(1 - G_1H_1 + G_1G_2H_2 + G_1G_3H_2 + G_1G_2G_3H_1)$

[2011 : 1 M]

2.9 The signal flow graph for a system is given below. The transfer function $Y(s)/U(s)$ for this system is given as

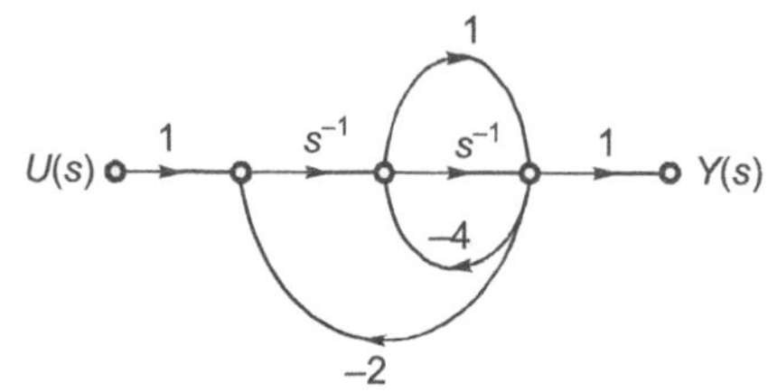

(a) $\dfrac{s+1}{5s^2+6s+2}$ (b) $\dfrac{s+1}{s^2+6s+2}$

(c) $\dfrac{s+1}{s^2+4s+2}$ (d) $\dfrac{1}{s^2+6s+2}$

[2013 : 2 M]

2.10 The signal flow graph of a system is shown. The expression for $Y(s)/X(s)$ is ________.

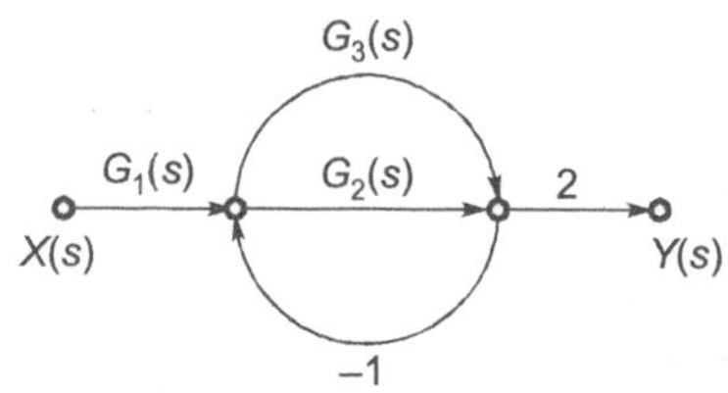

(a) $\dfrac{2G_1(s)G_2(s) + 2G_1(s)G_3(s)}{1 + G_2(s) + G_3(s)}$

(b) $2 + G_1(s) + G_3(s) + \dfrac{G_2(s)}{1 + G_2(s)}$

(c) $G_1(s) + G_3(s) - \dfrac{G_2(s)}{2 + G_2(s)}$

(d) $\dfrac{2G_1(s)G_2(s) + 2G_1(s)G_3(s) - G_1(s)}{1 + G_2(s) + G_3(s)}$

[2022 : 2 M]

■■■

Answers Block Diagram and Signal Flow Graph

2.1 (Sol.)	2.2 Sol.	2.3 (c)	2.4 (b)	2.5 (a)	2.6 (a)	2.7 (c)
2.8 (c)	2.9 (a)	2.10 (a)				

Explanations Block Diagram and Signal Flow Graph

2.1 Sol.

There is one forward path and two loops,

$$P_1 = G_1G_2G_3$$
$$L_1 = -H_1G_1G_2,$$
$$L_2 = -H_2G_2G_3$$

$$\text{T.F.} = \frac{P_1\Delta_1}{\Delta}$$
$$= \frac{G_1G_2G_3[1-0]}{1 + H_1G_1G_2 + H_2G_2G_3}$$
$$= \frac{G_1G_2G_3}{1 + G_1G_2H_1 + G_2G_3H_2}$$

2.2 Sol.

a-f, b-e, c-i, d-g

I[st] :

All pass filter $\Rightarrow \dfrac{1-s}{1+s}$,

$\because \quad |A| = 1$

and the poles and zeros are symmetric w.r.t. ($j\omega$) axis.

IInd :

Transport delay $\Rightarrow e^{-0.5s}$

delay time $\Rightarrow$ 0.5 sec

IIIrd :

Lead compensator $\Rightarrow \dfrac{10(1+0.3s)}{(1+0.1s)}$

the singularity near to origin is zero

$$\phi = \tan^{-1} 0.3 - \tan^{-1} 0.1 = 10.9°$$

$\phi \Rightarrow$ +ve

$\therefore$ Lead compensator

IVth :

Unstable $\Rightarrow \dfrac{1}{s^2+2s-3}$

$\because$ poles are lying at R.H.S.

$\dfrac{s}{s^2+s+100} \Rightarrow$ Band pass filter (B.P.F.)

2.3 (c)

Between y_1 and y_2

forward paths,

$$P_1 = a, \quad P_2 = b$$

Loop $\Rightarrow L_1 = c$

$$\text{T.F.} = \frac{P_1\Delta_1 + P_2\Delta_2}{\Delta}$$

$$= \frac{a(1-0)+b(1-0)}{1-c} = \frac{a+b}{1-c}$$

2.4 (b)

$$\frac{C(s)}{R(s)} = \frac{K_p K}{1+s\tau+K_p K}$$

$$\frac{E(s)}{R(s)} = \frac{R(s)-C(s)}{R(s)} = \frac{1+s\tau}{1+s\tau+K_p K}$$

So, steady state error, $e_{ss} = \lim_{s\to 0} s \cdot E(s)$

$$= \lim_{s\to 0} \frac{s \cdot R(s)\,(1+s\tau)}{1+s\tau+K_p K}$$

Input $R(s) = \dfrac{1}{s}$

$$e_{ss} = \lim_{s\to 0} \frac{1+s\tau}{1+s\tau+KK_p}$$

$$\boxed{e_{ss} = \frac{1}{1+KK_p}}$$

2.5 (a)

Forward path gain

$$P_1 = G_1 G_2$$

Loop $L_1 = -G_1 H_1$

$$L_2 = -G_2 H_2$$

No non-touching loop

$$\Delta = 1 - (L_1 + L_2)$$

$$= 1 + G_1 H_1 + G_2 H_2$$

$$\Delta_1 = 1$$

Therefore according to Mason's gain formula

$$T(s) = \frac{C(s)}{R(s)} = \frac{P_1\Delta_1}{\Delta}$$

$$= \frac{G_1 G_2}{1+G_1 H_1 + G_2 H_2}$$

2.6 (a)

Forward path gain, $P_1 = \dfrac{1}{s}$

$$P_2 = \frac{1}{s^3}$$

$$\Delta_1 = \Delta_2 = 1$$

Loop gain,

$$P_{11} = -\frac{1}{s^2}$$

$$P_{21} = -\frac{1}{s^2}$$

$$P_{31} = \frac{1}{s^2}$$

Transfer function $T(s) = \dfrac{P_1\Delta_1 + P_2\Delta_2}{1-P_{11}-P_{21}-P_{31}}$

$$= \frac{\dfrac{1}{s}+\dfrac{1}{s^3}}{1+\dfrac{1}{s^2}+\dfrac{1}{s^2}-\dfrac{1}{s^2}} = \frac{(s^2+1)}{s(s^2+1)} = \frac{1}{s}$$

2.7 (c)

Transfer function $T(s) = \dfrac{KG(s)}{1+KG(s)\,H(s)}$

$$S_G^T = \frac{\partial T}{\partial G}\frac{G}{T}$$

$$= \frac{G}{T} \times \frac{[1+KG(s)H(s)]K - KG(s)\,[KH(s)]}{[1+KG(s)\,H(s)]^2} \times KG(s)$$

$$= \frac{1 \times KG(s)\,H(s)}{K} \times \frac{K}{[1+KG(s)\,H(s)]^2}$$

$$S_G^T = \frac{1}{1+KG(s)\,H(s)}$$

$\because$ K is very high.

Therefore $S_G^T \to 0$ (less sensitive)

Now,

$$S_H^T = \frac{\partial T}{\partial H}\frac{H}{T}$$

$$= -\frac{KG(s)}{[1+KG(s)\,H(s)]^2}\;\frac{[1+KG(s)\,H(s)]\,H(s)}{KG(s)}$$

$$= -\frac{KG(s)\,H(s)}{1+KG(s)\,H(s)}$$

$K \to$ very high $\to \infty$,

then, $S_H^T \to -1$

or $\left|S_H^T\right| \to 1$ (highly sensitive)

2.8 (c)

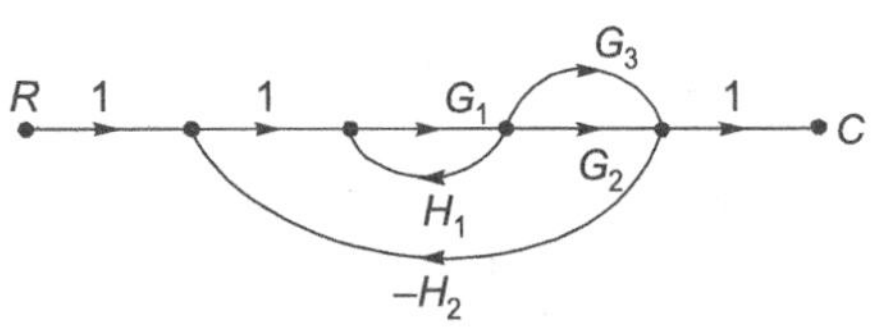

Step 1: There are two forward paths between input and output

$P_1 = G_1G_2$

$P_2 = G_1G_3$

Step 2: There are three loops, the loops gain are

$L_1 = G_1H_1$

$L_2 = -G_1\,G_3\,H_2$

$L_3 = -G_1G_2H_2$

Step 3: There are no pairs of non-touching loops.

$\Delta_1 = 1$ and $\Delta_2 = 1$

Step 4:

$$\Delta = 1-\left[L_1+L_2+L_3\right]$$

$$= 1 - G_1H_1 + G_1G_3H_2 + G_1G_2H_2$$

According to Mason's gain formula.

$$T(s) = \frac{\Sigma p_k\,\Delta_k}{\Delta}$$

$$= \frac{G_1G_2 \cdot 1 + G_1G_3 \cdot 1}{1-G_1H_1 + G_1G_3H_2 + G_1G_2H_2}$$

$$\therefore T(s) = \frac{G_1G_2 + G_1G_3}{1-G_1H_1 + G_1G_3H_2 + G_1G_2H_2}$$

2.9 (a)

Using Mason's gain formulae

$$\Delta = 1-\left[-2s^{-1} - 2s^{-2} - 4 - 4s^{-1}\right]$$

$$\Delta = 1+\frac{2}{s}+\frac{2}{s^2}+4+\frac{4}{s}$$

$$\Delta = \frac{5s^2+6s+2}{s^2}$$

$$P_1 = s^{-2} = \frac{1}{s^2}$$

$$P_2 = s^{-1} = \frac{1}{s}$$

$\Delta_1 = 1$

$\Delta_2 = 1$

$$\frac{Y(s)}{u(s)} = \frac{\sum P_k\,\Delta_k}{\Delta} = \frac{\frac{1}{s^2}\times 1+\frac{1}{s}\times 1}{\frac{5s^2+6s+2}{s^2}}$$

$$\frac{Y(s)}{u(s)} = \frac{s+1}{5s^2+6s+2}$$

2.10 (a)

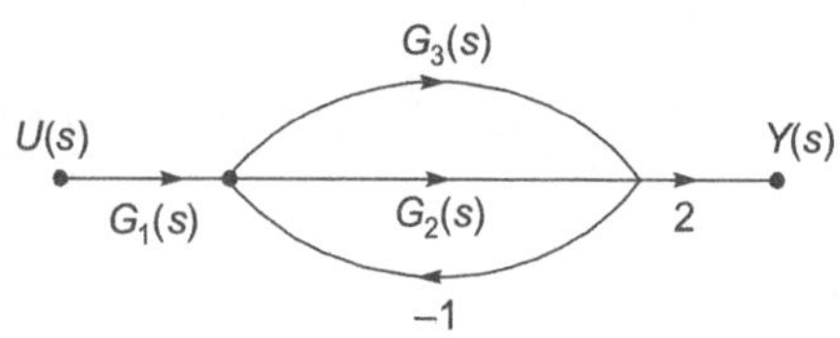

By Mason's gain formula

$$\frac{Y(s)}{U(s)} = \frac{P_1\Delta_1 + P_2\Delta_2}{\Delta}$$

$$\frac{Y(s)}{U(s)} = \frac{2G_1G_3 + 2G_2G_1}{1+G_2+G_3}$$

$$\frac{Y(s)}{U(s)} = \frac{2(G_2+G_3)G_1}{1+G_2+G_3}$$

3 Compensators and Controllers

3.1 The transfer function of a basic PD controller is given by (K's are real constants)

(a) $K_0 + \frac{K_1}{s} + K_2 s$ (b) $K_0 + K_2 s$

(c) $K_1 s + K_2 s$ (d) $K_0 + \frac{K_1}{s}$

[1997 : 1 M]

3.2 A first order system with a proportional controller in the negative feedback loop has an offset to a step input. This offset can be eliminated by

(a) adding a derivative mode to the controller
(b) adding an integral mode to the controller
(c) decreasing the magnitude of the gain of the proportional controller
(d) adding a delay in the controller loop

[1999 : 2 M]

3.3 A second order feedback system is found to be oscillating with a high frequency. The oscillations

(a) can be reduced by increasing the proportional action
(b) can be reduced by increasing the integral action
(c) can be reduced by increasing the derivative action
(d) cannot be reduced [2001 : 1 M]

3.4 An integral control is used to

(a) improve the transient response
(b) reduce the offset
(c) eliminate the offset
(d) reduce the settling time [2002 : 1 M]

3.5 The transfer function of a PID controller is given by $G(s) = 4\left(1 + \frac{1}{2s} + 0.5s\right)$, as ω tends to infinity

(a) magnitude of $G(j\omega)$ tends to zero and phase angle of $G(j\omega)$ tends to infinity
(b) magnitude of $G(j\omega)$ tends to infinity and phase angle of $G(j\omega)$ tends to zero
(c) magnitude of $G(j\omega)$ tends to infinity and phase angle of $G(j\omega)$ tends to +90°
(d) magnitude of $G(j\omega)$ tends to zero and phase angle of $G(j\omega)$ tends to +90°

[2002 : 2 M]

3.6 A PID controller has the transfer function $2 + \frac{0.4}{s}$ with the unit of time expressed in minutes. The parameters proportional band and reset time for the above controller are respectively

(a) 200% and 0.4 min (b) 50% and 0.4 min
(c) 200% and 5 min (d) 50% and 5 min

[2003 : 2 M]

3.7 The proportional band of a PI controller is 50% and the reset time is 0.2 s. The transfer function of the controller is given by

(a) $50\left(1 + \frac{1}{0.2s}\right)$ (b) $50\left(1 + \frac{1}{5s}\right)$

(c) $2\left(1 + \frac{1}{0.2s}\right)$ (d) $2\left(1 + \frac{1}{s}\right)$

[2004 : 1 M]

3.8 The transfer function of a position servo system is given as $G(s) = \frac{1}{s(s+1)}$.

A first order compensator is designed in a unity feedback configuration so that the poles of the compensated system are placed at $-1 \pm j1$ and -4. The transfer function of the compensated system is

(a) $\frac{s+3}{2(s+3)}$ (b) $\frac{2s+3}{s+5}$

(c) $\frac{5(s+1.6)}{s+5}$ (d) $\frac{3(2s+3)}{s+4}$

[2006 : 2 M]

Common Data Questions (3.9 to 3.10):

The following figure describes the block diagram of a closed loop process control system. The unit of time is given in minute.

3.9 The digital implementation of the controller with a sampling time of 0.1 minute using velocity algorithm is

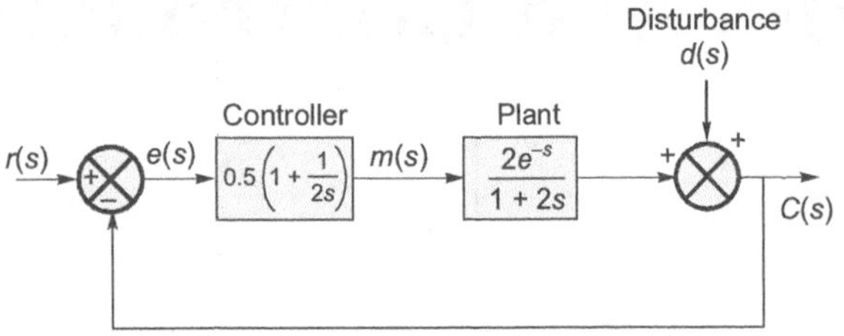

(a) $m(k) = 0.5\left[e(k) + 0.5\sum_{i=1}^{k} e(k)\right]$

(b) $m(k) = 2.0\left[e(k) + 2.0\sum_{i=1}^{k-1} e(k-1)\right]$

(c) $m(k) - m(k-1) = 0.5[e(k) - 0.85\, e(k-1)]$

(d) $m(k) - m(k-1) = 0.5[1.05\, e(k) - e(k-1)]$

[2006 : 2 M]

3.10 Suppose a disturbance signal $d(t) = \sin 0.2t$ units is applied. Then at steady state, the amplitude of the output $e(t)$ due to the effect of disturbance alone is

(a) 0.129 unit (b) 0.40 unit
(c) 0.528 unit (d) 2.102 unit

[2006 : 2 M]

3.11 The control action recommended for reducing the effect of disturbance at the output (provided that the disturbance signal is measurable) is

(a) cascade control
(b) P-D control
(c) ratio control
(d) feedback-feedforward control [2006 : 2 M]

3.12 A plant with a transfer function $\frac{2}{s(s+3)}$ is controlled by a PI controller with $K_p = 1$ and $K_i \geq 0$ in a unity feedback configuration. The lowest value of K_i that ensures zero steady state error for a step change in the reference input is

(a) 0 (b) 1/3
(c) 1/2 (d) 1 [2009 : 1 M]

Common Data for Questions (3.13 and 3.14):
Statement for Linked Answer Questions:

A disturbance input $d(t)$ is injected into the unity feedback control loop shown in the figure. Take the reference input $r(t)$ to be a unit step.

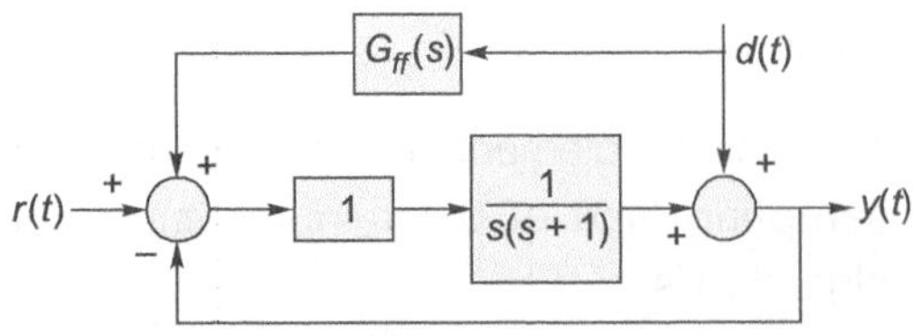

3.13 If the disturbance is measurable, its effect on the output can be minimized significantly using a feedforward controller $G_{ff}(s)$. To eliminate the component of the output due to $d(t) = \sin t \cdot G_{ff}(j\omega) \mid \omega = 1$ should be

(a) $\frac{1}{\sqrt{2}}\angle -\frac{3\pi}{4}$ (b) $\frac{1}{\sqrt{2}}\angle \frac{\pi}{4}$

(c) $\sqrt{2}\angle\pi$ (d) $\sqrt{2}\angle -\frac{\pi}{4}$

[2009 : 2 M]

3.14 Let $G_{ff}(s)$ be a PD controller. If $d(t) = \sin 2t$, the amplitude of the frequency component of $y(t)$ due to $d(t)$ is

(a) $\frac{5}{\sqrt{13}}$ (b) $\sqrt{\frac{9}{13}}$

(c) $\sqrt{\frac{17}{13}}$ (d) $\sqrt{\frac{20}{13}}$ [2009 : 2 M]

Statement for Linked Answer (3.15 and 3.16):

The transfer function of a compensator is given as

$$G_c(s) = \frac{s+a}{s+b}$$

3.15 $G_c(s)$ is a lead compensator if

(a) $a = 1, b = 2$ (b) $a = 3, b = 2$
(c) $a = -3, b = -1$ (d) $a = 3, b = 1$

[2012 : 2 M]

3.16 The phase of the above lead compensator is maximum at

(a) $\sqrt{2}$ rad/s (b) $\sqrt{3}$ rad/s
(c) $\sqrt{6}$ rad/s (d) $1/\sqrt{3}$ rad/s

[2012 : 2 M]

3.17 Consider the control system shown in figure with feed forward action for rejection of a measurable disturbance $d(t)$. The value of K, for which the disturbance response at the output $y(t)$ is zero mean, is

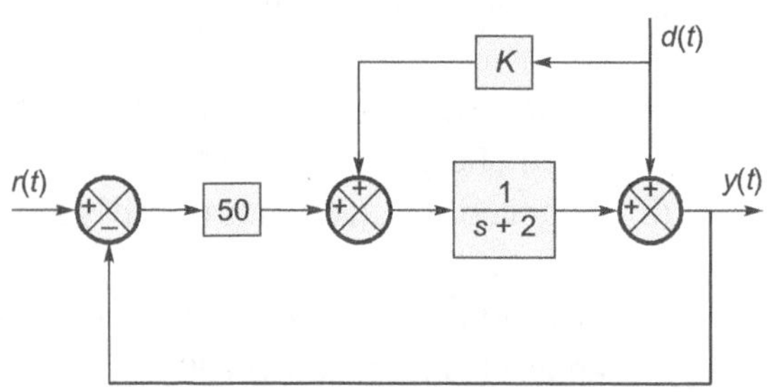

(a) 1 (b) –1
(c) 2 (d) –2

[2014 : 2 M]

3.18 A phase lead network has the transfer function $G(s) = \frac{1+0.2s}{1+0.05s}$. The angular frequency at which the maximum phase shift for the network occurs is ______.

(a) 20 rad/s
(b) 200 rad/s
(c) 100 rad/s
(d) 10 rad/s

[2020 : 1 M]

3.19 A differentiator has a transfer function whose

(a) magnitude decreases linearly with frequency
(b) magnitude remains constant
(c) phase increase linearly with frequency
(d) magnitude increases linearly with frequency

[2020 : 1 M]

3.20 Consider the unity feedback configuration with a plant and PID controller as shown in the figure

$$G(s) = \frac{1}{(s+1)(s+3)}$$

and $$C(s) = K\frac{(s+3-j)(s+3+j)}{s}$$

with K being scalar. The closed loop is

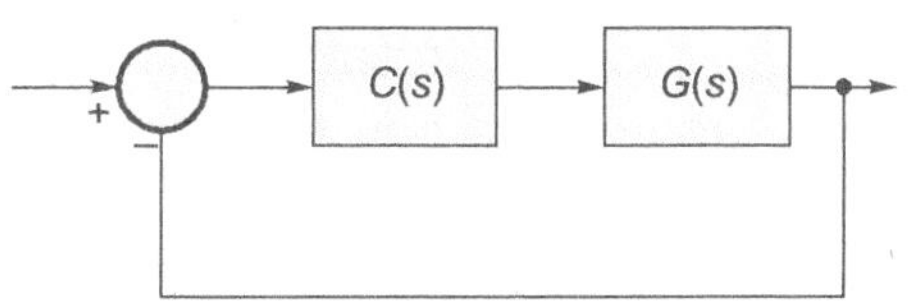

(a) stable for all values of K
(b) only stable for K between −1 and +1
(c) only stable for $K < 0$
(d) only stable for $K > 0$

[2021 : 2 M]

3.21 A proportional-integral-derivative (PID) controller is employed to stably control a plant with transfer function

$$P(s) = \frac{1}{(s+1)(s+2)}$$

Now, the proportional gain is increased by a factor of 2, the integral gain is increased by a factor of 3, and the derivative gain is left unchanged. Given that the closed-loop system continues to remain stable with the new gains, the steady-state error in tracking a ramp reference signal ________.

(a) Remains unchanged
(b) Decreases by a factor of 2
(c) Decreases by a factor of 3
(d) Decreases by a factor of 5

[2022 : 2 M]

3.22 In a unity-gain feedback control system, the plant

$$P(s) = \frac{0.001}{s(2s+1)(0.01s+1)}$$

is controlled by a lag compensator

$$C(s) = \frac{s+10}{s+0.1}$$

The slope (in dB/decade) of the *asymptotic* Bode magnitude plot of the loop gain at ω = 3 rad/s is ________ (in integer)

[2022 : 2 M]

■■■

Answers Compensators and Controllers

3.1 (b)	3.2 (b)	3.3 (c)	3.4 (c)	3.5 (c)	3.6 (d)	3.7 (c)
3.8 (c)	3.9 (a)	3.10 (b)	3.11 (b)	3.12 (a)	3.13 (d)	3.14 (a)
3.15 (a)	3.16 (a)	3.17 (d)	3.18 (d)	3.19 (d)	3.20 (d)	3.21 (c)
3.22 (−60)						

Explanations Compensators and Controllers

3.1 (b)

A proportional derivative (PD) controller has a gain term and a derivative term, as

$$y(t) = k_0 x(t) + k_2 \frac{d}{dt} x(t)$$

taking Laplace transform both sides, we get

$$Y(s) = k_0 X(s) + k_2 sX(s)$$

$$\therefore \quad \boxed{Y(s) = k_0 + k_2 s}$$

3.2 (b)

Adding an integral mode to a controller removes the offset present in the response of the process.

3.3 (c)

By adding derivative controller :

- Zero added to the system.
- Root locus shift to left.
- Stability increases.
- Oscillation decreases.

3.4 (c)

The integral control action reduces the steady state error to zero.

3.5 (c)

$$G(j\omega) = 4\left[1 + \frac{1}{2j\omega} + \frac{j\omega}{2}\right] = 4\left[\frac{2j\omega + 1 + (j\omega)^2}{2j\omega}\right]$$

$$= 4\left[\frac{1 + 2j\omega - \omega^2}{2j\omega}\right] = 4\left[\frac{1-\omega^2}{2j\omega} + \frac{2j\omega}{2j\omega}\right]$$

$$= 4\left[1 - j\frac{(1-\omega^2)}{2\omega}\right]$$

$$|G(j\omega)| = 4\sqrt{1 + \frac{(1-\omega^2)^2}{4\omega^2}} = 4\sqrt{1 + \frac{1}{4\omega^2} + \frac{\omega^4}{4\omega^2} - \frac{2\omega^2}{4\omega^2}}$$

as, $\omega \to 0$

$$|G(j\omega)| = 4\sqrt{1 + \left[\frac{1}{0} + 0 - \frac{1}{2}\right]} = 4\sqrt{1+\infty} = \infty$$

$$\angle G(j\omega) = \tan^{-1}\left(\frac{\omega^2 - 1}{2\omega}\right)$$

as, $\omega \to 0$

$$\angle G(j\omega) \to \tan^{-1}(-\infty) = 180° - \tan^{-1}\infty$$

$$= 180° - 90° = +90°$$

3.6 (d)

$$T(s) = 2 + \frac{0.4}{2} = 2\left(1 + \frac{0.2}{5}\right)$$

Gain = 2

$$\text{Proportional band} = \frac{100}{\text{Gain}} = \frac{100}{2} = 50\%$$

$$\text{Reset time} = \frac{1}{0.2} = 5 \text{ min.}$$

3.7 (c)

$$T(s) = K_p\left(1 + \frac{1}{T_r s}\right)$$

$$\text{Proportional band (PB)} = \frac{100}{\text{gain}}$$

or $\quad \text{gain} = \dfrac{100}{\text{PB}} = K_p$

and $\quad T_r = 0.2 \text{ sec}$

and $\quad \text{Integral (I)} = \dfrac{1}{\text{reset time}}$

So, the appropriate transfer function of the PI controller is $= 2\left(1 + \dfrac{1}{0.2s}\right)$.

3.8 (c)

Assume compensator T.F.

$$G_c(s) = \frac{k(s+z_1)}{(s+p_1)}$$

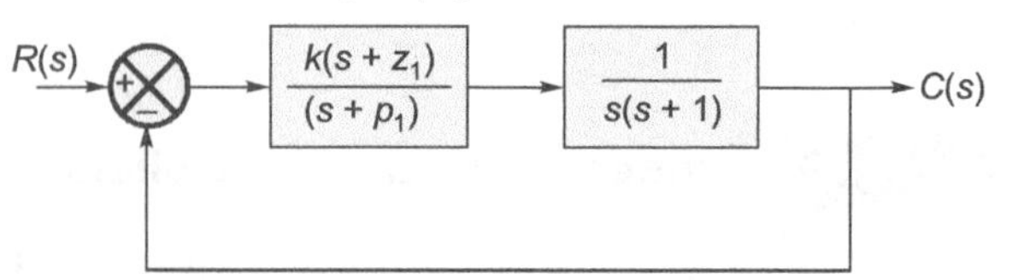

T.F. of closed loop,

$$\frac{C(s)}{R(s)} = \frac{k(s+z_1)}{s(s+p_1)(s+1) + k(s+z_1)}$$

$$= \frac{k(s+z_1)}{s^3 + p_1 s^2 + s^2 + p_1 s + ks + kz_1}$$

$$\frac{C(s)}{R(s)} = \frac{k(s+z_1)}{s^3 + (p_1+1)s^2 + (p_1+k)s + kz_1}$$

$\therefore$ characteristic equation,

$s^3 + (p_1 + 1)s^2 + (p_1 + k)s + kz_1 = 0$ ...(i)

and form given data,

$$[(s+1)^2 + 1](s+4) = 0$$

$$(s^2 + 2s + 2)(s + 4) = 0$$

$$s^3 + 2s^2 + 2s + 4s^2 + 8s + 8 = 0$$

$$\therefore \quad s^3 + 6s^2 + 10s + 8 = 0 \quad \text{...(ii)}$$

comparing equation (i) and (ii),

$$p_1 + 1 = 6 \Rightarrow p_1 = 5$$

$$p_1 + k = 10 \Rightarrow k = 10 - 5$$

$\Rightarrow k = 5$

$kz_1 = 8 \Rightarrow z_1 = 8/5 = z_1 = 1.6$

$$\therefore \quad G_c(s) = \frac{5(s+1.6)}{(s+5)}$$

3.9 (a)

$$m(s) = e(s)\left\{0.5\left(1+\frac{1}{2s}\right)\right\}$$

$$= 0.5\left\{e(s)+\frac{0.5}{s}e(s)\right\}$$

$$m(k) = 0.5\left[e(k) + 0.5\sum_{i=1}^{k} e(k)\right]$$

Because, Laplace transform of $\int i(t)\,dt = \frac{I(s)}{s}$. Here we use the same analogy in digital domain and integrator is substituted by summation.

3.10 (b)

For disturbance input only diagram will be reduced as

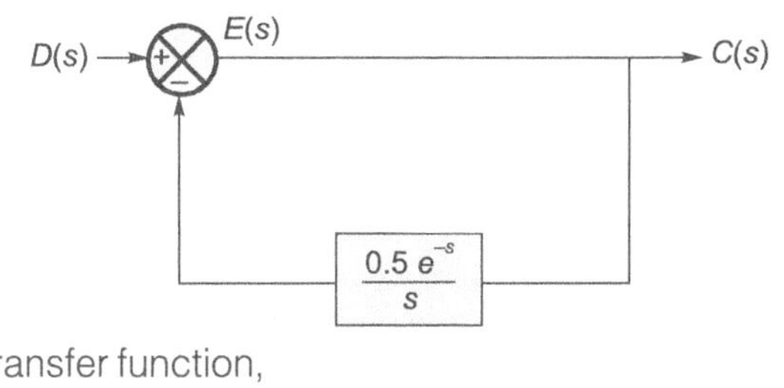

transfer function,

$$H(s) = \frac{C(s)}{D(s)} = \frac{E(s)}{D(s)} = \frac{1}{1+\frac{0.5e^{-s}}{s}}$$

$$\frac{E(s)}{D(s)} = \frac{s}{0.5+0.5s} = \frac{2s}{(s+1)}$$

$$E(s) = \frac{2s}{(s+1)}D(s)$$

$$d(t) = \sin 0.2t$$

$$\omega = 0.2$$

amplitude of $d(t)$ will multiplied by factor of $\left|\frac{2s}{s+1}\right|_{\omega=0.2}$

$$A = \left|\frac{2j\omega}{j\omega+1}\right| = \frac{2\times 0.2}{\sqrt{1+(0.2)^2}} = 0.3922$$

$$A = 0.40$$

So, amplitude of output $e(t)$

$= A \times$ initial amplitude

$= 0.40 \times 1 = 0.40$

3.11 (b)

P-D controller can be used.

3.12 (a)

Given, $K_P = 1$, transfer function for PI controller is

$$T(s) = K_P\left(1+\frac{1}{K_i s}\right) = \frac{K_i s + 1}{K_i s}$$

Open loop transfer function

$$G(s) = \frac{K_i s+1}{K_i s}\times\frac{2}{s(s+3)} = \frac{2(K_i s+1)}{K_i s^2(s+3)}$$

therefore characteristic equation is

$$1 + G(s) = 0$$

$$K_i s^3 + 3K_i s^2 + 2K_i s + 2 = 0$$

Routh array will be

s^3	K_i	$2K_i$
s^2	$3K_i$	2
s^1	$2K_i - 2/3$	
s^0	2	

for system to be stable $K_i > 0$

and $2K_i - 2/3 > 0$

$\Rightarrow \quad K_i > 0$ and $K_i > 1/3$

$\Rightarrow \quad K_i > 1/3$

Now $K_p \to$ position error constant $= \lim_{s\to 0} G(s) = \infty$

$$\therefore \quad e_{ss} = \frac{1}{1+K_p} = 0$$

So it is independent of K_i

therefore lowest value is $K_i = 0$

3.13 (d)

Output due to $D(s)$

$$= D(s) + D(s)\cdot G_{ff}(s)\cdot\frac{1}{s(s+1)}$$

$$= D(s)\left[1+\frac{G_{ff}(s)}{s(s+1)}\right]$$

$$0 = D(s)\left[1+\frac{G_{ff}(j\omega)}{j\omega(j\omega+1)}\right]$$

$$\Rightarrow \quad G_{ff}(j\omega)\big|_{\omega=1} = -j(1+j) = -j+1 = 1-j$$
$$= \sqrt{2}\angle -\pi/4$$

3.14 (a)

$$[D(s)\,G_{ff}(s) - Y(s)]\left[\frac{1}{s(s+1)}\right] + D(s) = Y(s)$$

$$D(s)\,G_{ff}(s) + (s^2+s)\,D(s) = (s^2+s+1)Y(s)$$

$$\frac{Y(s)}{D(s)} = H(s) = \frac{G_{ff}(s)+s^2+s}{(s^2+s+1)}$$

But, $G_{ff}(s) = (1+s)$

$$\therefore \quad H(s) = \frac{(s^2+2s+1)}{(s^2+s+1)}$$

$$|H(j\omega)|_{\omega=2} = \frac{5}{\sqrt{9+4}} = \frac{5}{\sqrt{13}}$$

3.15 (a)

$$G_c(s) = \frac{s+a}{s+b}$$

$$G_c(j\omega) = \frac{j\omega+a}{j\omega+b}$$

$$\angle G_c(j\omega) = \tan^{-1}\frac{\omega}{a} - \tan^{-1}\frac{\omega}{b}$$

$$= \tan^{-1}\left(\frac{\frac{\omega}{a}-\frac{\omega}{b}}{1+\frac{\omega^2}{ab}}\right)$$

For $G_c(s)$ to be a lead compensator

$$\angle G_c(j\omega) > 0$$

$$\frac{\omega}{a} > \frac{\omega}{b} \Rightarrow b > a$$

Option (a) satisfies the above equation.

3.16 (a)

For phase to be maximum

$$\frac{d\phi}{dt} = 0$$

$$\Rightarrow \left(1+\frac{\omega^2}{2}\right)\left(\frac{1}{a}-\frac{1}{b}\right) - \left(\frac{\omega}{a}-\frac{\omega}{b}\right)\left(\frac{2\omega}{ab}\right) = 0$$

$$\left(1+\frac{\omega^2}{2}\right)\left(\frac{1}{1}-\frac{1}{2}\right) = \left(\frac{\omega}{1}-\frac{\omega}{2}\right)\left(\frac{2\omega}{2}\right)$$

$$\frac{1}{2}+\frac{\omega^2}{4} = \frac{\omega^2}{2} \Rightarrow \frac{\omega^2}{4} = \frac{1}{2}$$

$$\Rightarrow \quad \omega = \sqrt{2}\text{ rad/sec.}$$

3.17 (d)

$$Y(s) = [-50\,Y(s) + kD(s)]\frac{1}{s+2} + D(s)$$

$$Y(s) = \left[1+\frac{50}{s+2}\right] = \left[\frac{K}{s+2}+1\right]D(s)$$

$$Y(s) = \frac{K+s+2}{s+2+50}D(s)$$

$$Y(j\omega) = \frac{K+j\omega+2}{j\omega+2+50}D(j\omega)$$

Zero mean,

at $\omega = 0$,

$$Y(j0) = 0$$

$$\frac{K+j0+2}{j0+52}D(j0) = 0$$

$$K = -2$$

3.18 (d)

$$G(s) = \frac{1+0.2s}{1+0.05s}$$

$$Z = 5$$

$$P = 20$$

$$\omega_n = \sqrt{Z.P} = 10 \text{ rad/s}$$

3.19 (d)

$$y(t) = \frac{dx(t)}{dt}$$

Taking Fourier Transform

$$Y(\omega) = j\omega\, X(\omega)$$

$$\Rightarrow \quad H(\omega) = \frac{Y(\omega)}{X(\omega)} = j\omega$$

$$\Rightarrow \quad |H(\omega)| = \omega$$

$$\angle H(\omega) = 90^\circ$$

Magnitude increases linearly and phase is constant.

3.20 (d)

$$q(s) = 1 + C(s)G(s) = 0$$

$$q(s) = s^3 + s^2(K+4) + s(6K+3) + 10K = 0$$

Necessary: $K > -4$; $K > -\frac{1}{2}$; $K > 0$

Sufficient : $bc > ad$

$$(K+4)(6K+3) > 10K$$

$$\Rightarrow \quad 6K^2 + 17K + 12 > 0$$

$$\Rightarrow \quad K > \frac{-4}{3} \text{ and } K > -\frac{3}{2} \text{ (or) } K < -\frac{4}{3} \text{ and } K < -3$$

Finally, $K > 0$ and $K > \frac{-4}{3}$

$\therefore$ The closed loop system is stable only for $K > 0$.

3.21 (c)

Transfer function of PID controller,

$$G_c(s) = K_p + \frac{K_I}{s} + sK_D$$

$$= \left(\frac{s^2K_D + sK_P + K_I}{s}\right)$$

Overall loop-transfer function,

$$L(s) = G_c(s) \cdot P(s)$$

$$L(s) = \frac{(s^2K_D + sK_P + K_I)}{s(s+1)(s+2)} \quad ...(1)$$

Steady state error due to ramp signal,

$$K_V = \lim_{s\to 0} sL(s)$$

$$K_V = \frac{K_I}{2} \Rightarrow e_{ss} = \frac{1}{K_V} = \frac{2}{K_I}$$

Now, $K'_P = 2K_P,\ K'_I = 3K_I$ and $K'_D = K_D$

Then $K'_V = \lim_{s\to 0} sL'(s)$

$$K'_V = \frac{3K_I}{2}$$

$$e'_{ss} = \frac{1}{K'_V} = \frac{2}{3K_I}$$

So, steady state error decreased by factor 3.

3.22 (–60)

Given, $P(s) = \frac{0.001}{s(2s+1)(0.01s+1)}$

and $C(s) = \frac{s+10}{(s+0.1)}$

$$\therefore \quad P(s)C(s) = \frac{0.001(s+10)}{s(2s+1)(0.01s+1)(s+0.1)}$$

Slope at ω = 3 rad/sec is

= –20 – 20 – 20 = –60 dB/dec

■■■

4 Time Response Analysis

4.1 For a unit step input, a system with a closed loop transfer function of

$G(s) = \frac{20}{s^2 + 2s + 5}$ has a steady-state output of

(a) 10 (b) 5

(c) 2 (d) 4 [1995 : 1 M]

4.2 A unity feedback system has an open loop gain of $G(s) = \frac{9}{s(s+3)}$, then the closed loop system will have a damping ratio of

(a) 0.33 (b) 0.5

(c) 1.0 (d) 3.0

[1996 : 1 M]

4.3 Match the roots in the s-plane with the impulse response shown in figure

Column I : Roots in the s-plane

(a) Two imaginary Roots

(b) Two complex roots in the left half plane

(c) A single root on the negative real axis

(d) A single root at the origin

Column II : Impulse Response

(i)

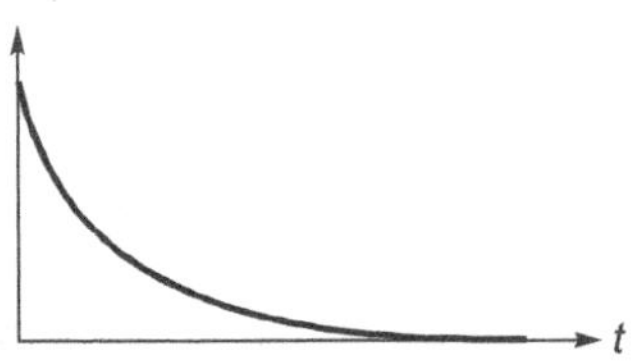

(ii)

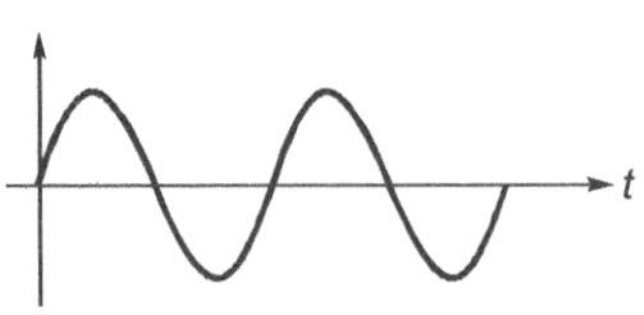

(iii)

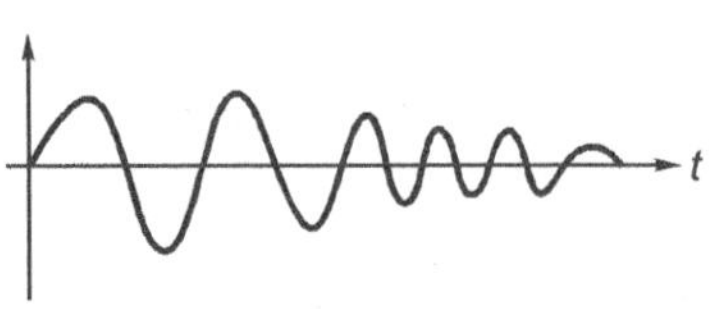

(iv)

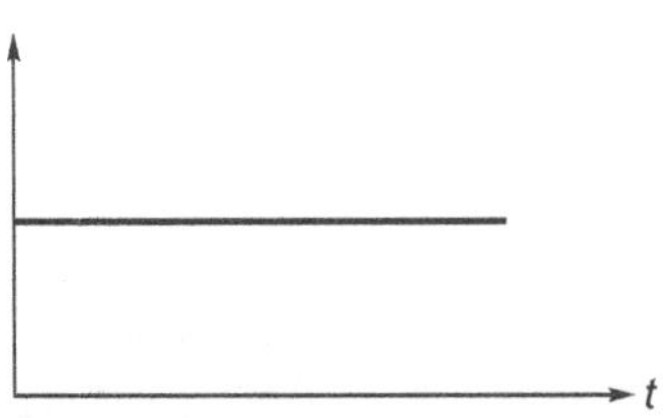

(v)

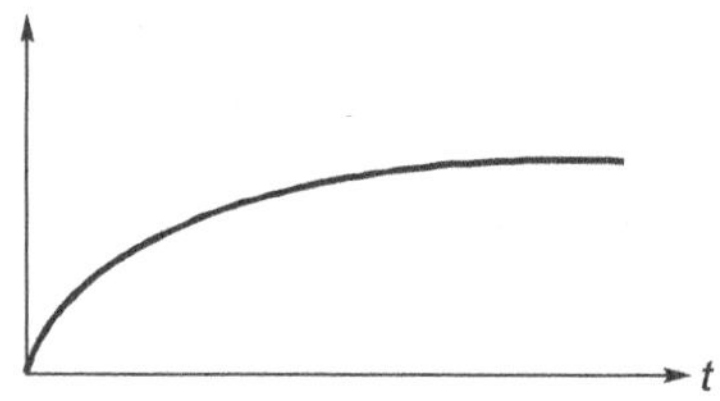

[1996 : 2 M]

4.4 The transfer function of a control system is given by:

$$\frac{C(s)}{R(s)} = \frac{25}{s^2 + 6s + 25}$$

The first maximum value of the response occurs at a time t_{max} given by

(a) $\frac{\pi}{8}$ (b) $\frac{\pi}{4}$

(c) $\frac{\pi}{2}$ (d) π

[1997 : 1 M]

4.5 A system with transfer function $\frac{1}{\tau s + 1}$, subjected to a step input takes 10 seconds to reach 50% of the step height. The value of τ is

(a) 6.9 s (b) 10 s

(c) 14.4 s (d) 20 s

[1999 : 1 M]

4.6 For a unity feedback system with open loop transfer function $G(s) = \frac{9}{s(s+2)}$, the damping ratio is

(a) 1/3 (b) 1/2

(c) 1 (d) 2

[2000 : 1 M]

4.7 The final value of a function $y(t)$ whose Laplace transform $Y(s) = \frac{4}{s^2 + 2s + 2}$ is

(a) 4 (b) 2

(c) 1 (d) 0 [2001 : 1 M]

4.8 The loop transfer function of a system is given by

$$G(s)\,H(s) = \frac{10e^{-0.1s}}{s(s+5)\,(s^2+4s+16)}$$

When the system is subjected to a parabolic input of $t^2/2$, the steady state error is

(a) zero (b) 0.1
(c) infinity (d) $e^{-0.1}/0.1$ [2002 : 2 M]

4.9 The transfer function of a system is $\frac{A}{s^2+\omega^2}$. The steady-state gain of the system to a unit-step input is

(a) $\frac{A}{\omega^2}$
(b) 0
(c) ∞
(d) Not possible to be determined [2003 : 2 M]

4.10 The response of a system to a unit impulse is $y(t) = e^{-5(t-1)}$. Which one of the following is the correct statement about the system?

(a) The system is nonlinear.
(b) The system is unstable in open loop.
(c) The steady-state gain of the system for a unit step input is 0.2.
(d) The steady-stage gain of the system for a unit step input is 1.0.

[2003 : 2 M]

4.11 For a first order instrument a 5% settling time is equal to

(a) three times the time constant
(b) two times the time constant
(c) the time constant
(d) time required for the output signal to reaches 5% of the final value

[2004 : 1 M]

4.12 The steady state error due to a unit-step disturbance input $D(s)$ in figure is

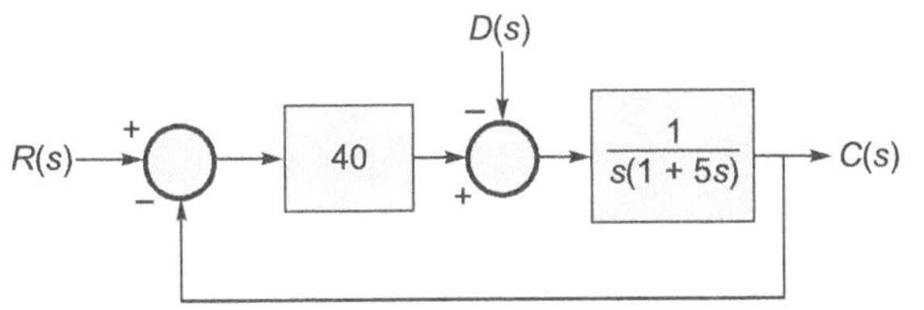

(a) 0 (b) 0.012
(c) 0.021 (d) 0.025

[2004 : 1 M]

4.13 A certain system exhibited an overshoot of 16% when subjected to an input of $2u(t)$, where $u(t)$ is a step input. The damping ratio and decay ratio respectively are

(a) (0.8, 0.0810) (b) (0.5, 0.0256)
(c) (1.0, 0.1626) (d) (1.1, 0.0089)

[2004 : 2 M]

4.14 Consider a unity feedback control system whose forward path transfer function $G(s) = K/s^2$. The steady state error for a step input is

(a) 1.0
(b) ∞
(c) 0.0
(d) does not exist

[2005 : 2 M]

Statement for Linked Answer Questions (4.15 & 4.16):

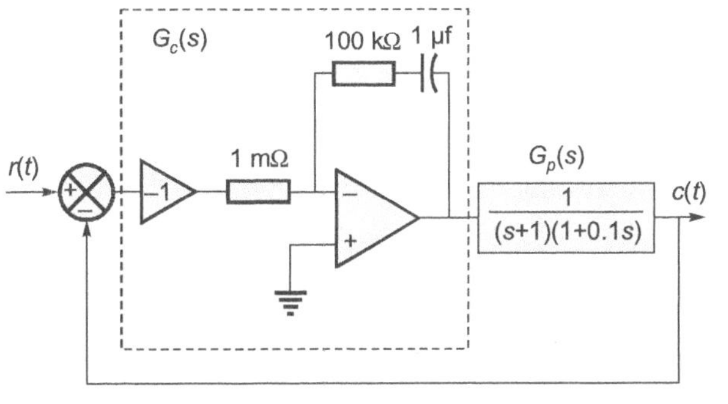

Given figure shows a closed loop unity feedback system. The controller block has a transfer function denoted by $G_c(s)$. The controller is cascaded to the plant, which is denoted by $G_p(s)$.

4.15 The loop transfer function $G_c(s)$ is

(a) $\frac{1+0.1s}{s}$ (b) $-\frac{1+0.1s}{s}$
(c) $-\frac{s}{s+1}$ (d) $\frac{s}{s+1}$

[2005 : 2 M]

4.16 The peak percentage overshoot of the closed loop system is

(a) 5.0% (b) 10.0%
(c) 16.3% (d) 1.63%

[2005 : 2 M]

4.17 If a first order system and its time response to a unit step input are as shown below, the gain K is

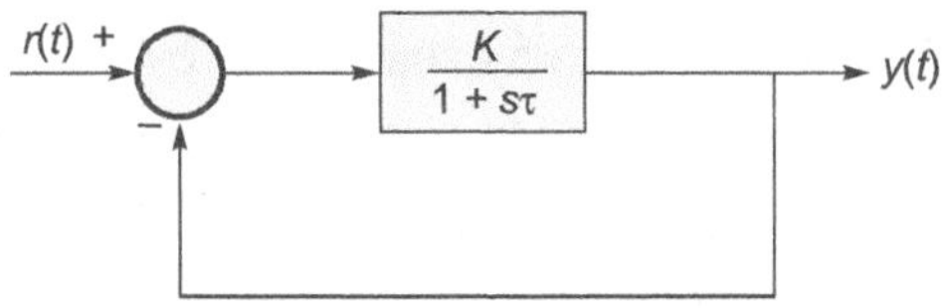

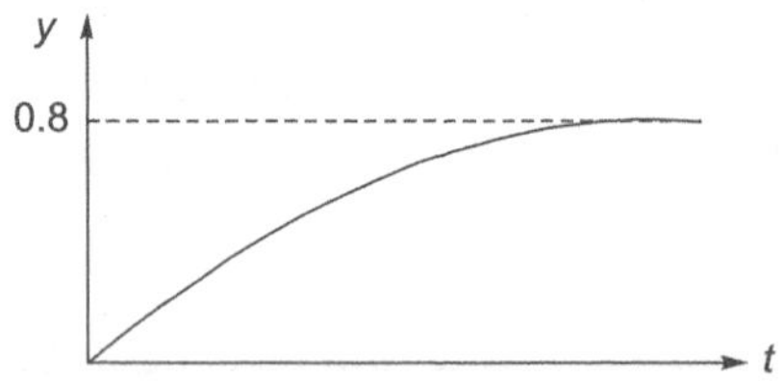

(a) 0.25 (b) 0.8
(c) 1 (d) 4 [2008 : 2 M]

Statement for Linked Answer Questions (4.18 & 4.19):
A unity feedback system has open loop transfer function $G(s) = \frac{100}{s(s+p)}$. The time at which the response to a unit step input reaches its peak is $\pi/8$ seconds.

4.18 The damping coefficient for the closed looped system is

(a) 0.4 (b) 0.6
(c) 0.8 (d) 1

[2008 : 2 M]

4.19 The value of p is

(a) 6 (b) 12
(c) 14 (d) 16 [2008 : 2 M]

4.20 A unity feedback system has the transfer function

$$\frac{K(s+b)}{s^2(s+20)}$$

The value of b for which the loci of all the three roots of the closed loop characteristic equation meet at a single point is

(a) $\frac{10}{9}$ (b) $\frac{20}{9}$
(c) $\frac{30}{9}$ (d) $\frac{40}{9}$ [2009 : 2 M]

4.21 A unit ramp input is applied to the system shown in the adjoining figure. The steady state error in its output is

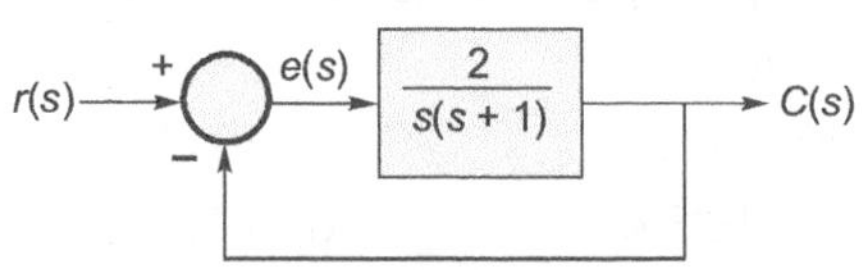

(a) 0 (b) 0.5
(c) 1 (d) 2

[2010 : 2 M]

4.22 A unity feedback system has an open loop transfer function $G(s) = \frac{k}{s(s+3)}$. The value of k that yields damping ratio of 0.5 for the closed loop system is

(a) 1 (b) 3
(c) 5 (d) 9 [2010 : 2 M]

4.23 The unit-step response of a negative unity feedback system with the open-loop transfer function $G(s) = \frac{6}{s+5}$ is

(a) $1 - e^{-5t}$ (b) $6 - 6e^{-5t}$
(c) $\frac{6}{5} - \frac{6}{5}e^{-5t}$ (d) $\frac{6}{11} - \frac{6}{11}e^{-11t}$

[2011 : 2 M]

4.24 The open-loop transfer function of a dc motor is given as $\frac{\omega(s)}{V_a(s)} = \frac{10}{1+10s}$. When connected in feedback as shown below, the approximate value of K_a that will reduce the time constant of the closed loop system by one hundred times as compared to that of the open-loop system is

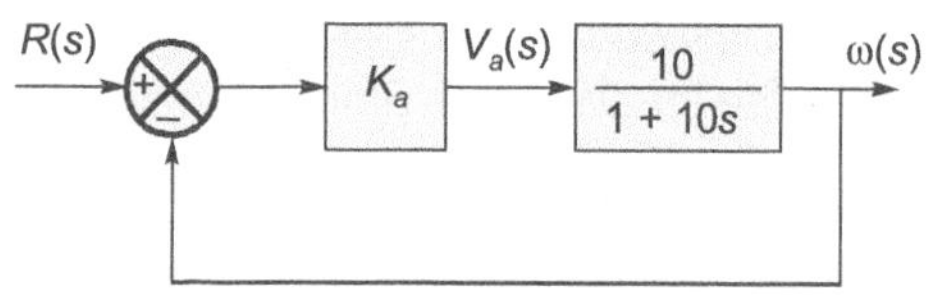

(a) 1 (b) 5
(c) 10 (d) 100 [2013 : 2 M]

4.25 A plant has an open-loop transfer function,

$$G_p(s) = \frac{20}{(s+0.1)(s+2)(s+100)}$$

The approximate model obtained by retaining only one of the above poles, which is closest to the frequency response of the original transfer function at low frequency is

(a) $\frac{0.1}{s+0.1}$ (b) $\frac{2}{s+2}$
(c) $\frac{100}{s+100}$ (d) $\frac{20}{s+0.1}$

[2014 : 1 M]

4.26 A system with transfer function $G(s) = \frac{1}{s^2+1}$ has zero initial conditions. The percentage overshoot in its step response is _____ %.

[2015 : 1 M]

4.27 Match the unit-step responses (1), (2) and (3) with the transfer functions $P(s)$, $Q(s)$ and $R(s)$, given below.

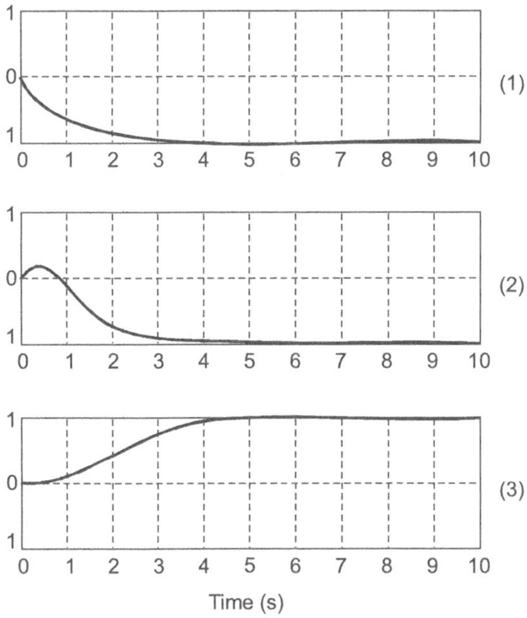

$$P(s) = \frac{-1}{(s+1)};\ Q(s) = \frac{2(s-1)}{(s+10)(s+2)};\ R(s) = \frac{1}{(s+1)^2}$$

(a) $P(s)-(3)$, $Q(s)-(2)$, $R(s)-(1)$
(b) $P(s)-(1)$, $Q(s)-(2)$, $R(s)-(3)$
(c) $P(s)-(2)$, $Q(s)-(1)$, $R(s)-(3)$
(d) $P(s)-(1)$, $Q(s)-(3)$, $R(s)-(2)$

[2016 : 2 M]

4.28 The block diagram of a closed-loop control system is shown in the figure. The values of k and k_p are such that the system has a damping ratio of 0.8 and an undamped natural frequency ω_n of 4 rad/s respectively. The value of k_p will be ______.

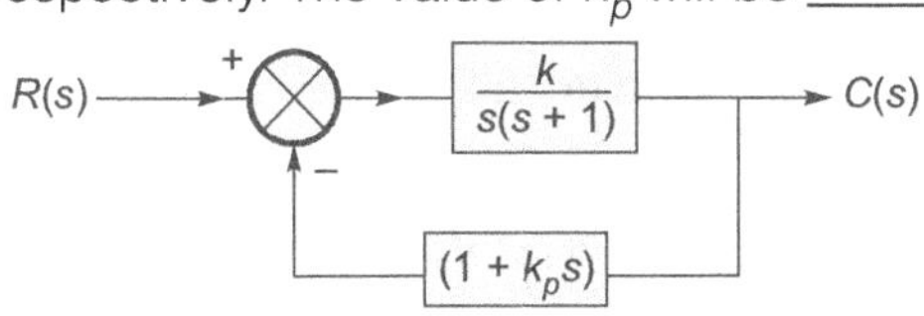

[2017 : 2 M]

4.29 Unit step response of a linear time invariant (LTI) system is given by $y(t) = (1 - e^{-2t})\,u(t)$. Assuming zero initial condition, the transfer function of the system is

(a) $\frac{1}{s+1}$ (b) $\frac{2}{(s+1)(s+2)}$

(c) $\frac{1}{s+2}$ (d) $\frac{2}{s+2}$

[2018 : 2 M]

4.30 In the control system shown in the figure below, a reference signal $r(t) = t^2$ is applied at time $t = 0$. The control system employs a PID controller $C(s) = K_p + \frac{K_I}{s} + K_D s$ and the plant has a transfer function $P(s) = \frac{3}{s}$. If $K_p = 10$, $K_I = 1$ and $K_D = 2$, the steady state value of e is

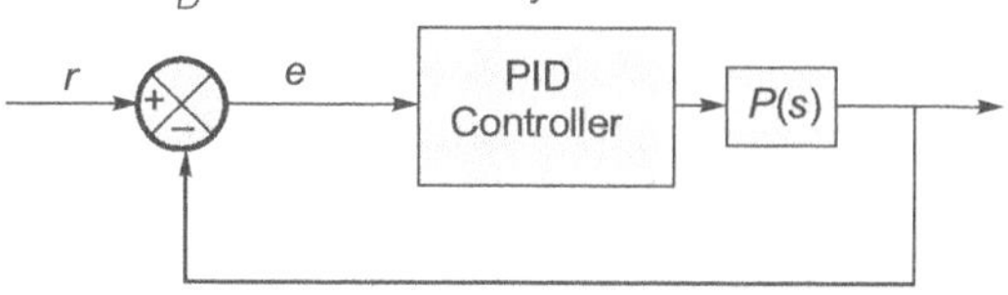

(a) 0 (b) 2/3
(c) 1 (d) ∞

[2019 : 2 M]

4.31 A second order system has closed loop poles located as $s = -3 \pm j4$. The time t at which the maximum value of the step response occurs (in seconds, rounded off to two decimal places) is ______.

[2020 : 1 M]

4.32 The system shown in figure (a) has a time response $y(t)$ to an input $r(t) = 10\,u(t)$ as shown in figure (b), $u(t)$ being the unit step input. Both K, τ are positive. The gain K of the system is ______.

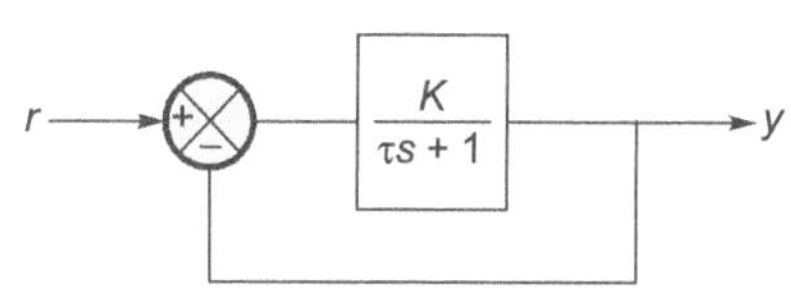

Fig. (a)

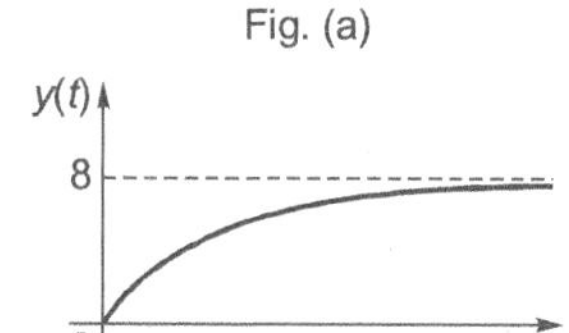

Fig. (b)

[2020 : 2 M]

4.33 Consider the system with transfer function $G(s) = \frac{2}{s+1}$. A unit step function $u(t)$ is applied to the system, which results in an output $y(t)$. If $e(t) = y(t) - u(t)$, then $\lim_{t\to\infty} e(t)$ is ______.

[2021 : 2 M]

4.34 The step response of a circuit is seen to have an oscillatory behaviour at the output with oscillations dying down after some time. The correct inference(s) regarding the transfer function from input to output is/are

(a) that it is of at least second order.
(b) that it has at least one pole-pair that is underdamped.
(c) that it does not have a real pole.
(d) that it is first order system. [2021 : 1 M]

4.35 The input $x(t)$ to a system is related to its output $y(t)$ as

$$\frac{dy(t)}{dt} + y(t) = 3x(t-3)u(t-3)$$

Here $u(t)$ represents a unit-step function.
The transfer function of this system is ________.

(a) $\frac{e^{-3s}}{s+3}$ (b) $\frac{3e^{-3s}}{s+1}$

(c) $\frac{3e^{-(s/3)}}{s+1}$ (d) $\frac{e^{-(s/3)}}{s+3}$ [2022 : 1 M]

4.36 A system has the transfer-function $\frac{Y(s)}{X(s)} = \frac{s-\pi}{s+\pi}$.

Let $u(t)$ be the unit-step function. The input $x(t)$ that results in a steady-state output $y(t) = \sin\pi t$ is ______.

(a) $x(t) = \sin(\pi t)u(t)$

(b) $x(t) = \sin\left(\pi t + \frac{\pi}{2}\right)u(t)$

(c) $x(t) = \sin\left(\pi t - \frac{\pi}{2}\right)u(t)$

(d) $x(t) = \cos\left(\pi t + \frac{\pi}{2}\right)u(t)$

[2023 : 1 M]

4.37 Consider the feedback control system shown in the figure. The steady-state error $e_{ss} = \lim_{t\to\infty}(r(t) - y(t))$ due to unit step reference r(t) is

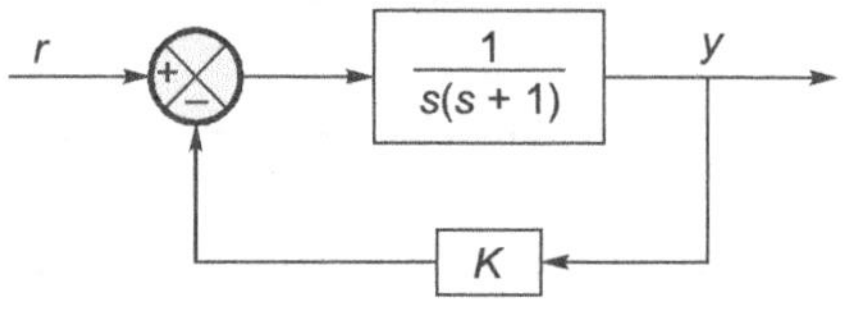

(a) $\frac{K-1}{K}$ (b) $\frac{1}{2}$

(c) 0 (d) $\frac{1-K}{K}$ [2024 : 2 M]

4.38 The transfer function of a system is

$$G(s) = \frac{\omega_n^2}{s^2 + 2\xi\omega_n s + \omega_n^2}$$

Choose the range of ξ and ω_n (in rad/s) from the following options such that the poles lie on the shaded region of the s-plane as shown in the figure.

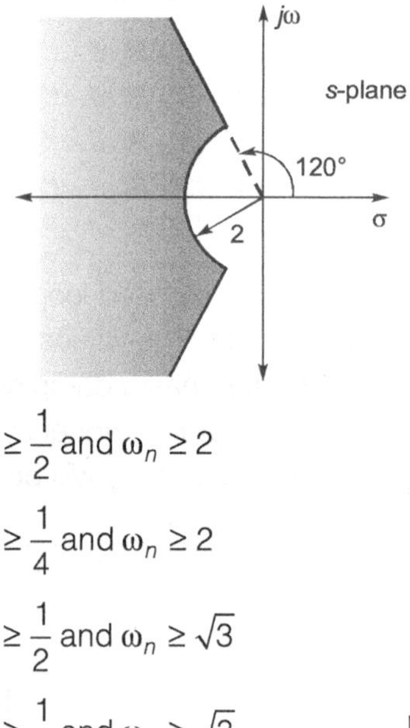

(a) $\xi \geq \frac{1}{2}$ and $\omega_n \geq 2$

(b) $\xi \geq \frac{1}{4}$ and $\omega_n \geq 2$

(c) $\xi \geq \frac{1}{2}$ and $\omega_n \geq \sqrt{3}$

(d) $\xi \geq \frac{1}{4}$ and $\omega_n \geq \sqrt{3}$ [2024 : 2 M]

■■■

Answers Time Response Analysis

4.1 (d)	4.2 (b)	4.3 (Sol.)	4.4 (b)	4.5 (c)	4.6 (a)	4.7 (d)
4.8 (c)	4.9 (a)	4.10 (a)	4.11 (a)	4.12 (a)	4.13 (b)	4.14 (c)
4.15 (a)	4.16 (c)	4.17 (d)	4.18 (b)	4.19 (b)	4.20 (b)	4.21 (b)
4.22 (d)	4.23 (d)	4.24 (c)	4.25 (a)	4.26 (100)	4.27 (b)	4.28 (0.337)
4.29 (d)	4.30 (b)	4.31 (0.78)	4.32 (4)	4.33 (1)	4.34 (a, b)	4.35 (b)
4.36 (c)	4.37 (a)	4.38 (a)				

Explanations Time Response Analysis

4.1 (d)

Output at steady state

$$= \lim_{s\to 0} s \cdot \frac{1}{s} \cdot \frac{20}{s^2+2s+5} = \frac{20}{5} = 4.0$$

4.2 (b)

Closed loop T.F.

$$\text{T.F.} = \frac{9}{s(s+3)+9} = \frac{9}{s^2+3s+9}$$

$\therefore \quad \omega_n = 3$

$2\zeta \times 3 = 3$

$\zeta = 0.5$

4.3 Sol.

(a)-(ii), (b)-(iii), (c)-(i), (d)-(iv)

4.4 (b)

t_{max} is the peak time (t_p)

$\omega_n = 5, \zeta = 0.6$

$$t_p = \frac{n\pi}{\omega_n(1-\zeta^2)}$$

first maxima is at $n = 1$

$$\therefore \quad t_{max} = \frac{\pi}{5\sqrt{1-(0.6)^2}} = \frac{\pi}{5\times 0.8} = \frac{\pi}{4}$$

4.5 (c)

The response of a first order system for unit step input is,

$$\theta(t) = \theta_0[1 - e^{-t/\tau}]$$

$$\frac{\theta_0}{2} = \theta_0[1 - e^{-10/\tau}]$$

$$\frac{1}{2} = 1 - e^{-10/t}$$

$$e^{-10/\tau} = \frac{1}{2}$$

$$\frac{-10}{\tau} = 0.693$$

$\therefore \quad \tau = 14.4$ sec

4.6 (a)

T.F. of closed loop system, is

$$\Rightarrow \frac{9}{s(s+2)+9} \Rightarrow \frac{9}{s^2+2s+9}$$

$\therefore \quad \omega_n = 3$

$2\zeta \times 3 = 2$

$\therefore \quad \zeta = \frac{1}{3}$

4.7 (d)

$$\text{Final value} = \lim_{s\to 0} sY(s) = \lim_{s\to 0} s \cdot \frac{4}{s^4+4s+2}$$

$$= \lim_{s\to 0} \frac{4}{\left(s+2+\frac{2}{s}\right)} = \frac{4}{0+2+\infty} = \frac{4}{\infty} = 0$$

4.8 (c)

$$e_{ss} = \lim_{s\to 0} \frac{s\cdot\frac{1}{s^3}}{1+s(s+5)(s^2+4s+16)}$$

$$= \lim_{s\to 0} \frac{1}{s^2[1+s(s+5)(s^2+4s+16)]}$$

$$= \frac{1}{0} = \infty \text{ (infinity)}$$

4.9 (a)

$$T(s) = \frac{A}{s^2+\omega^2};$$

Steady-state gain and steady state output is different

So, steady state gain in s-domain is

$$T(s) = \lim_{s \to 0} \frac{A}{0+\omega^2} = \frac{A}{\omega^2}$$

4.10 (a)

For a linear system, output and input are related as:

$y(t) = A\,x(t) + B\,x(t-1) + Cx(t-2) + ...$

So system is non-linear.

4.11 (a)

For 5%, $t_s = 3$ RC, and for 2%, $t_s = 4$ RC.

4.12 (a)

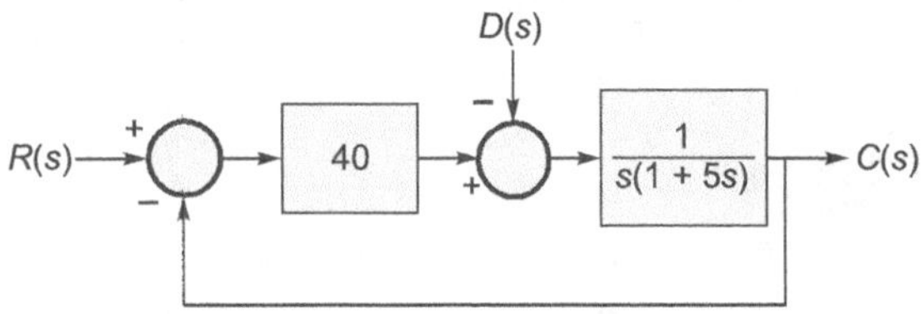

Now we have to find e_{ss} due to $D(s)$ input so, $R(s) = 0$. The reduced block diagram will be

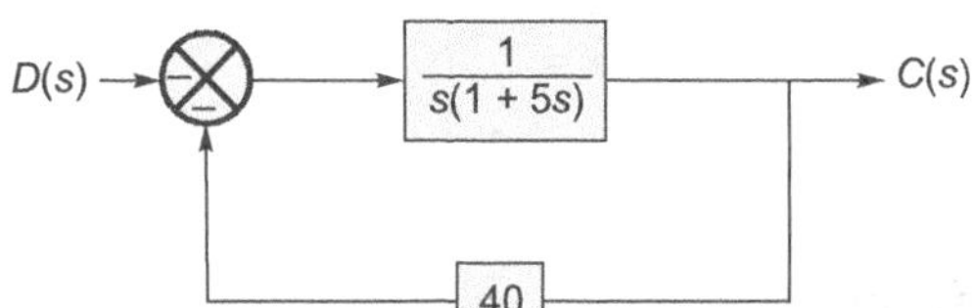

As we know the standard from of e_{ss}

$$e_{ss} = \lim_{s \to 0} \frac{s \cdot R(s)}{1+G(s)H(s)}$$

Here input is $= -D(s)$

$$e_{ss} = \lim_{s \to 0} \frac{s \cdot \left(-\frac{1}{s}\right)}{1+\frac{40}{s(1+5s)}}$$

$$= \lim_{s \to 0} \frac{-1}{\frac{40}{s(1+5s)}+1} = \frac{-1}{\frac{40}{0}+1} = \frac{-1}{\infty} = 0$$

4.13 (b)

$$\% \text{ of maximum overshoot} = 100e^{-\frac{\pi\zeta}{\sqrt{1-\zeta^2}}}$$

$$16 = 100e^{-\frac{\pi\zeta}{\sqrt{1-\zeta^2}}}$$

$$\therefore \quad \frac{-\pi\zeta}{\sqrt{1-\zeta^2}} = \ln(0.16)$$

$$\zeta^2 = \frac{1}{3.939}$$

$$\zeta = 0.5$$

$\therefore$ Damping ratio $(\zeta) = 0.5$

4.14 (c)

$$E(s) = \frac{R(s)}{1+G(s)H(s)}$$

$$e_{ss} = \underset{t \to \infty}{\text{Lt}}\, e(t) = \underset{s \to 0}{\text{Lt}} \frac{sR(s)}{1+G(s)}$$

$$= \underset{s \to 0}{\text{Lt}} \frac{s \cdot \frac{1}{s}}{1+\frac{k}{s^2}} = \underset{s \to 0}{\text{Lt}} \frac{1}{1+\frac{k}{s^2}} = \frac{1}{\infty} = 0$$

So, $\quad e_{ss} = \underset{t \to \infty}{\text{Lt}}\, e(t) = 0.$

4.15 (a)

The controller block is

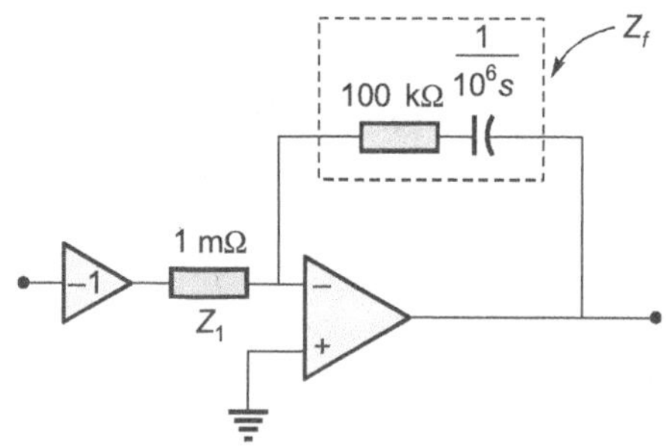

$$Z_f = 10^5 + \frac{1}{10^{-6}s}$$

$$G_c(s) = \frac{V_o}{V_{in}} = -\frac{Z_f}{Z_1}$$

$$= \frac{-\left(10^5 + \frac{1}{10^{-6}s}\right)}{-10^6}$$

$$= \frac{1}{10} + \frac{1}{s} = \frac{s+10}{10s} = \frac{0.1s+1}{s}$$

$$\therefore \quad \boxed{G_c(s) = \frac{0.1s+1}{s}}$$

4.16 (c)

The simplified block diagram is:

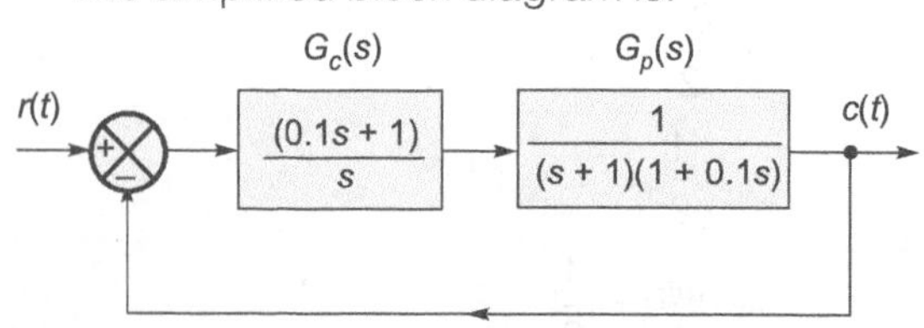

$$\frac{C(s)}{R(s)} = \frac{\frac{1}{s(s+1)}}{1+\frac{1}{s(s+1)}} = \frac{1}{s^2+s+1} \quad \ldots(i)$$

Standard second order equation is

$$\frac{C(s)}{R(s)} = \frac{\omega_n^2}{s^2+2\xi\omega_n s+\omega_n^2} \quad \ldots(ii)$$

Compare equation (i) with equation (ii), we get

$$\omega_n = 1 \text{ rad/sec}$$
$$2\xi\omega_n = 1$$
$$\therefore \quad \xi = 0.5$$

Percentage of peak overshoot

$$M_p = 100\times e^{\frac{-\pi\xi}{\sqrt{1-\xi^2}}}$$
$$= 100\times e^{\frac{-\pi 0.5}{\sqrt{1-(0.5)^2}}}$$
$$= 100 \times 0.163 = 16.3\%$$

4.17 (d)

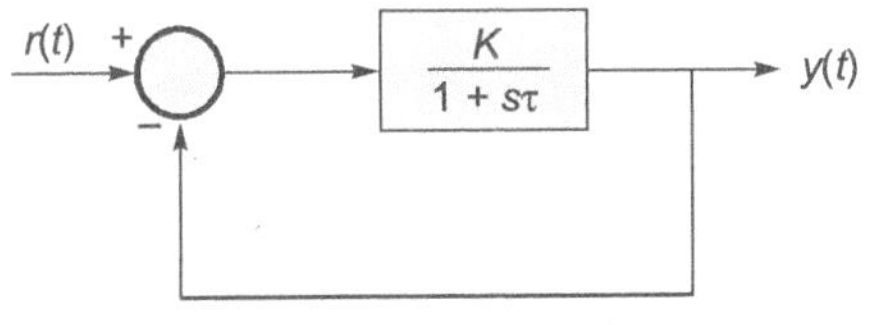

$$\frac{Y(s)}{R(s)} = \frac{k}{1+s\tau+k} = \frac{\frac{k}{\tau}}{s+\frac{(1+k)}{\tau}}$$

By the use of partial fraction:

$$\therefore \quad Y(s) = \frac{\frac{k}{\tau}}{s+\left(\frac{k+1}{\tau}\right)}\cdot\frac{1}{s}\frac{\frac{k+1}{\tau}}{\frac{k+1}{\tau}}$$
$$= \left[\frac{1}{s}-\frac{1}{s+\left(\frac{k+1}{\tau}\right)}\right]\cdot\frac{k}{k+1}$$
$$y(t) = \frac{k}{k+1}\left[1-e^{-\left(\frac{k+1}{\tau}\right)t}\right]$$

At steady state

$$y(\infty) = 0.8 = \frac{k}{k+1}$$
$$k = 0.8\,k + 0.8$$
$$0.2k = 0.8$$
$$k = \frac{0.8}{0.2} = 4$$

4.18 (b)

$$G(s) = \frac{100}{s(s+p)}$$

Closed loop transfer function is

$$\frac{C(s)}{R(s)} = \frac{100}{s^2+sp+100}$$

Compare with the 2nd order standard transfer function.

$$\frac{C(s)}{R(s)} = \frac{\omega_n^2}{s^2+2\xi\omega_n S+\omega_n^2}$$

Here $\quad \omega_n^2 = 100$

$\Rightarrow \quad \omega_n = 10$ rad/sec

We know that,

$$t_p(\text{peak time}) = \frac{\pi}{\omega_n\sqrt{1-\xi^2}} = \frac{\pi}{8}$$
$$\frac{\pi}{10\sqrt{1-\xi^2}} = \frac{\pi}{8}$$

After solving we get $\xi = 0.6$

4.19 (b)

$$2\xi\omega_n = P$$
$$\therefore \quad P = 2\times 10\times 0.6 = 12$$

4.20 (b)

$$G(s) = \frac{k(s+b)}{s^2(s+20)}$$

characteristic equation

$1 + G(s)\,H(s) = 0$

$$1+\frac{k(s+b)}{s^2(s+20)}\times 1 = 0$$

$\therefore\ s^2(s+20) + k(s+b) = 0$

$$\therefore \quad k = -\frac{(s^3+20s^2)}{(s+b)}$$
$$\frac{dk}{ds} = 0$$
$$\left[\frac{(s+b)(3s^2+40s)-(s^3+20s^2)(1)}{(s+b)^2}\right] = 0$$

$3s^3 + 3bs^2 + 40s^2 + 40bs - s^3 - 20s^2 = 0$

$\therefore\ 2s^3 + (3b+20)s^2 + 40bs = 0$

$\therefore\ s[2s^2 + (3b+20)s + 40b] = 0$

$\therefore\ s = 0$ and $2s^2 + (3b+20)s + 40b = 0$

For a single point,

$$B^2 = 4\,AC$$

$$(3b+20)^2 = 4\times 2\times 400$$
$$9b^2 + 120\,b + 400 = 3200$$
$$\therefore \quad 9b^2 - 200\,b + 400 = 0$$

$\therefore$ $b = 20$ and $b = 20/9$

If $b = 20$, pole and zero will be cancle out. It is not possible.

$$\therefore \quad b = 20/9$$

4.21 (b)

$$\frac{C(s)}{R(s)} = \frac{2}{s^2+s+2}$$

Given as input is unit ramp input

Hence, $R(s) = \frac{1}{s^2}$

$$e_{ss} = \lim_{s\to 0} \frac{s}{s^2}\cdot\frac{s^2+s}{s^2+s+2} = \frac{1}{2} = 0.5$$

4.22 (d)

$$\frac{C(s)}{R(s)} = \frac{k}{s^2+3s+k}$$

Compare with 2nd order standard equation.

$$\frac{C(s)}{R(s)} = \frac{\omega_n^2}{s^2+2\xi\omega_n s+\omega_n^2}$$

Here
$$\omega_n = \sqrt{k}$$
$$2\xi\omega_n = 3$$
$$2\times 0.5\times\sqrt{k} = 3 \qquad (\because \xi = 0.5)$$
$$\sqrt{k} = 3$$
$$k = 9$$

4.23 (d)

$$\frac{C(s)}{R(s)} = \frac{\frac{6}{s+5}}{1+\frac{6}{s+5}} = \frac{6}{s+11}$$

$$C(s) = \frac{1}{s}\cdot\frac{6}{s+11} \qquad \left(\because R(s) = \frac{1}{s}\right)$$

By partial fraction,

$$C(s) = \frac{6}{11}\cdot\frac{1}{s} - \frac{6}{11}\frac{1}{s+11}$$

$$\boxed{c(t) = \frac{6}{11} - \frac{6}{11}e^{-11t}}$$

4.24 (c)

$$\frac{\omega(s)}{V_a(s)} = \frac{10}{1+10s}$$
$$1 + Ts = 1 + 10s$$
$$Ts = 10s$$
$$T = 10$$
$$T' = \frac{T}{100} = \frac{10}{100} = \frac{1}{10}$$

$$\frac{\omega(s)}{R(s)} = \frac{K_a\times\frac{10}{1+10s}}{1+K_a\frac{10}{1+10s}}$$

$$\frac{\omega(s)}{R(s)} = \frac{10K_a}{1+10s+10K_a} = \frac{10K_a}{(1+10K_a)+10s}$$

$$\frac{\omega(s)}{R(s)} = \frac{10K_a}{(1+10K_a)\left[1+\frac{10s}{1+10K_a}\right]}$$

$$T' = \frac{10}{1+10K_a} = \frac{1}{10}$$
$$1 + 10\,K_a = 100$$
$$10\,K_a = 99$$
$$K_a = 9.9 \cong 10$$

4.25 (a)

At low frequency, $\omega \to 0$

$\therefore \quad s \to 0$

Retaining pole $(s+0.1)$ as it is nearest to the frequency response

$$= \frac{20}{(s+0.1)\times 2\times 100} = \frac{0.1}{(s+0.1)}$$

4.26 (100)

$$G(s) = \frac{1}{s^2+(0\times s)+1}$$

So, here clearly,

$$\omega = 1,\ \zeta = 0$$

$$\%M_p = e^{-\frac{\zeta\pi}{\sqrt{1-\xi^2}}\times 100}$$
$$= e^{-0}\times 100\% \qquad [\because \zeta = 0]$$
$$= 1\times 100\% = 100\%$$

4.27 (b)

$$P(s) = \frac{-1}{s+1}$$

Response, $C(s) = P(s) \cdot u(s) = \frac{-1}{s+1} \cdot \frac{1}{s}$

$$\therefore \quad c(\infty) = \underset{s \to 0}{Lt}\, s \cdot C(s)$$

$$= \underset{s \to 0}{Lt}\, s \cdot \left[\frac{-1}{s+1} \cdot \frac{1}{s}\right] = -1$$

$\therefore P(s) \Rightarrow 1$ Match

$$Q(s) = \frac{2(s-1)}{(s+10)(s+2)}$$

$$C(s) = Q(s) \cdot u(s)$$

$$= \frac{2(s-1)}{(s+10)(s+2)} \cdot \frac{1}{s}$$

$$c(\infty) = \underset{s \to 0}{Lt}\, sC(s)$$

$$= \underset{s \to 0}{Lt}\, S\left[\frac{2(s-1)}{(s+10)(s+2)} \cdot \frac{1}{s}\right] = \frac{-1}{2}$$

$Q(s) \Rightarrow 2$ Match

$$R(s) = \frac{1}{(s+1)^2}; \quad C(s) = \frac{1}{(s+1)^2} \cdot \frac{1}{s}$$

$$c(\infty) = \underset{s \to 0}{Lt}\, sC(s) = \underset{s \to 0}{Lt} \frac{1}{(s+1)^2} \cdot \frac{1}{s} = 1$$

$R(s) \Rightarrow 3$ Match

$P(s) \rightarrow 1,\ Q(s) \rightarrow 2,\ R(s) \rightarrow 3$

4.28 0.337 (0.32 to 0.4)

Characteristic equation,

$$q(s) = 1 + \frac{k(1 + k_p s)}{s(s+1)} = 0$$

$$q(s) = s^2 + s(1 + kk_p) + k = 0$$

$$\omega_n = \sqrt{k} = 4$$

$$\therefore \quad k = 16$$

$$2\xi\omega_n = 1 + kk_p = 2\,(0.8)(4)$$

$$\therefore \quad k_p = 0.337$$

4.29 (d)

$$x(t) = u(t) \xrightarrow{\ ?\ } y(t) = (1 - e^{-2t})u(t)$$

$$\frac{1}{s} \longrightarrow Y(s) = \frac{1}{s} - \frac{1}{s+2}$$

$$\frac{s+2-s}{s(s+2)} = \frac{2}{s(s+2)}$$

$$H(s) = \frac{2}{s+2}$$

4.30 (b)

$$\frac{E(s)}{R(s)} = \frac{1}{1+[(PID)(P)]} = \frac{s^2}{3\left[s^2 + K_P s + K_i + K_D s^2\right]}$$

$\therefore$ Steady-state value of E is

$$E(\infty) = \lim_{s \to 0} sE(s)$$

$$= \lim_{s \to 0} s \cdot \frac{s^2}{3\left[s^2 + K_P s + K_i + K_D s^2\right]} \times \frac{2}{s^3} = \frac{2}{3}$$

4.31 (0.78)

Given, Poles $= -3 \pm j4 = -\xi\omega_n \pm \omega_n\sqrt{1-\xi^2}$

So, $\omega_n\sqrt{1-\xi^2} = 4$

and since $t_p = \dfrac{\pi}{\omega_n\sqrt{1-\xi^2}} = 0.78$

4.32 (4)

$$\text{CLTF} = \frac{K}{\tau s + 1 + K}$$

$$\text{DC gain} = \frac{K}{1+K} = \frac{8}{10}$$

$$\therefore \quad K = 4$$

4.33 (1)

$$\text{TF} = \frac{2}{s+1}$$

$$\text{OLTF} = \frac{2}{s-1}$$

Type = 0

$e_{ss} = \dfrac{1}{1+k_p}$ for step input,

$$k_p = \lim_{s \to 0} \frac{2}{s-1} = -2$$

$$\therefore \quad e_{ss} = \frac{1}{1-2} = -1$$

$e_{ss} = +1$ ($\because$ In question, $e = Y - U$)

4.34 (a, b)

If system has oscillatory behaviour it should be underdamped.

For underdamped system, least order of system should be 2.

For 3rd order system, it can have a real system with underdamped in nature.

4.35 (b)

Given input-output relation $u(t-3)$

$$\frac{dy(t)}{dt} + y(t) = 3x(t-3)$$

Taking laplace both sides,

$$sY(s) + Y(s) = 3X(s)e^{-3s}$$

$$H(s) = \frac{Y(s)}{X(s)} = \frac{3e^{-3s}}{(s+1)}$$

4.36 (c)

We know,

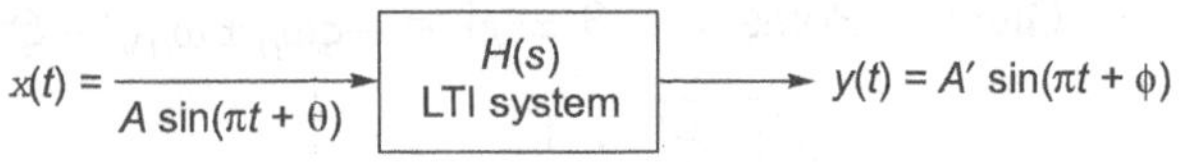

where, $A' = A \cdot |H(\omega)|_{\omega=\omega_o}$

$$\phi = \theta + \angle H(\omega)|_{\omega=\omega_o}$$

We have, $\omega_o = \pi$, $A' = 1$, $\phi = 0$

Now, $|H(\omega)|_{\omega=\pi} = \dfrac{\sqrt{\omega^2+\pi^2}}{\sqrt{\omega^2+\pi^2}} = 1$

$$\angle H(\omega)|_{\omega=\pi} = 180° - \tan^{-1}\left(\frac{\omega}{\pi}\right) - \tan^{-1}\left(\frac{\omega}{\pi}\right)$$

$$= 180° - 45° - 45°$$

$$= 90°$$

$\therefore$ $A = 1$

$\theta = -90°$

Now, $x(t) = \sin(\pi t - 90°) = \sin\left(\pi t - \frac{\pi}{2}\right)$

4.37 (a)

Given system

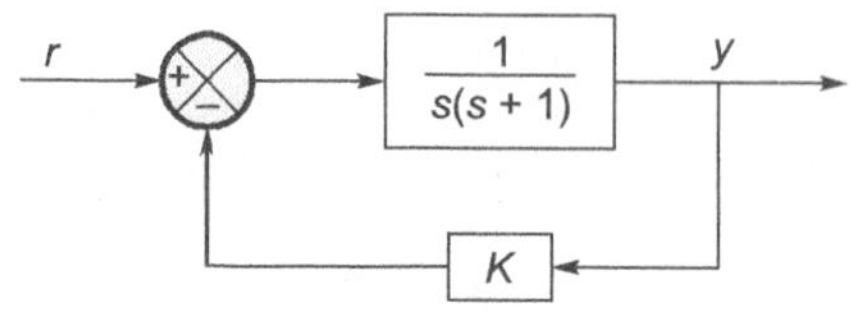

Transfer function $T(s) = \dfrac{1}{s^2+s+K}$

$$y(s) = \frac{1}{s(s^2+s+K)}$$

$$\lim_{s\to 0} sy(s) = \frac{1}{K}$$

$$e_{ss} = 1 - \frac{1}{K} = \frac{K-1}{K}$$

4.38 (a)

$\because$ $-60 < \theta < 60$

$\therefore$ $\xi \geq \dfrac{1}{2}$ and $\omega_n \geq 2$

■■■

5 Stability Analysis

5.1 The condition for stability of a closed-loop system with a characteristic equation $s^3 + bs^2 + cs + 1 = 0$, with positive coefficients is

(a) $b + c > 1$ (b) $bc > 1$

(c) $b = c$ (d) $b > c$ [1995 : 1 M]

5.2 The partial Routh array of the characteristic equation of a system is given by;

$$\begin{array}{cccc} s^4 & 1 & a & 8 \\ s^3 & 3 & 12 & \end{array}$$

The system oscillates with a frequency of 2 rad/s. The value of the parameter 'a' of the system is

(a) 6 (b) 2

(c) 8 (d) 12 [2002 : 2 M]

5.3 The closed loop control system shown in figure has $\tau_1 > 0$. The system will remain stable for all τ_1 in the range

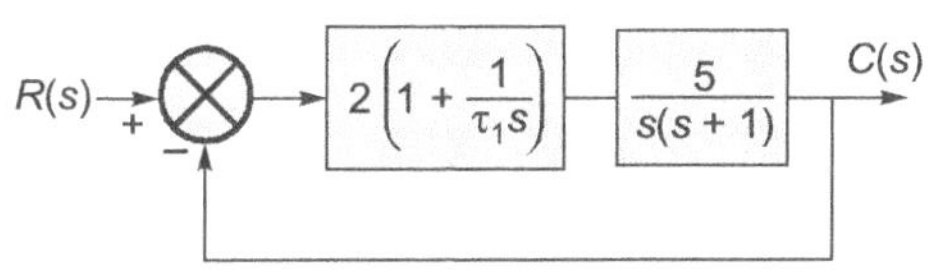

(a) $\tau_1 < 0.5$ (b) $0.5 < \tau_1 < 1.0$

(c) $\tau_1 < 1.0$ (d) $\tau_1 > 1.0$ [2003 : 2 M]

5.4 A certain closed loop system with unity feedback has the forward path transfer function given by

$$G(s) = \frac{K}{s(s+1)(s+5)}$$

With the gain set at the ultimate value, the system will oscillate at an angular frequency of

(a) $\sqrt{6}$ rad/sec (b) $\sqrt{5}$ rad/sec

(c) $\sqrt{2}$ rad/sec (d) $\sqrt{3}$ rad/sec

[2004 : 2 M]

5.5 The Routh-Hurwitz array is given for a third order characteristic equation,

$$\begin{array}{c|cccc} s^3 & 1 & b & 0 & 0 \\ s^2 & a & c & 0 & 0 \\ s^1 & \dfrac{(6-k)}{3} & 0 & & \\ s^0 & k & & & \end{array}$$

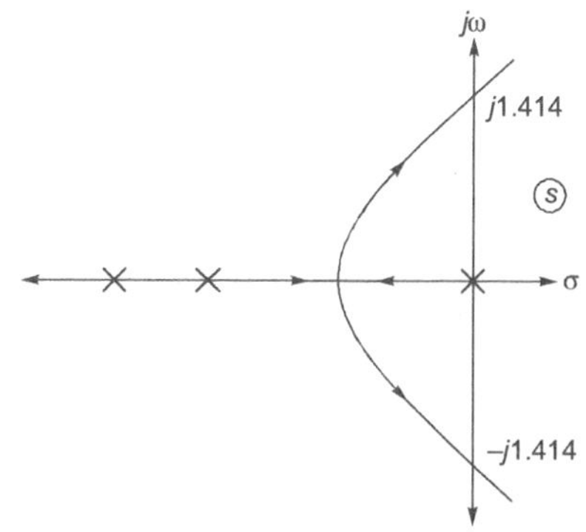

The coefficient a, b, c and k are such that $a = 3$, $b > 0$, $k > 0$, $c > 0$ and c is a function of k. The root locus for the corresponding characteristic equation is as shown in the given figure. The values of k and c for critical stability respectively are

(a) 6 and 6 (b) 6 and 2

(c) 3 and 2 (d) 3 and 3 [2005 : 2 M]

5.6 The range of the controller gains (K_p, K_i) that makes the closed loop control system (shown in the following figure) stable is given as

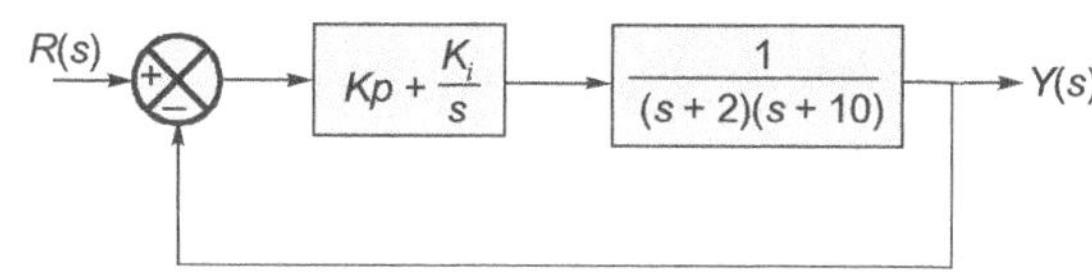

(a) $K_i < 0$ and $K_p < \frac{K_i}{12} - 20$

(b) $K_i > 0$ and $K_p > 0$

(c) $K_i < 0$ and $K_p > \frac{K_i}{12} - 20$

(d) $K_i > 0$ and $K_p > \frac{K_i}{12} - 20$ [2006 : 2 M]

5.7 A closed loop control system is shown below. The range of the controller gain K_C which will make the real parts of all the closed loop poles more negative than –1 is

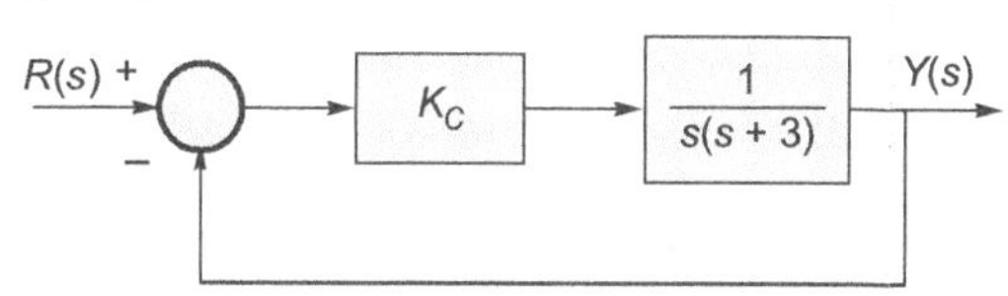

(a) $K_C > -4$ (b) $K_C > 0$

(c) $K_C > 2$ (d) $K_C < 2$ [2008 : 2 M]

5.8 The open loop transfer function of a unity gain feedback system is given by:

$$G(s) = \frac{k(s+3)}{(s+1)(s+2)}$$

The range of positive values of k for which the closed loop system will remain stable is

(a) $1 < k < 3$ (b) $0 < k < 10$
(c) $5 < k < \infty$ (d) $0 < k < \infty$

[2010 : 2 M]

5.9 The first two rows of Routh's table of a third-order characteristic equation are

s^3 3 3
s^2 4 4

It can be inferred that the system has

(a) one real pole in the right-half of s-plane
(b) a pair of complex conjugate poles in the right-half of s-plane
(c) a pair of real poles symmetrically placed around $s = 0$
(d) a pair of complex conjugate poles on the imaginary axis of the s-plane

[2011 : 2 M]

5.10 The value of a_0 which will ensure that the polynomial $s^3 + 3s^2 + 2s + a_0$ has roots on the left half of the s-plane is

(a) 11 (b) 9
(c) 7 (d) 5

[2016 : 1 M]

5.11 For the feedback system given below, the transfer function $G(s) = \frac{1}{(s+1)^2}$. The system **CANNOT** be stabilized with

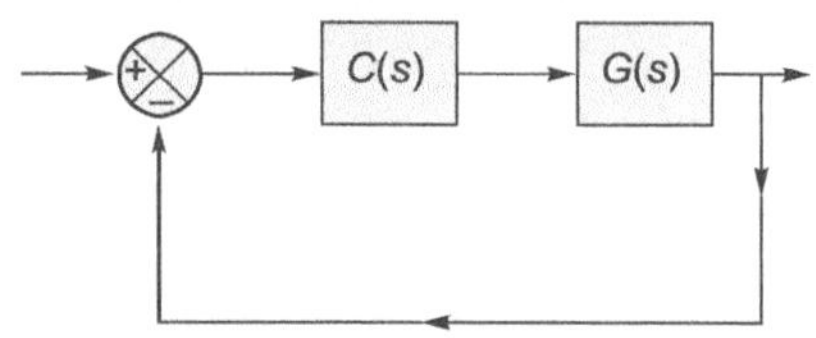

(a) $C(s) = 1 + \frac{3}{s}$ (b) $C(s) = 3 + \frac{7}{s}$

(c) $C(s) = 3 + \frac{9}{s}$ (d) $C(s) = \frac{1}{s}$

[2016 : 2 M]

5.12 A closed-loop system is shown in the figure. The system parameter α is not known. The condition for asymptotic stability of the closed loop system is

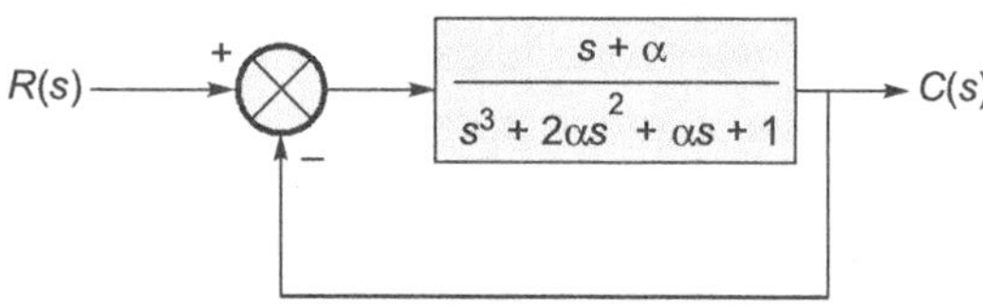

(a) $\alpha < -0.5$ (b) $-0.5 < \alpha < 0.5$
(c) $0 < \alpha < 0.5$ (d) $\alpha > 0.5$

[2017 : 2 M]

5.13 Consider a standard negative feedback configuration with $G(s) = \frac{1}{(s+1)(s+2)}$ and $H(s) = \frac{s+\alpha}{s}$. For the closed loop system to have poles on the imaginary axis, the value of α should be equal to (up to one decimal place) ________.

[2018 : 2 M]

5.14 The number of zeros of the polynomial $P(s) = s^3 + 2s^2 + 5s + 80$ in the right-half plane is ______.

[2023 : 1 M]

■■■

Answers Stability Analysis

5.1 (b)	5.2 (a)	5.3 (d)	5.4 (b)	5.5 (a)	5.6 (d)	5.7 (c)
5.8 (d)	5.9 (d)	5.10 (d)	5.11 (c)	5.12 (d)	5.13 (9)	5.14 (2)

Explanations Stability Analysis

5.1 (b)

For the system to be stable.
By Routh Hurwitz criteria

$$\begin{array}{c|cc} s^3 & 1 & c \\ s^2 & b & 1 \\ s^1 & \dfrac{bc-1}{b} & \\ s^0 & 1 & \end{array}$$

$\therefore$ for the system to be stable, all the elements of the column 1 should be > 0

$$\therefore \quad \frac{bc-1}{b} > 0$$

$$\therefore \quad bc - 1 > 0$$

$$\boxed{bc > 1}$$

5.2 (a)

$$\begin{array}{c|ccc} s^4 & 1 & a & 8 \\ s^3 & 3 & 12 & \end{array}$$

for the system to oscillate at $\omega = 2$ rad/sec

$$s^4 + as^2 + 8 = 0 \qquad \text{at } s = j\omega = 2j$$

$$(2j)^4 + a(2j)^2 + 8 = 0$$

$$16 - 4a + 8 = 0$$

$$\therefore \quad 4a = 24$$

$$a = \frac{24}{4} = 6$$

$$\boxed{a = 6}$$

5.3 (d)

The characteristics equation is,

$$1 + G(s)H(s) = 0$$

$$1 + \frac{10(\tau_1 s + 1)}{\tau_1 s^2 (s+1)} = 0$$

$$\tau_1 s^3 + \tau_1 s^2 + 10\tau_1 s + 10 = 0$$

According to R-H array

$$\begin{array}{c|cc} s^3 & \tau_1 & 10\tau_1 \\ s^2 & \tau_1 & 10 \\ s^1 & 10(\tau_1 - 1) & 0 \\ s^0 & 10 & 0 \end{array}$$

For stability $\tau_1 - 1 > 0$

$$\therefore \quad \tau_1 > 1.0$$

5.4 (b)

$$G(s)H(s) = \frac{K}{s(s+1)(s+5)}$$

As, we know in the root locus plot the system will oscillate when either the roots will lie on the $j\omega$ axis or near to origin.

Here ultimate value of gain means the value of K when loci intersects imaginary axis at the frequency of ω rad/sec.

Now, characteristic equation is

$$1 + G(s)H(s) = 0$$

$$1 + \frac{K}{s(s+1)(s+5)} = 0$$

$$s(s+1)(s+5) + K = 0$$

$$s^3 + 6s^2 + 5s + K = 0$$

According to R-H array

$$\begin{array}{c|cc} s^3 & 1 & 5 \\ s^2 & 6 & K \\ s^1 & \dfrac{(30-K)}{6} & 0 \\ s^0 & K & 0 \end{array}$$

The value of K that makes the s^1 term in the first column equal zero is

$$\frac{30-K}{6} = 0$$

$$\Rightarrow \quad K = 30$$

The crossing point on the imaginary axis can then be found by solving the auxiliary equation obtained from the s^2 row.

$$6s^2 + 30 = 0$$

$$\Rightarrow \quad s = \pm j\sqrt{5}$$

$\therefore$ Angular frequency $(\omega) = \sqrt{5}$ rad/sec

5.5 (a)

Given R-H array is

$$\begin{array}{c|cccc} s^3 & 1 & b & 0 & 0 \\ s^2 & a & c & 0 & 0 \\ s^1 & \dfrac{(6-k)}{3} & 0 & & \\ s^0 & k & & & \end{array}$$

Derived, R-H array is

$$\begin{array}{c|cc} s^3 & 1 & b \\ s^2 & a & c \\ s^1 & \dfrac{ab-c}{a} & 0 \\ s^0 & c & 0 \end{array}$$

Compare the given RH table with derived RH table we get,

$$c = k$$

$$\frac{ab-c}{a} = \frac{6-k}{3}$$

For critical stability

$$6 - k \geq 0$$

$$k = 6$$

Hence, $c = k = 6$

and $a = 3$

and $ab = 6$

$\Rightarrow$ $b = 2$

So, $k = 6$ and $c = 6$

5.6 (d)

Characteristic equation $1 + G(s)\,H(s) = 0$

$$1+\left(\frac{k_p s + k_I}{s}\right)\left(\frac{1}{(s+2)(s+10)}\right) = 0$$

$$1+\frac{(k_p s + k_I)}{s(s+2)(s+10)} = 0$$

$$s^3 + 12s^2 + (20 + k_p)\,s + k_I = 0$$

According to R-H array:

$$\begin{array}{c|cc} s^3 & 1 & (k_p+20) \\ s^2 & 12 & k_I \\ s^1 & \dfrac{12k_p+240-k_I}{12} & 0 \\ s^0 & k_I & \end{array}$$

For stability criteria to be satisfied,

$$\frac{12k_p+240-k_I}{12} > 0 \text{ and } k_I > 0$$

or $\quad k_p > \dfrac{k_I}{12} - 20$ and $k_I > 0$

5.7 (c)

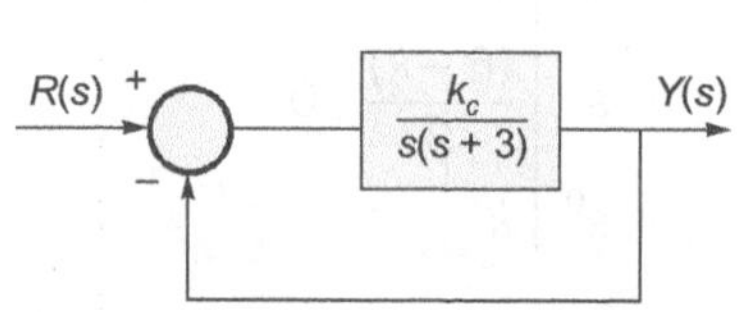

$$\frac{Y(s)}{R(s)} = \frac{k_c}{s(s+3)+k_c} = \frac{k_c}{s^2+3s+k_c}$$

Closed loop poles one given by

$$s_1,\, s_2 = \frac{-3 \pm \sqrt{9-4k_c}}{2}$$

For real parts more negative than −1

$$\frac{-3 \pm \sqrt{9-4k_c}}{2} < -1$$

$$-3 \pm \sqrt{9-4k_c} < -2$$

$$\pm\sqrt{9-4k_c} < 1$$

$$4k_c > 8$$

$$\Rightarrow \quad k_c > \frac{8}{4} = 2$$

So, according to the options k_c must be greater than 2.

5.8 (d)

The characteristic equation is

$$1+\frac{k(s+3)}{(s+1)(s+2)} = 0$$

$$(s + 1)(s + 2) + k(s + 3) = 0$$

$$s^2 + 3s + 2 + ks + 3k = 0$$

$$s^2 + (3 + k)\,s + (2 + 3k) = 0$$

According to R-H array:

$$\begin{array}{c|cc} s^2 & 1 & (2+3k) \\ s^1 & 3+k & 0 \\ s^0 & 2+3k & 0 \end{array}$$

For stability

$$3 + k > 0 \Rightarrow k > -3$$

and $2 + 3k > 0 \Rightarrow k > -\dfrac{2}{3}$

$$\Rightarrow \quad k > -\frac{2}{3}$$

From the options the appropriate of range k is $0 < k < \infty$.

5.9 (d)

Given first two rows of Routh's table of a third order characteristic equation are

$$\begin{array}{c|cc} s^3 & 3 & 3 \\ s^2 & 4 & 4 \\ s^1 & 8 & 0 \\ s^0 & 4 & \end{array}$$

This is the special case of Routh's array because first two row's having same elements so it will produce a row of zeros prematurely.

Since a row of zeros appear prematurely, we since a row equation using the form the auxiliary equation using the coefficient s^2 row.

$$A(s) = 4s^2 + 4 = 0 \quad \text{...(i)}$$

$$\Rightarrow \quad \frac{dA(s)}{ds} = 8s = 0$$

From which the coefficient 8 and 0 replace the zeros in the s'row of the original table.

Since these are no sign changes in the first column of the entire Routh's tabulation, the equation (i) does not have any root in the right-half s-plane. Solving the auxiliary equation, we get the two roots at $s = j$ and $s = -j$ which are also the roots of original characteristic equation. Thus, the equation has two roots on the $j\omega$-axis, and the system is marginally stable.

5.10 (d)

Characteristic equation

$$s^3 + 3s^2 + 2s + a_0$$

By RH criteria

s^3	1	2
s^2	3	a_0
s^1	$\frac{6-a_0}{3}$	0
s^0	a_0	0

$\therefore$ For roots to lie on LHP

$$a_0 > 0$$

$$\frac{6-a_0}{3} > 0$$

The a_0 value should be $0 < a_0 < 6$

$$\therefore \quad a_0 = 5$$

5.11 (c)

For $$C(s) = 3 + \frac{9}{s}$$

characteristic equation

$$1 + C(s)\,G(s) = 0$$

$$1 + \left(3 + \frac{9}{s}\right)\left(\frac{1}{(s+1)^2}\right) = 0$$

$$1 + \left(\frac{3s+9}{s}\right)\left(\frac{1}{s^2+2s+1}\right) = 0$$

$$s^3 + 2s^2 + 4s + 9 = 0$$

Using RH criteria

s^3	1	H
s^2	2	9
s^1	$\frac{8-9}{2}$	0
s^0	9	0

$\because$ 1st column of RH table does not contains positive sign only

$\therefore$ System become unstable.

5.12 (d)

Characteristics equation,

$$q(s) = s^3 + 2\alpha s^2 + s(\alpha + 1) + (\alpha + 1) = 0$$

By RH criteria,

s^3	1	$\alpha + 1$
s^2	2α	$\alpha + 1$
s^1	$\frac{2\alpha(\alpha+1)-(\alpha+1)}{2\alpha}$	0
s^0	$\alpha + 1$	0

Necessary condition: $\alpha > 0$

Sufficient condition: $\frac{2\alpha(\alpha+1)-(\alpha+1)}{2\alpha} > 0$

$\therefore \alpha > 0.5$

5.13 9 (8.9 to 9.1)

$$G(s) = \frac{1}{(s+1)(s+2)};\; H(s) = \frac{s+\alpha}{s}$$

$$q(s) = s^3 + 3s^2 + 3s + \alpha = 0$$

For poles on imaginary axis i.e. Marginal stability,

$$\alpha = 3 \times 3 = 9$$

5.14 (2)

$$s^3 + 2s^2 + 5s + 80$$

Forming RH array

s^3	1	5
s^2	2	80
s^1	$\frac{10-80}{2} = -35$	
s^0	80	

$\therefore$ There are two sign change.

Hence, 2 open loop zeros in RHS.

6 Root Locus

6.1 The break away point of the root locus on the real axis for a closed-loop system with a loop gain

$$G(s)\,H(s) = \frac{K(s+10)}{(s+2)(s+5)} \text{ lies}$$

(a) between –2 and origin
(b) between –2 and –5
(c) between – 10 and – ∞
(d) at – ∞ **[1995 : 1 M]**

Common data for Questions (6.3 and 6.4):

The root locus plot for a system, with transfer function $\frac{2}{s(s+1)(s+2)}$, is shown in below figure. A unity feedback proportional control system is built using this system.

6.2 The maximum possible controller gain, for which the unity feedback system is stable, is approximately

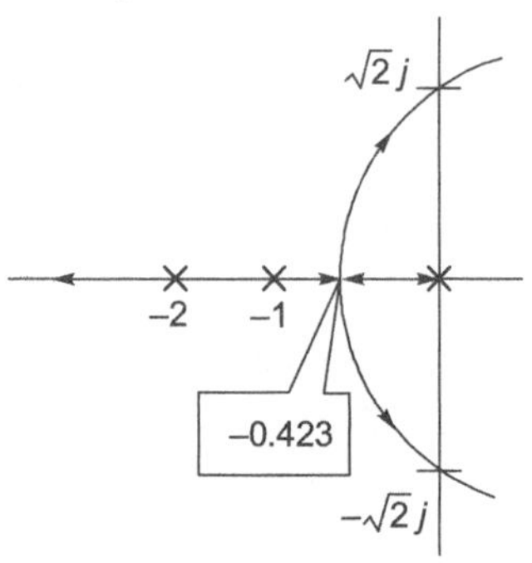

(a) 6.0 (b) 3.0
(c) 0.4 (d) 0.2

[2003 : 2 M]

6.3 The maximum possible controller gain, for which the unity feedback system exhibits a non-oscillatory response to a unit step input, is
(a) 6.0 (b) 3.0
(c) 0.4 (d) 0.2

[2003 : 2 M]

6.4 The root locus of a plant is given in the following figure. The root locus crosses imaginary axis at $\omega = 4\sqrt{2}$ rad/s with gain $K = 384$. It is observed that the point $s = -1.5 + j1.5$ lies in the root locus. The gain K at $s = -1.5 + j1.5$ is computed as

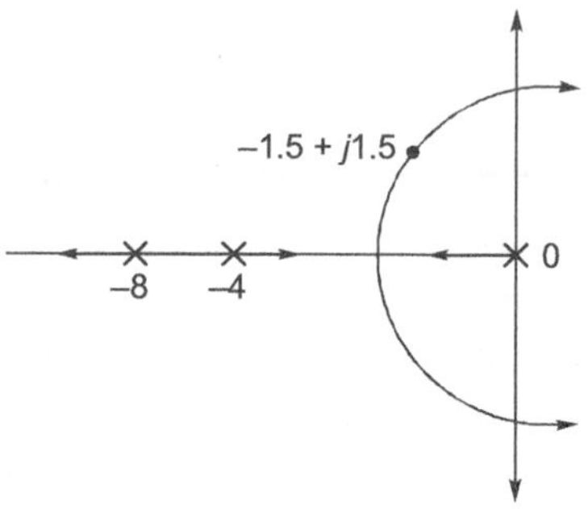

(a) 11.3 (b) 21.2
(c) 41.25 (d) 61.2 **[2006 : 2 M]**

Statement for Linked Answer Questions (6.5 and 6.6):

A transfer function with unity DC gain has three poles at –1, –2 and –3 and no finite zeros. A plant with this transfer function is connected with a proportional controller of gain K in the forward path, in a unity feedback configuration.

6.5 The transfer function is

(a) $\frac{s}{(s-1)(s-2)(s-3)}$

(b) $\frac{6}{(s+1)(s+2)(s+3)}$

(c) $\frac{s}{(s+1)(s+2)(s+3)}$

(d) $\frac{6}{(s-1)(s-2)(s-3)}$ **[2007 : 2 M]**

6.6 If the root locus plot of the closed loop system passes through the points $\pm j\sqrt{11}$, the maximum value of k for stability of the unity feedback closed loop system is

(a) $\sqrt{11}$ (b) 6
(c) 10 (d) $6\sqrt{11}$

[2007 : 2 M]

6.7 The open loop transfer function of a unity feedback system is

$$G(s) = \frac{K(s+2)}{(s+1+j1)(s+1-j1)}.$$

The root locus plot of the system has

(a) two breakaway points located at $s = -0.59$ and $s = -3.41$
(b) one breakaway point located at $s = -0.59$
(c) one breakaway point located at $s = -3.41$
(d) one breakaway point located at $s = -1.41$

[2008 : 2 M]

6.8 Consider the second-order system with the characteristic equation $s(s + 3) + K(s + 5) = 0$. Based on the properties of the root loci, it can be shown that the complex portion of the root loci of the given system for $0 < K < \infty$ is described by a circle, and the two break away points on the real axis are

(a) $-5 \pm \sqrt{5}/2$ (b) $-5 \pm \sqrt{5}$
(c) $-5 \pm \sqrt{10}$ (d) $-5 \pm 2\sqrt{5}$

[2011 : 2 M]

6.9 The open loop transfer function of a unity gain negative feedback control system is given by

$$G(s) = \frac{s^2 + 4s + 8}{s(s+2)(s+8)}.$$

The angle θ, at which the root locus approaches the zeros of the system, satisfies

(a) $|\theta| = \pi - \tan^{-1}\left(\frac{1}{4}\right)$

(b) $|\theta| = \frac{3\pi}{4} - \tan^{-1}\left(\frac{1}{3}\right)$

(c) $|\theta| = \frac{\pi}{2} - \tan^{-1}\left(\frac{1}{4}\right)$

(d) $|\theta| = \frac{\pi}{4} - \tan^{-1}\left(\frac{1}{3}\right)$

[2012 : 2 M]

6.10 The Bode plot of transfer function $G(s)$ is shown in the figure below.

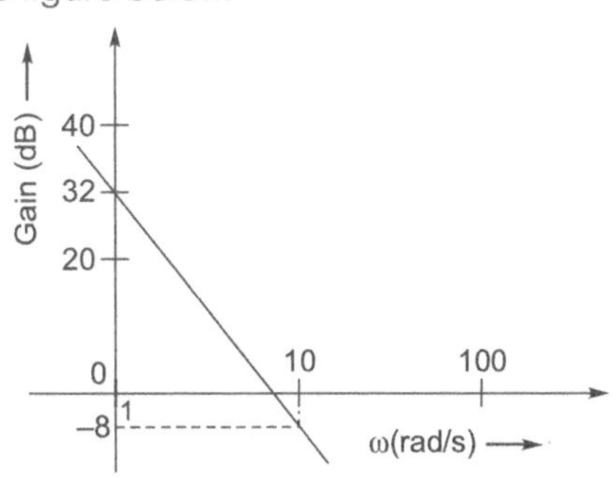

The gain $(20 \log|G(s)|)$ is 32 dB and −8 dB at 1 rad/s and 10 rad/s respectively. The phase is negative for all ω. Then $G(s)$ is

(a) $\frac{39.8}{s}$ (b) $\frac{39.8}{s^2}$
(c) $\frac{32}{s}$ (d) $\frac{32}{s^2}$

[2013 : 1 M]

6.11 A loop transfer function is given by

$$G(s)\,H(s) = \frac{K(s+2)}{s^2(s+10)}$$

The point of intersection of the asymptotes of $G(s)\,H(s)$ on the real axis in the s-plane is at ___.

[2014 : 1 M]

6.12 The open loop transfer function of a system is $G(s) = \frac{s^2 + 6s + 10}{s^2 + 2s + 2}$. The angles of arrival of its root loci are

(a) $\pm\frac{\pi}{4}$ (b) $\pm\frac{\pi}{3}$
(c) $\pm\frac{\pi}{2}$ (d) $\pm\frac{5\pi}{6}$

[2015 : 2 M]

6.13 The loop transfer function of a closed-loop system is given by $G(s)H(s) = \frac{K(s+6)}{s(s+2)}$. The breakaway point of the root-loci will be _______.

[2017 : 2 M]

6.14 In a control system with unity gain feedback, the plant has the transfer function $P(s) = \frac{3}{s}$. Assuming that a controller of the form $C(s) = \frac{K}{(s+p)}$ is used, where K is a positive constant, the value of p for which the root-locus of the closed-loop system passes through the points $-3 \pm j3\sqrt{3}$ where $j = \sqrt{-1}$, is

(a) 3 (b) $3\sqrt{3}$
(c) 6 (d) 9

[2019 : 2 M]

6.15 The loop transfer function of a negative feedback system is given by

$$G(s)H(s) = \frac{K}{s(s+2)(s+6)},$$ where K > 0. The value of K at the breakaway point of the root locus for the above system (rounded off to one decimal place) is _____.

[2020 : 2 M]

6.16 Consider a standard negative feedback configuration with $G(s) = \frac{1}{(s-2)(s-3)}$ and the controller $C(s) = K_P + \frac{K_1}{s} + K_D s$. The root-locus of $G(s)C(s)$ is presented in the figure below. The gain $C(j\omega) = 2$ at $\omega = 1$ rad/s. The value of K_D is ______ (rounded off to one decimal place).

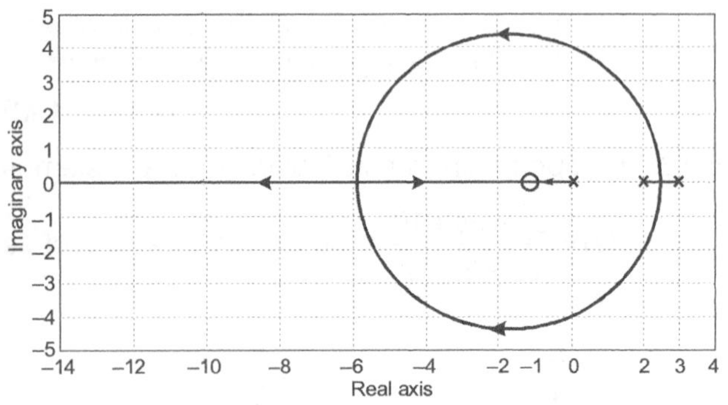

[2023 : 2 M]

6.17 Consider a unity negative feedback system with its open-loop pole-zero map as shown in the figure. If the point $s = j\alpha$, $\alpha > 0$, lies on the root locus, the value of α is ______ (rounded off to nearest integer).

Note: The poles are marked with × in the figure.

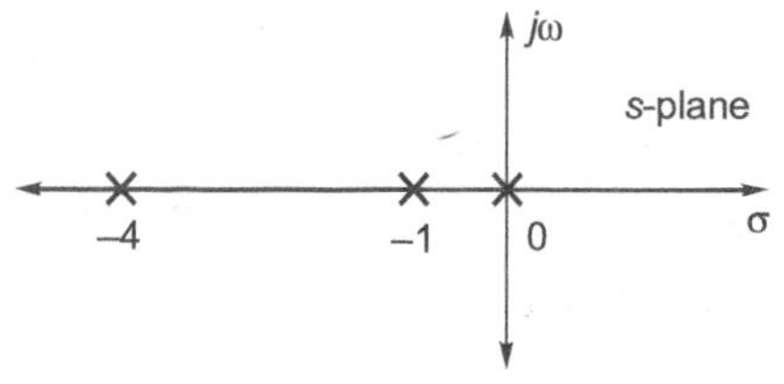

[2024 : 2 M]

Answers Root Locus

6.1	(b)	6.2	(b)	6.3	(d)	6.4	(c)	6.5	(b)	6.6	(c)	6.7	(c)
6.8	(c)	6.9	(d)	6.10	(b)	6.11	(–4)	6.12	(a)	6.13	(–1.10)	6.14	(c)
6.15	(5)	6.16	(1)	6.17	(2)								

Explanations Root Locus

6.1 (b)

The singularities are at,

$$s = -2, -5, -10$$

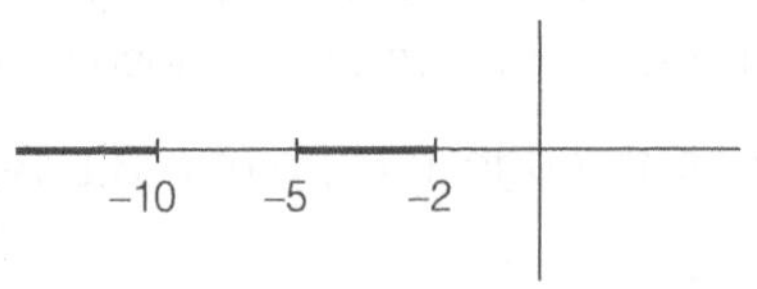

so, the break away point can lie only in the range between (–5, –2) or (–∞, –10)

$$1 + \frac{K(s+10)}{(s+2)(s+5)} = 0$$

$$\therefore \quad K = \frac{-(s+2)(s+5)}{(s+10)}$$

$$\frac{dK}{ds} = -\left\{\frac{(s+10)[s+2+s+5]-(s+2)(s+5)}{(s+10)^2}\right\} = 0$$

$\therefore$ $(s+10)(2s+7) = (s+2)(s+5)$

$2s^2 + 7s + 20s + 70 = s^2 + 7s + 10$

$\therefore$ $s^2 + 20s + 60 = 0$

$$s = \frac{-20 \pm \sqrt{400-240}}{2} = \frac{-20 \pm 12.64}{2}$$

$s = -3.68, -16.32$

$\therefore$ the breakaway point is –3.68 as the –16.32 do not lie in the set interval.

$\therefore$ the breakaway point lies in the interval –2 and –5.

6.2 (b)

At $\quad s = 0 \pm \sqrt{2}j$,

The system will have maximum controller gain and is

$$k = \left|\frac{s(s+1)(s+2)}{2}\right|_{s=\sqrt{2}j} = \left|\frac{\sqrt{2}j\,(1+\sqrt{2}j)\,(2+\sqrt{2}j)}{2}\right|$$

$$= 3.0$$

6.3 (d)

At $s = -0.423$, the system have maximum non-oscillatory response, because after that the system will become oscillatory since roots will become complex.

Now, according to magnitude conditions:

$$|T(s)| = \left|\frac{2k}{s(s+1)(s+2)}\right| = 1$$

$$\therefore \quad |k| = \left|\frac{s(s+1)(s+2)}{2}\right|_{s=-0.423}$$

$$= \left|\frac{-0.423 \times 0.577 \times 1.577}{2}\right|$$

$$= 0.1924 \simeq 0.2$$

6.4 (c)

Open-loop gain in pole-zero form

$$k = k'\left[\frac{\prod_{j=1}^{n}(s+p_j)}{\prod_{i=1}^{n}(s+z_i)}\right]$$

$$= K' s(s+4)(s+8)$$

At $\omega = 4\sqrt{2}$ rad/sec and $k = 384$

$$384 = k'(4\sqrt{2})(4+4\sqrt{2}j)(8+4\sqrt{2}j)$$

$\Rightarrow \quad K' = 1$

At, $\quad s = -1.5 + j1.5$

$$k = |(-1.5+j1.5)(2.5+j1.5)(6.5+j1.5)|$$

$$= 41.25$$

6.5 (b)

$$T(s) = \frac{k}{(s+1)(s+2)(s+3)}$$

For DC gain, put $s = 0$

$$1 = \frac{k}{6}$$

$\Rightarrow \quad k = 6$

$$\therefore \quad T(s) = \frac{6}{(s+1)(s+2)(s+3)}$$

6.6 (c)

$$T(s) = \frac{6k}{(s+1)(s+2)(s+3)}$$

Characteristic equation is

$(s + 1)(s + 2)(s + 3) + 6k = 0$

$s^3 + 6s^2 + 11s + 6(1 + k) = 0$

According to R-H array

s^3	1	11
s^2	6	$6+6k$
s^1	$\frac{66-(6+6k)}{6}$	0
s^0	$6+6k$	0

$$6s^2 + 6(1 + k) = 0$$

$$6\omega^2 + 6(1 + k) = 0$$

$$-6 \times 11 + 6(1 + k) = 0 \qquad (\because \omega = \pm\sqrt{11})$$

$$k = 10$$

6.7 (c)

Root locus of $G(s)\,H(s)$

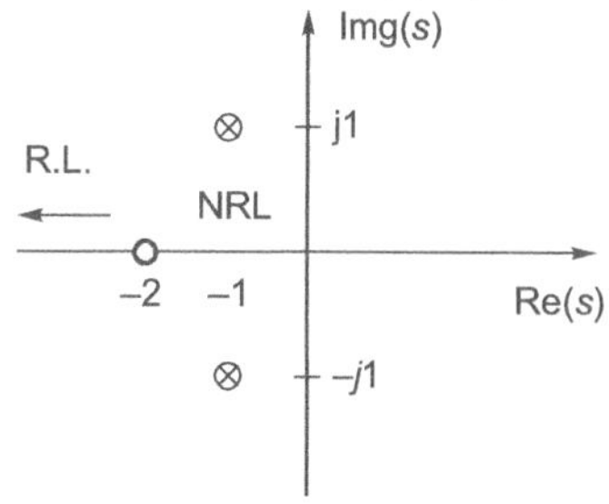

R.L. $\rightarrow$ Root locus part

NRL $\rightarrow$ Not root locus part

$$G(s)H(s) = \frac{K(s+2)}{s^2+2s+2}$$

For the break-away point

$$\frac{dG(s)H(s)}{ds} = 0$$

$$\frac{K(s^2+2s+2)\cdot 1 - K(s+2)(2s+2)}{(s^2+2s+2)^2} = 0$$

Solving this we get

$s^2 + 4s + 2 = 0$

$$s = \frac{-4 \pm 2\sqrt{2}}{2} = -2 \pm \sqrt{2}$$

$$s = -3.414, -0.59$$

From the root locus part it is obvious that the valid break-away point = –3.414

6.8 (c)

Characteristic equation is

$s(s + 3) + k(s + 5) = 0$

$$k = -\frac{s(s+3)}{s+5}$$

for break away points

$$\frac{dk}{ds} = 0$$

$$\Rightarrow \quad \frac{(2s+3)(s+5) - s(s+3)}{(s+5)^2} = 0$$

$\Rightarrow \quad s^2 + 10s + 15 = 0$

$$s = \frac{-10 \pm \sqrt{100 - 4 \times 15}}{2} = -5 \pm \sqrt{10}$$

6.9 (d)

$$G(s) = \frac{s^2+4s+8}{s(s+2)(s+8)}$$

Zeros: $\quad s^2 + 4s + 8$

$$s = \frac{-4 \pm \sqrt{4^2 - 4 \times 8}}{2} = \frac{-4 \pm \sqrt{16-32}}{2}$$

$$= \frac{-4 \pm 4j}{2} = -2 \pm 2j$$

Poles: $s = 0, -2, -8$

Pole zero plot:

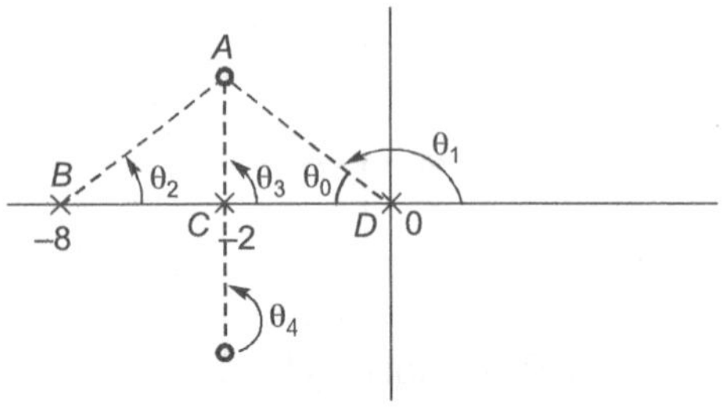

$$\theta_{arrival} = \pm\pi - \text{sum}(\theta_z) + \text{sum}(\theta_p)$$

$$= \pi - [\theta_1 + \theta_3 + \theta_2] - \theta_4$$

$$\theta_1 = \pi - \theta_a$$

$$= \pi - \tan^{-1}\left(\frac{2}{2}\right) = \pi - \frac{\pi}{4} = \frac{3\pi}{4}$$

$$\theta_2 = \tan^{-1}\frac{AC}{BC}$$

$$= \tan^{-1}\left(\frac{2}{6}\right) = \tan^{-1}\left(\frac{1}{3}\right)$$

$$\theta_3 = \frac{\pi}{2}$$

$$\theta_4 = \frac{\pi}{2}$$

$$\therefore \quad \theta_{arrival} = \pm\pi - \left[\frac{3\pi}{4} + \frac{\pi}{2} + \tan^{-1}\left(\frac{1}{3}\right)\right] - \frac{\pi}{2}$$

$$= \pi + \frac{3\pi}{4} + \tan^{-1}\left(\frac{1}{3}\right)$$

$$\text{or } -\pi + \frac{3\pi}{4} + \tan^{-1}\left(\frac{1}{3}\right)$$

$$= \frac{7\pi}{4} + \tan^{-1}\left(\frac{1}{3}\right)$$

$$\text{or } -\frac{\pi}{4} + \tan^{-1}\left(\frac{1}{3}\right)$$

$\therefore$ for $\theta_{arrival} = -\frac{\pi}{4} + \tan^{-1}\left(\frac{1}{3}\right)$

the $|\theta_{arrival}| = \frac{\pi}{4} - \tan^{-1}\left(\frac{1}{3}\right)$

6.10 (b)

Let, $G(s) = \frac{K}{s^n}$; $\quad G(j\omega) = \frac{K}{(j\omega)^n}$

$\because$ Gain = 32 dB at ω = 1 rad/s

$$\therefore \quad 20\log\left|\frac{K}{1^n}\right| = 32$$

$\Rightarrow \quad 20\log K = 32 \quad \Rightarrow K = 39.8$

Now,

$\because$ Gain = –8 dB at ω = 10 rad/s,

$$\therefore \quad 20\log\left|\frac{39.8}{10^n}\right| = -8$$

$\Rightarrow \quad 10^n = 100$

$\Rightarrow \quad n = 2$

6.11 (–4)

Point of intersection of asymptotes

$$\sigma = \frac{\sum(\text{poles}) - \sum(\text{zeros})}{\text{poles} - \text{zeros}} = \frac{-10-(-2)}{3-1} = -\frac{8}{2} = -4$$

6.12 (a)

$$\phi_a = 180 - \angle GH$$

$$G(s) = \frac{(s+3+i)(s+3-i)}{(s+1+i)(s+1-i)}$$

$$G(3+i) = \frac{[-3+i+3+i][-3+i+3-i]}{[-3-i+1+i][-3+i+1-i]}$$

$$= \frac{[2i]}{[-2+2i][-2]}$$

$$\angle G(-3+i) = 90° - \left[180 - \tan^{-1}\frac{2}{2}\right] - 180°$$

$$= 90° - 180° + 45° - 180°$$

$$= 135°$$

$$\phi_a = 180° - 135° = 45° = \frac{\pi}{4}$$

$\therefore$ Other angle, $\phi_d = -\frac{\pi}{4}$

$\therefore$ Angles of arrival = $\pm\frac{\pi}{4}$

6.13 –1.10 (–1.2 to –1.0)

$$GH(s) = \frac{k(s+6)}{s(s+2)}$$

Breakaway point and Break in point can be obtained from the solutions of $\frac{dk}{ds}$.

$$k = \frac{-s(s+2)}{s+6}$$

$$\frac{dk}{ds} = 0$$

$$\Rightarrow \quad s = -6 \pm 2\sqrt{6}$$

Breakaway point = $-6+2\sqrt{6} = -1.1$

Break in point = $-6-2\sqrt{6} = -10.9$

6.14 (c)

$$\text{OLTF} = \frac{3k}{s(s+P)}$$

Given root locus passes through the point $-3 \pm j3\sqrt{3}$

$$\therefore \angle G(s)H(s)_{s=-3+j3\sqrt{3}} = \pm 180°$$

$$\therefore \quad P = 6$$

6.15 (5)

$$G(s)H(s) = \frac{K}{s(s+2)(s+6)}$$

Breakaway point = -0.9

$$\therefore \left|\frac{K}{s(s+2)(s+6)}\right| = 1$$

$$S = -0.9$$

$$\therefore \quad K = 5$$

6.16 (1)

$$G(s) = \frac{1}{(s-2)(s-3)}$$

$$C(s) = K_P + \frac{K_i}{s} + K_D S$$

$$G(s)C(s) = \frac{sK_P + K_i + K_D s^2}{s(s-2)(s-3)}$$

Two zeros are at (-1)

$$(s+1)^2 = 0$$

$$s^2 + 2s + 1 = 0 \qquad \ldots(1)$$

Compared with

$$K_D s^2 + K_P s + K_i = 0 \qquad \ldots(2)$$

$$K_D = 1,\ K_P = 2 \text{ and } K_i = 1$$

$$K_D = 1$$

as per given

$$C(s) = K_P + \frac{K_i}{s} + K_D \cdot S$$

$$C(j\omega) = 2 + \frac{1}{j\omega} + j\omega$$

$$C(j1) = 2 - j + j = 2$$

6.17 (2)

If $GH = \dfrac{K}{s(s+a)(s+b)}$

Then it's root locus intersects $j\omega$ axis at $j\sqrt{ab}$.

$$\therefore \quad \alpha = \sqrt{ab} = 2$$

■■■

7 Frequency Response Analysis, Bode Plot and Nyquist Plot

7.1 The open-loop transfer function of a feedback control system is $\dfrac{1}{(s+1)^3}$.
The gain margin of the system is

(a) 16 (b) 8
(c) 4 (d) 2 [1997 : 1 M]

7.2 The phase margin of the system for which loop gain $GH(j\omega) = \dfrac{1}{(j\omega+1)^3}$ is

(a) $-\pi$ (b) π
(c) 0 (d) $\pi/2$ [1998 : 1 M]

7.3 The asymptotic Bode plot for the gain magnitude of a minimum phase system $G(s)$ is shown in the figure. The transfer function of the system $G(s)$ is

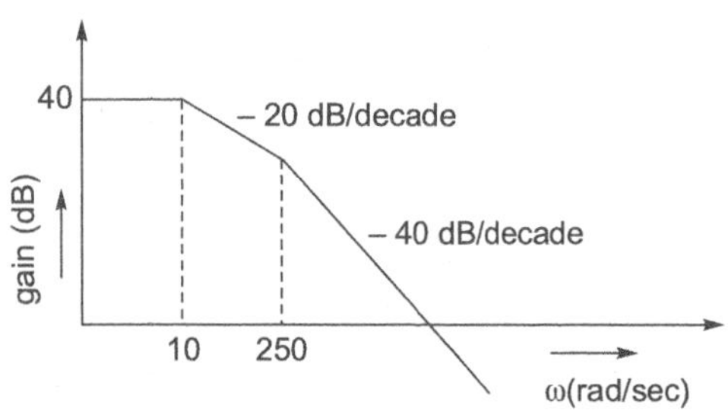

(a) $\dfrac{100}{(1+s/10)(1+s/250)}$

(b) $\dfrac{40}{s(s+250)}$

(c) $\dfrac{100}{(s+10)(s+250)}$

(d) $\dfrac{100s}{(s+10)(s+250)}$

[2000 : 1 M]

7.4 The loop transfer function of a system is given by

$$G(s)\,H(s) = \frac{10e^{-Ls}}{s}$$

The phase cross-over frequency is 5 rad/s. The value of the dead time L is

(a) $\pi/20$ (b) $\pi/10$
(c) $-\pi/20$ (d) zero

[2002 : 2 M]

7.5 The transfer function of a system is given by $\dfrac{Y(s)}{X(s)} = \dfrac{e^{-0.1s}}{1+s}$. If $x(t)$ is 0.5 sin t, then the phase angle between the output and the input will be

(a) –39.27° (b) –45°
(c) –50.73° (d) –90°

[2003 : 2 M]

7.6 Figure shows the Bode magnitude plot of a system. The minimum phase transfer function of the system is given by

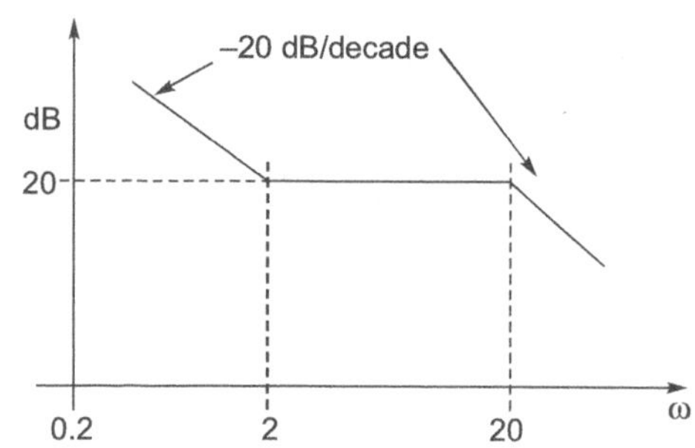

(a) $\dfrac{80\left(\dfrac{s}{2}+1\right)}{s\left(\dfrac{s}{20}+1\right)}$ (b) $\dfrac{20\left(\dfrac{s}{2}+1\right)}{s\left(\dfrac{s}{20}+1\right)}$

(c) $\dfrac{8(s+2)}{s(s+20)}$ (d) $\dfrac{8(s+2)}{s(s-20)}$

[2004 : 2 M]

7.7 Figure shows the polar plot of a system. The transfer function of the system is

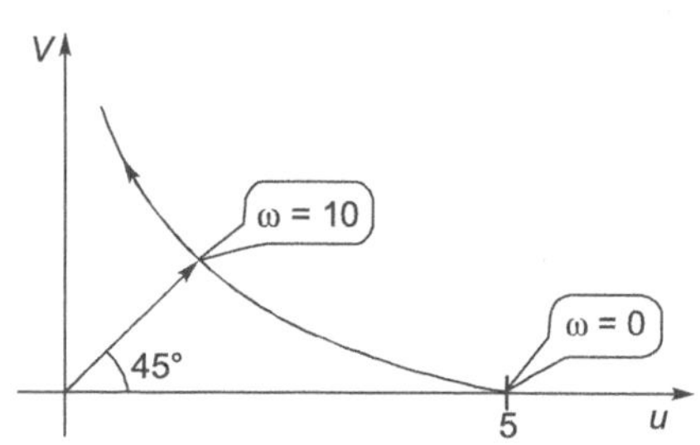

(a) $5(1+0.1s)$ (b) $(1+0.5s)$
(c) $5(1+10s)$ (d) $5(1+s)$

[2004 : 2 M]

7.8 The loop transfer function of a system is given by $G(s) = \frac{e^{-0.1s}}{s}$. The phase cross-over frequency is given by

(a) $\frac{\pi}{2}$ (b) $\frac{\pi}{10}$

(c) $\frac{\pi}{0.2}$ (d) $\frac{\pi}{4}$ **[2004 : 2 M]**

7.9 Consider the following systems:

System 1: $G(s) = \frac{1}{(2s+1)}$

System 2: $G(s) = \frac{1}{(5s+1)}$

The true statement regarding the system is

(a) Bandwidth of system 1 is greater than the bandwidth of system 2

(b) Bandwidth of system 1 is lower than the bandwidth of system 2

(c) Bandwidth of both the systems are the same

(d) Bandwidth of both the systems are infinite

[2004 : 2 M]

7.10 The asymptotic magnitude Bode plot of an open loop system $G(s)$ with $K > 0$ and all poles and zeros on the left hand side of the s-plane is shown in the figure. It is completely symmetric about ω_c. The minimum absolute phase angle contribution by $G(s)$ is given by

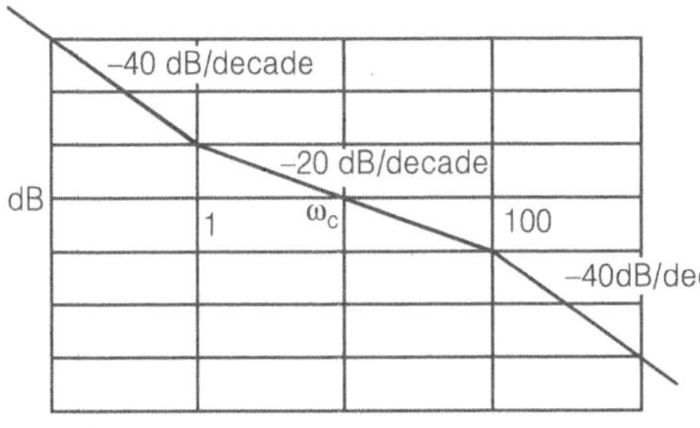

(a) 78.6° (b) 90°

(c) 101.4° (d) 180° **[2005 : 2 M]**

7.11 The open loop part of a unity feedback control system $G(s)$ is unstable and has two poles on the right hand side of the s-plane. However, the closed loop system is stable. The number of encirclements made by the Nyquist plot of the (–1, 0) point in the $G(s)$ plane is

(a) 2 (b) 0

(c) –3 (d) –1 **[2005 : 2 M]**

Statement for Linked Answer Questions (7.12 to 7.13):

The thickness of a metal sheet from a rolling mill is manipulated by controlling the roll gap $d(t)$ and measuring the sheet thickness $m(t)$ at a certain distance from the rolls as shown in the figure. The thickness of the sheet at the roll out point is equal to the roll gap $d(t)$ as shown and remains invariant during transit.

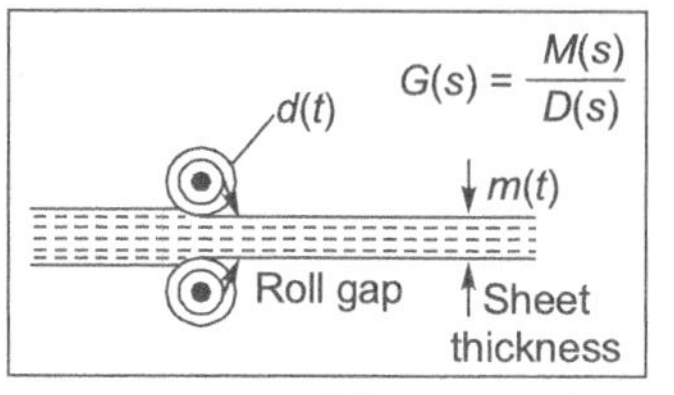

(a)

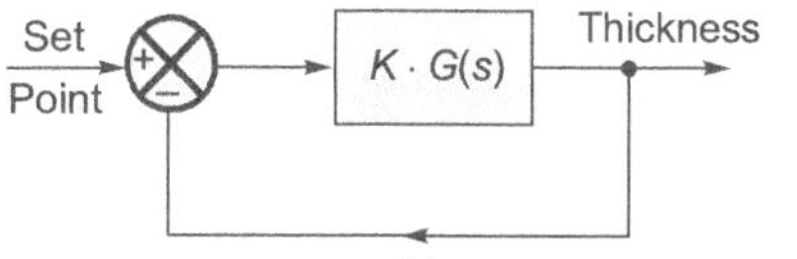

(b)

7.12 Given the distance at which the thickness is measured is 0.5 m from the roll gap and the velocity of the sheet 0.2 m/s, the transfer function system, $G(s) = M(s)/D(s)$, is

(a) exp(–2.5*s*) (b) exp(+2.5*s*)

(c) exp(+0.4*s*) (d) exp(–0.4*s*)

[2005 : 2 M]

7.13 The gap $d(t)$ is controlled by a proportional controller along with a hydraulic actuator of combined gain K and a unity feedback loop exists around K. $G(s)$ in figure (b) Increasing K beyond a certain value causes the system to become unstable. The frequency of oscillations at the critical value of K is

(a) $\omega = \pi$ (b) $\omega = 2.5\pi$

(c) $\omega = 0.4\pi$ (d) $\omega = \frac{\pi}{2.5}$

[2005 : 2 M]

7.14 A unity feedback system has the following open loop frequency response:

ω(rad/sec)	2	3	4	5	6	8	10
$\lvert G(j\omega)\rvert$	7.5	4.8	3.15	2.25	1.70	1.00	0.64
$\angle G(j\omega)$	–118°	–130°	–140°	–150°	–157°	–170°	–180°

The gain and phase margin of the system are

(a) 0 dB, –180° (b) 3.88 dB, –170°

(c) 0 dB, 10° (d) 3.88 dB, 10°

[2006 : 2 M]

Common Data Questions (7.15 and 7.16):

The following figure represents a proportional control scheme of a first order system with transportation lag.

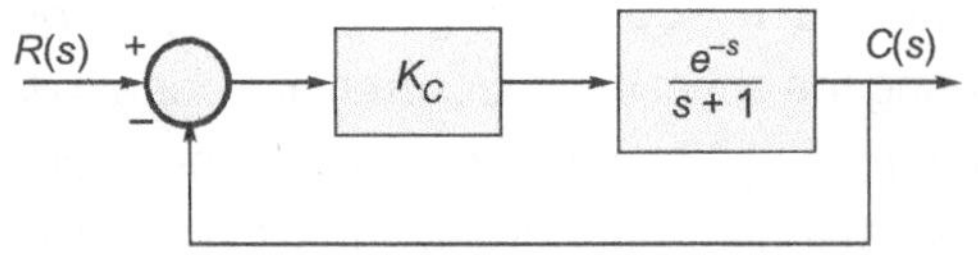

7.15 The angular frequency in radians/s at which the loop phase lag becomes 180° is

(a) 0.408 (b) 0.818
(c) 1.56 (d) 2.03

[2007 : 2 M]

7.16 The steady state error for a unit step input when the gain $K_C = 1$ is

(a) 1/4 (b) 1/2
(c) 1 (d) 2

[2007 : 2 M]

7.17 The Bode asymptotic plot of a transfer function is given below. In the frequency range shown, the transfer function has

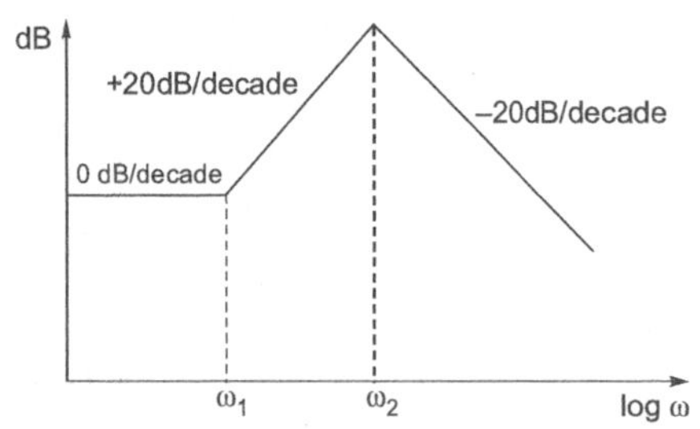

(a) 3 poles and 1 zero
(b) 1 pole and 2 zero
(c) 2 poles and 1 zero
(d) 2 poles and 2 zero

[2008 : 1 M]

Statement for Linked Answer Questions (7.18 & 7.19):

Consider a unity feedback system with open loop transfer function $G(s) = \dfrac{1+6s}{s^2(1+s)(1+2s)}$.

7.18 The phase cross-over frequency of the system in radians per second is

(a) 0.125 (b) 0.25
(c) 0.5 (d) 1

[2008 : 2 M]

7.19 The gain margin of the system is

(a) 0.125 (b) 0.25
(c) 0.5 (d) 1

[2008 : 2 M]

7.20 A unity feedback control loop with an open loop transfer function of the from $\dfrac{K}{s(s+\sigma)}$ has a gain cross-over frequency of 1 rad/s and a phase margin is 60°. If an element having a transfer function $\dfrac{s-\sqrt{3}}{s+\sqrt{3}}$ is inserted into the loop, the phase margin will become

(a) 0° (b) 30°
(c) 45° (d) 60°

[2009 : 2 M]

7.21 The asymptotic Bode magnitude plot of a lead network with its pole and zero on the left half of the s-plane is shown in the adjoining figure. The frequency at which the phase angle of the network is maximum (in rad/s) is

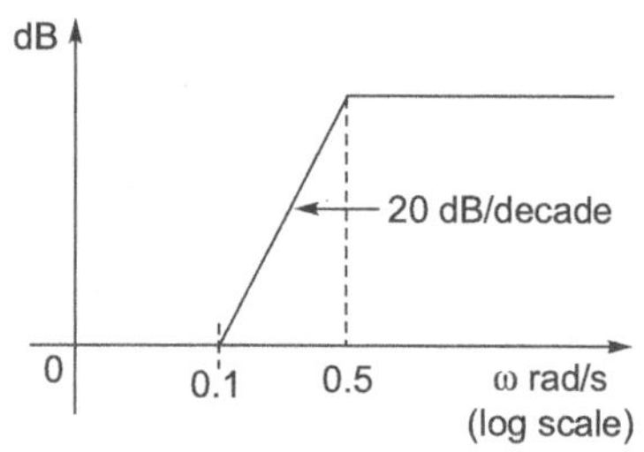

(a) $\dfrac{3}{\sqrt{10}}$ (b) $\dfrac{1}{\sqrt{20}}$
(c) $\dfrac{1}{20}$ (d) $\dfrac{1}{30}$

[2009 : 2 M]

Common data Question for (7.22 and 7.23):

The open-loop transfer function of a unity negative feedback control system is given by

$$G(s) = \frac{K}{(s+5)^3}$$

7.22 The value of K for the phase margin of the system to be 45° is

(a) $250\sqrt{5}$ (b) $250\sqrt{2}$
(c) $125\sqrt{5}$ (d) $125\sqrt{2}$

[2011 : 2 M]

7.23 The value of K for the damping ratio ζ to be 0.5, corresponding to the dominant closed-loop complex conjugate pole pair is

(a) 250 (b) 125
(c) 75 (d) 50

[2011 : 2 M]

7.24 A system with transfer function:

$$G(s) = \frac{(s^2+9)(s+2)}{(s+1)(s+3)(s+4)}$$

is excited by sin (ωt). The steady-state output of the system is zero at

(a) $\omega = 1$ rad/s (b) $\omega = 2$ rad/s
(c) $\omega = 3$ rad/s (d) $\omega = 4$ rad/s

[2012 : 1 M]

7.25 The feedback system shown below oscillates at 2 rad/s when

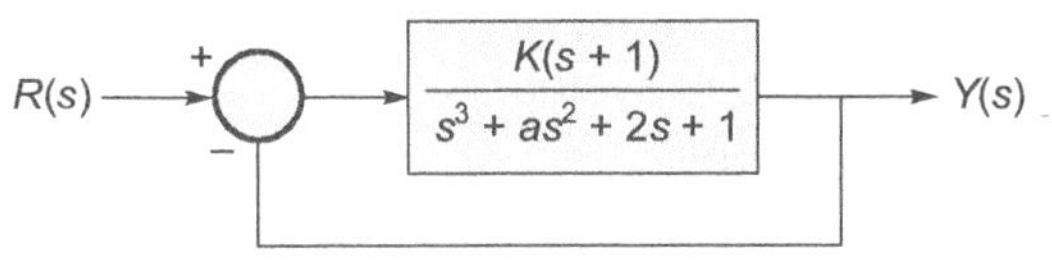

(a) $K = 2$ and $a = 0.75$
(b) $K = 3$ and $a = 0.75$
(c) $K = 4$ and $a = 0.5$
(d) $K = 2$ and $a = 0.5$ [2012 : 2 M]

7.26 The open loop transfer function of a unity negative feedback control system is given by:

$$G(s) = \frac{150}{s(s+9)(s+25)}$$

The gain margin of the system is

(a) 10.8 dB (b) 22.3 dB
(c) 34.1 dB (d) 45.6 dB

[2012 : 2 M]

7.27 The transfer function of a system is given by

$$G(s) = \frac{e^{-s/500}}{s+500}$$

The input to the system is $x(t) = \sin 100\pi t$. In periodic steady state the output of the system is found to be $y(t) = A\sin(100\pi t - \phi)$. The phase angle ($\phi$) in degree is ____.

[2014 : 2 M]

7.28 The loop transfer function of a feedback control system is given by

$$G(s)\,H(s) = \frac{1}{s(s+1)(9s+1)}$$

Its phase crossover frequency (in rad/s), approximated to two decimal places, is ____.

[2014 : 2 M]

7.29 A transfer function $G(s)$ with the degree of its numerator polynomial zero and the degree of its denominator polynomial two has a Nyquist plot shown in the figure. The transfer function represents

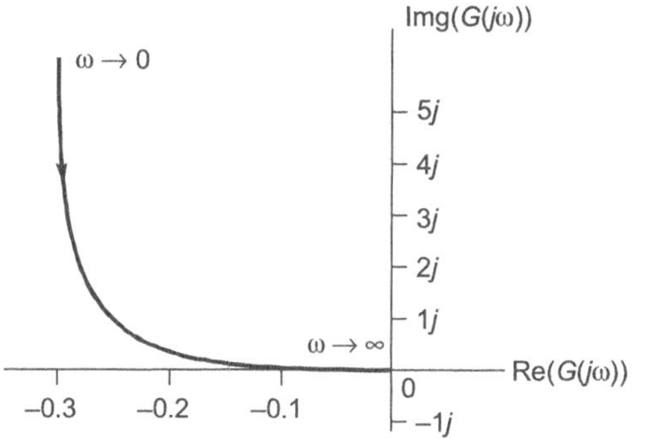

(a) a stable, type-0 system
(b) a stable, type-1 system
(c) an unstable, type-0 system
(d) an unstable, type-1 system [2015 : 2 M]

7.30 The number of times the Nyquist plot of $G(s) = \frac{s-1}{s+1}$ will encircle the origin clockwise is ____.

[2016 : 1 M]

7.31 The input $i(t) = 2\sin(3t+\pi)$ is applied to a system whose transfer function $G(s) = \frac{8}{(s+10)^2}$. The amplitude of the output of the system is____.

[2016 : 1 M]

7.32 The transfer function $G(s)$ of a system which has the asymptotic Bode plot shown below is:

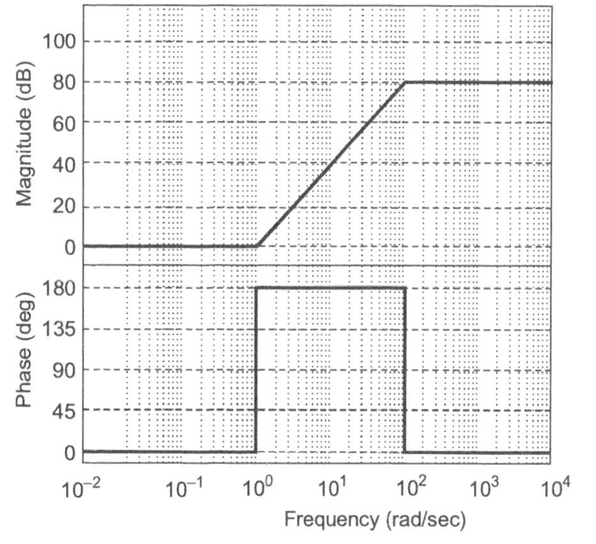

(a) $10^4 \frac{(s-1)^2}{(s+100)^2}$ (b) $10^4 \frac{(s+1)^2}{(s+100)^2}$

(c) $10^4 \frac{(s+1)}{(s+100)^2}$ (d) $10^4 \frac{(s-1)^2}{(s-100)^2}$

[2016 : 2 M]

7.33 An input $p(t) = \sin(t)$ is applied to the system $G(s) = \dfrac{s-1}{s+1}$. The corresponding steady state output is $y(t) = \sin(t + \phi)$, where the phase ϕ (in degrees). When restricted to $0° \le \phi \le 360°$, is _______ .

[2018 : 1 M]

7.34 The approximate phase response of $\dfrac{100^2 e^{-0.01s}}{s^2 + 0.2s + 100^2}$ is

(a)

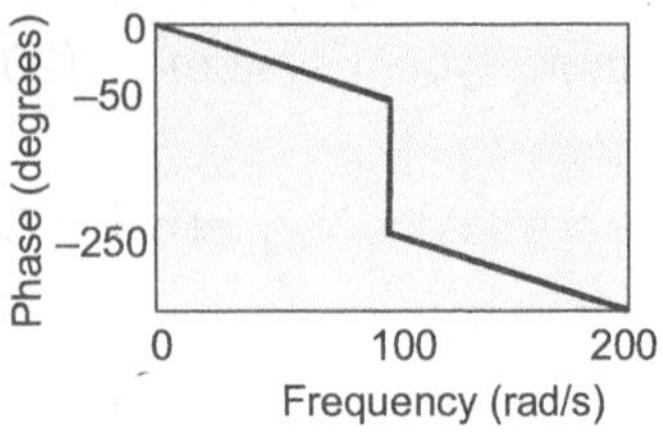

(b)

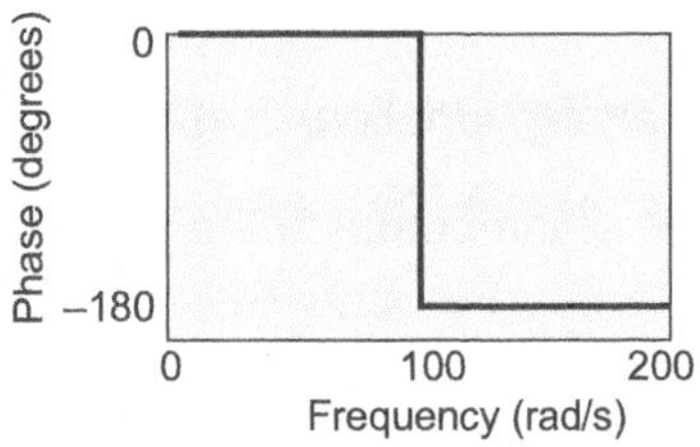

(c)

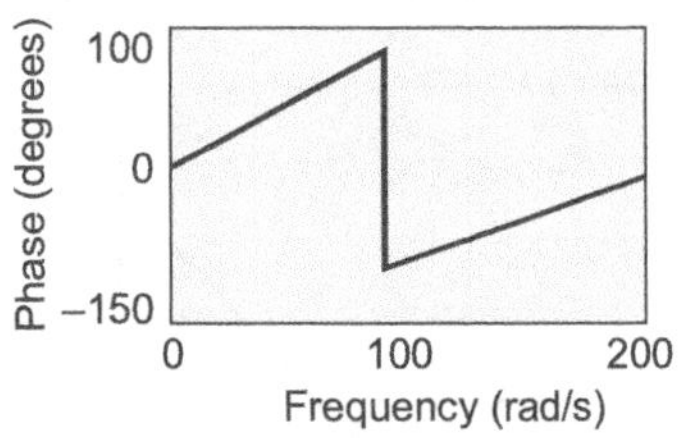

(d)

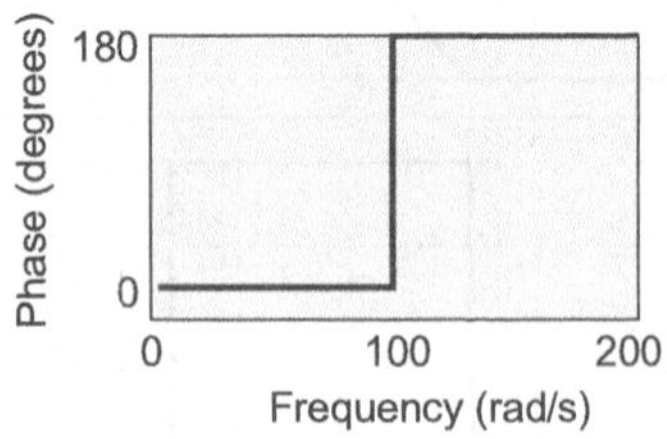

[2018 : 1 M]

7.35 Consider the transfer function $G(s) = \dfrac{2}{(s+1)(s+2)}$. The phase margin of $G(s)$ in degrees is ______.

[2018 : 1 M]

7.36 Consider the standard negative feedback configuration with $G(s) = \dfrac{s^2 + 0.2s + 100}{s^2 - 0.2s + 100}$ and $H(s) = \dfrac{1}{2}$. The number of clockwise encirclements of (–1, 0) in the Nyquist plot of the Loop transfer-function $G(s)\,H(s)$ is __________.

[2018 : 2 M]

7.37 The loop-gain function $L(s)$ of a control system with unity feedback is given to be $L(s) = \dfrac{k}{(s+1)(s+2)(s+3)}$, where $k > 0$. If the gain cross-over frequency of the loop-gain function is less than its phase cross-over frequency, the closed-loop system is

(a) unstable
(b) marginally stable
(c) conditionally stable
(d) stable

[2019 : 1 M]

7.38 The forward path transfer function $L(s)$ of the control system shown in figure (a) has the asymptotic Bode plot shown in figure (b). If the disturbance $d(t)$ is given by $d(t) = 0.1 \sin(\omega t)$, where $\omega = 5$ rad/s, the steady-state amplitude of the output $y(t)$ is

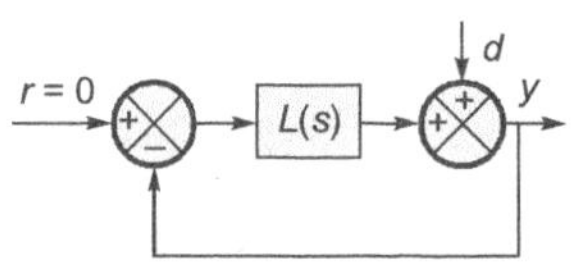

Figure (a)

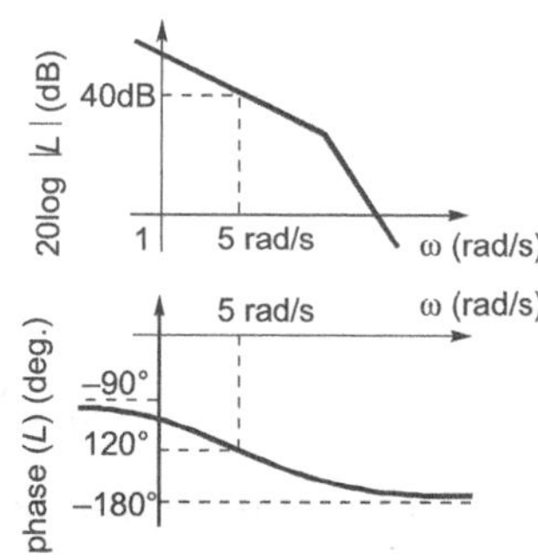

Figure (a)

(a) 1.00×10^{-3}
(b) 2.50×10^{-3}
(c) 5.00×10^{-3}
(d) 10.00×10^{-3}

[2019 : 2 M]

7.39 In a control system with unity gain feedback, the transfer function of the loop-gain function is $L(s) = \dfrac{9e^{-0.1s}}{s}$. The phase margin of the loop-gain function $L(s)$ is______degrees. [2019 : 2 M]

7.40 The loop transfer function of negative feedback system is $G(s)H(s) = \dfrac{2(s+1)}{s^2}$. The phase margin of the system (in degrees, rounded off to one decimal place) is _____. [2020 : 2 M]

7.41 The loop transfer function of a negative feedback system is $G(s)H(s) = \dfrac{1}{s(s-2)}$. The Nyquist plot for the above system _____.

(a) does not encircle $(-1 + j0)$ point
(b) encircles $(-1 + j0)$ point once in the clockwise direction
(c) encircles $(-1 + j0)$ point twice in the counterclockwise direction
(d) encircles $(-1 + j0)$ point once in the counterclockwise direction

[2020 : 2 M]

7.42 Taking N as positive for clockwise encirclement, otherwise negative, the number of encirclements N of $(-1, 0)$ in the Nyquist plot of $G(s) = \dfrac{3}{s-1}$ is _____.

[2021 : 1 M]

7.43 A sinusoid $\left(\sqrt{2}\sin t\right)\mu(t)$, where $\mu(t)$ is the step input is applied to a system with transfer function $G(s) = \dfrac{1}{s+1}$. The amplitude of the steady state output is _____.

[2021 : 2 M]

7.44 The Nyquist plot of a stable open-loop system $G(j\omega)$ is plotted in the frequency range $0 \le \omega < \infty$ as shown. It is found to intersect a unit circle with center at the origin at the point $P = -0.77 - 0.64j$. The points Q and R lie on $G(j\omega)$ and assume values $Q = 14.40 + 0.00j$ and $R = -0.21 + 0.00j$. The phase margin (PM) and the gain margin (GM) of the system are _______.

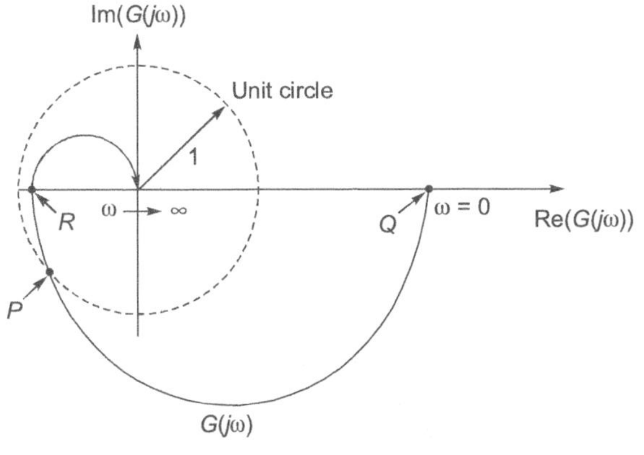

(a) PM = 39.7° and GM = 4.76
(b) PM = 39.7° and GM = 0.07
(c) PM = –39.7° and GM = 4.76
(d) PM = –39.7° and GM = 0.07

[2022 : 2 M]

7.45 The number of times the Nyquist plot of $G(s)H(s) = \dfrac{1(s-1)(s-2)}{2(s+1)(s+2)}$ encircles the origin is _________.

[2023 : 1 M]

7.46 The magnitude and phase plots shown in the figure match with the transfer-function _______.

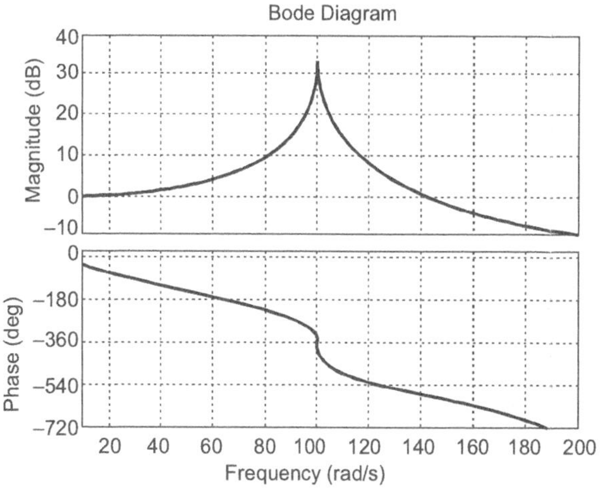

(a) $\dfrac{10000}{s^2 + 2s + 10000}$

(b) $\dfrac{10000}{s^2 + 2s + 10000}e^{-0.05s}$

(c) $\dfrac{10000}{s^2 + 2s + 10000}e^{-0.5\times10^{-12}s}$

(d) $\dfrac{100}{s^2 + 2s + 100}$

[2023 : 2 M]

7.47 The phase margin of the transfer function $G(s) = \dfrac{2(1-s)}{(1+s)^2}$ is _________degrees. (rounded off to the nearest integer).

[2023 : 2 M]

7.48 The sinusoidal transfer function corresponding to the polar plot shown in the figure, for $T > 0$, is

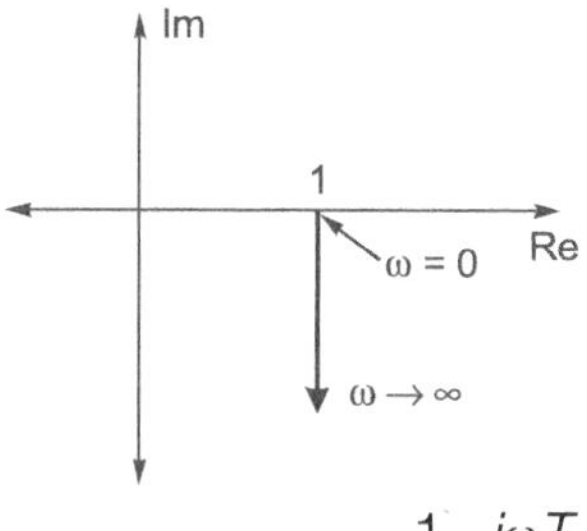

(a) $1 - j\omega T$
(b) $\dfrac{1 - j\omega T}{1 + j\omega T}$
(c) $1 + j\omega T$
(d) $\dfrac{1}{1 + j\omega T}$

[2024 : 1 M]

■■■

Answers Frequency Response Analysis, Bode Plot and Nyquist Plot

7.1 (b)	7.2 (b)	7.3 (a)	7.4 (b)	7.5 (c)	7.6 (b)	7.7 (a)
7.8 (c)	7.9 (a)	7.10 (d)	7.11 (a)	7.12 (a)	7.13 (c, d)	7.14 (d)
7.15 (d)	7.16 (b)	7.17 (c)	7.18 (c)	7.19 (a)	7.20 (a)	7.21 (b)
7.22 (b)	7.23 (b)	7.24 (c)	7.25 (a)	7.26 (c)	7.27 (–68.12)	7.28 (0.33)
7.29 (d)	7.30 (1)	7.31 (0.146)	7.32 (b)	7.33 (90)	7.34 (a)	7.35 (180)
7.36 (0)	7.37 (d)	7.38 (a)	7.39 (38.43)	7.40 (65.5)	7.41 (b)	7.42 (–1)
7.43 (1)	7.44 (a)	7.45 (2)	7.46 (b)	7.47 (0)	7.48 (a)	

Explanations Frequency Response Analysis, Bode Plot and Nyquist Plot

7.1 (b)

$$-3\tan^{-1}\omega = -180°$$

$$\therefore \quad \tan^{-1}\omega = 60°$$

$$\therefore \quad \omega = \sqrt{3} \quad \text{(phase crossover freq.)}$$

$$a = \frac{1}{(\sqrt{1+\omega^2})^3} = \frac{1}{(\sqrt{1+3})^3} = \frac{1}{2^3} = \frac{1}{8}$$

$$\text{Gain margin} = \frac{1}{a} = 8$$

7.2 (b)

PM is determined at gain cross over frequency

$$\frac{1}{(\sqrt{1+\omega^2})^3} = 1$$

$$1+\omega^2 = 1 \qquad \therefore \omega = 0$$

$$PM = \phi + 180°$$

$$= -3\tan^{-1}0 + 180° = 0 + 180°$$

$$= 180° = \pi$$

7.3 (a)

The initial slope is 0 dB/decade, thus it is a type zero system

$$\therefore \; 20\log k = 40$$

$$\log k = \frac{40}{20} \qquad \Rightarrow k = 10^2 = 100$$

$$\therefore \quad \text{T.F.} = \frac{100}{\left(1+\frac{s}{10}\right)\left(1+\frac{s}{250}\right)}$$

7.4 (b)

$$\frac{-\pi}{2} - \frac{180}{\pi}\times L\times 5 = -\pi$$

$$-\frac{180}{\pi}\times 5L = -\pi + \frac{\pi}{2} = -\frac{\pi}{2}$$

$$\frac{180}{\pi}\times 5L = \frac{\pi}{2} \qquad \text{(But } 180° \to \pi\text{)}$$

$$\therefore \quad 5L = \frac{\pi}{2}$$

$$L = \frac{\pi}{10}$$

7.5 (c)

$$\text{Phase angle} = -\tan^{-1}\omega - \frac{0.100\times 180°}{\pi}$$

For $\omega = 1$ $\quad (\because x(t) = 0.5\sin t;\ \omega = 1)$

$$\angle Y(j\omega) - \angle X(j\omega) = -\tan^{-1}(1) - \frac{0.1\times 180°}{\pi}$$

$$= -50.73°$$

7.6 (b)

Since at ω = 2 rad/sec, slope changes from –20 dB/decade to zero, then these would be zero at $\omega = 2$.

At ω = 20 rad/sec, slope changes from 0 dB/decade to –20 dB/decade, then there would be pole at $\omega = 20$.

The slope cut vertical axis at 40 dB.

$$\text{Then, } 20\log\frac{k}{0.2} = 40 \qquad \Rightarrow \quad k = 20$$

$$\text{Hence, transfer function } T(s) = \frac{20\left(\frac{s}{2}+1\right)}{s\left(\frac{s}{20}+1\right)}$$

7.7 (a)

$$T(s) = 5(1 + 0.1s)$$

$$T(j\omega) = 5(1 + 0.1\, j\omega)$$

$$\angle T(j\omega) = \tan^{-1} 0.1\omega$$

At $\omega = 10, \angle T(j\omega) = \tan^{-1}(1) = 45°$

At $\omega = 0, \quad \angle T(j\omega) = \tan^{-1}(0) = 0°$

7.8 (c)

$$-90° - \frac{180° \times \omega_1 t_d}{\pi} = -180°$$

$$\omega_1 = \frac{\pi}{2t_d} = \frac{\pi}{2 \times 0.1} = \frac{\pi}{0.2}$$

7.9 (a)

System 1: $G(s) = \frac{1}{2s+1}$

Time-constant $(\tau_1) = 2$ sec

System 2: $G(s) = \frac{1}{5s+1}$

Time constant $(\tau_2) = 5$ sec.

Bandwidth is inversely proportional to time constant.

For system 1 $\Rightarrow BW_1 \propto \frac{1}{2}$

For system 2 $\Rightarrow BW_2 \propto \frac{1}{5}$

Hence, system 1 is having more bandwidth compared to system 2.

7.10 (d)

Transfer function of the system,

$$T(s) = \frac{k(s+1)}{s^2\left(\frac{s}{100}+1\right)}$$

$$\angle T(j\omega) = -180° + \tan^{-1}\omega - \tan^{-1}\frac{\omega}{100}.$$

Put the minimum value of ω, then will get the minimum absolute phase being 180°.

7.11 (a)

$$N = P - Z$$

where, N = Number of encirclement of the origin mode by $\Delta(s)$ – phase locus $\sqrt{\Delta}$

Z = number of zeros of $\Delta(s)$ encircled by the s-plane locus $\sqrt{s}$ in the s-plane.

Here, $Z = 0$ and number of poles = 2

So, $N = 2 - 0 = 2$

7.12 (a)

The distance from roll gap where thickness is measured = 0.5 m and speed of sheet = 0.2 m/sec.

So, time taken to move this distance

$$= L/v = \frac{0.5}{0.2} = 2.5\,\text{sec}$$

$\therefore$ $m(t)$ is delayed by $d(t)$ by 2.5 sec

So, $m(t) = d(t - 2.5)$

So, $G(s) = \frac{M(s)}{D(s)} = \frac{e^{-2.5s}\cdot D(s)}{D(s)} = e^{-2.5s}$

7.13 (c, d)

At critical value of K, system will be marginally stable and the roll gap is oscillatory with time period of, $T_0 = 2T_{delay} = 5$ sec.

So, the frequency of oscillation will be,

$$\omega_0 = \frac{2\pi}{T_0} = \frac{2\pi}{5} = 0.4\,\pi$$

7.14 (d)

Gain margin: Gain margin are defined at the point of phase cross-over frequency that is the gain when the phase crosses-180°.

Phase margin are defined at the frequency when the gain becomes 1.0 that is 0 dB.

$$\text{Gain margin (GM)} = 20\log\frac{1}{|G(j\omega)|_{\omega=\omega_p}}$$

Where, ω_p = Phase cross-over frequency

$$= 20\log\frac{1}{0.64} = 3.88\text{ dB.}$$

$$\text{Phase margin } (P_m) = 180° + \angle G(j\omega)|_{\omega=\omega_g}$$

Hence, $P_m = 180° - 170° = 10°$

7.15 (d)

$$\angle G(j\omega) = \tan^{-1}\omega - 57.3\,\omega$$

At, $\omega = 2.03$ rad

$$\angle G(j\omega) = \tan^{-1}(2.03) - 57.3 \times 2.03$$

$$= 180°$$

7.16 (b)

$$\frac{C(s)}{R(s)} = \frac{k_c e^{-s}}{(s+1) + k_c e^{-s}}$$

$$E(s) = R(s) - C(s)$$

$$= R(s) - \frac{R(s)k_c e^{-s}}{(s+1)+k_c e^{-s}}$$

$$= R(s)\left\{1-\frac{k_c e^{-s}}{(s+1)+k_c e^{-s}}\right\}$$

$$= R(s)\left[\frac{(s+1)}{(s+1)+k_c e^{-s}}\right]$$

$$e_{ss} = \lim_{s\to 0} s\cdot E(s)$$

$$= \lim_{s\to 0} \frac{s\cdot R(s)\cdot(s+1)}{(s+1)+k_c e^{-s}}$$

$$= \lim_{s\to 0} \frac{s\cdot\frac{1}{s}\cdot(s+1)}{s+1+e^{-s}}$$

$$\left(\because R(s)=\frac{1}{s} \text{ and } k_c = 1\right)$$

$$= \lim_{s\to 0} \frac{(s+1)}{s+1+e^{-s}} = \frac{1}{2}$$

7.17 (c)

0 dB/decade → is a constant line.

+20 dB/decade → is a zero.

Again to reach –20 dB/decade, the +20 dB/decade line will go to 0 dB/decade and then –20 dB/decade.

So, there will be one zero and two poles.

7.18 (c)

$$G(s) = \frac{1+6s}{s^2(1+s)(1+2s)}$$

$$\angle G(j\omega) = -180^\circ + \tan^{-1}(6\omega) - \tan^{-1}\omega - \tan^{-1}2\omega = -180^\circ$$

or $$\omega = \frac{1}{2}\text{ rad/sec}$$

7.19 (a)

$$a = |G(j\omega)|_{\omega=\frac{1}{2}} = 8$$

$$G_m = \frac{1}{a} = \frac{1}{8} = 0.125$$

7.20 (a)

$$T(s) = \frac{k}{s(s+\sigma)}$$

$$P_m = 60^\circ = 180^\circ + \angle T(j\omega)\big|_{\omega=1}$$

$$60^\circ = 180^\circ + \left(-90^\circ - \tan^{-1}\frac{\omega}{\sigma}\right)$$

$$\Rightarrow \tan^{-1}\frac{\omega}{\sigma} = 30^\circ$$

$$\Rightarrow \frac{\omega}{\sigma} = \tan 30^\circ = \frac{1}{\sqrt{3}}$$

$$\Rightarrow \sigma = \sqrt{3}\omega$$

$$\therefore T'(s) = \frac{s-\sqrt{3}}{s+\sqrt{3}}\cdot\frac{k}{s(s+\sigma)}$$

$$\angle T'(j\omega) = -2\tan^{-1}\frac{\omega}{\sqrt{3}} - 90^\circ - \tan^{-1}\left(\frac{\omega}{\sigma}\right)$$

$$\therefore P_m = 180^\circ + \angle T'(j\omega)\big|_{\omega=1}$$

$$= 180^\circ - 2\tan^{-1}\frac{1}{\sqrt{3}} - 90^\circ - \tan^{-1}\left(\frac{1}{\sqrt{3}}\right) = 0^\circ$$

7.21 (b)

System bode plot starts with 0 dB/decade, thus it is a type zero system with transfer function of

$$\frac{k\left(1+\frac{s}{0.1}\right)}{\left(1+\frac{s}{0.5}\right)} = \frac{k[1+10s]}{\left(1+\frac{10s}{5}\right)} = \frac{k(1+10s)}{(1+2s)}$$

$$20\log k = 0$$

$$\therefore \log k = 0$$

$$\therefore k = 1$$

$$\therefore \text{Transfer function of system} = \frac{(1+10s)}{(1+2s)}$$

$$\phi = -\tan^{-1}(2\omega) + \tan^{-1}(10\omega)$$

for ϕ to be maximum, $\frac{d\phi}{d\omega} = 0$,

$$-\frac{2}{1+(2\omega)^2} + \frac{10}{1+(10\omega)^2} = 0$$

$$-\frac{2}{1+4\omega^2} + \frac{10}{1+100\omega^2} = 0$$

$$\frac{10}{1+100\omega^2} = \frac{2}{1+4\omega^2}$$

$$5(1+4\omega^2) = 1 + 100\omega^2$$

$$5 + 20\omega^2 = 1 + 100\omega^2$$

$$4 = 80\omega^2$$

$$\therefore \omega^2 = \frac{1}{20}$$

$$\omega = \frac{1}{\sqrt{20}}\text{ rad/sec}$$

7.22 (b)

Given, $$G(s) = \frac{K}{(s+5)^3}$$

$$\angle G(j\omega) = -3\tan^{-1}\frac{\omega}{5}$$

Phase margin = $\angle G(j\omega) + 180°$

$$45° = -3\tan^{-1}\frac{\omega}{5} + 180°$$

$$\tan^{-1}\left(\frac{\omega}{5}\right) = \frac{135}{3} = 45°$$

$$\frac{\omega}{5} = \tan 45°$$

$$\therefore \quad \omega = 5 \text{ rad/sec}$$

According to magnitude condition

$$|G(s)H(s)| = 1$$

$$\frac{K}{(\omega^2+5^2)^{3/2}} = 1$$

$$K = (\omega^2+5^2)^{3/2}$$

$$\boxed{K = 250\sqrt{2}} \qquad (\because \omega = 5 \text{ rad/sec})$$

7.23 (b)

Closed loop T.F.

$$= \frac{G(s)H(s)}{1+G(s)H(s)}$$

$$= \frac{\frac{k}{(s+5)^3}\times 1}{1+\frac{k}{(s+5)^3}\times 1} = \frac{k}{(s+5)^3+k}$$

Characteristic equation,

$(s+5)^3 + k = 0$

$\therefore s^3 + 15s^2 + 75s + k + 125 = 0$ ...(i)

Characteristic for the dominant

Conjugate pole pair,

$(s^2 + 2\zeta\omega_n s + \omega_n^2)(s+\alpha) = 0$

$\zeta = 0.5$ (is given)

$\therefore (s^2 + \omega_n s + \omega_n^2)(s+\alpha) = 0$

$\therefore s^3 + (\alpha+\omega_n)s + (2\omega_n + \omega_n^2)s + \alpha\omega_n^2 = 0$ ...(ii)

comparing equation (i) and (ii)

$\alpha + \omega_n = 15 \Rightarrow \alpha = 15 - \omega_n$

$$\alpha\omega_n + \omega_n^2 = 75$$

$$\therefore \quad (15-\omega_n)\,\omega_n + \omega_n^2 = 75$$

$$\therefore \quad 15\omega_n - \omega_n^2 + \omega_n^2 = 75$$

$$\therefore \quad 15\,\omega_n = 75$$

$$\therefore \quad \omega_n = 5 \text{ rad/sec}$$

and $\alpha = 15 - 5 = 10$

$$\therefore \quad \alpha = 10$$

$$k + 125 = \alpha\omega_n^2 = 10\times(5)^2 = 250$$

$$\therefore \quad k = 125$$

7.24 (c)

For sinusoidal excitation

$$s = j\omega$$

$$\therefore \quad G(j\omega) = \frac{(-\omega^2+9)(j\omega+2)}{(j\omega+1)(j\omega+3)(j\omega+4)}$$

For zero steady-state output

$$|G(j\omega)| = 0$$

$$= \frac{(-\omega^2+9)\sqrt{\omega^2+4}}{(\sqrt{\omega^2+1})(\sqrt{\omega^2+9})(\sqrt{\omega^2+16})}$$

For zero steady-state output

$\Rightarrow \quad \omega^2 = 9$

$\Rightarrow \quad \omega = 3$ rad/sec

7.25 (a)

The characteristic equation

$$1 + G(s)H(s) = 0$$

$\Rightarrow s^3 + as^2 + (k+2)s + (k+1) = 0$

Routh array for above equation

s^3	1	$(k+2)$
s^2	a	$(k+1)$
s^1	$\frac{a(k+2)-(k+1)}{a}$	0
s^0	$(k+1)$	0

$$as^2 + (k+1) = 0$$

For $s = j\omega$ and $\omega = 2$

$$-a\omega^2 + (k+1) = 0$$

$$k = 4a - 1$$

and $$a = \frac{k+1}{k+2}$$

$\Rightarrow \quad a = 0.75$ and $k = 2$

7.26 (c)

Phase cross over frequency is given as

$$-90° - \tan^{-1}\frac{\omega}{9} - \tan^{-1}\frac{\omega}{25} = -180°$$

$$\tan^{-1}\frac{\omega}{9} + \tan^{-1}\frac{\omega}{25} = 90°$$

$$\tan^{-1}\frac{\left(\frac{\omega}{9}+\frac{\omega}{25}\right)}{\left(1-\frac{\omega^2}{225}\right)} = 90°$$

$$\frac{\left(\frac{\omega}{9}+\frac{\omega}{25}\right)}{\left(1-\frac{\omega^2}{225}\right)} = \tan 90° = \infty$$

$$\therefore \quad 1-\frac{\omega^2}{225} = 0$$

$$\omega = 15$$

(Phase cross over frequency).

$$a = \frac{150}{15(9+15j)(25+15j)}$$

$$= \frac{150}{15\sqrt{9^2+15^2}\sqrt{25^2+15^2}}$$

$$= \frac{150}{15\sqrt{306}\sqrt{850}} = 0.0196$$

$$GM = 20\log\frac{1}{a}$$

$$= 20\log\frac{1}{0.0196} = 34.1 \text{ dB}$$

7.27 (-68.12)

$$G(s) = \frac{e^{-s/500}}{s+500}$$

$$G(j\omega) = \frac{e^{-j\omega/500}}{j\omega+500}$$

$$\phi = \angle G(j\omega) = -\left[\tan^{-1}\frac{\omega}{500}+\frac{\omega}{500}\right]$$

$$\omega = 100\pi$$

$$\phi = -\left[\tan^{-1}\frac{100\pi}{500}+\frac{100\pi}{500}\right]$$

$$= -\left[\tan^{-1}\frac{\pi}{5}+\frac{\pi}{5}\right] = -[32.12° + 36°]$$

$$= -68.12°$$

7.28 (0.33)

Phase cross-over frequency

$$-90° - \tan^{-1}\omega - \tan^{-1}9\omega = -180°$$

$$-\tan^{-1}\omega - \tan^{-1}9\omega = -90°$$

$$\tan^{-1}\omega + \tan^{-1}9\omega = 90°$$

$$\tan^{-1}\left(\frac{\omega+9\omega}{1-\omega\times 9\omega}\right) = 90°$$

$$\tan^{-1}\left(\frac{10\omega}{1-9\omega^2}\right) = 90°$$

$$\frac{10\omega}{1-9\omega^2} = \infty$$

$$1-9\omega^2 = 0$$

$$\therefore \quad \omega = \frac{1}{3} = 0.33 \text{ rad/sec}$$

7.29 (d)

According to above figure:

$-1 + j0$ point is on the right side of the given Nyquist plot. So system is unstable.

$\because$ Magnitude at $\omega = 0$ is infinite so, there is no constant part in the denominator of transfer function of the given system.

$$\text{Assume,} \quad G(s) = \frac{1}{s^2-s} = \frac{1}{s(s-1)}$$

$$\text{Magnitude} = \frac{1}{\sqrt{\omega^2(\omega^2+1)}} = \frac{1}{\omega\sqrt{\omega^2+1}}$$

$$\text{Phase } \phi = -270° + \tan^{-1}(\omega)$$

$$\omega = 0\,;\ M = \infty\,;\ \phi = -270°$$

$$\omega = \infty\,;\ M = 0\,;\ \phi = -270° + 90° = -180°$$

So, our assumption is true.

The system is type-1 and is unstable.

7.30 (1)

$$G(s) = \frac{s-1}{s+1} \quad G(j\omega) = \frac{j\omega-1}{j\omega+1}$$

$$= 1\angle 180° - 2\tan^{-1}\omega$$

on varying $\omega = 0^+$ to $+\infty$

Now on varying $\omega = 0^-$ to $-\infty$

$$G(j\omega)|_{\omega=0} = 1\angle 180°$$

$$G(j\omega)|_{\omega=0^-} = 1\angle 180°$$

$$G(j\omega)|_{\omega=1} = 1\angle 90°$$

$$G(j\omega)|_{\omega=-1} = 1\angle 270°$$

$$G(j\omega)|_{\omega=\infty} = 1\angle 0°$$

$$G(j\omega)|_{\omega=-\infty} = 1\angle 360°$$

On plotting,

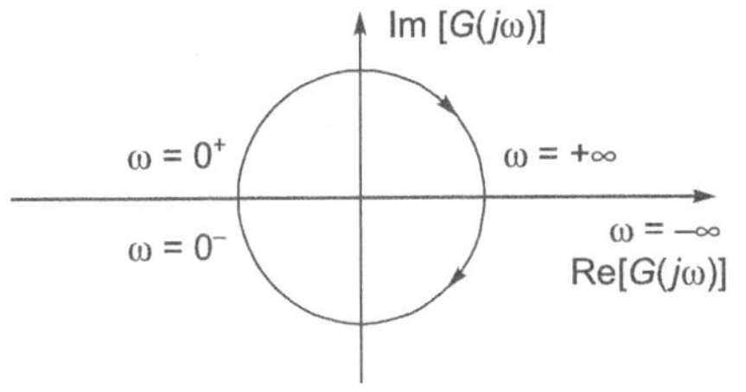

$\therefore$ No. of times the Nyquist plot encircle the origin clockwise = 1

7.31 0.146 (0.14 to 0.15)

$$G(s) = \frac{8}{(s+10)^2}$$
$$s = j\omega$$
$$G(j\omega) = \frac{8}{(j\omega+10)^2}$$
$$|G(j\omega)| = \frac{8}{\left(\sqrt{\omega^2+10^2}\right)^2}$$

output amplitude = input amplitude × $|G(\omega)|$

$$= 2\times\frac{8}{\left(\sqrt{\omega^2+10^2}\right)^2}$$

where ω = 3 rad/sec

$$\text{Output amplitude} = \frac{2\times 8}{\left(\sqrt{9+100}\right)^2} = \frac{16}{109}$$
$$= 0.146 \approx 0.15$$

7.32 (b)

Since the initial slope is 0 dB/decade, thus it a type zero system

$\therefore$ $20\log K = 0$

$K = 1$

now at $\omega_1 = 10^0$ rad/sec

Slope goes to 40 dB/decade

$\therefore$ Two zeros added at $\omega_1 = 10^0$ rad/sec

and at $\omega_2 = 10^2$ rad/sec

Slope goes to 0 dB/dec

$\therefore$ Two poles added at

$\omega_2 = 10^2$ rad/sec

Thus,

$$\text{T.F.} = \frac{K\left(\frac{s}{\omega_1}+1\right)^2}{\left(\frac{s}{\omega_2}+1\right)^2} = \frac{K\left(\frac{s}{1}+1\right)^2}{\left(\frac{s}{100}+1\right)^2}$$
$$= 10^4\frac{(s+1)^2}{(s+100)^2}$$

7.33 90 (90 to 90 or −270 to −270)

$$G(s) = \frac{s-1}{s+1}$$
$$p(t) = \sin t$$
$$y(t) = \sin(t+\phi)$$
$$\therefore \quad \phi = \angle p(t) + \angle G(j\omega) \quad \text{at } \omega = 1$$
$$= 0^\circ + 180^\circ - 90^\circ = 90^\circ$$

7.34 (a)

$$\text{TF} = \frac{100^2 e^{-0.01s}}{s^2+0.2s+100^2}$$

Here, phase of $\frac{100^2}{s^2+0.2s+100^2}$ varies from 0° to −180° as $\omega : 0 \to \infty$, therefore $e^{-0.01s}$ causes further phase lag, hence the correct answer is (a).

7.35 180 (180 to 180 or −180 to −180)

$$G(s) = \frac{2}{(s+1)(s+2)}$$
$$\text{PM} = 180^\circ + \angle G(j\omega)\Big|_{\omega=\omega_{gc}}$$

At, $|G(j\omega)|_{\omega=\omega_{gc}} = 1$

$$\therefore \quad \frac{2}{\sqrt{\omega_{gc}^2+1}\sqrt{\omega_{gc}^2+4}} = 1$$

$\therefore$ $\omega_{gc} = 0$ and PM = 180°

7.36 (0)

Given, $G(s) = \frac{s^2+0.2s+100}{s^2-0.2s+100}$, $H(s) = \frac{1}{2}$

$$q(s) = 1 + G(s)H(s) = 0$$
$$= 3s^2 - 0.2s + 300 = 0$$

This system is unstable with two roots in right side of s-plane.

$\therefore$ $P - N = 2$

$P = 2$ since OLTF has two poles in right side of s-plane.

$\therefore$ $N = 0$

Or,

Nyquist plot:

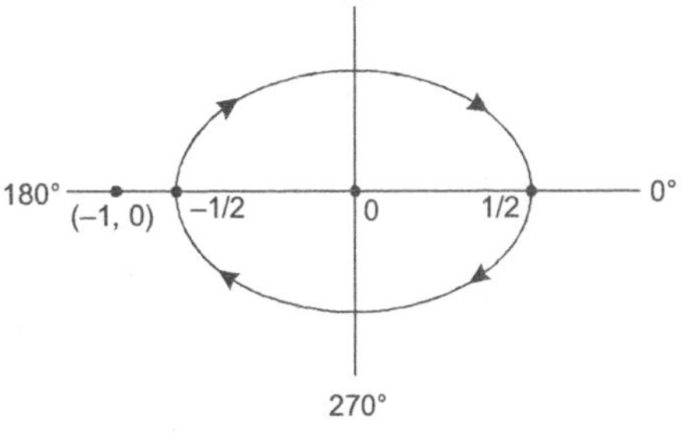

$\because$ (–1, 0) lies outside the circle,

$\therefore \quad N = 0$

7.37 (d)

$$\text{OLTF} = \frac{k}{(s+1)(s+2)(s+3)}$$

If the $\omega_{gc} < \omega_{pc}$ then that minimum phase system is stable.

Therefore, answer (d)

7.38 (a)

$$\frac{Y(s)}{D(s)} = \frac{1}{1+L(s)}$$

$$\left|\frac{Y(j\omega)}{D(j\omega)}\right|_{\omega=5} = \frac{1}{\sqrt{1+(100)^2}} \simeq \frac{1}{100}$$

$\therefore$ Steady-state amplitude of output

$$= 0.1 \times \frac{1}{100} = 1.00 \times 10^{-3}$$

7.39 38.43 (37 to 39)

$$G(s) = \frac{9e^{-0.1s}}{s}$$

$$PM = 180° + \left[-90° - 0.1\omega \times \frac{180°}{\pi}\right]_{\omega=\omega_{gc}}$$

$$\omega_{gc} = 9 \text{ r/s}$$

$\therefore$ Phase margin = 38.43°

7.40 (65.5)

$$G(s)H(s) = \frac{2(s+1)}{s^2}$$

$$\omega_{gc} = 2.2$$

$$\text{PM} = 180° + [-180° + \tan^{-1}(\omega_{gc})]$$

$$= 65.5°$$

7.41 (b)

$$G(s)H(s) = \frac{1}{s(s-2)}$$

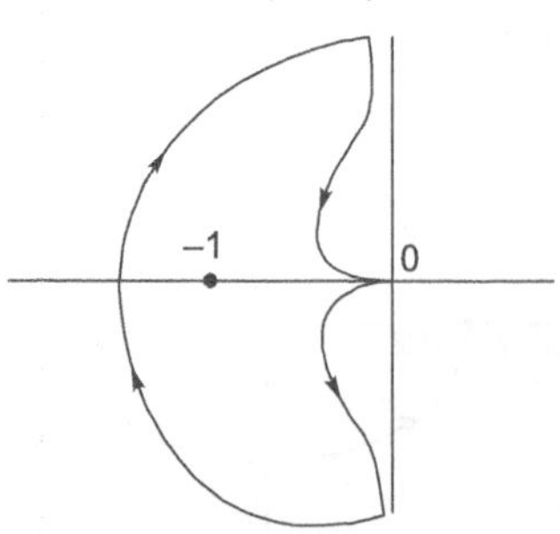

From Nyquist plot

$$N = -1$$

$$P = 1$$

$\therefore$ Nyquist plot encircles (–1, 0) once in clockwise direction.

7.42 (–1)

$$\text{OLTF} = \frac{3}{s-1}$$

Nyquist plot

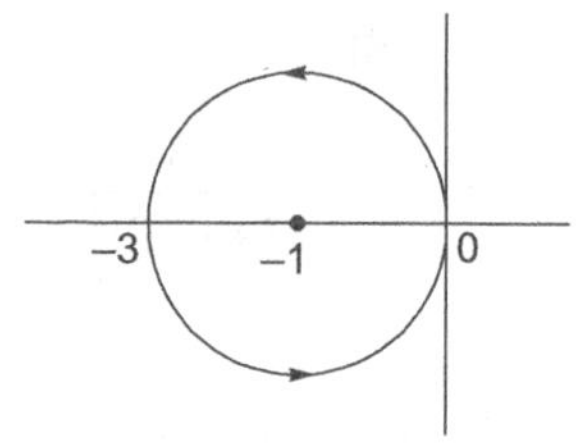

Since N is positive in CW and negative in ACW as in the question,

$\therefore \quad N = -1$

7.43 (1)

$$G(s) = \frac{1}{s+1}$$

$$r(t) = A \sin \omega t = \sqrt{2}\sin t$$

$$c(t) = B \sin (\omega t + \phi)$$

$$B = A|G(j\omega)|$$

$$= \sqrt{2}\left|\frac{1}{s+1}\right|_{s=j1} = 1$$

7.44 (a)

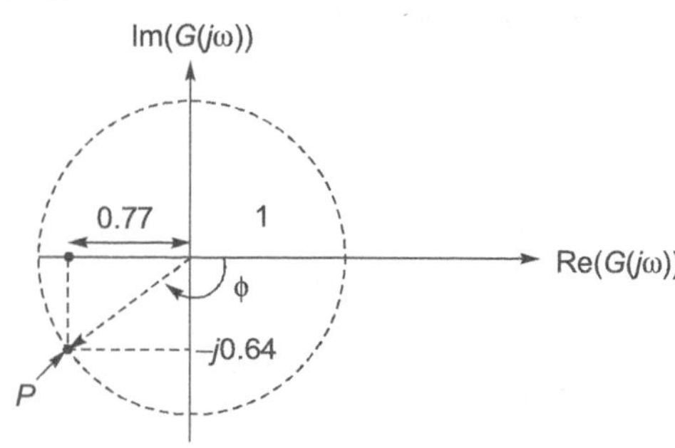

$$\phi = \left[90° + \tan^{-1}\left(\frac{0.77}{0.64}\right)\right]$$

$$\phi = -140.26°$$

$$\text{P.M} = 180° + \phi = 39.73°$$

$$\text{G.M} = \left|\frac{1}{R}\right| = \frac{1}{0.21} = 4.76$$

7.45 (2)

$$G(s)H(s) = \frac{(s-1)(s-2)}{2(s+1)(s+2)}$$

∴ Number of poles which are in RHS, $P = 0$

Number of zeros which are in RHS, $Z = 2$

(Here it is asking for number of encirclement around the origin).

∴ Encirclement,

$$N = P - Z = 0 - 2$$

$$N = -2$$

∴ 2 clockwise encirclement.

7.46 (b)

From the figure it is clear that there should be phase lag. Hence, option (b) and (c) could be answer.

From the figure, at $\omega = 100$ rad/sec, Phase is $-360°$.

Now considering option (b) only delay element

$$e^{-0.05s} = e^{-0.05\times 100j} = e^{-j5}$$

Hence, phase is = –5 radians = –286.5°

and from the rest of transfer function

$$\frac{1000}{s^2+2s+10000} \text{ at } \omega = \omega_n = 100 \text{ rad/sec phase}$$

is –90°.

$$= -286.5 - 90° = -376.5°$$

Hence, option (b) satisfy the graph.

7.47 (0)

Given transfer function,

$$G(s) = \frac{2(1-s)}{(1+s)^2}$$

Phase margin is calculated at gain crossover frequency ω_{gc}.

ω_{gc} is the frequency at which the magnitude of $G(j\omega)H(j\omega)$ is 1.

$$|G(j\omega)H(j\omega)|_{\omega=\omega_{gc}} = \frac{2\sqrt{1+\omega_{gc}^2}}{\left(1+\omega_{gc}^2\right)} = \frac{2}{\sqrt{1+\omega_{gc}^2}} = 1$$

$$2 = \sqrt{1+\omega_{gc}^2}$$

$$4 = 1+\omega_{gc}^2$$

$$\omega_{gc} = \sqrt{3} \text{ rad/sec}$$

Now ϕ at $\omega_{gc} = \tan^{-1}[-\omega_{gc}] - 2\tan^{-1}[\omega_{gc}]$

$$= -3\tan^{-1}[\omega_{gc}]$$

$$= -3\tan^{-1}\left(\sqrt{3}\right)$$

$$= -180°$$

∴Phase margin $= 180 + \phi$ at ω_{gc}

$$= 180° - 180° = 0°$$

■■■

8 State Space Analysis

8.1 For a dynamic system, various matrices for a state space model are given by $A = \begin{bmatrix} 0 & 1 \\ -2 & -3 \end{bmatrix}, B = \begin{bmatrix} 0 \\ 1 \end{bmatrix}$ and $C = [1 \ \ 0]$.

The system poles are located at

(a) −2, −1 (b) −2, −3

(c) −3, −1 (d) −1, 0 **[2003 : 2 M]**

8.2 A state space representation for the transfer function $\frac{Y(s)}{U(s)} = \frac{s+6}{s^2+5s+6}$ is $\dot{x} = Ax + Bu$, where

$A = \begin{bmatrix} 0 & 1 \\ -6 & -5 \end{bmatrix}, \ B = \begin{bmatrix} 0 \\ 1 \end{bmatrix}$, and the value of C is

(a) $\begin{bmatrix} 1 \\ 2 \end{bmatrix}$ (b) $[6 \ \ 1]$

(c) $\begin{bmatrix} 1 & 2 \\ -1 & 0 \end{bmatrix}$ (d) $[1 \ \ -5]$

[2004 : 2 M]

8.3 A single input single output linear system is a described by the equations $\frac{dx}{dt} = Ax + Bu$ and $y = cx + du$, where A is an $[n \times n]$ matrix and x and b are $[n \times 1]$ vectors. The transfer function is given by

(a) $c[sI - A]^{-1}b$ (b) $c[sI - A]^{-1}\, b + d$

(c) $c[sI - A]b$ (d) $c[sI - A]b + d$

[2005 : 2 M]

8.4 The state-variable representation of a plant is given by

$$\dot{X} = Ax + Bu$$
$$y = Cx$$

where x is the state, u is the input and y is the output. Assuming zero initial condition, the impulse response of the plant is given by

(a) $\exp(At)$ (b) $\int \exp[A(t-\tau)]Bu(\tau)d\tau$

(c) $C\exp(At)B$ (d) $C\int \exp[A(t-\tau)]Bu(\tau)d\tau$

[2006 : 1 M]

8.5 Consider the following standard state-space description of a linear time-invariant single input single output system: $\dot{x} = Ax + Bu$, $y = Cx + Du$. Which one of the following statements about the transfer function cannot be true if $D \neq 0$?

(a) The system is unstable

(b) The system is strictly proper

(c) The system is low pass

(d) The system is of type zero **[2007 : 1 M]**

8.6 The state space representation of a system is given by

$$\dot{x} = \begin{bmatrix} 0 & 1 \\ 0 & -3 \end{bmatrix} x + \begin{bmatrix} 1 \\ 0 \end{bmatrix} u, y = [1 0]x$$

The transfer function $\frac{Y(s)}{U(s)}$ of the system will be

(a) $\frac{1}{s}$ (b) $\frac{1}{s(s+3)}$

(c) $\frac{1}{s+3}$ (d) $\frac{1}{s^2}$ **[2008 : 2 M]**

8.7 A linear time-invariant single-input single-output system has a state space model given by

$\frac{dx}{dt} = Fx + Gu;\ y = Hz$

where, $F = \begin{bmatrix} 0 & 1 \\ -4 & -2 \end{bmatrix}; G = \begin{bmatrix} 0 \\ 1 \end{bmatrix};\ H = [1 \ \ 0]$.

Here, x is the state vector, u is the input, and y is the output.

The damping ratio of the system is

(a) 0.25 (b) 0.5

(c) 1 (d) 2 **[2009 : 2 M]**

8.8 The transfer function of the system described by the state-space equations

$$\begin{bmatrix} \dot{x}_1 \\ \dot{x}_2 \end{bmatrix} = \begin{bmatrix} -4 & -1 \\ -3 & -1 \end{bmatrix}\begin{bmatrix} x_1 \\ x_2 \end{bmatrix} + \begin{bmatrix} 1 \\ 1 \end{bmatrix} u, \quad y = [1 \ \ 0]\begin{bmatrix} x_1 \\ x_2 \end{bmatrix}$$

(a) $\frac{s}{s^2+5s+1}$ (b) $\frac{2s}{s^2+5s+1}$

(c) $\frac{3s}{s^2+5s+1}$ (d) $\frac{4s}{s^2+5s+1}$

[2011 : 2 M]

8.9 The state variable description of an LTI system is given by

$$\begin{pmatrix}\dot{x}_1\\ \dot{x}_2\\ \dot{x}_3\end{pmatrix} = \begin{pmatrix}0 & a_1 & 0\\ 0 & 0 & a_2\\ a_3 & 0 & 0\end{pmatrix}\begin{pmatrix}x_1\\ x_2\\ x_3\end{pmatrix} + \begin{pmatrix}0\\ 0\\ 1\end{pmatrix}u$$

$$y = (1\ \ 0\ \ 0)\begin{pmatrix}x_1\\ x_2\\ x_3\end{pmatrix}$$

where y is the output and u is the input. The system is controllable for

(a) $a_1 \neq 0,\ a_2 = 0,\ a_3 \neq 0$
(b) $a_1 = 0,\ a_2 \neq 0,\ a_3 \neq 0$
(c) $a_1 = 0,\ a_2 \neq 0,\ a_3 = 0$
(d) $a_1 \neq 0,\ a_2 \neq 0,\ a_3 = 0$

[2012 : 2 M]

8.10 A system is represented in state-space as $\dot{X} = Ax + Bu$, where $A = \begin{bmatrix}1 & 2\\ \alpha & 6\end{bmatrix}$ and $B = \begin{bmatrix}1\\ 1\end{bmatrix}$. The value of α for which the system is not controllable is _____.

[2015 : 2 M]

8.11 Consider the linear system $\dot{x} = \begin{bmatrix}-1 & 0\\ 0 & -2\end{bmatrix}x$, with initial condition $x(0) = \begin{bmatrix}1\\ 1\end{bmatrix}$. The solution $x(t)$ for this system is

(a) $x(t) = \begin{bmatrix}e^{-t} & te^{-2t}\\ 0 & e^{-2t}\end{bmatrix}\begin{bmatrix}1\\ 1\end{bmatrix}$

(b) $x(t) = \begin{bmatrix}e^{-t} & 0\\ 0 & e^{2t}\end{bmatrix}\begin{bmatrix}1\\ 1\end{bmatrix}$

(c) $x(t) = \begin{bmatrix}e^{-t} & -t^2e^{-2t}\\ 0 & e^{-2t}\end{bmatrix}\begin{bmatrix}1\\ 1\end{bmatrix}$

(d) $x(t) = \begin{bmatrix}e^{-t} & 0\\ 0 & e^{-2t}\end{bmatrix}\begin{bmatrix}1\\ 1\end{bmatrix}$

[2018 : 2 M]

8.12 The dynamics of the state $\begin{bmatrix}x_1\\ x_2\end{bmatrix}$ of a system is governed by the differential equation

$$\begin{bmatrix}\dot{x}_1\\ \dot{x}_2\end{bmatrix} = \begin{bmatrix}1 & 2\\ -3 & -4\end{bmatrix}\begin{bmatrix}x_1\\ x_2\end{bmatrix} + \begin{bmatrix}20\\ 10\end{bmatrix}$$

Given that the initial state is $\begin{bmatrix}0\\ 0\end{bmatrix}$, the steady state value of $\begin{bmatrix}x_1\\ x_2\end{bmatrix}$ is

(a) $\begin{bmatrix}-30\\ -40\end{bmatrix}$ (b) $\begin{bmatrix}-20\\ -10\end{bmatrix}$

(c) $\begin{bmatrix}5\\ -15\end{bmatrix}$ (d) $\begin{bmatrix}50\\ -35\end{bmatrix}$

[2019 : 2 M]

8.13 Consider the following state variable equations :

$\dot{x}_2(t) = x_2(t)$

$\dot{x}_2(t) = -6x_1(t) - 5x_2(t)$

The initial conditions are $x_1(0) = 0$ and $x_2(0) = 1$. At $t = 1$ second, the value of $x_2(1)$ (rounded off to two decimal places) is _____.

[2020 : 2 M]

8.14 The transfer function of a system is:

$$\frac{(s+1)(s+3)}{(s+5)(s+7)(s+9)}$$

In the state-space representation of the system, the minimum number of state variables (in integer) necessary is _______.

[2022 : 1 M]

8.15 Consider the state-space representation of a system

$$\dot{x} = Ax + Bu$$

where x is the state vector, u is the input, A is the system matrix and B is the input matrix. Choose the matrix A from the following options such that the system has a pole at the origin.

(a) $\begin{bmatrix}0 & 1\\ -2 & -3\end{bmatrix}$ (b) $\begin{bmatrix}1 & -1.5\\ -2 & 3\end{bmatrix}$

(c) $\begin{bmatrix}1 & 1.5\\ 2 & -3\end{bmatrix}$ (d) $\begin{bmatrix}0 & 1\\ -2 & 3\end{bmatrix}$

[2024 : 1 M]

■■■

Answers State Space Analysis

8.1 (a)	8.2 (b)	8.3 (b)	8.4 (d)	8.5 (a)	8.6 (a)	8.7 (b)
8.8 (a)	8.9 (d)	8.10 (–3)	8.11 (d)	8.12 (d)	8.13 (–0.12)	8.14 (3)
8.15 (b)						

Explanations State Space Analysis

8.1 (a)

$$T(s) = C|sI - A|^{-1} B$$

$$= [1 \quad 0]\begin{bmatrix} s & -1 \\ 2 & s+3 \end{bmatrix}^{-1}\begin{bmatrix} 0 \\ 1 \end{bmatrix}$$

$$= \frac{[1 \quad 0]\begin{bmatrix} s+3 & 1 \\ -2 & s \end{bmatrix}\begin{bmatrix} 0 \\ 1 \end{bmatrix}}{s^2+3s+2} = \frac{1}{s^2+3s+2}$$

The system poles are given by

$$s^2 + 3s + 2 = 0$$

$$(s + 2)(s + 1) = 0$$

or $\quad s = -1, s = -2$

8.2 (b)

$$\frac{Y(s)}{U(s)} = \frac{s+6}{s^2+5s+6}$$

or $\quad \frac{X_1(s)}{U(s)} \times \frac{Y(s)}{X_1(s)} = \frac{s+6}{s^2+5s+6}$

Hence, $\quad \frac{X_1(s)}{U(s)} = \frac{1}{s^2+5s+6}$

Let the state variable be defined as

$$x_1(t) = y(t)$$

$$\dot{x}_1 = x_2$$

$$\ddot{x}_1 = \dot{x}_2$$

then $\ddot{x}_1 + 5\dot{x}_1 + 6x_1 = u(t) \quad$ ...(i)

or $\quad \dot{x}_2 = -5x_2 - 6x_1 + u(t)$

and $\frac{Y(s)}{X_1(s)} = s + 6$

or $\quad Y(t) = \dot{x}_1 + 6x_1 = x_2 + 6x_1 \quad$ ...(ii)

Writing equation (i) and equation (ii) in matrix from, we have

$$y = [6 \quad 1]\begin{bmatrix} x_1 \\ x_2 \end{bmatrix}$$

and $\begin{bmatrix} \dot{x}_1 \\ \dot{x}_2 \end{bmatrix} = \begin{bmatrix} 0 & 1 \\ -6 & -5 \end{bmatrix}\begin{bmatrix} x_1 \\ x_2 \end{bmatrix} + \begin{bmatrix} 0 \\ 1 \end{bmatrix}u(t)$

Hence, the value of C is = [6 1]

8.4 (d)

This is the direct formula.

8.5 (a)

$\because \quad D \neq 0$

then, $\quad Y = CX + DU$

if input U is bounded, then output Y is bounded so can not be unstable.

8.6 (a)

$$T(s) = C[sI - A]^{-1}B$$

$$= [1 \quad 0]\begin{bmatrix} s & -1 \\ 0 & s+3 \end{bmatrix}^{-1}\begin{bmatrix} 1 \\ 0 \end{bmatrix}$$

$$= \frac{[1 \quad 0]\begin{bmatrix} s+3 & +1 \\ 0 & s \end{bmatrix}\begin{bmatrix} 1 \\ 0 \end{bmatrix}}{s(s+3)}$$

$$= \frac{(s+3)}{s(s+3)} = \frac{1}{s}$$

8.7 (b)

$$T(s) = C[sI - A]^{-1}B$$

Here, $\quad A = F = \begin{bmatrix} 0 & 1 \\ -4 & -2 \end{bmatrix};$

$$B = G = \begin{bmatrix} 0 \\ 1 \end{bmatrix} \text{ and } C = H = [1 \quad 0];$$

$$T(s) = [1 \quad 0]\begin{bmatrix} s & -1 \\ 4 & (s+2) \end{bmatrix}^{-1}\begin{bmatrix} 0 \\ 1 \end{bmatrix}$$

$$= \frac{[1 \quad 0]\begin{bmatrix} (s+2) & 1 \\ -4 & s \end{bmatrix}\begin{bmatrix} 0 \\ 1 \end{bmatrix}}{s^2+2s+4} = \frac{1}{s^2+2s+4}$$

$$2\xi\omega_n = 2$$

$$\xi = \frac{1}{2} = 0.5 \qquad (\because \omega_n = 2)$$

8.8 (a)

Here, $A = \begin{bmatrix} -4 & -1 \\ -3 & -1 \end{bmatrix}$

$B = \begin{bmatrix} 1 \\ 1 \end{bmatrix}$

$C = [1 \quad 0]$

$D = 0$

$$T(s) = C[sI - A]^{-1} B + D$$

$$= C[sI - A]^{-1}B \qquad (\because D = 0)$$

$$[sI - A] = \begin{bmatrix} (s+4) & 1 \\ 3 & (s+1) \end{bmatrix}$$

$$[sI - A]^{-1} = \begin{bmatrix} (s+1) & -1 \\ -3 & (s+4) \end{bmatrix} \cdot \frac{1}{s^2 + 5s + 1}$$

$$\therefore \quad T(s) = \frac{[1 \quad 0]\begin{bmatrix} (s+1) & -1 \\ -3 & (s+4) \end{bmatrix}\begin{bmatrix} 1 \\ 1 \end{bmatrix}}{(s^2 + 5s + 1)}$$

$$= \frac{[(s+1) \quad -1]\begin{bmatrix} 1 \\ 1 \end{bmatrix}}{s^2 + 5s + 1}$$

$$= \frac{s+1-1}{s^2+5s+1} = \frac{s}{s^2+5s+1}$$

8.9 (d)

$$A = \begin{bmatrix} 0 & a_1 & 0 \\ 0 & 0 & a_2 \\ a_3 & 0 & 0 \end{bmatrix},$$

$$B = \begin{bmatrix} 0 \\ 0 \\ 1 \end{bmatrix}$$

$$C = [1 \quad 0 \quad 0]^T$$

For system to be controllable, the matrix Q_c must be non singular.

$$Q_c = [B \quad AB \quad A^2B]$$

$$AB = [0 \quad a_2 \quad 0]^T$$

$$A^2 = \begin{bmatrix} 0 & 0 & a_1 a_2 \\ a_2 a_3 & 0 & 0 \\ 0 & a_1 a_3 & 0 \end{bmatrix}$$

$$A^2B = [a_1 a_2 \quad 0 \quad 0]^T$$

$$Q_c = \begin{bmatrix} 0 & 0 & a_1 a_2 \\ 0 & a_2 & 0 \\ 1 & 0 & 0 \end{bmatrix}$$

$$|Q_c| = -a_1 a_2^2$$

For Q_c to be non-singular

$$|Q_c| \neq 0$$

$$\therefore \quad a_1 \neq 0 \text{ and } a_2 \neq 0 \text{ and } a_3 \in R$$

8.10 (–3)

$$Q_c = |B \quad AB| = 0$$

$$AB = \begin{bmatrix} 1 & 2 \\ X & 6 \end{bmatrix}\begin{bmatrix} 1 \\ 1 \end{bmatrix} = \begin{bmatrix} 3 \\ \alpha + 6 \end{bmatrix}$$

$$Q_c = \begin{vmatrix} 1 & 3 \\ 1 & \alpha + 6 \end{vmatrix} = 0$$

$$\alpha + 6 - 3 = 0$$

$$\alpha = -3$$

8.11 (d)

$$\dot{X} = \begin{bmatrix} -1 & 0 \\ 0 & -2 \end{bmatrix}; \; x(0) = \begin{bmatrix} 1 \\ 1 \end{bmatrix}$$

$$X(t) = \phi(t) \cdot x(0)$$

$$\phi(t) = L^{-1}\left[(sI - A)^{-1}\right] = \begin{bmatrix} e^{-t} & 0 \\ 0 & e^{-2t} \end{bmatrix}$$

$$\therefore \quad X(t) = \begin{bmatrix} e^{-t} & 0 \\ 0 & e^{-2t} \end{bmatrix}\begin{bmatrix} 1 \\ 1 \end{bmatrix}$$

8.12 (d)

$$X(s) = \phi(s) \cdot x(0) + \phi(s) \cdot B \cdot U(s)$$

where, $\phi(s) = (sI - A)^{-1}$ Given $x(0) = \begin{bmatrix} 0 \\ 0 \end{bmatrix}$

$$\therefore \quad X(s) = \phi(s) \cdot B \cdot U(s)$$

$$= \frac{1}{(s-1)(s+4)+6} \cdot \begin{bmatrix} 20(s+4) + 20 \\ -60 + 10(s-1) \end{bmatrix}$$

Steady-state value of $X(t)$ for step input is

$$X(\infty) = \lim_{s \to 0} s\,X(s) = \begin{bmatrix} 50 \\ -35 \end{bmatrix}$$

8.13 (–0.12)

$$\dot{x}_1(t) = x_2(t)$$

$$Sx_1(s) - x_1(0)_0 = x_2(s)$$

$$\therefore \quad x_1(s) = \frac{x_2(s)}{s}$$

$$\dot{x}_2(t) = -6x_1(t) - 5x_2(t)$$

$$Sx_2(s) - x_2(0) = -6x_1(s) - 5x_2(s)$$

$$Sx_2(s) - 1 = -\frac{6x_2(s)}{s} - 5x_2(s)$$

$$x_2(s)\left(s + \frac{6}{s} + 5\right) = 1$$

$$x_2(s) = \frac{s}{s^2 + 5s + 6}$$

$$x_2(s) = \frac{-2}{s+2} + \frac{3}{s+3}$$

$$x_2(t) = -2e^{-2t} + 3e^{-3t}$$

$$\therefore \quad x_2(1) = -2e^{-2} + 3e^{-3}$$

$$= -0.12$$

8.14 (3)

The transfer function,

$$G(s) = \frac{(s+1)(s+3)}{(s+5)(s+7)(s+9)}$$

The transfer function is of order = 3.

So, minimum number of state variable = 3.

8.15 (b)

We know that,

characteristic equation is given as $|sI - A| = 0$

$$\left|\begin{bmatrix} s & 0 \\ 0 & s \end{bmatrix} - \begin{bmatrix} 1 & 1.5 \\ -2 & -3 \end{bmatrix}\right| = 0$$

$$\begin{vmatrix} s-1 & s-1.5 \\ 2 & s+3 \end{vmatrix} = 0$$

$$(s - 1)(s + 3) - 2(s - 1.5) = 0$$

$$s^3 + 3s - s - 3 - 2s + 3 = 0$$

$$s^2 = 0$$

$$s = 0, 0$$

Hence, option (b) is correct.

■■■

9 Process Control

9.1 The Laplace transform of a 4 second transportation lag element is

(a) $\frac{1}{s+4}$ (b) e^{4s}

(c) e^{-4s} (d) $e^{-s/4}$ [1995 : 1 M]

9.2 Match the transfer functions with the systems

(a) All pass filter (e) e^{-as}

(b) Transport delay (f) $\frac{1-s}{1+s}$

(c) Lag network (g) $\frac{1+as}{1+bs}, a < b$

(d) Servomotor (h) $\frac{K}{s(1+as)}$

[1998 : 2 M]

9.3 A process in a feedback loop with a proportional controller with ultimate gain $K_u = 10$ is oscillating at a frequency of $P_u = 8$ Hz. The Ziegler-Nichols setting for the proportional controller is

(a) 8 (b) 5

(c) 1.25 (d) 10 [2001 : 2 M]

9.4 A unity feed back system has a forward path transfer function $G(s) = \frac{10}{s(s+5)}$. It is subjected to a unity ramp input. The integral of the squared error (ISE) is

(a) zero (b) infinity

(c) always negative (d) 0.5 [2002 : 1 M]

9.5 The forward path transfer function of a unity feedback system is given by

$$G(s) = \frac{(1+5s)(1+10s)(1+2s)}{(1+s)(1+8s)(1+20s)}$$

If e(t) is the error to a unit impulse input, the value of performance index.

$$J = \int_0^\infty e(t)\,dt$$

(a) zero (b) infinity

(c) –12 (d) 0.5

[2002 : 2 M]

9.6 In figure K_1 and K_2 are proportional controllers. For the control system to remain stable, the upper limits of K_1 and K_2 are

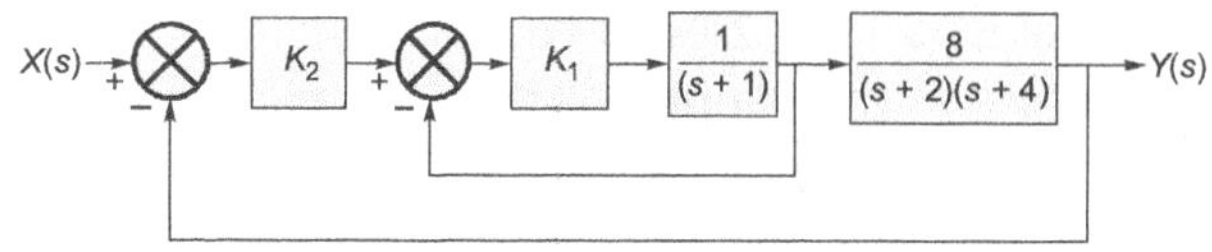

(a) ∞, ∞ (b) 1, ∞

(c) ∞, 10 (d) 10, 1

[2003 : 2 M]

9.7 A chamber is heated with a heater of maximum power rating of 1000 W. The process transfer relationship between the steady-state heater power and chamber temperature is assumed to be linear with a slope of 2 Watts/°C. The heater power is the sum of the output of a proportional controller having a proportional band of 200% and a constant offset of 500 W. The temperature achieved for a set point of 300°C is

(a) 260°C (b) 60°C

(c) 275°C (d) 150°C

[2007 : 2 M]

9.8 A cascade control system with proportional controllers is shown below:

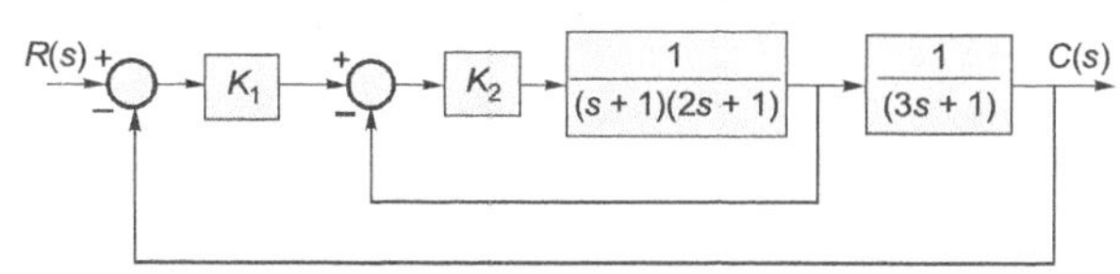

Theoretically, the largest values of the gains K_1 and K_2 that can be set without causing instability of the closed loop system are

(a) 10 and 100 (b) 100 and 10

(c) 10 and 10 (d) ∞ and ∞

[2007 : 2 M]

9.9 For the closed loop system shown below to be stable, the value of time delay T_D (in seconds) should be less than

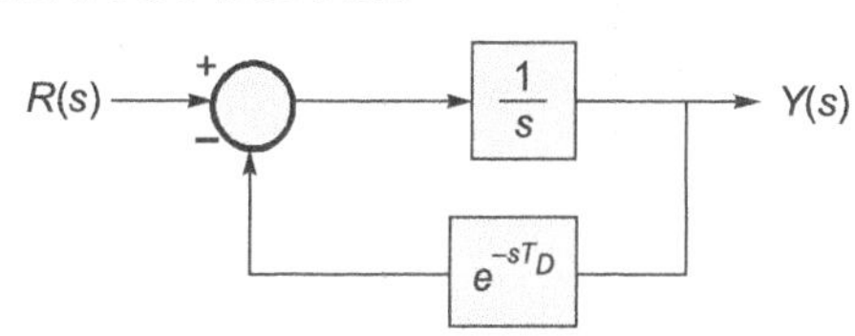

(a) $\frac{\pi}{4}$ (b) $\frac{\pi}{3}$

(c) $\frac{\pi}{2}$ (d) π

[2008 : 2 M]

Linked Answer Questions (9.10 and 9.11):

Consider a plant with the transfer function $G(s) = 1/(s+1)^3$. Let K_u and T_u be the ultimate gain and ultimate period corresponding to the frequency response based closed loop Ziegler-Nichols cycling method, respectively. The Ziegler-Nichols tuning rule for a P-controller is given as: $K = 0.5\,K_u$.

9.10 The values of K_u and T_u, respectively are

(a) $2\sqrt{2}$ and 2π (b) 8 and 2π

(c) 8 and $2\pi/\sqrt{3}$ (d) $2\sqrt{2}$ & $2\pi/\sqrt{3}$

[2013 : 2 M]

9.11 The gain of the transfer function between the plant output and an additive load disturbance input frequency $2\pi/T_u$ in closed loop with a P-controller designed according to the Ziegler-Nichols tuning rule as given above is

(a) −1.0 (b) 0.5

(c) 1.0 (d) 2.0

[2013 : 2 M]

9.12 Consider a transport lag process with a transfer function

$$G_p(s) = e^{-s}$$

The process is controlled by a purely integral controller with transfer function

$$G_c(s) = \frac{K_i}{s}$$

in a unity feedback configuration. The value of K_i for which the closed loop plant has a pole at $s = -1$, is ______.

[2014 : 2 M]

9.13 The standard for long distance analog signal transmission in process control industry is

(a) 4 - 20 mV (b) 0 - 20 mA

(c) 4 - 20 mA (d) 0 - 5 V

[2017 : 1 M]

9.14 The term hysteresis is associated with

(a) ON-OFF control

(b) P-I control

(c) Feed-forward control

(d) Ratio control

[2017 : 1 M]

9.15 In a cascade control system, the closed loop transfer function of the inner loop may be assumed to have a single time-constant τ_1. Similarly, the closed loop transfer function of the outer loop may be assumed to have a single time-constant τ_2. The desired relationship between τ_1 and τ_2 in a well-designed control system is

(a) τ_1 is much less than τ_2

(b) τ_1 is equal to τ_2

(c) τ_1 is much greater than τ_2

(d) τ_1 is independent of τ_2

[2019 : 1 M]

■■■

Answers Process Dynamics and Control

9.1 (c)	9.2 Sol.	9.3 (b)	9.4 (b)	9.5 (*)	9.6 (c)	9.7 (a)
9.8 (a)	9.9 (c)	9.10 (c)	9.11 (b)	9.12 (0.375)	9.13 (c)	9.14 (a)
9.15 (a)						

Explanations Process Dynamics and Control

9.1 (c)

For a system having a transportation lag of t_d sec. the transfer function is $e^{-t_d s}$

∴ for 4 sec. transportation lag.

Transfer function = e^{-4s}

9.2 Sol.

(a)-(f), (b)-(e), (c)-(g), (d)-(h)

9.3 (b)

Using the values of K_u and P_u, Ziegler and Nichols recommended the following settings for feedback controllers:

	K_c	$\tau_{I(\text{min})}$	$\tau_{D(\text{min})}$
P	$k_u/2$	–	–
P-I	$k_u/2.2$	$P_u/1.2$	–
P-I-D	$k_u/1.7$	$P_u/2$	$P_u/8$

then, the Ziegler-Nichols setting for the proportional controller is

$$\frac{k_u}{2} = \frac{10}{2} = 5$$

9.4 (b)

For a unit ramp input of magnitude k, the steady state output error in the response is

$$e_{ss} = -k\tau$$

$$\text{ISE} = \int_0^\infty \varepsilon^2(t)\,dt$$

$$= \int_0^\infty \underbrace{k^2\tau^2}_{\text{constant}}\,dt = t\Big|_0^\infty = \infty$$

9.6 (c)

We take the case of the primary loop. The open loop transfer function for the primary loop is given by

$$K_2\frac{1}{(s+1)}\cdot\frac{8}{(s+2)(s+4)}$$

Phase (ϕ) for the open loop can be find by

$$\phi = -\tan^{-1}\omega - \tan^{-1}\frac{\omega}{2} - \tan^{-1}\frac{\omega}{4}$$

Phase cross-over frequency is given by

$$-180° = -\tan^{-1}\omega_{c_o} - \tan^{-1}\frac{\omega_{c_o}}{2} - \tan^{-1}\frac{\omega_{c_o}}{4}$$

$$\tan^{-1}\omega_{c_o} + \tan^{-1}\frac{\omega_{c_o}}{2} + \tan^{-1}\frac{\omega_{c_o}}{4} = 180°$$

$$\tan^{-1}\left[\frac{\omega_{c_o} + \frac{\omega_{c_o}}{2}}{1-\omega_{c_o}\left(\frac{\omega_{c_o}}{2}\right)}\right] + \tan^{-1}\frac{\omega_{c_o}}{4} = 180°$$

$$\tan^{-1}\left[\frac{\frac{3\omega_{c_o}}{2}}{1-\frac{\omega_{c_o}^2}{2}}\right] + \tan^{-1}\frac{\omega_{c_o}}{4} = 180°$$

$$\tan^{-1}\left(\frac{3\omega_{c_o}}{2-\omega_{c_o}^2}\right) + \tan^{-1}\frac{\omega_{c_o}}{4} = 180°$$

$$\therefore \quad \frac{\frac{3\omega_{c_o}}{2-\omega_{c_o}^2} + \frac{\omega_{c_o}}{4}}{1-\left(\frac{3\omega_{c_o}}{2-\omega_{c_o}^2}\right)\left(\frac{\omega_{c_o}}{4}\right)} = \tan 180°$$

$$\therefore \quad \frac{3\omega_{c_o}}{2-\omega_{c_o}^2} + \frac{\omega_{c_o}}{4} = 0$$

$$\frac{3}{2-\omega_{c_o}^2} = -\frac{1}{4}$$

$$12 = \omega_{c_o}^2 - 2$$

$$\omega_{c_o} = \sqrt{14}$$

We know that magnitude = 1, at ω_{c_o}.

(By polar plot)

$$M = K_2 \frac{1}{\sqrt{1+\omega^2}} \frac{8}{\sqrt{\omega^2+4}\sqrt{\omega^2+16}}$$

$$I = K_2 \frac{1}{\sqrt{1+\omega_{c_o}^2}} \frac{8}{\sqrt{\omega_{c_o}^2}\sqrt{\omega_{c_o}^2+16}}$$

$$I = K_2 \frac{1}{\sqrt{1+14}} \frac{8}{\sqrt{14+4}\sqrt{14+16}}$$

$$K_2 = \frac{\sqrt{15}\sqrt{18}\sqrt{30}}{8} = 11.25$$

$$\boxed{K_2 = 11.25}$$

The open loop transfer function for the secondary loop is given by

$$G_{secondary} = K_1 \frac{1}{(s+1)}$$

Crossover frequency for secondary loop.

$$-\tan^{-1}\omega_{c_o} = -180°$$

$$\omega_{c_o} = 0.$$

Since these is no cross-over frequency for the secondary process, so we can use large gains for the secondary process.

So, we can use any value for K_1 or $K_1 = \infty$.

Hence option (c) is correct.

Method-2:

From block diagram, we have,

$$\frac{Y(s)}{X(s)} = \frac{8K_1K_2}{(s+2)(s+4)(s+1+K_1)+8K_1K_2}$$

⇒ Characteristic equation,

$(s + 2)(s + 4)(s + 1 + K_1) + 8K_1K_2$

$= s_3 + (7 + K_1)s_2 + (14 + 6K_1)s + 8 + 8K_1 + 8K_1K_2$

$= 0$

Applying Routh's stability criterion,

s^3	1	$(6K_1+14)$	0
s^2	$(7+K_1)$	$8(1+K_1+K_1K_2)$	0
s^1	$\frac{48K_1+6K_1^2+90-8K_1K_2}{7+K_1}$	0	
s	$8(1+K_1+K_1K_2)$		

For stability

$$K_1 + 7 > 0 \quad \text{...(i)}$$

$$\frac{48K_1+6K_1^2+90-8K_1K_2}{7+K_1} > 0$$

$$\Rightarrow 48K_1+6K_1^2+90-8K_1K_2 > 0 \quad \text{...(ii)}$$

Using (i),

and $8(1+K_1+K_1K_2) > 0$ ...(iii)

from option (i) and (iii) are always satisfied.

Now, checking options for equation (ii).

(a) $K_1 = K_2 \to \infty$

$$\lim_{K_1\to\infty} 48K_1+6K_1^2+90-8K_1^2 \to -\infty$$

(b) $K_1 = 1$, $K_2 \to \infty$

$$\lim_{K_2\to\infty} 48+6+90-6K_2 \to -\infty$$

(c) $K_2 = 10$, $K_1 = \infty$

$$\lim_{K_1\to\infty} 48K_1+6K_1^2+90-80K_1 > 0$$

So, answer is (c).

(d) $4.8 \times 10 + 6 \times 10^2 + 90 - 8 \times 10 \times 1 > 0$

But this not upper limit of K_1 and K_2.

9.7 (a)

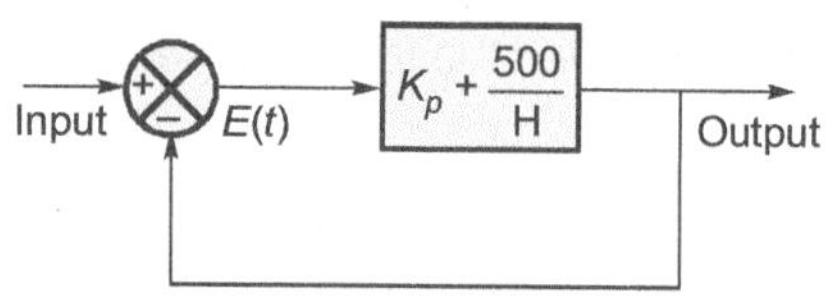

input = 300°C

Let, output = T°C

$E(t) = (300 - T)$°C

Corresponding power $P = 2 \times (300 - T)$

Proportional band $B = \frac{100}{K_P}$

$$\Rightarrow \quad K = \frac{100}{B} = \frac{100}{200} = 0.5$$

$$\text{output} = [2\times(300-T)]\times\left(K_P + \frac{500}{P}\right)$$

$$\text{output} = (300 - T) + 500$$

temperature equivalent is $\frac{(300-T)+500}{2}$

and this must be equal to T

$$\therefore \quad T = \frac{300 - T + 500}{2}$$

$$3T = 800$$

$$T = 266.66°C$$

from the options 260°C.

9.8 (a)

The open loop transfer function of the primary loop transfer is given by

$$K_1 \frac{1}{(s+1)(2s+1)} \cdot \frac{1}{(3s+1)}$$

Phase cross-over frequency for primary loop is given by

$$-\tan^{-1}\omega - \tan^{-1}2\omega - \tan^{-1}3\omega = -180°$$

$$\tan^{-1}\omega + \tan^{-1}2\omega + \tan^{-1}3\omega = 180°$$

$$\tan^{-1}\left(\frac{\omega+2\omega}{1-2\omega^2}\right) + \tan^{-1}3\omega = 180°$$

$$\tan^{-1}\left(\frac{3\omega}{1-2\omega^2}\right) + \tan^{-1}3\omega = 180°$$

$$\tan^{-1}\left[\frac{\frac{3\omega}{1-2\omega^2}+3\omega}{1-\left(\frac{3\omega}{1-2\omega^2}\right)3\omega}\right] = 180°$$

$$\frac{3\omega}{1-2\omega^2}+3\omega = 0$$

$$\frac{3\omega}{1-2\omega^2} = -3\omega$$

$$1 = -(1-2\omega^2)$$

$$1 = 2\omega^2 - 1$$

$$2\omega^2 = 2$$

$$\omega = 1$$

At phase cross-over frequency, $M = 1$ at $\omega = 1$.

$$M = K_1 \frac{1}{\sqrt{\omega^2+1}} \frac{1}{\sqrt{4\omega^2+1}} \frac{1}{\sqrt{9\omega^2+1}}$$

$$I = K_1 \frac{1}{\sqrt{2}} \frac{1}{\sqrt{5}} \frac{1}{\sqrt{10}}$$

$$\therefore \quad \boxed{K_1 = 0}$$

The open loop transfer function for the secondary loop is given by

$$G_{secondary} = K_2 \frac{1}{(s+1)(2s+1)}$$

Then, cross-over frequency for the secondary loop is given by

$$-\tan^{-1}\omega - \tan^{-1}2\omega = 180°$$

$$\tan^{-1}\frac{\omega+2\omega}{1-(\omega)(2\omega)} = 180°$$

$$3\omega = 180°$$

$$\omega_{c_o} = 0$$

Since there, is no cross-over frequency for the secondary loop, so we can use large gains for the secondary process.

So, we can use the higher values of the gain for the secondary loop, so as the have the faster response for the secondary process. For $K_1 = 10$, K_2 can be 100 (by selecting the higher gains).

9.9 (c)

For the system to be as the verge of instability phase margin (PM) = 0.

For a system to be stable PM should be positive. i.e. PM > 0 (for stable system)

for $\frac{1}{s} \times e^{-t_d s}$

$$-90° + \left(\frac{-180° \times t_d \times \omega}{\pi}\right) + 180° > 0$$

(for system to be stable)

$$\frac{-180\, t_d\, \omega}{\pi} + 90° > 0$$

$$\therefore \quad \frac{t_d\,\omega}{\pi} < \frac{1}{2}$$

$$\frac{t_d \times 1}{\pi} < \frac{1}{2}$$

$$\therefore \quad \boxed{t_d < \frac{\pi}{2}}$$

9.10 (c)

Finding ω_{co}

$$G(s) = \frac{1}{(s+1)^3}$$

$$-3\tan^{-1}\omega_{co} = -180°$$

$$\tan^{-1}\omega_{co} = +60°$$

$$\omega_{co} = \tan 60° = \sqrt{3}$$

$$\therefore \quad T_u = \frac{2\pi}{\omega_{co}} = \frac{2\pi}{\sqrt{3}}$$

$$\boxed{T_u = \frac{2\pi}{\sqrt{3}}}$$

$$M = \frac{1}{\left[\sqrt{1+(\sqrt{3})^2}\right]^3} = \frac{1}{(\sqrt{1+3})^3} = \frac{1}{2^3} = \frac{1}{8}$$

$$K_u = \frac{1}{M} = \frac{1}{1/8}$$

$$\boxed{K_u = 8}$$

9.11 (b)

$\because$ Input frequency $= \dfrac{2\pi}{T_u}$

and from above

$$T_U = \frac{2\pi}{\sqrt{3}}$$

So, input frequency $= \sqrt{3}$ rad/sec

Open loop gain

$$K_C = 0.5\,K_u$$

$\because$ $K_u = 8$

$\therefore$ $K_C = 0.5 \times 8 = 4$

So, gain of the transfer $f \times n$

$$\text{at } \omega = \sqrt{3} \text{ rad/sec} = \left|\frac{4}{(s+1)^3}\right|_{\omega=\sqrt{3}}$$

$$= \left.\frac{4}{\left[(j\omega)^2+1\right]^{3/2}}\right|_{\omega=\sqrt{3}} = \frac{4}{(\sqrt{3}^2+1^2)^{3/2}} = 0.5$$

9.12 (0.375)

$$\text{Closed loop } TF \Rightarrow \frac{e^{-s} \times \left(\dfrac{K_i}{s}\right)}{1+e^{-s}\left(\dfrac{K_i}{s}\right)}$$

Taking denominator equal to zero

$$1+e^{-s}\left(\frac{K_i}{s}\right) = 0$$

$$1+\left(1-s+\frac{s^2}{2!}-\frac{s^3}{3!}\right)\frac{K_i}{s} = 0 \qquad [\text{for } s = -1]$$

Putting $s = -1$

We get, $K_i = 0.375$

9.13 (c)

To differentiate between fault and zero input, lower value is kept at 4 mA.

9.14 (a)

ON-OFF controller is designed to tolerate a range of error to avoid self sustained oscillations; known as hysteresis characteristics.

9.15 (a)

In a Cascade control system, inner loop should function fast compared to the outer loop. Infact, it is the basic characteristic of cascade control strategy.

Generally speed of response of loop is inversely proportional to time constant of the transfer function.

So, time constant of inner loop τ_1 should be less compared to the time constant of outer loop τ_2. Option (a) is correct answer.

■■■

Digital Electronics and Microprocessors

UNIT

VII

CONTENTS

1. Number Systems 361
2. Boolean Algebra & Minimization of Boolean Expression 363
3. Logic Gates 369
4. Combinational Circuits 377
5. Sequential Circuits 381
6. Logic Families 394
7. Memories and Interfacing 397
8. ADC & DAC 400
9. Basics of Microprocessor and Instruction Set 408

Digital Electronics and Microprocessors

Syllabus

Combinational logic circuits, minimization of Boolean functions. IC families: TTL and CMOS. Arithmetic circuits, comparators, Schmitt trigger, multi-vibrators, sequential circuits, flipflops, shift registers, timers and counters; sample-and-hold circuit, multiplexer, analog-to-digital (successive approximation, integrating, flash and sigma-delta) and digital-to-analog converters (weighted R, R-2R ladder and current steering logic). Characteristics of ADC and DAC (resolution, quantization, significant bits, conversion/settling time); basics of number systems, Embedded Systems: Microprocessor and microcontroller applications, memory and input-output interfacing; basics of data acquisition systems, basics of distributed control systems (DCS) and programmable logic controllers (PLC).

Analysis of Previous GATE Papers

Exam Year	1 Mark Ques.	2 Marks Ques.	Total Marks
1994	3	6	**15**
1995	3	2	**7**
1996	2	–	**2**
1997	3	2	**7**
1998	7	1	**9**
1999	2	5	**12**
2000	2	2	**6**
2001	5	1	**7**
2002	3	3	**9**
2003	5	9	**23**
2004	4	7	**18**
2005	3	7	**17**
2006	1	9	**19**
2007	1	11	**23**
2008	2	8	**18**

Exam Year	1 Mark Ques.	2 Marks Ques.	Total Marks
2009	4	4	**12**
2010	2	3	**8**
2011	4	4	**12**
2012	3	1	**5**
2013	1	2	**5**
2014	1	4	**9**
2015	3	2	**7**
2016	2	4	**10**
2017	2	1	**4**
2018	3	4	**11**
2019	4	4	**12**
2020	3	2	**7**
2021	3	3	**9**
2022	3	1	**5**
2023	3	3	**9**
2024	1	1	**3**

1 Number Systems

1.1 A number in 4-bit two's complement representation is $X_3 X_2 X_1 X_0$. This number when stored using 8-bits will be

(a) 0000 $X_3 X_2 X_1 X_0$

(b) 1111 $X_3 X_2 X_1 X_0$

(c) $X_3 X_3 X_3 X_3 X_3 X_2 X_1 X_0$

(d) $\overline{X}_3\overline{X}_3\overline{X}_3\overline{X}_3 X_3 X_2 X_1 X_0$

[1999 : 1 M]

1.2 For a shaft encoder, the most appropriate 2-bit code is

(a) 11, 10, 01, 00 (b) 11, 10, 00, 01

(c) 01, 10, 11, 00 (d) 01, 00, 11, 10

[1999 : 1 M]

1.3 Two 4-bit 2's complement numbers 1011 and 0110 are added. The result expressed in 4-bit 2's complement notation is

(a) 0001

(b) 0010

(c) 1101

(d) cannot be expressed in 4-bit 2's complement

[2003 : 1 M]

1.4 7EH and 5FH are XORed. The output is multiplied by 10H. The result is

(a) 0210H (b) 7E5FH

(c) 5F7EH (d) 2100H

[2004 : 2 M]

1.5 A number N is stored in a 4-bit 2's complement representation as

a_3	a_2	a_1	a_0

It is copied into a 6-bit register and after a few operations, the final bit pattern is

a_3	a_3	a_2	a_1	a_0	1

The value of this bit pattern in 2's complement representation is given in terms of the original number N as

(a) $32\,a_3 + 2\,N + 1$ (b) $32\,a_3 - 2N - 1$

(c) $2\,N - 1$ (d) $2\,N + 1$

[2006 : 2 M]

1.6 The binary representation of the decimal number 1.375 is

(a) 1.111 (b) 1.010

(c) 1.011 (d) 1.001

[2009 : 1 M]

1.7 The base of the number system for the addition operation 24 +14 = 41 to be true is

(a) 8 (b) 7

(c) 6 (d) 5

[2011 : 1 M]

1.8 The result of $(45)_{10} - (45)_{16}$ expressed in 6-bit 2's complement representation is

(a) 011000 (b) 100111

(c) 101000 (d) 101001

[2011 : 2 M]

1.9 The representation of the decimal number $(27.625\}_{10}$ in base-2 number system is

(a) 11011.110 (b) 11101.101

(c) 11011.101 (d) 10111.110

[2018 : 1 M]

■■■

Answers Number Systems

1.1	(c)	1.2	(b)	1.3	(a)	1.4	(a)	1.5	(d)	1.6	(c)	1.7	(b)
1.8	(c)	1.9	(c)										

Explanations Number Systems

1.1 (c)

To extend number of bits in 1's compliment or 2's compliment copy MSB bit, so when the 4-bit number $X_3 X_2 X_1 X_0$ is stored using 8-bit, MSB is copied. So the required 8-bit number is $X_3 X_3 X_3 X_3 X_3 X_2 X_1 X_0$.

1.2 (b)

In shaft encoder one bit change is preferred to indicate the shaft position, hence Gray code is preferred as a code. So the most appropriate 2-bit code is 11, 10, 00, 01.

1.3 (a)

1011 ← 2's complement representation of +5
0110 ← 2's complement representation of 10
Addition of these two numbers will give 15 i.e. $(1111)_2$
15 in 2's complement form = $(0001)_2$

1.4 (a)

7EH ⟶ 0111 1110
5FH ⟶ 0101 1111
XORed 0010 0001 = 21H

21 H
× 10 H
Result ⟶ 0210 H

1.5 (d)

Given number is | a_3 | a_2 | a_1 | a_0 | in 2's complement form.

We know that in 2's complement form if we copy MSB at left of MSB any times the number remains unchanged.

So $a_3\, a_2\, a_1\, a_0 = a_3\, a_3\, a_2\, a_1\, a_0 = N$

When we left shift a number by 1 bit then it is multiplied by 2, so, $a_3\, a_3\, a_2\, a_1\, a_0 0 = 2\text{ N}$

Now,

$a_3\, a_3\, a_2\, a_1\, a_0 0 + 1 = a_3\, a_3\, a_2\, a_1\, a_0 1 = 2\text{ N} + 1$

1.6 (c)

	Integer	Remainder
0.375×2 =	0	0.750
0.750×2 =	1	0.5
0.5×2 =	1	0

Thus binary representation of $(1.375)_{10}$ is = 1.011

1.7 (b)

Let the base is x, then

$$(24)_x + (14)_x = (41)_x$$
$$(2x^2 + 4x) + (x^2 + 4x) = 4x^2 + x$$
$$3x^2 + 8x = 4x^2 + x$$
$$\Rightarrow \quad x^2 = 7x$$
$$\Rightarrow \quad x = 7$$

1.8 (c)

$$(45)_{10} - (45)_{16} = (45)_{10} - (69)_{10} = (-24)_{10}$$
$$(24)_{10} \rightarrow (011000)_2$$
$$(-24)_{10} \rightarrow (101000)_2$$

1.9 (c)

The given decimal number is 27.625.

2	27	
2	13	1
2	6	1
2	3	0
2	1	1
	0	1

Conversion of decimal part,

$$(27)_{10} = (11011)_2$$

Conversion of fractional part,

$$0.625 \times 2 = 1.250 \Rightarrow 1$$
$$0.250 \times 2 = 0.500 \Rightarrow 0$$
$$0.500 \times 2 = 1.000 \Rightarrow 1$$
$$(0.625)_{10} = (0.101)_2$$

Hence, $(27.625)_{10} = (11011.101)_2$

2 Boolean Algebra & Minimization of Boolean Expression

2.1 A combinatorial circuit has inputs *A*, *B* and *C* and its Karnaugh map is given in Fig. The output of the circuit is

C \ AB	00	01	11	10
0		①		①
1	①		①	

(a) $(\bar{A}B + A\bar{B})C$ (b) $(\bar{A}B + A\bar{B})\bar{C}$

(c) $A \oplus B \oplus C$ (d) $\bar{A}\bar{B}\bar{C}$ [1995 : 1 M]

2.2 The minimal sum of products form of $f = A\bar{B}CD + \bar{A}BC + BCD + \bar{A}\bar{B}C$ is

(a) $\bar{A}C + BD$ (b) $\bar{A}C + CD$

(c) $AC + \bar{B}D$ (d) $A\bar{B} + C\bar{D}$

[1998 : 1 M]

2.3 The expression $A = \overline{AB}$ is equivalent to

(a) $\bar{A} + \bar{B}$ (b) $AB + A$

(c) $A + B$ (d) AB [2000 : 1 M]

2.4 The Karnaugh map for a four variable Boolean function is given in below figure. The correct Boolean sum of product is

RS \ PQ	00	01	11	10
00	0	0	0	0
01	1	0	0	1
11	1	0	0	1
10	0	1	0	0

(a) $PQRS + \bar{Q}S$ (b) $\bar{P}QR\bar{S} + \bar{Q}S$

(c) $PQR + Q\bar{S}$ (d) $PQRS + \bar{Q}$

[2003 : 1 M]

2.5 The simplest form of the Boolean expression:

$AB\bar{C}\bar{D} + ABC\bar{D} + AB\bar{C}D + ABCD$ is

(a) AD (b) BC

(c) $\bar{A}B$ (d) AB [2004 : 1 M]

2.6 Min-term (Sum of products) expression for a Boolean function is given as follows:

$f(A, B, C) = \Sigma m\,(0, 1, 2, 3, 5, 6)$

where *A* is the *MSB* and *C* is the *LSB*. The minimized expression for the function is

(a) $A + (B \oplus C)$ (b) $(A \oplus B) + C$

(c) $\bar{A} + (B \oplus C)$ (d) $\overline{ABC}$ [2006 : 2 M]

2.7 A logic circuit implements the Boolean function, $F = \bar{X} \cdot Y + X \cdot \bar{Y} \cdot \bar{Z}$. It is found that the input combination $X = Y = 1$ can never occur. Taking this into account, a simplified expression for F is given by

(a) $\bar{X} + \bar{Y} \cdot \bar{Z}$ (b) $X + Z$

(c) $X - Z$ (d) $Y + X \cdot \bar{Z}$

[2007 : 1 M]

2.8 Let $X = X_1 X_0$ and $Y = Y_1 Y_0$ be unsigned 2-bit numbers. The function $F = 1$ if $X > Y$ and $F = 0$ otherwise. The minimized sum of products expression for F is

(a) $Y_1 \cdot Y_0 + X_0 \cdot Y_0 + \bar{X}_1 \cdot \bar{X}_0 \cdot \bar{Y}_1$

(b) $X_0 \cdot \bar{Y}_1 + Y_1 \cdot \bar{Y}_0 + X_1 \cdot \bar{X}_0$

(c) $Y_1 \cdot \bar{X}_1 + Y_0 \cdot \bar{X}_1 \cdot \bar{X}_0 + Y_1 \cdot Y_0 \cdot \bar{X}_0$

(d) $X_1 \cdot \bar{Y}_1 + X_0 \cdot \bar{Y}_0 \cdot \bar{Y}_1 + X_0 \cdot X_1 \cdot \bar{Y}_0$

[2007 : 2 M]

2.9 The minimum sum of product form of the Boolean expression is

$Y = \bar{P}\bar{Q}\bar{R}\bar{S} + P\bar{Q}\bar{R}\bar{S} + P\bar{Q}\bar{R}S + P\bar{Q}RS + P\bar{Q}R\bar{S} + \bar{P}\bar{Q}R\bar{S}$

(a) $Y = P\bar{Q} + \bar{Q}\bar{S}$ (b) $Y = P\bar{Q} + \bar{Q}R\bar{S}$

(c) $Y = P\bar{Q} + \bar{Q}\bar{R}\bar{S}$ (d) $Y = \bar{Q}\bar{S} + P\bar{Q}R$

[2008 : 2 M]

2.10 The minimal sum-of-products expression for the logic function f represented by the given Karnaugh map is

RS \ PQ	00	01	11	10
00	0	1	0	0
01	0	1	1	1
11	1	1	1	0
10	0	0	1	0

(a) $QS+P\bar{R}S+PQR+\bar{P}RS+\bar{P}Q\bar{R}$

(b) $\overline{QS}+\bar{P}R\bar{S}+\overline{PQ}R+P\bar{R}S+P\bar{Q}R$

(c) $\bar{P}R\bar{S}+\bar{P}\bar{Q}\bar{R}+P\bar{R}S+P\bar{Q}R$

(d) $P\bar{R}S+PQR+\bar{P}RS+\bar{P}Q\bar{R}$ [2009 : 1 M]

2.11 For the Boolean expression

$$f=\bar{a}\bar{b}\bar{c}+\bar{a}b\bar{c}+a\bar{b}\bar{c}+abc+ab\bar{c}$$

The minimized Product of Sum (PoS) expression is

(a) $f=(b+\bar{c})\cdot(a+\bar{c})$ (b) $f=(\bar{b}+c)\cdot(\bar{a}+c)$

(c) $f=(\bar{b}+c).(a+\bar{c})$ (d) $f=\bar{c}+abc$

[2011 : 1 M]

2.12 The output Y of a 2-bit comparator is logic 1 whenever the 2-bit input A is greater than the 2-bit input B. The number of combinations for which the output is logic 1, is

(a) 4 (b) 6

(c) 8 (d) 10 [2012 : 1 M]

2.13 In the sum of products function

$$f(X, Y, Z) = \sum(2, 3, 4, 5),$$

the prime implicants are

(a) $\bar{X}Y, X\bar{Y}$

(b) $\bar{X}Y, X\bar{Y}\bar{Z}, X\bar{Y}Z$

(c) $\bar{X}Y\bar{Z}, \bar{X}YZ, X\bar{Y}$

(d) $\bar{X}Y\bar{Z}, \bar{X}YZ, X\bar{Y}\bar{Z}, X\bar{Y}Z$ [2012 : 1 M]

2.14 A bulb in a staircase has two switches, one switch being at the ground floor and the other one at the first floor. The bulb can be turned ON and also can be turned OFF by anyone of the switches irrespective of the state of the other switch. The logic of switching of the bulb resembles.

(a) an AND gate (b) an OR gate

(c) an XOR gate (d) a NAND gate

[2013 : 1 M]

2.15 The Boolean expression $XY + (X' + Y')Z$ is equivalent to

(a) $XYZ' + X'Y'Z$ (b) $X'Y'Z' + XYZ$

(c) $(X + Z)(Y + Z)$ (d) $(X' + Z)(Y' + Z)$

[2016 : 1 M]

2.16 The product of sum expression of a Boolean function $F(A, B, C)$ of three variables is given by

$$F(A+B+C)=(A+B+\bar{C})\cdot(A+\bar{B}+\bar{C})\cdot(\bar{A}+B+C)\cdot(\bar{A}+\bar{B}+\bar{C})$$

The canonical sum of product expression of $F(A, B, C)$ is given by

(a) $\bar{A}\bar{B}C+\bar{A}BC+A\bar{B}\bar{C}+ABC$

(b) $\bar{A}\bar{B}\bar{C}+\bar{A}B\bar{C}+A\bar{B}C+AB\bar{C}$

(c) $AB\bar{C}+A\bar{B}\bar{C}+\bar{A}BC+\bar{A}\bar{B}\bar{C}$

(d) $\bar{A}\bar{B}\bar{C}+\bar{A}BC+AB\bar{C}+ABC$

[2018 : 2 M]

2.17 The total number of Boolean functions with distinct truth tables that can be defined over 3 Boolean variables is________. [2019 : 1 M]

2.18 $X = X_1X_0$ and $Y = Y_1Y_0$ are 2-bit binary numbers. The Boolean function S that satisfies the condition "If $X > Y$, then $S = 1$", in its minimized form, is

(a) $X_1Y_1 + X_0Y_0$

(b) $X_1\bar{Y}_1 + X_0\bar{Y}_0\bar{Y}_1 + X_0\bar{Y}_0X_1$

(c) $X_1\bar{Y}_1X_0\bar{Y}_0$

(d) $X_1\bar{Y}_1 + X_0\bar{Y}_0Y_1 + X_0\bar{Y}_0\bar{X}_1$ [2019 : 2 M]

2.19 Given $f(A, B, C, D) = \Sigma m(0, 1, 2, 6, 8, 9, 10, 11) + \Sigma d(3, 7, 14, 15)$ is a Boolean function, where m represents min-term and d represents don't-cares. The minimal sum of products expression for f is _____.

(a) $f=\bar{B}+C$ (b) $f=A\bar{B}+CB$

(c) $f=\bar{D}+A$ (d) $f=\bar{A}B+\bar{C}D$

[2020 : 1 M]

2.20 The Boolean expression for the shaded regions as shown in the figure is _____.

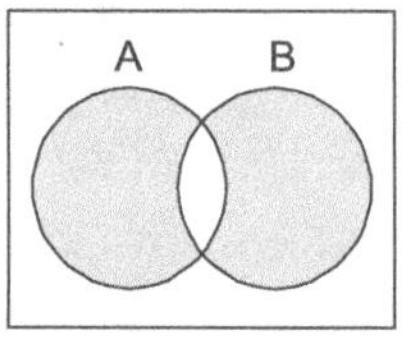

(a) $(A+\bar{B})\bullet(\bar{A}+B)$ (b) $(\bar{A}+B)\bullet(\bar{A}+B)$

(c) $(\bar{A}+\bar{B})\bullet(A+\bar{B})$ (d) $(A+B)\bullet(\bar{A}+\bar{B})$

[2020 : 1 M]

2.21 A Boolean function F of three variables X, Y and Z is given as

$$F(X, Y, Z) = (X' + Y + Z)\cdot(X + Y' + Z')\cdot(X' + Y + Z')\cdot(X'Y'Z' + X'YZ' + XYZ')$$

Which one of the following is true?

(a) $F(X, Y, Z) = (X + Y + Z')\cdot(X' + Y' + Z')$

(b) $F(X, Y, Z) = (X' + Y)\cdot(X + Y' + Z')$

(c) $F(X, Y, Z) = X'Z' + YZ'$

(d) $F(X, Y, Z) = X'Y'Z + XYZ$ [2021 : 2 M]

2.22 $A = a_1a_0$ and $B = b_1b_0$ are two 2-bit unsigned binary numbers. If $F(a_1, a_0, b_1, b_0)$ is a Boolean function such that $F = 1$ only when $A > B$, and $F = 0$ otherwise, then F can be minimized to the form _____.

(a) $a_1\bar{b}_1 + a_1a_0\bar{b}_0$

(b) $a_1\bar{b}_1 + a_1a_0\bar{b}_0 + a_0\bar{b}_0\bar{b}_1$

(c) $a_1a_0\bar{b}_0 + a_0\bar{b}_0\bar{b}_1$

(d) $a_1\bar{b}_1 + a_1a_0\bar{b}_0 + a_0\bar{b}_0b_1$ [2022 : 2 M]

2.23 The simplified form of the Boolean function $F(W, X, Y, Z) = \Sigma(4, 5, 10, 11, 12, 13, 14, 15)$ with the minimum number of terms and smallest number of literals in each term is________.

(a) $WX + \bar{W}X\bar{Y} + W\bar{X}Y$ (b) $WX + WY + X\bar{Y}$

(c) $X\bar{Y} + WY$ (d) $\bar{X}Y + \bar{W}\bar{Y}$

[2023 : 2 M]

2.24 Among the given options, the simplified form of the Boolean function $F = (A + \bar{A}\cdot B) + \bar{A}\cdot(A + \bar{B})\cdot C$ is

(a) $A + B + C$ (b) $A \cdot B \cdot C$

(c) $B + \bar{A}\cdot C$ (d) $\bar{A} + B\cdot C$

[2024 : 1 M]

Answers Boolean Algebra & Minimization of Boolean Expression

2.1	(c)	2.2	(b)	2.3	(a)	2.4	(b)	2.5	(d)	2.6	(c)	2.7	(d)
2.8	(d)	2.9	(a)	2.10	(d)	2.11	(a)	2.12	(b)	2.13	(a)	2.14	(c)
2.15	(c)	2.16	(b)	2.17	(256)	2.18	(b)	2.19	(a)	2.20	(d)	2.21	(c)
2.22	(b)	2.23	(c)	2.24	(a)								

Explanations Boolean Algebra & Minimization of Boolean Expression

2.1 (c)

From the *K*-map,

Output $F = \Sigma(1, 2, 4, 7)$ since, (as MSB)

$$= C\bar{A}\bar{B} + CAB + \bar{C}\bar{A}B + \bar{C}A\bar{B}$$

$$= C(\bar{A}\bar{B} + AB) + \bar{C}(\bar{A}B + A\bar{B}) = C(A \odot B) + \bar{C}(A \oplus B)$$

$$= C(\overline{A \oplus B}) + \bar{C}(A \oplus B) \quad \text{(using } A \odot B = \overline{A \oplus B}\text{)}$$

$$f = A \oplus B \oplus C$$

2.2 (b)

$$f = A\bar{B}CD + \bar{A}BC + BCD + \bar{A}\bar{B}C$$

The *K*-map of above function is

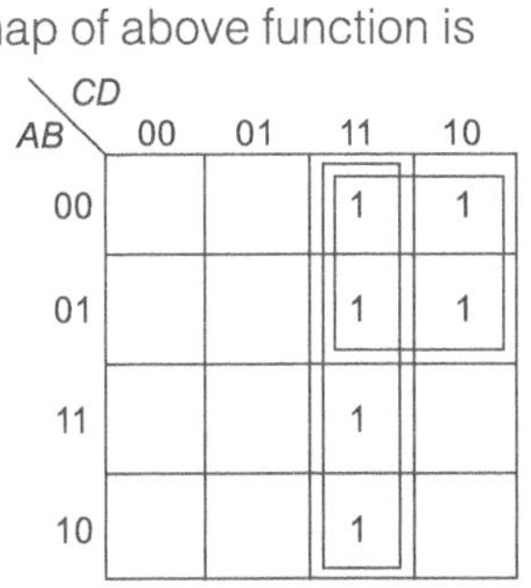

So, the required minimal SOP of *f* is

$$f = \bar{A}C + CD$$

2.3 (a)

$$A = \overline{AB}$$

By Demorgan's theorem

$$\overline{AB} = \bar{A} + \bar{B}$$

2.4 (b)

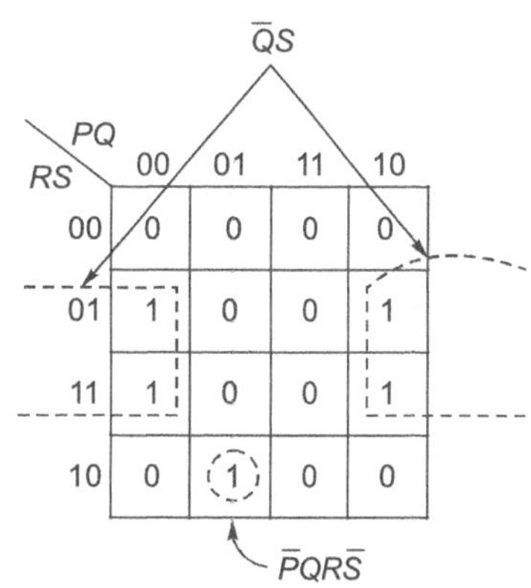

$$SOP = \bar{P}QR\bar{S} + \bar{Q}S$$

2.5 (d)

$$Y = AB\bar{C}\bar{D} + ABC\bar{D} + AB\bar{C}D + ABCD$$

$$= AB\bar{C}(\bar{D} + D) + ABC(\bar{D} + D)$$

$$= AB\bar{C}\cdot 1 + ABC\cdot 1 = AB \cdot 1 = AB$$

2.6 (c)

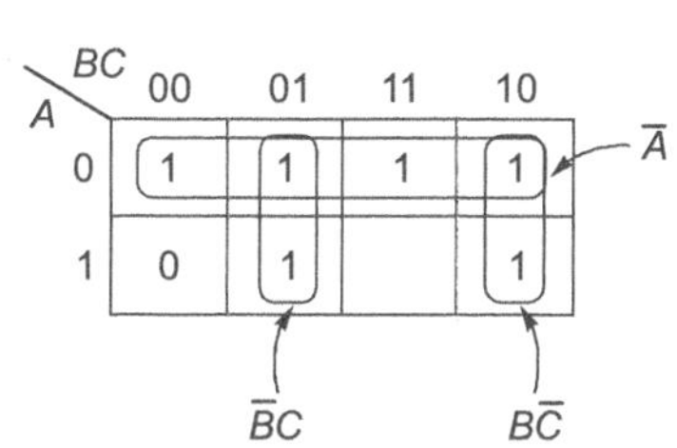

$$f(A, B, C) = \bar{A} + \bar{B}C + B\bar{C} = \bar{A} + (B \oplus C)$$

2.7 (d)

$$F = \bar{X}Y + X\bar{Y}\bar{Z} = \bar{X}Y(Z+\bar{Z}) + X\bar{Y}\bar{Z}$$
$$= \bar{X}YZ + \bar{X}Y\bar{Z} + X\bar{Y}\bar{Z}$$

Truth table:

X	Y	Z	F
0	0	0	0
0	0	1	0
0	1	0	1
0	1	1	1
1	0	0	1
1	0	1	0
1	1	0	x
1	1	1	x

K-map

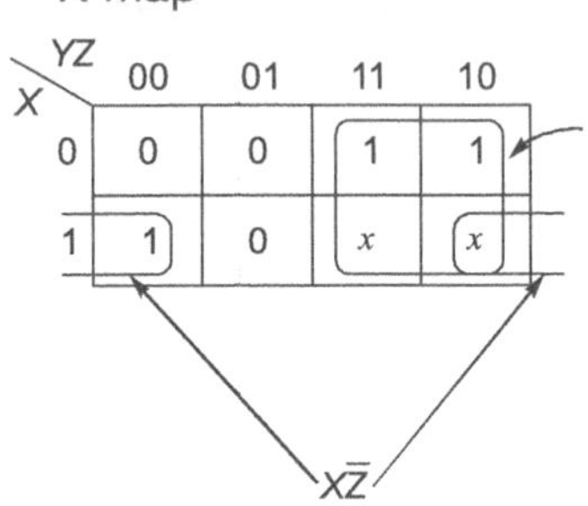

$$\boxed{F = Y + X\bar{Z}}$$

2.8 (d)

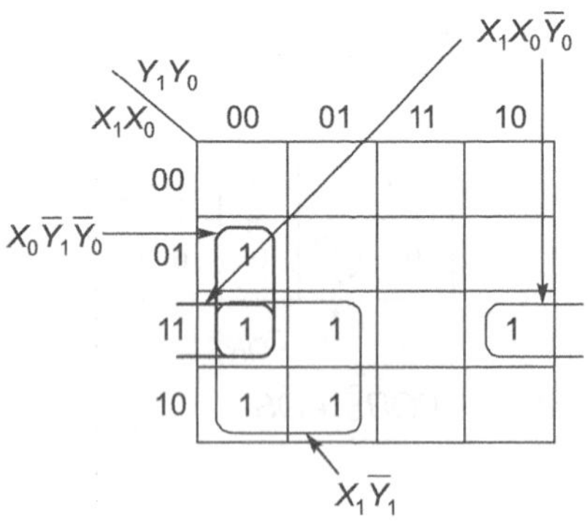

From the K-map

$$F = X_1\bar{Y}_1 + X_0\bar{Y}_0\bar{Y}_1 + X_1X_0\bar{Y}_0$$

2.9 (a)

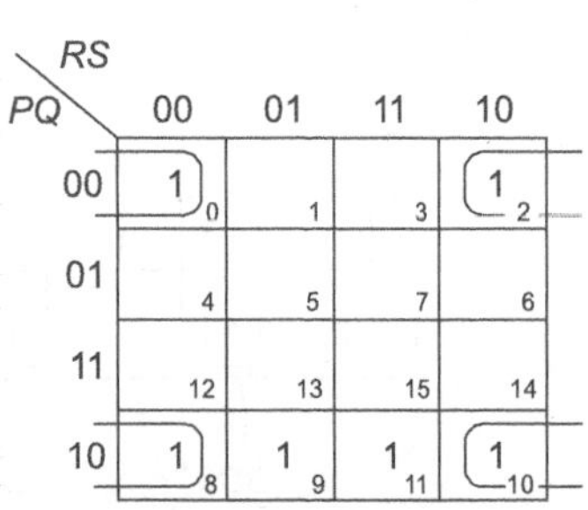

$$Y = P\bar{Q} + \bar{Q}\bar{S}$$

2.10 (d)

K-map

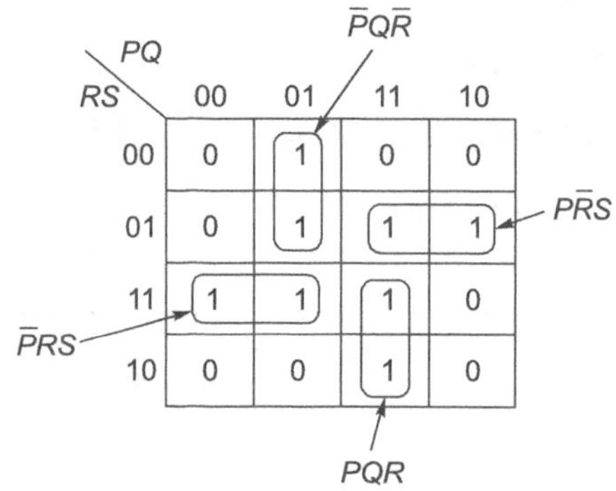

$$F = P\bar{R}S + PQR + \bar{P}RS + \bar{P}Q\bar{R}$$

2.11 (a)

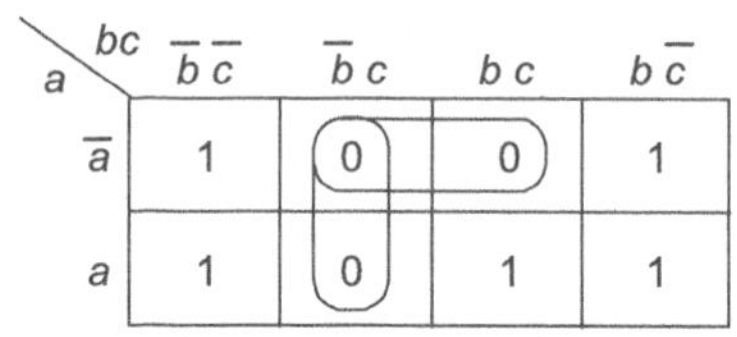

$$\bar{Y} = \bar{a}c + \bar{b}c \Rightarrow Y = (a+\bar{c})\cdot(b+\bar{c})$$

2.12 (b)

A	B	No. of cases
01	00	1
10	00, 01	2
11	00, 01, 10	3
		total = 6

2.13 (a)

$$f(X, Y, Z) = \Sigma(2, 3, 4, 5)$$

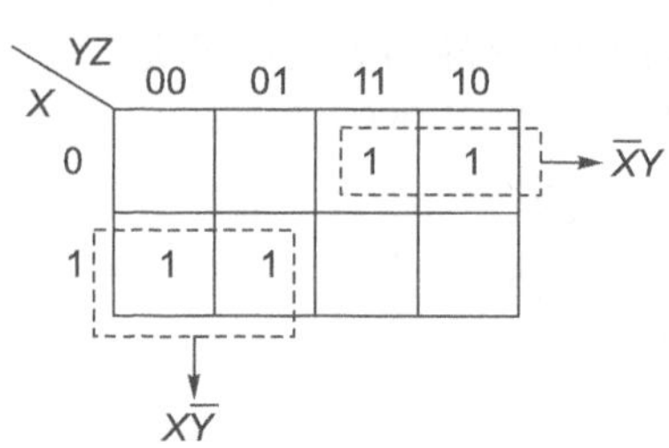

$$\therefore\ f(X, Y, Z) = X\bar{Y} + \bar{X}Y$$

2.14 (c)

Truth table of XOR gate is,

A	B	Y
0	0	0
0	1	1
1	0	1
1	1	0

So, from the XOR gate truth table it is clear that the bulb can be turned ON and also can be turned OFF by any one of the switches irrespective of the state of the other switch.

2.15 (c)

$$XY + (\bar{X} + \bar{Y})Z = XY + \overline{XY}Z$$
$$= (XY + \overline{XY})(XY + Z)$$
$$= XY + Z$$
$$= (X + Z)(Y + Z)$$

2.16 (b)

The function is given in POS form as,

$$F = (A + B + \bar{C})(A + \bar{B} + \bar{C})(\bar{A} + B + C)(\bar{A} + \bar{B} + \bar{C})$$

the above function can also be expressed as,

$$F(A, B, C) = \Pi M(1, 3, 4, 7)$$

the function can be expressed in SOP form as,

$$F(A, B, C) = \Sigma m(0, 2, 5, 6)$$
$$= \bar{A}\bar{B}\bar{C} + \bar{A}B\bar{C} + A\bar{B}C + AB\bar{C}$$

2.17 (256)

The total number of Boolean function with distinct truth table will be for n Boolean variables (2^{2^n}) so for 3 Boolean variables = $(2^{2^3}) = 256$

2.18 (b)

Given $X = X_1X_0$ and $Y = Y_1Y_0$

If $X > Y$

Then, $S = 1$

The truth table for S can be drawn as follows,

	X_1	X_0	y_1	y_0	S
	x		*y*		
0	0	0	0	0	0
1	0	0	0	1	0
2	0	0	1	0	0
3	0	0	1	1	0
4	0	1	0	0	1
5	0	1	0	1	0
6	0	1	1	0	0
7	0	1	1	1	0
8	1	0	0	0	1
9	1	0	0	1	1
10	1	0	1	0	0
11	1	0	1	1	0
12	1	1	0	0	1
13	1	1	0	1	1
14	1	1	1	0	1
15	1	1	1	1	0

The Boolean function S can be represented in the sum of minterms form as

$S(X_1, X_0, Y_1, Y_0) = \Sigma(4, 8, 9, 12, 13, 14)$

The above function can be simplified using *k*-Map as follows :

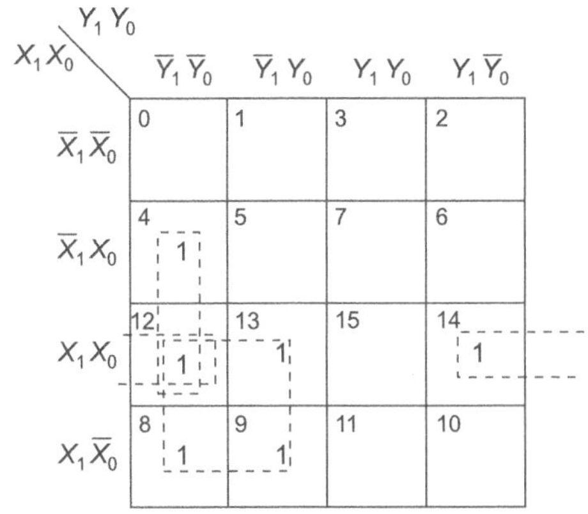

$$S = X_1\bar{Y}_1 + X_0\bar{Y}_1\bar{Y}_0 + X_1X_0\bar{Y}_0$$

2.19 (a)

$$f(A, B, C, D) = \Sigma m(0, 1, 2, 6, 8, 9, 10, 11) + \Sigma d(3, 7, 14, 15)$$

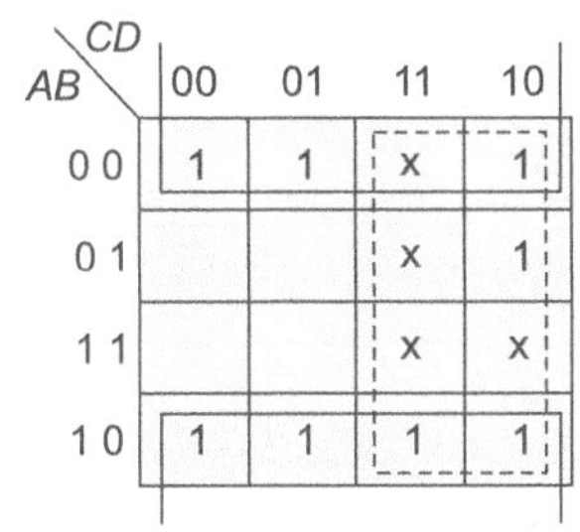

$$f = \bar{B} + C$$

2.20 (d)

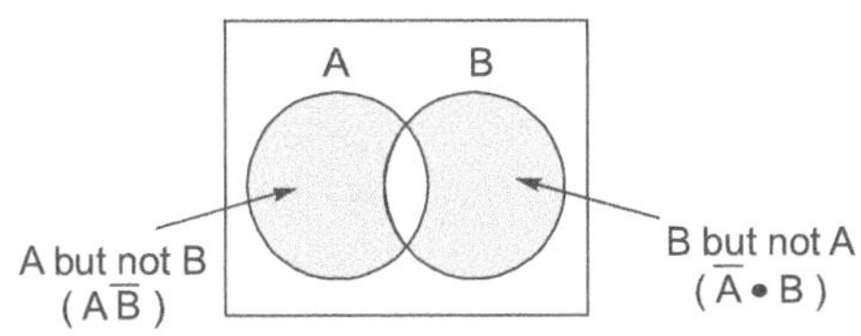

$$f = A\bar{B} + A\bar{B}$$
$$f = (\bar{A} + \bar{B}).(A + B)$$

2.21 (c)

(X'Y'Z' + X'YZ' + XYZ')

= X'Z' (Y' + Y) + XYZ'

= X'Z' + XYZ'

= Z' (X' + XY)

= Z' (X' + Y)

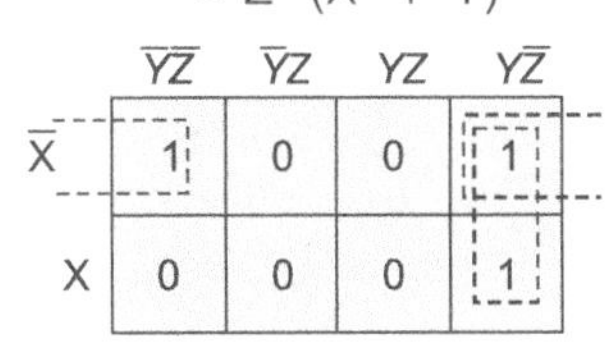

By using k-map,

$F = X'Z' + YZ'$

2.22 (b)

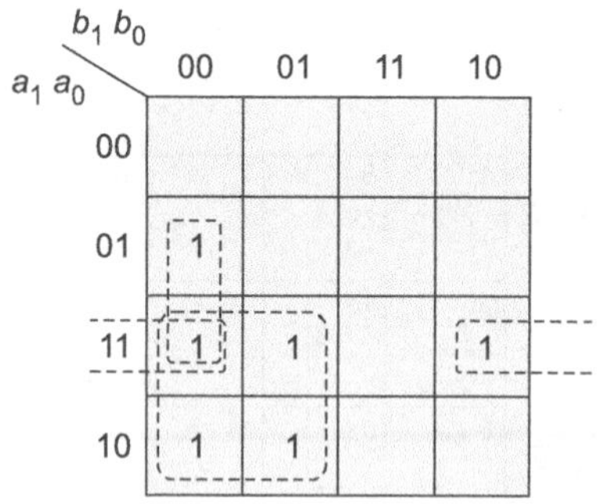

Minimized form of

$$F = a_1 a_0 \bar{b}_0 + a_1 \bar{b}_1 + a_0 \bar{b}_1 \bar{b}_0$$

2.23 (c)

Given:

$F(w, x, y, z) = \Sigma m(4, 5, 10, 11, 12, 13, 14, 15)$

K-Map:

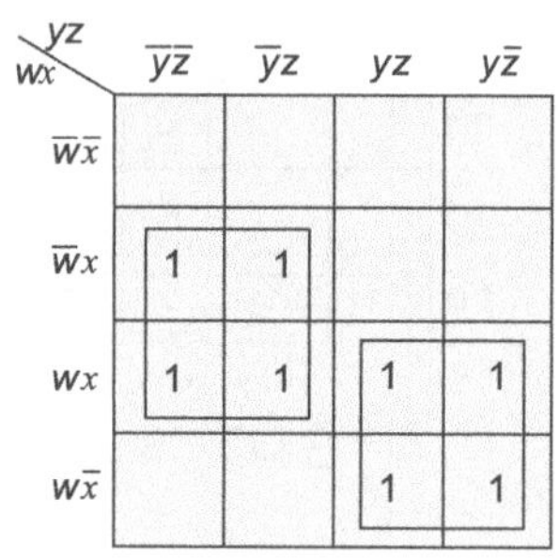

$$F(w, x, y, z) = x\bar{y} + wy$$

2.24 (a)

$$F = (A + \bar{A}B) + \bar{A}(A + \bar{B}) \cdot C$$

$$F = A + \bar{A}B + (A\bar{A} + \bar{A}\bar{B})C$$

$$F = A + \bar{A}B + \bar{A}\bar{B}C$$

$$F = (A + \bar{A})(A + B) + \bar{A}\bar{B}C$$

$$F = \bar{A} + B + \bar{A}\bar{B}C$$

$$F = (A + \bar{A})(A + \bar{B}C) + B$$

$$F = A + B + \bar{B}C$$

$$F = A + (B + \bar{B})(B + C)$$

$$F = A + B + C$$

■■■

3 Logic Gates

3.1 If A and B are the inputs to a logic gate, then match the logic with its output

(a) NAND	(i) $\overline{\overline{A}+\overline{B}}$
(b) NOR	(ii) $\overline{A}+\overline{B}$
(c) XNOR	(iii) $\overline{A}\overline{B}+AB$
(d) AND	(h) $\overline{AB}$

[1995 : 1 M]

3.2 For the logic circuit shown in figure write the truth stable of X, Y and Z.

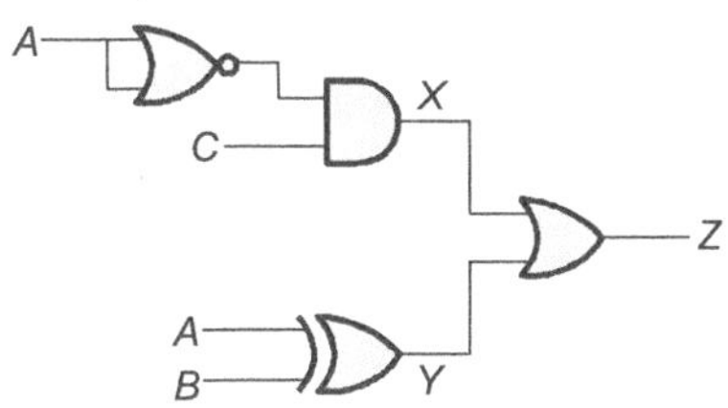

[1998 : 2 M]

3.3 In the logic circuit shown in figure the output X is

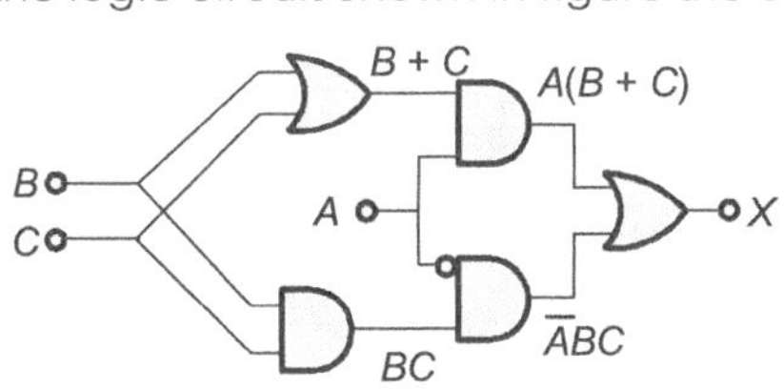

(a) $A\overline{B}+B\overline{C}+C\overline{A}$ (b) $A + B + C$
(c) $AB + BC + CA$ (d) $\overline{AB}+\overline{BC}+\overline{CA}$

[2001 : 1 M]

3.4 The logic circuit of figure is a

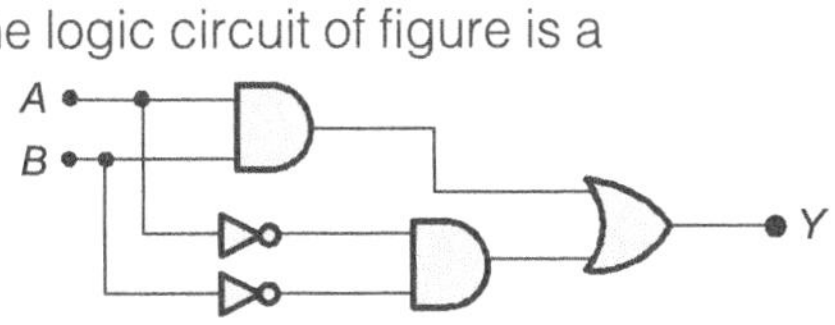

(a) half adder (b) XOR
(c) equality detector (d) full adder

[2003 : 2 M]

3.5 The two NAND gates before the latch circuit shown in figure are used to

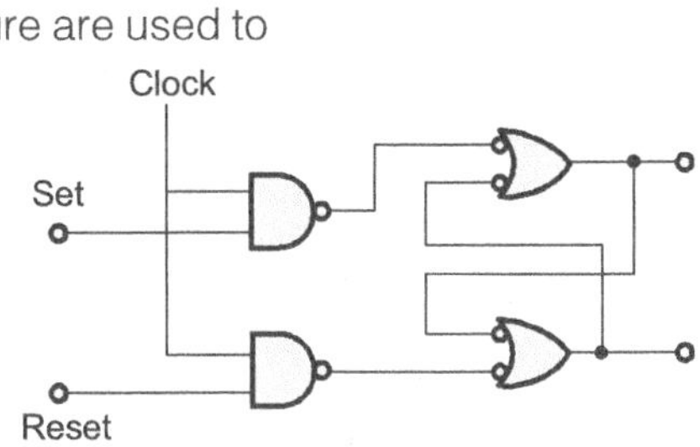

(a) act as buffers
(b) operate the latch faster
(c) avoid racing problem
(d) invert the latching action

[2004 : 1 M]

3.6 In the circuit of figure A, B, C are the inputs and P, Q are the two outputs. The circuit is a

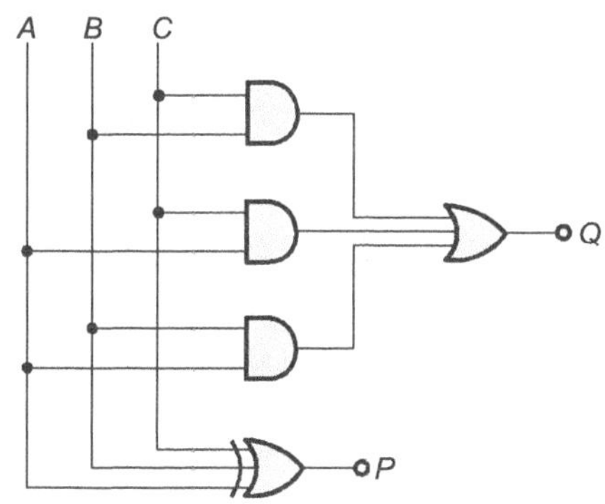

(a) half adder where P is the sum and Q is the carry
(b) half adder where P is the carry and Q is the sum
(c) full adder where P is the sum and Q is the carry
(d) full adder where P is the carry and Q is the sum

[2004 : 2 M]

3.7 The logic gate circuit shown in the adjoining figure realizes the function

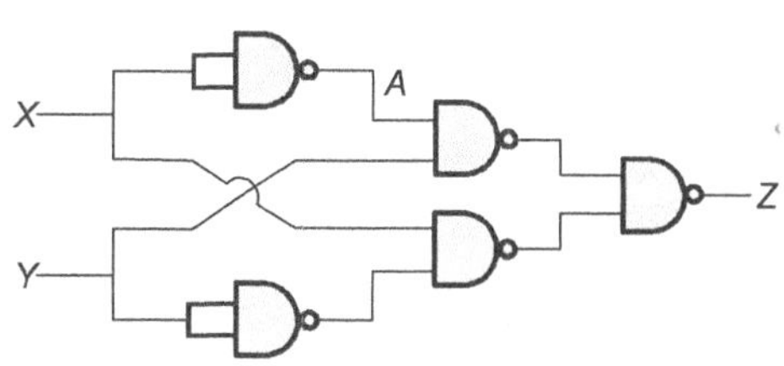

(a) XOR (b) XNOR
(c) Half adder (d) Full adder

[2010 : 2 M]

3.8 In the circuit shown, the switch is momentarily closed and then opened. Assuming the logic gates to have equal non-zero delay, at steady state, the logic states of X and Y are

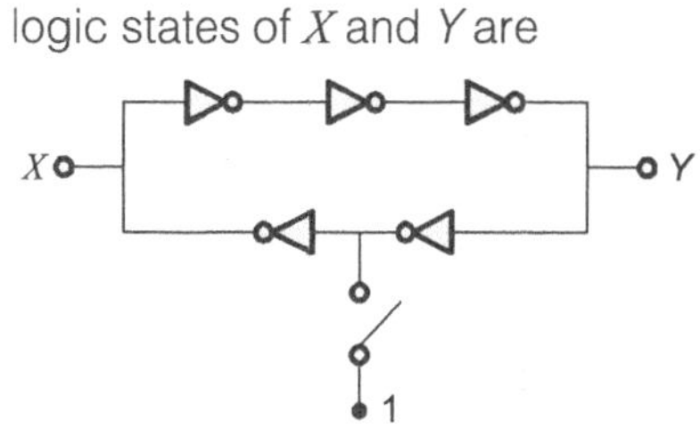

(a) X is latched, Y toggles continuously
(b) X and Y are both latched
(c) Y is latched, X toggles continuously
(d) X and Y both toggle continuously

[2015 : 1 M]

3.9 The logic evaluated by the circuit at the output is

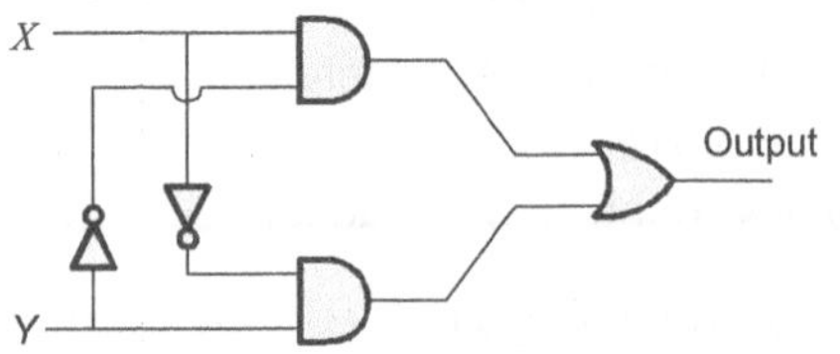

(a) $X\overline{Y}+Y\overline{X}$ (b) $(\overline{X+Y})XY$
(c) $\overline{XY}+XY$ (d) $\overline{X}Y+X\overline{Y}+X+Y$

[2015 : 1 M]

3.10 Consider the logic circuit with input signal TEST shown in the figure. All gates in the figure shown have identical non-zero delay. The signal TEST which was at logic LOW is switched to logic HIGH and maintained at logic HIGH. The output

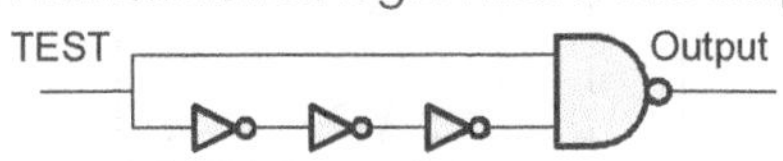

(a) stays HIGH throughout
(b) stays LOW throughout
(c) pulses from LOW to HIGH to LOW
(d) pulses from HIGH to LOW to HIGH

[2015 : 1 M]

3.11 In the digital circuit given below, F is:

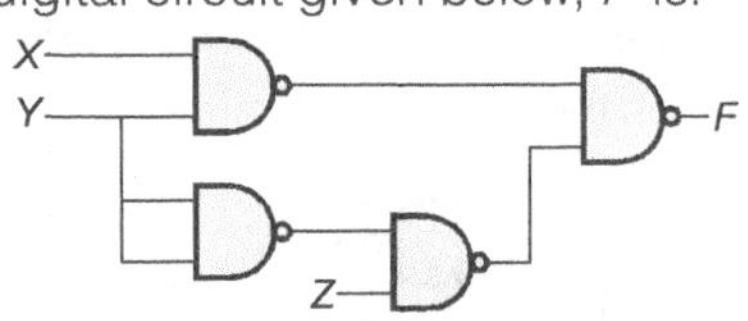

(a) $XY+Y\overline{Z}$ (b) $XY+\overline{Y}Z$
(c) $\overline{X}\overline{Y}+Y\overline{Z}$ (d) $XZ+\overline{Y}$

[2016 : 1 M]

3.12 The comparators (output = '1', when input ≥ 0 and output = '0', when input < 0), exclusive-OR gate and the unity gain low-pass filter given in the circuit are ideal. The logic output voltages of the exclusive-OR gate are 0 V and 5 V. The cutoff frequency of the low-pass filter is 0.1 Hz. For $V_1 = 1\sin(3000\,t+36°)$ V and $V_2 = 1\sin(3000\,t)$ V, the value of V_0 in volt is ______.

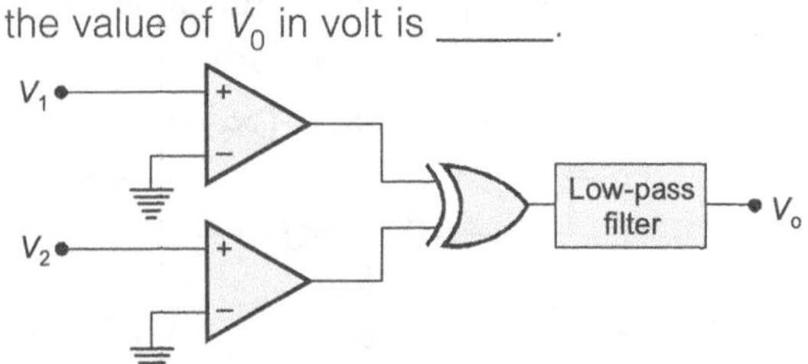

[2016 : 2 M]

3.13 A and B are the logical inputs and X is the logical output shown in the figure. The output X is related to A and B by

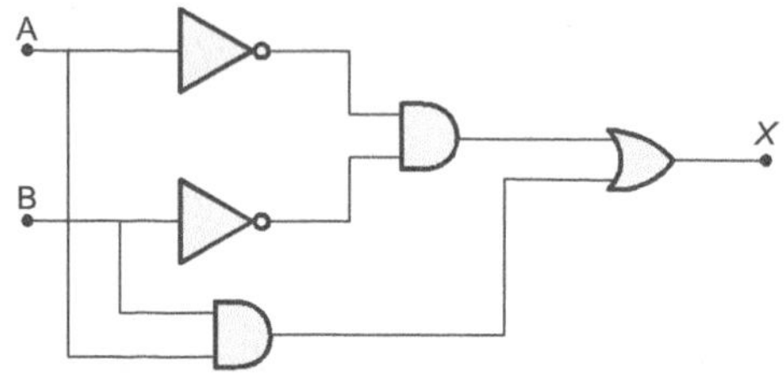

(a) $X=\overline{A}B+\overline{B}A$ (b) $X=AB+\overline{B}A$
(c) $X=AB+\overline{A}\overline{B}$ (d) $X=\overline{A}\overline{B}+\overline{B}A$

[2017 : 1 M]

3.14 The Boolean function $F(X, Y)$ realized by the given circuit is

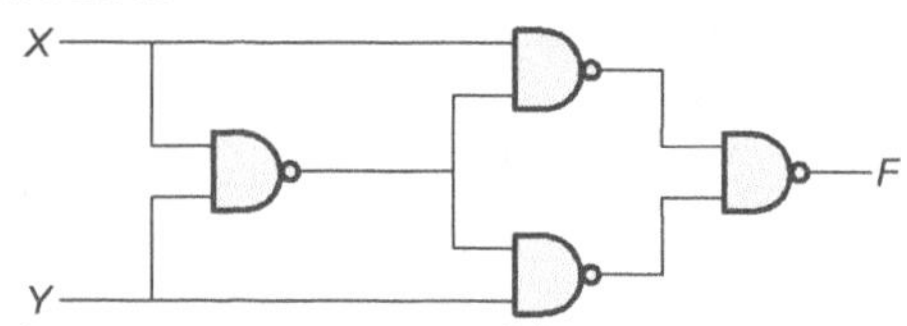

(a) $\overline{X}Y+X\overline{Y}$ (b) $\overline{X}\overline{Y}+XY$
(c) $\overline{X}+Y$ (d) $\overline{X}\cdot\overline{Y}$ [2018 : 2 M]

3.15 The logic block shown has an output F given by ____.

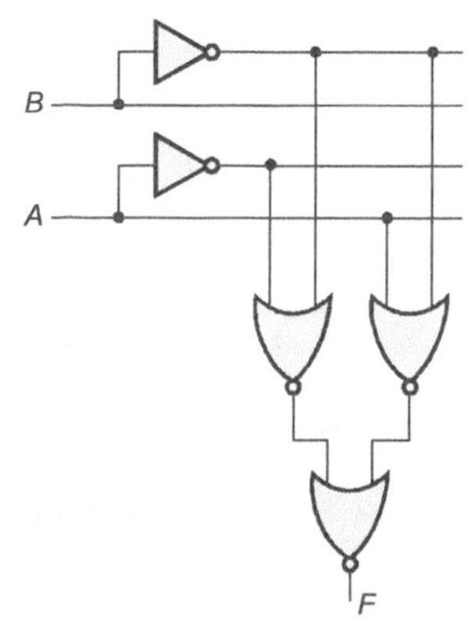

(a) $A+B$ (b) $A\cdot\overline{B}$
(c) $A+\overline{B}$ (d) $\overline{B}$ [2022 : 1 M]

3.16 Match the Exclusive-OR (XOR) operations (i) to (iv) with the results (p) to (s), where X is a Boolean input.

(i) $X\oplus X$ (p) 1
(ii) $X\oplus\overline{X}$ (q) 0
(iii) $X\oplus 0$ (r) $\overline{X}$
(iv) $X\oplus 1$ (s) X

(a) (i) - (q), (ii) - (r), (iii) - (s), (iv) - (p)
(b) (i) - (q), (ii) - (r), (iii) - (p), (iv) - (s)
(c) (i) - (p), (ii) - (s), (iii) - (q), (iv) - (r)
(d) (i) - (q), (ii) - (p), (iii) - (s), (iv) - (r)

[2023 : 1 M]

■■■

Answers Logic Gates

3.1	(Sol.)	3.2	Sol.	3.3	(c)	3.4	(c)	3.5	(d)	3.6	(c)	3.7	(a)
3.8	(d)	3.9	(a)	3.10	(d)	3.11	(b)	3.12	(1)	3.13	(c)	3.14	(a)
3.15	(d)	3.16	(d)										

Explanations Logic Gates

3.1 Sol.

(a-ii) (b-iv) (c-iii) (d-i)

NAND → $\overline{AB} = \bar{A}+\bar{B}$ (using Demorgan's theorem)

NOR → $\overline{A+B} = \bar{A}\cdot\bar{B}$

XNOR → $\bar{A}\bar{B}+AB$

AND → $A\cdot B = \overline{\overline{AB}} = \overline{\bar{A}+\bar{B}}$

3.2 Sol.

From the given digital circuit.

$X = \bar{A}C$, $Y = A\oplus B$, $Z = X+Y$

So the truth table of X, Y and Z is

A	B	C	X	Y	Z
0	0	0	0	0	0
0	0	1	1	0	1
0	1	0	0	1	1
0	1	1	1	1	1
1	0	0	0	1	1
1	0	1	0	1	1
1	1	0	0	0	0
1	1	1	0	0	0

3.3 (c)

From the give logic circuit

$$X = \bar{A}(BC)+A(B+C)$$
$$= AB+AC+\bar{A}BC$$
$$= B(A+\bar{A}C)+AC$$
$$= B(A+\bar{A})(A+C)+AC$$
$$= B(A+C)+AC$$
$$X = AB+BC+AC$$

3.4 (c)

A	B	Y
0	0	1
0	1	0
1	0	0
1	1	1

$$Y = AB+\bar{A}\bar{B}$$

Actually this circuit will behave as X-NOR logic gate which can be used as a 1-bit comparator or equality detector circuit.

3.5 (d)

When clock is high then NAND gates used inverts the *S* and *R*, so invert the latching action.

3.6 (c)

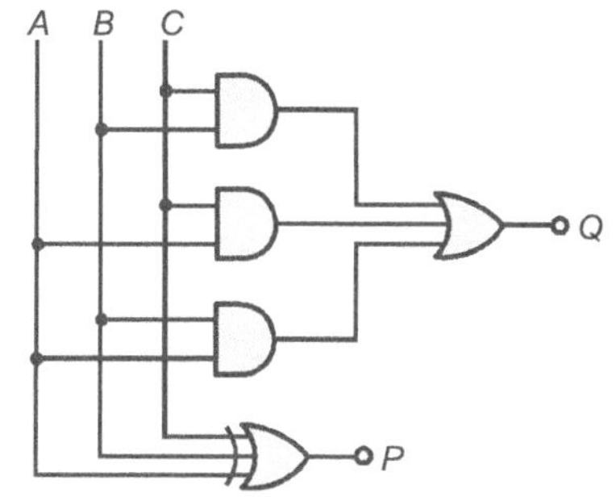

The above circuit is a full adder circuit where *P* is the sum and *Q* is the carry.

$$P = A\oplus B\oplus C$$

and $$Q = AB+BC+CA$$

3.7 (a)

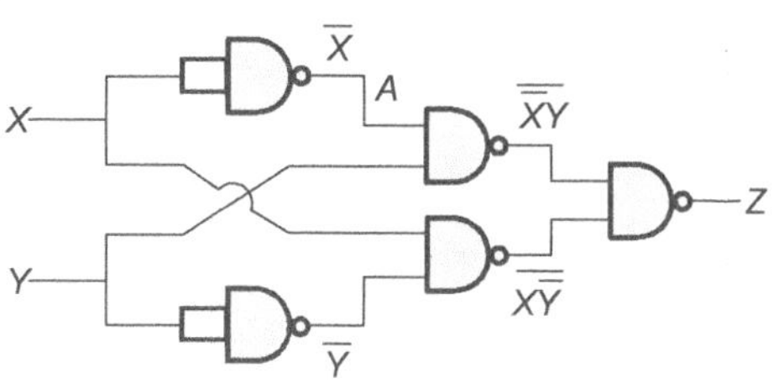

$$Z = \overline{(\overline{\bar{X}Y})(\overline{X\bar{Y}})} = (\overline{\overline{\bar{X}Y}})+(\overline{\overline{X\bar{Y}}})$$
$$= \bar{X}Y+X\bar{Y} = X\oplus Y$$

3.8 (d)

X	Y
0	1
1	0
0	1
1	0

Both *X* and *Y* will toggle continuously.

3.9 (a)

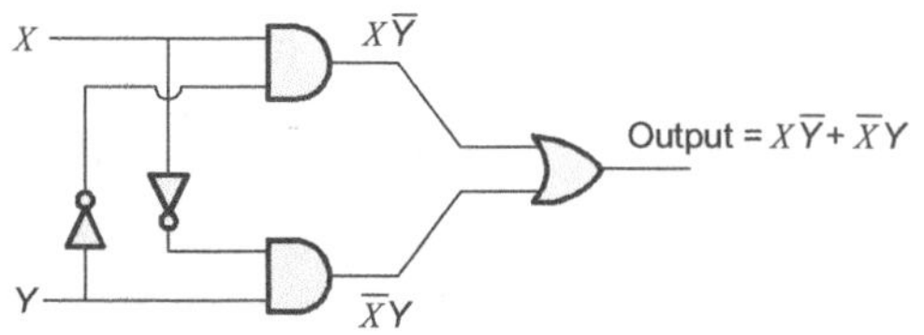

3.10 (d)

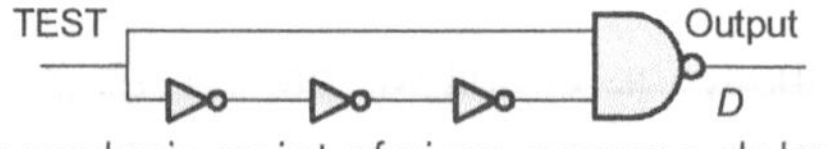

For analysis point of view, assume delay of each gate is 100 msec. However, we can take any value.

- By referring the circuit the upper input to the NAND gate is direct TEST signal.
- The lower input to NAND gate is TEST but with a delay of 300 msec.
- Assume the delay of NAND gate is 0, first draw output waveform (ideal case) then shift that by 100 msec i.e. introduce the delay.

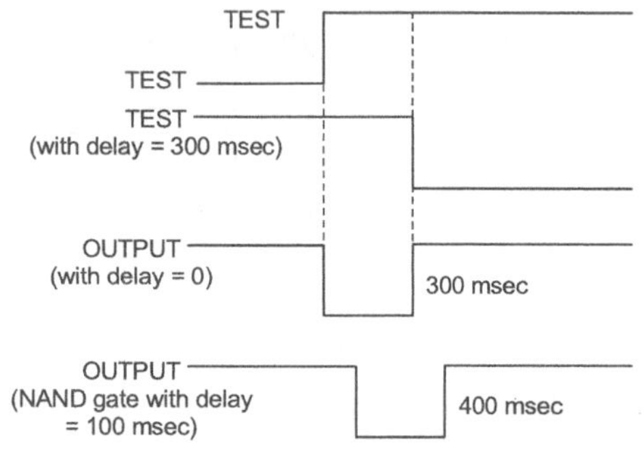

3.11 (b)

$$F = \overline{\overline{XY}\cdot\overline{Z\overline{Y}}} = XY + Z\overline{Y}$$

3.12 (1)

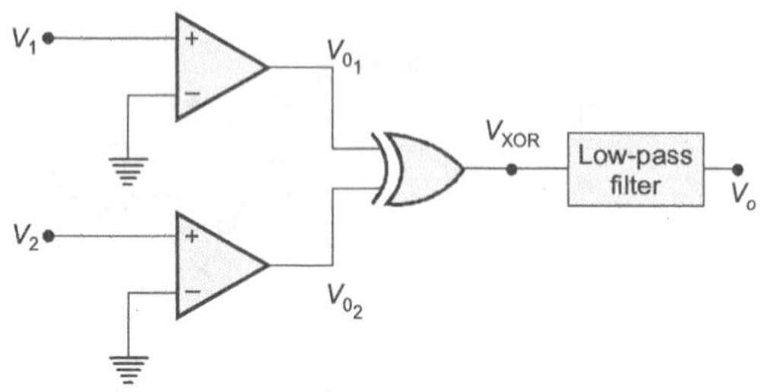

Output waveform

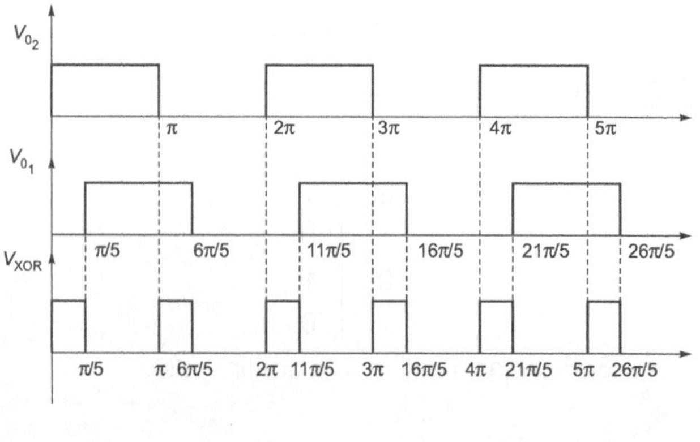

Output of low pass filter is the average value of output of XOR GATE

$$V_{DC} = \frac{1}{2\pi}\left[\int_0^{\pi/5} 5dt + \int_{\pi}^{6\pi/5} 5dt\right]$$

$$= \frac{1}{2\pi}\left[5\left(\frac{\pi}{5}\right) + 5\left(\frac{6\pi}{5} - \pi\right)\right]$$

$$= \frac{1}{2\pi}[\pi + \pi] = 1 \text{ volt}$$

3.14 (a)

The given circuit is the standard circuit. It is the implementation of EX-OR gate using NAND gates.

Hence, $F = X \oplus Y = \overline{X}Y + X\overline{Y}$

3.15 (d)

$$F = \left(\overline{\overline{A+\overline{B}} + \overline{\overline{A}+\overline{B}}}\right)$$

$$F = \left(\overline{\overline{A+\overline{B}}}\right)\cdot\left(\overline{\overline{\overline{A}+\overline{B}}}\right)$$

$$F = (A+\overline{B})\cdot(\overline{A}+\overline{B})$$

$$F = A\overline{A} + A\overline{B} + \overline{B}\overline{A} + \overline{B}\overline{B}$$

$$F = \overline{B}(A+\overline{A}) + \overline{B}$$

$$F = \overline{B} + \overline{B} = \overline{B}$$

3.16 (d)

Truth table for EXOR gate

Inputs		Output	
A	B	Y	
0	0	0	→ Same inputs
1	1	1	→ Different inputs
1	0	1	
1	1	0	→ Same inputs

Inputs		Output
A	B	Y
0	0 →	0
0	1 →	1
1	0	1
1	1	0

If one input of EXOR gate is 0, then the output is same as second input.

Inputs		Output
A	B	Y
0	0	0
0	1	1
1	0 →	1
1	1 →	0

If one output of EXOR gate is 1, then the output is the complement of second input.

(1) → (Q)
(2) → (P)
(3) → (S)
(4) → (R)

4 Combinational Circuits

4.1 The combinational circuit shown in the figure employs a 4 to 1 multiplexer. The output Q of the circuit is

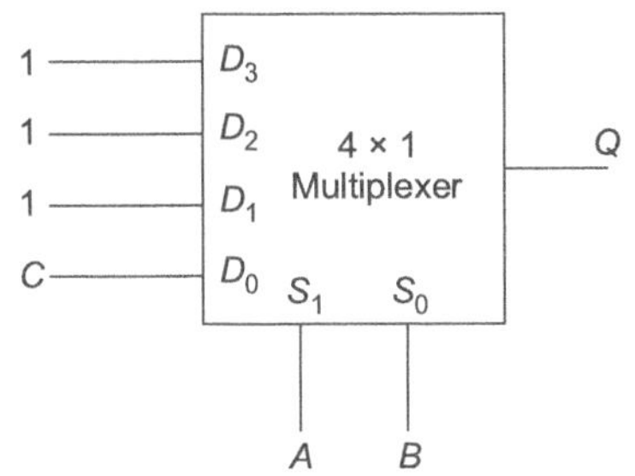

(a) $\bar{A}\bar{B}C$ (b) $A + B + C$

(c) $A \oplus B \oplus C$ (d) $\bar{A}\bar{B}\bar{C}$

[1995 : 1 M]

4.2 Design the logic system shown in figure, to satisfy the truth table given, using minimum number of gates.

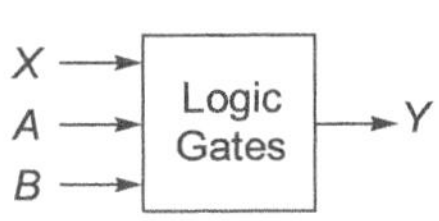

A	B	Y
0	0	$\bar{X}$
0	1	X
1	0	1
1	1	0

[1997 : 2 M]

4.3 Write the truth table for the following in Boolean expression, and realize it using only an 8 : 1 multiplexer. $f = \overline{AB} + BD + AB\bar{C}$

[1997 : 2 M]

4.4 A 2-to-1 digital multiplexer having a switching delay of 1 μs is connected as shown in figure. The output of the multiplexer is tied to its own select input S. The input which gets selected when $S = 0$ is tied to 1 and the input that gets selected when $S = 1$ is tied to 0. The output V_0 will be

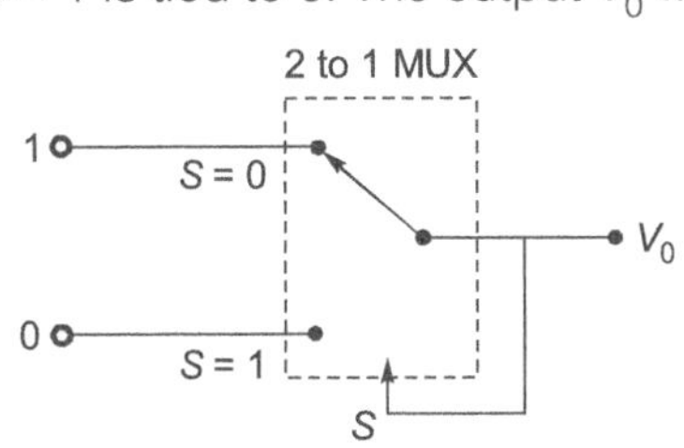

(a) 0

(b) 1

(c) pulse train of frequency 0.5 MHz

(d) pulse train of frequency 1.0 MHz

[2003 : 2 M]

4.5 The cell of a Field Programmable Gate Array is shown in the figure. It has three 2-to-1 multiplexers with their select lines G_0, G_1, G_2 and 4 digital signal input lines I_0, I_1, I_2 and I_3. The logical function that relates the output O to the select and signal input lines is

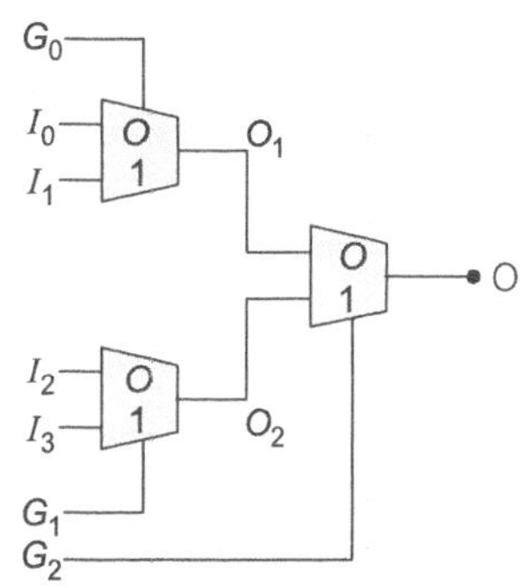

(a) $\bar{G}_0\bar{G}_1 I_2 + \bar{G}_0\bar{G}_1 I_3 + \bar{G}_2\bar{G}_1 I_0 + \bar{G}_2\bar{G}_1 I_1$

(b) $\bar{G}_0 I_2 + \bar{G}_0 G_1 + \bar{G}_2 I_0 + \bar{G}_2 G_1 I_1 + G_0$

(c) $\bar{G}_0\bar{G}_2 I_0 + G_0\bar{G}_2 I_1 + G_2\bar{G}_1 I_2 + G_2 G_1 I_3$

(d) $G_2 G_1 \bar{I}_2 + \bar{G}_2\bar{G}_1\bar{I}_3 + G_2\bar{G}_0 I_0 + G_0\bar{G}_2 I_1$

[2005 : 2 M]

4.6 A combinational logic circuit has three inputs A, B and C and one output Y. The output $Y = 1$ when at least two inputs are 1. Otherwise, $Y = 0$. In its minimized SOP realization, the maximum number of two-input terms is

(a) 1 (b) 2

(c) 3 (d) 4 **[2005 : 2 M]**

4.7 Not with standing overflow, the addition and subtraction of K-bit signed binary numbers can be realized using.

One K-bit full adder and

(a) 2 - Input AND Gates

(b) 2 - Input NOR Gates

(c) 2 - Input OR Gates

(d) 2 - Input XOR Gates **[2005 : 2 M]**

4.8 A combinational circuit using a 8-to-1 multiplexer is shown in the figure. The minimized expression for the output (Z) is

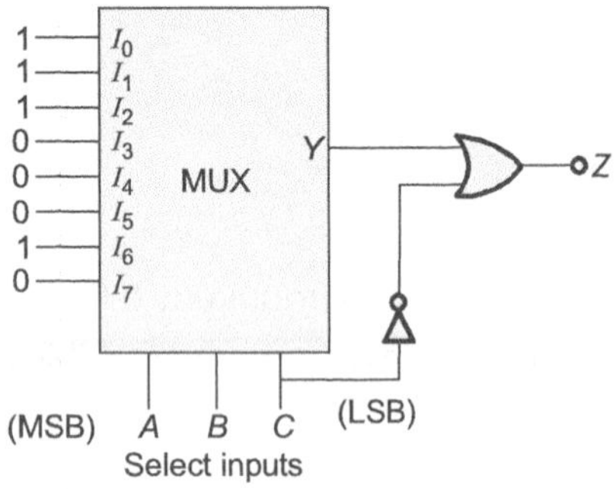

(a) $C(\overline{A}+\overline{B})$ (b) $C(A+B)$

(c) $\overline{C}+\overline{A}\,\overline{B}$ (d) $\overline{C}+AB$

[2006 : 2 M]

4.9 Two square waves of equal period T, but with a time delay τ are applied to a digital circuit whose truth table is shown in the following figure.

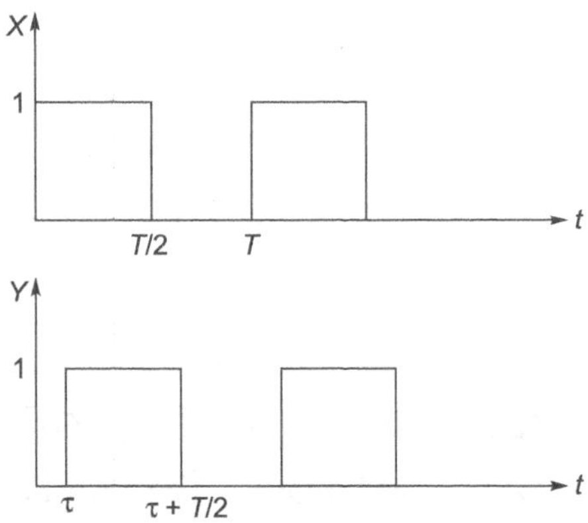

X	Y	Output
0	0	1
0	1	0
1	0	0
1	1	1

The high and the low levels of the output of the digital circuit are 5 V and 0 V, respectively. Which one of the following figures shows the correct variation of the average value of the output voltage as a function of τ for $0 \le t \le (T/2)$?

(a)

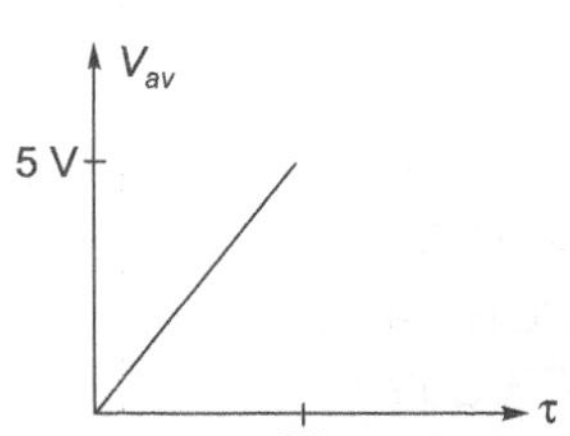

(b)

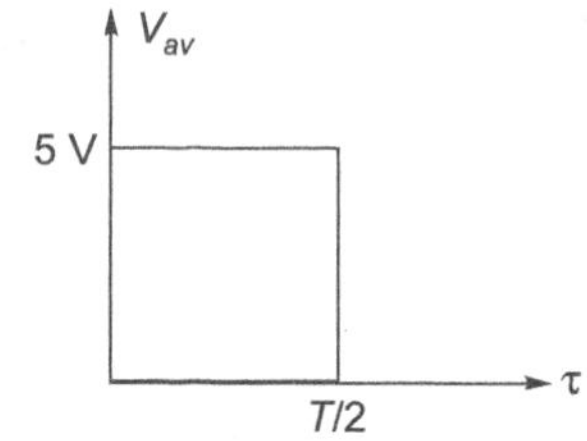

(c)

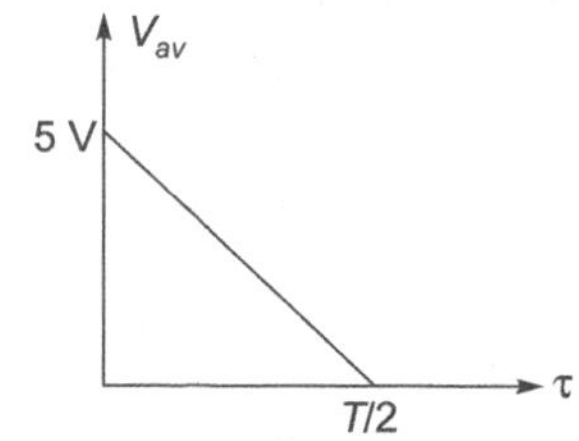

(d)

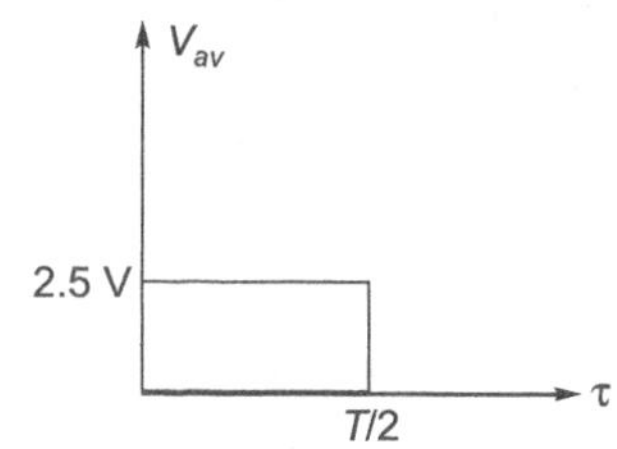

[2007 : 2 M]

4.10 A MUX circuit shown in the figure below implements a logic function F_1. The correct expression for F_1 is

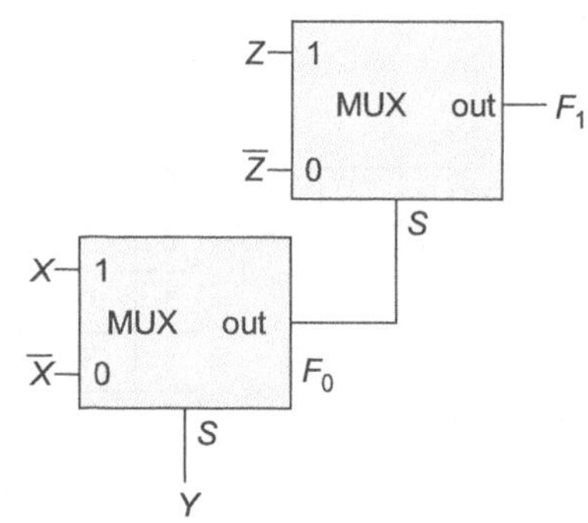

(a) $\overline{(X\oplus Y)}\oplus Z$ (b) $\overline{(\overline{X\oplus Y})\oplus Z}$

(c) $(X\oplus Y)\oplus\overline{Z}$ (d) $(X\oplus Y)+Z$

[2007 : 2 M]

4.11 The output F of the multiplexer circuit shown below expressed in terms of the inputs P, Q and R is

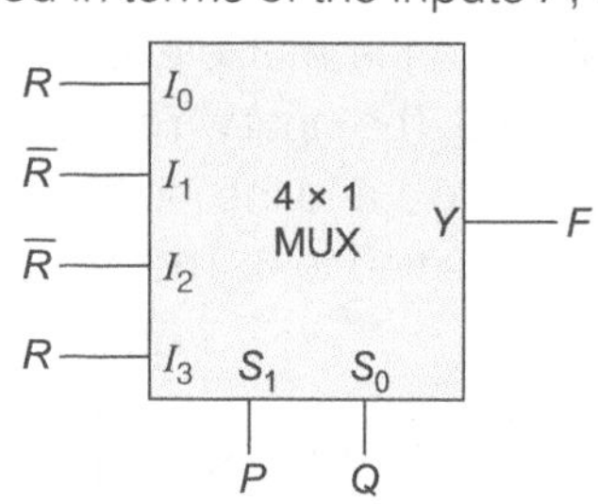

(a) $F = P\oplus Q\oplus R$ (b) $F = PQ + QR + RP$

(c) $F = (P\oplus Q)R$ (d) $F = (P\oplus Q)\overline{R}$

[2008 : 2 M]

4.12 A 4 to 1 multiplexer to realize a Boolean function $F(X, Y, Z)$ is shown in the figure below. The inputs Y and Z are connected to the selectors of the MUX (Y is more significant). The canonical sum-of-product expression for $F(X, Y, Z)$ is

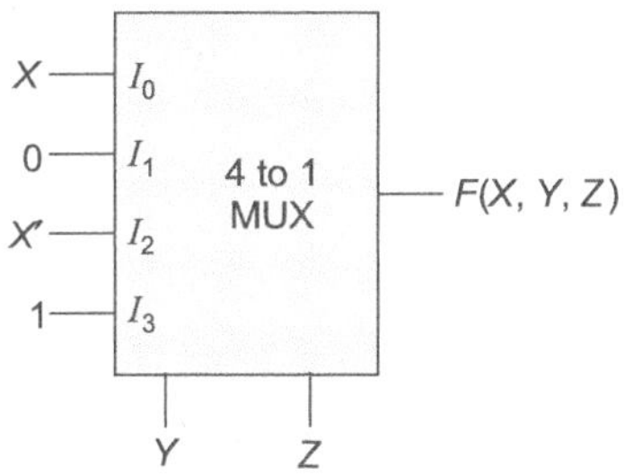

(a) $\Sigma m(2, 3, 4, 7)$ (b) $\Sigma m(1, 3, 5, 7)$
(c) $\Sigma m(0, 2, 4, 6$ (d) $\Sigma m(2, 3, 5, 6)$

[2016 : 2 M]

4.13 The figure below shows the ith full-adder block of a binary adder circuit. C_i is the input carry and C_{i+1} is the output carry of the circuit. Assume that each logic gate has a delay of 2 nanosecond, with no additional time delay due to the interconnecting wires. If the inputs A_i, B_i are available and stable throughout the carry propagation, the maximum time taken for an input C_i to produce a steady-state output C_{i+1} is____nanosecond.

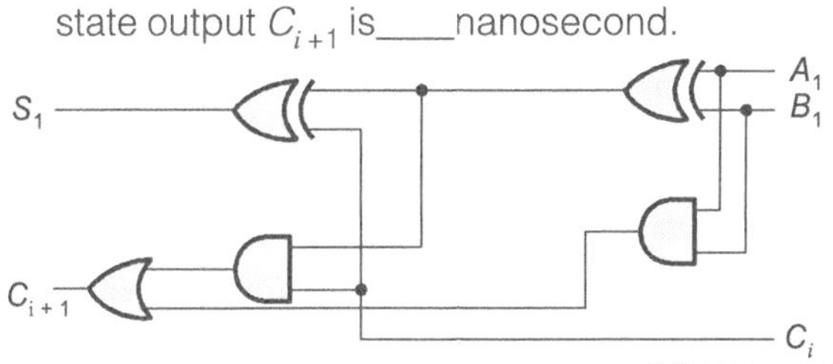

[2019 : 1 M]

4.14 In the circuit shown below, assume that the comparators are ideal and all components have zero propagation delay. In one period of the input signal $V_{in} = 6 \sin(\omega t)$, the fraction of the time for which the output OUT is in logic state HIGH is

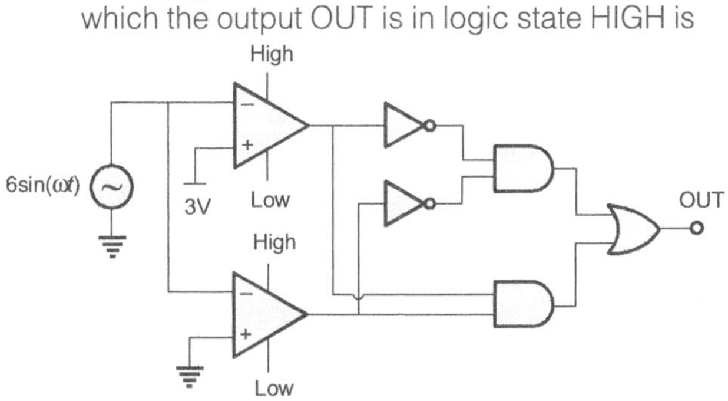

(a) $\frac{1}{12}$ (b) $\frac{1}{2}$
(c) $\frac{2}{3}$ (d) $\frac{5}{6}$

[2019 : 2 M]

4.15 The Boolean operation performed by the following circuit at the output O is _____.

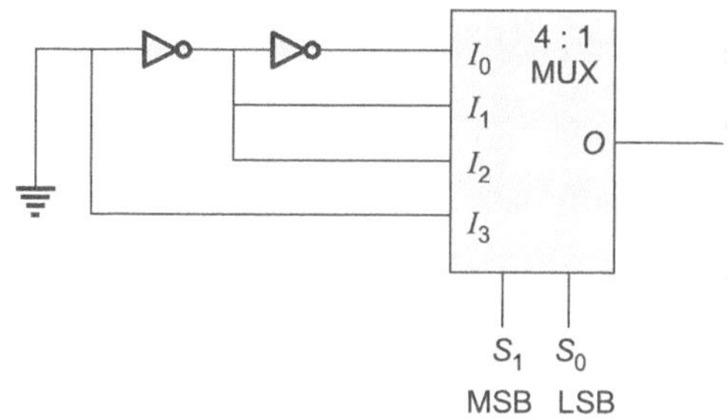

(a) $O = S_0 \bullet \bar{S}_1$ (b) $O = S_1 \bullet \bar{S}_0$
(c) $O = S_1 + S_0$ (d) $O = S_1 \oplus S_0$

[2020 : 1 M]

4.16 A 4 × 1 multiplexer with two selector lines is used to realize a Boolean function, F having four Boolean variables X, Y, Z and W as shown below. S_0 and S_1 denote the least significant bit (LSB) and most significant bit (MSB) of the selector lines of the multiplexer respectively. I_0, I_1, I_2, I_3 are the input lines of the multiplexer.

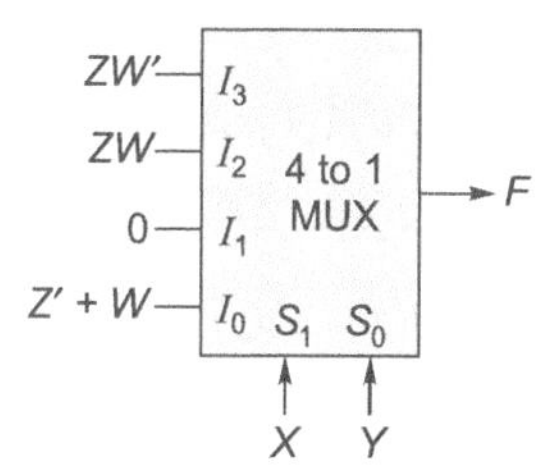

The canonical sum of product representation of F is
(a) $F(X, Y, Z, W) = \Sigma m(0, 1, 3, 14, 15)$
(b) $F(X, Y, Z, W) = \Sigma m(0, 1, 3, 11, 14)$
(c) $F(X, Y, Z, W) = \Sigma m(2, 5, 9, 11, 14)$
(d) $F(X, Y, Z, W) = \Sigma m(1, 3, 7, 9, 15)$

[2021 : 2 M]

4.17 The output F of the digital circuit shown can be written in the form(s) ________

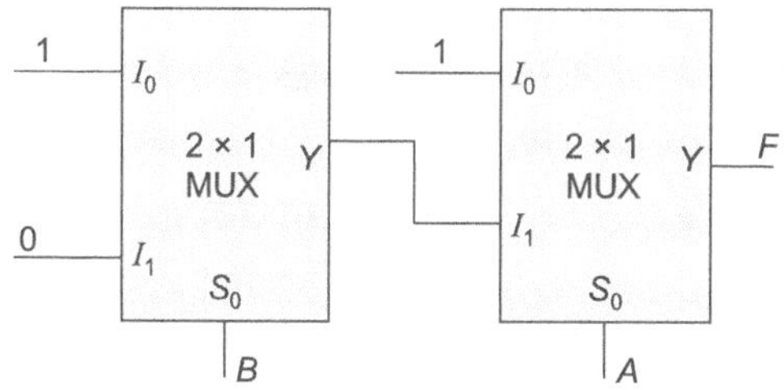

(a) $\overline{A \cdot B}$ (b) $\bar{A} + \bar{B}$
(c) $\overline{A + B}$ (d) $\bar{A} \cdot \bar{B}$

[2022 : 1 M]

4.18 For the given digital circuit, $A = B = 1$. Assume that AND, OR, and NOT gates have propagation delays of 10 ns, 10 ns, and 5 ns respectively. All lines have zero propagation delay. Given that $C = 1$ when the circuit is turned on, the frequency of steady-state oscillation of the output Y is ______.

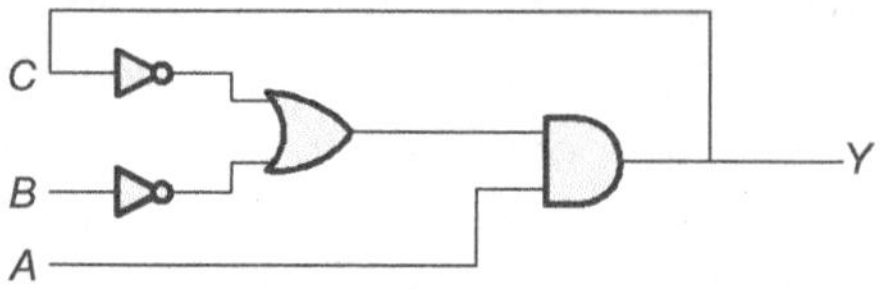

(a) 20 MHz (b) 15 MHz
(c) 40 MHz (d) 50 MHz

[2023 : 2 M]

Answers Combinational Circuits

4.1 (b)	4.2 (Sol.)	4.3 (Sol.)	4.4 (c)	4.5 (c)	4.6 (c)	4.7 (d)
4.8 (c)	4.9 (c)	4.10 (b)	4.11 (a)	4.12 (a)	4.13 (4)	4.14 (d)
4.15 (d)	4.16 (b)	4.17 (a, b)	4.18 (a)			

Explanations Combinational Circuits

4.1 (b)

The output of the MUX is

$$Q = \bar{S}_1\bar{S}_0 D_0 + \bar{S}_1 S_0 D_1 + S_1\bar{S}_0 D_2 + S_1 S_0 D_3$$

from given circuit,

$$Q = \bar{A}\bar{B}C + \bar{A}B + A\bar{B} + AB$$
$$= B(A+\bar{A}) + A\bar{B} + \bar{A}\bar{B}C$$
$$= B + \bar{B}(A + \bar{A}C)$$
$$= B + \bar{B}(A+\bar{A})(A+C)$$
$$= B + \bar{B}(A+C)$$
$$= B + \bar{B}A + \bar{B}C$$
$$= (B+\bar{B})(B+A) + \bar{B}C$$
$$= B + A + \bar{B}C$$
$$= A + (B+\bar{B})(B+C)$$
$$= A + B + C$$

Hence, option (b) is correct.

4.2 Sol.

$$Y = \bar{A}\bar{B}\bar{X} + \bar{A}BX + A\bar{B}$$

A \ BX	00	01	11	10
0	1		1	
1	1	1		

$$Y = \bar{B}\bar{X} + A\bar{B} + \bar{A}BX$$
$$= \bar{B}(A+\bar{X}) + B(\bar{A}X)$$
$$= \bar{B}(\overline{\bar{A}X}) + B(\bar{A}X)$$
$$= B \odot (\bar{A}X)$$

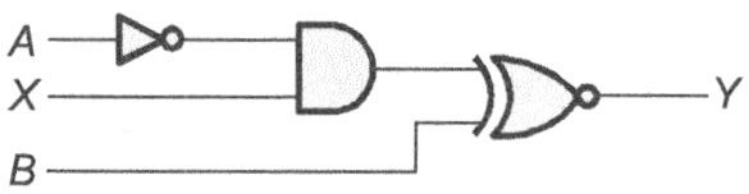

4.3 Sol.

Given function is

$$f = \overline{AB} + BD + AB\bar{C}$$
$$= \bar{A} + (\bar{B}+B)(\bar{B}+D) + AB\bar{C}$$
$$= \bar{A} + \bar{B} + D + AB\bar{C}$$

Truth table for the given function f is

A	B	C	D	f
0	0	0	0	1
0	0	0	1	1
0	0	1	0	1
0	0	1	1	1
0	1	0	0	1
0	1	0	1	1
0	1	1	0	1
0	1	1	1	1
1	0	0	0	1
1	0	0	1	1
1	0	1	0	1
1	0	1	1	1
1	1	0	0	1
1	1	0	1	1
1	1	1	0	0
1	1	1	1	1

k-map

AB \ CD	00	01	11	10
00	1_0	1_1	1_3	1_2
01	1_4	1_5	1_7	1_6
11	1_{12}	1_{13}	1_{15}	0_{14}
10	1_8	1_9	1_{11}	1_{10}

realization of given function using 8 × 1 MUX

A	I_0	I_1	I_2	I_3	I_4	I_5	I_6	I_7
	BCD							
0	(0)	(1)	(2)	(3)	(4)	(5)	(6)	(7)
1	(8)	(9)	(10)	(11)	(12)	(13)	14	(16)
	1	1	1	1	1	1	0 → $\bar{A}$	1

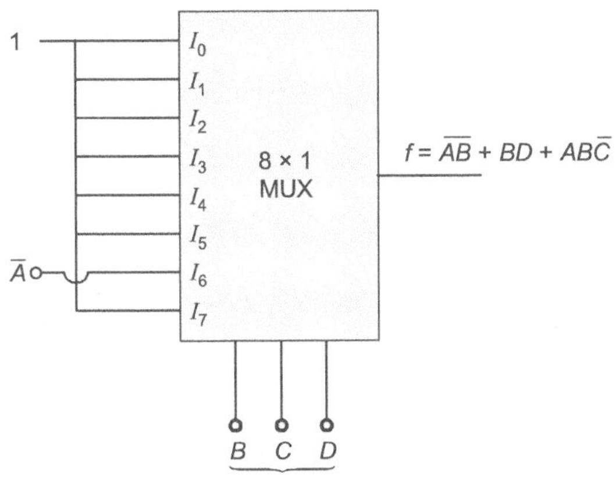

4.4 (c)

When output is 1, then s will be also 1 and s will go to 1 after 1 μsec and output become zero, again s becomes 0 and switch will go to 0 after 1 μsec and output goes to 1 this will continue and we will get the waveform at output as

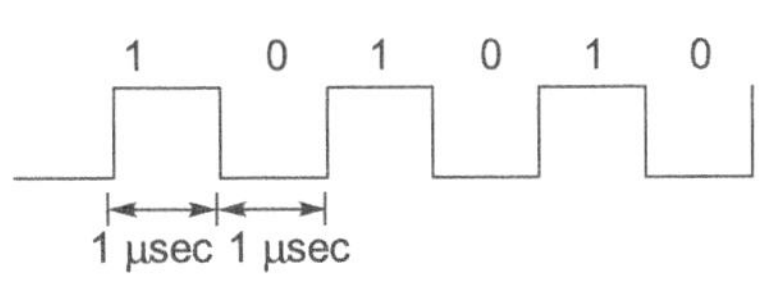

so, $T = 2\,\mu\text{sec}$

then, $f = \dfrac{1}{T} = \dfrac{1}{2\times10^{-6}} = 0.5 \times 10^{6}$

$= 0.5\text{ MHz}$

4.5 (c)

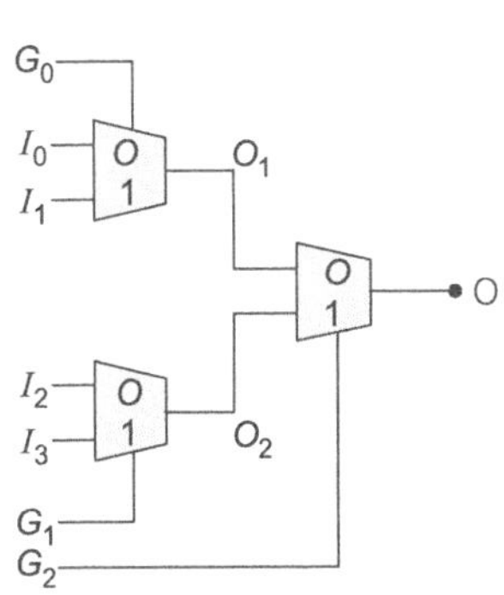

$$O_1 = \bar{G}_0 I_0 + G_0 I_1$$

$$O_2 = \bar{G}_1 I_2 + G_1 I_3$$

$$\begin{aligned} O &= \bar{G}_2 O_1 + G_2 O_2 \\ &= \bar{G}_2(\bar{G}_0 I_0 + G_0 I_1) + G_2(\bar{G}_1 I_2 + G_1 I_3) \\ &= \bar{G}_2 \bar{G}_0 \cdot I_0 + \bar{G}_2 G_0 I_1 + G_2 \bar{G}_1 I_2 + G_2 G_1 I_3 \\ &= \bar{G}_0 \bar{G}_2 I_0 + G_0 \bar{G}_2 I_1 + G_2 \bar{G}_1 I_2 + G_2 G_1 I_3 \end{aligned}$$

4.6 (c)

A	B	C	Y
0	0	0	0
0	0	1	0
0	1	0	0
0	1	1	1
1	0	0	0
1	0	1	1
1	1	0	1
1	1	1	1

K-map

A \ BC	00	01	11	10
0	0	0	1	0
1	0	1	1	1

$$Y = AB + BC + CA$$

Hence, maximum number of two input terms is = 3.

4.7 (d)

Condition for overflow is checked when both the numbers are of same sign.

There is no overflow if result will be also of same sign.

If it is not the case then this overflow condition will be checked by using 2-input XOR gate.

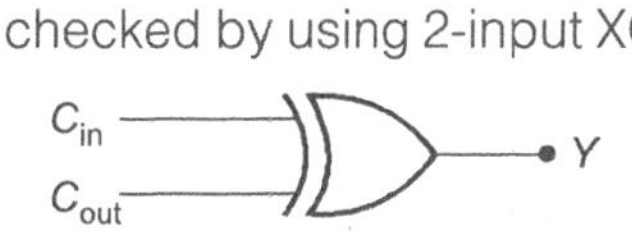

Where C_{in} → is carry to MSB.
C_{out} → Carry generated after addition is performed at MSB.

if both are same then no overflow i.e. $Y = 0$

if both are different then overflow i.e. $Y = 1$

We can understand it as suppose both are –ve numbers. Then MSB will be zero for both.

If carry is there at MSB then $C_{in} = 1$ and $C_{out} = 0$ so $Y = 1$ and MSB of result = 1 which is not zero so overflow.

If $C_{in} = 0$, then $Y = 0$, no overflow as also clear that MSB of result = 0.

4.8 (c)

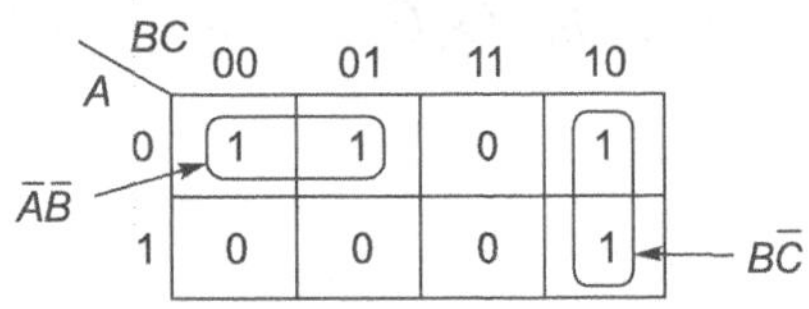

$$Y = \bar{A}\bar{B} + B\bar{C}$$

$$Z = Y + \bar{C}$$

$$Z = \bar{A}\bar{B} + B\bar{C} + \bar{C}$$

$$= \bar{A}\bar{B} + \bar{C}(B+1) = \bar{A}\bar{B} + \bar{C}\cdot 1 = \bar{A}\bar{B} + \bar{C} = \bar{C} + \bar{A}\bar{B}$$

4.9 (c)

For $\tau = 0$, output for $0 \le t \le T/2$ is

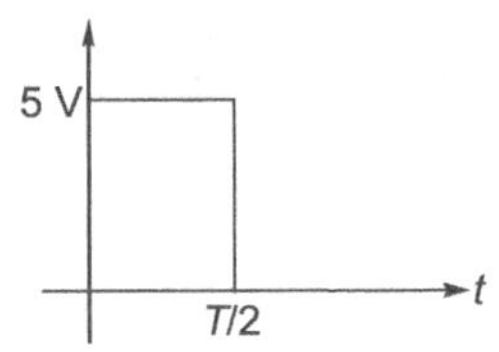

$$V_{avg} = 5\text{ V} \qquad (\text{for } \tau = 0)$$

For $\tau = T/2$ output is

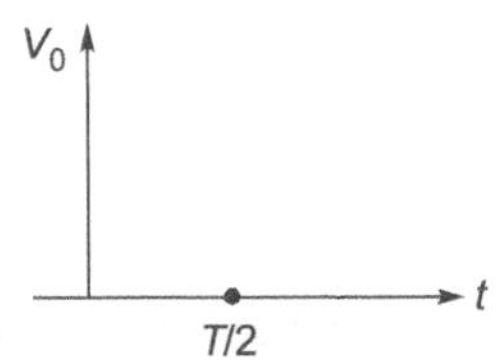

For τ in between 0 and $T/2$ output is

$$V_{avg} = \frac{5\times(T/2-\tau)}{T/2} = 5\left(1-\frac{2\tau}{T}\right)$$

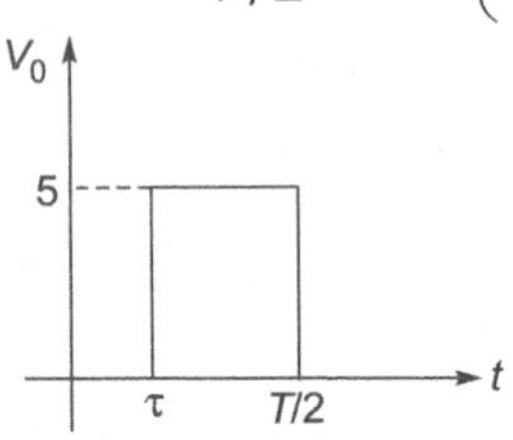

V_{avg} is varying linearly with τ with –ve slope and having slope of $-10/T$.

4.10 (b)

$$F = \overline{Y}\overline{X} + YX = \overline{X \oplus Y}$$

$$F_1 = \overline{S}\overline{Z} + SZ = \overline{S \oplus Z}$$

$$= \overline{\overline{X \oplus Y} \oplus Z} \quad (\because S = F_0 = \overline{X \oplus Y})$$

4.11 (a)

$$F = \bar{P}\bar{Q}R + \bar{P}Q\bar{R} + P\bar{Q}\bar{R} + PQR$$

$$= (\bar{P}\bar{Q} + PQ)R + \bar{R}(\bar{P}Q + P\bar{Q})$$

$$= (\overline{P \oplus Q})R + \bar{R}(P \oplus Q)$$

Let, $P \oplus Q = x$

$$\therefore \quad F = \bar{X}R + \bar{R}X$$

$$= X \oplus R = P \oplus Q \oplus R$$

4.12 (a)

$$F(X, Y, Z) = \overline{Y}\overline{Z}(X) + \overline{Y}Z(0) + Y\overline{Z}(\overline{X}) + YZ(1)$$

$$= \underset{(4)}{X\overline{Y}\overline{Z}} + \underset{(2)}{\overline{X}Y\overline{Z}} + \underset{(3,7)}{YZ}$$

$$= \Sigma m\,(2, 3, 4, 7)$$

4.13 (4)

Given that each logic gate has a delay of 2 ns

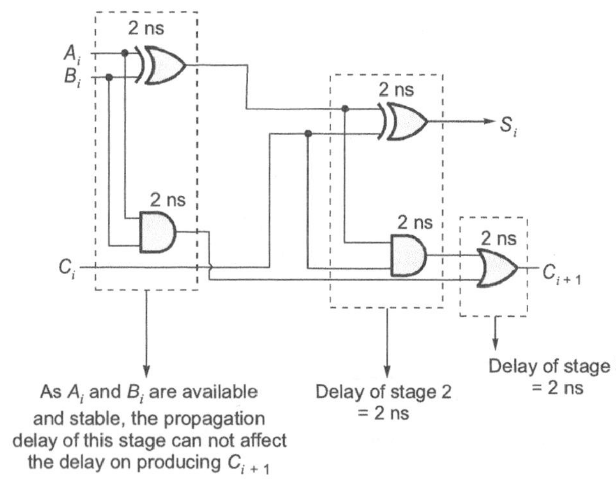

$\therefore$ total delay for an input C_i to produce $C_{i+1} = 2 + 2 = 4$ ns.

4.14 (d)

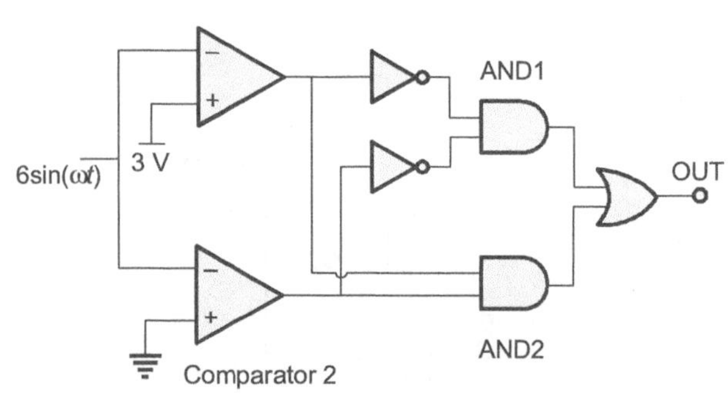

4.15 (d)

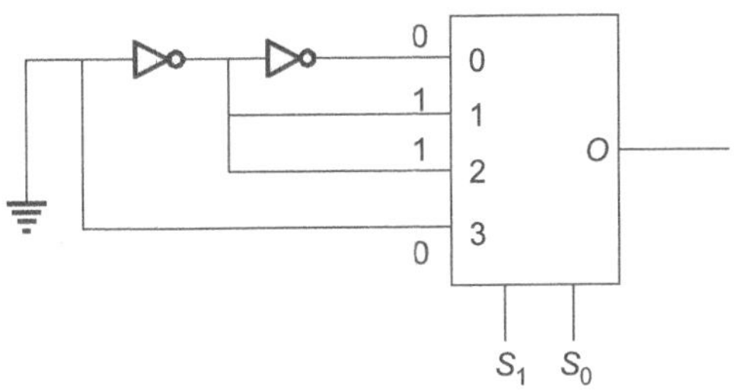

$$O = \bar{S}_1\bar{S}_0 . 0 + \bar{S}_1 S_0 . 1 + S_1\bar{S}_0 . 1 + S_1 S_0 . 0$$

$$= \bar{S}_1 S_0 + S_1 \bar{S}_0 = S_1 \oplus S_0$$

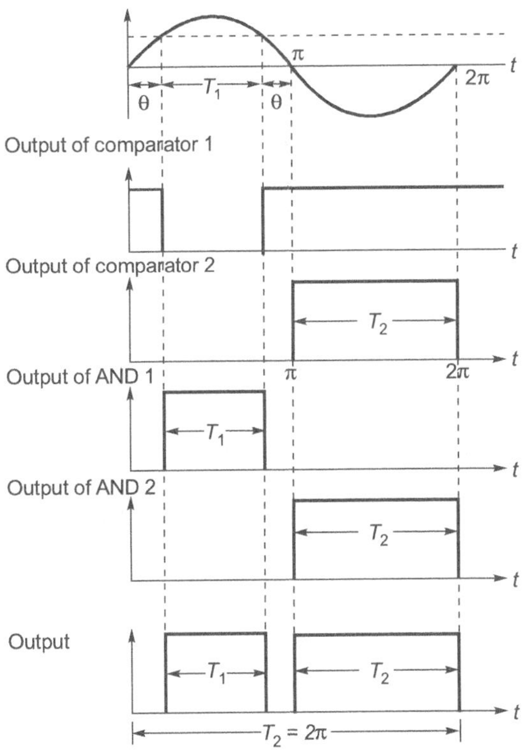

$$6\sin\theta = 3$$

$$\Rightarrow \qquad \theta = \sin^{-1}\left(\frac{3}{6}\right) = \frac{\pi}{6}$$

$$T_1 = \pi - 2\theta$$

$$= \pi - \frac{2\pi}{6} = \frac{2\pi}{3}$$

$$T_2 = \pi$$

$$\text{Fraction of time for which output is high} = \frac{T_1 + T_2}{T}$$

$$= \frac{\frac{2\pi}{3} + \pi}{2\pi} = \frac{5}{6}$$

Alternate method :

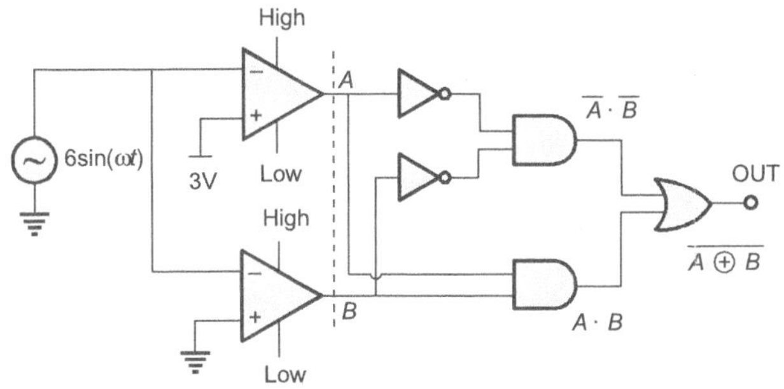

Above digital circuit behaves like XNOR gate.

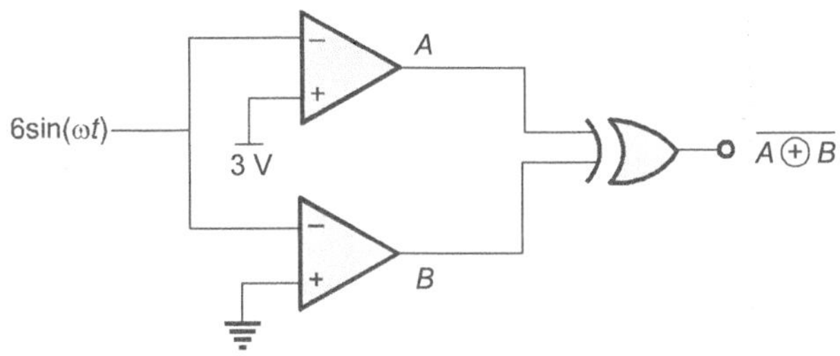

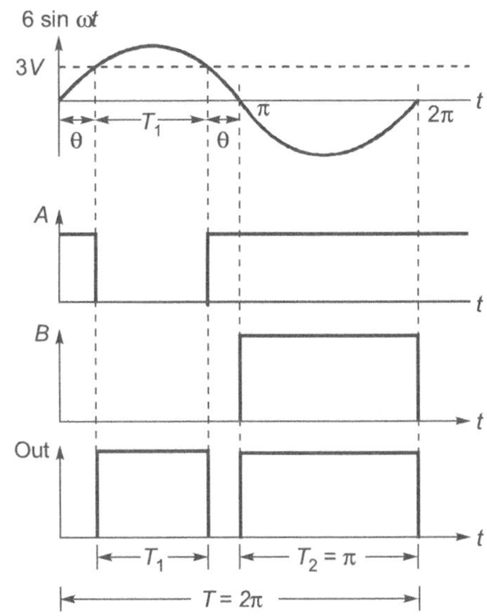

$$6\sin\theta = 3$$

$$\Rightarrow \qquad \theta = \sin^{-1}\left(\frac{3}{6}\right) = \frac{\pi}{6}$$

$$T_1 = \pi - 2\theta = \pi - \frac{2\pi}{6} = \frac{2\pi}{3}$$

$$T_2 = \pi$$

$$\text{Fraction of time for which output is high} = \frac{T_1 + T_2}{T}$$

$$= \frac{\frac{2\pi}{3} + \pi}{2\pi} = \frac{5}{6}$$

4.16 (b)

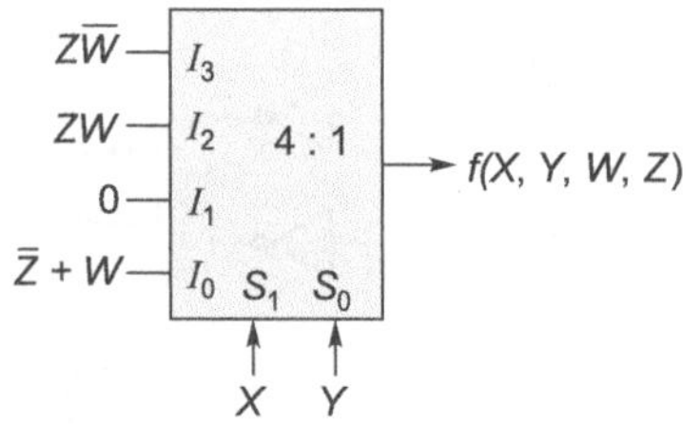

$$f = \overline{x}\overline{y}(\overline{z}+w)+\overline{x}y\cdot 0+x\overline{y}(zw)+xy(z\overline{w})$$

$\overline{X}$ $\overline{Y}$ $\overline{Z}$ _ + $\overline{X}$ $\overline{Y}$ _ W + X $\overline{Y}$ Z W + X Y Z $\overline{W}$

0 0 0 0   0 0 0 1   1 0 1 1   1 1 1 0

0 0 0 1   0 0 1 1

$$f(X, Y, Z, W) = \Sigma m(0, 1, 3, 11, 14)$$

4.17 (a, b)

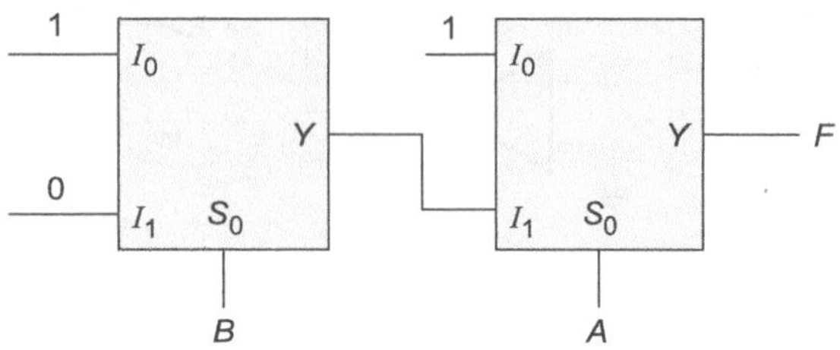

$$F_1 = I_0\overline{B}+I_1B$$

$$F_1 = (1)\overline{B}+(0)B=\overline{B}$$

So, $F = I_0\overline{A}+I_1A$

Here, $I_0 = 1,\ I_1 = \overline{B}$

So, $F = (1)\overline{A}+\overline{B}A$

$$F = \overline{A}+A\overline{B}=\overline{A}+\overline{B}$$

$$F = \overline{A\cdot B}$$

4.18 (a)

Given $A = B = 1$

$C = 1$ (Initially)

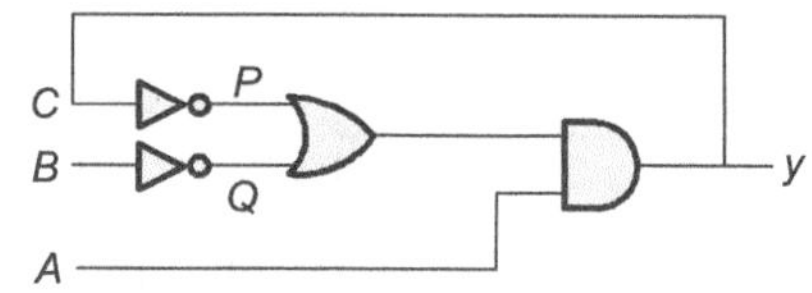

From circuit $P = \overline{C} = \overline{y}$

$$Q = \overline{B}$$

Next state of the output

$$y^+ = A\cdot(\overline{B}+\overline{C})$$

$A = 1$ and $B = 1$

$$y^+ = 1\cdot(0+\overline{C}) = 0+\overline{y}=\overline{y}$$

$$y^+ = \overline{y}$$

Overall propagation delay of circuit,

$$t_{Pd} = 5 + 10 + 10 = 25 \text{ nsec}$$

So, after every 25 nsec the output will toggle.

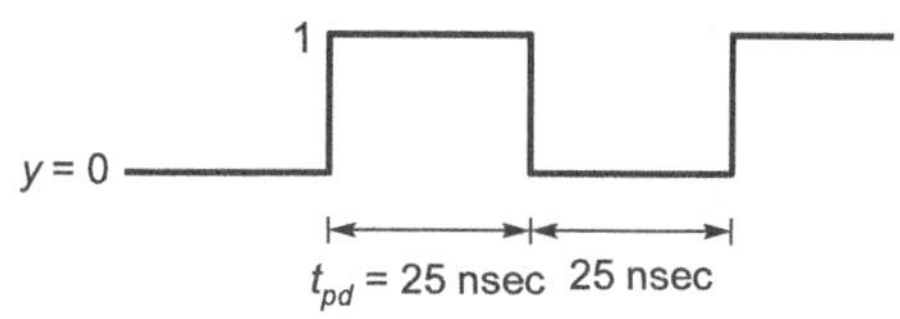

Frequency of steady state oscillations,

$$f = \frac{1}{50\times10^{-9}} = 20\times10^{6} \text{ Hz}$$

$$f = 20 \text{ MHz}$$

5 Sequential Circuits

5.1 The circuit shown in figure is that of an R–S latch.

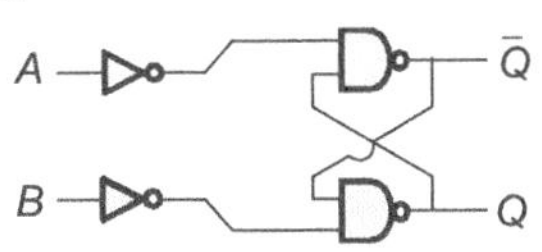

[1995 : 2 M]

5.2 Figure shows a sequential circuit with four J-K flip-flops. Generate a table of output (Q_3 Q_2 Q_1 Q_0) changes with each clock pulse. Start with Q_3 Q_2 Q_1 Q_0 = 0001 and complete a full cycle.

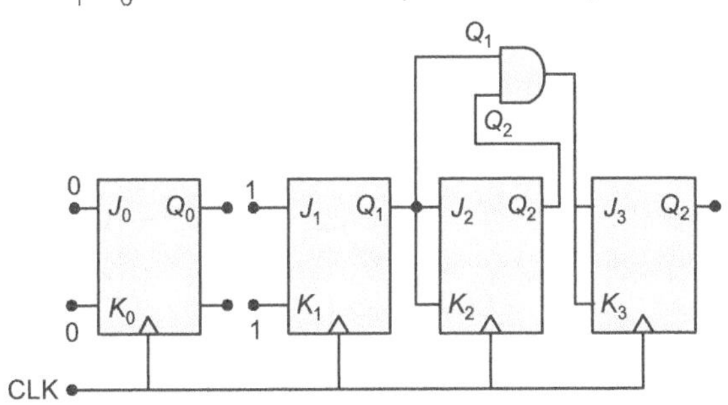

[1995 : 2 M]

5.3 A 4-bit synchronous counter with a series carry, uses flip-flop and AND gates, having a propagation delay of 30 ns and 10 ns respectively. The maximum time interval required between two successive clock pulses for reliable operation of the counter is

(a) 10 ns
(b) 30 ns
(c) 40 ns
(d) 50 ns [1997 : 1 M]

5.4 The minimum number of flip-flops needed to make mod-2 counter is

(a) 1
(b) 2
(c) 3
(d) 4 [1998 : 1 M]

5.5 (a) For the sequential logic circuit shown in figure express S_1 and R_1 as function of Q_2, R and T.

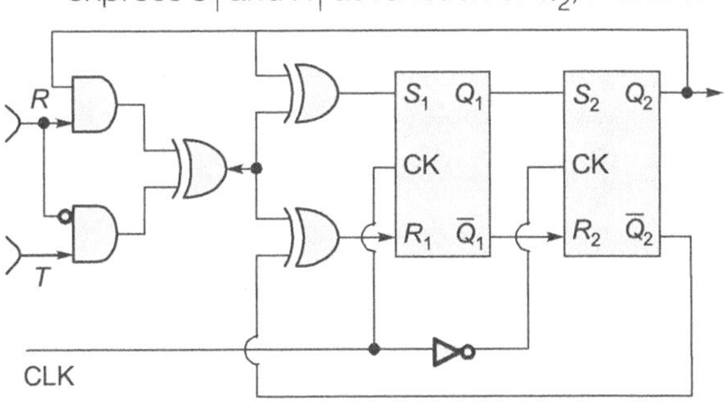

(b) Give the truth table with inputs Q_2, R and T and outputs S_1 and S_2. Assume Q_2 to be the most significant bit and T the least significant bit.

[1999 : 2 M]

5.6 In the circuit shown in figure when inputs $A = B = 0$, the possible logic states of C and D are

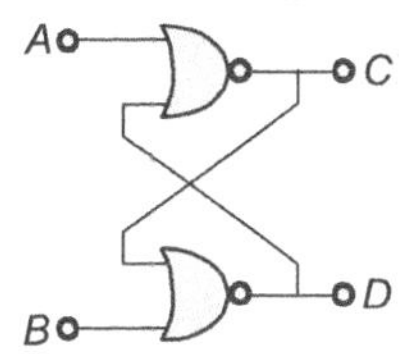

(a) $C = 0$, $D = 1$ or $C = 1$, $D = 0$
(b) $C = 1$, $D = 1$ or $C = 0$, $D = 0$
(c) $C = 1$, $D = 0$
(d) $C = 0$, $D = 1$

[2001 : 2 M]

5.7 A counter timer has a basic clock of 16 MHz. The count value displayed is in error by 1 count. The frequency at which the error in the displayed value is the same whether the counter-time is used in the frequency mode of operation or period mode of operation is

(a) 15 MHz
(b) 10 MHz
(c) 8 MHz
(d) 4 MHz [2002 : 1 M]

5.8 The 14-bit timer is loaded with the counter value of 07DOH. The timer input is connected to a clock with a frequency of 800 kHz. The timer is programmed to produce a continuous signature wave output. The frequency of the square wave output is

(a) 400 kHz
(b) 800 kHz
(c) 400 Hz
(d) 2000 kHz

[2002 : 2 M]

5.9 The clock frequency of a timer-counter is 10 MHz. The timer-counter is used in the period mode and the input to the timer-counter is a square wave of frequency 2 kHz. The display of the timer-counter will show a value

(a) 200
(b) 2000
(c) 5000
(d) 50000 [2003 : 1 M]

5.10 The correct cyclic sequence of the outputs (Q_0, Q_1) for the JK master-slave flip-flop circuit shown in figure when the input clock applied is

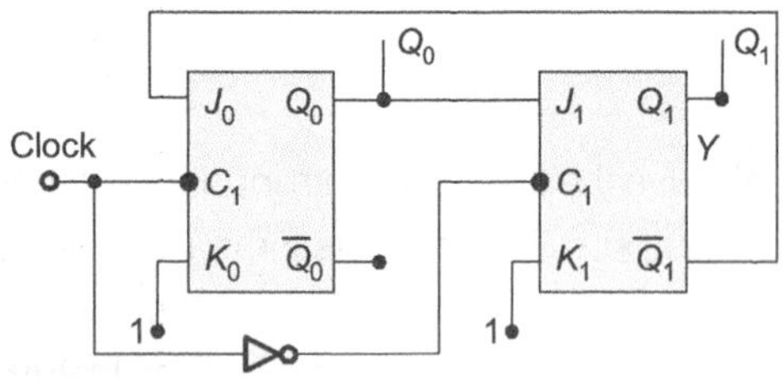

(a) 00, 01, 10, 11, 00, 01, ...
(b) 00, 10, 01, 00, 10, 01, ...
(c) 00, 01, 10, 00, 01, 10, ...
(d) 00, 11, 00, 11, 00, 11, ... **[2003 : 2 M]**

5.11 The square wave C_1 shown in figure is given to the clock input of a 4-bit binary up/down counter whose $UP/\overline{DN}$ input is fed with the pulse train P_u. The counter is a negative edge triggered one. The counter starts with 0000 and will reach 0000 again at the

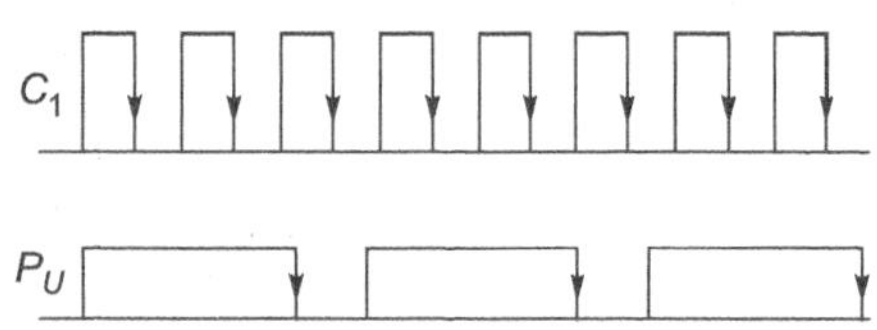

(a) 15th clock pulse
(b) 16th clock pulse
(c) 44th clock pulse
(d) 48th clock pulse **[2003 : 2 M]**

5.12 In figure, initially $Q = A = B = 0$. After three clock triggers, the states of Q, A and B will be respectively

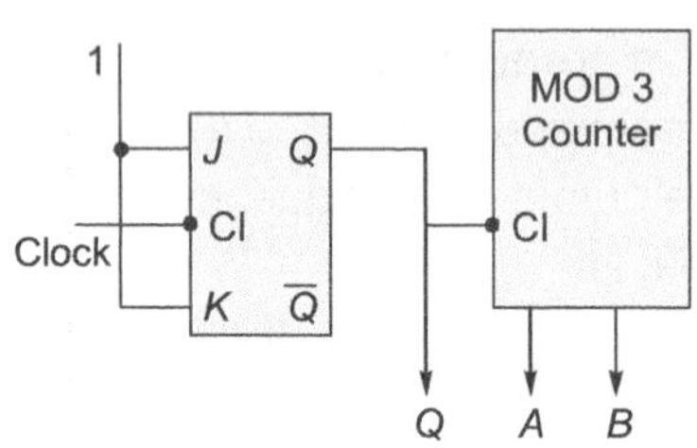

(a) 110 (b) 011
(c) 001 (d) 101
[2004 : 2 M]

5.13 In the digital circuit shown in figure the flip-flops have set time of 5 ns and a worst case delay of 15 ns. The AND gate has a delay of 5 ns. Maximum possible clock rate for the circuit to operate faithfully is

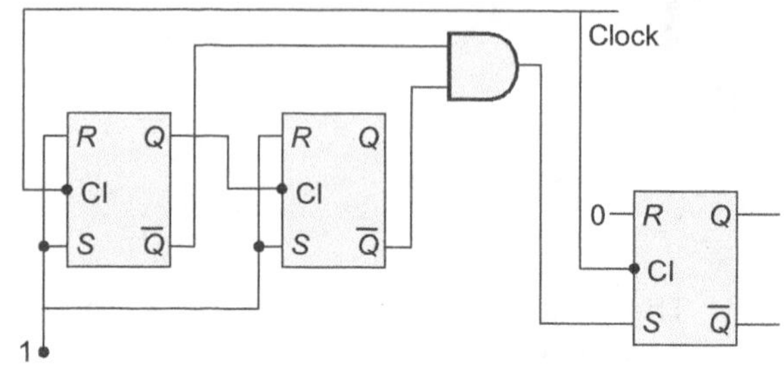

(a) 21 MHz (b) 22 MHz
(c) 25 MHz (d) 30 MHz
[2004 : 2 M]

5.14 For the digital counter shown in the figure with outputs $Q_0, Q_1, Q_2, \ldots\ldots$, where Q_0 indicates the LSB of the count value, the correct statement when the switch Sw is closed is

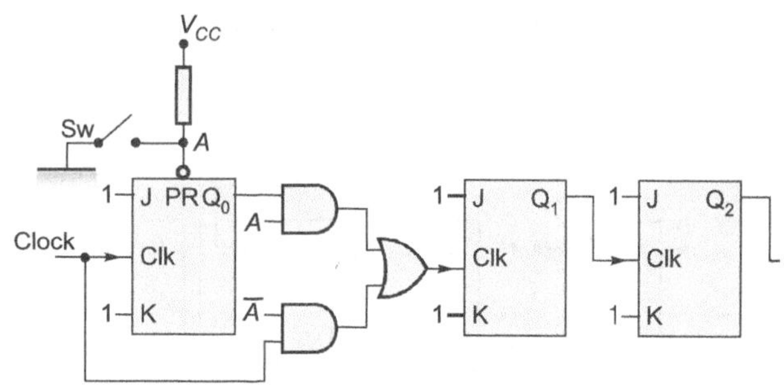

(a) counter outputs are both even and odd numbers.
(b) counter outputs are only odd numbers.
(c) counter outputs are only even numbers.
(d) counter stops counter.
[2005 : 2 M]

5.15 Given that the initial state (Q_1, Q_0) is 00, the counting sequence of the counter shown in the following figure is, $Q_1 Q_0$ =

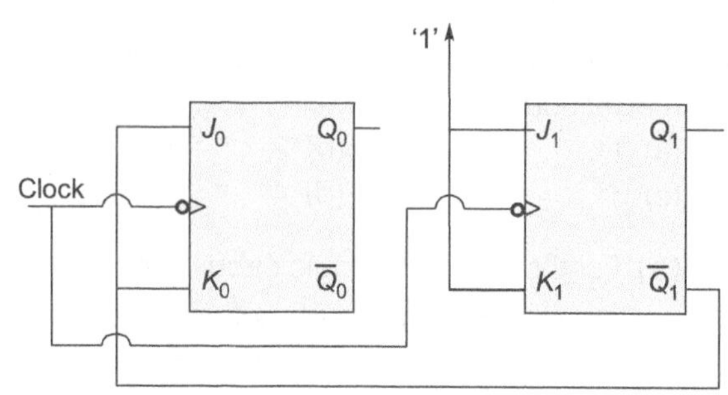

(a) 00-11-01-10-00 (b) 00-01-11-10-00
(c) 00-11-10-01-00 (d) 00-10-01-11-00
[2006 : 2 M]

5.16 All the logic gates in the circuit shown below have finite propagation delay. The circuit can be used as a clock generator, if

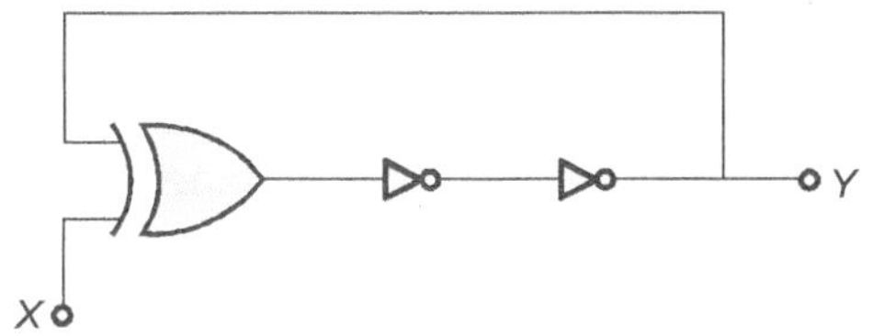

(a) $X = 0$ (b) $X = 1$

(c) $X = 0$ or 1 (d) $X = Y$

[2006 : 2 M]

5.17 A sequential circuit is shown in the figure below. Let the state of the circuit be encoded as $Q_A Q_B$. The notation $X \rightarrow Y$ implies that state Y is reachable from state X in a finite number of clock transitions.

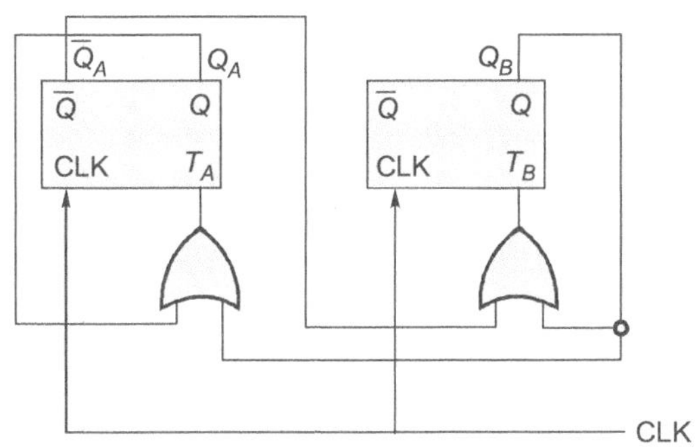

Identify the incorrect statements.

(a) $01 \rightarrow 00$ (b) $11 \rightarrow 01$

(c) $01 \rightarrow 11$ (d) $01 \rightarrow 10$

[2007 : 2 M]

Statement for Linked Answer Questions (5.18 to 5.19):

Consider the circuit shown in the following figure:

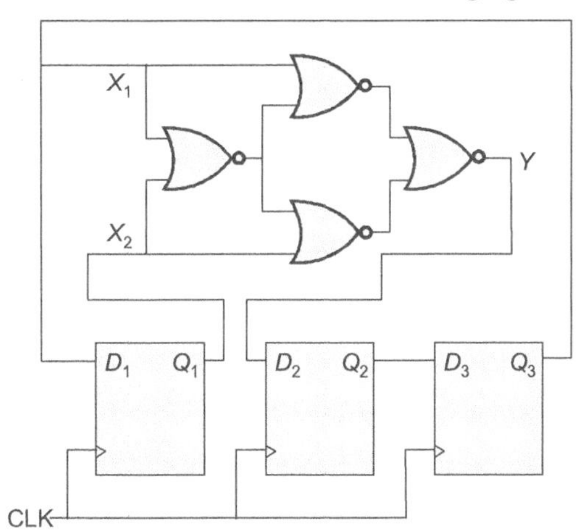

5.18 The correct input/output relationship between Y and (X_1, X_2) is

(a) $Y_1 = X_1 + X_2$ (b) $Y = X_1 X_2$

(c) $Y = X_1 \oplus X_2$ (d) $Y = \overline{X_1 \oplus X_2}$

[2007 : 2 M]

5.19 The D flip-flop are initialized to $Q_1Q_2Q_3 = 000$. After 1 clock cycle, $Q_1Q_2Q_3$ is equal to

(a) 011 (b) 010

(c) 100 (d) 101 [2007 : 2 M]

5.20 The inverters in the ring oscillator circuit shown below are identical. If the output waveform has a frequency of 10 MHz, the propagation delay of each inverter is

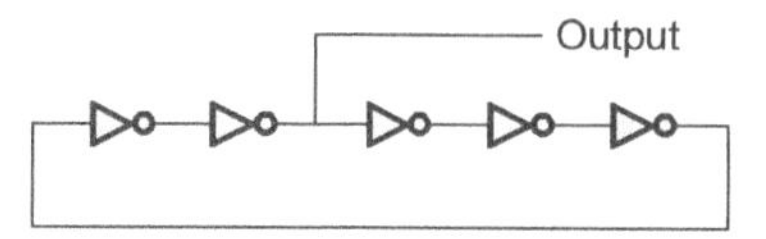

(a) 5 ns (b) 10 ns

(c) 20 ns (d) 50 ns

[2008 : 1 M]

Statement for Linked Answer Questions (5.21 & 5.22):

Consider the counter circuit shown below:

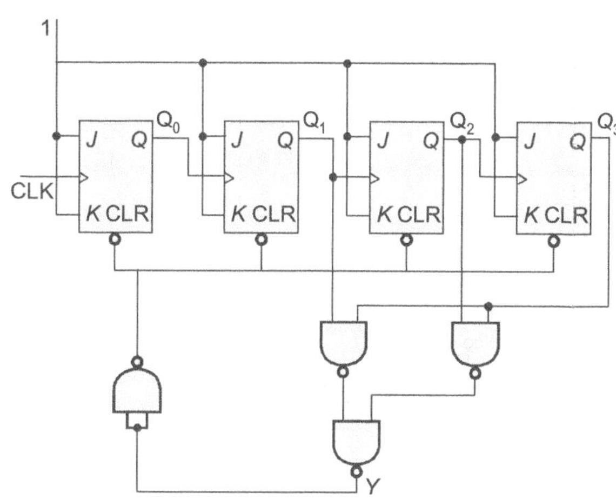

5.21 In the above figure, Y can be expressed as

(a) $Q_3(Q_2 + Q_1)$ (b) $Q_3 + Q_2Q_1$

(c) $\overline{Q_3(Q_2 + Q_1)}$ (d) $\overline{Q_3 + Q_2Q_1}$

[2008 : 2 M]

5.22 The above circuit is a

(a) Mod-8 counter (b) Mod-9 counter

(c) Mod-10 counter (d) Mod-11 counter

[2008 : 2 M]

5.23 In the figure shown, the initial state of Q is 0. The output is observed after the application of each clock pulse. The output sequence at Q is

(a) 0 0 0 0...

(b) 1 0 1 0...

(c) 1 1 1 1...

(d) 1 0 0 0...

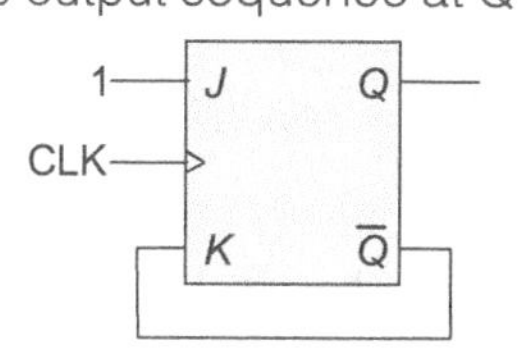

[2009 : 1 M]

5.24 The figure below shows a 3-bit ripple counter, with Q_2 as the MSB. The flip-flop are rising-edge triggered. The counting direction is

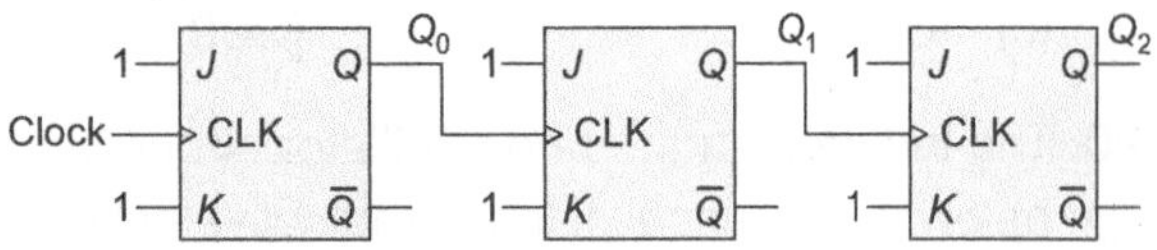

(a) always down
(b) always up
(c) up or down depending on the initial state of Q_0 only
(d) up or down depending on the initial states of Q_2, Q_1 and Q_0 [2009 : 2 M]

5.25 The circuit below shows an up/down counter working with a decoder and a flip-flop. Preset and Clear of the flip-flop are asynchronous active-low inputs.

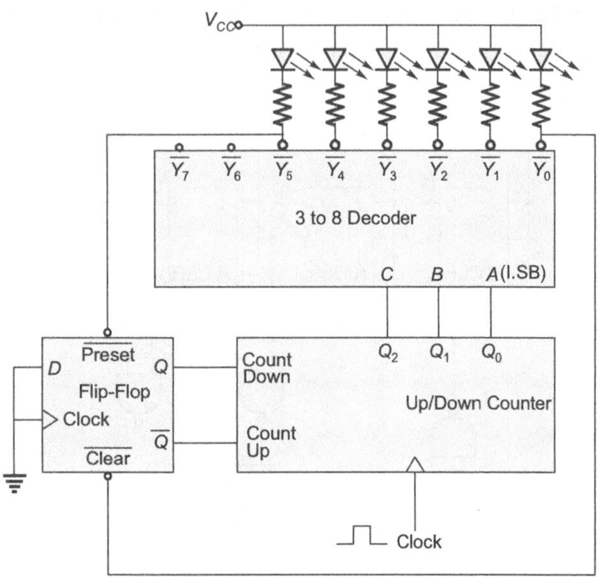

Assuming that the initial value of counter output ($Q_2\,Q_1\,Q_0$) as zero, the counter outputs in decimal for 12 clock cycles are
(a) 0, 1, 2, 3, 4, 4, 3, 2, 1, 1, 2, 3, 4
(b) 0, 1, 2, 3, 4, 5, 0, 1, 2, 3, 4, 5, 0
(c) 0, 1, 2, 3, 4, 5, 5, 4, 3, 2, 1, 0, 1
(d) 0, 1, 2, 3, 4, 5, 4, 3, 2, 1, 0, 1, 2

[2011 : 2 M]

5.26 Consider the given circuit

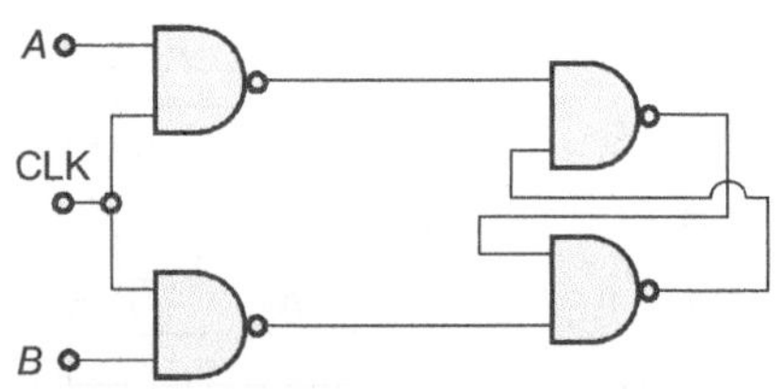

In this circuit, the race around
(a) does not occur
(b) occurs when CLK = 0
(c) occurs when CLK = 1 and $A = B = 1$
(d) occurs when CLK = 1 and $A = B = 0$

[2012 : 1 M]

5.27 The state transition diagram for the logic circuit shown is

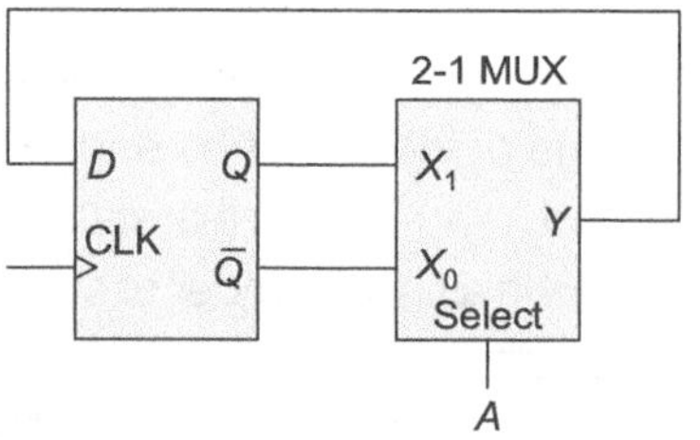

(a)
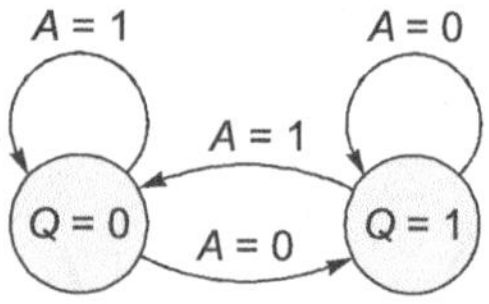

(b)
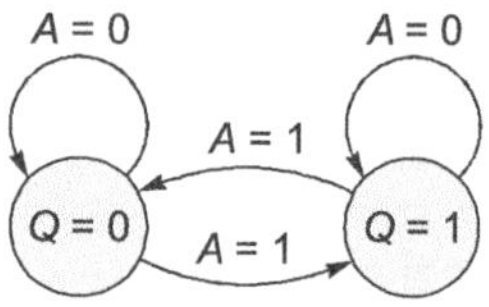

(c)
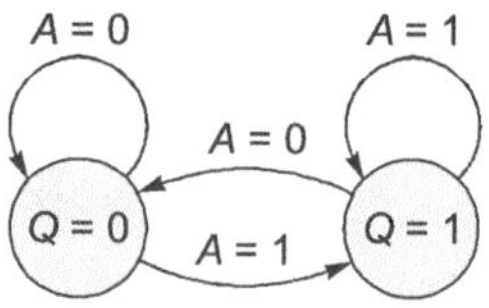

(d)
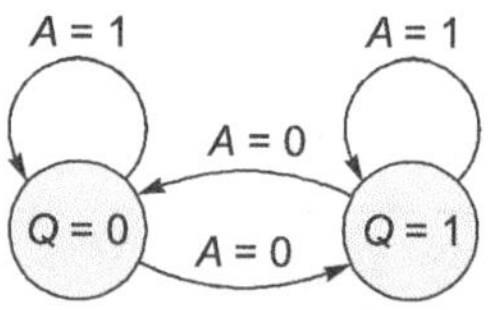

[2012 : 2 M]

5.28 The digital circuit shown below uses two negative edge-triggered D-flip-flops. Assuming initial condition of Q_1 and Q_0 as zero, the output Q_1, Q_0 of this circuit is

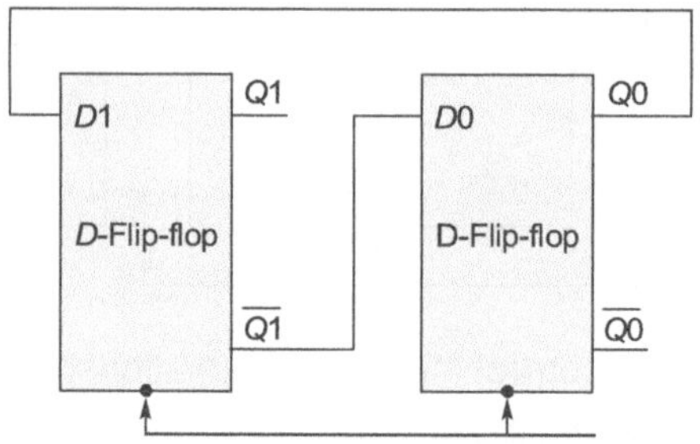

(a) 00, 01, 10, 11, 00 ...
(b) 00, 01, 11, 10, 00 ...
(c) 00, 11, 10, 01, 00 ...
(d) 00, 01, 11, 11, 00 ... [2013 : 2 M]

5.29 Frequency of an analog periodic signal in the range of 5kHz-10kHz is to be measured with a resolution of 100Hz by measuring its period with a counter. Assuming negligible signal and transition delays the minimum clock frequency and minimum number of bits in the counter needed, respectively, are:

(a) 1 MHz, 10-bits (b) 10 MHz, 10-bits
(c) 1 MHz, 8-bits (d) 10MHz, 8-bits

[2014 : 2 M]

5.30 The number of clock cycles for the duration of an input pulse is counted using a cascade of N decade counters (DC 1 to DC N) as shown in the figure. If the clock frequency in mega hertz is f, the resolution and range of measurement of input pulse width, both in μs, are respectively,

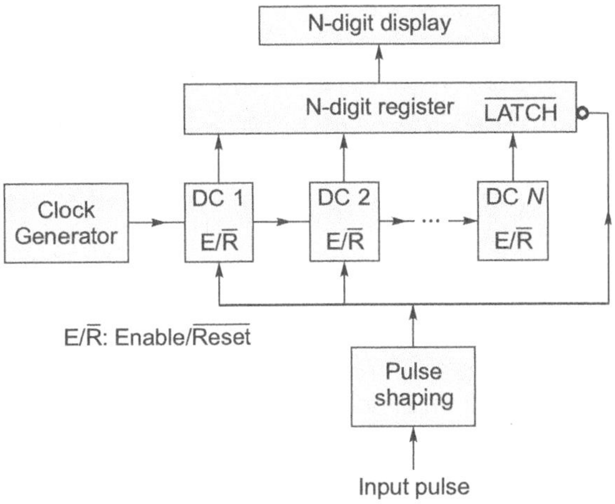

(a) $\frac{1}{f}$ and $\frac{(2^N - 1)}{f}$ (b) $\frac{1}{f}$ and $\frac{(10^N - 1)}{f}$

(c) $\frac{10^N}{f}$ and $\frac{(10^N - 1)}{f}$ (d) $\frac{2^N}{f}$ and $\frac{(2^N - 1)}{f}$

[2015 : 2 M]

5.31 A synchronous counter using two J - K flip flops that goes through the sequence of states: $Q_1Q_2 = 00 \rightarrow 10 \rightarrow 01 \rightarrow 11 \rightarrow 00$ ... is required. To achieve this, the inputs to the flip flops are:

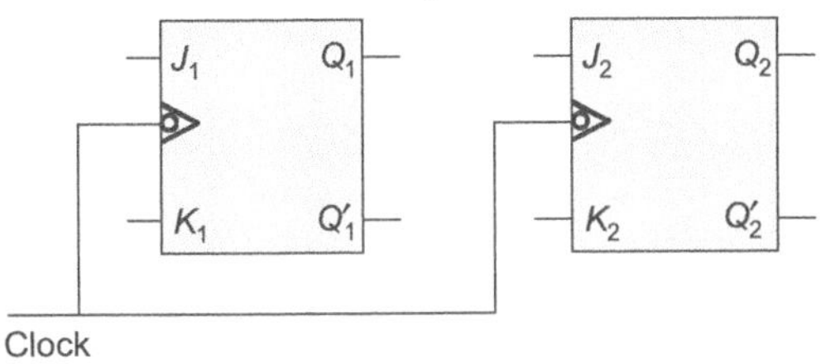

(a) $J_1 = Q_2, K_1 = 0$; $J_2 = Q_1', K_2 = Q_1$
(b) $J_1 = 1, K_1 = 1$; $J_2 = Q_1, K_2 = Q_1$
(c) $J_1 = Q_2, K_1 = Q_2'$; $J_2 = 1, K_2 = 1$
(d) $J_1 = Q_2', K_1 = Q_2, J_2 = Q_1, K_2 = Q_1'$

[2016 : 2 M]

5.32 The two inputs A and B are connected to an R-S latch via two AND gates as shown in the figure. If $A = 1$ and ($B = 0$), the output $Q\overline{Q}$ is

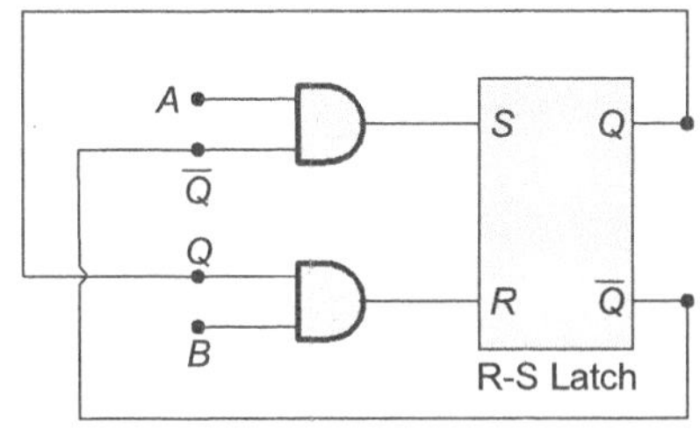

(a) 00 (b) 10
(c) 01 (d) 11 [2017 : 2 M]

5.33 A 2-bit synchronous counter using two J-K flip flops is shown. The expressions for the inputs to the J-K flip flops are also shown in the figure. The output sequence of the counter starting from $Q_1Q_2 = 00$ is

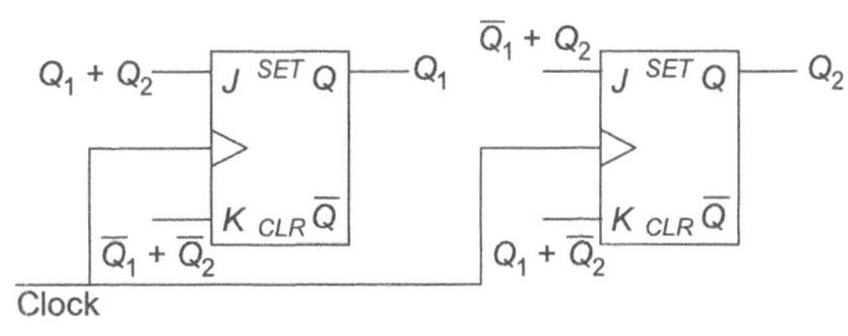

(a) $00 \rightarrow 11 \rightarrow 10 \rightarrow 01 \rightarrow 00$ ...
(b) $00 \rightarrow 01 \rightarrow 10 \rightarrow 11 \rightarrow 00$ ...
(c) $00 \rightarrow 01 \rightarrow 11 \rightarrow 10 \rightarrow 00$ ...
(d) $00 \rightarrow 10 \rightarrow 11 \rightarrow 01 \rightarrow 00$ ... [2018 : 2 M]

5.34 For the 3-bit binary counter shown in the figure, the output increments at every positive transition in the clock (CLK). Assume ideal diodes and the starting state of the counter as 000. If output high is 1 V and output low is 0 V, the current I (in mA) flowing through the 50 Ω resistor during the 5th clock cycle is (up to one decimal place) _______.

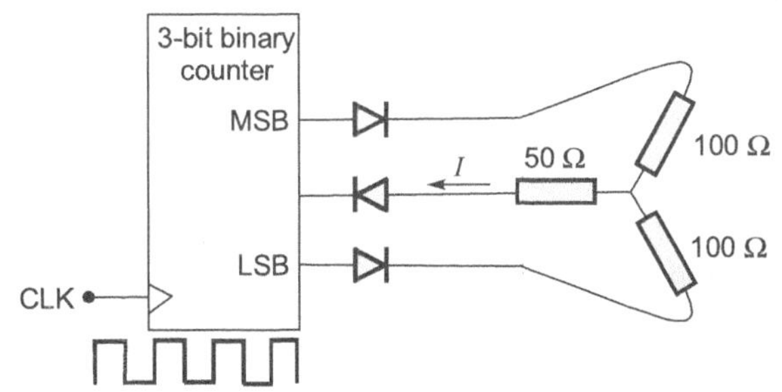

[2018 : 1 M]

5.35 The circuit shown in the figure below uses ideal positive edge-triggered synchronous J-K flip flops with outputs X and Y. If the initial state of the output is X = 0 and Y = 0 just before the arrival of the first clock pulse, the state of the output just before the arrival of the second clock pulse is

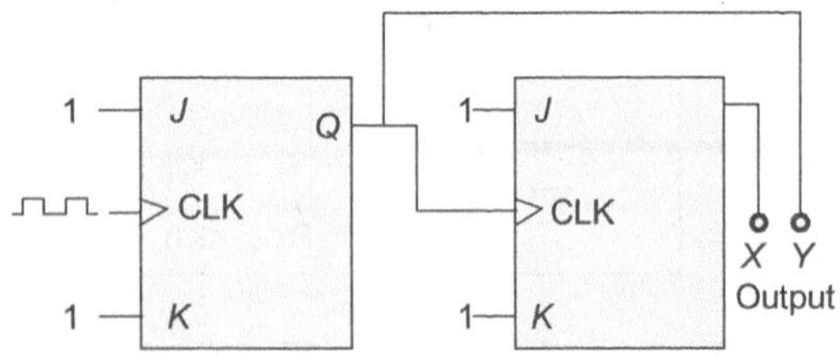

(a) $X = 0, Y = 0$ (b) $X = 0, Y = 1$
(c) $X = 1, Y = 0$ (d) $X = 1, Y = 1$

[2019 : 1 M]

5.36 In the circuit below, the light dependent resistor (LDR) receives light from the LED. The LDR has resistances of 5 kΩ and 500 Ω under dark and illuminated conditions, respectively. The LED is OFF at time $t < 0$. At time $t = 0$ s, the switch S_1 is closed for 1 ms and then kept open thereafter. Assuming zero propagation delay in the devices, the LED

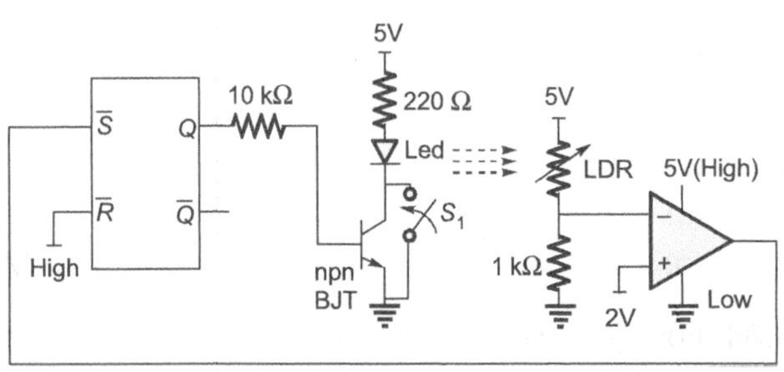

(a) turns ON when S_1 is closed and remains ON after S_1 is opened
(b) turns ON when S_1 is closed and turns OFF after S_1 is opened
(c) turns ON when S_1 is closed and toggles periodically from ON to OFF after S_1 is opened
(d) remains OFF when S_1 is closed and continues to remain OFF after S_1 is opened

[2019 : 2 M]

5.37 Two T-flip flops are interconnected as shown in the figure. The present state of the flip flops are : $A = 1$, $B = 1$. The input x is given as 1, 0, 1 in the next three clock cycles. The decimal equivalent of $(ABy)_2$ with A being the MSB and y being the LSB, after the 3rd clock cycle is _____.

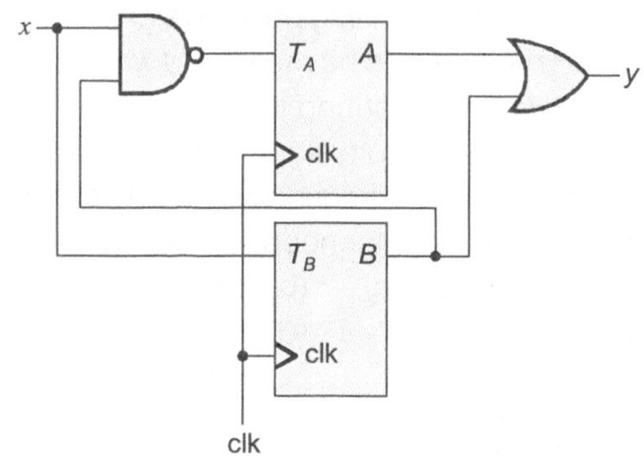

[2020 : 2 M]

5.38 Given below is the diagram of a synchronous sequential circuit with one J-K flip-flop and one T flip-flop with their outputs denoted as A and B respectively, with $J_A = (A' + B')$, $K_A = (A + B)$ and $T_B = A$.

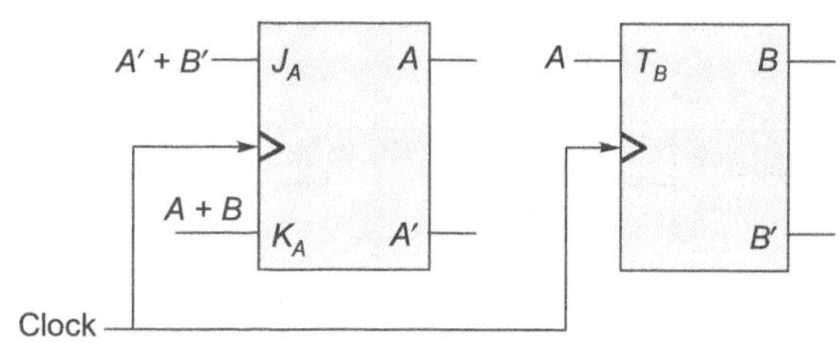

Starting from the initial state (AB = 00), the sequence of states (AB) visited by the circuit is

(a) $00 \rightarrow 10 \rightarrow 01 \rightarrow 11 \rightarrow 00$ .....
(b) $00 \rightarrow 01 \rightarrow 11 \rightarrow 00$ .....
(c) $00 \rightarrow 10 \rightarrow 11 \rightarrow 01 \rightarrow 00$ .....
(d) $00 \rightarrow 01 \rightarrow 10 \rightarrow 11 \rightarrow 00$ .....

[2021 : 2 M]

5.39 The digital circuit shown _______.

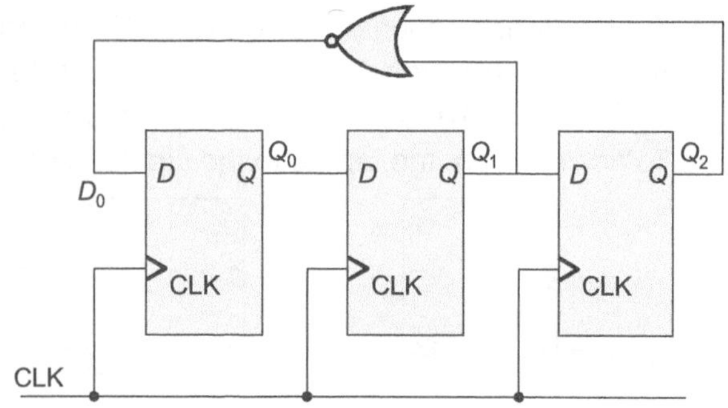

(a) is a divide-by-5 counter
(b) is a divide-by-7 counter
(c) is a divide-by-8 counter
(d) does not function as a counter due to disjoint cycles of states

[2022 : 2 M]

5.40 The table shows the present state $Q(t)$, next state $Q(t+1)$, and the control input in a flip-flop. Identify the flip-flop.

$Q(t)$	$Q(t+1)$	Input
0	0	0
0	1	1
1	0	1
1	1	0

(a) T flip-flop (b) D flip-flop
(c) SR flip-flop (d) JK flip-flop

[2023 : 1 M]

5.41 In the circuit shown, the initial binary content of shift register A is 1101 and that of shift register B is 1010. The shift registers are positive-edge triggered, and the gates have no delay.
When the shift control is high, what will be the binary content of the shift registers A and B after four clock pulses?

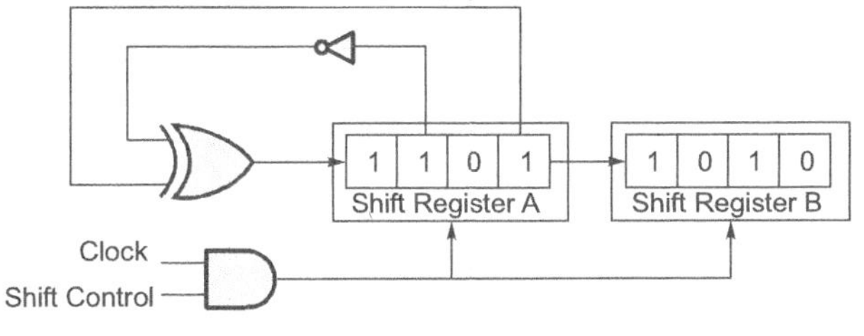

(a) $A = 1101, B = 1101$
(b) $A = 1110, B = 1001$
(c) $A = 0101, B = 1101$
(d) $A = 1010, B = 1111$

[2023 : 2 M]

5.42 In the figure shown, the positive edge triggered D flip-flops are initially reset to $Q = 0$. The logic gates and the multiplexers have no propagation delay. After reset, a train of clock pulses (CLK) are applied. The logic-states of the inputs DIN, S and the clock pulses are also shown in the figure. Assuming no timing violations, the sequence of output Y from the 3rd clock to the 5th clock, $Y_3Y_4Y_5$ is

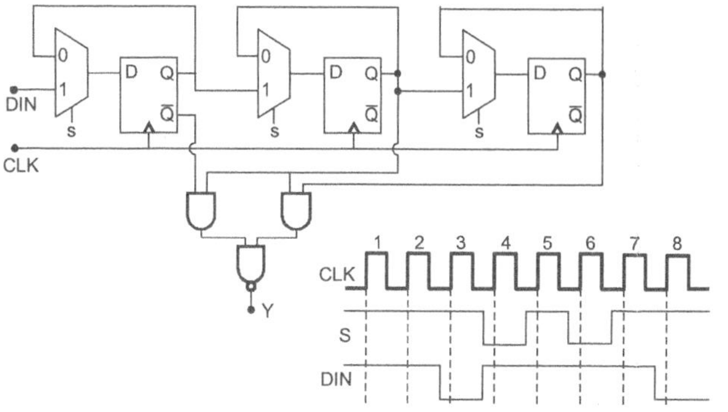

(a) 001 (b) 010
(c) 000 (d) 011

[2024 : 2 M]

Answers Sequential Circuits

5.1 (sol.)	5.2 (sol.)	5.3 (d)	5.4 (a)	5.5 (a)	5.6 (a)	5.7 (*)
5.8 (c)	5.9 (c)	5.10 (b)	5.11 (c)	5.12 (a)	5.13 (c)	5.14 (b)
5.15 (a)	5.16 (b)	5.17 (c)	5.18 (d)	5.19 (b)	5.20 (b)	5.21 (a)
5.22 (c)	5.23 (c)	5.24 (a)	5.25 (d)	5.26 (a)	5.27 (d)	5.28 (b)
5.29 (c)	5.30 (b)	5.31 (b)	5.32 (b)	5.33 (c)	5.34 (10)	5.35 (d)
5.36 (a)	5.37 (7)	5.38 (a)	5.39 (a)	5.40 (a)	5.41 (c)	5.42 (a)

Explanations Sequential Circuits

5.1 Sol.

True, the truth table of above circuit is

A	B	Q	$\overline{Q}$
0	0	previous state	
0	1	0	1
1	0	1	0
1	1	intermediate	

5.2 Sol.

From the given sequential circuit

$J_3 = Q_2\, Q_1 = K_3,\ J_2 = K_2 = Q_1,\ J_1 = K_1 = 1,$

$J_0 = Q_0 = 0$

the table of the given sequential circuit is

CLK	Q_0	Q_1	Q_2	Q_3	J_3	K_3	J_2	K_2	J_1	K_1	J_0	Q_0
Initial content	0	0	0	1	0	0	0	0	1	1	0	0
1st →	0	0	1	1	0	0	1	1	1	1	0	0
2nd →	0	1	0	1	0	0	0	0	1	1	0	0
3rd →	0	1	1	1	1	1	1	1	1	1	0	0
4th →	1	0	0	1	0	0	0	0	1	1	0	0
5th →	1	0	1	1	0	0	1	1	1	1	0	0
6th →	1	1	0	1	0	0	0	0	1	1	0	0
7th →	1	1	1	1	1	0	1	1	1	1	0	0
8th →	0	0	0	1								

5.3 (d)

For a 'n' bit synchronous counter with a series carry

$T_{CLK} \geq t_{pdff} + (n-2) \times 10$ ns

$T_{CLK} \geq 50$ ns

so option (d) is correct.

5.4 (a)

For mod-2 counter there are only two state, so required number of flip-flops are

$N \leq 2^n$

$2 \leq 2^n$

so, $n = 1$

one FF is required.

5.5 (a)

From the given sequential circuit

$$S_1 = Q_2(\overline{RQ_2 \oplus \overline{R}T}) + \overline{Q}_2(RQ_2 \oplus \overline{R}T)$$

$$= Q_2(RQ_2 \odot \overline{R}T) + \overline{Q}_2(RQ_2 \oplus \overline{R}T)$$

$$= Q_2(RQ_2\overline{R}T + \overline{RQ_2}\,\overline{\overline{R}T}) + \overline{Q}_2(\overline{RQ_2}\,\overline{R}T + RQ_2\overline{\overline{R}T})$$

$$= Q_2[RQ_2\overline{R}T + (\overline{R} + \overline{Q}_2)(R + \overline{T})]$$

$$+\overline{Q}_2((\overline{R} + \overline{Q}_2)\overline{R}T + RQ_2(R + \overline{T})]$$

$$= Q_2[0 + \overline{R}\,\overline{T} + RQ_2 + \overline{T}\,\overline{Q}_2]$$

$$+\overline{Q}_2[\overline{R}T + \overline{Q}_2\overline{R}T + RQ_2 + RQ_2\overline{T}]$$

$$= \overline{R}\,\overline{T}Q_2 + \overline{Q}_2\overline{R}T$$

$$\Rightarrow S_1 = Q_2\overline{R}\,\overline{T} + \overline{Q}_2\overline{R}T$$

$$R_1 = \overline{Q}_2(\overline{RQ_2 \oplus \overline{R}T}) + Q_2(RQ_2 \oplus \overline{R}T)$$

$$= \overline{Q}_2(RQ_2 \odot \overline{R}T) + Q_2(RQ_2 \oplus \overline{R}T)$$

$$= \overline{Q}_2[RQ_2\overline{R}T + \overline{RQ_2}\,\overline{\overline{R}T}]$$

$$+ Q_2[(\overline{RQ_2}\,\overline{R}T + RQ_2\overline{\overline{R}T}]$$

$$= \overline{Q}_2(\overline{R} + \overline{Q}_2)\,(R + \overline{T})$$

$$+ Q_2(\overline{R} + \overline{Q}_2)\overline{R}T + RQ_2(R + \overline{T})]$$

$$= \overline{R}\,\overline{T}\,\overline{Q}_2 + R\overline{Q}_2 + \overline{Q}_2\overline{T} + \overline{R}TQ_2$$

$$+RQ_2 + RQ_2\overline{T}$$

$$\overline{T}\,\overline{Q}_2(R + 1) + R + \overline{R}TQ_2 + RQ_2\overline{T}$$

$$\overline{T}\,\overline{Q}_2 + R(1 + Q_2\overline{T}) + \overline{R}TQ_2$$

$$= \overline{T}\,\overline{Q}_2 + R + \overline{R}TQ_2$$

5.6 (a)

For $A = B = 0$

If previous value of C is 1 and D is 0 then for $A = B = 0$

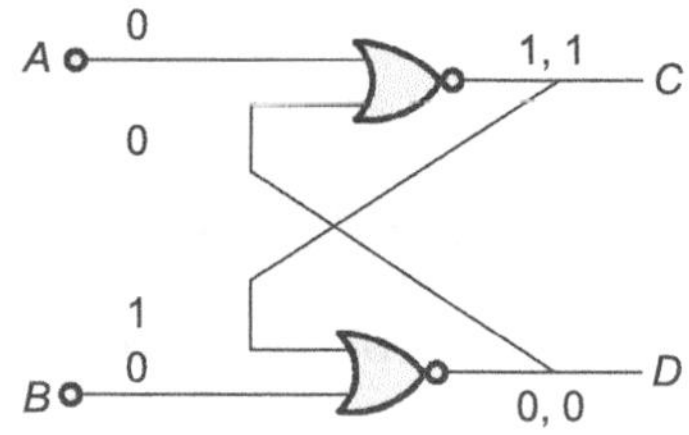

if C is 0 and D is 1

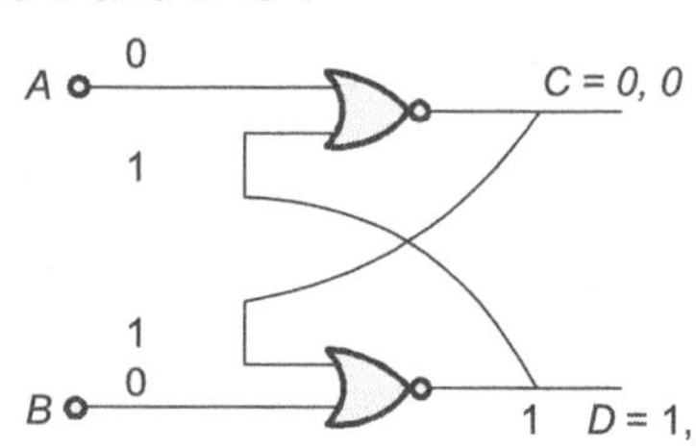

so for $A = B = 0$, circuit hold its previous state.

5.8 (c)

$(07D0)H \rightarrow (0 \times 16^3 + 7 \times 16^2 + 13 \times 16^1 + 0 \times 16^0)$

$= (2000)_{10}$

frequency of square wave generated

$$= \frac{800 \times 10^3}{2000} = 400 \text{ Hz}$$

5.9 (c)

$$\text{Display} = \frac{\text{clock frequency}}{\text{input frequency}}$$

$$= \frac{10 \times 10^6}{2 \times 10^3} = 5000$$

5.10 (b)

$J_0(\bar{Q}_1)$	K_0	$J_1(Q_0)$	K_1	Q_0	Q_1	
				0	0	← Initially
1	1	0	1	1	0	
1	1	1	1	0	1	
0	1	0	1	0	0	

So, the correct sequence is
00, 10, 01, 00, 10, 01, ...

5.11 (c)

Here in every P_u pulse there are 3 clock pulses. For initial two clock pulses counter will work as up counter as P_u is high and for 3rd clock pulse it will work as down counter so in every three clock pulse it will be incremented by only 1 (2 increment and one decrement), so after $14 \times 3 = 42$ clock pulses counter goes to 14. After that P_u goes high for two clock pulses and counter will be incremented by 2 is 14 + 2 = 16 i.e. equivalent to 0000. So after 42 + 2 = 44 clock pulses counter will be go to 0000 again.

5.12 (a)

J-K flip-flop is in toggle mode so after every clock pulse output Q toggles so output Q will be as

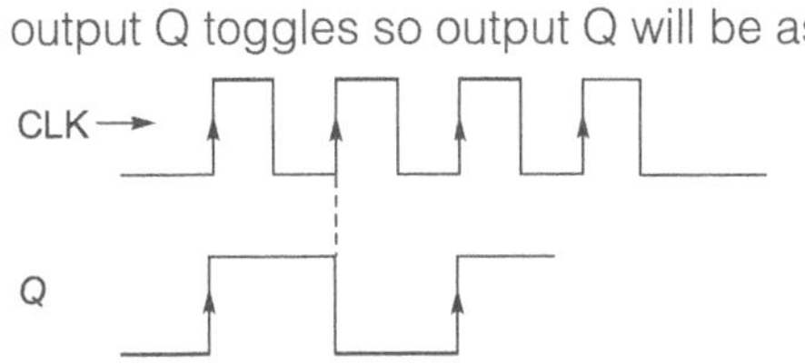

and Q is input to MOD-3 counter then

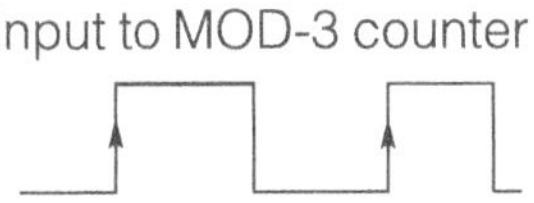

after 3 clock pulses of input CLK there are 2 +ve edge of clock input Q so output of counter goes to $2 = (10)_2$

so, $AB = 10$

so, Q, A, B respectively are 110.

5.13 (c)

Initially two flip-flops have fix inputs so no requirement of setup time.

⇒ two inputs to AND gate will, reach, one after 15 nsec and other 30 nsec, so complete input to AND will be available to AND after 30 nsec after arrival of clock.

⇒ delay of AND gate is 5 nsec so it will give output after 35 nsec after arrival of clock.

⇒ Now output of AND gate is S input of flip-flop which must be set atleast 5 nsec (setup time) before arrival of clock. So T must be greater than 35 nsec (delay) + 5 nsec (setup time) to avoid distortion.

So, $T \geq 40$ nsec

$$f \leq \frac{1}{40 \times 10^{-9}} \quad ; \quad f \leq 25 \text{ MHz}$$

$\therefore \quad f_{max} = 25$ MHz

5.14 (b)

When switch is closed A becomes zero and low active preset pin PR gets active and Q_0 become equal to 1.

And if LSB of any number is 1 then it will be definitely an odd number.

5.15 (a)

J_1	K_1	$J_0(\bar{Q}_1)$	$K_0(\bar{Q}_1)$	Q_1	Q_0
				0	0
1	1	1	1	1	1
1	1	0	0	0	1
1	1	1	1	1	0
1	1	0	0	0	0

So, the correct sequence is
00 – 11 – 01 – 10 – 00.

5.16 (b)

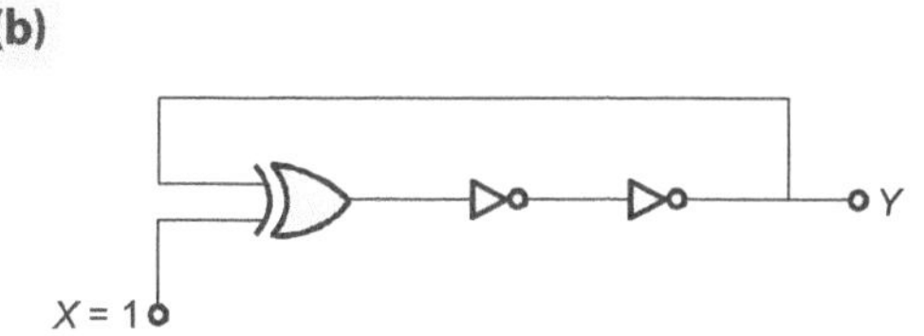

Clock generator means alternate 0 and 1 with a fix time period as shown below.

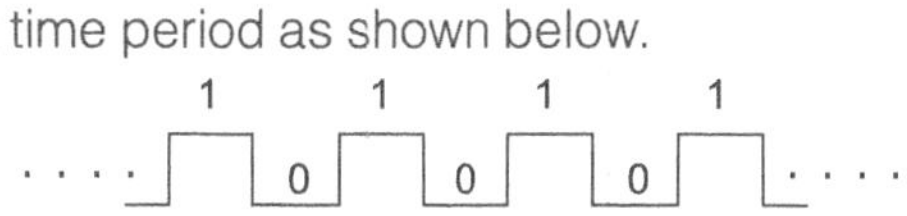

We will get this if output of XOR gate is inserted every time and final output will be getting after some delay due to 2 NOT gates. That is XOR gate must work as inverter for Y. And this is possible if $X = 1$, $Y \oplus 1 = \bar{Y}$.

5.17 (c)

Here, $T_B = \overline{Q}_A + Q_B$; $T_A = Q_A + Q_B$

and we know that output of T flip-flop $= T \oplus Q$.

So, started with $Q_A Q_B = 01$

$$T_B = 1$$
$$T_A = 1$$

then, $Q_A Q_B = 10$

Now, $T_B = 0$
$$T_A = 1$$

then, $Q_A Q_B = 01$

so sequence is repeated as 01, 10, 01

so it will never reach to 11.

5.18 (d)

The output of NOR gate start from left side to the right side.

The output of first NOR gate

$$= \overline{X_1 + X_2}$$

The output of upper NOR gate in second stage

$$= \overline{X_1 + (\overline{X_1 + X_2})} = \overline{X}_1 \cdot (X_1 + X_2) = \overline{X}_1 X_2$$

The output of lower NOR gate in second stage

$$= \overline{X_2 + (\overline{X_1 + X_2})} = \overline{X}_2 \cdot (X_1 + X_2) = X_1 \overline{X}_2$$

The output of right side NOR gate is

$$Y = \overline{X}_1 X_2 + X_1 \overline{X}_2 = X_1 \oplus X_2$$

5.19 (b)

Q_1	Q_2	Q_3	
0	0	0	Q_n
0	1	0	Q_{n+1}

$Q_1 = X_2,\ Q_2 = Y,\ Q_3 = D_1,\ Y = D_2$

D-flip truth table:

D_n	Q_{n+1}
0	0
1	1

5.20 (b)

We know if there are n odd numbers of inverters, then we get output wave of period $T = 2\,nt_d$

where t_d is delay of each inverter.

So, $$2\,nt_d = \frac{1}{f}$$

$$2 \times 5\,t_d = \frac{1}{10 \times 10^6}$$

$$t_d = \frac{1}{10^8} = 10^{-8} = 10 \text{ nsec}$$

5.21 (a)

$$Y = \overline{(\overline{Q_1\,Q_3}) \cdot (\overline{Q_2\,Q_3})} = (\overline{\overline{Q_1\,Q_3}}) + (\overline{\overline{Q_2\,Q_3}})$$

$$Y = Q_1\,Q_3 + Q_2\,Q_3 = Q_3(Q_1 + Q_2)$$

5.22 (c)

To reset the counter output Y must be one.

$$Y = Q_3(Q_1 + Q_2)$$

Q_3	Q_2	Q_1	Q_0
1	0	1	0
1	1	0	0
1	1	1	0

When the counter output $Q_3\,Q_2\,Q_1\,Q_0 = 1010$ then the output Y will give output 1 and it will be given to inverter, so it will reset the counter and again counting will start and it will not go for further combinations.

5.23 (c)

J	$K(\overline{Q})$	Q_{n+1}
1	1	0
1	0	1
1	0	1
		1

5.24 (a)

Since clock is the +ve edge triggered and Q_0, Q_1, Q_2 are connected to clock inputs. So it is down counter always. We can check always in case of ripple counter as:

If clk is (–ve edge triggered) then

– up counter, if Q is connected to clk.

– down counter, if $\overline{Q}$ is connected to clk.

If clk is +ve edge triggered then

– up counter, if $\overline{Q}$ is connected to clk.

– down counter, if Q is connected to clk.

It does not depend on initial values of Q_2, Q_1, Q_0.

5.25 (d)

Initially $Q = 0$ and count up $(\overline{Q} = 1)$ is active so it started counting up and when it reaches to 5 then decoder output at pin 5 becomes 0 and preset will be active and it will set Q and it will make the counter mode down and count becomes 4, then 3 then 2 then 1 then 0, as soon as it reaches 0, decoder output at pin 0 is low and clear is active and Q goes to 0 and $\overline{Q} = 1$ so up is active and it counts 1, 2,...... so sequence is 0, 1, 2, 3, 4, 5, 4, 3, 2, 1, 0, 1, 2.

5.26 (a)

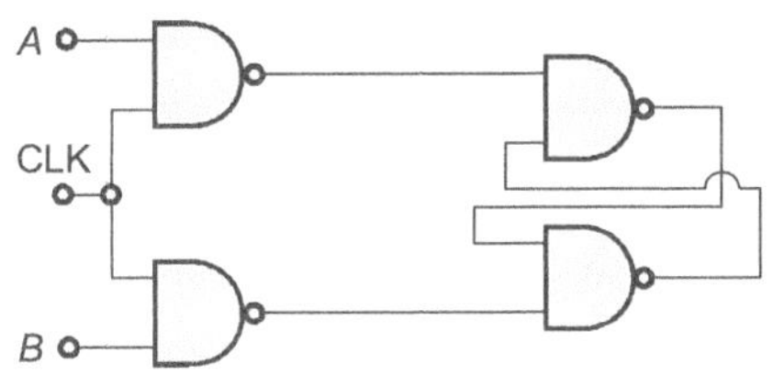

The above circuit is SR flip-flop. The race around condition occurs in J-K flip-flop.

5.27 (d)

When $A = 1$, then Q will be selected by MUX and feedback to D flip-flop which gives output Q again.

So at $A = 1$ it holds its state. When $A = 0$, then $\bar{Q}$ will be selected by MUX and feedback to D flip-flop and output will be inverted.

So for $A = 1$ it holds the state and for $A = 0$ it interchange the state i.e. if $Q = 0$ then it will go to $Q = 1$ and if $Q = 1$ then it will go to $Q = 0$.

So option (d) is correct option.

5.28 (b)

the initial condition $Q_1 Q_0$ is 00

	Q_1	Q_0
	0	0
Ist CLK →	0	1
IInd CLK →	1	1
IIIrd CLK →	1	0

5.29 (c)

Clock frequency = 10 k × 100

$= 10^6$ Hz = 1 MHz

$$500 \geq \frac{10^6}{2^N}$$

$\therefore \quad N = 8$ (minimum)

5.30 (b)

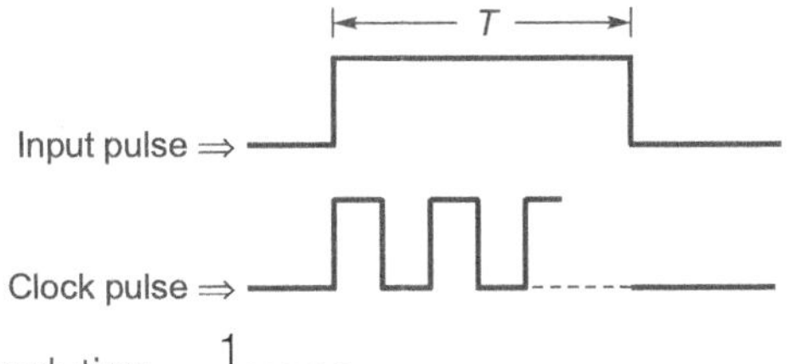

Resolution $= \frac{1}{f}\,\mu$ sec.

Range of measurement of input width = T

$$T = (10^N - 1)T_{clk} = \frac{(10^N - 1)}{f}$$

5.31 (b)

Present State		Next State		Flip-flop Input			
Q_1	Q_2	Q_1	Q_2	J_1	K_1	J_2	K_2
0	0	1	0	1	x	0	x
0	1	1	1	1	x	x	0
1	0	0	1	x	1	1	x
1	1	0	0	x	1	x	1

$$J_1 = K_1 = 1$$
$$J_2 = Q_1$$
$$K_2 = Q_1$$

5.32 (b)

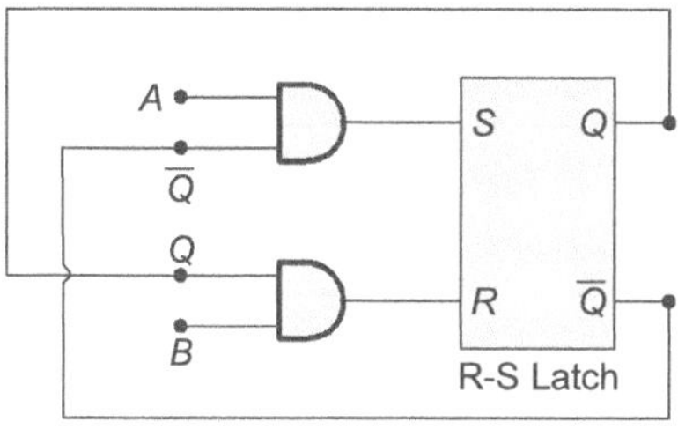

From the figure shown above

$$S = A\bar{Q}$$
$$R = BQ$$

Since Q and $\bar{Q}$ are complement of each other thus S and R can not be 1 at the same time but can be zero at the same time. Thus R-S latch have 1, 1 as a invalid state, which is the possible case only in S-R NOR latch but not in S-R NAND latch.

S-R NOR latch truth table

S	R	Q_{n+1}
0	0	No change
0	1	0
1	0	1
1	1	Invalid state

Now For S-R NOR latch

$$Q_{n+1} = S + \bar{R}.Q_n$$
$$= \bar{Q}_n + \bar{0}.Q_n$$
$$= \bar{Q}_n + Q_n = 1$$

When $AB = 10$, then irrespective of present state, the next state will be set state $Q\bar{Q} = 10$

5.33 (c)

The state table of the counter circuit can be drawn as follows,

Present state		Flip-Flop Inputs				Next state	
Q_1	Q_2	J_1 $(Q_1 + Q_2)$	K_1 $(\bar{Q}_1 + \bar{Q}_2)$	J_2 $(\bar{Q}_1 + Q_2)$	K_2 $(Q_1 + \bar{Q}_2)$	Q_1	Q_2
0	0	0	1	1	1	0	1
0	1	1	1	1	0	1	1
1	1	1	0	1	1	1	0
1	0	1	1	0	1	0	0

Hence, the sequence of states followed by the counter is,

00, 01, 11, 10, 00, 01 ...

5.34 (10)

During the 5th clock cycle, the combination of output bits of the counter is (101).

For this combination, the equivalent of the given circuit can be given as,

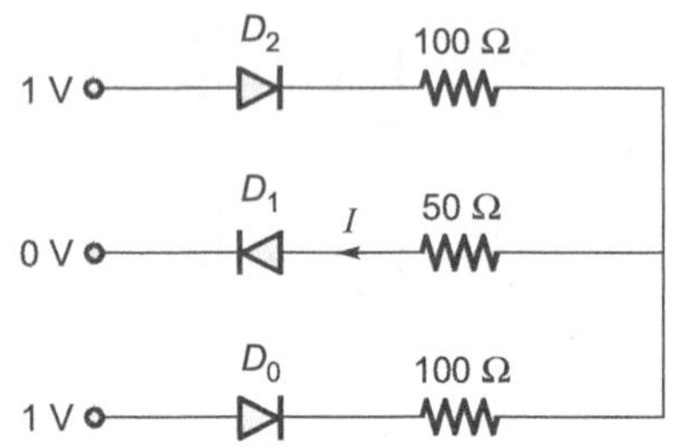

Let us assume all the three diodes are in ON state. So, for this assumption, the circuit can be equivalently drawn as shown below:

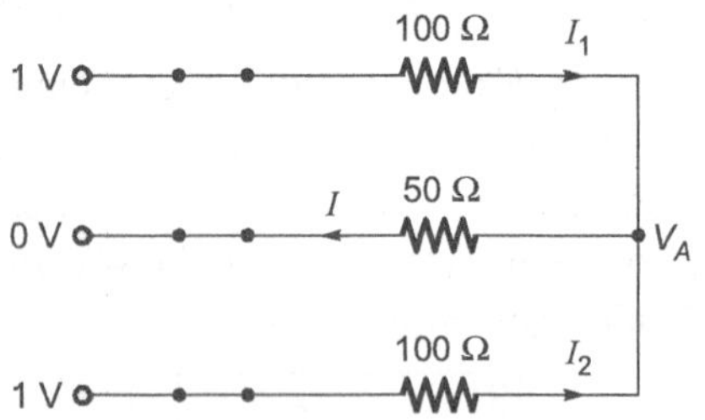

By applying KCL at node A, we get,

$$\frac{1-V_A}{100}+\frac{1-V_A}{100}=\frac{V_A-0}{50}$$

$$\frac{1-V_A}{50}=\frac{V_A}{50}$$

$$V_A=\frac{1}{2}\text{ V}$$

$$I_1=\frac{1-V_A}{100}=\frac{1}{200}\text{ A}=5\text{ mA}$$

$$I_2=\frac{1-V_A}{100}=\frac{1}{200}\text{ A}=5\text{ mA}$$

$$I=\frac{V_A}{50}=\frac{1}{100}\text{ A}=10\text{ mA}$$

$I_1 > 0, I_2 > 0, I > 0$. So, our assumption is correct and the correct answer is,

$$I = 10\text{ mA}$$

5.35 (d)

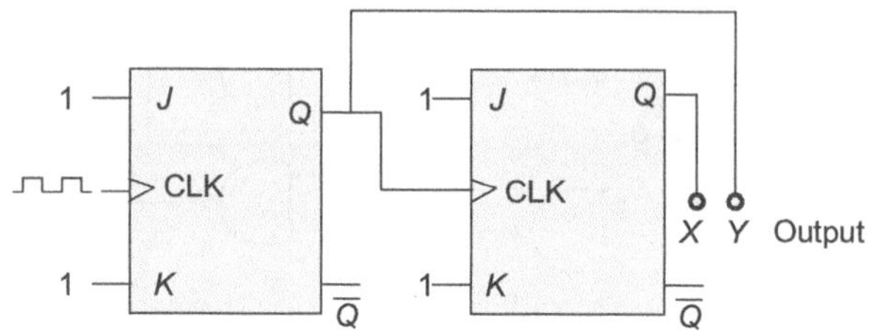

Initial state of output is $X = 0$ and $Y = 0$

After 1 Clock pulse,

$$y = 1$$

$$\Rightarrow \quad x = 1$$

So option (d) is the correct answer.

5.36 (a)

When switch S_1 is closed, a conducting path will be formed from V_{cc} to ground irrespective of the operating condition of the BJT. So, LED turns ON. In this situation, LDR will have a resistance of 500 Ω. Voltage at the inverting terminal of the op-amp is,

$$V^- = \frac{1000}{1000+500}\times 5V = 3.33\text{V}$$

$$V^+ = 2\text{V}$$

$$V^- > V^+$$

$$\Rightarrow \quad V_{out} = 0\text{ V (logic – LOW)}$$

So, $\bar{S}$ will be at logic - LOW and $\bar{R}$ will be at logic -HIGH. It pulses Q to logic-HIGH, which cause some current to flow into the base terminal of the BJT.

When switch S_1 is opened after 1 ms:

Now, let us assume that LED is in OFF state. To check the validity of this assumption, we have to check whether current is flowing into the base terminal of the BJT.

When LED is in OFF state, LDR will have a resistance of 5 kΩ.

$$V^- = \frac{1}{1+5}\times 5V = 0.83\text{V}$$

$$V^+ = 2\text{V}$$

$$V^- < V^+$$

$$\Rightarrow \quad V_{out} = 5\text{ V (logic – HIGH)}$$

So, $\bar{S}$ will be at logic - HIGH and $\bar{R}$ will be at logic -HIGH. It pulses Q to be in previous state (i.e., logic -HIGH). This makes the transistor to be in ON state and in turn LED will be in ON state.

So, our assumption is wrong, i.e., the LED will not be in OFF state after S_1 is opened. Hence, LED will be in ON state both during the switch S_1 is closed and after it is opened.

5.37 (7)

$$T_A = \overline{X \cdot B}$$
$$A^+ = T_A \oplus A$$
$$= \overline{X \cdot B} \oplus A$$
$$T_B = X$$
$$B^+ = T_B \oplus B$$
$$= X \oplus B$$
$$y = A + B$$

X	A	B	A^+	B^+	y
1	1	1	1	0	1
0	1	0	0	0	0
1	0	0	1	1	1

So $(ABy)_2 = (111)_2 = 7$

5.38 (a)

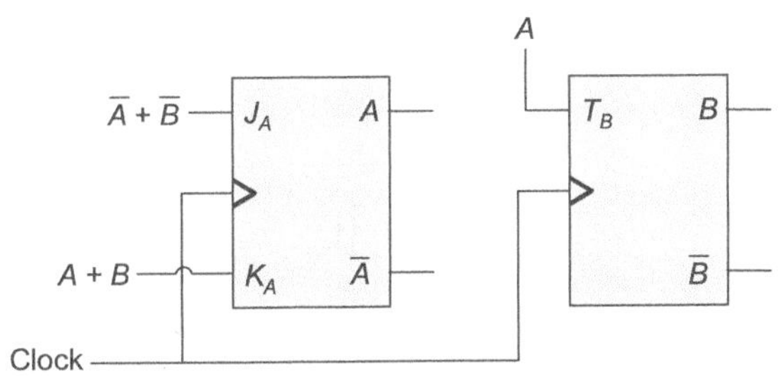

$$A^+ = J\bar{A} + \bar{K}A$$
$$B^+ = T_B \oplus B$$
$$A^+ = (\bar{A}+\bar{B}) \cdot \bar{A} + (\overline{A+B}) \cdot A$$
$$B^+ = A \oplus B$$
$$A^+ = \bar{A}$$

	$\bar{A}$	$A \oplus B$
AB	A^+	B^+
00	1	0
10	0	1
01	1	1
11	0	0

00 → 10 → 01 → 11 → 00....

5.39 (a)

For given digital circuit

Next state, $Q_0^+ = \bar{Q}_1 \cdot \bar{Q}_2$

$$Q_1^+ = Q_0$$
$$Q_2^+ = Q_1$$

Present State			Next State		
Q_2	Q_1	Q_0	Q_2^+	Q_1^+	Q_0^+
0	0	0	0	0	1
0	0	1	0	1	1
0	1	1	1	1	0
1	1	0	1	0	0
1	0	0	0	0	0

0 – 4 – 6 – 4 – 0

So, total states are 5. So, frequency divides by 5.

5.40 (a)

When, Input = 0

$$Q(t+1) = Q(t)$$

When, Input = 1

$$Q(t+1) = \bar{Q}(t)$$

Here, the output toggle for input = 1

So, it is a T-flip-flop.

5.41 (c)

	Register A	Register B
	1 1 0 1	1 0 1 0
Clock 1:	1 1 1 0	1 1 0 1
Clock 2:	0 1 1 1	0 1 1 0
Clock 3:	1 0 1 1	1 0 1 1
Clock 4:	0 1 0 1	1 1 0 1

After 4th clock pulse,

$A = 0101$

$B = 1101$

■■■

6 Logic Families

6.1 Introducing a Schottky diode between the base and collector of the output transistor in a TTL circuit
(a) increases the speed of operation by inhibiting saturation
(b) decrease the speed of operation by inhibiting saturation
(c) increases the FAN OUT by enabling saturation
(d) increases the speed of operation by enabling saturation

[2003 : 1 M]

6.2 For the CMOS analog switch shown in figure, the positive supply is $+V_{DD}$ and the negative supply is $-V_{EE}$. The input V_i is bipolar. The switch will be ON and V_0 will be equal to V_i provided

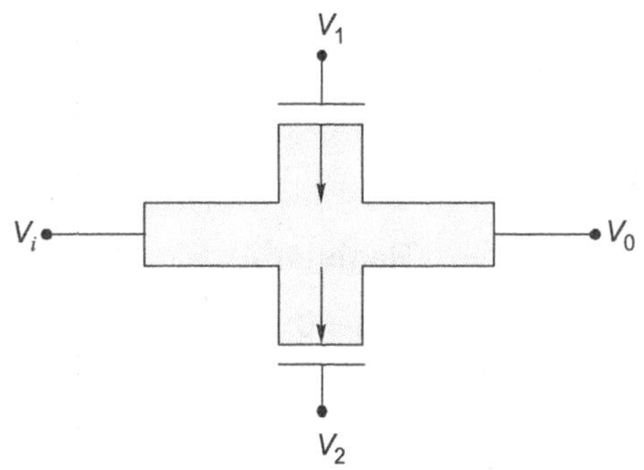

(a) $V_1 = V_2 = +V_{DD}$
(b) $V_1 = V_2 = -V_{EE}$
(c) $V_1 = +V_{DD}$ and $V_2 = -V_{EE}$
(d) $V_1 = -V_{EE}$ and $V_2 = +V_{DD}$

[2003 : 2 M]

6.3 The sink current of the TTL NAND gate shown in figure is 16 mA and the input capacitance of the CMOS NAND gate is 10 pF. The minimum pull-up time constant in ns for the CMOS gate with these specifications is

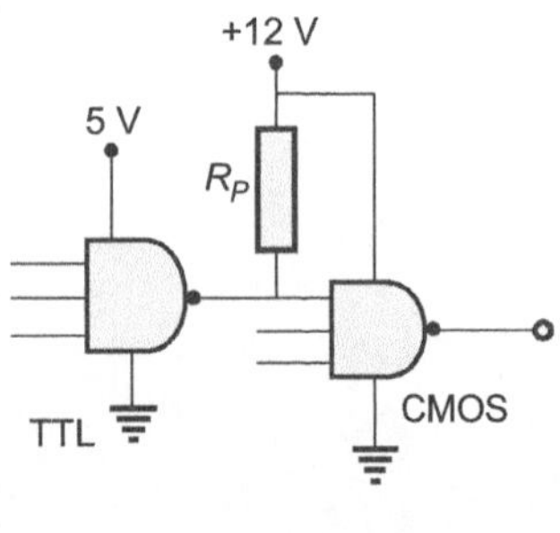

(a) 10.2 (b) 7.5
(c) 8.5 (d) 6.5 [2004 : 2 M]

6.4 In figure, if the input $C = 0$ and $D = 0$, the output Y can be made 0 by making

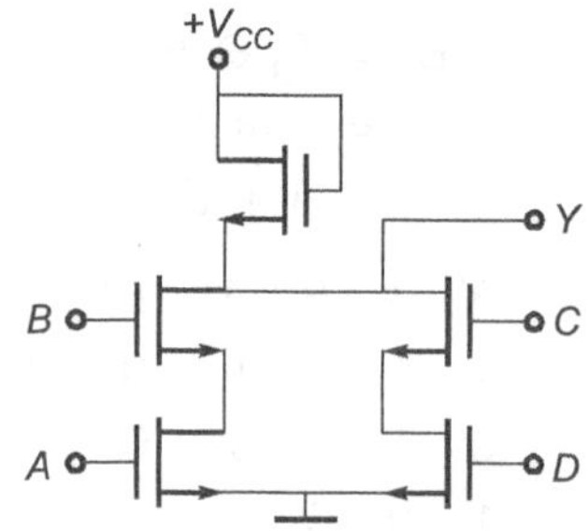

(a) $A = 0, B = 1$ (b) $A = 0, B = 0$
(c) $A = 1, B = 1$ (d) $A = 1, B = 0$

[2004 : 2 M]

6.5 Identify the logic gate given in the figure.

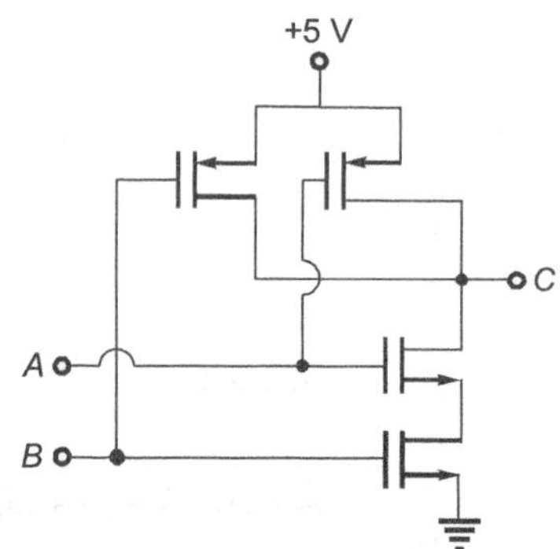

(a) NOR (b) NAND
(c) AND (d) OR

[2005 : 2 M]

6.6 A CMOS implementation of a logic gate is shown in the following figure:

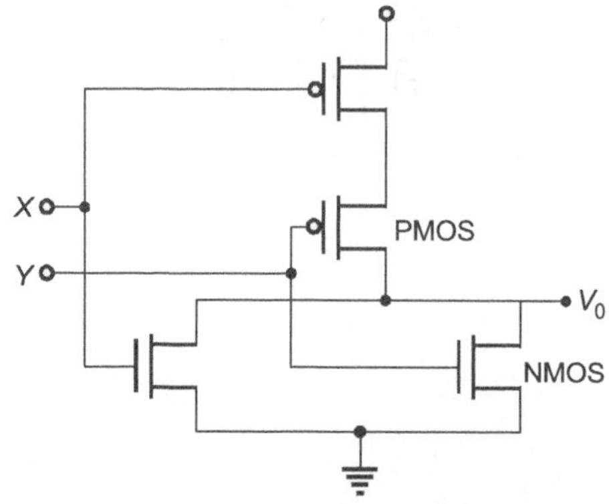

The Boolean logic function realized by the circuit is

(a) AND (b) NAND
(c) NOR (d) OR

[2007 : 1 M]

6.7 The diodes in the circuit shown are ideal. A voltage of 0 V represents logic 0 and +5 V represents logic. The logic function Z realized by the circuit for logic inputs X and Y is

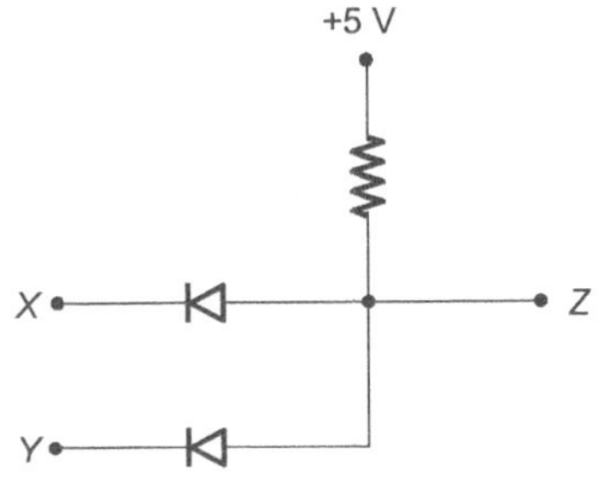

(a) $Z = X = Y$ (b) $Z = X\,Y$
(c) $Z = \overline{X+Y}$ (d) $Z = \overline{XY}$

[2009 : 1 M]

6.8 In the circuit shown below, Q_t has negligible collector-to-emitter saturation voltage and the diode drops negligible voltage across it under forward bias. If V_{cc} is +5 V, X and Y are digital signals with 0 V as logic 0 and V_{cc} as logic 1, the Boolean expression for Z is

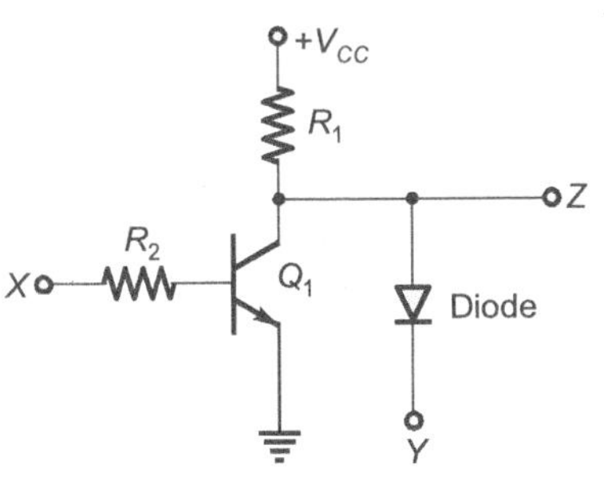

(a) XY (b) $\overline{X}Y$
(c) $X\overline{Y}$ (d) $\overline{XY}$

[2013 : 2 M]

6.9 The figure is a logic circuit with inputs A and B and output Y. V_{ss} = 5 V. The circuit is of type

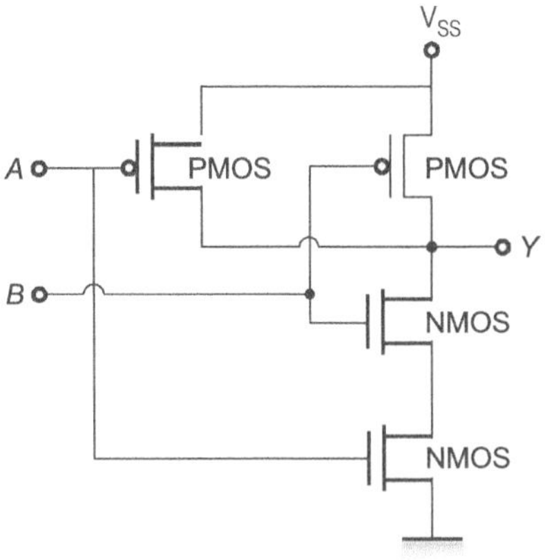

(a) NOR (b) AND
(c) OR (d) NAND

[2014 : 1 M]

6.10 Choose the fastest logic family among the following:

(a) Transistor-Transistor Logic
(b) Emitter-Coupled Logic
(c) CMOS Logic
(d) Resistor-Transistor Logic

[2023 : 1 M]

■■■

Answers Logic Families

6.1 (a)	6.2 (d)	6.3 (b)	6.4 (c)	6.5 (b)	6.6 (c)	6.7 (b)
6.8 (b)	6.9 (d)	6.10 (b)				

Explanations Logic Families

6.1 (a)

Schottky diode is used to speed up the operation by inhibiting the saturation (Because a part of current passes through it and so base current decreases so transistor does not go in deep saturation).

6.2 (d)

Here MOS connected to V_1 is PMOS and MOS connected to V_2 is NMOS as seen from the arrows directions. Since V_i is bipolar so if it is +ve NMOS has to be ON so V_2 must be high. And if V_i is –ve then PMOS has to be ON to pass this to V_0 for this V_1 must be –ve i.e. $V_1 = -V_{EE}$.

6.3 (b)

When sink current present TTL NAND gate output is zero.

Hence, $R_p = \frac{12\text{V}}{16\text{mA}} = \frac{3}{4}\text{k}\Omega.$

Time constant $= CR_p$

$= 10\times10^{-12}\times\frac{3}{4}\times10^{3} = 7.5\text{ ns}$

6.4 (c)

$$Y = \overline{AB+CD}$$

Given $C = 0$ and $D = 0$

So, only $A = 1$ and $B = 1$ will give output 0

$$Y = \overline{1.1+0.0} = \overline{1+0} = \overline{1} = 0.$$

6.5 (b)

It is CMOS gate where 2 PMOS are parallel and in series with 2 NMOS (series combination of NMOS). It is equivalent to NAND gate.

Series combination of NMOS equivalent to parallel combination of PMOS.

6.6 (c)

Truth table:

X	Y	V_0
0	0	1
0	1	0
1	0	0
1	1	0

$$V_0 = \overline{X+Y}$$

6.7 (b)

X	Y	Z
0	0	0
0	1	0
1	0	0
1	1	1

6.8 (b)

X	Y	Z
0	0	0
0	1	1
1	0	0
1	1	0

X	Y	Q	Z
0	0	OFF	0
0	+5V	OFF	+5V
+5V	0	ON	0
+5V	+5V	ON	0

6.9 (d)

The given circuit realize to,

A	B	Y
0	0	1
0	1	1
1	0	1
1	1	0

■■■

7 Memories and Interfacing

7.1 A computer has a memory space of 2^{16} and word length of 24 bits. The memory chips available have 10 address and 8 data lines. The number of chips required for the computer memory space is
(a) 192 (b) 256
(c) 512 (d) 1024
[1999 : 1 M]

7.2 The important feature of a microcontroller is that it has on-chip
(a) math co-processor
(b) program memory
(c) interface for I/O devices
(d) hardware multiplier
[1999 : 1 M]

7.3 Indicate which one of the following statements is correct
(a) Static RAM are faster than dynamic RAMs
(b) RAMs cannot be used to realize Read Write Memory (RWM)
(c) ROMs are not Random Access Devices
(d) RAMs are generally non-volatile.
[2000 : 1 M]

7.4 Find the correct match among the following pair in the context of an 8085 microprocessor
(a) DAA (i) Program control instruction
(b) LXI (ii) Data movement Instruction
(c) RST (iii) Interrupt instruction
(d) JMP (iv) Arithmetic instruction

Codes:
(a) (a) – (i), (b) – (ii), (c) – (iii), (d) – (iv)
(b) (a) – (iv), (b) – (ii), (c) – (iii), (d) – (i)
(c) (a) – (iv), (b) – (iii), (c) – (ii), (d) – (i)
(d) (a) – (ii), (b) – (iv), (c) – (iii), (d) – (i)
[2000 : 1 M]

7.5 In a microprocessor with 16 address and 12 data lines, the maximum number of opcodes is
(a) 2^6 (b) 2^8
(c) 2^{12} (d) 2^{16}
[2001 : 1 M]

7.6 A minimal microcomputer system is constructed using INTEL 8085 microprocessor, an 8156 RAM and an 8355 ROM. The chip enable *CE* of 8355 are connected to the address line A_{12} of 8085 the address of port *A* of the 8156 chip is
(a) 2 IH (b) 12 H
(c) 11 H (d) 20 H
[2002 : 2 M]

7.7 A ROM is interfaced to an 8085 CPU as indicated in figure. The address range occupied by the ROM is

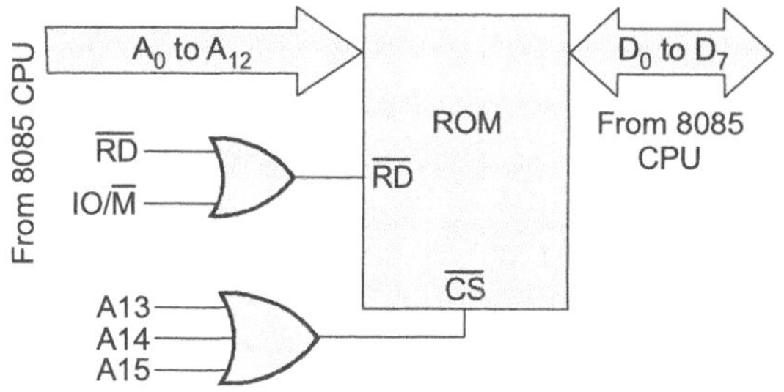

(a) 0000 - 0FFF (b) 0000 - 1FFF
(c) 0000 - 2FFF (d) 8000 - 9FFF
[2003 : 2 M]

7.8 An 8-bit microcontroller has an external RAM is the memory map from 8000H to 9FFFH. The number of bytes this RAM can store is
(a) 8193 (b) 8191
(c) 8192 (d) 8000
[2005 : 1 M]

7.9 A 2K × 8 bit RAM is interfaced to an 8-bit microprocessor. If the address of the memory location in the RAM is 0800H, the address of the last memory location will be
(a) 1000 H (b) 0FFFH
(c) 4800H (d) 47FFH
[2008 : 1 M]

7.10 An 8-bit DAC is interfaced with a microprocessor having 16 address lines ($A_0 ... A_{15}$) as shown in the adjoining figure. A possible valid address for this DAC is

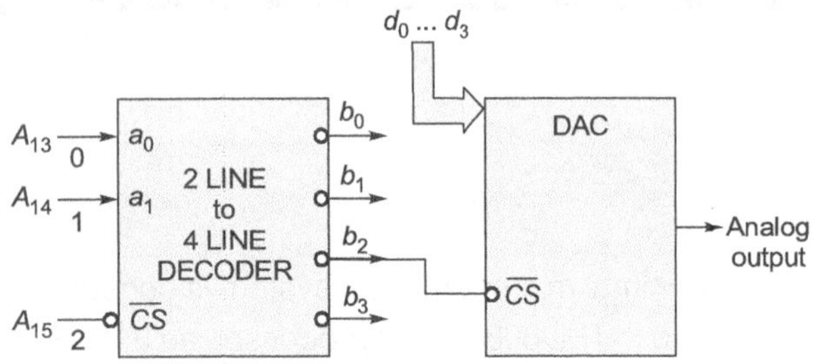

(a) 3000H (b) 4FFFH
(c) AFFFH (d) C000H

[2010 : 1 M]

7.11 An 8K x 8 bit RAM is interfaced to an 8085 microprocessor. In a fully decoded scheme if the address of the last memory location of this RAM is 4FFFH, the address of the first memory location of the RAM will be

(a) 1000 H (b) 2000 H
(c) 3000 H (d) 4000 H

[2011 : 1 M]

7.12 An ADC is interfaced with a microprocessor as shown in the figure. All signals have been indicated with typical notations. Acquisition of one new sample of the analog input signal by the microprocessor involves

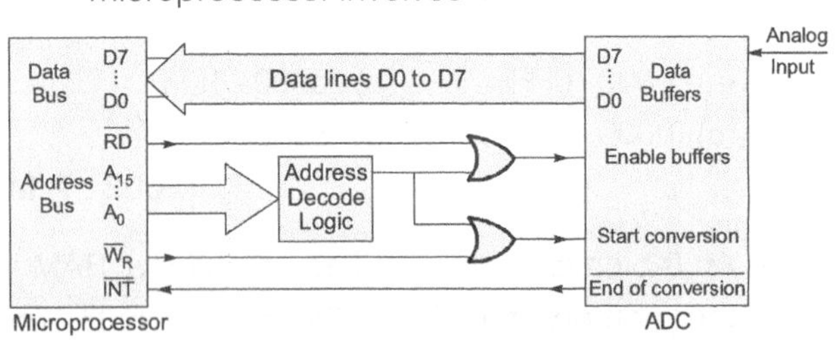

(a) one READ cycle only
(b) one WRITE cycle only
(c) one WRITE cycle followed by one READ cycle
(d) one READ cycle followed by one WRITE cycle

[2015 : 2 M]

7.13 A 1 Kbyte memory module has to be interfaced with an 8-bit microprocessor that has 16 address lines. The address lines A_0 to A_9 of the processor are connected to the corresponding address lines of the memory module. The active low chip select $\overline{CS}$ of the memory module is connected to the y_5 output of a 3 to 8 decoder with active low outputs. S_0, S_1, and S_2 are the input lines to the decoder, with S_2 as the MSB. The decoder has one active low $\overline{EN_1}$ and one active high EN_2 enable lines as shown below. The address range(s) that gets mapped onto this memory module is (are)

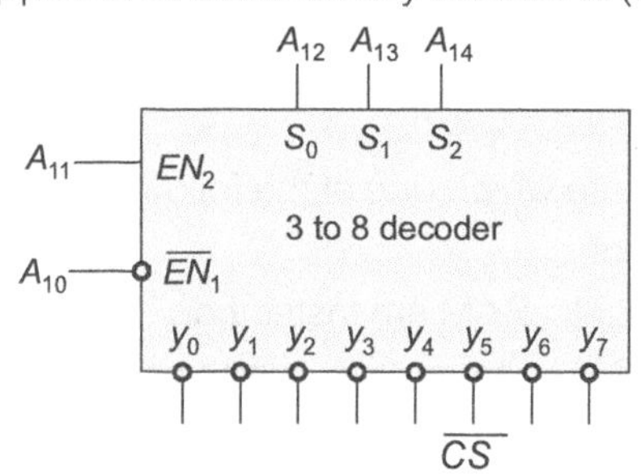

(a) 3000_H to $33FF_H$ and $E000_H$ to $E3FF_H$
(b) 1400_H to $17FF_H$
(c) 5300_H to $53FF_H$ and $A300_H$ to $A3FF_H$
(d) 5800_H to $5BFF_H$ and $D800_H$ to $DBFF_H$

[2016 : 2 M]

■■■

Answers Memories & Interfacing

7.1	(a)	7.2	(c)	7.3	(a)	7.4	(b)	7.5	(c)	7.6	(*)	7.7	(b)
7.8	(c)	7.9	(b)	7.10	(b)	7.11	(c)	7.12	(c)	7.13	(d)		

Explanations Memories & Interfacing

7.1 (a)

$$\text{Memory space of computer} = 2^{16} \times 24$$

[Memory space = $2^{(\text{Address lines})} \times$ (Data lines)]

$$\text{Memory space of chip} = 2^{10} \times 8$$

$$\text{Number of chips required} = \frac{2^{16} \times 24}{2^{10} \times 8} = 192$$

7.7 (b)

Chip select lines			Register Select lines													Range
A_{15}	A_{14}	A_{13}	A_{12}	A_{11}	A_{10}	A_9	A_8	A_7	A_6	A_5	A_4	A_3	A_2	A_1	A_0	
0	0	0	0	0	0	0	0	0	0	0	0	0	0	0	0	← 0000*H*
0	0	0	1	1	1	1	1	1	1	1	1	1	1	1	1	← 1*FFFH*

Address range → 0000H – 1FFFH

7.8 (c)

Number of bytes = Higher address – lower address + 1

= 9FFFH – 8000H + 0001H = 1FFFH + 0001H = 2000H = $(8192)_d$

So, the total number of bytes stored is = 8192.

7.9 (b)

The address of last memory location

= Starting address + number of bytes – $(0001)_H$

= 0800H + 2048 bytes – $(0001)_H$

= 0800H + 0800H – $(0001)_H$ = 1000H – $(0001)_H$ = $(0FFF)_H$

7.10 (b)

A_{15}	A_{14}	A_{13}	A_{12}	A_{11}	A_{10}	A_9	A_8	A_7	A_6	A_5	A_4	A_3	A_2	A_1	A_0	
0	1	0	0	0	0	0	0	0	0	0	0	0	0	0	0	= $(4000)_H$
fix																
0	1	0	1	1	1	1	1	1	1	1	1	1	1	1	1	= $(5\text{FFF})_H$

So, address can be lie in range of $(4000)_H$ to $(5\text{ FFF})_H$.

So, from the options (b) is correct.

7.11 (c)

Capacity of chip = last memory address-first memory address + 1

∴ Address of first memory address

= last memory address-capacity of chip + 1 = 4FFFH – 2000H + 1 = 3000H

7.13 (d)

1 KB memory is interfaced

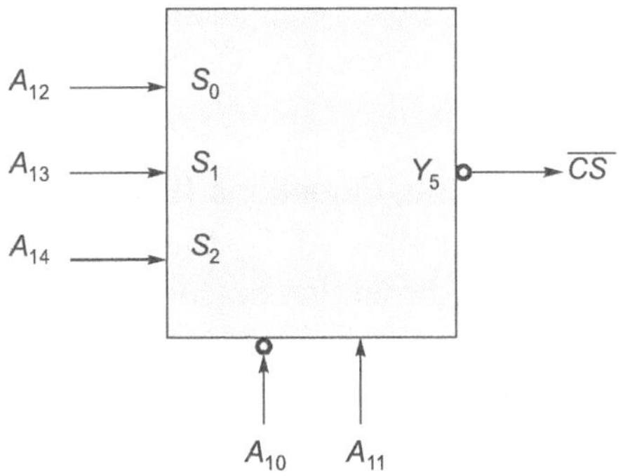

A_{15}	A_{14}	A_{13}	A_{12}	A_{11}	A_{10}	A_9.....A_0
X	1	0	1	1	0	0.......0
X	1	0	1	1	0	1.......1

If $A_{15} = 0 \Rightarrow$ 5800 H to 5 BFFH ; If $A_{15} = 1, \Rightarrow$ D800H to DBFFH

■■■

8 ADC & DAC

8.1 An 8 bit ADC outputs all 1's when V_{in} = 1.275 V. The quantization error is

(a) + 5 mV (b) –5 mV
(c) 10 mV (d) ± 2.5 mV

[1996 : 1 M]

8.2 A 3-bit resistance ladder *D*/*A* (*R*-2R network) has resistor values of R = 10 kΩ and 20 kΩ V_{REF} equals 8 volts. What is I_{out} for a digital input of 111?

(a) 0.7 mA (b) 0.2 mA
(c) 0.1 mA (d) 1 mA

[1996 : 1 M]

8.3 A sample-and-hold circuit is normally required before the following type of *A*/*D* converter

(a) Successive approximation
(b) Flash (parallel) converter
(c) Voltage - to - frequency converter
(d) Dual slope integrator

[1997 : 1 M]

8.4 Integrated output waveform for the dual slope ADC is shown in the figure. The time *T* for an 8 bit counter with 4 MHz clock will be

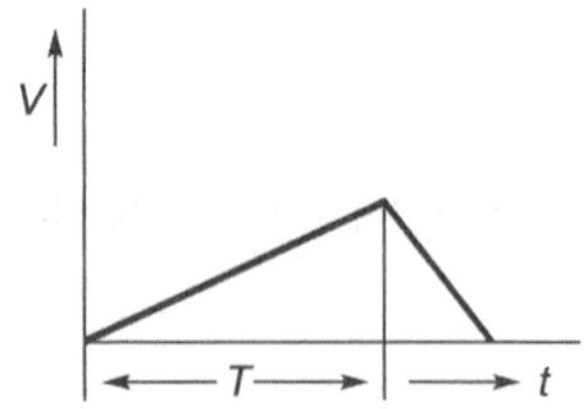

(a) 0.032 ms (b) 0.064 ms
(c) 0.64 ms (d) 0.024 ms

[1998 : 1 M]

8.5 The full scale input voltage to an ADC is 10 V. The resolution required is 5 mV. The minimum number of bits required for ADC is

(a) 8 (b) 10
(c) 11 (d) 12 [1998 : 1 M]

8.6 Percent resolution of an 8 bit *D*/*A* converter is

(a) 0.39 (b) 0.78
(c) 2.56 (d) None of these

[1998 : 1 M]

8.7 The advantage of a dual slope converter over successive approximation converter is that the dual slope converter ______.

(a) is faster
(b) eliminates error due to drift
(c) can reduce errors due to power supply
(d) does not require a stable voltage source

[1998 : 1 M]

8.8 The advantage of a dual slope converter over successive approximation converter is that the dual slope converter

(a) is faster
(b) eliminates error due to drift
(c) can reduce the errors due to power supply
(d) does not require a stable voltage reference

[1999 : 2 M]

8.9 The conversion time of an 8-bits successive approximation converter with a 1 MHz clock is nearly

(a) 512 μs (b) 256 μs
(c) 128 μs (d) 8 μs

[1999 : 2 M]

8.10 Majority of digital voltmeters are built with a dual slope ADC because

(a) Dual slope ADCs are less complex than other type of ADCs
(b) Dual slope ADCs are faster than other types of ADCs
(c) Dual slope ADCs can be designed to be insensitive to noise and interference
(d) Dual slope ADCs provide BCD outputs.

[2002 : 1 M]

Common Data Questions (8.11 to 8.12):

In a dual slope ADC, the reference voltage is 100 mV and the first integration period is set as 50 msec. The input resistor of the integrator is 1 kΩ and the integrating capacitor 0.047 μF.

8.11 For an input voltage of 120 mV, the second integration (de-integration) period will be

(a) 50 ms (b) 60 ms
(c) 100 ms (d) 120 ms [2003 : 2 M]

8.12 If the input of 120 mV is corrupted by power supply interference at 50 Hz having peak amplitude of 3 π mV, the worst-case error introduced by the interference in the reading is

(a) 0% (b) 1%
(c) 3% (d) π%

[2003 : 2 M]

8.13 The number of comparators required in an 8-bit flash-type A/D converter is

(a) 256 (b) 255
(c) (8 + 2) (d) 5

[2004 : 1 M]

8.14 A single channel signal acquisition system with 0 - 10 V range consists of a sample and hold circuit with worst case drop rate of 100 μV/ms and 10-bit ADC. The maximum conversion time for the ADC is

(a) 49 μs (b) 0.49 ms
(c) 4.9 ms (d) 49 ms

[2006 : 2 M]

Common Data for Questions (8.15 to 8.16):

An R-2R ladder type DAC is shown below. If a switch status is '0', 0 V is applied and if a switch status is '1', 5 V is applied to the corresponding terminal of the DAC.

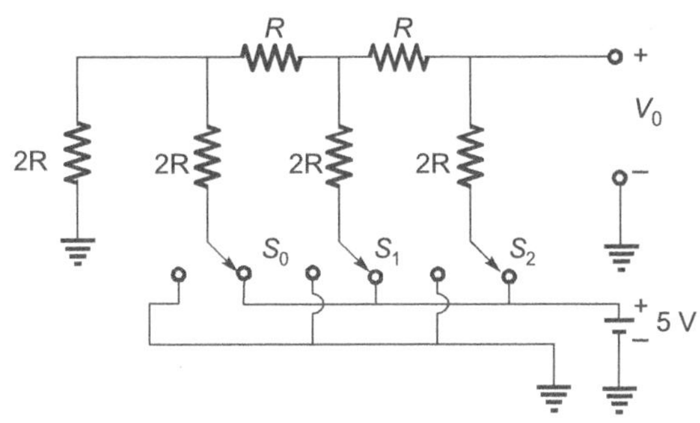

8.15 What is the output voltage (v_0) for the switch status $S_0 = 0$, $S_1 = 1$, $S_2 = 1$?

(a) $\frac{5}{4}$ V (b) $\frac{15}{4}$ V

(c) $\frac{17.5}{4}$ V (d) $\frac{22.5}{4}$ V

[2006 : 2 M]

8.16 What is the step size of the DAC?

(a) 0.125 V (b) 0.525 V
(c) 0.625 V (d) 0.75 V

[2006 : 2 M]

8.17 The circuit shown in the figure below works as a 2-bit analog to digital converter for $0 \le V_{in} \le 3$ V.

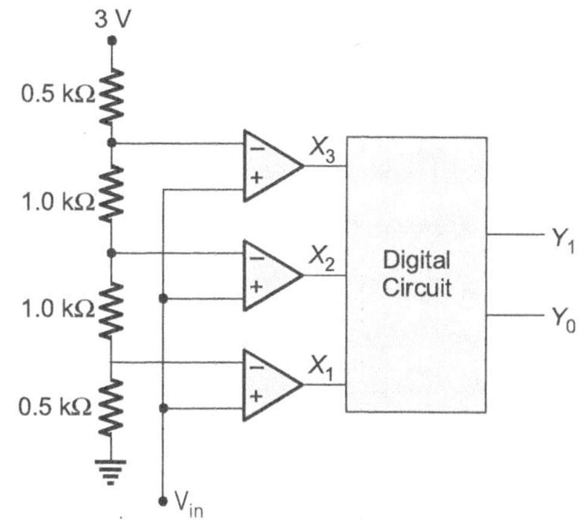

The MSB of the output Y_1, expressed as a Boolean function of the inputs X_1, X_2, X_3 is given by

(a) X_1 (b) X_2
(c) X_3 (d) $X_1 + X_2$

[2007 : 2 M]

8.18 The pulse width T of an asynchronous pulse is measured by a counter with an edge-triggered clock of known frequency f_c as shown in the figure below:

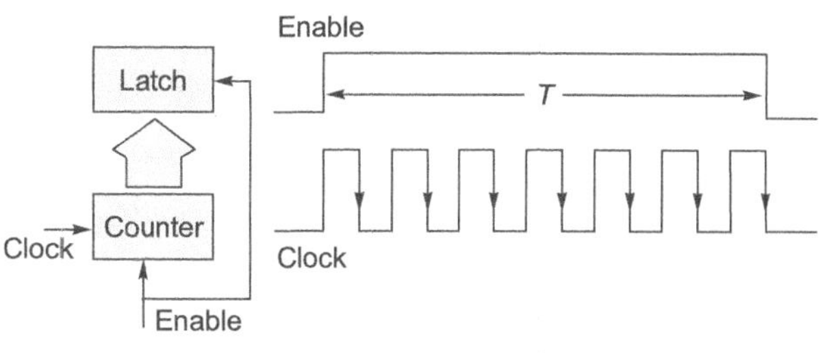

The pulse width to be measured is applied to the Enable pin of the counter. The counter counts while the Enable is high and is held reset to zero otherwise. The counter output is latched by the negative edge of the Enable signal. The measured pulse width is taken to be N times the clock period, where N is the count reached at the end of a count cycle. Assuming no overflow, the measurement error will be limited to x% of T if

(a) $T > \frac{100}{xf_c}$ (b) $T < \frac{100}{xf_c}$

(c) $T > \frac{200}{xf_c}$ (d) $T < \frac{200}{xf_c}$

[2007 : 2 M]

Common Data Questions (8.19 to 8.20):

A data acquisition system (DAS) shown below employs a successive approximation type 12-bit ADC having a conversion time of 5 μs.

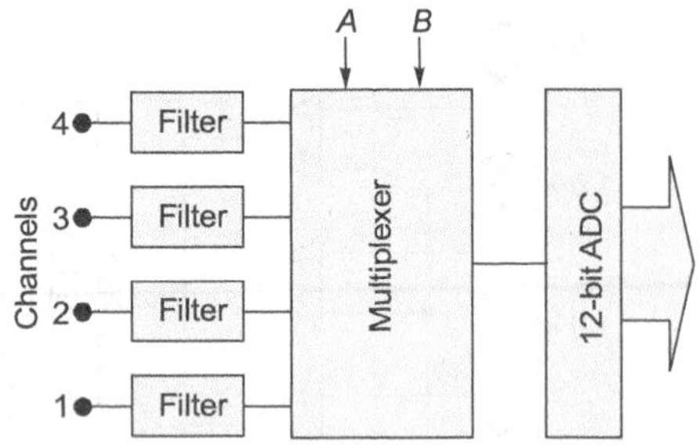

8.19 The quantization error of the ADC is

(a) 0% (b) ±0.012%
(c) ±0.024% (d) ±0.048%

[2008 : 2 M]

8.20 The system is used as a single channel DAS with channel 1 selected as input to the ADC which is set in the continuous conversion mode. For avoiding aliasing error, the cut-off frequency f_c of the filter in channel 1 should be

(a) f_c < 100 kHz
(b) f_c = 100 kHz
(c) 100 kHz < f_c < 200 kHz
(d) f_c = 200 kHz

[2008 : 2 M]

8.21 If the multiplexer is controlled such that the channels one sequenced every 5 μs as 1, 2, 1, 3, 1, 4, 1, 2, 1, 3, 1, 4, 1, ... , the input connected to channel 1 will be sampled at the rate of

(a) 25 k samples/sec (b) 50 k samples/sec
(c) 100 k samples/sec (d) 200 k samples/sec

[2008 : 2 M]

Common Data Questions (8.22 to 8.23):

The figure shows a sample and hold circuit using a MOSFET as a switch. The threshold voltage of the MOSFET is +2 V. It has zero leakage current in the off state. Assume that the capacitor is ideal.

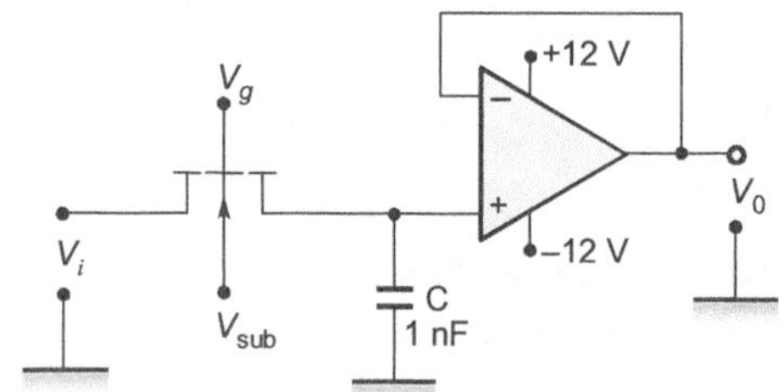

8.22 The input voltage v_i ranges from –5 V to +5 V. Appropriate values of V_{sub}, of V_g during sampling, and of V_g during hold are respectively

(a) 12 V, ≥ +7 V and ≤ –3 V
(b) –12 V, ≥ +3 V and ≤ –7 V
(c) +12 V, ≥ +3 V and ≤ –7 V
(d) –12 V, +7 V and ≤ –3 V

[2009 : 2 M]

8.23 The circuit is used at a sampling rate of 1 kHz, with an A/D converter having a conversion time of 200 μs. The opamp has an input bias current of 10 nA. The maximum hold error is

(a) 1 mV (b) 2 mV
(c) 5 mV (d) 10 mV

[2009 : 2 M]

8.24 A 4-bit successive approximation type ADC has a full scale value of 15 V. The sequence of the states, the SAR will traverse, for the conversion of an input of 8.15 V is

(a) Start Conversion → 1000 → 1100 → 1010 → 1001 → 1000 → End Conversion
(b) Start Conversion → 1000 → 0100 → 0010 → 0001 → 0000 → End Conversion
(c) Start Conversion → 1000 → 0100 → 0110 → 0111 → 1000 → End Conversion
(d) Start Conversion → 1000 → 1100 → 1110 → 1111 → 1111 → End Conversion

[2010 : 2 M]

8.25 Figure below shows a circuit for implementing an 8-bit Digital-to-Analog converter (DAC) using two identical 4-bit DAC's with equal reference voltage. Assume that b_0 represents LSB, b_7 MSB and the opamp is ideal. To obtain correct analog values corresponding to an 8-bit DAC at the output V_0, the value of resistor R is

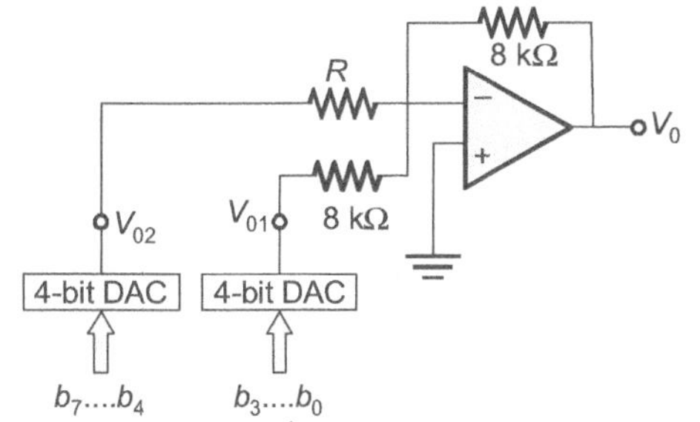

(a) 0.25 kΩ (b) 0.5 kΩ
(c) 1 kΩ (d) 8 kΩ

[2011 : 1 M]

8.26 A 4-bit successive approximation type of A/D converter has an input range of 0 to 15 volts. The output bit b_1 next to the LSB has a stuck-at-zero fault. The pair of input voltages, that produces the same output code word is

(a) 2 V and 4 V (b) 4 V and 6 V
(c) 1 V and 2V (d) 8 V and 9 V

[2011 : 2 M]

8.27 The circuit in the figure represents a counter-based unipolar ADC. When SOC is asserted the counter is reset and clock is enabled so that the counter counts up and the DAC output grows. When the DAC output exceeds the input sample value, the comparator switches from logic 0 to logic 1, disabling the clock and enabling the output buffer by asserting EOC. Assuming all components to be ideal, V_{ref}, DAC output and input to be positive, the maximum error in conversion of the analog sample value is

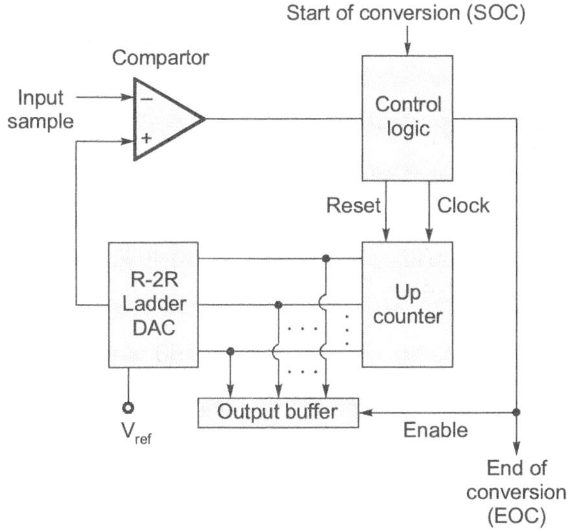

(a) directly proportional to V_{ref}
(b) inversely proportional to V_{ref}
(c) independent of V_{ref}
(d) directly proportional to clock frequency

[2014 : 2 M]

8.28 An N-bit ADC has an analog reference voltage V. Assuming zero mean and uniform distribution of the quantization error, the quantization noise power will be

(a) $\frac{V^2}{12(2^N-1)^2}$ (b) $\frac{V^2}{12(2^N-1)}$

(c) $\frac{V}{12(2^N-1)}$ (d) $\frac{V^2}{\sqrt{12}}$

[2014 : 2 M]

8.29 The number of comparators required for implementing an 8-bit flash analog-to-digital converter is

(a) 8 (b) 128
(c) 255 (d) 256 [2018 : 1 M]

8.30 An 8-bit weighted resistor digital-to-analog converter (DAC) has the smallest resistance of 500 Ω. The largest resistance has a value________kΩ. [2019 : 1 M]

8.31 The address lines A_9 .... A_2 of a 10 bit, 1.023 V full-scale digital to analog converter (DAC) is connected to the data lines D_7 to D_0 of an 8-bit microprocessor, with A_1 and A_0 of the DAC grounded. Now, D_7 .... D_0 is changed from 1010 1010 to 1010 1011. The corresponding change in the output of the DAC (in mV, rounded off to one decimal places) is _____. [2020 : 2 M]

8.32 A 10-bit ADC has a full-scale of 10.230 V, when the digital output is $(11\ 1111\ 1111)_2$. The quantization error of the ADC in millivolt is _____.

[2021 : 1 M]

8.33 For a 4-bit Flash type Analog to Digital Converter (ADC) with full scale input voltage range "V". Which of the following statement(s) is/are true?

(a) A change in the input voltage by $\frac{V}{16}$ will always flip MSB of the output.
(b) A change in the input voltage by $\frac{V}{16}$ will always flip the LSB of the output.
(c) The ADC requires 15 comparators.
(d) The ADC requires one 4 to 2 priority encoder and 4 comparators.

[2021 : 1 M]

8.34 An analog-to-digital converter with resolution 0.01 V converts analog signals between 0 V to +10 V to an unsigned binary output. The minimum number of bits (in integer) in the output is _______

[2022 : 1 M]

■■■

Answers ADC & DAC

8.1 (a)	8.2 (b)	8.3 (a)	8.4 (b)	8.5 (c)	8.6 (a)	8.7 (c)
8.8 (c)	8.9 (d)	8.10 (c)	8.11 (b)	8.12 (b)	8.13 (b)	8.14 (d)
8.15 (b)	8.16 (c)	8.17 (b)	8.18 (a)	8.19 (b)	8.20 (d)	8.21 (c)
8.22 (a)	8.23 (b)	8.24 (a)	8.25 (b)	8.26 (c)	8.27 (a)	8.28 (a)
8.29 (c)	8.30 (64)	8.31 (4)	8.32 (5)	8.33 (b, c)	8.34 (10)	

Explanations ADC & DAC

8.1 (a)

Quantization error

$$= \frac{FSD}{2^n - 1} = \frac{1.275}{(2^n - 1)} = \frac{1.275}{2^8 - 1} = 5 \text{ mV}$$

8.2 (b)

I_{out} for resistance ladder D/A (R-2R network)

$$I_{out} = \frac{V_{ref}}{2^n}\sum_{i=0}^{n-1}(2^i b_i)\left(\frac{1}{R+2R}\right)$$

$$= \frac{8}{2^3}(4+2+1)\frac{1}{3R}$$

$$= \frac{7}{3R} = \frac{7}{3\times 10} = 233 \text{ mA} \approx 0.2 \text{ mA}$$

8.3 (a)

For successive approximation A/D converter a sample and hold circuit is normally required.

8.4 (b)

Time $T = 2^n T_{CLK}$

$$= \frac{2^n}{f} = \frac{2^8}{4\times 10^6} = 64\ \mu s = 0.064 \text{ ms}$$

8.5 (c)

$$\text{Resolution} = \frac{\text{Full scale voltage}}{2^n - 1}$$

$$5\times 10^{-3} = \frac{10}{2^n - 1}$$

$$\Rightarrow \quad 2^n - 1 = \frac{10^4}{5}$$

$$\Rightarrow \quad 2^n = 2001$$

$$n = 10.96$$

$$\therefore \quad n = 11 \text{ bits required for ADC}$$

8.6 (a)

$$\%\ \text{Resolution} = \frac{1}{2^n - 1}\times 100$$

$$= \frac{1}{2^8 - 1}\times 100 = 0.39$$

8.7 (c)

Dual-slope integrating type A/D converter rejects noise by averaging due to integration.

8.9 (d)

Conversion time for SAR is

$$nt_{CLK} = \frac{8\times 1}{f_{CLK}} = \frac{8}{1\times 10^6} = 8\times 1\times 10^{-6} = 8\ \mu s$$

8.11 (b)

De-integration period,

$$T_2 - T_1 = \frac{V_a}{V_R}\cdot T_1 = \frac{120}{100}\cdot 50 = 60 \text{ ms}$$

8.12 (b)

Due to interference maximum change in voltage can be = 3π mV due to this additional voltage integrating time (second integration) will be increased by,

$$t = \frac{3\pi}{100}\times T_1 = \frac{3\pi}{100}\times 50 = 4.71 \text{ msec}$$

so, error due to this period

$$= \frac{RC}{t}\times 100$$

$$= \frac{4.7\times 10^{-5}}{4.71\times 10^{-3}}\times 100 = 1\%$$

8.13 (b)

Number of comparators $= 2^n - 1 = 2^8 - 1$

$= 256 - 1 = 255$

Where, n = number of bits of an ADC

8.14 (d)

In a ADC along with S-H circuit (sample and hold) circuit, for avoiding error at output, voltage of capacitor should not drop by more than $\pm\Delta/2$, where Δ is step size.

Here, $\Delta = \dfrac{10-0}{(2^{10}-1)} = 9.775\times10^{-3}$ V

so, $\dfrac{\Delta}{2} = 4.8875\times10^{-3}$ V

So conversion time (maximum) should be such that the drop across capacitor voltage must reach maximum value $\Delta/2$.

so, time taken for this will be

$$t = \frac{\Delta/2}{\text{drop rate}} = \frac{4.8875\times10^{-3}}{10^{-4}\ \text{V/msec}}$$

t = 48.875 msec

$t \simeq 49$ msec

8.15 (b)

At switch states, $S_0 = 0$, $S_1 = 1$, $S_2 = 1$

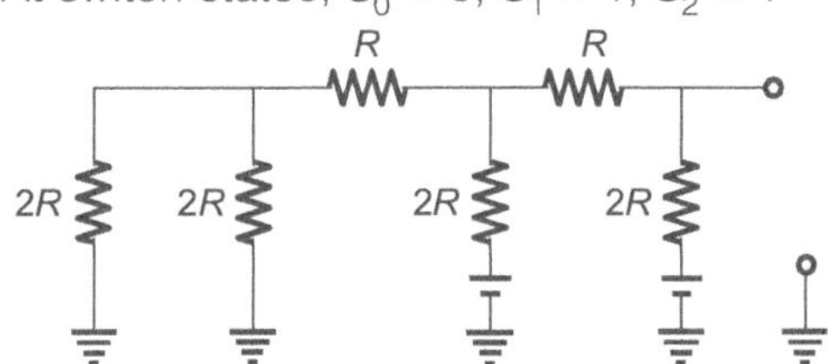

Thevenin's equivalent

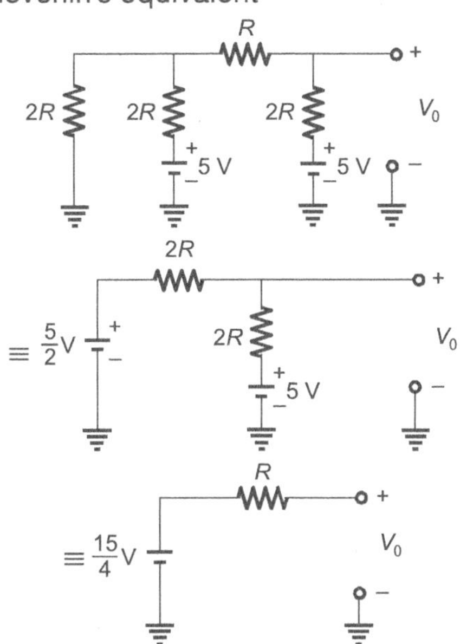

$\therefore \quad V_0 = \dfrac{15}{4}$ V

8.16 (c)

$$\text{Step size} = \frac{V_{FS}}{\text{no. of steps}} = \frac{5}{2^3} = \frac{5}{8} = 0.625\ \text{V}$$

8.17 (b)

It is encoder (digital circuit block).

⇒ so when X_2, X_1, X_0 all are zero then output Y_1 Y_0 = 00

⇒ when X_1 becomes 1 and other are zero then Y_1 Y_0 = 01

⇒ when X_2 becomes 1 and X_3 equal to 0 and X_1 becomes 1 then output Y_1 Y_0 = 10

⇒ when X_3 becomes 1 and X_2, X_1 are also 1, then Y_1 Y_0 = 11

⇒ it is a priority encoder

So,

X_3	X_2	X_1	Y_1	Y_0
0	0	0	0	0
0	0	1	0	1
0	1	1	1	0
1	1	1	1	1
0	1	0	1	0
1	0	0	1	1
1	1	0	1	1
1	0	1	1	1

last four are actually don't care conditions and will not occur

so, $\overline{Y}_1 = \overline{X}_2$

$$\boxed{Y_1 = X_2}$$

8.18 (a)

Maximum error possible will be one clock period of clock, this occurs when –ve edge of enable occurs just before –ve edge of clock so that count is 1 less than actual count.

Since error should be limited to x% of T.

So maximum error should not be greater than x% of T.

i.e. $\dfrac{1}{f} < \dfrac{xT}{100}$ (1/f = T' clock period)

$$\boxed{T > \frac{100}{xf}}$$

8.19 (b)

Quantization error lies between $-\dfrac{\delta}{2} < Q_e \leq \dfrac{\delta}{2}$ (for uniform quantizer).

Where δ is step size.

So, maximum quantization error = $\dfrac{\delta}{2}$

So, % error = $\dfrac{1}{2L}\times100 = \dfrac{100}{2\times2^{12}} = 0.012\%$

8.20 (d)

$$f_c = \frac{1}{5\times10^{-6}} = 200\text{ kHz}$$

8.21 (c)

Sampling period,

$$T_s = 2 \times 5\ \mu s = 10\ \mu s$$

Sampling rate,

$$f_c = \frac{1}{T_s} = \frac{1}{10\times10^{-6}} = 100\text{ k samples/sec.}$$

8.22 (a)

It is NMOS so V_{sub} is +ve and equal to supply voltage V_{sub} = 12 V

Now during sampling MOSFET must be on, so V_{GS} must be greater than threshold voltage.

$$V_{GS} \geq 2$$

$\Rightarrow$ $V_g - V_s \geq 2$

$\Rightarrow$ $V_g - V_i \geq 2$

for V_i maximum it must be on.

So, $V_g - 5 \geq 2$

$\Rightarrow$ $V_g \geq 7$ V (so it will be on for every value of V_i)

$\Rightarrow$ during hold MOSFET must be turned off. For this

$$V_{GS} \leq 2$$

$$V_g - V_i \leq 2$$

$\Rightarrow$ when V_i is minimum then V_{GS} is maximum so for this value also it must be off.

That is $V_g - (-5) \leq 2$

$$V_g \leq -3\text{ V}$$

8.23 (b)

Sampling frequency f_s = 1 kHz.

Sampling period, T_s = 1 msec

So we get the samples after every T_s seconds.

Now ADC convert it after 200 μsec, so capacitor should hold this value for 200 μsec, but due to bias current discharging takes place and voltage drop in 200 μsec will be error in output.

$$C\frac{dV_c}{dt} = i$$

$$10^{-9}\times\frac{dV}{200\times10^{-6}} = 10^{-8}$$

$$dV = \frac{200\times10^{-8}\times10^{-6}}{10^{-9}}$$

$$= 2 \times 10^{-3} = 2\text{ mV}$$

8.24 (a)

It is weighted type of ADC having weights 8, 4, 2, 1 (for four bits).

$$\text{Step size} = \frac{V_{FS}}{2^n-1} = \frac{15}{15} = 1\text{ V}$$

$\Rightarrow$ First of all MSB is set and analog voltage is compared $(1000)_2 = (8)_{10}$ equivalent to 8 V which is less than 8.15 V, so this bit remains set.

$\Rightarrow$ In next step next higher bit is set i.e. $(1100)_2 = (12)_{10}$ and this is equivalent to 12 V which is greater than 8 V so this bit is reset i.e. 1000.

$\Rightarrow$ Now next higher bit is set i.e. $(1010)_2 = (10)_{10}$ and it is equivalent to 10 V which is greater than 8.15 V so it is reset again i.e. 1000.

$\Rightarrow$ Now LSB is set i.e. $(1001)_2 = (9)_{10}$ and it is equivalent to 9 V which is again greater than 8.15 V so it is again reset i.e. 1000.

$\Rightarrow$ So sequence of operation is 1000 $\rightarrow$ 1100 $\rightarrow$ 1010 $\rightarrow$ 1001 $\rightarrow$ 1000.

8.25 (b)

Since the DAC output of 2nd DAC (higher byte) V_{02} must be amplified 16 times more than V_{01} is amplified in order to get correct output so R must be 1/16 times 8 kΩ.

$$R = \frac{8}{16} = 0.5\text{ k}\Omega$$

8.26 (c)

Fault is $\rightarrow$ $b_3\ b_2\ b_1\ b_0$ (b_1 = 0)

$$\text{Resolution} = \frac{15}{2^4-1} = \frac{15}{16-1} = 1\text{ V}$$

for input V_i = 2 V

	b_3	b_2	b_1	b_0		
I	1	0	0	0	$\Rightarrow$	8 V
II	0	1	0	0	$\Rightarrow$	4 V
III	0	0	0 (zero fault)	0	$\Rightarrow$	0 V
IV	0	0	0	1	$\Rightarrow$	1 V

output code word (IV)

for input V_i = 1 V

	b_3	b_2	b_1	b_0		
I	1	0	0	0	$\Rightarrow$	8 V
II	0	1	0	0	$\Rightarrow$	4 V
III	0	0	0 (zero fault)	0	$\Rightarrow$	0 V
IV	0	0	0	1	$\Rightarrow$	1 V

output code word (IV)

8.27 (a)

Maximum error in analog to digital conversion is equal to 1 bit only i.e. 1 LSB change which is equal to resolution i.e.

$$\text{Resolution or error} = \frac{V_{ref}}{2^n}$$

$$\text{error} \propto V_{ref}$$

8.28 (a)

$$\text{Quantization noise power} = \frac{\Delta^2}{12} = x$$

then,
$$\Delta = \frac{V}{2^N - 1} = \text{step size in } N\text{-Bit}$$

$$x = \left(\frac{V}{2^N - 1}\right)^2 \times \frac{1}{12} = \frac{V^2}{12(2^N - 1)^2}$$

8.29 (c)

In general, in n-bit flash type ADC, the number of comparators needed is equal to, $2^n - 1$. Hence, for 8-bit ADC, the number of comparators needed $= 2^8 - 1 = 255$.

8.30 (64)

In general, in a binary weighted resistor DAC, the relationship between LSB and MSB resistance is given as :

LSB Resistance = (2^{n-1}) MSB resistance where, n is the number of bits in the DAC.

Given, $n = 8$

smallest resistance = 500 Ω

MSB resistance = 500 Ω

$\therefore$ LSB resistance $= (2^{8-1}) \times 500\ \Omega$

$= 128 \times 500\ \Omega$

$= 64000\ \Omega = 64\ k\Omega$

8.31 (4)

V_{FS} = 1.023 10 bit DAC

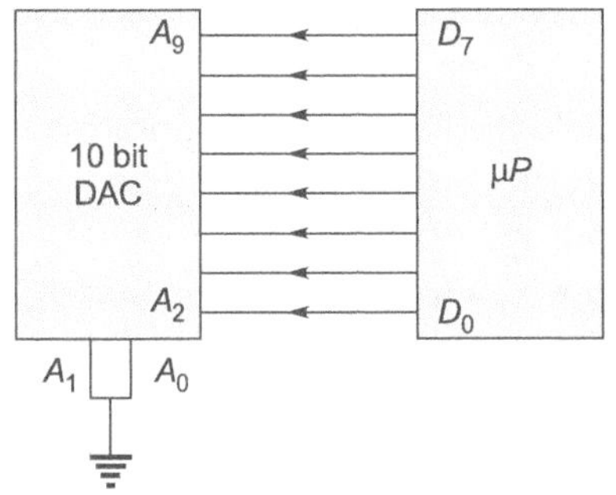

1010101000 = initial input of DAC

1010101100 = changed input of DAC

0000000100 = change in input

Output voltage of DAC,

$$V_{DAC} = \frac{V_{FS}}{2^{10} - 1} \times \text{(Decimal equivalent of binary input)}$$

Change in output of DAC for corresponding input change in

$$\Delta V_{DAC} = \frac{1.023}{1023} \times 4 = 4 \text{ mV}$$

8.32 (5)

Given, 10 bit ADC, Full scale = 10.23 V

$$\text{Resolution} = \frac{V_{FS}}{2^n - 1} = \frac{10.23}{2^{10} - 1} = 10 \text{ mV}$$

$$\text{Quantization error} = \frac{\text{Step size}}{2} = 5 \text{ mV}$$

(Step size = resolution of ADC)

8.33 (b, c)

Given, 4 bit flash ADC, $n = 4$

$$\text{Resolution} = \frac{V_R}{2^n} = \frac{V_R}{2^4} = \frac{V_R}{16}$$

Any change $\frac{V_R}{16}$ changes LSB.

Example,

Input : 0 V $\rightarrow$ 0000

$\frac{V_R}{16} \rightarrow 0001$

No. of comparators required for 4-bit flash type ADC $= 2^4 - 1 = 15$

8.34 (10)

Given resolution, $R = 0.01$ V

$$\text{Resolution} = \frac{V_{in}}{2^n - 1} = 0.01$$

$$\frac{10}{2^n - 1} = 0.01$$

$$2^n = 1001$$

$$n = \log_2 1001 \cong 9.96$$

$n = 10$ number of bits

■■■

9 Basics of Microprocessor and Instruction Set

9.1 The stack pointer in a microprocessor is a register containing
(a) the address of the next operand
(b) the current size of the stack
(c) the address of the top of the stack
(d) the address for storing the result of arithmetic operations
[1997 : 1 M]

9.2 Identify the software interrupt instruction in 8085 microprocessor
(a) INT (b) RST 5
(c) RST 7.5 (d) RST 6.5
[1998 : 1 M]

9.3 For an 8085 microprocessor, the Stack Pointer and Program Counter registers contain the number F000 and 2400 in Hex respectively. The contents of the register after execution of the instruction CALL E000 would be
(a) PC : F003 SP : 2400
(b) PC : 2400 SP : 2400
(c) PC : E000 SP : 2401
(d) PC : E000 SP : 23FE
[2000 : 1 M]

9.4 An m-bit microprocessor has an m-bit
(a) flag register
(b) instruction register
(c) data counter
(d) program counter
[2001 : 1 M]

9.5 In 8085 microprocessor, CY flag may be set by the instruction
(a) SUB (b) INX
(c) CMA (d) ANA
[2001 : 1 M]

9.6 Microprocessor 8085 regains control of the bus
(a) immediately after HOLD goes low
(b) immediately after HOLD goes high
(c) after half-clock cycle after HLDS goes low
(d) after half-clock cycle after HLDA goes high
[2001 : 1 M]

9.7 In an INTEL 8085 microprocessor the ADDRESS-DATA bus and the DATA bus are
(a) Non multiplexed
(b) Multiplexed
(c) Duplicated
(d) Same as CONTROL bus **[2002 : 1 M]**

9.8 In an 8085 based system the subroutine TEST given below is called by another program. When the processor returns from the subroutine TEST, the value in the accumulator will be

```
TEST    MOV A,#00H
        CALL TEST
TEST    INR A
        RET
```

(a) 02 (b) 00
(c) FF (d) 20 **[2002 : 2 M]**

9.9 Some of the pins of an 8085 CPU and their functions are listed below. Identify the correct answer that matches the pins to their respective functions.

Group-I
P. RST 7.5
Q. HOLD
R. $IO/\overline{M}$
S. ALE

Group-II
1. Selects I/O or memory
2. Demultiplexes the address and data bus
3. Is a vectored interrupt
4. Facilitates direct memory access
5. Is a clock
6. Selects BCD mode of operation

(a) P-3, Q-2, R-1, S-4
(b) P-4, Q-1, R-5, S-3
(c) P-3, Q-4, R-1, S-2
(d) P-2, Q-3, R-6, S-1 **[2003 : 1 M]**

9.10 In an 8085 system containing 8 KB of ROM and 8 KB of RAM, the ROM is selected when A_{15} is 0 and the RAM is selected when A_{15} is 1. A_{13} and A_{14} are unused. The CPU executes the following program

```
Prog 1  MVI A,00H
        STA 8080H
        DCR A
        STA C080H
        RET
```

The content of memory location 8080H after the execution of the return instruction is

(a) FFH (b) FEH
(c) 01H (d) 00H [2003 : 2 M]

9.11 The vectored address corresponding to the software interrupt command RST7 in 8085 Microprocessor is

(a) 0017H (b) 0027H
(c) 0038H (d) 0700H [2004 : 1 M]

9.12 The following 8085 instructions are executed sequentially.

```
Prog:  XRA   A
       MOV   L, A
       MOV   H, L
       INX   H
       DAD   H
```

After execution, the content of HL register pair is

(a) 0000H (b) 0101H
(c) 0001H (d) 0002H

[2004 : 2 M]

9.13 The time period of a square wave in the audio frequency range is measured using an 8085 Microprocessor by feeding the square wave to one of the four interrupts, namely, RST 7.5, RST 6.5, RST 5.5, or INT. The algorithm used starts a timer at the beginning of a time period, stops the timer at the beginning of the next time period and reads the time values for time measurement. Which of the following interrupts should be selected for this applications?

(a) INTR (b) RST 5.5
(c) RST 6.5 (d) RST 7.5

[2005 : 1 M]

9.14 In an 8085 microprocessor, which one of the following is the correct sequence of the machine cycles for the execution of the DCR M instruction?

(a) op-code fetch
(b) op-code fetch, memory read, memory write
(c) op-code fetch, memory read
(d) op-code fetch, memory write, memory write

[2005 : 1 M]

9.15 A microprocessor has an instruction XOR (r_1, r_2), which performs an Exclusive OR operation of register r_1, r_2 and stores the result in r_1. After the following instructions are executed

XOR (r_2, r_1)
XOR (r_1, r_2)
XOR (r_2, r_1)

Which one of the following is true?

(a) Content of register r_1 is half sum of r_1 and r_2
(b) Content of register r_2 is half sum of r_1 and r_2
(c) Contents of registers r_1 and r_2 remain unaltered
(d) Contents of registers r_1 and r_2 are swapped

[2005 : 2 M]

9.16 In a 8085 Microprocessor the value of the stack pointer (SP) is 2010H and that of DE register pair is 1234H before the following code is executed. The value of the DE register pair after the following code is executed is

```
LXI H,   0000H
PUSH,    H
PUSH     H
POP      B
DAD      SP
XCHG
```

(a) 200EH (b) 200CH
(c) 2010H (d) 1232H [2005 : 2 M]

9.17 A memory mapped I/O device has an address of 00F0H. Which of the following 8085 instructions output the content of the accumulator to the I/O device?

(a) LXI H, 00F0H
MOV M, A
(b) LXI H, 00F0H
OUT M
(c) LXI H, 00F0H
OUT F0H
(d) LXI H, 00F0H
MOV A, M

[2006 : 1 M]

9.18 An 8085 assembly language program is given as follows. The execution time of each instruction is given against the instruction in terms of T-state.

	Instruction	T-states
	....	....
	MVI B, 0AH	7T
LOOP:	MVI C, 05H	7T
	DCR C	4T
	DCR B	4T
	JNZ LOOP	10T/7T

The execution time of the program in terms of T-states is

(a) 247 T (b) 250 T
(c) 254 T (d) 257 T [2006 : 2 M]

9.19 A snapshot of the address, data and control buses of an 8085 microprocessor executing a program is given below:

Address	2020H
Data	24H
IO/$\overline{M}$	Logic High
$\overline{RD}$	Logic High
$\overline{WR}$	Logic Low

The assembly language instruction being executed is

(a) IN 24H
(b) IN 20H
(c) OUT 24H
(d) OUT 20H

[2007 : 2 M]

9.20 8-bit signed integers in 2's complement form are read into the accumulator of an 8085 Microprocessor from an I/O port using the following assembly language program segment with symbolic addresses.

```
BEGIN:   IN PORT
         RAL
         JNC      BEGIN
         RAR
END:     HLT
```

This program

(a) halts upon reading a negative number
(b) halts upon reading a positive number
(c) halts upon reading a zero
(d) never halts

[2007 : 2 M]

9.21 A part of a program written for an 8085 microprocessor is shown below. When the program execution reaches LOOP2, the value of register C will be

```
          SUB A
          MOV C, A
LOOP 1:   INR A
          DAA
          JC LOOP 2
          INR C
          JNC LOOP 1
LOOP 2:   NOP
```

(a) 63 H
(b) 64 H
(c) 99 H
(d) 100 H

[2008 : 2 M]

9.22 The following is an assembly language program for 8085 microprocessors:

When this program halts, the accumulator contains

Address	Instruction Code	Mnemonic
1000H	3E, 06	MVI A, 06H
1002H	C6, 70	ADI 70H
1004H	32, 07, 10	STA 1007H
1007H	AF	XRA A
1008H	76	HLT

(a) 00H
(b) 06H
(c) 70H
(d) 76H

[2009 : 2 M]

9.23 The subroutine SBX given below is executed by an 8085 processor. The value in the accumulator immediately after the execution of the subroutine will be

```
SBX     MVI A, 99H
        ADI 11H
        MOV C, A
        RET
```

(a) 00H
(b) 11H
(c) 99H
(d) AAH

[2010 : 1 M]

9.24 In an 8085 processor, the main program calls the subroutine SUB1 given below. When the program returns to the main program after executing SUB1, the value in the accumulator is

Address	Opcode	level	Mnemonic
2000	3E, 00	SUB1:	MVI A, 00H
2002	CD, 05, 20		CALL SUB2
2005	3C	SUB2:	INR A
2006	C9		RET

Accumulator for A = 00H increases and finally A = 14H

(a) 00
(b) 01
(c) 02
(d) 03

[2010 : 2 M]

9.25 The number of objects crossing a window sequentially at variable speed is to be counted using an interrupt in the 8085 microprocessor. The objects are sensed by an optical source and a detector with associated signal conditioning circuit. The circuit produces a logic high output as long as the object is in front of the window, and this output is used to interrupt the processor. The duration of an object being in front of the window is in the range of 100 ms to 2 s. The processor takes 1 ms to process the input after an interrupt. The best choice of interrupt for an error free counting is

(a) RST 5.5
(b) RST 6.5
(c) RST 7.5
(d) INTR

[2011 : 2 M]

9.26 A microprocessor accepts external interrupts (Ext INT) through a Programmable Interrupt Controller as shown in the figure.

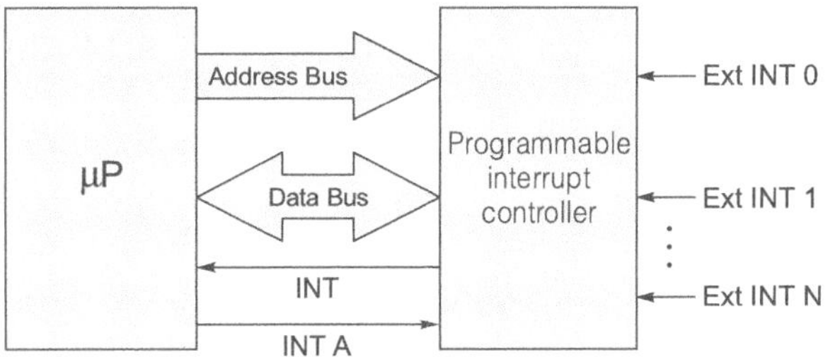

Assuming vectored interrupt, a correct sequence of operations when a single external interrupt (Ext INT1) is received will be

(a) Ext INT1 → INT → Data Read → INT
(b) Ext INT1 → INT → INT → Data Read
(c) Ext INT1 → INT → INTA → Address write
(d) Ext INT1 → INT → Data Read → Address write

[2014 : 2 M]

9.27 An 8-bit microcontroller with 16 address lines has 3 fixed interrupts i.e. Int 1, Int2 and Int3 with corresponding interrupt vector addresses as 0008 H, 0010 H and 0018H. To execute a 32-byte long interrupt service subroutine for Int1 starting at the address ISS1, the location 0008H onwards should ideally contain

(a) a CALL to ISS1
(b) an unconditional JUMP to ISS1
(c) a conditional JUMP to ISS1
(d) Only ISS1

[2017 : 1 M]

9.28 A portion of an assembly language program written for an 8-bit microprocessor is given below along with explanations. The code is intended to introduce a software time delay. The processor is driven by a 5 MHz clock. The time delay (in μs) introduced by the program is ________.

```
MVI B, 64 H    ; Move immediate the given byte
                 into register B. Takes 7 clock
                 periods.
LOOP : DCR B   ; Decrement register B. Affects
                 flags. Takes 4 clock periods.
JNZ LOOP       ; Jump to address with Label
                 LOOP if zero flag is not set.
                 Takes 10 clock periods when
                 jump is performed and 7 clock
                 periods when jump is not
                 performed.
```

[2018 : 2 M]

9.29 In a microprocessor with a 16 bit address bus. the most significant address lines A15 to A12 are used to select a 4096 word memory unit, while lines A0 to A11 are used to address a particular word in the memory unit. If the 3 least significant lines of the address bus A0 to A2 are short-circuited to ground, the addressable number of words in the memory unit is________.

[2019 : 2 M]

9.30 A 16-bit microprocessor has twenty address lines (A_0 to A_{19}) and 16 data lines. The higher eight significant lines of the data bus of the processor are tied to the 8-data lines of a 16 Kbyte memory that can store one byte in each of its 16K address locations. The memory chip should map onto contiguous memory locations and occupy only 16 Kbyte of memory space. Which of the following statement(s) is/are correct with respect to the above design?

(a) The 16 Kbyte memory cannot be mapped with contiguous address locations with a starting address as 0F000H using only A_{19} to A_{14} for generating chip select.
(b) The above chip cannot be interfaced as the width of the data bus of the processor and the memory chip differs.
(c) If the 16 Kbyte of memory chip is mapped with a starting address of 80000H, then the ending address will be 83FFFH.
(d) The active high chip-select needed to map the 16 Kbyte memory with a starting address at F0000H is given by the logic expression $(A_{19} \cdot A_{18} \cdot A_{17} \cdot A_{16})$.

[2021 : 1 M]

■■■

Answers Basics of Microprocessor and Instruction Set

9.1 (c)	9.2 (b)	9.3 (c)	9.4 (c)	9.5 (a)	9.6 (b)	9.7 (b)
9.8 (a)	9.9 (c)	9.10 (d)	9.11 (c)	9.12 (d)	9.13 (a)	9.14 (b)
9.15 (d)	9.16 (a)	9.17 (a)	9.18 (c)	9.19 (d)	9.20 (a)	9.21 (c)
9.22 (a)	9.23 (d)	9.24 (b)	9.25 (d)	9.26 (b)	9.27 (b)	9.28 (280.80)
9.29 (512)	9.30 (a, c)					

Explanations Basics of Microprocessor and Instruction Set

9.1 (c)

∵ Stack pointer handles address of the stack memory.

9.2 (b)

∵ RST_n is a software interrupt if n is an integer.

9.3 (c)

Initially SP ← F000H

PC ← 2400 CALL E000H
2401 ———
2402 ———

So, after executing CALL E000H, the PC has to take the address E000H and after executing CALL instruction it should come back to 2401H and for the mean time it was executing CALL, the address of the next instruction at 2401H was stored in the stack points (SP)

∴ $SP \leftarrow$ 2401H
$PC \leftarrow$ E000H

9.7 (b)

The lower 8-bits of address register carry both data and address bits, they are time multiplexed whereas the upper 8-bit carry only address bits.

9.9 (c)

RST 7.5 → is a vectored interrupt.
HOLD signal is used for direct memory access (DMA). ALE is used to demultiplex lower order address bus and data bus.

$IO/\overline{M}$ signal is used to select I/O or memory.

If $IO/\overline{M} = 1$ then it selects I/O and if $IO/\overline{M} = 0$ it selects memory.

9.10 (d)

Prog 1: MVI A, 00H ← Accumulator contains: A = 00H
STA 8080H ← Store accumulator content at 8080H → 00H
DCR ← Decrement accumulator content by one so A → FFH
STA C080H ← Store accumulator content at C080H → FFH
RET ← Return to the main program

So, after execution of RET instruction then also 8080H will having the content as 00H.

9.11 (c)

Vectored address corresponding to RST 7 is Hexadecimal equivalent of,

$$8 \times 7 = (56)_{10}$$
$$= 0038H$$

9.12 (d)

Prog: XRA A ← A = 00H
MOV L, A ← L = 00H, A = 00
MOV, H, L ← H = 00H, L = 00H
INX H ← HL = 0001H
DAD H ← Add content of HL register pair to HL itself.
HL = 0002H

So, content of HL register pair is 0002H.

9.13 (a)

INTR is best suited for measurement of frequency of square wave.

9.14 (b)

DCR M ← This instruction will reduce the content of memory by one, whose address is pointed by the content of HL register pair. By default the first operation is opcode fetching so, the first operation is opcode fetch, then second the content of memory is read by the microprocessor, so second machine cycle is memory read and again it is written back to the same memory location after decrementing the content of memory by one, so third machine cycle is memory write.

9.15 (d)

$XOR(r_2, r_1) \rightarrow r_2 \leftarrow r_2 \oplus r_1$

$XOR(r_1, r_2) \rightarrow r_1 \leftarrow r_1 \oplus (r_2 \oplus r_1)$

or $r_1 \leftarrow r_2$

$XOR(r_2, r_1) \rightarrow r_2 \leftarrow (r_2 \oplus r_1) \oplus r_2$

or $r_2 \leftarrow r_1$

It is clear that content of r_1 and r_2 are swapped.

9.16 (a)

LXI H, 0000H ← Load HL register pair with value 0000H.

PUSH H ← Store the content of HL register pair in stack memory

PUSH H ← Store the content of HL register pair in stack memory

POP B ← Retrieve the content of BC register pair from stack memory

DAD SP ← Add content of HL register pair with the content of SP register and store in HL.

XCHG ← Exchange content of DE register pair with HL register pair

Memory layout:

	Stack	
200CH	00H	
200DH	00H	
200EH	00H	← SP
200FH	00H	
SP→2010H	X X	

After DAD SP instruction

HL → 0000 H

SP → 200E H

HL → 200E H After addition

After exchange instruction

D | 20 | 0EH | E

H | 12 | 34H | L

9.17 (a)

Since I/O device is memory mapped I/O, therefore memory related instructions will be used for data transfer.

LXI H, 00F0H → load the address of memory in HL pair.

MOV M, A → load the memory by content of A.

9.18 (c)

	Instruction	T-state
	MVI B, 0AH	7T
Loop:	MVI C, 05H	7T
	DCR C	4T
	DCR B	4T
	JNL Loop	10T/7T

Execution time = $(7 + (9 \times 25) + 22)T$

$= 254\,T$

The loop will run for 9 times and at the last when the instruction JNZ loop will be false then microprocessor will come out from the loop and this instruction will take only 7T state instead of 10T state.

9.19 (d)

Address 2020H: The contents of lower and higher address lines are same i.e. 20H. So, this indicates that the technique is I/O mapped I/O.

Data 24H: This indicate that the content of accumulator is 24 H, which have to sent to I/O port.

$IO/\overline{M}$ (Logic High): $IO/\overline{M}$ high indicates that this is input/output operation.

$\overline{RD}$ (Logic High): $\overline{RD}$ logic high indicates that read operation is inactive.

$\overline{WR}$ (Logic Low): $\overline{WR}$ logic low indicates that write operation is active.

So, by observing all the operations the appropriate instruction is OUT20H.

9.20 (a)

e.g. 01H positive number → $(0000\ 0001)_b$.

After RAL

CY	Accumulator
0	0000 0010 or 00000011

JNC is TRUE and again IN instruction will load accumulator with same port address, and microprocessor will go into infinite loop.

Negative number → $(1000\ 0000)_b$ after RAL

CY	Accumulator
1	0000 0001 or 00000000

JNC is false and microprocessor will come out from the loop.

So, this program halts upon reading a negative number.

9.21 (c)

```
SUB A       → A = 00H and
              CY = 0
MOV C, A    → C = 00H and
              A = 00H and
              CY = 0
LOOP 1: INR A → Increment A by 1.
DAA         → decimal adjust
              accumulator if any of
              the bytes are greater
              than 9.
JC LOOP 2   → go to LOOP 2 if carry is
              set.
INR C       → increment C but there is
              no effect on carry flag.
```

The only way to go to LOOP 2 is when carry get set.

And it get set when carry is generated due to DAA instruction.

When A and C become 99H and control goes to LOOP 1 then INRA will make, A = 9AH.

Now DAA will adjust the lower order byte of accumulator since it is greater than 9.

Now,
```
  9 AH
+ 06 H
------
  A 0 H
```

Now again it will adjust the higher order byte by adding 6 to higher data byte, that is

```
   A0H
 + 60H
 -----
  (1)00H
```
Carry generated (the circled 1)

therefore CY = 1 and control goes to LOOP 2.

So, content of C = 99H and content of A = 00H.

9.22 (a)

```
MVI A, 06H  →  A = 06H
ADI 70H     →  A = 06H + 70H = 76H
STA 1007H   →  A=76H, (1007H)→76H
XRA A       →  A = 00H
HLT
```

So, when the above mentioned program will halt then accumulator contains 00H.

9.23 (d)

```
SBX: MVI A, 99H  →  A = 99H
     ADI 11H     →  A = 99H + 11H = AAH
     MOV C, A    →  A = AAH, C = AAH
     RET
```

9.24 (b)

```
SUB1:  MVI A, 00H  →  A = 00H
       CALL SUB2   →  transfer to subroutine
                      SUB2
SUB2:  INR A       →  A = 00H + 01H = 001H
       RET         →  Return to main
                      program
```

So, before returning to main program accumulator contains 01H.

9.25 (d)

The best choice is INTR.

INTR is also suited for measurement of square wave frequency by counting the number of cycles completed in a given interval of time.

So, normally for counting purposes INTR is best suited.

9.27 (b)

As the service routine length is more than 8 bytes. It should not disturb vector locations of other interrupts.

9.28 280.80 (279 to 282)

$$64H = (6 \times 16 + 4)_{10} = (100)_{10}$$

Instruction	Number of times executed	Clock cylces required for one time execution
MVI B, 64H	1	7
DCR B	100	4
JNZ Loop	99 (jump performed) + 1 (jump not performed)	10 (jump performed) and 7(jump not performed)

Total number of clock cycles

$$= 7 + 400 + (99 \times 10 + 1 \times 7)$$
$$= 407 + 990 + 7 = 1404$$

$$T_{clock} = \frac{1}{f_{clock}} = \frac{1}{5\,\text{MHz}} = \frac{1}{5}\,\mu\text{sec}$$

Time delay introduced

$$= \frac{1404}{5}\,\mu\text{sec} = 280.80\,\mu\text{sec}$$

9.29 (512)

Total address line = 16

3 lines grounded

4 lines for chip select

Remaining lines 9

$2^9 = 512$ words.

9.30 (a, c)

The type of microprocessor → 16 bit

Data bus length → 16 bit

Number of address lines→ 20 ($A_{19} - A_0$)

Memory chip connected → 16 kB

Given condition is memory chip of 16 kB should be mapped onto contiguous memory locations.

For a 16 kB memory chip, number of address lines required,

$$2^n = N = 16\text{ kB} = 2^{14}$$
$$n = 14 \text{ i.e. } A_{13} - A_0$$

Remaining 6 lines can be used for chip select i.e. $A_{19} - A_{14}$.

As per option 153, if starting address is 0F000H i.e.

$A_{19}A_{18}A_{17}A_{16}A_{15}A_{14}A_{13}A_{12}$ .... A_0

0 0 0 0 1 1 1 1 0 0 0 0 0 0

(Chip select lines: A_{19} – A_{13})

The ending for starting address of 0F000H is 12FFFH but the chip select lines get modified. So the statement is correct.

If starting addressing 80000H than ending address would be 83FFFH which does not change A_{19} to B_{14} lines i.e. 100000.

For 16 kB to find ending address 3FFF H should be added to starting address.

■■■

Measurements

UNIT

VIII

CONTENTS

1. Digital Voltmeter (DVM) 418
2. Galvanometers, Voltmeters and Ammeters 421
3. Measurement of Resistance and Potentiometer, DC Bridge 431
4. Measurement of Inductance Capacitance, AC Bridges & Q-meter 442
5. Measurement of Energy and Power 452
6. Measurement of Phase and Frequency 458
7. CRO & Electronic Measurements 459
8. Static and Dynamic Characteristics, Error Analysis and Statistical Analysis of Data 466
9. Miscellaneous 476

Measurements

Syllabus

SI units, standards (R, L, C, voltage, current and frequency), systematic and random errors in measurement, expression of uncertainty - accuracy and precision, propagation of errors, linear and weighted regression. Bridges: Wheatstone, Kelvin, Megohm, Maxwell, Anderson, Schering and Wien for measurement of R, L, C and frequency, Q-meter. Measurement of voltage, current and power in single and three phase circuits; ac and dc current probes; true rms meters, voltage and current scaling, instrument transformers, timer/counter, time, phase and frequency measurements, digital voltmeter, digital multimeter; oscilloscope, shielding and grounding.

Analysis of Previous GATE Papers

Exam Year	1 Mark Ques.	2 Marks Ques.	Total Marks
1995	3	1	**5**
1996	6	1	**8**
1997	2	1	**4**
1998	3	1	**5**
1999	5	–	**5**
2000	3	–	**3**
2001	1	4	**9**
2002	4	4	**12**
2003	5	4	**13**
2004	5	9	**23**
2005	3	7	**17**
2006	2	5	**12**
2007	3	4	**11**
2008	2	6	**14**
2009	3	3	**9**

Exam Year	1 Mark Ques.	2 Marks Ques.	Total Marks
2010	4	4	**12**
2011	–	4	**8**
2012	3	1	**5**
2013	–	2	**4**
2014	1	1	**3**
2015	2	3	**8**
2016	4	3	**10**
2017	2	3	**8**
2018	2	3	**8**
2019	1	2	**5**
2020	3	5	**13**
2021	2	5	**12**
2022	2	5	**12**
2023	4	4	**12**
2024	1	3	**7**

1 Digital Voltmeter (DVM)

1.1 The type of A/D converter normally used in a $3\frac{1}{2}$ digit multi meter is
(a) Dual-slope integrating type
(b) Voltage to frequency converter type
(c) Flash (or parallel) type
(d) Successive approximation type
[1995 : 1 M]

1.2 A $3\frac{1}{2}$ digit digital multimeter can display
(a) 999.5 (b) 5.999
(c) 555.5 (d) 1.999
[1999 : 1 M]

1.3 A $3\frac{1}{2}$ digit multi meter has an accuracy specification of 0.5% of reading plus 5 counts. The value of an unknown resistance is read as 50.0 Ω on the 200 Ω scale of the meter, the value of the resistance is
(a) 50.0 ± 0.25 Ω (b) 50.0 ± 0.5 Ω
(c) 50.0 ± 0.75 Ω (d) 50.0 ± 100 Ω
[2002 : 1 M]

1.4 A $3\frac{1}{2}$ digit digital voltmeter has an internal reference of 200 mV. It's full scale count and resolution are
(a) 2000 and 1 mV (b) 4000 and 0.1 mV
(c) 2000 and 0.1 mV (d) 4000 and 1 mV
[2002 : 2 M]

1.5 A $3\frac{1}{2}$ digit multimeter has an accuracy specification of (±0.5% of reading ±5 counts). If the meter reads 2.00 mA on a full scale of 20 mA, the worst-case error in the reading is
(a) 0.5% (b) 2.5%
(c) 3.0% (d) 5.5% [2003 : 1 M]

1.6 A $3\frac{1}{2}$ digit, 200 mV full scale DVM has an accuracy specification of ±0.5% of reading plus 5 counts. When the meter reads 100 mV, the voltage being measured as
(a) any value between 99.5 mV and 100.5 mV
(b) any value between 99.0 mV and 101.0 mV
(c) exactly 99.5 mV
(d) exactly 100 mV [2008 : 2 M]

1.7 A $3\frac{1}{2}$ digit DMM has an accuracy specification of ±1% of full scale (accuracy class 1). A reading of 100.0 mA is obtained on its 200 mA full scale range. The worst case error in the reading in milliampere is ± ______.
[2016 : 1 M]

1.8 A 200 mV full scale dual-slope $3\frac{1}{2}$ digit DMM has a reference voltage of 100 mV and a first integration time of 100 ms. For an input of $[100 + 10\cos(100\pi t)]$ mV, the conversion time (without taking the auto-zero phase time into consideration) in millisecond is______.
[2016 : 2 M]

1.9 A $3\frac{1}{2}$ digit, rectifier type digital meter is set to read in its 2000 V range. A symmetrical square wave of frequency 50 Hz and amplitude ±100 V is measured using the meter. The meter will read ______.
[2021 : 2 M]

1.10 A 1.999 V True RMS 3-1/2 digit multimeter has an accuracy of ±0.1% of reading ±2 digits.
It is used to measure 100 A (RMS) current flowing through a line using a 100:5 ratio, Class-1 current transformer with a burden of 0.1 Ω ±0.5 %.
The worst-case absolute error in the multimeter output is ________ V (rounded off to three decimal places).
[2023 : 2 M]

■■■

Answers Digital Voltmeter (DVM)

1.1 (a) 1.2 (d) 1.3 (c) 1.4 (c) 1.5 (c) 1.6 (b) 1.7 (2)

1.8 (200) 1.9 (111) 1.10 (0.011)

Explanations Digital Voltmeter (DVM)

1.1 (a)

Because of highest accuracy, highest stability and highest noise rejection. Dual-slope integrating type A/D converter is normally used in $3\frac{1}{2}$ digit multimeter.

1.2 (d)

The DMM has $3\frac{1}{2}$ digit display the $\frac{1}{2}$ digit at most significant side of digital readout can display 0 or 1. It can not display any other digit except 0 or 1.

Therefore, only 1.999 can be displayed by $3\frac{1}{2}$ digit DMM.

1.3 (c)

50 Ω will be displayed as: $\underline{0}\ \ \underline{5}\ \ \underline{0}\cdot\underline{0}\ \ \Omega$

error = ±(0.5% of reading + 5 counts)

number of counts from 0 to 1999 are 2000

$$\therefore \quad \text{error} = \pm\left[\left(50\times\frac{0.5}{100}\right)+5\times\frac{200}{2000}\right]\Omega$$

$$= \pm\, 0.75\ \Omega$$

1.4 (c)

Count range for $3\frac{1}{2}$ digit DVM is from 0 to 1999

i.e. 2000 counts

and $\quad \text{Resolution} = \dfrac{200\text{ mV}}{2000} = 0.1\text{ mV}$

1.5 (c)

Error due to (± 0.5%) variation of the measured value

$$= \frac{0.5}{100}\times 2 = 0.01\text{ mA}$$

the meter can read from 0 to 1999 = 2000 counts

$\therefore$ error due to ± 5 counts

$$= \frac{20}{2000}\times 5 = 0.05\text{ mA}$$

% age error in the measured value

$$= \frac{0.01+0.05}{2} = \frac{0.06}{2}\times 100 = 3\%$$

1.6 (b)

Error on reading 100 mV

$$= 100\text{ mV}\times\frac{0.5}{100} = 0.5\text{ mV}$$

also, error due to ±5 counts

$$= \frac{200}{1999}\times 5\text{ mV} = 0.5\text{ mV}$$

Hence,

Total error = 0.5 + 0.5 = 1.0 mV

$\therefore$ Meter will read between (100 – 1) mV and (100 + 1) mV.

i.e. between (99.0 mV and 101 mV).

1.7 (2)

Fundamentally the accuracy specification of a DMM is never specified in terms of its full scale.

It is always specified in terms of its reading.

Neglecting the above and applying the rules of normal error analysis

$$\text{DMM} = 200\text{ mA} \pm 1\%\text{ of FSD}$$

i.e. $\quad \dfrac{\delta A}{A_s}\times 100 = 1$

when $\quad A_s = 200\text{ mA}$

$$\delta A = \frac{200\times 1}{100} = \pm\, 2\text{mA}$$

As δA is always constant,

$$\epsilon_r = \frac{\delta A}{A_s}\times 100 \text{ when}$$

$$A_s = 100\text{ mA}$$

$$\epsilon_r = \frac{2}{100}\times 100 = \pm\, 2\text{ mA}$$

1.8 200 (199 to 201)

In dual slope converter total conversion time

$$T = \text{1st integration period} + \text{2nd integration period}$$

$$= T_1 + T_2 = 100 \text{ msec} + T_2$$

To obtain T_2 we can use $V_{in} T_1 = V_{ref} T_2$

$\Rightarrow$ 100 mV × 100 msec = 100 mV × T_2

$\Rightarrow \quad T_2 = 100$ msec

$\Rightarrow \quad T = 100 + 100 = 200$ msec.

1.9 (111)

As given meter is $3\frac{1}{2}$ digit meter reading will be displayed as 100.0

Input is symmetrical square wave of amplitude 100 V.

Multiplying factor for correction = 1.11

Meter reading = 1.11 × 100 = 111

1.10 (0.011)

Given that $3\frac{1}{2}$ DMM, V_{FSD} = 1.999 Volts

1 = 100 Amps, Burden $\Rightarrow Z_e = 0.1\ \Omega \pm 0.5\%$

Output voltage = $I * Z_e$ = 100 Amp × 0.1 Ω,

Reading = 10 Volt

$$\text{Error} = \pm 0.1\% \text{ reading} + 2 \text{ digits}$$

$$= \pm\left[\frac{0.1}{100} \times 10 \text{ volt} + 2 \text{ digits}\right]$$

$$= \pm[0.01 + 2 \text{ digits}]$$

$$1 \text{ digit} = \frac{V_{FSD}}{2000} = \frac{1.999}{2000} = 0.0009995$$

$$2 \text{ digits} = 2 \times 0.0009995 = 0.00199$$

$$\text{Error} = \pm[0.01 + 0.00199] = \pm 0.0119$$

Output voltage = 10 Volt ± 0.0119

Absolute Error = ±0.011

■■■

2 Galvanometers, Voltmeters and Ammeters

2.1 An electronic voltmeter consisting of a full wave precision rectifier and an average circuit gives correct R.M.S value for square wave inputs. It's reading for a 2 V peak to peak sinusoidal input will be

(a) $\frac{2}{\pi}$ V
(b) $\frac{1}{\pi}$ V
(c) $\frac{2}{\sqrt{2}\pi}$ V
(d) $\frac{\sqrt{2}}{\pi}$ V

[1995 : 1 M]

2.2 The instrument that does not have any restoring torque is
(a) D'Arsonval galvanometer
(b) Flux meter
(c) Ballistic galvanometer
(d) Moving iron instrument

[1996 : 1 M]

2.3 An ammeter with input resistance of 50 Ω gives full scale deflection for 50 μA current. The input resistance of a 0 - 1 mA ammeter obtained by connecting a shunt across the 0 - 50 μA meter will be
(a) 50 Ω
(b) 25 Ω
(c) 5 Ω
(d) 2.5 Ω

[1997 : 1 M]

2.4 For the two square inputs in figure, the PMMC meter will read, maximum when the phase difference between them is _____.

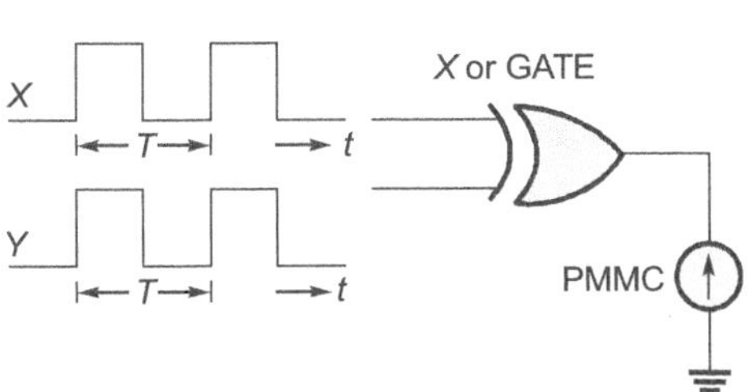

(a) 0
(b) $T/4$
(c) $T/2$
(d) T

[1998 : 1 M]

2.5 A voltmeter connected across the 10 kΩ resistor as shown in figure reads 5 V. The voltmeter is rated at 1000 ohm/volt and has a full scale reading of 10 V. The supply voltage V_S in volt is____.

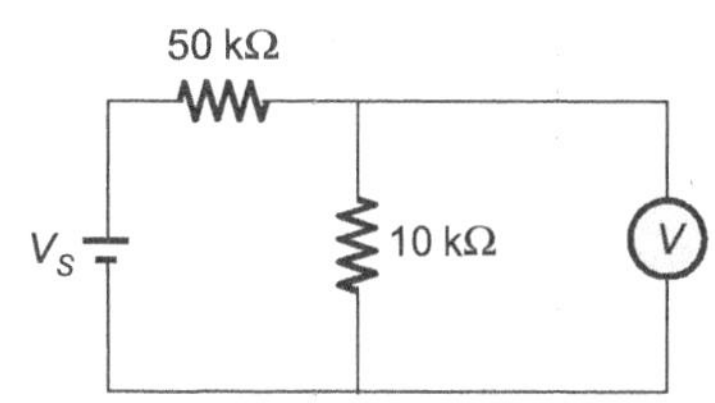

(a) 30
(b) 50
(c) 55
(d) 80

[1999 : 1 M]

2.6 The scale of a full wave rectifier type average sensing meter is calibrated to read the r.m.s value for ideal sine wave signal (form factor 1.11). Determine the multiplying factor of the meter if it is used to measure the r.m.s value of square waves.
(a) 1.11
(b) 1/1.11
(c) 1.0
(d) depend on the frequency of the square wave

[2000 : 1 M]

2.7 In the DC milli voltmeter circuit shown in figure, the input voltage for full scale deflection is ____.

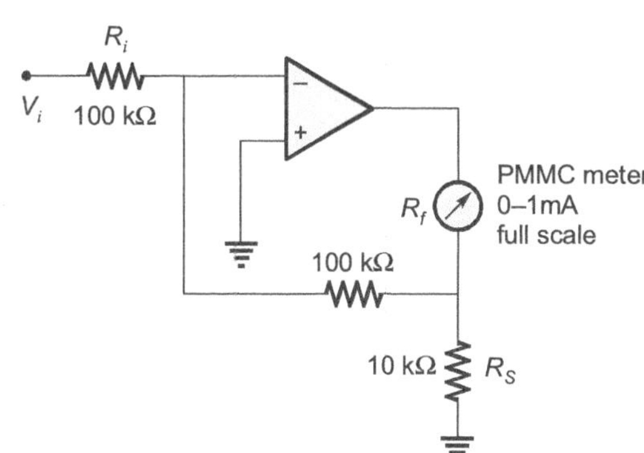

(a) 10 V
(b) 1V
(c) 100 mV
(d) 10 mV

[2000 : 1 M]

2.8 In the context of a PMMC instrument, identify the correct matches:

List-I
(i) a pair of springs
(ii) aluminium former
List-II
(p) to provide controlling torque
(q) to provide damping torque
(r) to act as a base for the coil
(s) to provide current into and out of the moving coil

(a) (i) – (p) and (q), (ii) – (r) and (s)
(b) (i) – (r) and (s), (ii) – (p) and (q)
(c) (i) – (p) and (s), (ii) – (q) and (r)
(d) (ii) – (q) and (r), (ii) – (p) and (s)

[2001 : 2 M]

2.9 A rectifier type PMMC voltmeter has a sensitivity of 20 kΩ/V. A reading of 4.5 V is obtained when measuring a voltage source with an internal resistance on it's 5 V scale. When the scale is changed to 10 V full scale, a reading of 6 V is obtained. The value of the voltage and it's internal resistance R are

(a) 10 V and 100 kΩ (b) 9 V and 100 kΩ
(c) 10 V and 200 kΩ (d) 9 V and 200 kΩ

[2002 : 2 M]

2.10 The insulation level for a 5 A full scale rectifier type ammeter is specified as 2.5 kV. It is then safe to use the meter with a potential difference of upto 2.5 kV between

(a) the terminals of the meter
(b) case and the ground
(c) both the terminals and the case
(d) one of the terminals and the ground

[2003 : 1 M]

2.11 A dc voltage of magnitude 11 kV is to be measured without loading the source under steady condition. The best-suited instrument is

(a) a rectifier type voltmeter
(b) an electrodynamic type voltmeter
(c) a megger
(d) an electrostatic voltmeter

[2004 : 1 M]

2.12 An ideal diode has been connected across a 10 Ω, 100 mA, center-zero PMMC meter as shown in figure. The meter will read

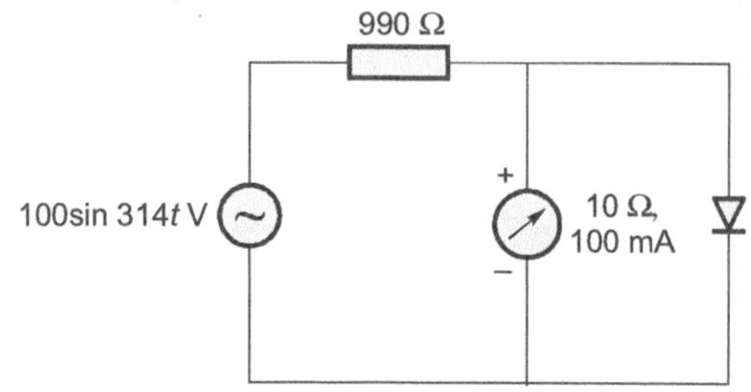

(a) +100 mA (b) +31.8 mA
(c) –31.8 mA (d) –67.3 mA

[2004 : 2 M]

2.13 A voltmeter has a range of [4 V, 20 V] and a resolution of 1 mV. The dynamic range is

(a) 24 dB (b) 60 dB
(c) 72 dB (d) 84 dB

[2006 : 1 M]

2.14 An ac voltmeter uses the circuit shown below, where the PMMC meter has an internal resistance of 100 Ω and requires a dc current of 1 mA for full scale deflection.

Assuming the diodes to be ideal, the value of R_s to obtain full scale deflection with 100 V (ac rms) applied to the input terminal would be

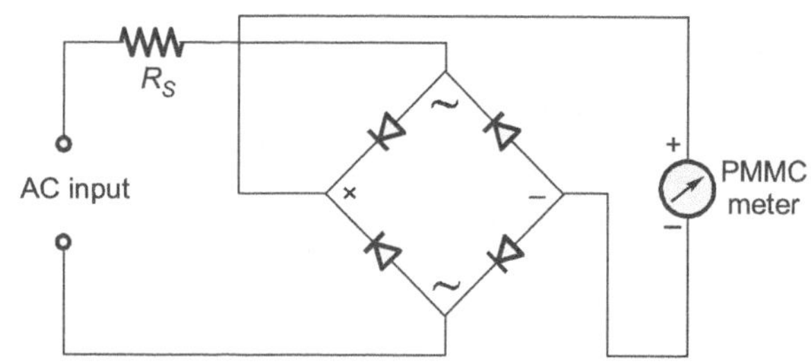

(a) 80 kΩ (b) 89 kΩ
(c) 89.9 kΩ (d) 90 kΩ

[2006 : 2 M]

2.15 If a current

$$\left[-6\sqrt{2}\sin(100\pi t)+6\sqrt{2}\cos\left(300\pi t+\frac{\pi}{4}\right)+6\sqrt{2}\right]\text{Amp.}$$

is passed through a true RMS ammeter, the meter reading will be

(a) $6\sqrt{2}$ A (b) $\sqrt{126}$ A
(c) 12 A (d) $\sqrt{216}$ A

[2008 : 1 M]

2.16 A 2 A full-scale PMMC type dc ammeter has a voltage drop of 100 mV at 2 A. The meter can be converted into a 10 A full-scale dc ammeter by connecting a

(a) 12.5 mΩ resistor in parallel with the meter
(b) 12.5 mΩ resistor in series with the meter
(c) 50.0 mΩ resistor in parallel with the meter
(d) 50.0 mΩ resistor in series with the meter

[2008 : 2 M]

2.17 The op-amp based circuit of a half wave rectifier electronic voltmeter shown below uses a PMMC ammeter with a full scale deflection (FSD) current of 1 mA and a coil resistance of 1 kΩ. The value of *R* that gives FSD for a sinusoidal input voltage of 100 mV (RMS) is

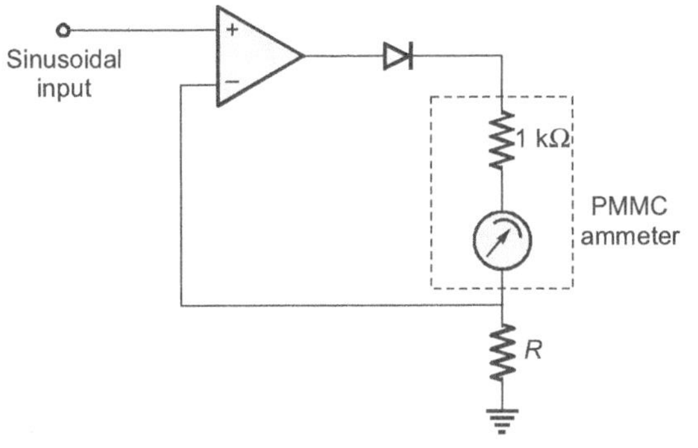

(a) 45 Ω (b) 67.5 Ω
(c) 100 Ω (d) 144.4 Ω

[2008 : 2 M]

2.18 A galvanometer with internal resistance 100 Ω and full-scale current 1 mA is used to realize a dc voltmeter with a full scale range of 1 V. The full scale range of this voltmeter can be extended to 10 V by connecting an external resistance of value
(a) 9 kΩ (b) 9.9 kΩ
(c) 10 kΩ (d) 11 kΩ

[2009 : 1 M]

2.19 The PMMC ammeter A in the adjoining figure has a range of 0 to 3 mA. When switch S_1 is opened, the pointer of the ammeter swings to the 1 mA mark, returns and settles at 0.9 mA. The meter is

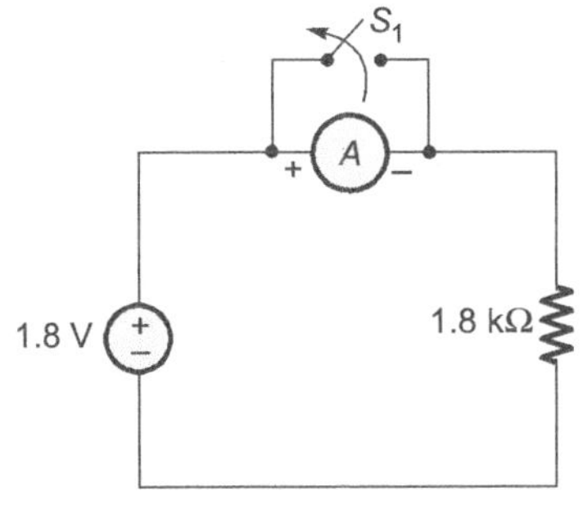

(a) critically damped and has a coil resistance of 100 Ω
(b) critically damped and has a coil resistance of 200 Ω
(c) under damped and has a coil resistance of 100 Ω
(d) under damped and has a coil resistance of 200 Ω

[2010 : 1 M]

2.20 The deflection angle of the pointer of an ideal moving iron ammeter is 20° for 1.0 ampere dc current. If a current of 3 sin (314 *t*) amperes is passed through the ammeter then the deflection angle is
(a) 0° (b) 42°
(c) 60° (d) 90°

[2010 : 1 M]

Common Data Questions (2.21 to 2.22):

A PMMC type ammeter has full scale current of 100 μA and a coil resistance of 100 Ω.

2.21 The resistance required to convert the 100 μA ammeter into a 1 A full scale dc ammeter is
(a) 10 mΩ in series with the meter
(b) 10 mΩ in parallel with the meter
(c) 1 mΩ in series with the meter
(d) 1 mΩ in parallel with the meter

[2010 : 2 M]

2.22 The above PMMC meter is connected in the circuit shown in the adjoining figure. The opamp is ideal. The voltage $v_i(t)$ = 1.0 sin 314 *t* V. Assuming the source impedance of $v_i(t)$ to be zero, the ammeter will indicate a current of

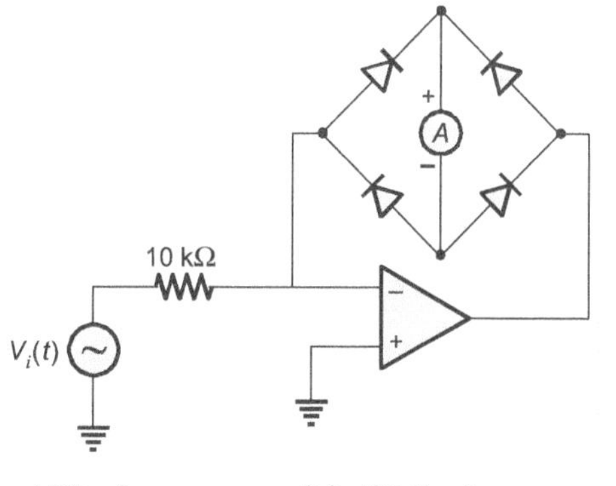

(a) 100 μA (b) 70.7 μA
(c) 63.7 μA (d) 31.8 μA

[2010 : 2 M]

2.23 A periodic voltage waveform observed on an oscilloscope across a load is shown. A permanent magnet moving coil (PMMC) meter connected across the same load reads

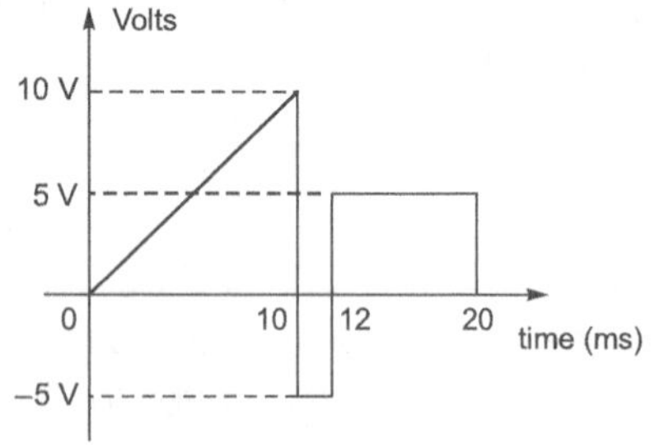

(a) 4 V (b) 5 V
(c) 8 V (d) 10 V

[2012 : 1 M]

2.24 An analog voltmeter uses external multiplier settings. With a multiplier setting of 20 kΩ, it reads 440 V and with a multiplier setting of 80 kΩ, it reads 352 V. For a multiplier setting of 40 kΩ, the voltmeter reads

(a) 371 V (b) 383 A
(c) 394 V (d) 406 V

[2012 : 2 M]

2.25 The circuit below incorporates a permanent magnet coil milli-ammeter of range 1 mA having a series resistance of 10 kΩ. Assuming constant diode forward resistance of 50 Ω, a forward diode drop of 0.7 V and infinite reverse diode resistance for each diode, the reading of the meter in mA is

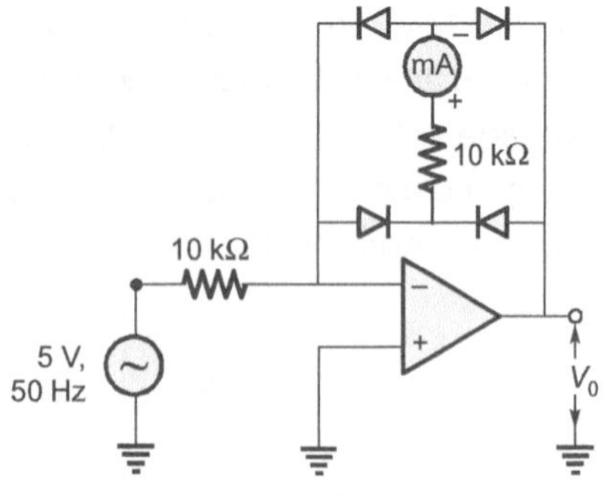

(a) 0.45 (b) 0.5
(c) 0.7 (d) 0.9

[2013 : 2 M]

2.26 Identify the instrument that does not exist:

(a) Dynamometer-type ammeter
(b) Dynamometer-type wattmeter
(c) Moving-iron voltmeter
(d) Moving-iron wattmeter

[2017 : 1 M]

2.27 A current waveform $i(t)$, shown in the figure, is passed through a Permanent Magnet Moving Coil (PMMC) type ammeter. The reading of the ammeter up to two decimal place is

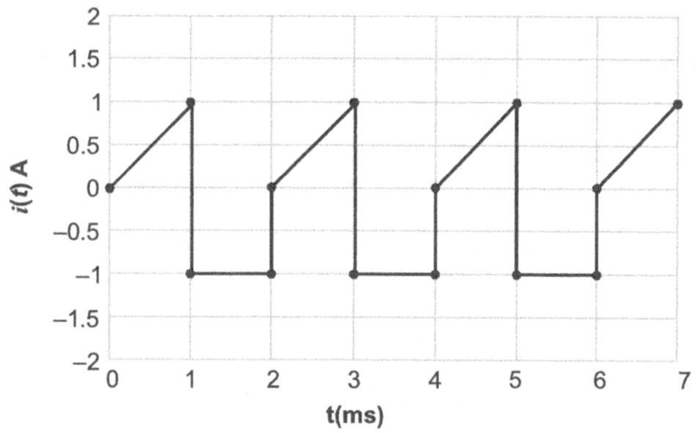

(a) –0.25 A (b) –0.12 A
(c) 0.37 A (d) 0.5 A

[2017 : 1 M]

2.28 When the voltage across a battery is measured using a dc potentiometer, the reading shows 1.08 V. But when the same voltage is measured using a Permanent Magnet Moving Coil (PMMC) voltmeter, the voltmeter reading shows 0.99 V. If the resistance of the voltmeter is 1100 Ω, the internal resistance of the battery, in Ω is ______.

[2017 : 2 M]

2.29 Assuming that the opamp used in the circuit shown is ideal, the reading of the 1 Hz bandwidth, permanent magnet moving coil (PMMC) type voltmeter (in volts) is _____.

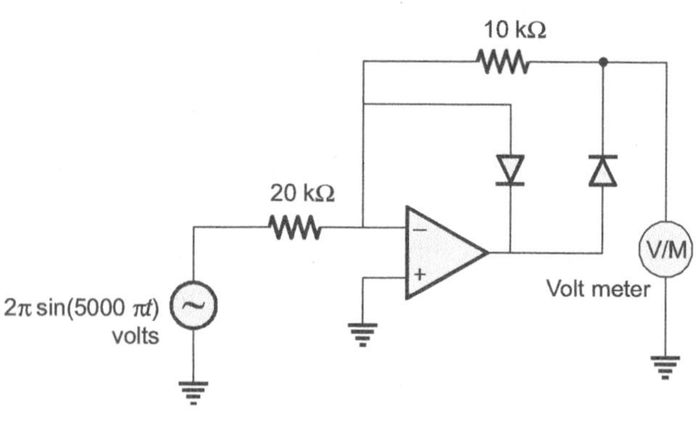

[2020 : 2 M]

2.30 In the diagram shown, the frequency of the sinusoidal source voltage V_S is 50 Hz. The load voltage is 230 V (RMS), and the load impedance is $\frac{230}{\sqrt{2}} + j\frac{230}{\sqrt{2}}\,\Omega$. The value of attenuator

$A_1 = \frac{1}{50\sqrt{2}}$. The multiplier output voltage $V_0 = \frac{V_x V_y}{1\text{ V}}$, where V_x and V_y are the inputs. The magnitude of the average value of the multiplier output V_0 is ________V (rounded off to one decimal place).

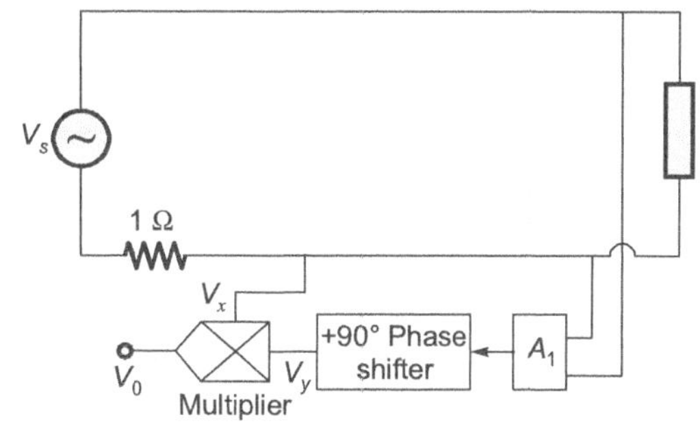

[2023 : 2 M]

Answers Galvanometers, Voltmeters and Ammeters

2.1 (a)	2.2 (b)	2.3 (d)	2.4 (c)	2.5 (c)	2.6 (b)	2.7 (a)
2.8 (c)	2.9 (b)	2.10 (a)	2.11 (d)	2.12 (c)	2.13 (d)	2.14 (c)
2.15 (c)	2.16 (a)	2.17 (a)	2.18 (a)	2.19 (d)	2.20 (d)	2.21 (b)
2.22 (c)	2.23 (a)	2.24 (d)	2.25 (a)	2.26 (d)	2.27 (a)	2.28 (100)
2.29 (1)	2.30 (2.4)					

Explanations Galvanometers, Voltmeters and Ammeters

2.1 (a)

Given : $(V_{pp})_{input} = 2\text{ V}$

$$\Rightarrow \quad (V_p)_{input} = \frac{2}{2} = 1\text{ V}$$

For full wave rectifier

$$(V_{dc})_{ouput} = \frac{2}{\pi}(V_p)_{input}$$

$$= \frac{2}{\pi} \times 1 = \frac{2}{\pi}\text{ V}$$

Now,

$$V_{out} = (\text{Form Factor of Square wave}) \times \frac{2}{\pi}$$

$$= 1 \times \frac{2}{\pi} = \frac{2}{\pi}\text{ V}$$

2.3 (d)

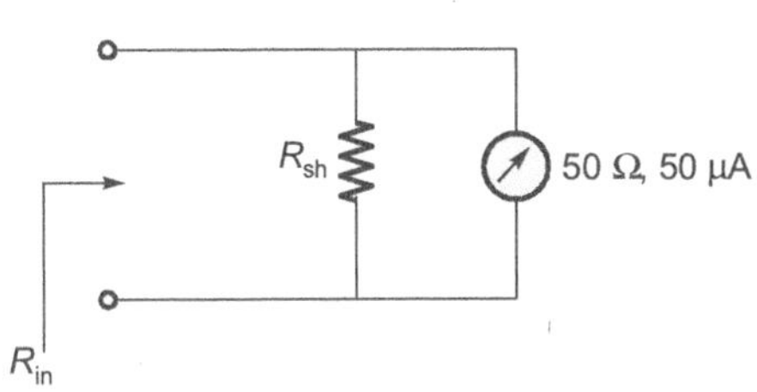

Shunt resistance required so that ammeter can read 0 – 1 mA is,

$$R_{sh} = \frac{R_m}{\frac{I}{I_m} - 1} = \frac{50\ \Omega}{\frac{1\text{ mA}}{50\ \mu\text{A}} - 1} = 2.63\ \Omega$$

Now, input resistance of meter is,

$$R_{in} = R_m || R_{sh} = 50 || 2.63$$

$$= \frac{50 \times 2.63}{50 + 2.63} = 2.5\ \Omega$$

2.4 (c)

PMMC meter will read average value of the output of XOR Gate which will be present only when inputs of the XOR Gate are different or have highest possible phase difference viz. $T/2$.

2.5 (c)

Method fo find internal resistance of voltmeter when sensitivity in ohms/volt is given.

(i) Sensitivity, S_v Ω/volt means voltmeter need S_v Ω of resistance to read 1 volt extra.

(ii) Now if full scale voltage is V_{fs} volts then voltmeter reads V_{fs} volts. So resistance required is $R_v = V_{fs} \times S_v$.

Note:
$R_v = V_{fs} \cdot S_v$ where V_{fs} is FULL SCALE READING.
One must not multiply sensitivity, S_v with measured value to get R_v. i.e., $R_v \neq V_m \cdot S_v$.
Here,

$$R_v = 1000 \times 10 = 10\ \text{k}\Omega$$

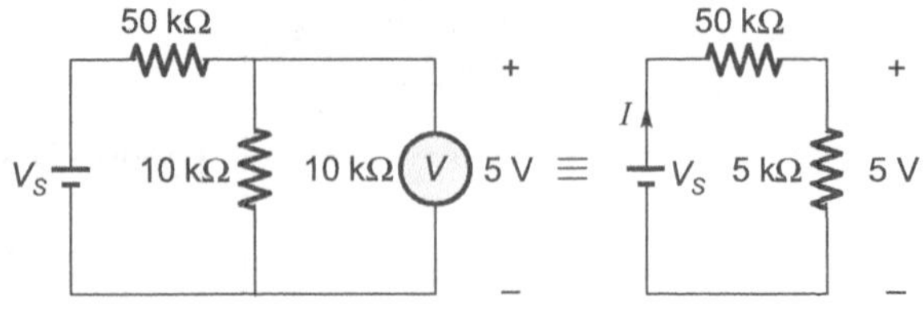

$$I = \frac{5\ \text{V}}{5\ \text{k}\Omega} = 1\ \text{mA}$$

$$\Rightarrow\ V_s = (50 + 5) \times 1 = 55\ \text{V}$$

2.6 (b)

Given: Meter reads rms value viz. = 1.11 $V_{avg.}$ (for sinusoidal wave)

$\because$ For a square wave input, full wave rectifier will give DC output.

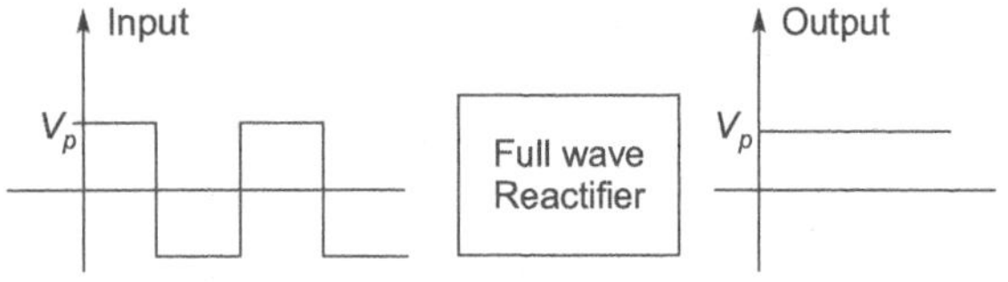

For DC output,

$$V_{avg} = V_p$$

$\therefore$ meter will read 1.11 V_{avg} or 1.11 V_p

But the true value of rectifier output which will be read by this meter = V_p

Hence, correction factor

$$= \frac{\text{True value}}{\text{Measured value}} = \frac{1}{1.11}$$

2.7 (a)

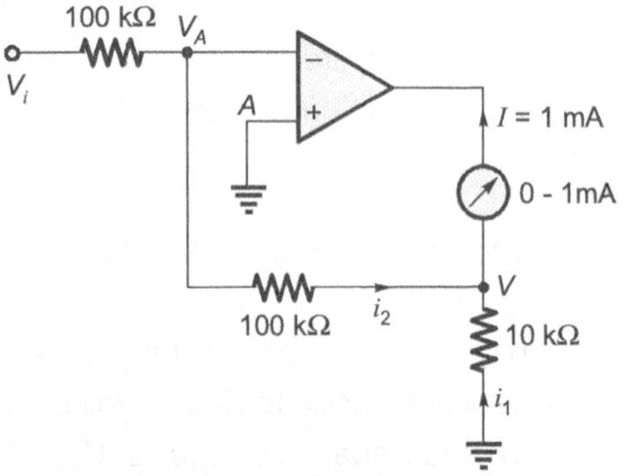

For full scale deflection, $I_m = 1$ mA

Also, using virtual ground concept, $V_A = 0$ V

$$\Rightarrow\ V = 10i_1 = 100i_2$$

Also, $i_1 + i_2 = 1$ mA

$$\Rightarrow\ i_1 = \frac{10}{11}\ \text{mA}$$

$$i_2 = \frac{1}{11}\ \text{mA}$$

Applying KVL in input Loop,

$$V_i - V_A = 100i_1$$

$$\Rightarrow\ V_i = \frac{100}{11}\ \text{V} = 9.09\ \text{V} \approx 10\ \text{V}$$

2.8 (c)

In PMMC, spring is used to

(i) Provide controlling torque.

(ii) Provide current into and out of moving coil and this is the reason why pointer points to zero (null) if spring is snapped (broken).

Aluminium former is used in PMMC to,

(i) Provide damping torque by producing eddy currents.

(ii) Also acts as base of coil.

2.9 (b)

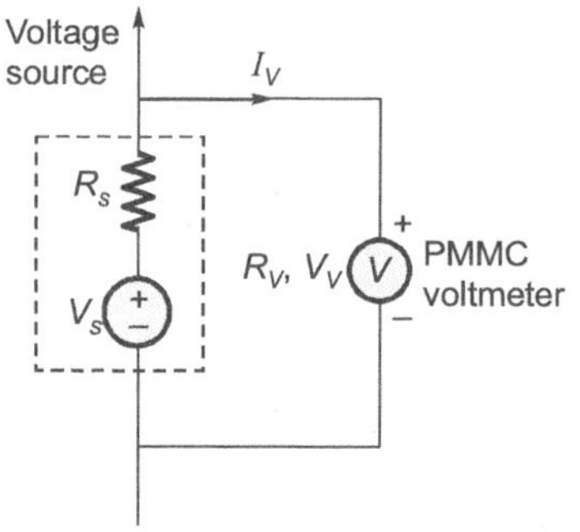

$$V_v = V_s \times \frac{R_v}{R_s + R_v}$$

For 5 V scale:

$$R_v = V \cdot S_v = 5 \times 20 = 100\ \text{k}\Omega$$

using above written formula,

$$4.5 = V_s \times \frac{100}{R_s + 100} \quad \text{...(i)}$$

For 10 V scale:

$$R_v = 10 \times 20 = 200\ \text{k}\Omega$$

using same formula

$$6 = V_s \times \frac{200}{R_s + 200} \quad \text{...(ii)}$$

on solving equations (i) and (ii) simultaneously

we get, $R_s = 100$ kΩ

and $V_s = 9$ V

2.10 (a)

∵ Maximum potential difference between the terminal of the meter equal to 2.5 kV.

2.11 (d)

For measurement of high voltages of the order of K volts, electrostatic voltmeters are used.

2.12 (c)

In +ve half cycle of input diode D gets on and no current flows through PMMC.
In –ve half cycle diode gets off and current flows through PMMC meters.

$$I_{max} = \frac{V_m}{1000} = 100 \text{ mA}$$

So current through PMMC meter will be as

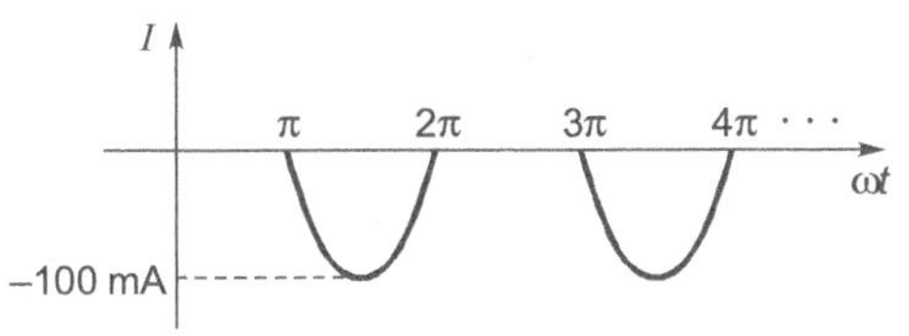

So PMMC meter will read the average value of current

i.e. $$I_{avg} = -\frac{I_{max}}{\pi} = -31.83 \text{ mA}$$

2.13 (d)

$$\text{Dynamic range, } 2^n = \frac{20-4}{1 \text{ mV}} = 16000$$

Dynamic range (dB) = $20\log_{10}16000 = 84$ dB

2.14 (c)

It is a full wave rectifier.
So current output across PMMC meter will be as

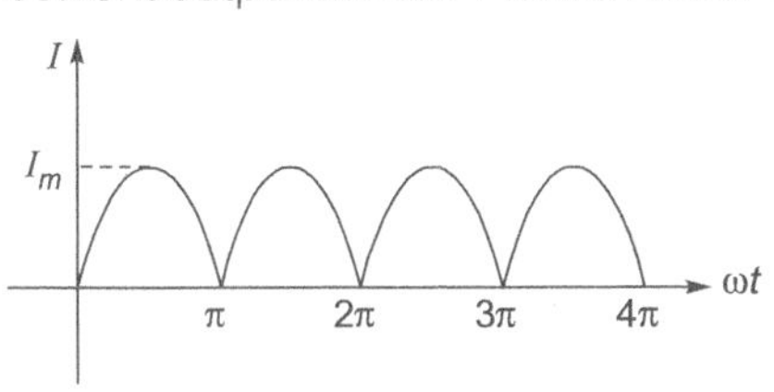

for full wave bridge rectifier,

$$R_s = 0.9 \times [S_{DC}] \times V_{rms}(AC) - 2R_D - R_M$$

where, S_{DC} = DC sensitivity

$$= \frac{1}{I_{DC(FSD)}} = \frac{1}{10^{-3}} = 10^3$$

R_D = Forward diode resistance = 0 Ω
R_M = Internal meter resistance = 100 Ω
$R_S = 0.9 \times 10^3 \times 100 \text{ V} - 100\ \Omega = 89.9 \text{ k}\Omega$

2.15 (c)

Since given current signal has components of different frequencies so,

$$I = \sqrt{I_{dc}^2 + \frac{1}{2}\sum_{k=1}^{n}(I_{mk})^2}$$

$$\Rightarrow \quad I = \sqrt{(6\sqrt{2})^2 + \frac{1}{2}\left[(6\sqrt{2})^2 + (-6\sqrt{2})^2\right]}$$

$$\Rightarrow \quad I = \sqrt{72+36+36} = 12 \text{ A}$$

2.16 (a)

PMMC resistance,

$$R_m = \frac{100 \text{ mV}}{2 \text{ A}} = 50 \text{ m}\Omega$$

Now, Multiplying factor, $m = \frac{10}{2} = 5$

⇒ Shunt resistance required,

$$R_{sh} = \frac{R_m}{m-1} = \frac{50}{5-1} = 12.5 \text{ m}\Omega$$

2.17 (a)

For half wave rectifier
Average voltage,

$$V_B = V_{avg} = \frac{100\sqrt{2}}{\pi} = 45 \text{ mV}$$

$$V_A = V_B$$

(By the concept of virtual ground)

∴ $V_A = 45$ mV

also, it can be observed that $V_C = V_B$

∴ $V_C = 45$ mV

at point 'c',

$$V_C = IR$$

$$45 \times 10^{-3} = 1 \times 10^{-3} \times R$$

$$\therefore \quad R = \frac{45 \times 10^{-3}}{1 \times 10^{-3}} = 45\ \Omega$$

2.18 (a)

When galvanometer measures full scale voltage of 1 V, its resistance is,

$$R = \frac{1 \text{ V}}{I_{fs}} = \frac{1 \text{ V}}{1 \text{ mV}} = 1 \text{ k}\Omega$$

Now, to measure full scale voltage of 10 V, its resistance is,

$$R = \frac{10}{I_{fs}} = \frac{10}{1 \text{ mA}} = 10 \text{ k}\Omega$$

⇒ External resistance required = 10 – 1 = 9 kΩ

2.19 (d)

Current through the PMMC ammeter is noted as 0.9 mA.

At steady state,

$$\text{Current} = \frac{\text{Voltage}}{\text{Total resistance}}$$

$$0.9 \times 10^{-3} = \frac{1.8}{(1.8\times10^3 + R_a)}$$

where, R_a = Ammeter resistance

$$(0.9 \times 10^{-3})(1.8 \times 10^3 + R_a) = 1.8$$

$$0.9 \times 1.8 + 0.9 \times 10^{-3} R_a = 1.8$$

$$0.9 \times 10^{-3} R_a = 1.8 - 0.9 \times 1.8$$

$$= 1.8 \times 0.1$$

$$\therefore \quad R_a = 200\ \Omega$$

Since, the points of the ammeter first swings around 1 mA and then settles at 0.9 mA, hence the ammeter is under-damped.

2.20 (d)

In the moving iron ammeter,

Deflection, $\theta \propto i_{rms}^2$

$$\frac{20}{\theta} = \frac{(1)^2}{\left(\frac{3}{\sqrt{2}}\right)^2} = \frac{2}{9}$$

$$\therefore \quad \theta = 90°$$

2.21 (b)

Voltage in parallel is same

$$\therefore (1 - 100 \times 10^{-6}) R_{sh} = (100 \times 10^{-6} \times 100)$$

$$R_{sh} = \frac{10^{-2}}{0.9999}\ \Omega$$

$$\simeq 1 \times 10^{-2}\ \Omega$$

$$= 10 \times 10^{-3}\ \Omega = 10\ \text{m}\Omega$$

2.22 (c)

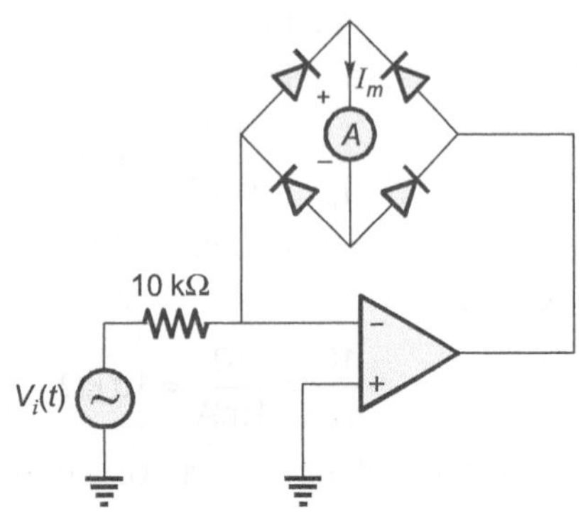

During both positive and negative cycle of supply voltage, current through PMMC is same as shown in figure, so PMMC current I_m is

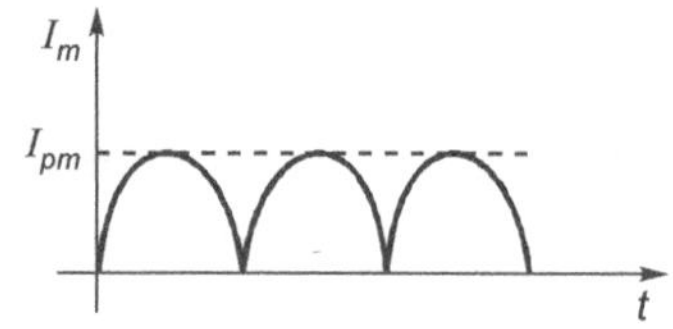

Also PMMC measures DC so,

Average current,

$$I = \frac{2 \times I_{pm}}{\pi} = \frac{2 \times 1}{\pi \times 10\ \text{k}\Omega} = 63.7\ \mu\text{A}$$

2.23 (a)

As PMMC meter reads only DC value or average value and average value is equal to

$$V_{avg} = \frac{\text{Area under the curve}}{\text{Total time}}$$

$$V_{avg} = \frac{\left(\frac{1}{2} \times 10 \times 10\right) + (-5 \times 2) + (5 \times 8)}{20}$$

$$= \frac{80}{20} = 4\ \text{V}$$

2.24 (d)

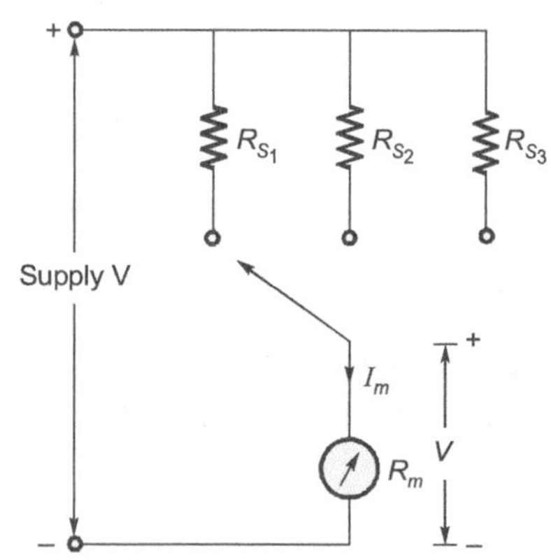

When, $R_{S_1} = 20\ \text{k}\Omega$, $V = 440$ V

$$V = \frac{440}{R_m} \times 20 + 440 \qquad \text{...(i)}$$

When, $R_{S_2} = 80\ \text{k}\Omega$, $V = 352$ V

$$V = \frac{352}{R_m} \times 80 + 352 \qquad \text{...(ii)}$$

Equating equations (i) and (ii),

$$\Rightarrow \frac{352 \times 80 - 440 \times 20}{88} = R_m$$

$R_m = 220\ k\Omega,\ V = 480\ V$

When $R_{S3} = 40\ k\Omega$ and $V = 480\ V$

$$480 = \frac{V}{R_m}\times 40 + V = \frac{13}{11}V$$

$\Rightarrow\ V = 406.15\ V$

2.25 (a)

Current in the inverting arm

$$I_m = \frac{5\sqrt{2}-0}{10^4} = 0.707\ mA$$

PMMC measures the average value

$$= \frac{2I_m}{\pi} = \frac{2\times 0.707}{\pi} = 0.45\ mA$$

2.27 (a)

The PMMC ammeter in this case measures the average value of the current

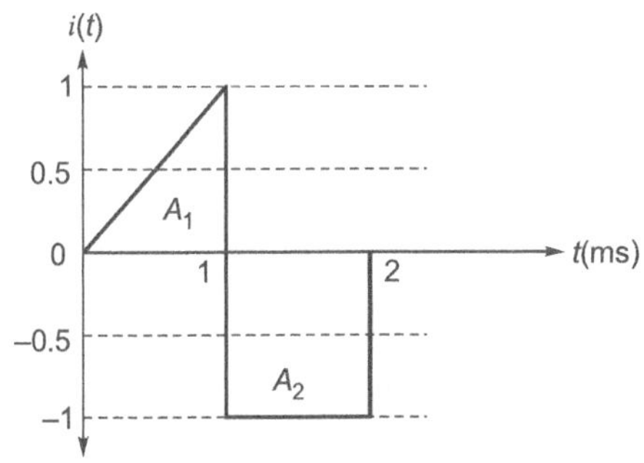

$$\therefore\ I_{average} = \frac{A_1 + A_2}{T}$$

$$= \frac{\frac{1}{2}\times 1\times 1 + (-1\times 1)}{2} = \frac{0.5-1}{2}$$

$= -0.25\ A$

2.28 (100)

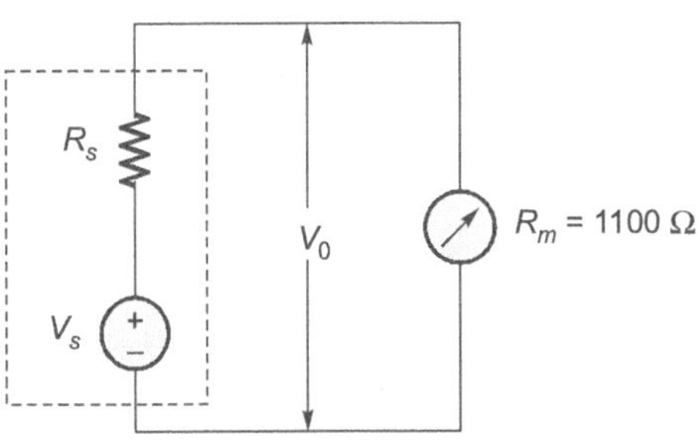

Open circuit voltage,

$V_s = 1.08\ V$

$R_m = 1100\ \Omega$

$$V_0 = V_s\frac{R_m}{R_m + R_s}$$

$$0.99 = (1.08)\frac{(1100)}{(1100)+R_s}$$

$1089 + 0.99\ R_s = 1188$

$$R_s = \frac{1188-1089}{0.99} = 100\ \Omega$$

2.29 (1)

The

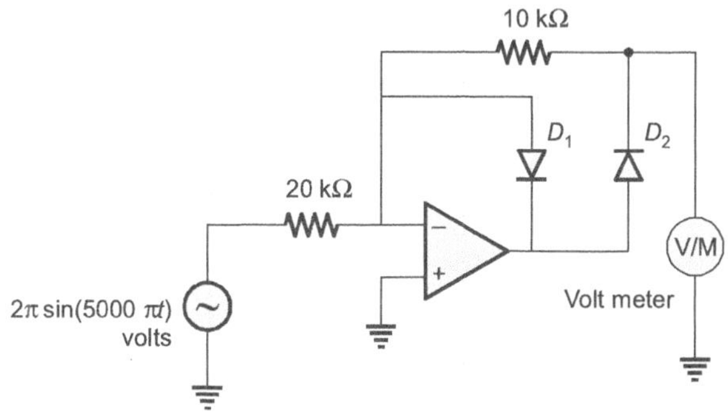

Case-1: For +ve half cycle

V_0 = –ve as opamp is in inverting mode

D_1 is FB and D_2 is RB.

From the concept of virtual ground

$V_0 = 0$

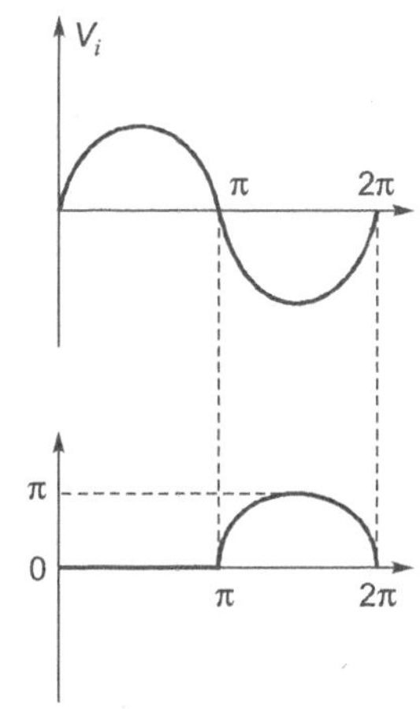

Case-2 : For –ve half cycle

V_0 = +ve

D_1 is reverse biased and D_2 is forward biased.

$$\therefore\ V_0 = -(-2\pi)\times\frac{10}{20} = +\pi$$

Here V_0 is the peak voltage

From the waveform

As a PMMC reads the average value

$$\therefore V_{avg} = \frac{V_P}{\pi} = \frac{\pi}{\pi} = 1\ V$$

2.30 (2.4)

Total Impedance = 1 Ω + Z_L

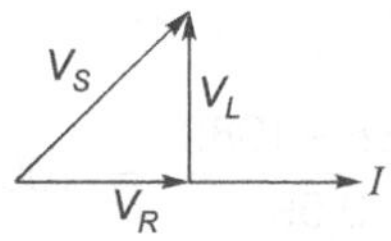

$$Z = 1 + \frac{230}{\sqrt{2}} + j\frac{230}{\sqrt{2}}$$

$$Z = \frac{1}{\sqrt{2}}\left[\sqrt{2} + 230 + j230\right]$$

$$= \frac{1}{\sqrt{2}}[231.414 + j230]$$

$$= \frac{326.2705}{\sqrt{2}}\angle 44.82^\circ$$

Given; $Z_L = 230\left[\frac{1}{\sqrt{2}} + j\frac{1}{\sqrt{2}}\right]$

So, Current in circuit,

$$I = \frac{V_S}{Z} = \frac{V_s}{\frac{326.2705}{\sqrt{2}}}\angle 44.82^\circ$$

$$= \frac{\sqrt{2} \cdot V_S}{326.2705}\angle -44.82^\circ$$

Given; V_L(rms) = 230 and let angle is θ.

So, $V_L \Rightarrow |V_L| \angle\theta_{V_L} = I \cdot Z_L$

$$= \frac{\sqrt{2} \cdot V_S}{326.2705}\angle -44.85\left[\frac{230}{\sqrt{2}}(1+j)\right]$$

$$= \frac{V_S \cdot 230 \cdot \sqrt{2}}{326.2705}\angle -44.82 + 45^\circ$$

$$|V_L| \cdot \angle\theta_{V_L} = 0.9969 \cdot V_S\angle +0.18^\circ$$

$$230 = 0.9969\ V_S \Rightarrow V_S = 230.708 \text{ V(rms)}$$

$$\theta_{V_L} = +0.18^\circ$$

$$V_X = \text{I.R} = \frac{\sqrt{2} \cdot (230.708)}{326.2705} \times 1\angle -44.82^\circ$$

$$= 0.9999\angle -44.82^\circ$$

$$V_y = [\angle +90^\circ] \times A_1 \cdot \bar{V}_L$$

$$= \frac{1}{50\sqrt{2}} 230\angle +0.18^\circ + 90^\circ$$

$$= \frac{230}{50\sqrt{2}}\angle 90.18^\circ$$

$$V_x = 0.9999\angle -44.82^\circ$$

$$V_0 = \frac{V_x V_y}{1V}$$

$$= 0.9999 \times \frac{230}{50\sqrt{2}}\angle -44.82^\circ + 90.18^\circ$$

$$V_0 = 3.252\angle 45.36^\circ \text{ V [RMS]}$$

$$V_0\ (\text{avg.}) = \frac{2\sqrt{2}}{\pi} V(\text{rms})$$

$$= 2 \times 1.4639 \text{ V (avg.)}$$

$$= 2.9278 \text{ V (Average)}$$

■■■

3 Measurement of Resistance and Potentiometer, DC Bridge

3.1 In the potentiometer circuit shown in figure balance is obtained. The unknown E_x is then

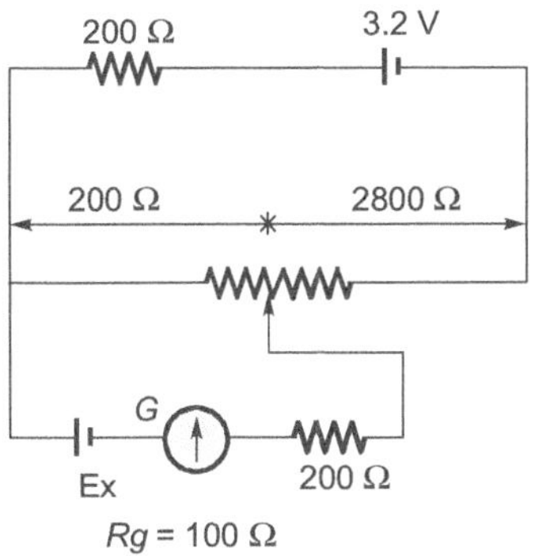

(a) 200 mV (b) 2.8 V
(c) 3.0 V (d) 400 mV

[1996 : 1 M]

3.2 For the circuit given in figure the unbalance voltage e_0 is

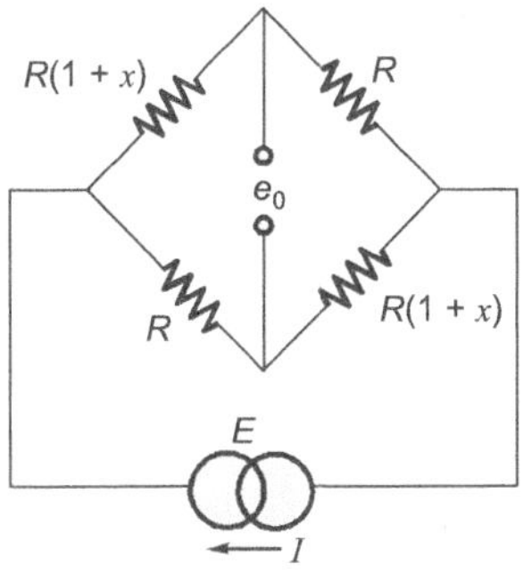

(a) $e_0 = IRx$ (b) $e_0 = \dfrac{IRx}{2}$
(c) $e_0 = 2IRx$ (d) $e_0 = IR(1 + x)$

[1998 : 1 M]

3.3 The bridge circuit in figure is balanced. The magnitude of current I is

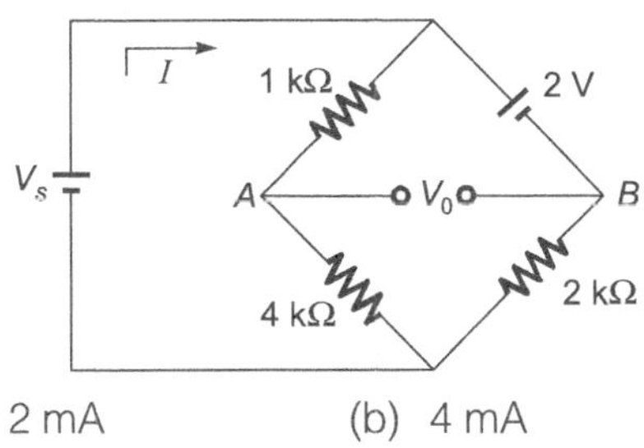

(a) 2 mA (b) 4 mA
(c) 5 mA (d) 6 mA

[1999 : 1 M]

3.4 A resistance potentiometer has a total resistance of 10000 Ω and is rated 4 W. If the range of potentiometer is 0 to 100 mm, then its sensitivity in V/mm

(a) 1.0 (b) 2.0
(c) 2.5 (d) 25

[1999 : 1 M]

3.5 In the context of low resistances, identify the INCORRECT statement

(a) standard low resistances have four terminal construction.
(b) Kelvin bridge is used for measuring low resistance, precision depends on the detector sensitivity.
(c) A pair of ratio arms in Kelvin bridge for measuring the low resistance eliminates the error due to thermo emf.
(d) low resistances used for ammeter shunts are usually made with a suitable number of plates of large area and all the plates connected in parallel.

[2001 : 2 M]

3.6 Kelvin double bridge is best suited for the measurement of

(a) Resistances of very low value
(b) Low value capacitance
(c) Resistance of very high value
(d) High value capacitance

[2002 : 1 M]

3.7 A potentiometer circuit is shown in figure. The jockey of the potentiometer is kept exactly at the mid-point and the ammeter reads the current, I, as 0.2 mA. When V_x is reversed the ammeter reads 3.8 mA. The internal resistance of the ammeter and the unknown potential V_x are respectively

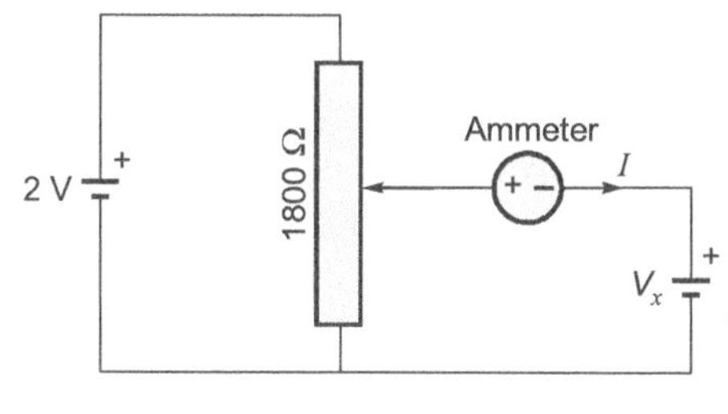

(a) 500 Ω, 1.1 V (b) 50 Ω, 1.1 V
(c) 500 Ω, 0.9 V (d) 50 Ω, 0.9 V

[2003 : 2 M]

3.8 The wiper of the 20 kΩ potentiometer in figure is positioned half way. Then the voltage V_0 of the circuit is

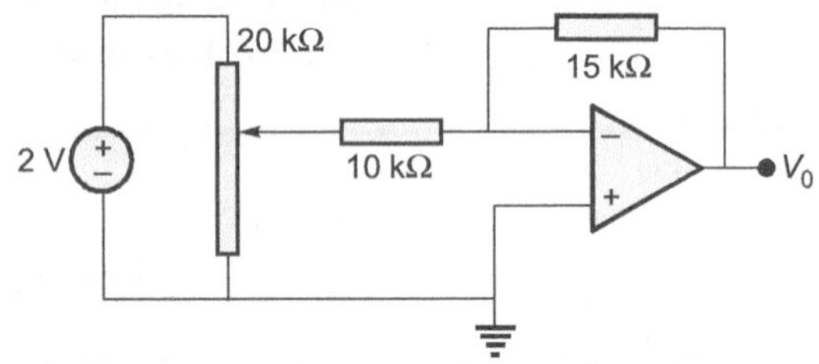

(a) –1.5 V (b) –1.0 V
(c) –0.75 V (d) +2.5 V

[2004 : 2 M]

3.9 A wire potentiometer of length 11 m and resistance 1 Ω/m balances a standard cell voltage of 1.018 V at a length of 10 m 18 cm. If the voltage of the battery supplying the current through the potentiometer is 2.0 V, then the value of the series resistance connected to the potentiometer is

(a) 9 Ω (b) 90 Ω
(c) 900 Ω (d) 990 Ω

[2004 : 2 M]

3.10 A 4 kΩ, 0.02 W potentiometer is used in the circuit shown below. The minimum value of the resistance R_s in order to protect the potentiometer is

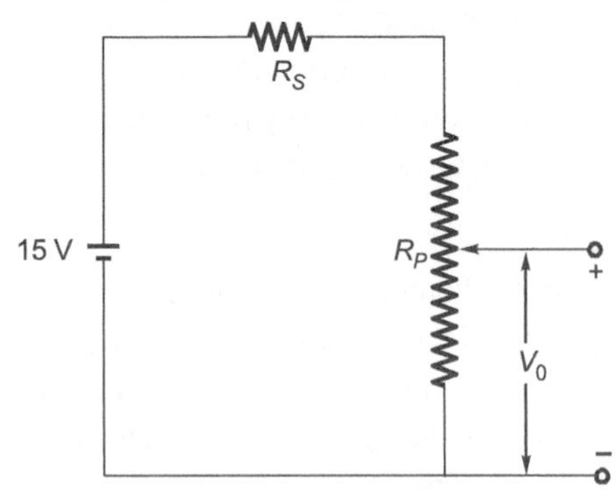

(a) 2.23 kΩ (b) 2.71 kΩ
(c) 3.82 kΩ (d) 8.92 kΩ

[2006 : 1 M]

3.11 An unknown voltage source E_u (with negligible internal resistance) is connected to a potentiometer circuit as shown in the following figure. If the galvanometer current is 10 μA with the direction as indicated, then value of E_u is

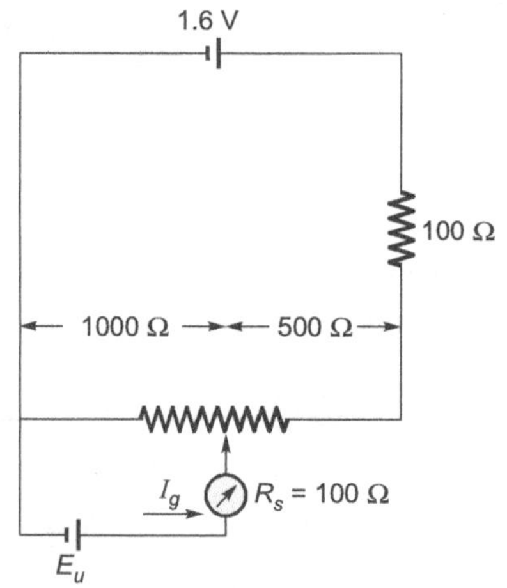

(a) 0.9910 V (b) 0.9991 V
(c) 1.0012 V (d) 1.0048 V

[2006 : 2 M]

3.12 The voltmeter shown in the following figure has a sensitivity of 500 Ω/V and a full scale of 100 V. When connected in the circuit as shown, then meter reads 20 V.

Then R_x is

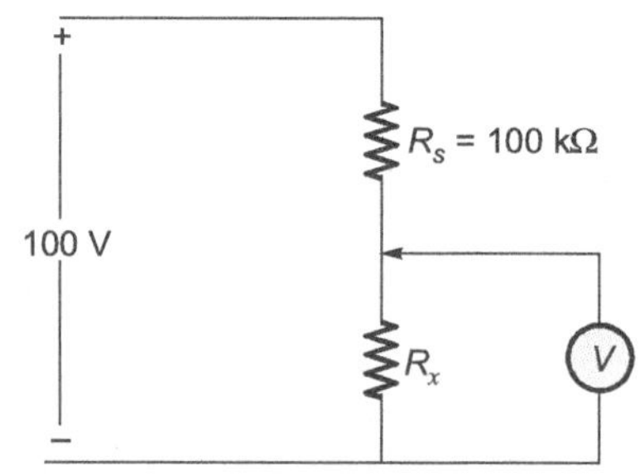

(a) 75 kΩ (b) 50 kΩ
(c) 25 kΩ (d) 10 kΩ

[2006 : 2 M]

3.13 The figure shows a potentiometer of total resistance R_T with a sliding contact.

The resistance between the points P and Q of the potentiometer at the position of the contact shown is R_C and the voltage ratio V_0/V_s at this point is 0.5. If the ratio $R_L/R_T = 1$, the ratio R_C/R_T is

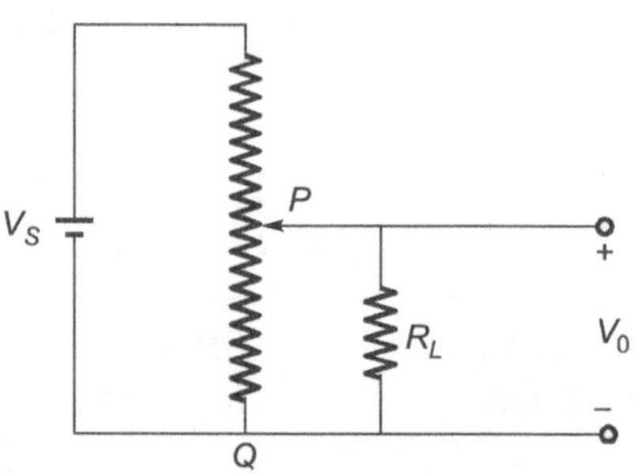

(a) $\frac{-1+\sqrt{5}}{2}$ (b) $\frac{1+\sqrt{5}}{2}$
(c) $-1+\sqrt{5}$ (d) $1+\sqrt{5}$

[2007 : 2 M]

Statement for Linked Answer Questions (3.14 to 3.15):

In the Wheatstone bridge shown below the galvanometer G has a current sensitivity of 1 μA/mm, a resistance of 2.5 kΩ and a scale resolution of 1 mm. Let ΔR be the minimum increase in R from its nominal value of 2 kΩ that can be detected by this bridge.

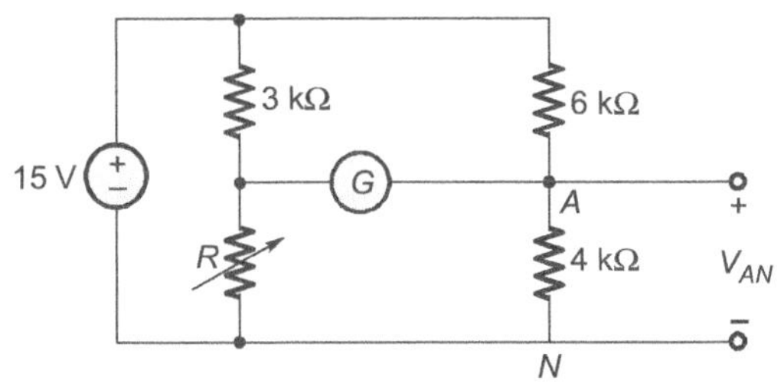

3.14 When R is 2 kΩ + ΔR, V_{AN} is
(a) 6 V (b) 6.0024 V
(c) 6.0038 V (d) 6.005 V

[2008 : 2 M]

3.15 The value of ΔR is approximately
(a) 2.8 Ω (b) 3.4 Ω
(c) 5.2 Ω (d) 12 Ω

[2008 : 2 M]

3.16 The dc potentiometer shown in the figure has a working current of 10 mA with switch S open. Let $R_g + R_1$ = 100 Ω. The galvanometer G can only detect currents greater than 10 μA. The maximum percentage error in the measurement of the unknown emf E_x as calculated from the slider position shown is closest to

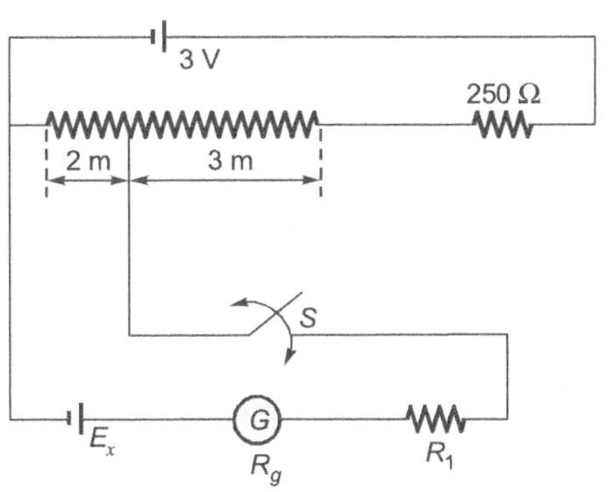

(a) 0.3 (b) 0.5
(c) 0.6 (d) 1.0

[2009 : 2 M]

3.17 In the Wheatstone bridge shown below, when the resistance R_1 increases by 1 Ω, the current through the galvanometer is

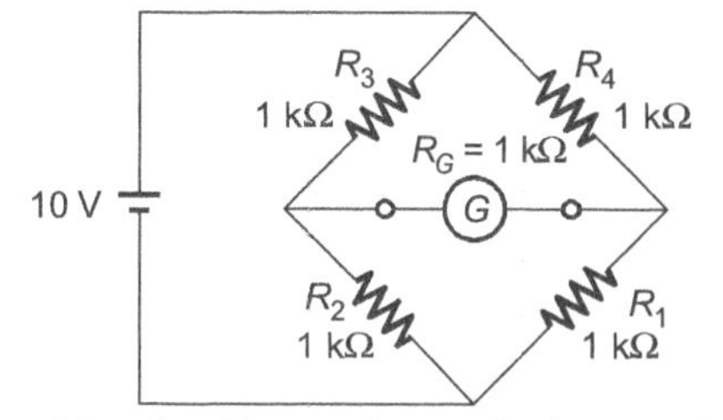

(consider the Thevenin equivalent resistance of the bridge in the calculations)
(a) 1.25 μA (b) 2.5 μA
(c) 12.5 μA (d) 25 μA

[2011 : 2 M]

3.18 The bridge most suited for measurement of a four-terminal resistance in range of 0.001 Ω to 0.1 Ω is
(a) Wien's bridge (b) Kelvin double bridge
(c) Maxwell's bridge (d) Schering bridge

[2015 : 1 M]

3.19 In the potentiometer circuit shown in the figure, the expression for V_x is

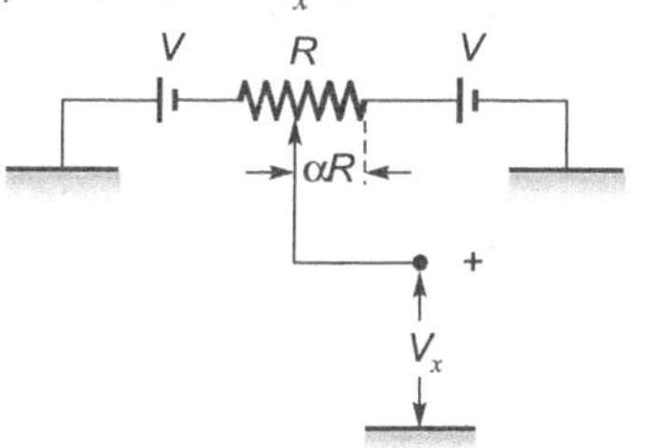

(a) $(1-2\alpha)V$ (b) $(1-\alpha)V$
(c) $(\alpha-1)V$ (d) αV

[2015 : 2 M]

3.20 If the deflection of the galvanometer in the bridge circuit shown in the figure is zero, then the value of R_x in ohms is ____ Ω.

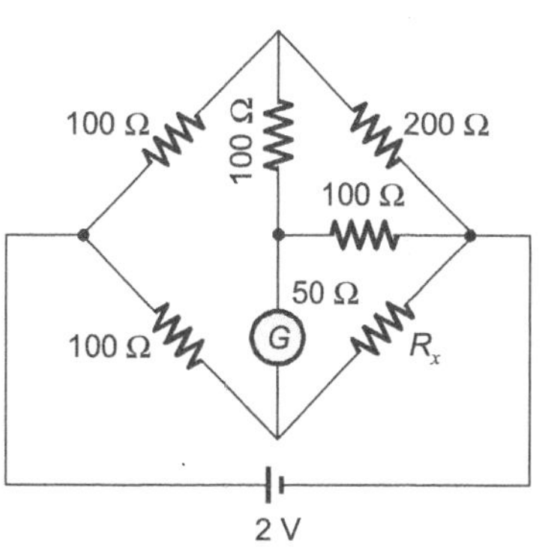

[2015 : 2 M]

3.21 A DC potentiometer, shown in figure below, is made by connecting fifteen 10 Ω resistors and a 10 Ω slide wire of length 1000 mm in series. The potentiometer is standardized with the current I_p = 10.0000 mA. Balance for an unknown voltage is obtained when the dial is in position 11 (11 numbers of the fixed 10 Ω resistor are included) and the slide wire is on the 234th mm position. The unknown voltage (up to four decimal places) in volt is _______.

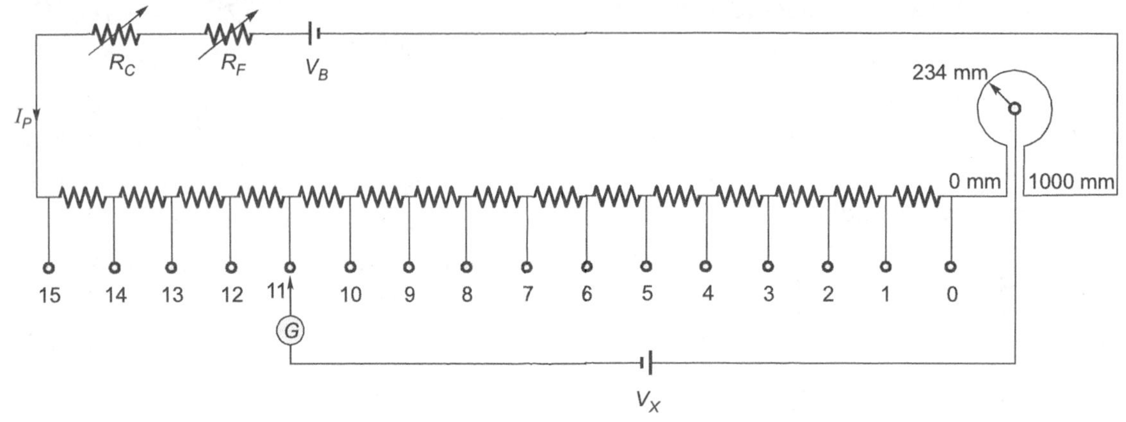

[2016 : 1 M]

3.22 A resistance-meter has five measurement range-settings between 200 Ω and 2 MΩ in multiples of 10. The meter measures resistance of a device by measuring a full-range voltage of 2 V across the device by passing an appropriate constant current for each range-setting. If a device having a resistance value in the range 8 kΩ to 12 kΩ and a maximum power rating of 100 μW is to be measured safely with this meter, the choice for range-setting on the meter for best resolution in measurement, in kΩ, is

(a) 2 (b) 20
(c) 200 (d) 2000

[2019 : 2 M]

3.23 In the bridge circuit shown, the voltmeter V showed zero when the value of the resistors are: R_1 = 100 Ω. R_2 = 110 Ω. and R_3 = 90 Ω. If $(R_1/R_2) = (R_A/R_B)$, the value of R_4 in ohm is _____.

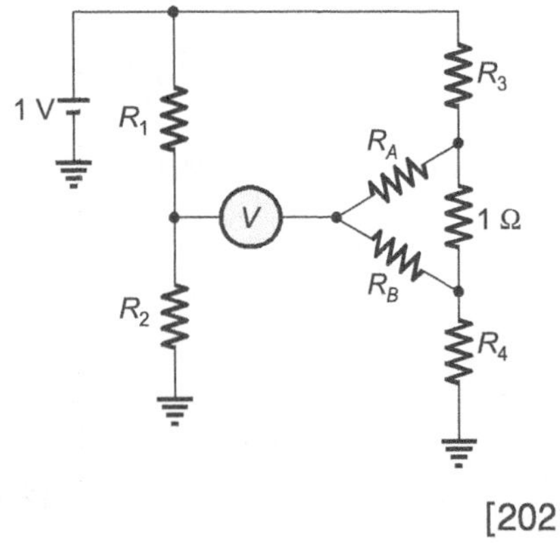

[2021 : 2 M]

3.24 In the Wheatstone bridge circuit shown, R_1 = 1.5 kΩ and $R_2 = R_3 = R_4$ = 1 kΩ. The switch is initially open and the voltage between the points C and D is V_{CD}. Upon closing the switch at t = 0, the resistance in the arm AD changes by an amount δR_1, and the voltage between C and D changes by δV_{CD}. The sensitivity of the bridge (in volt/kiloohm), defined as $\left|\frac{\delta V_{CD}}{\delta R_1}\right|$, is _______ (round off to two decimal places)

[2022 : 2 M]

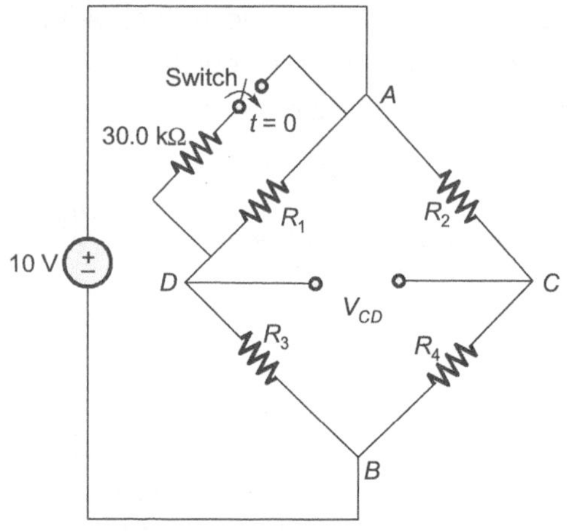

3.25 When the bridge given below is balanced, the current through the resistor R_a is ________ mA (rounded off to two decimal places).

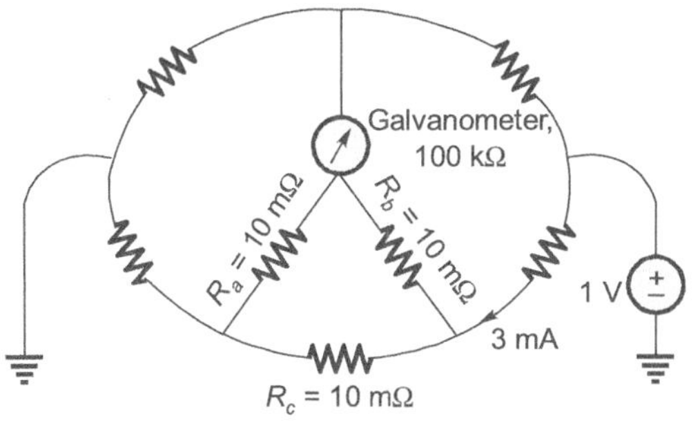

[2023 : 1 M]

3.26 A wire-wound 'resistive potentiometer type' angle sensor with 72 turns is used in an application. The first turn of the potentiometer is connected to ground while its last turn is connected to 3.6 V. The width of the wiper covers two turns ensuring make-before-break. The output (wiper) voltage when the wiper is on top of both the turns 35 and 36 is ______ V (rounded off to two decimal places).

[2023 : 2 M]

3.27 Consider the figure shown. For zero deflection in the galvanometer, the required value of resistor R_x is ____ Ω (rounded off to nearest integer).

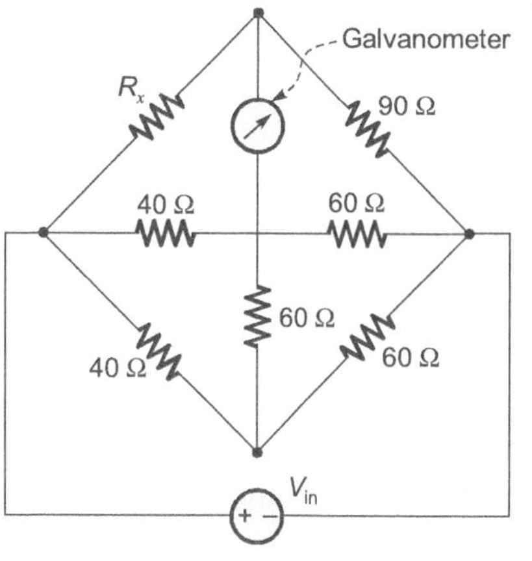

[2024 : 2 M]

Answers Measurement of Resistance and Potentiometer, DC Bridge

3.1 (a)	3.2 (b)	3.3 (d)	3.4 (b)	3.5 (c)	3.6 (a)	3.7 (d)
3.8 (b)	3.9 (a)	3.10 (b)	3.11 (d)	3.12 (b)	3.13 (a)	3.14 (b)
3.15 (b)	3.16 (b)	3.17 (a)	3.18 (b)	3.19 (a)	3.20 (33.33)	3.21 (1.1234)
3.22 (c)	3.23 (99)	3.24 (1.65)	3.25 (1)	3.26 (1.78)	3.27 (60)	

Explanations Measurement of Resistance and Potentiometer, DC Bridge

3.1 (a)

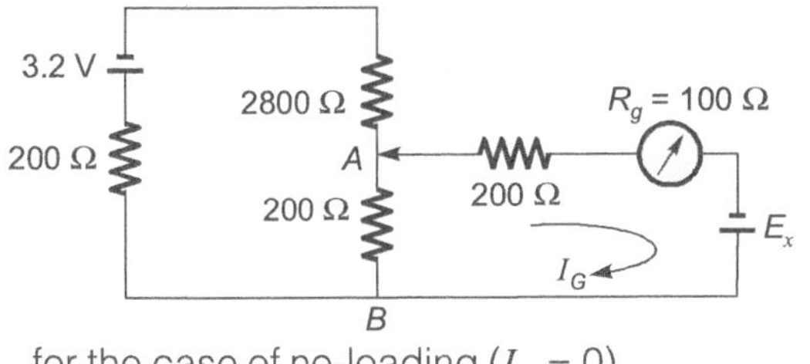

for the case of no-loading ($I_G = 0$)

$\therefore \; V_{AB} = E_x$

$$V_{AB} = \frac{200}{200+2800+200} \times 3.2$$
$$= 0.2 \text{ V}$$
$$= 200 \text{ mV}$$

$\therefore \quad \boxed{E_x = 200 \text{ mV}}$

3.2 (b)

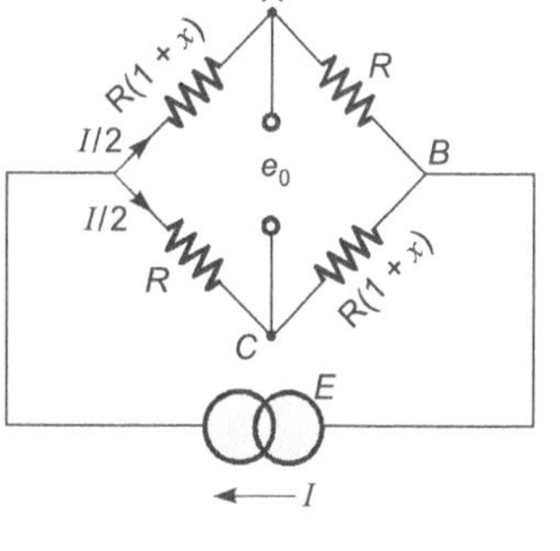

$$e_0 = V_{CB} - V_{AB}$$
$$= R(1+x)\frac{I}{2} - R\frac{I}{2} = \frac{IRx}{2}$$

3.3 (d)

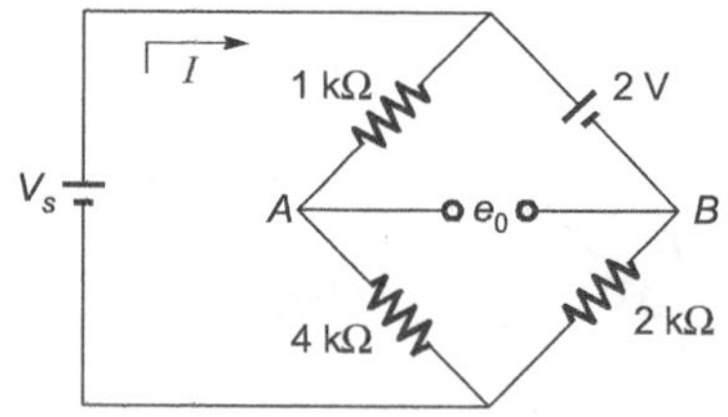

In case of balanced bridge

$$V_A = V_B$$

$$V_A = \frac{4}{4+1}V_s = \frac{4}{5}V_s$$

also, $V_B = V_s - 2$

$$\therefore \quad \frac{4}{5}V_s = V_s - 2$$

$$4\,V_s = 5\,V_s - 10$$

$$V_s = 10\text{ V}$$

$$I = \frac{V_S}{5k} + \frac{V_S - 2}{2k} = \left(\frac{10}{5} + \frac{10-2}{2}\right)\text{mA}$$

$$= (2 + 4)\text{mA} = 6\text{ mA}$$

Method 2:

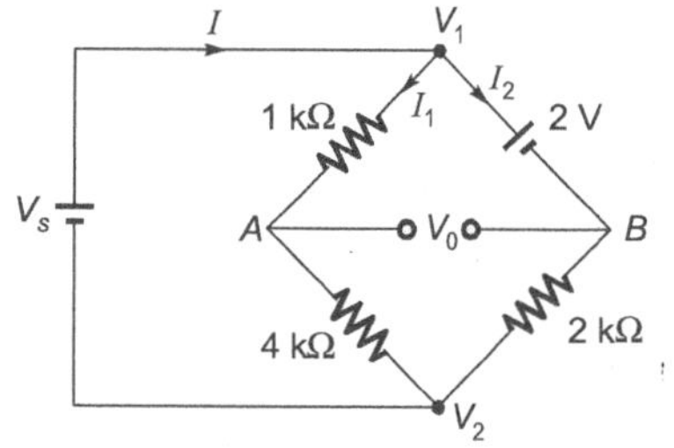

Since bridge is balanced,

$$V_A = V_B$$

$\Rightarrow \quad 2 = I_1 \times 1\text{ k}\Omega$

$\Rightarrow \quad I_1 = 2\text{ mA} \quad$ ...(i)

Also, $\quad 4I_1 = 2I_2$

$\Rightarrow \quad 4 \times 2 = 2I_2$

$\Rightarrow \quad I_2 = 4\text{ mA} \quad$ ...(ii)

from (i) and (ii),

$$I = I_1 + I_2 = 2 + 4 = 6\text{ mA}$$

3.4 (b)

$$R = 10000\,\Omega$$

$$P = 4\text{ W}$$

$$\frac{V^2}{R} = 4$$

$$\frac{V^2}{10000} = 4, \quad V = 200\text{ V}$$

Range 0 to 100 mm

$$\therefore \quad \text{Sensitivity} = \frac{200}{100} = 2\text{ V/mm}$$

3.5 (c)

(i) Low resistance has four terminals to remove effect of lead resistance.

(ii) Kelvin double bridge is used to measure low resistance whose precision depends on the detector sensitivity.

(iii) Pair of ratio arms in Kelvin bridge for measuring the low resistance eliminates error due to lead resistance.

Note:

Low resistance → Four terminal to remove effect to lead resistance

Medium resistance → Two terminals

High resistance → Three terminals to remove insulation resistance.

3.7 (d)

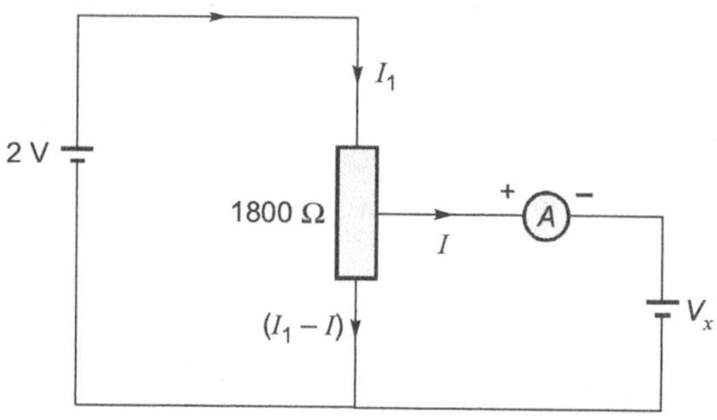

Writing KVL in outer loop,

$2 - I_1 \times 900 - IR - V_X = 0 \quad$ ...(i)

(Here R is resistance of ammeter)

In inner loop,

$2 - 900\,I_1 - 900(I_1 - I) = 0$

$1 + 450\,I = 900\,I_1 \quad$ ...(ii)

from (i) and (ii) we have

$2 - (1 + 450\,I) - IR - V_X = 0$

$1 - 450\,I - IR - V_X = 0 \quad$ ...(iii)

I is equal = 0.2 mA

So from equation (iii),

$1 - 450 \times 0.2 \times 10^{-3} - 0.2 \times 10^{-3}\,R - V_X = 0$

$1000 - 90 - 0.2\,R - 1000\,V_X = 0$

$0.2\,R + 1000\,V_X = 910 \quad$ ...(iv)

When direction of V_X is reversed then from equation (iii),

$1 - 450\,I - IR + V_X = 0 \quad$ ...(v)

Here, $\quad I = 3.8\text{ mA}$

so from equation (v),

$1 - 450 \times 3.8 \times 10^{-3} - 3.8 \times 10^{-3}\,R + V_X = 0$

$1000 - 1710 - 3.8\,R + 1000\,V_X = 0$

$-3.8\,R + 1000\,V_X = 710 \quad$ ...(vi)

Solving, (iv) and (v) we get,

$$R = 50\,\Omega$$

$$V_X = 0.9\text{ V}$$

3.8 (b)

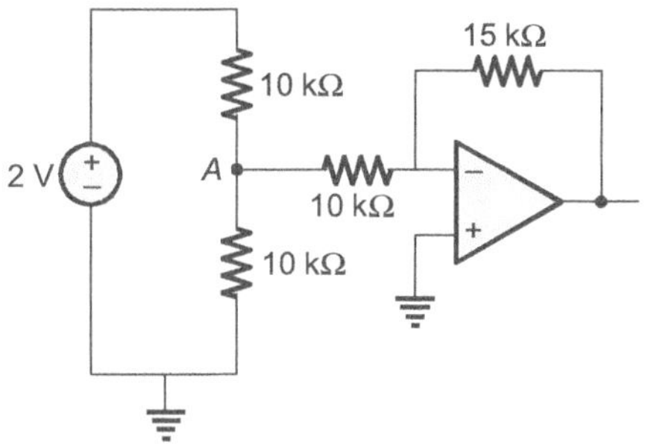

This is the modified circuit.

Due to virtual ground concept $V_- = 0$ V

Writing KCL at node A, we get

$$\frac{V_A}{10}+\frac{V_A}{10}+\frac{V_A-2}{10} = 0$$

$$3V_A - 2 = 0$$

$$V_A = 2/3 \text{ Volts}$$

Now KCL at –ve terminal of OPAMP

$$\frac{V_- - V_A}{10}+\frac{V_- - V_0}{15} = 0$$

$$\frac{-2}{3\times 10}-\frac{V_0}{15} = 0$$

$$V_0 = \frac{-2}{3}\times\frac{15}{10} = -1 \text{ Volt}$$

3.9 (a)

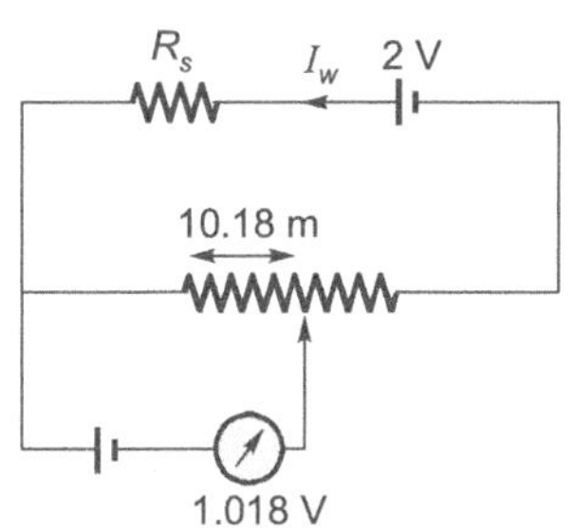

Working current,

$$I_w = \frac{1.018}{10.18} = 0.1\text{ A}$$

Now, applying KVL,

$$2 = (R_s + 11)0.1$$

$$\Rightarrow \quad R_s = 9\ \Omega$$

3.10 (b)

Current capacity (I) of the potentiometer is

$$I^2 = \frac{0.02}{4000} = 5\times 10^{-6}$$

$$I = \sqrt{5\times 10^{-6}} = 2.23\times 10^{-3}\text{ A}$$

$$\therefore \quad I = \frac{V}{R_s + R_{pot.}}$$

$$2.23\times 10^{-3} = \frac{15}{R_s + 4000}$$

$$(2.23\times 10^{-3})(R_s + 4000) = 15$$

$$2.23\times 10^{-3} R_s + 8.92 = 15$$

$$2.23\times 10^{-3} R_s = 6.08$$

$$R_s = \frac{6.08}{2.23\times 10^{-3}} = 2.726\text{ k}\Omega$$

$$\simeq 2.71\text{ k}\Omega$$

Method 2:

Maximum allowable drop across potentiometer is,

$$V_{max} = \sqrt{PR_p}$$

$$\Rightarrow \quad V_{max} = \sqrt{0.02\times 4\times 1000} = \sqrt{80}\text{ V}$$

$$\Rightarrow \quad I_{max} = \frac{\sqrt{80}}{4\text{ k}\Omega} = \sqrt{5}\text{ mA}$$

Now applying KVL in loop when maximum current flows,

$$15 - \sqrt{5}(R_s)_{min} - 4\sqrt{5} = 0$$

$$\Rightarrow \quad (R_s)_{min} = 2.71\text{ k}\Omega$$

3.11 (d)

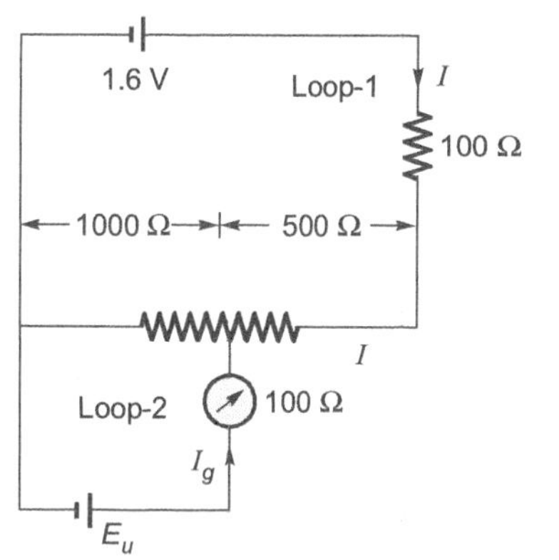

$$I_g = 10\ \mu\text{A} = 10^{-5}\text{ A}$$

writing KVL in Loop 1, we get

$$1.6 - 100I - 500I - 1000(I + I_g) = 0$$

$$1.6 - 1000I_g = 1600I$$

$$1.6 - 1000\times 10^{-5} = 1600I$$

$$I = \frac{1.6-0.01}{1600} = 9.9375\times 10^{-4}\text{ A}$$

Now in loop 2,

$$E_u - I_g\times 100 - 1000(I_g + I) = 0$$

$$E_u = 10^{-5}\times 100 + 1000\times 10^{-5} + 1000 \times 9.9375\times 10^{-4}$$

$$= 1100\times 10^{-5} + 0.99375$$

$$= 0.011 + 0.99375 = 1.00475 = 1.0048\text{ V}$$

3.12 (b)

Voltmeter resistance,

$$= S_V \times V_{fs}$$

$$= 500\ \Omega/V \times 100\ V = 50\ k\Omega$$

Since voltmeter reads 20 V

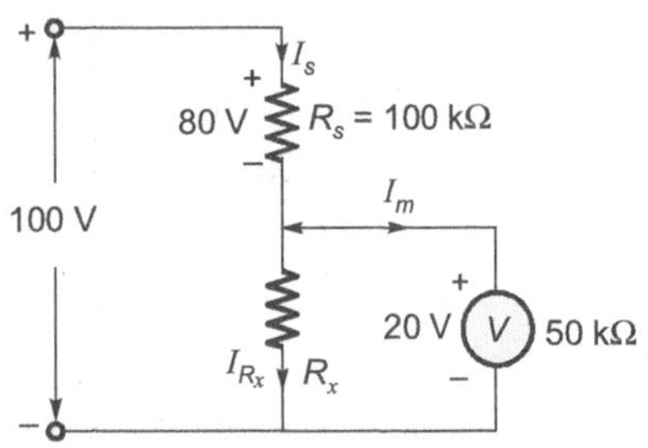

$\Rightarrow$ Voltage drop across $R_s = 100 - 20 = 80$ V

$$\Rightarrow \quad I_s = \frac{80}{100\text{ k}} = 0.8\text{ mA}$$

Now, $\quad I_{R_x} = I_s - I_m$

$$\Rightarrow \quad I_{R_x} = 0.8 - \frac{20}{50} = 0.4\text{ mA}$$

$$\Rightarrow \quad R_x = \frac{20}{0.4\text{ mA}} = 50\ k\Omega$$

3.13 (a)

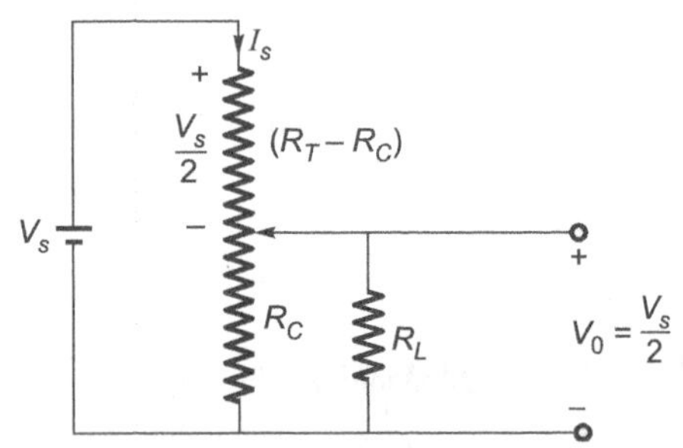

Given, $\quad \frac{R_L}{R_T} = 1$

$$\Rightarrow \quad R_L = R_T \quad \text{...(i)}$$

Also,

$$\frac{V_s}{2} = (R_T - R_C)I_s = (R_T \,||\, R_C)I_s$$

using (i),

$$R_T - R_C = \frac{R_C \cdot R_T}{R_C + R_T}$$

$$\Rightarrow \quad R_T^2 - R_C^2 = R_C R_T$$

$$\Rightarrow \quad \frac{R_T}{R_C} - \frac{R_C}{R_T} = 1$$

Let $\quad \frac{R_C}{R_T} = x$

$$\Rightarrow \quad \frac{1}{x} - x = 1$$

$$\Rightarrow \quad x^2 + x - 1 = 0$$

$$\Rightarrow \quad x = \frac{-1 \pm \sqrt{5}}{2}$$

Since, $\quad x = \frac{R_C}{R_T} > 0$

$$\Rightarrow \quad \frac{R_C}{R_T} = \frac{\sqrt{5} - 1}{2}$$

3.14 (b)

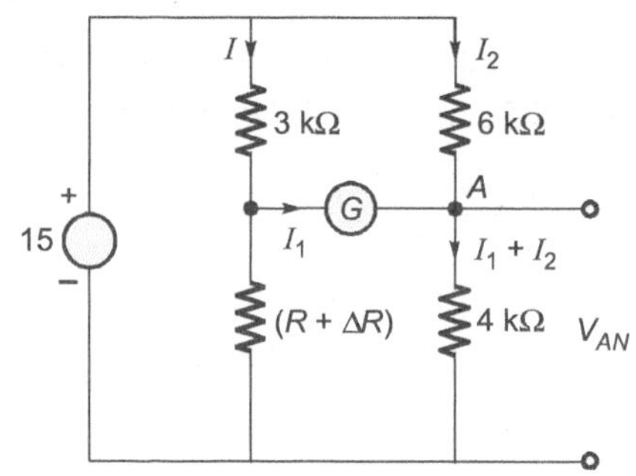

- $\Rightarrow$ Bridge can detect a change in R if galvanometer deflects.
- $\Rightarrow$ If change in R is minimum, then minimum deflection in galvanometer should be 1 mm for detection as 1 mm is it's resolution. So for this 1 μA current must flow through it. i.e. $I_1 = 1\ \mu A = 10^{-6}$ A.

Now in outer loop,

$$15 - 6 \times 10^3 I_2 - 4 \times 10^3 (I_1 + I_2) = 0$$

$$15 - 4 \times 10^3 I_1 = 10^4 I_2$$

$$15 - 4 \times 10^3 \times 10^{-6} = 10^4 I_2$$

$$15 - 0.004 = 10^4 I_2$$

$$14.996 = 10^4 I_2$$

So, $\quad V_{AN} = 4 \times 10^3 (I_1 + I_2)$

$$= 4 \times 10^3 (10^{-6} + 1.4996 \times 10^{-3})$$

$$= 0.004 + 5.9984 = 6.0024\text{ V}$$

3.15 (b)

Writing KVL in inner loop we get,

$$3 \times 10^3 + 0.0025 - 6 \times 10^3 I_2 = 0$$

$$3 \times 10^3 I = -0.0025 + 8.9976$$

$$I = 2.9983\text{ mA}$$

Writing KVL in another inner loop we get,

$$(R + \Delta R) \times (I - I_1) - 0.0025 - 6.0024 = 0$$

$$(R + \Delta R) = \frac{6.0049}{(2.9983 - 0.001) \times 10^{-3}}$$

$$R + \Delta R = 2003.436$$

$$\Delta R = 3.436\ \Omega$$

3.16 (b)

When the switch 's' is open,

$$I = 10 \text{ mA}$$

$$I = \frac{3}{R_{pot} + 250}$$

$$(10 \times 10^{-3})(R_{pot} + 250) = 3$$

$$R_{pot} + 250 = 300$$

$$\therefore \quad R_{pot} = 50$$

Length of pot = 5 m

$$5 \text{ m} = 50\ \Omega$$

$$\therefore \quad 1 \text{ m} = \frac{50}{5}\ \Omega$$

$$2 \text{ m} = \frac{50}{5} \times 2 = 20\ \Omega$$

$$V_{2m} = (20\ \Omega) \times (10 \text{ mA}) = 200 \text{ mV}$$

But the errors in the reading are encountered due to the presence of extra resistance due to the galvanometer resistance (R_g) and R_1

$$\therefore \quad \text{error} = \text{drop across } (R_g + R_1)$$

$$= I_g(R_g + R_1) = 10 \times 10^{-6}(100)$$

$$= 10^{-3} \text{ volts} = 1 \text{ mV}$$

$$\%\text{age error} = \frac{1 \text{mV}}{200 \text{ mV}} \times 100 = 0.5\%$$

3.17 (a)

When all the resistance R_1, R_2, R_3 and R_4 are equal (= 1 kΩ) the bridge is balanced and hence current I_G flowing through the galvanometer is zero.

When R_1 increases by 1 Ω i.e. R_1 = 1001 Ω, current $I_G (\neq 0)$ flows through the galvanometer I_G is calculated by obtaining the Thevenin equivalent across the terminal *A*, *B* of the galvanometer shown in below figure.

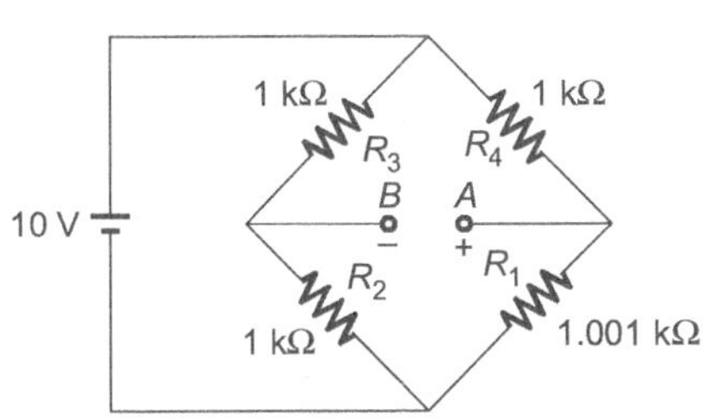

$$V_{Th} = V_A - V_B$$

$$= 10\left[\frac{1.001}{1+1.001}\right] - 10\left[\frac{1}{1+1}\right] = 2.5 \text{ mV}$$

$$R_{Th} = \frac{1 \times 1}{1+1} + \frac{1.001 \times 1}{1.001+1} = 1.00025 \text{ k}\Omega$$

The Thevenin equivalent circuit is shown below.

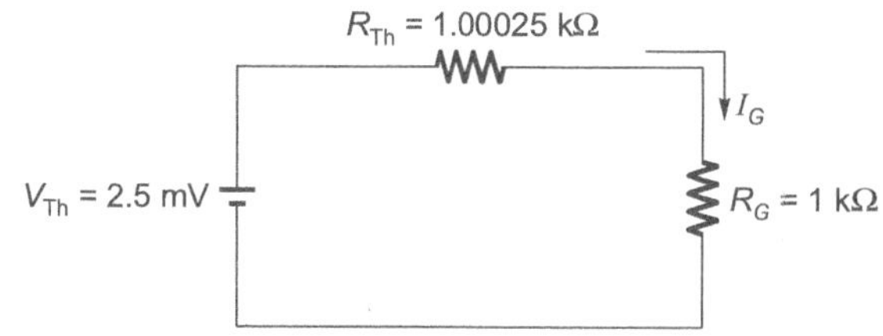

Galvanometer current,

$$I_G = \frac{2.5 \text{mV}}{1 \text{k}\Omega + 1.00025 \text{ k}\Omega} = 1.25\ \mu\text{A}$$

3.19 (a)

$$+V + IR + V = 0$$

$$\therefore \quad I = -\frac{2V}{R}$$

$$V_x = (\alpha R)\, I + V$$

$$= \alpha R \frac{(-2 \text{ V})}{R} + V$$

$$= -2 \times \alpha \times V + V$$

$$= (1 - 2\alpha) \text{ V}$$

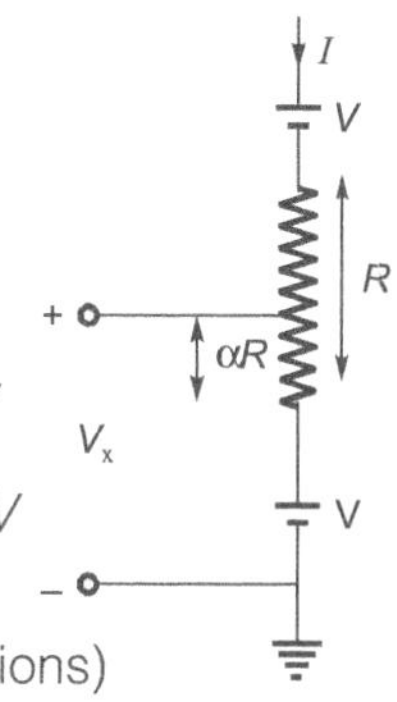

Method-2 (Eliminating options)

At $\alpha = 0$, $\quad V_x = V$

So, answer is either (a) or (b)

At $\alpha = 1$, $\quad V_x = -V$

⇒ Answer is (a).

3.20 33.33 (33 to 34)

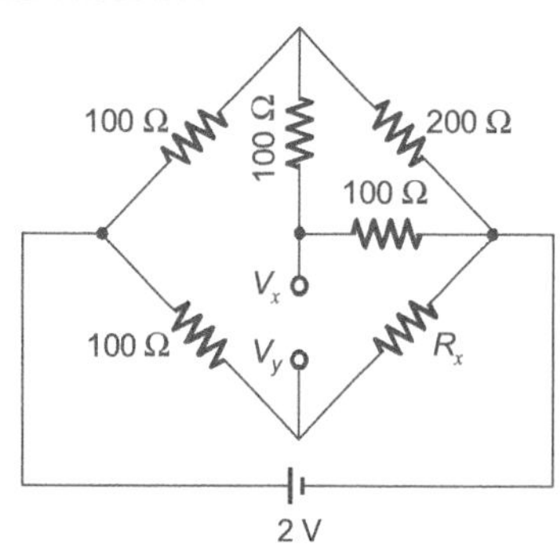

Since current through the galvanometer is zero, thus $V_x = V_y$, (no voltage drop so net current flow through galvanometer) is zero, circuit could be further simplified as

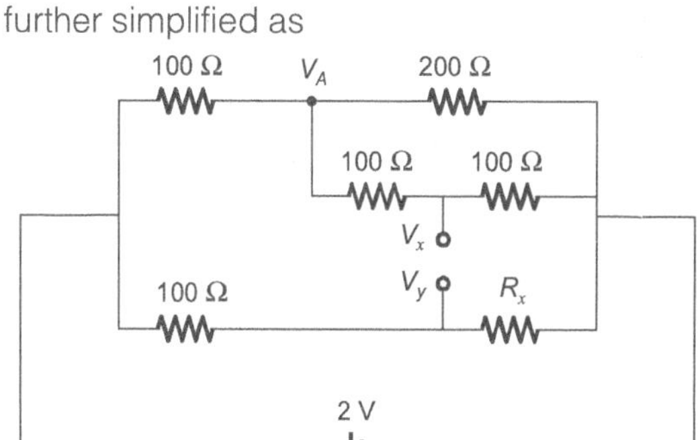

$$V_A = \frac{(200 \| 200)}{(200 \| 200) + 100} \times 2 = 1 \text{ V}$$

$$V_x = \frac{100}{100 + 100} V_A = 0.5 \text{ V}$$

$$V_y = \frac{R_x}{R_x + 100} \times 2$$

$$V_x = V_y$$

Thus, $$\frac{1}{2} = \frac{2Rx}{100 + Rx}$$

$\therefore$ $$R_x = 33.33\ \Omega$$

3.21 (1.1234)

The unknown voltage V_x will be

$$V_x = I_P[11 \text{ resistors of } 10\ \Omega + \text{Slide wire resistance}]$$

$$= 10 \times 10^{-3}\left(11 \times 10 + \frac{234}{1000} \times 10\right)$$

$$= 1.1234 \text{ V}$$

3.22 (c)

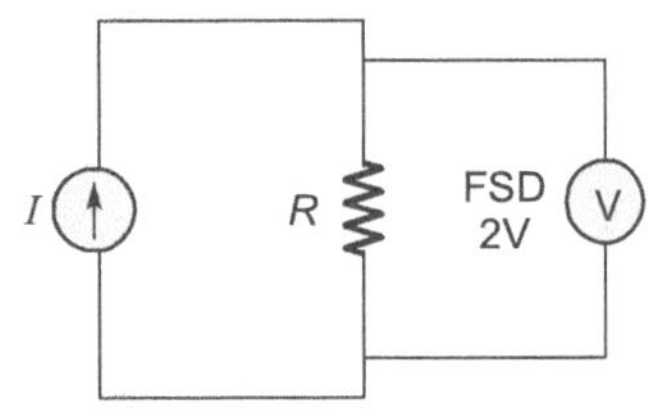

Possible ranges
200 Ω
2 kΩ
20 kΩ
200 kΩ
2 MΩ

Resistance to be measured ⇒ 8 kΩ to 12 kΩ

So, minimum range should be 20 kΩ.

With 20 kΩ range:

$$I = \frac{2V}{20\text{k}\Omega} = \frac{1}{10}\text{mA}$$

$$P_{max} = I^2 R_{max}$$

$$= \left(\frac{1}{10} \times 10^{-3}\right)^2 \times 12 \times 10^3 \text{ W}$$

$$= 120\ \mu\text{W} > 100\ \mu\text{W}$$

With 200 kΩ range:

$$I = \frac{2V}{200\text{k}\Omega} = \frac{1}{100}\text{mA}$$

$$P_{max} = I^2 R_{max}$$

$$= \left(\frac{1}{100} \times 10^{-3}\right)^2 \times 12 \times 10^3 \text{ W}$$

$$= 1.2\ \mu\text{W}$$

3.23 (99)

By redrawing circuit, we can get, "KELVIN DOUBLE BRIDGE"

Given that, $R_1 = 100\ \Omega$, $R_2 = 110\ \Omega$, $R_3 = 90\ \Omega$

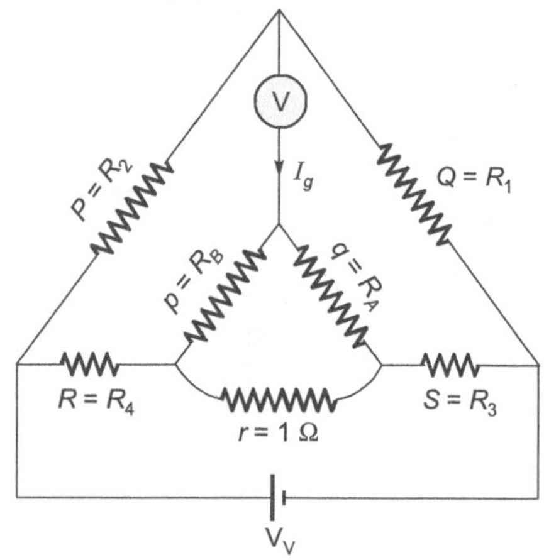

Balance condition, $$R = \frac{P}{Q} \times S + \frac{qr}{p+q+r}\left[\frac{P}{Q} - \frac{p}{q}\right]$$

By comparing, $$\frac{R_1}{R_2} = \frac{R_A}{R_B} \Rightarrow \frac{Q}{P} = \frac{q}{p} \Rightarrow R = \frac{P}{Q} \times S$$

$\therefore$ $$R_H = \frac{R_2}{R_1} \times R_3 = \frac{110}{100} \times 90$$

$\Rightarrow$ $$R_4 = 99\ \Omega$$

3.24 (1.65)

Case-I when switch open,

$$V_{CD} = V_C - V_D$$

$$= 10\left[\frac{1}{2} - \frac{1}{2.5}\right]$$

$$= 10[0.5 - 0.4] = 1 \text{ V}$$

Case-II when switch closed

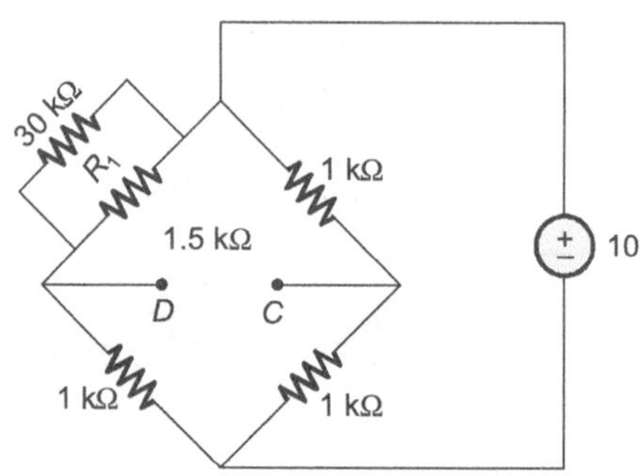

$$V_{CD} = 10\left[\frac{1}{2} - \frac{1}{1 + 1.4285}\right] = 0.882$$

$$\text{Sensitivity} = \frac{\Delta V_{CD}}{\Delta R} = \frac{0.882 - 1}{1.4285 - 1.5} \text{ V/k}\Omega$$

$$= 1.645 \text{ V/k}\Omega$$

3.25 (1)

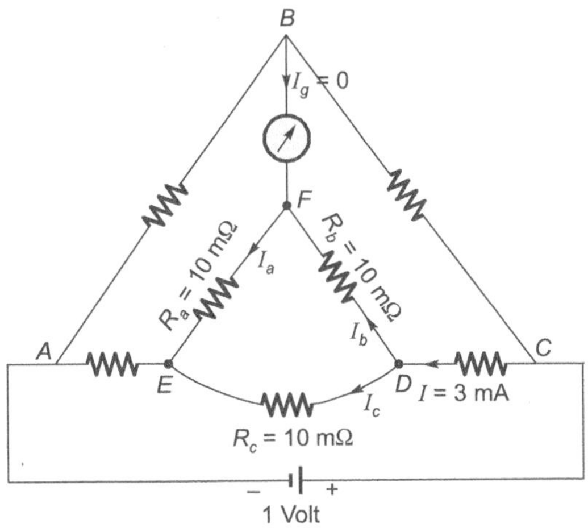

This is a Kelvin double bridge.

Under balanced condition $\Rightarrow$ $I_g = 0$

We can redraw the bridge $\Rightarrow$

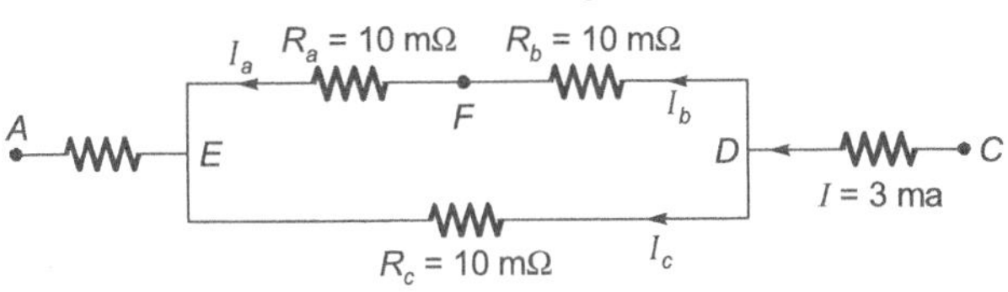

$$I_a = I_b = 3\text{ mA} \times \frac{10}{10+10+10} = 1\text{ mA}$$

3.26 (1.78)

Total turns in angle sensor: 72

Given width of wiper covers two turns.

So,

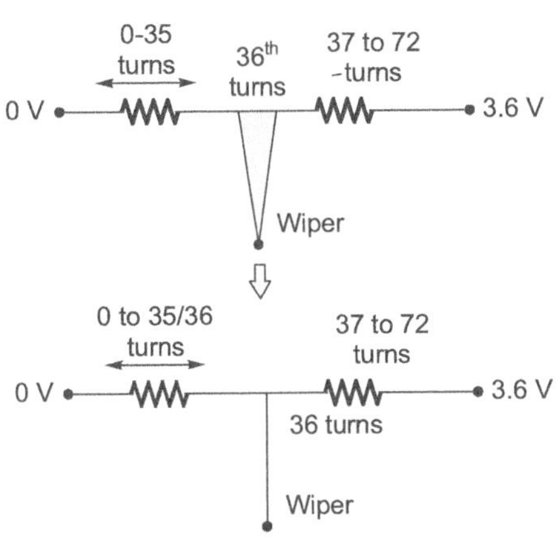

$$\text{So output voltage} = 3.6 \times \frac{35x}{71x} = 1.774$$

Output voltage = 1.77 V

3.27 (60)

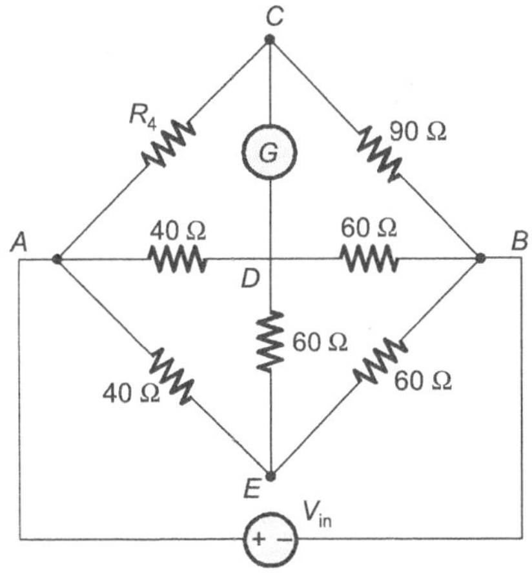

$$40 \times 60 = 40 \times 60$$
$$240 = 240$$

Bridge is balanced.

Remove the 60 Ω resistor between D and E.

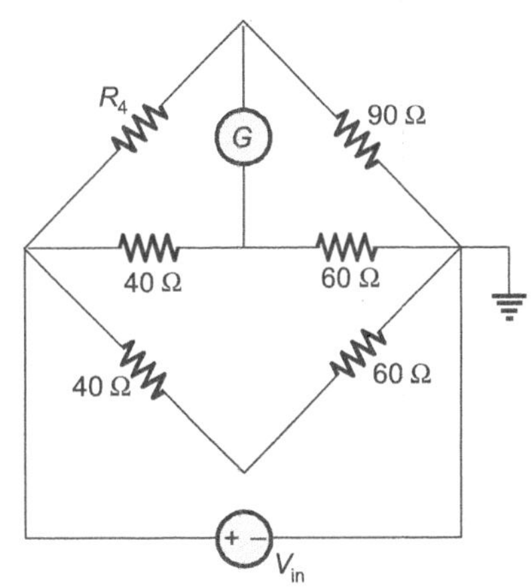

The galvanometer reads zero Amp. i.e., potential difference is zero volts.

$$V_{in} \times \frac{R_4}{R_4 + 90} = V_{in} \times \frac{60}{60+40}$$
$$R_4 = 60\ \Omega$$

4 Measurement of Inductance Capacitance, AC Bridges & Q-meter

4.1 Find the excitation frequency (Hz) and the value of the resistance R_2 in the ac bridge shown in figure under balanced condition. The circuit component values are given as

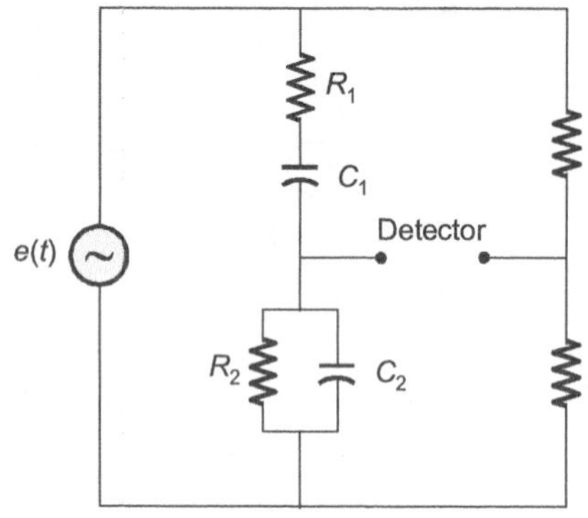

$R_1 = 100\ k\Omega$, $R_3 = R_4 = 100\ k\Omega$, $C_1 = 2C_2 = 10nF$.

[1995 : 2 M]

4.2 In the a.c. bridge circuit shown in figure, at balance the parameters of the coil are calculated to be 50 mH and 20 Ω. Another coil is connected in parallel to the previous coil and balance is restored by changing R_2 to 2.5 kΩ. The parameters of the new coil are then

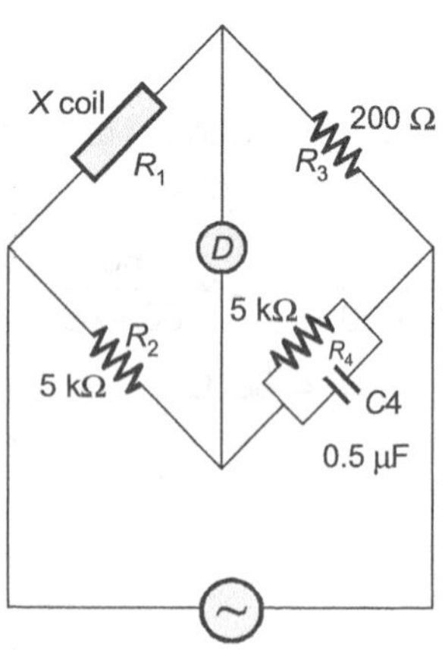

(a) 25 mH, 20 Ω (b) 50 mH, 20 Ω
(c) 50 mH, 10 Ω (d) 100 mH, 40 Ω

[1996 : 1 M]

4.3 The equations under balanced conditions for a bridge are :

$R_1 = \dfrac{R_2R_3}{R_4}$ and $L_1 = R_2R_3C_4$ where R_1 and L_1 are respectively unknown resistance and inductance.

In order to achieve faster balance which of the following variables are chosen?

(a) R_2 and R_3 (b) R_2 and C_4
(c) R_4 and C_4 (d) R_3 and C_4

[1996 : 1 M]

4.4 In the bridge circuit shown in figure when $\dfrac{X_c}{R} = 1$, then the voltmeter reads.

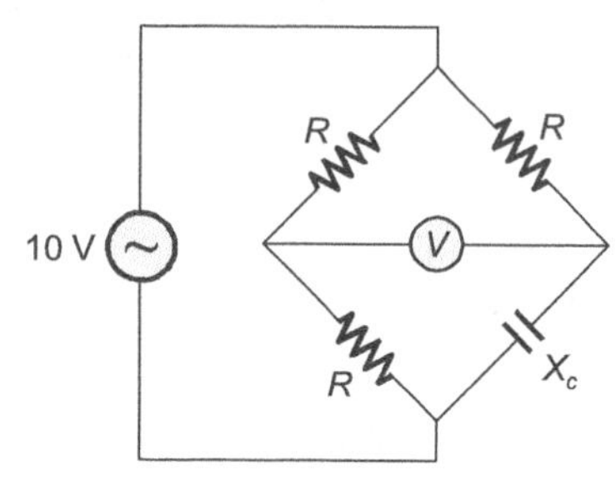

(a) 5 V (b) 0.0 V
(c) 2.5 V (d) 10 V

[1996 : 1 M]

4.5 Match the parameter with the type of bridge used for measurement

List-I
(a) low value of R
(b) high value of Q
(c) Inductance L
(d) capacitance C

List-II
(p) Schering bridge
(q) Maxwell-wien bridge
(r) Kelvin double bridge
(s) Hay bridge

[1997 : 2 M]

4.6 The a.c bridge in figure remains balance if Z comprises of

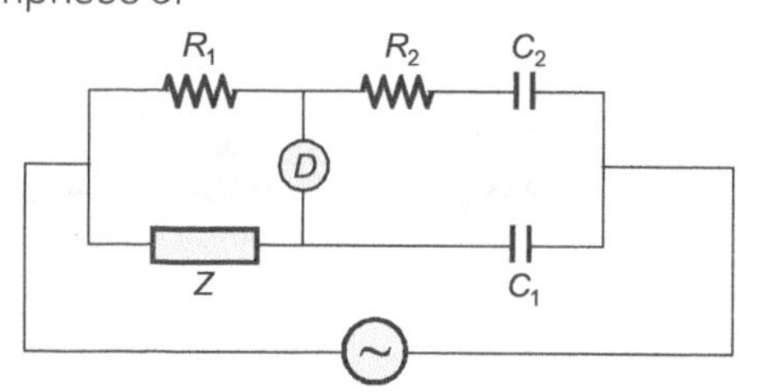

(a) Resistance only
(b) Capacitance only
(c) Resistance and capacitance in parallel
(d) Resistance and inductor in series

[1998 : 1 M]

4.7 Match the measured variables with the bridge

List-I	List-II
(a) Self inductance	(p) Anderson
(b) Capacitance	(q) Wien-Robinson
(c) Mutual inductance	(r) Campbell
(d) Frequency	(s) Schering

[1998 : 2 M]

4.8 Wien bridge is best suited for the measurement of

(a) frequency (b) capacitor
(c) inductor (d) resistor

[2002 : 1 M]

4.9 For the bridge circuit in figure, the voltage are:

$\underline{V}_1 = \sqrt{2}\cos 1000\,t$ V,

$\underline{V}_2 = 2\cos(1000\,t + 45°)$ and

$V_d = 0$.

The value of $R = 100\ \Omega$. Then Z_x will be

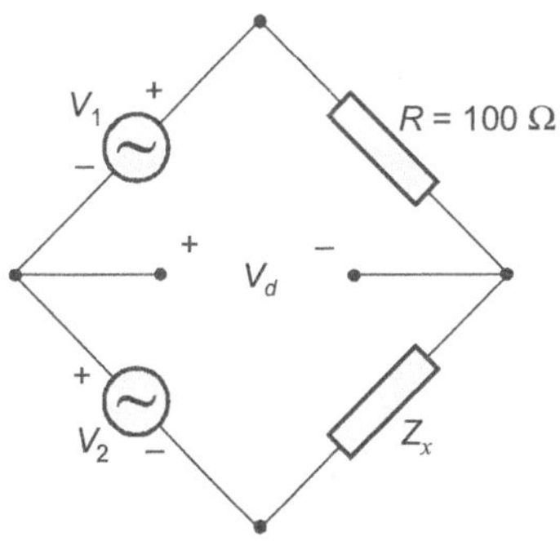

(a) 100 Ω resistor in series with 100 mH inductor
(b) 100 Ω resistor in series with 100 μF capacitor
(c) 100 Ω resistance in parallel with 100 mH inductor
(d) 100 Ω resistance in parallel with 100 μF capacitor

[2003 : 2 M]

4.10 A very low-loss coil is tested with a Q-meter and the distributed (self) capacitance of the coil is found to be 820 pF. Resonance occurred at an angular frequency ω of 10^6 rad/s with a capacitance of 9.18 nF. The inductance of the coil is

(a) 100 pH (b) 100 μH
(c) 100 nH (d) 100 mH

[2003 : 2 M]

4.11 The resistance values of the bridge circuit shown in figure are $R_1 = R_2 = R_3 = R$ and $R_4 = R + \Delta R$. The bridge is balanced by introducing a small voltage v. The value of ΔR is

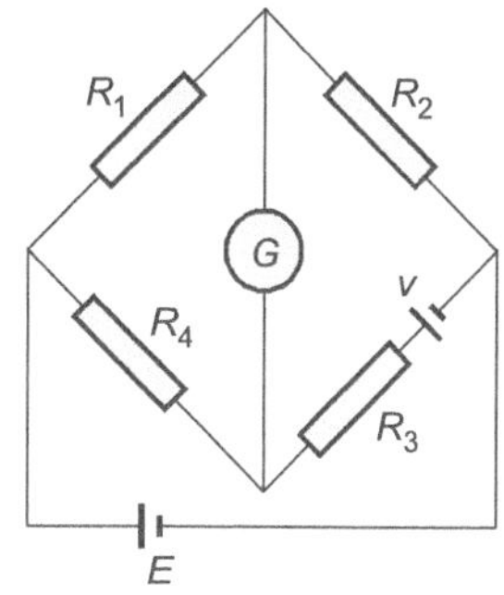

(a) $\dfrac{Rv}{E}$ (b) $\dfrac{Rv}{E-v}$

(c) $\dfrac{R2v}{E-2v}$ (d) $\dfrac{Rv}{E+2v}$

[2004 : 2 M]

4.12 In the ac bridge shown in figure the detector D reads zero. Then Z is made of

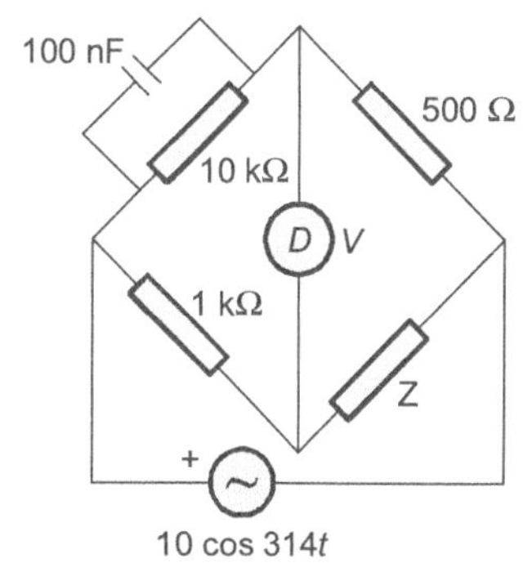

(a) 50 mH in parallel with 50 Ω
(b) 50 mH is series with 50 Ω
(c) 50 nF in series with 10 kΩ
(d) 50 nF in parallel with 5 kΩ

[2004 : 2 M]

4.13 A Q-meter having an insertion resistance of 0.02 Ω is used to measure the inductance of a coil. Resonance occurs at an angular frequency of 10^6 rad/sec. with a capacitance of 40 pF. The inductance of the coil is

(a) 250 μH (b) 2.5 mH
(c) 25 mH (d) 0.25 H

[2004 : 2 M]

4.14 The bridge shown in figure is balanced. The dissipation factor for the unknown lossy capacitor

$R_x + \dfrac{1}{j\omega C_x}$ is

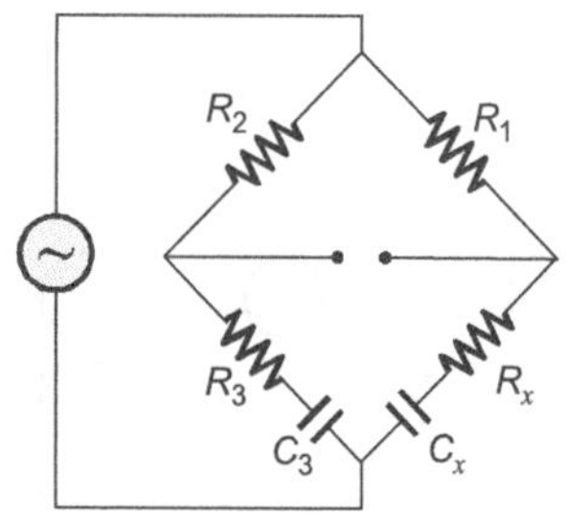

(a) $\omega R_2 C_3$ (b) $\frac{\omega R_1^2 C_3}{R_2}$

(c) $\omega R_3 C_3$ (d) $\frac{R_1}{R_2}$

[2005 : 2 M]

4.15 Consider the AC bridge shown below. If $\omega RC = 1$ and $\Delta C/C < 0.01$, then ratio $\frac{|V_0|}{|V_s|}$ is approximately equal to

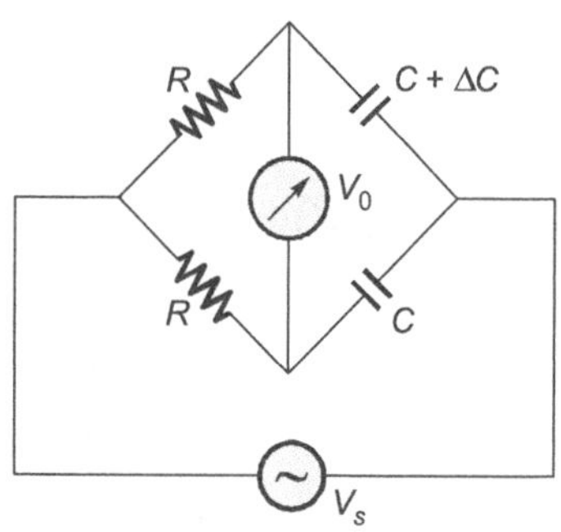

(a) $\sqrt{1+\frac{\Delta C}{C}}$ (b) $\frac{2\Delta C}{C}$

(c) $\frac{\Delta C}{C}$ (d) $\frac{\Delta C}{2C}$

[2006 : 2 M]

4.16 Consider the AC bridge shown in the figure below, with *R*, *L* and *C* having positive finite values.

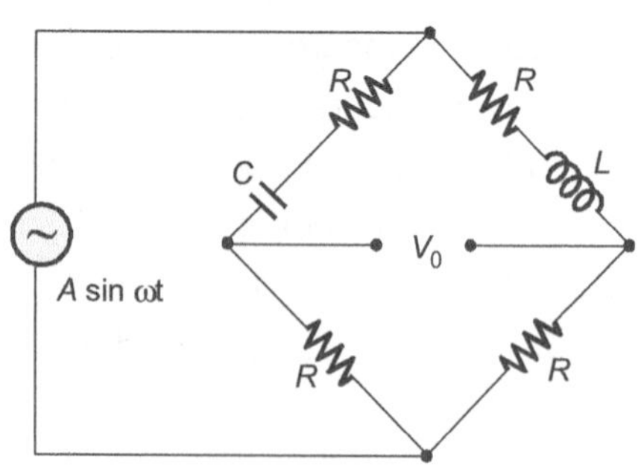

(a) $V_0 = 0$ if $\omega L = 1/\omega C$

(b) $V_0 = 0$ if $L = C$

(c) $V_0 = 0$ if $R = \frac{1}{\omega\sqrt{LC}}$

(d) V_0 cannot be made zero [2007 : 1 M]

4.17 If the ac bridge circuit shown below is balanced the element *Z* can be a

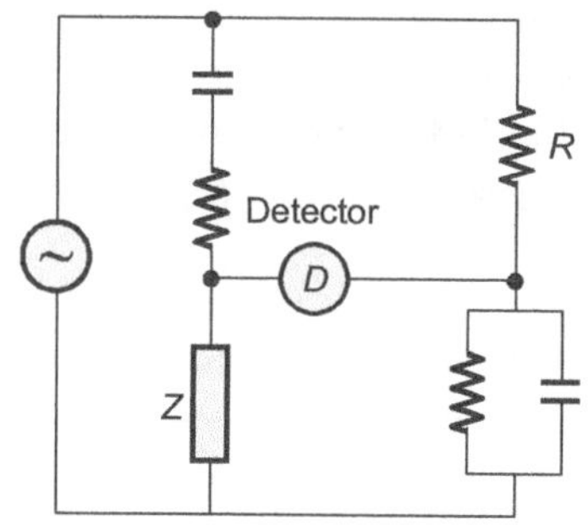

(a) pure capacitor
(b) pure inductor
(c) *R-L* series combination
(d) *R-L* parallel combination [2008 : 1 M]

4.18 The bridge method commonly used for finding mutual inductance is
(a) Heaviside Campbell bridge
(b) Schering bridge
(c) De Sauty bridge
(d) Wien bridge [2012 : 1 M]

4.19 The resistance and inductance of an inductive coil are measured using an AC bridge as shown in the figure. The bridge is to be balanced by varying the impedance Z_2.

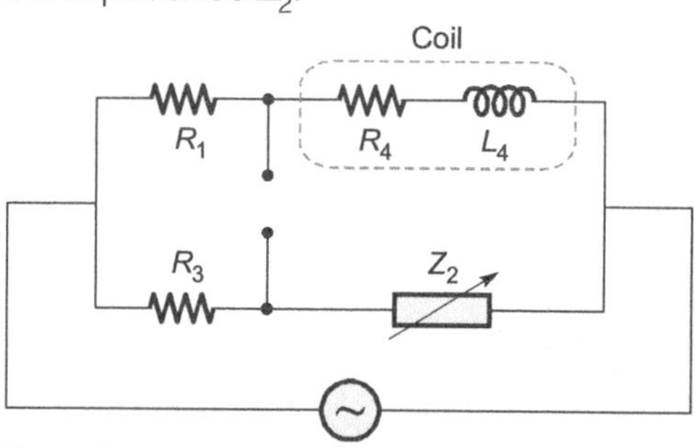

For obtaining balance, Z_2 should consist of element(s)

(a) *R* and *C* (b) *R* and *L*

(c) *L* and *C* (d) Only *C* [2014 : 1 M]

4.20 An air cored coil has a *Q* of 5 at a frequency of 100 kHz. The *Q* of the coil at 20 kHz (neglecting skin effect) will be________.

[2016 : 1 M]

4.21 A coil is tested with a series type *Q*-meter. Resonance at a particular frequency is obtained with a capacitance of 110 pF. When the frequency is doubled, the capacitance required for resonance is 20 pF. The distributed capacitance of the coil in Pico farad is______.

[2016 : 2 M]

4.22 In the AC bridge, shown in the figure, $R = 10^3\ \Omega$ and $C = 10^{-7}$ F. If the bridge is balanced at a frequency ω_0, the value of ω_0 in rad/s is _____.

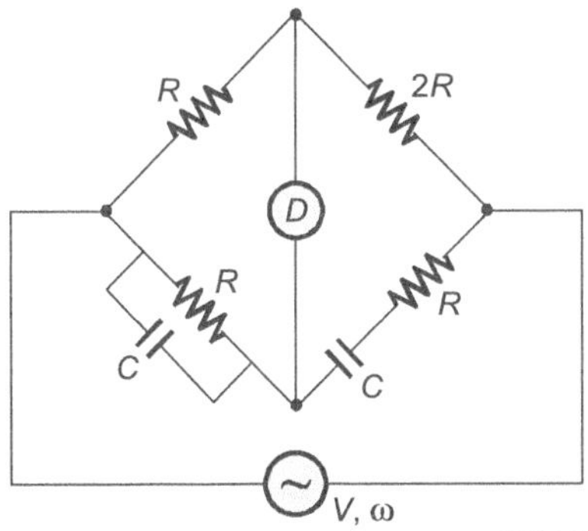

[2017 : 2 M]

4.23 A high Q coil having distributed (self) capacitance is tested with a Q-meter. First resonance at $\omega_1 = 10^6$ rad/s is obtained with a capacitance of 990 pF. The second resonance at $\omega_2 = 2 \times 10^6$ rad/s is obtained with a 240 pF capacitance. The value of the inductance (in mH) of the coil is (up to one decimal place) ________.

[2018 : 2 M]

4.24 The inductance of a coil is measured using the bridge shown in the figure. Balance ($D = 0$) is obtained with $C_1 = 1$ nF, $R_1 = 2.2$ MΩ, $R_2 = 22.2$ kΩ, $R_4 = 10$ kΩ. The value of the inductance L_x (in mH) is __________.

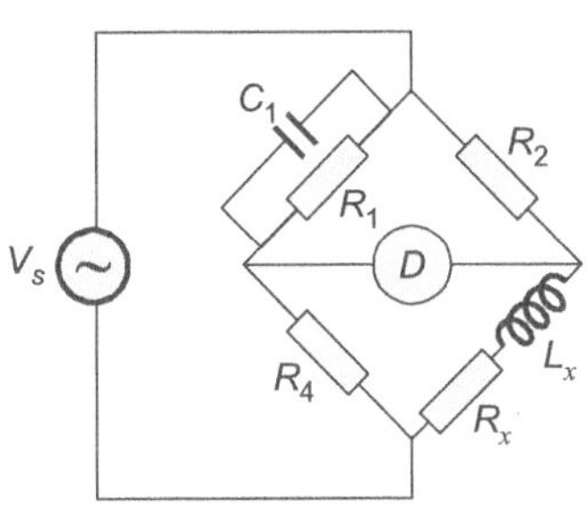

[2018 : 2 M]

4.25 The parallel resistance-capacitance bridge shown below has a standard capacitance value of $C_1 = 0.1\ \mu$F and a resistance value of $R_3 = 10$ kΩ. The bridge is balanced at a supply frequency of 100 Hz for $R_1 = 375$ kΩ, $R_3 = 10$ kΩ and $R_4 = 14.7$ kΩ. The value of the dissipation factor $D = \dfrac{1}{(\omega R_p C_p)}$ of the parallel combination of C_p and R_p is_________. (Answer should be rounded off to Three decimal places).

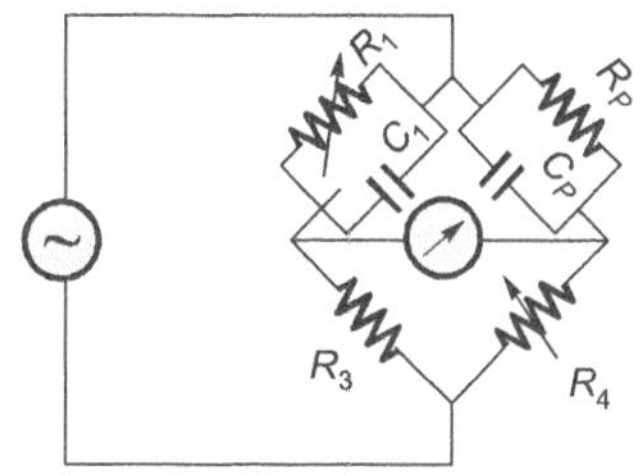

[2019 : 2 M]

4.26 A Q meter is best suited for the measurement of the _____.

(a) Quality factor of a capacitance.
(b) Distributed capacitance of a coil.
(c) Turns-ratio of a transformer
(d) Quality factor of piezoelectric sensor.

[2020 : 1 M]

4.27 In the Maxwell-Wien bridge shown, the detector D reads zero when $C_1 = 100$ nF and $R_1 = 100$ kΩ. The Q factor of the coil is _____.

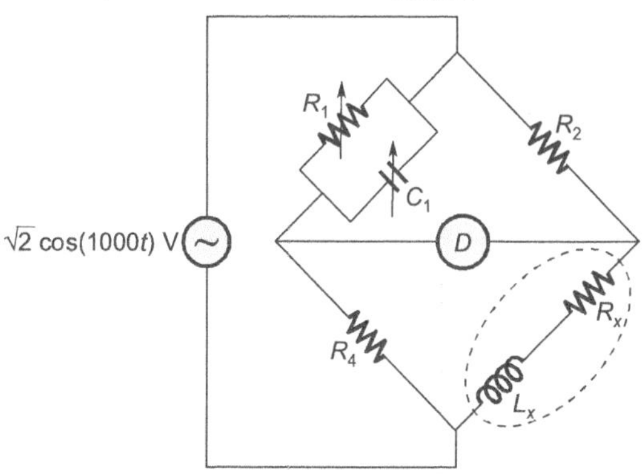

[2020 : 2 M]

4.28 In which of the following bridge(s) is the balancing condition frequency-independent?

(a) Maxwell bridge (b) Wien bridge
(c) Schering bridge (d) Wheatstone bridge

[2022 : 1 M]

4.29 The bridge shown is balanced when $R_1 = 100\ \Omega$, $R_2 = 210\ \Omega$, $C_2 = 2.9\ \mu$F, and $R_4 = 50\ \Omega$. The 2 kHz sine-wave generator supplies a voltage of 10 $V_{p\text{-}p}$. The value of L_3 (in millihenry) is ______ (round off to two decimal places)

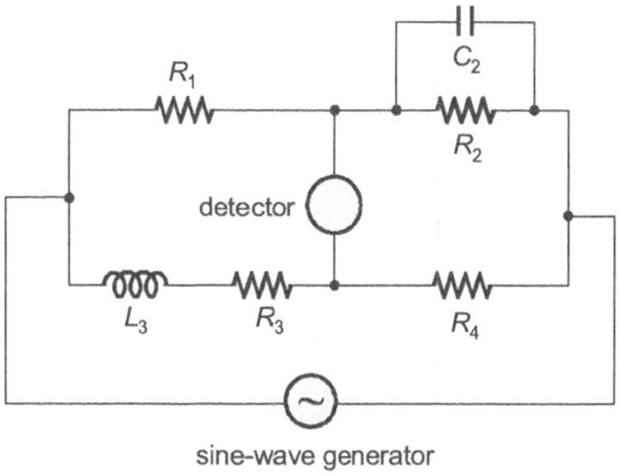

[2022 : 2 M]

4.30 Inductance of a coil is measured as 10 mH, using an LCR meter, when no other objects are present near the coil. The LCR meter uses a sinusoidal excitation at 10 kHz. If a pure copper sheet is brought near the coil, the same LCR meter will read ___________.

(a) less than 10 mH
(b) 10 mH
(c) more than 10 mH
(d) less than 10 mH initially and then stabilizes to more than 10 mH

[2023 : 1 M]

4.31 Capacitance '*C*' of a parallel plate structure is calculated as 20 pF using $C = \frac{\varepsilon_0 \varepsilon_r A}{d}$ where ε_0 is the permittivity of free space, ε_r is the relative permittivity of the dielectric, *A* is the overlapping area of the electrodes and *d* is the distance between them. The value of *C* is then measured using an LCR meter. If the meter is assumed to be ideal and it introduces no error due to cable capacitance, which one of the following readings is likely to be correct?

(a) 20.5 pF (b) 20 pF
(c) 19.5 pF (d) 10 pF

[2023 : 1 M]

4.32 In the bridge circuit shown in the figure, under balanced condition, the values of *R* and *C* respectively, are

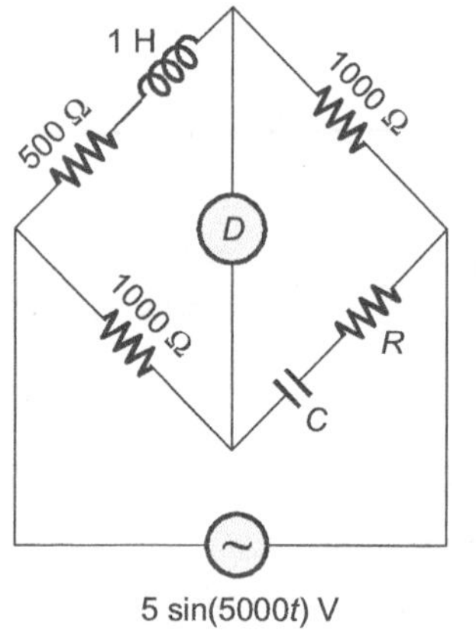

(a) 1.010 Ω and 19.802 Ω
(b) 9.901 Ω and 0.505 μf
(c) 19.802 Ω and 1.01 μf
(d) 39.604 Ω and 2.02 μf

[2024 : 2 M]

■■■

Answers Measurement of Inductance Capacitance, AC Bridges & Q-meter

4.1 (159.15)	4.2 (b)	4.3 (c)	4.4 (a)	4.5 (Sol.)	4.6 (c)	4.7 Sol.
4.8 (a)	4.9 (a)	4.10 (b)	4.11 (c)	4.12 (b)	4.13 (c)	4.14 (c)
4.15 (d)	4.16 (d)	4.17 (a)	4.18 (a)	4.19 (b)	4.20 (1)	4.21 (10)
4.22 (10000)	4.23 (1.0)	4.24 (222)	4.25 (0.042)	4.26 (b)	4.27 (10)	4.28 (a, c, d)
4.29 (14.50)	4.30 (a)	4.31 (a)	4.32 (c)			

Explanations Measurement of Inductance Capacitance, AC Bridges & Q-meter

4.1 159.15

At balance,

$$\left[R_1 + \frac{1}{j\omega C_1}\right]R_4 = R_3\left[\frac{R_2}{1 + j\omega C_2 R_2}\right]$$

Given, $R_3 = R_4$

$$\Rightarrow [j\omega R_1 C_1 + 1] = j\omega C_1\left[\frac{R_2}{1 + j\omega C_2 R_2}\right]$$

$$\Rightarrow [j\omega R_1 C_1 + 1]\,[j\omega R_2 C_2 + 1] = j\omega C_1 R_2$$

$$\Rightarrow [1 - \omega^2 R_1 R_2 C_1 C_2] + j[\omega(R_1 C_1 + R_2 C_2)] = j\omega C_1 R_2$$

Equating real and imaginary parts at LHS and RHS.

$$\Rightarrow 1 - \omega^2 R_1 R_2 C_1 C_2 = 0$$

$$\Rightarrow \quad 1 = \omega^2 R_1 R_2 C_1 C_2 \quad \text{...(i)}$$

and $\omega R_1 C_1 + \omega R_2 C_2 = \omega C_1 R_2$

$$\Rightarrow R_2 = \frac{R_1 C_1}{(C_1 - C_2)}$$

Given,

$$C_1 = 2C_2$$

$$\Rightarrow R_2 = \frac{R_1 \cdot 2C_1}{C_2} \Rightarrow R_2 = 2R_1 = 200\ \text{k}\Omega$$

Substituting R_2 in (i),

$$1 = \omega^2 \times 100\text{ k}\Omega \times 200\text{ k}\Omega \times 10\text{ nF} \times 5\text{ nF}$$

$$\Rightarrow \quad \omega = 1000\text{ rad/sec}$$

$$\Rightarrow \quad f = \frac{1000}{2\pi} = 159.15\text{ Hz}$$

4.2 (b)

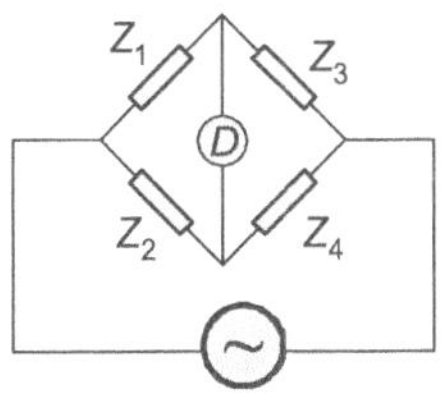

At balance,

$$Z_1 Z_4 = Z_2 Z_3$$

(Z_4 fixed, Z_3 fixed)

Since, Z_4 and Z_3 are fixed.

So, if Z_2 is made half its value.

In order to maintain balance, Z_1 should also be made half.

Z_1 can be make half by connecting the coil of the exact value as Z_1 is parallel to Z_1 is arm (1).

So, (b) is the correct option.

4.3 (c)

By choosing R_4 and C_4 as the variables, both the equation i.e. for R_1 and L_1 can be made independent and hence independent balance for R_1 and L_1 can be obtained.

4.4 (a)

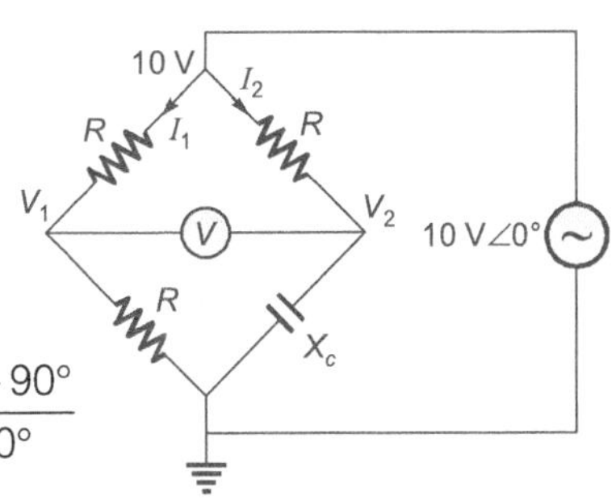

From figure,

$$\vec{V}_1 = \frac{10\angle 0° \times R}{R+R} = 5\angle 0°\text{ V}$$

$$\vec{V}_2 = \frac{10\angle 0° \times X_C\angle -90°}{R + X_C\angle -90°}$$

Given,

$$X_C = R$$

$$\Rightarrow \vec{V}_2 = \frac{10\angle 0° \times R\angle -90°}{R(1+\angle -90°)} = 5\sqrt{2}\angle -45°\text{ V} \quad \text{...(ii)}$$

From (i) and (ii),

Voltmeter reading,

$$|V| = \left|5\sqrt{2}\angle -45° - 5\angle 0°\right| = 5\text{ V}$$

Method - 2:

$$\vec{V}_1 = \frac{10 \times R}{R+R} = 5\text{ V} \quad \text{...(i)}$$

$$\vec{V}_2 = \frac{10 \times (-jX_C)}{R - jX_C}$$

Given, $X_C = R$

$$\Rightarrow \vec{V}_2 = \frac{-j10R}{C(1-j)} = 5 - 5i \quad \text{...(ii)}$$

$\Rightarrow$ Voltmeter reading,

$$V = \left|\vec{V}_2 - \vec{V}_1\right| = |5 - 5i - 5| = 5\text{ V}$$

4.5 Sol.

(a)-(r), (b)-(s), (c)-(q), (d)-(p)

4.6 (c)

At balance,

$$Z\left[R_2 + \frac{1}{j\omega C_2}\right] = R_1\left(\frac{1}{j\omega C_1}\right)$$

$$\therefore \quad Z = \frac{R_1\left(\frac{1}{j\omega C_1}\right)}{\left(R_2 + \frac{1}{j\omega C_2}\right)}$$

It is equivalent of a resistor and a capacitor connected in parallel.

4.7 Sol.

(a)-(p), (b)-(s), (c)-(r), (d)-(q)

4.8 (a)

Wein's bridge is best suited for the measurement of frequency. Although sometimes it can also be used for the measurement of capacitance.

4.9 (a)

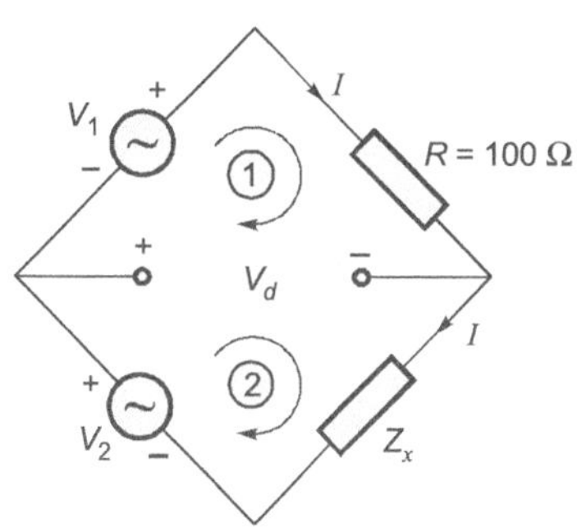

$$V_1 = \sqrt{2}\cos 1000t$$

$$V_2 = 2\cos(1000\,t + 45°)$$

$$V_d = 0$$

Here, $\omega = 1000$

Now in loop 1, writing KVL

$$V_1 - IR + V_d = 0$$

$$V_1 = IR$$

$$I = \frac{V_1}{R} = \frac{\sqrt{2}}{100}\cos 1000\,t \quad \text{...(i)}$$

Now in loop 2,

$$V_2 - V_d - IZ_x = 0$$

$$2\cos(1000\,t + 45°) = \frac{\sqrt{2}}{100}(\cos 1000\,t) \times Z_x$$

So both will be equal if Z_x provide 45° phase and magnitude $100\sqrt{2}$

i.e. $\quad Z_x = 100 + 100\,j$

So a resistance of 100 Ω in series with inductor providing reactance = 100 Ω

$$\omega L = 100\ \Omega$$

$$L = \frac{100}{\omega} = 0.1\text{H} = 100\text{ mH}$$

4.10 (b)

$$\omega = \frac{1}{\sqrt{L(c+c_d)}}$$

$$\therefore \quad \omega^2 = \frac{1}{L(c+c_d)}$$

$$L = \frac{1}{\omega^2(c+c_d)}$$

$$= \frac{1}{(10^6)^2(0.820+9.18)\times 10^{-9}} = 100\ \mu\text{H}$$

4.11 (c)

Since, the bridge becomes balanced, by introducing 'V' in the arm R_3, so the current flowing through the galvanometer becomes zero, and it can be shown as a open circuit.

$$I_1 = \frac{E}{R_1+R_2} = \frac{E}{R+R} = \frac{E}{2R}$$

$$I_2 = \frac{E-V}{R_3+R_4} = \frac{E-V}{R+\Delta R+R}$$

$$= \frac{E-V}{2R+\Delta R}$$

For bridge to be balanced,

$$E_{ab} = E_{ad}$$

$$\therefore \quad I_1 R_1 = I_2 R_4$$

$$\frac{E}{2R}(R) = \frac{E-v}{2R+\Delta R}(R+\Delta R)$$

$$\frac{E}{2} = \frac{(E-V)(R+\Delta R)}{(2R+\Delta R)}$$

$$E(2R+\Delta R) = 2(E-V)(R+\Delta R)$$

$$2\,ER + E(\Delta R) = 2[ER + (\Delta R)\,E - VR - V\Delta R]$$

$$\therefore \quad E(\Delta R) = 2E(\Delta R) - 2\,VR - 2\,V(\Delta R)$$

$$2\,VR + 2\,V(\Delta R) = E(\Delta R)$$

$$2\,VR = E(\Delta R) - 2\,V(\Delta R)$$

$$(\Delta R)\,[E - 2\,V] = 2\,VR$$

$$\Delta R = \frac{R2V}{E-2V}$$

4.12 (b)

It is a case of Maxwell's inductance bridge where Z is the (inductance with series resistance).

$$L_4 = C_1 R_3 R_2 = 100\times 10^{-9}\times 10^3\times 500$$

$$= 5\times 10^{-2}\ H = 50\text{ mH}$$

$$R_4 R_1 = R_2 R_3$$

$$\therefore \quad R_4 = \frac{R_2 R_3}{R_1} = \frac{500\times 1000}{10\times 10^3} = 50\ \Omega$$

4.13 (c)

At resonance, $\quad \omega_c = \dfrac{1}{\sqrt{LC}}$

$$\therefore \quad 10^6 = \frac{1}{\sqrt{L\times 40\times 10^{-12}}}$$

$$(10^6)^2 = \frac{1}{L\times 40\times 10^{-12}}$$

$$L = \frac{1}{40\times 10^{+12}\times 10^{-12}} = \frac{1}{40} = 0.025\text{ H} = 25\text{ mH}$$

4.14 (c)

At balance:

$$R_2\left[R_x + \frac{1}{j\omega C_x}\right] = R_1\left[R_3 + \frac{1}{j\omega C_3}\right]$$

$$R_2 R_x + \frac{R_2}{j\omega C_x} = R_1 R_3 + \frac{R_1}{j\omega C_3}$$

$$\frac{R_2 R_x\, j\omega C_x + R_2}{j\omega C_x} = \frac{R_1 R_3\, j\omega C_3 + R_1}{j\omega C_3}$$

$$R_2 R_x C_x C_3 \omega j + R_2 C_3 = R_1 R_3 C_3 C_x \omega j + R_1 C_x$$

Equating the real and the imaginary parts we get,

$$R_2 C_3 = R_1 C_x$$

$$C_x = \frac{R_2 C_3}{R_1}$$

$$R_2 R_x C_x C_3 = R_1 R_3 C_3 C_x$$

$$R_x = \frac{R_1 R_3}{R_2}$$

$$D = \frac{1}{Q} = \frac{1}{\dfrac{1}{\omega C_x R_x}} = \omega C_x R_x$$

$$= \omega \frac{C_3 R_2}{R_1} \times \frac{R_1 R_3}{R_2} = \omega C_3 R_3$$

4.15 (d)

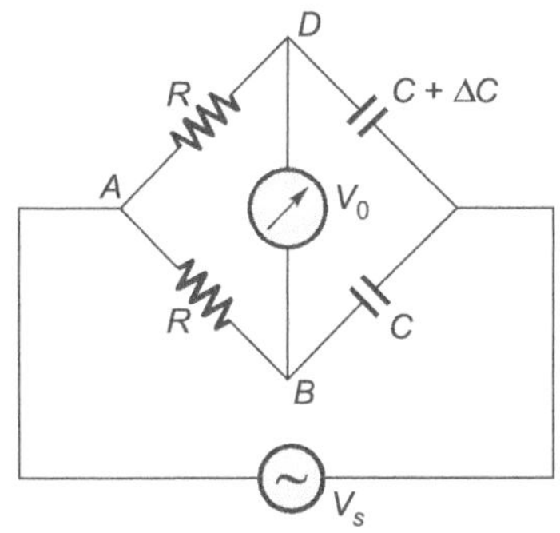

Let, $V_o = V_D - V_C$

$$V_D = \frac{1}{1 + j\omega R(C + \Delta C)} V_s$$

$$V_D = \frac{1}{1 + j\left[1 + \frac{\Delta C}{C}\right]} \quad [\because \omega RC = 1]$$

$$V_C = \frac{1}{1 + j} V_s$$

$$\frac{V_o}{V_s} = \frac{1}{1 + j\left[1 + \frac{\Delta C}{C}\right]} - \frac{1}{1 + j}$$

$$= \frac{1 + j - 1 + j\left[1 + \frac{\Delta C}{C}\right]}{[1 + j] - \left[1 + j\left(1 + \frac{\Delta C}{C}\right)\right]}$$

$$= \frac{(1 + j) - (1 + j) - j\left(\frac{\Delta C}{C}\right)}{(1 + j)^2} \quad \left[\because \left[1 + \frac{\Delta C}{C}\right] \simeq 1\right]$$

$$\left|\frac{V_o}{V_s}\right| = \frac{\left[\frac{\Delta C}{C}\right]}{(1 + j)^2} \simeq \frac{\Delta C}{2C}$$

4.16 (d)

for $V_0 = 0$, bridge must be balanced

$$R\left(R + \frac{1}{j\omega C}\right) = R(R + j\omega L)$$

$$R\left(R - \frac{j}{\omega C}\right) = R(R + j\omega L)$$

$$R^2 - \frac{R}{\omega C} j = R^2 + R\omega Lj$$

Comparing the imaginary parts as real parts are already equal

$$-\frac{R}{\omega C} = R\omega L$$

$$\omega^2 = -\frac{1}{LC} = \frac{\sqrt{-1}}{LC} = \frac{j}{\sqrt{LC}}$$

∴ Frequency is imaginary which is impossible. Hence, V_0 cannot be made zero.

4.17 (a)

At balance,

$$\left(R - \frac{j}{\omega C}\right)\left(\frac{R\left(\frac{-j}{\omega C}\right)}{\left(R - \frac{j}{\omega C}\right)}\right) = RZ$$

$$R\left(\frac{-j}{\omega C}\right) = R(X + jY) \qquad [\text{Let } Z = X + jY]$$

Comparing the real and imaginary parts we get,

$$RX = 0$$

$$X = 0$$

and $$R\left(\frac{-j}{\omega C}\right) = R\,jY$$

$$Y = \frac{-1}{\omega C} = \frac{-1}{\omega C}$$

∴ 'Z' is purely capacitive.

4.18 (a)

Heaviside Campbell bridge method is commonly used for finding mutual inductance.

4.19 (b)

At balance,

$$R_1 \times Z_2 = R_3\left[R_4 + j\omega L_4\right]$$

$$\therefore \quad Z_2 = \frac{R_3 R_4}{R_1} + \frac{j\omega L_4}{R_1}$$

∴ Z_2 should consist of R and L in series.

4.20 (1)

$$Q = \frac{\omega L}{R}$$

$$Q \propto \omega \propto f$$

$$\therefore \quad \frac{Q_1}{Q_2} = \frac{f_1}{f_2}$$

∴ Q of the coil at 20 kHz will be

$$Q_2 = \frac{Q_1 f_2}{f_1} = \frac{5 \times 20}{100} = 1$$

4.21 (10)

$$C_1 = 110\text{ pF},$$

$$C_2 = 20\text{ pF}$$

$$n = \frac{f_2}{f_1} = \frac{2f}{f} = 2$$

$$C_d = \frac{C_1 - n^2 C_2}{n^2 - 1} = \frac{110 - 4 \times 20}{3}$$

$$= 10 \text{ pF}$$

4.22 (10000)

Redraw the bridge:

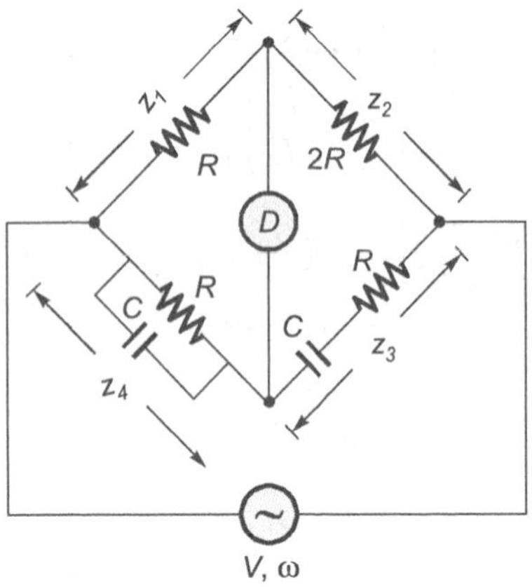

At balance condition,

$$Z_1 Z_3 = Z_2 Z_4$$

$$R\left[R + \frac{1}{j\omega C}\right] = 2R\left[\frac{R}{1 + j\omega RC}\right]$$

$$\left[R + \frac{1}{j\omega C}\right](1 + j\omega RC) = 2R$$

On comparing Real and Imaginary part

$$R + R = 2R \qquad \text{...(i)}$$

and, $$\frac{1}{j\omega C} + j\omega R^2 C = 0$$

$$\frac{1}{j\omega C} = -j\omega R^2 C$$

$$\omega = \frac{1}{RC} = \frac{1}{10^3 \times 10^{-7}}$$

$$= 10000 \text{ rad/sec.}$$

4.23 1.0 (0.9 to 1.1)

Given, $\omega_1 = 10^6$ rad/sec $\quad C_1 = 990$ pF

$\omega_2 = 2 \times 10^6$ rad/sec $\quad C_2 = 240$ pF

Here, $\omega_2 = 2\omega_1$

$$\therefore \quad C_d = \frac{C_1 - 4C_2}{3}$$

$$= \frac{990 - 4 \times 240}{3} = 10 \text{ pF}$$

We know at resonance

$$f = \frac{1}{2\pi\sqrt{L(C + C_d)}}$$

or $$L = \frac{1}{4\pi^2 f^2 (C + C_d)}$$

Total $C = C_1 = 990$ pF

$$L = \frac{1}{\omega_1^2 (C + C_d)}$$

$$= \frac{1}{10^6 \times 10^6 \times (990 + 10) \times 10^{-12}}$$

$$= 1.0 \text{ mH}$$

4.24 222 (221 to 223)

In the given bridge circuit.

$$Z_1 = \frac{R_1}{1 + j\omega C_1 R_1}$$

$$Z_2 = R_2$$

$$Z_3 = R_x + j\omega L_x$$

$$Z_4 = R_4$$

(2) balance product of impedance of opposite arms are equal.

i.e. $$Z_1 Z_3 = Z_2 Z_4$$

$$\frac{R_1}{1 + j\omega C_1 R_1}(R_x + j\omega L_x) = R_2 R_4$$

$$R_1 R_x + j\omega R_1 L_x = R_2 R_4 + j\omega C_1 R_1 R_2 R_4$$

Equating the imaginary parts

$$L_x = C_1 R_2 R_4$$

Substituting the values we have

$$L_x = 1 \times 10^{-9} \times 22.2 \times 10^3 \times 10 \times 10^3$$

$$= 222 \text{ mH}$$

4.25 (0.042) 0.035 to 0.045

In the given bridge circuit,

$$Z_1 = \frac{R_1}{1 + j\omega C_1 R_1}$$

$$Z_2 = \frac{R_P}{1 + j\omega C_P R_P}$$

$$Z_3 = R_3$$

$$Z_4 = R_4$$

At balance condition,

$$Z_1 Z_4 = Z_2 Z_3$$

$$\frac{R_1 R_4}{1 + j\omega C_1 R_1} = \frac{R_P R_3}{1 + j\omega C_P R_P}$$

$$R_1 R_4 + j\omega C_P R_P R_1 R_4 = R_P R_3 + j\omega C_1 R_1 R_P R_3$$

On comparing the real and imaginary components

$$R_1 R_4 = R_P R_3 \text{ or } R_P = \frac{R_1 R_4}{R_3}$$

$$C_P R_4 = C_1 R_3$$

or $$C_P = \frac{C_1R_3}{R_4}$$

Given, $$D = \frac{1}{\omega C_P R_P} = \frac{1}{\omega}\times\frac{R_4}{C_1R_3}\times\frac{R_3}{R_1R_4} = \frac{1}{\omega C_1R_1}$$

Given, $C_1 = 0.1 \times 10^{-6}$F

$R_1 = 375 \times 10^3\Omega$

$\omega = 2 \times \pi \times 100 = 628.318$

$$\therefore D = \frac{1}{628.318\times0.1\times10^{-6}\times375\times10^3} \simeq 0.042$$

4.27 (10)

Given, $C_1 = 100$ nF

$R_1 = 100$ kΩ

$\omega = 1000$

Bridge at balance condition,

$$Q = \omega C_1R_1$$

$= 1000 \times 100 \times 10^{-9} \times 100 \times 10^3 = 10$

4.28 (a, c, d)

Bridge in balanced conditions:

- Maxwell bridge at balance :

$$R_1 = \frac{R_2R_3}{R_4}$$

$$L_1 = R_2R_3C_4$$

- Schering bridge at balance :

$$r_1 = \frac{R_3C_4}{C_2}$$

$$C_1 = C_2\left(\frac{R_4}{R_3}\right)$$

- Wheat stone bridge is DC bridge used for measurement of resistance.
- Wein bridge at balance, for a given frequency has fixed values of circuit parameters which vary with change in frequency (dependent on frequency).

4.29 (14.50)

At balance condition,

$$R_1R_4 = (R_2 + j\omega L_3)\times\left[R_2 \parallel \frac{1}{jg\omega C_2}\right]$$

$$R_2 + j\omega L_3 = [R_1 R_4]\times\left[\frac{1}{R_2} + j\omega C_2\right]$$

$$R_2 + j\omega L_3 = \frac{R_1R_4}{R_2} + j\omega C_2R_1R_4$$

$L_3 = R_1 R_4 C_2 = 100 \times 50 \times 2.9 \times 10^{-6} = 14.50$ mH

4.30 (a)

For LCR meter,

frequency, $$f = \frac{1}{2\pi\sqrt{LC}}$$

$$L = \frac{1}{(2\pi)^2f^2C}$$

where, $$C = \frac{\epsilon A}{d} = \frac{\epsilon_0\epsilon_r A}{d}$$

$$L \propto \frac{1}{\epsilon_r}$$

by introducing copper sheet near the coil, inductor L reduces

$$\frac{L_1}{L_2} = \frac{\epsilon_{r_1}}{\epsilon_{r_2}}$$

$$L_2 = \frac{L_1}{\epsilon_{r_2}} = \frac{10\text{ mH}}{\epsilon_{r_2}}$$

Since, $\epsilon_{r_2} > 1$ for copper.

$\Rightarrow$ $L_2 < 10$ mH

4.32 (c)

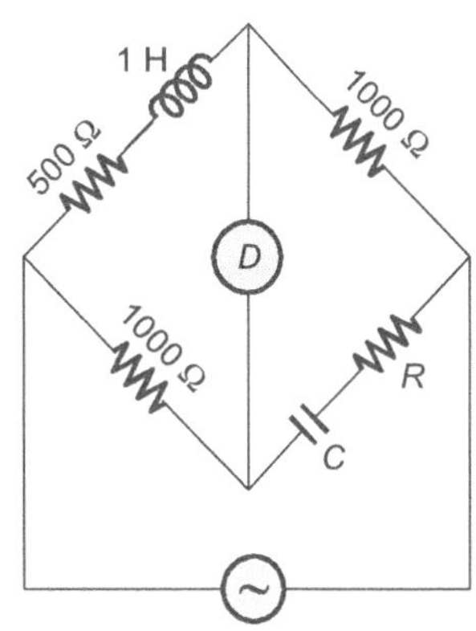

On balance condition

$$R + \frac{1}{j5000C} = \frac{1000\times1000}{500+j5000} = \frac{10^6}{500(j+j10)}$$

$$R + \frac{1}{j5000} = \frac{10^6(1-j10)}{500\times101}$$

$$R = \frac{10^6}{500\times101} = 19.80\ \Omega$$

$$\frac{1}{j5000C} = \frac{10^6\times10}{j(500\times101)}$$

$C = 1.01$ μf

5 Measurement of Energy and Power

5.1 The voltage drop across a non-linear element is given as $V = I + 2I^2$. The power loss across it for a current of sint is

(a) 1.0 W (b) 2.0 W
(c) 3.0 W (d) 0.5 W

[1996 : 1 M]

5.2 A 5 A, 110 V electrodynamic type wattmeter has a scale having 110 divisions. Its pressure coil is fed by a voltage of

$[110\sqrt{2}\cos(314t) + \sqrt{2}\sin(942t)\ V$

and its current coil carries a current of

$[5\sqrt{2}\cos(314t+60) + 2\sqrt{2}\sin(628t+90)$
$+\sqrt{2}\cos(642t+90)]\ A.$

The needle of the wattmeter will move to

(a) 110 divisions (b) 50 divisions
(c) 54 divisions (d) 55 divisions

[2002 : 2 M]

5.3 The current coil of a 200 V, 5 A, electrodynamometer type LPF wattmeter carries a current of $\sqrt{2}\cos(100\pi t)A$. The voltage across the pressure coil is $\sqrt{2}\sin(100\pi t)V$. The meter will indicate

(a) 0 W (b) 100 W
(c) 200 W (d) 400 W

[2003 : 1 M]

5.4 The brake magnet of an error free electromechanical, induction type energy meter is shifted from its position and moved a small distance towards the edge of the disc. Then the meter will

(a) continue to work correctly without any change in its characteristics
(b) continue to work, but registers a lower kWhr
(c) continue to work, but registers a higher kWhr
(d) stop working

[2004 : 1 M]

5.5 A 3-phase balanced R-L load is connected to a 3-phase balanced symmetrical 400 V, 50 Hz, RYB sequence mains as shown in figure. The wattmeter W reads 1.732 kW. Then,

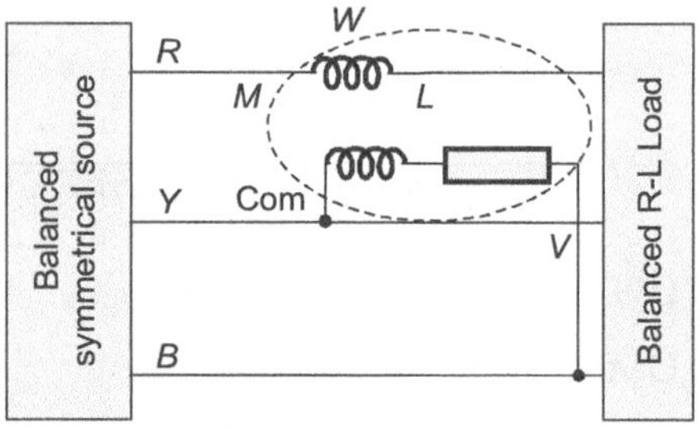

(a) the active power in the load is 1.732 kW
(b) the reactive power of the load is 1.732 kVAr
(c) the active power in the load is 3 kW
(d) the reactive power of the load is 3 kVAr

[2004 : 2 M]

5.6 The two-wattmeter method is used to measure power in a balanced 3-phase circuit, drawing lagging current. The power factor if one of the wattmeters reads zero is

(a) 0.5 (b) 0.8
(c) 1.0 (d) 0.0

[2005 : 2 M]

5.7 The two-wattmeter method is used to measure power in 3-phase, 3-wire balanced inductive circuit. The line voltage and line current are 400 V and 10 A respectively. If the load power factor is 0.866 lagging, then readings of the two wattmeters are

(a) 6000 W and 0 W
(b) 5000 W and 1000 W
(c) 4500 W and 1500 W
(d) 4000 W and 2000 W

[2006 : 2 M]

5.8 A dynamometer type wattmeter with a single scale marked for the smallest power range, has two current ranges, namely, 0- 5 A and 0-10 A as well as two voltage ranges, namely, 0-150 V and 0-300 V. To carry out a load test on a 230 V/115 V, 1 kVA, single phase transformer, the wattmeter is used on the high voltage side. The voltage and current ranges are chosen for maximum utilization of the scale. The multiplying factor to be used in this case is

(a) 0.5 (b) 1.0
(c) 2.0 (d) 4.0

[2007 : 1 M]

5.9 Power in a three phase star connected balanced inductive load is measured by two wattmeter method. The phase voltage and phase current are 230 V and 5 A, respectively. The power factor of the load is 0.707. The readings P_1 and P_2 of the two wattmeters are

(a) P_1 = 298 W and P_2 = 1111 W
(b) P_1 = 516 W and P_2 = 1924 W
(c) P_1 = 1220 W and P_2 = 1220 W
(d) P_1 = 1111 W and P_2 = – 516 W

[2011 : 2 M]

5.10 For the circuit in the figure, the voltage and current expressions are:

$$v(t) = E_1 \sin(\omega t) + E_3 \sin(3\omega t) \text{ and}$$
$$i(t) = I_1 \sin(\omega t - \phi_1) + I_3 \sin(3\omega t - \phi_3) + I_5 \sin(5\omega t)$$

The average power measured by the Wattmeter is

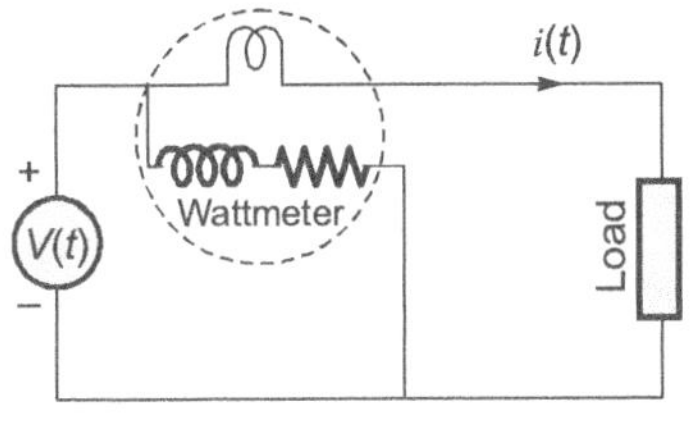

(a) $\frac{1}{2} E_1 I_1 \cos\phi_1$

(b) $\frac{1}{2}[E_1 I_1 \cos\phi_1 + E_1 I_3 \cos\phi_3 + E_1 I_5]$

(c) $\frac{1}{2}[E_1 I_1 \cos\phi_1 + E_3 I_3 \cos\phi_3]$

(d) $\frac{1}{2}[E_1 I_1 \cos\phi_1 + E_3 I_1 \cos\phi_1]$

[2012 : 1 M]

5.11 A symmetrical three-phase three-wire RYB system is connected to a balanced delta-connected load. The RMS values of the line current and line-to-line voltage are 10 A and 400 V respectively. The power in the system is measured using the two wattmeter method. The first wattmeter connected between *R*-line and *Y*-line reads zero. The reading of the second wattmeter (connected between *B*-line and *Y*-line) in watt is______.

[2016 : 2 M]

5.12 The power delivered to a single phase inductive load is measured with a dynamometer type wattmeter using a potential transformer (PT) of turns ratio 200:1 and the current transformer (CT) of turns ratio 1:5. Assume both the transformers to be ideal. The power factor of the load is 0.8. If the wattmeter reading is 200 W, then the apparent power of the load in kVA is _______.

[2017 : 2 M]

5.13 A 300 V, 5 A, 0.2 pf low power factor wattmeter is used to measure the power consumed by a load. The wattmeter scale has 150 divisions and the pointer is on the 100th division. The power consumed by the load (in Watts) is _______.

[2018 : 1 M]

5.14 The real power drawn by a balanced load connected to a 400 V, 50 Hz, balanced, symmetrical 3-phase, 3-wire, RYB sequence main is measured using the two-wattmeter method. Wattmeter W_1 is connected in the *R* line and wattmeter W_2 is connected in the *B* line. The line current is measured as $\frac{1}{\sqrt{3}}$ A. If the wattmeter W_1 reads zero, the reading on W_1 (in watts) is _____.

[2020 : 2 M]

5.15 A 300 V, 5 A, LPF wattmeter has a full scale of 300 W. The wattmeter can be used for loads supplied by 300 V ac mains with a maximum power factor of _____.
(rounded off to one decimal place)

[2021 : 1 M]

5.16 The power in a 400 V (rms, line-line) three-phase, three-wire RYB sequence system is measured using the two wattmeters, as shown. The R-line current is 5∠60°A Wattmeter W_1 in the R-line will read (in watt) _____.

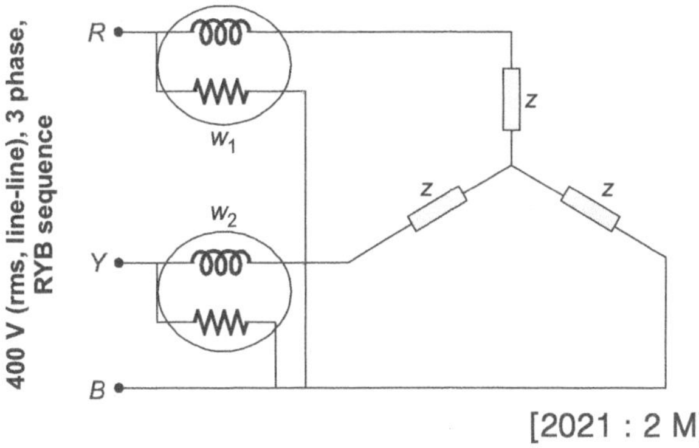

[2021 : 2 M]

5.17 The resistance of a pure copper wire of length 10 cm and diameter 1 mm is to be measured. The most suitable method from amongst the choices given below is ________

(a) Two wire method (b) Three wire method
(c) Four wire method (d) Ellipsometry

[2022 : 1 M]

5.18 In the balanced three-phase circuit shown, C_0 = 8.2 μF and the line-to-line r.m.s. voltage is 440 V at 50 Hz. The reading on the wattmeter (in watts) is ______ (round off to two decimal places)

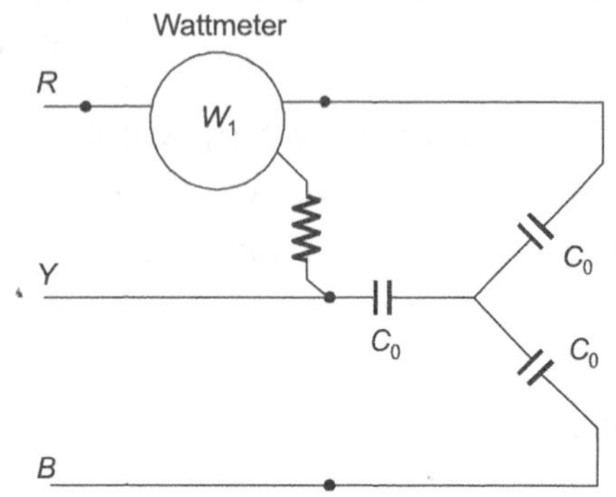

[2022 : 2 M]

5.19 The full-scale range of the wattmeter shown in the circuit is 100 W. The turns ratio of the individual transformers are indicated in the figure. The RMS value of the ac source voltage V_s is 200 V. The wattmeter reading will be ________ W (rounded off to the nearest integer).

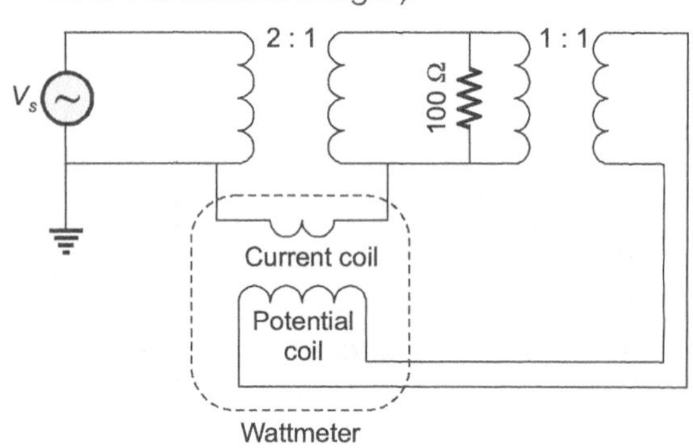

[2023 : 1 M]

5.20 In the figure shown, assume

- α is the phase angle between the load current and the load voltage
- β is the phase angle by which pressure coil current lags the pressure coil voltage of the wattmeter
- γ is the phase angle between currents in the pressure coil and the current coil of the wattmeter
- δ is the phase angle of the voltage transformer
- θ is the phase angle of the current transformer

When the load has a lagging phase angle of α, which one of the following options is correct?

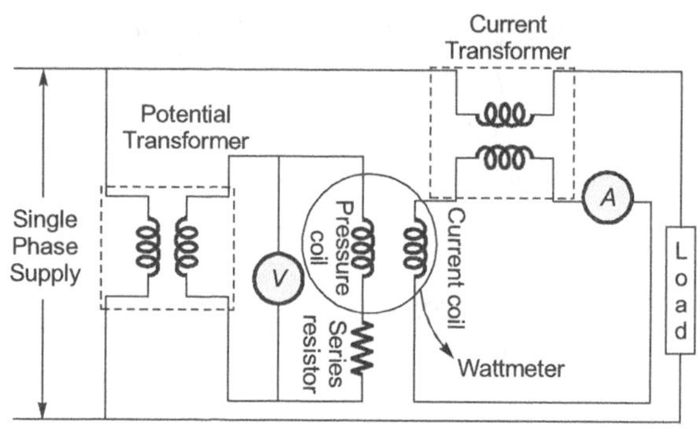

(a) $\alpha = -\gamma \pm \delta \pm \theta - \beta$
(b) $\alpha = -\gamma \pm \delta \pm \theta + \beta$
(c) $\alpha = -\gamma \pm \delta \pm \theta - \beta$
(d) $\alpha = -\gamma \pm \delta \pm \theta - \beta$

[2024 : 2 M]

■■■

Answers Measurement of Energy & Power

5.1 (*)	5.2 (d)	5.3 (a)	5.4 (b)	5.5 (b)	5.6 (a)	5.7 (d)
5.8 (d)	5.9 (b)	5.10 (c)	5.11 (3464.1)	5.12 (250)	5.13 (200)	5.14 (200)
5.15 (0.2)	5.16 (0)	5.17 (c)	5.18 (143.97)	5.19 (0)	5.20 (c)	

Explanations Measurement of Energy & Power

5.2 (d)

$$P_{avg} = V_{RMS} I_{RMS} \cos\phi$$

In V_{RMS} and I_{RMS} we consider frequency components which are present in both V and I.

$$P_{avg} = \frac{110\sqrt{2}}{\sqrt{2}} \cdot \frac{5\sqrt{2}}{\sqrt{2}} \cos(60°)$$

$$P_{avg} = 110 \times 5 \times \frac{1}{2} = 275 \text{ W}$$

Also at full scale, $P = 110 \times 5 = 550$ W which require 110 divisions.

So, for 275 W, number of divisions needed

$$= \frac{275}{550} \times 110 = 55$$

5.3 (a)

Reading of wattmeter = $V_{PC} I_{CC} \cos\phi$

where, f is angle between V_{PC} and I_{CC}.

Now, given,

$$I_{CC} = \sqrt{2}\cos(100\pi t)\text{ A}$$
$$= \sqrt{2}\sin(100\pi t + 90°)\text{ A}$$
$$V_{PC} = \sqrt{2}\sin(100\pi t)\text{ V}$$
$$P = VI\cos(90°) = 0 \text{ W}$$

5.4 (b)

Speed of the energy meter,

$$N \propto \frac{r}{\phi^2 R}$$

where, $r \Rightarrow$ resistance

$\phi \Rightarrow$ flux

$R \Rightarrow$ radius

So, if $R\uparrow$, $N\downarrow$ $\therefore$ it reads a lower kWhr.

5.5 (b)

Wattmeter reading,

$$B = |V_{PC}||I_{CC}|\cos\phi$$

Now, $V_{PC} = V_B = V_Y - V_B$

$I_{CC} = I_R$

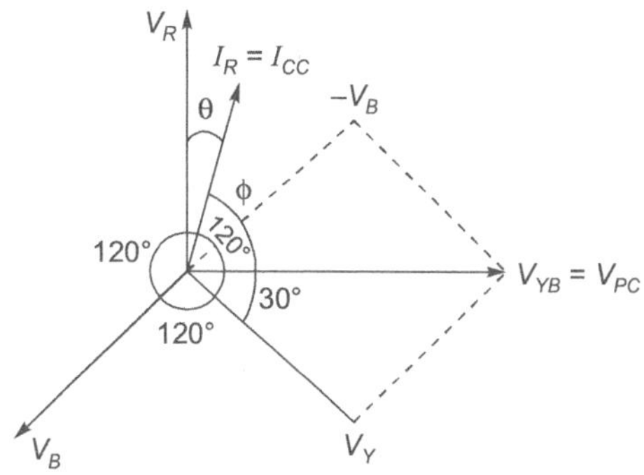

For phasor diagram,

$$\phi = (90 - \theta)$$

So, Wattmeter reading,

$$W = |V_{YB}||I_R|\cos\phi$$

$$\Rightarrow \quad W = \sqrt{3}VI\cos(90-\theta)$$

$$\Rightarrow \quad W = \sqrt{3}VI\sin\theta = Q$$

So, meter reads reactive power of the load.

5.6 (a)

Power factor $= \cos\phi$

$$= \cos\tan^{-1}\sqrt{3}\frac{(W_1 - W_2)}{(W_1 + W_2)}$$

When one of the two wattmeters reads zero.

i.e. say $W_2 = 0$

then, $$\cos\phi = \cos\tan^{-1}\sqrt{3}\frac{(W_1 - 0)}{(W_1 + 0)}$$

$$= \cos\tan^{-1}\sqrt{3}$$

$$= \cos\tan^{-1}(1.73) = 0.5$$

5.7 (d)

It is given that,

$$V_L = 400 \text{ V}$$
$$I_L = 10 \text{ A}$$
$$\cos\phi = 0.866$$
$$\therefore \quad \phi = \cos^{-1}(0.866) = 30°$$

$$W_1 = V_L I_L \cos(30° - \phi)$$
$$= 400 \times 10 \times \cos 0°$$
$$= 4000 \times 1 = 4000 \text{ W}$$

$$W_2 = V_L I_L \cos(30° + \phi)$$
$$= 400 \times 10 \times \cos(60°)$$
$$= 4000 \times \left(\frac{1}{2}\right) = 2000 \text{ W}$$

5.8 (d)

Upper voltage of a 230/115 V, 1 kVA transformer is 230 V.

Power of transformer = 1 kVA = 1000 kVA

$$\therefore \quad I = \frac{1000}{230} = 4.34 \text{ A}$$

∴ the current comes in the (0 – 5 A) range

∴ I, for wattmeter = 5 A

V, for wattmeter = 230 V

Full scale deflection for 230 V = 300 V (FDS)

$$\text{Multiplying factor (MF)} = \frac{V \times I \times \text{p.f.}}{FSD}$$
$$= \frac{230 \times 5 \times 1}{300} \quad \text{(for unity p.f)}$$
$$= 3.83 \simeq 4.0$$

5.9 (b)

Given,

$V_{ph} = 230$ V, $I_{ph} = 5$ A and $\cos\phi = 0.707$

For star connection,

$$V_L = \sqrt{3} V_{ph} \text{ and } I_L = I_{pH}$$

in two wattmeter method,

$$W_1 = V_L I_L \cos(30 + \phi)$$
$$= \sqrt{3} \times 230 \times 5 \times \cos 75° = 515.53 \text{ W}$$

and $$W_2 = V_L I_L \cos(30 - \phi)$$
$$= \sqrt{3} \times 230 \times 5 \times \cos 15° = 1923.99 \text{ W}$$

Method-2:

Total active power,

$$P = 3V_{ph} I_{ph} \cos\phi$$
$$= 3 \times 230 \times 5 \times 0.707 = 2439.15 \text{ W}$$

Also, $\cos\phi \neq 1$

$\Rightarrow \quad W_1 \neq W_2$

So, from option (b).

5.10 (c)

$$v(t) = E_1 \sin \omega t + E_3 \sin 3\omega t$$
$$i(t) = I_1 \sin(\omega t - \phi_1) + I_3 \sin(3\omega t - \phi_3) + I_5 \sin 5\omega t$$

Average power

$$P_{avg} = \frac{1}{2\pi} \int_0^{2\pi} vi \, d(\omega t)$$

The product of different frequency terms have zero average value.

$$\therefore P_{avg} = \frac{1}{2} E_1 I_1 \cos\phi_1 + \frac{1}{2} E_3 I_3 \cos\phi_3$$

5.11 (3464.1)

As the load is balanced and one of the watt-meter read zero power factor is (0.5) reading of the other watt-meter will be

$$W_1 = V_L I_L \cos(30 - \phi)$$
$$W_2 = V_L I_L \cos(30 + \phi)$$

when, $\cos\phi = 0.5$

$$W_2 = V_L I_L \cos(30 + 60)$$
$$= V_L I_L \cos 90 = 0$$
$$W_1 = 10 \times 400 \cos(30 - 60) = 3464.1 \text{ W}$$

5.12 (250)

As both the transformers are ideal the apparent power in KVA will be

$$\frac{V}{200} \times \frac{I}{5} \times \cos\phi = 200 \text{ W}$$
$$\Rightarrow \quad VI = \frac{200 \times 200 \times 5}{0.8} = 250 \text{ kVA}$$

5.13 (200)

At full scale the wattmeter will indicate

300 × 5 × 0.2 = 300 Watts

As the scale of the wattmeter is uniform.

300 W → 150 divisions

? → 100 divisions

$$= \frac{300 \times 100}{150} = 200 \text{ Watts}$$

5.14 (200)

In a 2 wattmeter method is one of the wattmeter reads '0' then $\phi = 60°$ and the second wattmeter indicates the total power consumed by the load and

$$\therefore \quad P = P_1 + P_2 = \frac{3}{2} V_{PH} I_{PH}$$

As $$I_{PH} = I_L = \frac{1}{\sqrt{3}}$$

$$V_{PH} = \frac{400}{\sqrt{3}}$$

$$\therefore \quad P = \frac{3}{2} \times \frac{400}{\sqrt{3}} \times \frac{1}{\sqrt{3}} = \frac{400}{2}$$
$$= 200 \text{ Watts}$$

5.15 (0.2)

For low power factor wattmeter.

$$VI\cos\phi = 300$$
$$300 \times 5 \cos\phi = 300$$
$$\cos\phi = 0.2$$

5.16 (0)

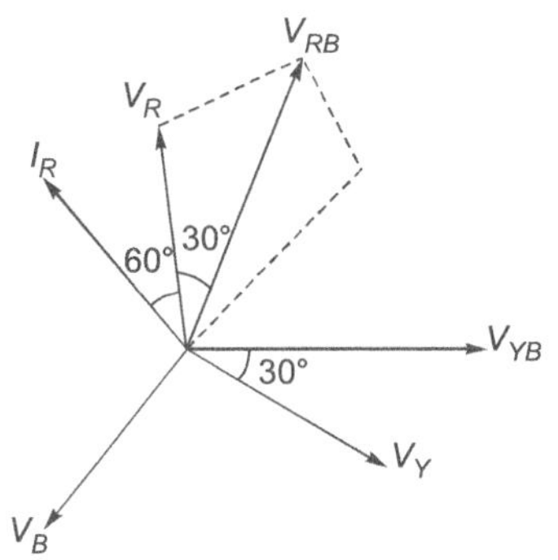

As wattmeter W_1 is connected with pressure coil in R-B phase and current coil in R-Phase.

So, Power measured in wattmeter W_1 is $= V_{RB} \times I_R\cos\theta = 400 \times 5 \times \cos 90° = 0$ Watt

5.17 (c)

For very low (in mΩ) resistance 4-wire method is used.

In this case,

$$R = \frac{\rho L}{A} = \frac{1.6\times10^{-8}\times10\times10^{-2}\times4}{3.14\times(1\times10^{-3})^2} = 8.03\ \text{m}\Omega$$

5.18 (143.97)

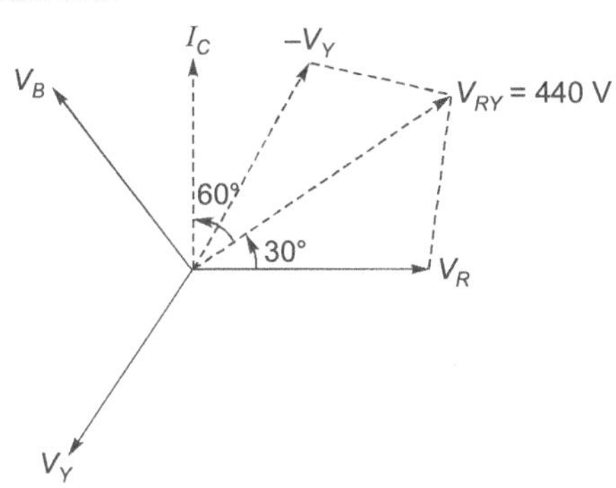

Reading of wattmeter :

$$W_1 = V_{RY} I_C \cos\angle V_{RY} I_C$$
$$= 440\times\frac{440}{\sqrt{3}}\times100\pi\times8.2\times10^{-6}\times\cos(60°)$$
$$= \frac{(440)^2\times100\pi\times8.2\times10^{-6}}{2\sqrt{3}}$$
$$W_1 = 143.97 \text{ Watt}$$

5.19 (0)

Since transformer is act as isolation device, there is no current flowing through secondary winding of 1 : 1 transformer hence the current coil draws no current. Hence, the wattmeter reads 0 Watt.

5.20 (c)

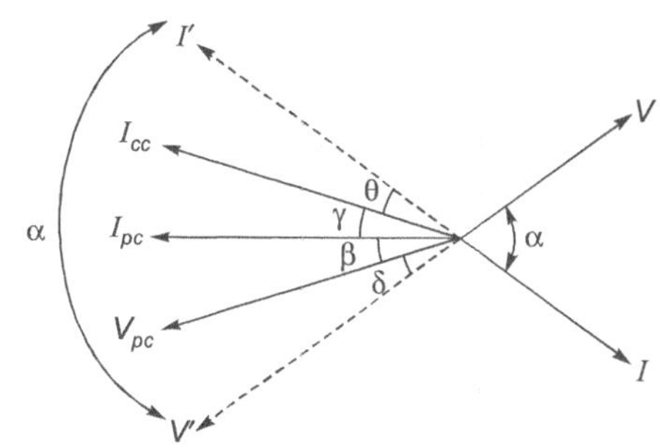

For Lagging leads $\theta = +ve$

$$\alpha = \gamma + \beta \pm \delta \pm \theta$$

■■■

6 Measurement of Phase and Frequency

6.1 In a dual slope type digital voltmeter, an unknown signal voltage is integrated over 100 cycles of the clock. If the signal has a 50 Hz pick up, the maximum clock frequency can be

(a) 50 Hz (b) 5 kHz
(c) 10 kHz (d) 50 kHz

[2001 : 2 M]

6.2 A counter timer has a basic clock of 16 MHz. The count value displayed is in error by 1 count. The frequency at which the error in the displayed value is the same whether the counter-timer is used in the frequency mode of operation or period mode of operation is_______ .

(a) 16 MHz (b) 10 MHz
(c) 8 MHz (d) 4 MHz

[2002 : 1 M]

■■■

Answers Measurement of Phase & Frequency

6.1 (b) 6.2 (a)

Explanations Measurement of Phase & Frequency

6.1 (b)

Clock frequency,

$$f_{CLK} = \frac{100}{n} \times f_s$$

For make clock frequency frequency,

$$n = 1$$

$$(f_{CLK})_{max} = \frac{100}{1} \times 50 \text{ Hz} \qquad [(n)_{minimum} = 1]$$

$$= 5 \text{ kHz}$$

6.2 (a)

Pulse frequency

$$= \frac{\text{Clock frequency}}{\text{Count value}}$$

$$= \frac{16 \text{ MHz}}{1} = 16 \text{ MHz}$$

■■■

7 CRO & Electronic Measurements

7.1 An amplitude modulated signal can be observed on a CRO by applying the following waveforms to the external trigger input:
(a) the modulated waveform itself
(b) derivative of the modulated waveform
(c) the modulating waveform
(d) the modulated waveform **[1995 : 1 M]**

7.2 The Post-Deflection Acceleration (PDA) in a CRT is used in high frequency Oscilloscopes mainly to ____.
(a) increase deflection sensitivity
(b) improve beam focusing
(c) improve deflection linearity
(d) improve the spot brightness

[1997 : 1 M]

7.3 An oscilloscope input impedance consists of 1 MΩ resistance connected in parallel with a 100 pF capacitor. A compensated 20 : 1 attenuator is obtained by connecting a parallel combination of _____.

(a) 19 MΩ and $\frac{100}{19}$ pF (b) 200 MΩ and $\frac{100}{20}$ pF

(c) 19 MΩ and 1900 pF (d) 19 MΩ and 2000 pF

[1999 : 1 M]

7.4 The elliptical Lissajous pattern shown in figure was observed from two sinusoidal signal sources of same frequency on a CRO.

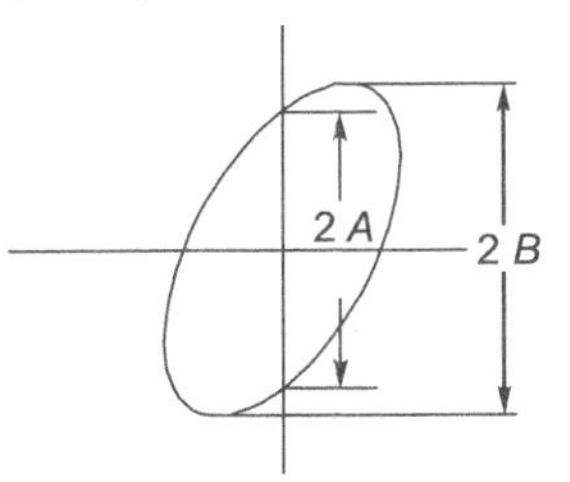

The phase angle between the signal is ____.
(a) $\cos^{-1}(A/B)$ (b) $\tan^{-1}(A/B)$
(c) $\sin^{-1}(A/B)$ (d) (A/B)

[2000 : 1 M]

7.5 A double beam CRO has a bandwidth of 10 MHz. Two signals $V_1(t) = 10 \sin(2\pi \times 20 \times 10^3\, t)$ and $V_2(t) = 10 \sin(2\pi \times 20 \times 10^6\, t)$ are applied to the two channels. If P_1 and P_2 are the peak-to-peak values of the signals displayed, then
(a) $P_1 = 20, P_2 = 20$
(b) $P_1 < 20, P_2 > 20$
(c) $P_1 = 20, P_2 < 20$
(d) $P_1 > 20, P_2 < 20$

[2001 : 2 M]

7.6 An oscilloscope screen displays a line inclined at 45°. Its *Y*-inputs is a sine wave of frequency *f* then the *X*-input should be
(a) sine wave of frequency *f* and 0° phase shift with the *Y*-input
(b) sine wave of frequency *f* and 45° phase shift with the *Y*-input
(c) sine wave of frequency *f* and 90° phase shift with the *Y*-input
(d) saw-tooth wave of frequency *f*

[2003 : 1 M]

7.7 An oscilloscope is powered from 230 V mains having a nominal frequency of nearly 50 Hz. The triggering mode of the oscilloscope is set to LINE. A square wave from a function generator is fed to the *Y* input and a stable display (properly triggered) is obtained when the frequency of the function generator is 297 Hz. The actual frequency of the power supply is
(a) 50 Hz (b) 49.5 Hz
(c) 49.0 Hz (d) 48.5 Hz

[2003 : 2 M]

7.8 In a single beam, single trace analog oscilloscope, a delay line is inserted between
(a) *X*-input and *X*-plates
(b) *Y*-input and *Y*-plates
(c) *X*-input and the trigger circuit
(d) *Y*-input and the trigger circuit

[2004 : 1 M]

7.9 A symmetrical square wave of frequency 25 kHz and a peak-to-peak amplitude of 10 V is fed to *Y*-input of a single channel oscilloscope. The screen appears as shown in figure. Then the *X* and *Y* sensitivities and the trigger settings respectively are

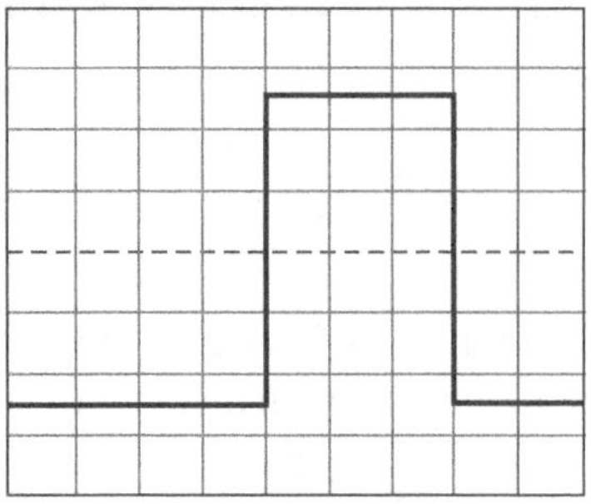

(a) 5 μs/cm, 2 V/cm and +ve slope
(b) 5 μs/cm, 2 V/cm and –ve slope
(c) 10 μs/cm, 1 V/cm and +ve slope
(d) 20 μs/cm, 2 V/cm and –ve slope

[2004 : 2 M]

7.10 A voltage signal $V(t) = A + B \sin \omega t$ is fed to an oscilloscope with the coupling mode set to DC, other settings being appropriate. The displayed wave can be expressed as

(a) $A + B \sin \omega t$
(b) $B \sin \omega t$
(c) A
(d) $|A + B \sin \omega t|$

[2005 : 1 M]

7.11 The linear sweep for the time base in an oscilloscope has deviation from its nominal waveform. The nominal (dashed line) and actual (solid line) sweep waveforms are shown in the following figure.

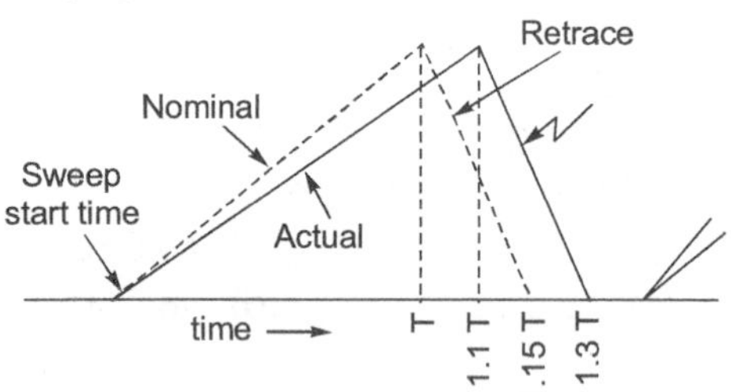

A 5 V p-p sine wave with a frequency of 1 kHz will be measured on the oscilloscope as a sine wave with

(a) 4.45 V p-p and 1 kHz frequency
(b) 5 V p-p and 1 kHz frequency
(c) 5 V p-p and 1.1 kHz frequency
(d) 5 V p-p and 1.15 kHz frequency

[2007 : 2 M]

7.12 The x and y sensitivities of an analog oscilloscope are set as 2 ms/cm and 1 V/cm respectively. The trigger is set at 0 V with negative slope. An input of $2 \cos(100 \pi t + 30°)$ V is fed to the y input of the oscilloscope. The wave form seen on the oscilloscope will be

(a)
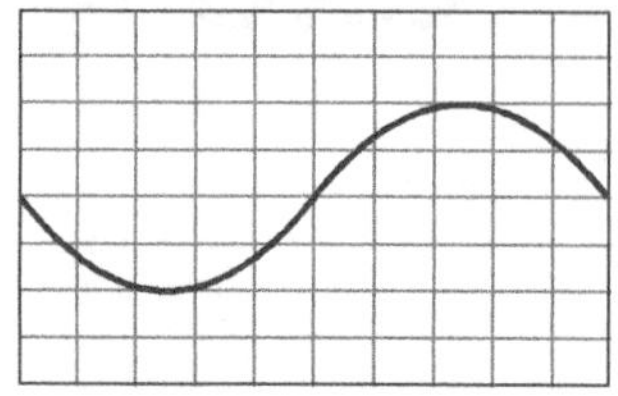

(b)
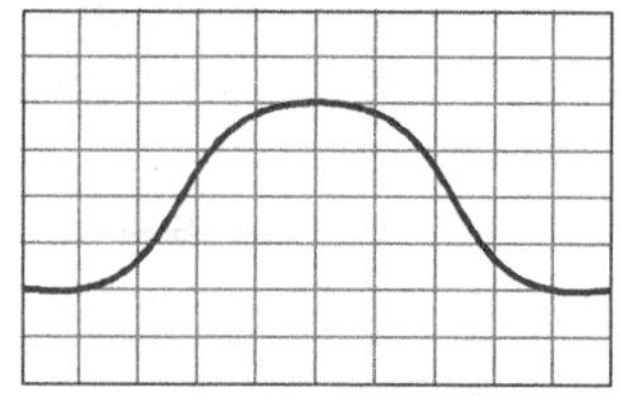

(c)
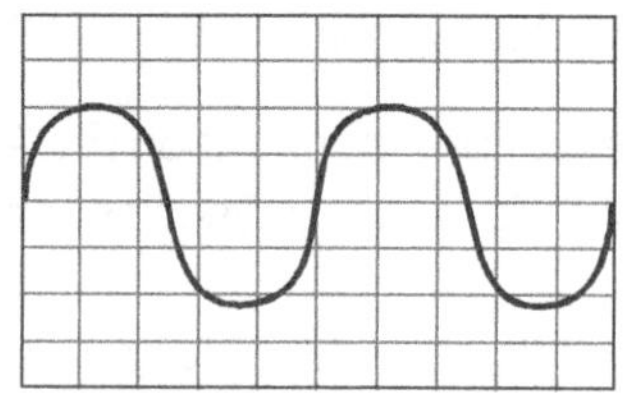

(d)
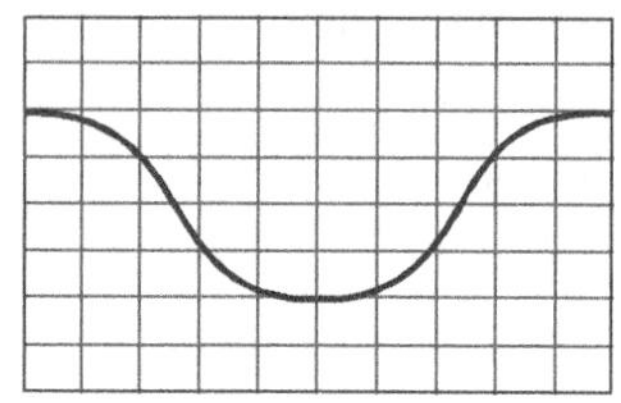

[2008 : 2 M]

7.13 The figure shows a periodic waveform to be displayed on a CRO. A trigger setting which ensures a stationary display is

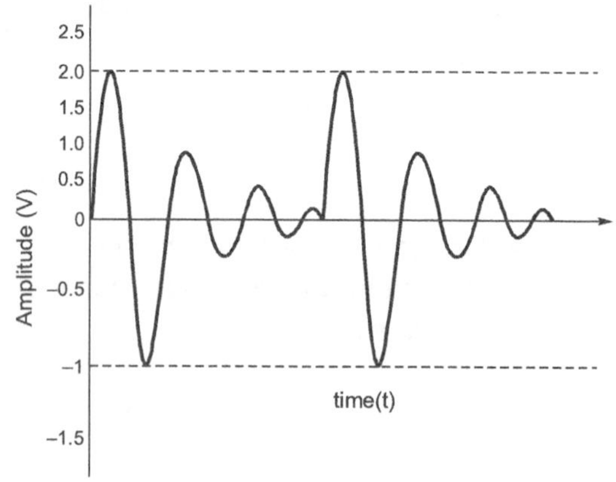

(a) level : 0.2 V, slope: –ve
(b) level: 0.5 V, slop: –ve
(c) level: 0.2 V, slope: +ve
(d) level: 1.8 V, slope: –ve

[2009 : 1 M]

7.14 The input impedance of a CRO is equivalent to a 1 MΩ resistance in parallel with a 45 pF capacitance. It is used with a compensated 10 to 1 attenuation probe. The effective input capacitance at the probe tip is

(a) 4.5 pF
(b) 5 pF
(c) 45 pF
(d) 450 pF

[2009 : 1 M]

7.15 In an analog single channel cathode ray oscilloscope (CRO), the x and y sensitivity are set as 1 ms/div. and 1 V/div. respectively. The y-input is connected to a voltage signal 4cos (200 πt – 45°) V. The trigger source is internal, level chosen is zero and the slope is positive. The display seen on the CRO screen is

(a) 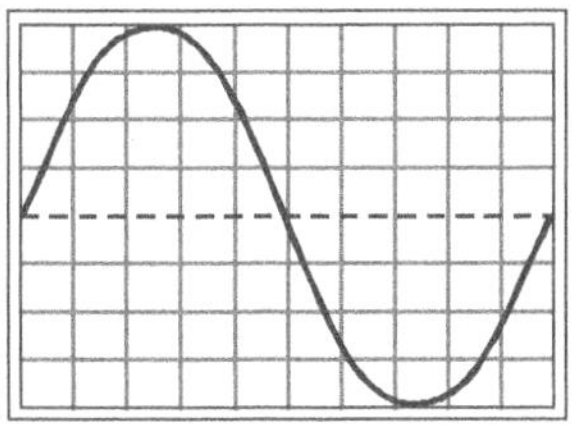

(b) 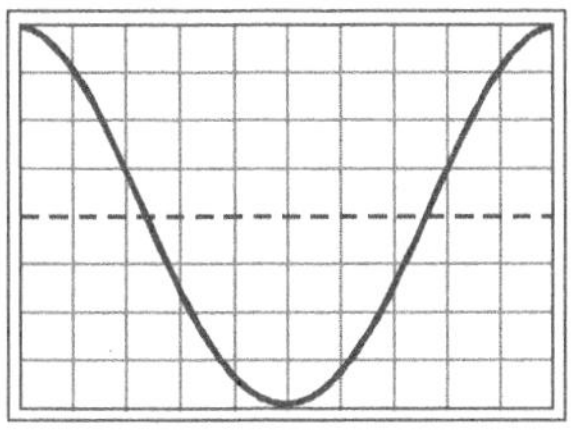

(c) 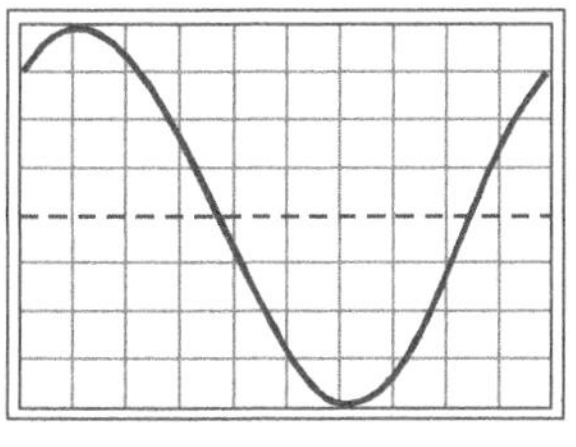

(d) 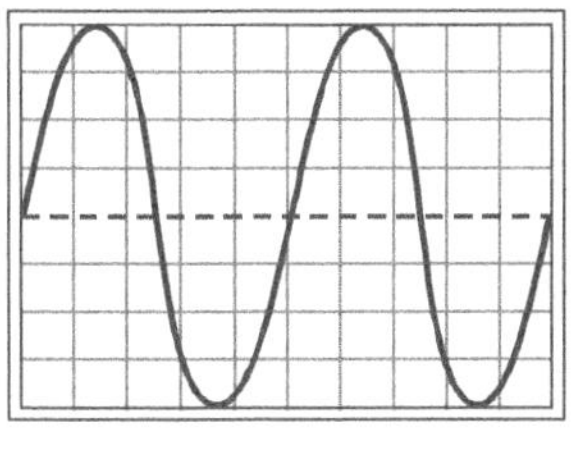

[2010 : 2 M]

7.16 A transfer characteristics of the circuit drawn below is observed on an oscilloscope used in XY mode. The display on the oscilloscope is shown on the right hand side. V_i is connected to the X input with a setting of 0.5 V/div, and V_0 is connected to the Y input with a setting of 2 V/div. The beam is positioned at the origin when V_i is zero.

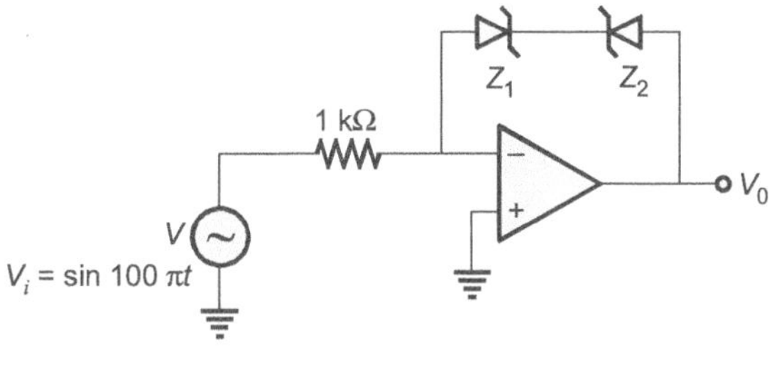

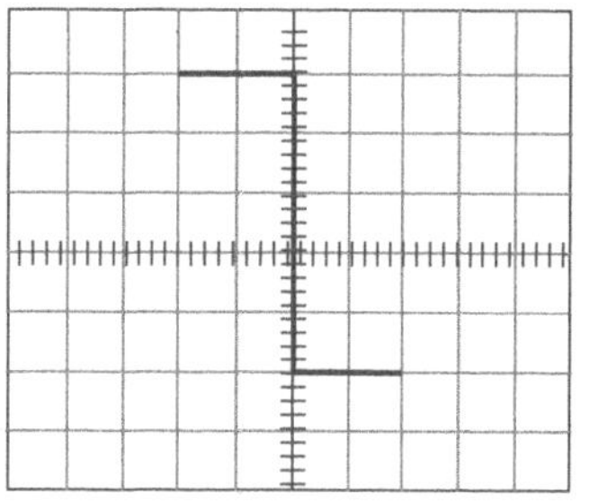

Assuming that the opamp is ideal and Zener diodes have forward biased voltage drop of 0.7 V, the values of reverse break-down voltages of Z_1 and Z_2 are respectively,

(a) 3.3 V and 5.3 V (b) 4.7 V and 6.7 V
(c) 6.7 V and 4.7 V (d) 5.3 V and 3.3 V

[2011 : 2 M]

7.17 A voltage of 6cos($100\pi t$) V is fed as y-input to a CRO. The waveform seen on the screen of the CRO is shown in the figure. The Y and X axes settings for the CRO are respectively

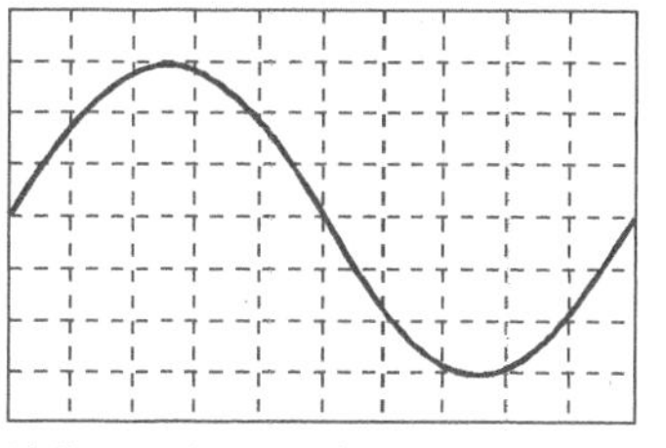

(a) 1 V/div and 1 ms/div
(b) 1 V/div and 2 ms/div
(c) 2 V/div and 1 ms/div
(d) 2 V/div and 2 ms/div

[2018 : 1 M]

7.18 A 200 mV full-scale dual-slope analog to digital converter (DS-ADC) has a reference voltage of 100 mV. The first integration time is set as 100 ms. The DS-ADC is operated in the continuous conversion mode. The conversion time of the DS-ADC for an input voltage of 123.4 mV (in ms, rounded off to one decimal place) is _____.

[2020 : 1 M]

7.19 A $6\frac{1}{2}$ digit timer-counter is set in the 'time period' mode of operation and the range is set as 'ns'. For an input signal, the timer-counter displays 1000000. With the same input signal, the timer-counter is changed to 'frequency' mode of operation and the range is set as 'Hz'. The display will show the number _____.

[2020 : 2 M]

7.20 A $10\frac{1}{2}$ digit Counter-timer is set in the 'frequency mode' of operation (with $T_s = 1s$). For a specific input, the reading obtained is 1000. Without disconnecting this input, the Counter-timer is changed to operate in the 'Period mode' and the range selected is microseconds (μs, with $f_s = 1$ MHz). The counter will then display.

(a) 0 (b) 100

(c) 10 (d) 1000

[2021 : 2 M]

7.21 A resistor ladder digital-to-analog converter (DAC) receives a digital input that results in the circuit having the state as shown in the figure. For this digital input, the Thevenin voltage, V_{th}, and Thevenin resistance, R_{th}, as seen at the output node are ________.

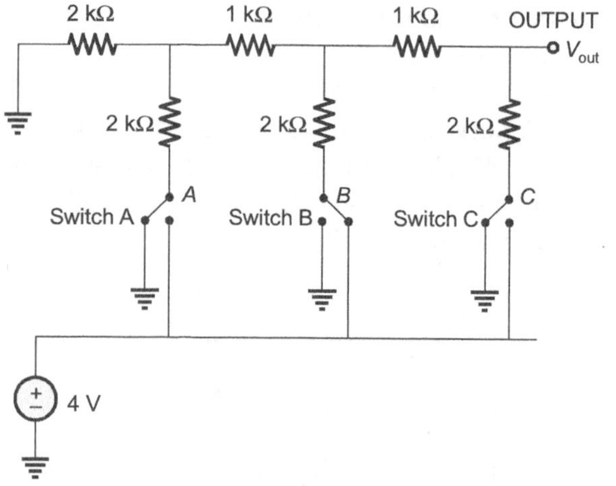

(a) $V_{th} = 0.5$ V, $R_{th} = 1$ kΩ

(b) $V_{th} = 0.5$ V, $R_{th} = 2$ kΩ

(c) $V_{th} = 1$ V, $R_{th} = 1$ kΩ

(d) $V_{th} = 1$ V, $R_{th} = 2$ kΩ

[2022 : 2 M]

■■■

Answers CRO & Electronic Measurements

7.1 (c)	7.2 (d)	7.3 (a)	7.4 (c)	7.5 (a)	7.6 (a)	7.7 (b)
7.8 (b)	7.9 (b)	7.10 (a)	7.11 (b)	7.12 (a)	7.13 (d)	7.14 (a)
7.15 (a)	7.16 (d)	7.17 (d)	7.18 (223.4)	7.19 (1000)	7.20 (d)	7.21 (c)

Explanations CRO & Electronic Measurements

7.2 (d)

Post deflecting acceleration (PDA) is used in high frequency CRO to improve/control spot brightness.

7.3 (a)

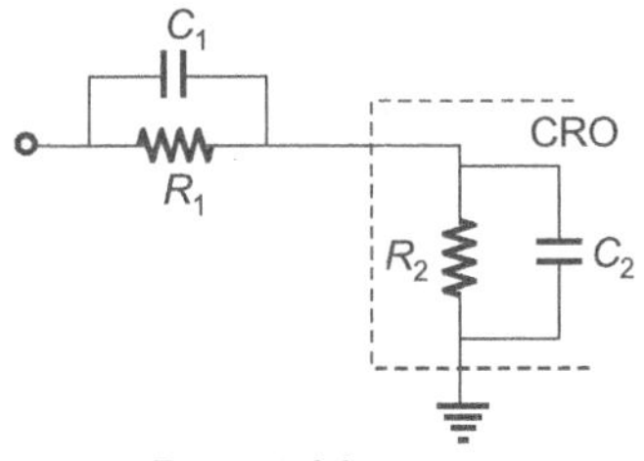

$$R_2 = 1\text{ M}\Omega$$
$$C_2 = 100\text{ pF}$$
$$\frac{R_2}{R_1+R_2} = \frac{1}{20}$$
$$\frac{1}{R_1+1} = \frac{1}{20}$$
$$R_1 + 1 = 20$$
$$\therefore \quad \boxed{R_1 = 19\text{ M}\Omega}$$
$$\frac{C_1}{C_1+C_2} = \frac{1}{20}$$
$$\frac{C_1}{C_1+100} = \frac{1}{20}$$
$$20\,C_1 = C_1 + 100$$
$$19\,C_1 = 100$$
$$\boxed{C_1 = \frac{100}{19}\text{pF}}$$

7.5 (a)

Given bandwidth of CRO = 10 MHz
and frequency of V_1 = 20 kHz
and frequency of V_2 = 20 MHz
So, CRO passes V_1 as it is but compresses V_2.
This compression in V_2 do not effect peak values and hence peak to peak value will be unchanged.
For V_1, peak to peak value is

$$P_1 = 10-(-10) = 20\text{ V}$$

Similarly for V_2, peak to peak value = 120 V

7.6 (a)

When we apply sinusoidal waves on both *XX* and *YY* plates, then pattern is Lissajous figure.
When phase difference is zero between them then we get a straight line inclined at 45° in other cases ellipse is formed.
Since oscilloscope displays a line
$\Rightarrow \phi = 0$ and frequency is same for X and Y and inclination is 45° $\Rightarrow V_X = V_Y$.

7.7 (b)

Frequency of squarewave = 297 Hz
So fundamental frequency of sinusoidals, when we express the square wave in terms of Fourier series = 297 Hz = f_1
So if *XX* input has sine wave with no phase difference then only we get a line.
This is possible if *YY* sine wave frequency is integral multiple of *XX* sine wave.
So, frequency of *XX* wave = 49.5 Hz

then, $\quad \frac{f_1}{f} = 6$

7.8 (b)

Delay line is introduced between *Y*-inputs and *Y*-plate to synchronize the output of both horizontal and vertical deflection system.

7.9 (b)

Here, in *Y* – direction – amplitude is there
in *X* – direction time axis
$\Rightarrow$ 2.5 cm on *Y*-axis correspond to 5 V

So, sensitivity $\frac{5\text{ V}}{2.5\text{ cm}} = 2\text{ V/cm}$

$\Rightarrow$ 4 cm on *x*-axis correspond to $T/2$.

$$\frac{T}{2} = \frac{1}{2f} = \frac{1}{2\times25\times10^3}$$
$$= 0.02\text{ msec} = 20\,\mu\text{sec}$$

so, sensitivity in *X*-direction

$$= \frac{T/2}{4\text{ cm}} = \frac{20\,\mu\text{sec}}{4\text{ cm}} = \frac{5\,\mu\text{sec}}{\text{cm}}$$

Trigger circuit triggers the UJT oscillator which is a –ve resistance device so slope will be –ve.

7.10 (a)

When a voltage signal is fed to an oscilloscope with DC coupling, entire waveform will be satisfied including the DC component.
Whereas, AC coupling removes the DC portion of the signal and displayed waveform will be centred at zero volts.

7.11 (b)

It is clear from the figure that there is no change in amplitude.
So it will show 5 V peak.
⇒ Also amplitude is function of Y-plates so no change.
The frequency of the wave will seem to be same as 1 kHz.

7.12 (a)

X-sensitivity = 2 msec/cm = S_X
Y-sensitivity = 1 V/cm = S_Y
for input peak to peak voltage, $V_{P-P} = 4$ V
so length on Y-axis from peak to peak,

$$l_1 = \frac{V_{PP}}{S_Y} = 4 \text{ cm}$$

$$\Rightarrow \quad T = \frac{2\pi}{\omega} = \frac{2\pi}{100\pi} = \frac{1}{50} = 20 \text{ msec}$$

$$T/2 = 10 \text{ msec}$$

Now half cycle on X-axis correspond to length,

$$l_2 = \frac{T/2}{S_x} = \frac{10 \text{ msec}}{2 \text{ msec/cm}} = 5 \text{ cm}$$

This is shown in figure (a) only.

7.14 (a)

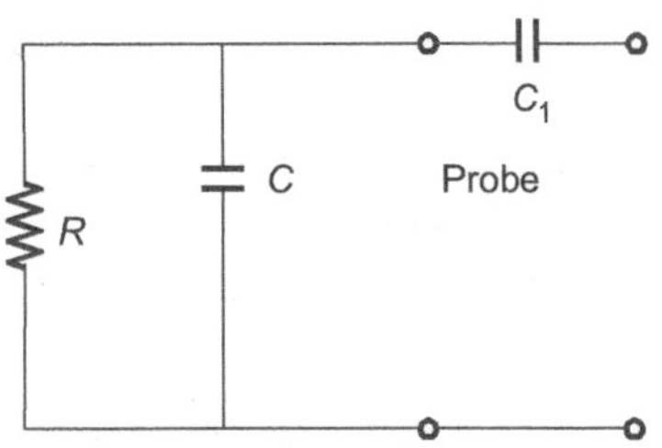

So effective capacitance

$$\frac{C_1 \cdot C}{C_1 + C} = \frac{C}{10}$$

$$10\,C_1 = C_1 + C\,;\, 9\,C_1 = C$$

$$C_1 = 5 \text{ pF}$$

but effective input capacitance at probe tip

$$= \frac{C_1 C}{C_1 + C} = 4.5 \text{ pF}.$$

7.15 (a)

For peak voltage to be 4 V on Y-axis, number of divisions on Y-axis,

$$d_1 = \frac{4}{S_Y} = \frac{4}{1 \text{ V/div}} = 4 \text{ div}$$

for $\quad T = \dfrac{2\pi}{200\pi} = 10$ msec

and $\quad T/2 = 5$ msc,

number of divisions on x-axis is

$$d_2 = \frac{T/2}{S_X} = \frac{5 \text{ msec}}{1 \text{ msec/div}} = 5 \text{ div}$$

The trigger is set at zero. So, it will start from zero level

7.16 (d)

When V_i is +ve,

$$V_{out} = -4 \text{ V}$$

[2 Division × 2 V/division = –4 V]

$$0 - V_{Z1} - V_{Z2} = V_0 \quad \therefore \; -V_{Z1} - V_{Z2} = -4$$

When V_i is +ve, Z_1 is forward bias and Z_2 is reverse bias

$$\therefore \quad V_{Z1} + V_{Z2} = 4$$

$$\therefore \quad 0.7 + V_{Z2} = 4$$

$$\therefore \quad \boxed{V_{Z2} = 3.3 \text{ V}}$$

When V_i is negative,

$$V_{out} = +6 \text{ V}$$

[3 Division × 2 V/Division = +6 V]

$$V_0 = V_{Z2} + V_{Z1}$$

$$6 = 0.7 + V_{Z1}$$

$$\therefore \quad \boxed{V_{Z1} = 5.3 \text{ V}}$$

7.17 (d)

Upon observing the waveform 3 divisions on the Y axis means that the Y setting is 2 V/div.
An angular frequency of $100\pi t$ would corresponds to frequency of 50 Hz.

i.e. Time period $= \dfrac{1}{50} = 20$ msec

∴ 10 correspond to a X setting of

$$\frac{20}{10} = 2 \text{ msec/div}$$

7.18 (223.4)

Dual Slope ADC

Reference voltage,

$$V_r = 100 \text{ mV}$$

For $V_A = 123.4$ mV

Disintegration time

$$= \frac{V_A}{V_r} \times \text{integration time}$$

$$= \frac{123.4}{100} \times 100 \text{ ms}$$

So, Total conversion time

$$= 100 \text{ ms} + 123.4 \text{ ms}$$

$$= 223.4 \text{ ms}$$

7.19 (1000)

$6\frac{1}{2}$ digital voltmeter time period

1	0	0	0	0	0	0
10^6	10^5	10^4	10^3	10^2	10^1	10^0

$$10^6 \text{ns} = 10^{-3} \text{ sec}$$

Frequency $\frac{1}{10^{-3}}$ Hz $= 1000$ Hz

7.20 (d)

When scale set on 1s, reading = 1000

$$\text{Time period} = \frac{1}{1000} = 10^{-3} \text{ sec}$$

If new range is 1 MHz, Reading $= \frac{10^{-3}}{10^{-6}} = 1000$

7.21 (c)

Given circuit can be redrawn as

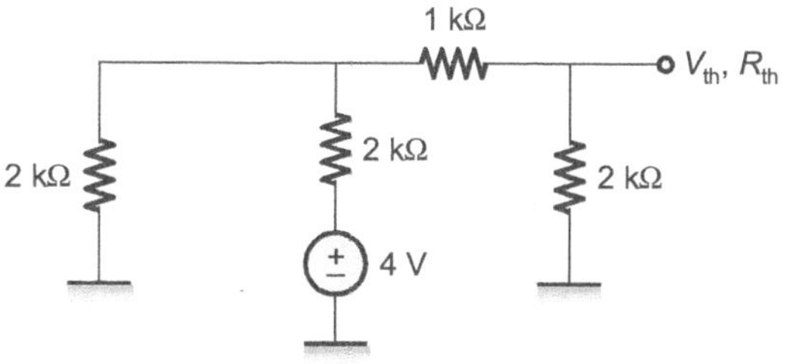

$$R_{th} = \{[(2 \parallel 2) + 1] \parallel 2\} \text{ k}\Omega$$

$$= 1 \text{ k}\Omega$$

By KCL at Node V,

$$\frac{V-4}{2} + \frac{V}{2} + \frac{V}{3} = 0$$

$$3V - 12 + 3V + 2V = 0$$

$$8V = 12$$

$$V = \frac{12}{8} \text{ Volt}$$

$$V_{Th} = \frac{12}{8} \times \frac{2}{1+2} = \frac{12}{8} \times \frac{2}{3} = 1 \text{ Volt}$$

■■■

8 Static and Dynamic Characteristics, Error Analysis and Statistical Analysis of Data

8.1 Match the terms and their definitions

List-I

A. Accuracy
B. Sensitivity
C. Repeatability
D. Resolution

List-II

1. The ability to display the same reading while measuring a quantity
2. Smallest perceptible change in the output
3. Closeness of the measurement to the mean value
4. Ratio of change in the instrument reading to the change in he measured quantity.

Codes:

	A	B	C	D
(a)	1	3	4	2
(b)	2	3	4	1
(c)	4	3	1	2
(d)	3	4	1	2

[1996 : 2 M]

8.2 Threshold of a measurement system is

(a) the smallest change in input which can be detected
(b) a measure of linearity of the system
(c) the smallest input which can be detected
(d) a measure of precision of the system

[2001 : 1 M]

8.3 Three DC currents, I_1, I_2 and I_3 meet at a node with I_1 entering and I_2 and I_3 leaving the node. I_1 and I_2 are measured as 100 mA and 99 mA with ±1% accuracy. Then the value I_3 and the accuracy of I_3 are

(a) 10 mA ± 2% (b) 199 mA ± 2%
(c) 1 mA ± 2% (d) 1 mA ± 199%

[2002 : 2 M]

8.4 A large number of 230 Ω resistors are obtained by combining 120 Ω resistors with α standard deviation of 4.0 Ω and 110 Ω resistors with a standard deviation of 3.0 Ω. The standard deviation of the 230 Ω resistors thus formed will be

(a) 3.5 Ω (b) 5.0 Ω
(c) 3.0 Ω (d) 12.0 Ω

[2003 : 1 M]

8.5 For a proving ring used in a press, the value of force and its uncertainty are given as 12355 ± 2354 N without any rounding off. If the error should be rounded off at the maximum to two significant figure only, the value of the reading is rounded off as

(a) 12300 ± 2300 N (b) 1230 ± 2350 N
(c) 12400 ± 2300 N (d) 12400 ± 2400 N

[2004 : 1 M]

8.6 The following terms used in the context of an instrument are numbered as shown:

1. accuracy 2. sensitivity
3. precision and 4. resolution

Match these with their possible definitions listed below:

P. Repeatability of readings on successive observations.
Q. Smallest perceptible change in the output
R. Deviation of the output from the true value
S. Minimum value of the input from the true value
T. Ratio of the change in the instrument reading to the change in the measured variable.

(a) 1-P, 2-Q, 3-R, 4-S (b) 1-S, 2-Q, 3-P, 4-T
(c) 1-R, 2-T, 3-P, 4-Q (d) 1-T, 2-Q, 3-P, 4-R

[2004 : 1 M]

8.7 The calibration data for a pressure compensator of pump is given below:

Input (x)	0	1	2	3	4	5	6	7	8	9	10
Output (y)	9.5	8.4	7.8	7.4	6.1	5.4	5.2	4.6	3.2	1.9	1.1

For the given data, the slope of the best-fit line applying least square method is

(a) 0.921 (b) –0.803
(c) 0.819 (d) –0.945 [2004 : 2 M]

8.8 A pressure sensor has the following specifications:
Sensitivity at the design temperature = 10 V/MPa,
Zero drift = 0.01 V/°C
Sensitivity drift = 0.01 (V/MPa)/°C
When the sensor is used in an ambient 20°C above the design temperature, the output from the device is 7.4 V. The true value of the pressure will be

(a) 0.71 MPa (b) 0.68 MPa
(c) 0.5 MPa (d) 0.61 MPa

[2004 : 2 M]

8.9 A mass of 10 kg is measured with an instrument and the readings are normally distributed with respect to the mean of 10 kg. Given that $\frac{2}{\sqrt{2\pi}}\int_0^{0.84}\exp(-\eta^2/2)d\eta = 0.6$ and that 60 per cent of the readings are found to be within 0.05 kg from the mean, the standard deviation of the data is

(a) 0.02 (b) 0.04
(c) 0.06 (d) 0.08 [2005 : 1 M]

8.10 The measurements of a source voltage are 5.9 V, 5.7 V and 6.1 V. The sample standard deviation of the reading is

(a) 0.013 (b) 0.04
(c) 0.115 (d) 0.2 [2005 : 1 M]

8.11 Position sensor units having a constant sensitivity of 1 V/mm are used for feedback in number of position controlled system units, each having an overall forward path dc gain of 50. If the random dc bias errors in outputs of various sensor units are characterised as normal with mean 0 and standard deviation (σ) 0.01 V, for a constant set point of 5 V, the true position outputs of the various controlled system units can be characterized as normal with

(a) mean of 5.0, σ of 0.01
(b) mean of 0.0, σ of 0.01
(c) mean of 5.0, σ of 0.0098
(d) mean of 4.902, σ of 0.0098 [2005 : 2 M]

8.12 A surface height profile is given by,

$$y = a + b\sin\frac{2\pi x}{l} \quad (0 \le x \le l)$$

where a, b, l are constants and x is the horizontal coordinate.

The roughness of the surface, based on absolute deviation, is

(a) $\frac{2b}{\pi}$ (b) $\frac{4b}{\pi}$
(c) $\frac{6b}{\pi}$ (d) $\frac{8b}{\pi}$ [2005 : 2 M]

8.13 A thermometer at room temperature 30°C is dipped suddenly into a bath of boiling water at 100°C. It takes 30 seconds to reach 96.5°C. The time required to reach a temperature of 98°C is

(a) 32.5 s (b) 34.6 s
(c) 35.6 s (d) 38.6 s [2005 : 2 M]

8.14 A temperature sensor having a range of 0-100°C has a worst case accuracy specification of ±1.0% of true value or ±0.3% of full scale, whichever is higher. The sensitivity of the sensor is experimentally obtained by taking its readings at true temperatures 25°C and 75°C. The maximum error in sensitivity, expressed as percentage of its ideal value is

(a) ± 0.105% (b) ± 0.21%
(c) ± 1.05% (d) ± 2.1% [2005 : 2 M]

8.15 Using the given data points tabulated below, a straight line passing through the origin is fitted using least squares method. The slope of the line is

x	1.0	2.0	3.0
y	1.5	2.2	2.7

(a) 0.9 (b) 1.0
(c) 1.1 (d) 1.5 [2005 : 2 M]

8.16 Two sensors have measurement errors that are Gaussian distributed with zero means and variances σ_1^2 and σ_2^2, respectively. The two sensor measurements x_1 and x_2 are combined to form the weighted average $x = \alpha x_1 + (1-\alpha)x_2$, $0 \le \alpha \le 1$. Assuming that the measurement errors of the two sensors are uncorrelated, the weighting factor α that yields the smallest error variance of x is

(a) $\frac{\sigma_2^2}{\sigma_1^2+\sigma_2^2}$ (b) $\frac{\sigma_1^2}{\sigma_1^2+\sigma_2^2}$
(c) $\frac{\sigma^2}{\sigma_1+\sigma_2}$ (d) 0.5 [2007 : 2 M]

8.17 A thermometer with time constant τ, initially at the ambient temperature, is used to measure the temperature of a liquid in a bath. The excess temperature of the thermometer and the liquid over the ambient are $\theta(t)$, $\theta_l(t)$ respectively, where t denotes the time. If $\theta_l(t) = kt$, where k is a constant, the steady state error, defined as $\lim_{t\to\infty}[\theta(t) - \theta_l(t)]$, is

(a) ∞ (b) 0
(c) $-k$ (d) $-k\tau$

[2007 : 2 M]

8.18 Consider a non-ideal voltage source whose output voltage is measured by a non-ideal voltmeter as shown below.

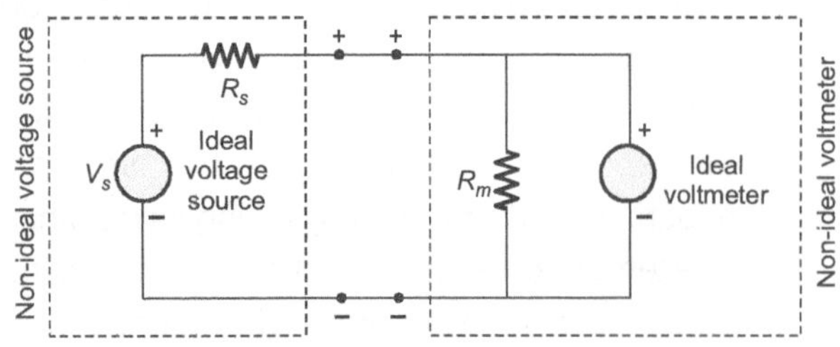

Let V_e be the difference between V_s and the measured voltage. V_e/V_s is function of

(a) R_m only (b) R_s only
(c) R_s/R_m (d) $R_m - R_s$

[2007 : 1 M]

8.19 A quantity x is calculated by using the formula

$$x = \frac{(p-q)}{r}$$

The measured values are $p = 9$, $q = 6$, $r = 0.5$

Assume that the measurement errors in p, q and r are independent. The absolute maximum error in the measurement of each of the three quantities is ε. The absolute maximum error in thc calculated value of x is

(a) ε (b) 2ε
(c) 3ε (d) 16ε [2009 : 2 M]

8.20 The response of a first order measurement system to a unit step input is $1 - e^{-0.5t}$, where t is in seconds. A ramp of 0.1 units per second is given as the input to this system. The error in the measured value after transients have died down is

(a) 0.02 units (b) 0.1 units
(c) 0.2 units (d) 1 unit [2009 : 2 M]

8.21 A measurement system with input $x(t)$ and output $y(t)$ is described by the differential equation $3\frac{dy}{dt} + 5y = 8x$. The static sensitivity of the system is

(a) 0.60 (b) 1.60
(c) 1.67 (d) 2.67 [2010 : 1 M]

8.22 The volume of a cylinder is computed from measurements of its height (h) and diameter (d). A set of several measurements of height has an average value of 0.2 m and a standard deviation of 1%. The average value obtained for the diameter is 0.1 m and the standard deviation is 1%. Assuming the errors in the measurements of height and diameter are uncorrelated, the standard deviation of the computed volume is

(a) 1.00% (b) 1.73%
(c) 2.23% (d) 2.41% [2010 : 2 M]

8.23 The output voltage of a transducer with an output resistance of 10 kΩ is connected to an amplifier. The minimum input resistance of the amplifier so that the error in recording the transducer output does not exceed 2% is

(a) 10 kΩ (b) 49 kΩ
(c) 490 kΩ (d) 1.2 kΩ [2010 : 1 M]

8.24 The characteristics of a thermometer measuring ambient temperature is $2\frac{dT_i}{dt} + T_i - T_a = 0$, where T_i and T_a are the indicated and ambient temperatures, respectively, both in °C, and time is in seconds. the –3 dB cut-off frequency in the frequency response of the thermometer is

(a) $1/4\pi$ Hz (b) 1/2 Hz
(c) 1 Hz (d) 2π Hz [2011 : 2 M]

8.25 Two ammeters A_1 and A_2 measure the same current and provide readings I_1 and I_2, respectively. The ammeter errors can be characterized as independent zero mean Gaussian random variable of standard deviations σ_1 and σ_2, respectively. The value of the current is computed as:

$$I = \mu I_1 + (1-\mu)I_2$$

The value of μ which gives the lowest standard deviation of I is

(a) $\frac{\sigma_2^2}{\sigma_1^2 + \sigma_2^2}$ (b) $\frac{\sigma_1^2}{\sigma_1^2 + \sigma_2^2}$

(c) $\frac{\sigma_2}{\sigma_1 + \sigma_2}$ (d) $\frac{\sigma_1}{\sigma_1 + \sigma_2}$

[2013 : 2 M]

8.26 The resistance of a wire is given by the expression

$$R = \frac{4\rho L}{\pi D^2},$$

where, ρ is the resistivity (Ω-meter), L is the length (meter) and D (meter) is the diameter of the wire. The error in measurement of each of the parameters ρ, L, and D is ±1.0%. Assuming that the errors are independent random variables, the percent error in measurement of R is ______.

[2014 : 2 M]

8.27 Consider the ammeter-voltmeter method of determining the value of the resistance R using the circuit shown in the figure. The maximum possible errors of the voltmeter and ammeter are known to be 1% and 2% of their readings, respectively. Neglecting the effects of meter resistances, the maximum possible percentage error in the value of R determined from the measurements, is ______ %.

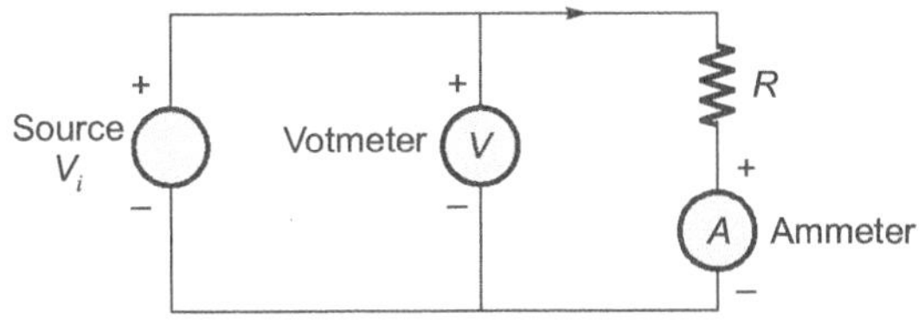

[2015 : 1 M]

8.28 The output frequency of an LC tank oscillator employing a capacitive sensor acting as the capacitor of the tank is 100 kHz. If the sensor capacitance increases by 10%, the output frequency in kilo-hertz becomes ____ kHz.

[2015 : 2 M]

8.29 A voltage V_1 is measured 100 times and its average and standard deviation are 100 V and 1.5 V respectively. A second voltage V_2, which is independent of V_1, is measured 200 times and its average and standard deviation are 150 V and 2 V respectively. V_3 is computed as: $V_3 = V_1 + V_2$. Then the standard deviation of V_3 in volt is_______.

[2016 : 1 M]

8.30 The voltage and current drawn by a resistive load are measured with a 300 V voltmeter of accuracy ± 1% of full scale and a 5 A ammeter of accuracy ± 0.5% of full scale. The readings obtained are 200 V and 2.5 A. The limiting error (in %) in computing the load resistance is (up to two decimal places)______.

[2018 : 2 M]

8.31 The resistance of a resistor is measured using a voltmeter and an ammeter. The voltage measurements nave a mean value of 1V and standard deviation of 0.12 V while current measurements have a mean value of 1 mA with standard deviation of 0.05 mA. Assuming that the errors in voltage and current measurements are independent, the standard deviation of the calculated resistance value is___________Ω.

[2019 : 1 M]

8.32 Two 100 Ω resistors having tolerance 3% and 4% are connected in series. The effective tolerance of the series combination (in %, rounded off to one decimal place) is _____.

[2020 : 1 M]

8.33 In an AC main, the rms voltage V_{ac}, rms current I_{ac} and power w_{ac} are measured as: $V_{ac} = 100\text{ V} \pm 1\%$. $I_{ac} = 1\text{ A} \pm 1\%$ and $W_{ac} = 50\text{ W} \pm 2\%$ (errors are with respect to readings). The percentage error in calculating the power factor using these readings is

(a) 1%
(b) 2%
(c) 3%
(d) 4%

[2021 : 1 M]

8.34 Five measurements are made using a weighing machine, and the readings are 80 kg, 79 kg, 81 kg, 79 kg and 81 kg. The sample standard deviation of the measurement is ______________ kg (rounded off to two decimal places).

[2023 : 2 M]

8.35 The resistance of a 20 kΩ resistor is measured six consecutive times using an LCR meter. The first five readings are 19 kΩ, 18 kΩ, 23 kΩ, 21 kΩ and 17 kΩ. If the mean of the measurements and the true value are equal, the last reading is ________ kΩ (rounded off to nearest integer).

[2024 : 1 M]

■■■

Answers Static and Dynamic characteristics .....

8.1 (d)	8.2 (c)	8.3 (d)	8.4 (b)	8.5 (d)	8.6 (c)	8.7 (b)
8.8 (a)	8.9 (c)	8.10 (d)	8.11 (a)	8.12 (*)	8.13 (c)	8.14 (c)
8.15 (a)	8.16 (a)	8.17 (d)	8.18 (c)	8.19 (d)	8.20 (c)	8.21 (b)
8.22 (c)	8.23 (c)	8.24 (a)	8.25 (a)	8.26 (2.45)	8.27 (3)	8.28 (95.35)
8.29 (2.5)	8.30 (2.50)	8.31 (130)	8.32 (3.5)	8.33 (d)	8.34 (1)	8.35 (22)

Explanations Static and Dynamic characteristics .....

8.3 (d)

Sum of all the currents meeting of a node = 0
incoming currents ⇒ +ve
outgoing currents ⇒ –ve

$$\therefore \quad I_1 - I_2 - I_3 = 0$$
$$I_1 = I_2 + I_3$$
$$I_3 = I_1 - I_2$$
$$= 100 - 99 = 1 \text{ mA}$$

Percentage accuracy

$$\Rightarrow \quad \frac{\partial I_3}{I_3} = \frac{\partial I_1}{I_1}\frac{I_1}{I_3} + \frac{\partial I_2}{I_2}\frac{I_2}{I_3}$$
$$= 1\left(\frac{100}{1}\right)\% + 1\left(\frac{99}{100}\right)\% = 199\%$$
$$\therefore \quad I_3 = 1 \text{ mA} + 199\%$$

8.4 (b)

Standard deviation of sum of two random variable

$$\text{stdev } (X+Y) = \sqrt{\text{Var}(X) + \text{Var}(Y) + 2\,\text{Cov}(X,Y)}$$

as X and Y in the given problem are independent, so their covariance will be zero

$\therefore$ Cov(X, Y) = 0

as variance = (standard deviation)^2 = σ^2

$$\therefore \quad \sigma_{(x+y)} = \sqrt{\sigma_x^2 + \sigma_x^2} = \sqrt{3^2 + 4^2} = 5$$

8.5 (d)

Given, F = 12355 ± 2354 N

Now rounding off at the maximum to two significant figure,

$$F = 12400 \pm 2400 \text{ N}$$

8.6 (c)

These all are the corresponding definitions.

8.7 (b)

Let the best-fit straight line

$$y = a + bx$$

$\therefore$ Slope of the line = b

By applying least square method the value of a and b

$$\begin{cases} a = \dfrac{(\Sigma y)(\Sigma x^2) - (\Sigma x)(\Sigma xy)}{n(\Sigma x^2) - (\Sigma x)^2} \\ b = \dfrac{n(\Sigma xy) - (\Sigma x)(\Sigma y)}{n(\Sigma x^2) - (\Sigma x)^2} \end{cases}$$

From the given data

n = 1, Σxy = 214.7

Σx = 55, Σy = 6.6

Σx^2 = 385

$$\therefore \quad b = \frac{(11 \times 214.7) - (55 \times 60.6)}{(11 \times 385) - (55)^2}$$
$$= -0.8027$$

8.8 (a)

Sensitivity = 10 V/MPa

$\therefore$ 10 V corresponds to 1 MPa

7.4 V corresponds to $\frac{1}{10} \times 7.4$ = 0.74 MPa

Errors:

Due to zero drift

Zero drift = 0.01 V/°C

$\therefore$ 1°C corresponds to 0.01 V

20° corresponds to 0.01 × 20 = 0.2 V

Due to sensitivity drift

Sensitivity drift = 0.01 (V/MPa)/°C

for 0.74 MPa and 20°C, the voltage error is

0.01 × 0.74 × 20 = 0.148 v

Total error voltage

= 0.2 + 0.148 = 0.348 V

$\therefore$ true output voltage should be

7.4 – 0.348 = 7.052 V

Sensitivity of instrument = 10 V/MPa

$\therefore \quad 10\,V \Rightarrow 1\,MPa$

$$\therefore \quad 7.052 \Rightarrow \frac{1}{10} \times 7.052 = 0.7052 \text{ MPa}$$

$$\simeq 0.71 \text{ MPa}$$

8.9 (c)

$$t = \frac{x}{\sigma}$$

x = given deviations

σ = standard deviation

$t = 0.84$

$$\sigma = \frac{x}{t} = \frac{0.05}{0.84} \quad \therefore \sigma = 0.06$$

8.10 (d)

$$\text{Arithmetic mean } \bar{x} = \frac{5.9 + 5.7 + 6.1}{3} = \frac{17.7}{3} = 5.9$$

$$\text{Deviation } d_1 = \bar{x}_1 - \bar{x} = 5.9 - 5.9 = 0$$

$$\text{Deviation } d_2 = \bar{x}_2 - \bar{x} = 5.7 - 5.9 = -0.2$$

$$\text{Deviation } d_3 = \bar{x}_3 - \bar{x} = 6.1 - 5.9 = 0.2$$

Standard deviation

$$\sigma = \sqrt{\frac{\Sigma d^2}{n-1}} \quad \text{(as } n < 20\text{)}$$

$$\sigma = \sqrt{\frac{(0)^2 + (-0.2)^2 + (0.2)^2}{(3-1)}} = 0.2$$

8.13 (c)

Thermometer formula, considering the initial conditions,

$$\theta(t) = \theta_f(1 - e^{-t/\tau}) + \theta_i e^{-t/\tau}$$

where, θ_f = final temperature (°C)

$\theta(t)$ = temperature at time (t°C)

θ_i = initial temperature (°C)

$$96.5 = 100(1 - e^{-30/\tau}) + 30 e^{-30/\tau}$$

$$= 100 - 100 e^{-30/\tau} + 30 e^{-30/\tau}$$

$$= 100 - 70 e^{-30/\tau}$$

$$70 e^{-30/\tau} = 3.5$$

$$e^{-30/\tau} = \frac{3.5}{70} = 0.05$$

Taking log both sides we get,

$$\log e^{-30/\tau} = \log 0.05$$

$$-\frac{30}{\tau} = -2.99 = -3.00$$

$$\therefore \quad t = \frac{30}{3} = 10 \text{ sec.}$$

for $\theta(t) = 98°C$

$$98 = 100(1 - e^{-t/10}) + 30 e^{-t/10}$$

$$98 = 100 - 70 e^{-t/10}$$

$$70 e^{-t/10} = 2$$

$$\therefore \quad e^{-t/10} = 2/70$$

Taking log both sides we get,

$$-\frac{t}{10} = \log\frac{2}{70} = -35.55$$

$$-t = -3.555\,5 \times 10$$

$$t = 35.55 = 35.6 \text{ sec}$$

8.14 (c)

In the given sensor, there are two sources of error, one expressed as % true value and other as % full scale.

$$\text{Also sensitivity of transducer} = \frac{\text{output}}{\text{input}}$$

$$\Rightarrow \text{Sensitivity} = \frac{x}{T}$$

where x is output of sensor at input temperature T.

Now, i.e., error in sensitivity,

(i)

At 25°C/75°C due to ±1 % of true value erro in input only,

$$e_1 = \frac{\frac{x}{24.75} - \frac{x}{25}}{\frac{x}{25}} \times 100 = \pm 1.01\%$$

$$e_2 = \frac{\frac{x}{74.25} - \frac{x}{75}}{\frac{x}{75}} \times 100 = \pm 1.01\%$$

(ii)

At 25°C/75°C due to ±0.3 % of full scale error in input only,

$$\text{error in input (in °C)} = \pm\frac{100 \times 0.3}{100} = 0.3°C$$

$$e_3 = \frac{\frac{x}{24.70} - \frac{x}{25}}{\frac{x}{25}} \times 100 = \pm 1.21\%$$

$$e_4 = \frac{\frac{x}{74.7} - \frac{x}{75}}{\frac{x}{75}} \times 100 = \pm 0.4\%$$

So, $e_{max} = \max(e_1, e_2, e_3, e_4)$

$\Rightarrow \quad e_{max} = \pm 1.21\%$

Note: While calculating maximum error, we have considered negative deviation. This is because error in above case will be more if negative deviation is taken.

For instance,

$$e_3' = \frac{\frac{x}{25.30} - \frac{x}{25}}{\frac{x}{25}} \times 100$$

$$= \pm 1.186\% < e_3$$

8.15 (a)

$$nb + a\sum x_i = \sum y_i$$

$$b\sum x_i + a\sum x_i^2 = \sum x_i y_i$$

here, n = 3

	x_i	y_i	x_i^2	$x_i y_i$
	1.0	1.5	1	1.5
	2.	2.2	4	4.4
	3.0	2.7	9	8.1
Sum	6.0	6.4	14	14

$$3b + 6a = 6.4 \quad \text{(i)}$$

$$6b + 14a = 14 \quad \text{...(ii)}$$

solving equation (i) and (ii) we get,

$$\therefore \quad b = 0.933 \approx 0.9$$

8.17 (d)

Steady state error for ramp input of magnitude $k\tau$ and first order system.

$$e_{ss} = \lim_{t \to \infty} [\theta_i(t) - \theta(t)] = k\tau$$

Here, $\lim_{t \to \infty} [\theta(t) - \theta_i(t)] = -k\tau$

8.18 (c)

Measured voltage = V_0

$$V_0 = \frac{R_m}{R_m + R_s} V_s$$

$$V_e = V_s - \frac{R_m}{R_m + R_s} V_s$$

$$= \frac{V_s(R_m + R_s) - V_s(R_m)}{R_m + R_s} = \frac{V_s R_s}{R_m + R_s}$$

$$\therefore \frac{V_e}{V_s} = \frac{R_s}{R_m + R_s} = \frac{R_s}{R_s\left(\frac{R_m}{R_s} + 1\right)}$$

$$= \frac{1}{\frac{R_m}{R_s} + 1} = \frac{1}{\frac{1}{R_s / R_m} + 1}$$

$$\therefore \text{Clearly, } \frac{V_e}{V_s} = f\left(\frac{R_s}{R_m}\right)$$

8.19 (d)

Given, $\delta p = \delta q = \delta r = \in$

and $p = 9,\ q = 6,\ r = 0.5$

$$x = \frac{(p - q)}{r}$$

let, $(p - q) = y$

$$\therefore \quad x = \frac{y}{r}$$

maximum absolute error in $y = \delta y$

$$\therefore \quad \frac{\delta y}{y} = \frac{\delta(p-q)}{(p-q)} = \left(\frac{\delta p}{p-q} + \frac{\delta q}{p-q}\right)$$

$$= \left(\frac{\in}{3} + \frac{\in}{3}\right) = \frac{2\in}{3}$$

maximum absolute error in $x = \delta x$

$$\therefore \quad \frac{\delta x}{x} = \left(\frac{\delta y}{y} + \frac{\delta r}{r}\right) = \left(\frac{2\in}{3} + \frac{\in}{0.5}\right) = \frac{8\in}{3}$$

$$\therefore \quad \delta x = \left(\frac{8\in}{3} \cdot x\right) = \left(\frac{8\in}{3} \times 6\right) = 16\in$$

8.20 (c)

$y(t) = 1 - e^{-0.5t}$ (unit step response)

therefore, impulse response is

$$h(t) = 0.5\, e^{-0.5t}$$

$$\Rightarrow H(s) = \frac{0.5}{(s+5)}$$

Let us assume unity feedback system

$$\Rightarrow H(s) = \frac{G(s)}{1+G(s)} = \frac{0.5}{s+0.5}$$

$$\Rightarrow \quad \frac{1+G(s)}{G(s)} = \frac{s}{0.5}$$

$$\Rightarrow \quad \frac{1}{G(s)} = \frac{s}{0.5} = G(s) = \frac{0.5}{s}$$

$$\Rightarrow \quad G(s) = \frac{1}{2s}$$

so velocity error constant,

$$K_v = \lim_{s \to 0} SG(s) = \frac{1}{2}$$

Now for ramp input $A\, r(t)$, steady state error

$e_{ss} = \frac{A}{K_v}$, so for $0.1\, r(t)$,

$$e_{ss} = \frac{0.1}{1/2} = 0.2$$

8.21 (b)

Differential equation,

$$3\frac{dy}{dt}+5y = 8x \quad ...(i)$$

This is a first-order system.

First order standard equation

$$\tau\frac{dy(t)}{dt}+y(t) = kx(t)$$

where k is the static sensitivity.

From equation (i),

$$\frac{3}{5}\frac{dy}{dt}+y = \frac{8}{5}x$$

$$\therefore \quad k = \frac{8}{5} = 1.6$$

8.22 (c)

Volume of cylinder, $V = \pi r^2 h$

$$\Rightarrow \quad \frac{\partial V}{\partial r} = 2\pi rh$$

$$\frac{\partial V}{\partial h} = \pi r^2$$

Given, $h = 0.2$ m and $d = 0.1$ m

$$\Rightarrow r = 0.05 \text{ m} ; \sigma_r = \frac{1}{100}\times 0.05 = 5\times 10^{-4}$$

$$\sigma_h = 0.2\times\frac{1}{100} = 2\times 10^{-3}$$

Standard deviation of volume,

$$\sigma_V = \sqrt{\left(\frac{\partial V}{\partial r}\right)^2 \sigma_r^2 + \left(\frac{\partial V}{\partial h}\right)^2 \sigma_h^2}$$

$$\Rightarrow \sigma_V = \sqrt{(2\pi rh)^2\cdot(5\times 10^{-4})^2 + (\pi r^2)^2 (2\times 10^{-3})^2}$$

$$\Rightarrow \sigma_V = 3.512\times 10^{-4}$$

$$\Rightarrow \%\sigma_V = \frac{3.5\times 10^{-4}}{\pi\times(0.05)^2\times 0.2}\times 100\% = 2.23\%$$

8.23 (c)

When the reading instrument have zero input impedance, it reads full. V volts, as is being supplied by the source.

When the reading instrument has resistance 'R' then the measured voltage is

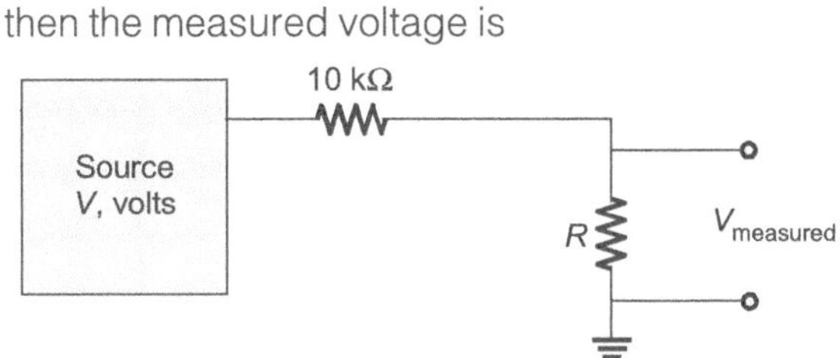

$\therefore$ Error due to the meter resistance R is 2% (given)

$$\therefore \quad \frac{V - \frac{R}{R+10}\times V}{V}\times 100 = 2$$

$$\frac{[(R+10)V - RV]\times 100}{V(R+10)} = 2$$

$$\frac{100(R+10-R)}{R+10} = 2$$

$$1000 = 2(R+10)$$

$$1000 = 2R+20$$

$$\therefore \quad 2R = 1000-20 = 980$$

$$\therefore \quad R = \frac{980}{2} = 490 \text{ k}\Omega$$

8.24 (a)

Clearly it is a first order equation

$$\frac{A_0}{A_1} = \frac{1}{\sqrt{1+(\omega\tau)^2}}$$

Standard first order equation is

$$\tau\frac{dy}{dt}+y = kx$$

where, τ = time constant

k = sensitivity of the system

$\therefore \tau = 2$ here

also, at 3 dB, $\frac{A_0}{A_1} = \frac{1}{\sqrt{2}}$

$$\therefore \quad \frac{1}{\sqrt{2}} = \frac{1}{\sqrt{1+(2\omega)^2}}, \quad \frac{1}{2} = \frac{1}{1+4\omega^2}$$

$$1+4\omega^2 = 2$$

$$\omega = \frac{1}{2} \quad \text{and} \quad 2\pi f = \frac{1}{2}$$

$$\therefore \quad \boxed{f = \frac{1}{4\pi}}$$

8.26 2.45 (2.3 to 2.5)

$$R = 4\frac{\rho L}{\pi D^2}$$

Assuming errors are in standard deviations.

$\%\sigma\rho$, $\%\sigma L$ and $\%\sigma D = \pm 1\%$,

$$\Delta R = \sqrt{\left(\frac{\partial R}{\partial \rho}\cdot\Delta\rho\right)^2 + \left(\frac{\partial R}{\partial L}\cdot\Delta L\right)^2 + \left(\frac{\partial R}{\partial D}\cdot\Delta D\right)^2}$$

$$\%\sigma_R = \sqrt{(\%\sigma\rho)^2 + (\%\sigma L)^2 + (2\times\%\sigma D)^2}$$

$\Rightarrow \quad \%\sigma_R = \sqrt{(1)^2+(1)^2+(2\times 1)^2} = \sqrt{6}$

$\Rightarrow \quad \%\sigma_R = \pm\sqrt{6} = \pm 2.45\%$

8.27 (3)

Given, $\dfrac{\partial V}{V} = 1\%$

and $\dfrac{\partial I}{I} = 2\%$

Since, $V = IR$

$$\Rightarrow \quad \frac{\partial R}{R} = \frac{\partial V}{V} + \frac{\partial I}{I} = \pm 2\% \pm 1\% = \pm 3\%$$

8.28 95.35 (95 to 96)

$$f_1 = \frac{1}{2\pi\sqrt{LC}}$$

if $\quad C' = C + 0.1\,C$

(as sensor capacitance increases by 10%)

then, $f_2 = \dfrac{1}{2\pi\sqrt{L(1.1C)}} = \dfrac{f}{\sqrt{1.1}}$

$$= \frac{100}{\sqrt{1.1}} = 95.35\,\text{kHz}$$

8.29 (2.5)

$V_1 = 100 \pm 1.5$ V

$V_2 = 150 \pm 2$ V

$V_3 = V_1 + V_2$

Standard deviation in V_3 will be

$$\sigma_{V_3} = \sqrt{\left(\frac{\partial V_3}{\partial V_1}\right)^2 \sigma_{V_1}^2 + \left(\frac{\partial V_3}{\partial V_2}\right)^2 \sigma_{V_2}^2}$$

$$\frac{\partial V_3}{\partial V_1} = \frac{\partial}{\partial V_1}(V_1 + V_2) = 1$$

$$\frac{\partial V_3}{\partial V_2} = \frac{\partial}{\partial V_2}(V_1 + V_2) = 1$$

$$\therefore \quad \sigma_{V_3} = \sqrt{1^2\times(1.5)^2 + 1^2\times(2)^2}$$

$$= \sqrt{6.25} = 2.5\text{ V}$$

8.30 2.50 (2.49 to 2.53)

1% of 300 V = 3 V

0.5% of 5 A = $\dfrac{1}{40}$ A

$V_L = 200\text{ V} \pm 3\text{ V} = 200\text{ V} \pm 1.5\%$

$I_L = 2.5\text{ A} \pm \dfrac{1}{40}\text{ A} = 2.5\text{ A} \pm 1\%$

$$R_L = \frac{V_L}{I_L} = \frac{200}{2.5} \pm (2.5\%)$$

$$= 80 \pm 2.5\%\ \Omega$$

8.31 (130)

Given, $\quad R = \dfrac{V}{I}$

where, $V = 1 \pm 0.12$ V, $I = 1 \pm 0.05$ mA

Here the standard deviation in R (σ_R) will be

$$\sigma_R = \pm\sqrt{\left(\frac{\partial R}{\partial V}\right)^2 \sigma_V^2 + \left(\frac{\partial R}{\partial I}\right)^2 \sigma_I^2}$$

$$\frac{\partial R}{\partial V} = \frac{\partial}{\partial V}\left(\frac{V}{I}\right) = \frac{1}{I} = \frac{1}{10^{-3}} = 10^3$$

$$\frac{\partial R}{\partial I} = \frac{\partial}{\partial I}\left(\frac{V}{I}\right) = -\frac{V}{I^2} = \frac{-1}{10^{-6}} = -10^6$$

$\sigma_V = 0.12V$ and $\sigma_I = 0.05$ mA

Substituting, we have

$$\sigma_R = \sqrt{\left(10^3\right)^2(0.12)^2 + \left(-10^6\right)^2\left(0.05\times 10^{-3}\right)^2}$$

$$= \sqrt{10^6\times 0.0144 + 10^{12}\times 2.5\times 10^{-9}}$$

$$= \sqrt{14400 + 2500} = \pm 130\,\Omega$$

8.32 (3.5)

Given, $\quad R_1 = 100\ \Omega \pm 3\%$

$R_2 = 100\ \Omega \pm 4\%$

$R = R_1 + R_2 = 100 + 100 = 200\ \Omega$

Here, $\dfrac{\delta R_1}{R_1}\times 100 = 3$

$$\delta R_1 = \frac{100\times 3}{100} = 3\ \Omega$$

$$\frac{\delta R_2}{R_2}\times 100 = 4$$

$$\delta R_2 = \frac{100\times 4}{100} = 4\ \Omega$$

We know, $\delta R = \delta R_1 + \delta R_2 = 3 + 4 = 7\ \Omega$

$$\therefore \quad \%\,\epsilon_r \text{ in } R = \frac{\delta R}{R}\times 100 = \frac{7}{200}\times 100 = 3.5\%$$

8.33 (d)

$V_{ac} = 100\pm 1\%,\ I_{ac} = \pm 1\%$

$W_{ac} = 50\text{W}\pm 2\%$

$P = VI\cos(\phi)$

$$\Rightarrow \quad \text{pf} = \cos(\phi) = \frac{P}{VI} = \frac{50}{100 \times 1} = 0.5$$

$$\%\text{Error} = \pm\left[\frac{\delta P}{P} + \frac{\delta V}{V} + \frac{\delta I}{I}\right]$$

$$= \pm[2\% + 1\% + 1\%] = \pm 4\%$$

8.34 (1)

Given,

X(kg)	80	79	81	79	81
$\bar{X}$	80	80	80	80	80

where, $\bar{X} = \frac{\text{Sum}}{N}; \ N = 5$

$$= \frac{80 + 79 + 81 + 79 + 81}{5} = \frac{400}{5} = 80$$

$$\text{S.D} = \sqrt{\frac{\sum_{i=1}^{5}(X - \bar{X})^2}{N-1}} = \sqrt{\frac{(1)^2 + (1)^2 + (1)^2 + (1)^2}{4}}$$

S.D = 1 kg

8.35 (22)

True value of resistance = 20 kΩ

Given mean of six readings = true value

let last reading is R'

$$\text{Now } \frac{19 + 18 + 23 + 21 + 17 + R'}{6} = 20$$

$$R' = 22 \text{ k}\Omega$$

■■■

9 Miscellaneous

9.1 The primary of a 200/1 A, 5 VA, 0.8 PF current transformer (CT) carries 100 A. The secondary current is 0.495 A. The ratio error of the CT is

(a) –1.0% (b) –0.5%
(c) +0.5% (d) +1.0%

[2003 : 2 M]

9.2 A 1000/1 A, 5 VA, UPF bar-primary measuring current transformer has 1000 secondary turns. The current transformer exhibits a ratio error of – 0.1% and a phase error of 3.438 minutes when the primary current is 1000 A. At this operating condition, the rms value of the magnetization current of the current transformer (in amperes, rounded off to two decimal places) is _____.

[2020 : 2 M]

9.3 A bar primary current transformer of rating 1000/1 A, 5VA; UPF has 995 secondary turns. It exhibits zero ratio error and phase error of 30 minutes at 1000 A with rated burden. The watt loss component of the primary excitation current in ampere is(rounded off to one decimal place).

[2021 : 2 M]

■■■

Answers Miscellaneous

9.1 (a) 9.2 (1.00) 9.3 (5)

Explanations Miscellaneous

9.1 (a)

Nominal ratio,

$$k_n = \frac{200}{1} = 200$$

Actual transformation ratio,

$$R = \frac{I_P}{I_S} = \frac{100}{0.495} = 202.020$$

$$\text{Ratio error} = \frac{\text{Nominal ratio} - \text{Actual ratio}}{\text{Actual ratio}} \times 100$$

$$= \frac{200 - 202.020}{202.020} \times 100 = -1.0\%$$

9.2 (1.00)

In a current transformer we know that the phase angle error

$$\beta = \frac{180}{\pi}\frac{I_m}{nI_s}$$

Here $\beta = 3.438 \text{ mins} = \frac{3.438}{60}\text{deg}$

$n = 1000$

$I_s = 1$

$$\therefore \frac{3.438}{60} = \frac{180}{\pi} \times \frac{I_m}{1000 \times 1}$$

$$I_m = \frac{3.438 \times 1000 \times \pi}{60 \times 180}$$

$$= \frac{108.0079}{108} = 1.00 \text{ A}$$

9.3 (5)

$n = 995$

Ratio error $= 0$

$k_n - R = 0$

$k_n = R$

$$k_n = \frac{1000}{1} = 1000$$

So, $R = \frac{1000}{R}$

We know that,

$$R = n + \frac{I_o \sin(\alpha + \delta)}{I_s}$$

Here, $\delta = 0$ as UPF

$$1000 = 995 + \frac{I_w}{1}$$

$I_w = 5 \text{ A}$

■■■

Optical Instrumentation

UNIT

IX

CONTENTS

1. Basics of Fibre Optics (Optical) 479
2. Interferometer, Applications in Metrology (Optical) 484
3. LED, Lasers & Photodiode (Optical) 487
4. Miscellaneous (Optical) 494
5. Mass Spectrometry, X-Ray and Nuclear Measurement (Analytical) 498
6. UV, Visible and IR Spectrometery Chromatography (Analytical) 502
7. Biomedical Instrumentation, ECG, EEG & EMG (Biomedical) 507
8. Clinical Measurement (Biomedical) 509
9. Ultrasonic Transducers and Ultrasonography (Biomedical) 512

Optical Instrumentation

Syllabus

Optical sources and detectors: LED, laser, photo-diode, light dependent resistor, square law detectors and their characteristics; interferometer: applications in metrology; basics of fiber optic sensing. UV-VIS Spectro photometers, Mass spectrometer.

Analysis of Previous GATE Papers

Exam Year	1 Mark Ques.	2 Marks Ques.	Total Marks
1995	–	3	**6**
1996	1	5	**11**
1997	1	2	**5**
1998	–	2	**4**
1999	–	10	**20**
2000	–	6	**12**
2001	4	–	**4**
2002	5	3	**11**
2003	3	7	**17**
2004	3	8	**19**
2005	2	5	**12**
2006	4	4	**12**
2007	4	8	**20**
2008	3	5	**13**
2009	1	5	**11**

Exam Year	1 Mark Ques.	2 Marks Ques.	Total Marks
2010	2	2	**6**
2011	2	2	**6**
2012	3	1	**5**
2013	1	1	**3**
2014	2	3	**8**
2015	2	4	**10**
2016	–	2	**4**
2017	–	1	**2**
2018	1	2	**5**
2019	2	3	**8**
2020	–	2	**4**
2021	1	1	**3**
2022	1	1	**3**
2023	–	1	**2**
2024	–	–	–

1 Basics of Fibre Optics (Optical)

1.1 A step index optical fiber, whose refractive indices of the core and cladding are 1.44 and 1.40 respectively, is surrounded by air. Its numerical aperture is
(a) 0.12 (b) 0.75
(c) 0.06 (d) 0.34 [1996 : 1 M]

1.2 An optical fibre has a refractive index of 1.641 for the core and 1.422 for the cladding. The critical angle above which a ray will be totally internally reflected is
(a) 37° (b) 41°
(c) 45° (d) 60° [1999 : 2 M]

1.3 The refractive indices of the core and cladding of an optical fibre are 1.46 and 1.455 respectively. The diameter of the core is 4 μm. When operated at a wavelength of 1.31 μm, the fibre functions as a
(a) single-mode fibre with a numerical aperture of 0.1
(b) multimode fibre with a numerical aperture of 0.15
(c) single-mode fibre with a numerical aperture of 0.12
(d) ten-mode fibre with a numerical aperture of 0.12 [2003 : 2 M]

1.4 The refractive index of the core of an optical fibre is n_1 and that of the cladding is n_2. If $(n_1 - n_2) = \Delta n$, then the fibre can be made single mode with numerical aperture unchanged by
(a) reducing core diameter and increasing Δn
(b) reducing both core diameter and Δn
(c) reducing core diameter alone
(d) reducing Δn alone [2004 : 1 M]

1.5 A step index fibre has a core refractive index of 1.46 and a cladding refractive index of 1.45. The pulse dispersion in ns/km due to the effect of numerical aperture alone is
(a) 33.3 (b) 31.5
(c) 29.5 (d) 25.2 [2004 : 2 M]

1.6 The maximum solid angle of acceptance of light coupled into a step index fibre having core and cladding refractive indices of 1.48 and 1.45 respectively, is
(a) 0.28 steradians (b) 0.30 steradians
(c) 0.32 steradians (d) 0.34 steradians
[2005 : 2 M]

Statement for Linked Answer Questions (1.7 to 1.8):

In a step-index optical fiber, the refractive indices of the core and the cladding are equal to 1.501 and 1.499, respectively, at the wavelength of 850 nm. The fiber is surrounded by air.

1.7 The numerical aperture of the optical fibre is
(a) 0.077 (b) 0.501
(c) 1.499 (d) 1.501 [2006 : 2 M]

1.8 An LED emitting light around wavelength of 850 nm is butt-coupled to the optical fiber. If the LED is assumed to be a Lambertial source, then efficiency of coupling LED light to the fiber is
(a) 7.70% (b) 1.02%
(c) 0.60% (d) 0.57% [2006 : 2 M]

Statement for Linked Answer Questions (1.9 to 1.10):
The numerical aperture of a step index fiber in (refractive index = 1), is 0.39. The diameter of the core is 200 μm.

1.9 The angle of acceptance when the fiber is used in water (refractive index = 1.33) is closest to
(a) 15° (b) 16°
(c) 17° (d) 18° [2007 : 2 M]

1.10 Two experiments are conducted in which light is launched into the fiber from a uniformly distributed planar source kept 5 mm away from the tip. In the first experiment (E_1) both the source and the fiber are in air. In the second experiment (E_2) both the source and the fiber are in water. Neglecting absorption in the medium between the source and the tip, the ratio of the amount of light coupled into the fiber in E_1 to the amount of light coupled into the fiber in E_2 is closest to
(a) 0.5 (b) 1.0
(c) 1.4 (d) 1.9 [2007 : 2 M]

1.11 Light coming out of an optical fiber is incident on a plane perpendicular to the fiber axis and 50 mm away from the end of the fiber. The light coming out creates a circular spot that can at most be of 20 mm diameter. Neglecting the diameter of the fiber, the numerical aperture of the fiber is, approximately,
(a) 0.14 (b) 0.20
(c) 0.34 (d) 0.40
[2010 : 2 M]

1.12 The core/cladding index difference of a single-mode optical fiber cable is 0.01. The refractive index of the material of the core is 1.5. The maximum angle of acceptance of the fiber is approximately equal to

(a) 17.5° (b) 12.1°
(c) 8.6° (d) 2.0° [2011 : 1 M]

1.13 A double convex lens is used to couple a laser beam of diameter 5 mm into an optical fiber with a numerical aperture of 0.5. The minimum focal length of the lens that should be used in order to focus the entire beam into the fiber is

(a) 1.44 mm (b) 2.50 mm
(c) 4.33 mm (d) 5.00 mm

[2012 : 2 M]

1.14 Monochromatic light from a step index (n_1 = 1.500 ; n_2 = 1.485), multimode optical fiber of core diameter 100 μm is incident through air (n = 1.000) onto a linear photo-detector array placed at 1 mm distance from the tip of the fiber. The tip of the fiber is polished and its exit plane is perpendicular to the axis of the fiber. The detector array is oriented parallel to the exit plane of the tip. The array consists of photo-detector elements each of 5 μm diameter. The distance between the edges of two adjacent elements can be assumed to be zero. The number of elements illuminated by the light coming out of the fiber is________.

[2014 : 2 M]

1.15 A beam of monochromatic light passes through two glass slabs of the same geometrical thickness at normal incidence. The refractive index of the first slab is 1.5 and that of the second, 2.0. The ratio of the time of passage of the beam through the first to the second slab is _____.

[2015 : 2 M]

1.16 A multi-mode optical fiber with a large core diameter has a core refractive index n_1 = 1.5 and cladding refractive index n_2 = 1.4142. The maximum value of θ_A (in degrees) for which the incident light from air will be guided in the optical fiber is ± ________.

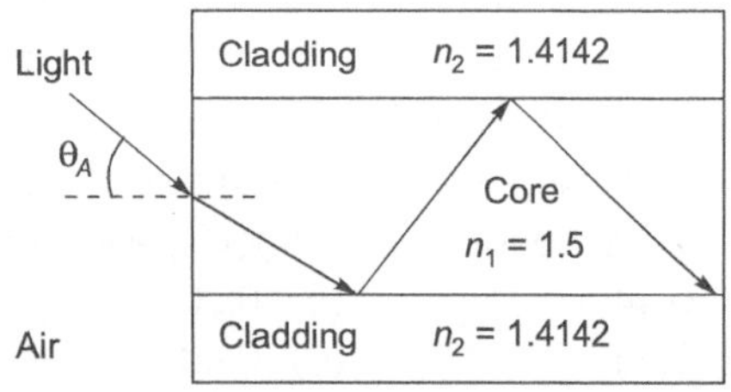

[2018 : 2 M]

1.17 In a single-mode optical fiber, the zero-dispersion wavelength refers to the wavelength at which the

(a) material dispersion is zero
(b) waveguide dispersion is zero.
(c) sum of material dispersion and waveguide dispersion is zero
(d) material dispersion and waveguide dispersion are simultaneously zero.

[2019 : 1 M]

1.18 A laser beam of 10 mm beam diameter is focused onto an optical fibre using a thin biconvex lens as shown in the figure. The refractive index of the lens is 1.5. The refractive indices of the core and cladding of the fibre are 1.55 and 1.54 respectively. The minimum value of the focal length of the lens to attain the maximum coupling to the fibre (in mm, rounded off to one decimal place) is _____.

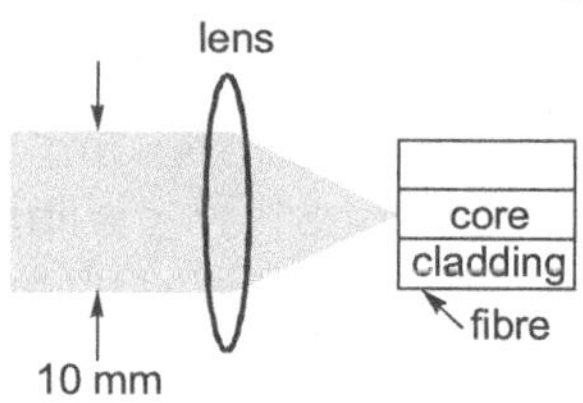

[2020 : 2 M]

1.19 In the figure shown, a large multimode fiber with n_{core} = 1.5 and n_{clad} = 1.2 is used for sensing. A portion with the cladding removed passes through a liquid with refractive index n_{liquid}. An LED is used to illuminate the fiber from one end and a paper is placed on the other end, 1 cm from the end of the fiber. The refractive index n_{liquid} of the liquid (rounded off to two decimal places) is_______.

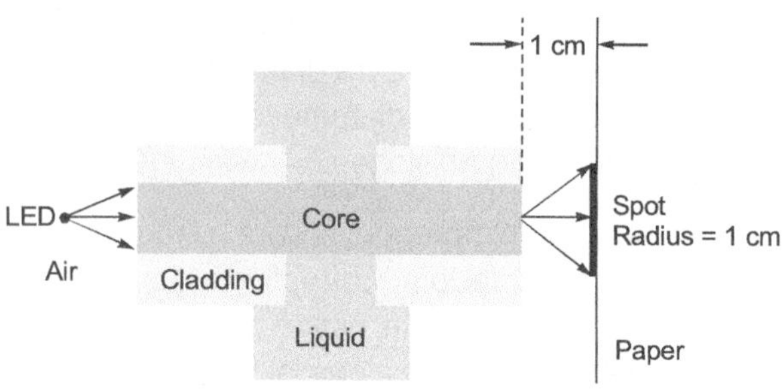

[2021 : 2 M]

1.20 A silica-glass fiber has a core refractive index of 1.47 and a cladding refractive index of 1.44. If the cladding is completely stripped out and the core is dipped in water having a refractive index of 1.33, the numerical aperture of the modified fiber is _____ (rounded off to three decimal places)

[2023 : 2 M]

■■■

Answers Basics of Fibre Optics (Optical)

1.1	(d)	1.2	(d)	1.3	(c)	1.4	(c)	1.5	(a)	1.6	(a)	1.7	(a)
1.8	(c)	1.9	(c)	1.10	(b)	1.11	(b)	1.12	(b)	1.13	(c)	1.14	(106)
1.15	(0.75)	1.16	(30)	1.17	(c)	1.18	(28)	1.19	(1.32)	1.20	(0.626)		

Explanations Basics of Fibre Optics (Optical)

1.1 (d)

$$NA = \sqrt{n_1^2 - n_2^2}$$
$$= \sqrt{(1.44)^2 - (1.40)^2} = \sqrt{0.1136}$$
$$= 0.337 \approx 0.34$$

1.2 (d)

$$n_1 \sin\theta_c = n_2 \sin 90°$$
$$\therefore \sin\theta_c = \left(\frac{n_2}{n_1}\right) = \frac{1.422}{1.641} = 0.866$$
$$\therefore \theta_c = \sin^{-1}(0.866) \approx 60°$$

1.3 (c)

If fibre core diameter is exceedingly small (2 - 10 μm) i.e if fibre core diameter is close to wavelength of propagated light wave then it is single mode transmission.

Now, numerical aperture,

$$N.A. = \sqrt{n_1^2 - n_2^2}$$
$$= \sqrt{(1.46)^2 - (1.455)^2} = 0.12$$

1.4 (c)

A multimode fiber optic cable can be made single made by either varying the Numerical Aperture (NA) or by reducing the diameter of core, and if we vary Δn NA will very as $NA = n\sqrt{2\Delta}$

Here, since NA is made fixed the fiber can then be made single mode fiber by reducing the core diameter alone, hence option (c) is correct.

1.5 (a)

$$\Delta = \frac{n_1 - n_2}{n_1} = \frac{1.46 - 1.45}{1.46} = \frac{0.01}{1.46} = 6.89 \times 10^{-3}$$

$$\Delta T = \frac{n_1 L}{C}\Delta \quad \text{(for step index fibers)}$$
$$= \frac{1.46 \times 1000}{3 \times 10^8} \times 6.89 \times 10^{-3}$$
$$= 3.33 \times 10^{-8} = 33.3 \text{ n-sec.}$$

1.6 (a)

$$\text{Solid angle} = \pi(\sin\theta_a)^2$$
$$\because \quad \sin\theta_A = \sqrt{n_1^2 - n_2^2} = \sqrt{1.48^2 - 1.45^2}$$
$$= \sqrt{0.0879} = 0.296$$
$$\text{Solid angle} = 0.28 \text{ steradians}$$

1.7 (a)

$$N.A. = \sqrt{n_1^2 - n_2^2} = \sqrt{(1.501)^2 - (1.499)^2} = 0.077$$

1.8 (c)

$$\text{Coupling Efficiency} = (NA)^2 = \left(\sqrt{n_1^2 - n_2^2}\right)^2$$

Substituting values,

Coupling Efficiency = $(1.501^2 - 1.499^2) = 0.006$

$\Rightarrow$ Coupling efficiency = $0.006 \times 100\% = 0.60\%$

1.9 (c)

$$A.A = \sin^{-1}\left(\frac{N.A}{\mu_w}\right) \quad (\mu_w = \text{R.I. of water})$$
$$= \sin^{-1}\left(\frac{0.39}{1.33}\right) \simeq 17°$$

1.10 (b)

The amount of light that can be gathered is depicted by the numerical aperture (N.A.) of the fiber, and N.A. is independent of the surrounding medium so the answer is 1.0.

1.11 (b)

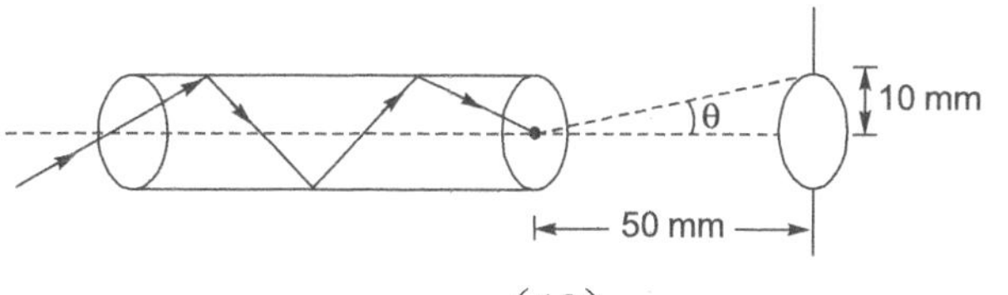

$$\text{Clearly,} \quad \theta = \tan^{-1}\left(\frac{10}{50}\right) = 11.45°$$
$$N.A. = \sin\theta = \sin(11.45°) = 0.19 \simeq 0.2$$

1.12 (b)

$\Delta = 0.01$; $n_1 = 1.5$

$$\text{N.A.} = n_1\sqrt{2\Delta}$$

$$= 1.5\sqrt{2\times 0.01} = 1.5\sqrt{0.02} = 0.212$$

$$\sin\theta_a = \text{N.A.}$$

$$\theta_a = \sin^{-1}(\text{N.A.}) = \sin^{-1}(0.212) \simeq 12.1^\circ$$

1.13 (c)

$$\text{N.A.} = \sin\theta_a = 0.5$$

$$\therefore \quad \theta_0 = \sin^{-1} 0.5 = 30^\circ$$

$$D = 5 \text{ mm}$$

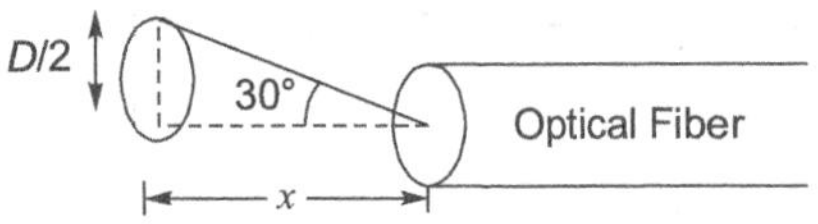

We have to find this distance 'X', as it is the focal of the lens

$$\tan 30^\circ = \frac{D/2}{x}$$

$$\frac{1}{\sqrt{3}} = \frac{5}{2x}$$

$$x = \frac{5\sqrt{3}}{2} = 4.33 \text{ mm}$$

1.14 106 (105 to 107)

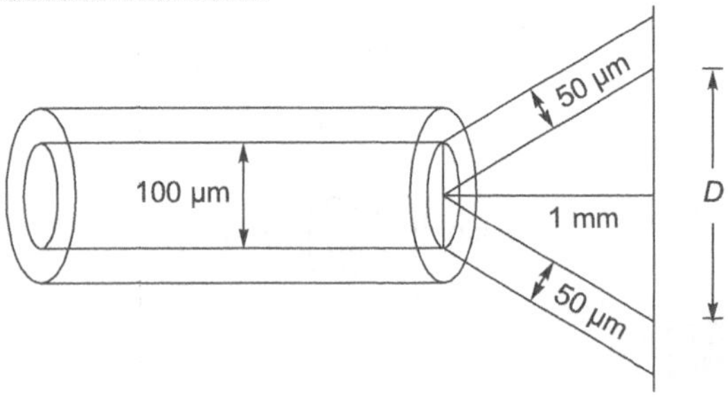

$\because$ Spot diameter $= 2r\theta_a$

here, $\theta_a = \sin^{-1}$ Na

$$= \sin^{-1}\sqrt{1.5^2 - 1.485^2}$$

$$= 12.216^\circ = 0.213 \text{ rad}$$

$$D = 2\times 1000\times 0.213 = 427\ \mu\text{m}$$

Total length of photo-detector array

$427 + 50 + 50\ \mu$m $= 527\ \mu$m

Diameter of one photodetector $= 5\ \mu$m

So, total number of photo detector in array

$$= \frac{527}{5} \simeq 106$$

1.15 (0.75)

$$v \propto \frac{1}{\mu}$$

but $$v \propto \frac{1}{t}$$

$$\therefore \quad t \propto \mu$$

$$\therefore \quad \frac{t_1}{t_2} = \frac{1.5}{2} = 0.75$$

1.16 30 (29 to 31)

From the given figure

$$\text{N.A.} = n_o \sin\theta_A = \sqrt{n_1^2 - n_2^2}$$

where, $n_0 \rightarrow$ Refractive index of air $[n_0 = 1]$
$n_1 \rightarrow$ Refractive index of core $[n_1 = 1.5]$
$n_2 \rightarrow$ Refractive index of cladding $[n_2 = 1.4142]$
$\theta_A \rightarrow$ Maximum acceptance angle

$$\theta_A = \sin^{-1}\left[\sqrt{n_1^2 - n_2^2}\right]$$

$$\theta_A = \sin^{-1}\left[\sqrt{(1.5)^2 - (1.4142)^2}\right]$$

$$\theta_A = 30^\circ$$

1.17 (c)

In single mode optical fiber 'zero-dispersion' wave length means, sum of material dispersion and waveguide dispersion is zero. So, option (c) is correct answer.

1.18 (28)

Given, the diameter of Laser beam = 10 mm
Refractive index of lens = 1.5
Refractive index of core (n_1) = 1.55
Refractive index of cladding (n_2) = 1.54

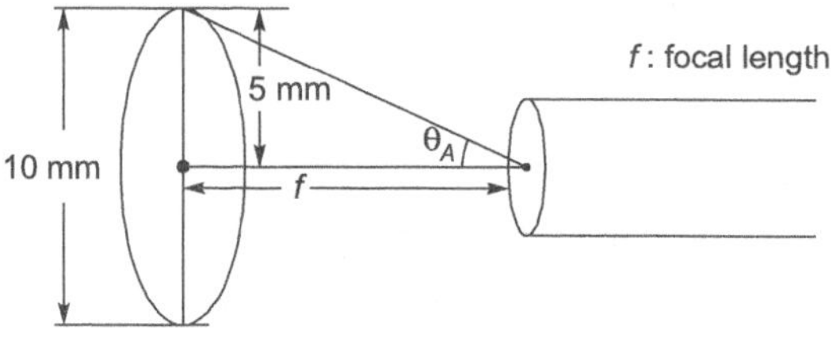

$$\text{Acceptance angle } (\theta_a) = \sin^{-1}\left[\frac{NA}{N_m}\right]$$

($NA \rightarrow$ Numerical aperture;
$N_m \rightarrow$ Surrounding medium (air))

$$\theta_A = \sin^{-1}\left(\frac{NA}{n_0}\right)$$

$$\theta_A = \sin^{-1}\left[\frac{\sqrt{n_1^2 - n_2^2}}{n_0}\right]$$

$$= \sin^{-1}\left[\sqrt{\frac{(1.55)^2 - (1.54)^2}{1}}\right]$$

$$\theta_A = \sin^{-1}(0.1757839) = 10.124°$$

From the figure,

$$\tan\theta_A = \frac{5\text{ mm}}{f}$$

$$f = \frac{5\text{ mm}}{\tan\theta_A}$$

$$= \frac{5\text{ mm}}{\tan(10.124°)} = 28\text{ mm}$$

1.19 (1.32)

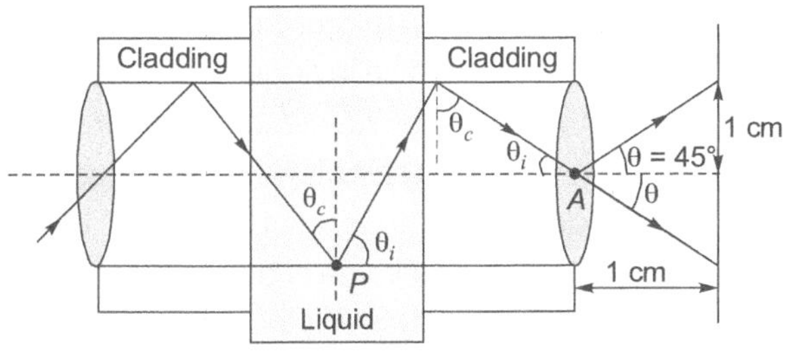

Applying Snell's law at point A,

$$n_1 \sin\theta_i = n_0 \sin\theta$$
$$1.5 \sin\theta_i = 1 \times \sin 45°$$
$$\sin\theta_i = 0.47$$
$$\theta_i = 28.12°$$
$$\therefore \quad \theta_c = 90° - \theta_i$$
$$\theta_c = 61.87°$$

Applying Snell's law at point P,

$$n_1 \sin\theta_c = n_{liquid}$$
$$1.5 \sin 61.87 = n_{liquid}$$
$$\therefore \quad n_{liquid} = 1.32$$

1.20 (0.626)

Numerical Aperture,

$$N.A = \sqrt{n_{core}^2 - n_{water}^2} = \sqrt{1.47^2 - 1.33^2}$$
$$N.A = 0.626$$

■■■

2 Interferometer, Applications in Metrology (Optical)

2.1 In a wave propagation, a path difference of x_0 corresponds to a phase difference of

(a) λx_0 (b) πx_0

(c) $\left(\frac{2\pi}{\lambda}\right)x_0$ (d) $\frac{x_0}{\lambda}$

[1995 : 2 M]

2.2 Match the instrument to the characteristic

List-I

(a) Interferometer

(b) Monochromator

List-II

(i) Wavelength dispersion Magnifying power

(ii) Flatness measurement

[1997 : 2 M]

2.3 For a change in elevation level of 1.5 μm between an optical flat and a surface, a light wave of wavelength 0.5 μm produces fringes. The number of fringes is

(a) 3 (b) 6

(c) 9 (d) 12

[1998 : 2 M]

2.4 An interference filter has been made with a dielectric of magnesium fluoride (refractive index $\mu = 1.38$) for a nominal wavelength of 455 nm. The required thickness of the dielectric for first order interference is

(a) 8.2 nm (b) 16.5 nm

(c) 82 nm (d) 165 nm

[2000 : 2 M]

2.5 An optical flat is tested against a standard reference flat in a Newton's interferoscope if the optical flat is as good as the reference, the fringes observed are a set of

(a) concentric rings

(b) elliptical rings

(c) hyperbolic fringes

(d) straight and parallel fringes

[2001 : 1 M]

2.6 Two narrow straight slits, 0.25 mm apart are illuminated by an unknown monochromatic source of light. If the fifth bright fringe is 12 mm away from the central fringe when the screen is at a distance of 0.8 m from the slits. Then the wave length of the unknown source is

(a) 540 nm (b) 600 nm

(c) 700 nm (d) 750 nm

[2002 : 2 M]

2.7 Koster's modification of Michelson interferometer is used to determine the thickness of a gauge G. The gauge G is placed on the mirror M_2 as shown in figure (a). The field-of-view observed through the lens L is shown in figure (b). The fractional-fringe shift e, as shown in figure (b), is measured to be 0.15 when the wavelength of light used is 0.63 μm. A quick check with a micrometer shows that the gauge length is approximately 100 μm. Then an accurate estimate of gauge length, in μm, is

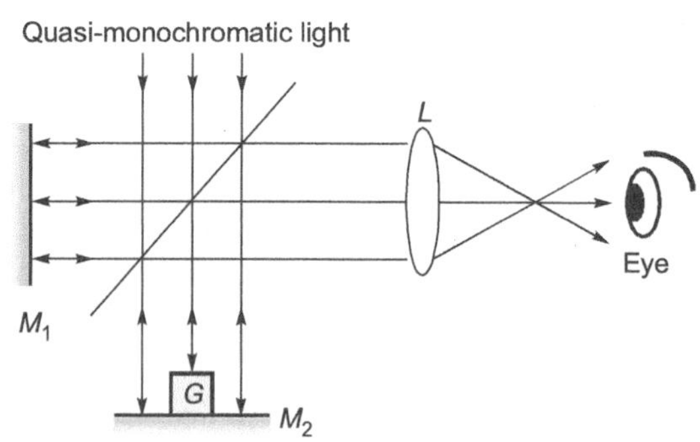

(a)

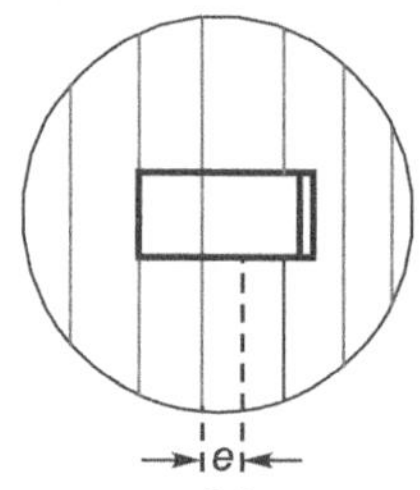

(b)

(a) 99.902 (b) 100.804

(c) 100.000 (d) 101.706

[2003 : 2 M]

2.8 In a Michelson's interferometer illuminated by a coherent and quasi-monochromatic light source, a very small frequency shift is introduced in the light returning from one of the arms. The effect on the observed fringe pattern is
(a) a decrease in contrast
(b) an increase in contrast
(c) appearance of two sets of the same pattern
(d) uniform movement across the observation plane

[2004 : 2 M]

2.9 A tissue with a refractive index 1.33 is introduced in one of the light paths of a Michelson interferometer operating with a monochromatic coherent light source of wavelength 589 nm. After the introduction of a tissue sample of thickness Δt, the fringe pattern is observed to shift by 50 fringes. If thickness is 2 Δt, the fringe pattern will shift by
(a) 25 fringes (b) 50 fringes
(c) 100 fringes (d) 200 fringes

[2008 : 2 M]

Statement for Linked Answer Questions (2.10 to 2.11):
A Michelson interferometer illuminated with a source of central wavelength λ_0 and spectral width Δl is adjusted for equal path difference for the beams returning from the two mirrors. When one of the mirrors is moved by a distance of 0.1 mm from this position, 300 fringes move past the field of view. When the mirror is moved further, the fringes completely disappear when the mirror is approximately 4 cm from the initial position.

2.10 The central wavelength of the source is
(a) 540 nm (b) 632.8 nm
(c) 667 nm (d) 720 nm

[2009 : 2 M]

2.11 The spectral width of the source $\Delta\lambda$ is approximately
(a) 0.0056 nm (b) 0.0100 nm
(c) 0.0500 nm (d) 0.1000 nm

[2009 : 2 M]

2.12 A Michelson Interferometer using a laser source of wavelength λ_0 = 500 nm with both the mirrors (M_1 and M_2) fixed and positioned equidistant from the splitter/combiner is shown in the figure. When a dielectric plate of refractive index n = 1.5, of thickness t, is placed in front of the mirror M_2, a dark fringe is observed on the detector. When the dielectric plate is removed without changing the position of the mirrors M_1 & M_2, a bright fringe is observed on the detector. The minimum thickness t (in nm} of the dielectric is ________ .

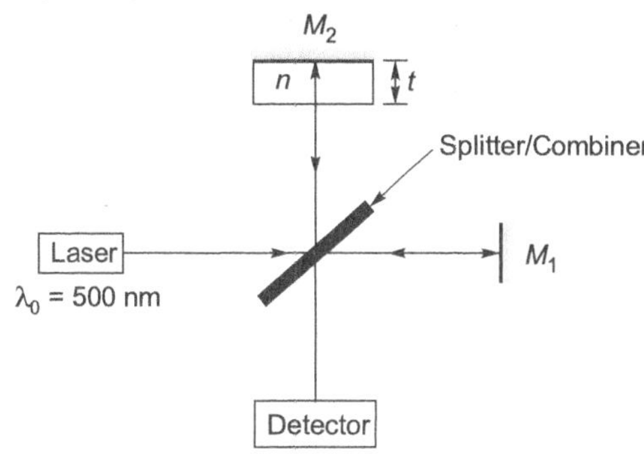

[2018 : 2 M]

2.13 Consider a Michelson interferometer as shown in the figure below. When the wavelength of the laser light source is switched from 400 nanometer to 500 nanometer, it is observed that the intensity measured at the output port P goes from a minimum to a maximum. This observation is possible when the smallest path difference between the two arms of the interferometer is____nanometer.

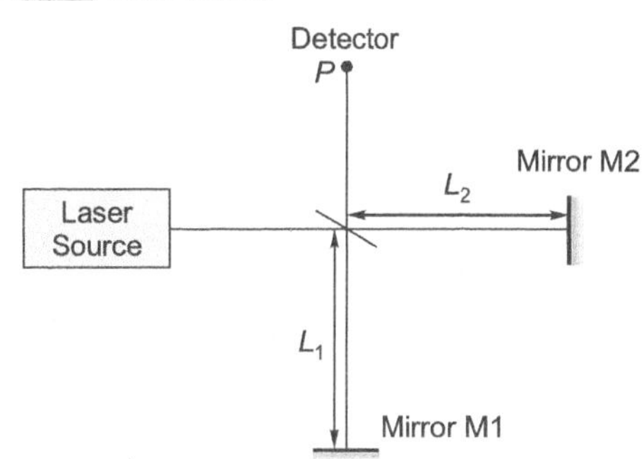

[2019 : 2 M]

Answers **Interferometer, Applications in Metrology (Optical)**

2.1 (c) 2.2 (a)-(ii), (b)-(i) 2.3 (b) 2.4 (d) 2.5 (a) 2.6 (d)
2.7 (a) 2.8 (c) 2.9 (c) 2.10 (c) 2.11 (a) 2.12 (250)
2.13 (500 or 1000)

Explanations Interferometer, Applications in Metrology (Optical)

2.1 (c)

$$\frac{\text{Phase difference}}{2\pi}\lambda = \text{Path difference}$$

i.e. $$\frac{\phi_0}{2\pi}\lambda = x_0$$

$$\Rightarrow \qquad \phi_0 = \frac{2\pi}{\lambda}x_0$$

2.3 (b)

$$2(\Delta x) = n\lambda$$

$$\therefore\ 2\times(1.5\times10^{-6}) = n\times0.5\times10^{-6}$$

$$\therefore \qquad n = \frac{2\times1.5\times10^{-6}}{0.5\times10^{-6}} = 6 \text{ fringes}$$

2.4 (d)

$$n\lambda = 2\mu t$$

for first order interference, $n = 1$

$$\therefore \quad \lambda = 2\mu t$$

$$t = \frac{\lambda}{2\mu} = \frac{455\times10^{-9}}{2\times1.38} = 164.8\times10^{-9} \approx 165 \text{ nm}$$

2.6 (d)

$d = 0.25$ mm ; $X_5 = 12$ mm ; $D = 0.8$ m

$$X_5 = 5\left(\frac{\lambda D}{d}\right)$$

$$\lambda = \frac{X_5\times d}{5D} = \frac{12\times10^{-3}\times0.25\times10^{-3}}{5\times0.8}$$

$$= 0.75\times10^{-6} = 750\times10^{-9}\text{ m} = 750\text{ nm}$$

2.7 (a)

Given fractional fringe shift, $e = 0.15$

and λ of light used = 0.63 μm

$\Rightarrow$ Error in length of gauge = $e\cdot\lambda$ = 0.0945 μm

So, Actual length = 100 − 0.0945 $\simeq$ 99.9055

$\simeq$ 99.902 μm

2.9 (c)

$$n\lambda = 2\Delta x = 2t(\mu - 1) \quad [\therefore n \propto t]$$

as Δt is replaced to $2(\Delta t)$.

$\therefore$ n will also gets doubled.

$$n' = 2n = 2\times50 = 100 \text{ fringes}$$

2.10 (c)

$$n\lambda = 2(\mu x)$$

$$\lambda = \frac{2(\Delta x)}{n} = \frac{2\times0.1\times10^{-3}}{300} = 0.66666\times10^{-6}$$

$$= 666.66 \text{ nm} \approx 667 \text{ nm}$$

2.11 (a)

The spectral width of the source ($\Delta\lambda$) is approximately

$$\Delta\lambda = \frac{(\lambda_{av})^2}{2x} = \frac{(667\times10^{-a})^2}{2\times4\times10^{-2}} = 0.0056 \text{ nm}$$

2.12 250 (249 to 251)

From the given explanation, we can understand that, "when a tissue of thickness 't' of refractive index 'N', is placed then a dark fringe is observed"

So,

$$\text{Path difference} = \frac{\lambda_0}{2}$$

$$\text{Path difference} = 2(N-1)t$$

$$2(N-1)t = \frac{\lambda_0}{2}$$

$$2(1.5-1)t = \frac{500\text{ nm}}{2}$$

$$t = 250 \text{ nm}$$

2.13 (500 or 1000)

As per the question the optical path difference in both the cases (400 nm or 500 nm) should be same but for 400 nm wavelength we should get destructive interference whereas for 500 nm wavelength we should get constructive interference.

$\therefore$ For constructive interference optical path difference $(2x) = n\lambda_c$

For distructive interference optical path difference

$$(2x) = \frac{(2m+1)\lambda_D}{2}$$

As per question,

$$h\lambda_c = \frac{(2m+1)\lambda_D}{2}$$

$$n(500) = \frac{(2m+1)400}{2}$$

$\therefore$ For $n = 2$ or $m = 2$ condition satisfied

$\therefore$ Optical path difference

$$2x = n\lambda = 2(500\text{ nm}) = 1000 \text{ nm}$$

■■■

3 LED, Lasers & Photodiode (Optical)

3.1 A laser beam (λ = 632.8 nm) having a divergence of 1 milli-radian is directed towards the moon at a distance of approximately 4×10^5 km. The beam would have spread to a diameter of

(a) 400 km (b) 200 km
(c) 100 km (d) 4 km

[1996 : 2 M]

3.2 Two crossed length of the He-Ne laser beam is 120 cm. Its coherence time in seconds is

(a) 4×10^{-1} (b) 4×10^{-3}
(c) 4×10^{-5} (d) 4×10^{-9}

[1997 : 1 M]

3.3 A frequency stabilized He-Ne laser with a wavelength of 6328 Å has a bandwidth of 1 MHz. Its coherence length is

(a) 0.3 m (b) 3 m
(c) 30 m (d) 300 m

[1999 : 2 M]

3.4 In a He-Ne laser operating at a wavelength of 632.8 nm. the actual laser transition takes place between the energy levels of

(a) Helium (b) Neon
(c) Helium and Neon (d) None of these

[2002 : 1 M]

3.5 A He-Ne laser beam having a wavelength of 632.8 mm is expanded to a diameter of 1 meter and then pointed towards the moon. If the distance of the moon from earth is 3.8×10^5 km, the diameter of the central maxima on the surface of the moon is.

(a) 1 m (b) 100 m
(c) 587 m (d) 720 m

[2002 : 2 M]

3.6 An unpolarized He-Ne laser beam of wavelength 632.8 nm is reflected from the flat surface of a dielectric. A detector analyzes the reflected beam. It is found that when the angle between the incident and the reflected beams is 115°-34′ the light is completely plane polarized. The refractive index of the dielectric at this wavelength is

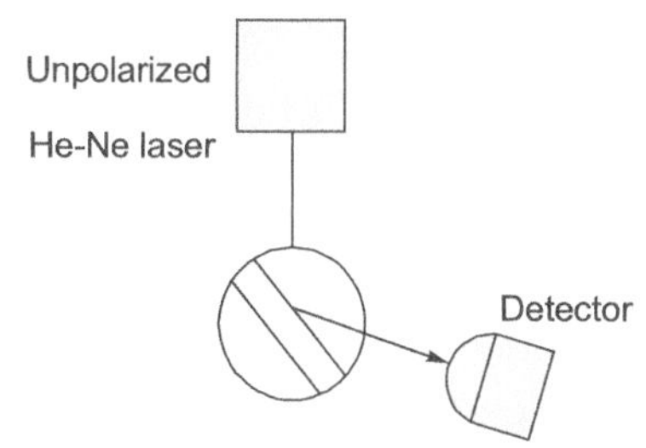

(a) 1.521 (b) 1.587
(c) 1.621 (d) 1.571

[2002 : 2 M]

3.7 A photo-detector circuit is given in figure. The photo-diode has an active area the photo-diode has an active area of 10^{-2} cm^2 and sensitivity of 0.55 A/W. When light of intensity 10 mW/cm^2 falls on the photo-diode, the output V_0 is

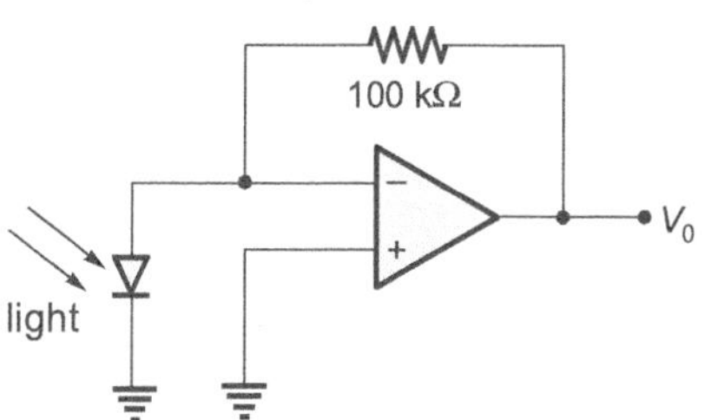

(a) 10.1 V (b) 5.5 V
(c) –5.5 V (d) –10.1 V

[2003 : 2 M]

3.8 A PIN photodetector has a junction capacitance of 16 pF, active area of 10 mm^2 and a responsivity of 0.5 A/W. It is illuminated by a light of intensity 1 mW/cm^2. A voltage output is obtained by connecting a 100 kΩ load resistor R_L in series with the detector. The steady state voltage across R_L will be

(a) 0.5 V (b) 1.5 V
(c) 5.0 V (d) 7.0 V

[2004 : 2 M]

3.9 A He-Ne laser of cavity length 500 mm has an oscillating bandwidth of 1500 MHz. The maximum number of longitudinal oscillating modes that is accommodated within the bandwidth is

(a) 4 (b) 5
(c) 30 (d) 40

[2004 : 2 M]

3.10 The longitudinal mode separation in a gas laser with a cavity length of 0.4 m (given velocity of light = 3×10^8 m/s) is

(a) 374 MHz (b) 375 MHz
(c) 748 MHz (d) 750 MHz

[2005 : 2 M]

3.11 Collimated light beams from a He-Ne laser and a Sodium vapour lamp (used for street lighting) are focused using a lens. The size of the spot at the focal point due to the laser is relatively smaller because

(a) laser light is relatively more monochromatic
(b) the limiting divergence of laser light is relatively smaller
(c) the output power of the laser is relatively larger
(d) the wavelength of the laser is relatively longer

[2006 : 1 M]

3.12 A measurement system for the rotational speed of a motor is shown in the figure below. The system consists of an opaque disk attached to the motor shaft with a hole as shown. A light source and a photodetector are placed on two sides of the disk so that whenever the hole cross the light path, the photodetector receives light through the hole. The photodetector circuit is also shown below. Assume sufficient light intensity and TTL logic levels for the inverter.

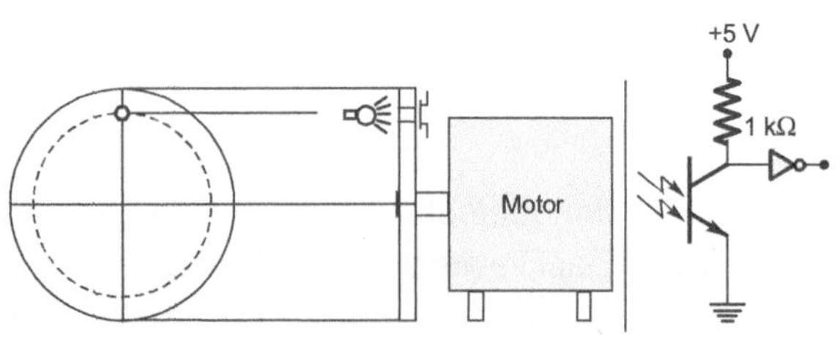

The output the photodetector is a

(a) triangular wave
(b) square wave with 50% duty cycle
(c) rectangular wave with duty cycle close to unity
(d) rectangular wave with duty cycle close to zero

[2007 : 2 M]

Common Data Questions (3.13 to 3.14):

Laser light is generated by energizing helium-neon gas in chamber. The ground and metastable states of helium are 0 eV and 20.61 eV respectively. The ground, higher and metastable energies of neon are 0 eV, 18.70 eV and 20.66 eV respectively. The values of speed of light, Planck constant and charge of electron are 3×10^8 m/s, 6.625×10^{-34} Js and 1.6×10^{-19} C respectively.

3.13 In this process, helium molecules

(a) play no role
(b) produce laser light
(c) give energy to neon molecules
(d) absorb energy from neon molecules

[2008 : 2 M]

3.14 Wave length of laser light generated in this process is

(a) 61.6 nm (b) 66.4 nm
(c) 633.8 nm (d) 650.3 nm

[2008 : 2 M]

3.15 An LED emitting at 1 μm with a spectral width of 50 nm is used in a Michelson interferometer. To obtain a sustained interference, the maximum optical path difference between the two arms of the interferometer is

(a) 200 μm (b) 20 μm
(c) 1 μm (d) 50 nm

[2012 : 1 M]

3.16 The responsivity of the PIN photodiode shown is 0.9 A/W. To obtain V_{out} of − 1 V for an incident optical power of 1 mW, the value of R to be used is

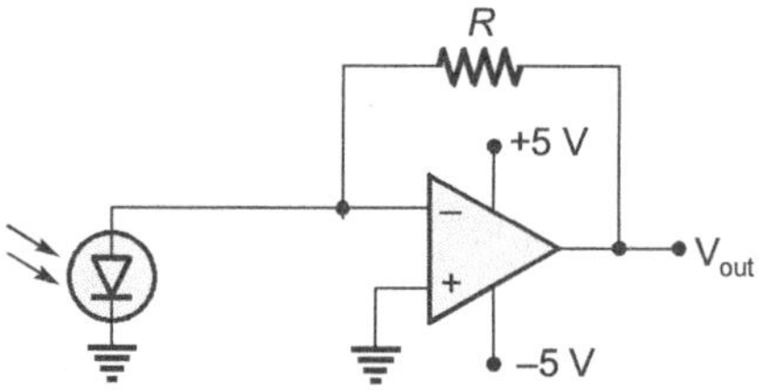

(a) 0.9 Ω (b) 1.1 Ω
(c) 0.9 kΩ (d) 1.1 kΩ

[2012 : 1 M]

3.17 A light detector circuit using an ideal photo-diode is shown in the figure. The sensitivity of the photo-diode is 0.5 μA/μW. With $V_r = 6$ V, the output voltage $V_o = -1.0$ V for 10 μW of incident light. If V_r is changed to 3 V, keeping all other parameters the same, the value of V_o in volts is ______ V.

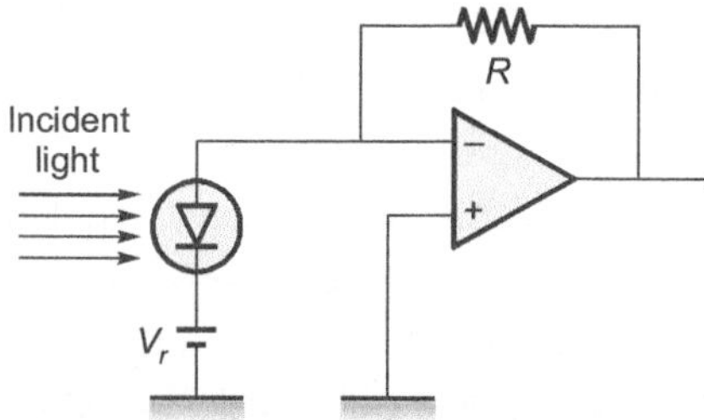

[2015 : 1 M]

3.18 The figure shows a spot of light of uniform intensity 50 W/m^2 and size 10 mm × 10 mm incident at the exact center of a photo-detector, comprising two identical photo-diodes D_1 and D_2. Each diode has sensitivity of 0.4 A/W and is operated in the photoconductive mode. If the spot of light is displaced upwards by 100 μm, the resulting difference between the photocurrents generated by D_1 and D_2 in micro amperes, is ______ μA.

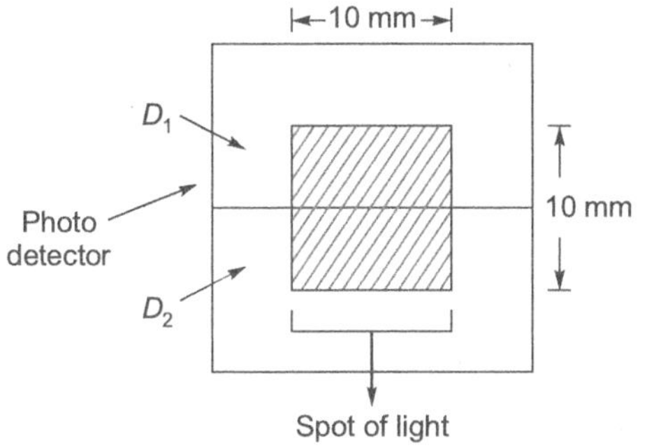

[2015 : 2 M]

3.19 The photo diode in the figure below has an active sensing area of 10 mm^2, a sensitivity of 0.5 A/W and a dark current of 1 μA. The i-to-v converter has a sensitivity of 100 mV/μA. For an input light intensity of 4 W/m^2, the output V_0 in volt is______

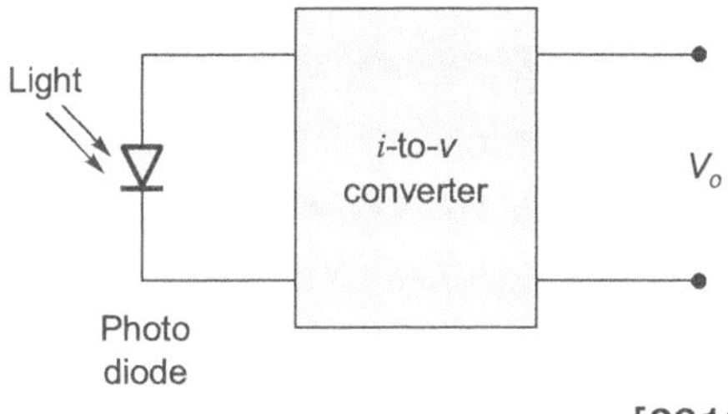

[2016 : 2 M]

3.20 The bandgap in eV of a semiconductor material required to construct an LED that emits peak power at the wavelength of 620 nm is______.
(Plank constant $h = 4.13567 \times 10^{-15}$ eV s and speed of light $c = 2.998 \times 10^8$ m/s)

[2016 : 2 M]

3.21 Quantum efficiency of a photodiode (ratio between the number of liberated electrons and the number of incident photons) is 0.75 at 830 nm. Given Plank's constant $h = 6.624 \times 10^{-34}$ J, the charge of an electron $e = 1.6 \times 10^{-19}$ C and the velocity of light in the photodiode $C_m = 2 \times 10^8$ m/s. For an incident optical power of 100 μW at 830 nm, the photocurrent in μA is ______.

[2017 : 2 M]

3.22 An optical pulse containing 6×10^6 photons is incident on a photodiode and 4.5×10^6 electron-hole pairs are created. The maximum possible quantum efficiency (in %) of the photodiode is ___.

[2018 : 1 M]

3.23 The correct biasing conditions for typical operation of light emitting diodes, photodiodes, Zener diodes are, respectively
(a) forward bias, reverse bias, reverse bias
(b) reverse bias, reverse bias, forward bias
(c) forward bias, forward bias, reverse bias
(d) reverse bias, forward bias, reverse bias

[2019 : 1 M]

3.24 A pulsed laser emits rectangular pulses of width 1 nanosecond at a repetition rate of 1 kHz. If the average power output is 1 mW, the average power over a single pulse duration, in Watts, is
(a) 1 (b) 10
(c) 100 (d) 1000

[2019 : 2 M]

3.25 A 100 W light source emits uniformly in all directions. A photodetector having a circular active area whose diameter is 2 cm is placed 1 m away from the source, normal to the incident light. If the responsivity of the photodetector is 0.4 A/W, the photo-current generated in the detector, in units of mA, is
(a) 1 (b) 4
(c) 100 (d) 400

[2019 : 2 M]

3.26 When the movable arm of a Michelson interferometer in vacuum ($n = 1$) is moved by 325 μm. The number of fringe crossings is 1000. The wavelength of the laser used in nanometers is ______.

[2021 : 1 M]

3.27 A photodiode is made of a semiconductor with a bandgap of 1.42 eV. Given that Planck's constant is 6.626×10^{-34} Js, the speed of light in vacuum is 3×10^8 m/s, and 1 eV = 1.6×10^{-19} J, the cut-off wavelength (in nanometers) of the photodiode is _______ (round off to one decimal place)

[2022 : 1 M]

■■■

Answers LED, Lasers & Photodiode (Optical)

3.1 (*)	3.2 (d)	3.3 (d)	3.4 (b)	3.5 (c)	3.6 (b)	3.7 (c)
3.8 (c)	3.9 (b)	3.10 (b)	3.11 (b)	3.12 (d)	3.13 (c)	3.14 (c)
3.15 (b)	3.16 (d)	3.17 (–1)	3.18 (40)	3.19 (2.1)	3.20 (2)	3.21 (75.18)
3.22 (75)	3.23 (a)	3.24 (d)	3.25 (a)	3.26 (650)	3.27 (874.9)	

Explanations LED, Lasers & Photodiode (Optical)

3.1 (*)

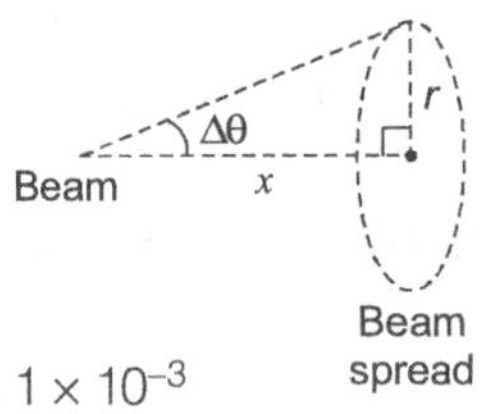

From figure,

Radius, $r = x\tan\Delta\theta$

For small θ, tanq $\simeq \theta$

$\Rightarrow \quad r = x\Delta\theta$

$\Rightarrow \quad r = 4 \times 10^5 \times 1 \times 10^{-3}$

$= 400$ km

So diameter, $d = 2 \times 400 = 800$ km

Alternate solution :

$\Delta\theta = 1$ milliradian

$= 1 \times 10^{-3}$

$r = 4 \times 10^5$ km

Spread diameter $(D) = 2r(\Delta\theta)$

$= 2 \times 4 \times 10^5 \times 1 \times 10^{-3}$

$= 8 \times 10^2 = 800$ km

3.2 (d)

$$t = \frac{L}{c} = \frac{120\times10^{-2}}{3\times10^{8}} = 4 \times 10^{-9}$$

3.3 (d)

Given bandwidth, $B = 1 \times 10^6$ sec^{-1}

$$\Rightarrow \text{Coherence length} = \frac{\text{Speed of light (m/sec)}}{\text{Bandwidth (sec}^{-1})}$$

$$= \frac{3\times10^8}{1\times10^6} = 300 \text{ m}$$

3.4 (b)

In He-Ne laser, He molecule gives energy to Ne molecule which leads laser transition of Ne.

3.5 (c)

$\lambda = 632.8$ mm ; $d = 1$ m

angle of divergence, Δθ,

$$\Delta\theta = \frac{1.22\lambda}{d} = \frac{1.22\times632.8\times10^{-9}}{1}$$

$= 772 \times 10^{-9}$ m

Diameter of central maxima $(D) = 2r(\Delta\theta)$

$= 2 \times 3.8 \times 10^5 \times 772 \times 10^{-9}$

$= 5867.2 \times 10^{-4} = 0.5867$ km

$= 586.7$ m ≈ 587 m

3.6 (b)

$2\,i = 115° - 34' = 115.566°$

$$\therefore \quad i = \frac{115.566}{2} = 57.783°$$

Refractive index (μ) = tan i

(By Brewster's law)

$= \tan(57.783°) = 1.587$

3.7 (c)

I = Sensitivity × Intensity × Area

$I = 0.55 \times 10 \times 10^{-3} \times 10^{-2}$

$= 5.5 \times 10^{-5}$ Amp

and o/p voltage,

$V_0 = -IR_L = -5.5 \times 10^{-5} \times 100 \times 10^3$

$V_0 = -5.5$ V

3.8 (c)

Light beam sensed by the photodiode of area 10 mm² when illuminated by a surge of 1 mW/cm².

$$1 \times 10^{-3} = \frac{\text{Power}}{10\text{ mm}^2} = \frac{\text{Power}}{10\times10^{-2}\text{ cm}^2}$$

$\therefore$ Power $= 10 \times 10^{-3} \times 10^{-2} = 10^{-4}$ watts

Responsivity = 0.5 A/watts

$\therefore$ amount of current flow corresponding to 10^{-4} watts is

$$0.5 = \frac{\text{Current}}{10^{-4}}$$

Current $= 0.5 \times 10^{-4}$ A

$\therefore$ Voltage induced across R_L

$= R_L \times 0.5 \times 10^{-4}$ V

$= 100 \times 10^3 \times 0.5 \times 10^{-4}$ V $= 5$ V

3.9 (b)

$$\Delta\nu = \frac{c}{2L} = \frac{3\times10^{8}}{2\times500\times10^{-3}}$$

$$\text{Number of modes} = \frac{\text{B.W}}{\Delta\nu} = \frac{1500\times10^{6}}{3\times10^{8}}$$

$$= 5 \text{ modes}$$

3.10 (b)

$$\Delta\nu = \frac{C}{2L} = \frac{3\times10^{8}}{2\times0.4}$$

$$= 3.75 \times 10^{8} \text{ Hz} = 375 \text{ MHz}$$

3.11 (b)

The limiting divergence of the laser light is relatively small.

Since the divergence is small so the spread is less hence the spot diameter is less.

3.12 (d)

Since the disk has only a single hole on its periphery. So the light falls on the detector (i.e. the base of the transistor) for a very small duration in one revolution. The transistor is the npn transistor. The npn transistor comes in saturation when a high is applied on its base and is cut off when a low is applied to its base.

When light falls:

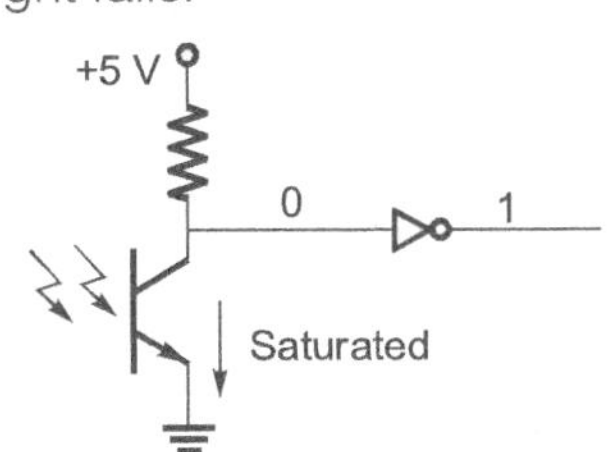

When light do not fall:

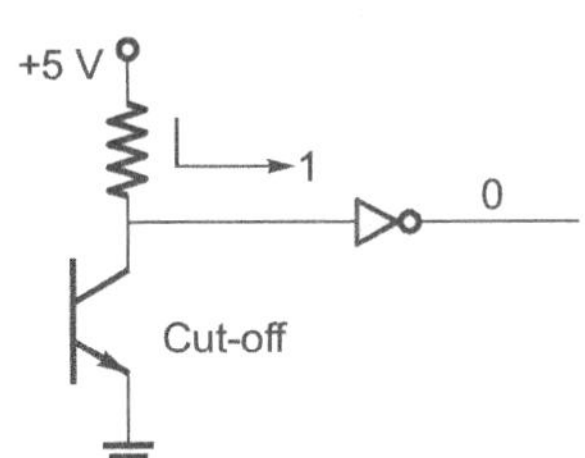

$$\text{Duty cycle} = \frac{T_{ON}}{T_{ON} + T_{OFF}}$$

Output is

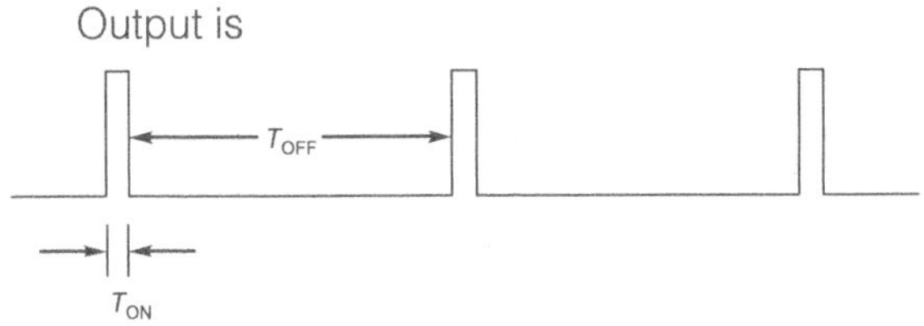

Hence the output is the rectangular wave, with a duty cycle

$$= \frac{T_{ON}}{T_{ON} + T_{OFF}} = 1 + \frac{1}{\frac{T_{OFF}}{T_{ON}}}$$

$$= \frac{1}{1+\infty} \simeq 0 \quad [\text{Since, } T_{ON} << T_{OFF}]$$

3.13 (c)

Give energy to Ne molecules.

3.14 (c)

The transition for the major element Ne in case of He-Ne laser is to be considered, since He is present just to excite the Ne atoms. And the transition between higher and the metastable state is kept on account as the atom attains stability in the metastable state.

So, Energy difference

$$= (20.66 - 18.70)\text{eV} = 1.96 \text{ eV}$$

We know, $E = h\nu$

$$1.96 \times 1.6 \times 10^{-19} = \frac{6.62\times10^{-34}\times3\times10^{8}}{\lambda}$$

$$\lambda = \frac{6.62\times10^{-34}\times3\times10^{8}}{1.96\times1.6\times10^{-19}}$$

$$= 6.3329 \times 10^{-7} = 633.29 \times 10^{-9}$$

$$\simeq 633.8 \text{ nm}$$

3.15 (b)

$$\because \ \Delta\lambda = \frac{(\lambda_{avg.})^2}{2re}$$

$$x = \frac{(\lambda_{avg.})^2}{2\Delta\lambda} = \frac{(1\times10^{-6})^2}{2\times50\times10^{-9}} = 10\ \mu\text{m}$$

Maximum distance = 2 × 10 μ m = 20 μm

3.16 (d)

Responsivity = 0.09 A/W

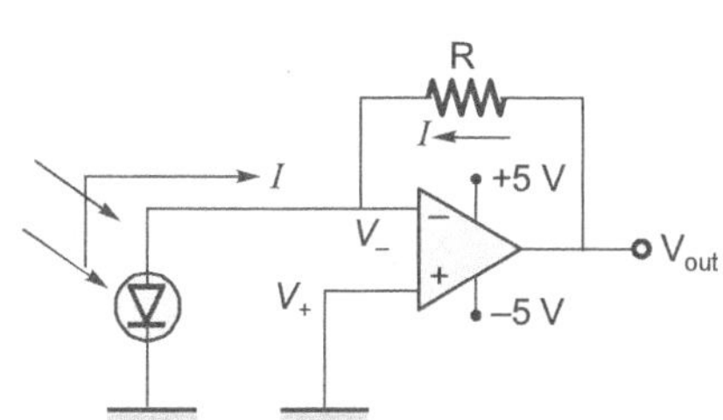

For incident power of 1 mW

$$I = 0.9 \times 10^{-3} \text{ A}$$

By virtual ground

$$V_- = V_+ \doteq 0\text{ V}$$

$$R = -\frac{V_{out}}{I} = \frac{1}{0.9\times10^{-3}} = 1.1\text{ k}\Omega$$

3.17 (-1)

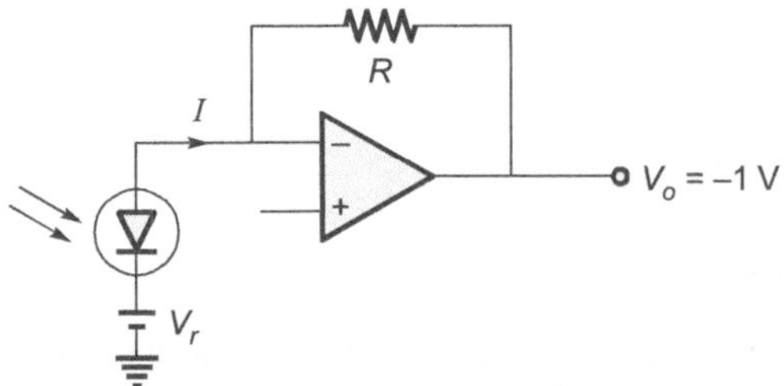

Since, incident light is constant so, photo current, I is also constant.

Applying KVL, $V_o = -I \cdot R$

Now as only V_r changes and since I and R are constant so V_o remains constant at –1 V.

3.18 (40)

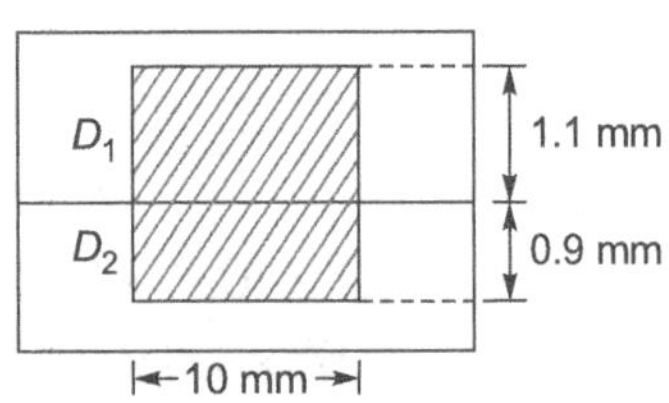

Photo current I_1 through diode D_1 is,

$$I_1 = 50\frac{W}{m^2}\times0.4\frac{A}{W}\times10\text{ mm}\times10.1\text{mm}$$

$$\Rightarrow \quad I_1 = 2020\,\mu\text{A} \quad ...(i)$$

Photo current I_2 through diode D_2 is,

$$I_2 = 50\frac{W}{m^2}\times0.4\frac{A}{W}\times10\text{ mm}\times9.9\text{ mm}$$

$$\Rightarrow \quad I_2 = 1980\,\mu\text{A} \quad ...(ii)$$

Using equations (i) and (ii),

Difference = $I_1 - I_2 = 40\,\mu\text{A}$

Note: Use unit/dimensional analysis to get formula.

3.19 (2.1)

OUTPUT current through photo diode (V_0)

= Intensity × Sensitivity × Area

$$V_0 = \frac{4\,W}{m^2}\times0.5\frac{A}{W}\times10\times10^{-6}m^2$$

$$= 20\,\mu\text{A}$$

Total current flow through photo diode

$= I_P + I_D$ (Dark current)

OUTPUT voltage i to v converter

$$V_0 = (20+1)\,\mu\text{A}\times\frac{100\text{mV}}{\mu\text{A}}$$

$$= 100\text{ mV} = 2.1\text{ V}$$

3.20 2 (1.92 to 2.1)

$$E_g = \frac{hc}{\lambda} = \frac{4.13567\times10^{-15}\times2.998\times10^8}{620\times10^{-9}}$$

$$= 2\text{ eV}$$

3.21 (75.18)

$$\text{Responsivity, }(R) = \frac{I}{P} = \frac{\eta e\lambda}{hC}\text{ A/W}$$

$$\therefore \quad I = \frac{\eta e\lambda}{hC}P$$

$$= \frac{0.75\times1.6\times10^{-19}\times830\times10^{-9}\times100\times10^{-6}}{6.624\times10^{-34}\times2\times10^8}$$

$$= 75.18\,\mu\text{A}$$

3.22 (75)

Number of incident photons 'N' = 6×10^6

Number of electron-hole pairs generated

$n = 4.5\times10^6$

Quantum efficiency of photo-diode

$$= \frac{n}{N}\times100 = \frac{4.5\times10^6}{6\times10^6} = 75$$

3.23 (a)

Light emitting diode

LED - Forward bias

Photo diode - Reverse bias

Zener diode - Reverse bias

3.24 (d)

Given that Laser emits rectangular pulse width of 1 nsec.

Repetition rate of 1 kHz.

If the average output power is 1 mW.

Energy of each pulse is given by

$$E = \frac{P_{av}}{R_{rate}} = \frac{1\times10^{-3}}{1\times10^3} = 1\times10^{-6}\text{ Joules}$$

$$\text{Power of each pulse } P_{pulse} = \frac{E}{\text{Duration of pulse}}$$

$$= \frac{1\times10^{-6}}{1\times10^{-9}} = 1000$$

3.25 (a)

Intensity of light at photo detector

$= \left(\frac{\text{Power}}{4\pi d^2}\right) \text{W/m}^2$

where d is distance in meters

$= \frac{100}{4\pi(1)^2} = \frac{100}{4\pi} \text{ W/m}^2$

Power received by photodetector = Intensity of light × Active area of photo detector

$$= \frac{100}{4\pi} \times \pi \times (0.01)^2 = \frac{1}{400} \text{W}$$

$$\text{Responsibility} = \frac{\text{Output current}}{\text{Power}}$$

Output current = Responsivity × Power

$$= \frac{1}{400} \times 0.4 = \frac{1}{1000} A = 1 \text{ mA}$$

3.26 (650)

Given, $d = 325$ mm

Path difference,

$$2d = 2 \times 325 \text{ mm}$$

Path difference,

$$n\lambda = 2\ d$$

$$\Rightarrow \quad n = \frac{2d}{\lambda} = \frac{2 \times 325 \mu m}{1000}$$

$$\therefore \quad n = 650 \text{ nm}$$

3.27 (874.9)

As we know,

$$E_g = hv = \frac{hC}{\lambda}$$

Cut-off wavelength, $\lambda = \frac{hC}{E_g}$

$$\lambda = \frac{6.626 \times 10^{-34} \times 3 \times 10^8}{1.42 \times 1.6 \times 10^{-19}}$$

$$\lambda = 874.91 \text{ nm}$$

■■■

4 Miscellaneous (Optical)

4.1 For a glass of refractive index 1.732, the angle of incidence of ordinary light, for which the reflected beam is completely polarized is
(a) 30 degrees (b) 60 degrees
(c) 45 degrees (d) 75 degrees
[1996 : 2 M]

4.2 When light passes from a medium of refractive index n_1 into a medium of refractive index n_2, the reflection at the boundary is expressed as
(a) $(n_2 - n_1)^2$ (b) $\left(\frac{n_2 - n_1}{n_2 + n_1}\right)^2$
(c) $\frac{1}{(n_2 - n_1)^2}$ (d) $(n_2 + n_1)^2$
[1999 : 2 M]

4.3 For alignment and testing of two surfaces at right angles, a constant deviation prism, also called an optical square, is used. It is a
(a) right angled prism (b) square prism
(c) pentagonal prism (d) hexagonal prism
[1999 : 2 M]

4.4 The relative permittivity of an optical medium is 2.5. Its refractive index is
(a) 0.4
(b) 1.25
(c) 1.58
(d) There is no interrelation between the two parameters [2000 : 2 M]

4.5 Brewster angle is the angle of incidence of monochromatic light at the interface of two media, for which
(a) the incidence beam would be totally internally reflected
(b) a polarized beam would become unpolarised after reflected
(c) an unpolarised beam would become polarized after reflected
[2000 : 2 M]

4.6 The refractive indices of glass and water are 1.50 and 1.33, respectively. If the glass is immersed in water, its relative refractive index is
(a) 0.89 (b) 1.13
(c) 1.33 (d) 2.83 [2001 : 1 M]

4.7 Light is incident on a glass plate of refractive index 1.732. It is plane polarized on reflection when the incidence angle is
(a) 30° (b) 45°
(c) 60° (d) 75° [2001 : 1 M]

4.8 The minimum number of lines that must be drawn on a grating needed to resolve the Sodium Doublet having nominal wavelengths of 589.0 nm and 589.6 nm is
(a) 600 (b) 1200
(c) 982 (d) 300 [2002 : 1 M]

4.9 To a swimmer viewing a Sodium lamp beside the pool. The colour of the lamp seen from underneath the water appears:
(a) Yellow
(b) Yellow with a reddish
(c) Yellow with a greenish tinge
(d) White [2002 : 1 M]

4.10 A parallel beam is incident normally on a air glass interface. If the refractive index of the glass is 1.720, then the loss of intensity through reflection is
(a) 7% (b) 4%
(c) 2% (d) 1% [2002 : 1 M]

4.11 A grating that has 1200 lines/mm is irradiated with a narrow beam of white light at an incident angle of 25° to the grating normal. The first order peak observed at an angle of 15° belongs to the wavelength
(a) 567.9 nm (b) 587.8 nm
(c) 597.1 nm (d) 600.7 nm
[2003 : 1 M]

4.12 A light beam is incident normally on a glass-air interface. The glass has refractive index 1.53. Air is assumed to be of refractive index 1.0. The Fresnel, intensity reflection coefficient is
(a) 0.53 (b) 0.25
(c) 0.044 (d) 0.022 [2003 : 1 M]

4.13 A light source is formed by putting a pinhole to let in sunlight. The above light source has
(a) NO spatial-or temporal coherence
(b) NO spatial coherence but has temporal coherence

(c) spatial coherence but NO temporal coherence
(d) quasi-monochromaticity

[2003 : 1 M]

4.14 A quasi-monochromatic light source with average central wavelength of 630 nm has a spectral width of 10 GHz. The coherence length of the source is
(a) 3 μm (b) 3 mm
(c) 3 cm (d) 3 m [2003 : 2 M]

4.15 The time taken by an ionized atom, of mass m kg and charge e coulombs, pulsed into a field-free region with V volts, to reach a detector L meters away is

(a) $\frac{1}{L}\sqrt{\frac{m}{2eV}}$ (b) $L\sqrt{\frac{m}{2eV}}$

(c) $m\sqrt{\frac{L}{2eV}}$ (d) $\frac{2}{L}\sqrt{\frac{m}{2eV}}$

[2004 : 2 M]

4.16 A glass plate is over coated with a thin film as shown below. The reflectance of the combination is monitored with a radiation of 600 nm wavelength. If n_g and n_f are refractive indices of the glass plate and film material respectively with $n_g > n_f$, then reflectance reaches a minimum when the geometrical film thickness equals

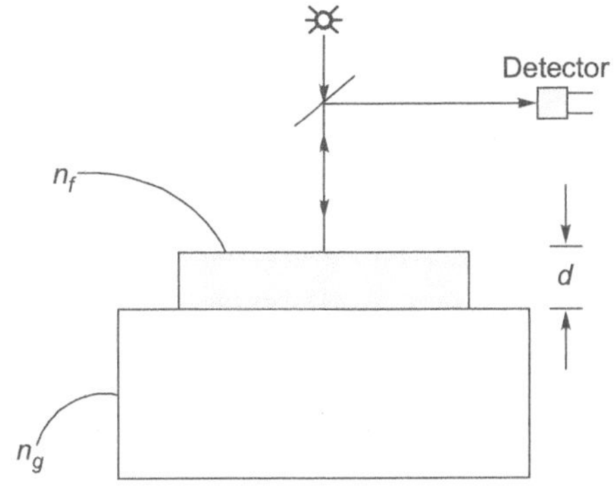

(a) 600 nm (b) 300 nm
(c) 150 nm (d) 100 nm [2006 : 2 M]

4.17 The dispersion in an X-ray diffractometer, $d\theta/d\lambda$ is given by the expression

(a) $\frac{m}{2d\cos\theta}$ (b) $\frac{m}{2d\sin\theta}$
(c) $2\,d\sin\theta$ (d) $2d\sin\theta$

[2007 : 1 M]

4.18 In the process of non-destructive testing of a 10 cm diameter cylinder, a cross-sectional (transaxial) image of the cylinder is reconstructed with the help of parallel beam computer tomography technique. To realize a spatial resolution of 1 mm in the image, the minimum number of ray samples in each projection set and the minimum set and the minimum number of projection sets required are
(a) 200 and 315 respectively
(b) 100 and 315 respectively
(c) 200 and 629 respectively
(d) 100 and 629 respectively [2008 : 2 M]

4.19 A beam of unpolarized light is first passed through a linear polarizer and then through a quarterwave plate. The emergent beam is
(a) unpolarized
(b) linearly polarized
(c) circularly polarized
(d) elliptically polarized [2010 : 1 M]

4.20 Light of wavelength 630 nm in vacuum, falling normally on a biological specimen of thickness 10 μm, splits into two beams that are polarized at right angles. The refractive index of the tissue for the two polarizations are 1.32 and 1.333. When the two beams emerge, they are out of phase by
(a) 0.13° (b) 74.3°
(c) 90.0° (d) 128.6°

[2012 : 1 M]

4.21 As shown in figure, a slab of finite thickness t with refractive index n_2 = 1.5, has air (n_1 = 1) above and below it. Light of free space wavelength 600 mm is incident normally from air as shown. For a destructive interference to be observed at R, the minimum value of thickness of the slab t (in nm) is _____.

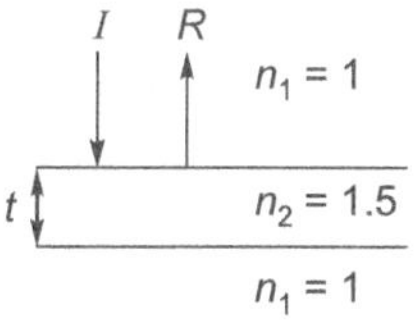

[2020 : 2 M]

Answers Miscellaneous (Optical)

4.1 (b)	4.2 (b)	4.3 (a)	4.4 (c)	4.5 (c)	4.6 (b)	4.7 (c)
4.8 (c)	4.9 (a)	4.10 (a)	4.11 (a)	4.12 (c)	4.13 (*)	4.14 (c)
4.15 (b)	4.16 (c)	4.17 (a)	4.18 (b)	4.19 (c)	4.20 (a)	4.21 (300)

Explanations Miscellaneous (Optical)

4.1 (b)

For the light beam to be completely polarized, according to Brewster's law.

Angle of incidence (i) = $\tan^{-1}(\mu)$

$= \tan^{-1}(1.732) = 60°$

4.4 (c)

$\because \quad \mu = \sqrt{\epsilon_o \epsilon_r}$

Here, $\epsilon_o = 1$

$\epsilon_r = 2.5$

$\mu = \sqrt{1 \times 2.5} = 1.58$

4.6 (b)

$\mu_g = 1.50$

$\mu_w = 1.33$

$$\frac{\mu_g}{\mu_w} = \frac{1.50}{1.33} = 1.13$$

4.7 (c)

By Brewster's law,

$\mu = \tan(i)$

$\therefore \quad i = \tan^{-1}(\mu)$

$= \tan^{-1}(1.732) = 60°$

4.8 (c)

the maximum number of lines required are

$$\frac{\text{mean}(\lambda_1, \lambda_2)}{\lambda_1 - \lambda_2} = \frac{\frac{589.6 + 589}{2}}{(589.6 - 589)}$$

$$= \frac{589.3}{0.6} = 982.16 \approx 982$$

4.10 (a)

$$\text{Reflection loss} = \left(\frac{1.720 - 1}{1.720 + 1}\right)^2 \times 100$$

$$= \left(\frac{0.72}{1.72}\right)^2 \times 100 = 0.0700 \times 100 = 7\%$$

4.11 (a)

$n\lambda = d(\sin i + \sin r)$

$$d = \frac{1}{200} = 8.33 \times 10^{-4} \text{m}$$

$= 8.33 \times 10^{-7}$ mm

$\therefore \quad \lambda = 8.33 \times 10^{-7} \times (\sin 25° + \sin 15°)$

$= 567.9$ nm

4.12 (c)

$$R = \left(\frac{n_2 - n_1}{n_2 + n_1}\right)^2 = \left(\frac{1.53 - 1.00}{1.53 + 1.00}\right)^2$$

$\therefore \quad R = 0.044$

4.14 (c)

$\Delta\nu = 10$ GHz

$$\Delta\nu = \frac{1}{\tau_c}$$

$$\therefore \quad \tau_c = \frac{1}{\Delta\nu} = \frac{1}{10 \times 10^9}$$

$= 10^{-10}$ sec

length of coherence (l) can be find as:

$l = c\tau_c$

$= 3 \times 10^8 \times 10^{-10}$

$= 3 \times 10^{-2}$ m $= 3$ cm

4.15 (b)

Distance of detector (L) = Velocity of atom (v) $\times$ time taken to reach the detector (t)

$\therefore \quad L = v \times t$

$$t = \frac{L}{v} = \frac{L}{\sqrt{\frac{2eV}{m}}} \qquad \left[\because v = \sqrt{\frac{2eV}{m}}\right]$$

$$\therefore \quad t = L\sqrt{\frac{m}{2eV}}$$

Hence, option (b) is correct.

4.16 (c)

$$t = \frac{\lambda}{4} \quad \text{for } n_g > n_f$$

$$= \frac{600}{4} \text{ nm}$$

$$t = 150 \text{ nm}$$

4.17 (a)

$$2\,d \sin\theta = m\lambda \quad \text{(Bragg's law)}$$

differentiating with respect to λ

$$2d\cos\theta\frac{d\theta}{d\lambda} = m$$

$$\boxed{\frac{d\theta}{d\lambda} = \frac{m}{2d\cos\theta}}$$

4.18 (b)

Circumference of cylinder = $2\pi r$

$$= 2\times3.14\times\frac{10}{2} = 31.4 \text{ cm}$$

Spatial resolution = 1 mm

$$\text{So, minimum projection set} = \frac{31.4\,\text{cm}}{1\,\text{mm}}$$

$$= 314 \simeq 315$$

Diameter of cylinder = 10 cm

Spatial resolution = 1 mm

$$\text{So, minimum number of rays} = \frac{10\text{ cm}}{1\text{ mm}} = 100$$

4.20 (a)

By Brewster's Law

$$\theta_1 = \tan^{-1} 1.32 = 52.85$$

$$\theta_2 = \tan^{-1} 1.333 = 53.12$$

$$\Delta\theta = \theta_2 - \theta_1 = 53.12 - 52.85$$

$$= 0.27°$$

So, the most nearby answers is (a).

4.21 (300)

Given; Wavelength (λ) = 600 nm;

Refractive index of air (N_1) = 1

Refractive Index of slab (N_2) = 1.5

Thickness of slab = t mm

Path difference of light ray in given setup is

$$PD = (N_2 - N_1)\times t$$

for the minimum thickness to be found, central destructive fringe should be observed.

So, $$PD = (2n+1)l/2$$

$$2\times(N_2 - N_1)\times t = l/2 \qquad (\because n = 0)$$

$$2\times(1.5-1)\times t = l/2$$

$$t = l/2$$

$$t = 600/2 = 300 \text{ nm}$$

■■■

5 Mass Spectrometry, X-Ray and Nuclear Measurement (Analytical)

5.1 The Geiger Counter has high quantum efficiency in the wavelength (A) range

(a) 0.2 to 0.3 (b) 0.4 to 0.5
(c) 1.5 to 2.1 (d) 6 to 7 [1995 : 2 M]

5.2 The Duane-Hunt law represents the minimum wavelength λ(Å) of X-rays generated for an applied voltage V(volts) and given by

(a) $V = \frac{12400}{\lambda}$ (b) $V = 12400\,\lambda$

(c) $V = \frac{12400}{\lambda^2}$ (d) $V = 12400\,\lambda^2$

[1995 : 2 M]

5.3 Match the and instruments with parameters they measure:

List-I	List-II
(a) Autocollimator	(i) Radiation scattered
(b) Hygrometer	(ii) Angular error
(c) Nephelometer	(iii) Molecular weight
(d) Mass spectrometer	(iv) Humidity
	(v) Density

[1996 : 2 M]

5.4 Match the name of the equation with the formula of the equation.

List-I	List-II
(a) Bragg equation	(i) $A = abc$
(b) Duane Hunt equation	(ii) $N = N_0\, e^{-\lambda t}$
(c) Radioactive decay equation	(iii) $\lambda = \frac{hc}{eV}$
(d) Beer Lambert Law	(iv) $(\mu - 1)t = n\lambda$
	(v) $2d \sin\theta = n\lambda$

[1996 : 2 M]

5.5 If the half-life of a radiation source is 1 year, its decay constant per year is

(a) 1 (b) 2
(c) 0.693 (d) 63.2

[1998 : 2 M]

5.6 The activity of a certain radioactive material decreases 2.5 times after 7 days. Find its half life.

[1999 : 2 M]

5.7 Attenuation of a narrow monochromatic X-ray beam in a metal plate of thickness 'd' is given by the equation

(a) $I = I_0 \exp(-\mu d)$ (b) $I = I_0 \exp(-\mu d^2)$
(c) $I = I_0 \exp(-\mu/d)$ (d) $I = I_0 \exp(-\mu/d^2)$

[1999 : 2 M]

5.8 For nuclear radiation measurement, a proportional counter has, in comparison to a Geiger counter;

(a) lower dead time and higher count rate
(b) higher dead time and higher count rate
(c) higher internal amplification
(d) higher internal amplification and lower count rate

[2004 : 1 M]

5.9 In an ideal X-Ray tube operated at 40 kV, the short wavelength limit of X-Ray emission in Å is (Planck's constant $h = 6.626 \times 10^{-34}$ Js, Charge of an electron $e = 1.6 \times 10^{-19}$ Coulomb velocity of light $c = 3 \times 10^8$ m/s)

(a) 0.155 (b) 0.31
(c) 0.62 (d) 0.93

[2005 : 2 M]

5.10 Two ionic species A and B of masses in the ratio 1 : 2 and of equal charge are simultaneously accelerated in a time of flight mass-spectrometer. When ion A reaches the end of a 1 m drift tube, ion B would be at a distance of

(a) $\sqrt{2}$ m (b) $\sqrt{2} - 1$m
(c) $\frac{1}{\sqrt{2}}$ m (d) $\frac{1}{2}$ m [2005 : 2 M]

5.11 An X-ray source radiates two characteristic wavelengths : λ_1 and λ_2, where $\lambda_1 < \lambda_2$.
The shorter wavelength can be filtered out by passing the radiation through

(a) a narrow slit
(b) an interference filter
(c) a crystalline material whose absorption edge lies between λ_1 and λ_2
(d) a polarizer that rotates the polarization state of radiation at λ_2

[2006 : 1 M]

5.12 An X-ray tube is operated at 80 kV anode voltage. In order to filter the low intensity X-rays, a 2.5 mm thick aluminium filter is used. It is given that at 80 kV anode voltage, the mass attenuation coefficients and densities of aluminium are 0.02 $m^2 kg^{-1}$ and 2699 kgm^{-3} respectively and for copper these are 0.075 $m^2 kg^{-1}$ and 8960 kgm^{-3} respectively. If a copper filter is to replace the aluminium filter with the same filtering effect, the thickness of the copper filter should be

(a) 0.2 mm (b) 0.66 mm
(c) 1.5 mm (d) 5 mm

[2008 : 2 M]

5.13 A mass spectrometer is to be used to resolve peaks corresponding to CO^+ and N_2^+. The atomic masses are ^{12}C = 12.0000, ^{16}O = 15.9949, and ^{14}N = 14.0031 amu. The resolving power of the mass spectrometer should be at least

(a) 250 (b) 350
(c) 2500 (d) 3500

[2009 : 1 M]

5.14 The operating voltage of an X-ray tube is changed from 40 kV to 50 kV. The resulting change in the shortest wavelength generated is

(a) +20% (b) –20%
(c) +25% (d) –36%

[2009 : 2 M]

5.15 A person weighing 60 kg receives radiation energy of 0.3 J over the entire body. The does of radiation absorbed (in rad) is

(a) 0.005 rad (b) 0.1 rad
(c) 0.3 rad (d) 0.5 rad

[2010 : 1 M]

5.16 A psychrometric chart is used to determine

(a) pH
(b) Sound velocity in glasses
(c) CO_2 concentration
(d) Relative humidity [2012 : 1 M]

5.17 In a time-of-flight mass spectrometer if q is the charge and m the mass of the ionized species, then the time of flight is proportional to

(a) $\frac{\sqrt{m}}{\sqrt{q}}$ (b) $\frac{\sqrt{q}}{\sqrt{m}}$

(c) $\frac{m}{\sqrt{q}}$ (d) $\frac{q}{\sqrt{m}}$

[2014 : 1 M]

5.18 An image of the chest of a patient is taken with an X-ray machine on a photographic film. The Hurter-Driffield (HD) curve of the film is shown in the figure. The highly absorbing parts of the body (e.g. bones), show up as low exposure regions (mapped near A) and the less absorbing parts (e.g. muscles) show up as high exposure regions (mapped near B).

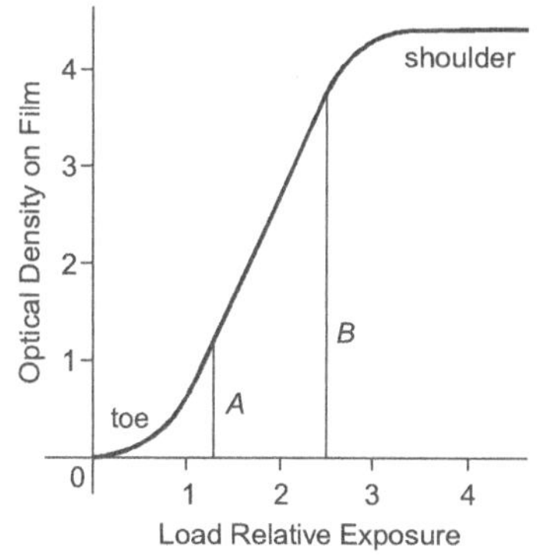

If the exposure time is increased 10 times, while keeping the voltages and currents in the X-ray machine constant, in the image,

(a) contrast decreases since B moves into the shoulder region
(b) contrast decreases since both A and B move into the shoulder region
(c) contrast increases since A moves into the toe region
(d) contrast decreases since both A and B move into the toe region

[2014 : 2 M]

5.19 The resolving power of a spectrometer consisting of a collimator, a grating and a telescope can be increased by

(a) increasing the angular magnification of the telescope
(b) increasing the period of the grating
(c) decreasing the period of the grating
(d) decreasing the slit-width of the collimator

[2015 : 2 M]

■■■

Answers Mass Spectrometry, X-Ray & Nuclear Measurement (Analytical)

5.1	(d)	5.2	(a)	5.3	(Sol.)	5.4	Sol.	5.5	(c)	5.6	(5.3)	5.7	(a)
5.8	(c)	5.9	(b)	5.10	(c)	5.11	(d)	5.12	(a)	5.13	(c)	5.14	(b)
5.15	(d)	5.16	(d)	5.17	(a)	5.18	(a)	5.19	(c)				

Explanations Mass Spectrometry, X-Ray & Nuclear Measurement (Analytical)

5.1 (d)

The quantum efficiency of a Geiger counter is highest in the wavelength range 6 to 7 Å range.

5.3 Sol.

(a)-(ii), (b)-(iv), (c)-(v), (d)-(iii)

5.4 Sol.

(a)-(v), (b)-(iii), (c)-(ii), (d)-(i)

5.5 (c)

Half life= 1 year

$$N(t) = N_0 e^{-\lambda t}$$

$$\frac{N(t)}{N_0} = e^{-\lambda t}$$

$$\frac{1}{2} = e^{-\lambda t}$$

$$-\lambda \times 1 = ln(0.5)$$

$$-\lambda = -0.693$$

$$\therefore \quad \boxed{\lambda = 0.693 \text{ year}^{-1}}$$

5.6 (5.3)

$$N(t) = N_0 e^{-\lambda t}$$

$$\frac{1}{2.5} = e^{-\lambda t}$$

$$-\lambda t = ln\left(\frac{1}{2.5}\right)$$

$$-\lambda \times 7 = -0.916$$

$$\lambda = 0.1308$$

$$\text{Half life} = \frac{0.693}{\lambda} = \frac{0.693}{0.131} = 5.3 \text{ days}$$

5.7 (a)

$$I = I_0 e^{-\mu d}$$

5.8 (c)

For nuclear radiation measurement, a proportional counter has higher internal amplification in comparison to a Geiger counter.

5.9 (b)

$$\lambda = \frac{hc}{E} = \frac{hc}{eV}$$

$$= \frac{6.626 \times 10^{-34} \times 3 \times 10^8}{1.6 \times 10^{-19} \times 40 \times 10^3}$$

$$= 0.316 \times 10^{-10} = 0.31 \text{ Å}$$

5.10 (c)

$$t = L\sqrt{\frac{m}{2ev}}$$

For A: $$t_A = L_A\sqrt{\frac{m_A}{2ev_A}}$$

For B: $$t_B = L_B\sqrt{\frac{m_B}{2ev_B}}$$

$$t_A = t_B$$

$$\therefore \quad L_A\sqrt{\frac{m_A}{2ev_A}} = L_B\sqrt{\frac{m_B}{2ev_B}}$$

$$L_A = 1 \text{ m}, \frac{m_A}{m_B} = \frac{1}{2}$$

$$v_A = v_B$$

$$\therefore \quad 1 \times \sqrt{\frac{m_A}{m_B}} = L_B$$

$$L_B = \frac{1}{\sqrt{2}} \text{ m}$$

5.12 (a)

For X-Ray filters.

$$\frac{I}{I_0} = e^{-\mu \rho x}$$

Case-I: Aluminium

$$\frac{I}{I_0} = e^{-\mu_a \rho_a t_a} \quad \text{(i)}$$

Case-II: Copper

$$\frac{I}{I_0} = e^{-\mu_c \rho_c t_c} \quad \text{(ii)}$$

equating (i) and (ii) we get,

$$e^{-\mu_a \rho_a t_a} = e^{-\mu_c \rho_c t_c}$$

also the effect of thickness of filter is taken:

$$\therefore \quad t_a \mu_a \rho_a = \mu_c \rho_c t_c$$

$$\therefore \quad t_c = \frac{t_a \mu_a \rho_a}{\mu_c \rho_c}$$

$$= \frac{2.5 \times 10^{-3} \times 0.02 \times 2699}{0.075 \times 8960}$$

$$= 0.2 \text{ mm}$$

5.13 (c)

$$\text{mass of CO} = 12 \text{ amu} + 15.9949 \text{ amu}$$
$$= 27.9949 \text{ amu}$$
$$\text{mass of } N_2 = 2 \times 14.0031 \text{ amu}$$
$$= 28.0062 \text{ amu}$$

Reading power of mass spectrometer

$$= \frac{\frac{m_1 + m_2}{2}}{(m_1 - m_2)}$$

where,

m_1 = mass of the substance which has greater atomic mass

Here, $m_1 \Rightarrow N_2$, $m_2 \Rightarrow O_2$

$\therefore$ Resolving power

$$= \frac{28.0062}{(28.0062 - 27.9949)}$$

$$= \frac{28.0062}{0.0113} = 2479.42 \simeq 2500$$

5.14 (b)

$$E = \frac{hc}{\lambda}$$

$$\therefore \quad \lambda = \frac{hc}{E}$$

so, we can see that $\lambda \propto \frac{1}{E}$

Since the voltage is increased from 40 to 50 kV and $\lambda \propto \frac{1}{E}$, therefore the wavelength will decrease, so the percentage change in wavelength will be negative

$$\therefore \text{Change in } \lambda = \frac{\lambda_2 - \lambda_1}{\lambda_1} \times 100$$

$$= \frac{\frac{hc}{E_2} - \frac{hc}{E_1}}{\lambda_1} \times 100 = \frac{\frac{1}{50} - \frac{1}{40}}{\frac{1}{40}} \times 100$$

$$= \frac{\frac{40-50}{40 \times 50}}{\frac{1}{40}} \times 100 = -20\%$$

5.15 (d)

Energy absorbed per unit mass $= \frac{0.3}{60}$ J/kg

We knot that,

$$1 \text{ Rad} = 0.01 \text{ J/kg}$$

Hence, $\quad 1 \text{ Rad} = 0.01 \text{ J/kg}$

or $\quad 0.01 \text{ J/kg} = 1 \text{ Rad}$

$$1 \text{ J/kg} = \frac{1}{0.01} \text{Rad}$$

$$\therefore \quad \frac{0.3}{60} \text{J/kg} = \frac{1}{0.01} \times \frac{0.3}{60}$$

$$= \frac{100 \times 3}{600} = 0.5 \text{ rad}$$

5.16 (d)

Psychrometric chart is used to determine the Relative Humidity.

5.17 (a)

For time-of-flight mass spectrometer time of flight,

$$t = L\sqrt{\frac{m}{2vq}}$$

$$\therefore \quad t \propto \frac{\sqrt{m}}{\sqrt{q}}$$

5.19 (c)

$\because$ As the number of slits increases. Resolving power of a spectrometer increase but period of grating decreases.

■■■

6 UV, Visible and IR Spectrometery Chromatography (Analytical)

6.1 In a spectrophotometer, the monochromator must be able to resolve two wavelengths 599.9 nm and 600.01 nm. The required resolution is

(a) 100 (b) 1000
(c) 3000 (d) 5000

[1996 : 2 M]

6.2 In a particular sample, the absorbance is 0.6. For a molar concentration of the solute of 10×10^{-4} M and a 2.0 cm path length, the molar absorptivity (litre per mol per cm) is

(a) 1.2 (b) 1200
(c) 300 (d) 3×10^{-4}

[1997 : 2 M]

6.3 Identify the correct set of matches from the following

	List-I		List-II
(a)	Beer's law	(i)	X-ray diffraction
(b)	Bragg's law	(ii)	Thermo-emf
(c)	Peltier effect	(iii)	Electro-chemical cell
(d)	Nernst equation	(iv)	Absorption of light

[2000 : 2 M]

6.4 The transmittance of a coloured solution is 0.5. The absorbance of the solution is

(a) 0.3 (b) 0.69
(c) 3.16 (d) –1.5

[2000 : 2 M]

6.5 A gas chromatograph is used for

(a) measuring flow rate of a gas
(b) measuring the temperature of a gas
(c) measuring the pressure of a gas
(d) analyzing the composition of a gas

[2001 : 1 M]

6.6 The following method is widely accepted to determine oxides of Nitrogen in an automobile emission:

(a) Orsat analysis
(b) Gas-chromatography
(c) Chemi-luminescence
(d) Flame-ionization detection

[2002 : 1 M]

6.7 Water vapour absorbs electromagnetic radiation primarily in the range of

(a) *X*-ray (b) *IR*
(c) UV (d) γ-ray

[2004 : 1 M]

6.8 The transmittance of a particular solution measured is *T*. The concentration of the solution is now doubled. Assuming that Beer-Lambert's law holds good for both the cases, the transmittance for the second would be

(a) $T/2$ (b) $2T$
(c) T^2 (d) $\sqrt{T}$

[2004 : 2 M]

6.9 A spectrophotometer is used to measure concentration of a molecule in a solution based on Beer's law using a cuvette of path length 10 mm and detectable absorbance of 0.01. The molar absorptivity of the solution is 10^4 litres/mol/cm. The minimum concentration of the solution that can detected is

(a) 1 μ mol/litre (b) 2 μ mol/litre
(c) 5 μ mol/litre (d) 10 μ mol/litre

[2005 : 2 M]

6.10 The transmittance of a sample solution is monitored using 500 nm wavelength. The transmittance is reduced from 100% to 10% due to addition of pollutants. As a result, the absorbance recorded by the spectrophotometer changes from

(a) $0 \to 1$ (b) $2 \to 1$
(c) $0 \to 2$ (d) $0 \to 10$

[2006 : 1 M]

6.11 Light of intensity I_0 is equally divided and passed through 2 cuvettes P_1 and P_2 containing an analyte at concentrations *c* and 0.5 c, respectively. The corresponding path lengths in P_1 and P_2 are 4 cm and 1 cm. The cross-sectional areas are 1 cm^2 and 3 cm^2, respectively. The ratio of the absorbances in P_1 and P_2 is

(a) 1.5 (b) 8/3
(c) 3 (d) 8

[2007 : 2 M]

6.12 The *H-D* curves of two *X*-ray films, F_1 and F_2, are shown below:

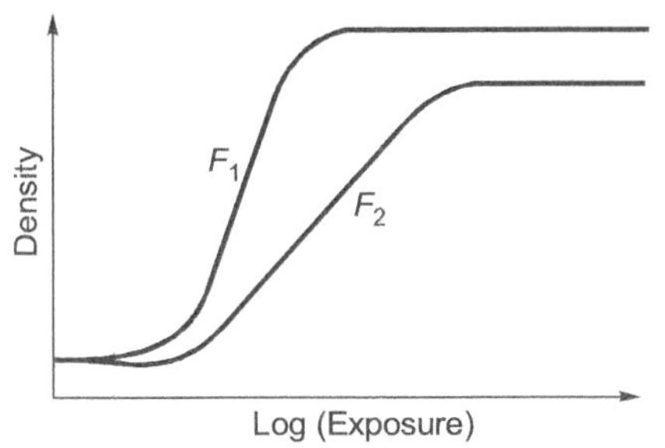

The curves show that, for the same change in exposure, F_1 has

(a) higher contrast than F_2 but the same fog level
(b) lower contrast than F_2 but the same fog level
(c) higher contrast and higher fog level than F_2
(d) lower contrast than F_2 but higher fog level

[2007 : 2 M]

6.13 For radioscope imaging, an Anger camera is fitted with a parallel hole collimator. If the thickness of the collimator is increased, the camera

(a) resolution and sensitivity will increase
(b) resolution and sensitivity will decrease
(c) resolution will increase and sensitivity will decrease
(d) resolution will decrease and sensitivity will increase

[2008 : 1 M]

6.14 A solution "*P*" is put in a spectrophotometer cuvette of optical path length 1 cm. The transmittance is found to be 10%. Another solution "*Q*" has a transmittance of 40% under the same circumstances. If equal volumes of *P* and *Q* are mixed together, the transmittance of the resulting solution (assuming the constituents of *P* and *Q* do not react with each other) is, approximately,

(a) 15% (b) 20%
(c) 25% (d) 30%

[2010 : 2 M]

Statement for Linked Answer Questions (6.15 & 6.16):

The level of water, stored in a truncated conical bath, is measured by a gamma-ray radiation sensor. The initial level of water 1 m, and the level is increasing due to water inflow at the constant rate of 0.125 m^3/s. Assume mass absorption coefficient of water is 77×10^{-4} m^2/kg and density of water is 1000 kg/m^3.

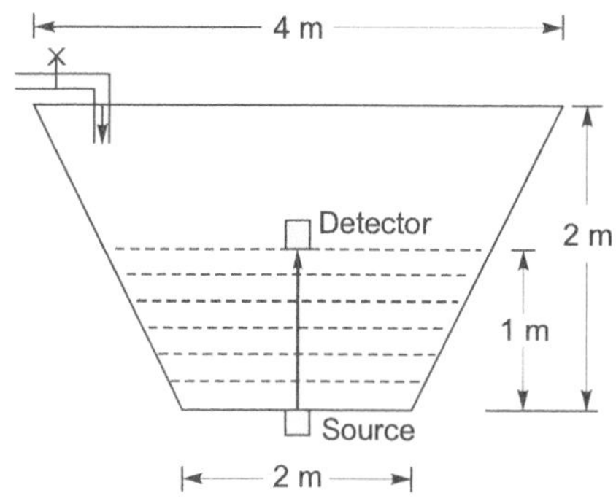

6.15 When the intensity of radiation received by the floating detector is half of the intensity detected initially, the level of water is

(a) 1.09 m (b) 1.5 m
(c) 1.8 m (d) 1.9 m

[2011 : 2 M]

6.16 When the floating detector is at the level calculated in **(Q. 6.16)**, the time elapsed is

(a) 4.1 s (b) 5.23 s
(c) 10.52 s (d) 50.63 s

[2011 : 2 M]

6.17 Measurement of optical absorption of a solution is disturbed by the additional stray light falling at the photo-detector. For estimation of the error caused by stray light the following data could be obtained from controlled experiments.

- Photo-detector output without solution and without stray light is 500 μW.
- Photo-detector output without solution and with stray light is 600 μW.
- Photo-detector output with solution and with stray light is 200 μW.

The percent error in computing absorption coefficient due to stray light is

(a) 12.50 (b) 31.66
(c) 33.33 (d) 94.98

[2013 : 2 M]

6.18 A mixture contains two mutually inert solutions '*X*' and '*Y*' in equal volumes. The mixture is examined in a spectrophotometer using a cuvette. It is observed that the transmittance is 0.40. With only the solution '*X*' in the same cuvette, the transmittance is 0.20. With only solution 'Y' in the cuvette the transmittance is________.

[2014 : 2 M]

6.19 A liquid level measurement system employing a radio-isotope is mounted on a tank as shown in the figure. The absorption coefficient of water for the radiation is 7.7 m^{-1}. If the height of water in the tank is reduced from 100 mm to 90 mm, the percentage change in the radiation intensity received by the detector, neglecting absorption of the radiation by air, is ______ %.

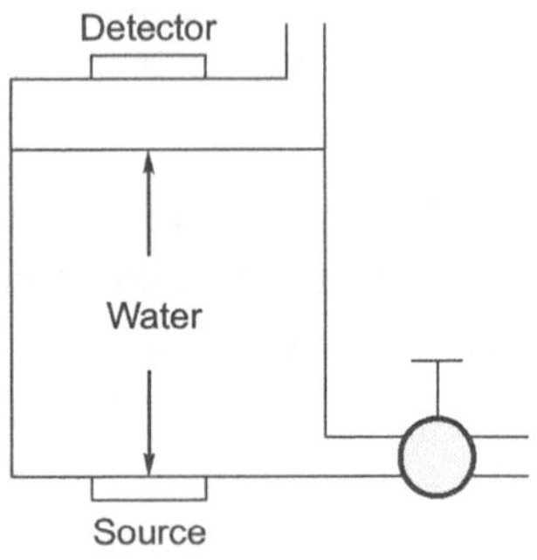

[2015 : 2 M]

6.20 Monochromatic light of wavelength 532 nm is used to measure the absorption coefficient of a material in a UV-Visible Spectrophotometer. The measured light intensity after transmission through a 1 cm thick sample of the material is 0.414 mW/cm^2. For a sample of thickness 2 cm, the measured light intensity is 0.186 mW/cm^2. The absorption coefficient (in cm^{-1}) of the material is _______ (round off to two decimal places)

[2022 : 2 M]

■■■

Answers UV, Visible & IR Spectrometery Chromatography (Analytical)

6.1 (c)	6.2 (c)	6.3 (Sol.)	6.4 (a)	6.5 (d)	6.6 (a)	6.7 (b)
6.8 (c)	6.9 (a)	6.10 (a)	6.11 (b)	6.12 (c)	6.13 (c)	6.14 (c)
6.15 (a)	6.16 (b)	6.17 (b)	6.18 Sol.	6.19 (8)	6.20 (0.80)	

Explanations UV, Visible & IR Spectrometery Chromatography (Analytical)

6.1 (c)

Resolution of the monochromator required

$$= \frac{\frac{600.1+599.9}{2}}{600.1-599.9} = \frac{600}{0.2} = 3000$$

6.2 (c)

$$A = abc \quad ; \quad 0.6 = 10\times10^{-4}\times2\times a$$

$$a = \frac{0.6}{20\times10^{-4}} = 300$$

6.3 Sol.

(a)-(iv), (b)-(i), (c)-(ii), (d)-(iii)

6.4 (a)

$$A = \log\left(\frac{1}{T}\right) = \log\left(\frac{1}{0.5}\right) = 0.3010$$

6.8 (c)

$$A = \log\frac{1}{T}$$

$$A = aC\varepsilon$$

when C is double, $A \to 2A$

$$\frac{A}{2A} = \frac{\log\frac{1}{T}}{\log\frac{1}{T'}}$$

$$\left[\log\frac{1}{T'}\right]\frac{1}{2} = \left[\log\frac{1}{T}\right]$$

$$\log\left(\frac{1}{T'}\right) = 2\log\left(\frac{1}{T}\right)$$

$$-\log T' = -\log T^2$$

$$\therefore \quad T' = T^2$$

6.9 (a)

Here,
$$A = 0.01$$
$$b = 10 \text{ mm} = 10\times10 = 1 \text{ cm}$$
$$\epsilon = 10^4 \text{ litres/mol/cm}$$
$$A = \epsilon\, b\, C$$
$$0.01 = 10^4\times C\times1$$
$$\therefore\ C = 0.01\times10^{-4} = 1\times10^{-6} = 1\ \mu\text{ mol/litre}$$

6.10 (a)

$$A = \log\frac{1}{T}$$

$$A_1 = \log\frac{100}{100} = \log 1 = 0$$

$$A_2 = \log\frac{100}{10} = \log 10 = 1$$

$\therefore$ the absorbance changes from $0 \to 1$.

6.11 (b)

Absorbance $\propto$ Concentration

$\propto$ Length of the cuvette

$$\therefore \quad \frac{A_1}{A_2} = \frac{C\times4\times1}{0.5c\times1\times3} = \frac{8}{3}$$

6.12 (c)

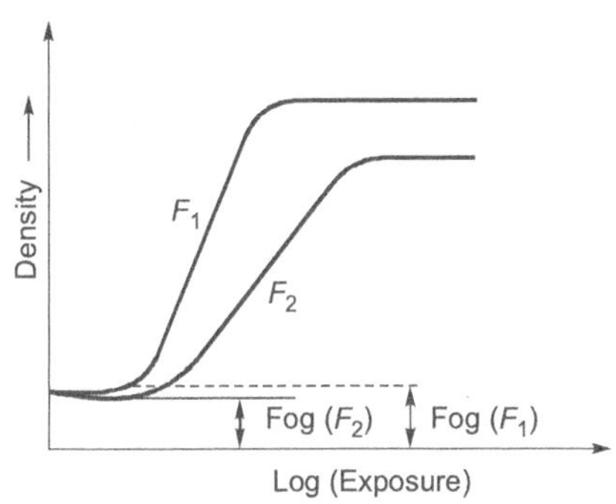

(i) Clearly the slope of F_1 is more steep than the F_2 (i.e. slope F_1 is more than compared to slope F_2)
Hence, $(\text{Contrast})_{F_1} > (\text{Contrast})_{F_2}$

(ii) also it can be observed from the figure above that the $(\text{fog})_{F_1} > (\text{fog})_{F_2}$.

6.14 (c)

$$T = \frac{T_1 + T_2}{2} \text{ for equal volume mixing}$$

$$= \frac{10+40}{2} = 25\%$$

6.15 (a)

$$I(t) = I_0\, e^{-\mu\rho x}$$

$$\frac{I_0}{2} = I_0 e^{-77\times10^{-4}\times1000\times x}$$

$$\frac{1}{2} = e^{-7.7x}$$

taking lag both sides we get,

$$ln(0.5) = ln(e^{-7.7x})$$

$$-0.693 = -7.7x$$

$$x = \frac{0.693}{7.7} = 0.09$$

I_0 was the initially received intensity when water level was at 1 m.

So, level of water when the radiation received by the floating detector is half of the intensity detected initially = (1 + 0.09) m = 1.09 m.

6.16 (b)

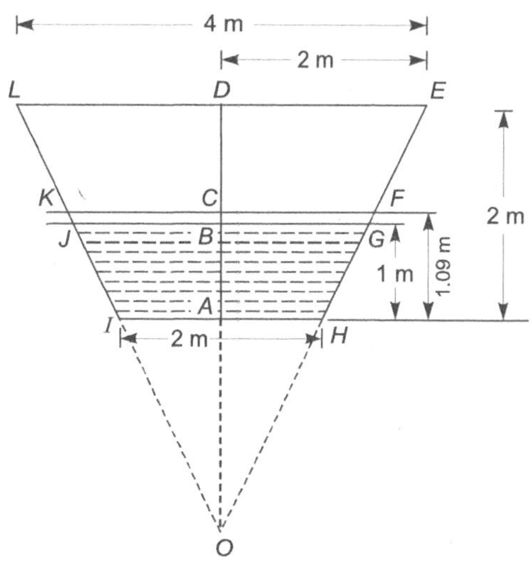

AD = 2 m, AB = 1 m (given), AD = 2 m

$$\therefore \quad BD = AD - AB$$

$$= 2 - 1 = 1 \text{ m}$$

$$HI = 2 \text{ m}$$

also it can be observed that

$$AH = \frac{1}{2}HI = \frac{1}{2}\times2 = 1 \text{ m}$$

(ΔOAH) is similar to (ΔODE)

$$\therefore \quad \frac{OA}{OD} = \frac{AH}{DE}$$

$$\frac{OA}{OA+AD} = \frac{AH}{DE}$$

$$\frac{OA}{OA+2} = \frac{1}{2}$$

$$2(OA) = OA + 2$$

$$\therefore \quad \boxed{OA = 2 \text{ m}}$$

$$\therefore \quad OC = OA + AC$$

$$= 2 + 1.09 = 3.09$$

and (ΔOAH) is similar to (ΔOBG)

$$\therefore \quad \frac{OA}{OB} = \frac{AH}{BG}$$

$$\frac{OA}{OA+AB} = \frac{AH}{BG}$$

$$\frac{2}{2+1} = \frac{1}{BG}$$

also, (ΔOAH) is similar to (ΔOCF)

$$\frac{OA}{OC} = \frac{AH}{CF}$$

$$\frac{OA}{OA+AC} = \frac{AH}{CF}$$

$$\therefore \quad \frac{2}{2+1.09} = \frac{1}{CF}$$

$$CF = \frac{3.09}{2}\text{ m} = 1.545\text{ m}$$

Volume of water filled in making half the intensity received

$$= \frac{1}{3}\pi(CF)^2(OC) - \frac{1}{3}\pi(BG)^2(OB)$$

$$= \frac{1}{3}\pi[(1.545)^2(3.09) - (1.5)^2(3)]$$

$$= \frac{1}{3}\pi \times 0.6259 = 0.655\text{ m}^3$$

Water flow rate = 0.125 m^3/s

$\therefore$ time elapsed in filling 0.635m^3

$$= \frac{0.655}{0.125} = 5.24\text{ sec}$$

6.17 (b)

Case-I (Measured value)

$$I_0 = 600\ \mu W$$
$$I(t) = 200\ \mu W$$
$$I(t) = I_0\, e^{-\mu_{MV}x}$$
$$\mu_{MV} = -ln\left(\frac{2}{6}\right) = -ln(0.33)$$

Case-II (True value)

$$I_0 = 500\ \mu W$$
$$I(t) = 100\ \mu W$$
$$I(t) = I0\, e^{-\mu_{TV}x}$$
$$\mu_{TV} = -ln\left(\frac{1}{5}\right) = -ln(0.2)$$

$$\%\text{ error} = \frac{MV - TV}{TV} \times 100$$

$$= \frac{\{[-ln(0.33)] - [-ln(0.2)]\} \times 100}{-ln(0.2)}$$

$$= 31.7\%$$

6.18 Sol.

(0.795 to 0.805)

6.19 (8)

$$I(t) = I_o\, e^{-\mu t}$$
$$I_{100} = I_o\, e^{-0.1 \times 7.7}$$
$$I_{90} = I_o\, e^{-0.09 \times 7.7}$$

Percentage charge in radiation intensity

$$= \frac{I_{90} - I_{100}}{I_{100}} \times 100$$

$$= \frac{e^{-0.09 \times 7.7} - e^{-0.1 \times 7.7}}{e^{-0.1 \times 7.7}} \times 100$$

$$= \frac{0.5 - 0.463}{0.463} \times 100 = 7.98\%$$

6.20 (0.80)

$$I_t = I_i\, e^{-\mu x}$$

$$0.414\text{ mW/cm}^2 = I_i\, e^{-\mu(1 \times 10^{-2}\text{ m})} \quad \text{...(i)}$$

$$0.186\text{ mW/cm}^2 = I_i\, e^{-\mu(2 \times 10^{-2}\text{ m})} \quad \text{...(ii)}$$

On dividing equation (i) by equation (ii),

$$\frac{0.414}{0.186} = e^{-\mu + 2\mu}$$
$$e^{\mu} = 2.225$$
$$\mu = \ln(2.225)$$

Absorption coefficient,

$$\mu = 0.80\text{ cm}^{-1}$$

7 Biomedical Instrumentation, ECG, EEG & EMG (Biomedical)

7.1 The bandwidth of an electrocardiogram (ECG) amplifier is

(a) dc to 0.01 Hz (b) 0.05 to 500 Hz
(c) 550 to 1500 Hz (d) 2000 to 10000 Hz

[1999 : 2 M]

7.2 Korotkoff sounds are used

(a) as a reference for sound level measurement
(b) for studying heart muscle functioning
(c) for blood pressure measurement
(d) for study of heart valve functioning

[1999 : 2 M]

7.3 Among the electrodes which are used to record ECG signals, the one used for ground referencing is placed at,

(a) left arm (b) right leg
(c) left leg

[1999 : 2 M]

7.4 For an ECG amplifier the input signal is 100 μV corrupted by a common mode noise of 5 mV. The output of the amplifier contains 50 mV of signal and 0.001 mV of noise. The CMRR of the amplifier is

(a) 40 dB (b) 60 dB
(c) 80 dB (d) 100 dB

[2003 : 2 M]

7.5 The *P* wave, *QRS* complex and *T* wave in one heart cycle of the *ECG* waveform correspond respectively to

(a) polarization of atria; depolarization of ventricles and repolarization of atria
(b) polarization of atria; polarization of ventricles and repolarization of atria
(c) depolarization of atria; depolarization of ventricles and repolarization of atria
(d) depolarization of atria; repolarization of ventricles and depolarization of atria

[2005 : 1 M]

7.6 Match **List-I** with **List-II** and select the correct answer using the codes given below the lists:

List-I	List-II
A. EEG	1. Eye movement
B. EOG	2. Gastric motility
C. EMG	3. Muscle activity
D. EGG	4. Brain activity
	5. Heart activity
	6. Bladder activity

Codes:

	A	B	C	D
(a)	4	5	4	3
(b)	4	1	3	2
(c)	3	2	6	1
(d)	2	3	1	6

[2005 : 1 M]

7.7 In the driven-right-leg system for reducing the common mode interference in *ECG* recording, the common mode voltage is sensed and connected to the right leg electrode. The desirable condition for the amplifier gain is

(a) a large positive gain
(b) a large negative gain
(c) a gain of 1
(d) a gain of –1

[2006 : 2 M]

7.8 The *ECG* of a patient is being recorded using the three standard frontal plane leads. If the cardiac vector is oriented at an angle of 45 degrees to Lead *I* and has a magnitude of 3 mV, the voltages seen on Leads *I*, *II* and *III* are

(a) 1.50, 2.57 and 0.77 mV
(b) 2.12, 0.77 and 2.89 mV
(c) 2.12, 3.15 and 1.03 mV
(d) 2.12, 2.89 and 0.77 mV

[2007 : 2 M]

7.9 In the stand are 12-lead ECG recording system, the minimum number of electrodes required to be attached to a human subject for recording any one of the unipolar chest lead signals is

(a) 1 (b) 2
(c) 4 (d) 5

[2008 : 1 M]

7.10 A standard three-lead frontal plane *ECG* is taken of a person with a normal heart. The peak amplitude of the R-wave is
(a) greatest in lead I
(b) greatest in lead II
(c) greatest in lead III
(d) equal in all the leads

[2009 : 2 M]

7.11 In order to remove respiration related motion artifacts from an *ECG* signal, the following filter should be used
(a) low-pass filter with $f_c = 0.5$ Hz
(b) high-pass filter with $f_c = 0.5$ Hz
(c) high-pass filter with $f_c = 49.5$ Hz
(d) band-pass filter with passband between 0.1 Hz and 0.5 Hz

[2014 : 1 M]

7.12 An apparatus to capture *ECG* signals has a filter followed by a data acquisition system. The filter best suited for this application is
(a) low pass with cut-off frequency 200 Hz
(b) high pass with cut-off frequency 200 Hz
(c) band pass with lower and upper cut-off frequencies 100 Hz and 200 Hz for its pass band
(d) band reject with lower and upper cut-off frequencies 1 Hz and 200 Hz for its stop band

[2015 : 1 M]

■■■

Answers Biomedical Instrumentation, ECG, EEG & EMG (Biomedical)

7.1 (b)	7.2 (c)	7.3 (b)	7.4 (d)	7.5 (c)	7.6 (b)	7.7 (b)
7.8 (d)	7.9 (c)	7.10 (b)	7.11 (b)	7.12 (a)		

Explanations Biomedical Instrumentation, ECG, EEG & EMG (Biomedical)

7.4 (d)

$$V_{in} = 100 \text{ mV}$$
$$V_c = 5 \text{ mV}$$
$$A_d = \frac{50 \text{ mV}}{100 \, \mu\text{V}} = 500$$
$$A_c = \frac{0.001 \text{ mV}}{5 \text{ mV}} = 0.2 \times 10^{-3}$$
$$\text{CMRR} = 20 \log \frac{A_d}{A_c}$$
$$= 20 \log \frac{500}{0.2 \times 10^{-3}}$$
$$= 20 \log 2.5 \times 10^6$$
$$= 127.9 \text{ db}$$

$\therefore$ most nearby answer is 100 db.

7.8 (d)

$$V_I = p \cos \alpha = 3 \cos 45° \text{ mV}$$
$$= 2.12 \text{ mV}$$

$$V_{II} = -\frac{1}{2} p\cos\alpha - \frac{\sqrt{3}}{2} p\sin\alpha$$
$$= -\frac{1}{2}(3)\cos 45° - \frac{\sqrt{3}}{2}(3)\sin 45° \text{ mV}$$
$$= -1.06 - 1.86 \text{ mV} = -2.89 \text{ mV}$$
$$|V_{II}| = 2.89 \text{ mV}$$
$$V_{III} = \frac{1}{2} p\cos\alpha - \frac{\sqrt{3}}{2} p\sin\alpha$$
$$= \frac{1}{2}(3)\cos 45° - \frac{\sqrt{3}}{2}(3)\sin 45° \text{ mV}$$
$$= 1.06 - 1.83 \text{ mV}$$
$$= -0.77 \text{ mV}$$
$$|V_{III}| = 0.77 \text{ mV}$$

7.12 (a)

Frequency range of *ECG* signal
= 0.5 Hz to 200 Hz

So, low pass filter can be used with cut-off frequency 200 Hz.

■■■

8 Clinical Measurement (Biomedical)

8.1 A cardiac output measurement set-up uses the thermo-diluation principle. The change in the output temperature is 7°C for 5.60 litres per minute of cardiac output with an integration time of 10s. For a second measurement, with all other parameters remaining unchanged, the change in the temperature is 10°C with an integration time of 20s. The cardiac output in litres per minute for this will be

(a) 5.60 (b) 4.21
(c) 3.92 (d) 1.96

[2003 : 2 M]

8.2 In Fick's method of cardiac output measurement, oxygen consumption is 200 ml/min., arterial oxygen content is 0.30 ml/ml and venous oxygen content is 0.20 ml/ml. The cardiac output in *l*/min. is

(a) 1 (b) 2
(c) 0.5 (d) 1.5

[2004 : 2 M]

8.3 For bio-potential recording, the impedance of an Ag/AgCl electrode, as compared to that of an Ag electrode of the same size, is

(a) lower at all frequencies
(b) lower for frequencies less than 10 Hz and higher for frequencies greater than 300 Hz
(c) higher for frequencies less than 10 Hz and lower for frequencies greater than 300 Hz
(d) higher at all frequencies

[2006 : 1 M]

8.4 A Clark oxygen sensor is

(a) potentiometric sensor
(b) amperometric sensor
(c) conductometric sensor
(d) magnetostrictive sensor

[2007 : 1 M]

8.5 During intravascular measurement of arterial blood pressure, catheters may be introduced in different configurations, as shown in the figures below:

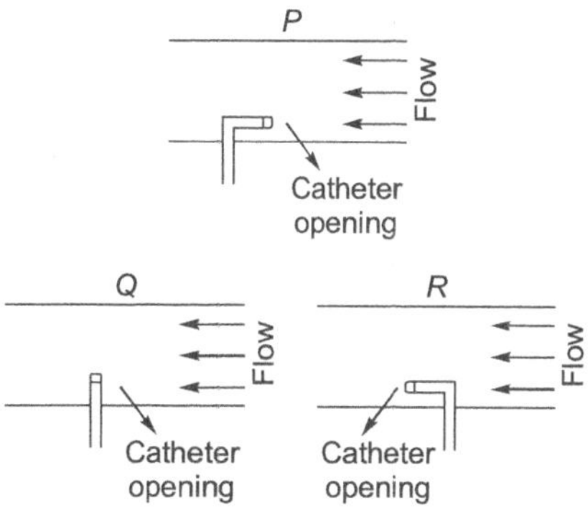

The static pressure will be measured correctly in the configuration (*s*)

(a) *P* and *Q* but not in *R*
(b) *R* only
(c) *Q* only
(d) *P* and *R* but not in *Q*

[2007 : 1 M]

8.6 In N_2-washout estimation of lung volume using spirometry, the lung volumes at the beginning and the end of the washout are the same. Let *T*, *V* and *F* denote temperature, volume and molar fraction (of N_2) respectively; subscripts *S* and *L* denote the spirometer and the lung; and t_1 and t_2 the beginning time and the end time of the experiment, respectively. Then,

(a) $V_L = \frac{T_L}{T_S}\left[\frac{F_S(t_2)\,V_S(t_2)}{F_L(t_1) - F_L(t_2)}\right]$

(b) $V_L = \frac{T_L}{T_S}\left[\frac{F_L(t_2)\,V_S(t_2)}{F_L(t_1) - F_L(t_2)}\right]$

(c) $V_L = F_L(t_1)\,V_L(t_2)$

(d) $V_L = T_S F_S(t_2)\,V_S(t_2)$

[2007 : 1 M]

8.7 A laser light with a wave length of 633 nm is passed through 1 cm length of tissue and 2 cm length of glass. The refractive indices of tissue and glass are 1.33 and 1.5 respectively. The velocities of laser light in the tissue and in the glass are in the ratio of

(a) 1.33 : 0.75 (b) 1.33 : 3.0
(c) 1.33 : 1.5 (d) 1.5 : 1.33

[2008 : 1 M]

8.8 The Treadmill Test is used to diagnose
(a) the balancing style during walk of the patient
(b) the auditory activity of the patient
(c) the visual activity of the patient
(d) the cardiac activity of the patient

[2011 : 1 M]

8.9 Match the following biomedical instrumentation techniques with their applications:

P : Otoscopy	U : Respiratory volume measurement
Q : Ultrasound technique	V : Ear diagnostics
R : Spirometry	W : Echo-cardiography
S : Thermodilution technique	X : Heart volume measurement

(a) P-U, Q-V, R-X, S-W
(b) P-V, Q-U, R-X, S-W
(c) P-V, Q-W, R-U, S-X
(d) P-V, Q-W, R-X, S-U

[2013 : 1 M]

Answers Biomedical Instrumentation, ECG, EEG & EMG (Biomedical)

8.1 (d) 8.2 (b) 8.3 (c) 8.4 (b) 8.5 (c) 8.6 (d) 8.7 (d)
8.8 (d) 8.9 (c)

Explanations Biomedical Instrumentation, ECG, EEG & EMG (Biomedical)

8.1 (d)

Case-I:

Cardiac output = 5.6 litres/min.
$\Delta T = 7°C$
Integration time = 10 sec.

Case-II:

Cardiac output = Q (say)
$\Delta T = 10°C$
Integration time = 20 sec.

We know that,

$$\text{Cardiac output} = \frac{(1.8)(C)(60)(V)(T_i - T_b)}{\int \Delta T\, dt}$$

taking everything constant (k) except $\int \Delta T\, dt$ which is different (given).
We get,

$$\text{Cardiac output} = \frac{K}{\int \Delta T\, dt}$$

For Case-I,

$$Q = \frac{K}{7 \times 10} \quad \text{...(i)}$$

For Case-II

$$Q = \frac{K}{10 \times 20} \quad \text{...(ii)}$$

Dividing (i) and (ii) we get,

$$\frac{5.6}{Q} = \frac{10 \times 20}{7 \times 10} = \frac{20}{7}$$

$$\therefore \quad Q = \frac{7 \times 5.6}{20} = 1.96 \text{ litre/min.}$$

8.2 (b)

$VO_2 = 200$ ml/min.
$C_A = 0.30$ ml/ml
$C_V = 0.20$ ml/ml

We know that, for Fick's method:

$$\text{Cardiac output} = \frac{VO_2}{(C_A - C_V)} \times 100$$

$$= \frac{200}{(0.30 - 0.20)} \times 100$$

$= 2$ l/min.

8.5 **(c)**

Clearly the configurations P and Q are valid, configuration R is not having any significance.
Here P is measuring total pressure, and
Q is measuring static pressure, so
Option (c) is correct.

8.7 **(d)**

$$\mu_t = 1.33$$
$$\mu_g = 1.5$$
$$\lambda = 633 \times 10^{-9}$$
$$d_t = 1 \text{ cm}$$
$$d_g = 2 \text{ cm}$$
$$\frac{V_t}{V_g} = \frac{\mu_g}{\mu_t} = \frac{1.5}{1.33}$$
$$\boxed{V_t : V_g = 1.5 : 1.33}$$

Here, the distance (d) has no significance in this example.

8.9 **(c)**

Otoscopy :
Ear-diagnostics
Ultrasound :
Echo-cardiography
Spirometry :
Respiratory volume measurement
Thermodilution technique :
Heart volume measurement

■■■

9 Ultrasonic Transducers and Ultrasonography (Biomedical)

9.1 In an electromagnetic blood flow meter, the induced voltages is directly proportional to the
(a) blood flow rate
(b) square root of the blood flow rate
(c) square of the blood flow rate
(d) logarithm of the blood flow rate

[1999 : 2 M]

9.2 An ultrasonic beam of frequency 1 MHz and intensity 0.5 W/cm^2 passes through a layer of soft tissue of thickness t with an attenuation coefficient of 1.18 cm^{-1}. The ratio of output to input power is $1/e^2$. The thickness of the tissue is
(a) 1 cm (b) 1.695 cm
(c) 2.408 cm (d) 3.721 cm

[2003 : 2 M]

9.3 A single-sensor, contact-type ultrasonic flaw detector uses a frequency of 330 kHz. When testing a specimen, an echo from a flaw is recorded 0.05 ms after the transmitted pulse. If the velocity of sound in the test object is 6.0 km/s then the flaw is at a depth of
(a) 120 cm (b) 60 cm
(c) 30 cm (d) 15 cm

[2004 : 2 M]

Statement for Linked Answer Questions (9.4 to 9.5):

Blood flow through a straight segment of an artery is measured using a Doppler flow meter. The probe is oriented at an angle of 45 degrees to the longitudinal axis of the artery. It is known that Doppler shift is proportional to the velocity of the blood in the direction of propagation of sound from the probe and the frequency generated from the probe. The velocity of sound in blood is 1500 m/s and the maximum blood flow velocity to be measured is 110 cm/s.

9.4 Signal processing limitations constrain the maximum Doppler shift to be less than 3 kHz. The maximum source frequency (probe output), to the nearest MHz, should be
(a) 2 MHz (b) 3 MHz
(c) 5 MHz (d) 10 MHz

[2007 : 2 M]

9.5 In (Q.9.4), if the piezoelectric crystal of the probe is made of quartz (Young's modulus, $Y = 80 \times 10^9$ N.M^{-2}, $\rho = 2.65 \times 10^3$ kg.m^{-3}), the maximum thickness of the crystal required to resonate at the frequency determined above, to the nearest mm, is
(a) 1 mm (b) 2 mm
(c) 3 mm (d) 4 mm

[2007 : 2 M]

9.6 In a pulsed ultrasound imaging system, a single 5 MHz crystal is used both as source and as detector. Bursts of at least 20 cycles are needed for acceptable image quality. The velocity of sound in the tissue being imaged is 1500 m/s. The minimum distance of the objects to be imaged should be
(a) 12 mm (b) 6 mm
(c) 1 mm (d) 3 mm

[2009 : 2 M]

■■■

Answers Ultrasonic Transducers and Ultrasonography (Biomedical)

9.1 (a) 9.2 (b) 9.3 (d) 9.4 (b) 9.5 (a) 9.6 (d)

Explanations Ultrasonic Transducers and Ultrasonography (Biomedical)

9.1 (a)

induced voltage (e) = Blv

$$\therefore \quad \boxed{e \propto v}$$

9.2 (b)

$$x = 1.695 \text{ cm}$$
$$P = P_0\, e^{-\mu x}$$
$$\therefore \quad \frac{P}{P_0} = e^{-\mu x}$$
$$\frac{P}{P_0} = \frac{1}{e^2}$$
$$\therefore \quad e^{-\mu x} = e^{-2}$$
$$\mu x = 2$$
$$x = \frac{2}{1.18}$$
$$\therefore \quad = 1.695 \text{ cm}$$

9.3 (d)

$$v = 60 \text{ km/sec.}$$
$$= 6 \times 10^5 \text{ cm/sec.}$$
$$t = 0.05 \text{ m-sec.}$$
$$= 0.05 \times 10^{-3} \text{ sec.}$$

We know,

$$\text{Depth of penetration} = \frac{v \times t}{2}$$
$$= \frac{6\times10^5\times0.05\times10^{-3}}{2} = 15 \text{ cm}$$

9.4 (b)

Maximum source frequency

$$f = \frac{\Delta f \times C}{2v\cos\theta}$$

Δf = Doppler frequency

C = Velocity of sound in blood

v = Velocity of blood

θ = Angle

$$\therefore \quad f = \frac{1500\times3\times10^3}{2\times110\times10^{-2}\times\cos45°}$$
$$\therefore \quad f = 2.89 \times 10^6 \text{ Hz}$$
$$\therefore \quad f \simeq 3 \text{ MHz}$$

9.5 (a)

Thickness,

$$t = \frac{1}{2f}\sqrt{\frac{Y}{\rho}}$$
$$\therefore \quad t = \frac{1}{2\times3\times10^6}\sqrt{\frac{80\times10^9}{2.65\times10^3}}$$
$$t = 0.9157 \text{ mm} = 1 \text{ mm}$$

9.6 (d)

$$\because \quad \Delta f = \frac{2f_t\cos\theta}{C}d$$
$$d = \frac{\Delta f\times C}{2f_t\cos\theta} = \frac{20\times1500}{2\times5\times10^6\times\cos0°}$$
$$= 3 \text{ mm}$$

■■■

Electricity & Magnetism

UNIT

X

CONTENTS

1. Electricity & Magnetism 516

Electricity & Magnetism

Syllabus

Coulomb's Law, Electric Field Intensity, Electric Flux Density, Gauss's Law, Divergence, Electric field and potential due to point, line, plane and spherical charge distributions, Effect of dielectric medium, Capacitance of simple configurations, Biot•]Savart•fs law, Ampere•fs law, Curl, Faraday•fs law, Lorentz force, Inductance, Magnetomotive force, Reluctance, Magnetic circuits, Self and Mutual inductance of simple configurations.

Analysis of Previous GATE Papers

Exam Year	1 Mark Ques.	2 Marks Ques.	Total Marks
2021	1	1	3
2022	0	2	4
2023	–	–	–
2024	2	1	4

1 Electricity & Magnetism

1.1 An infinitely long line, with uniform positive charge density, lies along the z-axis. In cylindrical coordinates $(\hat{r}, \hat{\phi}, \hat{z})$. at any point $\vec{P}$ not on the z-axis. The direction of the electric field is:

(a) $\hat{\phi}$ (b) $\frac{(\hat{r}+\hat{z})}{\sqrt{2}}$

(c) $\hat{z}$ (d) $\hat{r}$

[2021 : 1 M]

1.2 The figure below shows an electrically conductive bar of square cross-section resting on a plane surface. The bar of mass of 1 kg has a depth of 0.5 m along the y-direction. The coefficient of friction between the bar and the surface is 0.1. Assume the acceleration due to gravity to be 10 m/s^2. The system faces a uniform flux density $B = -1\hat{z}T$. At time, t = 0, a current of 10 A is switched onto the bar and is maintained.

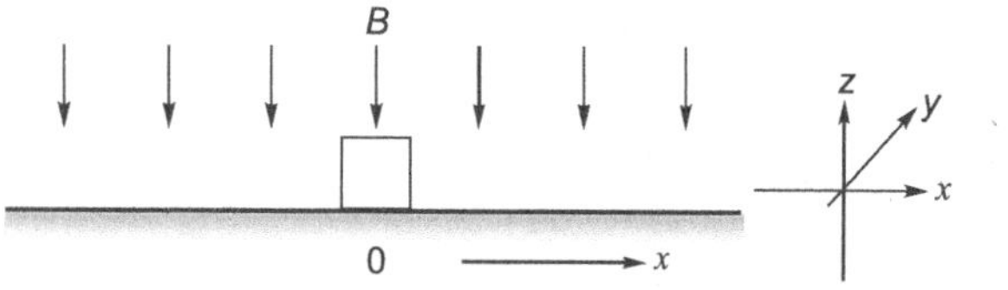

When the bar has moved by 1 m, its speed in metre per second is _____ (rounded off to one decimal place).

[2021 : 2 M]

1.3 A conducting semi-circular loop of radius R = 0.1 m, with its diameter centered at the origin, rotates in the x-y plane about the origin with a constant angular velocity, ω = 20 rad/s, as shown. A magnetic field of magnitude B = 2 T and normal to x-y plane exists in the region $x \geq 0$ as shown. If the loop has a resistance of 2 Ω, and negligible inductance, the peak-to-peak current (in milliamperes) in the loop is _________ (round off to one decimal place)

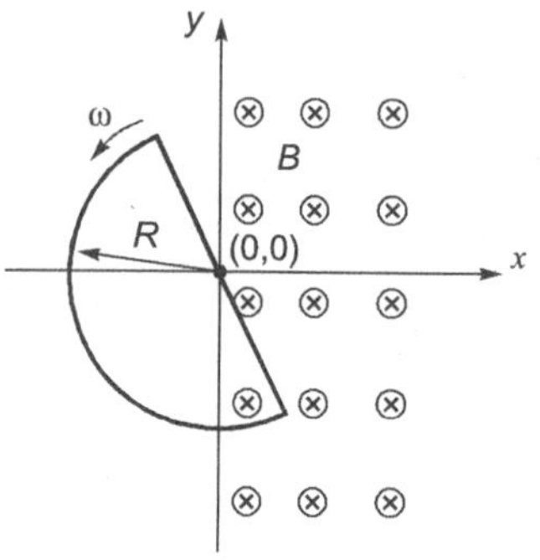

[2022 : 2 M]

1.4 A capacitor is constructed using two concentric spheres and air as the dielectric medium (permittivity of air = 8.854 × 10^{-12} F/m). The radii of the inner and outer spheres are a =10 cm and b =15 cm, respectively. The capacitance (in picofarads) is ________ (round off to 2 decimal places)

[2022 : 2 M]

1.5 The capacitor shown in the figure has parallel plates, with each plate having an area A. The thickness of the dielectric materials are d_1 and d_2 and their relative permittivities are ε_1 and ε_2, respectively. Assume that the fringing field effects are negligible and ε_0 is the permittivity of free space.

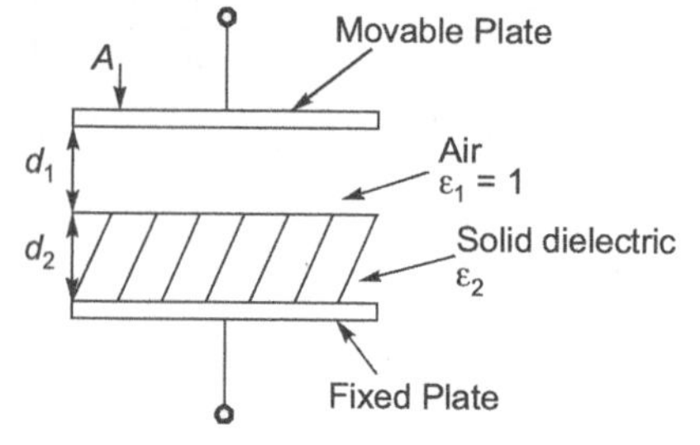

If d_1 is decreased by δd_1, the resultant capacitance becomes

(a) $\frac{\varepsilon_0 A}{d_1 - \delta d_1 + \frac{d_2}{\varepsilon_2}}$ (b) $\frac{\varepsilon_0 A}{d_2 + \frac{d_1}{\varepsilon_2}}$

(c) $\frac{\varepsilon_0 A}{d_2 - \delta d_2 + \frac{d_1}{\varepsilon_2}}$ (d) $\frac{\varepsilon_0 A}{d_1 - \delta d_1 + \frac{d_2}{\varepsilon_2}}$

[2024 : 1 M]

1.6 The capacitance formed between two concentric spherical metal shells having radii x and y with $y > x$ is

Note: $\in$ is the permittivity of the medium between the shells.

(a) $4\pi \in \left(\dfrac{xy}{y-x}\right)$ (b) $4\pi \in \left(\dfrac{x^2}{y-x}\right)$

(c) $4\pi \in \left(\dfrac{y^2}{y-x}\right)$ (d) $4\pi \in \left(\dfrac{y^2-xy}{x}\right)$

[2024 : 1 M]

1.7 A wire of circular cross section with radius a is shown in the figure. The current density is given by $J = ks^2$, where k is a constant, s is the radial distance from the axis and $0 \le s \le a$. The total current I in the wire is

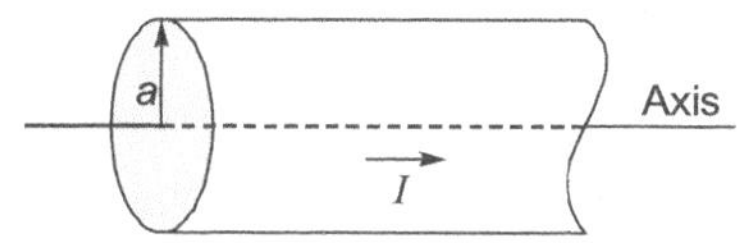

(a) $\dfrac{\pi k a^4}{2}$ (b) $\dfrac{2\pi k a^3}{3}$

(c) $\dfrac{\pi k a^3}{2}$ (d) $\dfrac{\pi k a^4}{4}$ [2024 : 2 M]

■■■

Answers Electricity & Magnetism

1.1 (d) 1.2 (2.8) 1.3 (200) 1.4 (33.36) 1.5 (a) 1.6 (a) 1.7 (a)

Explanations Electricity & Magnetism

1.2 (2.8)

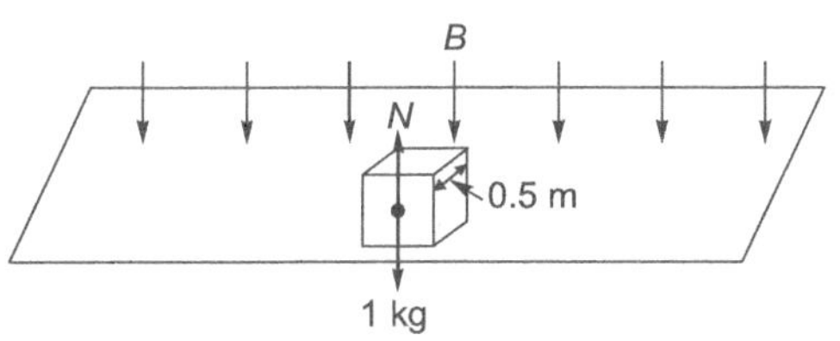

Weight of the body = 1 kg

Normal reaction from surface = 1 × 10 = 10 N

Friction force,

$$F_f = \mu N$$

Where μ is coefficient of friction of surface = 0.1

$\therefore$ $F_f = 10 \times 0.1 = 1$ N (opposing the motion)

Motion of bar caused by force,

$$|F| = |BIL| = |(-1)(10)(0.5)| = 5\,\text{N}$$

Net force on body

$F_n = 5 - 1 = 4$ N (in direction of motion)

Force, $F_n = ma$

$$4 = 1 \times a$$

$$a = 4 \text{ m/s}^2$$

Using motion equation: $v^2 = u^2 + 2aS$

$u = 0$, $S = 1$ m

$$v^2 = 2 \times 4 \times 1 = 8$$

$$v = \sqrt{8}\text{ m/sec} = 2.8\text{ m/sec}$$

1.3 (200)

Induced emf in semi-circular loop is given as

$$E_{\text{ind}} = \frac{1}{2}B\omega R^2$$

$$= \frac{1}{2} \times 2 \times 20 \times (0.1)^2 = 0.2\text{ V}$$

Peak current in loop $= \dfrac{E_{ind}}{R} = \dfrac{0.2}{2} = 0.1\text{ A}$

Peak to peak current in loop,

$$I_{p-p} = 2 \times 0.1$$

$$= 0.24$$

$$= 200 \text{ mA}$$

1.4 (33.36)

The capacitance,

$$C = \frac{4\pi\varepsilon}{\frac{1}{a} - \frac{1}{b}} = \frac{4\pi \times 8.85 \times 10^{-12}}{\frac{100}{10} - \frac{100}{15}} = 33.36\text{ pF}$$

1.5 (a)

Original capacitance with distance d_1 is

$$\frac{1}{C_1} = \frac{d_1}{\epsilon_0 A} + \frac{d_2}{\epsilon_2 A}$$

$$\frac{1}{C_1} = \frac{d_1(\epsilon_2 A) + d_2(\epsilon_0 A)}{(\epsilon_0 A)(\epsilon_2 A)}$$

$$C_1 = \frac{\epsilon_0 \epsilon_2 A^2}{d_1 \epsilon_2 A + d_2 \epsilon_0 A}$$

$$C_1 = \frac{\epsilon_0 \epsilon_2 A}{d_1 \epsilon_2 + d_2 \epsilon_0}$$

New capacitance, when d_1 is changed by δd_1

$$C_{new} = C_2 = \frac{\epsilon_0 \epsilon_2 A}{(d_1 - \delta d_1)\epsilon_2 + d_2 \epsilon_0}$$

$$C_2 = \frac{\epsilon_0 A}{(d_1 - \delta d_1) + \frac{d_2 \epsilon_0}{\epsilon_2}}$$

Hence, most close answer is option (a).

1.6 (a)

We know that the capacitance formed between two concentric spherical metal shells having radii x and y is given as,

$$C = \frac{4\pi \in xy}{y - x}$$

Hence, option (a) is correct.

1.7 (a)

Area of cylindrical cross section = $A = \pi a^2$

$$a = \sqrt{\frac{A}{\pi}}$$

$$\text{Current} \rightarrow i = \iint J \cdot ds = \iint Ks^2 \cdot s \cdot ds \cdot d\phi$$

$$= K\int_0^a s^3 ds \cdot \int_0^{2\pi} d\phi$$

$$= K \cdot \frac{\left[s^4\right]_0^a}{4} \cdot [2\pi]$$

$$i = \frac{\pi K \cdot a^4}{2}$$

■■■

General Aptitude

UNIT

XI

CONTENTS

1. Numerical Ability 521

2. Verbal Ability 546

General Aptitude *Syllabus*

Verbal Aptitude : Basic English grammar: tenses, articles, adjectives, prepositions, conjunctions, verb-noun agreement, and other parts of speech Basic vocabulary: words, idioms, and phrases in context Reading and comprehension Narrative sequencing.

Quantitative Aptitude : Data interpretation: data graphs (bar graphs, pie charts, and other graphs representing data), 2 and 3-dimensional plots, maps, and tables Numerical computation and estimation: ratios, percentages, powers, exponents and logarithms, permutations and combinations, and series Mensuration and geometry Elementary statistics and probability

Analytical Aptitude : Logic: deduction and induction ; Analogy : Numerical relations and reasoning

Spatial Aptitude : Transformation of shapes: translation, rotation, scaling, mirroring, assembling, and grouping Paper folding, cutting, and patterns in 2 and 3 dimensions

Analysis of Previous GATE Papers

Exam Year	1 Mark Ques.	2 Marks Ques.	Total Marks
2010	5	5	**15**
2011	5	5	**15**
2012	5	5	**15**
2013	5	5	**15**
2014	5	5	**15**
2015	5	5	**15**
2016	5	5	**15**
2017	5	5	**15**
2018	5	5	**15**
2019	5	5	**15**
2020	5	5	**15**
2021	5	5	**15**
2022	5	5	**15**
2023	5	5	**15**
2024	5	5	**15**

1 Numerical Ability

1.1 25 persons are in a room. 15 of them play hockey, 17 of them play football and 10 of them play both hockey and football. Then the number of persons playing neither hockey nor football is

(a) 2 (b) 17
(c) 13 (d) 3

[2010 : 1 M]

1.2 If 137 + 276 = 435 how much is 731 + 672?

(a) 534 (b) 1403
(c) 1623 (d) 1531

[2010 : 2 M]

1.3 5 skilled workers can build a wall in 20 days; 8 semiskilled workers can build a wall in 25 days; 10 unskilled workers can build a wall in 30 days. If a team has 2 skilled, 6 semiskilled and 5 unskilled workers, how long will it take to build the wall?

(a) 20 days (b) 18 days
(c) 16 days (d) 15 days

[2010 : 2 M]

1.4 Given digits 2, 2, 3, 3, 3, 4, 4, 4, 4 how many distinct 4 digit numbers greater than 3000 can be formed?

(a) 50 (b) 51
(c) 52 (d) 54

[2010 : 2 M]

1.5 Hari (*H*), Gita (*G*), Irfan (*I*) and Saira (*S*) are siblings (i.e. brothers and sisters). All were born on Ist January. The age difference between any two successive siblings (that is born one after another) is less than 3 years. Given the following facts:

1. Hari's age + Gita's age > Irfan's age + Saira's age.
2. The age difference between Gita and Saira is 1 year. However, Gita is not the oldest and Saira is not the youngest.
3. There are no twins.

In what order were they born (oldest first)?

(a) HSIG (b) SGHI
(c) IGSH (d) IHSG

[2010 : 2 M]

1.6 There are two candidates *P* and *Q* in an election. During the campaign 40% of the voters promised to vote for *P*, and rest for *Q*. However, on the day of election 15% of the voters went back on their promise to vote for *P* and instead voted for *Q*. 25% of the voters went back on their promise to vote for *Q* and instead voted for *P*. Suppose, *P* lost by 2 votes, then what was the total number of voters?

(a) 100 (b) 110
(c) 90 (d) 95

[2011 : 1 M]

1.7 The fuel consumed by a motorcycle during a journey while travelling at various speeds is indicated in the graph below

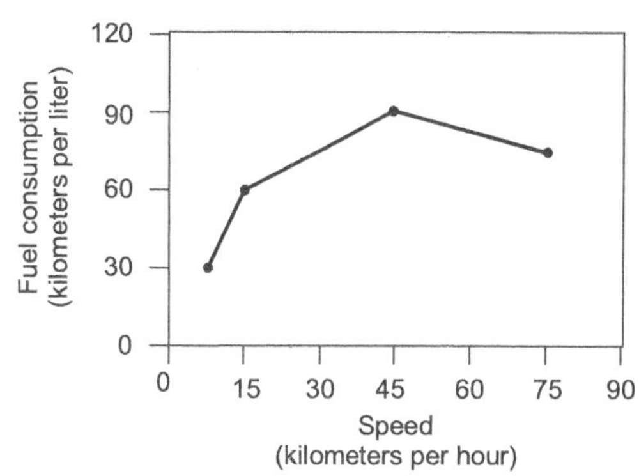

The distance covered during four laps of the journey are listed in the table below

Lap	Distance (kilometers)	Average speed (kilometers per hour)
P	15	15
Q	75	45
R	40	75
S	10	10

From the given data, we can conclude that the fuel consumed per kilometre was least during the lap

(a) *P* (b) *Q*
(c) *R* (d) *S*

[2011 : 2 M]

1.8 Three friends, R, S and T shared toffee from a bowl. R took 1/3rd of the toffees, but returned to the bowl. S took 1/4th of what was left but returned three toffees to the bowl. T took half of the remainder but returned two back into the bowl. If the bowl had 17 toffees left, how many toffees were originally there in the bowl?

(a) 38 (b) 31
(c) 48 (d) 41

[2011 : 2 M]

1.9 Given that $f(y) = |y|$, y and q is any non-zero real number, the value of $|f(q) - f(-q)|$ is

(a) 0 (b) –1
(c) 1 (d) 2

[2011 : 2 M]

1.10 The sum of n terms of the series 4 + 44 + 444 + ... is

(a) $(4/81)[10^{n+1} - 9n - 1]$
(b) $(4/81)[10^{n-1} - 9n - 1]$
(c) $(4/81)[10^{n+1} - 9n - 10]$
(d) $(4/81)[10^{n} - 9n - 10]$

[2011 : 2 M]

1.11 If $(1.001)^{1259} = 3.52$, $(1.001)^{2062} = 7.85$, then $(1.001)^{3321}$

(a) 2.23 (b) 4.33
(c) 11.37 (d) 27.64

[2012 : 1 M]

1.12 The data given in the following table summarizes the monthly budget of an average household.

Category	Amount
Food	4000
Clothing	1200
Rent	2000
Savings	1500
Others	1800

The approximate percentage of the monthly budget **NOT** spent on savings is

(a) 10% (b) 14%
(c) 81% (d) 86%

[2012 : 2 M]

1.13 A and B friends. They decide to meet between 1 PM and 2 PM on a given day. There is a condition that whoever arrives first will not wait for the other for more than 15 minutes. The probability that they will meet on that day is

(a) 1/4 (b) 1/16
(c) 7/16 (d) 9/16

[2012 : 2 M]

1.14 Raju has 14 currency notes in his pocket consisting of only Rs.20 notes and Rs.10 notes. The total money value of the notes is Rs.230. The number of Rs.10 notes that Raju has is

(a) 5 (b) 6
(c) 9 (d) 10

[2012 : 2 M]

1.15 There are eight bags of rice looking alike, seven of which have equal and one is slightly heavier. The weighing balance is of unlimited capacity. Using this balance the minimum number of weighting required to identify the heavier bag is

(a) 2 (b) 3
(c) 4 (d) 8

[2012 : 2 M]

1.16 In the summer of 2012, in New Delhi, the mean temperature of Monday to Wednesday was 41°C and of Tuesday to Thursday was 43°C. If the temperature on Thursday was 15% higher than that of Monday, then the temperature in °C on Thursday was

(a) 40 (b) 43
(c) 46 (d) 49

[2013 : 1 M]

1.17 A car travels 8 km in the first quarter of an hour, 6 km in the second quarter and 16 km in the third quarter. The average speed of the car in km per hour over the entire journey is

(a) 30 (b) 36
(c) 40 (d) 24

[2013 : 2 M]

1.18 Find the sum to n terms of the series 10 + 84 + 734 +....

(a) $\frac{9(9^n+1)}{10}+1$ (b) $\frac{9(9^n-1)}{8}+1$

(c) $\frac{9(9^n-1)}{8}+n$ (d) $\frac{9(9^n-1)}{8}+n^2$

[2013 : 2 M]

1.19 The set of values of p for which the roots of the equation $3x^2 + 2x + p(p-1) = 0$ are of opposite sign is

(a) $(-\infty, 0)$ (b) $(0, 1)$
(c) $(1, \infty)$ (d) $(0, \infty)$

[2013 : 2 M]

1.20 What is the change that a leap year, selected at random, will contain 53 Saturdays?

(a) $\frac{2}{7}$ (b) $\frac{3}{7}$
(c) $\frac{1}{7}$ (d) $\frac{5}{7}$

[2013 : 2 M]

1.21 In any given year, the probability of an earthquake greater than Magnitude 6 occurring in the Garhwal Himalayas is 0.04. The average time between successive occurrences of such earthquakes is ____ years.

[2014 : 1 M]

1.22 The population of a new city is 5 million and is growing at 20% annually. How many years would it take to double at this growth rate?

(a) 3-4 years (b) 4-5 years
(c) 5-6 years (d) 6-7 years

[2014 : 1 M]

1.23 In a group of four children, Som is younger to Riaz. Shiv is elder to Ansu. Ansu is youngest in the group. Which of the following statements is/are required to find the eldest child in the group?

Statements

1. Shiv is younger to Riaz.
2. Shiv is elder to Som.

(a) Statement 1 by itself determines the eldest child.
(b) Statement 2 by itself determines the eldest child.
(c) Statements 1 and 2 are both required to determine the eldest child.
(d) Statements 1 and 2 are not sufficient to determine the eldest child.

[2014 : 2 M]

1.24 The total exports and revenues from the exports of a country are given in the two pie charts below. The pie chart for exports shows the quantity of each item as a percentage of the total quantity of exports. The pie chart for the revenues shows the percentage of the total revenue generated through export of each item. The total quantity of exports of all the items is 5 lakh tonnes and the total revenues are 250 crore rupees. What is the ratio of the revenue generated through export of Item 1 per kilogram to the revenue generated through export of Item 4 per kilogram?

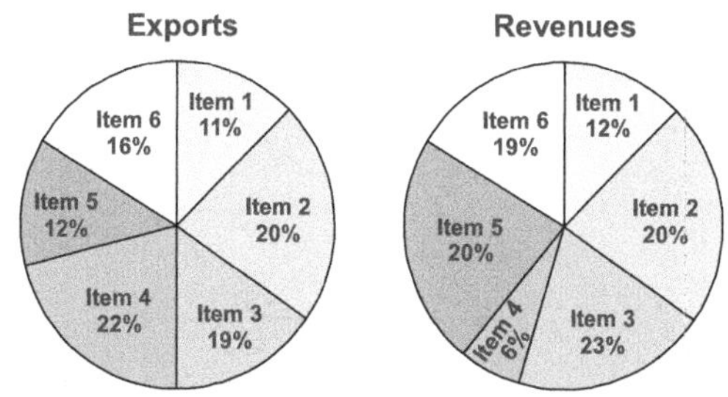

(a) 1 : 2 (b) 2 : 1
(c) 1 : 4 (d) 4 : 1

[2014 : 2 M]

1.25 X is 1 km northeast of Y. Y is 1 km southeast of Z. W is 1 km west of Z. P is 1 km south of W. Q is 1 km east of P. What is the distance between X and Q in km?

(A) 1 (b) $\sqrt{2}$
(c) $\sqrt{3}$ (d) 2

[2014 : 2 M]

1.26 10% of the population in a town is HIV^+. A new diagnostic kit for HIV detection is available; this kit correctly identifies HIV^+ individuals 95% of the time, and HIV^- individuals 89% of the time. A particular patient is tested using this kit and is found to be positive. The probability that the individual is actually positive is ______.

[2014 : 2 M]

1.27 Five teams have to compete in a league, with every team playing every other team exactly once, before going to the next round. How many matches will have to held to complete the league round of matches?

(a) 20 (b) 10
(c) 8 (d) 5

[2015 : 1 M]

1.28 Tanya is older than Eric.
Cliff is older than Tanya.
Eric is older than Cliff.

If the first two statements are true, then the third statement is

(a) True (b) False
(c) Uncertain (d) Data insufficient

[2015 : 1 M]

1.29 In the given angle Q is a right angle, $PS : QS = 3 : 1$, $RT : QT = 5 : 2$ and $PU : UR = 1 : 1$. If area of triangle QTS is 20 cm^2, then the area of triangle PQR in cm^2 is ____.

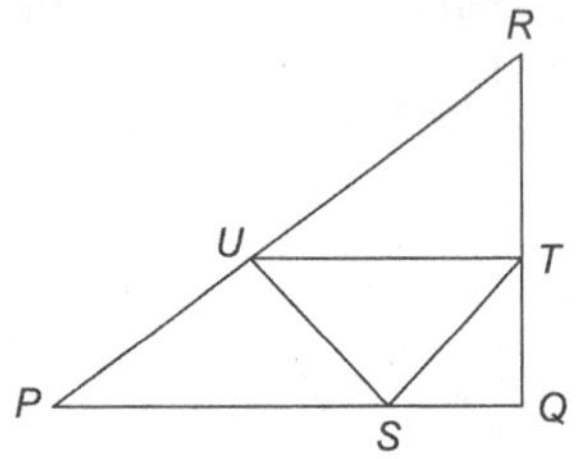

[2015 : 2 M]

1.30 Given below are two statements followed by two conclusions. Assuming these statements to be true, decide which one logically follows:

Statements:

I. No manager is a leader.

II. All leaders are executives.

Conclusions:

I. No manager is an executive.

II. No executive is a manager

(a) Only conclusion I follows.
(b) Only conclusion II follows.
(c) Neither conclusion I nor II follows.
(d) Both conclusions I and II follow

[2015 : 2 M]

1.31 A coin is tossed thrice. Let X be the even that head occurs in each of the first two tosses. Let Y be the event that a tail occurs on the third toss. Let Z be the event that two tails occur in three tosses. Based on the above information, which one of the following statements is TRUE?

(a) X and Y are not independent
(b) Y and Z are dependent
(c) Y and Z are independent
(d) X and Z are independent

[2015 : 2 M]

1.32 Right triangle PQR is to be constructed in the xy-plane so that the right angle is at P and line PR is parallel to the x-axis. The x and y coordinates of P, Q, and R are to be integers that satisfy the inequalities:

$-4 \le x \le 5$ and $6 \le y \le 16$. How many different triangle could be constructed with these properties?

(a) 110 (b) 1100
(c) 9900 (d) 10000

[2015 : 2 M]

1.33 M has a son Q and a daughter R. He has no other children. E is the mother of P and daughter-in-law of M. How is P related to M?

(a) P is the son-in-law of M
(b) P is the grandchild of M
(c) P is the daughter-in law of M
(d) P is the grandfather of M

[2016 : 1 M]

1.34 The number that least fits this set: (324, 441, 97 and 64) is ______.

(a) 324 (b) 441
(c) 97 (d) 64

[2016 : 1 M]

1.35 It takes 10s and 15s, respectively, for two trains travelling at different constant speeds to completely pass a telegraph post. The length of the first train is 120 m and that of the second train is 150 m. The magnitude of the difference in the speeds of the two trains (in m/s) is _____.

(a) 2.0 (b) 10.0
(c) 12.0 (d) 22.0

[2016 : 1 M]

1.36 The velocity V of a vehicle along a straight line is measured in m/s and plotted as shown with respect to time in seconds. At the end of the 7 seconds, how much will the odometer reading increase by (in m)?

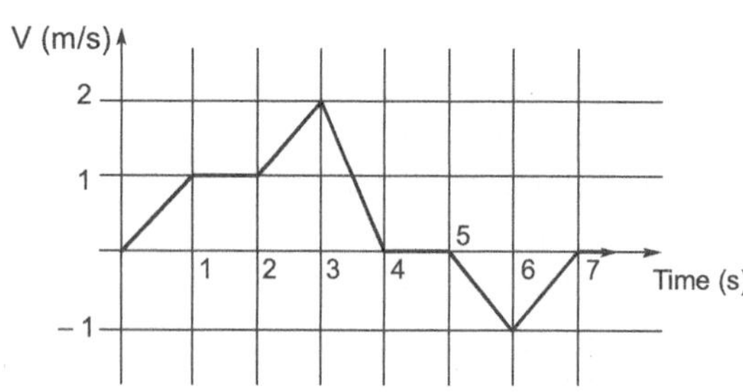

(a) 0 (b) 3
(c) 4 (d) 5

[2016 : 2 M]

1.37 A flat is shared by four first year undergraduate students. They agreed to allow the oldest of them to enjoy some extra space in the flat. Manu is two months older than Sravan, who is three months younger than Trideep. Pavan is one month older than Sravan. Who should occupy the extra space in the flat?

(a) Manu (b) Sravan
(c) Trideep (d) Pavan

[2016 : 2 M]

1.38 Find the area bounded by the lines $3x + 2y = 14$, $2x - 3y = 5$ in the first quadrant.

(a) 14.95 (b) 15.25
(c) 15.70 (d) 20.35

[2016 : 2 M]

1.39 A straight line is fit to a data set ($ln\ x$, y). This line intercepts the abscissa at $\ln x = 0.1$ and has a slope of -0.02. What is the value of y at $x = 5$ from the fit?

(a) −0.030 (b) −0.014
(c) 0.014 (d) 0.030

[2016 : 2 M]

1.40 Two dice are thrown simultaneously. The probability that the product of the numbers appearing on the top faces of the dice is a perfect square is

(a) $\frac{1}{9}$ (b) $\frac{2}{9}$
(c) $\frac{1}{3}$ (d) $\frac{4}{9}$

[2017 : 1 M]

1.41 Four cards lie on a table. Each card has a number printed on one side and a colour on the other. The faces visible on the cards are 2, 3, red and blue.

Proposition: If a card has an even value on one side, then its opposite face is red.

The cards which MUST be turned over to verify the above proposition are

(a) 2, red (b) 2, 3, red
(c) 2, blue (d) 2, red, blue

[2017 : 1 M]

1.42 What is the value of x when $81 \times \left(\frac{16}{25}\right)^{x+2} \div \left(\frac{3}{5}\right)^{2x+4} = 144$?

(a) 1 (b) −1
(c) −2 (d) Cannot be determined

[2017 : 1 M]

1.43 The points in the graph below represent the halts of a lift for durations of 1 minute, over a period of 1 hour.

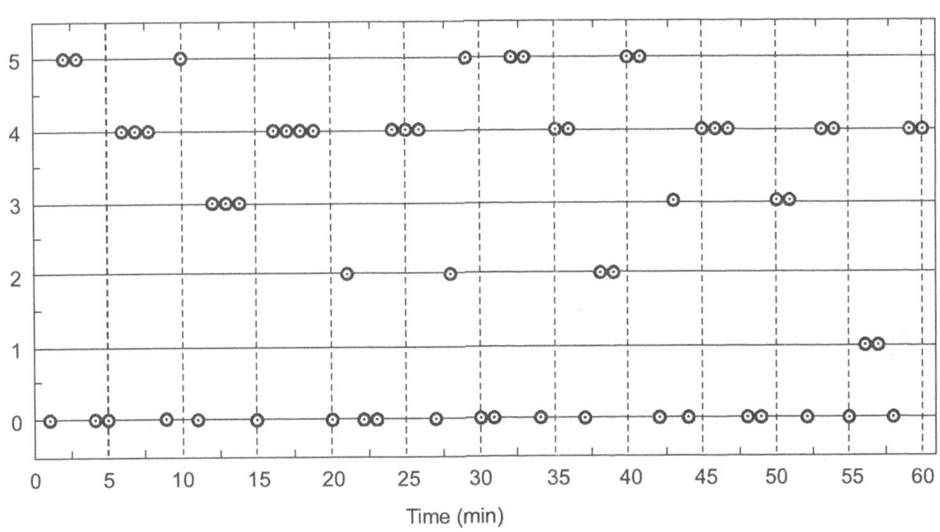

Which of the following statements are correct?

(i) The elevator never moves directly from any non-ground floor to another non-ground floor over the one hour period.

(ii) The elevator stays on the fourth floor for the longest duration over the one hour period.

(a) Only (i) (b) Only (ii)

(c) Both (i) and (ii) (d) Neither (i) nor (ii)

[2017 : 1 M]

1.44 Budhan covers a distance of 19 km in 2 hours by cycling one fourth of the time and walking the rest. The next day he cycles (at the same speed as before) for half the time and walks the rest (at the same speed as before) and covers 26 km in 2 hours. The speed in km/h at which Budhan walks is

(a) 1 (b) 4

(c) 5 (d) 6 [2017 : 1 M]

1.45 A map shows the elevations of Darjeeling. Gangtok, Kalimpong, Pelling and Siliguri. Kalimpong is at a lower elevation than Gangtok. Pelling is at a lower elevation than Gangtok. Pelling is at a higher elevation than Siliguri. Darjeeling is at a higher elevation than Gangtok.

Which of the following statements can be inferred from the paragraph above?

(i) Pelling is at a higher elevation that Kalimpong.

(ii) Kalimpong is at a lower elevation than Darjeeling.

(iii) Kalimpong is at a higher elevation than Siliguri.

(iv) Siliguri is at a lower elevation than Gangtok.

(a) Only (ii) (b) Only (ii) and (iii)

(c) Only (ii) and (iv) (d) Only (iii) and (iv)

[2017 : 1 M]

1.46 P, Q, R, S, T and U are seated around a circular table. R is seated two places to the right of Q, P is seated three places to the left R. S is seated opposite U. If P and U now switch seats, which of the following must necessarily be true?

(a) P is immediately to the right of R

(b) T is immediately to the left of P

(c) T is immediately to the left of P or P is immediately to the right of Q

(d) U is immediately to the right of R or P is immediately to the left of T

[2017 : 1 M]

1.47 For $0 \le x \le 2\pi$, $\sin x$ and $\cos x$ are both decreasing functions in the interval ______.

(a) $\left(0, \frac{\pi}{2}\right)$ (b) $\left(\frac{\pi}{2}, \pi\right)$

(c) $\left(\pi, \frac{\pi}{2}\right)$ (d) $\left(\frac{3\pi}{2}, 2\pi\right)$

[2018 : 1 M]

1.48 Arrange the following three-dimensional objects in the descending order of their volumes:

(i) A cuboid with dimensions 10 cm, 8 cm and 6 cm.

(ii) A cube of side 8 cm.

(iii) A cylinder with base radius 7 cm and height 7 cm.

(iv) A sphere of radius 7 cm.

(a) (i), (ii), (iii), (iv) (b) (ii), (i), (iv), (iii)

(c) (iii), (ii), (i), (iv) (d) (iv), (iii), (ii), (i)

[2018 : 1 M]

1.49 The area of an equilateral triangle is $\sqrt{3}$. What is the perimeter of the triangle?

(a) 2 (b) 4

(c) 6 (d) 8 [2018 : 1 M]

1.50 An automobile travels from city A to city B and returns to city A by the same route. The speed of the vehicle during the onward and return journeys were constant at 60 km/h and 90 km/h, respectively. What is the average speed in km/h for the entire journey?

(a) 72 (b) 73

(c) 74 (d) 75

[2018 : 2 M]

1.51 In a detailed study of annual crow births in India, it was found that there was relatively no growth during the period 2002 to 2004 and a sudden spike from 2004 to 2005. In another unrelated study, it was found that the revenue from cracker sales in India which remained fairly flat from 2002 to 2004, saw a sudden spike in 2005 before declining again in 2006. The solid line in the graph below refers to annual sale of crackers and the dashed line refers to the annual crow births in India. Choose the most appropriate inference from the above data.

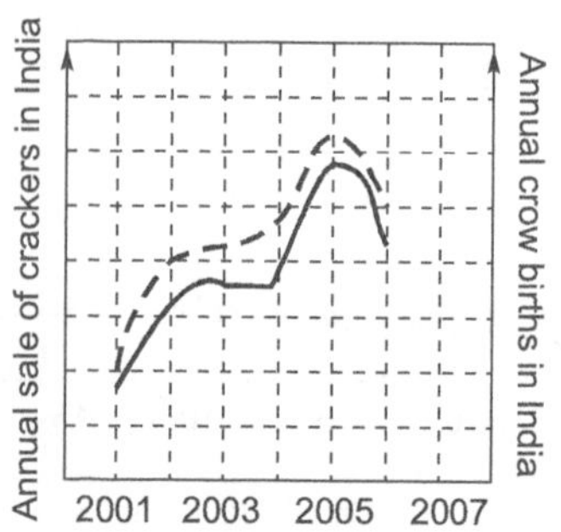

(a) There is a strong correlation between crow birth and cracker sales.
(b) Cracker usage increases crow birth rate.
(c) If cracker sale declines, crow birth will decline.
(d) Increased birth rate of crows will cause an increase in the sale of crackers.

[2018 : 2 M]

1.52 If $x^2 + x - 1 = 0$ what is the value of $x^4 + \frac{1}{x^4}$?

(a) 1 (b) 5
(c) 7 (d) 9

[2018 : 2 M]

1.53 To pass a test, a candidate needs to answer at least 2 out of 3 questions correctly. A total of 6,30,000 candidates appeared for the test. Question *A* was correctly answered by 3,30,000 candidates. Question *B* was answered correctly by 2,50,000 candidates. Question *C* was answered correctly by 2,60,000 candidates. Both questions *A* and *B* were answered correctly by 1,00,000 candidates. Both questions *B* and *C* were answered correctly by 90,000 candidates. Both questions *A* and *C* were answered correctly by 80,000 candidates. If the number of students answering all questions correctly is the same as the number answering none, how many candidates failed to clear the test?

(a) 30,000 (b) 2,70,000
(c) 3,90,000 (d) 4,20,000

[2018 : 2 M]

1.54 A set of 4 parallel lines intersect with another set of 5 parallel lines. How many parallelograms are formed?

(a) 20 (b) 48
(c) 60 (d) 72

[2018 : 2 M]

1.55 Five numbers 10, 7, 5,4 and 2 are to be arranged in a sequence from left to right following the directions given below:

1. No two odd or even numbers are next to each other.
2. The second number from the left is exactly half of the left-most number.
3. The middle number is exactly twice the right-most number.

Which is the second number from the right?

(a) 2 (b) 4
(c) 7 (d) 10 [2019 : 1 M]

1.56 The radius as well as the height of a circular cone increases by 10%. The percentage increase in its volume is______.

(a) 17.1 (b) 21.0
(c) 33.1 (d) 72.8 **[2019 : 1 M]**

1.57 In a country of 1400 million population 70% own mobile phones. Among the mobile phone owners, only 294 million access the Internet. Among these Internet users, only half buy goods from e-commerce portals. What is the percentage of these buyers in the country?

(a) 10.50 (b) 14.70
(c) 15.00 (d) 50.00

[2019 : 2 M]

1.58 Two trains started at 7 AM from the same point. The first train travelled north at a speed of 80 km/h and the second train travelled south at a speed of 100 km/h. The time at which they were 540 km apart is ______AM.

(a) 9 (b) 10
(c) 11 (d) 11.30

[2019 : 2 M]

1.59 *P*, *Q*, *R* and *S* are to be uniquely coded using α and β. If *P* is coded as $\alpha\alpha$ and *Q* as $\alpha\beta$, then *R* and *S*, respectively, can be coded as _____.

(a) $\beta\alpha$ and $\alpha\beta$ (b) $\alpha\beta$ and $\beta\beta$
(c) $\beta\alpha$ and $\beta\beta$ (d) $\beta\beta$ and $\alpha\alpha$

[2020 : 2 M]

1.60 The bar graph shows the data of students who appeared and passed in an examination for four schools P, Q, R and S. The average of success rates (in percentage) of these four schools is ____.

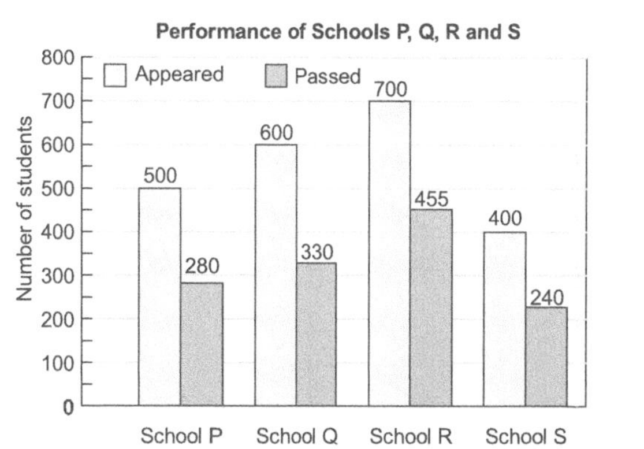

(a) 58.8% (b) 58.5%
(c) 59.0% (d) 59.3%

[2020 : 2 M]

1.61 Define $[x]$ as the greatest integer less than or equal to x, for each $x \in (-\infty, \infty)$. If $y = [x]$ then area under y for $x \in [1, 4]$ is _____.

(a) 4 (b) 1
(c) 6 (d) 3

[2020 : 1 M]

1.62 The sum of the first n terms in the sequence 8, 88, 888, 8888, ... is _____.

(a) $\frac{80}{81}(10^n - 1) - \frac{8}{9}n$ (b) $\frac{81}{80}(10^n - 1) + \frac{9}{8}n$

(c) $\frac{81}{80}(10^n - 1) - \frac{9}{8}n$ (d) $\frac{80}{81}(10^n - 1) + \frac{8}{9}n$

[2020 : 2 M]

1.63 Select the graph that schematically represents BOTH $y = x^m$ and $y = x^{1/m}$ properly in the interval $0 \le x \le 1$, for integer values of m, where $m > 1$.

(a) 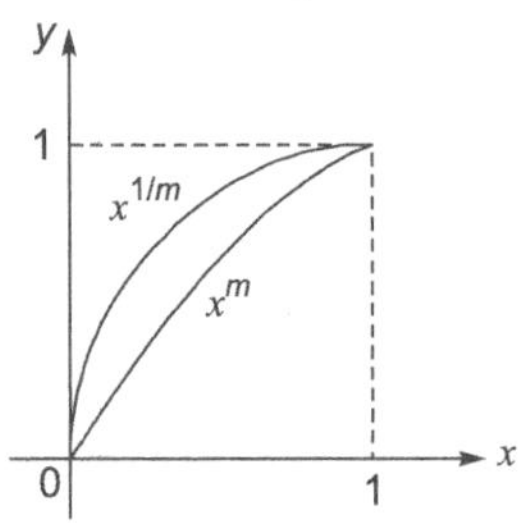

(b) 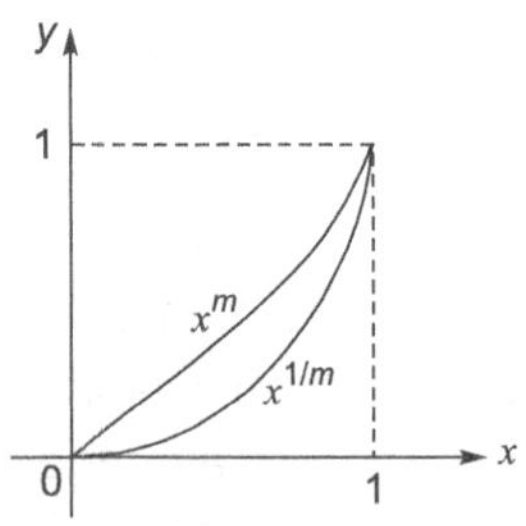

(c) 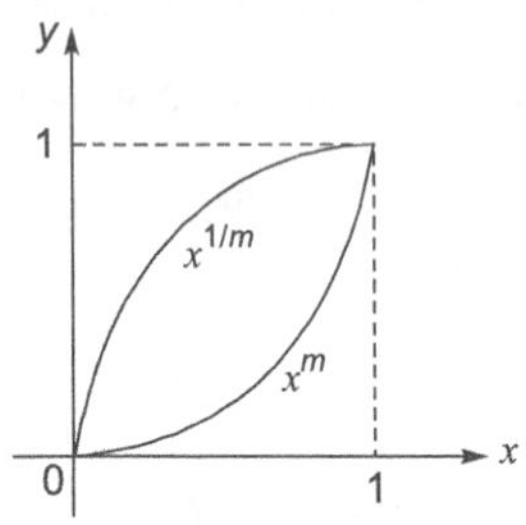

(d) 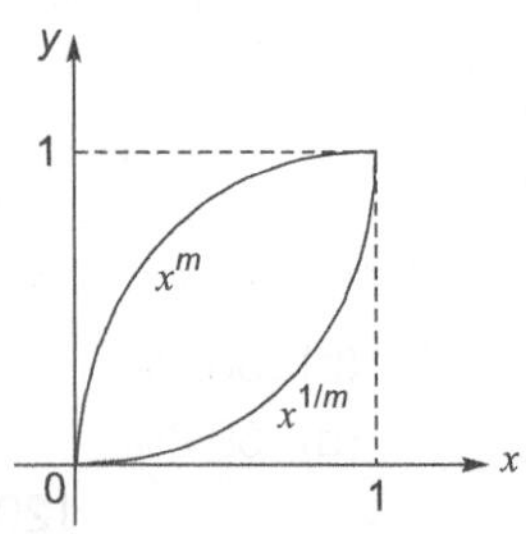

[2020 : 2 M]

1.64 The mirror image of the about text about the X-axis is

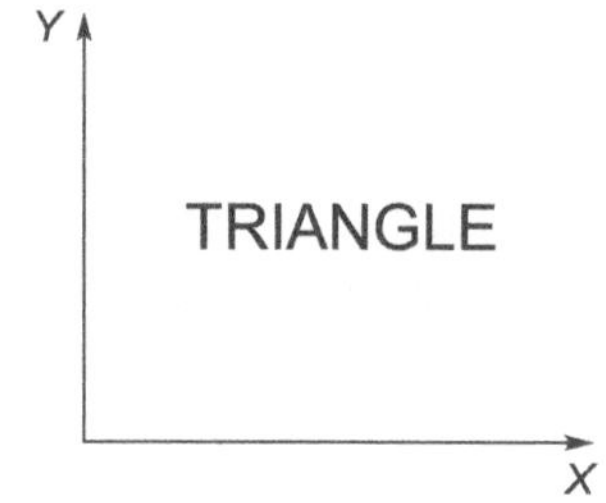

(a)  (b) 
(c)  (d) 

[2021 : 1 M]

1.65 In a company, 35% of the employees drink coffee, 40% of the employees drink tea and 10% of the employees drink both tea and coffee. What percentage of employees drink neither tea nor coffee?

(a) 15 (b) 25
(c) 35 (d) 40

[2021 : 1 M]

1.66 Four persons P, Q, R and S are to be seated in a row. All facing the same direction, but not necessarily in the same order. P and R cannot sit adjacent to each other. S should be seated to the right of Q. The number of distinct seating arrangements possible is:

(a) 4 (b) 2
(c) 8 (d) 6 [2021 : 1 M]

1.67 **Statement:** Either P marries Q or X marries Y. Among the options below, the logical NEGATION of the above statement is:

(a) P does not marry Q and X marries Y.
(b) P marries Q and X marries Y.
(c) X does not marry Y and P marries Q.
(d) Neither P marries Q nor X marries Y.

[2021 : 2 M]

1.68 A function λ, is defined by

$$\lambda(p,q) = \begin{cases} (p-q)^2, & \text{if } p \ge q \\ p+q, & \text{if } p < q \end{cases}$$

The value of expression $\frac{\lambda[-(-3+2),(-2,+3)]}{[-(-2+1)]}$ is:

(a) −1 (b) 0
(c) $\frac{16}{3}$ (d) 16

[2021 : 2 M]

1.69 ⊕ and ⊙ are two operators on numbers p and q such that

$$p \oplus Q = \frac{p^2 + q^2}{pq} \text{ and } p \odot q = \frac{p^2}{q};$$

If $x \oplus y = 2 \odot 2$, then x =

(a) y (b) $\frac{y}{2}$

(c) $\frac{3y}{2}$ (d) $2y$ [2021 : 1 M]

1.70 Five line segments of equal lengths, *PR*, *PS*, *QS*, *QT* and *RT* as used to form a star shown in the figure above.

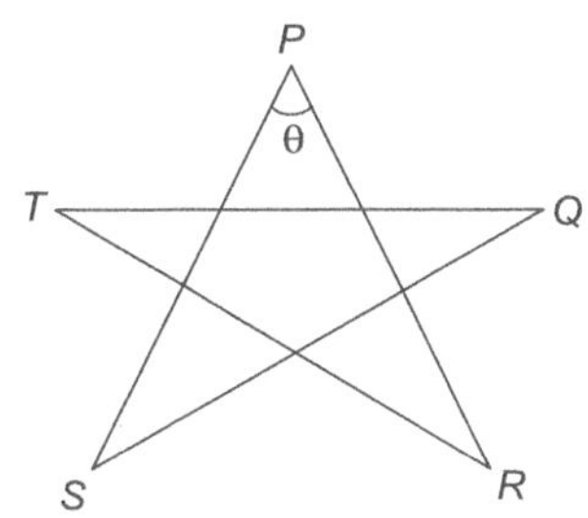

The value of θ, in degrees, is______.

(a) 72 (b) 108

(c) 36 (d) 45 [2021 : 2 M]

1.71 Consider two rectangular sheets, Sheet *M* and Sheet *N* of dimensions 6 cm × 4 cm each.

Folding operation 1: The sheet is folded into half by joining the short edges of the current shape,

Folding operation 2: The sheet is folded into half by joining the long edges of the current shape.

Folding operation 1 is carried out on Sheet *M* three times,

Folding operation 2 is carried out on Sheet *N* three times,

The ratio of perimeters of the final folded shape of Sheet *N* to the final folded shape of Sheet *M* is_____.

(a) 13 : 7 (b) 7 : 5

(c) 5 : 13 (d) 3 : 2

[2021 : 2 M]

1.72 A sphere of radius *r* cm is packed in a box of cubical shape.

What should be the minimum volume (in cm^3) of the box that can enclose the sphere?

(a) $\frac{r^3}{8}$ (b) r^3

(c) $2r^3$ (d) $8r^3$

[2022 : 1 M]

1.73 Pipes P and Q can fill a storage tank in full with water in 10 and 6 minutes, respectively. Pipe R draws the water out from the storage tank at a rate of 34 litres per minute. P, Q and R operate at a constant rate.

If it takes one hour to completely empty a full storage tank with all the pipes operating simultaneously, what is the capacity of the storage tank (in litres)?

(a) 26.8 (b) 60.0

(c) 120.0 (d) 127.5

[2022 : 1 M]

1.74 Six persons P, Q, R, S, T and U are sitting around a circular table facing the center not necessarily in the same order. Consider the following statements:

- P sits next to S and T.
- Q sits diametrically opposite to P.
- The shortest distance between S and R is equal to the shortest distance between T and U.

Based on the above statements, Q is a neighbor of

(a) U and S (b) R and T

(c) R and U (d) P and S

[2022 : 1 M]

1.75 A building has several rooms and doors as shown in the top view of the building given below. The doors are closed initially.

What is the minimum number of doors that need to be opened in order to go from the point P to the point Q?

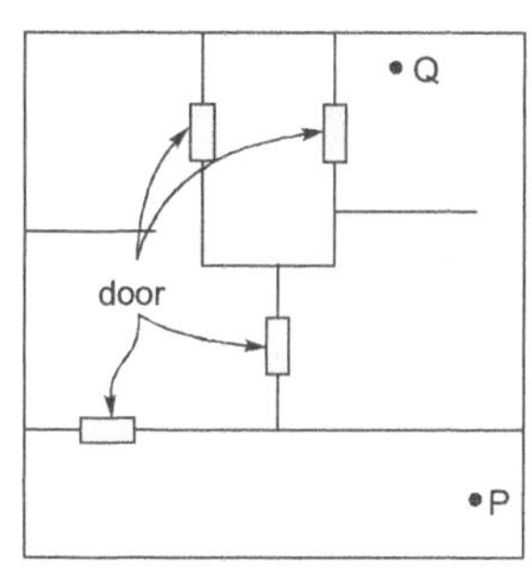

(a) 4 (b) 3

(c) 2 (d) 1 [2022 : 1 M]

1.76 A game consists of spinning an arrow around a stationary disk as shown below. When the arrow comes to rest, there are eight equally likely outcomes. It could come to rest in any one of the sectors numbered 1, 2, 3, 4, 5, 6, 7 or 8 as shown. Two such disks are used in a game where their arrows are independently spun.

What is the probability that the sum of the numbers on the resulting sectors upon spinning the two disks is equal to 8 after the arrows come to rest?

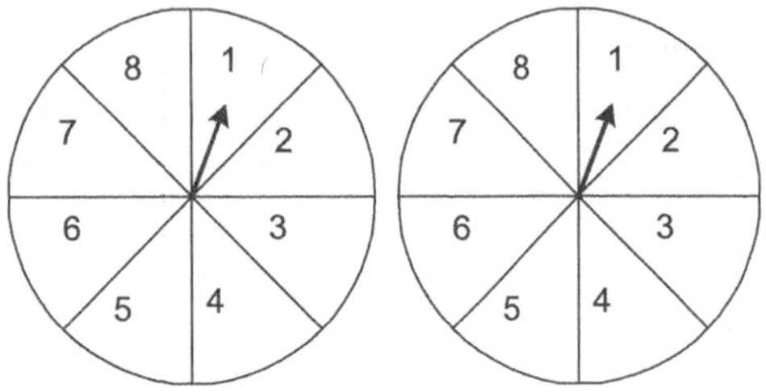

(a) $\frac{1}{16}$ (b) $\frac{5}{64}$

(c) $\frac{3}{32}$ (d) $\frac{7}{64}$

[2022 : 2 M]

1.77 Consider the following inequalities.

(i) $3p - q < 4$

(ii) $3q - p < 12$

Which one of the following expressions below satisfies the above two inequalities?

(a) $p + q < 8$ (b) $p + q = 8$

(c) $8 \leq p + q < 16$ (d) $p + q \geq 16$

[2022 : 2 M]

1.78 Given below are three statements and four conclusions drawn based on the statements.

Statement 1: Some engineers are writers.
Statement 2: No writer is an actor.
Statement 3: All actors are engineers.

Conclusion I: Some writers are engineers.
Conclusion II: All engineers are actors.
Conclusion III: No actor is a writer.
Conclusion IV: Some actors are writers.

Which one of the following options can be logically inferred?

(a) Only conclusion I is correct
(b) Only conclusion II and conclusion III are correct
(c) Only conclusion I and conclusion III are correct
(d) Either conclusion III or conclusion IV is correct

[2022 : 2 M]

1.79 Which one of the following sets of pieces can be assembled to form a square with a single round hole near the center? Pieces cannot overlap.

(a)

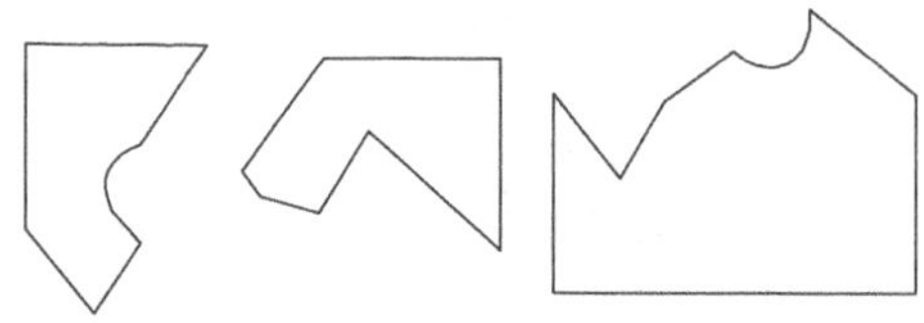

(b)

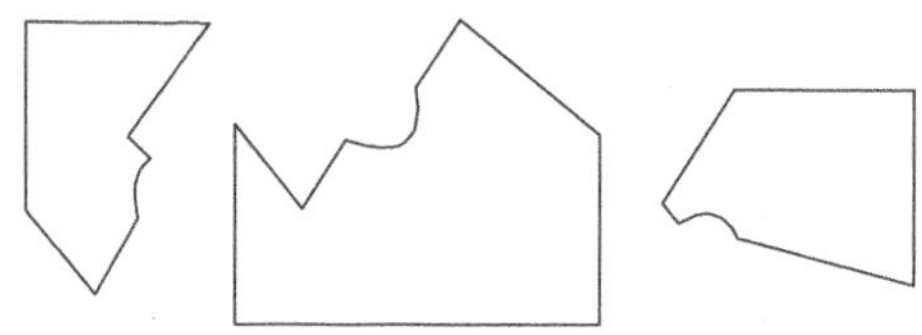

(c)

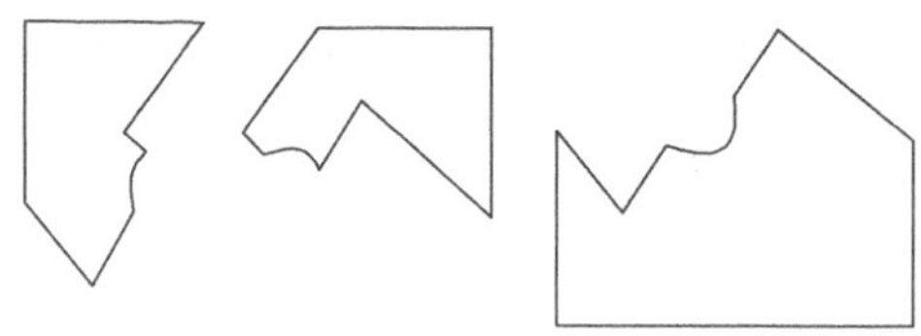

(d)

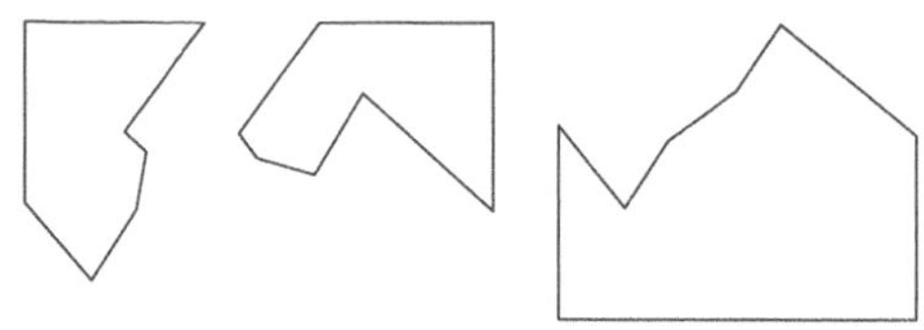

[2022 : 2 M]

1.80 A 'frabjous' number is defined as a 3 digit number with all digits odd, and no two adjacent digits being the same. For example, 137 is a frabjous number, while 133 is not. How many such frabjous numbers exist?

(a) 125 (b) 720
(c) 60 (d) 80

[2023 : 1 M]

1.81 Which one among the following statements must be TRUE about the mean and the median of the scores of all candidates appearing for GATE 2023?

(a) The median is at least as large as the mean.
(b) The mean is at least as large as the median.
(c) At most half the candidates have a score that is larger than the median.
(d) At most half the candidates have a score that is larger than the mean.

[2023 : 1 M]

1.82 Ankita has to climb 5 stairs starting at the ground, while respecting the following rules:

1. At any stage, Ankita can move either one or two stairs up.
2. At any stage, Ankita cannot move to a lower step.

Let $F(N)$ denote the number of possible ways in which Ankita can reach the Nth stair For example, $F(1)=1$, $F(2)= 2$, $F(3) = 3$.

The value of $F(5)$ is _______.

(a) 8 (b) 7
(c) 6 (d) 5 [2023 : 2 M]

1.83 Which one of the given figures P, Q, R and S represents the graph of the following function?

$$f(x) = \left|\,|x+2| - |x-1|\,\right|$$

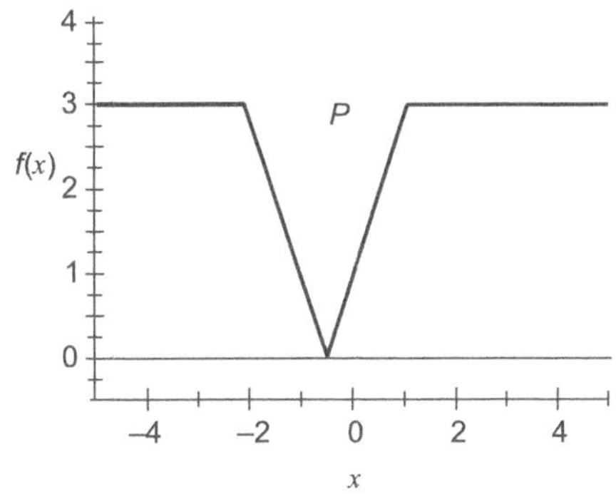

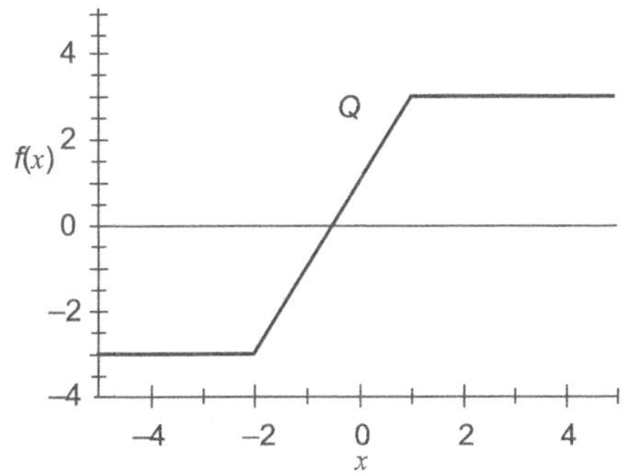

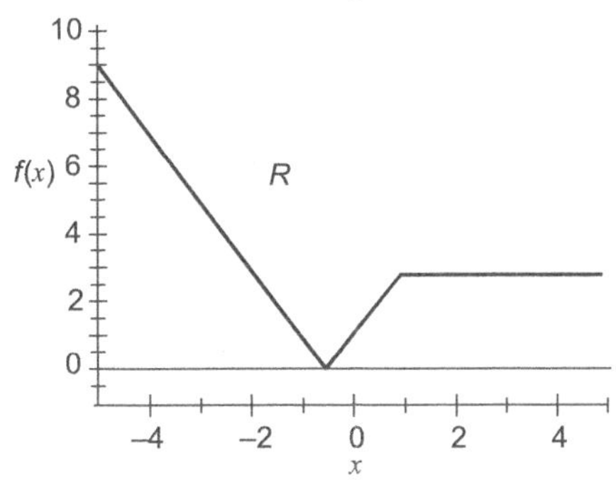

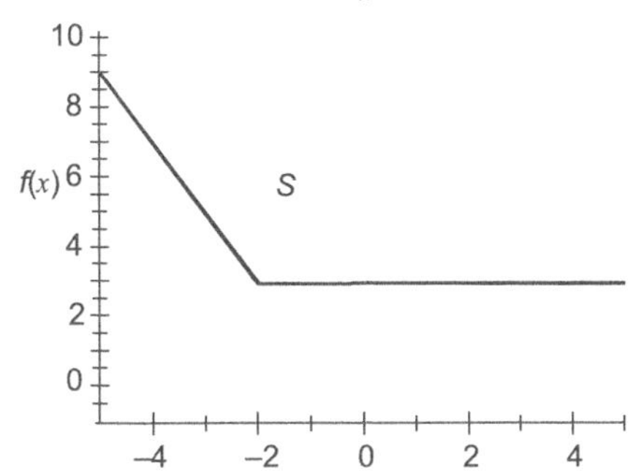

(a) P (b) Q
(c) R (d) S [2023 : 2 M]

1.84 Statements:

1. All heroes are winners.
2. All winners are lucky people.

Inferences:

I. All lucky people are heroes.
II. Some lucky people are heroes.
III. Some winners are heroes.

Which of the above inferences can be logically deduced from statements 1 and 2?

(a) Only I and II (b) Only II and III
(c) Only I and III (d) Only III [2024 : 1 M]

1.85 A student was supposed to **multiply** a positive real number p with another positive real number q. Instead, the student **divided** p by q. If the percentage error in the student's answer is 80%, the value of q is

(a) 5 (b) $\sqrt{2}$
(c) 2 (d) $\sqrt{5}$ [2024 : 1 M]

1.86 If the sum of the first 20 consecutive positive odd numbers is divided by 20^2, the result is

(a) 1 (b) 20
(c) 2 (d) 1/2 [2024 : 1 M]

1.87 The ratio of the number of girls to boys in class VIII is the same as the ratio of the number of boys to girls in class IX. The total number of students (boys and girls) in classes VIII and IX is 450 and 360, respectively. If the number of girls in classes VIII and IX is the same, then the number of girls in each class is

(a) 150 (b) 200
(c) 250 (d) 175 [2024 : 1 M]

1.88 Seven identical cylindrical chalk-sticks are fitted tightly in a cylindrical container. The figure below shows the arrangement of the chalk-sticks inside the cylinder.

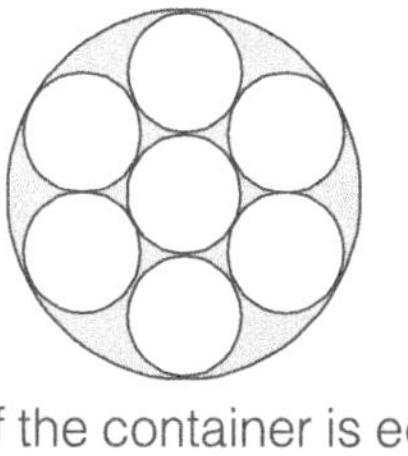

The length of the container is equal to the length of the chalk-sticks. The ratio of the occupied space to the empty space of the container is

(a) $\frac{5}{2}$ (b) $\frac{7}{2}$
(c) $\frac{9}{2}$ (d) 3 [2024 : 2 M]

1.89 The plot below shows the relationship between the mortality risk of cardiovascular disease and the number of steps a person walks per day. Based on the data, which one of the following options is true?

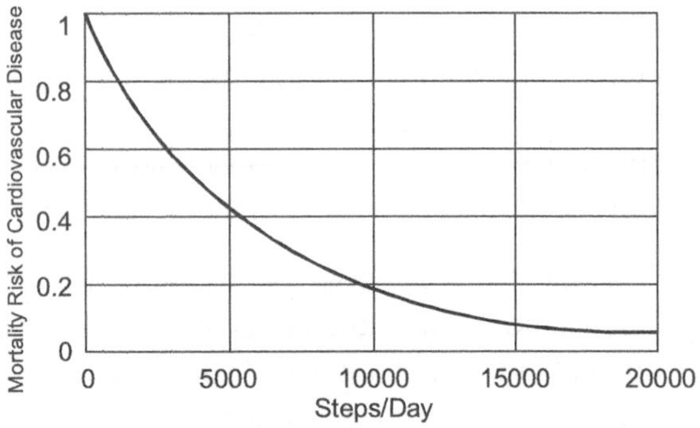

(a) The risk reduction on increasing the steps/day from 0 to 10000 is less than the risk reduction on increasing the steps/day from 10000 to 20000.
(b) The risk reduction on increasing the steps/day from 0 to 5000 is less than the risk reduction on increasing the steps/day from 15000 to 20000.
(c) For any 5000 increment in steps/day the largest risk reduction occurs on going from 0 to 5000.
(d) For any 5000 increment in steps/day the largest risk reduction occurs on going from 15000 to 20000.

[2024 : 2 M]

1.90 Five cubes of identical size and another smaller cube are assembled as shown in Figure A. If viewed from direction X, the planar image of the assembly appears as Figure B.

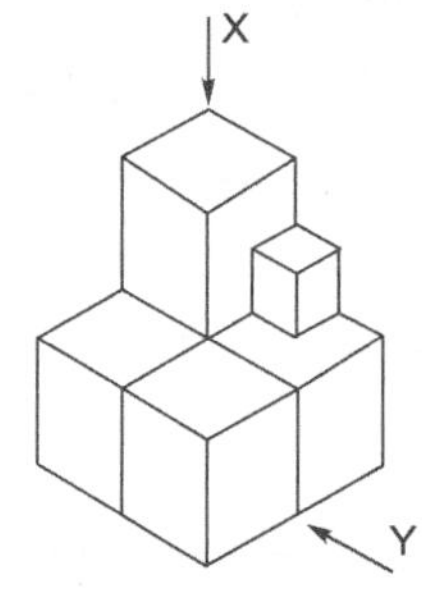

Figure A

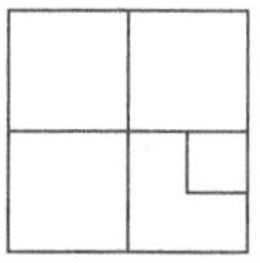

Figure B

If viewed from direction Y, the planar image of the assembly (Figure A) will appear as

(a) 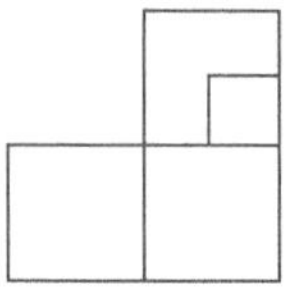

(b) 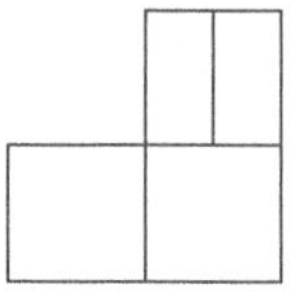

(c) 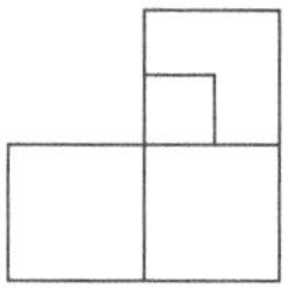

(d) 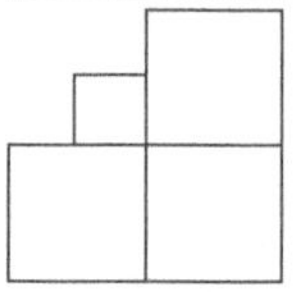

[2024 : 2 M]

1.91 Visualize a cube that is held with one of the four body diagonals aligned to the vertical axis. Rotate the cube about this axis such that its view remains unchanged. The magnitude of the minimum angle of rotation is
(a) 120°
(b) 60°
(c) 90°
(d) 180° [2024 : 2 M]

■■■

Answers Numerical Ability

1.1	(d)	1.2	(c)	1.3	(d)	1.4	(b)	1.5	(b)	1.6	(a)	1.7	(b)
1.8	(c)	1.9	(d)	1.10	(c)	1.11	(d)	1.12	(d)	1.13	(c)	1.14	(a)
1.15	(a)	1.16	(c)	1.17	(c)	1.18	(d)	1.19	(b)	1.20	(a)	1.21	(25)
1.22	(a)	1.23	(a)	1.24	(d)	1.25	(c)	1.26	(0.48)	1.27	(b)	1.28	(b)
1.29	(280)	1.30	(c)	1.31	(b)	1.32	(c)	1.33	(b)	1.34	(c)	1.35	(a)
1.36	(d)	1.37	(c)	1.38	(b)	1.39	(a)	1.40	(b)	1.41	(c)	1.42	(b)
1.43	(d)	1.44	(d)	1.45	(c)	1.46	(c)	1.47	(b)	1.48	(d)	1.49	(c)
1.50	(a)	1.51	(a)	1.52	(c)	1.53	(d)	1.54	(c)	1.55	(c)	1.56	(c)
1.57	(a)	1.58	(b)	1.59	(c)	1.60	(c)	1.61	(c)	1.62	(a)	1.63	(c)
16.4	(b)	1.65	(c)	1.66	(d)	1.67	(d)	1.68	(b)	1.69	(a)	1.70	(c)
1.71	(a)	1.72	(d)	1.73	(c)	1.74	(c)	1.75	(c)	1.76	(d)	1.77	(a)
1.78	(c)	1.79	(c)	1.80	(d)	1.81	(c)	1.82	(a)	1.83	(a)	1.84	(b)
1.85	(d)	1.86	(a)	1.87	(b)	1.88	(b)	1.89	(c)	1.90	(a)	1.91	(a)

Explanations Numerical Ability

1.1 (d)

Using the set theory formula

$n(A)$: Number of people who play hockey = 15

$n(B)$: Number of people who play football = 17

$n(A \cap B)$: Persons who play both hockey and football = 10

$n(A \cup B)$: Persons who play either hockey or football or both

Using the formula

$$n(A \cup B) = n(A) + n(B) - n(A \cap B)$$
$$n(A \cup B) = 15 + 17 - 10 = 22$$

Thus people who play neither hockey nor football = 25 – 22 = 3

Alternative method:

Refer to Venn diagram given below:

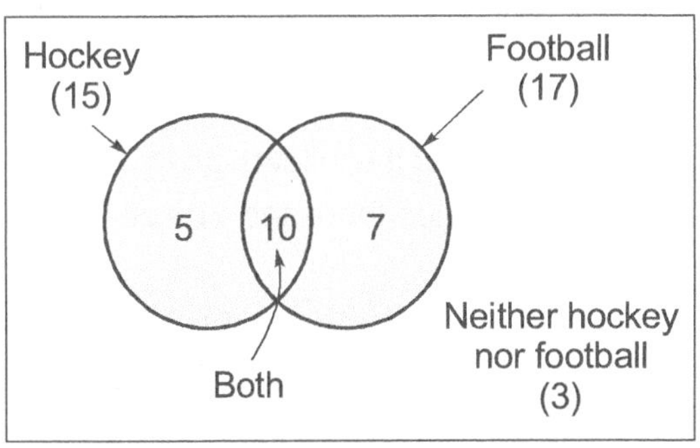

1.2 (c)

7 and 6 added is becoming five means the given two numbers are added on base 8.

$$\begin{array}{r} (137)_8 \\ +(276)_8 \\ \hline (435)_8 \end{array}$$

Hence we have to add the another two given set of numbers also on base 8.

$$\begin{array}{r} (731)_8 \\ +(672)_8 \\ \hline (1623)_8 \end{array}$$

Hence the overall problem was based on identifying base which was 8 and adding number on base 8.

1.3 (d)

Per day work or rate of 5 skilled workers = $\frac{1}{20}$

⇒Per day work or rate of one skill worker

$$= \frac{1}{5 \times 20} = \frac{1}{100}$$

Similarly Per day work or rate of 8 semiskilled workers = $\frac{1}{25}$

⇒Per day work or rate of one semi-skill worker

$$= \frac{1}{8 \times 25} = \frac{1}{200}$$

And per day work or rate of 10 unskilled workers

$$= \frac{1}{30}$$

⇒ Per day work or rate of one semi-skill worker

$$= \frac{1}{10 \times 30} = \frac{1}{300}$$

Thus total per day work of 2 skilled, 6 semiskilled and 5 unskilled workers

$$= \frac{2}{100} + \frac{6}{200} + \frac{5}{300} = \frac{12 + 18 + 10}{600}$$
$$= \frac{40}{600} = \frac{1}{15}$$

Thus time to complete the work is 15 days.

Alternative method:

Let one day work of skilled semi-skilled and unskilled worker be a, b, c units respectively.

$5a \times 20 = 8b + 25 = 10c \times 30$ = Total unit of work

$$100a = 200b = 300c$$
$$a = 2b = 3c$$
$$\Rightarrow \quad b = \frac{a}{2} \text{ and } c = \frac{a}{3}$$

Given that 2 skilled, 6 semi-skilled and 5 unskilled workers are working. Let they finish the work in 'x' days.

$(2a + 6b + 5c)x = 5a + 20$ = Total units of work

$$\left(2a + 3a + \frac{5}{3}a\right)x = 5a \times 20$$
$$\frac{20a}{3}x = 5a \times 20$$
$$x = 15 \text{ days}$$

1.4 (b)

We have to make 4 digit numbers, so the number should be start with 3 or 4, two cases possible;

Case (1) thousands digit is 3

Now other three digits may be any of 2, 2, 3, 3, 4, 4, 4, 4.

(a) Using 2, 2, 3
⇒223, 232, 322 -------

$\left(\frac{3!}{2!} = 3 \text{ numbers are possible}\right)$

(b) Using 2, 2, 4 ⇒ 224, 242, 422

$\left(\frac{3!}{2!} = 3 \text{ numbers are possible}\right)$

(c) Using 2, 3, 3 ⇒ 233, 323, 332

$\left(\frac{3!}{2!} = 3 \text{ numbers are possible}\right)$

(d) Using 2, 3, 4 ⇒ 234, 243, 324, 342, 423, 432
(3! = 6 numbers are possible)

(e) Using 2, 4, 4 ⇒ 244, 424, 442

$\left(\frac{3!}{2!} = 3 \text{ numbers are possible}\right)$

(f) Using 3, 3, 4 ⇒ 334, 343, 433

$\left(\frac{3!}{2!} = 3 \text{ numbers are possible}\right)$

(g) Using 3, 4, 4 ⇒ 344, 434, 443

$\left(\frac{3!}{2!} = 3 \text{ numbers are possible}\right)$

(h) Using 4, 4, 4 ⇒ 444

$\left(\frac{3!}{3!} = 1 \text{ numbers are possible}\right)$

Total 4 digit numbers in case 1 = 3 + 3 + 3 + 6 + 3 + 3 + 3 + 1 = 25

Case (2) thousands digit is 4 ; Now other three digits may be any of 2, 2, 3, 3, 3, 4, 4, 4.

(a) Using 2, 2, 3 ⇒ 223, 232, 322

$\left(\frac{3!}{2!} = 3 \text{ numbers are possible}\right)$

(b) Using 2, 2, 4 ⇒ 224, 242, 422

$\left(\frac{3!}{2!} = 3 \text{ numbers are possible}\right)$

(c) Using 2, 3, 3 ⇒ 233, 323, 332

$\left(\frac{3!}{2!} = 3 \text{ numbers are possible}\right)$

(d) Using 2, 3, 4 ⇒ 234, 243, 324, 342,
423, 432

$\left(\frac{3!}{2!} = 3 \text{ numbers are possible}\right)$

(e) Using 2, 4, 4 ⇒ 244, 424, 442

$\left(\frac{3!}{2!} = 3 \text{ numbers are possible}\right)$

(f) Using 3, 3, 3 ⇒ 333

$\left(\frac{3!}{3!} = 1 \text{ numbers are possible}\right)$

(g) Using 3, 3, 4 ⇒ 334, 343, 433

$\left(\frac{3!}{2!} = 3 \text{ numbers are possible}\right)$

(h) Using 3, 4, 4 ⇒ 344, 434, 443

$\left(\frac{3!}{2!} = 3 \text{ numbers are possible}\right)$

(i) Using 4, 4, 4 ⇒ 444

$\left(\frac{3!}{3!} = 1 \text{ numbers are possible}\right)$

Total 4 digit numbers in case 2 = 3 + 3 + 3 + 6 + 3 + 3 = 1 + 3 + 1 = 26

Thus total 4 digits numbers using case (1) and case (2)
= 25 + 26 = 51

Alternative Method / Shortcut method

As the number is greater than 3000. So thousand's place can be their 3 or 4.

Let's consider the following two cases

Case (I) When thousand's place is 3.

<u>3</u> <u>a</u> <u>b</u> <u>c</u>

If there is no restriction on number of two's, three's and four's. Then each of *a*, *b*, *c* can be filled with 2 or 3 or 4 each in 3 ways.

So 3 × 3 × 3 = 27 numbers are there. Out of which 3222, 3333 are invalid as 2 can be used twice & three thrice only so number of such valid numbers beginning with 3 are 27 – 2 = 25.

Case (II) When thousand's place is 4

<u>4</u> <u>a</u> <u>b</u> <u>c</u>

Without restriction on number of 2's, 3's and 4's *a*, *b*, *c* (as explained in case I) can be filled in 27 ways.

Out of these 27 numbers, 4 2 2 2 is only invalid as two have to be used twice only.

So valid numbers are 27 – 1 = 26.

Total numbers from Case (I) & Case (II)
25 + 26 = 51.

1.5 (b)

Suppose: Hari's age : *H*, Gita's age : *G*,
Saira's age : *S*, Irfan's age : *I*

- $H + G > I + S$
- Using Statement (2) both $G - S = 1$ or

$S - G = 1$; G can't be oldest and S can't be youngest.

- There are no twins thus using statement (2) either *GS* or *SG* possible.

(A) *HSIG*: not possible as there is I between S and G which is not possible using statement (3)

(B) *SGHI*: *SG* order is possible, $S > G > H > I$ and $G + H > S + I$ (possible)
Because if { $S = G + 1$; and $G = H + 1$ and $H = I + 2$ then $G + (I + 2) > (G + 1) + I$ }

(C) *IGSH*: according to this $I > G$ and $S > H$ thus adding these both inequalities we get $I + S > G + H$ which is opposite of statement (2) thus not possible.

(D) *IHSG*: according to this $I > H$ and $S > G$ thus adding both inequalities $I + S > H + G$ which is opposite of statement (2) thus not possible.

1.6 (a)

Let there be overall 'x' candidates. Initially those who decided to vote for *P* and *Q* are $0.4x$ and $0.6x$ respectively. On the day of election 15% of $0.4x = 0.06x$ went from *P*'s side to *Q*'s side and 25% of $0.6x = 0.15x$ went from *Q*'s side to *P*'s side.

Now, after transfer '*p*' has $0.4x - 0.06x + 0.15x = 0.49x$

and after transfer '*q*' has $0.6x - 0.15x + 0.06x = 0.51x$

Given, in the question that *P* lost by 2 votes

$$Q - P = 2 \text{ votes}$$

$$0.51x - 0.49 = 0.02x = 2 \text{ votes}$$

Hence, ($x = 100$)

Total number of votes are 100.

Alternate Solution:

Using options:

Lets take option (a) 100 as total number of voters. Initially *P* and *Q* has 40 and 60 voters on their sides respectively.

15% went from *P*' s side to *Q* and 25% went from *Q*'s side to *P*.

P	*Q*
(40)	(60)
(–) 15% of 40	– 25% of (60)
+ 25% of 60	+ 15% of (40)
40 – 6 + 15	60 – 15 + 6
= 49	= 51

Hence *Q* and *P* has 51 and 49 respectively difference in number of votes = 2. Hence answer (a).

1.7 (b)

Fuel consumed per km will be least when mileage (kilometres per litres) mentioned on '*y*' axis of graph will be maximum irrespective of number of kilometres travelled.

From the graph ('*y*' axis) we can observe that mileage (kilometres per litres) is maximum when vehicle is driven at 45 kilometres per hour. Hence the stretch which '*Q*' covered at 45 kmph, mileage was highest and fuel consumption per kilometre was lowest.

1.8 (c)

Shortcut method:

The question can be done orally as *R* took 1/3rd of toffees initially. Hence total number of toffees have to be multiple of 3. There is only one option (c) as 48 which is multiple of 3.

Alternate Solution:

Using options:

If we use options for option (c), if total number of toffees are 48. R took 1/3rd if 48 and returned 4. So R has 16 – 4 = 12 toffees. Out of left over 36 toffees S took 1/4th of 36 and returned 3. So, S has 9 – 3 = 6 and remaining 30 will be left in bowl. Now T took 1/2 of 30 and returned two back. So, the bowl has 17 left.

*Always solve such questions by options.

1.9 (d)

$$f(y) = \frac{|y|}{y}$$

As we know,

$$|y| = \begin{cases} y & y \geq 0 \\ -y & y < 0 \end{cases} \quad \ldots(i)$$

So from (i),

$$f(y) = \frac{y}{y} = 1 \quad \text{If } y \geq 0$$

$$f(y) = \frac{-y}{y} = -1 \quad \text{If } y < 0$$

Hence, $|f(q) - f(-q)| = |1 - (-1)| = 2$

Alternate Solution:

Otherwise if $f(y) = \dfrac{|y|}{y}$

$$f(1) = \frac{|1|}{1} = 1$$

$$f(-1) = \frac{|-1|}{-1} = -1$$

For calculating $|f(q)-f(-q)|$

Let, $q = 1$, $|f(q)-f(-q)|$

$|f(1)-f(-1)| = 1-(-1) = 2$

1.10 (c)

4 + 44 + 444..............

If we see first term of series = 4

Hence, sum upto 1st term is also = 4

Put, $n = 1$ (first term) only option (c) satisfies.

$$\frac{4}{81}\left[10^{n+1}-9n-10\right] \in = \frac{4}{81}\left[10^{1+1}-9\times 1-10\right]$$

$$= \frac{4}{81}\times 81 = 4$$

Hence, option (c) is correct answer. Always solve these questions by putting values.

Alternative Solution:

4 + 44 + 444......................

$$\frac{4}{9}[9+99+999.............]$$

$$= \frac{4}{9}\left[(10-1)+(10^2-1)+(10^3-1).......\right]$$

$$= \frac{4}{9}\left[(10+10^2+10^3....10^n)-(1+1+1.....\right]$$

$$= \frac{4}{81}\left[10^{n+1}-9n-10\right]$$

1.11 (d)

$(1.001)^{1259} = 3.52$ and $(1.001)^{2062} = 7.85$

$$(1.001)^{1259}\times(1.001)^{2062} = (1.001)^{3321}$$

$$\downarrow \qquad \downarrow$$

$$3.52 \times 7.85 = (1.001)^{3321} \quad (\text{as } a^m \times a^n = a^{m+n})$$

Hence $(1.001)^{3321} = 27.632 \approx 27.64$

1.12 (d)

Category	Amount
Food	4000
Clothing	1200
Rent	2000
Savings	1500
Others	1800
Total	10500

So total expenses are 10,500.

Expenses excluding savings are 9,000.

Hence required percentage

$$= \frac{9,000}{10,500}\times 100 = 85.714\% \approx 86\%$$

1.13 (c)

Probability that A and B will meet will be given by the graphical representation where shaded region represents favorable area.

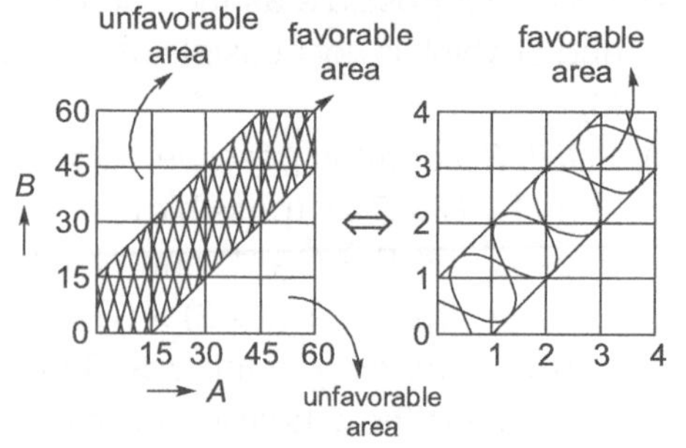

$$\text{Required probability} = \frac{\text{fabourable area}}{\text{total area}} = \frac{7}{16}$$

1.14 (a)

Let the number of Rs.20 notes be 'a' and Rs.10 notes be 'b' respectively.

So, $a + b = 14$

$20a + 10b = 230$

on solving $a = 9, b = 5$

Hence number of Rs.10 notes are 5.

Alternative Solution:

If we use option (a)

Number of ten Rs. notes are 5.

Rs.10 × 5 = Rs.50

The remaining nine notes have to be 20 Rs. notes

Rs. 20 × 9 = 180

and hence 50 + 180 = 230 (total money value)

Hence option (a) number of ten Rs. notes are 5.

1.15 (a)

Divide 8 bags into three parts.

3, 3 and 2 respectively. Now two cases arise:

Case (i):

Firstly, if we compare the 3, 3 bags on pans of a balance. We can identify on which side is the lighter bag placed. And then we will need only one more weighing for identifying faulty bag. So only two weighings are required.

Case (ii):

Secondly, if the 3, 3 compared bags are equal, then to be sure faulty bag is one among the remaining two bags. So one more weighing is required to find out faulty bag from these two bags. Hence overall two weighings are required. Thus in either cases only two weighings are required.

Alternative Solution:

In case of a weighing balance (i.e. beam balance) the following model can be observed.

$1-3 \rightarrow$ 1 weighing required

$4-9 \rightarrow$ 2 weighing required

$10-27 \rightarrow$ 3 weighing required

$28-81 \rightarrow$ 4 weighing required

and the process will go so on.

As there are eight objects.

Minimum number of weighings required will be only 2.

1.16 (c)

$$\frac{M + Tue + W}{3} = 41$$

$$\Rightarrow \quad M + Tue + W = 123 \quad \text{...(i)}$$

$$\frac{Tue + W + Th}{3} = 43$$

$$\Rightarrow \quad Tue + W + Th = 129 \quad \text{...(ii)}$$

From equation (ii) and equation (i),

$$Th - M = 6$$

Also $\quad Th = 1.15\,M$

$$1.15\,M - M = 6$$

$$\Rightarrow \quad 0.15\,M = 6$$

$$\Rightarrow \quad M = 40$$

$$Th = 1.15\,m$$

$$\Rightarrow \quad Th = 1.15 \times 40$$

$$\Rightarrow \quad Th = 46$$

1.17 (c)

$$\text{Average speed} = \frac{\text{Total distance}}{\text{Total time}}$$

$$\frac{\text{Total distance}}{\text{Total time}} = \frac{(8+6+16)\text{km}}{\left(\frac{1}{4}+\frac{1}{4}+\frac{1}{4}\right)\text{hr}}$$

$$\frac{30\text{km}}{\frac{3}{4}\text{hr}} = 40 \text{ km/hr}$$

1.18 (d)

$S = 10 + 84 + 73 + \ldots$

$S = (9 + 1) + (9^2 + 3) + (9^3 + 5) \ldots$

$S = \{9 + 9^2 + 9^3 + \ldots 9^n\} + \{1 + 3 + 5 \ldots (2n-1)\}$

$$S = \frac{9(9^n - 1)}{(9-1)} + n^2$$

$$\left(\begin{array}{l}\text{As } a + ar + ar^2 \text{ .... } ar^{n-1} \\ = \dfrac{a(r^n - 1)}{(r-1)} \quad \text{for} \quad r > 1\end{array}\right)$$

$$\left(\begin{array}{l}\text{and } 1 + 3 + 5 \text{ .... } (2n-1) = n^2 \\ \text{Sum of odd natural numbers } = n^2\end{array}\right)$$

So, $\quad S = \dfrac{9(9^n - 1)}{8} + n^2$

Alternate Method:

Put $n = 2$ in all options and check whether sum = 10 + 84 = 94 is in which option.

1.19 (b)

$3x^2 + 2x + p(p-1) = 0$

If roots are of opposite signs product of the roots will be negative.

$$\frac{p(p-1)}{3} < 0$$

$$p(p-1) < 0$$

$$0 < p < 1$$

$$P \in (0, 1)$$

1.20 (a)

Every leap year will have 366 days.

1 leap year = 366 days

= 52 × 7 + 2 days

Every leap year will have 52 complete weeks. For 53rd Saturday there are two favourable cases from 7 possible cases

Fri	Sat
Sat	Sun
Sun	Mon
Mon	Tue
Tue	Wed
Wed	Thu
Thu	Fri

(Fri–Sat, Sat–Sun) → 2 favourable cases

$$\text{Required probability} = \frac{2}{7}$$

1.21 (25)

The average time between successive occurrence of such earthquakes is

$$= \frac{1}{0.04} = 25 \text{ years}$$

1.22 (a)

$$P_{new} = P_{old}\left(1+\frac{r}{100}\right)^n$$

$$10 = 5\left(1+\frac{20}{100}\right)^n$$

$$2 = (1+0.2)^n$$

$$2 = 1.2^n$$

taking log both sides, we get

$$ln\,2 = n\,ln(1.2)$$

$$\therefore \quad n = \frac{ln2}{ln1.2} = 3.8 \text{ years}$$

1.23 (a)

↑ Riaz / Som   ↑ Shiv / Ansu

Ansu is youngest,

thus, ↑ Riaz, Som, Ansu

Only statement 1 (i.e. Shiv is younger to Riaz) justifies that Shiv is younger and not eldest, so clearly form previous statements its is clear that Riaz is the eldest.

1.24 (d)

For item 1

$$\frac{11}{100}\times 5\times 10^8 \longrightarrow \frac{12}{100}\times 250$$

$\therefore$ Revenue per kg $= 54.54 \times 10^{-3}$

For item 4

$$\frac{22}{100}\times 5\times 10^8 \longrightarrow \frac{6}{100}\times 250$$

$\therefore$ Revenue per kg $= 13.63 \times 10^{-8}$

$$\text{Ratio of } \frac{\text{Revenue by item 1/kg}}{\text{Revenue by item 2/kg}} = \frac{54.54}{13.63}$$

$$= 4 : 1$$

1.25 (c)

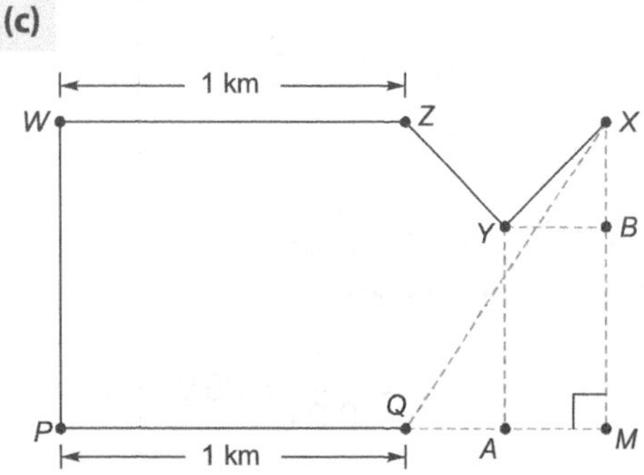

$$\angle XYB = 45^\circ$$

$$\frac{XB}{XY} = 1 \times \sin 45^\circ = \frac{1}{\sqrt{2}}$$

$$\tan\angle XYB = \frac{XB}{YB}$$

$$\frac{\frac{1}{\sqrt{2}}}{YB} = 1, \quad \therefore\ YB = \frac{1}{\sqrt{2}}$$

$$QM = 2YB = \frac{2}{\sqrt{2}} = \sqrt{2}\text{ km}$$

$$XM = WB = 1\text{ km}$$

In right angled Δ, XQM

$$(XQ)^2 = (XM)^2 + (QM)^2 = 1^2 + (\sqrt{2})^2 = 3$$

$$\therefore \quad XQ = \sqrt{3}\text{ km}$$

1.26 0.48 (0.48 to 0.49)

10% H^+ → 95% correct; → 5% incorrect

90% H^+ → 89% correct; → 11% incorrect

Probability that the individual is actually positive is

$$= \frac{0.1\times 0.95}{(0.1\times 0.95)+(0.9\times 0.11)} = 0.4896$$

1.27 (b)

Total number of matches $^5C_2 = 10$.

1.28 (b)

$T > E$, $C > T$, $\Rightarrow$ $C > T > E$. So Eric is actually order older than Cliff so third statement is false.

1.29 (280)

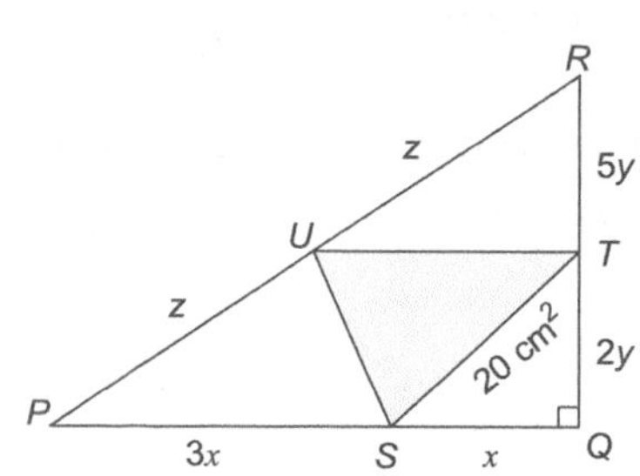

$$\text{Area}\,\Delta QTS = \frac{1}{2}SQ + QT\sin 90^\circ$$

$$20 = \frac{1}{2}\times x \times 2y \times 1$$

$$\Rightarrow \quad xy = 20$$

$$\text{Now Area for }\Delta PQR = \frac{1}{2}PQ\times QR\times \sin 90^\circ$$

$$= \frac{1}{2}\times 4x\times 7y\times \sin 90^\circ = 14\,xy = 280\text{ cm}^2$$

1.30 (c)

S1: 

S2: 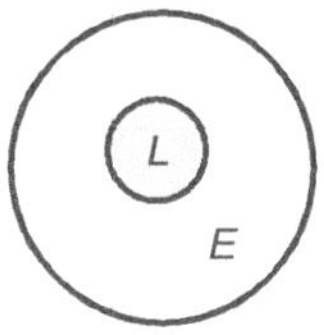

Therefore concluding diagram can be

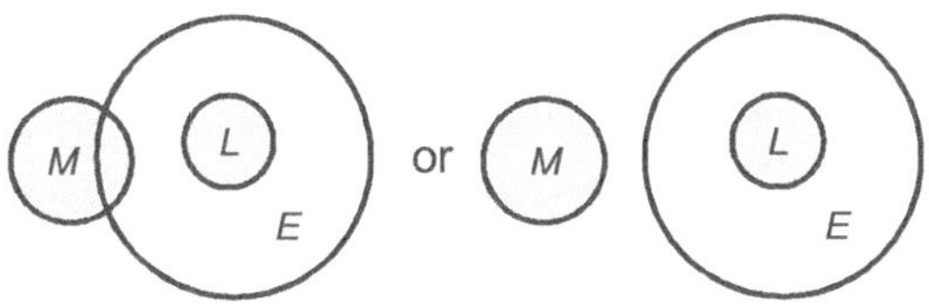

Fig. (a) Fig. (b)

1. Some manager can be executive.
2. Some executive can be manager.

1.32 (c)

PR is parallel to x axis

$-4 \le x \le 5$

$6 \le y \le 16$

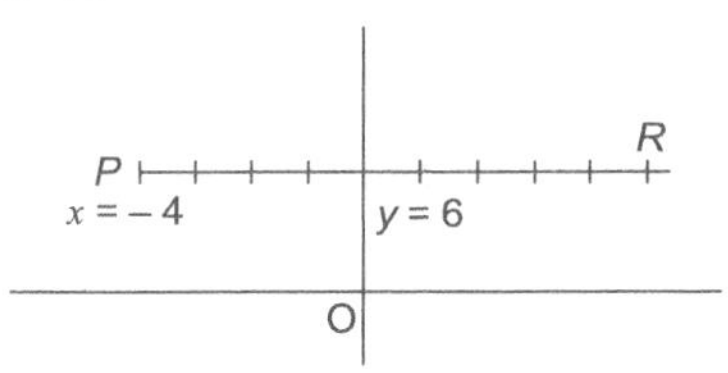

P at $x = -4$

Δ's = with base 1 unit *y* can take values (7, 8, ..., 16) : 10 values

Δ's = with base 2 unit *y* varies (7.... 16) : 10 values

Δ's = with base 9 unit *y* varies (7.... 16) : 10 values

Similarly *P* at $x = -3$ we will get 80 and 50 on

So [90 + 80 + ... 10] = 9[1 + 2 .... 10]

$$= 10\sum_{n=1}^{9} n$$

Similarly changing ordinate of *P* we will respectively get

$$[81 + 725 \ldots. 9]\ \Delta's = 9\ [1 + \ldots.9] = 9\sum_{n=1}^{9} n$$

and further (64 + 56 .... 8) $\Delta's = 8\sum_{n=1}^{8} n$ and so on

If we sum all of the above

$$10\sum_{n=1}^{9} n + 9\sum_{n=1}^{9} n + 8\sum_{n=1}^{8} n.\Sigma n.......$$

$$9\sum_{n=1}^{9} n[1+2+3.....10] = \sum_{n=1}^{9} n \times \sum_{n=1}^{10} n$$

But there are overall four orientation for rt. ____ as shown , so actual answer will be $4 \times \sum_{n=1}^{9} n \times \sum_{n=1}^{10} n$

$$4\times\left(\frac{9\times 10}{2}\right)\times\left(\frac{10\times 11}{2}\right) = 9900.$$

Alternate Method :

$-4 \le x \le 5$, $6 \le y \le 16$

PQ ll y axis and *PR* ll x axis $\angle QPR$ is 90°

As all co-ordinates of *P*,*Q*,*R*, are integers so, we can choose (x, y) integer co-ordinates for vertex *P* in 10 × 11 = 110 ways.

Now with respect to fixed y co-ordinate for *R* we can choose its x-coordinate in only 9 ways as one point is already occupied (two points of cannot coincide) overall 9 ways of doing it.

Now choosing *y*-coordinate for *Q* because x-coordinate of *Q* is already fixed (it has same x-coordinate as *R*) it can take 10 different values. So, overall 110 × 9 × 10 = 9900 ways 9900 such Δ's will be possible.

1.34 (c)

97 is not a perfect square.

1.36 (d)

The odometer reading increases from starting point to end point

Magnitude of area of the given diagram = Odometer reading

Magnitude of area of the velocity and time graph per second

1st sec $\Rightarrow$ triangle $= \frac{1}{2}\times 1\times 1 = \frac{1}{2}$

2nd sec $\Rightarrow$ square = 1 × 1 = 1

3rd sec $\Rightarrow$ square + triangle

$$= 1\times 1 + \frac{1}{2}\times 1\times 1 = 1\frac{1}{2}$$

4th sec $\Rightarrow$ triangle $= \frac{1}{2}\times 1\times 2 = 1$

5th sec $\Rightarrow$ straight line = 0

6th sec $\Rightarrow$ triangle $= \frac{1}{2}\times 1\times 1 = \frac{1}{2}$

7th sec $\Rightarrow$ triangle $= \frac{1}{2}\times 1\times 1 = \frac{1}{2}$

Total Odometer reading at 7 seconds

$$= \frac{1}{2} + 1 + 1\frac{1}{2} + 1 + 0 + \frac{1}{2} + \frac{1}{2} = 5$$

1.37 (c)

Manu's age = Sravan's age + 2 months
Manu's age = Trideep's age – 3 months
Pavan's age = Sravan's age + 1 month
From this Trideep's age > Manu > Pavan > Sravan
∴ Trideep can occupy the extra space in the flat.

1.38 (b)

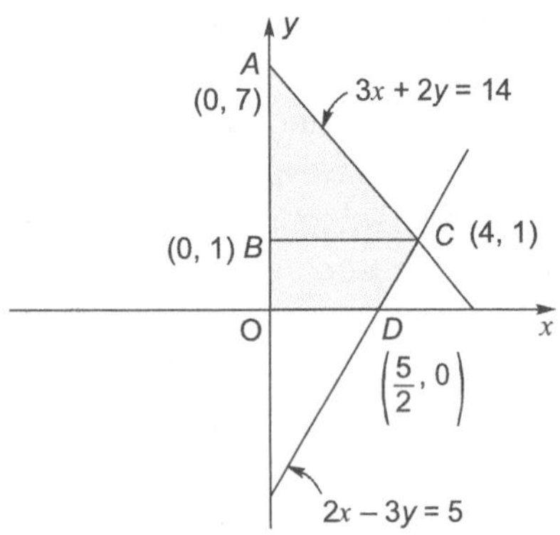

Required area = Area of ΔABC + Area of trapezoid $BCDO$

$$= \frac{1}{2}\times 4\times 6+\frac{1}{2}\times(4+2.5)\times 1$$
$$= 12 + 3.25 = 15.25 \text{ sq. units}$$

1.39 (a)

Straight line equation, $y = mx + c$

m = slope = –0.02 $(ln\ x,\ y)$

If $ln\ x = X$, then set $\begin{matrix}(x, & y)\\ \downarrow & \downarrow \\ 0.1 & 0\end{matrix}$

$$y = mX + C$$
$$0 = -0.02 \times 0.1 + C$$
$$\therefore \quad C = 0.002$$
$$y = mX + C$$
$$y = -0.02 \times \ln x + C$$

At $x = 5$

$$y = -0.02 \times \ln 5 + 0.002 = -0.030$$

1.40 (b)

Sample space = 6 × 6 = 36 events = Total chances

Product of numbers on 2 dice have to perfect square = Favourable chances

(1, 1) (2, 2) (3, 3) (4, 4) (5, 5) (6, 6) (1, 4) (4, 1) ⇒ 8 events

$$\text{Probability} = \frac{\text{Favourable chances}}{\text{Total chances}} = \frac{8}{36} = \frac{2}{9}$$

1.41 (c)

In order to verify proposition we have to turn to card 2 and blue from given four cards as proposition says it has even an one side opposite is red. Vice-versa might or might not be true so, ans. (c) as all other options are eliminated.

1.42 (b)

$$\text{Put, } (x = -1) \quad \frac{81\left[\frac{16}{25}\right]^{1}}{\left[\frac{3}{5}\right]^{2}} = \frac{\frac{3^4\times 2^4}{5^2}}{\frac{3^2}{5^2}} = 144$$

Hence option (b) is correct.

1.43 (d)

(i) is incorrect as it has move directly.
(ii) is incorrect as it stayed for maximum duration on ground floor.

1.44 (d)

Let cycling speed = C
and walking speed = W

$$C\left(\frac{1}{2}\right)+W\left(\frac{3}{2}\right) = 19 \qquad ...(i)$$
$$C + W = 26 \qquad ...(ii)$$

On solving equation (i) and (ii) we get,

$$W = 6 \text{ km/hr}$$

1.45 (c)

Following information is given
$K < G$, $P < G$, $S < P$ and $G < D$
From these it can be inferred
$S < P < G < D$ and also $(K < G)$
So, $K < D$ as well but no comparison between K & S or K & P is given
So (ii) and (iv) follows.

1.46 (c)

Following two possibilities can be drawn

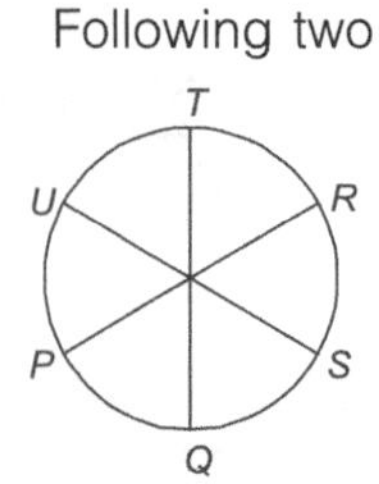

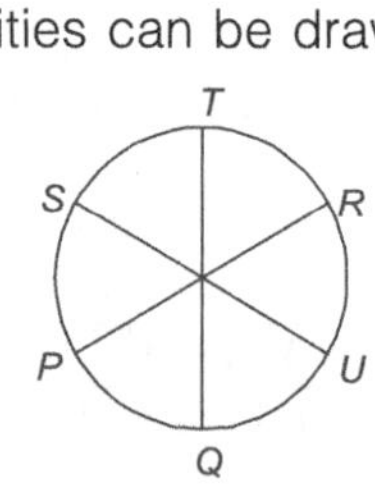

Given in question P & U interchange then new diagram can be drawn.

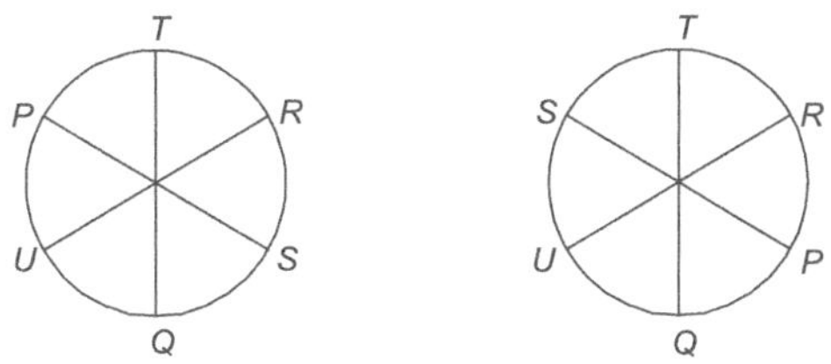

Now option (c) only follows.

1.47 (b)

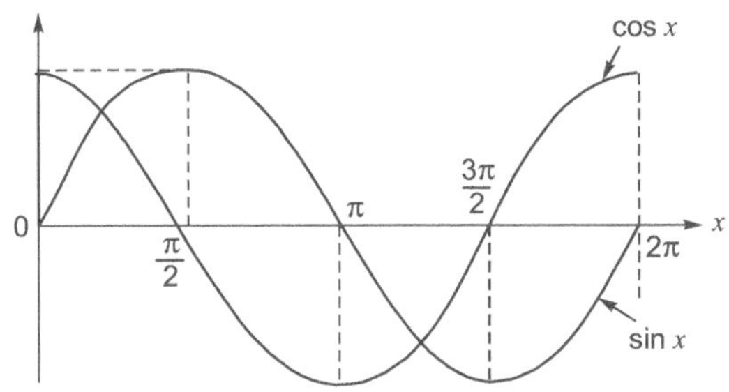

From the curve it is clear that sin$x$ and cos$x$ both are decreasing in the interval $\left(\frac{\pi}{2}, \pi\right)$.

1.48 (d)

(i) Cuboid volume = $8 \times 10 \times 6 = 480$ cm^3

(ii) Cube volume = $8 \times 8 \times 8 = 512$ cm^3

(iii) Cylinder volume = $\pi r^2 h$

$$= \frac{22}{7} \times 7 \times 7 \times 7 = 1078 \text{ cm}^3$$

(iv) Sphere volume = $\frac{4}{3}\pi r^3 = \frac{4}{3}\pi \times (7)^3$

$$= 1436.75 \text{ cm}^3$$

Hence, the descending order of their volumes is (iv), (iii), (ii), (i).

1.49 (c)

Area of equilateral triangle = $\sqrt{3}$

$$\frac{\sqrt{3}}{4}a^2 = \sqrt{3}$$

$$a^2 = 4 \quad \Rightarrow a = 2$$

$$\text{perimeter} = 3a = 3 \times 2 = 6$$

1.50 (a)

Total distance = x

(Onward journey) S_1 = 60 km/h

(Return journey) S_2 = 90 km/h

$$\text{Average speed} = \frac{\text{Total distance}}{\text{Total time}}$$

$$= \frac{x + x}{\frac{x}{60} + \frac{x}{90}} = \frac{2x}{x\left[\frac{3+2}{90\times 2}\right]} = \frac{2\times 90\times 2}{5} = 72 \text{ km/h}$$

1.52 (c)

$$x^2 + x = 1$$

$$x(x+1) = 1$$

$$x + 1 = \frac{1}{x}$$

$$\left(x - \frac{1}{x}\right) = (-1)$$

Squaring above equation,

$$x^2 + \frac{1}{x^2} - 2 = (+1)$$

$$x^2 + \frac{1}{x^2} = 3$$

Squaring above equation,

$$x^4 + \frac{1}{x^4} + 2 = 9$$

$$x^4 + \frac{1}{x^4} = 7$$

1.53 (d)

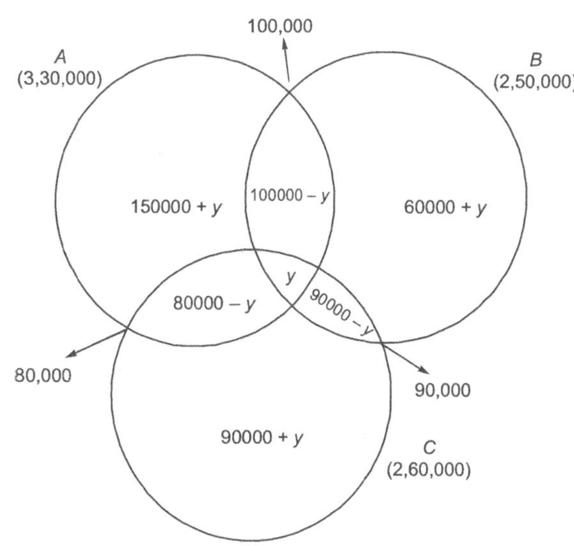

$$6,30,000 = 2y + 1,50,000 + 100000 + 80000 + 60000 + 90000 + 90000$$

$$6,30,000 - 5,70,000 = 2y$$

$$y = 30000$$

Students who failed to clear the test

$$= 150000 + 60000 + 90000 + 4y$$

$$= 300000 + 4 \times 30000$$

Students who failed to clear the test = 420000

1.54 (c)

Number of parallelogram

$$= {}^4C_2 \times {}^5C_2 = 6 \times 10 = 60$$

1.55 (c)

According to given data the only possible arrangement is 10 5 4 7 2

So, second from right will be 7.

1.56 (c)

We know formula for volume of a (right circular) cone is $\frac{1}{3}\pi r^2 h$

Original volume, $V_0 = \frac{1}{3}\pi r_1^2 h_1$ ... (i)

Now we know radius and height both are increased by 10%. So after increase the new volume is

$$V_n = \frac{1}{3}\pi(1.1r_1)^2(1.1)h_1$$

$$= 1.331\left(\frac{1}{3}\pi r_1^2 h_1\right) = 1.331\ (V_0)$$

% change in volume

$$= \frac{V_n - V_0}{V_0}\times 100$$

$$= \frac{1.331V_0 - V_0}{V_0}\times 100 = 33.1\%$$

1.57 (a)

Total population= 1400 million

According to the question 70% of mobile phone users data is an unnecessary data. Its not necessary to calculate tabing the relevant data into consideration,

Number of internet users = 294 million

Half of which (internet) use e-commerce portal

$$\frac{294}{2} = 147$$

Percentage of these (e-commerce portals) in the country (out of 1400 million)

$$= \frac{147}{1400}\times 100 = 10.5\%$$

1.58 (b)

According to the concept of relative speed in opposite direction speeds should be added

$$\text{Time of activity} = \frac{\text{Sum of distance}}{\text{Sum of speeds}}$$

$$= \frac{540}{100+80} = \frac{540}{180}$$

= 3 hours from 7 AM = 10 AM

1.59 (c)

Given: Code of

$P = \alpha\alpha$; Code of $Q = \alpha\beta$

Then, Code of

$R = \beta\alpha$

and Code of $S = \beta\beta$

1.60 (c)

Average percentage of passing students

$$= \frac{\dfrac{\text{Total no. of passed students}}{\text{Total no. of students}}}{4}\times 100$$

$$= \frac{\frac{280}{500}+\frac{330}{600}+\frac{455}{700}+\frac{240}{400}}{4}\times 100$$

$$= \frac{0.56+0.55+0.65+0.60}{4}\times 100 = 59\%$$

1.61 (c)

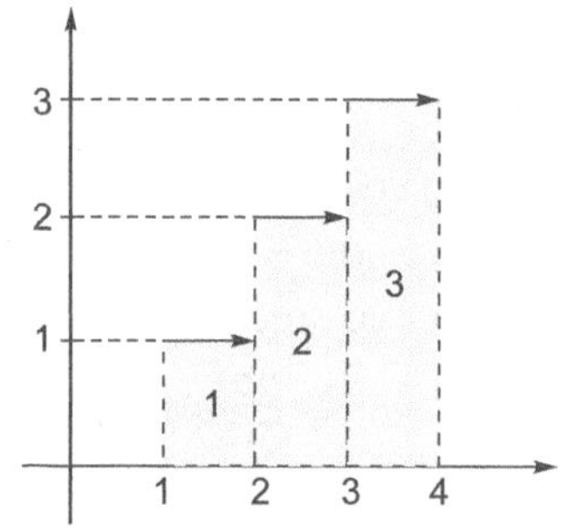

Area = 1 × 1 + 1 × 2 + 1 × 3

= 1 + 2 + 3 = 6

1.62 (a)

Put n = 1 in all options.

$$\frac{80}{81}\left[10^1-1\right]-\frac{8}{9}[1] = \frac{80}{9}-\frac{8}{9} = \frac{72}{9}=8$$

1.63 (c)

$y = x^m$ and $y = x^{1/m}$

for $m = 2$ in $0 \le x \le 1$

$y = x^2$ and $y = \sqrt{x}$

∴ graphs are shown as

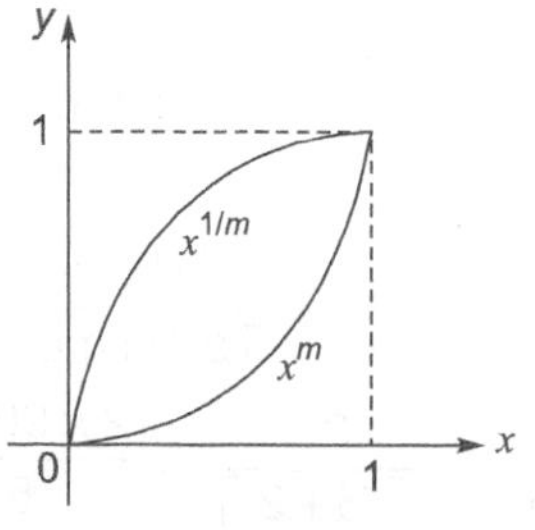

16.4 (b)

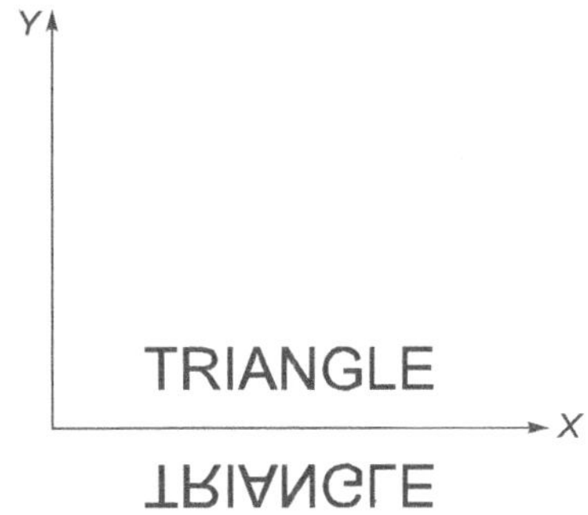

1.65 (c)

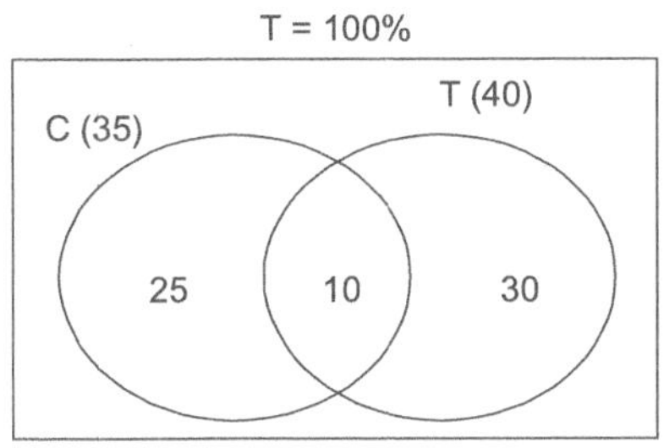

Percentage of employees who drink neither tea nor coffee = 100 – 25 – 10 – 35 = 35

1.66 (d)

Following cases can be

PQSR, RQSP, QPSR, QRSP, RQPS and PQRS

and $\begin{bmatrix} P & R \\ R & P \end{bmatrix}$ not possible.

1.68 (b)

$$\frac{\lambda(-(-3+2),(-2+3))}{(-(-2+1))} = \lambda\frac{(1,1)}{1} = \lambda(1,1)$$

So, 1st definition will be applicable as $p = q$.

Hence, $\lambda(1, 1) = (1-1)^2 = 0$

1.69 (a)

$$x \oplus y = 2 \odot 2$$

$$\frac{x^2+y^2}{xy} = \frac{2^2}{2}$$

$$x^2 + y^2 = 2xy$$

$$(x - y)^2 = 0$$

$$x = y$$

1.70 (c)

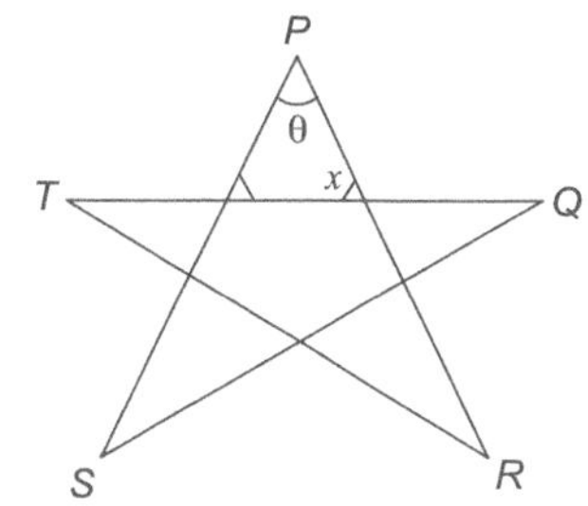

Sum of angle formed at the pentagon = 540°

Each angle of pentagon = $\frac{540}{5} = 108°$

$\angle x = 180 - 108 = 72°$

Sum of angle of triangle = 180°

$72° + 72° + \theta = 180°$

$\theta = 36°$

1.71 (a)

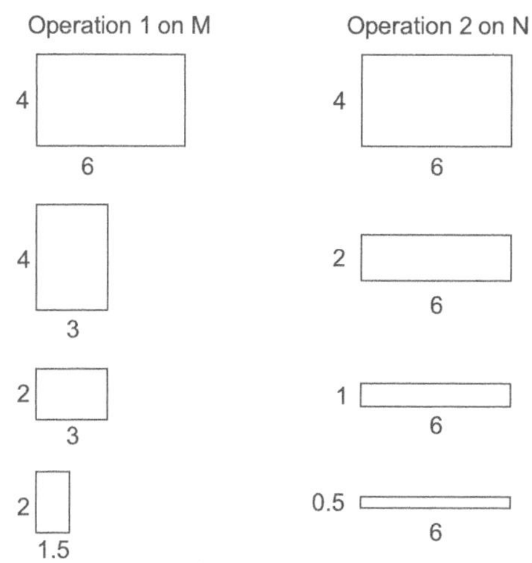

$(\text{Perimeter})_M = 2(2 + 1.5) = 7$

$(\text{Perimeter})_N = 2(0.5 + 6) = 13$

Required ratio = $\frac{13}{7}$

1.72 (d)

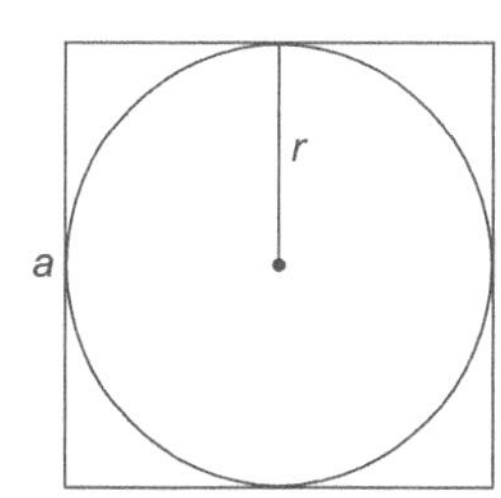

Diameter of sphere = Side of cube

$a = 2r$

Volume of cube = $(2r)^3$

$= 8r^3$

1.73 (c)

Let the capacity of storage tank is x liters. Amount of water drawn only by pipe R is in one minute.

$$\frac{x}{60}+\left(\frac{x}{6}+\frac{x}{10}\right) = \frac{17x}{60}$$

Total time taken R to draw a storage tank

$$= \frac{x}{\frac{17x}{60}} = \frac{60}{17}\text{min}$$

$$\text{Capacity of tank} = \frac{60}{17}\times 34 = 120 \text{ liters}$$

1.74 (c)

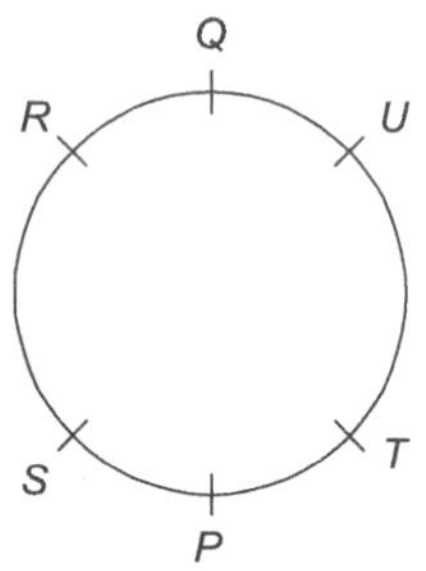

1.76 (d)

Probability of getting only one number $= \frac{1}{8}$

As they are independent spins probability of getting one number for each disc is

$$P = \frac{1}{8}\times\frac{1}{8}$$

Possible outcomes to the sum of 8

= (1, 7), (2, 6), (3, 5), (4, 4), (5, 3), (6, 2), (7, 1)

$$\text{Required probability} = \frac{7}{64}$$

1.77 (a)

Adding both equation

$$2p + 2q < 16$$

$$p + q < 8$$

1.78 (c)

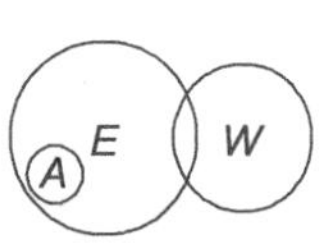

1.80 (d)

As we know that there are 5 odd digits.

So, total frabjous number = 5 × 4 × 4 = 80

1.82 (a)

Use (1) if Ankita move one stairs up

Use (2) if Ankita move two stairs up

Possible ways in which Ankita can reach the 5th stair.

F(5) ⇒ ① ① ① ① ① | ① ② ① ①
① ② ② | ② ① ① ①
② ② ①
② ① ②
① ① ① ②
① ① ② ①

1.84 (b)

Conclusion 1 ×
Conclusion 2 ✓
Conclusion 3 ✓

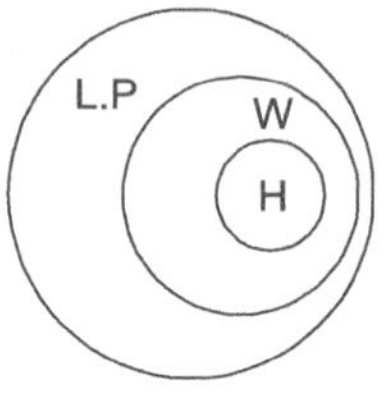

1.85 (d)

$$\frac{R\times q-\frac{R}{q}}{R\times q} = \frac{80}{100} = \frac{4}{5}$$

$$\frac{Rq-\frac{R}{q}}{Rq} = \frac{4}{5}$$

$$1-\frac{1}{q^2} = \frac{4}{5}$$

$$\frac{1}{q^2} = \frac{1}{5}$$

$$q = \sqrt{5}$$

1.86 (a)

The sum of first n odd natural number $= n^2$

$$\therefore \text{ Result is } \frac{20^2}{20^2} = 1$$

1.87 (b)

	Class VIII	Class IX
Total students	450	360
Number of girls	G	G
Number of boys	450 – G	360 – G

According to question,

$$\frac{G}{450-G} = \frac{360-G}{G}$$

$$G^2 = 360 \times 450 - 450\,G - 360\,G + G^2$$

$$810\,G = 360 - 450$$

$$G = \frac{360 \times 450}{810} = 200$$

1.88 (b)

Volume of outer cylinder $= \pi R^2 h$

Volume of smaller cylinder $= \pi r^2 h$

$R = 3r$

Ratio of occupied space to empty space

$$= \frac{7 \times \pi r^2 h}{9\pi r^2 h - 7 \times \pi r^2 h} = \frac{7}{2}$$

■■■

2 Verbal Ability

2.1 Which of the following options is the closest in meaning to the word below?

Circuitous

(a) Cyclic (b) Indirect
(c) Confusing (d) Crooked

[2010 : 1 M]

2.2 The question below consists of a pair of related words followed by four pairs of words. Select the pair that best expresses the relation in the original pair.

Unemployed: Worker

(a) Fallow : Land
(b) Unaware : Sleeper
(c) Wit : Jester
(d) Renovated : House

[2010 : 1 M]

2.3 Choose the most appropriate word from the options given below to complete the following sentence:

If we manage to.............our natural resources, we would leave a better planet for our children.

(a) uphold (b) restrain
(c) cherish (d) conserve

[2010 : 1 M]

2.4 Choose the most appropriate word from the options given below to complete the following sentence:

His rather casual remarks on politics .... his lack of seriousness about the subject.

(a) masked (b) belied
(c) betrayed (d) suppressed

[2010 : 1 M]

2.5 Modern warfare has changed from large scale clashes of armies to suppression of civilian populations. Chemical agents that do their work silently appear to be suited to such warfare; and regretfully, there exist people in military establishments who think that chemical agents are useful fools for their cause.

Which of the following statements best sums up the meaning of the above passage?

(a) Modern warfare has resulted in civil strife.
(b) Chemical agents are useful in modern warfare.
(c) Use of chemical agents in warfare would be undesirable.
(d) People in military establishments like to use chemical agents in war.

[2010 : 2 M]

2.6 The question below consists of a pair of related words followed by four pairs of words. Select the pair that best expresses the relation in the original pair:

Gladiator : Arena

(a) dancer : stage
(b) commuter : train
(c) teacher : classroom
(d) lawyer : courtroom

[2011 : 1 M]

2.7 Choose the most appropriate word from the options given below to complete the following sentence:

It was her view that the country's problems had been _______ by foreign technocrats, so that to invite them to come back would be counter-productive.

(a) identified (b) ascertained
(c) exacerbated (d) analysed

[2011 : 1 M]

2.8 Choose the word from the options given below that is most nearly opposite in meaning to the given word:

Frequency

(a) periodicity (b) rarity
(c) gradualness (d) persistency

[2011 : 1 M]

2.9 Choose the most appropriate word from the options given below to complete the following sentence:

Under ethical guidelines recently adopted by the Indian Medical Association, human genes are to be manipulated only to correct diseases for which ______ treatments are unsatisfactory.

(a) similar (b) most
(c) uncommon (d) available

[2011 : 1 M]

2.10 The horse has played a little known but very important role in the field of medicine. Horses were injected with toxins of diseases until their blood built up immunities. Then a serum was made from their blood. Serums to fight with diphtheria and tetanus were developed this way.

It can be inferred from the passage, that horses were

(a) given immunity to diseases
(b) generally quite immune to diseases
(c) given medicines to fight toxins
(d) given diphtheria and tetanus serums

[2011 : 2 M]

2.11 Which one of the following options is the closest in meaning to the word given below?

Latitude

(a) Eligibility (b) Freedom
(c) Coercion (d) Meticulousness

[2012 : 1 M]

2.12 One of the parts (A, B, C, D) in the sentence given below contains an ERROR. Which one of the following is INCORRECT?

I requested that he should be given the driving test today instead of tomorrow.

(a) requested that (b) should be given
(c) the driving test (d) instead of tomorrow

[2012 : 1 M]

2.13 Choose the most appropriate alternative from the options given below to complete the following sentence:

If the tired soldier wanted to lie down, he ____ the mattress out on the balcony.

(a) should take (b) shall take
(c) should have taken (d) will have taken

[2012 : 1 M]

2.14 Choose the most appropriate word from the options given below to complete the following sentence:

Given the seriousness of the situation that he had to face, his ___ was impressive.

(a) beggary (b) nomenclature
(c) jealousy (d) nonchalance

[2012 : 1 M]

2.15 One of the legacies of the Roman legions was discipline. IN the legions, military law prevailed and discipline was brutal. Discipline on the battlefield kept units obedient, intact and fighting, even when the odds and conditions were against them.

Which one of the following statements best sums up the meaning of the above passage?

(a) Thorough regimentation was the main reason for the efficiency of the Roman legions even in adverse circumstances.
(b) The legions were treated inhumanly as if the men were animals.
(c) Discipline was the armies inheritance from their seniors.
(d) The harsh discipline to which the legions were subjected to led to the odds and conditions being against them.

[2012 : 2 M]

2.16 Choose the grammatically **CORRECT** sentence:

(a) Two and two add four.
(b) Two and two become four.
(c) Two and two are four.
(d) Two and two make four.

[2013 : 1 M]

2.17 **Statement:** You can always give me a ring whenever you need.

Which one of the following is the best inference from the above statement?

(a) Because I have a nice caller tune.
(b) Because I have a better telephone facility.
(c) Because a friend in need is a friend indeed.
(d) Because you need not pay towards the telephone bills when you give me a ring.

[2013 : 1 M]

2.18 Complete the sentence:

Dare ________ mistakes.

(a) commit (b) to commit
(c) committed (d) committing

[2013 : 1 M]

2.19 They were requested not to **quarrel** with others. Which one of the following options is the closest in meaning to the word **quarrel**?

(a) make out (b) call out
(c) dig out (d) fall out

[2013 : 1 M]

2.20 **Statement:** There were different streams of freedom movements in colonial India carried out by the moderates, liberals, radicals, socialists, and so on.

Which one of the following is the best inference from the above statement?

(a) The emergence of nationalism in colonial India led to our Independence.
(b) Nationalism in India emerged in the context of colonialism.
(c) Nationalism in India is homogeneous.
(d) Nationalism in India is heterogeneous.

[2013 : 2 M]

2.21 Choose the most appropriate word from the options given below to complete the following sentence.

A person suffering from Alzheimer's disease _________ short-term memory loss.

(a) experienced (b) has experienced
(c) is experiencing (d) experiences

[2014 : 1 M]

2.22 Choose the most appropriate word from the options given below to complete the following sentence.

___________ is the key to their happiness; they are satisfied with what they have.

(a) Contentment (b) Ambition
(c) Perseverance (d) Hunger

[2014 : 1 M]

2.23 Which of the following options is the closest in meaning to the sentence below?

"As a woman, I have no country."

(a) Women have no country.
(b) Women are not citizens of any country.
(c) Women's solidarity knows no national boundaries.
(d) Women of all countries have equal legal rights.

[2014 : 1 M]

2.24 Moving into a world of big data will require us to change our thinking about the merits of exactitude. To apply the conventional mindset of measurement to the digital, connected world of the twenty-first century is to miss a crucial point. As mentioned earlier, the obsession with exactness is an artefact of the information-deprived analog era. When data was sparse, every data point was critical, and thus great care was taken to avoid letting any point bias the analysis. *From "BIG DATA" Viktor Mayer-Schonberger and Kenneth Cukier*

The **main** point of the paragraph is:

(a) The twenty-first century is a digital world
(b) Big data is obsessed with exactness
(c) Exactitude is not critical in dealing with big data
(d) Sparse data leads to a bias in the analysis

[2014 : 2 M]

2.25 Choose the appropriate word/phrase, out of the four options given below, to complete the following sentence:

Apparent lifelessness ______ dormant life.

(a) harbours (b) leads to
(c) supports (d) affects

[2015 : 1 M]

2.26 Choose the statement where underlined word is used correctly:

(a) When the teacher eludes to different authors, he is being <u>elusive</u>.
(b) When the thief keeps eluding the police, he is being <u>elusive</u>.
(c) Matters the are difficult to understand, identify or remember are <u>allusive</u>.
(d) Mirages can be <u>allusive</u>, but a better way to express them is illusory.

[2015 : 1 M]

2.27 Fill in the blank with the correct idiom/phrase.

That boy from the two was a ______ in the sleepy village.

(a) dog out of herd
(b) sheep from the heap
(c) fish out of water
(d) bird from the flock

[2015 : 1 M]

2.28 Select the appropriate option in place of underlined part of the sentence:

<u>Increased productivity necessary</u> reflects greater efforts made by the employees.

(a) Increase in productivity necessary
(b) Increase productivity is necessary
(c) Increase in productivity necessarily
(d) No improvement required

[2015 : 2 M]

2.29 An apple costs ₹ 10. An onion costs ₹ 8. Select the most suitable sentence with respect to grammar and usage.

(a) The price of an apple is greater than an onion.
(b) The price of an apple is more than onion.
(c) The price of an apple is greater than of an onion.
(d) Apples are more costlier than onions.

[2016 : 1 M]

2.30 The Buddha said, "Holding on to anger is like <u>grasping</u> a hot coal with the intent of throwing it at someone else; you are the one who gets burnt."

Select the word below which is closest in meaning to the word underlined above.

(a) burning (b) igniting
(c) clutching (d) flinging

[2016 : 1 M]

2.31 The overwhelming number of people infected with rabies in India has been flagged by the World Health Organization as a source of concern. It is estimated that inoculating 70% of pets and stray dogs against rabies can lead to a significant reduction in the number of people infected with rabies.

Which of the following can be logically inferred from the above sentences?

(a) The number of people in India infected with rabies is high.
(b) The number of people in other parts of the world who are infected with rabies is low.
(c) Rabies can be eradicated in India by vaccinating 70% of stray dogs.
(d) Stray dogs are the main source of rabies worldwide.

[2016 : 2 M]

2.32 The event would have been successful if you ________ able to come.

(a) are (b) had been
(c) have been (d) would have been

[2017 : 1 M]

2.33 There was no doubt that their work was thorough.

Which of the words below is closest in meaning to the underlined word above?

(a) pretty (b) complete
(c) sloppy (d) haphazard

[2017 : 1 M]

2.34 Bhaichung was observing the pattern of people entering and leaving a car service centre. There was a single window where customers were being served. He saw that people inevitably came out of the centre in the order that they went in. However, the time they spent inside seemed to vary a lot: Some people came out in a matter of minutes while for others it took much longer.

From this, what can one conclude?

(a) The centre operates on a first-come-first-served basis, but with variable service times, depending on specific customer needs.
(b) Customers were served in an arbitrary order, since they took varying amounts of time for service completion in the centre.
(c) Since some people came out within a few minutes of entering the centre, the system is likely to operate on a last-come-first-served basis.
(d) Entering the centre early ensured that one would have shorter service times and most people attempted to do this.

[2017 : 1 M]

2.35 "In spite of being warned repeatedly, he failed to correct his ________ behaviour." The word that best fills that blank in the above sentence is

(a) rational (b) reasonable
(c) errant (d) good

[2018 : 1 M]

2.36 "When she fell down the ________. She received many ________ but little help." The words that best fill the blanks in the above sentence are

(a) stairs, stares (b) stairs, stairs
(c) stares, stairs (d) stares, stares

[2018 : 1 M]

2.37 The fishermen, ________ the flood victims owed their lives, were rewarded by the government.

(a) whom (b) to which
(c) to whom (d) that

[2019 : 1 M]

2.38 Some students were not involved in the strike.
If the above statement is true, which of the following conclusions is/are logically necessary?
1. Some who were involved in the strike were students.
2. No student was involved in the strike.
3. At least one student was involved in the strike.
4. Some who were not involved in the strike were students.

(a) 1 and 2 (b) only 3
(c) only 4 (d) 2 and 3

[2019 : 1 M]

2.39 Until Iran came along, India had never been_______in kabaddi
(a) defeated (b) defeating
(c) defeat (d) defeatist

[2019 : 1 M]

2.40 Since the last one year, after a 125 basis point reduction in repo rate by the Reserve Bank of India, banking institutions have been making a demand to reduce interest rates on small saving schemes. Finally, the government announced yesterday a reduction in interest rates on small saving schemes to bring them on par with fixed deposit interest rates.
Which one of the following statements can be inferred from the given passage?
(a) Whenever the Reserve Bank of India reduces the repo rate, the interest rates on small saving schemes are also reduced
(b) Interest rates on small saving schemes are always maintained on par with fixed deposit interest rates
(c) The government sometimes takes into consideration the demands of banking institutions before reducing the interest rates on small saving schemes
(d) A reduction in interest rates on small saving schemes follow only after a reduction in repo rate by the Reserve Bank of India

[2019 : 2 M]

2.41 "I read somewhere that in ancient times the prestige of a kingdom depended upon the number of taxes that it was able to levy on its people. It was very much like the prestige of a head-hunter in his own community."
Based on the paragraph above, the prestige of a head-hunter depended upon_____
(a) the prestige of the kingdom
(b) the prestige of the heads
(c) the number of taxes he could levy
(d) the number of heads he could gather

[2019 : 2 M]

2.42 The nomenclature of Hindustani music has changed over the centuries. Since the medieval period *dhrupad* styles were identified as *baanis*. Terms like *gayaki* and *baaj* were used to refer to vocal and instrumental styles, respectively. With the institutionalization of music education the term *gharana* became acceptable. *Gharana* originally referred to hereditary musicians from a particular lineage, including disciples and grand disciples.
Which one of the following pairings is NOT correct?
(a) *dhrupad, baani* (b) *gayaki*, vocal
(c) *baaj*, institution (d) *gharana*, lineage

[2019 : 2 M]

2.43 I do not think you know the case well enough to opinions. Having said that, I agree with your other point. What does the phrase "having said that" means in the given text?
(a) contrary to what I have said
(b) in addition to what I have said
(c) as opposed to what I have said
(d) despite what I have said

[2020 : 1 M]

2.44 He is known for his unscrupulous ways. He always sheds ____ tears to deceive people.
(a) fox (b) fox's
(c) crocodile's (d) crocodile

[2020 : 1 M]

2.45 Select the word that fits the analogy:
Build : Building :: Grow : _____
(a) Grown (b) Grew
(c) Growed (d) Growth

[2020 : 1 M]

2.46 Crowd funding deals with mobilization of funds for a project from a large number of people, who would be willing to invest smaller amounts through web-based platforms in the project.
Based on the above paragraph, which of the following is correct about crowd funding?

(a) Funds raised through unwilling contributions on web-based platforms.
(b) Funds raised through coerced contributions on web-based platforms.
(c) Funds raised through large contributions on web-based platforms.
(d) Funds raised through voluntary contributions on web-based platforms.

[2020 : 2 M]

2.47 Jofra Archer, the England fast bowler, is _____ than accurate.

(a) more faster (b) faster
(c) less fast (d) more fast

[2020 : 1 M]

2.48 Humans have the ability to construct worlds entirely in their minds, which don't exist in the physical world. So far as we know, no other species possesses this ability. This skill is so important that we have different words to refer to its different flavors, such as imagination, invention and innovation.

Based on the above passage, which one of the following is TRUE?

(a) The terms imagination, invention and innovation refer to unrelated skills.
(b) Imagination, invention and innovation are unrelated to the ability to construct mental worlds.
(c) We do not know of any species other than humans who possess the ability to construct mental worlds.
(d) No species possess the ability to construct worlds in their minds.

[2021 : 2 M]

2.49 Getting to the top is_____ than staying on top.

(a) easiest (b) more easy
(c) easier (d) much easy

[2021 : 1 M]

2.50 Inhaling the smoke from a burning ________ could ________ you quickly.

(a) tire / tier (b) tire / tyre
(c) tyre / tire (d) tyre / tier

[2022 : 1 M]

2.51 Rice, a versatile and inexpensive source of carbohydrate, is a critical component of diet worldwide. Climate change, causing extreme weather, poses a threat to sustained availability of rice. Scientists are working on developing Green Super Rice (GSR), which is resilient under extreme weather conditions yet gives higher yields sustainably.

Which one of the following is the CORRECT logical inference based on the information given in the above passage?

(a) GSR is an alternative to regular rice, but it grows only in an extreme weather.
(b) GSR may be used in future in response to adverse effects of climate change.
(c) GSR grows in an extreme weather, but the quantity of produce is lesser than regular rice.
(d) Regular rice will continue to provide good yields even in extreme weather.

[2022 : 2 M]

2.52 The village was nestled in a green spot, ______ the ocean and the hills.

(a) through (b) in
(c) at (d) between

[2023 : 1 M]

2.53 Disagree : Protest : : Agree : _______.

(By word meaning)

(a) Refuse (b) Pretext
(c) Recommend (d) Refute

[2023 : 1 M]

2.54 In the given diagram, ovals are marked at different heights (h) of a hill. Which one of the following options P, Q, R, and S depicts the top view of the hill?

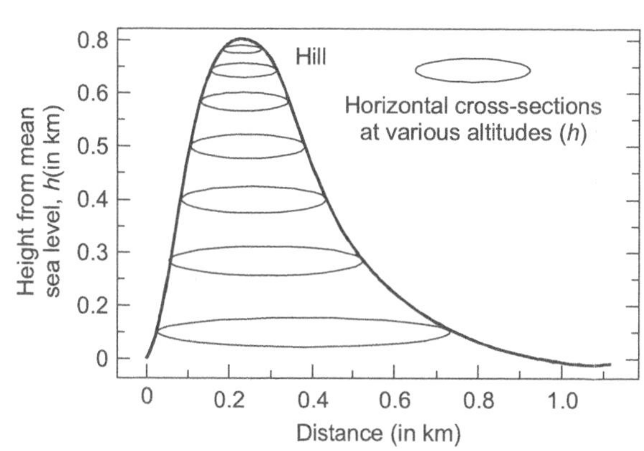

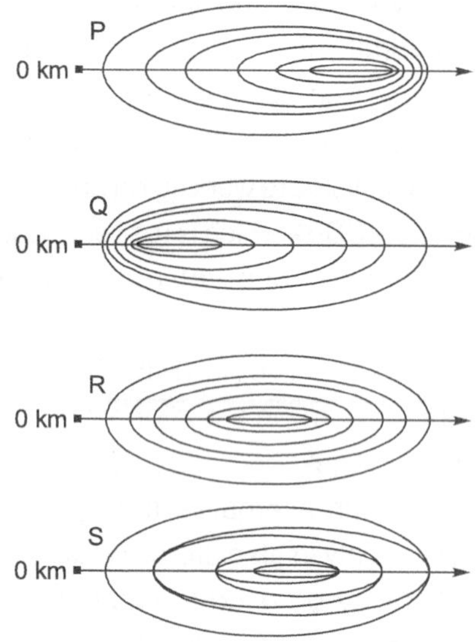

(a) P (b) Q
(c) R (d) S

[2023 : 1 M]

2.55 Residency is a famous housing complex with many well-established individuals among its residents. A recent survey conducted among the residents of the complex revealed that all of those residents who are well established in their respective fields happen to be academicians. The survey also revealed that most of these academicians are authors of some best-selling books.

Based only on the information provided above, which one of the following statements can be logically inferred with certainty?

(a) Some residents of the complex who are well established in their fields are also authors of some best-selling books.
(b) All academicians residing in the complex are well established in their fields.
(c) Some authors of best-selling books are residents of the complex who are well established in their fields.
(d) Some academicians residing in the complex are well established in their fields.

[2023 : 2 M]

2.56 The information contained in DNA is used to synthesize proteins that are necessary for the functioning of life. DNA is composed of four nucleotides: Adenine (A), Thymine (T), Cytosine (C), and Guanine (G). The information contained in DNA can then be thought of as a sequence of these four nucleotides: A, T, C, and G. DNA has coding and non-coding regions. Coding regions—where the sequence of these nucleotides are read in groups of three to produce individual amino acids—constitute only about 2% of human DNA. For example, the triplet of nucleotides CCG codes for the amino acid glycine, while the triplet GGA codes for the amino acid proline. Multiple amino acids are then assembled to form a protein.

Based only on the information provided above, which of the following statements can be logically inferred with certainty?

(i) The majority of human DNA has no role in the synthesis of proteins.
(ii) The function of about 98% of human DNA is not understood.

(a) only (i) (b) only (ii)
(c) both (i) and (ii) (d) neither (i) nor (ii)

[2023 : 2 M]

2.57 An opaque cylinder (shown below) is suspended in the path of a parallel beam of light, such that its shadow is cast on a screen oriented perpendicular to the direction of the light beam. The cylinder can be reoriented in any direction within the light beam. Under these conditions, which one of the shadows *P*, *Q*, *R*, and *S* is NOT possible?

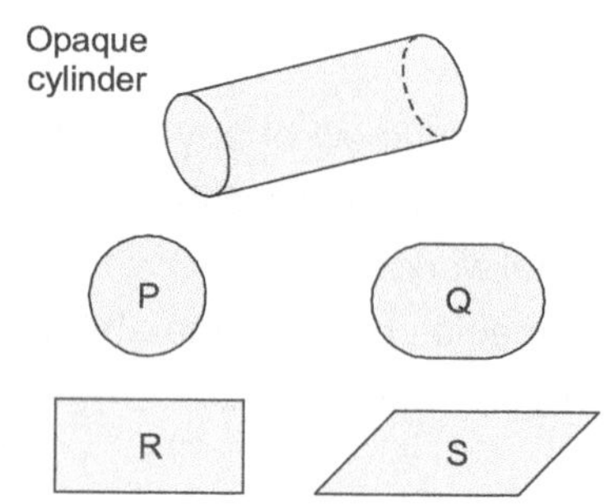

(a) P (b) Q
(c) R (d) S

[2023 : 2 M]

2.58 In the given text, the blanks are numbered (i)-(iv). Select the best match for all the blanks.

Yoko Roi stands ___(i)___ as an author for standing ___(ii)___ as an honorary fellow, after she stood ___(iii)___ her writings that stand ___(iv)___ the freedom of speech.

(a) (i) out (ii) down (iii) in (iv) for
(b) (i) down (ii) out (iii) by (iv) in
(c) (i) down (ii) out (iii) for (iv) in
(d) (i) out (ii) down (iii) by (iv) for
[2024 : 2 M]

2.59 If '→' denotes increasing order of intensity, then the meaning of the words [drizzle → rain → downpour] is analogous to [______ → quarrel → feud]. Which one of the given options is appropriate to fill the blank?
(a) bicker (b) bog
(c) dither (d) dodge [2024 : 1 M]

■■■

Answers Verbal Ability

2.1 (b)	2.2 (a)	2.3 (d)	2.4 (c)	2.5 (d)	2.6 (a)	2.7 (c)
2.8 (c)	2.9 (c)	2.10 (c)	2.11 (b)	2.12 (a)	2.13 (c)	2.14 (d)
2.15 (a)	2.16 (d)	2.17 (c)	2.18 (a)	2.19 (d)	2.20 (d)	2.21 (d)
2.22 (a)	2.23 (c)	2.24 (c)	2.25 (a)	2.26 (b)	2.27 (c)	2.28 (c)
2.29 (c)	2.30 (c)	2.31 (a)	2.32 (b)	2.33 (b)	2.34 (a)	2.35 (c)
2.36 (a)	2.37 (c)	2.38 (c)	2.39 (a)	2.40 (c)	2.41 (d)	2.42 (c)
2.43 (d)	2.44 (d)	2.45 (d)	2.46 (d)	2.47 (d)	2.48 (c)	2.49 (c)
2.50 (c)	2.51 (b)	2.52 (d)	2.53 (c)	2.54 (b)	2.55 (*)	2.56 (d)
2.57 (d)	2.58 (d)	2.59 (a)				

Explanations Verbal Ability

2.1 (b)

Circuitous: Deviating from a straight course ⇒ Indirect
(A) Cyclic: Recurring in cycle
(B) Indirect: Not leading by straight line
(C) Confusing: Lacking clarity
(D) Crooked: For shapes (irregular in shape)

2.2 (a)

Unemployed: Worker ⇒ Here one is opposite to other.
(A) Fallow: Land ⇒ Fallow means undeveloped land.
(B) Unaware: sleeper ⇒ Both are same unaware or asleep.
(C) Wit: Jester ⇒ Wit means ability to make jokes and jester is a joker.
(D) Renovated : House ⇒ Renovate means to make better and house can be renovated.

2.3 (d)

(A) Uphold: cause to remain ⇒ not appropriate
(B) Restrain: keep under control ⇒ not appropriate
(C) Cherish: be fond of ⇒ not related
(D) Conserve: Keep in safety and protect from harm, decay, loss, or destruction ⇒ most appropriate.

2.4 (c)

(A) Masked: Hide under a false appearance ⇒ opposite
(B) Belied: Be in contradiction with ⇒ not appropriate
(C) Betrayed: Reveal unintentionally⇒ most appropriate
(D) Suppressed: To put down by force or authority ⇒ irrelevant

2.5 (d)

(A) Modern warfare has resulted in civil strife: There is no direct consequence of warfare given, so it is not appropriate.
(B) Chemical agents are useful in modern warfare: Passage does not say whether chemical agents are useful or not, so not appropriate.

(C) Use of chemical agents in warfare would be undesirable: Given that people in military think these are useful, undesirable is wrong

(D) People in military establishments like to use chemical agents in war; Correct choice as last statement tells that military people think that chemical agents are useful tools for their cause (work silently in warfare).

2.11 (b)

Latitude refers to freedom of action, freedom of expression from restrictions etc. For example, he allowed his children a fair amount of latitude.
Coercion refers to force which is an opposite of the word Latitude. Meticulousness refers to being extremely careful and conscientious.

2.12 (a)

'Requested that' part should be modified. The sentence is in Reported speech so 'should be given' part is correct. Instead of Request, words like 'suggested', 'recommended' can be introduced.

2.13 (c)

If the tired soldier wanted to lie down, he should have taken the mattress out on the balcony.

2.14 (d)

Given the seriousness of the situation that he had to face, his nonchalance was impressive.
Nonchalance refers to casual / non-serious/ indifferent attitude. The author wants to convey that in spite of the seriousness of the situation, he was very casual.
No other word fits in the blank. Beggary is word that is related to extreme poverty (Beggar). Nomenclature refers to the art of naming.

2.15 (a)

Thorough regimentation was the main reason for the efficiency of the Roman legions even in adverse circumstances. As the author says, 'Discipline on the battlefield kept units obedient, intact and fighting, even when the odds and conditions were against them.' So choice (A) emerges as the best answer option.
We have been asked to summarize the paragraph in the best possible manner. Choice (C) presents one-sided picture of the paragraph. Summary is the essence or the gist of the paragraph.
Choice (B) and Choice (D) are wrong as per the information mentioned in the paragraph.

2.16 (d)

This is a famous idiom in English.

2.17 (c)

'You can always give me a ring whenever you need' implies that I am available even at a single call (whenever you need me). 'Ring' has been used as a metaphor here and not literally.

2.20 (d)

As the given argument states that there were different streams of freedom movement during colonialism, it clearly implies that nationalism in India is heterogeneous.
Option (c) is ruled out as it states the contrary.
Options (a) and (b) fail to establish a direct link with the argument.

2.25 (a)

Harbours means to give shelter.

2.27 (c)

Fish out of water : implies uncomfortable situation.

2.28 (c)

Increase in productivity necessarily reflects greater efforts made, by the employees.

2.32 (b)

Use of conditional sentence based on past perfect structure. Conditional clause uses (had + V_3), main clause is based on would / should / could + have + V_3.

2.33 (b)

Thorough refers to somethings that is exhaustively complete, accurate, careful or absolute.

2.34 (a)

People coming out in the same order in which they enter indicates that the centre operates on a first-come-first-serve basis.

2.35 (c)

Errant means misbehaving, exhibiting inappropriate behavior / offending conduct.

2.36 (a)

Stairs - A construction designed to bridge a large vertical distance by dividing it into smaller vertical distances, called steps

Stares - To look at someone for a long time

2.38 (c)

Given statements: Some students were not involved in the strike.

Venn diagram:

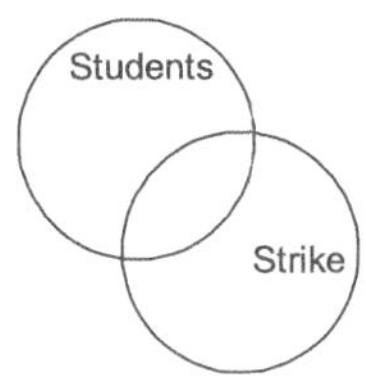

Only conclusion 4 is logically necessary. Hence, option (c) is correct.

2.39 (a)

Until Iran came along, India had never been defeated in Kabaddi.

If two events occur in past one after another, the event completing first takes past perfect tense. And another event is expressed in Simple past tense.

2.40 (c)

The argument says that banking institutions had been demanding for a reduction in interest rates for the last one year. Finally, the government decided to reduce the interest rates of small saving schemes thus implying that the govt. does consider the demands of banking institutions before making any such policy decision.

2.41 (d)

The way prestige of a kingdom depended upon the number of taxes, the prestige of a head-hunter depended upon the number of heads he could collect.

2.42 (c)

Baaj, institution

Following are the correct pairings:

- Dhrupad style - baani
- gayaki - vocal style
- baaj - instrumental style
- gharana - musicians from a particular lineage

2.48 (c)

Option (a) and (b) are weakened by the work 'UNRELATED SKILLS'. Option (d) is weakened by the expression no species posses the ability. Hence, answer is (c) which reflects the information given in the passage.

2.49 (c)

When the comparison is between two things we use the second degree of the adjective. The degree form of easy are-(easy-easier-easiest).

2.59 (a)

Bicker → To argue about un-important thing or issue

Quarrel → Angry, argument or disagreement

Feud → Angry or serious argument between two people that continues for long period of time.

■■■